Graduate Texts in Mathematics

continued after Index

Graduate Texts in Mathematics 96

Editorial Board
F. W. Gehring   P. R. Halmos (Managing Edito
C. C. Moore

John B. Conway

A Course
in Functional Analysis

Springer-Verlag
New York Berlin Heidelberg Tokyo

John B. Conway
Department of Mathematics
Indiana University
Bloomington, IN 47405
U.S.A.

AMS Classifications: 46-01, 45B05

Library of Congress Cataloging in Publication Data
Conway, John B.
    A course in functional analysis.
    (Graduate texts in mathematics: 96)
    Bibliography: p.
    Includes index.
    1. Functional analysis. I. Title. II. Series.
QA320.C658 1985   515.7   84-10568

With 1 illustration

Typeset by Science Typographers, Medford, New York.
Printed and bound by R. R. Donnelley & Sons, Harrisonburg, Virginia.
Printed in the United States of America.

9 8 7 6 5 4 3 2 1

ISBN 0-387-96042-2 Springer-Verlag New York Berlin Heidelberg Tokyo
ISBN 3-540-96042-2 Springer-Verlag Berlin Heidelberg New York Tokyo

For Ann (of course)

Preface

Functional analysis has become a sufficiently large area of mathematics that it is possible to find two research mathematicians, both of whom call themselves functional analysts, who have great difficulty understanding the work of the other. The common thread is the existence of a linear space with a topology or two (or more). Here the paths diverge in the choice of how that topology is defined and in whether to study the geometry of the linear space, or the linear operators on the space, or both.

In this book I have tried to follow the common thread rather than any special topic. I have included some topics that a few years ago might have been thought of as specialized but which impress me as interesting and basic. Near the end of this work I gave into my natural temptation and included some operator theory that, though basic for operator theory, might be considered specialized by some functional analysts.

The word "course" in the title of this book has two meanings. The first is obvious. This book was meant as a text for a graduate course in functional analysis. The second meaning is that the book attempts to take an excursion through many of the territories that comprise functional analysis. For this purpose, a choice of several tours is offered the reader—whether he is a tourist or a student looking for a place of residence. The sections marked with an asterisk are not (strictly speaking) necessary for the rest of the book, but will offer the reader an opportunity to get more deeply involved in the subject at hand, or to see some applications to other parts of mathematics, or, perhaps, just to see some local color. Unlike many tours, it is possible to retrace your steps and cover a starred section after the chapter has been left.

There are some parts of functional analysis that are not on the tour. Most authors have to make choices due to time and space limitations, to say nothing of the financial resources of our graduate students. Two areas that

are only briefly touched here, but which constitute entire areas by themselves, are topological vector spaces and ordered linear spaces. Both are
beautiful theories and both have books which do them justice.

The prerequisites for this book are a thoroughly good course in measure
and integration—together with some knowledge of point set topology. The
appendices contain some of this material, including a discussion of nets in
Appendix A. In addition, the reader should at least be taking a course in
analytic function theory at the same time that he is reading this book. From
the beginning, analytic functions are used to furnish some examples, but it
is only in the last half of this text that analytic functions are used in the
proofs of the results.

It has been traditional that a mathematics book begin with the most
general set of axioms and develop the theory, with additional axioms added
as the exposition progresses. To a large extent I have abandoned tradition.
Thus the first two chapters are on Hilbert space, the third is on Banach
spaces, and the fourth is on locally convex spaces. To be sure, this causes
some repetition (though not as much as I first thought it would) and the
phrase "the proof is just like the proof of ..." appears several times. But I
firmly believe that this order of things develops a better intuition in the
student. Historically, mathematics has gone from the particular to the
general—not the reverse. There are many reasons for this, but certainly one
reason is that the human mind resists abstraction unless it first sees the need
to abstract.

I have tried to include as many examples as possible, even if this means
introducing without explanation some other branches of mathematics (like
analytic functions, Fourier series, or topological groups). There are, at the
end of every section, several exercises of varying degrees of difficulty with
different purposes in mind. Some exercises just remind the reader that he is
to supply a proof of a result in the text; others are routine, and seek to fix
some of the ideas in the reader's mind; yet others develop more examples;
and some extend the theory. Examples emphasize my idea about the nature
of mathematics and exercises stress my belief that doing mathematics is the
way to learn mathematics.

Chapter I discusses the geometry of Hilbert spaces and Chapter II begins
the theory of operators on a Hilbert space. In Sections 5–8 of Chapter II,
the complete spectral theory of normal compact operators, together with a
discussion of multiplicity, is worked out. This material is presented again in
Chapter IX, when the Spectral Theorem for bounded normal operators is
proved. The reason for this repetition is twofold. First, I wanted to design
the book to be usable as a text for a one-semester course. Second, if the
reader understands the Spectral Theorem for compact operators, there will
be less difficulty in understanding the general case and, perhaps, this will
lead to a greater appreciation of the complete theorem.

Chapter III is on Banach spaces. It has become standard to do some of
this material in courses on Real Variables. In particular, the three basic

principles, the Hahn–Banach Theorem, the Open Mapping Theorem, and the Principle of Uniform Boundedness, are proved. For this reason I contemplated not proving these results here, but in the end decided that they should be proved. I did bring myself to relegate to the appendices the proofs of the representation of the dual of L^p (Appendix B) and the dual of $C_0(X)$ (Appendix C).

Chapter IV hits the bare essentials of the theory of locally convex spaces —enough to rationally discuss weak topologies. It is shown in Section 5 that the distributions are the dual of a locally convex space.

Chapter V treats the weak and weak-star topologies. This is one of my favorite topics because of the numerous uses these ideas have.

Chapter VI looks at bounded linear operators on a Banach space. Chapter VII introduces the reader to Banach algebras and spectral theory and applies this to the study of operators on a Banach space. It is in Chapter VII that the reader needs to know the elements of analytic function theory, including Liouville's Theorem and Runge's Theorem. (The latter is proved using the Hahn–Banach Theorem in Section III.8.)

When in Chapter VIII the notion of a C^*-algebra is explored, the emphasis of the book becomes the theory of operators on a Hilbert space.

Chapter IX presents the Spectral Theorem and its ramifications. This is done in the framework of a C^*-algebra. Classically, the Spectral Theorem has been thought of as a theorem about a single normal operator. This it is, but it is more. This theorem really tells us about the functional calculus for a normal operator and, hence, about the weakly closed C^*-algebra generated by the normal operator. In Section IX.8 this approach culminates in the complete description of the functional calculus for a normal operator. In Section IX.10 the multiplicity theory (a complete set of unitary invariants) for normal operators is worked out. This topic is too often ignored in books on operator theory. The ultimate goal of any branch of mathematics is to classify and characterize, and multiplicity theory achieves this goal for normal operators.

In Chapter X unbounded operators on Hilbert space are examined. The distinction between symmetric and self-adjoint operators is carefully delineated and the Spectral Theorem for unbounded normal operators is obtained as a consequence of the bounded case. Stone's Theorem on one parameter unitary groups is proved and the role of the Fourier transform in relating differentiation and multiplication is exhibited.

Chapter XI, which does not depend on Chapter X, proves the basic properties of the Fredholm index. Though it is possible to do this in the context of unbounded operators between two Banach spaces, this material is presented for bounded operators on a Hilbert space.

There are a few notational oddities. The empty set is denoted by □. A reference number such as (8.10) means item number 10 in Section 8 of the present chapter. The reference (IX.8.10) is to (8.10) in Chapter IX. The reference (A.1.1) is to the first item in the first section of Appendix A.

There are many people who deserve my gratitude in connection with writing this book. In three separate years I gave a course based on an evolving set of notes that eventually became transfigured into this book. The students in those courses were a big help. My colleague Grahame Bennett gave me several pointers in Banach spaces. My ex-student Marc Raphael read final versions of the manuscript, pointing out mistakes and making suggestions for improvement. Two current students, Alp Eden and Paul McGuire, read the galley proofs and were extremely helpful. Elena Fraboschi typed the final manuscript.

John B. Conway

Contents

CHAPTER IV
Locally Convex Spaces

CHAPTER V
Weak Topologies

CHAPTER VI
Linear Operators on a Banach Space

CHAPTER VII
Banach Algebras and Spectral Theory for Operators on a Banach Space

CHAPTER VIII
C^*-Algebras

CHAPTER IX
Normal Operators on Hilbert Space

CHAPTER X
Unbounded Operators

CHAPTER XI
Fredholm Theory

APPENDIX A
Preliminaries

APPENDIX B
The Dual of $L^p(\mu)$

APPENDIX C
The Dual of $C_0(X)$

Bibliography

List of Symbols

Index

CHAPTER I

Hilbert Spaces

A Hilbert space is the abstraction of the finite-dimensional Euclidean spaces of geometry. Its properties are very regular and contain few surprises, though the presence of an infinity of dimensions guarantees a certain amount of surprise. Historically, it was the properties of Hilbert spaces that guided mathematicians when they began to generalize. Some of the properties and results seen in this chapter and the next will be encountered in more general settings later in this book, or we shall see results that come close to these but fail to achieve the full power possible in the setting of Hilbert space.

§1. Elementary Properties and Examples

Throughout this book $\mathbb{F}$ will denote either the real field, $\mathbb{R}$, or the complex field, $\mathbb{C}$.

1.1. Definition. If $\mathscr{X}$ is a vector space over $\mathbb{F}$, a *semi-inner product* on $\mathscr{X}$ is a function $u: \mathscr{X} \times \mathscr{X} \to \mathbb{F}$ such that for all α, β in $\mathbb{F}$ and x, y, z in $\mathscr{X}$, the following are satisfied:

(a) $u(\alpha x + \beta y, z) = \alpha u(x, z) + \beta u(y, z)$,
(b) $u(x, \alpha y + \beta z) = \bar{\alpha} u(x, y) + \bar{\beta} u(x, z)$,
(c) $u(x, x) \geq 0$,
(d) $u(x, y) = \overline{u(y, x)}$.

Here, for α in $\mathbb{F}$, $\bar{\alpha} = \alpha$ if $\mathbb{F} = \mathbb{R}$ and $\bar{\alpha}$ is the complex conjugate of α if $\mathbb{F} = \mathbb{C}$. If $\alpha \in \mathbb{C}$, the statement that $\alpha \geq 0$ means that $\alpha \in \mathbb{R}$ and α is non-negative.

Note that if $\alpha = 0$, then property (a) implies that $u(0, y) = u(\alpha \cdot 0, y) = \alpha u(0, y) = 0$ for all y in $\mathscr{X}$. This and similar reasoning shows that for a semi-inner product u,

(e) $u(x, 0) = u(0, y) = 0$ for all x, y in $\mathscr{X}$.

In particular, $u(0, 0) = 0$.

An *inner product* on $\mathscr{X}$ is a semi-inner product that also satisfies the following:

(f) If $u(x, x) = 0$, then $x = 0$.

An inner product in this book will be denoted by

$$\langle x, y \rangle = u(x, y).$$

There is no universally accepted notation for an inner product and the reader will often see (x, y) and $(x|y)$ used in the literature.

1.2. Example. Let $\mathscr{X}$ be the collection of all sequences $\{\alpha_n \colon n \geq 1\}$ of scalars α_n from $\mathbb{F}$ such that $\alpha_n = 0$ for all but a finite number of values of n. If addition and scalar multiplication are defined on $\mathscr{X}$ by

$$\{\alpha_n\} + \{\beta_n\} \equiv \{\alpha_n + \beta_n\},$$
$$\alpha\{\alpha_n\} \equiv \{\alpha\alpha_n\},$$

then $\mathscr{X}$ is a vector space over $\mathbb{F}$.

If $u(\{\alpha_n\}, \{\beta_n\}) \equiv \sum_{n=1}^{\infty} \alpha_{2n}\bar{\beta}_{2n}$, then u is a semi-inner product that is not an inner product. On the other hand,

$$\langle \{\alpha_n\}, \{\beta_n\} \rangle = \sum_{n=1}^{\infty} \alpha_n \bar{\beta}_n,$$

$$\langle \{\alpha_n\}, \{\beta_n\} \rangle = \sum_{n=1}^{\infty} \frac{1}{n} \alpha_n \bar{\beta}_n,$$

$$\langle \{\alpha_n\}, \{\beta_n\} \rangle = \sum_{n=1}^{\infty} n^5 \alpha_n \bar{\beta}_n,$$

all define inner products on $\mathscr{X}$.

1.3. Example. Let (X, Ω, μ) be a measure space consisting of a set X, a σ-algebra Ω of subsets of X, and a countably additive $\mathbb{R} \cup \{\infty\}$ valued measure μ defined on Ω. If f and $g \in L^2(\mu) \equiv L^2(X, \Omega, \mu)$, then Hölder's inequality implies $f\bar{g} \in L^1(\mu)$. If

$$\langle f, g \rangle = \int f\bar{g}\, d\mu,$$

then this defines an inner product on $L^2(\mu)$.

Note that Hölder's inequality also states that $|\int f\bar{g}\, d\mu| \leq [\int |f|^2\, d\mu]^{1/2} \cdot [\int |g|^2\, d\mu]^{1/2}$. This is, in fact, a consequence of the following result on semi-inner products.

1.4.   The Cauchy–Bunyakowsky–Schwarz Inequality. *If* $\langle \cdot \cdot \rangle$ *is a semi-inner product on* $\mathcal{X}$, *then*

$$|\langle x, y \rangle|^2 \le \langle x, x \rangle \langle y, y \rangle$$

for all x and y in $\mathcal{X}$.

PROOF. If $\alpha \in \mathbb{F}$ and x and $y \in \mathcal{X}$, then

$$0 \le \langle x - \alpha y, x - \alpha y \rangle$$
$$= \langle x, x \rangle - \alpha \langle y, x \rangle - \bar{\alpha} \langle x, y \rangle + |\alpha|^2 \langle y, y \rangle.$$

Suppose $\langle y, x \rangle = be^{i\theta}$, $b \ge 0$, and let $\alpha = e^{-i\theta}t$, t in $\mathbb{R}$. The above inequality becomes

$$0 \le \langle x, x \rangle - e^{-i\theta}tbe^{i\theta} - e^{i\theta}tbe^{-i\theta} + t^2 \langle y, y \rangle$$
$$= \langle x, x \rangle - 2bt + t^2 \langle y, y \rangle$$
$$= c - 2bt + at^2 \equiv q(t),$$

where $c = \langle x, x \rangle$ and $a = \langle y, y \rangle$. Thus $q(t)$ is a quadratic polynomial in the real variable t and $q(t) \ge 0$ for all t. This implies that the equation $q(t) = 0$ has at most one real solution t. From the quadratic formula we find that the discriminant is not positive; that is, $0 \ge 4b^2 - 4ac$. Hence

$$0 \ge b^2 - ac = |\langle x, y \rangle|^2 - \langle x, x \rangle \langle y, y \rangle,$$

proving the inequality.   ■

The inequality in (1.4) will be referred to as the CBS inequality.

1.5.   Corollary. *If* $\langle \cdot, \cdot \rangle$ *is a semi-inner product on* $\mathcal{X}$ *and* $\|x\| \equiv \langle x, x \rangle^{1/2}$ *for all x in* $\mathcal{X}$, *then*

(a) $\|x + y\| \le \|x\| + \|y\|$ *for x, y in* $\mathcal{X}$,
(b) $\|\alpha x\| = |\alpha| \|x\|$ *for* α *in* $\mathbb{F}$ *and x in* $\mathcal{X}$.

If $\langle \cdot, \cdot \rangle$ *is an inner product, then*

(c) $\|x\| = 0$ *implies* $x = 0$.

PROOF. The proofs of (b) and (c) are left as an exercise. To see (a), note that for x and y in $\mathcal{X}$,

$$\|x + y\|^2 = \langle x + y, x + y \rangle$$
$$= \|x\|^2 + \langle y, x \rangle + \langle x, y \rangle + \|y\|^2$$
$$= \|x\|^2 + 2\,\mathrm{Re}\langle x, y \rangle + \|y\|^2.$$

By the CBS inequality, $\mathrm{Re}\langle x, y \rangle \le |\langle x, y \rangle| \le \|x\| \|y\|$. Hence,

$$\|x + y\|^2 \le \|x\|^2 + 2\|x\| \|y\| + \|y\|^2$$
$$= (\|x\| + \|y\|)^2.$$

The inequality now follows by taking square roots.   ■

If $\langle \cdot, \cdot \rangle$ is a semi-inner product on $\mathcal{X}$ and if $x, y \in \mathcal{X}$, then as was shown in the preceding proof,

$$\|x + y\|^2 = \|x\|^2 + 2 \operatorname{Re}\langle x, y \rangle + \|y\|^2.$$

This identity is often called the *polar identity*.

The quantity $\|x\| = \langle x, x \rangle^{1/2}$ for an inner product $\langle \cdot, \cdot \rangle$ is called the *norm* of x. If $\mathcal{X} = \mathbb{F}^d$ ($\mathbb{R}^d$ or $\mathbb{C}^d$) and $\langle \{\alpha_n\}, \{\beta_n\} \rangle \equiv \sum_{n=1}^d \alpha_n \bar{\beta}_n$, then the corresponding norm is $\|\{\alpha_n\}\| = [\sum_{n=1}^d |\alpha_n|^2]^{1/2}$.

The virtue of the norm on a vector space $\mathcal{X}$ is that $d(x, y) = \|x - y\|$ defines a metric on $\mathcal{X}$ [by (1.5)] so that $\mathcal{X}$ becomes a metric space. In fact, $d(x, y) = \|x - y\| = \|(x - z) + (z - y)\| \le \|x - z\| + \|z - y\| = d(x, z) + d(z, y)$. The other properties of a metric follow similarly. If $\mathcal{X} = \mathbb{F}^d$ and the norm is defined as above, this distance function is the usual Euclidean metric.

1.6.   Definition. A *Hilbert space* is a vector space $\mathcal{H}$ over $\mathbb{F}$ together with an inner product $\langle \cdot, \cdot \rangle$ such that relative to the metric $d(x, y) = \|x - y\|$ induced by the norm, $\mathcal{H}$ is a complete metric space.

If $\mathcal{H} = L^2(\mu)$ and $\langle f, g \rangle = \int f\bar{g} \, d\mu$, then the associated norm is $\|f\| = [\int |f|^2 \, d\mu]^{1/2}$. It is a standard result of measure theory that $L^2(\mu)$ is a Hilbert space. It is also easy to see that $\mathbb{F}^d$ is a Hilbert space.

REMARK. The inner products defined on $L^2(\mu)$ and $\mathbb{F}^d$ are the "usual" ones. Whenever these spaces are discussed these are the inner products referred to. The same is true of the next space.

1.7.   Example. Let I be any set and let $l^2(I)$ denote the set of all functions $x: I \to \mathbb{F}$ such that $x(i) = 0$ for all but a countable number of i and $\sum_{i \in I} |x(i)|^2 < \infty$. For x and y in $l^2(I)$ define

$$\langle x, y \rangle = \sum_i x(i)\overline{y(i)}.$$

Then $l^2(I)$ is a Hilbert space (Exercise 2).

If $I = \mathbb{N}$, $l^2(I)$ is usually denoted by l^2. Note that if $\Omega = $ the set of all subsets of I and for E in Ω, $\mu(E) \equiv \infty$ if E is infinite and $\mu(E) = $ the cardinality of E if E is finite, then $l^2(I)$ and $L^2(I, \Omega, \mu)$ are equal.

Recall that an absolutely continuous function on the unit interval $[0, 1]$ has a derivative a.e. on $[0, 1]$.

1.8.   Example. Let $\mathcal{H} = $ the collection of all absolutely continuous functions $f: [0, 1] \to \mathbb{F}$ such that $f(0) = 0$ and $f' \in L^2(0, 1)$. If $\langle f, g \rangle = \int_0^1 f'(t)g'(t) \, dt$ for f and g in $\mathcal{H}$, then $\mathcal{H}$ is a Hilbert space (Exercise 3).

Suppose $\mathcal{X}$ is a vector space with an inner product $\langle \cdot, \cdot \rangle$ and the norm is defined by the inner product. What happens if $(\mathcal{X}, d)$ ($d(x, y) \equiv \|x - y\|$) is not complete?

1.9. Proposition. *If $\mathscr{X}$ is a vector space and $\langle \cdot , \cdot \rangle_{\mathscr{X}}$ is an inner product on $\mathscr{X}$ and if $\mathscr{H}$ is the completion of $\mathscr{X}$ with respect to the metric induced by the norm on $\mathscr{X}$, then there is an inner product $\langle \cdot , \cdot \rangle_{\mathscr{H}}$ on $\mathscr{H}$ such that $\langle x, y \rangle_{\mathscr{H}} = \langle x, y \rangle_{\mathscr{X}}$ for x and y in $\mathscr{X}$ and the metric on $\mathscr{H}$ is induced by this inner product. That is, the completion of $\mathscr{X}$ is a Hilbert space.*

The preceding result says that an incomplete inner product space can be completed to a Hilbert space. It is also true that a Hilbert space over $\mathbb{R}$ can be imbedded in a complex Hilbert space (see Exercise 7).

This section closes with an example of a Hilbert space from analytic function theory.

1.10. Definition. If G is an open subset of the complex plane $\mathbb{C}$, then $L^2_a(G)$ denotes the collection of all analytic functions $f: G \to \mathbb{C}$ such that

$$\int\int_G |f(x + iy)|^2 \, dx \, dy < \infty.$$

$L^2_a(G)$ is called the *Bergman space* for G.

Several alternatives for the integral with respect to two-dimensional Lebesgue measure will be used. In addition to $\int\int_G f(x + iy) \, dx \, dy$ we will also see

$$\int\int_G f \quad \text{and} \quad \int_G f \, d\text{Area}.$$

Note that $L^2_a(G) \subseteq L^2(\mu)$, where $\mu = \text{Area}|G$, so that $L^2_a(G)$ has a natural inner product and norm from $L^2(\mu)$.

1.11. Lemma. *If f is analytic in a neighborhood of $\overline{B}(a; r)$, then*

$$f(a) = \frac{1}{\pi r^2} \int\int_{B(a; r)} f.$$

[Here $B(a; r) \equiv \{z: |z - a| < r\}$ and $\overline{B}(a; r) \equiv \{z: |z - a| \leq r\}$.]

PROOF. By the mean value property, if $0 < t \leq r$, $f(a) = (1/2\pi)\int_{-\pi}^{\pi} f(a + te^{i\theta}) \, d\theta$. Hence

$$(\pi r^2)^{-1}\int\int_{B(a; r)} f = (\pi r^2)^{-1}\int_0^r t\left[\int_{-\pi}^{\pi} f(a + te^{i\theta}) \, d\theta\right] dt$$

$$= (2/r^2)\int_0^r tf(a) \, dt = f(a). \quad \blacksquare$$

1.12. Corollary. *If $f \in L^2_a(G)$, $a \in G$, and $0 < r < \text{dist}(a, \partial G)$, then*

$$|f(a)| \leq \frac{1}{r\sqrt{\pi}} \|f\|_2.$$

PROOF. Since $\bar{B}(a;r) \subseteq G$, the preceding lemma and the CBS inequality imply

$$
\begin{aligned}
|f(a)| &= \frac{1}{\pi r^2} \left| \int\int_{B(a;r)} f \cdot 1 \right| \\
&\leq \frac{1}{\pi r^2} \left[\int\int_{B(a;r)} |f|^2 \right]^{1/2} \left[\int\int_{B(a;r)} 1^2 \right]^{1/2} \\
&\leq \frac{1}{\pi r^2} \|f\|_2 r\sqrt{\pi} . \quad \blacksquare
\end{aligned}
$$

1.13. Proposition. $L_a^2(G)$ *is a Hilbert space.*

PROOF. If μ = area measure on G, then $L^2(\mu)$ is a Hilbert space and $L_a^2(G) \subseteq L^2(\mu)$. So it suffices to show that $L_a^2(G)$ is closed in $L^2(\mu)$. Let $\{f_n\}$ be a sequence in $L_a^2(G)$ and let $f \in L^2(\mu)$ such that $\int |f_n - f|^2 \, d\mu \to 0$ as $n \to \infty$.

Suppose $\bar{B}(a;r) \subseteq G$ and let $0 < \rho < \text{dist}(B(a;r), \partial G)$. By the preceding corollary there is a constant C such that $|f_n(z) - f_m(z)| \leq C\|f_n - f_m\|_2$ for all n, m and for $|z - a| \leq \rho$. Thus $\{f_n\}$ is a uniformly Cauchy sequence on any closed disk in G. By standard results from analytic function theory (Montel's Theorem or Morera's Theorem, for example), there is an analytic function g on G such that $f_n(z) \to g(z)$ uniformly on compact subsets of G. But since $\int |f_n - f|^2 \, d\mu \to 0$, a result of Riesz implies there is a subsequence $\{f_{n_k}\}$ such that $f_{n_k}(z) \to f(z)$ a.e. $[\mu]$. Thus $f = g$ a.e. $[\mu]$ and so $f \in L_a^2(\mu)$. $\quad \blacksquare$

EXERCISES

1. Verify the statements made in Example 1.2.

2. Verify that $l^2(I)$ (Example 1.7) is a Hilbert space.

3. Show that the space $\mathscr{H}$ in Example 1.8 is a Hilbert space.

4. Describe the Hilbert spaces obtained by completing the space $\mathscr{X}$ in Example 1.2 with respect to the norm defined by each of the inner products given there.

5. (A variation on Example 1.8) Let $n \geq 2$ and let $\mathscr{H}$ = the collection of all functions $f: [0,1] \to \mathbb{F}$ such that (a) $f(0) = 0$; (b) for $1 \leq k \leq n-1$, $f^{(k)}(t)$ exists for all t in $[0,1]$ and $f^{(k)}$ is continuous on $[0,1]$; (c) $f^{(n-1)}$ is absolutely continuous and $f^{(n)} \in L^2(0,1)$. For f and g in $\mathscr{H}$, define

$$
\langle f, g \rangle = \sum_{k=1}^{n} \int_0^1 f^{(k)}(t) \overline{g^{(k)}(t)} \, dt.
$$

Show that $\mathscr{H}$ is a Hilbert space.

6. Let u be a semi-inner product on $\mathscr{X}$ and put $\mathscr{N} = \{x \in \mathscr{X} : u(x,x) = 0\}$.

   (a) Show that $\mathscr{N}$ is a linear subspace of $\mathscr{X}$.

(b) Show that if

$$\langle x + \mathcal{N}, y + \mathcal{N} \rangle \equiv u(x, y)$$

for all $x + \mathcal{N}$ and $y + \mathcal{N}$ in the quotient space $\mathcal{X}/\mathcal{N}$, then $\langle \cdot, \cdot \rangle$ is a well-defined inner product on $\mathcal{X}/\mathcal{N}$.

7. Let $\mathcal{H}$ be a Hilbert space over $\mathbb{R}$ and show that there is a Hilbert space $\mathcal{X}$ over $\mathbb{C}$ and a map $U: \mathcal{H} \to \mathcal{X}$ such that (a) U is linear; (b) $\langle Uh_1, Uh_2 \rangle = \langle h_1, h_2 \rangle$ for all h_1, h_2 in $\mathcal{H}$; (c) for any k in $\mathcal{X}$ there are unique h_1, h_2 in $\mathcal{H}$ such that $k = Uh_1 + iUh_2$. ($\mathcal{X}$ is called the *complexification* of $\mathcal{H}$.)

8. If $G = \{z \in \mathbb{C}: 0 < |z| < 1\}$ show that every f in $L_a^2(G)$ has a removable singularity at $z = 0$.

9. Which functions are in $L_a^2(\mathbb{C})$?

10. Let G be an open subset of $\mathbb{C}$ and show that if $a \in G$, then $\{f \in L_a^2(G): f(a) = 0\}$ is closed in $L_a^2(G)$.

11. If $\{h_n\}$ is a sequence in a Hilbert space $\mathcal{H}$ such that $\sum_n \|h_n\| < \infty$, then show that $\sum_{n=1}^{\infty} h_n$ converges in $\mathcal{H}$.

§2. Orthogonality

The greatest advantage of a Hilbert space is its underlying concept of orthogonality.

2.1. Definition. If $\mathcal{H}$ is a Hilbert space and $f, g \in \mathcal{H}$, then f and g are *orthogonal* if $\langle f, g \rangle = 0$. In symbols, $f \perp g$. If $A, B \subseteq \mathcal{H}$, then $A \perp B$ if $f \perp g$ for every f in A and g in B.

If $\mathcal{H} = \mathbb{R}^2$, this is the correct concept. Two non-zero vectors in $\mathbb{R}^2$ are orthogonal precisely when the angle between them is $\pi/2$.

2.2. The Pythagorean Theorem. *If $f_1, f_2, \ldots, f_n$ are pairwise orthogonal vectors in $\mathcal{H}$, then*

$$\|f_1 + f_2 + \cdots + f_n\|^2 = \|f_1\|^2 + \|f_2\|^2 + \cdots + \|f_n\|^2.$$

PROOF. If $f_1 \perp f_2$, then

$$\|f_1 + f_2\|^2 = \langle f_1 + f_2, f_1 + f_2 \rangle = \|f_1\|^2 + 2\operatorname{Re}\langle f_1, f_2 \rangle + \|f_2\|^2$$

by the polar identity. Since $f_1 \perp f_2$, this implies the result for $n = 2$. The remainder of the proof proceeds by induction and is left to the reader. ∎

Note that if $f \perp g$, then $f \perp -g$, so $\|f - g\|^2 = \|f\|^2 + \|g\|^2$. The next result is an easy consequence of the Pythagorean Theorem if f and g are orthogonal, but this assumption is not needed for its conclusion.

2.3.   Parallelogram Law. *If $\mathscr{H}$ is a Hilbert space and f and $g \in \mathscr{H}$, then*

$$\|f + g\|^2 + \|f - g\|^2 = 2(\|f\|^2 + \|g\|^2).$$

PROOF. For any f and g in $\mathscr{H}$ the polar identity implies

$$\|f + g\|^2 = \|f\|^2 + 2\operatorname{Re}\langle f, g \rangle + \|g\|^2,$$
$$\|f - g\|^2 = \|f\|^2 - 2\operatorname{Re}\langle f, g \rangle + \|g\|^2.$$

Now add.   ∎

The next property of a Hilbert space is truly pivotal. But first we need a geometric concept valid for any vector space over $\mathbb{F}$.

2.4.   Definition. If $\mathscr{X}$ is any vector space over $\mathbb{F}$ and $A \subseteq \mathscr{X}$, then A is a *convex set* if for any x and y in A and $0 \le t \le 1$, $tx + (1 - t)y \in A$.

Note that $\{tx + (1 - t)y: 0 \le t \le 1\}$ is the straight-line segment joining x and y. So a convex set is a set A such that if x and $y \in A$, the entire line segment joining x and y is contained in A.

If $\mathscr{X}$ is a vector space, then any linear subspace in $\mathscr{X}$ is a convex set. A singleton set is convex. The intersection of any collection of convex sets is convex. If $\mathscr{H}$ is a Hilbert space, then every open ball $B(f; r) = \{g \in \mathscr{H}: \|f - g\| < r\}$ is convex, as is every closed ball.

2.5.   Theorem. *If $\mathscr{H}$ is a Hilbert space, K is a closed convex nonempty subset of $\mathscr{H}$, and $h \in \mathscr{H}$, then there is a unique point k_0 in K such that*

$$\|h - k_0\| = \operatorname{dist}(h, K) \equiv \inf\{\|h - k\|: k \in K\}.$$

PROOF. By considering $K - h \equiv \{k - h: k \in K\}$ instead of K, it suffices to assume that $h = 0$. (Verify!) So we want to show that there is a unique vector k_0 in K such that

$$\|k_0\| = \operatorname{dist}(0, K) \equiv \inf\{\|k\|: k \in K\}.$$

Let $d = \operatorname{dist}(0, K)$. By definition, there is a sequence $\{k_n\}$ in K such that $\|k_n\| \to d$. Now the Parallelogram Law implies that

$$\left\|\frac{k_n - k_m}{2}\right\|^2 = \tfrac{1}{2}(\|k_n\|^2 + \|k_m\|^2) - \left\|\frac{k_n + k_m}{2}\right\|^2.$$

Since K is convex, $\tfrac{1}{2}(k_n + k_m) \in K$. Hence, $\|\tfrac{1}{2}(k_n + k_m)\|^2 \ge d^2$. If $\varepsilon > 0$, choose N such that for $n \ge N$, $\|k_n\|^2 < d^2 + \tfrac{1}{4}\varepsilon^2$. By the equation above, if $n, m \ge N$, then

$$\left\|\frac{k_n - k_m}{2}\right\|^2 < \tfrac{1}{2}(2d^2 + \tfrac{1}{2}\varepsilon^2) - d^2 = \tfrac{1}{4}\varepsilon^2.$$

Thus, $\|k_n - k_m\| < \varepsilon$ for $n, m \ge N$ and $\{k_n\}$ is a Cauchy sequence. Since $\mathscr{H}$ is complete and K is closed, there is a k_0 in K such that $\|k_n - k_0\| \to 0$.

Also for all k_n,

$$d \leq \|k_0\| = \|k_0 - k_n + k_n\|$$
$$\leq \|k_0 - k_n\| + \|k_n\| \to d.$$

Thus $\|k_0\| = d$.

To prove that k_0 is unique, suppose $h_0 \in K$ such that $\|h_0\| = d$. By convexity, $\frac{1}{2}(k_0 + h_0) \in K$. Hence,

$$d \leq \|\tfrac{1}{2}(h_0 + k_0)\| \leq \tfrac{1}{2}(\|h_0\| + \|k_0\|) \leq d.$$

So $\|\frac{1}{2}(h_0 + k_0)\| = d$. The Parallelogram Law implies

$$d^2 = \left\| \frac{h_0 + k_0}{2} \right\|^2 = d^2 - \left\| \frac{h_0 - k_0}{2} \right\|^2;$$

hence $h_0 = k_0$. ∎

If the convex set in the preceding theorem is in fact a closed linear subspace of $\mathcal{H}$, more can be said.

2.6. Theorem. *If $\mathcal{M}$ is a closed linear subspace of $\mathcal{H}$, $h \in \mathcal{H}$, and f_0 is the unique element of $\mathcal{M}$ such that $\|h - f_0\| = \mathrm{dist}(h, \mathcal{M})$, then $h - f_0 \perp \mathcal{M}$. Conversely, if $f_0 \in \mathcal{M}$ such that $h - f_0 \perp \mathcal{M}$, then $\|h - f_0\| = \mathrm{dist}(h, \mathcal{M})$.*

PROOF. Suppose $f_0 \in \mathcal{M}$ and $\|h - f_0\| = \mathrm{dist}(h, \mathcal{M})$. If $f \in \mathcal{M}$, then $f_0 + f \in \mathcal{M}$ and so $\|h - f_0\|^2 \leq \|h - (f_0 + f)\|^2 = \|(h - f_0) - f\|^2 = \|h - f_0\|^2 - 2\,\mathrm{Re}\langle h - f_0, f \rangle + \|f\|^2$. Thus

$$2\,\mathrm{Re}\langle h - f_0, f \rangle \leq \|f\|^2$$

for any f in $\mathcal{M}$. Fix f in $\mathcal{M}$ and substitute $te^{i\theta}f$ for f in the preceding inequality, where $\langle h - f_0, f \rangle = re^{i\theta}$, $r \geq 0$. This yields $2\,\mathrm{Re}\{te^{-i\theta}re^{i\theta}\} \leq t^2\|f\|^2$, or $2tr \leq t^2\|f\|$. Letting $t \to 0$, we see that $r = 0$; that is, $h - f_0 \perp f$.

For the converse, suppose $f_0 \in \mathcal{M}$ such that $h - f_0 \perp \mathcal{M}$. If $f \in \mathcal{M}$, then $h - f_0 \perp f_0 - f$ so that

$$\|h - f\|^2 = \|(h - f_0) + (f_0 - f)\|^2$$
$$= \|h - f_0\|^2 + \|f_0 - f\|^2$$
$$\geq \|h - f_0\|^2.$$

Thus $\|h - f_0\| = \mathrm{dist}(h, \mathcal{M})$. ∎

If $A \subseteq \mathcal{H}$, let $A^\perp \equiv \{f \in \mathcal{H}: f \perp g \text{ for all } g \text{ in } A\}$. It is easy to see that $A^\perp$ is a closed linear subspace of $\mathcal{H}$.

Note that Theorem 2.6, together with the uniqueness statement in Theorem 2.5, shows that if $\mathcal{M}$ is a closed linear subspace of $\mathcal{H}$ and $h \in \mathcal{H}$, then there is a unique element f_0 in $\mathcal{M}$ such that $h - f_0 \in \mathcal{M}^\perp$. Thus a function $P: \mathcal{H} \to \mathcal{M}$ can be defined by $Ph = f_0$.

2.7.   Theorem. *If $\mathcal{M}$ is a closed linear subspace of $\mathcal{H}$ and $h \in \mathcal{H}$, let Ph be the unique point in $\mathcal{M}$ such that $h - Ph \perp \mathcal{M}$. Then*

(a) *P is a linear transformation on $\mathcal{H}$,*
(b) *$\|Ph\| \leq \|h\|$ for every h in $\mathcal{H}$,*
(c) *$P^2 = P$ (here P^2 means the composition of P with itself),*
(d) *$\ker P = \mathcal{M}^{\perp}$ and $\operatorname{ran} P = \mathcal{M}$.*

PROOF. Keep in mind that for every h in $\mathcal{H}$, $h - Ph \in \mathcal{M}^{\perp}$ and $\|h - Ph\| = \operatorname{dist}(h, \mathcal{M})$.

(a) Let $h_1, h_2 \in \mathcal{H}$ and $\alpha_1, \alpha_2 \in \mathbb{F}$. If $f \in \mathcal{M}$, then $\langle [\alpha_1 h_1 + \alpha_2 h_2] - [\alpha_1 Ph_1 + \alpha_2 Ph_2], f \rangle = \alpha_1 \langle h_1 - Ph_1, f \rangle + \alpha_2 \langle h_2 - Ph_2, f \rangle = 0$. By the uniqueness statement of (2.6), $P(\alpha h_1 + \alpha_2 h_2) = \alpha_1 Ph_1 + \alpha_2 Ph_2$.
(b) If $h \in \mathcal{H}$, then $h = (h - Ph) + Ph$, $Ph \in \mathcal{M}$, and $h - Ph \in \mathcal{M}^{\perp}$. Thus $\|h\|^2 = \|h - Ph\|^2 + \|Ph\|^2 \geq \|Ph\|^2$.
(c) If $f \in \mathcal{M}$, then $Pf = f$. For any h in $\mathcal{H}$, $Ph \in \mathcal{M}$; hence $P^2 h \equiv P(Ph) = Ph$. That is, $P^2 = P$.
(d) If $Ph = 0$, then $h = h - Ph \in \mathcal{M}^{\perp}$. Conversely, if $h \in \mathcal{M}^{\perp}$, then 0 is the unique vector in $\mathcal{M}$ such that $h - 0 = h \perp \mathcal{M}$. Therefore $Ph = 0$. That $\operatorname{ran} P = \mathcal{M}$ is clear.   ■

2.8.   Definition. If $\mathcal{M}$ is a closed linear subspace of $\mathcal{H}$ and P is the linear map defined in the preceding theorem, then P is called the *orthogonal projection* of $\mathcal{H}$ onto $\mathcal{M}$. If we wish to show this dependence of P on $\mathcal{M}$, we will denote the orthogonal projection of $\mathcal{H}$ onto $\mathcal{M}$ by $P_{\mathcal{M}}$.

It also seems appropriate to introduce the notation $\mathcal{M} \leq \mathcal{H}$ to signify that $\mathcal{M}$ is a closed linear subspace of $\mathcal{H}$. We will use the term *linear manifold* to designate a linear subspace of $\mathcal{H}$ that is not necessarily closed. A *linear subspace* of $\mathcal{H}$ will always mean a closed linear subspace.

2.9.   Corollary. *If $\mathcal{M} \leq \mathcal{H}$, then $(\mathcal{M}^{\perp})^{\perp} = \mathcal{M}$.*

PROOF. If I is used to designate the identity operator on $\mathcal{H}$ (viz., $Ih = h$) and $P = P_{\mathcal{M}}$, then $I - P$ is the orthogonal projection of $\mathcal{H}$ onto $\mathcal{M}^{\perp}$ (Exercise 2). By part (d) of the preceding theorem, $(\mathcal{M}^{\perp})^{\perp} = \ker(I - P)$. But $0 = (I - P)h$ iff $h = Ph$. Thus $(\mathcal{M}^{\perp})^{\perp} = \ker(I - P) = \operatorname{ran} P = \mathcal{M}$.
                                                                                          ■

2.10.   Corollary. *If $A \subseteq \mathcal{H}$, then $(A^{\perp})^{\perp}$ is the closed linear span of A in $\mathcal{H}$.*

The proof is left to the reader; see Exercise 4 for a discussion of the term "closed linear span."

2.11.   Corollary. *If $\mathcal{Y}$ is a linear manifold in $\mathcal{H}$, then $\mathcal{Y}$ is dense in $\mathcal{H}$ iff $\mathcal{Y}^{\perp} = (0)$.*

PROOF. Exercise.

EXERCISES

1. Let $\mathcal{H}$ be a Hilbert space and suppose $f, g \in \mathcal{H}$ with $\|f\| = \|g\| = 1$. Show that $\|tf + (1 - t)g\| < 1$ for $0 < t < 1$. What does this say about $\{h \in \mathcal{H}: \|h\| \leq 1\}$?

2. If $\mathcal{M} \leq \mathcal{H}$ and $P = P_{\mathcal{M}}$, show that $I - P$ is the orthogonal projection of $\mathcal{H}$ onto $\mathcal{M}^{\perp}$.

3. If $\mathcal{M} \leq \mathcal{H}$, show that $\mathcal{M} \cap \mathcal{M}^{\perp} = (0)$ and every h in $\mathcal{H}$ can be written as $h = f + g$ where $f \in \mathcal{M}$ and $g \in \mathcal{M}^{\perp}$. If $\mathcal{M} \dotplus \mathcal{M}^{\perp} \equiv \{(f, g): f \in \mathcal{M}, g \in \mathcal{M}^{\perp}\}$ and $T: \mathcal{M} \dotplus \mathcal{M}^{\perp} \to \mathcal{H}$ is defined by $T(f, g) = f + g$, show that T is a linear bijection and a homeomorphism if $\mathcal{M} \dotplus \mathcal{M}^{\perp}$ is given the product topology. (This is usually phrased by stating that $\mathcal{M}$ and $\mathcal{M}^{\perp}$ are *topologically complementary* in $\mathcal{H}$.)

4. If $A \subseteq \mathcal{H}$, let $\bigvee A \equiv$ the intersection of all closed linear subspaces of $\mathcal{H}$ that contain A. $\bigvee A$ is called the *closed linear span* of A. Prove the following:

   (a) $\bigvee A \leq \mathcal{H}$ and $\bigvee A$ is the smallest closed linear subspace of $\mathcal{H}$ that contains A.
   (b) $\bigvee A = $ the closure of $\{\sum_{k=1}^{n} \alpha_k f_k: n \geq 1, \quad \alpha_k \in \mathbb{F}, \quad f_k \in A\}$.

5. Prove Corollary 2.10.

6. Prove Corollary 2.11.

§3. The Riesz Representation Theorem

The title of this section is somewhat ambiguous as there are at least two Riesz Representation Theorems. There is one so-called theorem that represents bounded linear functionals on the space of continuous functions on a compact Hausdorff space. That theorem will be discussed later in this book. The present section deals with the representation of certain linear functionals on Hilbert space. But first we have a few preliminaries to dispose of.

3.1. Proposition. *Let $\mathcal{H}$ be a Hilbert space and $L: \mathcal{H} \to \mathbb{F}$ a linear functional. The following statements are equivalent.*

(a) *L is continuous.*
(b) *L is continuous at 0.*
(c) *L is continuous at some point.*
(d) *There is a constant $c > 0$ such that $|L(h)| \leq c\|h\|$ for every h in $\mathcal{H}$.*

PROOF. It is clear that (a) $\Rightarrow$ (b) $\Rightarrow$ (c) and (d) $\Rightarrow$ (b). Let's show that (c) $\Rightarrow$ (a) and (b) $\Rightarrow$ (d).
    (c) $\Rightarrow$ (a): Suppose L is continuous at h_0 and h is any point in $\mathcal{H}$. If $h_n \to h$ in $\mathcal{H}$, then $h_n - h + h_0 \to h_0$. By assumption, $L(h_0) = \lim[L(h_n - h + h_0)] = \lim[L(h_n) - L(h) + L(h_0)] = \lim L(h_n) - L(h) + L(h_0)$. Hence $L(h) = \lim L(h_n)$.

(b) $\Rightarrow$ (d): The definition of continuity at 0 implies that $L^{-1}(\{\alpha \in \mathbb{F}:$ $|\alpha| < 1\})$ contains an open ball about 0. So there is a $\delta > 0$ such that $B(0; \delta) \subseteq L^{-1}(\{\alpha \in \mathbb{F}: |\alpha| < 1\})$. That is, $\|h\| < \delta$ implies $|L(h)| < 1$. If h is an arbitrary element of $\mathcal{H}$ and $\varepsilon > 0$, then $\|\delta(\|h\| + \varepsilon)^{-1}h\| < \delta$. Hence

$$1 > \left| L\left[\frac{\delta h}{\|h\| + \varepsilon} \right] \right| = \frac{\delta}{\|h\| + \varepsilon}|L(h)|;$$

thus,

$$|L(h)| < \frac{1}{\delta}(\|h\| + \varepsilon).$$

Letting $\varepsilon \to 0$ we see that (d) holds with $c = 1/\delta$. ∎

3.2.   Definition. A *bounded linear functional* L on $\mathcal{H}$ is a linear functional for which there is a constant $c > 0$ such that $|L(h)| \le c\|h\|$ for all h in $\mathcal{H}$. In light of the preceding proposition, a linear functional is bounded if and only if it is continuous.

For a bounded linear functional $L: \mathcal{H} \to \mathbb{F}$, define

$$\|L\| = \sup\{|L(h)|: \|h\| \le 1\}.$$

Note that by definition, $\|L\| < \infty$; $\|L\|$ is called the *norm* of L.

3.3.   Proposition. *If L is a linear functional, then*
$$\begin{aligned}
\|L\| &= \sup\{|L(h)|: \|h\| = 1\} \\
&= \sup\{|L(h)|/\|h\|: h \in \mathcal{H}, h \ne 0\} \\
&= \inf\{c > 0: |L(h)| \le c\|h\|, h \text{ in } \mathcal{H}\}.
\end{aligned}$$
Also, $|L(h)| \le \|L\| \|h\|$ for every h in $\mathcal{H}$.

PROOF. Let $\alpha = \inf\{c > 0: \|L(h)\| \le c\|h\|, h \text{ in } \mathcal{H}\}$. It will be shown that $\|L\| = \alpha$; the remaining equalities are left as an exercise. If $\varepsilon > 0$, then the definition of $\|L\|$ shows that $|L((\|h\| + \varepsilon)^{-1}h)| \le \|L\|$. Hence $|L(h)| \le \|L\|(\|h\| + \varepsilon)$. Letting $\varepsilon \to 0$ shows that $|L(h)| \le \|L\| \|h\|$ for all h. So the definition of α shows that $\alpha \le \|L\|$. On the other hand, if $|L(h)| \le c\|h\|$ for all h, then $\|L\| \le c$. Hence $\|L\| \le \alpha$. ∎

Fix an h_0 in $\mathcal{H}$ and define $L: \mathcal{H} \to \mathbb{F}$ by $L(h) = \langle h, h_0 \rangle$. It is easy to see that L is linear. Also, the CBS inequality gives that $|L(h)| = |\langle h, h_0 \rangle| \le \|h\| \|h_0\|$. So L is bounded and $\|L\| \le \|h_0\|$. In fact, $L(h_0/\|h_0\|) = \langle h_0/\|h_0\|, h_0 \rangle = \|h_0\|$, so that $\|L\| = \|h_0\|$. The main result of this section provides a converse to these observations.

3.4.   The Riesz Representation Theorem. *If $L: \mathcal{H} \to \mathbb{F}$ is a bounded linear functional, then there is a unique vector h_0 in $\mathcal{H}$ such that $L(h) = \langle h, h_0 \rangle$ for every h in $\mathcal{H}$. Moreover, $\|L\| = \|h_0\|$.*

PROOF. Let $\mathcal{M} = \ker L$. Because L is continuous, $\mathcal{M}$ is a closed linear subspace of $\mathcal{H}$. Since we may assume that $\mathcal{M} \neq \mathcal{H}$, $\mathcal{M}^{\perp} \neq (0)$. Hence there is a vector f_0 in $\mathcal{M}^{\perp}$ such that $L(f_0) = 1$. Now if $h \in \mathcal{H}$ and $\alpha = L(h)$, then $L(h - \alpha f_0) = L(h) - \alpha = 0$; so $h - L(h)f_0 \in \mathcal{M}$. Thus

$$0 = \langle h - L(h)f_0, f_0 \rangle$$
$$= \langle h, f_0 \rangle - L(h)\|f_0\|^2.$$

So if $h_0 = \|f_0\|^{-2}f_0$, $L(h) = \langle h, h_0 \rangle$ for all h in $\mathcal{H}$.

If $h_0' \in \mathcal{H}$ such that $\langle h, h_0 \rangle = \langle h, h_0' \rangle$ for all h, then $h_0 - h_0' \perp \mathcal{H}$. In particular, $h_0 - h_0' \perp h_0 - h_0'$ and so $h_0' = h_0$. The fact that $\|L\| = \|h_0\|$ was shown in the discussion preceding the theorem. ∎

3.5. Corollary. *If (X, Ω, μ) is a measure space and $F: L^2(\mu) \to \mathbb{F}$ is a bounded linear functional, then there is a unique h_0 in $L^2(\mu)$ such that*

$$F(h) = \int h\overline{h_0}\, d\mu$$

for every h in $L^2(\mu)$.

Of course the preceding corollary is a special case of the theorem on representing bounded linear functionals on $L^p(\mu)$, $1 \leq p < \infty$. But it is interesting to note that it is a consequence of the result for Hilbert space [and the result that $L^2(\mu)$ is a Hilbert space].

EXERCISES

1. Prove Proposition 3.3.

2. Let $\mathcal{H} = l^2(\mathbb{N})$. If $N \geq 1$ and $L: \mathcal{H} \to \mathbb{F}$ is defined by $L(\{\alpha_n\}) = \alpha_N$, find the vector h_0 in $\mathcal{H}$ such that $L(h) = \langle h, h_0 \rangle$ for every h in $\mathcal{H}$.

3. Let $\mathcal{H} = l^2(\mathbb{N} \cup \{0\})$. (a) Show that if $\{\alpha_n\} \in l^2$, then the power series $\sum_{n=0}^{\infty} \alpha_n z^n$ has radius of convergence ≥ 1. (b) If $|\lambda| < 1$ and $L: \mathcal{H} \to \mathbb{C}$ is defined by $L(\{\alpha_n\}) = \sum_{n=0}^{\infty} \alpha_n \lambda^n$, find the vector h_0 in $\mathcal{H}$ such that $L(h) = \langle h, h_0 \rangle$ for every h in $\mathcal{H}$. (c) What is the norm of the linear functional L defined in (b)?

4. With the notation as in Exercise 3, define $L: \mathcal{H} \to \mathbb{C}$ by $L(\{\alpha_n\}) = \sum_{n=1}^{\infty} n\alpha_n \lambda^{n-1}$, where $|\lambda| < 1$. Find a vector h_0 in $\mathcal{H}$ such that $L(h) = \langle h, h_0 \rangle$ for every h in $\mathcal{H}$.

5. Let $\mathcal{H}$ be the Hilbert space described in Example 1.8. If $0 < t \leq 1$, define $L: \mathcal{H} \to \mathbb{F}$ by $L(h) = h(t)$. Show that L is a bounded linear functional, find $\|L\|$, and find the vector h_0 in $\mathcal{H}$ such that $L(h) = \langle h, h_0 \rangle$ for all h in $\mathcal{H}$.

6. Let $\mathcal{H} = L^2(0, 1)$ and let $C^{(1)}$ be the set of all continuous functions on $[0, 1]$ that have a continuous derivative. Let $t \in [0, 1]$ and define $L: C^{(1)} \to \mathbb{F}$ by $L(h) = h'(t)$. Show that there is no bounded linear functional on $\mathcal{H}$ that agrees with L on $C^{(1)}$.

§4. Orthonormal Sets of Vectors and Bases

It will be shown in this section that, as in Euclidean space, each Hilbert space can be coordinatized. The vehicle for introducing the coordinates is an orthonormal basis. The corresponding vectors in $\mathbb{F}^d$ are the vectors $\{e_1, e_2, \ldots, e_d\}$, where e_k is the d-tuple having a 1 in the kth place and zeros elsewhere.

4.1. Definition. An *orthonormal* subset of a Hilbert space $\mathscr{H}$ is a subset $\mathscr{E}$ having the properties: (a) for e in $\mathscr{E}$, $\|e\| = 1$; (b) if $e_1, e_2 \in \mathscr{E}$ and $e_1 \neq e_2$, then $e_1 \perp e_2$.
A *basis* for $\mathscr{H}$ is a maximal orthonormal set.

Every vector space has a Hamel basis (a maximal linearly independent set). The term "basis" for a Hilbert space is defined as above and it relates to the inner product on $\mathscr{H}$. For an infinite-dimensional Hilbert space, a basis is never a Hamel basis. This is not obvious, but the reader will be able to see this after understanding several facts about bases.

4.2. Proposition. *If $\mathscr{E}$ is an orthonormal set in $\mathscr{H}$, then there is a basis for $\mathscr{H}$ that contains $\mathscr{E}$.*

The proof of this proposition is a straightforward application of Zorn's Lemma and is left to the reader.

4.3. Example. Let $\mathscr{H} = L^2_{\mathbb{C}}[0, 2\pi]$ and for n in $\mathbb{Z}$ define e_n in $\mathscr{H}$ by $e_n(t) = (2\pi)^{-1/2}\exp(\text{int})$. Then $\{e_n : n \in \mathbb{Z}\}$ is an orthonormal set in $\mathscr{H}$. (Here $L^2_{\mathbb{C}}[0, 2\pi]$ is the space of complex-valued square integrable functions.)

It is also true that the set in (4.3) is a basis, but this is best proved after a bit of theory.

4.4. Example. If $\mathscr{H} = \mathbb{F}^d$ and for $1 \leq k \leq d$, $e_k =$ the d-tuple with 1 in the kth place and zeros elsewhere, then $\{e_1, \ldots, e_d\}$ is a basis for $\mathscr{H}$.

4.5. Example. Let $\mathscr{H} = l^2(I)$ as in Example 1.7. For each i in I define e_i in $\mathscr{H}$ by $e_i(i) = 1$ and $e_i(j) = 0$ for $j \neq i$. Then $\{e_i : i \in I\}$ is a basis.

The proof of the next result is left as an exercise (see Exercise 5). It is very useful but the proof is not difficult.

4.6. The Gram–Schmidt Orthogonalization Process. *If $\mathscr{H}$ is a Hilbert space and $\{h_n : n \in \mathbb{N}\}$ is a linearly independent subset of $\mathscr{H}$, then there is an orthonormal set $\{e_n : n \in \mathbb{N}\}$ such that for every n, the linear span of $\{e_1, \ldots, e_n\}$ equals the linear span of $\{h_1, \ldots, h_n\}$.*

Remember that $\bigvee A$ is the closed linear span of A (Exercise 2.4).

4.7. Proposition. *Let* $\{e_1, \ldots, e_n\}$ *be an orthonormal set in* $\mathcal{H}$ *and let* $\mathcal{M} = \bigvee\{e_1, \ldots, e_n\}$. *If* P *is the orthogonal projection of* $\mathcal{H}$ *onto* $\mathcal{M}$, *then*

$$Ph = \sum_{k=1}^{n} \langle h, e_k \rangle e_k$$

for all h *in* $\mathcal{H}$.

PROOF. Let $Qh = \sum_{k=1}^{n} \langle h, e_k \rangle e_k$. If $1 \le j \le n$, then $\langle Qh, e_j \rangle = \sum_{k=1}^{n} \langle h, e_k \rangle \langle e_k, e_j \rangle = \langle h, e_j \rangle$ since $e_k \perp e_j$ for $k \ne j$. Thus $\langle h - Qh, e_j \rangle = 0$ for $1 \le j \le n$. That is, $h - Qh \perp \mathcal{M}$ for every h in $\mathcal{H}$. Since Qh is clearly a vector in $\mathcal{M}$, Qh is the unique vector h_0 in $\mathcal{M}$ such that $h - h_0 \perp \mathcal{M}$ (2.6). Hence $Qh = Ph$ for every h in $\mathcal{H}$. ∎

4.8. Bessel's Inequality. *If* $\{e_n : n \in \mathbb{N}\}$ *is an orthonormal set and* $h \in \mathcal{H}$, *then*

$$\sum_{n=1}^{\infty} |\langle h, e_n \rangle|^2 \le \|h\|^2.$$

PROOF. Let $h_n = h - \sum_{k=1}^{n} \langle h, e_k \rangle e_k$. Then $h_n \perp e_k$ for $1 \le k \le n$ (Why?). By the Pythagorean Theorem,

$$\|h\|^2 = \|h_n\|^2 + \left\| \sum_{k=1}^{n} \langle h, e_n \rangle e_k \right\|^2$$

$$= \|h_n\|^2 + \sum_{k=1}^{n} |\langle h, e_n \rangle|^2$$

$$\ge \sum_{k=1}^{n} |\langle h, e_k \rangle|^2.$$

Since n was arbitrary, the result is proved. ∎

4.9. Corollary. *If* $\mathscr{E}$ *is an orthonormal set in* $\mathcal{H}$ *and* $h \in \mathcal{H}$, *then* $\langle h, e \rangle \ne 0$ *for at most a countable number of vectors* e *in* $\mathscr{E}$.

PROOF. For each $n \ge 1$ let $\mathscr{E}_n = \{e \in \mathscr{E} : |\langle h, e \rangle| \ge 1/n\}$. By Bessel's Inequality, $\mathscr{E}_n$ is finite. But $\bigcup_{n=1}^{\infty} \mathscr{E}_n = \{e \in \mathscr{E} : \langle h, e_n \rangle \ne 0\}$. ∎

4.10. Corollary. *If* $\mathscr{E}$ *is an orthonormal set and* $h \in \mathcal{H}$, *then*

$$\sum_{e \in \mathscr{E}} |\langle h, e \rangle|^2 \le \|h\|^2.$$

This last corollary is just Bessel's Inequality together with the fact (4.9) that at most a countable number of the terms in the sum differ from zero.

Actually, the sum that appears in (4.10) can be given a better interpretation—a mathematically precise one that will be useful later. The question is, what is meant by $\Sigma\{h_i\colon i \in I\}$ if $h_i \in \mathcal{H}$ and I is an infinite, possibly uncountable, set? Let $\mathcal{F}$ be the collection of all finite subsets of I and order $\mathcal{F}$ by inclusion, so $\mathcal{F}$ becomes a directed set. For each F in $\mathcal{F}$, define

$$h_F = \sum\{h_i\colon i \in F\}.$$

Since this is a finite sum, h_F is a well-defined element of $\mathcal{H}$. Now $\{h_F\colon F \in \mathcal{F}\}$ is a net in $\mathcal{H}$.

4.11. Definition. With the notation above, the sum $\Sigma\{h_i\colon i \in I\}$ *converges* if the net $\{h_F\colon F \in \mathcal{F}\}$ converges; the value of the sum is the limit of the net.

If $\mathcal{H} = \mathbb{F}$, the definition above gives meaning to an uncountable sum of scalars. Now Corollary 4.10 can be given its precise meaning; namely, $\Sigma\{|\langle h, e \rangle|^2\colon e \in \mathcal{E}\}$ converges and the value $\leq \|h\|^2$ (Exercise 9).

If the set I in Definition 4.11 is countable, then this definition of convergent sum is not the usual one. That is, if $\{h_n\}$ is a sequence in $\mathcal{H}$, then the convergence of $\Sigma\{h_n\colon n \in \mathbb{N}\}$ is not equivalent to the convergence of $\sum_{n=1}^{\infty} h_n$. The former concept of convergence is that defined in (4.11) while the latter means that the sequence $\{\sum_{k=1}^{n} h_k\}_{n=1}^{\infty}$ converges. Even if $\mathcal{H} = \mathbb{F}$, these concepts do not coincide (see Exercise 12). If, however, $\Sigma\{h_n\colon n \in \mathbb{N}\}$ converges, then $\sum_{n=1}^{\infty} h_n$ converges (Exercise 10). Also see Exercise 11.

4.12. Lemma. *If $\mathcal{E}$ is an orthonormal set and $h \in \mathcal{H}$, then*

$$\sum\{\langle h, e \rangle e\colon e \in \mathcal{E}\}$$

converges in $\mathcal{H}$.

PROOF. By (4.9), there are vectors $e_1, e_2, \ldots$ in $\mathcal{E}$ such that $\{e \in \mathcal{E}\colon \langle h, e \rangle \neq 0\} = \{e_1, e_2, \ldots\}$. We also know that $\sum_{n=1}^{\infty}|\langle h, e_n \rangle|^2 \leq \|h\|^2 < \infty$. So if $\varepsilon > 0$, there is an N such that $\sum_{n=N}^{\infty}|\langle h, e_n \rangle|^2 < \varepsilon^2$. Let $F_0 = \{e_1, \ldots, e_{N-1}\}$ and let $\mathcal{F} = $ all the finite subsets of $\mathcal{E}$. For F in $\mathcal{F}$ define $h_F \equiv \Sigma\{\langle h, e \rangle e\colon e \in F\}$. If F and $G \in \mathcal{F}$ and both contain F_0, then

$$\|h_F - h_G\|^2 = \sum\{|\langle h, e \rangle|^2\colon e \in (F \backslash G) \cup (G \backslash F)\}$$
$$\leq \sum_{n=N}^{\infty}|\langle h, e_n \rangle|^2$$
$$< \varepsilon^2.$$

So $\{h_F\colon F \in \mathcal{F}\}$ is a Cauchy net in $\mathcal{H}$. Because $\mathcal{H}$ is complete, this net converges. In fact, it converges to $\sum_{n=1}^{\infty}\langle h, e_n \rangle e_n$. ∎

4.13. Theorem. *If $\mathcal{E}$ is an orthonormal set in $\mathcal{H}$, then the following statements are equivalent.*

(a) *$\mathcal{E}$ is a basis for $\mathcal{H}$.*
(b) *If $h \in \mathcal{H}$ and $h \perp \mathcal{E}$, then $h = 0$.*
(c) *$\bigvee \mathcal{E} = \mathcal{H}$.*

(d) *If $h \in \mathcal{H}$, then $h = \Sigma\{\langle h, e \rangle e: e \in \mathcal{E}\}$.*
(e) *If g and $h \in \mathcal{H}$, then*

$$\langle g, h \rangle = \Sigma\{\langle g, e \rangle \langle e, h \rangle: e \in \mathcal{E}\}.$$

(f) *If $h \in \mathcal{H}$, then $\|h\|^2 = \Sigma\{|\langle h, e \rangle|^2: e \in \mathcal{E}\}$ (Parseval's Identity).*

PROOF. (a) $\Rightarrow$ (b): Suppose $h \perp \mathcal{E}$ and $h \neq 0$; then $\mathcal{E} \cup \{h/\|h\|\}$ is an orthonormal set that properly contains $\mathcal{E}$, contradicting maximality.

(b) $\Leftrightarrow$ (c): By Corollary 2.11, $\bigvee \mathcal{E} = \mathcal{H}$ if and only if $\mathcal{E}^\perp = (0)$.

(b) $\Rightarrow$ (d): If $h \in \mathcal{H}$, then $f = h - \Sigma\{\langle h, e \rangle e: e \in \mathcal{E}\}$ is a well-defined vector by Lemma 4.12. If $e_1 \in \mathcal{E}$, then $\langle f, e_1 \rangle = \langle h, e_1 \rangle - \Sigma\{\langle h, e \rangle \langle e, e_1 \rangle: e \in \mathcal{E}\} = \langle h, e_1 \rangle - \langle h, e_1 \rangle = 0$. That is, $f \in \mathcal{E}^\perp$. Hence $f = 0$. (Is everything legitimate in that string of equalities? We don't want any illegitimate equalities.)

(d) $\Rightarrow$ (e): This is left as an exercise for the reader.

(e) $\Rightarrow$ (f): Since $\|h\|^2 = \langle h, h \rangle$, this is immediate.

(f) $\Rightarrow$ (a): If $\mathcal{E}$ is not a basis, then there is a unit vector e_0 ($\|e_0\| = 1$) in $\mathcal{H}$ such that $e_0 \perp \mathcal{E}$. Hence, $0 = \Sigma\{|\langle e_0, e \rangle|^2: e \in \mathcal{E}\}$, contradicting (f). ∎

Just as in finite-dimensional spaces, a basis in Hilbert space can be used to define a concept of dimension. For this purpose the next result is pivotal.

4.14. Proposition. *If $\mathcal{H}$ is a Hilbert space, any two bases have the same cardinality.*

PROOF. Let $\mathcal{E}$ and $\mathcal{F}$ be two bases for $\mathcal{H}$ and put $\varepsilon =$ the cardinality of $\mathcal{E}$, $\eta =$ the cardinality of $\mathcal{F}$. If ε or η is finite, then $\varepsilon = \eta$ (Exercise 15). Suppose both ε and η are infinite. For e in $\mathcal{E}$, let $\mathcal{F}_e \equiv \{f \in \mathcal{F}: \langle e, f \rangle \neq 0\}$; so $\mathcal{F}_e$ is countable. By (4.13b), each f in $\mathcal{F}$ belongs to at least one set $\mathcal{F}_e$, e in $\mathcal{E}$. That is, $\mathcal{F} = \bigcup\{\mathcal{F}_e: e \in \mathcal{E}\}$. Hence $\eta \leq \varepsilon \cdot \aleph_0 = \varepsilon$. Similarly, $\varepsilon \leq \eta$. ∎

4.15. Definition. The *dimension* of a Hilbert space is the cardinality of a basis and is denoted by $\dim \mathcal{H}$.

If (X, d) is a metric space that is separable and $\{B_i = B(x_i; \varepsilon_i): i \in I\}$ is a collection of pairwise disjoint open balls in X, then I must be countable. Indeed, if D is a countable dense subset of X, $B_i \cap D \neq \square$ for each i in I. Thus there is a point x_i in $B_i \cap D$. So $\{x_i: i \in I\}$ is a subset of D having the cardinality of I; thus I must be countable.

4.16. Proposition. *If $\mathcal{H}$ is an infinite-dimensional Hilbert space, then $\mathcal{H}$ is separable if and only if $\dim \mathcal{H} = \aleph_0$.*

PROOF. Let $\mathscr{E}$ be a basis for $\mathscr{H}$. If $e_1, e_2 \in \mathscr{E}$, then $\|e_1 - e_2\|^2 = \|e_1\|^2 + \|e_2\|^2 = 2$. Hence $\{ B(e; 1/\sqrt{2}): e \in \mathscr{E} \}$ is a collection of pairwise disjoint open balls in $\mathscr{H}$. From the discussion preceding this proposition, the assumption that $\mathscr{H}$ is separable implies $\mathscr{E}$ is countable. The converse is an exercise. ∎

EXERCISES

1. Verify the statements in Example 4.3.

2. Verify the statements in Example 4.4.

3. Verify the statements in Example 4.5.

4. Find an infinite orthonormal set in the Hilbert space of Example 1.8.

5. Using the notation of the Gram–Schmidt Orthogonalization Process, show that up to scalar multiple $e_1 = h_1/\|h_1\|$ and for $n \geq 2$, $e_n = \|h_n - f_n\|^{-1}(h_n - f_n)$, where f_n is the vector defined formally by

$$
f_n = \frac{-1}{\det\left[\langle h_i, h_j\rangle\right]_{i,j=1}^n} \det
\begin{bmatrix}
\langle h_1, h_1\rangle & \cdots & \langle h_{n-1}, h_1\rangle & \langle h_n, h_1\rangle \\
\vdots & & \vdots & \vdots \\
\langle h_1, h_{n-1}\rangle & \cdots & \langle h_{n-1}, h_{n-1}\rangle & \langle h_n, h_{n-1}\rangle \\
h_1 & \cdots & h_{n-1} & 0
\end{bmatrix}
$$

In the next three exercises, the reader is asked to apply the Gram–Schmidt Orthogonalization Process to a given sequence in a Hilbert space. A reference for this material is pp. 82–96 of Courant and Hilbert [1953].

6. If the sequence $1, x, x^2, \ldots$ is orthogonalized in $L^2(-1, 1)$, the sequence $e_n(x) = [\frac{1}{2}(2n + 1)]^{1/2} P_n(x)$ is obtained, where

$$
P_n(x) = \frac{1}{2^n n!}\left(\frac{d}{dx}\right)^n (x^2 - 1)^n.
$$

The functions $P_n(x)$ are called *Legendre polynomials*.

7. If the sequence $e^{-x^2/2}, xe^{-x^2/2}, x^2 e^{-x^2/2}, \ldots$ is orthogonalized in $L^2(-\infty, \infty)$, the sequence $e_n(x) = [2^n n! \sqrt{\pi}]^{-1/2} H_n(x) e^{-x^2/2}$ is obtained, where

$$
H_n(x) = (-1)^n e^{x^2}\left(\frac{d}{dx}\right)^n e^{-x^2}.
$$

The functions H_n are *Hermite polynomials* and satisfy $H_n'(x) = 2nH_{n-1}(x)$.

8. If the sequence $e^{-x/2}, xe^{-x/2}, x^2 e^{-x/2}, \ldots$ is orthogonalized in $L^2(0, \infty)$, the sequence $e_n(x) = e^{-x/2} L_n(x)/n!$ is obtained, where

$$
L_n(x) = e^x\left(\frac{d}{dx}\right)^n (x^n e^{-x}).
$$

The functions L_n are called *Laguerre polynomials*.

9. Prove Corollary 4.10 using Definition 4.11.

10. If $\{h_n\}$ is a sequence in Hilbert space and $\Sigma\{h_n: n \in \mathbb{N}\}$ converges to h (Definition 4.11), then $\lim_n \Sigma_{k=1}^n h_k = h$. Show that the converse is false.

11. If $\{h_n\}$ is a sequence in a Hilbert space and $\sum_{n=1}^{\infty}\|h_n\| < \infty$, show that $\Sigma\{h_n:$ $n \in \mathbb{N}\}$ converges in the sense of Definition 4.11.

12. Let $\{\alpha_n\}$ be a sequence in $\mathbb{F}$ and prove that the following statements are equivalent: (a) $\Sigma\{\alpha_n: n \in \mathbb{N}\}$ converges in the sense of Definition 4.11. (b) If π is any permutation of $\mathbb{N}$, then $\sum_{n=1}^{\infty}\alpha_{\pi(n)}$ converges (*unconditional convergence*). (c) $\sum_{n=1}^{\infty}|\alpha_n| < \infty$.

13. Let $\mathcal{E}$ be an orthonormal subset of $\mathcal{H}$ and let $\mathcal{M} = \vee\mathcal{E}$. If P is the orthogonal projection of $\mathcal{H}$ onto $\mathcal{M}$, show that $Ph = \Sigma\{\langle h,e\rangle e: e \in \mathcal{E}\}$ for every h in $\mathcal{H}$.

14. Let λ = Area measure on $\mathbb{D}$ and show that $1, z, z^2, \dots$ are orthogonal vectors in $L^2(\lambda)$. Find $\|z^n\|$, $n \geq 0$. If $e_n = \|z^n\|^{-1}z^n$, $n \geq 0$, is $\{e_0, e_1, \dots\}$ a basis for $L^2(\lambda)$?

15. In the proof of (4.14), show that if either ε or η is finite, then $\varepsilon = \eta$.

16. If $\mathcal{H}$ is an infinite-dimensional Hilbert space, show that no orthonormal basis for $\mathcal{H}$ is a Hamel basis. Show that a Hamel basis is uncountable.

17. Let $d \geq 1$ and let μ be a regular Borel measure on $\mathbb{R}^d$. Show that $L^2(\mu)$ is separable.

18. Suppose $L^2(X, \Omega, \mu)$ is separable and $\{E_i: i \in I\}$ is a collection of pairwise disjoint subsets of X, $E_i \in \Omega$, and $0 < \mu(E_i) < \infty$ for all i. Show that I is countable. Can you allow $\mu(E_i) = \infty$?

19. If $\{h \in \mathcal{H}: \|h\| \leq 1\}$ is compact, show that dim $\mathcal{H} < \infty$.

20. What is the cardinality of a Hamel basis for l^2?

§5. Isomorphic Hilbert Spaces and the Fourier Transform for the Circle

Every mathematical theory has its concept of isomorphism. In topology there is homeomorphism and homotopy equivalence; algebra calls them isomorphisms. The basic idea is to define a map which preserves the basic structure of the spaces in the category.

5.1. Definition. If $\mathcal{H}$ and $\mathcal{K}$ are Hilbert spaces, an *isomorphism* between $\mathcal{H}$ and $\mathcal{K}$ is a linear surjection $U: \mathcal{H} \to \mathcal{K}$ such that

$$\langle Uh, Ug\rangle = \langle h, g\rangle$$

for all h, g in $\mathcal{H}$. In this case $\mathcal{H}$ and $\mathcal{K}$ are said to be *isomorphic*.

It is easy to see that if $U: \mathcal{H} \to \mathcal{K}$ is an isomorphism, then so is $U^{-1}:$ $\mathcal{K} \to \mathcal{H}$. Similar such arguments show that the concept of "isomorphic" is an equivalence relation on Hilbert spaces. It is also certain that this is the correct equivalence relation since an inner product is the essential ingredient for a Hilbert space and isomorphic Hilbert spaces have the "same" inner product. One might object that completeness is another essential ingredient in the definition of a Hilbert space. So it is! However, this too is preserved by an isomorphism. An *isometry* between metric spaces is a map that preserves distance.

5.2. Proposition. *If $V: \mathscr{H} \to \mathscr{K}$ is a linear map between Hilbert spaces, then V is an isometry if and only if $\langle Vh, Vg \rangle = \langle h, g \rangle$ for all h, g in $\mathscr{H}$.*

PROOF. Assume $\langle Vh, Vg \rangle = \langle h, g \rangle$ for all h, g in $\mathscr{H}$. Then $\|Vh\|^2 = \langle Vh, Vh \rangle = \langle h, h \rangle = \|h\|^2$ and V is an isometry.

Now assume that V is an isometry. If $h, g \in \mathscr{H}$ and $\lambda \in \mathbb{F}$, then $\|h + \lambda g\|^2 = \|Vh + \lambda Vg\|^2$. Using the polar identity on both sides of this equation gives

$$\|h\|^2 + 2\operatorname{Re}\bar{\lambda}\langle h, g \rangle + |\lambda|^2 \|g\|^2 = \|Vh\|^2 + 2\operatorname{Re}\bar{\lambda}\langle Vh, Vg \rangle + |\lambda|^2 \|Vg\|^2.$$

But $\|Vh\| = \|h\|$ and $\|Vg\| = \|g\|$, so this equation becomes

$$\operatorname{Re}\bar{\lambda}\langle h, g \rangle = \operatorname{Re}\bar{\lambda}\langle Vh, Vg \rangle$$

for any λ in $\mathbb{F}$. If $\mathbb{F} = \mathbb{R}$, take $\lambda = 1$. If $\mathbb{F} = \mathbb{C}$, first take $\lambda = 1$ and then take $\lambda = i$ to find that $\langle h, g \rangle$ and $\langle Vh, Vg \rangle$ have the same real and imaginary parts. ∎

Note that an isometry between metric spaces maps Cauchy sequences into Cauchy sequences. Thus an isomorphism also preserves completeness. That is, if an inner product space is isomorphic to a Hilbert space, then it must be complete.

5.3. Example. Define $S: l^2 \to l^2$ by $S(\alpha_1, \alpha_2, \dots) = (0, \alpha_1, \alpha_2, \dots)$. Then S is an isometry that is not surjective.

The preceding example shows that isometries need not be isomorphisms.

A word about terminology. Many call what we call an isomorphism a *unitary operator*. We shall define a unitary operator as a linear transformation $U: \mathscr{H} \to \mathscr{H}$ that is a surjective isometry. That is, a unitary operator is an isomorphism whose range coincides with its domain. This may seem to be a minor distinction, and in many ways it is. But experience has taught me that there is some benefit in making such a distinction, or at least in being aware of it.

5.4. Theorem. *Two Hilbert spaces are isomorphic if and only if they have the same dimension.*

PROOF. If $U: \mathscr{H} \to \mathscr{K}$ is an isomorphism and $\mathscr{E}$ is a basis for $\mathscr{H}$, then it is easy to see that $U\mathscr{E} \equiv \{Ue: e \in \mathscr{E}\}$ is a basis for $\mathscr{K}$. Hence, dim $\mathscr{H} =$ dim $\mathscr{K}$.

Let $\mathcal{H}$ be a Hilbert space and let $\mathcal{E}$ be a basis for $\mathcal{H}$. Consider the Hilbert space $l^2(\mathcal{E})$. If $h \in \mathcal{H}$, define $\hat{h}: \mathcal{E} \to \mathbb{F}$ by $\hat{h}(e) = \langle h, e \rangle$. By Parseval's Identity $\hat{h} \in l^2(\mathcal{E})$ and $\|h\| = \|\hat{h}\|$. Define $U: \mathcal{H} \to l^2(\mathcal{E})$ by $Uh = \hat{h}$. Thus U is linear and an isometry. It is easy to see that ran U contains all the functions f in $l^2(\mathcal{E})$ such that $f(e) = 0$ for all but a finite number of e; that is, ran U is dense. But U, being an isometry, must have closed range. Hence $U: \mathcal{H} \to l^2(\mathcal{E})$ is an isomorphism.

If $\mathcal{K}$ is a Hilbert space with a basis $\mathcal{F}$, $\mathcal{K}$ is isomorphic to $l^2(\mathcal{F})$. If $\dim \mathcal{H} = \dim \mathcal{K}$, $\mathcal{E}$ and $\mathcal{F}$ have the same cardinality; it is easy to see that $l^2(\mathcal{E})$ and $l^2(\mathcal{F})$ must be isomorphic. Therefore $\mathcal{H}$ and $\mathcal{K}$ are isomorphic. ∎

5.5.   Corollary. *All separable infinite dimensional Hilbert spaces are isomorphic.*

This section concludes with a rather important example of an isomorphism, the Fourier transform on the circle.

The proof of the next result can be found as an Exercise on p. 263 of Conway [1978]. Another proof will be given later in this book after the Stone–Weierstrass Theorem is proved. So the reader can choose to assume this for the moment. Let $\mathbb{D} = \{z \in \mathbb{C}: |z| < 1\}$.

5.6.   Theorem. *If $f: \partial\mathbb{D} \to \mathbb{C}$ is a continuous function, then there is a sequence $\{p_n(z, \bar{z})\}$ of polynomials in z and $\bar{z}$ such that $p_n(z, \bar{z}) \to f(z)$ uniformly on $\partial\mathbb{D}$.*

Note that if $z \in \partial\mathbb{D}$, $\bar{z} = z^{-1}$. Thus a polynomial in z and $\bar{z}$ on $\partial\mathbb{D}$ becomes a function of the form

$$\sum_{k=-m}^{n} \alpha_k z^k.$$

If we put $z = e^{i\theta}$, this becomes a function of the form

$$\sum_{k=-m}^{n} \alpha_k e^{ik\theta}.$$

Such functions are called *trigonometric polynomials*.

We can now show that the orthonormal set in Example 4.3 is a basis for $L^2_\mathbb{C}[0, 2\pi]$. This is a rather important result.

5.7.   Theorem. *If for each n in $\mathbb{Z}$, $e_n(t) \equiv (2\pi)^{-1/2}\exp(int)$, then $\{e_n: n \in \mathbb{Z}\}$ is a basis for $L^2_\mathbb{C}[0, 2\pi]$.*

PROOF. Let $\mathcal{T} = \{\sum_{k=-n}^{n} \alpha_k e_k: \alpha_k \in \mathbb{C}, n \geq 0\}$. Then $\mathcal{T}$ is a subalgebra of $C_\mathbb{C}[0, 2\pi]$, the algebra of all continuous $\mathbb{C}$-valued functions on $[0, 2\pi]$. Note that if $f \in \mathcal{T}$, $f(0) = f(2\pi)$. We want to show that the uniform closure of $\mathcal{T}$

is $\mathscr{C} \equiv \{f \in C_{\mathbb{C}}[0, 2\pi]: f(0) = f(2\pi)\}$. To do this, let $f \in \mathscr{C}$ and define F: $\partial\mathbb{D} \to \mathbb{C}$ by $F(e^{it}) = f(t)$. F is continuous. (Why?) By (5.6) there is a sequence of polynomials in z and $\bar{z}$, $\{p_n(z, \bar{z})\}$, such that $p_n(z, \bar{z}) \to F(z)$ uniformly on $\partial\mathbb{D}$. Thus $p_n(e^{it}, e^{-it}) \to f(t)$ uniformly on $[0, 2\pi]$. But $p_n(e^{it}, e^{-it}) \in \mathscr{T}$.

Now the closure of $\mathscr{C}$ in $L^2_{\mathbb{C}}[0, 2\pi]$ is all of $L^2_{\mathbb{C}}[0, 2\pi]$ (Exercise 6). Hence $\bigvee\{e_n: n \in \mathbb{Z}\} = L^2_{\mathbb{C}}[0, 2\pi]$ and $\{e_n\}$ is thus a basis (4.13). ∎

Actually, it is usually preferred to normalize the measure on $[0, 2\pi]$. That is, replace dt by $(2\pi)^{-1} dt$, so that the total measure of $[0, 2\pi]$ is 1. Now define $e_n(t) = \exp(int)$. Hence $\{e_n: n \in \mathbb{Z}\}$ is a basis for $\mathscr{H} \equiv L^2_{\mathbb{C}}([0, 2\pi], (2\pi)^{-1} dt)$. If $f \in \mathscr{H}$, then

5.8
$$\hat{f}(n) \equiv \langle f, e_n \rangle = \frac{1}{2\pi} \int_0^{2\pi} f(t) e^{-int} \, dt$$

is called the nth *Fourier coefficient* of f, n in $\mathbb{Z}$. By (5.7) and (4.13d),

5.9
$$f = \sum_{n=-\infty}^{\infty} \hat{f}(n) e_n,$$

where this infinite series converges to f in the metric defined by the norm of $\mathscr{H}$. This is called the *Fourier series* of f. This terminology is classical and has been adopted for a general Hilbert space.

If $\mathscr{H}$ is any Hilbert space and $\mathscr{E}$ is a basis, the scalars $\{\langle h, e \rangle; e \in \mathscr{E}\}$ are called the *Fourier coefficients* of h (relative to $\mathscr{E}$) and the series in (4.13d) is called the *Fourier expansion* of h (relative to $\mathscr{E}$).

Note that Parseval's Identity applied to (5.9) gives that $\sum_{n=-\infty}^{\infty} |\hat{f}(n)|^2 < \infty$. This proves a classical result.

5.10. The Riemann–Lebesgue Lemma. *If $f \in L^2[0, 2\pi]$, then $\int_0^{2\pi} f(t) e^{-int} \, dt \to 0$ as $n \to \pm\infty$.*

If $f \in L^2_{\mathbb{C}}[0, 2\pi]$, then the Fourier series of f converges to f in L^2-norm. It was conjectured by Lusin that the series converges to f almost every-where. This was proved in Carleson [1966]. Hunt [1967] showed that if $f \in L^p_{\mathbb{C}}[0, 2\pi]$, $1 < p \leq \infty$, then the Fourier series also converges to f a.e. Long before that, Kolmogoroff had furnished an example of a function f in $L^1_{\mathbb{C}}[0, 2\pi]$ whose Fourier series does not converge to f a.e.

For f in $L^2_{\mathbb{C}}[0, 2\pi]$, the function $\hat{f}: \mathbb{Z} \to \mathbb{C}$ is called the *Fourier transform* of f; the map $U: L^2_{\mathbb{C}}[0, 2\pi] \to l^2(\mathbb{Z})$ defined by $Uf = \hat{f}$ is the *Fourier transform*. The results obtained so far can be applied to this situation to yield the following.

5.11. Theorem. *The Fourier transform is a linear isometry from $L^2_{\mathbb{C}}[0, 2\pi]$ onto $l^2(\mathbb{Z})$.*

PROOF. Let U: $L_{\mathbb{C}}^2[0, 2\pi] \to l^2(\mathbb{Z})$ be the Fourier transform. That U maps $L^2 \equiv L_{\mathbb{C}}^2[0, 2\pi]$ into $l^2(\mathbb{Z})$ and satisfies $\|Uf\| = \|f\|$ is a consequence of Parseval's Identity. That U is linear is an exercise. If $\{\alpha_n\} \in l^2(\mathbb{Z})$ and $\alpha_n = 0$ for all but a finite number of n, then $f = \sum_{n=-\infty}^{\infty} \alpha_n e_n \in L^2$. It is easy to check that $\hat{f}(n) = \alpha_n$ for all n, so $Uf = \{\alpha_n\}$. Thus ran U is dense in l^2. But U is an isometry, so ran U is closed; hence U is surjective. ∎

Note that functions in $L_{\mathbb{C}}^2[0, 2\pi]$ can be defined on $\partial \mathbb{D}$ by letting $f(e^{i\theta}) = f(\theta)$. The ambiguity for $\theta = 0$ and 2π (or $e^{i\theta} = 1$) might cause us to pause, but remember that elements of $L_{\mathbb{C}}^2[0, 2\pi]$ are equivalence classes of functions—not really functions. Since $\{0, 2\pi\}$ has zero measure, there is really no ambiguity. In this way $L_{\mathbb{C}}^2[0, 2\pi]$ can be identified with $L_{\mathbb{C}}^2(\partial \mathbb{D})$, where the measure on $\partial \mathbb{D}$ is normalized arc-length measure (normalized so that the total measure of $\partial \mathbb{D}$ is 1). So $L_{\mathbb{C}}^2[0, 2\pi]$ and $L_{\mathbb{C}}^2(\partial \mathbb{D})$ are (naturally) isomorphic). Thus, Theorem 5.11 is a theorem about the Fourier transform of the circle.

The importance of Theorem 5.11 is not the fact that $L^2[0, 2\pi]$ and $l^2(\mathbb{Z})$ are isomorphic, but that the Fourier transform is an isomorphism. The fact that these two spaces are isomorphic follows from the abstract result that all separable infinite dimensional Hilbert spaces are isomorphic (5.5).

EXERCISES

1. Verify the statements in Example 5.3.

2. Define V: $L^2(0, \infty) \to L^2(0, \infty)$ by $(Vf)(t) = f(t + 1)$. Show that V is an isometry that is not surjective.

3. Define V: $L^2(\mathbb{R}) \to L^2(\mathbb{R})$ by $(Vf)(t) = f(t + 1)$ and show that V is an isomorphism (a unitary operator).

4. Let $\mathcal{H}$ be the Hilbert space of Example 1.8 and define U: $\mathcal{H} \to L^2(0, 1)$ by $Uf = f'$. Show that U is an isomorphism and find a formula for U^{-1}.

5. Let (X, Ω, μ) be a σ-finite measure space and let u: $X \to \mathbb{F}$ be an Ω-measurable function such that $\sup\{|u(x)|: x \in X\} < \infty$. Show that U: $L^2(X, \Omega, \mu) \to L^2(X, \Omega, \mu)$ defined by $Uf = uf$ is an isometry if and only if $|u(x)| = 1$ a.e. $[\mu]$, in which case U is surjective.

6. Let $\mathscr{C} = \{f \in C[0, 2\pi]: f(0) = f(2\pi)\}$ and show that $\mathscr{C}$ is dense in $L^2[0, 2\pi]$.

7. Show that $\{(1/\sqrt{2\pi}), (1/\sqrt{\pi})\cos nt, (1/\sqrt{\pi})\sin nt: 1 \le n < \infty\}$ is a basis for $L^2[-\pi, \pi]$.

8. Let (X, Ω) be a measurable space and let μ, ν be two measures defined on (X, Ω). Suppose $\nu \ll \mu$ and ϕ is the Radon–Nikodym derivative of ν with respect to μ ($\phi = d\nu/d\mu$). Define V: $L^2(\nu) \to L^2(\mu)$ by $Vf = \sqrt{\phi}f$. Show that V is a well-defined linear isometry and V is an isomorphism if and only if $\mu \ll \nu$ (that is, μ and ν are mutually absolutely continuous).

§6. The Direct Sum of Hilbert Spaces

Suppose $\mathcal{H}$ and $\mathcal{K}$ are Hilbert spaces. We want to define $\mathcal{H} \oplus \mathcal{K}$ so that it becomes a Hilbert space. This is not a difficult assignment. For any vector spaces $\mathcal{X}$ and $\mathcal{Y}$, $\mathcal{X} \oplus \mathcal{Y}$ is defined as the Cartesian product $\mathcal{X} \times \mathcal{Y}$ where the operations are defined on $\mathcal{X} \times \mathcal{Y}$ coordinatewise. That is, if elements of $\mathcal{X} \oplus \mathcal{Y}$ are defined as $\{x \oplus y: x \in \mathcal{X}, y \in \mathcal{Y}\}$, then $(x_1 \oplus y_1) + (x_2 \oplus y_2) \equiv (x_1 + x_2) \oplus (y_1 + y_2)$, and so on.

6.1.   Definition. If $\mathcal{H}$ and $\mathcal{K}$ are Hilbert spaces, $\mathcal{H} \oplus \mathcal{K} = \{h \oplus k: h \in \mathcal{H}, k \in \mathcal{K}\}$ and

$$\langle h_1 \oplus k_1, h_2 \oplus k_2 \rangle \equiv \langle h_1, h_2 \rangle + \langle k_1, k_2 \rangle.$$

It must be shown that this defines an inner product on $\mathcal{H} \oplus \mathcal{K}$ and that $\mathcal{H} \oplus \mathcal{K}$ is complete (Exercise).

Now what happens if we want to define $\mathcal{H}_1 \oplus \mathcal{H}_2 \oplus \cdots$ for a sequence of Hilbert spaces $\mathcal{H}_1, \mathcal{H}_2, \ldots$? There is a problem about the completeness of this infinite direct sum, but this can be overcome as follows.

6.2.   Proposition. *If $\mathcal{H}_1, \mathcal{H}_2, \ldots$ are Hilbert spaces, let $\mathcal{H} = \{(h_n)_{n=1}^{\infty}: h_n \in \mathcal{H}_n$ for all n and $\sum_{n=1}^{\infty} \|h_n\|^2 < \infty\}$. For $h = (h_n)$ and $g = (g_n)$ in $\mathcal{H}$, define*

6.3
$$\langle h, g \rangle = \sum_{n=1}^{\infty} \langle h_n, g_n \rangle.$$

Then $\langle \cdot, \cdot \rangle$ is an inner product on $\mathcal{H}$ and the norm relative to this inner product is $\|h\| = [\sum_{n=1}^{\infty} \|h_n\|^2]^{1/2}$. With this inner product $\mathcal{H}$ is a Hilbert space.

PROOF. If $h = (h_n)$ and $g = (g_n) \in \mathcal{H}$, then the CBS inequality implies $\sum |\langle h_n, g_n \rangle| \leq \sum \|h_n\| \|g_n\| \leq (\sum \|h_n\|^2)^{1/2} (\sum \|g_n\|^2)^{1/2} < \infty$. Hence the series in (6.3) converges absolutely. The remainder of the proof is left to the reader.   ∎

6.4.   Definition. If $\mathcal{H}_1, \mathcal{H}_2, \ldots$ are Hilbert spaces, the space $\mathcal{H}$ of Proposition 6.2 is called the *direct sum* of $\mathcal{H}_1, \mathcal{H}_2, \ldots$ and is denoted by $\mathcal{H} \equiv \mathcal{H}_1 \oplus \mathcal{H}_2 \oplus \cdots$.

This is part of a more general process. If $\{\mathcal{H}_i: i \in I\}$ is a collection of Hilbert spaces, $\mathcal{H} \equiv \oplus \{\mathcal{H}_i: i \in I\}$ is defined as the collection of functions $h: I \to \bigcup \{\mathcal{H}_i: i \in I\}$ such that $h(i) \in \mathcal{H}_i$ for all i and $\sum \{\|h(i)\|^2: i \in I\} < \infty$. If $h, g \in \mathcal{H}$, $\langle h, g \rangle \equiv \sum \{\langle h(i), g(i) \rangle: i \in I\}$; $\mathcal{H}$ is a Hilbert space.

The main reason for considering direct sums is that they provide a way of manufacturing operators on Hilbert space. In fact, Hilbert space is a rather dull subject, except for the fact that there are numerous interesting questions about the linear operators on them that are as yet unresolved. This subject is introduced in the next chapter.

EXERCISES

1. Let $\{(X_i, \Omega_i, \mu_i): i \in I\}$ be a collection of measure spaces and define X, Ω, and μ as follows. Let $X =$ the disjoint union of $\{X_i: i \in I\}$ and let $\Omega = \{\Delta \subseteq X: \Delta \cap X_i \in \Omega_i$ for all $i\}$. For Δ in Ω put $\mu(\Delta) = \Sigma_i \mu_i(\Delta \cap X_i)$. Show that (X, Ω, μ) is a measure space and $L^2(X, \Omega, \mu)$ is isomorphic to $\oplus \{L^2(X_i, \Omega_i, \mu_i): i \in I\}$.

2. Let (X, Ω) be a measurable space, let μ_1, μ_2 be measures defined on (X, Ω), and put $\mu = \mu_1 + \mu_2$. Show that the map $V: L^2(X, \Omega, \mu) \to L^2(X, \Omega, \mu_1) \oplus L^2(X, \Omega, \mu_2)$ defined by $Vf = f_1 \oplus f_2$, where f_j is the equivalence class of $L^2(X, \Omega, \mu_j)$ corresponding to f, is well defined, linear, and injective. Show that U is an isomorphism iff μ_1 and μ_2 are mutually singular.

CHAPTER II

Operators on Hilbert Space

A large area of current research interest is centered around the theory of operators on Hilbert space. Several other chapters in this book will be devoted to this topic.

There is a marked contrast here between Hilbert spaces and the Banach spaces that are studied in the next chapter. Essentially all of the information about the geometry of Hilbert space is contained in the preceding chapter. The geometry of Banach space lies in darkness and has attracted the attention of many talented research mathematicians. However, the theory of linear operators (linear transformations) on a Banach space has very few general results, whereas Hilbert space operators have an elegant and well-developed general theory. Indeed, the reason for this dichotomy is related to the opposite status of the geometric considerations. Questions concerning operators on Hilbert space don't necessitate or imply any geometric difficulties.

In addition to the fundamentals of operators, this chapter will also present an interesting application to differential equations in Section 6.

§1. Elementary Properties and Examples

The proof of the next proposition is similar to that of Proposition I.3.1 and is left to the reader.

1.1. Proposition. *Let $\mathscr{H}$ and $\mathscr{K}$ be Hilbert spaces and $A: \mathscr{H} \to \mathscr{K}$ a linear transformation. The following statements are equivalent.*

(a) *A is continuous.*
(b) *A is continuous at 0.*

(c) *A is continuous at some point.*
(d) *There is a constant $c > 0$ such that $\|Ah\| \leq c\|h\|$ for all h in $\mathscr{H}$.*

As in (I.3.3), if

$$\|A\| = \sup\{\|Ah\|: h \in \mathscr{H}, \|h\| \leq 1\},$$

then

$$\|A\| = \sup\{\|Ah\|: \|h\| = 1\}$$
$$= \sup\{\|Ah\|/\|h\|: h \neq 0\}$$
$$= \inf\{c > 0: \|Ah\| \leq c\|h\|, h \text{ in } \mathscr{H}\}.$$

Also, $\|Ah\| \leq \|A\|\,\|h\|$. $\|A\|$ is called the *norm* of A and a linear transformation with finite norm is called *bounded*. Let $\mathscr{B}(\mathscr{H}, \mathscr{K})$ be the set of bounded linear transformations from $\mathscr{H}$ into $\mathscr{K}$. For $\mathscr{H} = \mathscr{K}$, $\mathscr{B}(\mathscr{H}, \mathscr{H}) \equiv \mathscr{B}(\mathscr{H})$. Note that $\mathscr{B}(\mathscr{H}, \mathbb{F}) = $ all the bounded linear functionals on $\mathscr{H}$.

1.2. Proposition. (a) *If A and $B \in \mathscr{B}(\mathscr{H}, \mathscr{K})$, then $A + B \in \mathscr{B}(\mathscr{H}, \mathscr{K})$, and $\|A + B\| \leq \|A\| + \|B\|$.*
(b) *If $\alpha \in \mathbb{F}$ and $A \in \mathscr{B}(\mathscr{H}, \mathscr{K})$, then $\alpha A \in \mathscr{B}(\mathscr{H}, \mathscr{K})$ and $\|\alpha A\| = |\alpha|\,\|A\|$.*
(c) *If $A \in \mathscr{B}(\mathscr{H}, \mathscr{K})$ and $B \in \mathscr{B}(\mathscr{K}, \mathscr{L})$, then $BA \in \mathscr{B}(\mathscr{H}, \mathscr{L})$ and $\|BA\| \leq \|B\|\,\|A\|$.*

PROOF. Only (c) will be proved; the rest of the proof is left to the reader. If $k \in \mathscr{K}$, then $\|Bk\| \leq \|B\|\,\|k\|$. Hence, if $h \in \mathscr{H}$, $k = Ah \in \mathscr{K}$ and so $\|BAh\| \leq \|B\|\,\|Ah\| \leq \|B\|\,\|A\|\,\|h\|$. ∎

By virtue of the preceding proposition, $d(A, B) = \|A - B\|$ defines a metric on $\mathscr{B}(\mathscr{H}, \mathscr{K})$. So it makes sense to consider $\mathscr{B}(\mathscr{H}, \mathscr{K})$ as a metric space. This will not be examined closely until later in the book, but later in this chapter the idea of the convergence of a sequence of operators will be used.

1.3. Example. If $\dim \mathscr{H} = n < \infty$ and $\dim \mathscr{K} = m < \infty$, let $\{e_1, \ldots, e_n\}$ be an orthonormal basis for $\mathscr{H}$ and let $\{\varepsilon_1, \ldots, \varepsilon_m\}$ be an orthonormal basis for $\mathscr{K}$. It can be shown that every linear transformation from $\mathscr{H}$ into $\mathscr{K}$ is bounded (Exercise 3). If $1 \leq j \leq n$, $1 \leq i \leq m$, let $\alpha_{ij} = \langle Ae_j, \varepsilon_i \rangle$. Then the $m \times n$ matrix (α_{ij}) represents A and every such matrix represents an element of $\mathscr{B}(\mathscr{H}, \mathscr{K})$.

1.4. Example. Let $l^2 \equiv l^2(\mathbb{N})$ and let $e_1, e_2, \ldots$ be its usual basis. If $A \in \mathscr{B}(l^2)$, form $\alpha_{ij} = \langle Ae_j, e_i \rangle$. The infinite matrix (α_{ij}) represents A as finite matrices represent operators on finite dimensional spaces. However, this representation has limited value unless the matrix has a special form. One difficulty is that it is unknown how to find the norm of A in terms of the entries in the matrix. In fact, if $2 < n < \infty$, there is no known formula for the norm of a matrix in terms of its entries. A sufficient condition that is useful is known, however (see Exercise 11).

1.5.   Theorem. *Let (X, Ω, μ) be a σ-finite measure space and put $\mathcal{H} = L^2(X, \Omega, \mu) \equiv L^2(\mu)$. If $\phi \in L^\infty(\mu)$, define $M_\phi \colon L^2(\mu) \to L^2(\mu)$ by $M_\phi f = \phi f$. Then $M_\phi \in \mathcal{B}(L^2(\mu))$ and $\|M_\phi\| = \|\phi\|_\infty$.*

PROOF. Here $\|\phi\|_\infty$ is the μ-*essential supremum norm.* That is,

$$\|\phi\|_\infty \equiv \inf\{\sup\{|\phi(x)| \colon x \notin N\} \colon N \in \Omega, \mu(N) = 0\}$$

$$= \inf\{c > 0 \colon \mu(\{x \in X \colon |\phi(x)| > c\}) = 0\}.$$

Thus $\|\phi\|_\infty$ is the infimum of all $c > 0$ such that $|\phi(x)| \le c$ a.e. $[\mu]$ and, moreover, $|\phi(x)| \le \|\phi\|_\infty$ a.e. $[\mu]$. Thus we can, and do, assume that ϕ is a bounded measurable function and $|\phi(x)| \le \|\phi\|_\infty$ for all x. So if $f \in L^2(\mu)$, then $\int |\phi f|^2 \, d\mu \le \|\phi\|_\infty \int |f|^2 \, d\mu$. That is, $M_\phi \in \mathcal{B}(L^2(\mu))$ and $\|M_\phi\| \le \|\phi\|_\infty$. If $\varepsilon > 0$, the σ-finiteness of the measure space implies that there is a set Δ in Ω, $0 < \mu(\Delta) < \infty$, such that $|\phi(x)| \ge \|\phi\|_\infty - \varepsilon$ on Δ. (Why?) If $f = (\mu(\Delta))^{-1/2}\chi_\Delta$, then $f \in L^2(\mu)$ and $\|f\|_2 = 1$. So $\|M_\phi\|^2 \ge \|\phi f\|_2^2 = (\mu(\Delta))^{-1} \int_\Delta |\phi|^2 \, d\mu \ge (\|\phi\|_\infty - \varepsilon)^2$. Letting $\varepsilon \to 0$, we get that $\|M_\phi\| \ge \|\phi\|_\infty$. ∎

The operator M_ϕ is called a *multiplication operator.* The function ϕ is its *symbol.*

If the measure space (X, Ω, μ) is not σ-finite, then the conclusion of Theorem 1.5 is not necessarily valid. Indeed, let $\Omega = $ the Borel subsets of $[0, 1]$ and define μ on Ω by $\mu(\Delta) = $ the Lebesgue measure of Δ if $0 \notin \Delta$ and $\mu(\Delta) = \infty$ if $0 \in \Delta$. This measure has an infinite atom at 0 and, therefore, is not σ-finite. Let $\phi = \chi_{\{0\}}$. Then $\phi \in L^\infty(\mu)$ and $\|\phi\|_\infty = 1$. If $f \in L^2(\mu)$, then $\infty > \int |f|^2 \, d\mu \ge |f(0)|^2 \mu(\{0\})$. Hence every function in $L^2(\mu)$ vanishes at 0. Therefore $M_\phi = 0$ and $\|M_\phi\| < \|\phi\|_\infty$.

There are more general measure spaces for which (1.5) is valid—the decomposable measure spaces (see Kelley [1966]).

1.6.   Theorem. *Let (X, Ω, μ) be a measure space and suppose $k \colon X \times X \to \mathbb{F}$ is an $\Omega \times \Omega$-measurable function for which there are constants c_1 and c_2 such that*

$$\int_X |k(x, y)| \, d\mu(y) \le c_1 \qquad \text{a.e. } [\mu],$$

$$\int_X |k(x, y)| \, d\mu(x) \le c_2 \qquad \text{a.e. } [\mu].$$

If $K: L^2(\mu) \to L^2(\mu)$ is defined by

$$(Kf)(x) = \int k(x, y) f(y) \, d\mu(y),$$

then K is a bounded linear operator and $\|K\| \leq (c_1 c_2)^{1/2}$.

PROOF. Actually it must be shown that $Kf \in L^2(\mu)$, but this will follow from the argument that demonstrates the boundedness of K. If $f \in L^2(\mu)$,

$$|Kf(x)| \leq \int |k(x, y)| \, |f(y)| \, d\mu(y)$$

$$= \int |k(x, y)|^{1/2} |k(x, y)|^{1/2} |f(y)| \, d\mu(y)$$

$$\leq \left[\int |k(x, y)| \, d\mu(y) \right]^{1/2} \left[\int |k(x, y)| \, |f(y)|^2 \, d\mu(y) \right]^{1/2}$$

$$\leq c_1^{1/2} \left[\int |k(x, y)| \, |f(y)|^2 \, d\mu(y) \right]^{1/2}.$$

Hence

$$\int |Kf(x)|^2 \, d\mu(x) \leq c_1 \int \int |k(x, y)| \, |f(y)|^2 \, d\mu(y) \, d\mu(x)$$

$$= c_1 \int |f(y)|^2 \int |k(x, y)| \, d\mu(x) \, d\mu(y)$$

$$\leq c_1 c_2 \|f\|^2.$$

Now this shows that the formula used to define Kf is finite a.e. $[\mu]$, $Kf \in L^2(\mu)$, and $\|Kf\|^2 \leq c_1 c_2 \|f\|^2$. ∎

The operator described above is called an *integral operator* and the function k is called its *kernel*. There are conditions on the kernel other than the one in (1.6) that will imply that K is bounded.

A particular example of an integral operator is the *Volterra operator* defined below.

1.7. Example. Let $k: [0, 1] \times [0, 1] \to \mathbb{R}$ be the characteristic function of $\{(x, y): y < x\}$. The corresponding operator $V: L^2(0, 1) \to L^2(0, 1)$ defined by $Vf(x) = \int_0^1 k(x, y) f(y) \, dy$ is called the *Volterra operator*. Note that

$$Vf(x) = \int_0^x f(y) \, dy.$$

Another example of an operator was defined in Example I.5.3. The nonsurjective isometry defined there is called the *unilateral shift*. It will be studied in more detail later in this book. Note that any isometry is a bounded operator with norm 1.

EXERCISES

1. Prove Proposition 1.1.

2. Prove Proposition 1.2.

3. Suppose $\{e_n\}$ is an orthonormal basis for $\mathscr{H}$ and $A\colon \mathscr{H} \to \mathscr{K}$ is a linear transformation such that $\Sigma \|Ae_n\| < \infty$. Show that A is bounded.

4. Proposition 1.2 says that $d(A, B) = \|A - B\|$ is a metric on $\mathscr{B}(\mathscr{H}, \mathscr{K})$. Show that $\mathscr{B}(\mathscr{H}, \mathscr{K})$ is complete relative to this metric.

5. Show that a multiplication operator M_ϕ (1.5) satisfies $M_\phi^2 = M_\phi$ if and only if ϕ is a characteristic function.

6. Let (X, Ω, μ) be a measure space and let k_1, k_2 be two kernels satisfying the hypothesis of (1.6). Define

$$k\colon X \times X \to \mathbb{F} \quad \text{by } k(x, y) = \int k_1(x, z) k_2(z, y)\, d\mu(z).$$

(a) Show that k also satisfies the hypothesis of (1.6). (b) If K, K_1, K_2 are the integral operators with kernels k, k_1, k_2, show that $K = K_1 K_2$. What does this remind you of? Is more going on than an analogy?

7. If (X, Ω, μ) is a measure space and $k \in L^2(\mu \times \mu)$, show that k defines a bounded integral operator.

8. Let $\{e_n\}$ be the usual basis for l^2 and let $\{\alpha_n\}$ be a sequence of scalars. Show that there is a bounded operator A on l^2 such that $Ae_n = \alpha_n e_n$ for all n if and only if $\{\alpha_n\}$ is uniformly bounded, in which case $\|A\| = \sup\{|\alpha_n|\colon n \geq 1\}$. This type of operator is called a *diagonal operator* or is said to be *diagonalizable*.

9. (Schur test) Let $\{\alpha_{ij}\}_{i,j=1}^\infty$ be an infinite matrix such that $\alpha_{ij} \geq 0$ for all i, j and such that there are scalars $p_i > 0$ and $\beta, \gamma > 0$ with

$$\sum_{i=1}^\infty \alpha_{ij} p_i \leq \beta p_j,$$

$$\sum_{j=1}^\infty \alpha_{ij} p_j \leq \gamma p_i$$

for all $i, j \geq 1$. Show that there is an operator A on $l^2(\mathbb{N})$ with $\langle Ae_j, e_i \rangle = \alpha_{ij}$ and $\|A\|^2 \leq \beta\gamma$.

10. (Hilbert matrix) Show that $\langle Ae_j, e_i \rangle = (i + j + 1)^{-1}$ for $0 \leq i, j < \infty$ defines a bounded operator on $l^2(\mathbb{N} \cup \{0\})$ with $\|A\| \leq \pi$. (See also Choi [1983].)

11. Find the operator norm of a 2×2 matrix in terms of its entries.

12. (Direct sum of operators) Let $\{\mathscr{H}_i\}$ be a collection of Hilbert spaces and let $\mathscr{H} = \oplus_i \mathscr{H}_i$. Suppose $A_i \in \mathscr{B}(\mathscr{H}_i)$ for all i. Show that there is a bounded operator A on $\mathscr{H}$ such that $A|\mathscr{H}_i = A_i$ for all i if and only if $\sup_i \|A_i\| < \infty$. In this case, $\|A\| = \sup_i \|A_i\|$.

§2. The Adjoint of an Operator

2.1.   Definition. If $\mathscr{H}$ and $\mathscr{K}$ are Hilbert spaces, a function $u\colon \mathscr{H} \times \mathscr{K} \to \mathbb{F}$ is a *sesquilinear form* if for h, g in $\mathscr{H}$, k, f in $\mathscr{K}$, and α, β in $\mathbb{F}$,

(a) $u(\alpha h + \beta g, k) = \alpha u(h, k) + \beta u(g, k)$;
(b) $u(h, \alpha k + \beta f) = \bar{\alpha} u(h, k) + \bar{\beta} u(h, f)$.

   The prefix "sesqui" is used because the function is linear in one variable but (for $\mathbb{F} = \mathbb{C}$) only conjugate linear in the other. ("Sesqui" means "one-and-a-half.")

   A sesquilinear form is *bounded* if there is a constant M such that $|u(h, k)| \le M\|h\|\,\|k\|$ for all h in $\mathscr{H}$ and k in $\mathscr{K}$. The constant M is called a bound for u.

   Sesquilinear forms are used to study operators. If $A \in \mathscr{B}(\mathscr{H}, \mathscr{K})$, then $u(h, k) \equiv \langle Ah, k \rangle$ is a bounded sesquilinear form. Also, if $B \in \mathscr{B}(\mathscr{K}, \mathscr{H})$, $u(h, k) \equiv \langle h, Bk \rangle$ is a bounded sesquilinear form. Are there any more? Are these two forms related?

2.2.   Theorem. *If $u\colon \mathscr{H} \times \mathscr{K} \to \mathbb{F}$ is a bounded sesquilinear form with bound M, then there are unique operators A in $\mathscr{B}(\mathscr{H}, \mathscr{K})$ and B in $\mathscr{B}(\mathscr{K}, \mathscr{H})$ such that*

2.3 $$u(h, k) = \langle Ah, k \rangle = \langle h, Bk \rangle$$

for all h in $\mathscr{H}$ and k in $\mathscr{K}$ and $\|A\|, \|B\| \le M$.

PROOF. Only the existence of A will be shown. For each h in $\mathscr{H}$, define $L_h\colon \mathscr{K} \to \mathbb{F}$ by $L_h(k) = \overline{u(h, k)}$. Then L_h is linear and $|L_h(k)| \le M\|h\|\,\|k\|$. By the Riesz Representation Theorem there is a unique vector f in $\mathscr{K}$ such that $\langle k, f \rangle = L_h(k) = \overline{u(h, k)}$ and $\|f\| \le M\|h\|$. Let $Ah = f$. It is left as an exercise to show that A is linear (use the uniqueness part of the Riesz Theorem). Also, $\langle Ah, k \rangle = \overline{\langle k, Ah \rangle} = \overline{\langle k, f \rangle} = u(h, k)$.

   If $A_1 \in \mathscr{B}(\mathscr{H}, \mathscr{K})$ and $u(h, k) = \langle A_1 h, k \rangle$, then $\langle Ah - A_1 h, k \rangle = 0$ for all k; thus $Ah - A_1 h = 0$ for all h. Thus, A is unique.   ■

2.4.   Definition. If $A \in \mathscr{B}(\mathscr{H}, \mathscr{K})$, then the unique operator B in $\mathscr{B}(\mathscr{K}, \mathscr{H})$ satisfying (2.3) is called the *adjoint* of A and is denoted by $B = A^*$.

   The adjoint of an operator will usually be used for operators in $\mathscr{B}(\mathscr{H})$, rather than $\mathscr{B}(\mathscr{H}, \mathscr{K})$. There is one notable exception.

2.5.   Proposition. *If $U \in \mathscr{B}(\mathscr{H}, \mathscr{K})$, then U is an isomorphism if and only if U is invertible and $U^{-1} = U^*$.*

PROOF. Exercise.

From now on we will examine and prove results for the adjoint of operators in $\mathscr{B}(\mathscr{H})$. Often, as in the next proposition, there are analogous results for the adjoint of operators in $\mathscr{B}(\mathscr{H}, \mathscr{K})$. This simplification is justified, however, by the cleaner statements that result. Also, the interested reader will have no trouble formulating the more general statement when it is needed.

2.6. Proposition. *If* $A, B \in \mathscr{B}(\mathscr{H})$ *and* $\alpha \in \mathbb{F}$, *then*:

(a) $(\alpha A + B)^* = \bar{\alpha} A^* + B^*$.
(b) $(AB)^* = B^* A^*$.
(c) $A^{**} \equiv (A^*)^* = A$.
(d) *If* A *is invertible in* $\mathscr{B}(\mathscr{H})$ *and* A^{-1} *is its inverse, then* A^* *is invertible and* $(A^*)^{-1} = (A^{-1})^*$.

The proof of the preceding proposition is left as an exercise, but a word about part (d) might be helpful. The hypothesis that A is invertible in $\mathscr{B}(\mathscr{H})$ means that there is an operator A^{-1} in $\mathscr{B}(\mathscr{H})$ such that $AA^{-1} = A^{-1}A = I$. It is a remarkable fact that if A is only assumed to be bijective, then A is invertible in $\mathscr{B}(\mathscr{H})$. This is a consequence of the Open Mapping Theorem, which will be proved later.

2.7. Proposition. *If* $A \in \mathscr{B}(\mathscr{H})$, $\|A\| = \|A^*\| = \|A^*A\|^{1/2}$.

PROOF. For h in $\mathscr{H}$, $\|h\| \leq 1$, $\|Ah\|^2 = \langle Ah, Ah \rangle = \langle A^*Ah, h \rangle \leq \|A^*Ah\| \|h\| \leq \|A^*A\| \leq \|A^*\| \|A\|$. Hence $\|A\|^2 \leq \|A^*A\| \leq \|A^*\| \|A\|$. Using the two ends of this string of inequalities gives $\|A\| \leq \|A^*\|$ when $\|A\|$ is cancelled. But $A = A^{**}$ and so if A^* is substituted for A, we get $\|A^*\| \leq \|A^{**}\| = \|A\|$. Hence $\|A\| = \|A^*\|$. Thus the string of inequalities becomes a string of equalities and the proof is complete. ■

2.8. Example. Let (X, Ω, μ) be a σ-finite measure space and let M_ϕ be the multiplication operator with symbol ϕ (1.5). Then M_ϕ^* is $M_{\bar{\phi}}$, the multiplication operator with symbol $\bar{\phi}$.

If an operator on $\mathbb{F}^d$ is represented by a matrix, then its adjoint is represented by the conjugate transpose of the matrix.

2.9. Example. If K is the integral operator with kernel k as in (1.6), then K^* is the integral operator with kernel $k^*(x, y) \equiv \overline{k(y, x)}$.

2.10. Proposition. *If* $S: l^2 \to l^2$ *is defined by* $S(\alpha_1, \alpha_2, \ldots) = (0, \alpha_1, \alpha_2, \ldots)$, *then* S *is an isometry and* $S^*(\alpha_1, \alpha_2, \ldots) = (\alpha_2, \alpha_3, \ldots)$.

PROOF. It has already been mentioned that S is an isometry (I.5.3). For (α_n) and (β_n) in l^2, $\langle S^*(\alpha_n), (\beta_n) \rangle = \langle (\alpha_n), S(\beta_n) \rangle = \langle (\alpha_1, \alpha_2, \ldots), (0, \beta_1, \beta_2,$

$\dots$)) $= \alpha_2 \bar{\beta}_1 + \alpha_3 \bar{\beta}_2 + \cdots = \langle (\alpha_2, \alpha_3, \dots), (\beta_1, \beta_2, \dots) \rangle$. Since this holds for every (β_n), the result is proved. ∎

The operator S in (2.10) is called the *unilateral shift* and the operator S^* is called the *backward shift*.

The operation of taking the adjoint of an operator is, as the reader may have seen from the examples above, analogous to taking the conjugate of a complex number. It is good to keep the analogy in mind, but do not become too religious about it.

2.11.   Definition. If $A \in \mathscr{B}(\mathscr{H})$, then: (a) A is *hermitian* or *self-adjoint* if $A^* = A$; (b) A is *normal* if $AA^* = A^*A$.

In the analogy between the adjoint and the complex conjugate, hermitian operators become the analogues of real numbers and, by (2.5), unitaries are the analogues of complex numbers of modulus 1. Normal operators, as we shall see, are the true analogues of complex numbers. Notice that hermitian and unitary operators are normal.

In light of (2.8), every multiplication operator M_ϕ is normal; M_ϕ is hermitian if and only if ϕ is real-valued; M_ϕ is unitary if and only if $|\phi| \equiv 1$ a.e. $[\mu]$. By (2.9), an integral operator K with kernel k is hermitian if and only if $k(x, y) = \overline{k(y, x)}$ a.e. $[\mu \times \mu]$. The unilateral shift is not normal (Exercise 6).

2.12.   Proposition. *If $\mathscr{H}$ is a $\mathbb{C}$-Hilbert space and $A \in \mathscr{B}(\mathscr{H})$, then A is hermitian if and only if $\langle Ah, h \rangle \in \mathbb{R}$ for all h in $\mathscr{H}$.*

PROOF. If $A = A^*$, then $\langle Ah, h \rangle = \langle h, Ah \rangle = \overline{\langle Ah, h \rangle}$; hence $\langle Ah, h \rangle \in \mathbb{R}$.

For the converse, assume $\langle Ah, h \rangle$ is real for every h in $\mathscr{H}$. If $\alpha \in \mathbb{C}$ and $h, g \in \mathscr{H}$, then $\langle A(h + \alpha g), h + \alpha g \rangle = \langle Ah, h \rangle + \bar{\alpha} \langle Ah, g \rangle + \alpha \langle Ag, h \rangle + |\alpha|^2 \langle Ag, g \rangle \in \mathbb{R}$. So this expression equals its complex conjugate. Using the fact that $\langle Ah, h \rangle$ and $\langle Ag, g \rangle \in \mathbb{R}$ yields

$$\alpha \langle Ag, h \rangle + \bar{\alpha} \langle Ah, g \rangle = \bar{\alpha} \langle h, Ag \rangle + \alpha \langle g, Ah \rangle$$
$$= \bar{\alpha} \langle A^*h, g \rangle + \alpha \langle A^*g, h \rangle.$$

By first taking $\alpha = 1$ and then $\alpha = i$, we obtain the two equations

$$\langle Ag, h \rangle + \langle Ah, g \rangle = \langle A^*h, g \rangle + \langle A^*g, h \rangle,$$
$$i \langle Ag, h \rangle - i \langle Ah, g \rangle = -i \langle A^*h, g \rangle + i \langle A^*g, h \rangle.$$

A little arithmetic implies $\langle Ag, h \rangle = \langle A^*g, h \rangle$, so $A = A^*$. ∎

The preceding proposition is false if it is only assumed that $\mathscr{H}$ is an $\mathbb{R}$-Hilbert space. For example, if $A = \begin{bmatrix} 0 & 1 \\ -1 & 0 \end{bmatrix}$ on $\mathbb{R}^2$, then $\langle Ah, h \rangle = 0$

for all h in $\mathbb{R}^2$. However, A^* is the transpose of A and so $A^* \neq A$. Indeed, for any operator A on an $\mathbb{R}$-Hilbert space, $\langle Ah, g \rangle \in \mathbb{R}$.

2.13. Proposition. *If $A = A^*$, then*

$$\|A\| = \sup\{|\langle Ah, h \rangle| : \|h\| = 1\}.$$

PROOF. Put $M = \sup\{|\langle Ah, h \rangle| : \|h\| = 1\}$. If $\|h\| = 1$, then $|\langle Ah, h \rangle| \leq \|A\|$; hence $M \leq \|A\|$. On the other hand, if $\|h\| = \|g\| = 1$, then

$$\langle A(h \pm g), h \pm g \rangle = \langle Ah, h \rangle \pm \langle Ah, g \rangle \pm \langle Ag, h \rangle + \langle Ag, g \rangle$$

$$= \langle Ah, h \rangle \pm \langle Ah, g \rangle \pm \langle g, A^*h \rangle + \langle Ag, g \rangle.$$

Since $A = A^*$, this implies

$$\langle A(h \pm g), h \pm g \rangle = \langle Ah, h \rangle \pm 2\operatorname{Re}\langle Ah, g \rangle + \langle Ag, g \rangle.$$

Subtracting one of these two equations from the other gives

$$4\operatorname{Re}\langle Ah, g \rangle = \langle A(h + g), h + g \rangle - \langle A(h - g), h - g \rangle.$$

Now it is easy to verify that $|\langle Af, f \rangle| \leq M\|f\|^2$ for any f in $\mathcal{H}$. Hence using the parallelogram law we get

$$4\operatorname{Re}\langle Ah, g \rangle \leq M\left(\|h + g\|^2 + \|h - g\|^2\right)$$

$$= 2M\left(\|h\|^2 + \|g\|^2\right)$$

$$= 4M$$

since h and g are unit vectors. Now suppose $\langle Ah, g \rangle = e^{i\theta}|\langle Ah, g \rangle|$. Replacing h in the inequality above with $e^{-i\theta}h$ gives $|\langle Ah, g \rangle| \leq M$ if $\|h\| = \|g\| = 1$. Taking the supremum over all g gives $\|Ah\| \leq M$ when $\|h\| = 1$. Thus $\|A\| \leq M$. ∎

2.14. Corollary. *If $A = A^*$ and $\langle Ah, h \rangle = 0$ for all h, then $A = 0$.*

The preceding corollary is not true unless $A = A^*$, as the example given after Proposition 2.12 shows. However, if a complex Hilbert space is present, this hypothesis can be deleted.

2.15. Proposition. *If $\mathcal{H}$ is a $\mathbb{C}$-Hilbert space and $A \in \mathcal{B}(\mathcal{H})$ such that $\langle Ah, h \rangle = 0$ for all h in $\mathcal{H}$, then $A = 0$.*

The proof of (2.15) is left to the reader.

If $\mathcal{H}$ is a $\mathbb{C}$-Hilbert space and $A \in \mathcal{B}(\mathcal{H})$, then $B = (A + A^*)/2$ and $C = (A - A^*)/2i$ are self-adjoint and $A = B + iC$. The operators B and C are called, respectively, the *real and imaginary parts* of A.

2.16.   Proposition. *If $A \in \mathcal{B}(\mathcal{H})$, the following statements are equivalent.*

(a) *A is normal.*
(b) *$\|Ah\| = \|A^*h\|$ for all h.*

   If $\mathcal{H}$ is a $\mathbb{C}$-Hilbert space, then these statements are also equivalent to:

(c) *The real and imaginary parts of A commute.*

PROOF. If $h \in \mathcal{H}$, then $\|Ah\|^2 - \|A^*h\|^2 = \langle Ah, Ah \rangle - \langle A^*h, A^*h \rangle = \langle (A^*A - AA^*)h, h \rangle$. Since $A^*A - AA^*$ is hermitian, the equivalence of (a) and (b) follows from Corollary 2.14.
   If B, C are the real and imaginary parts of A, then a calculation yields

$$A^*A = B^2 - iCB + iBC + C^2,$$
$$AA^* = B^2 + iCB - iBC + C^2.$$

Hence $A^*A = AA^*$ if and only if $CB = BC$, and so (a) and (c) are equivalent. ■

2.17.   Proposition. *If $A \in \mathcal{B}(\mathcal{H})$, the following statements are equivalent.*

(a) *A is an isometry.*
(b) *$A^*A = I$.*
(c) *$\langle Ah, Ag \rangle = \langle h, g \rangle$ for all h, g in $\mathcal{H}$.*

PROOF. The proof that (a) and (c) are equivalent was seen in Proposition I.5.2. Note that if $h, g \in \mathcal{H}$, then $\langle A^*Ah, g \rangle = \langle Ah, Ag \rangle$. Hence (b) and (c) are easily seen to be equivalent. ■

2.18.   Proposition. *If $A \in \mathcal{B}(\mathcal{H})$, then the following statements are equivalent.*

(a) *A is unitary.*
(b) *A is a surjective isometry.*
(c) *A is a normal isometry.*

PROOF. (a) $\Rightarrow$ (b): Proposition I.5.2.
   (b) $\Rightarrow$ (c): By (2.17), $A^*A = I$. But it is easy to see that the fact that A is a surjective isometry implies that A^{-1} is also. Hence by (2.17) $I = (A^{-1})^*A^{-1} = (A^*)^{-1}A^{-1} = (AA^*)^{-1}$; this implies that $A^*A = AA^* = I$.
   (c) $\Rightarrow$ (a): By (2.17), $A^*A = I$. Since A is also normal, $AA^* = A^*A = I$ and so A is surjective. ■

   We conclude with a very important, though easily proved, result.

2.19.   Theorem. *If $A \in \mathcal{B}(\mathcal{H})$, then $\ker A = (\operatorname{ran} A^*)^{\perp}$.*

PROOF. If $h \in \ker A$ and $g \in \mathcal{H}$, then $\langle h, A^*g \rangle = \langle Ah, g \rangle = 0$, so $\ker A \subseteq (\operatorname{ran} A^*)^{\perp}$. On the other hand, if $h \perp \operatorname{ran} A^*$ and $g \in \mathcal{H}$, then $\langle Ah, g \rangle = \langle h, A^*g \rangle = 0$; so $(\operatorname{ran} A^*)^{\perp} \subseteq \ker A$. ■

Two facts should be noted. Since $A^{**} = A$, it also holds that $\ker A^* = (\operatorname{ran} A)^{\perp}$. Second, it is not true that $(\ker A)^{\perp} = \operatorname{ran} A^*$ since $\operatorname{ran} A^*$ may not be closed. All that can be said is that $(\ker A)^{\perp} = \operatorname{cl}(\operatorname{ran} A^*)$ and $(\ker A^*)^{\perp} = \operatorname{cl}(\operatorname{ran} A)$.

EXERCISES

1. Prove Proposition 2.5.

2. Prove Proposition 2.6.

3. Verify the statement in Example 2.8.

4. Verify the statement in Example 2.9.

5. Find the adjoint of a diagonal operator (Exercise 1.8).

6. Let S be the unilateral shift and compute SS^* and S^*S. Also compute S^nS^{*n} and $S^{*n}S^n$.

7. Compute the adjoint of the Volterra operator V (1.7) and $V + V^*$. What is $\operatorname{ran}(V + V^*)$?

8. Where was the hypothesis that $\mathscr{H}$ is a Hilbert space over $\mathbb{C}$ used in the proof of Proposition 2.12?

9. Suppose $A = B + iC$, where B and C are hermitian and prove that $B = (A + A^*)/2$, $C = (A - A^*)/2i$.

10. Prove Proposition 2.15.

11. If A and B are self-adjoint, show that AB is self-adjoint if and only if $AB = BA$.

12. Let $\sum_{n=0}^{\infty} \alpha_n z^n$ be a power series with radius of convergence R, $0 < R \le \infty$. If $A \in \mathscr{B}(\mathscr{H})$ and $\|A\| < R$, show that there is an operator T in $\mathscr{B}(\mathscr{H})$ such that for any h, g in $\mathscr{H}$, $\langle Th, g \rangle = \sum_{n=0}^{\infty} \alpha_n \langle A^n h, g \rangle$. [If $f(z) = \sum \alpha_n z^n$, the operator T is usually denoted by $f(A)$.]

13. Let A and T be as in Exercise 12 and show that $\|T - \sum_{k=0}^{n} \alpha_k A^k\| \to 0$ as $n \to \infty$. If $BA = AB$, show that $BT = TB$.

14. If $f(z) = \exp z = \sum_{n=0}^{\infty} z^n/n!$ and A is hermitian, show that $f(iA)$ is unitary.

15. If A is a normal operator on $\mathscr{H}$, show that A is injective if and only if A has dense range. Give an example of an operator B such that $\ker B = (0)$ but $\operatorname{ran} B$ is not dense. Give an example of an operator C such that C is surjective but $\ker C \ne (0)$.

16. Let M_{ϕ} be a multiplication operator (1.5) and show that $\ker M_{\phi} = (0)$ if and only if $\mu(\{x: \phi(x) = 0\}) = 0$. Give necessary and sufficient conditions on ϕ that $\operatorname{ran} M_{\phi}$ is closed.

§3. Projections and Idempotents; Invariant and Reducing Subspaces

3.1.   Definition. An *idempotent* on $\mathscr{H}$ is a bounded linear operator E on $\mathscr{H}$ such that $E^2 = E$. A *projection* is an idempotent P such that $\ker P = (\operatorname{ran} P)^{\perp}$.

If $\mathscr{M} \leq \mathscr{H}$, then $P_{\mathscr{M}}$ is a projection (Theorem I.2.7). It is not difficult to construct an idempotent that is not a projection (Exercise 1).

Let E be any idempotent and set $\mathscr{M} = \operatorname{ran} E$ and $\mathscr{N} = \ker E$. Since E is continuous, $\mathscr{N}$ is a closed subspace of $\mathscr{H}$. Notice that $(I - E)^2 = I - 2E + E^2 = I - 2E + E = I - E$; thus $I - E$ is also an idempotent. Also, $0 = (I - E)h = h - Eh$, if and only if $Eh = h$. So $\operatorname{ran} E \supseteq \ker(I - E)$. On the other hand, if $h \in \operatorname{ran} E$, $h = Eg$ and so $Eh = E^2g = Eg = h$; hence $\operatorname{ran} E = \ker(I - E)$. Similarly, $\operatorname{ran}(I - E) = \ker E$. These facts are recorded here.

3.2.   Proposition. (a) *E is an idempotent if and only if $I - E$ is an idempotent.* (b) $\operatorname{ran} E = \ker(I - E)$, $\ker E = \operatorname{ran}(I - E)$, *and both* $\operatorname{ran} E$ *and* $\ker E$ *are closed linear subspaces of* $\mathscr{H}$. (c) *If* $\mathscr{M} = \operatorname{ran} E$ *and* $\mathscr{N} = \ker E$, *then* $\mathscr{M} \cap \mathscr{N} = (0)$ *and* $\mathscr{M} + \mathscr{N} = \mathscr{H}$.

The proof of part (c) is left as an exercise. There is also a converse to (c). If $\mathscr{M}, \mathscr{N} \leq \mathscr{H}$, $\mathscr{M} \cap \mathscr{N} = (0)$, and $\mathscr{M} + \mathscr{N} = \mathscr{H}$, then there is an idempotent E such that $\mathscr{M} = \operatorname{ran} E$ and $\mathscr{N} = \ker E$; moreover, E is unique. The difficult part in proving this converse is to show that E is bounded. The same fact is true in more generality (for Banach spaces) and so this proof will be postponed.

Now we turn our attention to projections, which are peculiar to Hilbert space.

3.3.   Proposition. *If E is an idempotent on $\mathscr{H}$ and $E \neq 0$, the following statements are equivalent.*

(a) *E is a projection.*
(b) *E is the orthogonal projection of $\mathscr{H}$ onto* $\operatorname{ran} E$.
(c) $\|E\| = 1$.
(d) *E is hermitian.*
(e) *E is normal.*
(f) $\langle Eh, h \rangle \geq 0$ *for all h in* $\mathscr{H}$.

PROOF. (a) $\Rightarrow$ (b): Let $\mathscr{M} = \operatorname{ran} E$ and $P = P_{\mathscr{M}}$. If $h \in \mathscr{H}$, $Ph =$ the unique vector in $\mathscr{M}$ such that $h - Ph \in \mathscr{M}^{\perp} = (\operatorname{ran} E)^{\perp} = \ker E$ by (a). But $h - Eh = (I - E)h \in \ker E$. Hence $Eh = Ph$ by uniqueness.
   (b) $\Rightarrow$ (c): By (I.2.7), $\|E\| \leq 1$. But $Eh = h$ for h in $\operatorname{ran} E$, so $\|E\| = 1$.

(c) $\Rightarrow$ (a): Let $h \in (\ker E)^\perp$. Now $\operatorname{ran}(I - E) = \ker E$, so $h - Eh \in \ker E$. Hence $0 = \langle h - Eh, h \rangle = \|h\|^2 - \langle Eh, h \rangle$. Hence $\|h\|^2 = \langle Eh, h \rangle \le \|Eh\| \, \|h\| \le \|h\|^2$. So for h in $(\ker E)^\perp$, $\|Eh\| = \|h\| = \langle Eh, h \rangle^{1/2}$. But then for h in $(\ker E)^\perp$,

$$\|h - Eh\|^2 = \|h\|^2 - 2\operatorname{Re}\langle Eh, h \rangle + \|Eh\|^2 = 0.$$

That is, $(\ker E)^\perp \subseteq \ker(I - E) = \operatorname{ran} E$. On the other hand, if $g \in \operatorname{ran} E$, $g = g_1 + g_2$, where $g_1 \in \ker E$ and $g_2 \in (\ker E)^\perp$. Thus $g = Eg = Eg_2 = g_2$; that is, $\operatorname{ran} E \subseteq (\ker E)^\perp$. Therefore $\operatorname{ran} E = (\ker E)^\perp$ and E is a projection.

(b) $\Rightarrow$ (f): If $h \in \mathcal{H}$, write $h = h_1 + h_2$, $h_1 \in \operatorname{ran} E$, $h_2 \in \ker E = (\operatorname{ran} E)^\perp$. Hence $\langle Eh, h \rangle = \langle E(h_1 + h_2), h_1 + h_2 \rangle = \langle Eh_1, h_1 \rangle = \langle h_1, h_1 \rangle = \|h_1\|^2 \ge 0$.

(f) $\Rightarrow$ (a): Let $h_1 \in \operatorname{ran} E$ and $h_2 \in \ker E$. Then by (f), $0 \le \langle E(h_1 + h_2), h_1 + h_2 \rangle = \langle h_1, h_1 \rangle + \langle h_1, h_2 \rangle$. Hence $-\|h_1\|^2 \le \langle h_1, h_2 \rangle$ for all h_1 in $\operatorname{ran} E$ and h_2 in $\ker E$. If there are such h_1 and h_2 with $\langle h_1, h_2 \rangle = \bar{\alpha} \ne 0$, then substituting $k_2 = -2\alpha^{-1}\|h_1\|^2 h_2$ for h_2 in this inequality, we obtain $-\|h_1\|^2 \le -2\|h_1\|^2$, a contradiction. Hence $\langle h_1, h_2 \rangle = 0$ whenever $h_1 \in \operatorname{ran} E$ and $h_2 \in \ker E$. That is, E is a projection.

(a) $\Rightarrow$ (d): Let $h, g \in \mathcal{H}$ and put $h = h_1 + h_2$ and $g = g_1 + g_2$, where $h_1, g_1 \in \operatorname{ran} E$ and $h_2, g_2 \in \ker E = (\operatorname{ran} E)^\perp$. Hence $\langle Eh, g \rangle = \langle h_1, g_1 \rangle$. Also, $\langle E^*h, g \rangle = \langle h, Eg \rangle = \langle h_1, g_1 \rangle = \langle Eh, g \rangle$. Thus $E = E^*$.

(d) $\Rightarrow$ (e): clear.

(e) $\Rightarrow$ (a): By (2.16), $\|Eh\| = \|E^*h\|$ for every h. Hence $\ker E = \ker E^*$. But by (2.19), $\ker E^* = (\operatorname{ran} E)^\perp$, so E is a projection. ∎

Note that by part (b) of the preceding proposition, if E is a projection and $\mathcal{M} = \operatorname{ran} E$, then $E = P_{\mathcal{M}}$.

Let P be a projection with $\operatorname{ran} P = \mathcal{M}$ and $\ker P = \mathcal{N}$. So both $\mathcal{M}$ and $\mathcal{N}$ are closed subspaces of $\mathcal{H}$ and, hence, are also Hilbert spaces. As in (I.6.1), we can form $\mathcal{M} \oplus \mathcal{N}$. If $U: \mathcal{M} \oplus \mathcal{N} \to \mathcal{H}$ is defined by $U(h \oplus g) = h + g$ for h in $\mathcal{M}$ and g in $\mathcal{N}$, then it is easy to see that U is an isomorphism. Making this identification, we will often write $\mathcal{H} = \mathcal{M} \oplus \mathcal{N}$.

More generally, the following will be used.

3.4. Definition. If $\{\mathcal{M}_i\}$ is a collection of pairwise orthogonal subspaces of $\mathcal{H}$, then

$$\oplus_i \mathcal{M}_i \equiv \bigvee_i \mathcal{M}_i.$$

If $\mathcal{M}$ and $\mathcal{N}$ are two closed linear subspaces of $\mathcal{H}$, then

$$\mathcal{M} \ominus \mathcal{N} \equiv \mathcal{M} \cap \mathcal{N}^\perp.$$

This is called the *orthogonal difference* of $\mathcal{M}$ and $\mathcal{N}$.

Note that if $\mathcal{M}, \mathcal{N} \le \mathcal{H}$ and $\mathcal{M} \perp \mathcal{N}$, then $\mathcal{M} + \mathcal{N}$ is closed. (Why?) Hence $\mathcal{M} \oplus \mathcal{N} = \mathcal{M} + \mathcal{N}$. The same is true, of course, for any finite collection of pairwise orthogonal subspaces but not for infinite collections.

3.5.   Definition. If $A \in \mathcal{B}(\mathcal{H})$ and $\mathcal{M} \le \mathcal{H}$, say that $\mathcal{M}$ is an *invariant subspace* for A if $Ah \in \mathcal{M}$ whenever $h \in \mathcal{M}$. In other words, if $A\mathcal{M} \subseteq \mathcal{M}$. Say that $\mathcal{M}$ is a *reducing subspace* for A if $A\mathcal{M} \subseteq \mathcal{M}$ and $A\mathcal{M}^{\perp} \subseteq \mathcal{M}^{\perp}$.

If $\mathcal{M} \le \mathcal{H}$, then $\mathcal{H} = \mathcal{M} \oplus \mathcal{M}^{\perp}$. If $A \in \mathcal{B}(\mathcal{H})$, then A can be written as a 2×2 matrix with operator entries,

$$3.6 \qquad\qquad A = \begin{bmatrix} W & X \\ Y & Z \end{bmatrix},$$

where $W \in \mathcal{B}(\mathcal{M})$, $X \in \mathcal{B}(\mathcal{M}^{\perp}, \mathcal{M})$, $Y \in \mathcal{B}(\mathcal{M}, \mathcal{M}^{\perp})$, and $Z \in \mathcal{B}(\mathcal{M}^{\perp})$.

3.7.   Proposition. *If $A \in \mathcal{B}(\mathcal{H})$, $\mathcal{M} \le \mathcal{H}$, and $P = P_{\mathcal{M}}$, then statements (a) through (c) are equivalent.*

(a) $\mathcal{M}$ is invariant for A.
(b) $PAP = AP$.
(c) In (3.6), $Y = 0$.

Also, statements (d) through (g) are equivalent.

(d) $\mathcal{M}$ reduces A.
(e) $PA = AP$.
(f) In (3.6), Y and X are 0.
(g) $\mathcal{M}$ is invariant for both A and A^*.

PROOF. (a) $\Rightarrow$ (b): If $h \in \mathcal{H}$, $Ph \in \mathcal{M}$. So $APh \in \mathcal{M}$. Hence, $P(APh) = APh$. That is, $PAP = AP$.

(b) $\Rightarrow$ (c): If P is represented as a 2×2 operator matrix relative to $\mathcal{H} = \mathcal{M} \oplus \mathcal{M}^{\perp}$, then

$$P = \begin{bmatrix} I & 0 \\ 0 & 0 \end{bmatrix}.$$

Hence,

$$PAP = \begin{bmatrix} W & 0 \\ 0 & 0 \end{bmatrix} = AP = \begin{bmatrix} W & 0 \\ Y & 0 \end{bmatrix}.$$

So $Y = 0$.

(c) $\Rightarrow$ (a): If $Y = 0$ and $h \in \mathcal{M}$, then

$$Ah = \begin{bmatrix} W & X \\ 0 & Z \end{bmatrix} \begin{bmatrix} h \\ 0 \end{bmatrix} = \begin{bmatrix} Wh \\ 0 \end{bmatrix} \in \mathcal{M}.$$

(d) $\Rightarrow$ (e): Since both $\mathcal{M}$ and $\mathcal{M}^{\perp}$ are invariant for A, (b) implies that $AP = PAP$ and $A(I - P) = (I - P)A(I - P)$. Multiplying this second equation gives $A - AP = A - AP - PA + PAP$. Thus $PA = PAP = AP$.

(e) $\Rightarrow$ (f): Exercise.

(f) $\Rightarrow$ (g): If $X = Y = 0$, then

$$A = \begin{bmatrix} W & 0 \\ 0 & Z \end{bmatrix} \quad \text{and} \quad A^* = \begin{bmatrix} W^* & 0 \\ 0 & Z^* \end{bmatrix}.$$

By (c), $\mathcal{M}$ is invariant for both A and A^*.

(g) $\Rightarrow$ (d): If $h \in \mathcal{M}^\perp$ and $g \in \mathcal{M}$, then $\langle g, Ah \rangle = \langle A^*g, h \rangle = 0$ since $A^*g \in \mathcal{M}$. Since g was an arbitrary vector in $\mathcal{M}$, $Ah \in \mathcal{M}^\perp$. That is, $A\mathcal{M}^\perp \subseteq \mathcal{M}^\perp$. ∎

If $\mathcal{M}$ reduces A, then $X = Y = 0$ in (3.6). This says that a study of A is reduced to the study of the smaller operators W and Z. This is the reason for the terminology.

If $A \in \mathcal{B}(\mathcal{H})$ and $\mathcal{M}$ is an invariant subspace for A, then $A|\mathcal{M}$ is used to denote the restriction of A to $\mathcal{M}$. That is, $A|\mathcal{M}$ is the operator on $\mathcal{M}$ defined by $(A|\mathcal{M})h = Ah$ whenever $h \in \mathcal{M}$. Note that $A|\mathcal{M} \in \mathcal{B}(\mathcal{M})$ and $\|A|\mathcal{M}\| \leq \|A\|$. Also, if $\mathcal{M}$ is invariant for A and A has the representation (3.6) with $Y = 0$, then $W = A|\mathcal{M}$.

Exercises

1. Let $\mathcal{H}$ be the two-dimensional real Hilbert space $\mathbb{R}^2$, let $\mathcal{M} \equiv \{(x,0) \in \mathbb{R}^2 : x \in \mathbb{R}\}$ and let $\mathcal{N} \equiv \{(x, x\tan\theta) : x \in \mathbb{R}\}$, where $0 < \theta < \frac{1}{2}\pi$. Find a formula for the idempotent E_θ with $\operatorname{ran} E_\theta = \mathcal{M}$ and $\ker E_\theta = \mathcal{N}$. Show that $\|E_\theta\| = (\sin\theta)^{-1}$.

2. Prove Proposition 3.2 (c).

3. Let $\{\mathcal{M}_i : i \in I\}$ be a collection of closed subspaces of $\mathcal{H}$ and show that $\cap\{\mathcal{M}_i : i \in I\} = [\vee\{\mathcal{M}_i : i \in I\}]^\perp$ and $[\cap\{\mathcal{M}_i : i \in I\}]^\perp = \vee\{\mathcal{M}_i^\perp : i \in I\}$.

4. Let P and Q be projections. Show: (a) $P + Q$ is a projection if and only if $\operatorname{ran} P \perp \operatorname{ran} Q$. If $P + Q$ is a projection, then $\operatorname{ran}(P + Q) = \operatorname{ran} P + \operatorname{ran} Q$ and $\ker(P + Q) = \ker P \cap \ker Q$. (b) PQ is a projection if and only if $PQ = QP$. If PQ is a projection, then $\operatorname{ran} PQ = \operatorname{ran} P \cap \operatorname{ran} Q$ and $\ker PQ = \operatorname{cl}(\ker P + \ker Q)$.

5. Generalize Exercise 4 as follows. Suppose $\{\mathcal{M}_i : i \in I\}$ is a collection of subspaces of $\mathcal{H}$ such that $\mathcal{M}_i \perp \mathcal{M}_j$ if $i \neq j$. Let P_i be the projection of $\mathcal{H}$ onto $\mathcal{M}_i$ and show that for all h in $\mathcal{H}$, $\Sigma\{P_ih : i \in I\}$ converges to Ph, where P is the projection of $\mathcal{H}$ onto $\vee\{\mathcal{M}_i : i \in I\}$.

6. If P and Q are projections, then the following statements are equivalent. (a) $P - Q$ is a projection. (b) $\operatorname{ran} Q \subseteq \operatorname{ran} P$. (c) $PQ = Q$. (d) $QP = Q$. If $P - Q$ is a projection, then $\operatorname{ran}(P - Q) = (\operatorname{ran} P) \ominus (\operatorname{ran} Q)$ and $\ker(P - Q) = \operatorname{ran} Q + \ker P$.

7. Let P and Q be projections. Show that $PQ = QP$ if and only if $P + Q - PQ$ is a projection. If this is the case, then $\operatorname{ran}(P + Q - PQ) = \operatorname{cl}(\operatorname{ran} P + \operatorname{ran} Q)$ and $\ker(P + Q - PQ) = \ker P \cap \ker Q$.

8. Give an example of two noncommuting projections.

9. Let $A \in \mathcal{B}(\mathcal{H})$ and let $\mathcal{N} =$ graph $A \subseteq \mathcal{H} \oplus \mathcal{H}$. That is, $\mathcal{N} = \{h \oplus Ah:$ $h \in \mathcal{H}\}$. Because A is continuous and linear, $\mathcal{N} \leq \mathcal{H} \oplus \mathcal{H}$. Let $\mathcal{M} = \mathcal{H} \oplus (0) \leq \mathcal{H} \oplus \mathcal{H}$. Prove the following statements. (a) $\mathcal{M} \cap \mathcal{N} = (0)$ if and only if ker $A = (0)$. (b) $\mathcal{M} + \mathcal{N}$ is dense in $\mathcal{H} \oplus \mathcal{H}$ if and only if ran A is dense in $\mathcal{H}$. (c) $\mathcal{M} + \mathcal{N} = \mathcal{H} \oplus \mathcal{H}$ if and only if A is surjective.

10. Find two closed linear subspaces $\mathcal{M}, \mathcal{N}$ of an infinite-dimensional Hilbert space $\mathcal{H}$ such that $\mathcal{M} \cap \mathcal{N} = (0)$ and $\mathcal{M} + \mathcal{N}$ is dense in $\mathcal{H}$, but $\mathcal{M} + \mathcal{N} \neq \mathcal{H}$.

11. Define $A: l^2(\mathbb{Z}) \to l^2(\mathbb{Z})$ by $A(\ldots, \alpha_{-1}, \hat{\alpha}_0, \alpha_1, \ldots) = (\ldots, \hat{\alpha}_{-1}, \alpha_0, \alpha_1, \ldots)$, where $\land$ sits above the coefficient in the 0-place. Find an invariant subspace of A that does not reduce A.

12. Let $\mu =$ Area measure on $\mathbb{D} \equiv \{z \in \mathbb{C}: |z| < 1\}$ and define $A: L^2(\mu) \to L^2(\mu)$ by $(Af)(z) = zf(z)$ for $|z| < 1$ and f in $L^2(\mu)$. Find a nontrivial reducing subspace for A and an invariant subspace that does not reduce A.

§4. Compact Operators

It turns out that most of the statements about linear transformations on finite-dimensional spaces have nice generalizations to a certain class of operators on infinite-dimensional spaces—namely, to the compact operators.

4.1. Definition. A linear transformation $T: \mathcal{H} \to \mathcal{K}$ is *compact* if $T(\text{ball } \mathcal{H})$ has compact closure in $\mathcal{K}$. The set of compact operators from $\mathcal{H}$ into $\mathcal{K}$ is denoted by $\mathcal{B}_0(\mathcal{H}, \mathcal{K})$, and $\mathcal{B}_0(\mathcal{H}) = \mathcal{B}_0(\mathcal{H}, \mathcal{H})$.

4.2. Proposition. (a) $\mathcal{B}_0(\mathcal{H}, \mathcal{K}) \subseteq \mathcal{B}(\mathcal{H}, \mathcal{K})$.
(b) $\mathcal{B}_0(\mathcal{H}, \mathcal{K})$ is a linear space and if $\{T_n\} \subseteq \mathcal{B}_0(H, K)$ and $T \in \mathcal{B}(H, K)$ such that $\|T_n - T\| \to 0$, then $T \in \mathcal{B}_0(\mathcal{H}, \mathcal{K})$.
(c) If $A \in \mathcal{B}(\mathcal{H})$, $B \in \mathcal{B}(\mathcal{K})$, and $T \in \mathcal{B}_0(\mathcal{H}, \mathcal{K})$, then TA and $BT \in \mathcal{B}_0(\mathcal{H}, \mathcal{K})$.

PROOF. (a) If $T \in \mathcal{B}_0(\mathcal{H}, \mathcal{K})$, then cl$[T(\text{ball } \mathcal{H})]$ is compact in $\mathcal{K}$. Hence there is a constant $C > 0$ such that $T(\text{ball } \mathcal{H}) \subseteq \{k \in \mathcal{K}: \|k\| \leq C\}$. Thus $\|T\| \leq C$.

(b) It is left to the reader to show that $\mathcal{B}_0(\mathcal{H}, \mathcal{K})$ is a linear space. For the second part of (b), it will be shown that $T(\text{ball } \mathcal{H})$ is totally bounded. Since $\mathcal{K}$ is a complete metric space, this is equivalent to showing that $T(\text{ball } \mathcal{H})$ has compact closure. Let $\varepsilon > 0$ and choose n such that $\|T - T_n\| < \varepsilon/3$. Since T_n is compact, there are vectors $h_1, \ldots, h_m$ in ball $\mathcal{H}$ such that $T_n(\text{ball } \mathcal{H}) \subseteq \bigcup_{j=1}^m B(T_n h_j; \varepsilon/3)$. So if $\|h\| \leq 1$, there is an h_j with

$\|T_n h_j - T_n h\| < \varepsilon/3$. Thus

$$\|Th_j - Th\| \leq \|Th_j - T_n h_j\| + \|T_n h_j - T_n h\| + \|T_n h - Th\|$$
$$< 2\|T - T_n\| + \varepsilon/3$$
$$< \varepsilon.$$

Hence $T(\text{ball } \mathcal{H}) \subseteq \bigcup_{j=1}^{m} B(Th_j; \varepsilon)$.

The proof of (c) is left to the reader. ∎

4.3. Definition. An operator T on $\mathcal{H}$ has *finite rank* if $\operatorname{ran} T$ is finite dimensional. The set of finite-rank operators is denoted by $\mathscr{B}_{00}(\mathcal{H}, \mathcal{K})$; $\mathscr{B}_{00}(\mathcal{H}) = \mathscr{B}_{00}(\mathcal{H}, \mathcal{H})$.

It is easy to see that $\mathscr{B}_{00}(\mathcal{H}, \mathcal{K})$ is a linear space and $\mathscr{B}_{00}(\mathcal{H}, \mathcal{K}) \subseteq \mathscr{B}_0(\mathcal{H}, \mathcal{K})$ (Exercise 2). Before giving other examples of compact operators, however, the next result should be proved.

4.4. Theorem. *If $T \in \mathscr{B}(\mathcal{H}, \mathcal{K})$, the following statements are equivalent.*

(a) *T is compact.*
(b) *T^* is compact.*
(c) *There is a sequence $\{T_n\}$ of operators of finite rank such that $\|T - T_n\| \to 0$.*

PROOF. (c) ⇒ (a): This is immediate from (4.2b) and the fact that $\mathscr{B}_{00}(\mathcal{H}, \mathcal{K}) \subseteq \mathscr{B}_0(\mathcal{H}, \mathcal{K})$.

(a) ⇒ (c): Since $\text{cl}[T(\text{ball } \mathcal{H})]$ is compact, it is separable. Therefore $\text{cl}(\operatorname{ran} T) = \mathscr{L}$ is a separable subspace of $\mathcal{K}$. Let $\{e_1, e_2, \ldots\}$ be a basis for $\mathscr{L}$ and let P_n be the orthogonal projection of $\mathcal{K}$ onto $\bigvee\{e_j: 1 \leq j \leq n\}$. Put $T_n = P_n T$; note that each T_n has finite rank. It will be shown that $\|T_n - T\| \to 0$, but first we prove the following:

Claim. If $h \in \mathcal{H}$, $\|T_n h - Th\| \to 0$.

In fact, $k = Th \in \mathscr{L}$, so $\|P_n k - k\| \to 0$ by (I.4.13d) and (I.4.7). That is, $\|P_n Th - Th\| \to 0$ and the claim is proved.

Since T is compact, if $\varepsilon > 0$, there are vectors $h_1, \ldots, h_m$ in ball $\mathcal{H}$ such that $T(\text{ball } \mathcal{H}) \subseteq \bigcup_{j=1}^{m} B(Th_j; \varepsilon/3)$. So if $\|h\| \leq 1$, choose h_j with $\|Th - Th_j\| < \varepsilon/3$. Thus for any integer n,

$$\|Th - T_n h\| \leq \|Th - Th_j\| + \|Th_j - T_n h_j\| + \|P_n(Th_j - Th)\|$$
$$\leq 2\|Th - Th_j\| + \|Th_j - T_n h_j\|$$
$$\leq 2\varepsilon/3 + \|Th_j - T_n h_j\|.$$

Using the claim we can find an integer n_0 such that $\|Th_j - T_n h_j\| < \varepsilon/3$ for $1 \leq j \leq m$ and $n \geq n_0$. So $\|Th - T_n h\| < \varepsilon$ uniformly for h in ball $\mathcal{H}$. Therefore $\|T - T_n\| < \varepsilon$ for $n \geq n_0$.

(c) $\Rightarrow$ (b): If $\{T_n\}$ is a sequence in $\mathcal{B}_{00}(\mathcal{H}, \mathcal{K})$ such that $\|T_n - T\| \to 0$, then $\|T_n^* - T^*\| = \|T_n - T\| \to 0$. But $T_n^* \in \mathcal{B}_{00}(\mathcal{H}, \mathcal{K})$ (Exercise 3). Since (c) implies (a), T^* is compact.

(b) $\Rightarrow$ (a): Exercise. ∎

A fact emerged in the proof that (a) implies (c) in the preceding theorem that is worth recording.

4.5.   Corollary. *If $T \in \mathcal{B}_0(\mathcal{H}, \mathcal{K})$, then $\mathrm{cl}(\mathrm{ran}\, T)$ is separable and if $\{e_n\}$ is a basis for $\mathrm{cl}(\mathrm{ran}\, T)$ and P_n is the projection of $\mathcal{K}$ onto $\bigvee\{e_j : 1 \le j \le n\}$, then $\|P_n T - T\| \to 0$.*

4.6.   Proposition. *Let $\mathcal{H}$ be a separable Hilbert space with basis $\{e_n\}$; let $\{\alpha_n\} \subseteq \mathbb{F}$ with $M = \sup\{|\alpha_n| : n \ge 1\} < \infty$. If $Ae_n = \alpha_n e_n$ for all n, then A extends by linearity to a bounded operator on $\mathcal{H}$ with $\|A\| = M$. The operator A is compact if and only if $\alpha_n \to 0$ as $n \to \infty$.*

PROOF. The fact that A is bounded and $\|A\| = M$ is an exercise; such an operator is said to be *diagonalizable* (see Exercise 1.8). Let P_n be the projection of $\mathcal{H}$ onto $\bigvee\{e_1, \ldots, e_n\}$. Then $A_n = A - AP_n$ is seen to be diagonalizable with $A_n e_j = \alpha_j e_j$ if $j > n$ and $A_n e_j = 0$ if $j \le n$. So $AP_n \in \mathcal{B}_{00}(\mathcal{H})$ and $\|A_n\| = \sup\{|\alpha_j| : j > n\}$. If $\alpha_n \to 0$, then $\|A_n\| \to 0$ and so A is compact since it is the limit of a sequence of finite-rank operators. Conversely, if A is compact, then Corollary 4.5 implies $\|A_n\| \to 0$; hence $\alpha_n \to 0$. ∎

4.7.   Proposition. *If (X, Ω, μ) is a measure space and $k \in L^2(X \times X, \Omega \times \Omega, \mu \times \mu)$, then*

$$(Kf)(x) = \int k(x, y) f(y)\, d\mu(y)$$

is a compact operator and $\|K\| \le \|k\|_2$.

The following lemma is useful for proving this proposition.

4.8.   Lemma. *If $\{e_i : i \in I\}$ is a basis for $L^2(X, \Omega, \mu)$ and*

$$\phi_{ij}(x, y) = e_i(x)\overline{e_j(y)}$$

for i, j in I and x, y in X, then $\{\phi_{ij} : i, j \in I\}$ is a basis for $L^2(X \times X, \Omega \times \Omega, \mu \times \mu)$.

PROOF. Since $\int\int |\phi_{ij}|^2\, d\mu\, d\mu = \|e_i\|^2 \|e_j\|^2 = 1$, $\phi_{ij} \in L^2(\mu \times \mu)$. If $(i, j) \ne (\alpha, \beta)$, then

$$\langle \phi_{\alpha\beta}, \phi_{ij} \rangle = \int\int \phi_\alpha(x)\overline{\phi_\beta(y)}\,\overline{\phi_i(x)}\phi_j(y)\, d\mu(x)\, d\mu(y)$$

$$= \langle \phi_\alpha, \phi_i \rangle \langle \phi_j, \phi_\beta \rangle = 0.$$

So $\{\phi_{ij}\}$ is an orthonormal family.

If $\phi \in L^2(\mu \times \mu)$, then the fact that $\iint |\phi(x, y)|^2 \, d\mu(x) \, d\mu(y) < \infty$ implies that $\int |\phi(x, y)|^2 \, d\mu(x) < \infty$ for almost all y in X. That is, if $\phi_y(x) = \phi(x, y)$, then $\phi_y \in L^2(\mu)$ for almost all y. Thus $f_i(y) = \langle e_i, \phi_y \rangle = \int \phi(x, y) e_i(x) \, d\mu(x)$ is well defined. Moreover, $f_i \in L^2(\mu)$ (Exercise). But

$$\|f_i\|^2 = \sum_j |\langle e_j, f_i \rangle|^2 = \sum_j \left| \int \overline{f_i(y)} e_j(y) \, d\mu(y) \right|^2$$

$$= \sum_j \left| \int \int \phi(x, y) \overline{e_i(x)} e_j(y) \, d\mu(x) \, d\mu(y) \right|^2$$

$$= \sum_j |\langle \phi, \phi_{ij} \rangle|^2.$$

So if $\phi \perp \phi_{ij}$ for all i, j, then $f_i = 0$ for all i. Thus $\phi_y = 0$ in $L^2(\mu)$ for almost all y. That is, $\phi = 0$. Therefore, $\{\phi_{ij}\}$ is a basis. ∎

PROOF OF PROPOSITION 4.7. Let $\{e_i\}$ and $\{\phi_{ij}\}$ be as in Lemma 4.8. Since $k \in L^2(\mu \times \mu)$,

$$\|k\|_2^2 = \sum_{i, j} |\langle k, \phi_{ij} \rangle|^2$$

$$= \sum_{i, j} \left| \int \int k(x, y) \overline{e_i(x)} e_j(y) \, d\mu(x) \, d\mu(y) \right|^2$$

$$= \sum_{i, j} \left| \int \left[\int k(x, y) e_j(y) \, d\mu(y) \right] \overline{e_i(x)} \, d\mu(x) \right|^2$$

$$= \sum_{i, j} |\langle Ke_j, e_i \rangle|^2.$$

Thus, if $f = \sum_j \alpha_j e_j \in L^2(\mu)$, $\sum_j |\alpha_j|^2 < \infty$, then

$$|\langle Kf, e_i \rangle|^2 = \left| \sum_j \alpha_j \langle Ke_j, e_i \rangle \right|^2$$

$$\leq \left(\sum_j |\alpha_j|^2 \right) \left(\sum_j |\langle Ke_j, e_i \rangle|^2 \right).$$

Therefore,

$$\|Kf\|^2 = \sum_i |\langle Kf, e_i \rangle|^2$$

$$\leq \|k\|_2^2 \|f\|^2.$$

This shows that K is bounded and $\|K\| \leq \|k\|_2$.

Now assume that k is a linear combination of a finite number of the $\{\phi_{ij}\}$. It is left to the reader to show that in this case K has finite rank. If k is an arbitrary element of $L^2(\mu \times \mu)$, then k is in the linear span of a countable number of ϕ_{ij}. Say that $k = \sum_{n, m=1}^{\infty} \alpha_{nm} \phi_{nm}$, $\phi_{nm}(x, y) =$

$e_n(x)\overline{e_m(y)}$. If $k_N = \sum_{n,m=1}^{N} \alpha_{n,m}\phi_{n,m}$, then $\|k_N - k\|_2 \to 0$ as $N \to \infty$. If K_N is the integral operator corresponding to k_N, K_N has finite rank and $\|K_N - K\| \le \|k_N - k\|_2 \to 0$. Thus K is compact. ∎

In particular, note that the preceding proposition shows that the Volterra operator (1.7) is compact.

One of the dominant tools in the study of linear transformations on finite-dimensional spaces is the concept of eigenvalue.

4.9.   Definition. If $A \in \mathcal{B}(\mathcal{H})$, a scalar α is an *eigenvalue* of A if $\ker(A - \alpha) \ne (0)$. If h is a nonzero vector in $\ker(A - \alpha)$, h is called an *eigenvector* for α; thus $Ah = \alpha h$. Let $\sigma_p(A)$ denote the set of eigenvalues of A.

4.10.   Example. Let A be the diagonalizable operator in Proposition 4.6. Then $\sigma_p(A) = \{\alpha_1, \alpha_2, \ldots\}$. If $\alpha \in \sigma_p(A)$, let $J_\alpha = \{j \in \mathbb{N}: \alpha_j = \alpha\}$. Then h is an eigenvector for α if and only if $h \in \bigvee\{e_j: j \in J_\alpha\}$.

4.11.   Example. The Volterra operator has no eigenvalues.

4.12.   Example. Let $h \in \mathcal{H} = L_{\mathbb{C}}^2(-\pi, \pi)$ and define $K: \mathcal{H} \to \mathcal{H}$ by $(Kf)(x) = \int_{-\pi}^{\pi} h(x - y)f(y)\,dy$. If $\lambda_n = (2\pi)^{-1/2}\int_{-\pi}^{\pi} h(x)\exp(-inx)\,dx = \hat{h}(n)$, the nth Fourier coefficient of h, then $Ke_n = \lambda_n e_n$, where $e_n(x) = (2\pi)^{-1/2}\exp(inx)$.

The way to see this is to extend functions in $L_{\mathbb{C}}^2(-\pi, \pi)$ to $\mathbb{R}$ by periodicity and perform a change of variables in the formula for $(Ke_n)(x)$. The details are left to the reader.

Operators on finite-dimensional spaces always have eigenvalues. As the Volterra operator illustrates, the analogy between operators on finite-dimensional spaces and compact operators breaks down here. If, however, a compact operator has an eigenvalue, several nice things can be said if the eigenvalue is not zero.

4.13.   Proposition. *If $T \in \mathcal{B}_0(\mathcal{H})$, $\lambda \in \sigma_p(T)$, and $\lambda \ne 0$, then the eigenspace $\ker(T - \lambda)$ is finite dimensional.*

PROOF. Suppose there is an infinite orthonormal sequence $\{e_n\}$ in $\ker(T - \lambda)$. Since T is compact, there is a subsequence $\{e_{n_k}\}$ such that $\{Te_{n_k}\}$ converges. Thus, $\{Te_{n_k}\}$ is a Cauchy sequence. But for $n_k \ne n_j$, $\|Te_{n_k} - Te_{n_j}\|^2 = \|\lambda e_{n_k} - \lambda e_{n_j}\|^2 = 2|\lambda|^2 > 0$ since $\lambda \ne 0$. This contradiction shows that $\ker(T - \lambda)$ must be finite dimensional. ∎

The next result on the existence of eigenvalues is not a practical way to show that a specific example has a nonzero eigenvalue, but it is a good theoretical tool that will be used later in this book (in particular, in the next section).

4.14. Proposition. *If T is a compact operator on $\mathcal{H}$, $\lambda \neq 0$, and $\inf\{\|(T - \lambda)h\| : \|h\| = 1\} = 0$, then $\lambda \in \sigma_p(T)$.*

PROOF. By hypothesis, there is a sequence of unit vectors $\{h_n\}$ such that $\|(T - \lambda)h_n\| \to 0$. Since T is compact, there is a vector f in $\mathcal{H}$ and a subsequence $\{h_{n_k}\}$ such that $\|Th_{n_k} - f\| \to 0$. But $h_{n_k} = \lambda^{-1}[(\lambda - T)h_{n_k} + Th_{n_k}] \to \lambda^{-1}f$. So $1 = \|\lambda^{-1}f\| = |\lambda|^{-1}\|f\|$ and $f \neq 0$. Also, it must be that $Th_{n_k} \to \lambda^{-1}Tf$. Since $Th_{n_k} \to f$, $f = \lambda^{-1}Tf$, or $Tf = \lambda f$. That is, $f \in \ker(T - \lambda)$ and $f \neq 0$, so $\lambda \in \sigma_p(T)$. ∎

4.15. Corollary. *If T is a compact operator on $\mathcal{H}$, $\lambda \neq 0$, $\lambda \notin \sigma_p(T)$, and $\bar{\lambda} \notin \sigma_p(T^*)$, then $\operatorname{ran}(T - \lambda) = \mathcal{H}$ and $(T - \lambda)^{-1}$ is a bounded operator on $\mathcal{H}$.*

PROOF. Since $\lambda \notin \sigma_p(T)$, the preceding proposition implies that there is a constant $c > 0$ such that $\|(T - \lambda)h\| \geq c\|h\|$ for all h in $\mathcal{H}$. If $f \in \operatorname{cl} \operatorname{ran}(T - \lambda)$, then there is a sequence $\{h_n\}$ in $\mathcal{H}$ such that $(T - \lambda)h_n \to f$. Thus $\|h_n - h_m\| \leq c^{-1}\|(T - \lambda)h_n - (T - \lambda)h_m\|$ and so $\{h_n\}$ is a Cauchy sequence. Hence $h_n \to h$ for some h in $\mathcal{H}$. Thus $(T - \lambda)h = f$. So $\operatorname{ran}(T - \lambda)$ is closed and, by (2.19), $\operatorname{ran}(T - \lambda) = [\ker(T - \lambda)^*]^{\perp} = \mathcal{H}$, by hypothesis.

So for f in $\mathcal{H}$ let $Af =$ the unique vector h such that $(T - \lambda)h = f$. Thus $(T - \lambda)Af = f$ for all f in $\mathcal{H}$. From the inequality above, $c\|Af\| \leq \|(T - \lambda)Af\| = \|f\|$. So $\|Af\| \leq c^{-1}\|f\|$ and A is bounded. Also, $(T - \lambda)A(T - \lambda)h = (T - \lambda)h$, so $0 = (T - \lambda)[A(T - \lambda)h - h]$. Since $\lambda \notin \sigma_p(T)$, $A(T - \lambda)h = h$. That is, $A = (T - \lambda)^{-1}$. ∎

It will be proved in a later chapter that if $\lambda \notin \sigma_p(T)$ and $\lambda \neq 0$, then $\bar{\lambda} \notin \sigma_p(T^*)$.

More will be shown about arbitrary compact operators in Chapter VI. In the next section the theory of compact self-adjoint operators will be explored.

EXERCISES

1. Prove Proposition 4.2(c).

2. Show that every operator of finite rank is compact.

3. If $T \in \mathcal{B}_{00}(\mathcal{H}, \mathcal{K})$, show that $T^* \in \mathcal{B}_{00}(\mathcal{K}, \mathcal{H})$ and $\dim(\operatorname{ran} T) = \dim(\operatorname{ran} T^*)$.

4. Show that an idempotent is compact if and only if it has finite rank.

5. Show that no nonzero multiplication operator on $L^2(0, 1)$ is compact.

6. Show that if $T: \mathcal{H} \to \mathcal{K}$ is a compact operator and $\{e_n\}$ is any orthonormal sequence in $\mathcal{H}$, then $\|Te_n\| \to 0$. Is the converse true?

7. If T is compact and $\mathcal{M}$ is an invariant subspace for T, show that $T|\mathcal{M}$ is compact.

8. If $h, g \in \mathcal{H}$, define $T: \mathcal{H} \to \mathcal{H}$ by $Tf = \langle f, h \rangle g$. Show that T has rank 1 [that is, $\dim(\operatorname{ran} T) = 1$]. Moreover, every rank 1 operator can be so represented. Show that if T is a finite-rank operator, then there are orthonormal vectors $e_1, \ldots, e_n$ and vectors $g_1, \ldots, g_n$ such that $Th = \sum_{j=1}^{n} \langle h, e_j \rangle g_j$ for all h in $\mathcal{H}$. In this case show that T is normal if and only if $g_j = \lambda_j e_j$ for some scalars $\lambda_1, \ldots, \lambda_n$. Find $\sigma_p(T)$.

9. Show that a diagonalizable operator is normal.

10. Verify the statements in Example 4.10.

11. Verify the statement in Example 4.11.

12. Verify the statement in Example 4.12. (Note that the operator K in this example is diagonalizable.)

13. If $T_n \in \mathcal{B}(\mathcal{H}_n)$, $n \geq 1$, with $\sup_n \|T_n\| < \infty$ and $T = \bigoplus_{n=1}^{\infty} T_n$ on $\mathcal{H} = \bigoplus_{n=1}^{\infty} \mathcal{H}_n$, show that T is compact if and only if each T_n is compact and $\|T_n\| \to 0$.

§5*. The Diagonalization of Compact Self-Adjoint Operators

This section and the remaining ones in this chapter may be omitted if the reader intends to continue through to the end of this book, as the material in these sections (save for Section 6) will be obtained in greater generality in Chapter IX. It is worthwhile, however, to examine this material even if Chapter IX is to be read, since the intuition provided by this special case is valuable.

The main result of this section is the following.

5.1. Theorem. *If T is a compact self-adjoint operator on $\mathcal{H}$, $\{\lambda_1, \lambda_2, \ldots\}$ are the distinct nonzero eigenvalues of T, and P_n is the projection of $\mathcal{H}$ onto $\ker(T - \lambda_n)$, then $P_n P_m = P_m P_n = 0$ if $n \neq m$, each λ_n is real, and*

5.2
$$T = \sum_{n=1}^{\infty} \lambda_n P_n,$$

where the series converges to T in the metric defined by the norm of $\mathcal{B}(\mathcal{H})$. [Of course, (5.2) may be only a finite sum.]

The proof of Theorem 5.1 requires a few preliminary results. Before beginning this process, let's look at a few consequences.

5.3. Corollary. *With the notation of* (5.1):

(a) $\ker T = [\vee\{P_n\mathcal{H}: n \geq 1\}]^{\perp} = (\operatorname{ran} T)^{\perp}$;
(b) *each* P_n *has finite rank*;
(c) $\|T\| = \sup\{|\lambda_n|: n \geq 1\}$ *and* $\lambda_n \to 0$ *as* $n \to \infty$.

PROOF. Since $P_n \perp P_m$ for $n \neq m$, if $h \in \mathcal{H}$, then (5.2) implies $\|Th\|^2 = \sum_{n=1}^{\infty}\|\lambda_n P_n h\|^2 = \sum_{n=1}^{\infty}|\lambda_n|^2\|P_n h\|^2$. Hence $Th = 0$ if and only if $P_n h = 0$ for all n. That is, $h \in \ker T$ if and only if $h \perp P_n\mathcal{H}$ for all n, whence (a).
  Part (b) follows by Proposition 4.13.
  For part (c), if $\mathcal{L} = \operatorname{cl}[\operatorname{ran} T]$, $\mathcal{L}$ is invariant for T. Since $T = T^*$, $\mathcal{L} = (\ker T)^{\perp}$ and $\mathcal{L}$ reduces T. So we can consider the restriction of T to $\mathcal{L}$, $T|\mathcal{L}$. Now $\mathcal{L} = \vee\{P_n\mathcal{H}: n \geq 1\}$ by (a). Let $\{e_j^{(n)}: 1 \leq j \leq N_n\}$ be a basis for $P_n\mathcal{H} = \ker(T - \lambda_n)$, so $Te_j^{(n)} = \lambda_n e_j^{(n)}$ for $1 \leq j \leq N_n$. Thus $\{e_j^{(n)}: 1 \leq j \leq N_n, n \geq 1\}$ is a basis for $\mathcal{L}$ and $T|\mathcal{L}$ is diagonalizable with respect to this basis. Part (c) now follows by (4.6). ∎

  The proof of (c) in the preceding corollary revealed an interesting fact that deserves a statement of its own.

5.4. Corollary. *If* T *is a compact self-adjoint operator, then there is a sequence* $\{\mu_n\}$ *of real numbers and an orthonormal basis* $\{e_n\}$ *for* $(\ker T)^{\perp}$ *such that for all* h,

$$Th = \sum_{n=1}^{\infty} \mu_n \langle h, e_n\rangle e_n.$$

  Note that there may be repetitions in the sequence $\{\mu_n\}$ in (5.4). How many repetitions?

5.5. Corollary. *If* $T \in \mathcal{B}_0(\mathcal{H})$, $T = T^*$, *and* $\ker T = (0)$, *then* $\mathcal{H}$ *is separable*.

  Also note that by (4.6), if (5.2) holds, $T \in \mathcal{B}_0(\mathcal{H})$.
  To begin the proof of Theorem 5.1, we prove a few results about not necessarily compact operators.

5.6. Proposition. *If* A *is a normal operator and* $\lambda \in \mathbb{F}$, *then* $\ker(A - \lambda) = \ker(A - \lambda)^*$ *and* $\ker(A - \lambda)$ *is a reducing subspace for* A.

PROOF. Since A is normal, so is $A - \lambda$. Hence $\|(A - \lambda)h\| = \|(A - \lambda)^*h\|$ (2.16). Thus $\ker(A - \lambda) = \ker(A - \lambda)^*$. If $h \in \ker(A - \lambda)$, $Ah = \lambda h \in \ker(A - \lambda)$. Also $A^*h = \bar{\lambda}h \in \ker(A - \lambda)$. Therefore $\ker(A - \lambda)$ reduces A. ∎

5.7. Proposition. *If* A *is a normal operator and* λ, μ *are distinct eigenvalues of* A, *then* $\ker(A - \lambda) \perp \ker(A - \mu)$.

PROOF. If $h \in \ker(A - \lambda)$ and $g \in \ker(A - \mu)$, then the fact (5.6) that $A^*g = \bar{\mu}g$ implies that $\lambda\langle h, g \rangle = \langle Ah, g \rangle = \langle h, A^*g \rangle = \langle h, \bar{\mu}g \rangle = \mu\langle h, g \rangle$. Thus $(\lambda - \mu)\langle h, g \rangle = 0$. Since $\lambda - \mu \neq 0$, $h \perp g$. ∎

5.8. Proposition. *If $A = A^*$ and $\lambda \in \sigma_p(A)$, then λ is a real number.*

PROOF. If $Ah = \lambda h$, then $Ah = A^*h = \bar{\lambda}h$ by (5.6). So $(\lambda - \bar{\lambda})h = 0$. Since h can be chosen different from 0, $\lambda = \bar{\lambda}$. ∎

The main result prior to entering the proof of Theorem 5.1 is to show that a compact self-adjoint operator has nonzero eigenvalues. If (5.3c) is examined, we see that there is a λ_n in $\sigma_p(T)$ with $|\lambda_n| = \|T\|$. Since the preceding proposition says that $\lambda_n \in \mathbb{R}$, it must be that $\lambda_n = \pm\|T\|$. That is, either $\pm\|T\| \in \sigma_p(T)$. This is the key to showing that $\sigma_p(T)$ is nonvoid.

5.9. Lemma. *If T is a compact self-adjoint operator, then either $\pm\|T\|$ is an eigenvalue of T.*

PROOF. If $T = 0$, the result is clear. So suppose $T \neq 0$. By Proposition 2.13 there is a sequence $\{h_n\}$ of unit vectors such that $|\langle Th_n, h_n \rangle| \to \|T\|$. By passing to a subsequence if necessary, we may assume that $\langle Th_n, h_n \rangle \to \lambda$, where $|\lambda| = \|T\|$. It will be show that $\lambda \in \sigma_p(T)$. Since $|\lambda| = \|T\|$, $0 \leq \|(T - \lambda)h_n\|^2 = \|Th_n\|^2 - 2\lambda\langle Th_n, h_n \rangle + \lambda^2 \leq 2\lambda^2 - 2\lambda\langle Th_n, h_n \rangle \to 0$. Hence $\|(T - \lambda)h_n\| \to 0$. By (4.14), $\lambda \in \sigma_p(T)$. ∎

PROOF OF THEOREM 5.1. By Lemma 5.9 there is a real number λ_1 in $\sigma_p(T)$ with $|\lambda_1| = \|T\|$. Let $\mathcal{E}_1 = \ker(T - \lambda_1)$, $P_1 =$ the projection onto $\mathcal{E}_1$, $\mathcal{H}_2 = \mathcal{E}_1^\perp$. By (5.6) $\mathcal{E}_1$ reduces T, so $\mathcal{H}_2$ reduces T. Let $T_2 = T|\mathcal{H}_2$; then T_2 is a self-adjoint compact operator on $\mathcal{H}_2$. (Why?)

By (5.9) there is an eigenvalue λ_2 for T_2 such that $|\lambda_2| = \|T_2\|$. Let $\mathcal{E}_2 = \ker(T_2 - \lambda_2)$. It is easy to check that $\mathcal{E}_2 = \ker(T - \lambda_2)$ and so $\lambda_2 \neq \lambda_1$. Let $P_2 =$ the projection of $\mathcal{H}$ onto $\mathcal{E}_2$ and put $\mathcal{H}_3 = (\mathcal{E}_1 \oplus \mathcal{E}_2)^\perp$. Note that $\|T_2\| \leq \|T\|$ so that $|\lambda_2| \leq |\lambda_1|$.

Using induction (give the details) we obtain a sequence $\{\lambda_n\}$ of real eigenvalues of T such that

(i) $|\lambda_1| \geq |\lambda_2| \geq \cdots$;

(ii) If $\mathcal{E}_n = \ker(T - \lambda_n)$, $|\lambda_{n+1}| = \|T|(\mathcal{E}_1 \oplus \cdots \oplus \mathcal{E}_n)^\perp\|$.

By (i) there is a nonnegative number α such that $|\lambda_n| \to \alpha$.

Claim. $\alpha = 0$; that is, $\lim \lambda_n = 0$.

In fact, let $e_n \in \mathcal{E}_n$, $\|e_n\| = 1$. Since T is compact, there is an h in $\mathcal{H}$ and a subsequence $\{e_{n_j}\}$ such that $\|Te_{n_j} - h\| \to 0$. But $e_n \perp e_m$ for $n \neq m$ and

$Te_{n_j} = \lambda_{n_j} e_{n_j}$. Hence $\|Te_{n_j} - Te_{n_i}\|^2 = \lambda_{n_j}^2 + \lambda_{n_i}^2 \geq 2\alpha^2$. Since $\{Te_{n_j}\}$ is a Cauchy sequence, $\alpha = 0$.

Now put P_n = the projection of $\mathcal{H}$ onto $\mathcal{E}_n$ and examine $T - \sum_{j=1}^n \lambda_j P_j$. If $h \in \mathcal{E}_k$, $1 \leq k \leq n$, then $(T - \sum_{j=1}^n \lambda_j P_j)h = Th - \lambda_k h = 0$. Hence $\mathcal{E}_1 \oplus \cdots \oplus \mathcal{E}_n \subseteq \ker(T - \sum_{j=1}^n \lambda_j P_j)$. If $h \in (\mathcal{E}_1 \oplus \cdots \oplus \mathcal{E}_n)^\perp$, then $P_j h = 0$ for $1 \leq j \leq n$; so $(T - \sum_{j=1}^n \lambda_j P_j)h = Th$. These two statements, together with the fact that $(\mathcal{E}_1 \oplus \cdots \oplus \mathcal{E}_n)^\perp$ reduces T, imply that

$$\left\| T - \sum_{j=1}^n \lambda_j P_j \right\| = \| T |(\mathcal{E}_1 \oplus \cdots \oplus \mathcal{E}_n)^\perp \|$$

$$= |\lambda_{n+1}| \to 0.$$

Therefore the series $\sum_{n=1}^\infty \lambda_n P_n$ converges in the metric of $\mathcal{B}(\mathcal{H})$ to T. ∎

Theorem 5.1 is called the *Spectral Theorem* for compact self-adjoint operators. Using it, one can answer virtually every question about compact hermitian operators, as will be seen before the end of this chapter.

If in Theorem 5.1 it is assumed that T is normal and compact, then the same conclusion, except for the statement that each λ_n is real, is true provided that $\mathcal{H}$ is a $\mathbb{C}$-Hilbert space. The proof of this will be given in Section 7.

EXERCISES

1. Prove Corollary 5.4.

2. Prove Corollary 5.5

3. Let K and k be as in Proposition 4.7 and suppose that $k(x, y) = \overline{k(y, x)}$. Show that K is self-adjoint and if $\{\mu_n\}$ are the eigenvalues of K, each repeated $\dim(K - \mu_n)$ times, then $\sum_1^\infty |\mu_n|^2 < \infty$.

4. If T is a compact self-adjoint operator and $\{e_n\}$ and $\{\mu_n\}$ are as in (5.4) and if h is a given vector in $\mathcal{H}$, show that there is a vector f in $\mathcal{H}$ such that $Tf = h$ if and only if $h \perp \ker T$ and $\sum_n \mu_n^{-2} |\langle h, e_n \rangle|^2 < \infty$. Find the form of the general vector f such that $Tf = h$.

5. Let T, $\{\mu_n\}$, and $\{e_n\}$ be as in (5.4). If $\lambda \neq 0$ and $\lambda \neq \mu_n$ for any μ_n, then for every h in $\mathcal{H}$ there is a unique f in $\mathcal{H}$ such that $(T - \lambda)f = h$. Moreover, $f = \lambda^{-1}[h + \sum_{n=1}^\infty \lambda_n (\lambda - \lambda_n)^{-1} \langle h, e_n \rangle e_n]$. Interpret this when T is an integral operator.

§6*. An Application: Sturm–Liouville Systems

In this section, $[a, b]$ will be a proper interval with $-\infty < a < b < \infty$. $C[a, b]$ denotes the continuous functions $f: [a, b] \to \mathbb{R}$ and for $n \geq 1$, $C^{(n)}[a, b]$ denotes those functions in $C[a, b]$ that have n continuous derivatives. $C_{\mathbb{C}}^{(n)}[a, b]$ denotes the corresponding spaces of complex-valued functions. We want to consider the differential equation

6.1 $-h'' + qh - \lambda h = f,$

where λ is a given complex number, $q \in C[a, b]$, and $f \in L^2[a, b]$, together with the boundary conditions

6.2
$$\begin{cases} \text{(a) } \alpha h(a) + \alpha_1 h'(a) = 0 \\ \text{(b) } \beta h(b) + \beta_1 h'(b) = 0 \end{cases},$$

where α, α_1, β, and β_1 are real numbers and $\alpha^2 + \alpha_1^2 > 0$, $\beta^2 + \beta_1^2 > 0$.

Equation (6.1) together with the boundary conditions (6.2) is called a (*regular*) *Sturm–Liouville system*. Such systems arise in a number of physical problems, including the description of the motion of a vibrating string. In this section we will discuss solutions of the Sturm–Liouville system by relating the system to a certain compact self-adjoint integral operator.

Recall that an absolutely continuous function h on $[a, b]$ has a derivative a.e. and $h(x) = \int_a^x h'(t)\, dt + h(a)$ for all x.

Define

$$\mathscr{D}_a \equiv \{ h \in C^{(1)}\mathbb{C}[a, b] : h' \text{ is absolutely continuous,}$$

$$h'' \in L^2[a, b], \text{ and } h \text{ satisfies } (6.2a) \}.$$

$\mathscr{D}_b$ is defined similarly but each h in $\mathscr{D}_b$ satisfies (6.2b) instead of (6.2a). The space $\mathscr{D} = \mathscr{D}_a \cap \mathscr{D}_b$.

Define $L: \mathscr{D} \to L^2[a, b]$ by

6.3 $Lh = -h'' + qh.$

L is called a *Sturm–Liouville operator*.

Note that $\mathscr{D}$ is a linear space and L is a linear transformation. The Sturm–Liouville problem thus becomes: if $\lambda \in \mathbb{C}$ and $f \in L^2[a, b]$, is there an h in $\mathscr{D}$ with $(L - \lambda)h = f$. Equivalently, for which λ is f in ran$(L - \lambda)$?

By placing a suitable norm on $\mathscr{D}$, L can be made into a bounded operator. This does not help much. The best procedure is to consider $(L - \lambda)^{-1}$. Integration is the inverse of differentiation, and it turns out that $(L - \lambda)^{-1}$ (when we can define it) is an integral operator.

Begin by considering the case when $\lambda = 0$. (Equivalently, replace q by $q - \lambda$.) To define L^{-1} (even if only on its range), we need that L is injective. Thus we make an assumption:

6.4   If $h \in \mathscr{D}$ and $Lh = 0$, then $h = 0$.

The first lemma is from ordinary differential equations and says that certain initial-value problems have nontrivial (nonzero) solutions.

6.5. Lemma. *If* $\alpha, \alpha_1, \beta, \beta_1 \in \mathbb{R}$, $\alpha^2 + \alpha_1^2 > 0$, *and* $\beta^2 + \beta_1^2 > 0$, *then there are functions* h_a, h_b *in* $\mathscr{D}_a, \mathscr{D}_b$, *respectively, such that* $L(h_a) = 0$ *and* $L(h_b) = 0$ *and* h_a, h_b *are real-valued and not identically zero.*

The *Wronskian* of h_a and h_b is the function

$$W = \det \begin{bmatrix} h_a & h_b \\ h'_a & h'_b \end{bmatrix} = h_a h'_b - h'_a h_b.$$

Note that $W' = h_a h''_b - h''_a h_b = h_a(qh_b) - (qh_a)h_b = 0$. Hence $W(x) \equiv W(a)$ for all x.

6.6. Lemma. *Assuming* (6.4), $W(a) \neq 0$ *and so* h_a *and* h_b *are linearly independent.*

PROOF. If $W(a) = 0$, then linear algebra tells us that the column vectors in the matrix used to define $W(a)$ are linearly dependent. Thus there is a λ in $\mathbb{R}$ such that $h_b(a) = \lambda h_a(a)$ and $h'_b(a) = \lambda h'_a(a)$. Thus $h_b \in \mathcal{D}$ and $L(h_b) = 0$. By (6.4), $h_b \equiv 0$, a contradiction. ∎

Put $c = W(a)$ and define $g: [a, b] \times [a, b] \to \mathbb{R}$ by

6.7
$$g(x, y) = \begin{cases} c^{-1} h_a(x) h_b(y) & \text{if } a \leq x \leq y \leq b \\ c^{-1} h_a(y) h_b(x) & \text{if } a \leq y \leq x \leq b. \end{cases}$$

The function g is the *Green function* for L.

6.8. Lemma. *The function g defined in* (6.7) *is real-valued, continuous, and* $g(x, y) = g(y, x)$.

PROOF. Exercise.

6.9. Theorem. *Assume* (6.4). *If g is the Green function for L defined in* (6.7) *and* $G: L^2[a, b] \to L^2[a, b]$ *is the integral operator defined by*

$$(Gf)(x) = \int_a^b g(x, y) f(y) \, dy,$$

then G is a compact self-adjoint operator, $\operatorname{ran} G = \mathcal{D}$, $LGf = f$ *for all f in* $L^2[a, b]$, *and* $GLh = h$ *for all h in* $\mathcal{D}$.

PROOF. That G is self-adjoint follows from the fact that g is real-valued and $g(x, y) = g(y, x)$; G is compact by (4.7). Fix f in $L^2[a, b]$ and put $h = Gf$. It must be shown that $h \in \mathcal{D}$.

Put

$$H_a(x) = c^{-1} \int_a^x h_a(y) f(y) \, dy \quad \text{and} \quad H_b(x) = c^{-1} \int_x^b h_b(y) f(y) \, dy.$$

Then

$$h(x) = \int_a^b g(x, y) f(y) \, dy$$

$$= c^{-1} \int_a^x h_a(y) h_b(x) f(y) \, dy + c^{-1} \int_x^b h_a(x) h_b(y) f(y) \, dy.$$

That is, $h = H_a h_b + h_a H_b$. Differentiating this equation gives $h' = (c^{-1}h_a f)h_b + H_a h_b' + h_a' H_b + h_a(-c^{-1}h_b f) = H_a h_b' + h_a' H_b$ a.e. Since $H_a h_b' + h_a' H_b$ is absolutely continuous, as part of showing that $h \in \mathcal{D}$ we want to show the following.

Claim. $h' = H_a h'_b + h'_a H_b$ everywhere.

Put $\phi = H_a h_b' + h_a' H_b$ and put $\psi(x) = h(a) + \int_a^x \phi(y)\,dy$. So h and ψ are absolutely continuous, $h(a) = \psi(a)$, and $h' = \psi'$ a.e. Thus $h = \psi$ everywhere. But ψ has a continuous derivative ϕ, so h does too. That is, the claim is proved.

Differentiating $h' = H_a h_b' + h_a' H_b$ gives that a.e., $h'' = (c^{-1}h_a f)h_b' + H_a h_b'' + h_a'' H_b + h_a'(-c^{-1}h_b f)$; since each of these summands belongs to $L^2[a, b]$, $h'' \in L^2[a, b]$.

Because $H_a(a) = 0$ and $h_a \in \mathcal{D}_a$, $\alpha h(a) + \alpha_1 h'(a) = \alpha h_a(a)H_b(a) + \alpha_1 h_a'(a)H_b(a) = [\alpha h_a(a) + \alpha_1 h_a'(a)]H_b(a) = 0$. Hence $h \in \mathcal{D}_a$. Similarly, $h \in \mathcal{D}_b$. Thus $h \in \mathcal{D}$. Hence $\operatorname{ran} G \subseteq \mathcal{D}$.

Now to show that $LGf = f$. If $h = Gf$, $L(h) = -h'' + qh = -[c^{-1}h_a h_b' f + H_a h_b'' + h_a'' H_b - c^{-1}h_a' h_b f] + q(H_a h_b + h_a H_b) = (-h_b'' + qh_b)H_a + (-h_a'' + qh_a)H_b + c^{-1}(h_a' h_b - h_a h_b')f = f$ since $L(h_a) = L(h_b) = 0$ and $h_a' h_b - h_a h_b' = W = c$.

If $h \in \mathcal{D}$, then $Lh \in L^2[a, b]$. So by the first part of the proof, $LGLh = Lh$. Thus $0 = L(GLh - h)$. Since $\ker L = (0)$, $h = GLh$ and so $h \in \operatorname{ran} G$. ∎

6.10. Corollary. *Assume* (6.4). *If* $h \in \mathcal{D}$, $\lambda \in \mathbb{C}$, *and* $Lh = \lambda h$, *then* $Gh = \lambda^{-1}h$. *If* $h \in L^2[a, b]$ *and* $Gh = \lambda^{-1}h$, *then* $h \in \mathcal{D}$ *and* $Lh = \lambda h$.

PROOF. This is immediate from the theorem. ∎

6.11. Lemma. *Assume* (6.4). *If* $\alpha \in \sigma_p(G)$, *then* $\dim \ker(G - \alpha) = 1$.

PROOF. Suppose there are linearly independent functions h_1, h_2 in $\ker(G - \alpha)$. By (6.10), h_1, h_2 are solutions of the equation

$$-h'' + (q - \alpha^{-1})h = 0.$$

Since this is a second-order linear differential equation, every solution of it must be a linear combination of h_1 and h_2. But $h_1, h_2 \in \mathcal{D}$ so they satisfy (6.2). But a solution can be found to this equation satisfying any initial conditions at a—and thus not satisfying (6.2). This contradiction shows that linearly independent h_1, h_2 in $\ker(G - \alpha)$ cannot be found. ∎

6.12. Theorem. *Assume* (6.4). *Then there is a sequence* $\{\lambda_1, \lambda_2, \ldots\}$ *of real numbers and a basis* $\{e_1, e_2, \ldots\}$ *for* $L^2[a, b]$ *such that*

(a) $0 < |\lambda_1| < |\lambda_2| < \cdots$ *and* $\lambda_n \to \infty$.
(b) $e_n \in \mathcal{D}$ *and* $Le_n = \lambda_n e_n$ *for all* n.

(c) If $\lambda \neq \lambda_n$ for any λ_n and $f \in L^2[a, b]$, then there is a unique h in $\mathcal{D}$ with $Lh - \lambda h = f$.

(d) If $\lambda = \lambda_n$ for some n and $f \in L^2[a, b]$, then there is an h in $\mathcal{D}$ with $Lh - \lambda h = f$ if and only if $\langle f, e_n \rangle = 0$. If $\langle f, e_n \rangle = 0$, any two solutions of $Lh - \lambda h = f$ differ by a multiple of e_n.

PROOF. Parts (a) and (b) follow by Theorem 5.1, Corollary 6.10, and Lemma 6.11. For parts (c) and (d), first note that

6.13 $Lh - \lambda h = f$ if and only if $h - \lambda Gh = Gf$.

This is, in fact, a straightforward consequence of Theorem 6.9.

(c) If $\lambda \neq \lambda_n$ for any n, $\lambda^{-1} \notin \sigma_p(G)$. Since $G = G^*$, Corollary 4.15 implies $G - \lambda^{-1}$ is bijective. So if $f \in L^2[a, b]$, there is a unique h in $L^2[a, b]$ with $Gf = (\lambda^{-1} - G)h$. Thus $h \in \mathcal{D}$ and (6.13) implies $L(h/\lambda) - \lambda(h/\lambda) = f$.

(d) Suppose $\lambda = \lambda_n$ for some n. If $Lh - \lambda_n h = f$, then $h - \lambda_n Gh = Gf$. Hence $\langle Gf, e_n \rangle = \langle h, e_n \rangle - \lambda_n \langle Gh, e_n \rangle = \langle h, e_n \rangle - \lambda_n \langle h, Ge_n \rangle = \langle h, e_n \rangle - \lambda_n \lambda_n^{-1} \langle h, e_n \rangle = 0$. So $0 = \langle Gf, e_n \rangle = \langle f, Ge_n \rangle = \lambda_n \langle f, e_n \rangle$. Hence $f \perp e_n$.

Since $\mathbb{C} e_n = \ker(G - \lambda_n^{-1})$, $[e_n]^\perp \equiv \mathcal{N}$ reduces G. Let $G_1 = G|\mathcal{N}$. So G_1 is a compact self-adjoint operator on $\mathcal{N}$ and $\lambda_n \notin \sigma_p(G_1)$. By (4.15), $\text{ran}(G_1 - \lambda_n) = \mathcal{N}$. As in the proof of (c), if $f \perp e_n$, there is a unique h in $\mathcal{N}$ such that $Lh - \lambda_n h = f$. Note that $h + \alpha e_n$ is also a solution. If h_1, h_2 are two solutions, $h_1 - h_2 \in \ker(L - \lambda_n)$, so $h_1 - h_2 = \alpha e_n$. ∎

What happens if $\ker L \neq (0)$? In this case it is possible to find a real number μ such that $\ker(L - \mu) = (0)$ (Exercise 6). Replacing q by $q - \mu$, Theorem 6.12 now applies. More information on this problem can be found in Exercises 2 through 5.

EXERCISES

1. Consider the Sturm–Liouville operator $Lh = -h''$ with $a = 0$, $b = 1$, and for each of the following boundary conditions find the eigenvalues $\{\lambda_n\}$, the eigenvectors $\{e_n\}$, and the Green function $g(x, y)$: (a) $h(0) = h(1) = 0$; (b) $h'(0) = h'(1) = 0$; (c) $h(0) = 0$ and $h'(1) = 0$; (d) $h(0) = h'(0)$ and $h(1) = -h'(1)$.

2. In Theorem 6.12 show that $\sum_{n=1}^{\infty} \lambda_n^{-2} < \infty$ (see Exercise 5.3).

3. In Theorem 6.12 show that $h \in \mathcal{D}$ if and only if $h \in L^2[a, b]$ and $\sum_{n=1}^{\infty} \lambda_n^2 |\langle h, e_n \rangle|^2 < \infty$. If $h \in \mathcal{D}$, show that $h(x) = \sum_{n=1}^{\infty} \langle h, e_n \rangle e_n(x)$, where this series converges uniformly and absolutely on $[a, b]$.

4. In Theorem 6.12(c), show that $h(x) = \sum_{n=1}^{\infty} (\lambda_n - \lambda)^{-1} \langle f, e_n \rangle e_n(x)$ and this series converges uniformly and absolutely on $[a, b]$.

5. In Theorem 6.12(d), show that if $f \perp e_n$ and $Lh - \lambda_n h = f$, then $h(x) = \sum_{j \neq n} (\lambda_j - \lambda_n)^{-1} \langle f, e_j \rangle e_j(x) + \alpha e_n(x)$ for some α, where the series converges uniformly and absolutely on $[a, b]$.

6. This exercise demonstrates how to handle the case in which $\ker L \neq (0)$. (a) If $h, g \in C^{(1)}[a, b]$ with h', g' absolutely continuous and $h'', g'' \in L^2[a, b]$, show that

$$\int_a^b (h''g - hg'') = [h'(b)g(b) - h(b)g'(b)] - [h'(a)g(a) - h(a)g'(a)].$$

(b) If $h, g \in \mathcal{D}$, show that $\langle Lh, g \rangle = \langle h, Lg \rangle$. (The inner product is in $L^2[a, b]$.)
(c) If $h, g \in \mathcal{D}$ and $\lambda, \mu \in \mathbb{R}$, $\lambda \neq \mu$, and if $h \in \ker(L - \lambda)$, $g \in \ker(L - \mu)$, then $h \perp g$. (d) Show that there is a real number μ with $\ker(L - \mu) = (0)$.

§7*. The Spectral Theorem and Functional Calculus for Compact Normal Operators

We begin by characterizing the operators that commute with a diagonalizable operator. If one considers the definition of a diagonalizable operator (4.6), it is possible to reformulate it in a way that is more tractable for the present purpose and closer to the form of a compact self-adjoint operator given in (5.2). Unlike (4.6), it will not be assumed that the underlying Hilbert space is separable.

7.1. Proposition. *Let $\{P_i: i \in I\}$ be a family of pairwise orthogonal projections in $\mathcal{B}(\mathcal{H})$. (That is, $P_iP_j = P_jP_i = 0$ for $i \neq j$.) If $h \in \mathcal{H}$, then $\sum_i \{P_ih: i \in I\}$ converges in $\mathcal{H}$ to Ph, where P is the projection of $\mathcal{H}$ onto $\bigvee\{P_i\mathcal{H}: i \in I\}$.*

This appeared as Exercise 3.5 and its proof is left to the reader.
If $\{P_i: i \in I\}$ is as in the preceding proposition and $\mathcal{M}_i = P_i\mathcal{H}$, then with the notation of Definition 3.4, P is the projection of $\mathcal{H}$ onto $\oplus_i \mathcal{M}_i$. Write $P = \sum_i P_i$. A word of caution here: $Ph = \sum_i P_ih$, where the convergence is in the norm of $\mathcal{H}$. However, $\sum_i P_i$ does not converge to P in the norm of $\mathcal{B}(\mathcal{H})$. In fact, it never does unless I is finite (Exercise 1).

7.2. Definition. A *partition of the identity* on $\mathcal{H}$ is a family $\{P_i: i \in I\}$ of pairwise orthogonal projections on $\mathcal{H}$ such that $\bigvee_i P_i\mathcal{H} = \mathcal{H}$. This might be indicated by $1 = \sum_i P_i$ or $1 = \oplus_i P_i$. [Note that 1 is often used to denote the operator on $\mathcal{H}$ defined by $1(h) = h$ for all h. Similarly if $\alpha \in \mathbb{F}$, α is the operator defined by $\alpha(h) = \alpha h$ for all h.]

7.3. Definition. An operator A on $\mathcal{H}$ is *diagonalizable* if there is a partition of the identity on $\mathcal{H}$, $\{P_i: i \in I\}$, and a family of scalars $\{\alpha_i: i \in I\}$ such that $\sup_i |\alpha_i| < \infty$ and $Ah = \alpha_i h$ whenever $h \in \operatorname{ran} P_i$.

It is easy to see that this is equivalent to the definition given in (4.6) when $\mathcal{H}$ is separable (Exercise 2). Also, $\|A\| = \sup_i |\alpha_i|$.

To denote a diagonalizable operator satisfying the conditions of (7.3), write

$$A = \sum_i \alpha_i P_i \quad \text{or} \quad A = \oplus_i \alpha_i P_i.$$

Note that it was not assumed that the scalars α_i in (7.3) are distinct. There is no loss in generality in assuming this, however. In fact, if $\alpha_i = \alpha_j$, then we can replace P_i and P_j with $P_i + P_j$.

7.4. Proposition. *An operator A on $\mathscr{H}$ is diagonalizable if and only if there is an orthonormal basis for $\mathscr{H}$ consisting of eigenvectors for A.*

PROOF. Exercise.

Also note that if $A = \oplus_i \alpha_i P_i$, then $A^* = \oplus_i \bar{\alpha}_i P_i$ and A is normal (Exercise 5).

7.5. Theorem. *If $A = \oplus_i \alpha_i P_i$ is diagonalizable and all the α_i are distinct, then an operator B in $\mathscr{B}(\mathscr{H})$ satisfies $AB = BA$ if and only if for each i, ran P_i reduces B.*

PROOF. If all the α_i are distinct, then ran $P_i = \ker(A - \alpha_i)$. If $AB = BA$ and $Ah = \alpha_i h$, then $ABh = BAh = B(\alpha_i h) = \alpha_i Bh$; hence $Bh \in \text{ran } P_i$ whenever $h \in \text{ran } P_i$. Thus ran P_i is left invariant by B. Therefore B leaves $\bigvee\{\text{ran } P_j: j \neq i\} = \mathscr{N}_i$ invariant. But since $\oplus_i P_i = 1$, $\mathscr{N}_i = (\text{ran } P_i)^\perp$. Thus ran P_i reduces B.

Now assume that B is reduced by each ran P_i. Thus $BP_i = P_i B$ for all i. If $h \in \mathscr{H}$, then $Ah = \sum_i \alpha_i P_i h$. Hence $BAh = \sum_i \alpha_i BPh = \sum_i \alpha_i PBh = ABh$. (Why is the first equality valid?) ∎

Using the notation of the preceding theorem, if $AB = BA$, let $B_i = B|\text{ran } P_i$. Then it is appropriate to write $B = \oplus_i B_i$ on $\mathscr{H} = \oplus_i (P_i \mathscr{H})$. One might paraphrase Theorem 7.5 by saying that B commutes with a diagonalizable operator if and only if B can be "diagonalized with operator entries."

7.6. Spectral Theorem for Compact Normal Operators. *If T is a compact normal operator on the complex Hilbert space $\mathscr{H}$, $\{\lambda_1, \lambda_2, \dots\}$ are the distinct nonzero eigenvalues of T, and P_n is the projection of $\mathscr{H}$ onto $\ker(T - \lambda_n)$, then $P_n P_m = P_m P_n = 0$ if $n \neq m$ and*

7.7
$$T = \sum_{n=1}^{\infty} \lambda_n P_n,$$

where this series converges to T in the metric defined by the norm on $\mathscr{B}(\mathscr{H})$.

PROOF. Let $A = (T + T^*)/2$, $B = (T - T^*)/2i$. So A, B are compact self-adjoint operators, $T = A + iB$, and $AB = BA$ since T is normal. The idea of the proof is rather simple. We'll get started in this proof together but the reader will have to complete the details.

By Theorem 5.1, $A = \sum_1^\infty \alpha_n E_n$, where $\alpha_n \in \mathbb{R}$, $\alpha_n \neq \alpha_m$ if $n \neq m$, and E_n is the projection of $\mathcal{H}$ onto $\ker(A - \alpha_n)$. Since $AB = BA$, the idea is to use Theorem 7.5 and Theorem 5.1 applied to B to diagonalize A and B simultaneously; that is, to find an orthonormal basis for $\mathcal{H}$ consisting of vectors that are simultaneously eigenvectors of A and B.

Since $BA = AB$, $E_n \mathcal{H} = \mathcal{L}_n$ reduces B for every n (7.5). Let $B_n = B|\mathcal{L}_n$; then $B_n = B_n^*$ and $\dim \mathcal{L}_n < \infty$. Applying (5.1) to B_n (or, rather, the corresponding theorem from linear algebra) there is a basis $\{e_j^{(n)}: 1 \leq j \leq d_n\}$ for $\mathcal{L}_n$ and real numbers $\{\beta_j^{(n)}: 1 \leq j \leq d_n\}$ such that $B_n e_j^{(n)} = \beta_j^{(n)} e_j^{(n)}$. Thus $T e_j^{(n)} = A e_j^{(n)} + i B e_j^{(n)} = (\alpha_n + i\beta_j^{(n)}) e_j^{(n)}$.

Therefore $\{e_j^{(n)}: 1 \leq j \leq d_n, n \geq 1\}$ is a basis for $\mathrm{cl}(\mathrm{ran}\, A)$ consisting of eigenvectors for T. It may be that $\mathrm{cl}(\mathrm{ran}\, A) \neq \mathrm{cl}(\mathrm{ran}\, T)$. Since B is reduced by $\ker A = (\mathrm{ran}\, A)^\perp$ and $B_0 = B|\ker A$ is a compact self-adjoint operator, there is an orthonormal basis $\{e_j^{(0)}: j \geq 1\}$ for $\mathrm{cl}(\mathrm{ran}\, B_0)$ and scalars $\{\beta_j^{(0)}: j \geq 1\}$ such that $B e_j^{(0)} = \beta_j^{(0)} e_j^{(0)}$. It follows that $T e_j^{(0)} = i\beta_j^{(0)} e_j^{(0)}$. Moreover, $\ker T \subseteq \ker A \cap \ker B$, so $\mathrm{cl}(\mathrm{ran}\, T) \subseteq \mathrm{cl}(\mathrm{ran}\, A) \oplus \mathrm{cl}(\mathrm{ran}\, B_0)$.

The remainder of the proof now consists in a certain amount of bookkeeping to gather together the eigenvectors belonging to the same eigenvalues of T and the performing of some light housekeeping chores to obtain the convergence of the series (7.7)   ∎

7.8. Corollary. *With the notation of* (7.6):

(a) $\ker T = [\vee\{P_n \mathcal{H}: n \geq 1\}]^\perp$;
(b) *each* P_n *has finite rank*;
(c) $\|T\| = \sup\{|\lambda_n|: n \geq 1\}$ *and* $\lambda_n \to 0$ *as* $n \to \infty$.

The proof of (7.8) is similar to the proof of (5.3).

7.9. Corollary. *If T is a compact operator, then T is normal if and only if T is diagonalizable.*

If T is a normal operator which is not necessarily compact, there is a spectral theorem for T which has a somewhat different form. This theorem states that T can be represented as an integral with respect to a measure whose values are not numbers but projections on a Hilbert space. Theorem 7.6 will be a consequence of this more general theorem and correspond to the case in which this projection valued measure is "atomic."

The approach to this more general spectral theorem will be to develop a functional calculus for normal operators T. That is, an operator $\phi(T)$ will be defined for every bounded Borel function ϕ on $\mathbb{C}$ and certain properties of the map $\phi \mapsto \phi(T)$ will be deduced. The projection valued measure will then be obtained by letting $\mu(\Delta) = \chi_\Delta(T)$. These matters are taken up in Chapter IX.

At this point, Theorem 7.6 will be used to develop a functional calculus for compact normal operators. For the remainder of this section $\mathscr{H}$ is a complex Hilbert space.

7.10.   Definition. Denote by $l^\infty(\mathbb{C})$ all the bounded functions $\phi: \mathbb{C} \to \mathbb{C}$. If T is a compact normal operator satisfying (7.7), define $\phi(T)$: $\mathscr{H} \to \mathscr{H}$ by

$$\phi(T) = \sum_{n=1}^\infty \phi(\lambda_n) P_n + \phi(0) P_0,$$

where $P_0 = $ the projection of $\mathscr{H}$ onto $\ker T$.

Note that $\phi(T)$ is a diagonalizable operator and $\|\phi(T)\| = \sup\{|\phi(0)|, |\phi(\lambda_1)|, \ldots\}$ (4.6). Much more can be said.

7.11. Functional Calculus for Compact Normal Operators. *If T is a compact normal operator on a $\mathbb{C}$-Hilbert space $\mathscr{H}$, then the map $\phi \mapsto \phi(T)$ of $l^\infty(\mathbb{C}) \to \mathscr{B}(\mathscr{H})$ has the following properties*:

(a) $\phi \mapsto \phi(T)$ *is a multiplicative linear map of $l^\infty(\mathbb{C})$ into $\mathscr{B}(\mathscr{H})$. If $\phi \equiv 1$, $\phi(T) = 1$; if $\phi(z) = z$, then $\phi(T) = T$.*
(b) $\|\phi(T)\| = \sup\{|\phi(\lambda)|: \lambda \in \sigma_p(T)\}.$
(c) $\phi(T)^* = \phi^*(T)$, *where ϕ^* is the function defined by $\phi^*(z) = \overline{\phi(z)}$.*
(d) *If $A \in \mathscr{B}(\mathscr{H})$ and $AT = TA$, then $A\phi(T) = \phi(T)A$ for all ϕ in $l^\infty(\mathbb{C})$.*

PROOF. Adopt the notation of Theorem 7.6 and (7.10).

(a) If $\phi, \psi \in l^\infty(\mathbb{C})$, then $(\phi\psi)(z) = \phi(z)\psi(z)$ for z in $\mathbb{C}$. Also, $\phi(T)\psi(T)h = [\phi(0)P_0 + \Sigma\phi(\lambda_n)P_n][\psi(0)P_0 + \Sigma\psi(\lambda_m)P_m]h = [\phi(0)P_0 + \Sigma_n\phi(\lambda_n)P_n][\psi(0)P_0 h + \Sigma_m\psi(\lambda_m)P_m h]$. Since $P_n P_m = 0$ when $n \neq m$, this gives that $\phi(T)\psi(T)h = \phi(0)\psi(0)P_0 h + \Sigma_n\phi(\lambda_n)\psi(\lambda_n)P_n h = (\phi\psi)(T)h$. Thus $\phi \mapsto \phi(T)$ is multiplicative. The linearity of the map is left to the reader. If $\phi(z) \equiv 1$, then $\phi(T) = 1(T) = P_0 + \Sigma_{n=1}^\infty P_n = 1$ since $\{P_0, P_1, \ldots\}$ is a partition of the identity. If $\phi(z) = z$, $\phi(\lambda_n) = \lambda_n$ and so $\phi(T) = T$.

Parts (b) and (c) follow from Exercise 5.

(d) If $AT = TA$, Theorem 7.5 implies that $P_0\mathscr{H}, P_1\mathscr{H}, \ldots$ all reduce A. Fix h_n in $P_n\mathscr{H}$, $n \geq 0$. If $\phi \in l^\infty(\mathbb{C})$, then $Ah_n \in P_n\mathscr{H}$ and so $\phi(T)Ah_n = \phi(\lambda_n)Ah_n = A(\phi(\lambda_n)h_n) = A\phi(T)h_n$. If $h \in \mathscr{H}$, then $h = \Sigma_{n=0}^\infty h_n$, where $h_n \in P_n$. Hence $\phi(T)Ah = \Sigma_{n=0}^\infty\phi(T)Ah_n = \Sigma_{n=0}^\infty A\phi(T)h_n = A\phi(T)h$. (Justify the first equality.)   ∎

Which operators on $\mathscr{H}$ can be expressed as $\phi(T)$ for some ϕ in $l^\infty(\mathbb{C})$? Part (d) of the preceding theorem provides the answer.

7.12.   Theorem. *If T is a compact normal operator on a $\mathbb{C}$-Hilbert space, then $\{\phi(T): \phi \in l^\infty(\mathbb{C})\}$ is equal to*

$$\{B \in \mathcal{B}(\mathcal{H}): BA = AB \text{ whenever } AT = TA\}.$$

PROOF. Half of the desired equality is obtained from (7.11d). So let $B \in \mathcal{B}(\mathcal{H})$ and assume that $BA = AB$ whenever $AT = TA$. Thus, B must commute with T itself. By (7.5), B is reduced by each $P_n\mathcal{H} \equiv \mathcal{H}_n$, $n \geq 0$; put $B_n = B|\mathcal{H}_n$. Fix $n \geq 0$ for the moment and let A_n be any bounded operator in $\mathcal{B}(\mathcal{H}_n)$. Define $Ah = A_nh$ if $h \in \mathcal{H}_n$ and $Ah = 0$ if $h \in \mathcal{H}_m$, $m \neq n$, and extend A to $\mathcal{H}$ by linearity; so $A = \oplus_{m=0}^\infty A_m$ where $A_m = 0$ if $m \neq n$. By (7.5), $AT = TA$; hence $BA = AB$. This implies that $B_nA_n = A_nB_n$. Since A_n was arbitrarily chosen from $\mathcal{B}(\mathcal{H}_n)$, $B_n = \beta_n$ for some β_n (Exercise 7). If $\phi: \mathbb{C} \to \mathbb{C}$ is defined by $\phi(0) = \beta_0$ and $\phi(\lambda_n) = \beta_n$ for $n \geq 1$, then $B = \phi(T)$.   ∎

7.13.   Definition. If $A \in \mathcal{B}(\mathcal{H})$, then A is *positive* if $\langle Ah, h \rangle \geq 0$ for all h in $\mathcal{H}$. In symbols this is denoted by $A \geq 0$.

Note that by Proposition 2.12 every positive operator is self-adjoint.

7.14.   Proposition. *If T is a compact normal operator, then T is positive if and only if all its eigenvalues are positive real numbers.*

PROOF. Let $T = \sum_1^\infty \lambda_n P_n$. If $T \geq 0$ and $h \in P_n\mathcal{H}$ with $\|h\| = 1$, then $Th = \lambda_n h$. Hence $\lambda_n = \langle Th, h \rangle \geq 0$. Conversely, assume each $\lambda_n \geq 0$. If $h \in \mathcal{H}$, $h = h_0 + \sum_{n=1}^\infty h_n$, where $h_0 \in \ker T$ and $h_n \in P_n\mathcal{H}$ for $n \geq 1$. Then $Th = \sum_1^\infty \lambda_n h_n$. Hence $\langle Th, h \rangle = \langle \sum_{n=1}^\infty \lambda_n h_n, h_0 + \sum_{m=1}^\infty h_m \rangle = \sum_{n=1}^\infty \sum_{m=0}^\infty \lambda_n \langle h_n, h_m \rangle = \sum_{n=1}^\infty \lambda_n \|h_n\|^2 \geq 0$ since $\langle h_n, h_m \rangle = 0$ when $n \neq m$.   ∎

7.15.   Theorem. *If T is a compact self-adjoint operator, then there are unique positive compact operators A, B such that $T = A - B$ and $AB = BA = 0$.*

PROOF. Let $T = \sum_{n=1}^\infty \lambda_n P_n$ as in (7.6). Define $\phi, \psi: \mathbb{C} \to \mathbb{C}$ by $\phi(\lambda_n) = \lambda_n$ if $\lambda_n > 0$, $\phi(z) = 0$ otherwise; $\psi(\lambda_n) = -\lambda_n$ if $\lambda_n < 0$, $\psi(z) = 0$ otherwise. Put $A = \phi(T)$ and $B = \psi(T)$. Then $A = \sum\{\lambda_n P_n: \lambda_n > 0\}$ and $B = \sum\{-\lambda_n P_n: \lambda_n < 0\}$. Thus $T = A - B$. Since $\phi\psi = 0$, $AB = BA = 0$ by (7.11a). Since $\phi, \psi \geq 0$, $A, B \geq 0$ by the preceding proposition. It remains to show that A, B are unique.

Suppose $T = C - D$ where C, D are compact positive operators and $CD = DC = 0$. It is easy to check that C and D commute with T. Put $\lambda_0 = 0$ and $P_0 =$ the projection of $\mathcal{H}$ onto $\ker T$. Thus C and D are reduced by $P_n\mathcal{H} \equiv \mathcal{H}_n$ for all $n \geq 0$. Let $C_n = C|\mathcal{H}_n$ and $D_n = D|\mathcal{H}_n$. So $C_nD_n = D_nC_n = 0$, $\lambda_n P_n = T|\mathcal{H}_n = C_n - D_n$, and C_n, D_n are positive. Suppose $\lambda_n > 0$ and let $h \in \mathcal{H}_n$. Since $C_nD_n = 0$, $\ker C_n \supseteq \text{cl}[\text{ran } D_n] =$

$(\ker D_n)^{\perp}$. So if $h \in (\ker D_n)^{\perp}$, then $\lambda_n h = -D_n h$. Hence $\lambda_n \|h\|^2 = -\langle D_n h, h \rangle \leq 0$. Thus $h = 0$ since $\lambda_n > 0$. That is, $\ker D_n = \mathcal{H}_n$. Thus $D_n = 0 = B|\mathcal{H}_n$ and $C_n = \lambda_n P_n = A|\mathcal{H}_n$. Similarly, if $\lambda_n < 0$, $C_n = 0 = A|\mathcal{H}_n$ and $D_n = -\lambda_n P_n = B|\mathcal{H}_n$. On $\mathcal{H}_0$, $T|\mathcal{H}_0 = 0 = C_0 - D_0$. Thus $C_0 = D_0$. But $0 = C_0 D_0 = C_0^2$. Thus $0 = \langle C_0^2 h, h \rangle = \|C_0 h\|^2$, so $C_0 = 0 = A|\mathcal{H}_0$ and $D_0 = 0 = B|\mathcal{H}_0$. Therefore $C = A$ and $D = B$. ∎

Positive operators are analogous to positive numbers. With this in mind, the next result seems reasonable.

7.16. Theorem. *If T is a positive compact operator, then there is a unique positive compact operator A such that $A^2 = T$.*

PROOF. Let $T = \sum_{n=1}^{\infty} \lambda_n P_n$ as in the Spectral Theorem. Since $T \geq 0$, $\lambda_n > 0$ for all n (7.14). Let $\phi(\lambda_n) = \lambda_n^{1/2}$ and $\phi(z) = 0$ otherwise; put $A = \phi(T)$. It is easy to check that $A \geq 0$; $A = \sum_1^{\infty} \lambda_n^{1/2} P_n$ so that A is compact; and $A^2 = T$.

The proof of uniqueness is left to the reader. ∎

EXERCISES

1. If $\{P_n\}$ is a sequence of pairwise orthogonal nonzero projections and $P = \sum P_n$, show that $\|P - \sum_{j=1}^{n} P_j\| = 1$ for all n.

2. If $\mathcal{H}$ is separable, show that the definitions of a diagonalizable operator in (4.6) and (7.3) are equivalent.

3. If $A = \sum \alpha_i P_i$ as in (7.3), show that A is compact if and only if: (a) $\alpha_i = 0$ for all but a countable number of i; (b) P_i has finite rank whenever $\alpha_i \neq 0$; (c) if $\{\alpha_1, \alpha_2, \ldots\} = \{\alpha_i : \alpha_i \neq 0\}$, then $\alpha_n \to 0$ as $n \to \infty$.

4. Prove Proposition 7.4.

5. If $A = \oplus_i \alpha_i P_i$, show that $A^* = \oplus_i \bar{\alpha}_i P_i$, A is normal, and $\|A\| = \sup\{|\alpha_i| : i \in I\}$.

6. Give the remaining details in the proof of (7.6).

7. If $A \in \mathcal{B}(\mathcal{H})$ and $AT = TA$ for every compact operator T, show that A is a multiple of the identity operator.

8. Suppose T is a compact normal operator on a $\mathbb{C}$-Hilbert space such that $\dim \ker(T - \lambda) \leq 1$ for all λ in $\mathbb{C}$. Show that if $A \in \mathcal{B}(\mathcal{H})$ and $AT = TA$, then $A = \phi(T)$ for some ϕ in $l^{\infty}(\mathbb{C})$.

9. Prove a converse to Exercise 8: if T is a compact normal operator such that $\{A \in \mathcal{B}(\mathcal{H}) : AT = TA\} = \{\phi(T) : \phi \in l^{\infty}(\mathbb{C})\}$, then $\dim \ker(T - \lambda) \leq 1$ for all λ in $\mathbb{C}$.

10. Let T be a compact normal operator and show that $\ker(T - \lambda) \leq 1$ for all λ in $\mathbb{C}$ if and only if there is a vector h in $\mathcal{H}$ such that $\{p(T)h : p$ is a polynomial in one variable$\}$ is dense in $\mathcal{H}$. (Such a vector h is called a *cyclic vector* for T.)

11. If $\lambda \in \mathbb{C}$, let δ_λ be the unit point mass at λ; that is, δ_λ is the measure on $\mathbb{C}$ such that $\delta_\lambda(\Delta) = 1$ if $\lambda \in \Delta$ and $\delta_\lambda(\Delta) = 0$ if $\lambda \notin \Delta$. If $\{\lambda_1, \lambda_2, \ldots\}$ are distinct complex numbers and $\{\alpha_n\}$ is a sequence of real numbers with $\alpha_n > 0$ and $\sum_n \alpha_n < \infty$, let $\mu = \sum_{n=1}^\infty \alpha_n \delta_{\lambda_n}$; so μ is a finite measure. If $\phi \in l^\infty(\mathbb{C})$, let M_ϕ be the multiplication operator on $L^2(\mu)$. Define $T: L^2(\mu) \to L^2(\mu)$ by $(Tf)(\lambda_n) = \lambda_n f(\lambda_n)$. Prove: (a) T is a normal operator; (b) T has a cyclic vector (see Exercise 10); (c) if $A \in \mathscr{B}(\mathscr{H})$ and $AT = TA$, then $A = M_\phi$ for some ϕ in $l^\infty(\mathbb{C})$; (d) T is compact if and only if $\lambda_n \to 0$. (e) Find all of the cyclic vectors for T. (f) If T is compact, find the decomposition (7.7) for T.

12. Using the notation of Theorem 7.11, give necessary and sufficient conditions on T and ϕ that $\phi(T)$ be compact. (Hint: consider separately the cases where $\ker T$ is finite or infinite dimensional.)

13. Prove the uniqueness part of Theorem 7.16.

14. If $T \in \mathscr{B}(\mathscr{H})$, show that $T^*T \geq 0$.

15. Let T be a compact normal operator and show that there is a compact positive operator A and a unitary operator U such that $T = UA = AU$. Discuss the uniqueness of A and U.

16. (*Polar decomposition* of compact operators.) Let $T \in \mathscr{B}_0(\mathscr{H})$ and let A be the unique positive square root of T^*T [(7.16) and Exercise 14]. (a) Show that $\|Ah\| = \|Th\|$ for all h in $\mathscr{H}$. (b) Show that there is a unique operator U such that $\|Uh\| = \|h\|$ when $h \perp \ker T$, $Uh = 0$ when $h \in \ker T$, and $UA = T$. (c) If U and A are as in (a) and (b), show that $T = AU$ if and only if T is normal.

17. Prove the following uniqueness statement for the functional calculus (7.11). If T is a compact normal operator on a $\mathbb{C}$-Hilbert space $\mathscr{H}$ and $\tau: l^\infty(\mathbb{C}) \to \mathscr{B}(\mathscr{H})$ is a multiplicative linear map such that $\|\tau(\phi)\| = \sup\{|\phi(\lambda)\|: \lambda \in \sigma_p(T)\}$, $\tau(1) = 1$, and $\tau(z) = T$, then $\tau(\phi) = \phi(T)$ for every ϕ in $l^\infty(\mathbb{C})$.

§8*. Unitary Equivalence for Compact Normal Operators

In Section I.5 the concept of an isomorphism between Hilbert spaces was defined as the natural equivalence relation on Hilbert spaces. This equivalence relation between the spaces induces a natural equivalence relation between the operators on the spaces.

8.1. Definition. If A, B are bounded operators on Hilbert spaces $\mathscr{H}, \mathscr{K}$, then A and B are *unitarily equivalent* if there is an isomorphism $U: \mathscr{H} \to \mathscr{K}$ such that $UAU^{-1} = B$. In symbols this is denoted by $A \cong B$.

Some of the elementary properties of unitary equivalence are contained in Exercises 1 and 2. Note that if $UAU^{-1} = B$, then $UA = BU$.

The purpose of this section is to give necessary and sufficient conditions that two compact normal operators are unitarily equivalent. Later, in Section IX.10, necessary and sufficient conditions that any two normal operators be unitarily equivalent are given and the results of this section are subsumed by those of that section.

8.2.   Definition. If T is a compact operator, the *multiplicity function* for T is the function $m_T: \mathbb{C} \to \mathbb{C} \cup \{\infty\}$ defined by $m_T(\lambda) = \dim \ker(T - \lambda)$.

Hence $m_T(\lambda) \geq 0$ for all λ and $m_T(\lambda) > 0$ if and only if λ is an eigenvalue for T. Note that by Proposition 4.13, $m_T(\lambda) < \infty$ if $\lambda \neq 0$.

If T, S are compact operators on Hilbert spaces and $U: \mathcal{H} \to \mathcal{K}$ is an isomorphism with $UTU^{-1} = S$, then $U \ker(T - \lambda) = \ker(S - \lambda)$ for every λ in $\mathbb{C}$. In fact, if $Th = \lambda h$, then $SUh = UTh = \lambda Uh$ and so $Uh \in \ker(S - \lambda)$. Conversely, if $k \in \ker(S - \lambda)$ and $h = U^{-1}k$, then $Th = TU^{-1}k = U^{-1}Sk = \lambda h$. In particular, it must be that $m_T = m_S$. If S and T are normal, this condition is also sufficient for unitary equivalence.

8.3.   Theorem. *Two compact normal operators are unitarily equivalent if and only if they have the same multiplicity function.*

PROOF. Let T, S be compact normal operators on Hilbert spaces $\mathcal{H}, \mathcal{K}$. If $T \cong S$, then it has already been shown that $m_T = m_S$. Suppose now that $m_T = m_S$. We must manufacture a unitary operator $U: \mathcal{H} \to \mathcal{K}$ such that $UTU^{-1} = S$.

Let $T = \sum_{n=1}^{\infty} \lambda_n P_n$ and let $S = \sum_{n=1}^{\infty} \mu_n Q_n$ as in the Spectral Theorem (7.6). So if $n \neq m$, then $\lambda_n \neq \lambda_m$ and $\mu_n \neq \mu_m$, and each of the projections P_n and Q_n has finite rank. Let P_0, Q_0 be the projections of $\mathcal{H}, \mathcal{K}$ onto $\ker T, \ker S$; so $P_0 = (\sum_1^{\infty} P_n)^{\perp}$ and $Q_0 = (\sum_1^{\infty} Q_n)^{\perp}$. Put $\lambda_0 = \mu_0 = 0$.

Since $m_T = m_S$, $0 < m_T(\lambda_n) = m_S(\lambda_n)$. Hence there is a unique μ_j such that $\mu_j = \lambda_n$. Define $\pi: \mathbb{N} \to \mathbb{N}$ by letting $\mu_{\pi(n)} = \lambda_n$. Let $\pi(0) = 0$. Note that π is one-to-one. Also, since $0 < m_S(\mu_n) = m_T(\mu_n)$, for every n there is a j such that $\pi(j) = n$. Thus $\pi: \mathbb{N} \cup \{0\} \to \mathbb{N} \cup \{0\}$ is a bijection or permutation. Since $\dim P_n = m_T(\lambda_n) = m_S(\mu_{\pi(n)}) = \dim Q_{\pi(n)}$, there is an isomorphism $U_n: P_n \mathcal{H} \to Q_{\pi(n)} \mathcal{K}$ for $n \geq 0$. Define $U: \mathcal{H} \to \mathcal{K}$ by letting $U = U_n$ on $P_n \mathcal{H}$ and extending by linearity. Hence $U = \oplus_{n=0}^{\infty} U_n$. It is easy to check that U is an isomorphism. Also, if $h \in P_n \mathcal{H}$, $n \geq 0$, then $UTh = \lambda_n Uh = \mu_{\pi(n)} Uh = SUh$. Hence $UTU^{-1} = S$. ∎

If V is the Volterra operator, then $m_V \equiv 0$ (4.11) and V and the zero operator are definitely not unitarily equivalent, so the preceding theorem only applies to compact normal operators. There are no known necessary and sufficient conditions for two arbitrary compact operators to be unitarily equivalent. In fact, there are no known necessray and sufficient conditions that two arbitrary operators on a finite-dimensional space be unitarily equivalent.

EXERCISES

1. Show that "unitary equivalence" is an equivalence relation on $\mathscr{B}(\mathscr{H})$.

2. Let $U: \mathscr{H} \to \mathscr{K}$ be an isomorphism and define $\rho: \mathscr{B}(\mathscr{H}) \to \mathscr{B}(\mathscr{K})$ by $\rho(A) = UAU^{-1}$. Prove: (a) $\|\rho(A)\| = \|A\|$, $\rho(A^*) = \rho(A)^*$, and ρ is an isomorphism between the two algebras $\mathscr{B}(\mathscr{H})$ and $\mathscr{B}(\mathscr{K})$. (b) $\rho(A) \in \mathscr{B}_0(\mathscr{K})$ if and only if $A \in \mathscr{B}_0(\mathscr{H})$. (c) If $T \in \mathscr{B}(\mathscr{H})$, then $AT = TA$ if and only if $\rho(T)\rho(A) = \rho(A)\rho(T)$. (d) If $A \in \mathscr{B}(\mathscr{H})$ and $\mathscr{M} \leq \mathscr{H}$, then $\mathscr{M}$ is invariant (reducing) for A if and only if $U\mathscr{M}$ is invariant (reducing) for $\rho(A)$.

3. Say that an operator A on $\mathscr{H}$ is *irreducible* if the only reducing subspaces for A are (0) and $\mathscr{H}$. Prove: (a) The Volterra operator is irreducible. (b) The unilateral shift is irreducible.

4. Suppose $A = \oplus \{A_i: i \in I\}$ and $B = \oplus \{B_i: i \in I\}$ where each A_i and B_i is irreducible (Exercise 3). Show that $A \cong B$ if and only if there is a bijection $\pi: I \to I$ such that $A_i \cong B_{\pi(i)}$.

5. If T is a compact normal operator and $m_T = m$ is its multiplicity function, prove: (a) $\{\lambda: m(\lambda) > 0\}$ is countable and 0 is its only possible cluster point; (b) $m(\lambda) < \infty$ if $\lambda \neq 0$. Show that if $m: \mathbb{C} \to \mathbb{N} \cup \{0, \infty\}$ is any function satisfying (a) and (b), then there is a compact normal operator T such that $m_T = m$.

6. Show that two projections P and Q are unitarily equivalent if and only if $\dim(\operatorname{ran} P) = \dim(\operatorname{ran} Q)$ and $\dim(\ker P) = \dim(\ker Q)$.

7. Let $A: L^2(0,1) \to L^2(0,1)$ be defined by $(Af)(x) = xf(x)$ for f in $L^2(0,1)$ and x in $(0,1)$. Show that $A \cong A^2$.

8. Say that a compact normal operator T is *simple* if $m_T \leq 1$. (See Exercises 7.10 and 7.11.) Show that every compact normal operator T on a separable Hilbert space is unitarily equivalent to $\oplus_{n=1}^{\infty} T_n$, where each T_n is a simple compact normal operator and $m_{T_n} \geq m_{T_{n+1}}$ for all n. Show that $\|T_n\| \to 0$. (Of course, there may only be a finite number of T_n.)

9. Using the notation of Exercise 8, suppose also that S is a compact normal operator and $S \cong \oplus_{n=1}^{\infty} S_n$, where S_n is a simple compact normal operator and $m_{S_n} \geq m_{S_{n+1}}$ for all n. Show that $T \cong S$ if and only if $T_n \cong S_n$ for all n.

10. If T is a compact normal operator on a separable Hilbert space, show that there are simple compact normal operators $T_1, T_2, \ldots$ such that $T \cong 0 \oplus T_1 \oplus T_2^{(2)} \oplus T_3^{(3)} \oplus \cdots$, where: (a) for any operator A, $A^{(n)} \equiv A \oplus \cdots \oplus A$ (n times); (b) 0 is the zero operator on an infinite-dimensional space; (c) for $n \neq k$, $m_{T_n} m_{T_k} \equiv 0$; and (d) if $\ker T$ is infinite dimensional, then $\ker T_n = (0)$ for all n. (Of course not all of the summands need be present.) Show that $\|T_n\| \to 0$.

11. Using the notation of Exercise 10, let S be a compact normal operator and let $0 \oplus S_1 \oplus S_2^{(2)} \oplus \cdots$ be the corresponding decomposition. Show that $T \cong S$ if and only if $T_n \cong S_n$ and $\ker T$ and $\ker S$ have the same dimension.

12. If T is a compact normal operator, show that T and $T \oplus T$ are not unitarily equivalent.

13. Give an example of a nontrivial operator T such that $T \cong T \oplus T$. Show that if $T \cong T \oplus T$, then $T \cong T \oplus T \oplus \cdots$. Characterize the diagonalizable normal operators T such that $T \cong T \oplus T$.

14. Let $\mathcal{H}$ be the space defined in Example I.1.8 and let $U: \mathcal{H} \to L^2(0, 1)$ be the isomorphism defined by $Uf = f'$ (Exercise I.1.4). If $(Af)(x) = xf(x)$ for f in $\mathcal{H}$, what is UAU^{-1}?

CHAPTER III

Banach Spaces

The concept of a Banach space is a generalization of Hilbert space. A Banach space assumes that there is a norm on the space relative to which the space is complete, but it is not assumed that the norm is defined in terms of an inner product. There are many examples of Banach spaces that are not Hilbert spaces, so that the generalization is quite useful.

§1. Elementary Properties and Examples

1.1. Definition. If $\mathscr{X}$ is a vector space over $\mathbb{F}$, a *seminorm* is a function $p: \mathscr{X} \to [0, \infty)$ having the properties:

(a) $p(x + y) \le p(x) + p(y)$ for all x, y in $\mathscr{X}$.
(b) $p(\alpha x) = |\alpha| p(x)$ for all α in $\mathbb{F}$ and x in $\mathscr{X}$.

It follows from (b) that $p(0) = 0$. A *norm* is a seminorm p such that

(c) $x = 0$ if $p(x) = 0$.

Usually a norm is denoted by $\| \cdot \|$.
The norm on a Hilbert space is a norm. Also, the norm on $\mathscr{B}(\mathscr{H})$ is a norm.
If $\mathscr{X}$ has a norm, then $d(x, y) = \|x - y\|$ defines a metric on $\mathscr{X}$.

1.2. Definition. A *normed space* is a pair $(\mathscr{X}, \| \cdot \|)$, where $\mathscr{X}$ is a vector space and $\| \cdot \|$ is a norm on $\mathscr{X}$. A *Banach space* is a normed space that is complete with respect to the metric defined by the norm.

1.3. Proposition. *If $\mathscr{X}$ is a normed space, then*

(a) *the function $\mathscr{X} \times \mathscr{X} \to \mathscr{X}$ defined by $(x, y) \mapsto x + y$ is continuous;*
(b) *the function $\mathbb{F} \times \mathscr{X} \to \mathscr{X}$ defined by $(\alpha, x) \mapsto \alpha x$ is continuous.*

PROOF. If $x_n \to x$ and $y_n \to y$, then $\|(x_n + y_n) - (x + y)\| = \|(x_n - x) + (y_n - y)\| \leq \|x_n - x\| + \|y_n - y\| \to 0$ as $n \to \infty$. This proves (a). The proof of (b) is left to the reader. ∎

The next lemma is quite useful.

1.4. Lemma. *If p and q are seminorms on a vector space $\mathscr{X}$, then the following statements are equivalent.*

(a) $p(x) \leq q(x)$ *for all* x. *(That is, $p \leq q$.)*
(b) $\{x \in \mathscr{X} : q(x) < 1)\} \subseteq \{x \in \mathscr{X} : p(x) < 1\}$.
(b') $p(x) < 1$ *whenever* $q(x) < 1$.
(c) $\{x : q(x) \leq 1\} \subseteq \{x : p(x) \leq 1\}$.
(c') $p(x) \leq 1$ *whenever* $q(x) \leq 1$.
(d) $\{x : q(x) < 1\} \subseteq \{x : p(x) \leq 1\}$.
(d') $p(x) \leq 1$ *whenever* $q(x) < 1$.

PROOF. It is clear that (b) and (b'), (c) and (c'), and (d) and (d') are equivalent. It is also clear that (a) implies all of the remaining conditions and that (b) implies (d). It will be shown that (d) implies (a). The proof that (c) implies (a) is left as an exercise.

Assume that (d) holds and put $q(x) = \alpha$. If $\varepsilon > 0$, then $q((\alpha + \varepsilon)^{-1}x) = (\alpha + \varepsilon)^{-1}\alpha < 1$. By (d), $1 \geq p((\alpha + \varepsilon)^{-1}x) = (\alpha + \varepsilon)^{-1}p(x)$, so $p(x) \leq \alpha + \varepsilon = q(x) + \varepsilon$. Letting $\varepsilon \to 0$ shows (a). ∎

If $\| \cdot \|_1$ and $\| \cdot \|_2$ are two norms on $\mathscr{X}$, they are said to be *equivalent norms* if they define the same topology on $\mathscr{X}$.

1.5. Proposition. *If $\| \cdot \|_1$ and $\| \cdot \|_2$ are two norms on $\mathscr{X}$, then these norms are equivalent if and only if there are positive constants c and C such that*

$$c\|x\|_1 \leq \|x\|_2 \leq C\|x\|_1$$

for all x in $\mathscr{X}$.

PROOF. Suppose there are constants c and C such that $c\|x\|_1 \leq \|x\|_2 \leq C\|x\|_1$ for all x in $\mathscr{X}$. Fix x_0 in $\mathscr{X}$, $\varepsilon > 0$. Then

$$\{x \in \mathscr{X} : \|x - x_0\|_1 < \varepsilon/C\} \subseteq \{x \in \mathscr{X} : \|x - x_0\|_2 < \varepsilon\},$$

$$\{x \in \mathscr{X} : \|x - x_0\|_2 < c\varepsilon\} \subseteq \{x \in \mathscr{X} : \|x - x_0\|_1 < \varepsilon\}.$$

This shows that the two topologies are the same. Now assume that the two norms are equivalent. Hence $\{x : \|x\|_1 < 1\}$ is an open neighborhood of 0 in the topology defined by $\| \cdot \|_2$. Therefore there is an $r > 0$ such that $\{x:$

$\|x\|_2 < r\} \subseteq \{x: \|x\|_1 < 1\}$. If $q(x) = r^{-1}\|x\|_2$ and $p(x) = \|x\|_1$, the preceding lemma implies $\|x\|_1 \leq r^{-1}\|x\|_2$ or $c\|x\|_1 \leq \|x\|_2$, where $c = r$. The other inequality is left to the reader. ∎

There are two types of properties of a Banach space: those that are topological and those that are metric. The metric properties depend on the precise norm; the topological ones depend only on the equivalence class of norms (see Exercise 4).

1.6.   Example. Let X be any Hausdorff space (all spaces in this book are assumed to be Hausdorff unless the contrary is specified) and let $C_b(X) =$ all continuous functions $f: X \to \mathbb{F}$ such that $\|f\| \equiv \sup\{|f(x)|: x \in X\} < \infty$. For f, g in $C_b(X)$, define $(f + g): X \to \mathbb{F}$ by $(f + g)(x) = f(x) + g(x)$; for α in $\mathbb{F}$ define $(\alpha f)(x) = \alpha f(x)$. Then $C_b(X)$ is a Banach space.

The proofs of the statements in (1.6) are all routine except, perhaps, for the fact that $C_b(X)$ is complete. To see this, let $\{f_n\}$ be a Cauchy sequence in $C_b(X)$. So if $\varepsilon > 0$, there is an integer N_ε such that for $n, m \geq N_\varepsilon$, $\varepsilon > \|f_n - f_m\| = \sup\{|f_n(x) - f_m(x)|: x \in X\}$. In particular, for any x in X, $|f_n(x) - f_m(x)| \leq \|f_n - f_m\| < \varepsilon$ when $n, m \geq N_\varepsilon$. So $\{f_n(x)\}$ is a Cauchy sequence in $\mathbb{F}$. Let $f(x) = \lim f_n(x)$ if $x \in X$. Now fix x in X. If $n, m \geq N_\varepsilon$, then $|f(x) - f_n(x)| \leq |f(x) - f_m(x)| + \|f_m - f_n\| < |f(x) - f_m(x)| + \varepsilon$. Letting $m \to \infty$ gives that $|f(x) - f_n(x)| \leq \varepsilon$ when $n \geq N_\varepsilon$. This is independent of x. Hence $\|f - f_n\| \leq \varepsilon$ for $n \geq N_\varepsilon$.

What has been just shown is that $\|f - f_n\| \to 0$ as $n \to \infty$. Note that this implies that $f_n(x) \to f(x)$ uniformly on X. It is standard that f is continuous. Also, $\|f\| \leq \|f - f_n\| + \|f_n\| < \infty$. Hence $f \in C_b(X)$ and so $C_b(X)$ is complete.

Note that a linear subspace $\mathcal{Y}$ of a Banach space $\mathcal{X}$ that is topologically closed is also a Banach space if it has the norm of $\mathcal{X}$.

1.7.   Proposition. *If X is a locally compact space and $C_0(X) =$ all continuous functions $f: X \to \mathbb{F}$ such that for all $\varepsilon > 0$, $\{x \in X: |f(x)| \geq \varepsilon\}$ is compact, then $C_0(X)$ is a closed subspace of $C_b(X)$ and hence is a Banach space.*

PROOF. That $C_0(X)$ is a linear manifold in $C_b(X)$ is left as an exercise. It will only be shown that $C_0(X)$ is closed in $C_b(X)$. Let $\{f_n\} \subseteq C_0(X)$ and suppose $f_n \to f$ in $C_b(X)$. If $\varepsilon > 0$, there is an integer N such that $\|f_n - f\| < \varepsilon/2$; that is, $|f_n(x) - f(x)| < \varepsilon/2$ for all $n \geq N$ and x in X. If $|f(x)| \geq \varepsilon$, then $\varepsilon \leq |f(x) - f_n(x) + f_n(x)| \leq \varepsilon/2 + |f_n(x)|$ for $n \geq N$; so $|f_n(x)| \geq \varepsilon/2$ for $n \geq N$. Thus, $\{x \in X: |f(x)| \geq \varepsilon\} \subseteq \{x \in X: |f_N(x)| \geq \varepsilon/2\}$ so that $f \in C_0(X)$. ∎

The space $C_0(X)$ is the set of continuous functions on X that *vanish at infinity*. If $X = \mathbb{R}$, then $C_0(\mathbb{R}) =$ all of the continuous functions $f: \mathbb{R} \to \mathbb{F}$ such that $\lim_{x \to \pm\infty} f(x) = 0$. If X is compact, $C_0(X) = C_b(X) \equiv C(X)$.

If I is any set, then give I the discrete topology. Hence I becomes locally compact. Also any function on I is continuous. Rather than $C_b(I)$, the customary notation is $l^\infty(I)$. That is, $l^\infty(I) = $ all bounded functions f: $I \to \mathbb{F}$ with $\|f\| = \sup\{|f(i)|: i \in I\}$. $C_0(I)$ consists of all functions f: $I \to \mathbb{F}$ such that for every $\varepsilon > 0$, $\{i \in I: |f(i)| \geq \varepsilon\}$ is finite. If $I = \mathbb{N}$, the usual notation for these spaces is l^∞ and c_0. Note that l^∞ consists of all bounded sequences of scalars and c_0 consists of all sequences that converge to 0.

1.8.   Example. If (X, Ω, μ) is a measure space and $1 \leq p \leq \infty$, then $L^p(X, \Omega, \mu)$ is a Banach space.

The preceding example is usually proved in courses on integration and no proof is given here.

1.9.   Example. Let I be a set and $1 \leq p < \infty$. Define $l^p(I)$ to be the set of all functions f: $I \to \mathbb{F}$ such that $\Sigma\{|f(i)|^p: i \in I\} < \infty$; and define $\|f\|_p = (\Sigma\{|f(i)|^p: i \in I\})^{1/p}$. Then $l^p(I)$ is a Banach space. If $I = \mathbb{N}$, then $l^p(\mathbb{N}) = l^p$.

If $\Omega = $ all subsets of I and for each Δ in Ω, $\mu(\Delta) = $ the number of points in Δ if Δ is finite and $\mu(\Delta) = \infty$ otherwise, then $l^p(I) = L^p(I, \Omega, \mu)$. So the statement in (1.9) is a consequence of the one in (1.8).

1.10.   Example. Let $n \geq 1$ and let $C^{(n)}[0,1] = $ the collection of functions f: $[0,1] \to \mathbb{F}$ such that f has n continuous derivatives. Define $\|f\| = \sup_{0 \leq k \leq n}\{\sup\{|f^{(k)}(x)|: 0 \leq x \leq 1\}\}$. Then $C^{(n)}[0,1]$ is a Banach space.

1.11.   Example. Let $1 \leq p < \infty$ and $n \geq 1$ and let $W_p^n[0,1] = $ the functions f: $[0,1] \to \mathbb{F}$ such that f has $n - 1$ continuous derivatives, $f^{(n-1)}$ is absolutely continuous, and $f^{(n)} \in L^p[0,1]$. For f in $W_p^n[0,1]$, define

$$\|f\| = \sum_{k=0}^{n}\left[\int_0^1 |f^{(k)}(x)|^p\, dx\right]^{1/p}.$$

Then $W_p^n[0,1]$ is a Banach space.

The following is a useful fact about seminorms.

1.12.   Proposition. *If p is a seminorm on $\mathscr{X}$, $|p(x) - p(y)| \leq p(x - y)$ for all x, y in $\mathscr{X}$. If $\|\cdot\|$ is a norm, then $|\,\|x\| - \|y\|\,| \leq \|x - y\|$ for all x, y in $\mathscr{X}$.*

PROOF. Of course, the inequality for norms is a consequence of the one for seminorms. Note that if $x, y \in \mathscr{X}$, $p(x) = p(x - y + y) \leq p(x - y) + p(y)$, so $p(x) - p(y) \leq p(x - y)$. Similarly, $p(y) - p(x) \leq p(x - y)$. ∎

There is the concept of "isomorphism" for the category of Banach spaces.

1.13. **Definition.** If $\mathscr{X}$ and $\mathscr{Y}$ are normed spaces, $\mathscr{X}$ and $\mathscr{Y}$ are *isometrically isomorphic* if there is a surjective linear isometry from $\mathscr{X}$ onto $\mathscr{Y}$.

The term *isomorphism* in Banach space theory is reserved for linear bijections $T\colon \mathscr{X} \to \mathscr{Y}$ that are homeomorphisms.

EXERCISES

1. Complete the proof of Proposition 1.3.

2. Complete the proof of Proposition 1.5.

3. For $1 \le p < \infty$ and $x = (x_1, \ldots, x_d)$ in $\mathbb{F}^d$, define $\|x\|_p \equiv [\sum_{j=1}^{d} |x_j|^p]^{1/p}$; define $\|x\|_\infty \equiv \sup\{|x_j|\colon 1 \le j \le d\}$. Show that all of these norms are equivalent. For $1 \le p, q \le \infty$, what are the best constants c and C such that $c\|x\|_p \le \|x\|_q \le C\|x\|_p$ for all x in $\mathbb{F}^d$?

4. If $1 \le p \le \infty$ and $\|\cdot\|_p$ is defined on $\mathbb{R}^2$ as in Exercise 3, graph $\{x \in \mathbb{R}^2\colon \|x\|_p = 1\}$. Note that if $1 < p < \infty$, $\|x\|_p = \|y\|_p = 1$, and $x \ne y$, then for $0 < t < 1$, $\|tx + (1 - t)y\|_p < 1$. The same cannot be said for $p = 1, \infty$.

5. Let $c =$ the set of all sequences $\{\alpha_n\}_1^\infty$, α_n in $\mathbb{F}$, such that $\lim \alpha_n$ exists. Show that c is a closed subspace of l^∞ and hence is a Banach space.

6. Let $X = \{n^{-1}\colon n \ge 1\} \cup \{0\}$. Show that $C(X)$ and the space of c of Exercise 5 are isometrically isomorphic.

   (a) Show that if $1 \le p < \infty$ and I is an infinite set, then $l^p(I)$ has a dense set of the same cardinality as I.
   (b) Show that if $1 \le p < \infty$, $l^p(I)$ and $l^p(J)$ are isometrically isomorphic if and only if I and J have the same cardinality.

7. If $l^\infty(I)$ and $l^\infty(J)$ are isometrically isomorphic, do I and J have the same cardinality?

8. Show that l^∞ is not separable.

9. Complete the proof of Proposition 1.7.

10. Verify the statements in Example 1.10.

11. Verify the statements in Example 1.11.

12. Let X be locally compact and let $X_\infty = X \cup \{\infty\}$ be the one-point compactification of X. Show that $C_0(X)$ and $\{f \in C(X_\infty)\colon f(\infty) = 0\}$, with the norm it inherits as a subspace of $C(X_\infty)$, are isometrically isomorphic Banach spaces.

13. Let X be locally compact and define $C_c(X)$ to be the continuous functions $f\colon X \to \mathbb{F}$ such that $\operatorname{spt} f \equiv \operatorname{cl}\{x \in X\colon f(x) \ne 0\}$ is compact ($\operatorname{spt} f$ is the *support* of f). Show that $C_c(X)$ is dense in $C_0(X)$.

14. If X is a metrizable locally compact space that is σ-*compact*, then $C_0(X)$ is separable. (X is σ-*compact* if $X = \bigcup_{n=1}^{\infty} K_n$, where each K_n is compact.)

15. If $W_p^n[0,1]$ is defined as in Example 1.11 and $f \in W_p^n[0,1]$, let $\||f\|| \equiv [\int |f(x)|^p \, dx]^{1/p} + [\int |f^{(n)}(x)|^p \, dx]^{1/p}$. Show that $\|| \cdot \||$ is equivalent to the norm defined on $W_p^n[0,1]$.

16. Let $\mathscr{X}$ be a normed space and let $\hat{\mathscr{X}}$ be its completion as a metric space. Show that $\hat{\mathscr{X}}$ is a Banach space.

§2. Linear Operators on Normed Spaces

This section gathers together a few pertinent facts and examples concerning linear operators on normed spaces. A fuller study of operators on Banach spaces will be pursued later.

The proof of the first result is similar to that of Proposition I.3.1 and is left to the reader [Also see (II.1.1)]. $\mathscr{B}(\mathscr{X}, \mathscr{Y}) =$ all continuous linear transformations $A: \mathscr{X} \to \mathscr{Y}$.

2.1. Proposition. *If $\mathscr{X}$ and $\mathscr{Y}$ are normed spaces and $A: \mathscr{X} \to \mathscr{Y}$ is a linear transformation, the following statements are equivalent.*

(a) $A \in \mathscr{B}(\mathscr{X}, \mathscr{Y})$.
(b) *A is continuous at 0.*
(c) *A is continuous at some point.*
(d) *There is a positive constant c such that $\|Ax\| \leq c\|x\|$ for all x in $\mathscr{X}$.*
   If $A \in \mathscr{B}(\mathscr{X}, \mathscr{Y})$ and

$$\|A\| = \sup\{\|Ax\| : \|x\| \leq 1\},$$

   then

$$\|A\| = \sup\{\|Ax\| : \|x\| = 1\}$$
$$= \sup\{\|Ax\|/\|x\| : x \neq 0\}$$
$$= \inf\{c > 0 : \|Ax\| \leq c\|x\| \text{ for } x \text{ in } \mathscr{X}\}.$$

$\|A\|$ is called the *norm* of A and $\mathscr{B}(\mathscr{X}, \mathscr{Y})$ becomes a normed space if addition and scalar multiplication are defined pointwise. $\mathscr{B}(\mathscr{X}, \mathscr{Y})$ is Banach space if $\mathscr{Y}$ is a Banach space (Exercise 1). A continuous linear operator is also called a *bounded linear operator*.

The following examples are reminiscent of those that were given in Section II.1.

2.2. Example. If (X, Ω, μ) is a σ-finite measure space and $\phi \in L^{\infty}(X, \Omega, \mu)$, define $M_{\phi}: L^p(X, \Omega, \mu) \to L^p(X, \Omega, \mu)$, $1 \leq p \leq \infty$, by $M_{\phi}f = \phi f$ for all f in $L^p(X, \Omega, \mu)$. Then $M_{\phi} \in \mathscr{B}(L^p(X, \Omega, \mu))$ and $\|M_{\phi}\| = \|\phi\|_{\infty}$.

2.3. Example. If (X, Ω, μ), k, c_1, and c_2 are as in Example II.1.6 and $1 \le p \le \infty$, then $K: L^p(\mu) \to L^p(\mu)$, defined by

$$(Kf)(x) = \int k(x, y)f(y)\, d\mu(y)$$

for all f in $L^p(\mu)$ and x in X, is a bounded operator on $L^p(\mu)$ and $\|K\| \le c_1^{1/q} c_2^{1/p}$, where $1/p + 1/q = 1$.

2.4. Example. If X and Y are compact spaces and $\tau: Y \to X$ is a continuous map, define $A: C(X) \to C(Y)$ by $(Af)(y) = f(\tau(y))$. Then $A \in \mathcal{B}(C(X), C(Y))$ and $\|A\| = 1$.

EXERCISES

1. Show that $\mathcal{B}(\mathcal{X}, \mathcal{Y})$ is a Banach space if and only if $\mathcal{Y}$ is a Banach space.

2. Let $\mathcal{X}$ be a normed space, let $\mathcal{Y}$ be a Banach space, and let $\hat{\mathcal{X}}$ be the completion of $\mathcal{X}$. Show that if $\rho: \mathcal{B}(\hat{\mathcal{X}}, \mathcal{Y}) \to \mathcal{B}(\mathcal{X}, \mathcal{Y})$ is defined by $\rho(A) = A|\mathcal{X}$, then ρ is an isometric isomorphism.

3. If (X, Ω, μ) is a σ-finite measure space, $\phi: X \to \mathbb{F}$ is an Ω-measurable function, $1 \le p \le \infty$, and $\phi f \in L^p(\mu)$ whenever $f \in L^p(\mu)$, then show that $\phi \in L^\infty(\mu)$.

4. Verify the statements in Example 2.2.

5. Verify the statements in Example 2.3.

6. Verify the statements in Example 2.4.

7. Let A and τ be as in Example 2.4. (a) Give necessary and sufficient conditions on τ that A be injective. (b) Give such a condition that A be surjective. (c) Give such a condition that A be an isometry. (d) If $X = Y$, show that $A^2 = A$ if and only if τ is a retraction.

§3. Finite-Dimensional Normed Spaces

In functional analysis it it always good to see what significance a concept has for finite-dimensional spaces.

3.1. Theorem. *If $\mathcal{X}$ is a finite-dimensional vector space over $\mathbb{F}$, then any two norms on $\mathcal{X}$ are equivalent.*

PROOF. Let $\{e_1, \ldots, e_d\}$ be a Hamel basis for $\mathcal{X}$. For $x = \sum_{j=1}^{d} x_j e_j$, define $\|x\|_\infty \equiv \max\{|x_j|: 1 \le j \le d\}$. It is left to the reader to verify that $\|\cdot\|_\infty$ is a norm. Let $\|\cdot\|$ be any norm on $\mathcal{X}$. It will be shown that $\|\cdot\|$ and $\|\cdot\|_\infty$ are equivalent.

If $x = \sum_j x_j e_j$, then $\|x\| \le \sum_j |x_j|\, \|e_j\| \le C\|x\|_\infty$, when $C = \sum_j \|e_j\|$. To show the other inequality, let $\mathcal{T}$ be the topology defined on $\mathcal{X}$ by $\|\cdot\|_\infty$

and let $\mathcal{U}$ be the topology defined on $\mathcal{X}$ by $\|\cdot\|$. Put $B = \{x \in \mathcal{X}: \|x\|_\infty \leq 1\}$. The first part of the proof implies $\mathcal{T} \supseteq \mathcal{U}$. Since B is $\mathcal{T}$-compact and $\mathcal{T} \supseteq \mathcal{U}$, B is $\mathcal{U}$-compact and the relativizations of the two topologies to B agree. Let $A = \{x \in \mathcal{X}: \|x\|_\infty < 1\}$. Since A is $\mathcal{T}$-open, it is open in $(B, \mathcal{U})$. Hence there is a set U in $\mathcal{U}$ such that $U \cap B = A$. Thus $0 \in U$ and there is an $r > 0$ such that $\{x \in \mathcal{X}: \|x\| < r\} \subseteq U$. Hence

3.2 $\|x\| < r$ and $\|x\|_\infty \leq 1$ implies $\|x\|_\infty < 1$.

Claim. $\|x\| < r$ implies $\|x\|_\infty < 1$.

Let $\|x\| < r$ and put $x = \Sigma x_j e_j$, $\alpha = \|x\|_\infty$. So $\|x/\alpha\|_\infty = 1$ and $x/\alpha \in B$. If $\alpha \geq 1$, then $\|x/\alpha\| < r/\alpha \leq r$, and hence $\|x/\alpha\|_\infty < 1$ by (3.2), a contradiction. Thus $\|x\|_\infty = \alpha < 1$ and the claim is established.

By Lemma 1.4, $\|x\|_\infty < r^{-1}\|x\|$ for all x and so the proof is complete. ∎

3.3. Proposition. *If $\mathcal{X}$ is a normed space and $\mathcal{M}$ is a finite dimensional linear manifold in $\mathcal{X}$, then $\mathcal{M}$ is closed.*

PROOF. Let $x_0 \in \mathcal{X} \setminus \mathcal{M}$ and put $\mathcal{M}_1 =$ the linear span of $\mathcal{M}$ and $\{x_0\}$. Define a norm $\|\cdot\|_1$ on $\mathcal{M}_1$ by $\|x + \alpha_0 x_0\|_1 \equiv \|x\| + |\alpha_0|$, for x in $\mathcal{M}$ and α_0 in $\mathbb{F}$. It is left as an exercise to show that $\|\cdot\|_1$ is a norm on $\mathcal{M}_1$. By Theorem 3.1 and Proposition 1.5, there are constants c and C such that $c\|x + \alpha_0 x_0\| \leq \|x\| + |\alpha_0| \leq C\|x + \alpha_0 x_0\|$ for all x in $\mathcal{M}$ and α_0 in $\mathbb{F}$. Hence for all x in $\mathcal{M}$, $\|x_0 - x\| \geq C^{-1}(\|x\| + 1) \geq C^{-1}$. Thus $0 < C^{-1} \leq \inf\{\|x_0 - x\|: x \in \mathcal{M}\} \equiv \mathrm{dist}(x_0, \mathcal{M})$. That is, every point x_0 not in $\mathcal{M}$ is at a positive distance from $\mathcal{M}$. Hence $\mathcal{M}$ is closed. ∎

3.4. Proposition. *Let $\mathcal{X}$ be a finite-dimensional normed space and let $\mathcal{Y}$ be any normed space. If $T: \mathcal{X} \to \mathcal{Y}$ is a linear transformation, then T is continuous.*

PROOF. Since all norms on $\mathcal{X}$ are equivalent and $T: \mathcal{X} \to \mathcal{Y}$ is continuous with respect to one norm on $\mathcal{X}$ precisely when it is continuous with respect to any equivalent norm, we may assume that $\|\Sigma_{j=1}^d \xi_j e_j\| = \max\{|\xi_j|: 1 \leq j \leq d\}$, where $\{e_j\}$ is a Hamel basis for $\mathcal{X}$. Thus, for $x = \Sigma \xi_j e_j$, $\|Tx\| = \|\Sigma_j \xi_j T e_j\| \leq \Sigma_j |\xi_j| \|T e_j\| \leq C\|x\|$, where $C = \Sigma_j \|T e_j\|$. By (2.1), T is continuous. ∎

EXERCISES

1. Show that if $\mathcal{X}$ is a locally compact normed space, then $\mathcal{X}$ is finite dimensional.

2. Show that $\|\cdot\|_1$, defined on $\mathcal{M}_1$ in the proof of Proposition 3.3, is a norm.

§4. Quotients and Products of Normed Spaces

Let $\mathscr{X}$ be a normed space, let $\mathscr{M}$ be a linear manifold in $\mathscr{X}$, and let Q: $\mathscr{X} \to \mathscr{X}/\mathscr{M}$ be the natural map $Qx = x + \mathscr{M}$. We want to make $\mathscr{X}/\mathscr{M}$ into a normed space, so define

4.1
$$\|x + \mathscr{M}\| = \inf\{\|x + y\|: y \in \mathscr{M}\}.$$

Note that because $\mathscr{M}$ is a linear space, $\|x + \mathscr{M}\| = \inf\{\|x - y\|: y \in \mathscr{M}\}$ $= \operatorname{dist}(x, \mathscr{M})$, the distance from x to $\mathscr{M}$. It is left to the reader to show that (4.1) defines a seminorm on $\mathscr{X}/\mathscr{M}$. But if $\mathscr{M}$ is not closed in $\mathscr{X}$, (4.1) cannot define a norm. (Why?) If, however, $\mathscr{M}$ is closed, then (4.1) does define a norm.

4.2. Theorem. *If $\mathscr{M} \leq \mathscr{X}$ and $\|x + \mathscr{M}\|$ is defined as in (4.1), then $\|\cdot\|$ is a norm on $\mathscr{X}/\mathscr{M}$. Also:*

(a) $\|Q(x)\| \leq \|x\|$ *for all x in $\mathscr{X}$ and hence Q is continuous.*
(b) *If $\mathscr{X}$ is a Banach space, then so is $\mathscr{X}/\mathscr{M}$.*
(c) *A subset W of $\mathscr{X}/\mathscr{M}$ is open relative to the norm if and only if $Q^{-1}(W)$ is open in $\mathscr{X}$.*
(d) *If U is open in $\mathscr{X}$, then $Q(U)$ is open in $\mathscr{X}/\mathscr{M}$.*

PROOF. It is left as an exercise to show that (4.1) defines a norm on $\mathscr{X}/\mathscr{M}$. To show (a), $\|Q(x)\| = \|x + \mathscr{M}\| \leq \|x\|$ since $0 \in \mathscr{M}$; Q is therefore continuous by (2.1).

(b) Let $\{x_n + \mathscr{M}\}$ be a Cauchy sequence in $\mathscr{X}/\mathscr{M}$. There is a subsequence $\{x_{n_k} + \mathscr{M}\}$ such that

$$\|(x_{n_k} + \mathscr{M}) - (x_{n_{k+1}} + \mathscr{M})\| = \|x_{n_k} - x_{n_{k+1}} + \mathscr{M}\| < 2^{-k}.$$

Let $y_1 = 0$. Choose y_2 in $\mathscr{M}$ such that

$$\|x_{n_1} - x_{n_2} + y_2\| \leq \|x_{n_1} - x_{n_2} + \mathscr{M}\| + 2^{-1} < 2 \cdot 2^{-1}.$$

Choose y_3 in $\mathscr{M}$ such that

$$\|(x_{n_2} + y_2) - (x_{n_3} + y_3)\| \leq \|x_{n_2} - x_{n_3} + \mathscr{M}\| + 2^{-2} < 2 \cdot 2^{-2}.$$

Continuing, there is a sequence $\{y_k\}$ in $\mathscr{M}$ such that

$$\|(x_{n_k} + y_k) - (x_{n_{k+1}} + y_{k+1})\| < 2 \cdot 2^{-k}.$$

Thus $\{x_{n_k} + y_k\}$ is a Cauchy sequence in $\mathscr{X}$ (Why?). Since $\mathscr{X}$ is complete, there is an x_0 in $\mathscr{X}$ such that $x_{n_k} + y_k \to x_0$ in $\mathscr{X}$. By (a), $x_{n_k} + \mathscr{M} = Q(x_{n_k} + y_k) \to Qx_0 = x_0 + \mathscr{M}$. Since $\{x_n + \mathscr{M}\}$ is a Cauchy sequence, $x_n + \mathscr{M} \to x_0 + \mathscr{M}$ and $\mathscr{X}/\mathscr{M}$ is complete (Exercise 3).

(c) If W is open in $\mathscr{X}/\mathscr{M}$, then $Q^{-1}(W)$ is open in $\mathscr{X}$ because Q is continuous. Now assume that $W \subseteq \mathscr{X}/\mathscr{M}$ and $Q^{-1}(W)$ is open in $\mathscr{X}$. Let $r > 0$ and put $B_r \equiv \{x \in \mathscr{X}: \|x\| < r\}$. It will be shown that $Q(B_r) = \{x$

$+\mathcal{M}$: $\|x + \mathcal{M}\| < r\}$. In fact, if $\|x\| < r$, then $\|x + \mathcal{M}\| \leq \|x\| < r$. On the other hand, if $\|x + \mathcal{M}\| < r$, then there is a y in $\mathcal{M}$ such that $\|x + y\| < r$. Thus $x + \mathcal{M} = Q(x + y) \in Q(B_r)$. If $x_0 + \mathcal{M} \in W$, then $x_0 \in Q^{-1}(W)$. Since $Q^{-1}(W)$ is open, there is an $r > 0$ such that $x_0 + B_r = \{x: \|x - x_0\| < r\} \subseteq Q^{-1}(W)$. The preceding argument now implies that $W = QQ^{-1}(W) \supseteq Q(x_0 + B_r) = \{x + \mathcal{M}: \|x - x_0 + \mathcal{M}\| < r\}$. Hence W is open.

(d) If U is open in $\mathcal{X}$, then $Q^{-1}(Q(U)) = U + \mathcal{M} \equiv \{u + y: u \in U, y \in \mathcal{M}\} = \cup\{U + y: y \in \mathcal{M}\}$. Each $U + y$ is open, so $Q^{-1}(Q(U))$ is open in $\mathcal{X}$. By (c), $Q(U)$ is open in $\mathcal{X}/\mathcal{M}$. ∎

Because Q is an open map [part (d)], it does not follow that Q is a closed map (Exercise 4).

4.3. Proposition. *If $\mathcal{X}$ is a normed space, $\mathcal{M} \leq \mathcal{X}$, and $\mathcal{N}$ is a finite dimensional subspace of $\mathcal{X}$, then $\mathcal{M} + \mathcal{N}$ is a closed subspace of $\mathcal{X}$.*

PROOF. Consider $\mathcal{X}/\mathcal{M}$ and the quotient map $Q: \mathcal{X} \to \mathcal{X}/\mathcal{M}$. Since $\dim Q(\mathcal{N}) \leq \dim \mathcal{N} < \infty$, $Q(\mathcal{N})$ is closed in $\mathcal{X}/\mathcal{M}$. Since Q is continuous $Q^{-1}(Q(\mathcal{N}))$ is closed in $\mathcal{X}$; but $Q^{-1}(Q(\mathcal{N})) = \mathcal{M} + \mathcal{N}$. ∎

Now for the product or direct sum of normed spaces. Here there is a difficulty because, unlike Hilbert space, there is no canonical way to proceed. Suppose $\{\mathcal{X}_i: i \in I\}$ is a collection of normed spaces. Then $\prod\{\mathcal{X}_i: i \in I\}$ is a vector space if the linear operations are defined coordinatewise. The idea is to put a norm on a linear subspace of this product.

Let $\|\cdot\|$ denote the norm on each $\mathcal{X}_i$. For $1 \leq p < \infty$, define

$$\oplus_p \mathcal{X}_i \equiv \left\{ x \in \prod_i \mathcal{X}_i: \|x\| \equiv \left[\sum_i \|x(i)\|^p\right]^{1/p} < \infty\right\}.$$

Define

$$\oplus_\infty \mathcal{X}_i \equiv \left\{ x \in \prod_i \mathcal{X}_i: \|x\| \equiv \sup_i \|x(i)\| < \infty\right\}.$$

If $\{\mathcal{X}_1, \mathcal{X}_2, \ldots\}$ is a sequence of normed spaces, define

$$\oplus_0 \mathcal{X}_n \equiv \left\{ x \in \prod_{n=1}^\infty \mathcal{X}_n: \|x(n)\| \to 0\right\};$$

give $\oplus_0 \mathcal{X}_n$ the norm it has as a subspace of $\oplus_\infty \mathcal{X}_n$.

The proof of the next proposition is left as an exercise.

4.4. Proposition. *Let $\{\mathcal{X}_i: i \in I\}$ be a collection of normed spaces and let $\mathcal{X} = \oplus_p \mathcal{X}_i$, $1 \leq p \leq \infty$.*

(a) *$\mathcal{X}$ is a normed space and the projection $P_i: \mathcal{X} \to \mathcal{X}_i$ is a continuous linear map with $\|P_i(x)\| \leq \|x\|$ for each x in $\mathcal{X}$.*

(b) $\mathscr{X}$ is a Banach space if and only if each $\mathscr{X}_i$ is a Banach space.

(c) Each projection P_i is an open map of $\mathscr{X}$ onto $\mathscr{X}_i$.

A similar result holds for $\oplus_0 \mathscr{X}_n$, but the formulation and proof of this is left to the reader.

EXERCISES

1. Show that if $\mathscr{M} \leq \mathscr{X}$, then (4.1) defines a norm on $\mathscr{X}/\mathscr{M}$.

2. Prove that $\mathscr{X}$ is a Banach space if and only if whenever $\{x_n\}$ is a sequence in $\mathscr{X}$ such that $\Sigma \|x_n\| < \infty$, then $\Sigma_{n=1}^\infty x_n$ converges in $\mathscr{X}$.

3. Show that if (X, d) is a metric space and $\{x_n\}$ is a Cauchy sequence such that there is a subsequence $\{x_{n_k}\}$ that converges to x_0, then $x_n \to x_0$.

4. Find a Banach space $\mathscr{X}$ and a closed subspace $\mathscr{M}$ such that the natural map Q: $\mathscr{X} \to \mathscr{X}/\mathscr{M}$ is not a closed map. Can the natural map ever be a closed map?

5. Prove the converse of (4.2b): If $\mathscr{X}$ is a normed space, $\mathscr{M} \leq \mathscr{X}$, and both $\mathscr{M}$ and $\mathscr{X}/\mathscr{M}$ are complete, then $\mathscr{X}$ is complete. (This is an example of what is called a "two-out-of-three" result. If any two of $\mathscr{X}$, $\mathscr{M}$, and $\mathscr{X}/\mathscr{M}$ are complete, so is the third.)

6. Let $\mathscr{M} = \{x \in l^p: x(2n) = 0 \text{ for all } n\}$, $1 \leq p \leq \infty$. Show that $l^p/\mathscr{M}$ is isometrically isomorphic to l^p.

7. Let X be a normal locally compact space and F a closed subset of X. If $\mathscr{M} \equiv \{f \in C_0(X): f(x) = 0 \text{ for all } x \text{ in } F\}$, then $C_0(X)/\mathscr{M}$ is isometrically isomorphic to $C_0(F)$.

8. Prove Proposition 4.4.

9. Formulate and prove a version of Proposition 4.4 for $\oplus_0 \mathscr{X}_n$.

10. If $\{\mathscr{X}_1, \ldots, \mathscr{X}_n\}$ is a finite collection of normed spaces and $1 \leq p \leq \infty$, show that the norms on $\oplus_p \mathscr{X}_k$ are all equivalent.

11. Here is an abstraction of Proposition 4.4. Suppose $\{\mathscr{X}_i: i \in I\}$ is a collection of normed spaces and Y is a normed space contained in $\mathbb{F}^I$. Define $\mathscr{X} \equiv \{x \in \Pi_i \mathscr{X}_i:$ there is a y in Y with $\|x(i)\| \leq y(i)$ for all $i\}$. If $x \in \mathscr{X}$, define $\|x\| \equiv \inf\{\|y\|:$ $\|x(i)\| \leq y(i)$ for all $i\}$. Then $(\mathscr{X}, \|\cdot\|)$ is a normed space. Give necessary and sufficient conditions on Y that each of the parts of (4.4) are valid for $\mathscr{X}$.

12. Let $\mathscr{X}$ be a normed space and $\mathscr{M} \leq \mathscr{X}$. (a) If $\mathscr{X}$ is separable, so is $\mathscr{X}/\mathscr{M}$. (b) If $\mathscr{X}/\mathscr{M}$ and $\mathscr{M}$ are separable, then $\mathscr{X}$ is separable. (c) Give an example such that $\mathscr{X}/\mathscr{M}$ is separable but $\mathscr{X}$ is not.

13. Let $1 \leq p < \infty$ and put $\mathscr{X} = \oplus_p \mathscr{X}_i$. Show that $\mathscr{X}$ is separable if and only if I is countable and each $\mathscr{X}_i$ is separable. Show that $\oplus_\infty \mathscr{X}_i$ is separable if and only if I is finite and each $\mathscr{X}_i$ is separable.

14. Show that $\oplus_0 \mathscr{X}_n$ is separable if and only if each $\mathscr{X}_n$ is separable.

15. Let $J \subseteq I$, and $\mathscr{X} \equiv \oplus_p \{\mathscr{X}_i : i \in I\}$, $\mathscr{M} \equiv \{x \in \mathscr{X} : x(j) = 0 \text{ for } j \text{ in } J\}$. Show that $\mathscr{X}/\mathscr{M}$ is isometrically isomorphic to $\oplus_p \{\mathscr{X}_j : j \in J\}$.

16. Let $\mathscr{H}$ be a Hilbert space and suppose $\mathscr{M} \le \mathscr{H}$. Show that if $Q: \mathscr{H} \to \mathscr{H}/\mathscr{M}$ is the natural map, then $Q: \mathscr{M}^\perp \to \mathscr{H}/\mathscr{M}$ is an isometric isomorphism.

§5. Linear Functionals

Let $\mathscr{X}$ be a vector space over $\mathbb{F}$. A *hyperplane* in $\mathscr{X}$ is a linear manifold $\mathscr{M}$ in $\mathscr{X}$ such that $\dim(\mathscr{X}/\mathscr{M}) = 1$. If $f: \mathscr{X} \to \mathbb{F}$ is a linear functional and $f \not\equiv 0$, then $\ker f$ is a hyperplane. In fact, f induces an isomorphism between $\mathscr{X}/\ker f$ and $\mathbb{F}$. Conversely, if $\mathscr{M}$ is a hyperplane, let $Q: \mathscr{X} \to \mathscr{X}/\mathscr{M}$ be the natural map and let $T: \mathscr{X}/\mathscr{M} \to \mathbb{F}$ be an isomorphism. Then $f \equiv T \circ Q$ is a linear functional on $\mathscr{X}$ and $\ker f = \mathscr{M}$.

Suppose now that f and g are linear functionals on $\mathscr{X}$ such that $\ker f = \ker g$. Let $x_0 \in \mathscr{X}$ such that $f(x_0) = 1$; so $g(x_0) \ne 0$. If $x \in \mathscr{X}$ and $\alpha = f(x)$, then $x - \alpha x_0 \in \ker f = \ker g$. So $0 = g(x) - \alpha g(x_0)$, or $g(x) = (g(x_0))\alpha = (g(x_0))f(x)$. Thus $g = \beta f$ for a scalar β. This is summarized as follows.

5.1. Proposition. *A linear manifold in $\mathscr{X}$ is a hyperplane if and only if it is the kernel of a linear functional. Two linear functionals have the same kernel if and only if one is a nonzero multiple of the other.*

Hyperplanes in a normed space fall into one of two categories.

5.2. Proposition. *If $\mathscr{X}$ is a normed space and $\mathscr{M}$ is a hyperplane in $\mathscr{X}$, then either $\mathscr{M}$ is closed or $\mathscr{M}$ is dense.*

PROOF. Consider $\operatorname{cl} \mathscr{M}$, the closure of $\mathscr{M}$. By Proposition 1.3, $\operatorname{cl} \mathscr{M}$ is a linear manifold in $\mathscr{X}$. Since $\mathscr{M} \subseteq \operatorname{cl} \mathscr{M}$ and $\dim \mathscr{X}/\mathscr{M} = 1$, either $\operatorname{cl} \mathscr{M} = \mathscr{M}$ or $\operatorname{cl} \mathscr{M} = \mathscr{X}$. ∎

If $\mathscr{X} = c_0$ and $f: \mathscr{X} \to \mathbb{F}$ is defined by $f(\alpha_1, \alpha_2, \dots) = \alpha_1$, then $\ker f = \{(\alpha_n) \in c_0 : \alpha_1 = 0\}$ is closed in c_0. To get an example of a dense hyperplane, let $\mathscr{X} = c_0$ and let e_n be the element of c_0 such that $e_n(k) = 0$ if $k \ne n$ and $e_n(n) = 1$. (It is best to think of c_0 as a collection of functions on $\mathbb{N}$.) Let $x_0(n) = 1/n$ for all n; so $x_0 \in c_0$ and $\{x_0, e_1, e_2, \dots\}$ is a linearly independent set in c_0. Let $\mathscr{B} = a$ Hamel basis in c_0 which contains $\{x_0, e_1, e_2, \dots\}$. Put $\mathscr{B} = \{x_0, e_1, e_2, \dots\} \cup \{b_i : i \in I\}$, $b_i \ne x_0$ or e_n for any i or n. Define $f: c_0 \to \mathbb{F}$ by $f(\alpha_0 x_0 + \sum_{n=1}^\infty \alpha_n e_n + \sum_i \beta_i b_i) = \alpha_0$. (Remember that in the preceding expression at most a finite number of the α_n and β_i are not zero.) Since $e_n \in \ker f$ for all $n \ge 1$, $\ker f$ is dense but clearly $\ker f \ne c_0$.

The dichotomy that exists for hyperplanes should be reflected in a dichotomy for linear functionals.

5.3.   Theorem. *If $\mathcal{X}$ is a normed space and $f\colon \mathcal{X} \to \mathbb{F}$ is a linear functional, then f is continuous if and only if $\ker f$ is closed.*

PROOF. If f is continuous, $\ker f = f^{-1}(\{0\})$ and so $\ker f$ must be closed. Assume now that $\ker f$ is closed and let $Q\colon \mathcal{X} \to \mathcal{X}/\ker f$ be the natural map. By (4.2), Q is continuous. Let $T\colon \mathcal{X}/\ker f \to \mathbb{F}$ be an isomorphism; by (3.4), T is continuous. Thus, if $g = T \circ Q\colon \mathcal{X} \to \mathbb{F}$, g is continuous and $\ker f = \ker g$. Hence (5.1) $f = \alpha g$ for some α in $\mathbb{F}$ and so f is continuous. ∎

If $f\colon \mathcal{X} \to \mathbb{F}$ is a linear functional, then f is a linear transformation and so Proposition 2.1 applies. Continuous linear functionals are also called *bounded linear* functionals and

$$\|f\| \equiv \sup\{|f(x)|\colon \|x\| \le 1\}.$$

The other formulas for $\|f\|$ given in (2.1) are also valid here. Let $\mathcal{X}^* \equiv$ the collection of all bounded linear functionals on $\mathcal{X}$. If $f, g \in \mathcal{X}^*$ and $\alpha \in F$, define $(\alpha f + g)(x) = \alpha f(x) + g(x)$; $\mathcal{X}^*$ is called the *dual space* of $\mathcal{X}$. Note that $\mathcal{X}^* = \mathcal{B}(\mathcal{X}, \mathbb{F})$.

5.4.   Proposition. *If $\mathcal{X}$ is a normed space, $\mathcal{X}^*$ is a Banach space.*

PROOF. It is left as an exercise for the reader to show that $\mathcal{X}^*$ is a normed space. To show that $\mathcal{X}^*$ is complete, let $B = \{x \in \mathcal{X}\colon \|x\| \le 1\}$. If $f \in \mathcal{X}^*$, define $\rho(f)\colon B \to \mathbb{F}$ by $\rho(f)(x) = f(x)$; that is, $\rho(f)$ is the restriction of f to B. Note that $\rho\colon \mathcal{X}^* \to C_b(B)$ is a linear isometry. Thus to show that $\mathcal{X}^*$ is complete, it suffices, since $C_b(B)$ is complete (1.6), to show that $\rho(\mathcal{X}^*)$ is closed. Let $\{f_n\} \subseteq \mathcal{X}^*$ and suppose $g \in C_b(B)$ such that $\|\rho(f_n) - g\| \to 0$ as $n \to \infty$. Let $x \in \mathcal{X}$. If $\alpha, \beta \in \mathbb{F}$, $\alpha, \beta \ne 0$, such that $\alpha x, \beta x \in B$, then $\alpha^{-1}g(\alpha x) = \lim \alpha^{-1} f_n(\alpha x) = \lim \beta^{-1} f_n(\beta x) = \beta^{-1}g(\beta x)$. Define $f\colon \mathcal{X} \to \mathbb{F}$ by letting $f(x) = \alpha^{-1}g(\alpha x)$ for any $\alpha \ne 0$ such that $\alpha x \in B$. It is left as an exercise for the reader to show that $f \in \mathcal{X}^*$ and $\rho(f) = g$. ∎

Compare the preceding result with Exercise 2.1.

It should be emphasized that it is not assumed in the preceding proposition that $\mathcal{X}$ is complete. In fact, if $\mathcal{X}$ is a normed space and $\hat{\mathcal{X}}$ is its completion (Exercise 1.16), then $\mathcal{X}^*$ and $\hat{\mathcal{X}}^*$ are isometrically isomorphic (Exercise 2.2).

5.5.   Theorem. *Let (X, Ω, μ) be a measure space and let $1 < p < \infty$. If $1/p + 1/q = 1$ and $g \in L^q(X, \Omega, \mu)$, define $F_g\colon L^p(\mu) \to \mathbb{F}$ by*

$$F_g(f) = \int fg \, d\mu.$$

Then $F_g \in L^p(\mu)^*$ *and the map* $g \mapsto F_g$ *defines an isometric isomorphism of* $L^q(\mu)$ *onto* $L^p(\mu)^*$.

Since this theorem is often proved in courses in measure and integration, the proof of this result, as well as the next two, is contained in the Appendix. See Appendix B for the proofs of (5.5) and (5.6).

5.6.   Theorem. *If* (X, Ω, μ) *is a σ-finite measure space and* $g \in L^\infty(X, \Omega, \mu)$, *define* $F_g \colon L^1(\mu) \to \mathbb{F}$ *by*

$$F_g(f) = \int fg \, d\mu.$$

Then $F_g \in L^1(\mu)^*$ *and the map* $g \mapsto F_g$ *defines an isometric isomorphism of* $L^\infty(\mu)$ *onto* $L^1(\mu)^*$.

Note that when $p = 2$ in Theorem 5.5, there is a little difference between (5.5) and (I.3.5) owing to the absence of a complex conjugate in (5.5). Also, note that (5.6) is false if the measure space is not assumed to be σ-finite (Exercise 3).

If X is a locally compact space, $M(X)$ denotes the space of all $\mathbb{F}$-valued regular Borel measures on X with the total variation norm. See Appendix C for the definitions as well as the proof of the next theorem.

5.7. Riesz Representation Theorem. *If* X *is a locally compact space and* $\mu \in M(X)$, *define* $F_\mu \colon C_0(X) \to \mathbb{F}$ *by*

$$F_\mu(f) = \int f \, d\mu.$$

Then $F_\mu \in C_0(X)^*$ *and the map* $\mu \to F_\mu$ *is an isometric isomorphism of* $M(X)$ *onto* $C_0(X)^*$.

There are special cases of these theorems that deserve to be pointed out.

5.8.   Example. The dual of c_0 is isometrically isomorphic to l^1. In fact, $c_0 = C_0(\mathbb{N})$, if $\mathbb{N}$ is given the discrete topology, and $l^1 = M(\mathbb{N})$.

5.9.   Example. The dual of l^1 is isometrically isomorphic to l^∞. In fact, $l^1 = L^1(\mathbb{N}, 2^\mathbb{N}, \mu)$, where $\mu(\Delta) = $ the number of points in Δ. Also, $l^\infty = L^\infty(\mathbb{N}, 2^\mathbb{N}, \mu)$.

5.10.   Example. If $1 < p < \infty$, the dual of l^p is l^q, where $1 = 1/p + 1/q$.

What is the dual of $L^\infty(X, \Omega, \mu)$? There are two possible representations. One is to identify $L^\infty(X, \Omega, \mu)^*$ with the space of finitely additive measures defined on Ω that are "absolutely continuous" with respect to μ and have finite total variation (see Dunford and Schwartz [1958], p. 296). Another representation is to obtain a compact space Z such that $L^{\infty}(X, \Omega, \mu)$ is isometrically isomorphic to $C(Z)$ and then use the Riesz Representation Theorem. This will be done later in this book (VIII.2.1).

What is the dual of $M(X)$? For this, define $L^{\infty}(M(X))$ as the set of all F in $\prod\{L^{\infty}(\mu): \mu \in M(X)\}$ such that if $\mu \ll \nu$; then $F(\mu) = F(\nu)$ a.e. $[\mu]$. This is an inverse limit of the spaces $L^{\infty}(\mu)$, μ in $M(X)$.

5.11. Lemma. *If* $F \in L^{\infty}(M(X))$, *then*

$$\|F\| \equiv \sup_{\mu} \|F(\mu)\|_{\infty} < \infty.$$

PROOF. If $\|F\| = \infty$, then there is a sequence $\{\mu_n\}$ in $M(X)$ such that $\|F(\mu_n)\|_{\infty} \geq n$. Let $\mu = \sum_{n=1}^{\infty} 2^{-n} |\mu_n| / \|\mu_n\|$. Then $\mu_n \ll \mu$ for all n, so $F(\mu_n) = F(\mu)$ a.e. $[\mu_n]$ for each n. Hence $\|F(\mu)\|_{\infty} \geq \|F(\mu_n)\|_{\infty} \geq n$ for each n, a contradiction. ∎

5.12. Theorem. *If* X *is locally compact and* $F \in L^{\infty}(M(X))$, *define* Φ_F: $M(X) \to \mathbb{F}$ *by*

$$\Phi_F(\mu) = \int F(\mu)\, d\mu.$$

Then $\Phi_F \in M(X)^*$ *and the map* $F \mapsto \Phi_F$ *is an isometric isomorphism of* $L^{\infty}(M(X))$ *onto* $M(X)^*$.

PROOF. It is easy to see that Φ_F is linear. Also, $|\Phi_F(\mu)| \leq \int |F(\mu)|\, d|\mu| \leq \|F(\mu)\|_{\infty} \|\mu\| \leq \|F\| \|\mu\|$. Thus $\Phi_F \in M(X)^*$ and $\|\Phi_F\| \leq \|F\|$.

Now fix Φ in $M(X)^*$. If $\mu \in M(X)$ and $f \in L^1(|\mu|)$, then $\nu = f\mu \in M(X)$. (That is, $\nu(\Delta) = \int_{\Delta} f\, d\mu$ for every Borel set Δ.) Also $\|\nu\| = \int |f|\, d|\mu|$. In fact, the Radon–Nikodym Theorem can be interpreted as an identification (isometrically isomorphic) of $L^1(|\mu|)$ with $\{\eta \in M(X): \eta \ll |\mu|\}$. Thus $f \mapsto \Phi(f\mu)$ is a linear functional on $L^1(|\mu|)$ and $|\Phi(f\mu)| \leq \|\Phi\| \int |f|\, d|\mu|$. Hence there is an $F(\mu)$ in $L^{\infty}(|\mu|)$ such that $\Phi(f\mu) = \int fF(\mu)\, d\mu$ for every f in $L^1(|\mu|)$ and $\|F(\mu)\|_{\infty} \leq \|\Phi\|$. (We have been a little nonchalant about using μ or $|\mu|$, but what was said is perfectly correct. Fill in the details.) In particular, taking $f = 1$ gives $\Phi(\mu) = \int F(\mu)\, d\mu$. It must be shown that $F \in L^{\infty}(M(X))$; it then follows that $\Phi = \Phi_F$ and $\|\Phi_F\| \geq \|F\|_{\infty}$.

To show that $F \in L^{\infty}(M(X))$, let μ and ν be measures such that $\mu \ll \nu$. By the Radon–Nikodym Theorem, there is an f in $L^1(|\mu|)$ such that $\nu = f\mu$. Hence if $g \in L^1(|\nu|)$, then $gf \in L^1(|\mu|)$ and $\int g\, d\nu = \int gf\, d\mu$. Thus, $\int gF(\nu)\, d\nu = \Phi(g\nu) = \Phi(gf\mu) = \int gfF(\mu)\, d\mu = \int gF(\mu)\, d\nu$. So $F(\nu) = F(\mu)$ a.e. $[\nu]$ and $F \in L^{\infty}(M(X))$. ∎

EXERCISES

1. Complete the proof of Proposition 5.4.

2. Show that $\mathscr{X}^*$ is a normed space.

3. Give an example of a measure space (X, Ω, μ) that is not σ-finite for which the conclusion of Theorem 5.6 is false.

4. Let $\{\mathcal{X}_i : i \in I\}$ be a collection of normed spaces. If $1 \le p < \infty$, show that the dual space of $\oplus_p \mathcal{X}_i$ is isometrically isomorphic to $\oplus_q \mathcal{X}_i^*$, where $1/p + 1/q = 1$.

5. If $\mathcal{X}_1, \mathcal{X}_2, \ldots$ are normed spaces, show that $(\oplus_0 \mathcal{X}_n)^*$ is isometrically isomorphic to $\oplus_1 \mathcal{X}_n^*$.

6. Let $n \ge 1$ and let $C^{(n)}[0,1]$ be defined as in Example 1.10. Show that $\|f\| = \sum_{k=0}^{n-1} |f^{(k)}(0)| + \sup\{|f^{(n)}(x)| : 0 \le x \le 1\}$ is an equivalent norm on $C^{(n)}[0,1]$. Show that $L \in (C^{(n)}[0,1])^*$ if and only if there are scalars $\alpha_0, \alpha_1, \ldots, \alpha_{n-1}$ and a measure μ on $[0,1]$ such that $L(f) = \sum_{k=0}^{n-1} \alpha_k f^{(k)}(0) + \int f^{(n)} d\mu$. Is there a formula for $\|L\|$ in terms of $|\alpha_0|, |\alpha_1|, \ldots, |\alpha_{n-1}|$, and $\|\mu\|$?

§6. The Hahn–Banach Theorem

The Hahn–Banach Theorem is one of the most important results in mathematics. It is used so often it is rightly considered as a cornerstone of functional analysis. It is one of those theorems that when it or one of its immediate consequences is used, it is used without quotation or reference and the reader is assumed to realize that it is being invoked.

6.1. Definition. If $\mathcal{X}$ is a vector space, a *sublinear functional* is a function $q \colon \mathcal{X} \to \mathbb{R}$ such that

(a) $q(x + y) \le q(x) + q(y)$ for all x, y in $\mathcal{X}$;
(b) $q(\alpha x) = \alpha q(x)$ for x in $\mathcal{X}$ and $\alpha \ge 0$.

Note that every seminorm is a sublinear functional, but not conversely. In fact, it should be emphasized that a sublinear functional is allowed to assume negative values and that (b) in the definition only holds for $\alpha \ge 0$.

6.2. The Hahn–Banach Theorem. *Let $\mathcal{X}$ be a vector space over $\mathbb{R}$ and let q be a sublinear functional on $\mathcal{X}$. If $\mathcal{M}$ is a linear manifold in $\mathcal{X}$ and $f \colon \mathcal{M} \to \mathbb{R}$ is a linear functional such that $f(x) \le q(x)$ for all x in $\mathcal{M}$, then there is a linear functional $F \colon \mathcal{X} \to \mathbb{R}$ such that $F|\mathcal{M} = f$ and $F(x) \le q(x)$ for all x in $\mathcal{X}$.*

Note that the substance of the theorem is not that the extension exists but that an extension can be found that remains dominated by q. Just to find an extension, let $\{e_i\}$ be a Hamel basis for $\mathcal{M}$ and let $\{y_j\}$ be vectors in $\mathcal{X}$ such that $\{e_i\} \cup \{y_j\}$ is a Hamel basis for $\mathcal{X}$. Now define $F \colon \mathcal{X} \to \mathbb{R}$ by

$F(\Sigma_i \alpha_i e_i + \Sigma_j \beta_j y_j) = \Sigma_i \alpha_i f(e_i) = f(\Sigma_i \alpha_i e_i)$. This extends f. If $\{\gamma_j\}$ is any collection of real numbers, then $F(\Sigma_i \alpha_i e_i + \Sigma_j \beta_j y_j) = f(\Sigma_i \alpha_i e_i) + \Sigma_j \beta_j \gamma_j$ is also an extension of f. Moreover, any extension of f has this form. The difficulty is that we must find one of these extensions that is dominated by q.

Before proving the theorem, let's see some of its immediate corollaries. The first is an extension of the theorem to complex spaces. For this a lemma is needed. Note that if $\mathscr{X}$ is a vector space over $\mathbb{C}$, it is also a vector space over $\mathbb{R}$. Also, if $f: \mathscr{X} \to \mathbb{C}$ is $\mathbb{C}$-linear, then $\mathrm{Re}\, f: \mathscr{X} \to \mathbb{R}$ is $\mathbb{R}$-linear. The following lemma is the converse of this.

6.3. Lemma. *Let $\mathscr{X}$ be a vector space over $\mathbb{C}$.*

(a) *If $f: \mathscr{X} \to \mathbb{R}$ is an $\mathbb{R}$-linear functional, then $\tilde{f}(x) = f(x) - if(ix)$ is a $\mathbb{C}$-linear functional and $f = \mathrm{Re}\, \tilde{f}$.*
(b) *If $g: \mathscr{X} \to \mathbb{C}$ is $\mathbb{C}$-linear, $f = \mathrm{Re}\, g$, and $\tilde{f}$ is defined as in (a), then $\tilde{f} = g$.*
(c) *If p is a seminorm on $\mathscr{X}$ and f and $\tilde{f}$ are as in (a), then $|f(x)| \le p(x)$ for all x if and only if $|\tilde{f}(x)| \le p(x)$ for all x.*
(d) *If $\mathscr{X}$ is a normed space and f and $\tilde{f}$ are as in (a), then $\|f\| = \|\tilde{f}\|$.*

PROOF. The proofs of (a) and (b) are left as an exercise. To prove (c), suppose $|\tilde{f}(x)| \le p(x)$. Then $f(x) = \mathrm{Re}\, \tilde{f}(x) \le |\tilde{f}(x)| \le p(x)$. Also, $-f(x) = \mathrm{Re}\, \tilde{f}(-x) \le |\tilde{f}(-x)| \le p(x)$. Hence $|f(x)| \le p(x)$. Now assume that $|f(x)| \le p(x)$. Choose θ such that $\tilde{f}(x) = e^{i\theta}|\tilde{f}(x)|$. Hence $|\tilde{f}(x)| = \tilde{f}(e^{-i\theta}x) = \mathrm{Re}\, \tilde{f}(e^{-i\theta}x) = f(e^{-i\theta}x) \le p(e^{-i\theta}x) = p(x)$.

Part (d) is an easy application of (c). ■

6.4. Corollary. *Let $\mathscr{X}$ be a vector space, let $\mathscr{M}$ be a linear manifold in $\mathscr{X}$, and let $p: \mathscr{X} \to [0, \infty)$ be a seminorm. If $f: \mathscr{M} \to \mathbb{F}$ is a linear functional such that $|f(x)| \le p(x)$ for all x in $\mathscr{M}$, then there is a linear functional $F: \mathscr{X} \to \mathbb{F}$ such that $F|\mathscr{M} = f$ and $|F(x)| \le p(x)$ for all x in $\mathscr{X}$.*

PROOF. *Case 1:* $\mathbb{F} = \mathbb{R}$. Note that $f(x) \le |f(x)| \le p(x)$ for x in $\mathscr{M}$. By (5.2) there is an extension $F: \mathscr{X} \to \mathbb{R}$ of f such that $F(x) \le p(x)$ for all x. Hence $-F(x) = F(-x) \le p(-x) = p(x)$. Thus $|F| \le p$.

Case 2: $\mathbb{F} = \mathbb{C}$. Let $f_1 = \mathrm{Re}\, f$. By (6.3c), $|f_1| \le p$. By Case 1, there is an $\mathbb{R}$-linear functional $F_1: \mathscr{X} \to \mathbb{R}$ such that $F_1|\mathscr{M} = f_1$ and $|F_1| \le p$. Let $F(x) = F_1(x) - iF_1(ix)$ for all x in $\mathscr{X}$. By (6.3c), $|F| \le p$. Clearly, $F|\mathscr{M} = f$. ■

6.5. Corollary. *If $\mathscr{X}$ is a normed space, $\mathscr{M}$ is a linear manifold in $\mathscr{X}$, and $f: \mathscr{M} \to \mathbb{F}$ is a bounded linear functional, then there is an F in $\mathscr{X}^*$ such that $F|\mathscr{M} = f$ and $\|F\| = \|f\|$.*

PROOF. Use Corollary 6.4 with $p(x) = \|f\|\, \|x\|$. ■

6.6. Corollary. *If $\mathscr{X}$ is a normed space, $\{x_1, x_2, \ldots, x_d\}$ is a linearly independent subset of $\mathscr{X}$, and $\alpha_1, \alpha_2, \ldots, \alpha_d$ are arbitrary scalars, then there is an f in $\mathscr{X}^*$ such that $f(x_j) = \alpha_j$ for $1 \leq j \leq d$.*

PROOF. Let $\mathscr{M}$ = the linear span of $x_1, \ldots, x_d$ and define $g\colon \mathscr{M} \to \mathbb{F}$ by $g(\Sigma_j \beta_j x_j) = \Sigma_j \beta_j \alpha_j$. So g is linear. Since $\mathscr{M}$ is finite dimensional, g is continuous. Let f be a continuous extension of g to $\mathscr{X}$. ∎

6.7. Corollary. *If $\mathscr{X}$ is a normed space and $x \in \mathscr{X}$, then*

$$\|x\| = \sup\{|f(x)|\colon f \in \mathscr{X}^* \text{ and } \|f\| \leq 1\}.$$

Moreover, this supremum is attained.

PROOF. Let $\alpha = \sup\{|f(x)|\colon f \in \mathscr{X}^* \text{ and } \|f\| \leq 1\}$. If $f \in \mathscr{X}^*$ and $\|f\| \leq 1$, then $|f(x)| \leq \|f\| \, \|x\| \leq \|x\|$; hence $\alpha \leq \|x\|$. Now let $\mathscr{M} = \{\beta x\colon \beta \in \mathbb{F}\}$ define $g\colon \mathscr{M} \to \mathbb{F}$ by $g(\beta x) = \beta\|x\|$. Then $g \in \mathscr{M}^*$ and $\|g\| = 1$. By Corollary 6.5, there is an f in $\mathscr{X}^*$ such that $\|f\| = 1$ and $f(x) = g(x) = \|x\|$. ∎

This introduces a certain symmetry in the definitions of the norms in $\mathscr{X}$ and $\mathscr{X}^*$ that will be explored later (§11).

6.8. Corollary. *If $\mathscr{X}$ is a normed space, $\mathscr{M} \leq \mathscr{X}$, $x_0 \in \mathscr{X} \backslash \mathscr{M}$, and $d = \text{dist}(x_0, \mathscr{M})$, then there is an f in $\mathscr{X}^*$ such that $f(x_0) = 1$, $f(x) = 0$ for all x in $\mathscr{M}$, and $\|f\| = d^{-1}$.*

PROOF. Let $Q\colon \mathscr{X} \to \mathscr{X}/\mathscr{M}$ be the natural map. Since $\|x_0 + \mathscr{M}\| = d$, by the preceding corollary there is a g in $(\mathscr{X}/\mathscr{M})^*$ such that $g(x_0 + \mathscr{M}) = d$ and $\|g\| = 1$. Let $f = d^{-1}g \circ Q\colon \mathscr{X} \to \mathbb{F}$. Then f is continuous, $f(x) = 0$ for x in $\mathscr{M}$, and $f(x_0) = 1$. Also, $|f(x)| = d^{-1}|g(Q(x))| \leq d^{-1}\|Q(x)\| \leq d^{-1}\|x\|$; hence $\|f\| \leq d^{-1}$. On the other hand, $\|g\| = 1$ so there is a sequence $\{x_n\}$ such that $|g(x_n + \mathscr{M})| \to 1$ and $\|x_n + \mathscr{M}\| < 1$ for all n. Let $y_n \in \mathscr{M}$ such that $\|x_n + y_n\| < 1$. Then $|f(x_n + y_n)| = d^{-1}|g(x_n + \mathscr{M})| \to d^{-1}$, so $\|f\| = d^{-1}$. ∎

To prove the Hahn–Banach Theorem, we first show that we can extend the functional to a space of one dimension more.

6.9. Lemma. *Suppose the hypothesis of (6.2) is satisfied and, in addition, $\dim \mathscr{X}/\mathscr{M} = 1$. Then the conclusion of (6.2) is valid.*

PROOF. Fix x_0 in $\mathscr{X} \backslash \mathscr{M}$; so $\mathscr{X} = \mathscr{M} \vee \{x_0\} = \{tx_0 + y\colon t \in \mathbb{R}, y \in \mathscr{M}\}$. For the moment assume that the extension $F\colon \mathscr{X} \to \mathbb{R}$ of f exists with $F \leq q$. Let's see what F must look like. Put $\alpha_0 = F(x_0)$. If $t > 0$ and $y_1 \in \mathscr{M}$, then $F(tx_0 + y_1) = t\alpha_0 + f(y_1) \leq q(tx_0 + y_1)$. Hence $\alpha_0 \leq -t^{-1}f(y_1) + t^{-1}q(tx_0 + y_1) = -f(y_1/t) + q(x_0 + y_1/t)$ for every y_1 in

$\mathcal{M}$. Since $y_1/t \in \mathcal{M}$, this gives that

6.10 $\alpha_0 \le -f(y_1) + q(x_0 + y_1)$

for all y_1 in $\mathcal{M}$. Also note that if α_0 satisfies (6.10), then by reversing the inequalities in the preceding argument, it follows that $t\alpha_0 + f(y_1) \le q(tx_0 + y_1)$ whenever $t \ge 0$.

If $t \ge 0$ and $y_2 \in \mathcal{M}$ and if F exists, then $F(-tx_0 + y_2) = -t\alpha_0 + f(y_2) \le q(-tx_0 + y_2)$. As above, this implies that

6.11 $\alpha_0 \ge f(y_2) - q(-x_0 + y_2)$

for all y_2 in $\mathcal{M}$. Moreover, (6.11) is sufficient that $-t\alpha_0 + f(y_2) \le q(-tx_0 + y_2)$ for all $t \ge 0$ and y_2 in $\mathcal{M}$.

Combining (6.10) and (6.11) we see that we must show that α_0 can be chosen satisfying (6.10) and (6.11) simultaneously. Thus we must show that

6.12 $f(y_2) - q(-x_0 + y_2) \le -f(y_1) + q(x_0 + y_1)$

for all y_1, y_2 in $\mathcal{M}$. But this means we want to show that $f(y_1 + y_2) \le q(x_0 + y_1) + q(-x_0 + y_2)$. But

$$f(y_1 + y_2) \le q(y_1 + y_2) = q((y_1 + x_0) + (-x_0 + y_2))$$
$$\le q(y_1 + x_0) + q(-x_0 + y_2),$$

so (6.12) is satisfied. If α_0 is chosen with $\sup\{f(y_2) - q(-x_0 + y_2): y_2 \in \mathcal{M}\} \le \alpha_0 \le \inf\{-f(y_1) + q(x_0 + y_1): y_1 \in \mathcal{M}\}$ and $F(tx_0 + y) = t\alpha_0 + f(y_1)$, F satisfies the conclusion of (6.2). ∎

PROOF OF THE HAHN–BANACH THEOREM. Let $\mathcal{S}$ be the collection of all pairs $(\mathcal{M}_1, f_1)$, where $\mathcal{M}_1$ is a linear manifold in $\mathcal{X}$ such that $\mathcal{M}_1 \supseteq \mathcal{M}$ and $f_1: \mathcal{M}_1 \to \mathbb{R}$ is a linear functional with $f_1|\mathcal{M} = f$ and $f_1 \le q$ on $\mathcal{M}_1$. If $(\mathcal{M}_1, f_1)$ and $(\mathcal{M}_2, f_2) \in \mathcal{S}$, define $(\mathcal{M}_1, f_1) \le (\mathcal{M}_2, f_2)$ to mean that $\mathcal{M}_1 \subseteq \mathcal{M}_2$ and $f_2|\mathcal{M}_1 = f_1$. So $(\mathcal{S}, \le)$ is a partially ordered set. Suppose $\mathcal{C} = \{(\mathcal{M}_i, f_i): i \in I\}$ is a chain in $\mathcal{S}$. If $\mathcal{N} \equiv \cup\{\mathcal{M}_i: i \in I\}$, then the fact that $\mathcal{C}$ is a chain implies that $\mathcal{N}$ is a linear manifold. Define $F: \mathcal{N} \to \mathbb{R}$ by setting $F(x) = f_i(x)$ if $x \in \mathcal{M}_i$. It is easily checked that F is well defined, linear, and satisfies $F \le q$ on $\mathcal{N}$. So $(\mathcal{N}, F) \in \mathcal{S}$ and $(\mathcal{N}, F)$ is an upper bound for $\mathcal{C}$. By Zorn's Lemma, $\mathcal{S}$ has a maximal element $(\mathcal{Y}, F)$. But the preceding lemma implies that $\mathcal{Y} = \mathcal{X}$. Hence F is the desired extension. ∎

This section concludes with one important consequence of the Hahn–Banach Theorem. It will be generalized later (IV.3.11), but it is used so often it is worth singling out for consideration.

6.13.   Theorem. *If $\mathcal{X}$ is a normed space and $\mathcal{M}$ is a linear manifold in $\mathcal{X}$, then*

$$\text{cl } \mathcal{M} = \bigcap\{\ker f: f \in \mathcal{X}^* \text{ and } \mathcal{M} \subseteq \ker f\}.$$

PROOF. Let $\mathcal{N} = \cap \{\ker f\colon f \in \mathcal{X}^* \text{ and } \mathcal{M} \subseteq \ker f\}$. If $f \in \mathcal{X}^*$ and $\mathcal{M} \subseteq \ker f$, then the continuity of f implies that $\mathrm{cl}\, \mathcal{M} \subseteq \ker f$. Hence $\mathrm{cl}\, \mathcal{M} \subseteq \mathcal{N}$. If $x_0 \notin \mathrm{cl}\, \mathcal{M}$, then $d = \mathrm{dist}(x_0, \mathcal{M}) > 0$. By Corollary 6.8 there is an f in $\mathcal{X}^*$ such that $f(x_0) = 1$ and $f(x) = 0$ for every x in $\mathcal{M}$. Hence $x_0 \notin \mathcal{N}$. Thus $\mathcal{N} \subseteq \mathrm{cl}\, \mathcal{M}$ and the proof is complete. ∎

6.14. Corollary. *If $\mathcal{X}$ is a normed space and $\mathcal{M}$ is a linear manifold in $\mathcal{X}$, then $\mathcal{M}$ is dense in $\mathcal{X}$ if and only if the only bounded linear functional on $\mathcal{X}$ that annihilates $\mathcal{M}$ is the zero functional.*

EXERCISES

1. Complete the proof of Lemma 6.3.

2. Give the details of the proof of Corollary 6.5.

3. Show that c^* is isometrically isomorphic to l^1. Are c and c_0 isometrically isomorphic?

4. If μ is a measure on $[0,1]$ and $\int x^n \, d\mu(x) = 0$ for all $n \geq 0$, show that $\mu = 0$.

5. If $n \geq 1$, show that there is a measure μ on $[0,1]$ such that for every polynomial p of degree at most n,

$$\int p \, d\mu = \sum_{k=1}^{n} p^{(k)}(k/n).$$

6. If $n \geq 1$, does there exist a measure μ on $[0,1]$ such that $p'(0) = \int p \, d\mu$ for every polynomial of degree at most n?

7. Does there exist a measure μ on $[0,1]$ such that $\int p \, d\mu = p'(0)$ for every polynomial p?

8. Let K be a compact subset of $\mathbb{C}$ and define $A(K)$ to be $\{f \in C(K)\colon f$ is analytic on $\mathrm{int}\, K\}$. (Functions here are complex valued.) Show that if $a \in K$, then there is a probability measure μ supported on ∂K such that $f(a) = \int f \, d\mu$ for every f in $A(K)$. (A *probability measure* is a nonnegative measure μ such that $\|\mu\| = 1$.)

9. If $K = \mathrm{cl}\, \mathbb{D}$ ($\mathbb{D} = \{|z| < 1\}$) and $a \in K$, find the measure μ whose existence was proved in Exercise 8.

10. Let $P = \{p|\partial\mathbb{D}\colon p = $ an analytic polynomial$\}$ and consider P as a manifold in $C(\partial\mathbb{D})$. Show that if μ is a real-valued measure on $\partial\mathbb{D}$ such that $\int p \, d\mu = 0$ for every p in P, then $\mu = 0$. Give an example of a complex-valued measure μ such that $\mu \neq 0$ but $\int p \, d\mu = 0$ for every p in P.

§7*. An Application: Banach Limits

If $x = \{x(n)\} \in c$, define $L(x) = \lim x(n)$. Then L is a linear functional, $\|L\| = 1$, and, if for x in c, x' is defined by $x' = (x(2), x(3), \ldots)$, then $L(x) = L(x')$. Also, if $x \geq 0$ [that is, $x(n) \geq 0$ for all n], then $L(x) \geq 0$.

In this section it will be shown that these properties of the limit functional can be extended to l^∞. The proof uses the Hahn–Banach Theorem.

7.1.   Theorem. *There is a linear functional* $L: l^\infty \to \mathbb{F}$ *such that*

(a) $\|L\| = 1$.
(b) *If* $x \in c$, $L(x) = \lim x(n)$.
(c) *If* $x \in l^\infty$ *and* $x(n) \geq 0$ *for all* n, *then* $L(x) \geq 0$.
(d) *If* $x \in l^\infty$ *and* $x' \equiv (x(2), x(3), \ldots)$, *then* $L(x) = L(x')$.

PROOF. First assume $\mathbb{F} = \mathbb{R}$; that is, $l^\infty = l^\infty_\mathbb{R}$. If $x \in l^\infty$, let x' denote the element of l^∞ defined in part (d) above. Put $\mathscr{M} = \{x - x' : x \in l^\infty\}$. Note that $(x + \alpha y)' = x' + \alpha y'$ for any x, y in l^∞ and α in $\mathbb{R}$; hence $\mathscr{M}$ is a linear manifold in l^∞. Let 1 denote the sequence $(1, 1, 1, \ldots)$ in l^∞.

7.2.   Claim. $\operatorname{dist}(1, \mathscr{M}) = 1$.

Since $0 \in \mathscr{M}$, $\operatorname{dist}(1, \mathscr{M}) \leq 1$. Let $x \in l^\infty$; if $(x - x')(n) \leq 0$ for any n, then $\|1 - (x - x')\|_\infty \geq |1 - (x(n) - x'(n))| \geq 1$. Suppose $0 \leq (x - x')(n) = x(n) - x'(n) = x(n) - x(n + 1)$ for all n. Thus $x(n + 1) \leq x(n)$ for all n. Since $x \in l^\infty$, $\alpha = \lim x(n)$ exists. Thus $\lim(x - x')(n) = 0$ and $\|1 - (x - x')\|_\infty \geq 1$. This proves the claim.

By Corollary 6.8 there is a linear functional $L: l^\infty \to \mathbb{R}$ such that $\|L\| = 1$, $L(1) = 1$, and $L(\mathscr{M}) = 0$. So this functional satisfies (a) and (d) of the theorem. To prove (b), we establish the following.

7.3.   Claim. $c_0 \subseteq \ker L$.

If $x \in c_0$, let $x^{(1)} = x'$ and let $x^{(n+1)} = (x^{(n)})'$ for $n \geq 1$. Note that $x^{(n+1)} - x = [x^{(n+1)} - x^{(n)}] + \cdots + [x' - x] \in \mathscr{M}$. Hence $L(x) = L(x^{(n)})$ for all $n \geq 1$. If $\varepsilon > 0$, then let n be such that $|x(m)| < \varepsilon$ for $m > n$. Hence $|L(x)| = |L(x^{(n)})| \leq \|x^{(n)}\|_\infty = \sup\{|x(m)| : m > n\} < \varepsilon$. Thus $x \in \ker L$. Condition (b) is now clear.

To show (c), suppose there is an x in l^∞ such that $x(n) \geq 0$ for all n and $L(x) < 0$. If x is replaced by $x/\|x\|_\infty$, it remains true that $L(x) < 0$ and it is also true that $1 \geq x(n) \geq 0$ for all n. But then $\|1 - x\|_\infty \leq 1$ and $L(1 - x) = 1 - L(x) > 1$, contradicting (a). Thus (c) holds.

Now assume that $\mathbb{F} = \mathbb{C}$. Let L_1 be the functional obtained on $l^\infty_\mathbb{R}$. If $x \in l^\infty_\mathbb{C}$, then $x = x_1 + ix_2$ when $x_1, x_2 \in l^\infty_\mathbb{R}$. Define $L(x) = L_1(x_1) + iL_1(x_2)$. It is left as an exercise to show that L is $\mathbb{C}$-linear. It's clear that (b), (c), and (d) hold. It remains to show that $\|L\| = 1$.

Let $E_1, \ldots, E_m$ be pairwise disjoint subsets of $\mathbb{N}$ and let $\alpha_1, \ldots, \alpha_m \in \mathbb{C}$ with $|\alpha_k| \leq 1$ for all k. Put $x = \sum_{k=1}^m \alpha_k \chi_{E_k}$; so $x \in \ell^\infty$ and $\|x\|_\infty \leq 1$. Then $L(x) = \sum_k \alpha_k L(\chi_{E_k}) = \sum_k \alpha_k L_1(\chi_{E_k})$. But $L_1(\chi_{E_k}) \geq 0$ and $\sum_k L_1(\chi_{E_k}) = L_1(\chi_E)$, where $E = \bigcup_k E_k$. Hence $\sum_k L_1(\chi_{E_k}) \leq 1$. Because $|\alpha_k| \leq 1$ for all k, $|L(x)| \leq 1$. (This is a small convexity argument.) If x is an arbitrary element of l^∞, $\|x\|_\infty \leq 1$, then there is a sequence $\{x_n\}$ if elements of l^∞ such that $\|x_n - x\|_\infty \to 0$, $\|x_n\|_\infty \leq 1$, and each x_n is the type of element of l^∞ just discussed that takes on only a finite number of values (Exercise 3). Clearly, $\|L\| \leq 2$, so $L(x_n) \to L(x)$. Since $|L(x_n)| \leq 1$ for all n, $|L(x)| \leq 1$. Hence $\|L\| \leq 1$. Since $L(1) = 1$, $\|L\| = 1$.   ∎

A linear functional of the type described in Theorem 7.1 is called a *Banach limit*. They are useful for a variety of things, among which is the construction of representations of the algebra of bounded operators on a Hilbert space.

EXERCISES

1. If L is a Banach limit, show that there are x and y in l^∞ such that $L(xy) \neq L(x)L(y)$.

2. Let X be a set and Ω a σ-algebra of subsets of X. Suppose μ is a complex-valued countably additive measure defined on Ω such that $\|\mu\| = \mu(X) < \infty$. Show that $\mu(\Delta) \geq 0$ for every Δ in Ω. (Though it is difficult to see at this moment, this fact is related to the proof of (c) in Theorem 7.1 for the complex case.)

3. Show that if $x \in l^\infty$, $\|x\|_\infty \leq 1$, then there is a sequence $\{x_n\}$, x_n in l^∞ such that $\|x_n\|_\infty \leq 1$, $\|x_n - x\| \to 0$, and each x_n takes on only a finite number of values.

§8*. An Application: Runge's Theorem

8.1. Runge's Theorem. *Let K be a compact subset of $\mathbb{C}$ and let E be a subset of $\mathbb{C}_\infty \setminus K$ that meets each component of $\mathbb{C}_\infty \setminus K$. If f is analytic in a neighborhood of K, then there are rational functions f_n whose only poles lie in E such that $f_n \to f$ uniformly on K.*

The main tool in proving Runge's Theorem is Theorem 6.13. (A proof that does not use functional analysis can be found on p. 198 of Conway [1978].) To do this, let $R(K, E)$ be the closure in the space $C(K)$ of the rational functions with poles in E. By (6.13) and the Riesz Representation Theorem, it suffices to show that if $\mu \in M(K)$ and $\int g \, d\mu = 0$ for each g in $R(K, E)$, then $\int f \, d\mu = 0$.

Let $R > 0$ and let λ be area measure. Pick $\rho > 0$ such that $B(0; R) \subseteq B(z; \rho)$ for every z in K. Then for z in K,

$$\int_{B(0;\, R)} |z - w|^{-1} \, d\lambda(w) \leq \int_{B(z;\, \rho)} |z - w|^{-1} \, d\lambda(w)$$

$$= \int_0^{2\pi} \int_0^\rho dr \, d\theta = 2\pi\rho.$$

If $\mu \in M(K)$, define $\tilde{\mu} \colon \mathbb{C} \to [0, \infty]$ by

$$\tilde{\mu}(w) = \int \frac{d|\mu|(z)}{|z - w|}$$

when the integral is finite, and $\tilde{\mu}(w) = \infty$ otherwise. The inequalities above imply

$$\int_{B(0;\, R)} \tilde{\mu}(w)\, d\lambda(w) = \int_{B(0;\, R)} \int_K \frac{d|\mu|(z)}{|z - w|}\, d\lambda(w)$$

$$= \int_K \int_{B(0;\, R)} \frac{d\lambda(w)}{|z - w|}\, d|\mu|(z)$$

$$\leq 2\pi\rho\|\mu\|.$$

Thus $\tilde{\mu}(w) < \infty$ a.e. $[\lambda]$.

8.2. Lemma. *If $\mu \in M(K)$, then*

$$\hat{\mu}(w) = \int \frac{d\mu(z)}{z - w}$$

is in $L^1(B(0; R), \lambda)$ for any $R > 0$, $\hat{\mu}$ is analytic on $\mathbb{C}_\infty \setminus K$, and $\hat{\mu}(\infty) = 0$.

PROOF. The first statement follows from what came before the statement of this lemma. To show that $\hat{\mu}$ is analytic on $\mathbb{C}_\infty \setminus K$, let $w, w_0 \in \mathbb{C} \setminus K$ and note that

$$\frac{\hat{\mu}(w) - \hat{\mu}(w_0)}{w - w_0} = \int_K \frac{d\mu(z)}{(z - w)(z - w_0)}.$$

As $w \to w_0$, $[(z - w)(z - w_0)]^{-1} \to (z - w_0)^{-2}$ uniformly for z in K, so that $\hat{\mu}$ has a derivative at w_0 and

$$\frac{d\hat{\mu}}{dw}(w_0) = \int_K (z - w_0)^{-2}\, d\mu(z).$$

So $\hat{\mu}$ is analytic on $\mathbb{C} \setminus K$. To show that it is analytic at infinity, note that $\hat{\mu}(z) \to 0$ as $z \to \infty$, so infinity is a removable singularity.   ∎

It is not difficult to see that for w_0 in $\mathbb{C} \setminus K$,

8.3
$$\left(\frac{d}{dw} \right)^n \hat{\mu}(w_0) = n! \int (z - w_0)^{-n-1}\, d\mu(z)$$

Also, we can easily find the power series expansion of $\hat{\mu}$ at infinity. Indeed,

$$\hat{\mu}(w) = \int \frac{1}{z - w}\, d\mu(z) = -\frac{1}{w} \int \left(1 - \frac{z}{w} \right)^{-1} d\mu(z).$$

Choose w near enough to infinity that $|z/w| < 1$ for all z in K. Then

8.4
$$\hat{\mu}(w) = -\frac{1}{w} \sum_{n=0}^{\infty} \int \left(\frac{z}{w} \right)^n d\mu(z)$$

$$= -\sum_{n=0}^{\infty} \frac{a_n}{w^{n+1}},$$

where $a_n = \int z^n\, d\mu(z)$.

Now assume $\mu \in M(K)$ and $\int g \, d\mu = 0$ for every rational function g with poles in E. Let U be a component of $\mathbb{C}_\infty \setminus K$, and let $w_0 \in E \cap U$. If $w_0 \neq \infty$, then the hypothesis and (8.3) implies each derivative of $\hat{\mu}$ at w_0 vanishes. Hence $\hat{\mu} \equiv 0$ on U. If $w_0 = \infty$, then (8.4) implies $\hat{\mu} \equiv 0$ on U. Thus $\hat{\mu} \equiv 0$ on $\mathbb{C}_\infty \setminus K$.

If f is analytic on an open set G containing K, let $\gamma_1, \ldots, \gamma_n$ be straight-line segments in $G \setminus K$ such that

$$f(z) = \sum_{k=1}^{n} \frac{1}{2\pi i} \int_{\gamma_k} \frac{f(w)}{w - z} \, dw$$

for all z in K. (See p. 195 of Conway [1978].) Thus

$$\int_K f(z) \, d\mu(z) = \sum_{k=1}^{n} \frac{1}{2\pi i} \int_K \int_{\gamma_k} \frac{f(w)}{w - z} \, dw \, d\mu(z)$$

$$= -\sum_{k=1}^{n} \frac{1}{2\pi i} \int_{\gamma_k} f(w) \hat{\mu}(w) \, dw$$

by Fubini's Theorem. But $\hat{\mu}(w) = 0$ on γ_k $(\subseteq \mathbb{C} \setminus K)$, so $\int f \, d\mu = 0$. By (6.13), $f \in R(K, E)$. This proves Runge's Theorem. ∎

8.5.   Corollary. *If K is compact and $\mathbb{C} \setminus K$ is connected and if f is analytic in a neighborhood of K, then there is a sequence of polynomials that converges to f uniformly on K.*

EXERCISES

1. Let μ be a compactly supported measure on $\mathbb{C}$ that is absolutely continuous with respect to area measure. Show that $\hat{\mu}$ is continuous on $\mathbb{C}_\infty$.

2. Let $m =$ Lebesgue measure on $[0, 1]$. Show that $\hat{m}$ is not continuous at any point of $[0, 1]$.

§9*. An Application: Ordered Vector Spaces

In this section only vector spaces over $\mathbb{R}$ are considered.

There are numerous spaces in which there is a notion of $\leq$ in addition to the vector space structure. The L^p spaces and $C(X)$ are some that spring to mind. The concept of an ordered vector space is an attempt to study such spaces in an abstract setting. The first step is to abstract the notion of the positive elements.

9.1.   Definition. An *ordered vector space* is a pair $(\mathcal{X}, \leq)$ where $\mathcal{X}$ is a vector space over $\mathbb{R}$ and $\leq$ is a relation on $\mathcal{X}$ satisfying

(a) $x \leq x$ for all x;
(b) if $x \leq y$ and $y \leq z$, then $x \leq z$;

(c) if $x \leq y$ and $z \in \mathscr{X}$, then $x + z \leq y + z$;
(d) if $x \leq y$ and $\alpha \in [0, \infty)$, then $\alpha x \leq \alpha y$.

Note that it is not assumed that $\leq$ is *antisymmetric*. That is, it is not assumed that if $x \leq y$ and $y \leq x$, then $x = y$.

9.2. Definition. If $\mathscr{X}$ is a real vector space, a *wedge* is a nonempty subset P of $\mathscr{X}$ such that

(a) if $x, y \in P$, then $x + y \in P$;
(b) if $x \in P$ and $\alpha \in [0, \infty)$, then $\alpha x \in P$.

9.3. Proposition. (a) *If $(\mathscr{X}, \leq)$ is an ordered vector space and $P = \{ x \in \mathscr{X} : x \geq 0 \}$, then P is a wedge.* (b) *If P is a wedge in the real vector space $\mathscr{X}$ and $\leq$ is defined on $\mathscr{X}$ by declaring $x \leq y$ if and only if $y - x \in P$, then $(\mathscr{X}, \leq)$ is an ordered vector space.*

PROOF. Exercise.

If $(\mathscr{X}, \leq)$ is an ordered vector space, $P = \{ x \in \mathscr{X} : x \geq 0 \}$ is called the *wedge of positive elements*. The next result is also left as an exercise.

9.4. Proposition. *If $(\mathscr{X}, \leq)$ is an ordered vector space and P is the wedge of positive elements, $\leq$ is antisymmetric if and only if $P \cap (-P) = (0)$.*

9.5. Definition. A *cone* in $\mathscr{X}$ is a wedge P such that $P \cap (-P) = (0)$.

9.6. Definition. If $(\mathscr{X}, \leq)$ is an ordered vector space, a subset A of $\mathscr{X}$ is *cofinal* if for every $x \geq 0$ in $\mathscr{X}$ there is an a in A such that $a \geq x$. An element e of $\mathscr{X}$ is an *order unit* if for every x in $\mathscr{X}$ there is a positive integer n such that $-ne \leq x \leq ne$.

If X is a compact space and $\mathscr{X} = C(X)$, then any constant function is an order unit. ($f \leq g$ if and only if $f(x) \leq g(x)$ for all x). If $\mathscr{X} = C(\mathbb{R})$, all real-valued continuous functions on $\mathbb{R}$, then $\mathscr{X}$ has no order unit (Exercise 4). If e is an order unit, then $\{ ne: n \geq 1 \}$ is cofinal.

9.7. Definition. If $(\mathscr{X}, \leq)$ and $(\mathscr{Y}, \leq)$ are ordered vector spaces and T: $\mathscr{X} \to \mathscr{Y}$ is a linear map, then T is *positive* (in symbols $T \geq 0$) if $Tx \geq 0$ whenever $x \geq 0$.

The principal result of this section is the following.

9.8. Theorem. *Let $(\mathscr{X}, \leq)$ be an ordered vector space and let $\mathscr{Y}$ be a linear manifold in $\mathscr{X}$ that is cofinal. If $f: \mathscr{Y} \to \mathbb{R}$ is a positive linear functional, then there is a positive linear functional $\tilde{f}: \mathscr{X} \to \mathbb{R}$ such that $\tilde{f}|\mathscr{Y} = f$.*

PROOF. Let $P = \{ x \in \mathscr{X} : x \geq 0 \}$ and put $\mathscr{X}_1 = \mathscr{Y} + P - P$. It is easy to see that $\mathscr{X}_1$ is a linear manifold in $\mathscr{X}$. If there is a positive linear functional

$g: \mathcal{X}_1 \to \mathbb{R}$ that extends f, let $\tilde{f}$ be any linear functional on $\mathcal{X}$ that extends g (use a Hamel basis). If $x \geq 0$, then $x \in P \subseteq \mathcal{X}_1$ so that $\tilde{f}(x) = g(x) \geq 0$. Hence $\tilde{f}$ is positive. Thus, we may assume that $\mathcal{X} = \mathcal{Y} + P - P$.

9.9. Claim. $\mathcal{X} = \mathcal{Y} + P = \mathcal{Y} - P$.

Let $x \in \mathcal{X}$; so $x = y + p_1 - p_2$, y in $\mathcal{Y}$, p_1, p_2 in P. Since $\mathcal{Y}$ is cofinal there is a y_1 in $\mathcal{Y}$ such that $y_1 \geq p_1$. Hence $p_1 = y_1 - (y_1 - p_1) \in \mathcal{Y} - P$. Thus $x = y - p_2 + p_1 \in (\mathcal{Y} - P) + (\mathcal{Y} - P) \subseteq \mathcal{Y} - P$. So $\mathcal{X} = \mathcal{Y} - P$. Also, $\mathcal{X} = -\mathcal{X} = -\mathcal{Y} + P = \mathcal{Y} + P$.

9.10. Claim. If $x \in \mathcal{X}$, there are y_1, y_2 in $\mathcal{Y}$ such that $y_2 \leq x \leq y_1$.

In fact, Claim 9.9 states that we can write $x = y_1 - p_1 = y_2 + p_2$, $p_1, p_2 \in P$ and $y_1, y_2 \in \mathcal{Y}$. Thus $y_2 \leq x \leq y_1$.

By Claim 9.10, it is possible to define for each x in $\mathcal{X}$,

$$q(x) = \inf\{ f(y): y \in \mathcal{Y} \text{ and } y \geq x \}.$$

9.11. Claim. The function q is a sublinear functional on $\mathcal{X}$.

The proof of (9.11) is left as an exercise.

For y in $\mathcal{Y}$, let $y_1 \in \mathcal{Y}$ such that $y_1 \geq y$. Because f is positive, $f(y) \leq f(y_1)$. Hence $f(y) \leq q(y)$ for all y in $\mathcal{Y}$. The Hahn–Banach Theorem implies that there is a linear functional $\tilde{f}: \mathcal{X} \to \mathbb{R}$ such that $\tilde{f}|\mathcal{Y} = f$ and $\tilde{f} \leq q$ on $\mathcal{X}$. If $x \in P$, then $-x \leq 0$ (and $0 \in \mathcal{Y}$). Hence $q(-x) \leq f(0)$. Thus $-\tilde{f}(x) = \tilde{f}(-x) \leq q(-x) \leq 0$, or $\tilde{f}(x) \geq 0$. Therefore $\tilde{f}$ is positive. ∎

9.12. Corollary. *Let $(\mathcal{X}, \leq)$ be an ordered vector space with an order unit e. If $\mathcal{Y}$ is a linear manifold in $\mathcal{X}$ and $e \in \mathcal{Y}$, then any positive linear functional defined on $\mathcal{Y}$ has an extension to a positive linear functional defined on $\mathcal{X}$.*

EXERCISES

1. Prove Proposition 9.3.

2. Prove Proposition 9.4.

3. Show that e is an order unit for $(\mathcal{X}, \leq)$ if and only if for every x in $\mathcal{X}$ there is a $\delta > 0$ such that $e \pm tx \geq 0$ for $0 \leq t \leq \delta$.

4. Show that $C(\mathbb{R})$, the space of all continuous real-valued functions on $\mathbb{R}$, has no order unit.

5. Prove (9.11).

6. Characterize the order units of $C_b(X)$. Does $C_b(X)$ always have an order unit?

7. Characterize the order units of $C_0(X)$ if X is locally compact. Does $C_0(X)$ always have an order unit?

8. Let $\mathscr{X} = M_2(\mathbb{R})$, the 2×2 matrices over $\mathbb{R}$. Define A in $M_2(\mathbb{R})$ to be positive if $A = A^*$ and $\langle Ax, x \rangle \geq 0$ for all x in $\mathbb{R}^2$. Characterize the order units of $M_2(\mathbb{R})$.

9. If $1 \leq p < \infty$ and $\mathscr{X} = L^p(0,1)$, define $f \leq g$ to mean that $f(x) \leq g(x)$ a.e. Show that $\mathscr{X}$ is an ordered vector space that has no order unit.

§10. The Dual of a Quotient Space and a Subspace

Let $\mathscr{X}$ be a normed space and $\mathscr{M} \leq \mathscr{X}$. If $f \in \mathscr{X}^*$, then $f|\mathscr{M}$, the restriction of f to $\mathscr{M}$, belongs to $\mathscr{M}^*$ and $\|f|\mathscr{M}\| \leq \|f\|$. According to the Hahn–Banach Theorem, every bounded linear functional on $\mathscr{M}$ is obtainable as the restriction of a functional from $\mathscr{X}^*$. In fact, more can be said.

Note that if $\mathscr{M}^\perp \equiv \{g \in \mathscr{X}^*: g(\mathscr{M}) = 0\}$ (note the analogy with Hilbert space notation); then $\mathscr{M}^\perp$ is a closed subspace of the Banach space $\mathscr{X}^*$. Hence $\mathscr{X}^*/\mathscr{M}^\perp$ is a Banach space. Moreover, if $f + \mathscr{M}^\perp \in \mathscr{X}^*/\mathscr{M}^\perp$, then $f + \mathscr{M}^\perp$ induces a linear functional on $\mathscr{M}$, namely $f|\mathscr{M}$.

10.1. Theorem. *If $\mathscr{M} \leq \mathscr{X}$ and $\mathscr{M}^\perp \equiv \{g \in \mathscr{X}^*: g(\mathscr{M}) = 0\}$, then the map $\rho: \mathscr{X}^*/\mathscr{M}^\perp \to \mathscr{M}^*$ defined by*

$$\rho(f + \mathscr{M}^\perp) = f|\mathscr{M}$$

is an isometric isomorphism.

PROOF. It is easy to see that ρ is linear and injective. If $f \in \mathscr{X}^*$ and $g \in \mathscr{M}^\perp$, then $\|f|\mathscr{M}\| = \|(f + g)|\mathscr{M}\| \leq \|f + g\|$. Taking the infimum over all g we get that $\|f|\mathscr{M}\| \leq \|f + \mathscr{M}^\perp\|$. Suppose $\phi \in \mathscr{M}^*$. The Hahn–Banach Theorem implies that there is an f in $\mathscr{X}^*$ such that $f|\mathscr{M} = \phi$ and $\|f\| = \|\phi\|$. Hence $\phi = \rho(f + \mathscr{M}^\perp)$ and $\|\phi\| = \|f\| \geq \|f + \mathscr{M}^\perp\|$. ∎

Now consider $\mathscr{X}/\mathscr{M}$; what is $(\mathscr{X}/\mathscr{M})^*$? Let $Q: \mathscr{X} \to \mathscr{X}/\mathscr{M}$ be the natural map. If $f \in (\mathscr{X}/\mathscr{M})^*$, then $f \circ Q \in \mathscr{X}^*$ and $\|f \circ Q\| \leq \|f\|$. (Why?) This gives a way of mapping $(\mathscr{X}/\mathscr{M})^* \to \mathscr{X}^*$. What is its image? Is it an isometry?

10.2. Theorem. *If $\mathscr{M} \leq \mathscr{X}$ and $Q: \mathscr{X} \to \mathscr{X}/\mathscr{M}$ is the natural map, then $\rho(f) = f \circ Q$ defines an isometric isomorphism of $(\mathscr{X}/\mathscr{M})^*$ onto $\mathscr{M}^\perp$.*

PROOF. If $f \in (\mathscr{X}/\mathscr{M})^*$ and $y \in \mathscr{M}$, then $f \circ Q(y) = 0$, so $f \circ Q \in \mathscr{M}^\perp$. Again, it is easy to see that $\rho: (\mathscr{X}/\mathscr{M})^* \to \mathscr{M}^\perp$ is linear and, as was seen earlier, $\|\rho(f)\| \leq \|f\|$. Let $\{x_n + \mathscr{M}\}$ be a sequence in $\mathscr{X}/\mathscr{M}$ such that $\|x_n + \mathscr{M}\| < 1$ and $|f(x_n + \mathscr{M})| \to \|f\|$. For each n there is a y_n in $\mathscr{M}$ such that $\|x_n + y_n\| < 1$. Thus $\|\rho(f)\| \geq |\rho(f)(x_n + y_n)| = |f(x_n + \mathscr{M})| \to \|f\|$, so ρ is an isometry. ∎

To see that ρ is surjective, let $g \in \mathcal{M}^{\perp}$; then $g \in \mathcal{X}^*$ and $g(\mathcal{M}) = 0$. Define $f: \mathcal{X}/\mathcal{M} \to \mathbb{F}$ by $f(x + \mathcal{M}) = g(x)$. Because $g(\mathcal{M}) = 0$, f is well defined. Also, if $x \in \mathcal{X}$ and $y \in \mathcal{M}$, $|f(x + \mathcal{M})| = |g(x)| = |g(x + y)| \le \|g\| \|x + y\|$. Taking the infimum over all y gives $|f(x + \mathcal{M})| \le \|g\| \|x + \mathcal{M}\|$. Hence $f \in (\mathcal{X}/\mathcal{M})^*$, $\rho(f) = g$, and $\|f\| \le \|\rho(f)\|$. ∎

§11. Reflexive Spaces

If $\mathcal{X}$ is a normed space, then we have seen that $\mathcal{X}^*$ is a Banach space (5.4). Because $\mathcal{X}^*$ is a Banach space, it too has a dual space $(\mathcal{X}^*)^* \equiv \mathcal{X}^{**}$ and $\mathcal{X}^{**}$ is a Banach space. Hence $\mathcal{X}^{**}$ has a dual. Can this be kept up?

Before answering this question, let's examine a curious phenomenon. If $x \in \mathcal{X}$, then x defines an element $\hat{x}$ of $\mathcal{X}^{**}$; namely, define $\hat{x}: \mathcal{X}^* \to \mathbb{F}$ by

11.1 $\hat{x}(x^*) = x^*(x)$

for every x^* in $\mathcal{X}^*$. Note that Corollary 6.7 implies that $\|\hat{x}\| = \|x\|$ for all x in $\mathcal{X}$. The map $x \to \hat{x}$ of $\mathcal{X} \to \mathcal{X}^{**}$ is called the *natural map* of $\mathcal{X}$ into its *second dual*.

11.2.   Definition. A normed space $\mathcal{X}$ is *reflexive* if $\mathcal{X}^{**} = \{\hat{x}: x \in \mathcal{X}\}$, where $\hat{x}$ is defined in (11.1).

First note that a reflexive space $\mathcal{X}$ is isometrically isomorphic to $\mathcal{X}^{**}$, and hence must be a Banach space. It is not true, however, that a Banach space $\mathcal{X}$ that is isometric to $\mathcal{X}^{**}$ is reflexive. The definition of reflexivity stipulates that the isometry be the natural embedding of $\mathcal{X}$ into $\mathcal{X}^{**}$. In fact, James [1951] gives an example of a nonreflexive space $\mathcal{X}$ that is isometric to $\mathcal{X}^{**}$.

11.3.   Example. If $1 < p < \infty$, $L^p(X, \Omega, \mu)$ is reflexive.

11.4.   Example. c_0 is not reflexive. We know that $c_0^* = l^1$, so $c_0^{**} = (l^1)^* = l^{\infty}$. With these identifications, the natural map $c_0 \to c_0^{**}$ is precisely the inclusion map $c_0 \to l^{\infty}$.

A discussion of reflexivity is best pursued after the weak topology is understood (Chapter V). Until that time, we will say *adieu* to reflexivity.

EXERCISES

1. Show that $(\mathcal{X}^*)^{**}$ and $(\mathcal{X}^{**})^*$ are isometrically isomorphic.

2. Show that $C_b(X)$ is reflexive if and only if X is finite.

3. Let $\mathcal{M} \le \mathcal{X}$ and let $\rho_{\mathcal{X}}: \mathcal{X} \to \mathcal{X}^{**}$ and $\rho_{\mathcal{M}}: \mathcal{M} \to \mathcal{M}^{**}$ be the natural maps. If $i: \mathcal{M} \to \mathcal{X}$ is the inclusion map, show that there is an isometry $\phi: \mathcal{M}^{**} \to \mathcal{X}^{**}$

such that the diagram

$$\begin{array}{ccc} \mathscr{X} & \xrightarrow{\ \rho_{\mathscr{X}}\ } & \mathscr{X}^{**} \\ {\scriptstyle i}\uparrow & & \uparrow{\scriptstyle \phi} \\ \mathscr{M} & \xrightarrow{\ \rho_{\mathscr{M}}\ } & \mathscr{M}^{**} \end{array}$$

commutes. Prove that $\phi(\mathscr{M}^{**}) = (\mathscr{M}^{\perp})^{\perp} \equiv \{x^{**} \in \mathscr{X}^{**}: x^{**}(\mathscr{M}^{\perp}) = 0\}$.

4. Use Exercise 3 to show that if $\mathscr{X}$ is reflexive, then any closed subspace of $\mathscr{X}$ is also reflexive.

§12. The Open Mapping and Closed Graph Theorems

12.1. The Open Mapping Theorem. *If $\mathscr{X}, \mathscr{Y}$ are Banach spaces and A: $\mathscr{X} \to \mathscr{Y}$ is a continuous linear surjection, then $A(G)$ is open in $\mathscr{Y}$ whenever G is open in $\mathscr{X}$.*

PROOF. For $r > 0$, let $B(r) = \{x \in \mathscr{X}: \|x\| < r\}$.

12.2. Claim. $0 \in \text{int cl } A(B(r))$.

Note that because A is surjective, $\mathscr{Y} = \bigcup_{k=1}^{\infty} \text{cl}[A(B(kr/2))] = \bigcup_{k=1}^{\infty} k \, \text{cl}[A(B(r/2))]$. By the Baire Category Theorem, there is a $k \geq 1$ such that $k \, \text{cl}[A(B(r/2))]$ has nonempty interior. Thus $V = \text{int}\{\text{cl}[A(B(r/2))]\} \neq \square$. If $y_0 \in V$, let $s > 0$ such that $\{y \in \mathscr{Y}: \|y - y_0\| < s\} \subseteq V \subseteq \text{cl } A(B(r/2))$. Let $y \in \mathscr{Y}$, $\|y\| < s$. Since $y_0 \in \text{cl } A(B(r/2))$, there is a sequence $\{x_n\}$ in $B(r/2)$ such that $A(x_n) \to y_0$. There is also a sequence $\{z_n\}$ in $B(r/2)$ such that $A(z_n) \to y_0 + y$. Thus $A(z_n - x_n) \to y$ and $\{z_n - x_n\} \subseteq B(r)$; that is, $\{y \in \mathscr{Y}: \|y\| < s\} \subseteq \text{cl } A(B(r))$. This establishes Claim 12.2.

It will now be shown that

12.3 $$\text{cl } A(B(r/2)) \subseteq A(B(r)).$$

Note that if (12.3) is proved, then Claim 12.2 implies that $0 \in \text{int } A(B(r))$ for any $r > 0$. From here the theorem is easily proved. Indeed, if G is an open subset of $\mathscr{X}$, then for every x in G let $r_x > 0$ such that $B(x; r_x) \subseteq G$. But $0 \in \text{int } A(B(r_x))$ and so $A(x) \in \text{int } A(B(x; r_x))$. Thus there is an $s_x > 0$ such that $U_x \equiv \{y \in \mathscr{Y}: \|y - A(x)\| < s_x\} \subseteq A(B(x; r_x))$. Therefore $A(G) \supseteq \bigcup\{U_x: x \in G\}$. But $A(x) \in U_x$, so $A(G) = \bigcup\{U_x: x \in G\}$ and hence $A(G)$ is open.

To prove (12.3), fix y_1 in $\text{cl } A(B(r/2))$. By (12.2), $0 \in \text{int}[\text{cl } A(B(2^{-2}r))]$. Hence $[y_1 - \text{cl } A(B(2^{-2}r))] \cap A(B(r/2)) \neq \square$. Let $x_1 \in B(r/2)$ such that $A(x_1) \in [y_1 - \text{cl } A(B(2^{-2}r))]$; now $A(x_1) = y_1 - y_2$, where $y_2 \in \text{cl } A(B(2^{-2}r))$. Using induction, we obtain a sequence $\{x_n\}$ in $\mathscr{X}$ and a sequence $\{y_n\}$ in $\mathcal{Y}$ such that

$$12.4 \begin{cases} \text{(i)} & x_n \in B(2^{-n}r), \\ \text{(ii)} & y_n \in \operatorname{cl} A(B(2^{-n}r)), \\ \text{(iii)} & y_{n+1} = y_n - A(x_n). \end{cases}$$

But $\|x_n\| < 2^{-n}r$, so $\sum_1^\infty \|x_n\| < \infty$; hence $x = \sum_{n=1}^\infty x_n$ exists in $\mathcal{X}$ and $\|x\| < r$. Also,

$$\sum_{k=1}^n A(x_k) = \sum_{k=1}^n (y_k - y_{k+1}) = y_1 - y_{n+1}.$$

But (12.4ii) implies $\|y_n\| \le \|A\|2^{-n}r$; hence $y_n \to 0$. Therefore $y_1 = \sum_{k=1}^\infty A(x_k) = A(x) \in A(B(r))$, proving (12.3) and completing the proof of the theorem. ∎

The Open Mapping Theorem has several applications. Here are two important ones.

12.5. The Inverse Mapping Theorem. *If $\mathcal{X}$ and $\mathcal{Y}$ are Banach spaces and A: $\mathcal{X} \to \mathcal{Y}$ is a bounded linear transformation that is bijective, then A^{-1} is bounded.*

PROOF. Because A is continuous, bijective, and open by Theorem 12.1, A is a homeomorphism. ∎

12.6. The Closed Graph Theorem. *If $\mathcal{X}$ and $\mathcal{Y}$ are Banach spaces and A: $\mathcal{X} \to \mathcal{Y}$ is a linear transformation such that the graph of A,*

$$\operatorname{gra} A \equiv \{x \oplus Ax \in \mathcal{X} \oplus_1 \mathcal{Y} : x \in \mathcal{X}\}$$

is closed, then A is continuous.

PROOF. Let $\mathcal{G} = \operatorname{gra} A$. Since $\mathcal{X} \oplus_1 \mathcal{Y}$ is a Banach space and $\mathcal{G}$ is closed, $\mathcal{G}$ is a Banach space. Define P: $\mathcal{G} \to \mathcal{X}$ by $P(x \oplus Ax) = x$. It is easy to check that P is bounded and bijective. (Do it). By the Inverse Mapping Theorem, P^{-1}: $\mathcal{X} \to \mathcal{G}$ is continuous. Thus A: $\mathcal{X} \to \mathcal{Y}$ is the composition of the continuous map P^{-1}: $\mathcal{X} \to \mathcal{G}$ and the continuous map of $\mathcal{G} \to \mathcal{Y}$ defined by $x \oplus Ax \mapsto Ax$; A is therefore continuous. ∎

Let $\mathcal{X} = $ all functions f: $[0, 1] \to \mathbb{F}$ such that the derivative f' exists and is continuous on $[0, 1]$. Let $\mathcal{Y} = C[0, 1]$ and give both $\mathcal{X}$ and $\mathcal{Y}$ the supremum norm: $\|f\| = \sup\{|f(t)| : t \in [0, 1]\}$. So $\mathcal{X}$ is not a Banach space, though $\mathcal{Y}$ is. Define A: $\mathcal{X} \to \mathcal{Y}$ by $Af = f'$. Clearly, A is linear. If $\{f_n\} \subseteq \mathcal{X}$ and $(f_n, f_n') \to (f, g)$ in $\mathcal{X} \times \mathcal{Y}$, then $f_n' \to g$ uniformly on $[0, 1]$. Hence

$$f_n(t) - f_n(0) = \int_0^t f_n'(s)\, ds \to \int_0^t g(s)\, ds.$$

But $f_n(t) - f_n(0) \to f(t) - f(0)$, so

$$f(t) = f(0) + \int_0^t g(s)\, ds.$$

Thus $f' = g$ and gra A is closed. However, A is not bounded. (Why?)

The preceding example shows that the domain of the operator in the Closed Graph Theorem must be assumed to be complete. The next example (due to Alp Eden) shows that the range must also be assumed to be complete.

Let $\mathcal{X}$ be a separable infinite-dimensional Banach space and let $\{e_i : i \in I\}$ be a Hamel basis for $\mathcal{X}$ with $\|e_i\| = 1$ for all i. Note that a Baire Category argument shows that I is uncountable. If $x \in \mathcal{X}$, then $x = \Sigma_i \alpha_i e_i$, $\alpha_i \in \mathbb{F}$, and $\alpha_i = 0$ for all but a finite number of i in I. Define $\|x\|_1 \equiv \Sigma_i |\alpha_i|$. It is left as an exercise for the reader to show that $\| \cdot \|_1$ is a norm on $\mathcal{X}$. Since $\|e_i\| = 1$ for all i, $\|x\| \le \Sigma_i |\alpha_i| = \|x\|_1$. Let $\mathcal{Y} = \mathcal{X}$ with the norm $\| \cdot \|_1$ and let $T: \mathcal{Y} \to \mathcal{X}$ be defined by $T(x) = x$. Note that it was just shown that $T^{-1}: \mathcal{X} \to \mathcal{Y}$ is a contraction. Therefore gra T^{-1} is closed and hence so is gra T. But T is not continuous because if it were, then T would be a homeomorphism. Since $\mathcal{X}$ is separable, it would follow that $\mathcal{Y}$ is separable. But $\mathcal{Y}$ is not separable. To see this, note that $\|e_i - e_j\|_1 = 2$ for $i \ne j$ and since I is uncountable, $\mathcal{Y}$ cannot be separable.

When applying the Closed Graph Theorem, the following result is useful.

12.7. Proposition. *If $\mathcal{X}$ and $\mathcal{Y}$ are normed spaces and $A: \mathcal{X} \to \mathcal{Y}$ is a linear transformation, then gra A is closed if and only if whenever $x_n \to 0$ and $Ax_n \to y$, it must be that $y = 0$.*

PROOF. Exercise 3.

Note that (12.7) underlines the advantage of the Closed Graph Theorem. To show that A is continuous, it suffices to show that if $x_n \to 0$, then $Ax_n \to 0$. By (12.7) this is eased by allowing us to assume that $\{Ax_n\}$ is convergent.

It is possible to give a measure-theoretic solution to Exercise 2.3, but here is one using the Closed Graph Theorem. Let (X, Ω, μ) be a σ-finite measure space, $1 \le p \le \infty$, and $\phi: X \to \mathbb{F}$ an Ω-measurable function such that $\phi f \in L^p(\mu)$ whenever $f \in L^p(\mu)$. Define $A: L^p(\mu) \to L^p(\mu)$ by $Af = \phi f$. Thus A is linear and well defined. Suppose $f_n \to 0$ and $\phi f_n \to g$ in $L^p(\mu)$. If $1 \le p < \infty$, then $f_n \to 0$ in measure. By a theorem of Riesz, there is a subsequence $\{f_{n_k}\}$ such that $f_{n_k}(x) \to 0$ a.e. $[\mu]$. Hence $\phi(x) f_{n_k}(x) \to 0$ a.e. $[\mu]$. This implies $g = 0$ and so gra A is closed. If $p = \infty$, then $f_n(x) \to 0$ a.e. $[\mu]$ and the same argument implies gra A is closed. By the Closed Graph Theorem, A is bounded. Clearly, it may be assumed that $\|A\| = 1$. If $\delta > 0$, let $E = \{x: |\phi(x)| \ge 1 + \delta\}$. Now $\|A^n\| \le 1$, so $\|\phi^n f\|_p \le \|f\|_p$ for all

$n \geq 1$. Thus

$$\|f\|_p^p \geq \int |\phi|^{np} |f|^p \, d\mu \geq (1 + \delta)^{np} \int_E |f|^p \, d\mu.$$

But $(1 + \delta)^{np} \to \infty$ as $n \to \infty$. Hence $\int_E |f|^p \, d\mu = 0$ for each f in $L^p(\mu)$, and $\mu(E) = 0$. It follows that $\phi \in L^\infty(\mu)$ and $|\phi| \leq 1$ a.e. $[\mu]$.

12.8. Definition. If $\mathcal{X}, \mathcal{Y}$ are Banach spaces, an *isomorphism* of $\mathcal{X}$ and $\mathcal{Y}$ is a linear bijection $T: \mathcal{X} \to \mathcal{Y}$ that is a homeomorphism. Say that $\mathcal{X}$ and $\mathcal{Y}$ are *isomorphic* if there is an isomorphism of $\mathcal{X}$ onto $\mathcal{Y}$.

Note that the Inverse Mapping Theorem says that an isomorphism is a continuous bijection.

The use of the word "isomorphism" is counter to the spirit of category theory, but it is traditional in Banach space theory.

EXERCISES

1. Suppose $\mathcal{X}$ and $\mathcal{Y}$ are Banach spaces. If $A \in \mathcal{B}(\mathcal{X}, \mathcal{Y})$ and ran A is a second category space, show that ran A is closed.

2. Give both $C^{(1)}[0,1]$ and $C[0,1]$ the supremum norm. If $A: C^{(1)}[0,1] \to C[0,1]$ is defined by $Af = f'$, show that A is not bounded.

3. Prove Proposition 12.7.

4. Let $\mathcal{X}$ be a vector space and suppose $\| \cdot \|_1$ and $\| \cdot \|_2$ are two norms on $\mathcal{X}$ and that $\mathcal{I}_1$ and $\mathcal{I}_2$ are the corresponding topologies. Show that if $\mathcal{X}$ is complete in both norms and $\mathcal{I}_1 \supseteq \mathcal{I}_2$, then $\mathcal{I}_1 = \mathcal{I}_2$.

5. Let $\mathcal{X}$ and $\mathcal{Y}$ be Banach spaces and let $A \in \mathcal{B}(\mathcal{X}, \mathcal{Y})$. Show that there is a constant $c > 0$ such that $\|Ax\| \geq c\|x\|$ for all x in $\mathcal{X}$ if and only if $\ker A = (0)$ and ran A is closed.

6. Let X be compact and suppose that $\mathcal{X}$ is a Banach subspace of $C(X)$. If E is a closed subset of X such that for every g in $C(E)$ there is an f in $\mathcal{X}$ with $f|E = g$, show that there is a constant $c > 0$ such that for each g in $C(E)$ there is an f in $\mathcal{X}$ with $f|E = g$ and $\max\{|f(x)|: x \in X\} \leq c \max\{|g(x)|: x \in E\}$.

7. Let $1 \leq p \leq \infty$ and suppose (α_{ij}) is a matrix such that $(Af)(i) = \sum_{j=1}^\infty \alpha_{ij}(j)$ defines an element Af of l^p for every f in l^p. Show that $A \in \mathcal{B}(l^p)$.

8. Let (X, Ω, μ) be a σ-finite measure space, $1 \leq p < \infty$, and suppose that $k: X \times X \to \mathbb{F}$ is an $\Omega \times \Omega$ measurable function such that for f in $L^p(\mu)$ and a.e. x, $k(x, \cdot)f(\cdot) \in L^1(\mu)$ and $(Kf)(x) = \int k(x, y)f(y) \, d\mu(y)$ defines an element Kf of $L^p(\mu)$. Show that $K: L^p(\mu) \to L^p(\mu)$ is a bounded operator.

§13. Complemented Subspaces of a Banach Space

If $\mathscr{X}$ is a Banach space and $\mathscr{M} \leq \mathscr{X}$, say that $\mathscr{M}$ is *algebraically complemented* in $\mathscr{X}$ if there is an $\mathscr{N} \leq \mathscr{X}$ such that $\mathscr{M} \cap \mathscr{N} = (0)$ and $\mathscr{M} + \mathscr{N} = \mathscr{X}$. Of course, the definition makes sense in a purely algebraic setting, so the requirement that $\mathscr{M}$ and $\mathscr{N}$ be closed seems fatuous. Why is it made?

If $\mathscr{M}$ is a linear manifold in a vector space $\mathscr{X}$ (a Banach space or not), then a Hamel-basis argument can be fashioned to produce a linear manifold $\mathscr{N}$ such that $\mathscr{M} \cap \mathscr{N} = (0)$ and $\mathscr{M} + \mathscr{N} = \mathscr{X}$. So the requirement in the definition that $\mathscr{M}$ and $\mathscr{N}$ be closed subspaces of the Banach space $\mathscr{X}$ makes the existence problem more interesting. Also, since we are dealing with the category of Banach spaces, all definitions should involve only objects in that category.

If $\mathscr{M}$ and $\mathscr{N}$ are algebraically complemented closed subspaces of a normed space $\mathscr{X}$, then $A \colon \mathscr{M} \oplus_1 \mathscr{N} \to \mathscr{X}$ defined by $A(m \oplus n) = m + n$ is a linear bijection. Also, $\|A(m \oplus n)\| = \|m + n\| \leq \|m\| + \|n\| = \|m \oplus n\|$. Hence A is bounded. Say that $\mathscr{M}$ and $\mathscr{N}$ are *topologically complemented* if A is a homeomorphism; equivalently, if $\|\|m + n\|\| \equiv \|m\| + \|n\|$ is an equivalent norm. If $\mathscr{X}$ is a Banach space, then the Inverse Mapping Theorem implies A is a homeomorphism. This proves the following.

13.1.   Theorem. *If a pair of subspaces of a Banach space are algebraically complementary, then they are topologically complementary.*

This permits us to speak of *complementary subspaces* of a Banach space without modifying the term. The proof of the next result is left to the reader.

13.2.   Theorem. (a) *If $\mathscr{M}$ and $\mathscr{N}$ are complementary subspaces of a Banach space $\mathscr{X}$ and $E \colon \mathscr{X} \to \mathscr{X}$ is defined by $E(m + n) = m$ for m in $\mathscr{M}$ and n in $\mathscr{N}$, then E is a continuous linear operator such that $E^2 = E$, $\operatorname{ran} E = \mathscr{M}$, and $\ker E = \mathscr{N}$.* (b) *If $E \in \mathscr{B}(\mathscr{X})$ and $E^2 = E$, then $\mathscr{M} = \operatorname{ran} E$ and $\mathscr{N} = \ker E$ are complemented subspaces of $\mathscr{X}$.*

If $\mathscr{M} \leq \mathscr{X}$ and $\mathscr{M}$ is complemented in $\mathscr{X}$, its complementary subspace may not be unique. Indeed, finite-dimensional spaces furnish the necessary examples.

A result due to R. S. Phillips [1940] is that c_0 is not complemented in l^∞. A straightforward proof of this can be found in Whitely [1966]. Murray [1937] showed that l^p, $p \neq 2$, $p > 1$ has *uncomplemented* subspaces. This seems to be the first paper to exhibit uncomplemented subspaces of a Banach space.

Lindenstrauss [1967] showed that if $\mathscr{M}$ is an infinite-dimensional subspace of l^∞ that is complemented in l^∞, then $\mathscr{M}$ is isomorphic to l^∞. This same result holds if l^∞ is replaced by l^p, $1 \leq p < \infty$, c, or c_0.

Does there exist a Banach space $\mathcal{X}$ such that every closed subspace of $\mathcal{X}$ is complemented? Of course, if $\mathcal{X}$ is a Hilbert space, then this is true. But are there any Banach spaces that have this property and are not Hilbert spaces? Lindenstrauss [1971] proved that if $\mathcal{X}$ is a Banach space and every subspace of $\mathcal{X}$ is complemented, then $\mathcal{X}$ is isomorphic to a Hilbert space.

EXERCISES

1. If $\mathcal{X}$ is a vector space and $\mathcal{M}$ is a linear manifold in $\mathcal{X}$, show that there is a linear manifold $\mathcal{N}$ in $\mathcal{X}$ such that $\mathcal{M} \cap \mathcal{N} = (0)$ and $\mathcal{M} + \mathcal{N} = \mathcal{X}$.

2. Let $\mathcal{X}$ be a Banach space and let $E: \mathcal{X} \to \mathcal{X}$ be a linear map such that $E^2 = E$ and both ran E and ker E are closed. Show that E is continuous.

3. Prove Theorem 13.2.

4. Let $\mathcal{X}$ be a Banach space and show that if $\mathcal{M}$ is a complemented subspace of $\mathcal{X}$, then every complementary subspace is isomorphic to $\mathcal{X}/\mathcal{M}$.

5. Let X be a compact set and let Y be a closed subset of X. A *simultaneous extension* for Y is a bounded linear map $T: C(Y) \to C(X)$ such that for each g in $C(Y)$, $T(g)|Y = g$. Let $C_0(X \setminus Y) = \{ f \in C(X): f(y) = 0 \text{ for all } y \text{ in } Y \}$. Show that if there is a simultaneous extension for Y, then $C_0(X \setminus Y)$ is complemented in $C(X)$.

6. Show that if Y is a closed subset of $[0,1]$, then there is a simultaneous extension for Y (see Exercise 5). (Hint: Write $[0,1] \setminus Y$ as the union of disjoint intervals.)

7. Using the notation of Exercise 5, show that if Y is a retract of X, then $C_0(X \setminus Y)$ is complemented in $C(X)$.

§14. The Principle of Uniform Boundedness

There are several results that may be called the Principle of Uniform Boundedness (PUB) and all of these are called the PUB by various mathematicians. In this book the PUB will refer to any of the results of this section, though in a formal way the next result plays the role of the founder of the family.

14.1. Principle of Uniform Boundedness (PUB). *Let $\mathcal{X}$ be a Banach space and $\mathcal{Y}$ a normed space. If $\mathcal{A} \subseteq \mathcal{B}(\mathcal{X}, \mathcal{Y})$ such that for each x in $\mathcal{X}$, $\sup\{\|Ax\|: A \in \mathcal{A}\} < \infty$, then $\sup\{\|A\|: A \in \mathcal{A}\} < \infty$.*

PROOF. (Due to William R. Zame) For each x in $\mathcal{X}$ let $M(x) = \sup\{\|Ax\|: A \in \mathcal{A}\}$, so $\|Ax\| \leq M(x)$ for all x in $\mathcal{X}$. Suppose $\sup\{\|A\|: A \in \mathcal{A}\} = \infty$. Then there is a sequence $\{A_n\} \subseteq \mathcal{A}$ and a sequence $\{x_n\}$ of vectors in $\mathcal{X}$ such that $\|x_n\| = 1$ and $\|A_n x_n\| > 4^n$. Let $y_n = 2^{-n} x_n$; thus $\|y_n\| = 2^{-n}$ and $\|A_n y_n\| > 2^n$.

14.2. Claim. There is a subsequence $\{y_{n_k}\}$ such that for $k \geq 1$:

(a) $\|A_{n_{k+1}} y_{n_{k+1}}\| > 1 + k + \sum_{j=1}^{k} M(y_{n_j})$;

(b) $\|y_{n_{k+1}}\| < 2^{-k-1}[\sup\{\|A_{n_j}\|: 1 \leq j \leq k\}]^{-1}$.

The proof of (14.2) is by induction. Let $n_1 = 1$. The induction step is valid since $\|y_n\| \to 0$ and $\|A_n y_n\| \to \infty$. The details are left to the reader.

Since $\sum_k \|y_{n_k}\| < \infty$, $\sum_k y_{n_k} = y$ in $\mathscr{X}$ (here is where the completeness of $\mathscr{X}$ is used). Now for any $k \geq 1$,

$$
\|A_{n_{k+1}} y\| = \left\| \sum_{j=1}^{k} A_{n_{k+1}} y_{n_j} + A_{n_{k+1}} y_{n_{k+1}} + \sum_{j=k+2}^{\infty} A_{n_{k+1}} y_{n_j} \right\|
$$

$$
= \left\| A_{n_{k+1}} y_{n_{k+1}} - \left[-\sum_{j=1}^{k} A_{n_{k+1}} y_{n_j} - \sum_{j=k+2}^{\infty} A_{n_{k+1}} y_{n_j} \right] \right\|
$$

$$
\geq \left\| A_{n_{k+1}} y_{n_{k+1}} \right\| - \left\| \sum_{j=1}^{k} A_{n_{k+1}} y_{n_j} + \sum_{j=k+2}^{\infty} A_{n_{k+1}} y_{n_j} \right\|
$$

$$
\geq 1 + k + \sum_{j=1}^{k} M(y_{n_j}) - \left[\sum_{j=1}^{k} M(y_{n_j}) + \sum_{j=k+2}^{\infty} \|A_{n_{k+1}}\| \|y_{n_j}\| \right]
$$

$$
\geq 1 + k - \sum_{j=k+2}^{\infty} 2^{-j-1}
$$

$$
\geq k.
$$

That is, $M(y) \geq k$ for all k, a contradiction. ∎

14.3. Corollary. *If $\mathscr{X}$ is a normed space and $A \subseteq \mathscr{X}$, then A is a bounded set if and only if for every f in $\mathscr{X}^*$, $\sup\{|f(a)|: a \in A\} < \infty$.*

PROOF. Consider $\mathscr{X}$ as a subset of $\mathscr{B}(\mathscr{X}^*, \mathbb{F})$ ($= \mathscr{X}^{**}$) by letting $\hat{x}(f) = f(x)$ for every f in $\mathscr{X}^*$. Since $\mathscr{X}^*$ is a Banach space and $\|x\| = \|\hat{x}\|$ for all x, the corollary is a special case of the PUB. ∎

14.4. Corollary. *If $\mathscr{X}$ is a Banach space and $A \subseteq \mathscr{X}^*$, then A is a bounded set if and only if for every x in $\mathscr{X}$, $\sup\{|f(x)|: f \in A\} < \infty$.*

PROOF. Consider $\mathscr{X}^*$ as $\mathscr{B}(\mathscr{X}, \mathbb{F})$. ∎

Using Corollary 14.3, it is possible to prove the following improvement of (14.1).

14.5. Corollary. *If $\mathscr{X}$ is a Banach space and $\mathscr{Y}$ is a normed space and if $\mathscr{A} \subseteq \mathscr{B}(\mathscr{X}, \mathscr{Y})$ such that for every x in $\mathscr{X}$ and g in $\mathscr{Y}^*$,*

$$
\sup\{|g(A(x))|: A \in \mathscr{A}\} < \infty,
$$

then $\sup\{\|A\|: A \in \mathscr{A}\} < \infty$.

PROOF. Fix x in $\mathscr{X}$. By the hypothesis and Corollary 14.3, $\sup\{\|A(x)\|:$ $A \in \mathscr{A}\} < \infty$. By (14.1), $\sup\{\|A\|:\ A \in \mathscr{A}\} < \infty$. ∎

A special form of the PUB that is quite useful is the following.

14.6. The Banach–Steinhaus Theorem. *If $\mathscr{X}$ and $\mathscr{Y}$ are Banach spaces and $\{A_n\}$ is a sequence in $\mathscr{B}(\mathscr{X}, \mathscr{Y})$ with the property that for every x in $\mathscr{X}$ there is a y in $\mathscr{Y}$ such that $\|A_n x - y\| \to 0$, then there is an A in $\mathscr{B}(\mathscr{X}, \mathscr{Y})$ such that $\|A_n x - Ax\| \to 0$ for every x in $\mathscr{X}$ and $\sup\|A_n\| < \infty$.*

PROOF. If $x \in \mathscr{X}$, let $Ax = \lim A_n x$. By hypothesis $A: \mathscr{X} \to \mathscr{Y}$ is defined and it is easy to see that it is linear. To show that A is bounded, note that the PUB implies that there is a constant $M > 0$ such that $\|A_n\| \le M$ for all n. If $x \in \mathscr{X}$ and $\|x\| \le 1$, then for any $n \ge 1$, $\|Ax\| \le \|Ax - A_n x\| + \|A_n x\| \le \|Ax - A_n x\| + M$. Letting $n \to \infty$ shows that $\|Ax\| \le M$ whenever $\|x\| \le 1$. ∎

The Banach–Steinhaus Theorem is a result about sequences, not nets. Note that if I is the identity operator on $\mathscr{X}$ and for each $n \ge 1$, $A_n = n^{-1}I$ and for $n \le 0$, $A_n = nI$, then $\{A_n: n \in \mathbb{Z}\}$ is a countable net that converges in norm to 0, but the net is not bounded.

14.7. Proposition. *Let X be locally compact and let $\{f_n\}$ be a sequence in $C_0(X)$. Then $\int f_n\,d\mu \to \int f\,d\mu$ for every μ in $M(X)$ if and only if $\sup_n\|f_n\| < \infty$ and $f_n(x) \to f(x)$ for every x in X.*

PROOF. Suppose $\int f_n\,d\mu \to \int f\,d\mu$ for every μ in $M(X)$. Since $M(X) = C_0(X)^*$, (14.3) implies that $\sup_n\|f_n\| < \infty$. By letting $\mu = \delta_x$, the unit point mass at x, we see that $\int f_n\,d\delta_x = f_n(x) \to f(x)$. The converse follows by the Lebesgue Dominated Convergence Theorem. ∎

EXERCISES

1. Here is another proof of the PUB using the Baire Category Theorem. With the notation of (14.1), let $B_n \equiv \{x \in \mathscr{X}:\ \|Ax\| \le n \text{ for all } A \text{ in } \mathscr{A}\}$. By hypothesis, $\bigcup_{n=1}^{\infty} B_n = \mathscr{X}$. Now apply the Baire Category Theorem.

2. If $1 < p < \infty$ and $\{x_n\} \subseteq l^p$, then $\sum_{j=1}^{\infty} x_n(j)y(j) \to 0$ for every y in l^q, $1/p + 1/q = 1$, if and only if $\sup_n\|x_n\|_p < \infty$ and $x_n(j) \to 0$ for every $j \ge 1$.

3. If $\{x_n\} \subseteq l^1$, then $\sum_{j=1}^{\infty} x_n(j)y(j) \to 0$ for every y in c_0 if and only if $\sup_n\|x_n\|_1 < \infty$ and $x_n(j) \to 0$ for every $j \ge 1$.

4. If (X, Ω, μ) is a measure space, $1 < p < \infty$, and $\{f_n\} \subseteq L^p(X, \Omega, \mu)$, then $\int fg\,d\mu \to 0$ for every g in $L^q(\mu)$, $1/p + 1/q = 1$, if and only if $\sup\{\|f_n\|_p:\ n \ge 1\} < \infty$ and for every set E in Ω with $\mu(E) < \infty$, $\int_E f_n\,d\mu \to 0$ as $n \to \infty$.

5. If (X, Ω, μ) is a σ-finite measure space and $\{f_n\}$ is a sequence in $L^1(X, \Omega, \mu)$, then $\int f_n g\,d\mu \to 0$ for every g in $L^\infty(\mu)$ if and only if $\sup\{\|f_n\|_1:\ n \ge 1\} < \infty$ and $\int_E f_n\,d\mu \to 0$ for every E in Ω.

6. Let $\mathcal{H}$ be a Hilbert space and let $\mathcal{E}$ be an orthonormal basis for $\mathcal{H}$. Show that a sequence $\{h_n\}$ in $\mathcal{H}$ satisfies $\langle h_n, h \rangle \to 0$ for every h in $\mathcal{H}$ if and only if $\sup\{\|h_n\|: n \geq 1\} < \infty$ and $\langle h_n, e \rangle \to 0$ for every e in $\mathcal{E}$.

7. If X is locally compact and $\{\mu_n\}$ is a sequence in $M(X)$, then $L(\mu_n) \to 0$ for every L in $M(X)^*$ if and only if $\sup\{\|\mu_n\|: n \geq 1\} < \infty$ and $\mu_n(E) \to 0$ for every Borel set E.

8. In (14.6), show that $\|A\| \leq \limsup \|A_n\|$.

9. If (S, d) is a metric space and $\mathcal{X}$ is a normed space, say that a function $f: S \to \mathcal{X}$ is a *Lipschitz function* if there is a constant $M > 0$ such that $\|f(x) - f(t)\| \leq Md(s, t)$ for all s, t in S. Show that if $f: S \to \mathcal{X}$ is a function such that for all L in $\mathcal{X}^*$, $L \circ f: S \to \mathbb{F}$ is Lipschitz, then $f: S \to \mathcal{X}$ is a Lipschitz function.

10. Let $\mathcal{X}$ be a Banach space and suppose $\{x_n\}$ is a sequence in $\mathcal{X}$ that is linearly independent and such that for each x in $\mathcal{X}$ there are scalars $\{\alpha_n\}$ such that $\lim_{n \to \infty} \|x - \sum_{k=1}^{n} \alpha_k x_k\| = 0$. Such a sequence is called a *basis*. (a) Prove that $\mathcal{X}$ is separable. (b) Let $\mathcal{Y} = \{\{\alpha_n\} \in \mathbb{F}^{\mathbb{N}}: \sum_{n=1}^{\infty} \alpha_n x_n$ converges in $\mathcal{X}\}$ and for $y = \{\alpha_n\}$ in $\mathcal{Y}$ define $\|y\| = \sup_n \|\sum_{k=1}^{n} \alpha_k x_k\|$. Show that $\mathcal{Y}$ is a Banach space. (c) Show that there is a bounded bijection $T: \mathcal{X} \to \mathcal{Y}$. (d) If $n \geq 1$ and $f_n: \mathcal{X} \to \mathbb{F}$ is defined by $f_n(\sum_{k=1}^{\infty} \alpha_k x_k) = \alpha_n$, show that $f_n \in \mathcal{X}^*$. (e) Show that $x_n \notin$ the closed linear span of $\{x_k: k \neq n\}$.

CHAPTER IV

Locally Convex Spaces

A topological vector space is a generalization of the concept of a Banach space. The locally convex spaces are encountered repeatedly when discussing weak topologies on a Banach space, sets of operators on Hilbert space, or the theory of distributions. This book will only skim the surface of this theory, but it will treat locally convex spaces in sufficient detail as to enable the reader to understand the use of these spaces in the three areas of analysis just mentioned. For more details on this theory, see Bourbaki [1967], Robertson and Robertson [1966], or Schaefer [1971].

§1. Elementary Properties and Examples

A topological vector space is a vector space that is also a topological space such that the linear structure and the topological structure are vitally connected.

1.1. Definition. A *topological vector space* (TVS) is a vector space $\mathscr{X}$ together with a topology such that with respect to this topology

(a) the map of $\mathscr{X} \times \mathscr{X} \to \mathscr{X}$ defined by $(x, y) \mapsto x + y$ is continuous;
(b) the map of $\mathbb{F} \times \mathscr{X} \to \mathscr{X}$ defined by $(\alpha, x) \mapsto \alpha x$ is continuous.

It is easy to see that a normed space is a TVS (Proposition III.1.3).

Suppose $\mathscr{X}$ is a vector space and $\mathscr{P}$ is a family of seminorms on $\mathscr{X}$. Let $\mathscr{T}$ be the topology on $\mathscr{X}$ that has as a subbase the sets $\{x: p(x - x_0) < \varepsilon\}$, where $p \in \mathscr{P}$, $x_0 \in \mathscr{X}$, and $\varepsilon > 0$. Thus a subset U of $\mathscr{X}$ is open if and only if for every x_0 in U there are $p_1, \ldots, p_n$ in $\mathscr{P}$ and $\varepsilon_1, \ldots, \varepsilon_n > 0$ such that $\bigcap_{j=1}^{n} \{x \in \mathscr{X}: p_j(x - x_0) < \varepsilon_j\} \subseteq U$. It is not difficult to show that $\mathscr{X}$ with this topology is a TVS (Exercise 2).

1.2. Definition. A *locally convex space* (LCS) is a TVS whose topology is defined by a family of seminorms $\mathscr{P}$ such that $\bigcap_{p \in \mathscr{P}} \{x: p(x) = 0\} = (0)$.

The attitude that has been adopted in this book is that all topological spaces are Hausdorff. The condition in Definition 1.2 that $\bigcap_{p \in \mathscr{P}} \{x: p(x) = 0\} = (0)$ is imposed precisely so that the topology defined by $\mathscr{P}$ be Hausdorff. In fact, suppose that $x \neq y$. Then there is a p in $\mathscr{P}$ such that $p(x - y) \neq 0$; let $p(x - y) > \varepsilon > 0$. If $U = \{z: p(x - z) < \frac{1}{2}\varepsilon\}$ and $V = \{z: p(y - z) < \frac{1}{2}\varepsilon\}$, then $U \cap V = \square$ and U and V are neighborhoods of x and y, respectively.

If $\mathscr{X}$ is a TVS and $x_0 \in \mathscr{X}$, then $x \mapsto x + x_0$ is a homeomorphism of $\mathscr{X}$; also, if $\alpha \in \mathbb{F}$ and $\alpha \neq 0$, $x \mapsto \alpha x$ is a homeomorphism of $\mathscr{X}$ (Exercise 4). Thus the topology of $\mathscr{X}$ looks the same at any point. This might make the next statement less surprising.

1.3. Proposition. *Let $\mathscr{X}$ be a TVS and let p be a seminorm on $\mathscr{X}$. The following statements are equivalent.*

(a) *p is continuous.*
(b) *$\{x \in \mathscr{X}: p(x) < 1\}$ is open.*
(c) *$0 \in \operatorname{int}\{x \in \mathscr{X}: p(x) < 1\}$.*
(d) *$0 \in \operatorname{int}\{x \in \mathscr{X}: p(x) \leq 1\}$.*
(e) *p is continuous at 0.*
(f) *There is a continuous seminorm q on $\mathscr{X}$ such that $p \leq q$.*

PROOF. It is clear that (a) $\Rightarrow$ (b) $\Rightarrow$ (c) $\Rightarrow$ (d).

(d) *implies* (e): Clearly (d) implies that for every $\varepsilon > 0$, $0 \in \operatorname{int}\{x \in \mathscr{X}: p(x) \leq \varepsilon\}$; so if $\{x_i\}$ is a net in $\mathscr{X}$ that converges to 0 and $\varepsilon > 0$, there is an i_0 such that $x_i \in \{x: p(x) \leq \varepsilon\}$ for $i \geq i_0$; that is, $p(x_i) \leq \varepsilon$ for $i \geq i_0$. So p is continuous at 0.

(e) *implies* (a): If $x_i \rightarrow x$, then $|p(x) - p(x_i)| \leq p(x - x_i)$. Since $x - x_i \rightarrow 0$, (e) implies that $p(x - x_i) \rightarrow 0$. Hence $p(x_i) \rightarrow p(x)$.

Clearly (a) implies (f). So it remains to show that (f) implies (e). If $x_i \rightarrow 0$ in $\mathscr{X}$, then $q(x_i) \rightarrow 0$. But $0 \leq p(x_i) \leq q(x_i)$, so $p(x_i) \rightarrow 0$. ∎

1.4. Proposition. *If $\mathscr{X}$ is a TVS and $p_1, \ldots, p_n$ are continuous seminorms, then $p_1 + \cdots + p_n$ and $\max_i(p_i(x))$ are continuous seminorms. If $\{p_i\}$ is a family of continuous seminorms such that there is a continuous seminorm q with $p_i \leq q$ for all i, then $x \mapsto \sup_i\{p_i(x)\}$ defines a continuous seminorm.*

PROOF. Exercise.

If $\mathcal{P}$ is a family of seminorms on $\mathcal{X}$ that makes $\mathcal{X}$ into a LCS, it is often convenient to enlarge $\mathcal{P}$ by assuming that $\mathcal{P}$ is closed under the formation of finite sums and supremums of bounded families [as in (1.4)]. Sometimes it is convenient to assume that $\mathcal{P}$ consists of all continuous seminorms. In either case the resulting topology on $\mathcal{X}$ remains unchanged.

1.5.   Example. Let X be completely regular and let $C(X) =$ all continuous functions from X into $\mathbb{F}$. If K is a compact subset of X, define $p_K(f) = \sup\{|f(x)|: x \in K\}$. Then $\{p_K: K$ compact in $X\}$ is a family of seminorms that makes $C(X)$ into a LCS.

1.6.   Example. Let G be an open subset of $\mathbb{C}$ and let $H(G)$ be the subset of $C_{\mathbb{C}}(G)$ consisting of all analytic functions on G. Define the seminorms of (1.5) on $H(G)$. Then $H(G)$ is a LCS. Also, the topology defined on $H(G)$ by these seminorms is the topology of uniform convergence on compact subsets—the usual topology for discussing analytic functions.

1.7.   Example. Let $\mathcal{X}$ be a normed space. For each x^* in $\mathcal{X}^*$, define $p_{x^*}(x) = |x^*(x)|$. Then p_{x^*} is a seminorm and if $\mathcal{P} = \{p_{x^*}: x^* \in \mathcal{X}^*\}$, $\mathcal{P}$ makes $\mathcal{X}$ into a LCS. The topology defined on $\mathcal{X}$ by these seminorms is called the *weak topology* and is often denoted by $\sigma(\mathcal{X}, \mathcal{X}^*)$.

1.8.   Example. Let $\mathcal{X}$ be a normed space and for each x in $\mathcal{X}$ define p_x: $\mathcal{X}^* \to [0, \infty)$ by $p_x(x^*) = |x^*(x)|$. Then p_x is a seminorm and $\mathcal{P} = \{p_x: x \in \mathcal{X}\}$ makes $\mathcal{X}^*$ into a LCS. The topology defined by these seminorms is called the *weak-star* (or *weak** or *wk**) *topology* on $\mathcal{X}^*$. It is often denoted by $\sigma(\mathcal{X}^*, \mathcal{X})$.

The spaces $\mathcal{X}$ with its weak topology and $\mathcal{X}^*$ with its weak* topology are very important and will be explored in depth in Chapter V.

Recall the definition of convex set from (I.2.4). If $a, b \in \mathcal{X}$, then the *line segment* from a to b is defined as $[a, b] \equiv \{tb + (1 - t)a: 0 \le t \le 1\}$. So a set A is convex if and only if $[a, b] \subseteq A$ whenever $a, b \in A$. The proof of the next result is left to the reader.

1.9.   Proposition. (a) *A set A is convex if and only if whenever $x_1, \ldots, x_n \in A$ and $t_1, \ldots, t_n \in [0, 1]$ with $\sum_j t_j = 1$, then $\sum_j t_j x_j \in A$.* (b) *If $\{A_i: i \in I\}$ is a collection of convex sets, then $\cap_i A_i$ is convex.*

1.10.   Definition. If $A \subseteq \mathcal{X}$, the *convex hull* of A, denoted by $\mathrm{co}(A)$, is the intersection of all convex sets that contain A. If $\mathcal{X}$ is a TVS, then the *closed convex hull* of A is the intersection of all closed convex subsets of $\mathcal{X}$ that contain A; it is denoted by $\overline{\mathrm{co}}(A)$.

Since a vector space is itself convex, each subset of $\mathscr{X}$ is contained in a convex set. This fact and Proposition 1.9(b) imply that co(A) is well defined and convex. Also, $\overline{\text{co}}(A)$ is a closed convex set.

If $\mathscr{X}$ is a normed space, then $\{x: \|x\| \leq 1\}$ and $\{x: \|x\| < 1\}$ are both convex sets. If $f \in \mathscr{X}^*$, $\{x: |f(x)| \leq 1\}$, $\{x: \text{Re}\,f(x) \leq 1\}$, $\{x: \text{Re}\,f(x) > 1\}$ are all convex. In fact, if $T: \mathscr{X} \to \mathscr{Y}$ is a real linear map and C is a convex subset of $\mathscr{Y}$, then $T^{-1}(C)$ is convex in $\mathscr{X}$.

1.11. Proposition. *Let $\mathscr{X}$ be a TVS and let A be a convex subset of $\mathscr{X}$. Then (a) cl A is convex; (b) if $a \in$ int A and $b \in$ cl A, then $[a, b) \equiv \{tb + (1 - t)a: 0 \leq t < 1\} \subseteq$ int A.*

PROOF. Let $a \in A$, $b \in$ cl A, and $0 \leq t \leq 1$. Let $\{x_i\}$ be a net in A such that $x_i \to b$. Then $tx_i + (1 - t)a \to tb + (1 - t)a$. This shows that

1.12 b in cl A and a in A imply $[a, b] \subseteq$ cl A.

Using (1.12) it is easy to show that cl A is convex. To prove (b), fix t, $0 < t < 1$, and put $c = tb + (1 - t)a$, where $a \in$ int A and $b \in$ cl A. There is an open set V in $\mathscr{X}$ such that $0 \in V$ and $a + V \subseteq A$. (Why?) Hence for any d in A

$$A \supseteq td + (1 - t)(a + V)$$
$$= t(d - b) + tb + (1 - t)(a + V)$$
$$= [t(d - b) + (1 - t)V] + c.$$

If it can be shown that there is an element d in A such that $0 \in t(d - b) + (1 - t)V = U$, then the preceding inclusion shows that $c \in$ int A since U is open (Exercise 4). Note that the finding of such a d in A is equivalent to finding a d such that $0 \in t^{-1}(1 - t)V + (d - b)$ or $d \in b - t^{-1}(1 - t)V$. But $0 \in -t^{-1}(1 - t)V$ and this set is open. Since $b \in$ cl A, d can be found in A. ∎

1.13. Corollary. *If $A \subseteq \mathscr{X}$, then $\overline{\text{co}}(A)$ is the closure of co(A).*

A set $A \subseteq \mathscr{X}$ is *balanced* if $\alpha x \in A$ whenever $x \in A$ and $|\alpha| \leq 1$. A set A is *absorbing* if for each x in $\mathscr{X}$ there is an $\varepsilon > 0$ such that $tx \in A$ for $0 < t < \varepsilon$. Note that an absorbing set must contain the origin. If $a \in A$, then A is *absorbing at* a if the set $A - a$ is absorbing. Equivalently, A is absorbing at a if for every x in $\mathscr{X}$ there is an $\varepsilon > 0$ such that $a + tx \in A$ for $0 < t < \varepsilon$.

If $\mathscr{X}$ is a vector space and p is a seminorm, then $V = \{x: p(x) < 1\}$ is a convex balanced set that is absorbing at each of its points. It is rather remarkable that the converse of this is true. This fact will be used to give an abstract formulation of a LCS and also to explore some geometric consequences of the Hahn–Banach Theorem.

1.14. Proposition. *If $\mathscr{X}$ is a vector space over $\mathbb{F}$ and V is a convex, balanced set that is absorbing at each of its points, then there is a unique seminorm p on $\mathscr{X}$ such that $V = \{x \in \mathscr{X}: p(x) < 1\}$.*

PROOF. Define $p(x)$ by

$$p(x) = \inf\{t: t \geq 0 \text{ and } x \in tV\}.$$

Since V is absorbing, $\mathscr{X} = \bigcup_{n=1}^{\infty} nV$, so that the set whose infimum is $p(x)$ is nonempty. Clearly $p(0) = 0$. To see that $p(\alpha x) = |\alpha| p(x)$, we can suppose that $\alpha \neq 0$. Hence, because V is balanced,

$$p(\alpha x) = \inf\{t \geq 0: \alpha x \in tV\}$$
$$= \inf\left\{t \geq 0: x \in t\left(\frac{1}{\alpha}V\right)\right\}$$
$$= \inf\left\{t \geq 0: x \in t\left(\frac{1}{|\alpha|}V\right)\right\}$$
$$= |\alpha|\inf\left\{\frac{t}{|\alpha|}: x \in \frac{t}{|\alpha|}V\right\}$$
$$= |\alpha| p(x).$$

To complete the proof that p is a seminorm, note that if $\alpha, \beta \geq 0$ and $a, b \in V$, then

$$\alpha a + \beta b = (\alpha + \beta)\left(\frac{\alpha}{\alpha + \beta}a + \frac{\beta}{\alpha + \beta}b\right) \in (\alpha + \beta)V$$

by the convexity of V. If $x, y \in \mathscr{X}$, $p(x) = \alpha$, and $p(y) = \beta$, let $\delta > 0$. Then $x \in (\alpha + \delta)V$ and $y \in (\beta + \delta)V$. (Why?) Hence $x + y \in (\alpha + \delta)V + (\beta + \delta)V = (\alpha + \beta + 2\delta)V$ (Exercise 11). Letting $\delta \to 0$ shows that $p(x + y) \leq \alpha + \beta = p(x) + p(y)$.

It remains to show that $V = \{p(x) < 1\}$. If $p(x) = \alpha < 1$, then $\alpha < \beta < 1$ implies $x \in \beta V \subseteq V$ since V is balanced. Thus $V \supseteq \{x: p(x) < 1\}$. If $x \in V$, then $p(x) \leq 1$. Since V is absorbing at x, there is an $\varepsilon > 0$ such that for $0 < t < \varepsilon$, $x + tx = y \in V$. But $x = (1 + t)^{-1}y$, $y \in V$. Hence $p(x) = (1 + t)^{-1}p(y) \leq (1 + t)^{-1} < 1$.

Uniqueness follows by (III.1.4). ■

The seminorm p defined in the preceding proposition is called the *Minkowski functional* of V or the *gauge of V*.

Note that if $\mathscr{X}$ is a TVS space and V is an open set in $\mathscr{X}$, then V is absorbing at each of its points.

Using Proposition 1.14, the following characterization of a LCS can be obtained. The proof is left to the reader.

1.15. Proposition. *Let $\mathscr{X}$ be a TVS and let $\mathscr{U}$ be the collection of all open convex balanced subsets of $\mathscr{X}$. Then $\mathscr{X}$ is locally convex if and only if $\mathscr{U}$ is a basis for the neighborhood system at 0.*

Exercises

1. Let $\mathscr{X}$ be a TVS and let $\mathscr{U}$ be all the open sets containing 0. Prove the following. (a) If $U \in \mathscr{U}$, there is a V in $\mathscr{U}$ such that $V + V \subseteq U$. (b) If $U \in \mathscr{U}$, there is a V in $\mathscr{U}$ such that $V \subseteq U$ and $\alpha V \subseteq V$ for all $|\alpha| \leq 1$. (V is balanced.) (Hint: If $W \in \mathscr{U}$ and $\alpha W \subseteq U$ for $|\alpha| \leq \varepsilon$, then $\varepsilon W \subseteq \beta U$ for $|\beta| \geq 1$.)

2. Show that a LCS is a TVS.

3. Suppose that $\mathscr{X}$ is a TVS but do not assume that $\mathscr{X}$ is Hausdorff. (a) Show that $\mathscr{X}$ is Hausdorff if and only if the singleton set $\{0\}$ is closed. (b) If $\mathscr{X}$ is Hausdorff, show that $\mathscr{X}$ is a regular topological space.

4. Let $\mathscr{X}$ be a TVS. Show: (a) if $x_0 \in \mathscr{X}$, the map $x \mapsto x + x_0$ is a homeomorphism of $\mathscr{X}$ onto $\mathscr{X}$; (b) if $\alpha \in \mathbb{F}$ and $\alpha \neq 0$, the map $x \mapsto \alpha x$ is a homeomorphism.

5. Prove Proposition 1.4.

6. Verify the statements made in Example 1.5. Show that a net $\{f_i\}$ in $C(X)$ converges to f if and only if $f_i \to f$ uniformly on compact subsets of X.

7. Show that the space $H(G)$ defined in (1.6) is complete. (Every Cauchy net converges.)

8. Verify the statements made in Example 1.7. Give a basis for the neighborhood system at 0.

9. Verify the statements made in Example 1.8.

10. Prove Proposition 1.9.

11. Show that if A is a convex set and $\alpha, \beta > 0$, then $\alpha A + \beta A = (\alpha + \beta)A$. Give an example of a nonconvex set A for which this is untrue.

12. If $\mathscr{X}$ is a TVS and A is closed, show that A is convex if and only if $\frac{1}{2}(x + y) \in A$ whenever x and $y \in A$.

13. Let $s = $ the space of all sequences of scalars. Thus $s = $ all functions $x \colon \mathbb{N} \to \mathbb{F}$. Define addition and scalar multiplication in the usual way. If $x, y \in s$, define

$$d(x, y) = \sum_{n=1}^{\infty} 2^{-n} \frac{|x(n) - y(n)|}{1 + |x(n) - y(n)|}.$$

Show that d is a metric on s and that with this topology s is a TVS. Also show that s is complete.

14. Let (X, Ω, μ) be a finite measure space, let $\mathcal{M}$ be the space of Ω-measurable functions, and identify two functions that agree a.e. $[\mu]$. If $f, g \in \mathcal{M}$, define

$$d(f, g) = \int \frac{|f - g|}{1 + |f - g|} \, d\mu.$$

Then d is a metric on $\mathcal{M}$ and $(\mathcal{M}, d)$ is a complete TVS. Is there a relationship between this example and the space s of Exercise 13?

15. If $\mathscr{X}$ is a TVS and $A \subseteq \mathscr{X}$, then $\operatorname{cl} A = \bigcap \{A + V \colon 0 \in V \text{ and } V \text{ is open}\}$.

16. If $\mathcal{X}$ is a TVS and $\mathcal{M}$ is a closed linear space, then $\mathcal{X}/\mathcal{M}$ with the quotient topology is a TVS. If p is a seminorm on $\mathcal{X}$, define $\bar{p}$ on $\mathcal{X}/\mathcal{M}$ by $\bar{p}(x + \mathcal{M}) = \inf\{ p(x + y): y \in \mathcal{M} \}$. Show that $\bar{p}$ is a seminorm on $\mathcal{X}/\mathcal{M}$. Show that if $\mathcal{X}$ is a LCS, then so is $\mathcal{X}/\mathcal{M}$.

17. If $\{\mathcal{X}_i: i \in I\}$ is a family of TVS's, then $\mathcal{X} = \prod\{\mathcal{X}_i: i \in I\}$ with the product topology is a TVS. If each $\mathcal{X}_i$ is a LCS, then so is $\mathcal{X}$. If $\mathcal{X}$ is a LCS, must each $\mathcal{X}_i$ be a LCS?

18. If $\mathcal{X}$ is a finite-dimensional vector space and $\mathcal{T}_1, \mathcal{T}_2$ are two topologies on $\mathcal{X}$ that make $\mathcal{X}$ into a TVS, then $\mathcal{T}_1 = \mathcal{T}_2$.

19. If $\mathcal{X}$ is a TVS and $\mathcal{M}$ is a finite-dimensional linear manifold in $\mathcal{X}$, then $\mathcal{M}$ is closed and $\mathcal{Y} + \mathcal{M}$ is closed for any closed subspace $\mathcal{Y}$ of $\mathcal{X}$.

20. Let $\mathcal{X}$ be any infinite-dimensional vector space and let $\mathcal{T}$ be the collection of all subsets W of $\mathcal{X}$ such that if $x \in W$, then there is a convex balanced set U with $x + U \subseteq W$ and $U \cap \mathcal{M}$ open in $\mathcal{M}$ for every finite-dimensional linear manifold $\mathcal{M}$ in $\mathcal{X}$. (Each such $\mathcal{M}$ is given its usual topology.) Show: (a) $(\mathcal{X}, \mathcal{T})$ is a LCS; (b) a set F is closed in $\mathcal{X}$ if and only if $F \cap \mathcal{M}$ is closed for every finite-dimensional subspace $\mathcal{M}$ of $\mathcal{X}$; (c) if Y is a topological space and $f: \mathcal{X} \to Y$ (not necessarily linear), then f is continuous if and only if $f|\mathcal{M}$ is continuous for every finite-dimensional space $\mathcal{M}$; (d) if $\mathcal{Y}$ is a TVS and $T: \mathcal{X} \to \mathcal{Y}$ is a linear map, then T is continuous.

21. Let X be a locally compact space and for each ϕ in $C_0(X)$, define $p_\phi(f) = \|\phi f\|_\infty$ for f in $C_b(X)$. Show that p_ϕ is a seminorm on $C_b(X)$. Let $\beta = $ the topology defined by these seminorms. Show that $(C_b(X), \beta)$ is a LCS that is complete. β is called the *strict topology*.

22. For $0 < p < 1$, let $l^p = $ all sequences x such that $\sum_{n=1}^{\infty}|x(n)|^p < \infty$. Define $d(x, y) = \sum_{n=1}^{\infty}|x(n) - y(n)|^p$ (no pth root). Then d is a metric and (l^p, d) is a TVS that is not locally convex.

23. Let $\mathcal{X}$ and $\mathcal{Y}$ be locally convex spaces and let $T: \mathcal{X} \to \mathcal{Y}$ be a linear transformation. Show that T is continuous if and only if for every continuous seminorm p on $\mathcal{Y}$, $p \circ T$ is a continuous seminorm on $\mathcal{X}$.

24. Let $\mathcal{X}$ be a LCS and let G be an open connected subset of $\mathcal{X}$. Show that G is arcwise connected.

§2. Metrizable and Normable Locally Convex Spaces

Which LCS's are metrizable? That is, which have a topology which is defined by a metric? Which LCS's have a topology that is defined by a norm? Both are interesting questions and both answers could be useful.

If $\mathcal{P}$ is a family of seminorms on $\mathcal{X}$ and $\mathcal{X}$ is a TVS, say that $\mathcal{P}$ determines the topology on $\mathcal{X}$ if the topology of $\mathcal{X}$ is the same as the topology induced by $\mathcal{P}$.

2.1. Proposition. *Let* $\{p_1, p_2, \ldots\}$ *be a sequence of seminorms on* $\mathscr{X}$ *such that* $\cap_{n=1}^{\infty}\{x: p_n(x) = 0\} = (0)$. *For x and y in* $\mathscr{X}$, *define*

$$d(x, y) = \sum_{n=1}^{\infty} 2^{-n} \frac{p_n(x - y)}{1 + p_n(x - y)}.$$

Then d is a metric on $\mathscr{X}$ *and the topology on* $\mathscr{X}$ *defined by d is the topology on* $\mathscr{X}$ *defined by the seminorms* $\{p_1, p_2, \ldots\}$. *Thus a LCS is metrizable if and only if its topology is determined by a countable family of seminorms.*

PROOF. It is left as an exercise for the reader to show that d is a metric and induces the same topology as $\{p_n\}$. If $\mathscr{X}$ is a LCS and its topology is determined by a countable family of seminorms, it is immediate that $\mathscr{X}$ is metrizable. For the converse, assume that $\mathscr{X}$ is metrizable and its metric is ρ. Let $U_n = \{x: \rho(x, 0) < 1/n\}$. Because $\mathscr{X}$ is locally convex, there are continuous seminorms $q_1, \ldots, q_k$ and positive numbers $\varepsilon_1, \ldots, \varepsilon_k$ such that $\cap_{j=1}^{k}\{x: q_j(x) < \varepsilon_j\} \subseteq U_n$. If $p_n = \varepsilon_1^{-1}q_1 + \cdots + \varepsilon_k^{-1}q_k$, then $x \in U_n$ whenever $p_n(x) < 1$. Clearly, p_n is continuous for each n. Thus if $x_j \to 0$ in $\mathscr{X}$, then for each n, $p_n(x_j) \to 0$ as $j \to \infty$. Conversely, suppose that for each n, $p_n(x_j) \to 0$ as $j \to \infty$. If $\varepsilon > 0$, let $n > \varepsilon^{-1}$. Then there is a j_0 such that for $j \geq j_0$, $p_n(x_j) < 1$. Thus, for $j \geq j_0$, $x_j \in U_n \subseteq \{x: \rho(x, 0) < \varepsilon\}$. That is, $\rho(x_j, 0) < \varepsilon$ for $j \geq j_0$ and so $x_j \to 0$ in $\mathscr{X}$. This shows that $\{p_n\}$ determines the topology on $\mathscr{X}$. (Why?) ∎

2.2. Example. If $C(X)$ is as in Example 1.5, then $C(X)$ is metrizable if and only if $X = \cup_{n=1}^{\infty}K_n$, where each K_n is compact, $K_1 \subseteq K_2 \subseteq \cdots$, and if K is any compact subset of X, then $K \subseteq K_n$ for some n.

2.3. Example. If X is locally compact and $C(X)$ is as in Example 1.5, then $C(X)$ is metrizable if and only if X is σ-*compact* (that is, X is the union of a sequence of compact sets). If $H(G)$ is as in Example 1.6, then $H(G)$ is metrizable.

If $\mathscr{X}$ is a vector space and d is a metric on $\mathscr{X}$, say that d is *translation invariant* if $d(x + z, y + z) = d(x, y)$ for all x, y, z in $\mathscr{X}$. Note that the metric defined by a norm as well as the metric defined in (2.1) are translation invariant.

2.4. Definition. A *Frechet space* is a TVS $\mathscr{X}$ whose topology is defined by a translation invariant metric d and such that $(\mathscr{X}, d)$ is complete.

It should be pointed out that some authors include in the definition of a Frechet space the assumption that $\mathscr{X}$ is locally convex.

2.5. Definition. If $\mathscr{X}$ is a TVS and $B \subseteq \mathscr{X}$, then B is *bounded* if for every open set U containing 0, there is an $\varepsilon > 0$ such that $\varepsilon B \subseteq U$.

If $\mathcal{X}$ is a normed space, then it is easy to see that a set B is bounded if and only if $\sup\{\|b\|: b \in B\} < \infty$, so the definition is intuitively correct.

Also, notice that if $\| \cdot \|$ is a norm, $\{x: \|x\| < 1\}$ is itself bounded. This is not true for seminorms. For example, if $C(\mathbb{R})$ is topologized as in (1.5), let $p(f) = \sup\{|f(t)|: 0 \le t \le 1\}$. Then p is a continuous seminorm. However, $\{f: p(f) < 1\}$ is not bounded. In fact, if f_0 is any function in $C(\mathbb{R})$ that vanishes on $[0, 1]$, $\{\alpha f_0: \alpha \in \mathbb{R}\} \subseteq \{f: p(f) < 1\}$. The fact that a normed space possesses a bounded open set is characteristic.

2.6. Proposition. *If $\mathcal{X}$ is a LCS, then $\mathcal{X}$ is normable if and only if $\mathcal{X}$ has a bounded open set.*

PROOF. It has already been shown that a normed space has a bounded open set. So assume that $\mathcal{X}$ is a LCS that has a bounded open set U. It must be shown that there is norm on $\mathcal{X}$ that defines the same topology. By translation, it may be assumed that $0 \in U$ (see Exercise 4i). By local convexity, there is a continuous seminorm p such that $\{x: p(x) < 1\} \equiv V \subseteq U$ (Why?). It will be shown that p is a norm and defines the topology on $\mathcal{X}$.

To see that p is a norm, suppose that $x \in \mathcal{X}$, $x \ne 0$. Let W_0, W_x be disjoint open sets such that $0 \in W_0$ and $x \in W_x$. Then there is an $\varepsilon > 0$ such that $W_0 \supseteq \varepsilon U \supseteq \varepsilon V$. But $\varepsilon V = \{y: p(y) < \varepsilon\}$. Since $x \notin W_0$, $p(x) \ge \varepsilon$. Hence p is a norm.

Because p is continuous on $\mathcal{X}$, to show that p defines the topology of $\mathcal{X}$ it suffices to show that if q is any continuous seminorm on $\mathcal{X}$, there is an $\alpha > 0$ such that $q \le \alpha p$ (Why?). But because q is continuous, there is an $\varepsilon > 0$ such that $\{x: q(x) < 1\} \supseteq \varepsilon U \supseteq \varepsilon V$. That is, $p(x) < \varepsilon$ implies $q(x) < 1$. By Lemma III.1.4, $q \le \varepsilon^{-1} p$. ∎

EXERCISES

1. Supply the missing details in the proof of Proposition 2.1.

2. Verify the statements in Example 2.2.

3. Verify the statements in Example 2.3.

4. Let $\mathcal{X}$ be a TVS and prove the following: (a) If B is a bounded subset of $\mathcal{X}$, then so is cl B. (b) The finite union of bounded sets is bounded. (c) Every compact set is bounded. (d) If $B \subseteq \mathcal{X}$, then B is bounded if and only if for every sequence $\{x_n\}$ contained in B and for every $\{\alpha_n\}$ in c_0, $\alpha_n x_n \to 0$ in $\mathcal{X}$. (e) If $\mathcal{Y}$ is a TVS, $T: \mathcal{X} \to \mathcal{Y}$ is a continuous linear transformation, and B is a bounded subset of $\mathcal{X}$, then $T(B)$ is a bounded subset of $\mathcal{Y}$. (f) If $\mathcal{X}$ is a LCS and $B \subseteq \mathcal{X}$, then B is bounded if and only if for every continuous seminorm p, $\sup\{p(b): b \in B\} < \infty$. (g) If $\mathcal{X}$ is a normed space and $B \subseteq \mathcal{X}$, then B is bounded if and only if $\sup\{\|b\|: b \in B\} < \infty$. (h) If $\mathcal{X}$ is a Frechet space, then bounded sets have finite diameter, but not conversely. (i) The translate of a bounded set is bounded.

5. If $\mathscr{X}$ is a LCS, show that $\mathscr{X}$ is metrizable if and only if $\mathscr{X}$ is first countable. Is this equivalent to saying that $\{0\}$ is a G_δ set?

6. Let X be a locally compact space and give $C_b(X)$ the strict topology defined in Exercise 1.21. Show that a subset of $C_b(X)$ is β-bounded if and only if it is norm bounded.

7. With the notation of Exercise 6, show that $(C_b(X), \beta)$ is metrizable if and only if X is compact.

8. Prove the Open Mapping Theorem for Frechet spaces.

§3. Some Geometric Consequences of the Hahn–Banach Theorem

In order to exploit the Hahn–Banach Theorem in the setting of a LCS, it is necessary to establish some properties of continuous linear functionals. The proofs of the relevant propositions are similar to the proofs of the corresponding facts about linear functionals on normed spaces given in §III.5. For example, a hyperplane in a TVS is either closed or dense (see III.5.2). The proof of the next fact is similar to the proof of (III.2.1) and (III.5.3) and will not be given.

3.1.   Theorem. *If $\mathscr{X}$ is a TVS and $f: \mathscr{X} \to \mathbb{F}$ is a linear functional, then the following statements are equivalent.*

(a) *f is continuous.*
(b) *f is continuous at 0.*
(c) *f is continuous at some point.*
(d) *$\ker f$ is closed.*
(e) *$x \mapsto |f(x)|$ is a continuous seminorm.*

If $\mathscr{X}$ is a LCS and $\mathscr{P}$ is a family of seminorms that defines the topology on $\mathscr{X}$, then the statements above are equivalent to the following:

(f) *There are $p_1, \ldots, p_n$ in $\mathscr{P}$ and positive scalars $\alpha_1, \ldots, \alpha_n$ such that $|f(x)| \leq \sum_{k=1}^{n} \alpha_k p_k(x)$ for all x.*

The proof of the next proposition is similar to the proof of Proposition 1.14 and will not be given.

3.2.   Proposition. *Let $\mathscr{X}$ be a TVS and suppose that G is an open convex subset of $\mathscr{X}$ that contains the origin. If*

$$q(x) = \inf\{t: t \geq 0 \text{ and } x \in tG\},$$

then q is a non-negative continuous sublinear functional and $G = \{x: q(x) < 1\}$.

Note that the difference between the preceding proposition and (1.14) is that here G is not assumed to be balanced and the consequence is a sublinear functional $(q(\alpha x) = \alpha q(x)$ if $\alpha \geq 0)$ that is not necessarily a seminorm.

The geometric consequences of the Hahn–Banach Theorem are achieved by interpreting that theorem in light of the correspondence between linear functionals and hyperplanes and between sublinear functionals and open convex neighborhoods of the origin. The next result is typical.

3.3. Theorem. *If $\mathscr{X}$ is a TVS and G is an open convex nonempty subset of $\mathscr{X}$ that does not contain the origin, then there is a closed hyperplane $\mathscr{M}$ such that $\mathscr{M} \cap G = \square$.*

PROOF. *Case 1.* $\mathscr{X}$ is an $\mathbb{R}$-linear space. Pick any x_0 in G and let $H = x_0 - G$. Then H is an open convex set containing 0. (Verify). By (3.2) there is a continuous nonnegative sublinear functional $q: \mathscr{X} \to \mathbb{R}$ such that $H = \{x: q(x) < 1\}$. Since $x_0 \notin H$, $q(x_0) \geq 1$.

Let $\mathscr{Y} \equiv \{\alpha x_0: \alpha \in \mathbb{R}\}$ and define $f_0: \mathscr{Y} \to \mathbb{R}$ by $f_0(\alpha x_0) = \alpha q(x_0)$. If $\alpha \geq 0$, then $f(\alpha x_0) = \alpha q(x_0) = q(\alpha x_0)$; if $\alpha < 0$, then $f(\alpha x_0) = \alpha q(x_0) \leq \alpha < 0 \leq q(\alpha x_0)$. So $f_0 \leq q$ on $\mathscr{Y}$. Let $f: \mathscr{X} \to \mathbb{R}$ be a linear functional such that $f|\mathscr{Y} = f_0$ and $f \leq q$ on $\mathscr{X}$. Put $\mathscr{M} = \ker f$.

Now if $x \in G$, then $x_0 - x \in H$ and so $f(x_0) - f(x) = f(x_0 - x) \leq q(x_0 - x) < 1$. Therefore $f(x) > f(x_0) - 1 = q(x_0) - 1 \geq 0$ for all x in G. Thus $\mathscr{M} \cap G = \square$.

Case 2. $\mathscr{X}$ is a $\mathbb{C}$-linear space. Lemma III.6.3 will be used here. Using Case 1 and the fact that $\mathscr{X}$ is also an $\mathbb{R}$-linear space, there is a continuous $\mathbb{R}$-linear functional $f: \mathscr{X} \to \mathbb{R}$ such that $G \cap \ker f = \square$. If $F(x) = f(x) - if(ix)$, then F is a $\mathbb{C}$-linear functional and $f = \operatorname{Re} F$ (III.6.3). Hence $F(x) = 0$ if and only if $f(x) = f(ix) = 0$; that is, $\mathscr{M} = \ker F = \ker f \cap [i \ker f]$. So $\mathscr{M}$ is a closed hyperplane and $\mathscr{M} \cap G = \square$. ∎

An *affine hyperplane* in $\mathscr{X}$ is a set $\mathscr{M}$ such that for every x_0 in $\mathscr{M}$, $\mathscr{M} - x_0$ is a hyperplane. (See Exercise 3.) An *affine manifold* in $\mathscr{X}$ is a set $\mathscr{Y}$ such that for every x_0 in $\mathscr{Y}$, $\mathscr{Y} - x_0$ is a linear manifold in $\mathscr{X}$. An *affine subspace* of a TVS $\mathscr{X}$ is a closed affine manifold.

3.4. Corollary. *Let $\mathscr{X}$ be a TVS and let G be an open convex nomempty subset of $\mathscr{X}$. If $\mathscr{Y}$ is an affine subspace of $\mathscr{X}$ such that $\mathscr{Y} \cap G = \square$, then there is a closed affine hyperplane $\mathscr{M}$ in $\mathscr{X}$ such that $\mathscr{Y} \subseteq \mathscr{M}$ and $\mathscr{M} \cap G = \square$.*

PROOF. By considering $G - x_0$ and $\mathscr{Y} - x_0$ for any x_0 in $\mathscr{Y}$, it may be assumed that $\mathscr{Y}$ is a linear subspace of $\mathscr{X}$. Let $Q: \mathscr{X} \to \mathscr{X}/\mathscr{Y}$ be the natural map. Since $Q^{-1}(Q(G)) = \{y + G: y \in \mathscr{Y}\}$, $Q(G)$ is open in $\mathscr{X}/\mathscr{Y}$. It is easy to see that $Q(G)$ is also convex. Since $\mathscr{Y} \cap G = \square$, $0 \notin Q(G)$. By the preceding theorem, there is a closed hyperplane $\mathscr{N}$ in $\mathscr{X}/\mathscr{Y}$ such that $\mathscr{N} \cap Q(G) = \square$. Let $\mathscr{M} = Q^{-1}(\mathscr{N})$. It is easy to check that $\mathscr{M}$ has the desired properties. ∎

There is a great advantage inherent in a geometric discussion of real TVS's. Namely, if $f \colon \mathcal{X} \to \mathbb{R}$ is a nonzero continuous $\mathbb{R}$-linear functional, then the hyperplane $\ker f$ disconnects the space. That is, $\mathcal{X} \setminus \ker f$ has two components (see Exercises 4 and 5). It thus becomes convenient to make the following definitions.

3.5.   Definition. Let $\mathcal{X}$ be a real TVS. A subset S of $\mathcal{X}$ is called an *open half-space* if there is a continuous linear functional $f \colon \mathcal{X} \to \mathbb{R}$ such that $S = \{x \in \mathcal{X} \colon f(x) > \alpha\}$ for some α. S is a *closed half-space* if there is a continuous linear functional $f \colon \mathcal{X} \to \mathbb{R}$ such that $S = \{x \in \mathcal{X} \colon f(x) \geq \alpha\}$ for some α.

Two subsets A and B of $\mathcal{X}$ are said to be *strictly separated* if they are contained in disjoint open half-spaces; they are *separated* if they are contained in two closed half-spaces whose intersection is a closed affine hyperplane.

3.6.   Proposition. *Let $\mathcal{X}$ be a real TVS.*

(a) *The closure of an open half-space is a closed half-space and the interior of a closed half-space is an open half-space.*
(b) *If $A, B \subseteq \mathcal{X}$, then A and B are strictly separated (separated) if and only if there is a continuous linear functional $f \colon \mathcal{X} \to \mathbb{R}$ and a real scalar α such that $f(a) > \alpha$ for all a in A and $f(b) < \alpha$ for all b in B ($f(a) \geq \alpha$ for all a in A and $f(b) \leq \alpha$ for all b in B).*

PROOF. Exercise 6.

In many ways, the next result is the most important "separation" theorem as the other separation theorems follow from this one. However, the most used separation theorem is Theorem 3.9 below.

3.7.   Theorem. *If $\mathcal{X}$ is a real TVS and A and B are disjoint open convex subsets of $\mathcal{X}$, then A and B are strictly separated.*

PROOF. Let $G = A - B \equiv \{a - b \colon a \in A, b \in B\}$; it is easy to verify that G is convex (do it!). Also, $G = \bigcup\{A - b \colon b \in B\}$, so G is open. Moreover, because $A \cap B = \square$, $0 \notin G$. By Theorem 3.3 there is a closed hyperplane $\mathcal{M}$ in $\mathcal{X}$ such that $\mathcal{M} \cap G = \square$. Let $f \colon \mathcal{X} \to \mathbb{R}$ be a linear functional such that $\mathcal{M} = \ker f$. Now $f(G)$ is a convex subset of $\mathbb{R}$ and $0 \notin f(G)$. Hence either $f(x) > 0$ for all x in G or $f(x) < 0$ for all x in G; suppose $f(x) > 0$ for all x in G. Thus if $a \in A$ and $b \in B$, $0 < f(a - b) = f(a) - f(b)$; that is, $f(a) > f(b)$. Therefore there is a real number α such that

$$\sup\{f(b) \colon b \in B\} \leq \alpha \leq \inf\{f(a) \colon a \in A\}.$$

But $f(A)$ and $f(B)$ are open intervals (Exercise 7), so $f < \alpha$ on B and $f > \alpha$ on A.   ∎

3.8. **Lemma.** *If $\mathcal{X}$ is a TVS, K is a compact subset of $\mathcal{X}$, and V is an open subset of $\mathcal{X}$ such that $K \subseteq V$, then there is an open neighborhood of $0, U$, such that $K + U \subseteq V$.*

PROOF. Let $\mathcal{U}_0 = $ all of the open neighborhoods of 0. Suppose that for each U in $\mathcal{U}_0$, $K + U$ is not contained in V. Thus, for each U in $\mathcal{U}_0$ there is a vector x_U in K and a y_U in U such that $x_U + y_U \in \mathcal{X} \backslash V$. Order $\mathcal{U}_0$ by reverse inclusion; that is, $U_1 \geq U_2$ if $U_1 \subseteq U_2$. Then $\mathcal{U}_0$ is a directed set and $\{x_U\}$ and $\{y_U\}$ are nets. Now $y_U \to 0$ in $\mathcal{X}$. Because K is compact there is an x in K such that $x_U \xrightarrow{\text{cl}} x$ ($\{x_U\}$ clusters at x). Hence $x_U + y_U \xrightarrow{\text{cl}} x + 0 = x$. (Why?) Hence $x \in \text{cl}(\mathcal{X} \backslash V) = \mathcal{X} \backslash V$, a contradiction. ∎

The condition that K be compact in the preceding lemma is necessary; it is not enough to assume that K is closed. (What is counterexample?)

3.9. **Theorem.** *Let $\mathcal{X}$ be a real LCS and let A and B be two disjoint closed convex subsets of $\mathcal{X}$. If B is compact, then A and B are strictly separated.*

PROOF. By hypothesis, B is a compact subset of the open set $\mathcal{X} \backslash A$. The preceding lemma implies there is an open neighborhood U_1 of 0 such that $B + U_1 \subseteq \mathcal{X} \backslash A$. Because $\mathcal{X}$ is locally convex, there is a continuous seminorm p on $\mathcal{X}$ such that $\{x: p(x) < 1\} \subseteq U_1$. Put $U = \{x: p(x) < \frac{1}{2}\}$. Then $(B + U) \cap (A + U) = \square$ (Verify!), and $A + U$ and $B + U$ are open convex subsets of $\mathcal{X}$ that contain A and B, respectively. (Why?) So the result follows from Theorem 3.7. ∎

The fact that one of the two closed convex sets in the preceding theorem is assumed to be compact is necessary. In fact, if $\mathcal{X} = \mathbb{R}^2$, $A = \{(x, y) \in \mathbb{R}^2: y \leq 0\}$, and $B = \{(x, y) \in \mathbb{R}^2: y \geq x^{-1}\}$, then A and B are disjoint closed convex subsets of $\mathbb{R}^2$ that cannot be strictly separated.

The next result generalizes Corollary III.6.8, though, of course, the metric content of (III.6.8) is missing.

3.10. **Corollary.** *If $\mathcal{X}$ is a real LCS, A is a closed convex subset of $\mathcal{X}$, and $x \notin A$, then x is strictly separated from A.*

3.11. **Corollary.** *If $\mathcal{X}$ is a real LCS and $A \subseteq \mathcal{X}$, then $\overline{\text{co}}(A)$ is the intersection of the closed half-spaces containing A.*

PROOF. Let $\mathscr{H}$ be the collection of all closed half-spaces containing A. Since each set in $\mathscr{H}$ is closed and convex, $\overline{\text{co}}(A) \subseteq \cap\{H: H \in \mathscr{H}\}$. On the other hand, if $x_0 \notin \overline{\text{co}}(A)$, then (3.10) implies there is a continuous linear functional $f: \mathcal{X} \to \mathbb{R}$ and an α in $\mathbb{R}$ such that $f(x_0) > \alpha$ and $f(x) < \alpha$ for all x in $\overline{\text{co}}(A)$. Thus $H = \{x: f(x) \leq \alpha\}$ belongs to $\mathscr{H}$ and $x_0 \notin H$. ∎

The next result generalizes Theorem III.6.13.

3.12. Corollary. *If $\mathscr{X}$ is a real LCS and $A \subseteq \mathscr{X}$, then the closed linear span of A is the intersection of all closed hyperplanes containing A.*

If $\mathscr{X}$ is a complex LCS, it is also a real LCS. This can be used to formulate and prove versions of the preceding results. As a sample, the following complex version of Theorem 3.9 is presented.

3.13. Theorem. *Let $\mathscr{X}$ be a complex LCS and let A and B be two disjoint closed convex subsets of $\mathscr{X}$. If B is compact, then there is a continuous linear functional $f \colon \mathscr{X} \to \mathbb{C}$, an α in $\mathbb{R}$, and an $\varepsilon > 0$ such that for a in A and b in B,*

$$\operatorname{Re} f(a) \leq \alpha < \alpha + \varepsilon \leq \operatorname{Re} f(b).$$

3.14. Corollary. *If $\mathscr{X}$ is a LCS and $\mathscr{Y}$ is a linear manifold in $\mathscr{X}$, then $\mathscr{Y}$ is dense in $\mathscr{X}$ if and only if the only continuous linear functional on $\mathscr{X}$ that vanishes on $\mathscr{Y}$ is the identically zero functional.*

3.15. Corollary. *If $\mathscr{X}$ is a LCS, $\mathscr{Y}$ is a closed linear subspace of $\mathscr{X}$, and $x_0 \in \mathscr{X} \setminus \mathscr{Y}$, then there is a continuous linear functional $f \colon \mathscr{X} \to \mathbb{F}$ such that $f(y) = 0$ for all y in $\mathscr{Y}$ and $f(x_0) = 1$.*

These results imply that on a LCS there are many continuous linear functionals. Compare the results of this section with those of §III.6.

The hypothesis that $\mathscr{X}$ is locally convex does not appear in the results prior to Theorem 3.9. The reason for this is that in the preceding results the existence of an open convex subset of $\mathscr{X}$ is assumed. In Theorem 3.9 such a set must be manufactured. Without the hypothesis of local convexity it may be that the only open convex sets are the whole space itself and the empty set.

3.16. Example. For $0 < p < 1$, let $L^p(0,1)$ be the collection of equivalence classes of measurable functions $f \colon (0,1) \to \mathbb{R}$ such that

$$((f))_p = \int_0^1 |f(x)|^p \, dx < \infty.$$

It will be shown that $d(f, g) = ((f - g))_p$ is a metric on $L^p(0,1)$ and that with this metric $L^p(0,1)$ is a Frechet space. It will also be shown, however, that $L^p(0,1)$ has only one nonempty open convex set, namely itself. So $L^p(0,1)$, $0 < p < 1$, is most emphatically not locally convex. The proof of these facts begins with the following inequality.

3.17   For s, t in $[0, \infty)$ and $0 < p < 1$, $(s + t)^p \leq s^p + t^p$.

To see this, let $f(t) = s^p + t^p - (s + t)^p$ for $t \geq 0$, s fixed. Then $f'(t) = pt^{p-1} - p(s + t)^{p-1}$. Since $p - 1 < 0$ and $s + t \geq t$, $f'(t) \geq 0$. Thus $0 = f(0) \leq f(t)$. This proves (3.17).

If $d(f, g) = ((f - g))_p$ for f, g in $L^p(0, 1)$, then (3.17) implies that $d(f, g) \le d(f, h) + d(h, g)$ for all f, g, h in $L^p(0, 1)$. It follows that d is a metric on $L^p(0, 1)$. Clearly d is translation invariant.

3.18 $L^p(0, 1), 0 < p < 1$, is complete.

The proof of this is left as an exercise.

3.19 $L^p(0, 1)$ is a TVS.

The continuity of addition is a direct consequence of the translation invariance of d. If $f_n \to f$ and $\alpha_n \to \alpha$, α_n in $\mathbb{F}$, $d(\alpha_n f_n, \alpha f) = ((\alpha_n f_n - \alpha f))_p$ $\le ((\alpha_n f_n - \alpha_n f))_p + ((\alpha_n f - \alpha f))_p = |\alpha_n|^p ((f_n - f))_p + |\alpha_n - \alpha|^p ((f))_p$ $\le C((f_n - f))_p + |\alpha_n - \alpha|^p ((f))_p$, where C is a constant independent of n. Hence $\alpha_n f_n \to \alpha f$. Thus $L^p(0, 1)$ is a Frechet space.

3.20   If G is a nonempty open convex subset of $L^p(0, 1)$, then

$$G = L^p(0, 1).$$

To see this, first suppose $f \in L^p(0, 1)$ and $((f))_p = r < R$. As a function of t, $\int_0^t |f(x)|^p dx$ is continuous, assumes the value 0 at $t = 0$, and assumes the value r at $t = 1$. Let $0 < t < 1$ such that $\int_0^t |f(x)|^p dx = r/2$. Define $g, h: (0, 1) \to \mathbb{F}$ by $g(x) = f(x)$ for $x \le t$ and 0 otherwise; $h(x) = f(x)$ for $x \ge t$ and 0 otherwise. Now $f = g + h = \frac{1}{2}(2g + 2h)$ and $((2g))_p = ((2h))_p = 2^p(r/2) = r/2^{1-p}$. Hence $f \in \text{co } B(0; R/2^{1-p})$. This implies that $B(0; R) \subseteq \text{co } B(0; R/2^{1-p})$, or, equivalently, $B(0; 2^{1-p}R) \subseteq \text{co } B(0, R)$. Hence $B(0; 4^{1-p}R) \subseteq \text{co } B(0; 2^{1-p}R) \subseteq \text{co } B(0; R)$. Continuing we see that for all n, $B(0; 2^{n(1-p)}R) \subseteq \text{co } B(0; R)$.

So if G is a nonempty open convex subset of $L^p(0, 1)$, then by translation it may be assumed that $0 \in G$. Thus there is an $R > 0$ with $B(0; R) \subseteq G$. By the preceding paragraph, $B(0; 2^{n(1-p)}R) \subseteq \text{co } B(0; R) \subseteq G$ for all $n \ge 1$. Therefore $L^p(0, 1) \subseteq G$.

Also note that this says that the only continuous linear functional on $L^p(0, 1), 0 < p < 1$, is the identically zero functional.

EXERCISES

1. Prove Theorem 3.1.

2. Let p be a sublinear functional, $G \equiv \{x: p(x) < 1\}$, and define the sublinear functional q for the set G as in Proposition 3.2. Show that $q(x) = \max(p(x), 0)$ for all x in $\mathcal{X}$.

3. Let $\mathcal{M} \subseteq \mathcal{X}$, a TVS, and show that the following statements are equivalent: (a) $\mathcal{M}$ is an affine hyperplane; (b) there exists an x_0 in $\mathcal{M}$ such that $\mathcal{M} - x_0$ is a hyperplane; (c) there is a linear function $f: \mathcal{X} \to \mathbb{F}$ and an α in $\mathbb{F}$ such that $\mathcal{M} = \{x \in \mathcal{X}: f(x) = \alpha\}$.

4. Let $\mathcal{X}$ be a real TVS. Show: (a) if G is an open connected subset of $\mathcal{X}$, then G is arcwise connected; (b) if $f: \mathcal{X} \to \mathbb{R}$ is a continuous linear functional, then $\mathcal{X} \setminus \ker f$ has two components, $\{x: f(x) > 0\}$ and $\{x: f(x) < 0\}$.

5. If $\mathcal{X}$ is a complex TVS and $f: \mathcal{X} \to \mathbb{C}$ is a nonzero continuous linear function, show that $\mathcal{X} \setminus \ker f$ is connected.

6. Prove Proposition 3.6.

7. If $f: \mathscr{X} \to \mathbb{R}$ is a continuous $\mathbb{R}$-linear functional and A is an open convex subset of $\mathscr{X}$, then $f(A)$ is an open interval.

8. Prove Corollary 3.12.

9. Prove Theorem 3.13.

10. State and prove a version of Theorem 3.7 for a complex TVS.

11. State and prove a version of Corollary 3.11 for a complex LCS.

12. State and prove a version of Corollary 3.12 for a complex LCS.

13. Prove (3.18).

14. Give an example of a TVS $\mathscr{X}$ that is not locally convex and a subspace $\mathscr{Y}$ of $\mathscr{X}$ such that there is a continuous linear functional f on $\mathscr{Y}$ with no continuous extension to $\mathscr{X}$.

§4*. Some Examples of the Dual Space of a Locally Convex Space

As with a normed space, if $\mathscr{X}$ is a LCS, $\mathscr{X}^*$ denotes the space of all continuous linear functionals $f: \mathscr{X} \to \mathbb{F}$. $\mathscr{X}^*$ is called the *dual space* of $\mathscr{X}$.

4.1. Proposition. *Let X be completely regular and let $C(X)$ be topologized as in Example 1.5. If $L: C(X) \to \mathbb{F}$ is a continuous linear functional, then there is a compact set K and a regular Borel measure μ on K such that $L(f) = \int_K f\,d\mu$ for every f in $C(X)$. Conversely, each such measure defines an element of $C(X)^*$.*

PROOF. It is easy to see that each measure μ supported on a compact set K defines an element of $C(X)^*$. In fact, if $p_K(f) = \sup\{|f(x)|: x \in K\}$ and $L(f) = \int_K f\,d\mu$, then $|L(f)| \le \|\mu\|p_K(f)$, and so L is continuous.

Now assume $L \in C(X)^*$. There are compact sets $K_1, \ldots, K_n$ and positive numbers $\alpha_1, \ldots, \alpha_n$ such that $|L(f)| \le \sum_{j=1}^{n} \alpha_n p_{K_j}(f)$ (3.1f). Let $K = \bigcup_{j=1}^{n} K_j$ and $\alpha = \max\{|\alpha_j|: 1 \le j \le n\}$. Then $|L(f)| \le \alpha p_K(f)$. Hence if $f \in C(X)$ and $f|K \equiv 0$, then $L(f) = 0$.

Define $F: C(K) \to \mathbb{F}$ as follows. If $g \in C(K)$, let $\tilde{g}$ be any continuous extension of g to X and put $F(g) = L(\tilde{g})$. To check that F is well defined, suppose that $\tilde{g}_1$ and $\tilde{g}_2$ are both extensions of g to X. Then $\tilde{g}_1 - \tilde{g}_2 = 0$ on K, and hence $L(\tilde{g}_1) = L(\tilde{g}_2)$. Thus F is well defined. It is left as an exercise for the reader to show that $F: C(K) \to \mathbb{F}$ is linear. If $g \in C(K)$ and $\tilde{g}$ is an extension in $C(X)$, then $|F(g)| = |L(\tilde{g})| \le \alpha p_K(\tilde{g}) = \alpha\|g\|$, where the norm is the norm of $C(K)$. By (III.5.7) there is a measure μ in $M(K)$ such that $F(g) = \int_K g\,d\mu$. If $f \in C(X)$, then $g = f|K \in C(K)$ and so $L(f) = F(g) = \int_K f\,d\mu$. ∎

Let $\mathbb{C}_\infty$ denote the extended complex plane. Thus $\mathbb{C}_\infty = \mathbb{C} \cup \{\infty\}$ with the metric it obtains from its identification with the sphere. If $\gamma: [0, 1] \to \mathbb{C}$ is a rectifiable curve and f is a continuous function defined on the trace of γ, $\gamma([0,1])$, then $\int_\gamma f$ is the line integral of f over γ. That is, $\int_\gamma f \equiv \int_0^1 f(\gamma(t)) \, d\gamma(t)$. (See Conway [1978].) The next result generalizes to arbitrary regions in the plane, but for simplicity it is stated only for the disk $\mathbb{D}$. Recall the definition of $H(\mathbb{D})$ from Example 1.6.

4.2. Proposition. $L \in H(\mathbb{D})^*$ *if and only if there is an* $r < 1$ *and a unique function g analytic on* $\mathbb{C}_\infty \setminus \overline{B}(0; r)$ *with* $g(\infty) = 0$ *such that*

4.3
$$L(f) = \frac{1}{2\pi i} \int_\gamma fg$$

for every f in $H(\mathbb{D})$, *where* $\gamma(t) = \rho e^{it}$, $0 \le t \le 2\pi$, *and* $r < \rho < 1$.

PROOF. Let g be given and define L as in (4.3). If $K = \{z: |z| = \rho\}$, then

$$|L(f)| = \frac{1}{2\pi} \left| \int_0^{2\pi} f(\rho e^{it}) g(\rho e^{it}) i\rho e^{it} \, dt \right|$$

$$\le \frac{1}{2\pi} p_K(f) p_K(g) 2\pi\rho.$$

So if $c = \rho p_K(g)$, $|L(f)| \le c p_K(f)$, and $L \in H(\mathbb{D})^*$.

Now assume that $L \in H(\mathbb{D})^*$. The Hahn–Banach Theorem implies there is an F in $C(\mathbb{D})^*$ such that $F|H(\mathbb{D}) = L$. By Proposition 4.1 there is a compact set K contained in $\mathbb{D}$ and a measure μ on K such that $L(f) = \int_K f \, d\mu$ for every f in $H(\mathbb{D})$. Define $g: \mathbb{C}_\infty \setminus K \to \mathbb{C}$ by $g(\infty) = 0$ and $g(z) = -\int_K 1/(w - z) \, d\mu(w)$ for z in $\mathbb{C} \setminus K$. By Lemma III.8.2, g is analytic on $\mathbb{C}_\infty \setminus K$. Let $\rho < 1$ such that $K \subseteq B(0; \rho)$. If $\gamma(t) = \rho e^{it}$, $0 \le t \le 2\pi$, then Cauchy's Integral Formula implies

$$f(w) = \frac{1}{2\pi i} \int_\gamma \frac{f(z)}{z - w} \, dz$$

for $|w| < \rho$; in particular, this is true for w in K. Thus,

$$L(f) = \int_K f(w) \, d\mu(w)$$

$$= \int_K \left[\frac{\rho}{2\pi} \int_0^{2\pi} \frac{f(\rho e^{it})}{\rho e^{it} - w} e^{it} \, dt \right] d\mu(w)$$

$$= \frac{\rho}{2\pi} \int_0^{2\pi} f(\rho e^{it}) e^{it} \left[\int_K \frac{1}{\rho e^{it} - w} \, d\mu(w) \right] dt$$

$$= \frac{1}{2\pi i} \int_\gamma f(z) g(z) \, dz.$$

This completes the proof except for the uniqueness of g (Exercise 3). ∎

EXERCISES

1. Let $\{\mathcal{X}_i: i \in I\}$ be a family of LCS's and give $\mathcal{X} = \Pi\{\mathcal{X}_i: i \in I\}$ the product topology. (See Exercise 1.17.) Show that $L \in \mathcal{X}^*$ if and only if there is a finite subset F contained in I and there are x_j^* in $\mathcal{X}_j^*$ for j in F such that $L(x) = \sum_{j \in F} x_j^*(x(j))$ for each x in $\mathcal{X}$.

2. Show that the space s (Exercise 1.13) is linearly homeomorphic to $C(\mathbb{N})$ and describe s^*.

3. Show that the function g obtained in Proposition 4.2 is unique.

4. Show that $L \in H(\mathbb{D})^*$ if and only if there are scalars $b_0, b_1, \ldots$ in $\mathbb{C}$ such that $\limsup |b_n|^{1/n} < 1$ and $L(f) = \sum_{n=0}^{\infty} 1/(n!) f^{(n)}(0) b_n$.

5. If G is an annulus, describe $H(G)^*$.

6. (Buck [1958]). Let X be locally compact and let β be the strict topology on $C_b(X)$ defined in Exercise 1.21. (Also see Exercises 2.6 and 2.7.) Prove the following statements: (a) If $\mu \in M(X)$ and $\varepsilon_n \downarrow 0$, then there are compact sets $K_1, K_2, \ldots$ such that for each $n \geq 1$, $K_n \subseteq \text{int } K_{n+1}$ and $|\mu|(X \setminus K_n) < \varepsilon_n$. (b) If $\mu \in M(X)$, then there is a ϕ in $C_0(X)$ such that $\phi \geq 0$, $|\mu|(X \setminus \{x: \phi(x) > 0\}) = 0$, $1/\phi \in L^1(|\mu|)$, and $\int 1/\phi \, d|\mu| \leq 1$. (c) Show that if $\mu \in M(X)$ and $L(f) = \int f \, d\mu$ for f in $C_b(X)$, then $L \in (C_b(X), \beta)^*$. (d) Conversely, if $L \in (C_b(X), \beta)^*$, then there is a μ in $M(X)$ such that $L(f) = \int f \, d\mu$ for f in $C_b(X)$.

7. Let X be completely regular and let $\mathcal{M}$ be a linear manifold in $C(X)$. Show that if for every compact subset K of X, $\mathcal{M}|K \equiv \{f|K: f \in \mathcal{M}\}$ is dense in $C(K)$, then $\mathcal{M}$ is dense in $C(X)$.

§5*. Inductive Limits and the Space of Distributions

In this section the most general definition of an inductive limit will not be presented. Rather one that removes certain technicalities from the arguments and yet covers the most important examples will be given. For the more general definition see Köthe [1969], Robertson and Robertson [1966], or Schaefer [1971].

5.1. Definition. An *inductive system* is a pair $(\mathcal{X}, \{\mathcal{X}_i: i \in I\})$, where $\mathcal{X}$ is a vector space, $\mathcal{X}_i$ is a linear manifold in $\mathcal{X}$ that has a topology $\mathcal{T}_i$ such that $(\mathcal{X}_i, \mathcal{T}_i)$ is a LCS, and, moreover:

(a) I is a directed set and $\mathcal{X}_i \subseteq \mathcal{X}_j$ if $i \leq j$;
(b) if $i \leq j$ and $U_j \in \mathcal{T}_j$, then $U_j \cap \mathcal{X}_i \in \mathcal{T}_i$;
(c) $\mathcal{X} = \bigcup\{\mathcal{X}_i: i \in I\}$.

Note that condition (b) is equivalent to the condition that the inclusion map $\mathcal{X}_i \hookrightarrow \mathcal{X}_j$ is continuous.

5.2. Example. Let $d \geq 1$ and let Ω be an open subset of $\mathbb{R}^d$. Denote by $C_c^{(\infty)}(\Omega)$ all the functions $\phi: \Omega \to \mathbb{F}$ such that ϕ is infinitely differentiable and has compact support in Ω. (The *support* of ϕ is defined by spt $\phi \equiv \mathrm{cl}\{x: \phi(x) \neq 0\}$.) If K is a compact subset of Ω, define $\mathcal{D}(K) \equiv \{\phi \in C_c^\infty(\Omega): \mathrm{spt}\, \phi \subseteq K\}$. Let $\mathcal{D}(K)$ have the topology defined by the seminorms

$$p_{K,m}(\phi) = \sup\{|\phi^{(k)}(x)|: |k| \leq m,\, x \in K\},$$

where $k = (k_1, \ldots, k_d)$, $k_j \in \mathbb{N} \cup \{0\}$, $|k| = k_1 + \cdots + k_d$, and

$$\phi^{(k)} = \frac{\partial^{|k|}}{\partial x_1^{k_1} \cdots \partial x_d^{k_d}}.$$

Then $(C_c^\infty(\Omega), \{\mathcal{D}(K): K \text{ is compact in } \Omega\})$ is an inductive system. The space $C_c^\infty(\Omega)$ is often denoted in the literature by $\mathcal{D}(\Omega)$, as it will be in this book.

This example of an inductive system is the most important one as it is connected with the theory of distributions (below). In fact, this example was the inspiration for the definition of an inductive limit given now.

5.3. Proposition. *If $(\mathcal{X}, \{\mathcal{X}_i, \mathcal{T}_i\})$ is an inductive system, let $\mathcal{B} = $ all convex balanced sets V such that $V \cap \mathcal{X}_i \in \mathcal{T}_i$ for all i. Let $\mathcal{T} = $ the collection of all subsets U of $\mathcal{X}$ such that for every x_0 in U there is a V in $\mathcal{B}$ with $x_0 + V \subseteq U$. Then $(\mathcal{X}, \mathcal{T})$ is a (not necessarily Hausdorff) LCS.*

Before proving this proposition, it seems appropriate to make the following definition.

5.4. Definition. If $(\mathcal{X}, \{\mathcal{X}_i\})$ is an inductive system and $\mathcal{T}$ is the topology defined in (5.3), $\mathcal{T}$ is called the *inductive limit topology* and $(\mathcal{X}, \mathcal{T})$ is said to be the inductive limit of $\{\mathcal{X}_i\}$.

5.5. Lemma. *With the notation as in (5.3), $\mathcal{B} \subseteq \mathcal{T}$.*

PROOF. Fix V in $\mathcal{B}$. It will be shown that V is absorbing at each of its points. Indeed, if $x_0 \in V$ and $x \in \mathcal{X}$, then there is an $\mathcal{X}_i$ and an $\mathcal{X}_j$ such that $x_0 \in \mathcal{X}_i$ and $x \in \mathcal{X}_j$. Since I is directed. there is a k in I with $k \geq i, j$. Hence $x_0, x \in \mathcal{X}_k$. But $V \cap \mathcal{X}_k \in \mathcal{T}_k$. Thus there is an $\varepsilon > 0$ such that $x_0 + \alpha x \in V \cap \mathcal{X}_k \subseteq V$ for $|\alpha| < \varepsilon$.

Since V is convex, balanced, and absorbing at each of its points, there is a seminorm p on $\mathcal{X}$ such that $V = \{x \in \mathcal{X}: p(x) < 1\}$ (1.14). So if $x_0 \in V$, $p(x_0) = r_0 < 1$. Let $W = \{x \in \mathcal{X}: p(x) < \frac{1}{2}(1 - r_0)\}$. Then $W = \frac{1}{2}(1 - r_0)V$ and so $W \in \mathcal{B}$. Since $x_0 + W \subseteq V$, $V \in \mathcal{T}$. ∎

PROOF OF PROPOSITION 5.3. The proof that $\mathcal{T}$ is a topology is left as an exercise. To see that $(\mathcal{X}, \mathcal{T})$ is a LCS, note that Lemma 5.5 and Theorem 1.14 imply that $\mathcal{T}$ is defined by a family of seminorms. ∎

For all we know the inductive limit topology may be trivial. However, the fact that this topology has not been shown to be Hausdorff need not concern us, since we will concentrate on a particular type of inductive limit which will be shown to be Hausdorff. But for the moment we will continue at the present level of generality.

5.6. **Proposition.** *Let* $(\mathcal{X}, \{\mathcal{X}_i\})$ *be an inductive system and let* $\mathcal{T}$ *be the inductive limit topology. Then*

(a) *the relative topology on* $\mathcal{X}_i$ *induced by* $\mathcal{T}$ *(viz.,* $\mathcal{T}|\mathcal{X}_i$*) is smaller than* $\mathcal{T}_i$*;*
(b) *if* $\mathcal{U}$ *is a locally convex topology on* $\mathcal{X}$ *such that for every* i*,* $\mathcal{U}|\mathcal{X}_i \subseteq \mathcal{T}_i$*, then* $\mathcal{U} \subseteq \mathcal{T}$*;*
(c) *a seminorm* p *on* $\mathcal{X}$ *is continuous if and only if* $p|\mathcal{X}_i$ *is continuous for each* i*.*

PROOF. Exercise 3.

5.7. **Proposition.** *Let* $(\mathcal{X}, \mathcal{T})$ *be the inductive limit of the spaces* $\{(\mathcal{X}_i, \mathcal{T}_i): i \in I\}$*. If* $\mathcal{Y}$ *is a LCS and* $T: \mathcal{X} \to \mathcal{Y}$ *is a linear transformation, then* T *is continuous if and only if the restriction of* T *to each* $\mathcal{X}_i$ *is* $\mathcal{T}_i$*-continuous.*

PROOF. Suppose $T: \mathcal{X} \to \mathcal{Y}$ is continuous. By (5.6a), the inclusion map $(\mathcal{X}_i, \mathcal{T}_i) \to (\mathcal{X}, \mathcal{T})$ is continuous. Since the restriction of T to $\mathcal{X}_i$ is the composition of the inclusion map $\mathcal{X}_i \to \mathcal{X}$ and T, the restriction is continuous.

Now assume that each restriction is continuous. If p is a continuous seminorm on $\mathcal{Y}$, then $p \circ T|\mathcal{X}_i$ is a $\mathcal{T}_i$-continuous seminorm for every i. By (5.6c), $p \circ T$ is continuous on $\mathcal{X}$. By Exercise 1.23, T is continuous. ∎

It may have occurred to the reader that the definition of the inductive limit topology depends on the choice of the spaces $\mathcal{X}_i$ in more than the obvious way. That is, if $\mathcal{X} = \bigcup_j \mathcal{Y}_j$ and each $\mathcal{Y}_j$ has a topology that is "compatible" with that of the spaces $\{\mathcal{X}_i\}$, perhaps the inductive limit topology defined by the spaces $\{\mathcal{Y}_j\}$ will differ from that defined by the $\{\mathcal{X}_i\}$. This is not the case.

5.8. **Proposition.** *Let* $(\mathcal{X}, \{(\mathcal{X}_i, \mathcal{T}_i)\})$ *and* $(\mathcal{X}, \{(\mathcal{Y}_j, \mathcal{U}_j)\})$ *be two inductive systems and let* $\mathcal{T}$ *and* $\mathcal{U}$ *be the corresponding inductive limit topologies on* $\mathcal{X}$*. If for every* i *there is a* j *such that* $\mathcal{X}_i \subseteq \mathcal{Y}_j$ *and* $\mathcal{U}_j|\mathcal{X}_i \subseteq \mathcal{T}_i$*, then* $\mathcal{U} \subseteq \mathcal{T}$*.*

PROOF. Let V be a convex balanced subset of $\mathcal{X}$ such that for every j, $V \cap \mathcal{Y}_j \in \mathcal{U}_j$. If $\mathcal{X}_i$ is given, let j be such that $\mathcal{X}_i \subseteq \mathcal{Y}_j$ and $\mathcal{U}_j|\mathcal{X}_i \subseteq \mathcal{T}_i$. Hence $V \cap \mathcal{X}_i = (V \cap \mathcal{Y}_j) \cap \mathcal{X}_i \in \mathcal{T}_i$. Thus $V \in \mathcal{B}$ [as defined in (5.3)]. It now follows that $\mathcal{U} \subseteq \mathcal{T}$. ∎

5.9. **Example.** Let $\mathcal{X}$ be any vector space and let $\{\mathcal{X}_i: i \in I\}$ be all of the finite-dimensional subspaces of $\mathcal{X}$. Give each $\mathcal{X}_i$ the unique topology from its identification with a Euclidean space. Then $(\mathscr{X}, \{\mathscr{X}_i\})$ is an inductive system. Let $\mathscr{T}$ be the inductive limit topology. If $\mathscr{Y}$ is a LCS and $T: \mathscr{X} \to \mathscr{Y}$ is a linear transformation, then T is $\mathscr{T}$-continuous.

5.10. Example. Let X be a locally compact space and let $\{K_i: i \in I\}$ be the collection of all compact subsets of X. Let $\mathscr{X}_i =$ all f in $C(X)$ such that spt $f \subseteq K_i$. Then $\cup_i \mathscr{X}_i = C_c(X)$, the continuous functions on X with compact support. Topologize each $\mathscr{X}_i$ by giving it the supremum norm. Then $(C_c(X), \{\mathscr{X}_i\})$ is an inductive system.

Let U_j be the open subsets of X such that $\operatorname{cl} U_j$ is compact. Let $C_0(U_j)$ be the continuous functions on U_j vanishing at ∞ with the supremum norm. If $f \in C_0(U_j)$ and f is defined on X by letting it be identically 0 on $X \setminus U_j$, then $f \in C_c(X)$. Thus $(C_c(X), \{C_0(U_j)\})$ is an inductive system. Proposition 5.8 implies that these two inductive systems define the same inductive limit topology on $C_c(X)$.

5.11. Example. Let $d \geq 1$ and put $K_n = \{x \in \mathbb{R}^d: \|x\| \leq n\}$. Then $(\mathscr{D}(\mathbb{R}^d), \{\mathscr{D}(K_n)\}_{n=1}^\infty)$ is an inductive system. By (5.9), the inductive limit topology defined on $\mathscr{D}(\mathbb{R}^d)$ by this system equals the inductive limit topology defined by the system given in Example 5.2.

If Ω is any open subset of $\mathbb{R}^d$, then Ω can be written as the union of a sequence of compact subsets $\{K_n\}$ such that $K_n \subseteq \operatorname{int} K_{n+1}$. It follows by (5.9) that $\{\mathscr{D}(K_n)\}$ defines the same topology on $\mathscr{D}(\Omega)$ as was defined in Example 5.2.

The preceding example inspires the following definition.

5.12. Definition. A strict inductive system is an inductive system $(\mathscr{X}, \{\mathscr{X}_n, \mathscr{T}_n\}_{n=1}^\infty)$ such that for every $n \geq 1$, $\mathscr{X}_n \subseteq \mathscr{X}_{n+1}$, $\mathscr{T}_{n+1}|\mathscr{X}_n = \mathscr{T}_n$, and $\mathscr{X}_n$ is closed in $\mathscr{X}_{n+1}$. The inductive limit topology defined on $\mathscr{X}$ by such a system is called a *strict inductive limit topology* and $\mathscr{X}$ is said to be the *strict inductive limit* of $\{\mathscr{X}_n\}$.

Example 5.11 shows that $\mathscr{D}(\mathbb{R}^d)$, indeed $\mathscr{D}(\Omega)$, is a strict inductive limit. The following lemma is useful in the study of strict inductive limits as well as in other situations.

5.13. Proposition. *If $\mathscr{X}$ is a LCS, $\mathscr{Y} \leq \mathscr{X}$, and p is a continuous seminorm on $\mathscr{Y}$, then there is a continuous seminorm $\tilde{p}$ on $\mathscr{X}$ such that $\tilde{p}|\mathscr{Y} = p$.*

PROOF. Let $U = \{y \in \mathscr{Y}: p(y) < 1\}$. So U is open in $\mathscr{Y}$; hence there is an open subset V_1 of $\mathscr{X}$ such that $V_1 \cap \mathscr{Y} = U$. Since $0 \in V_1$ and $\mathscr{X}$ is a LCS, there is an open convex balanced set V in $\mathscr{X}$ such that $V \subseteq V_1$. Let $q =$ the gauge of V. So if $y \in \mathscr{Y}$ and $q(y) < 1$, then $p(y) < 1$. By Lemma III.1.4, $p \leq q|\mathscr{Y}$.

Let $W = \text{co}(U \cup V)$; it is easy to see that W is convex and balanced since both U and V are. It will be shown that W is open. First observe that $W = \{ tu + (1 - t)v : 0 \leq t \leq 1, u \in U, v \in V \}$ (verify). Hence $W = \bigcup \{ tU + (1 - t)V : 0 \leq t \leq 1 \}$. Put $W_t = tU + (1 - t)V$. So $W_0 = V$, which is open. If $0 < t < 1$, $W_t = \bigcup \{ tu + (1 - t)V : u \in V \}$, and hence is open. But $W_1 = U$, which is not open. However, if $u \in U$, then there is an $\varepsilon > 0$ such that $\varepsilon u \in V$. For $0 < t < 1$, let $y_t = t^{-1}[1 - \varepsilon + t\varepsilon]u$ $(\in \mathcal{Y})$. As $t \to 1$, $y_t \to u$. Since U is open in $\mathcal{Y}$, there is a t, $0 < t < 1$, with y_t in U. Thus $u = ty_t + (1 - t)(\varepsilon u) \in W_t$. Therefore $W = \bigcup \{ W_t : 0 \leq t < 1 \}$ and W is open.

5.14. Claim. $W \cap \mathcal{Y} = U$.

In fact, $U \subseteq W$, so $U \subset W \cap \mathcal{Y}$. If $w \in W \cap \mathcal{Y}$, then $w = tu + (1 - t)v$, u in U, v in V, $0 \leq t \leq 1$; it may be assumed that $0 < t < 1$. (Why?) Hence, $v = (1 - t)^{-1}(w - tu) \in \mathcal{Y}$. So $v \in V \cap \mathcal{Y} \subseteq U$; hence $w \in U$.

Let $\tilde{p}$ = the gauge of W. By Claim 5.14, $\{ y \in \mathcal{Y} : \tilde{p}(y) < 1 \} = \{ y \in \mathcal{Y} : p(y) < 1 \}$. By the uniqueness of the gauge, $\tilde{p}|\mathcal{Y} = p$. ∎

5.15. Corollary. *If $\mathcal{X}$ is the strict inductive limit of $\{ \mathcal{X}_n \}$, k is fixed, and p_k is a continuous seminorm on $\mathcal{X}_k$, then there is a continuous seminorm p on $\mathcal{X}$ such that $p|\mathcal{X}_k = p_k$. In particular, the inductive limit topology is Hausdorff.*

Proof. By (5.13) and induction, for every integer $n > k$, there is a continuous seminorm p_n such that $p_n|\mathcal{X}_{n-1} = p_{n-1}$. If $x \in \mathcal{X}$, define $p(x) = p_n(x)$ when $x \in \mathcal{X}_n$. Since $\mathcal{X}_n \subseteq \mathcal{X}_{n+1}$ for all n, the properties of $\{ p_n \}$ insure that p is well defined. Clearly p is a seminorm and by (5.6c) p is continuous.

If $x \in \mathcal{X}$ and $x \neq 0$, there is a $k \geq 1$ such that $x \in \mathcal{X}_k$. Thus there is a continuous seminorm p_k on $\mathcal{X}_k$ such that $p_k(x) \neq 0$. Using the first part of the corollary, we get a continuous seminorm p on $\mathcal{X}$ such that $p(x) \neq 0$. Thus $(\mathcal{X}, \mathcal{T})$ is Hausdorff. ∎

5.16. Proposition. *Let $\mathcal{X}$ be the strict inductive limit of $\{ \mathcal{X}_n \}$. A subset B of $\mathcal{X}$ is bounded if and only if there is an $n \geq 1$ such that $B \subseteq \mathcal{X}_n$ and B is bounded in $\mathcal{X}_n$.*

The proof will be accomplished only after a few preliminaries are settled. Before doing this, here are a few consequences of (5.16).

5.17. Corollary. *If $\mathcal{X}$ is the strict inductive limit of $\{ \mathcal{X}_n \}$, then a subset K of $\mathcal{X}$ is compact if and only if there is an $n \geq 1$ such that $K \subseteq \mathcal{X}_n$ and K is compact in $\mathcal{X}_n$.*

5.18. Corollary. *If $\mathcal{X}$ is the strict inductive limit of Frechet spaces $\{ \mathcal{X}_n \}$, $\mathcal{Y}$ is a LCS, and $T: \mathcal{X} \to \mathcal{Y}$ is a linear transformation, then T is continuous if and only if T is sequentially continuous.*

PROOF. By Proposition 5.7, T is continuous if and only if $T|\mathscr{X}_n$ is continuous for every n. Since each $\mathscr{X}_n$ is metrizable, the result follows. ∎

Note that using Example 5.11 it follows that for an open subset Ω of $\mathbb{R}^d$, $\mathscr{D}(\Omega)$ is the strict inductive limit of Frechet spaces [each $\mathscr{D}(K_n)$ is a Frechet space by Proposition 2.1]. So (5.18) applies.

5.19. Definition. If Ω is an open subset of $\mathbb{R}^d$, a *distribution* on Ω is a continuous linear functional on $\mathscr{D}(\Omega)$.

Distributions are, in a certain sense, generalizations of the concept of function as the following example illustrates.

5.20. Example. Let f be a Lebesgue measurable function on Ω that is locally integrable (that is, $\int_K |f| \, d\lambda < \infty$ for every compact subset K of Ω—here λ is d-dimensional Lebesgue measure). If $L_f: \mathscr{D}(\Omega) \to \mathbb{F}$ is defined by $L_f(\phi) = \int f\phi \, d\lambda$, L_f is a distribution.

From Corollary 5.18 we arrive at the following.

5.21. Proposition. *A linear functional* $L: \mathscr{D}(\Omega) \to \mathbb{F}$ *is a distribution if and only if for every sequence* $\{\phi_n\}$ *in* $\mathscr{D}(\Omega)$ *such that* $\mathrm{cl}[\bigcup_{n=1}^{\infty} \mathrm{spt}\, \phi_n] = K$ *is compact in* Ω *and* $\phi_n^{(k)}(x) \to 0$ *uniformly on* K *as* $n \to \infty$ *for every* $k = (k_1, \ldots, k_d)$, *it follows that* $L(\phi_n) \to 0$.

Proposition 5.21 is usually taken as the definition of a distribution in books on differential equations. There is the advantage that (5.21) can be understood with no knowledge of locally convex spaces and inductive limits. Moreover, most theorems on distributions can be proved by using (5.21). However, the realization that a distribution is precisely a continuous linear functional on a LCS contributes more than cultural edification. This knowledge brings power as it enables you to apply the theory of LCS's (including the Hahn-Banach Theorem).

The exercises contain more results on distributions, but now we must return to the proof of Proposition 5.16. To do this the idea of a topological complement is needed. We have seen this idea in Section III.13.

5.22. Proposition. *If* $\mathscr{X}$ *is a TVS and* $\mathscr{Y} \leq \mathscr{X}$, *the following statements are equivalent.*

(a) *There is a closed linear subspace* $\mathscr{Z}$ *of* $\mathscr{X}$ *such that* $\mathscr{Y} \cap \mathscr{Z} = (0)$, $\mathscr{Y} + \mathscr{Z} = \mathscr{X}$, *and the map of* $\mathscr{Y} \times \mathscr{Z} \to \mathscr{X}$ *given by* $(y, z) \mapsto y + z$ *is a homeomorphism.*

(b) *There is a continuous linear map* $P: \mathscr{X} \to \mathscr{X}$ *such that* $P\mathscr{X} = \mathscr{Y}$ *and* $P^2 = P$.

PROOF. (a) $\Rightarrow$ (b): Define $P: \mathcal{X} \to \mathcal{X}$ by $P(y + z) = y$, for y in $\mathcal{Y}$ and z in Z. It is easy to verify that P is linear and $P\mathcal{X} = \mathcal{Y}$. Also, $P^2(y + z) = PP(y + z) = Py = y = P(y + z)$; so $P^2 = P$. If $\{y_i + z_i\}$ is a net in $\mathcal{X}$ such that $y_i + z_i \to y + z$, then (a) implies that $y_i \to y$ (and $z_i \to z$). Hence $P(y_i + z_i) \to P(y + z)$ and P is continuous.

(b) $\Rightarrow$ (a): If P is given, let $\mathcal{Z} = \ker P$. So $\mathcal{Z} \leq \mathcal{X}$. Also, $x = Px + (x - Px)$ and $y = Px \in \mathcal{Y}$, and $z = x - Px$ has $Pz = Px - P^2x = Px - Px = 0$, so $z \in \mathcal{Z}$. Thus, $\mathcal{Y} + \mathcal{Z} = \mathcal{X}$. If $x \in \mathcal{Y} \cap \mathcal{Z}$, then $Px = 0$ since $x \in \mathcal{Z}$; but also $x = Pw$ for some w in $\mathcal{X}$ since $x \in \mathcal{Y} = P\mathcal{X}$. Therefore $0 = Px = P^2w = Pw = x$; that is, $\mathcal{Y} \cap \mathcal{Z} = (0)$. Now suppose that $\{y_i\}$ and $\{z_i\}$ are nets in $\mathcal{Y}$ and $\mathcal{Z}$. If $y_i \to y$ and $z_i \to z$, then $y_i + z_i \to y + z$ because addition is continuous. If, on the other hand, it is assumed that $y_i + z_i \to y + z$, then $y = P(y + z) = \lim P(y_i + z_i) = \lim y_i$ and $z_i = (y_i + z_i) - y_i \to z$. This proves (a). ■

5.23. Definition. If $\mathcal{X}$ is a TVS and $\mathcal{Y} \leq \mathcal{X}$, $\mathcal{Y}$ is *topologically complemented* in $\mathcal{X}$ if either (a) or (b) of (5.22) is satisfied.

5.24. Proposition. *If $\mathcal{X}$ is a LCS and $\mathcal{Y} \leq \mathcal{X}$ such that either* $\dim \mathcal{Y} < \infty$ *or* $\dim \mathcal{X}/\mathcal{Y} < \infty$, *then $\mathcal{Y}$ is topologically complemented in $\mathcal{X}$.*

PROOF. The proof will only be sketched. The reader is asked to supply the details (Exercise 9).

(a) Suppose $d = \dim \mathcal{Y} < \infty$ and let $y_1, \ldots, y_d$ be a basis for $\mathcal{Y}$. By the Hahn–Banach Theorem (III.6.6), there are $f_1, \ldots, f_d$ in $\mathcal{X}^*$ such that $f_i(y_j) = 1$ if $i = j$ and 0 otherwise. Define $Px = \sum_{j=1}^d f_j(x)y_j$.

(b) Suppose $d = \dim \mathcal{X}/\mathcal{Y} < \infty$, $Q: \mathcal{X} \to \mathcal{X}/\mathcal{Y}$ is the natural map, and $z_1, \ldots, z_d \in \mathcal{X}$ such that $Q(z_1), \ldots, Q(z_d)$ is a basis for $\mathcal{X}/\mathcal{Y}$. Let $\mathcal{Z} = \bigvee\{z_1, \ldots, z_d\}$. ■

PROOF OF PROPOSITION 5.16. Suppose $\mathcal{X}$ is the strict inductive limit of $\{(\mathcal{X}_n, \mathcal{T}_n)\}$ and B is a bounded subset of $\mathcal{X}$. It must be shown that there is an n such that $B \subseteq \mathcal{X}_n$ (the rest of the proof is easy). Suppose this is not the case. By replacing $\{\mathcal{X}_n\}$ by a subsequence if necessary, it follows that for each n there is an x_n in $B \setminus \mathcal{X}_n$. Let p_1 be a continuous seminorm on $\mathcal{X}_1$ such that $p_1(x_1) = 1$.

5.25. Claim. *For every $n \geq 2$ there is a continuous seminorm p_n on $\mathcal{X}_n$ such that $p_n(x_n) = n$ and $p_n|\mathcal{X}_{n-1} = p_{n-1}$.*

The proof of (5.25) is by induction. Suppose p_n is given and let $\mathcal{Y} = \mathcal{X}_n \vee \{x_{n+1}\}$. By (5.24), $\mathcal{X}_n$ and $\bigvee\{x_{n+1}\}$ are topologically complementary in $\mathcal{Y}$. Define $q: \mathcal{Y} \to [0, \infty)$ by $q(x + \alpha x_{n+1}) = p_n(x) + (n + 1)|\alpha|$, where $x \in \mathcal{X}_n$ and $\alpha \in \mathbb{F}$. Then q is a continuous seminorm on $(\mathcal{Y}, \mathcal{T}_{n+1}\mathcal{Y})$. (Verify!) By Proposition 5.13 there is a continuous seminorm p_{n+1} on $\mathcal{X}_{n+1}$ such that $p_{n+1}|\mathcal{Y} = q$. Thus $p_{n+1}|\mathcal{X}_n = p_n$ and $p_{n+1}(x_{n+1}) = n + 1$. This proves the claim.

Now define $p: \mathcal{X} \to [0, \infty)$ by $p(x) = p_n(x)$ if $x \in \mathcal{X}_n$. By (5.25), p is well defined. It is easy to see that p is a continuous seminorm. However, $\sup\{ p(x): x \in B \} = \infty$, so B is not bounded (Exercise 2.4f).   ∎

EXERCISES

1. Verify the statements made in Example 5.2.

2. Fill in the details of the proof of Proposition 5.3.

3. Prove Proposition 5.6.

4. Verify the statements made in Example 5.9.

5. Verify the statements made in Example 5.10.

6. Verify the statements made in Example 5.11.

7. With the notation of (5.10), show that if X is σ-compact, then the dual of $C(X)$ is the space of all extended $\mathbb{F}$-valued measures.

8. Is the inductive limit topology on $C(X)$ (5.10) different from the topology of uniform convergence on compact subsets of X (1.5)?

9. Prove Proposition 5.24.

10. Verify the statements made in Example 5.20.

    For the remaining exercises, Ω is always an open subset of $\mathbb{R}^d$, $d \geq 1$.

11. If μ is a measure on Ω, $\phi \mapsto \int \phi \, d\mu$ is a distribution Ω.

12. Let $f: \Omega \to \mathbb{F}$ be a function with continuous partial derivatives and let L_f be defined as in (5.20). Show that for every ϕ in $\mathcal{D}(\Omega)$ and $1 \leq j \leq d$, $L_f(\partial \phi / \partial x_j) = -L_g(\phi)$, where $g = \partial f / \partial x_j$. (Hint: Use integration by parts.)

13. Exercise 12 motivates the following definition. If L is a distribution on Ω, define $\partial L / \partial x_j: \mathcal{D}(\Omega) \to \mathbb{F}$ by $\partial L / \partial x_j(\phi) = -L(\partial \phi / \partial x_j)$ for all ϕ in $\mathcal{D}(\Omega)$. Show that $\partial L / \partial x_j$ is a distribution.

14. Using Example 5.20 and Exercise 13, one is justified to talk of the derivative of any locally integrable function as a *distribution*. By Exercise 11 we can differentiate measures. Let $f: \mathbb{R} \to \mathbb{R}$ be the characteristic function of $[0, \infty)$ and show that its derivative as a distribution is δ_0, the unit point mass at 0. [That is, δ_0 is the measure such that $\delta_0(\Delta) = 1$ if $0 \in \Delta$ and $\delta_0(\Delta) = 0$ if $0 \notin \Delta$.]

15. Let f be an absolutely continuous function on $\mathbb{R}$ and show that $(L_f)' = L_{f'}$.

16. Let f be a left continuous nondecreasing function on $\mathbb{R}$ and show that $(L_f)'$ is the distribution defined by the measure μ such that $\mu[a, b) = f(b) - f(a)$ for all $a < b$.

17. Let f be a C^∞ function on Ω and let L be a distribution on $\mathcal{D}(\Omega)$. Show that $M(\phi) \equiv L(\phi f)$, ϕ in $\mathcal{D}(\Omega)$, is a distribution. State and prove a product rule for finding the derivative of M.

CHAPTER V

Weak Topologies

The principal objects of study in this chapter are the weak topology on a Banach space and the weak-star topology on its dual. In order to carry out this study efficiently, the first two sections are devoted to the study of the weak topology on a locally convex space.

§1. Duality

As in §IV.4, for a LCS $\mathscr{X}$, let $\mathscr{X}^*$ denote the space of continuous linear functionals on $\mathscr{X}$. If $x^*, y^* \in \mathscr{X}^*$ and $\alpha \in \mathbb{F}$, then $(\alpha x^* + y^*)(x) \equiv \alpha x^*(x) + y^*(x)$, x in $\mathscr{X}$, defines an element $\alpha x^* + y^*$ in $\mathscr{X}^*$. Thus $\mathscr{X}^*$ has a natural vector-space structure.

It is convenient and, more importantly, helpful to introduce the notation

$$\langle x, x^* \rangle$$

to stand for $x^*(x)$, for x in $\mathscr{X}$ and x^* in $\mathscr{X}^*$. Also, because of a certain symmetry, we will use $\langle x^*, x \rangle$ to stand for $x^*(x)$. Thus

$$x^*(x) = \langle x, x^* \rangle = \langle x^*, x \rangle.$$

We begin by recalling two definitions (IV.1.7 and IV.1.8).

1.1. Definition. If $\mathscr{X}$ is a LCS, the *weak topology* on $\mathscr{X}$, denoted by "wk" or $\sigma(\mathscr{X}, \mathscr{X}^*)$, is the topology defined by the family of seminorms $\{ p_{x^*} : x^* \in \mathscr{X}^* \}$, where

$$p_{x^*}(x) = |\langle x, x^* \rangle|.$$

The *weak-star topology* on $\mathcal{X}^*$, denoted by "wk*" or $\sigma(\mathcal{X}^*, \mathcal{X})$, is the topology defined by the seminorms $\{p_x : x \in \mathcal{X}\}$, where

$$p_x(x^*) = |\langle x, x^* \rangle|.$$

So a subset U of $\mathcal{X}$ is weakly open if and only if for every x_0 in U there is an $\varepsilon > 0$ and there are $x_1^*, \ldots, x_n^*$ in $\mathcal{X}^*$ such that

$$\bigcap_{k=1}^{n} \{x \in \mathcal{X} : |\langle x - x_0, x_k^* \rangle| < \varepsilon\} \subseteq U.$$

A net $\{x_i\}$ in $\mathcal{X}$ converges weakly to x_0 if and only if $\langle x_i, x^* \rangle \to \langle x_0, x^* \rangle$ for every x^* in $\mathcal{X}^*$. (What are the analogous statements for the weak-star topology?)

Note that both $(\mathcal{X}, \text{wk})$ and $(\mathcal{X}^*, \text{wk}^*)$ are LCS's. Also, $\mathcal{X}$ already possesses a topology so that wk is a second topology on $\mathcal{X}$. However, $\mathcal{X}^*$ has no topology to begin with so that wk* is the only topology on $\mathcal{X}^*$. Of course if $\mathcal{X}$ is a normed space, this last statement is not correct since $\mathcal{X}^*$ is a Banach space (III.5.4). The reader should also be cautioned that some authors make no distinction between the weak and weak-star topologies. Finally, pay attention to the positions of $\mathcal{X}$ and $\mathcal{X}^*$ in the notation $\sigma(\mathcal{X}, \mathcal{X}^*) = \text{wk}$ and $\sigma(\mathcal{X}^*, \mathcal{X}) = \text{wk}^*$.

If $\{x_i\}$ is a net in $\mathcal{X}$ and $x_i \to 0$ in $\mathcal{X}$, then for every x^* in $\mathcal{X}^*$, $\langle x_i, x^* \rangle \to 0$. So if $\mathcal{T}$ is the topology on $\mathcal{X}$, wk $\subseteq \mathcal{T}$ (A.2.9) and each x^* in $\mathcal{X}^*$ is weakly continuous. The first result gives the converse of this.

1.2. Theorem. *If $\mathcal{X}$ is a LCS, $(\mathcal{X}, \text{wk})^* = \mathcal{X}^*$.*

PROOF. Let $f \in (\mathcal{X}, \text{wk})^*$; that is, f is a wk-continuous linear functional on $\mathcal{X}$. By (IV.3.1f) there are $x_1^*, x_2^*, \ldots, x_n^*$ in $\mathcal{X}^*$ such that $|f(x)| \leq \sum_{k=1}^{n} |\langle x, x_k^* \rangle|$ for all x in $\mathcal{X}$. This implies that $\bigcap_{k=1}^{n} \ker x_k^* \subseteq \ker f$. By (A.1.4), there are scalars $\alpha_1, \ldots, \alpha_n$ such that $f = \sum_{k=1}^{n} \alpha_k x_k^*$; hence $f \in \mathcal{X}^*$. ∎

There is a similar result for wk*; the proof is left for the reader.

1.3. Theorem. *If $\mathcal{X}$ is a LCS, $(\mathcal{X}^*, \text{wk}^*)^* = \mathcal{X}$.*

So $\mathcal{X}$ is the dual of a LCS—$(\mathcal{X}, \text{wk}^*)$—and hence has a weak-star topology—$\sigma((\mathcal{X}, \text{wk}^*)^*, \mathcal{X}^*)$. As an exercise in notational juggling, note that $\sigma((\mathcal{X}, \text{wk}^*)^*, \mathcal{X}^*) = \sigma(\mathcal{X}, \mathcal{X}^*)$.

All unmodified topological statements about $\mathcal{X}$ refer to its original topology. So if $A \subseteq \mathcal{X}$ and we say that it is closed, we mean that A is closed in the original topology of $\mathcal{X}$. To say that A is closed in the weak topology of $\mathcal{X}$ we say that A is weakly closed or wk-closed. Also cl A means the closure of A in the original topology while wk $-$ cl A means the closure of A in the weak topology. The next result shows that under certain circumstances this distinction is unnecessary.

1.4. Theorem. *If $\mathscr{X}$ is a LCS and A is a convex subset of $\mathscr{X}$, then* $\operatorname{cl} A = \operatorname{wk} - \operatorname{cl} A$.

PROOF. If $\mathscr{T}$ is the original topology of $\mathscr{X}$, then $\operatorname{wk} \subseteq \mathscr{T}$, hence $\operatorname{cl} A \subseteq \operatorname{wk} - \operatorname{cl} A$. Conversely, if $x \in \mathscr{X} \setminus \operatorname{cl} A$, then (IV.3.13) implies that there is an x^* in $\mathscr{X}^*$, an α in $\mathbb{R}$, and an $\varepsilon > 0$ such that

$$\operatorname{Re}\langle a, x^* \rangle \leq \alpha < \alpha + \varepsilon \leq \operatorname{Re}\langle x, x^* \rangle$$

for all a in $\operatorname{cl} A$. Hence $\operatorname{cl} A \subseteq B \equiv \{ y \in \mathscr{X} : \operatorname{Re}\langle y, x^* \rangle \leq \alpha \}$. But B is clearly wk-closed since x^* is wk-continuous. Thus $\operatorname{wk} - \operatorname{cl} A \subseteq B$. Since $x \notin B$, $x \notin \operatorname{wk} - \operatorname{cl} A$. ∎

1.5. Corollary. *A convex subset of $\mathscr{X}$ is closed if and only if it is weakly closed.*

There is a useful observation that can be made here. Because of (III.6.3) it can be shown that if $\mathscr{X}$ is a complex linear space, then the weak topology on $\mathscr{X}$ is the same as the weak topology it has if it is considered as a real linear space (Exercise 4). This will be used in the future.

1.6. Definition. If $A \subseteq \mathscr{X}$, the *polar* of A, denoted by A°, is the subset of $\mathscr{X}^*$ defined by

$$A^\circ \equiv \{ x^* \in \mathscr{X}^* : |\langle a, x^* \rangle| \leq 1 \text{ for all } a \text{ in } A \}.$$

If $B \subseteq \mathscr{X}^*$, the *prepolar* of B, denoted by $^\circ B$, is the subset of $\mathscr{X}$ defined by

$$^\circ B \equiv \{ x \in \mathscr{X} : |\langle x, b^* \rangle| \leq 1 \text{ for all } b^* \text{ in } B \}.$$

If $A \subseteq \mathscr{X}$ the *bipolar* of A is the set $^\circ(A^\circ)$. If there is no confusion, then it is also denoted by $^\circ A^\circ$.

The prototype for this idea is that if A is the unit ball in a normed space, A° is the unit ball in the dual space.

1.7. Proposition. *If $A \subseteq \mathscr{X}$, then*

(a) *A° is convex and balanced.*
(b) *If $A_1 \subseteq A$, then $A^\circ \subseteq A_1^\circ$.*
(c) *If $\alpha \in \mathbb{F}$ and $\alpha \neq 0$, $(\alpha A)^\circ = \alpha^{-1} A^\circ$.*
(d) *$A \subseteq {}^\circ A^\circ$.*
(e) *$A^\circ = ({}^\circ A^\circ)^\circ$.*

PROOF. The proofs of parts (a) through (d) are left as an exercise. To prove (e) note that $A \subseteq {}^\circ A^\circ$ by (d), so $({}^\circ A^\circ)^\circ \subseteq A^\circ$ by (b). But $A^\circ \subseteq {}^\circ (A^\circ)^\circ$ by an analog of (d) for prepolars. Also, $^\circ (A^\circ)^\circ = ({}^\circ A^\circ)^\circ$. ∎

There is an analogous result for prepolars. In fact, it is more than analogy that is at work here. By Theorem 1.3, $(\mathcal{X}^*, \mathrm{wk}^*)^* = \mathcal{X}$. Thus the result for prepolars is a consequence of the preceding proposition.

If A is a linear manifold in $\mathcal{X}$ and $x^* \in A^\circ$, then $ta \in A$ for all $t > 0$ and a in A. So $1 \geq |\langle ta, x^* \rangle| = t|\langle a, x^* \rangle|$. Letting $t \to \infty$ shows that $A^\circ = A^\perp$, where

$$A^\perp \equiv \{ x^* \text{ in } \mathcal{X}^* : \langle a, x^* \rangle = 0 \text{ for all } a \text{ in } A \}.$$

Similarly, if B is a linear manifold in $\mathcal{X}^*$, $^\circ B = {}^\perp B$, where

$$^\perp B \equiv \{ x \text{ in } \mathcal{X} : \langle x, b^* \rangle = 0 \text{ for all } b^* \text{ in } B \}.$$

The next result is a slight generalization of Corollary IV.3.12.

1.8.   Theorem. *If $\mathcal{X}$ is a LCS and $A \subseteq \mathcal{X}$, then $^\circ A^\circ$ is the closed convex balanced hull of A.*

PROOF. Let A_1 be the intersection of all closed convex balanced subsets of $\mathcal{X}$ that contain A. It must be shown that $A_1 = {}^\circ A^\circ$. Since $^\circ A^\circ$ is closed, convex, and balanced and $A \subseteq {}^\circ A^\circ$, it follows that $A_1 \subseteq {}^\circ A^\circ$.

Now assume that $x_0 \in \mathcal{X} \setminus A_1$. A_1 is a closed convex balanced set so by (IV.3.13) there is an x^* in $\mathcal{X}^*$, an α in $\mathbb{R}$, and an $\varepsilon > 0$ such that

$$\mathrm{Re}\langle a_1, x^* \rangle < \alpha < \alpha + \varepsilon < \mathrm{Re}\langle x_0, x^* \rangle$$

for all a_1 in A_1. Since $0 \in A_1$, $0 = \langle 0, x^* \rangle < \alpha$. By replacing x^* with $\alpha^{-1} x^*$ it follows that there is an $\varepsilon > 0$ (not the same as the first ε) such that

$$\mathrm{Re}\langle a_1, x^* \rangle < 1 < 1 + \varepsilon < \mathrm{Re}\langle x_0, x^* \rangle$$

for all a_1 in A_1. If $a_1 \in A_1$ and $\langle a_1, x^* \rangle = |\langle a_1, x^* \rangle| e^{i\theta}$, then $e^{-i\theta} a_1 \in A_1$ and so

$$|\langle a_1, x^* \rangle| = \mathrm{Re}\langle e^{-i\theta} a_1, x^* \rangle < 1 < \mathrm{Re}\langle x_0, x^* \rangle$$

for all a_1 in A_1. Hence $x^* \in A_1^\circ$, and $x_0 \notin {}^\circ A^\circ$. That is, $\mathcal{X} \setminus A_1 \subseteq \mathcal{X} \setminus {}^\circ A^\circ$. ∎

1.9.   Corollary. *If $\mathcal{X}$ is a LCS and $B \subseteq \mathcal{X}^*$, then $(^\circ B)^\circ$ is the wk^* closed convex balanced hull of B.*

Using the weak and weak* topologies and the concept of a bounded subset of a LCS (IV.2.5), it is possible to rephrase the results associated with the Principle of Uniform Boundedness (§III.14). As an example we offer the following reformulation of Corollary III.14.5 (which is, in fact, the most general form of the result).

1.10.   Theorem. *If $\mathcal{X}$ is a Banach space, $\mathcal{Y}$ is a normed space, and $\mathcal{A} \subseteq \mathcal{B}(\mathcal{X}, \mathcal{Y})$ such that for every x in $\mathcal{X}$, $\{ Ax: A \in \mathcal{A} \}$ is weakly bounded in $\mathcal{Y}$, then $\mathcal{A}$ is norm bounded in $\mathcal{B}(\mathcal{X}, \mathcal{Y})$.*

EXERCISES

1. Show that wk is the smallest topology on $\mathscr{X}$ such that each x^* in $\mathscr{X}^*$ is continuous.

2. Show that wk* is the smallest topology on $\mathscr{X}^*$ such that for each x in $\mathscr{X}$, $x^* \mapsto \langle x, x^* \rangle$ is continuous.

3. Prove Theorem 1.3.

4. Let $\mathscr{X}$ be a complex LCS and let $\mathscr{X}_{\mathbb{R}}^*$ denote the collection of all continuous real linear functionals on $\mathscr{X}$. Use the elements of $\mathscr{X}_{\mathbb{R}}^*$ to define seminorms on $\mathscr{X}$ and let $\sigma(\mathscr{X}, \mathscr{X}_{\mathbb{R}}^*)$ be the corresponding topology. Show that $\sigma(\mathscr{X}, \mathscr{X}^*) = \sigma(\mathscr{X}, \mathscr{X}_{\mathbb{R}}^*)$.

5. Prove the remainder of Proposition 1.7.

6. If $A \subseteq \mathscr{X}$, show that A is weakly bounded if and only if A° is absorbing in $\mathscr{X}^*$.

7. Let $\mathscr{X}$ be a normed space and let $\{x_n\}$ be a sequence in $\mathscr{X}$ such that $x_n \to x$ weakly. Show that there is a sequence $\{y_n\}$ such that $y_n \in \mathrm{co}\{x_1, x_2, \ldots, x_n\}$ and $\|y_n - x\| \to 0$. (Hint: use Theorem 1.4.)

8. If $\mathscr{H}$ is a Hilbert space and $\{h_n\}$ is a sequence in $\mathscr{H}$ such that $h_n \to h$ weakly and $\|h_n\| \to \|h\|$, then $\|h_n - h\| \to 0$. (The same type of result is true for L^p-spaces if $1 < p < \infty$.)

9. If $\mathscr{X}$ is a normed space show that the norm on $\mathscr{X}$ is lower semicontinuous for the weak topology and the norm on $\mathscr{X}^*$ is lower semicontinuous for the weak-star topology.

10. Suppose $\mathscr{X}$ is an infinite-dimensional normed space. If $S = \{x \in \mathscr{X}: \|x\| = 1\}$, then the weak closure of S is $\{x: \|x\| \leq 1\}$.

§2. The Dual of a Subspace and a Quotient Space

In §III.4 the quotient of a normed space $\mathscr{X}$ by a closed subspace $\mathscr{M}$ was defined and in (III.10.2) it was shown that the dual of a quotient space $\mathscr{X}/\mathscr{M}$ is isometrically isomorphic to $\mathscr{M}^\perp$. These results are generalized in this section to the setting of a LCS and, moreover, it is shown that when $(\mathscr{X}/\mathscr{M})^*$ and $\mathscr{M}^\perp$ are identified, the weak-star topology on $(\mathscr{X}/\mathscr{M})^*$ is precisely the relative weak-star topology that $\mathscr{M}^\perp$ receives as a subspace of $\mathscr{X}^*$.

The first result was presented in abbreviated form as Exercise IV.1.16.

2.1. Proposition. *Let $\mathscr{X}$ be a LCS and let $\mathscr{P}$ be a family of seminorms defining the topology of $\mathscr{X}$. If $\mathscr{M} \leq \mathscr{X}$ and $p \in \mathscr{P}$, define $\bar{p}$: $\mathscr{X}/\mathscr{M} \to [0, \infty)$ by*

$$\bar{p}(x + \mathscr{M}) = \inf\{p(x + y): y \in \mathscr{M}\}.$$

Then $\bar{p}$ is a seminorm on $\mathscr{X}/\mathscr{M}$, and the topology on $\mathscr{X}/\mathscr{M}$ defined by $\bar{\mathscr{P}} \equiv \{ \bar{p} \colon p \in \mathscr{P} \}$ is the quotient space topology.

PROOF. Exercise.

Thus if $\mathscr{X}$ is a LCS and $\mathscr{M} \leq \mathscr{X}$, then $\mathscr{X}/\mathscr{M}$ is a LCS. Let $f \in (\mathscr{X}/\mathscr{M})^*$. If $Q \colon \mathscr{X} \to \mathscr{X}/\mathscr{M}$ is the natural map, then $f \circ Q \in \mathscr{X}^*$. Moreover, $f \circ Q \in \mathscr{M}^\perp$. Hence $f \mapsto f \circ Q$ is a map of $(\mathscr{X}/\mathscr{M})^* \to \mathscr{M}^\perp \subseteq \mathscr{X}^*$.

2.2.   Theorem. *If $\mathscr{X}$ is a LCS, $\mathscr{M} \leq \mathscr{X}$, and $Q \colon \mathscr{X} \to \mathscr{X}/\mathscr{M}$ is the natural map, then $f \mapsto f \circ Q$ defines a linear bijection between $(\mathscr{X}/\mathscr{M})^*$ and $\mathscr{M}^\perp$. If $(\mathscr{X}/\mathscr{M})^*$ has its weak-star topology $\sigma((\mathscr{X}/\mathscr{M})^*, \mathscr{X}/\mathscr{M})$ and $\mathscr{M}^\perp$ has the relative weak-star topology $\sigma(\mathscr{X}^*, \mathscr{X})|\mathscr{M}^\perp$, then this bijection is a homeomorphism. If $\mathscr{X}$ is a normed space, then this bijection is an isometry.*

PROOF. Let $\rho \colon (\mathscr{X}/\mathscr{M})^* \to \mathscr{M}^\perp$ be defined by $\rho(f) = f \circ Q$. It was shown prior to the statement of the theorem that ρ is well defined and maps $(\mathscr{X}/\mathscr{M})^*$ into $\mathscr{M}^\perp$. It is easy to see that ρ is linear and if $0 = \rho(f) = f \circ Q$, then $f = 0$ since Q is surjective. So ρ is injective. Now let $x^* \in \mathscr{M}^\perp$ and define $f \colon \mathscr{X}/\mathscr{M} \to \mathbb{F}$ by $f(x + \mathscr{M}) = \langle x, x^* \rangle$. Because $\mathscr{M} \subseteq \ker x^*$, f is well defined and linear. Also, $Q^{-1}\{x + \mathscr{M} \colon |f(x + \mathscr{M})| < 1\} = \{x \in \mathscr{X} \colon |\langle x, x^* \rangle| < 1\}$ and this is open in $\mathscr{X}$ since x^* is continuous. Thus $\{x + \mathscr{M} \colon |f(x + \mathscr{M})| < 1\}$ is open in $\mathscr{X}/\mathscr{M}$ and so f is continuous. Clearly $\rho(f) = x^*$, so ρ is a bijection.

If $\mathscr{X}$ is a normed space, it was shown in (III.10.2) that ρ is an isometry. It remains to show that ρ is a weak-star homeomorphism. Let $\text{wk}^* = \sigma(\mathscr{X}^*, \mathscr{X})$ and let $\sigma^* = \sigma((\mathscr{X}/\mathscr{M})^*, \mathscr{X}/\mathscr{M})$. If $\{f_i\}$ is a net in $(\mathscr{X}/\mathscr{M})^*$ and $f_i \to 0(\sigma^*)$, then for each x in $\mathscr{X}$, $\langle x, \rho(f_i) \rangle = f_i(Q(x)) \to 0$. Hence $\rho(f_i) \to 0$ (wk*). Conversely, if $\rho(f_i) \to 0(\text{wk}^*)$, then for each x in $\mathscr{X}$, $f_i(x + \mathscr{M}) = \langle x, \rho(f_i) \rangle \to 0$; hence $f_i \to 0(\sigma^*)$. ∎

Once again let $\mathscr{M} \leq \mathscr{X}$. If $x^* \in \mathscr{X}^*$, then the restriction of x^* to $\mathscr{M}$, $x^*|\mathscr{M}$, belongs to $\mathscr{M}^*$. Also, the Hahn–Banach Theorem implies that the map $x^* \mapsto x^*|\mathscr{M}$ is surjective. If $\rho(x^*) = x^*|\mathscr{M}$, then $\rho \colon \mathscr{X}^* \to \mathscr{M}^*$ is clearly linear as well as surjective. It fails, however, to be injective. How does it fail? It's easy to see that $\ker \rho = \mathscr{M}^\perp$. Thus ρ induces a linear bijection $\tilde{\rho} \colon \mathscr{X}^*/\mathscr{M}^\perp \to \mathscr{M}^*$.

2.3.   Theorem. *If $\mathscr{X}$ is a LCS, $\mathscr{M} \leq \mathscr{X}$, and $\rho \colon \mathscr{X}^* \to \mathscr{M}^*$ is the restriction map, then ρ induces a linear bijection $\tilde{\rho} \colon \mathscr{X}^*/\mathscr{M}^\perp \to \mathscr{M}^*$. If $\mathscr{X}^*/\mathscr{M}^\perp$ has the quotient topology induced by $\sigma(\mathscr{X}^*, \mathscr{X})$ and $\mathscr{M}^*$ has its weak-star topology $\sigma(\mathscr{M}^*, \mathscr{M})$, then $\tilde{\rho}$ is a homeomorphism. If $\mathscr{X}$ is a normed space, then $\tilde{\rho}$ is an isometry.*

PROOF. The fact that $\tilde{\rho}$ is an isometry when $\mathscr{X}$ is a normed space was shown in (III.10.1). Let $\text{wk}^* = \sigma(\mathscr{M}^*, \mathscr{M})$ and let η^* be the quotient topology on $\mathcal{X}^*/\mathcal{M}^\perp$ defined by $\sigma(\mathcal{X}^*, \mathcal{X})$. Let $Q: \mathcal{X}^* \to \mathcal{X}^*/\mathcal{M}$ be the natural map. Therefore the diagram

$$\mathcal{X}^* \xrightarrow{\quad \rho \quad} \mathcal{M}^*$$

$$Q \searrow \qquad \nearrow \bar{\rho}$$

$$\mathcal{X}^*/\mathcal{M}^\perp$$

commutes. If $y \in \mathcal{M}$, then the commutativity of the diagram implies that

$$Q^{-1}\big(\bar{\rho}^{-1}\{y^* \in \mathcal{M}^*: |\langle y, y^*\rangle| < 1\}\big) = Q^{-1}\{x^* + \mathcal{M}: |\langle y, x^*\rangle| < 1\}$$

$$= \{x^* \in \mathcal{X}^*: |\langle y, x^*\rangle| < 1\},$$

which is weak-star open in $\mathcal{X}^*$. Hence $\bar{\rho}: (\mathcal{X}^*/\mathcal{M}^\perp, \eta^*) \to (\mathcal{M}^*, \mathrm{wk}^*)$ is continuous.

How is the topology on $\mathcal{X}^*/\mathcal{M}^\perp$ defined? If $x \in \mathcal{X}$, $p_x(x^*) = |\langle x, x^*\rangle|$ is a typical seminorm on $\mathcal{X}^*$. By Proposition 2.1, the topology on $\mathcal{X}^*/\mathcal{M}^\perp$ is defined by the seminorms $\{\bar{p}_x: x \in \mathcal{X}\}$, where

$$\bar{p}_x(x^* + \mathcal{M}) = \inf\{|\langle x, x^* + z^*\rangle|: z^* \in \mathcal{M}\}.$$

2.4. Claim. If $x \notin \mathcal{M}$, then $\bar{p}_x = 0$.

In fact, let $\mathcal{L} = \{\alpha x: \alpha \in \mathbb{F}\}$. If $x \notin \mathcal{M}$, then $\mathcal{L} \cap \mathcal{M} = (0)$. Since $\dim \mathcal{L} < \infty$, $\mathcal{M}$ is topologically complemented in $\mathcal{L} + \mathcal{M}$. Let $x^* \in \mathcal{X}^*$ and define $f: \mathcal{L} + \mathcal{M} \to \mathbb{F}$ by $f(\alpha x + y) = \langle y, x^*\rangle$ for y in $\mathcal{M}$ and α in $\mathbb{F}$. Because $\mathcal{M}$ is topologically complemented in $\mathcal{L} + \mathcal{M}$, if $\alpha_i x + y_i \to 0$, then $y_i \to 0$. Hence $f(\alpha_i x + y_i) = \langle y_i, x^*\rangle \to 0$. Thus f is continuous. By the Hahn–Banach Theorem, there is an x_1^* in $\mathcal{X}^*$ that extends f. Note that $x^* - x_1^* \in \mathcal{M}^\perp$. Thus $\bar{p}_x(x^* + \mathcal{M}^\perp) = \bar{p}_x(x_1^* + \mathcal{M}^\perp) \leq p_x(x_1^*) = |\langle x, x_1^*\rangle| = 0$. This proves (2.4).

Now suppose that $\{x_i^* + \mathcal{M}^\perp\}$ is a net in $\mathcal{X}^*/\mathcal{M}^\perp$ such that $\bar{\rho}(x_i^* + \mathcal{M}^\perp) = x_i^*|\mathcal{M} \to 0$ (wk*) in $\mathcal{M}^*$. If $x \in \mathcal{X}$ and $x \notin \mathcal{M}$, then Claim (2.4) implies that $\bar{p}_x(x_i^* + \mathcal{M}^\perp) = 0$. If $x \in \mathcal{M}$, then $\bar{p}_x(x_i^* + \mathcal{M}^\perp) \leq |\langle x, x_i^*\rangle| \to 0$. Thus $x_i^* + \mathcal{M}^\perp \to 0$ (η^*) and $\bar{\rho}$ is a weak-star homeomorphism. ∎

EXERCISES

1. In relation to Claim 2.4, show that if $\mathcal{L} \leq \mathcal{X}$, $\dim \mathcal{L} < \infty$, and $\mathcal{M} \leq \mathcal{X}$, then $\mathcal{L} + \mathcal{M}$ is closed.

2. Show that if $\mathcal{M} \leq \mathcal{X}$ and $\mathcal{M}$ is topologically complemented in $\mathcal{X}$, then $\mathcal{M}^\perp$ is topologically complemented in $\mathcal{X}^*$ and that its complement is weak-star and linearly homeomorphic to $\mathcal{X}^*/\mathcal{M}^\perp$.

§3. Alaoglu's Theorem

If $\mathscr{X}$ is any normed space, let's agree to denote by ball $\mathscr{X}$ the closed unit ball in $\mathscr{X}$. So ball $\mathscr{X} \equiv \{x \in \mathscr{X}: \|x\| \le 1\}$.

3.1. Alaoglu's Theorem. *If $\mathscr{X}$ is a normed space, then* ball $\mathscr{X}^*$ *is weak-star compact.*

PROOF. For each x in ball $\mathscr{X}$, let $D_x \equiv \{\alpha \in \mathbb{F}: |\alpha| \le 1\}$ and put $D = \prod\{D_x: x \in \text{ball } \mathscr{X}\}$. By Tychonoff's Theorem, D is compact. Define $\tau:$ ball $\mathscr{X}^* \to D$ by

$$\tau(x^*)(x) = \langle x, x^* \rangle.$$

That is, $\tau(x^*)$ is the element of the product space D whose x coordinate is $\langle x, x^* \rangle$. It will be shown that τ is a homeomorphism from (ball $\mathscr{X}^*$, wk*) onto $\tau(\text{ball } \mathscr{X}^*)$ with the relative topology from D, and that $\tau(\text{ball } \mathscr{X}^*)$ is closed in D. Thus it will follow that $\tau(\text{ball } \mathscr{X}^*)$, and hence ball $\mathscr{X}^*$, is compact.

To see that τ is injective, suppose that $\tau(x_1^*) = \tau(x_2^*)$. Then for each x in ball $\mathscr{X}$, $\langle x, x_1^* \rangle = \langle x, x_2^* \rangle$. It follows by definition that $x_1^* = x_2^*$.

Now let $\{x_i^*\}$ be a net in ball $\mathscr{X}^*$ such that $x_i^* \to x^*$. Then for each x in ball $\mathscr{X}$, $\tau(x_i^*)(x) = \langle x, x_i^* \rangle \to \langle x, x^* \rangle = \tau(x^*)(x)$. That is, each coordinate of $\{\tau(x_i^*)\}$ converges to $\tau(x^*)$. Hence $\tau(x_i^*) \to \tau(x^*)$ and τ is continuous.

Let x_i^* be a net in ball $\mathscr{X}^*$, let $f \in D$, and suppose $\tau(x_i^*) \to f$ in D. So $f(x) = \lim\langle x, x_1^* \rangle$ exists for every x in ball $\mathscr{X}$. If $x \in \mathscr{X}$, let $\alpha > 0$ such that $\|\alpha x\| \le 1$. Then define $f(x) = \alpha^{-1}f(\alpha x)$. If also $\beta > 0$ such that $\|\beta x\| \le 1$, then $\alpha^{-1}f(\alpha x) = \alpha^{-1}\lim\langle \alpha x, x_i^* \rangle = \beta^{-1}\lim\langle \beta x, x_i^* \rangle = \beta^{-1}f(\beta x)$. So $f(x)$ is well defined. It is left as an exercise for the reader to show that $f: \mathscr{X} \to \mathbb{F}$ is a linear functional. Also, if $\|x\| \le 1$, $f(x) \in D_x$ so $|f(x)| \le 1$. Thus $f = x^* \in \text{ball } \mathscr{X}^*$ and $\tau(x^*) = f$. Thus $\tau(\text{ball } \mathscr{X}^*)$ is closed in D. This implies that $\tau(\text{ball } \mathscr{X}^*)$ is compact and, hence, τ is a homeomorphism (A.2.8). ∎

EXERCISES

1. Show that the functional f occurring in the proof of Alaoglu's Theorem is linear.

2. Let $\mathscr{X}$ be a LCS and let V be an open neighborhood of 0. Show that V° is weak-star compact in $\mathscr{X}^*$.

3. If $\mathscr{X}$ is a Banach space, show that there is a compact space X such that $\mathscr{X}$ is isometrically isomorphic to a closed subspace of $C(X)$.

§4. Reflexivity Revisited

In §III.11 a Banach space $\mathscr{X}$ was defined to be reflexive if the natural embedding of $\mathscr{X}$ into its double dual, $\mathscr{X}^{**}$, is surjective. Recall that if $x \in \mathscr{X}$, then the image of x in $\mathscr{X}^{**}, \hat{x},$ is defined by (using our recent notation)

$$\langle x^*, \hat{x} \rangle = \langle x, x^* \rangle$$

for all x^* in $\mathscr{X}^*$. Also recall that the map $x \mapsto \hat{x}$ is an isometry.

To begin, note that $\mathscr{X}^{**}$, being the dual space of $\mathscr{X}^*$, has its weak-star topology $\sigma(\mathscr{X}^{**}, \mathscr{X}^*)$. Also note that if $\mathscr{X}$ is considered as a subspace of $\mathscr{X}^{**}$, then the topology $\sigma(\mathscr{X}^{**}, \mathscr{X}^*)$ when relativized to $\mathscr{X}$ is $\sigma(\mathscr{X}, \mathscr{X}^*)$, the weak topology on $\mathscr{X}$. This will be important later when it is combined with Alaoglu's Theorem applied to $\mathscr{X}^{**}$ in the discussion of reflexivity. But now the next result must occupy us.

4.1. Proposition. *If $\mathscr{X}$ is a normed space, then* ball $\mathscr{X}$ *is $\sigma(\mathscr{X}^{**}, \mathscr{X}^*)$ dense in* ball $\mathscr{X}^{**}$.

PROOF. Let $B =$ the $\sigma(\mathscr{X}^{**}, \mathscr{X}^*)$ closure of ball $\mathscr{X}$ in $\mathscr{X}^{**}$; clearly, $B \subseteq$ ball $\mathscr{X}^{**}$. If there is an x_0^{**} in ball $\mathscr{X}^{**} \setminus B$, then the Hahn–Banach Theorem implies there is an x^* in $\mathscr{X}^*$, an α in $\mathbb{R}$, and an $\varepsilon > 0$ such that

$$\mathrm{Re}\langle x, x^* \rangle < \alpha < \alpha + \varepsilon < \mathrm{Re}\langle x^*, x_0^{**} \rangle$$

for all x in ball $\mathscr{X}$. (Exactly how does the Hahn–Banach Theorem imply this?) Since $0 \in$ ball $\mathscr{X}$, $0 < \alpha$. Dividing by α and replacing x^* by $\alpha^{-1}x^*$, it may be assumed that there is an x^* in $\mathscr{X}^*$ and an $\varepsilon > 0$ such that

$$\mathrm{Re}\langle x, x^* \rangle < 1 < 1 + \varepsilon < \mathrm{Re}\langle x^*, x_0^{**} \rangle$$

for all x in ball $\mathscr{X}$. Since $e^{i\theta}x \in$ ball $\mathscr{X}$ whenever $x \in$ ball $\mathscr{X}$, this implies that $|\langle x, x^* \rangle| \leq 1$ if $\|x\| \leq 1$. Hence $x^* \in$ ball $\mathscr{X}^*$. But then $1 + \varepsilon < \mathrm{Re}\langle x^*, x_0^{**} \rangle \leq |\langle x^*, x_0^{**} \rangle| \leq \|x_0^{**}\| \leq 1$, a contradiction. ∎

4.2. Theorem. *If $\mathscr{X}$ is a Banach space, the following statements are equivalent.*

(a) $\mathscr{X}$ *is reflexive.*
(b) $\mathscr{X}^*$ *is reflexive.*
(c) $\sigma(\mathscr{X}^*, \mathscr{X}) = \sigma(\mathscr{X}^*, \mathscr{X}^{**})$.
(d) ball $\mathscr{X}$ *is weakly compact.*

PROOF. (a) $\Rightarrow$ (c): This is clear since $\mathscr{X} = \mathscr{X}^{**}$.

(d) $\Rightarrow$ (a): Note that $\sigma(\mathscr{X}^{**}, \mathscr{X}^*)|\mathscr{X} = \sigma(\mathscr{X}, \mathscr{X}^*)$. By (d), ball $\mathscr{X}$ is $\sigma(\mathscr{X}^{**}, \mathscr{X}^*)$ closed in ball $\mathscr{X}^{**}$. But the preceding proposition implies ball $\mathscr{X}$ is $\sigma(\mathscr{X}^{**}, \mathscr{X}^*)$ dense in ball $\mathscr{X}^{**}$. Hence ball $\mathscr{X} =$ ball $\mathscr{X}^{**}$ and so $\mathscr{X}$ is reflexive.

(c) $\Rightarrow$ (b): By Alaoglu's Theorem, ball $\mathscr{X}^*$ is $\sigma(\mathscr{X}^*, \mathscr{X})$-compact. By (c), ball $\mathscr{X}^*$ is $\sigma(\mathscr{X}^*, \mathscr{X}^{**})$ compact. Since it has already been shown that (d) implies (a), this implies that $\mathscr{X}^*$ is reflexive.

(b) $\Rightarrow$ (a): Now ball $\mathscr{X}$ is norm closed in $\mathscr{X}^{**}$; hence ball $\mathscr{X}$ is $\sigma(\mathscr{X}^{**}, \mathscr{X}^{***})$ closed in $\mathscr{X}^{**}$ (Corollary 1.5). Since $\mathscr{X}^* = \mathscr{X}^{***}$ by (b), this says that ball $\mathscr{X}$ is $\sigma(\mathscr{X}^{**}, \mathscr{X}^*)$ closed in $\mathscr{X}^{**}$. But, according to (4.1), ball $\mathscr{X}$ is $\sigma(\mathscr{X}^{**}, \mathscr{X}^*)$ dense in ball $\mathscr{X}^{**}$. Hence ball $\mathscr{X} =$ ball $\mathscr{X}^{**}$ and $\mathscr{X}$ is reflexive.

(a) $\Rightarrow$ (d): By Alaoglu's Theorem, ball $\mathscr{X}^{**}$ is $\sigma(\mathscr{X}^{**}, \mathscr{X}^*)$ compact. Since $\mathscr{X} = \mathscr{X}^{**}$, this says that ball $\mathscr{X}$ is $\sigma(\mathscr{X}, \mathscr{X}^*)$ compact. ∎

4.3. Corollary. *If $\mathscr{X}$ is a reflexive Banach space and $\mathscr{M} \leq \mathscr{X}$, then $\mathscr{M}$ is reflexive.*

PROOF. Note that ball $\mathscr{M} = \mathscr{M} \cap$ [ball $\mathscr{X}$], so ball $\mathscr{M}$ is $\sigma(\mathscr{X}, \mathscr{X}^{**})$ compact. It remains to show that $\sigma(\mathscr{X}, \mathscr{X}^*)|\mathscr{M} = \sigma(\mathscr{M}, \mathscr{M}^*)$. But this follows by (2.3). (How?) ∎

Call a sequence $\{x_n\}$ in $\mathscr{X}$ a *weakly Cauchy sequence* if for every x^* in $\mathscr{X}^*$, $\{\langle x_n, x^* \rangle\}$ is a Cauchy sequence in $\mathbb{F}$.

4.4. Corollary. *If $\mathscr{X}$ is reflexive, then every weakly Cauchy sequence in $\mathscr{X}$ converges weakly. That is, $\mathscr{X}$ is weakly sequentially complete.*

PROOF. Since $\{\langle x_n, x^* \rangle\}$ is a Cauchy sequence in $\mathbb{F}$ for each x^* in $\mathscr{X}^*$, $\{x_n\}$ is weakly bounded. By the PUB there is a constant M such that $\|x_n\| \leq M$ for all $n \geq 1$. But $\{x \in \mathscr{X}: \|x\| \leq M\}$ is weakly compact since $\mathscr{X}$ is reflexive. Thus there is an x in $\mathscr{X}$ such that $x_n \xrightarrow{\text{cl}} x$ weakly. But for each x^* in $\mathscr{X}^*$, $\lim \langle x_n, x^* \rangle$ exists. Hence $\langle x_n, x^* \rangle \rightarrow \langle x, x^* \rangle$, so $x_n \rightarrow x$ weakly. ∎

Not all Banach spaces are weakly sequentially complete.

4.5. Example. $C[0,1]$ is not weakly sequentially complete. In fact, let $f_n(t) = (1 - nt)$ if $0 \leq t \leq 1/n$ and $f_n(t) = 0$ if $1/n \leq t \leq 1$. If $\mu \in M[0,1]$, then $\int f_n \, d\mu \rightarrow \mu(\{0\})$ by the Monotone Convergence Theorem. Hence $\{f_n\}$ is a weakly Cauchy sequence. However, $\{f_n\}$ does not converge weakly to any continuous function on $[0,1]$.

4.6. Corollary. *If $\mathscr{X}$ is reflexive, $\mathscr{M} \leq \mathscr{X}$, and $x_0 \in \mathscr{X} \setminus \mathscr{M}$, then there is a point y_0 in $\mathscr{M}$ such that $\|x_0 - y_0\| = \text{dist}(x_0, \mathscr{M})$.*

PROOF. $x \mapsto \|x - x_0\|$ is weakly lower semicontinuous (Exercise 1.9). If $d = \text{dist}(x_0, \mathscr{M})$, then $\mathscr{M} \cap \{x: \|x - x_0\| \leq 2d\}$ is weakly compact and a lower semicontinuous function attains its minimum on a compact set. ∎

It is not generally true that the distance from a point to a linear subspace is attained. If $\mathcal{M} \subseteq \mathcal{X}$, call $\mathcal{M}$ *proximinal* if for every x in $\mathcal{X}$ there is a y in $\mathcal{M}$ such that $\|x - y\| = \text{dist}(x, \mathcal{M})$. So if $\mathcal{X}$ is reflexive, Corollary 4.6 implies that every closed linear subspace of $\mathcal{X}$ is proximinal. If $\mathcal{X}$ is any Banach space and $\mathcal{M}$ is a finite-dimensional subspace, then it is easy to see that $\mathcal{M}$ is proximinal. How about if $\dim(\mathcal{X}/\mathcal{M}) < \infty$?

4.7. Lemma. *If $\mathcal{X}$ is a Banach space and $x^* \in \mathcal{X}^*$, then $\ker x^*$ is proximinal if and only if there is an x in $\mathcal{X}$, $\|x\| = 1$, such that $\langle x, x^* \rangle = \|x^*\|$.*

PROOF. Let $\mathcal{M} = \ker x^*$ and suppose that $\mathcal{M}$ is proximinal. If $f: \mathcal{X}/\mathcal{M} \to \mathbb{F}$ is defined by $f(x + \mathcal{M}) = \langle x, x^* \rangle$, then f is a linear functional and $\|f\| = \|x^*\|$. Since $\dim \mathcal{X}/\mathcal{M} = 1$, there is an x in $\mathcal{X}$ such that $\|x + \mathcal{M}\| = 1$ and $f(x + \mathcal{M}) = \|f\|$. Because $\mathcal{M}$ is proximinal, there is a y in $\mathcal{M}$ such that $1 = \|x + \mathcal{M}\| = \|x + y\|$. Thus $\langle x + y, x^* \rangle = \langle x, x^* \rangle = f(x + \mathcal{M}) = \|f\| = \|x^*\|$.

Now assume that there is an x_0 in $\mathcal{X}$ such that $\|x_0\| = 1$ and $\langle x_0, x^* \rangle = \|x^*\|$. If $x \in \mathcal{X}$ and $\|x + \mathcal{M}\| = \alpha > 0$, then $\|\alpha^{-1}x + \mathcal{M}\| = 1$. But also $\|x_0 + \mathcal{M}\| = 1$. (Why?) Since $\dim \mathcal{X}/\mathcal{M} = 1$, there is a β in $\mathbb{F}$, $|\beta| = 1$, such that $\alpha^{-1}x + \mathcal{M} = \beta(x_0 + \mathcal{M})$. Hence $\alpha^{-1}x - \beta x_0 \in \mathcal{M}$, or, equivalently, $x - \alpha\beta x_0 \in \mathcal{M}$. However, $\|x - (x - \alpha\beta x_0)\| = \|\alpha\beta x_0\| = \alpha = \text{dist}(x, \mathcal{M})$. So the distance from x to $\mathcal{M}$ is attained at $x - \alpha\beta x_0$. ∎

4.8. Example. If $L: C[0,1] \to \mathbb{F}$ is defined by

$$L(f) = \int_0^{1/2} f(x)\, dx - \int_{1/2}^1 f(x)\, dx,$$

then $\ker L$ is not proximinal.

There is a result in James [1964b] that states that a Banach space is reflexive if and only if every closed hyperplane is proximinal. This result is very deep.

EXERCISES

1. Show that if $\mathcal{X}$ is reflexive and $\mathcal{M} \leq \mathcal{X}$, then $\mathcal{X}/\mathcal{M}$ is reflexive.

2. If $\mathcal{X}$ is a Banach space, $\mathcal{M} \leq \mathcal{X}$, and both $\mathcal{M}$ and $\mathcal{X}/\mathcal{M}$ are reflexive, must $\mathcal{X}$ be reflexive?

3. If X is compact, show that $C(X)$ is reflexive if and only if X is finite.

4. If (X, Ω, μ) is a σ-finite measure space, show that $L^1(X, \Omega, \mu)$ is reflexive if and only if it is finite dimensional.

5. Give the details of the proofs of the statements made in Example 4.5.

6. Verify the statement made in Example 4.8.

7. If (X, Ω, μ) is a σ-finite measure space, show that $L^\infty(\mu)$ is weakly sequentially complete but is reflexive if and only if it is finite dimensional.

8. Let X be compact and suppose there is a norm on $C(X)$ that is given by an inner product making $C(X)$ into a Hilbert space such that for every x in X the functional $f \mapsto f(x)$ on $C(X)$ is continuous with respect to the Hilbert space norm. Show that X is finite.

§5. Separability and Metrizability

The weak and weak-star topologies on an infinite-dimensional Banach space are never metrizable. It is possible, however, to show that under certain conditions these topologies are metrizable when restricted to bounded sets. In applications this is often sufficient.

5.1. Theorem. *If $\mathscr{X}$ is a Banach space, then* ball $\mathscr{X}^*$ *is weak-star metrizable if and only if $\mathscr{X}$ is separable.*

PROOF. Assume that $\mathscr{X}$ is separable and let $\{x_n\}$ be a countable dense subset of ball $\mathscr{X}$. For each n let $D_n = \{\alpha \in \mathbb{F}: |\alpha| \leq 1\}$. Put $X = \prod_{n=1}^\infty D_n$; X is a compact metric space. So if (ball $\mathscr{X}^*$, wk*) is homeomorphic to a subset of X, ball $\mathscr{X}^*$ is weak-star metrizable.

Define τ: ball $\mathscr{X}^* \to X$ by $\tau(x^*) = \{\langle x_n, x^* \rangle\}$. If $\{x_i^*\}$ is a net in ball $\mathscr{X}^*$ and $x_i^* \to x^*$ (wk*), then for each $n \geq 1$, $\langle x_n, x_i^* \rangle \to \langle x_n, x^* \rangle$; hence $\tau(x_i^*) \to \tau(x^*)$ and τ is continuous. If $\tau(x^*) = \tau(y^*)$, $\langle x_n, x^* - y^* \rangle = 0$ for all n. Since $\{x_n\}$ is dense, $x^* - y^* = 0$. Thus τ is injective. Since ball $\mathscr{X}^*$ is wk* compact, τ is a homeomorphism onto its image (A.2.8) and ball $\mathscr{X}^*$ is wk* metrizable.

Now assume that (ball $\mathscr{X}^*$, wk*) is metrizable. Thus there are open sets $\{U_n: n \geq 1\}$ in (ball $\mathscr{X}^*$, wk*) such that $0 \in U_n$ and $\bigcap_{n=1}^\infty U_n = (0)$. By the definition of the relative weak-star topology on ball $\mathscr{X}^*$, for each n there is a finite set F_n contained in $\mathscr{X}$ such that $\{x^* \in$ ball $\mathscr{X}^*: |\langle x, x^* \rangle| < 1$ for all x in $F_n\} \subseteq U_n$. Let $F = \bigcup_{n=1}^\infty F_n$; so F is countable. Also, $^\perp(F^\perp)$ is the closed linear span of F and this subspace of $\mathscr{X}$ is separable. But if $x^* \in F^\perp$, then for each $n \geq 1$ and for each x in F_n, $|\langle x, x^*/\|x^*\| \rangle| = 0 < 1$. Hence $x^*/\|x^*\| \in U_n$ for all $n \geq 1$; thus $x^* = 0$. Since $F^\perp = (0)$, $^\perp(F^\perp) = \mathscr{X}$ and $\mathscr{X}$ must be separable. ∎

Is there a corresponding result for the weak topology? If $\mathscr{X}^*$ is separable, then the weak topology on ball $\mathscr{X}$ is metrizable. In fact, this follows from Theorem 5.1 if the embedding of $\mathscr{X}$ into $\mathscr{X}^{**}$ is considered. This result is not very useful since there are few examples of Banach spaces $\mathscr{X}$ such that $\mathscr{X}^*$ is separable. Of course if $\mathscr{X}$ is separable and reflexive, then $\mathscr{X}^*$ is separable (Exercise 3), but in this case the weak topology on $\mathscr{X}$ is the same as its weak-star topology when $\mathscr{X}$ is identified with $\mathscr{X}^{**}$. Thus (5.1) is adequate for a discussion of the weak topology on the unit ball of a separable reflexive space. If $\mathcal{X} = c_0$, then $\mathcal{X}^* = l^1$ and this is separable but not reflexive. This is one of the few nonreflexive spaces with a separable dual space.

If $\mathcal{X}$ is separable, is (ball $\mathcal{X}$, wk) metrizable? The answer is no, as the following result of Schur demonstrates.

5.2. Proposition. *If a sequence in l^1 converges weakly, it converges in norm.*

PROOF. Recall that $l^\infty = (l^1)^*$. Since l^1 is separable, Theorem 5.1 implies that ball l^∞ is wk* metrizable. By Alaoglu's Theorem, ball l^∞ is wk* compact. Hence (ball l^∞, wk*) is a complete metric space and the Baire Category Theorem is applicable.

Let $\{f_n\}$ be a sequence of elements in l^1 such that $f_n \to 0$ weakly and let $\varepsilon > 0$. For each positive integer m let

$$F_m = \{\phi \in \text{ball } l^\infty : |\langle f_n, \phi \rangle| \le \varepsilon/3 \text{ for } n \ge m\}.$$

It is easy to see that F_m is wk* closed in ball l^∞ and, because $f_n \to 0$ (wk), $\bigcup_{m=1}^\infty F_m = \text{ball } l^\infty$. By the theorem of Baire, there is an F_m with nonempty weak interior.

An equivalent metric on (ball l^∞, wk) is given by

$$d(\phi, \psi) = \sum_{j=1}^\infty 2^{-j} |\phi(j) - \psi(j)|$$

(see Exercise 4). Since F_m has a nonempty weak interior, there is a ϕ in F_m and a $\delta > 0$ such that $\{\psi \in \text{ball } l^\infty : d(\phi, \psi) < \delta\} \subseteq F_m$. Let $J \ge 1$ such that $2^{-(J-1)} < \delta$. Fix $n \ge m$ and define ψ in l^∞ by $\psi(j) = \phi(j)$ for $1 \le j \le J$ and $\psi(j) = \text{sign}(f_n(j))$ for $j > J$. Thus $\psi(j)f_n(j) = |f_n(j)|$ for $j > J$. It is easy to see that $\psi \in \text{ball } l^\infty$. Also, $d(\phi, \psi) = \sum_{j=J+1}^\infty 2^{-j} |\phi(j) - \psi(j)| \le 2 \cdot 2^{-J} = 2^{-(J-1)} < \delta$. So $\psi \in F_m$ and hence $|\langle \psi, f_m \rangle| \le \varepsilon/3$. Thus

5.3
$$\left| \sum_{j=1}^J \phi(j)f_n(j) + \sum_{j=J+1}^\infty |f_n(j)| \right| \le \frac{\varepsilon}{3}$$

for $n \ge m$. But there is an $m_1 \ge m$ such that for $n \ge m_1$, $\sum_{j=1}^J |f_n(j)| < \varepsilon/3$. (Why?) Combining this with (5.3) gives that

$$\|f_n\| = \sum_{j=1}^\infty |f_n(j)|$$

$$< \frac{\varepsilon}{3} + \left| \sum_{j=J+1}^\infty |f_n(j)| + \sum_{j=1}^J \phi(j)f_n(j) \right| + \left| \sum_{j=1}^J \phi(j)f_n(j) \right|$$

$$< \frac{2\varepsilon}{3} + \sum_{j=1}^J |f_n(j)|$$

$$< \varepsilon$$

whenever $n \ge m_1$. ∎

So if (ball l^1, wk) were metrizable, the preceding proposition would say that the weak and norm topologies on l^1 agree. But this is not the case (Exercise 1.10).

Also, note that the preceding result demonstrates in a dramatic way that in discussions concerning the weak topology it is essential to consider nets and not just sequences.

A proof of (5.2) that avoids the Baire Category Theorem can be found in Banach [1955], p. 218.

EXERCISES

1. Let $B = $ ball $M[0,1]$ and for μ, ν in $M[0,1]$ define $d(\mu, \nu) = \sum_{n=0}^{\infty} 2^{-n} |\int_0^1 x^n \, d\mu - \int_0^1 x^n \, d\nu|$. Show that d is a metric on $M[0,1]$ that defines the weak-star topology on B but not on $M[0,1]$.

2. Let X be a compact space and let $\mathscr{U} = \{(U,V): U, V \text{ are open subsets of } X \text{ and } \operatorname{cl} U \subseteq V\}$. For $u = (U,V)$ in $\mathscr{U}$, let $f_u: X \to [0,1]$ be a continuous function such that $f_u \equiv 1$ on $\operatorname{cl} U$ and $f_u \equiv 0$ on $X \setminus V$. Show: (a) the linear span of $\{f_u: u \in \mathscr{U}\}$ is dense in $C(X)$; (b) if X is a metric space, then $C(X)$ is separable.

3. If $\mathscr{X}$ is a Banach space and $\mathscr{X}^*$ is separable, show that (a) $\mathscr{X}$ is separable; (b) if K is a weakly compact subset of $\mathscr{X}$, then K with the relative weak topology is metrizable.

4. If $B = $ ball l^∞, show that $d(\phi, \psi) = \sum_{j=1}^{\infty} 2^{-j} |\phi(j) - \psi(j)|$ defines a metric on B and that this metric defines the weak-star topology on B.

§6*. An Application: The Stone–Čech Compactification

Let X be any topological space and consider the Banach space $C_b(X)$. Unless some assumption is made regarding X, it may be that $C_b(X)$ is "very small." If, for example, it is assumed that X is completely regular, then $C_b(X)$ has many elements. The next result says that this assumption is also necessary in order for $C_b(X)$ to be "large." But first, here is some notation.

If $x \in X$, let $\delta_x: C_b(X) \to \mathbb{F}$ be defined by $\delta_x(f) = f(x)$ for every f in $C_b(X)$. It is easy to see that $\delta_x \in C_b(X)^*$ and $\|\delta_x\| = 1$. Let $\Delta: X \to C_b(X)^*$ be defined by $\Delta(x) = \delta_x$. If $\{x_i\}$ is a net in X and $x_i \to x$, then $f(x_i) \to f(x)$ for every f in $C_b(X)$. This says that $\delta_{x_i} \to \delta_x$ (wk*) in $C_b(X)^*$. Hence $\Delta: X \to (C_b(X)^*, \text{wk*})$ is continuous. Is Δ a homeomorphism of X onto $\Delta(X)$?

6.1. Proposition. *The map* $\Delta: X \to (\Delta(X), \text{wk*})$ *is a homeomorphism if and only if* X *is completely regular.*

PROOF. Assume X is completely regular. If $x_1 \neq x_2$, then there is an f in $C_b(X)$ such that $f(x_1) = 1$ and $f(x_2) = 0$; thus $\delta_{x_1}(f) \neq \delta_{x_2}(f)$. Hence Δ is injective. To show that $\Delta: X \to (\Delta(X), \mathrm{wk}^*)$ is an open map, let U be an open subset of X and let $x_0 \in U$. Since X is completely regular, there is an f in $C_b(X)$ such that $f(x_0) = 1$ and $f \equiv 0$ on $X \setminus U$. Let $V_1 = \{\mu \in C_b(X)^*: \langle f, \mu \rangle > 0\}$. Then V_1 is wk^* open in $C_b(X)^*$ and $V_1 \cap \Delta(X) = \{\delta_x: f(x) > 0\}$. So if $V = V_1 \cap \Delta(X)$, V is wk^* open in $\Delta(X)$ and $\delta_{x_0} \in V \subseteq \Delta(U)$. Since x_0 was arbitrary, $\Delta(U)$ is open in $\Delta(X)$. Therefore $\Delta: X \to (\Delta(X), \mathrm{wk}^*)$ is a homeomorphism.

Now assume that Δ is a homeomorphism onto its image. Since $(\mathrm{ball}\, C_b(X)^*, \mathrm{wk}^*)$ is a compact space, it is completely regular. Since $\Delta(X) \subseteq \mathrm{ball}\, C_b(X)^*$, $\Delta(X)$ is completely regular (Exercise 2). Thus X is completely regular. ■

6.2. Stone–Čech Compactification. *If X is completely regular, then there is a compact space βX such that:*

(a) *there is a continuous map $\Delta: X \to \beta X$ with the property that $\Delta: X \to \Delta(X)$ is a homeomorphism;*
(b) *$\Delta(X)$ is dense in βX;*
(c) *if $f \in C_b(X)$, then there is a continuous map $f^\beta: \beta X \to \mathbb{F}$ such that $f^\beta \circ \Delta = f$.*

Moreover, if Ω is a compact space having these properties, then Ω is homeomorphic to βX.

PROOF. Let $\Delta: X \to C_b(X)^*$ be the map defined by $\Delta(x) = \delta_x$ and let $\beta X =$ the weak-star closure of $\Delta(X)$ in $C_b(X)^*$. By Alaoglu's Theorem and the fact that $\|\delta_x\| = 1$ for all x, βX is compact. By the preceding proposition, (a) holds. Part (b) is true by definition. It remains to show (c).

Fix f in $C_b(X)$ and define $f^\beta: \beta X \to \mathbb{F}$ by $f^\beta(\tau) = \langle f, \tau \rangle$ for every τ in βX. [Remember that $\beta X \subseteq C_b(X)^*$, so this makes sense.] Clearly f^β is continuous and $f^\beta \circ \Delta(X) = f^\beta(\delta_x) = \langle f, \delta_x \rangle = f(x)$. So $f^\beta \circ \Delta = f$ and (c) holds.

To show that βX is unique, assume that Ω is a compact space and $\pi: X \to \Omega$ is a continuous map such that:

(a') *$\pi: X \to \pi(X)$ is a homeomorphism;*
(b') *$\pi(X)$ is dense in Ω;*
(c') *if $f \in C_b(X)$, there is an $\tilde{f}$ in $C(\Omega)$ such that $\tilde{f} \circ \pi = f$.*

Define $g: \Delta(X) \to \Omega$ by $g(\Delta(x)) = \pi(x)$. In other words, $g = \pi \circ \Delta^{-1}$. The idea is to extend g to a homeomorphism of βX onto Ω. If $\tau_0 \in \beta X$, then (b) implies that there is a net $\{x_i\}$ in X such that $\Delta(x_i) \to \tau_0$ in βX. Now $\{\pi(x_i)\}$ is a net in Ω and since Ω is compact, there is an ω_0 in Ω such that $\pi(x_i) \xrightarrow{\text{cl}} \omega_0$. If $F \in C(\Omega)$, let $f = F \circ \pi$; so $f \in C_b(X)$ (and $F = \tilde{f}$). Also, $f(x_i) = \langle f, \delta_{x_i} \rangle \to \langle f, \tau_0 \rangle = f^\beta(\tau_0)$. But it is also true that $f(x_i) =$

$F(\pi(x_i)) \xrightarrow[\text{cl}]{} F(\omega_0)$. Hence $F(\omega_0) = f^\beta(\tau_0)$ for any F in $C(\Omega)$. This implies that ω_0 is the unique cluster point of $\{\pi(x_i)\}$; thus $\pi(x_i) \to \omega_0$ (A.2.7). Let $g(\tau_0) = \omega_0$. It must be shown that the definition of $g(\tau_0)$ does not depend on the net $\{x_i\}$ in X such that $\Delta(x_i) \to \tau_0$. This is left as an exercise. To summarize, it has been shown that

6.3   There is a function $g \colon \beta X \to \Omega$

$$\text{such that if } f \in C_b(X), \text{ then } f^\beta = \tilde{f} \circ g.$$

To show that $g \colon \beta X \to \Omega$ is continuous, let $\{\tau_i\}$ be a net in βX such that $\tau_i \to \tau$. If $F \in C(\Omega)$, let $f = F \circ \pi$; so $f \in C_b(X)$ and $\tilde{f} = F$. Also, $f^\beta(\tau_i) \to f^\beta(\tau)$. But $F(g(\tau_i)) = f^\beta(\tau_i) \to f^\beta(\tau) = F(g(\tau))$. It follows (6.1) that $g(\tau_i) \to g(\tau)$ in Ω. Thus g is continuous.

It is left as an exercise for the reader to show that g is injective. Since $g(\beta X) \supseteq g(\Delta(X)) = \pi(X)$, $g(\beta X)$ is dense in Ω. But $g(\beta X)$ is compact, so g is bijective. By (A.2.8), g is a homeomorphism. ∎

The compact set βX obtained in the preceding theorem is called the *Stone–Čech compactification* of X. By properties (a) and (b), X can be considered as a dense subset of βX and the map Δ can be taken to be the inclusion map. With this convention, (c) can be interpreted as saying that every bounded continuous function on X has a continuous extension to βX.

The space βX is usually very much larger than X. In particular, it is almost never true that βX is the one-point compactification of X. For example, if $X = (0, 1]$, then the one-point compactification of X is $[0, 1]$. However, $\sin(1/x) \in C_b(X)$ but it has no continuous extension to $[0, 1]$, so $\beta X \neq [0, 1]$.

To obtain an idea of how large $\beta X \setminus X$ is, see Exercise 6, which indicates how to show that if $\mathbb{N}$ has the discrete topology, then $\beta \mathbb{N} \setminus \mathbb{N}$ has $2^{\aleph_0}$ pairwise disjoint open sets. The best source of information on the Stone-Čech compactification is the book by Gillman and Jerison [1960], though the approach to βX is somewhat different there than here.

6.4. Corollary. *If X is completely regular and $\mu \in M(\beta X)$, define $L_\mu \colon C_b(X) \to \mathbb{F}$ by*

$$L_\mu(f) = \int_{\beta X} f^\beta \, d\mu$$

for each f in $C_b(X)$. Then the map $\mu \mapsto L_\mu$ is an isometric isomorphism of $M(\beta X)$ onto $C_b(X)^*$.

PROOF. Define $V \colon C_b(X) \to C(\beta X)$ by $Vf = f^\beta$. It is easy to see that V is linear. Considering X as a subset of βX, the fact that $\beta X = \text{cl } X$ implies that V is an isometry. If $g \in C(\beta X)$ and $f = g|X$, then $g = f^\beta = Vf$; hence V is surjective.

If $\mu \in M(\beta X) = C(\beta X)^*$, it is easy to check that $L_\mu \in C_b(X)^*$ and $\|L_\mu\| = \|\mu\|$ since V is an isometry. Conversely, if $L \in C_b(X)^*$, then $L \circ V^{-1} \in C(\beta X)^*$ and $\|L \circ V^{-1}\| = \|L\|$. Hence there is a μ in $M(\beta X)$ such that $\int g\, d\mu = L \circ V^{-1}(g)$ for every g in $C(\beta X)$. Since $V^{-1}g = g|X$, it follows that $L = L_\mu$. ∎

The next result is from topology. It may be known to the reader, but it is presented here for the convenience of those to whom it is not.

6.5. Partition of Unity. *If X is normal and $\{U_1, \ldots, U_n\}$ is an open covering of X, then there are continuous functions $f_1, \ldots, f_n$ from X into $[0,1]$ such that*

(a) $\sum_{k=1}^n f_k(x) = 1$;
(b) $f_k(x) = 0$ for x in $X \setminus U_k$ and $1 \le k \le n$.

PROOF. First observe that it may be assumed that $\{U_1, \ldots, U_n\}$ has no proper subcover. The proof now proceeds by induction.

If $n = 1$, let $f_1 \equiv 1$. Suppose $n = 2$. Then $X \setminus U_1$ and $X \setminus U_2$ are disjoint closed subsets of X. By Urysohn's Lemma there is a continuous function f_1: $X \to [0,1]$ such that $f_1(x) = 0$ for x in $X \setminus U_1$ and $f_1(x) = 1$ for x in $X \setminus U_2$. Let $f_2 = 1 - f_1$ and the proof of this case is complete.

Now suppose the theorem has been proved for some $n \ge 2$ and $\{U_1, \ldots, U_{n+1}\}$ is an open cover of X that is minimal. Let $F = X \setminus U_{n+1}$; then F is closed, nonempty, and $F \subseteq \bigcup_{k=1}^n U_k$. Let V be an open subset of X such that $F \subseteq V \subseteq \mathrm{cl}\, V \subseteq \bigcup_{k=1}^n U_k$. Since $\mathrm{cl}\, V$ is normal and $\{U_1 \cap \mathrm{cl}\, V, \ldots, U_n \cap \mathrm{cl}\, V\}$ is an open cover of $\mathrm{cl}\, V$, the induction hypothesis implies that there are continuous functions $g_1, \ldots, g_n$ on $\mathrm{cl}\, V$ such that $\sum_{k=1}^n g_k = 1$ and for $1 \le k \le n$, $0 \le g_k \le 1$, and $g_k(\mathrm{cl}\, V \setminus U_k) = 0$. By Tietze's Extension Theorem there are continuous functions $\tilde{g}_1, \ldots, \tilde{g}_n$ on X such that $\tilde{g}_k = g_k$ on $\mathrm{cl}\, V$ and $0 \le \tilde{g}_k \le 1$ for $1 \le k \le n$.

Also, there is a continuous function h: $X \to [0,1]$ such that $h = 0$ on $X \setminus V$ and $h = 1$ on F. Put $f_k = \tilde{g}_k h$ for $1 \le k \le n$ and let $f_{n+1} = 1 - \sum_{k=1}^n f_k$. Clearly $0 \le f_k \le 1$ if $1 \le k \le n$. If $x \in \mathrm{cl}\, V$, then $f_{n+1}(x) = 1 - (\sum_{k=1}^n g_k(x))h(x) = 1 - h(x)$; so $0 \le f_{n+1}(x) \le 1$ on $\mathrm{cl}\, V$. If $x \in X \setminus V$, then $f_{n+1}(x) = 1$ since $h(x) = 0$. Hence $0 \le f_{n+1} \le 1$.

Clearly (a) holds. Let $1 \le k \le n$; if $x \in X \setminus U_k$, then either $x \in (\mathrm{cl}\, V) \setminus U_k$ or $x \in (X \setminus \mathrm{cl}\, V) \setminus U_k$. If the first alternative is the case, then $g_k(x) = 0$, so $f_k(x) = 0$. If the second alternative is true, then $h(x) = 0$ so that $f_k(x) = 0$. If $x \in X \setminus U_{n+1} = F$, then $h(x) = 1$ and so $f_{n+1}(x) = 1 - \sum_{k=1}^n g_k(x) = 0$. ∎

Partitions of unity are a standard way to put together local results to obtain global results. If $\{f_k\}$ is related to $\{U_k\}$ as in the statement of (6.5), then $\{f_k\}$ is said to be a partition of unity *subordinate to the cover* $\{U_k\}$.

6.6.   Theorem. *If X is completely regular, then $C_b(X)$ is separable if and only if X is a compact metric space.*

PROOF. Suppose X is a compact metric space with metric d. For each n, let $\{U_k^{(n)}: 1 \le k \le N_n\}$ be an open cover of X by balls of radius $1/n$. Let $\{f_k^{(n)}: 1 \le k \le N_n\}$ be a partition of unity subordinate to $\{U_k^{(n)}: 1 \le k \le N_n\}$. Let $\mathcal{Y}$ be the rational (or complex–rational) linear span of $\{f_k^{(n)}: n \ge 1, 1 \le k \le N_n\}$; thus $\mathcal{Y}$ is countable. It will be shown that $\mathcal{Y}$ is dense in $C(X)$.

Fix f in $C(X)$ and $\varepsilon > 0$. Since f is uniformly continuous there is a $\delta > 0$ such that $|f(x_1) - f(x_2)| < \varepsilon/2$ whenever $d(x_1, x_2) < \delta$. Choose $n > 2/\delta$ and consider the cover $\{U_k^{(n)}: 1 \le k \le N_n\}$. If $x_1, x_2 \in U_k^{(n)}$, $d(x_1, x_2) < 2/n < \delta$; hence $|f(x_1) - f(x_2)| < \varepsilon/2$. Pick x_k in $U_k^{(n)}$ and let $\alpha_k \in \mathbb{Q} + i\mathbb{Q}$ such that $|\alpha_k - f(x_k)| < \varepsilon/2$. Let $g = \sum_k \alpha_k f_k^{(n)}$, so $g \in \mathcal{Y}$. Therefore for every x in X,

$$|f(x) - g(x)| = \left| \sum_k f(x) f_k^{(n)}(x) - \sum_k \alpha_k f_k^{(n)}(x) \right|$$
$$\le \sum_k |f(x) - \alpha_k| f_k^{(n)}(x).$$

Examine each of these summands. If $x \in U_k^{(n)}$, then $|f(x) - \alpha_k| \le |f(x) - f(x_k)| + |f(x_k) - \alpha_k| < \varepsilon$. If $x \notin U_k^{(n)}$, then $f_k^{(n)}(x) = 0$. Hence $|f(x) - g(x)| < \sum_k \varepsilon f_k^{(n)}(x) = \varepsilon$. Thus $\|f - g\| < \varepsilon$ and $\mathcal{Y}$ is dense in $C(X)$. This shows that $C(X)$ is separable.

Now assume that $C_b(X)$ is separable. Thus (ball $C_b(X)^*$, wk*) is metrizable (5.1). Since X is homeomorphic to a subset of ball $C_b(X)^*$ (6.1), X is metrizable. It also follows that βX is metrizable. It must be shown that $X = \beta X$.

Suppose there is a τ in $\beta X \setminus X$. Let $\{x_n\}$ be a sequence in X such that $x_n \to \tau$. It can be assumed that $x_n \ne x_m$ for $n \ne m$. Let $A = \{x_n: n \text{ is even}\}$ and $B = \{x_n: n \text{ is odd}\}$. Then A and B are disjoint closed subsets of X (not closed in βX, but in X) since A and B contain all of their limit points in X. Since X is normal, there is a continuous function $f: X \to [0,1]$ such that $f = 0$ on A and $f = 1$ on B. But then $f^\beta(\tau) = \lim f(x_{2n}) = 0$ and $f^\beta(\tau) = \lim f(x_{2n+1}) = 1$, a contradiction. Thus $\beta X \setminus X = \square$.   ∎

EXERCISES

1. If $x \in X$ and $\delta_x(f) = f(x)$ for all f in $C_b(X)$, show that $\|\delta_x\| = 1$.

2. Prove that a subset of a completely regular space is completely regular.

3. Fill in the details of the proof of Theorem 6.2.

4. If X is completely regular, Ω is compact, and $f: X \to \Omega$ is continuous, show that there is a continuous map $f^\beta: \beta X \to \Omega$ such that $f^\beta | X = f$.

5. If X is completely regular, show that X is open in βX if and only if X is locally compact.

6. Let $\mathbb{N}$ have the discrete topology. Let $\{r_n: n \in \mathbb{N}\}$ be an enumeration of the rational numbers in $[0,1]$. Let S = the irrational numbers in $[0,1]$ and for each s in S let $\{r_n: n \in N_s\}$ be a subsequence of $\{r_n\}$ such that $s = \lim\{r_n: n \in N_s\}$. Show: (a) if s, $t \in S$ and $s \neq t$, $N_s \cap N_t$ is finite; (b) if for each s in S, cl N_s = the closure of N_s in $\beta\mathbb{N}$ and $A_s = (\text{cl } N_s) \backslash \mathbb{N}$, then $\{A_s: s \in S\}$ are pairwise disjoint subsets of $\beta\mathbb{N} \backslash \mathbb{N}$ that are both open and closed.

7. Show that if X is totally disconnected, then so is βX.

8. Show that if $\tau \in \beta X$ and there is a sequence $\{x_n\}$ in X such that $x_n \to \tau$ in βX, then $\tau \in X$.

9. Let X be the space of all ordinals less than the first uncountable ordinal and give X the order topology. Show that βX = the one point compactification of X. (You can find the pertinent definitions in Kelley [1955].)

§7. The Krein–Milman Theorem

7.1. Definition. If K is a convex subset of a vector space $\mathscr{X}$, then a point a in K is an *extreme point* of K if there is no proper open line segment that contains a and lies entirely in K. Let ext K be the set of extreme points of K.

Recall that an open line segment is a set of the form $(x_1, x_2) \equiv \{tx_2 + (1-t)x_1: 0 < t < 1\}$, and to say that this line segment is proper is to say that $x_1 \neq x_2$.

7.2. Examples.

(a) If $\mathscr{X} = \mathbb{R}^2$ and $K = \{(x,y) \in \mathbb{R}^2: x^2 + y^2 \leq 1\}$, then ext $K = \{(x,y): x^2 + y^2 = 1\}$.

(b) If $\mathscr{X} = \mathbb{R}^2$ and $K = \{(x,y) \in \mathbb{R}^2: x \leq 0\}$, then ext $K = \square$.

(c) If $\mathscr{X} = \mathbb{R}^2$ and $K = \{(x,y) \in \mathbb{R}^2: x < 0\} \cup \{(0,0)\}$, then ext $K = \{(0,0)\}$.

(d) If K = the closed region in $\mathbb{R}^2$ bordered by a regular polygon, then ext K = the vertices of the polygon.

(e) If $\mathscr{X}$ is any normed space and $K = \{x \in \mathscr{X}: \|x\| \leq 1\}$, then ext $K \subseteq \{x: \|x\| = 1\}$, though for all we know it may be that ext $K = \square$.

(f) If $\mathscr{X} = L^1[0,1]$ and $K = \{f \in L^1[0,1]: \|f\|_1 \leq 1\}$, then ext $K = \square$. This last statement requires a bit of proof. Let $f \in L^1[0,1]$ such that $\|f\|_1 = 1$. Choose x in $[0,1]$ such that $\int_0^x |f(t)| \, dt = \frac{1}{2}$. Let $h(t) = 2f(t)$ if $t \leq x$ and 0 otherwise; let $g(t) = 2f(t)$ if $t \geq x$ and 0 otherwise. Then $\|h\|_1 = \|g\|_1 = 1$ and $f = \frac{1}{2}(h + g)$. So ball $L^1[0,1]$ has no extreme points.

The next proposition is left as an exercise.

7.3. Proposition. *If K is a convex subset of a vector space $\mathscr{X}$ and $a \in K$, then the following statements are equivalent.*

(a) $a \in \text{ext } K$.

(b) *If $x_1, x_2 \in \mathscr{X}$ and $a = \frac{1}{2}(x_1 + x_2)$, then either $x_1 \notin K$ or $x_2 \notin K$ or $x_1 = x_2 = a$.*

(c) *If $x_1, x_2 \in \mathscr{X}$, $0 < t < 1$, and $a = tx_1 + (1-t)x_2$, then either $x_1 \notin K$, $x_2 \notin K$, or $x_1 = x_2 = a$.*

(d) *If $x_1, \ldots, x_n \in K$ and $a \in \text{co}\{x_1, \ldots, x_n\}$, then $a = x_k$ for some k.*

(e) *$K \setminus \{a\}$ is a convex set.*

7.4. The Krein–Milman Theorem. *If K is a nonempty compact convex subset of a LCS $\mathscr{X}$, then $\text{ext } K \neq \square$ and $K = \overline{\text{co}}(\text{ext } K)$.*

PROOF. (Léger [1968].) Note that (7.3c) says that a point a is an extreme point if and only if $K \setminus \{a\}$ is a relatively open convex subset. We thus look for a maximal proper relatively open convex subset of K. Let $\mathscr{U}$ be all the proper relatively open convex subsets of K. Since $\mathscr{X}$ is a LCS and $K \neq \square$ (and let's assume that K is not a singleton), $\mathscr{U} \neq \square$. Let $\mathscr{U}_0$ be a chain in $\mathscr{U}$ and put $U_0 = \bigcup\{U : U \in \mathscr{U}_0\}$. Clearly U is open, and since $\mathscr{U}_0$ is a chain, U_0 is convex. If $U_0 = K$, then the compactness of K implies that there is a U in $\mathscr{U}_0$ with $U = K$, a contradiction to the propriety of U. Thus $U_0 \in \mathscr{U}$. By Zorn's Lemma, $\mathscr{U}$ has a maximal element U.

If $x \in K$ and $0 \le \lambda \le 1$, let $T_{x,\lambda} \colon K \to K$ be defined by $T_{x,\lambda}(y) = \lambda y + (1-\lambda)x$. Note that $T_{x,\lambda}$ is continuous and $T_{x,\lambda}(\sum_{j=1}^{n}\alpha_j y_j) = \sum_{j=1}^{n}\alpha_j T_{x,\lambda}(y_j)$ whenever $y_1, \ldots, y_n \in K$, $\alpha_1, \ldots, \alpha_n \ge 0$, and $\sum_{j=1}^{n}\alpha_j = 1$. (This means that $T_{x,\lambda}$ is an *affine map* of K into K.) If $x \in U$ and $0 \le \lambda < 1$, then $T_{x,\lambda}(U) \subseteq U$. Thus $U \subseteq T_{x,\lambda}^{-1}(U)$ and $T_{x,\lambda}^{-1}(U)$ is an open convex subset of K. If $y \in (\text{cl } U) \setminus U$, $T_{x,\lambda}(y) \in [x, y) \subseteq U$ by Proposition IV.1.11. So $\text{cl } U \subseteq T_{x,\lambda}^{-1}(U)$ and hence the maximality of U implies $T_{x,\lambda}^{-1}(U) = K$. That is,

7.5 $$T_{x,\lambda}(K) \subseteq U \text{ if } x \in U \text{ and } 0 \le \lambda < 1.$$

Claim. *If V is any open convex subset of K, then either $V \cup U = U$ or $V \cup U = K$.*

In fact, (7.5) implies that $V \cup U$ is convex so that the claim follows from the maximality of U.

It now follows from the claim that $K \setminus U$ is a singleton. In fact, if $a, b \in K \setminus U$ and $a \neq b$, let V_a, V_b be disjoint open convex subsets of K such that $a \in V_a$ and $b \in V_b$. By the claim $V_a \cup U = K$ since $a \notin U$. But $b \notin V_a \cup U$, a contradiction. Thus $K \setminus U = \{a\}$ and $a \in \text{ext } K$ by (7.3e). Hence $\text{ext } K \neq \square$.

Note that we have actually proved the following.

7.6 If V is an open convex subset of $\mathscr{X}$ and $\operatorname{ext} K \subseteq V$, then $K \subseteq V$.

In fact, if V is open and convex, $V \cap K \in \mathscr{U}$ and is contained in a maximal element U of $\mathscr{U}$. Since $K \setminus U = \{a\}$ for some a in $\operatorname{ext} K$, this is a contradiction. Thus (7.6) holds.

Let $E = \overline{\operatorname{co}}(\operatorname{ext} K)$. If $x^* \in \mathscr{X}^*$, $\alpha \in \mathbb{R}$, and $E \subseteq \{x \in \mathscr{X}: \operatorname{Re}\langle x, x^*\rangle < \alpha\} = V$, then $K \subseteq V$ by (7.6). Thus the Hahn–Banach Theorem (IV.3.13) implies $E = K$. ∎

The Krein–Milman Theorem seems innocent enough, but it has widespread application. Two such applications will be seen in Sections 8 and 10; another will occur later when C^*-algebras are studied. Here a small application is given.

If $\mathscr{X}$ is a Banach space, then ball $\mathscr{X}^*$ is weak* compact by Alaoglu's Theorem. By the Krein–Milman Theorem, ball $\mathscr{X}^*$ has many extreme points. Keep this in mind.

7.7. Example. c_0 is not the dual of a Banach space. That is, c_0 is not isometrically isomorphic to the dual of a Banach space. In light of the preceding comments, in order to prove this statement, it suffices to show that ball c_0 has few extreme points. In fact, ball c_0 has no extreme points. Let $x \in$ ball c_0. It must be that $0 = \lim x(n)$. Let N be such that $|x(n)| < \frac{1}{2}$ for $n \geq N$. Define y_1, y_2 in c_0 by letting $y_1(n) = y_2(n) = x(n)$ for $n \leq N$, and for $n > N$ let $y_1(n) = x(n) + 2^{-n}$ and $y_2(n) = x(n) - 2^{-n}$. It is easy to check that y_1 and $y_2 \in$ ball c_0, $\frac{1}{2}(y_1 + y_2) = x$, and $y_1 \neq x$.

In light of Example 7.2(f), $L^1[0, 1]$ is not the dual of a Banach space.

The next two results are often useful in applying the Krein–Milman Theorem. Indeed, the first is often taken as part of that result.

7.8. Theorem. *If $\mathscr{X}$ is a LCS, K is a compact convex subset of $\mathscr{X}$, and $F \subset K$ such that $K = \overline{\operatorname{co}}(F)$, then $\operatorname{ext} K \subseteq \operatorname{cl} F$.*

PROOF. Clearly it suffices to assume that F is closed. Suppose that there is an extreme point x_0 of K such that $x_0 \notin F$. Let p be a continuous seminorm on $\mathscr{X}$ such that $F \cap \{x \in \mathscr{X}: p(x - x_0) < 1\} = \square$. Let $U_0 = \{x \in \mathscr{X}: p(x) < \frac{1}{3}\}$. So $(x_0 + U_0) \cap (F + U_0) = \square$; hence $x_0 \notin \operatorname{cl}(F + U_0)$.

Because F is compact, there are $y_1, \ldots, y_n$ in F such that $F \subseteq \bigcup_{k=1}^n (y_k + U_0)$. Let $K_k = \overline{\operatorname{co}}(F \cap (y_k + U_0))$. Thus $K_k \subseteq y_k + \operatorname{cl} U_0$ (Why?), and $K_k \subseteq K$. Now the fact that $K_1, \ldots, K_n$ are compact and convex implies that $\overline{\operatorname{co}}(K_1 \cup \cdots \cup K_n) = \operatorname{co}(K_1 \cup \cdots \cup K_n)$ (Exercise 8). Therefore

$$K = \overline{\operatorname{co}}(F) = \operatorname{co}(K_1 \cup \cdots \cup K_n).$$

Since $x_0 \in K$, $x_0 = \sum_{k=1}^n \alpha_k x_k$, $x_k \in K_k$, $\alpha_k \geq 0$, $\alpha_1 + \cdots + \alpha_n = 1$. But x_0 is an extreme point of K. Thus, $x_0 = x_k \in K_k$ for some k. But this implies that $x_0 \in K_k \subseteq y_k + \operatorname{cl} U_0 \subseteq \operatorname{cl}(F + U_0)$, a contradiction. ∎

You might think that the set of extreme points of a compact convex subset would have to be closed. This is untrue even if the LCS is finite dimensional, as Figure V-1 illustrates.

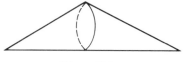

Figure V-1

7.9. Proposition. *If K is a compact convex subset of a LCS $\mathscr{X}$, $\mathscr{Y}$ is a LCS, and T: $K \to \mathscr{Y}$ is a continuous affine map, then $T(K)$ is a compact convex subset of $\mathscr{Y}$ and if y is an extreme point of $T(K)$, then there is an extreme point x of K such that $T(x) = y$.*

PROOF. Because T is affine, $T(K)$ is convex and it is compact by the continuity of T. Let y be an extreme point of $T(K)$. It is easy to see that $T^{-1}(y)$ is compact and convex. Let x be an extreme point of $T^{-1}(y)$. It now follows that $x \in \operatorname{ext} K$ (Exercise 9). ∎

Note that it is possible that there are extreme points x of K such that $T(x)$ is not an extreme point of $T(K)$. For example, let T be the orthogonal projection of $\mathbb{R}^3$ onto $\mathbb{R}^2$ and let $K = \operatorname{ball} \mathbb{R}^3$.

EXERCISES

1. If (X, Ω, μ) is a σ-finite measure space and $1 < p < \infty$, then the set of extreme points of ball $L^p(\mu)$ is $\{f \in L^p(\mu): \|f\|_p = 1\}$.

2. If (X, Ω, μ) is a σ-finite measure space, the set of extreme points of ball $L^1(\mu)$ is $\{\alpha\chi_E: E$ is an atom of μ, $\alpha \in \mathbb{F}$, and $|\alpha| = \mu(E)^{-1}\}$.

3. If (X, Ω, μ) is a σ-finite measure space, the set of extreme points of ball $L^\infty(\mu)$ is $\{f \in L^\infty(\mu): |f(x)| = 1$ a.e. $[\mu]\}$.

4. If X is completely regular, the set of extreme points of ball $C_b(X)$ is $\{f \in C_b(X): |f(x)| = 1$ for all $x\}$. So ball $C_{\mathbb{R}}[0,1]$ has only two extreme points.

5. Let X be a totally disconnected compact space. (That is, X is compact and if $x \in X$ and U is an open neighborhood of x, then there is a subset V of X that is both open and closed and such that $x \in V \subseteq U$. The Cantor set is an example of such a space.) Show that ball $C(X)$ is the norm closure of the convex hull of its extreme points.

6. Show that ball l^1 is the norm closure of the convex hull of its extreme points.

7. Show that if X is locally compact but not compact, then ball $C_0(X)$ has no extreme points.

8. If $\mathscr{X}$ is a LCS and $K_1, \ldots, K_n$ are compact convex subsets of $\mathscr{X}$, then $\overline{\mathrm{co}}(K_1 \cup \cdots \cup K_n) = \mathrm{co}(K_1 \cup \cdots \cup K_n)$.

9. Let K be convex and let $T: K \to \mathscr{Y}$ be an affine map. If y is an extreme point of $T(K)$ and x is an extreme point of $T^{-1}(y)$, then x is an extreme point of K.

10. If $\mathscr{H}$ is a Hilbert space, show that T is an extreme point of ball $\mathscr{B}(\mathscr{H})$ if and only if either T or T^* is an isometry.

§8. An Application: The Stone–Weierstrass Theorem

If $f: X \to \mathbb{C}$ is a function, then $\bar{f}$ denotes the function from X into $\mathbb{C}$ whose value at each x is the complex conjugate of $f(x), \overline{f(x)}$.

8.1. The Stone–Weierstrass Theorem. *If X is compact and $\mathscr{A}$ is a closed subalgebra of $C(X)$ such that:*

(a) $1 \in \mathscr{A}$;
(b) *if $x, y \in X$ and $x \neq y$, then there is an f in $\mathscr{A}$ such that $f(x) \neq f(y)$;*
(c) *if $f \in \mathscr{A}$, then $\bar{f} \in \mathscr{A}$;*
    then $\mathscr{A} = C(X)$.

If $C(X)$ is the algebra of continuous functions from X into $\mathbb{R}$, then condition (c) is not needed. Also, an algebra in $C(X)$ that has property (b) is said to *separate the points* of X (see Exercise 1).

The proof of this result makes use of the Krein–Milman Theorem and is due to L. de Branges [1959].

PROOF OF THE STONE–WEIERSTRASS THEOREM. To prove the theorem it suffices to show that $\mathscr{A}^{\perp} = (0)$ (III.6.14). Suppose $\mathscr{A}^{\perp} \neq (0)$. By Alaoglu's Theorem, ball $\mathscr{A}^{\perp}$ is weak* compact. By the Krein–Milman Theorem, there is an extreme point μ of ball $\mathscr{A}^{\perp}$. Let $K =$ the support of μ. Since $\mathscr{A}^{\perp} \neq (0)$, $\|\mu\| = 1$ and $K \neq \square$. Fix x_0 in K. It will be shown that $K = \{x_0\}$.

Let $x \in X$, $x \neq x_0$. By (b) there is an f_1 in $\mathscr{A}$ such that $f_1(x_0) \neq f_1(x) = \beta$. By (a), the function $\beta \in \mathscr{A}$. Hence $f_2 = f_1 - \beta \in \mathscr{A}$, $f_2(x_0) \neq 0 = f_2(x)$. By (c), $f_3 = |f_2|^2 = f_2 \bar{f}_2 \in \mathscr{A}$. Also, $f_3(x) = 0 < f_3(x_0)$ and $f_3 \geq 0$. Put $f = (\|f_3\| + 1)^{-1} f_3$. Then $f \in \mathscr{A}$, $f(x) = 0$, $f(x_0) > 0$, and $0 \leq f < 1$ on X. Moreover, because $\mathscr{A}$ is an algebra, gf and $g(1 - f) \in \mathscr{A}$ for every g in $\mathscr{A}$. Because $\mu \in \mathscr{A}^{\perp}$, $0 = \int gf \, d\mu = \int g(1 - f) \, d\mu$ for every g in $\mathscr{A}$. Therefore $f\mu$ and $(1 - f)\mu \in \mathscr{A}^{\perp}$.

(For any bounded Borel function h on X, $h\mu$ denotes the measure whose value at a Borel set Δ is $\int_{\Delta} h \, d\mu$. Note that $\|h\mu\| = \int |h| \, d|\mu|$.)

Put $\alpha = \|f\mu\| = \int f \, d|\mu|$. Since $f(x_0) > 0$, there is an open neighborhood U of x_0 and an $\varepsilon > 0$ such that $f(y) > \varepsilon$ for y in U. Thus, $\alpha = \int f \, d|\mu| \geq$

$\int_U f\, d|\mu| \geq \varepsilon|\mu|(U) > 0$ since $U \cap K \neq \square$. Similarly, since $f(x_0) < 1$, $\alpha < 1$. Therefore, $0 < \alpha < 1$. Also, $1 - \alpha = 1 - \int f\, d|\mu| = \int (1 - f)\, d|\mu| = \|(1 - f)\mu\|$. Since

$$\mu = \alpha \left[\frac{f\mu}{\|f\mu\|} \right] + (1 - \alpha) \left[\frac{(1 - f)\mu}{\|(1 - f)\mu\|} \right]$$

and μ is an extreme point of ball $\mathscr{A}^{\perp}$, $\mu = f\mu\|f\mu\|^{-1} = \alpha^{-1} f\mu$. But the only way that the measures μ and $\alpha^{-1} f\mu$ can be equal is if $\alpha^{-1} f = 1$ a.e. $[\mu]$. Since f is continuous, it must be that $f \equiv \alpha$ on K. Since $x_0 \in K$, $f(x_0) = \alpha$. But $f(x_0) > f(x) = 0$. Hence $x \notin K$. This establishes that $K = \{x_0\}$ and so $\mu = \gamma \delta_{x_0}$ where $|\gamma| = 1$. But $\mu \in \mathscr{A}^{\perp}$ and $1 \in \mathscr{A}$, so $0 = \int 1\, d\mu = \gamma$, a contradiction. Therefore $\mathscr{A}^{\perp} = (0)$ and $\mathscr{A} = C(X)$. ∎

With an important theorem it is good to ask what happens if part of the hypothesis is deleted. If $x_0 \in X$ and $\mathscr{A} = \{f \in C(X): f(x_0) = 0\}$, then $\mathscr{A}$ is a closed subalgebra of $C(X)$ that satisfies (b) and (c) but $\mathscr{A} \neq C(X)$. This is the worse that can happen.

8.2. Corollary. *If X is compact and $\mathscr{A}$ is a closed subalgebra of $C(X)$ that separates the points of X and is closed under complex conjugation, then either $\mathscr{A} = C(X)$ or there is a point x_0 in X such that $\mathscr{A} = \{f \in C(X): f(x_0) = 0\}$.*

PROOF. Identify $\mathbb{F}$ and the one-dimensional subspace of $C(X)$ consisting of the constant functions. Since $\mathscr{A}$ is closed, $\mathscr{A} + \mathbb{F}$ is closed (III.4.3). It is easy to see that $\mathscr{A} + \mathbb{F}$ is an algebra and satisfies the hypothesis of the Stone–Weierstrass Theorem; hence $\mathscr{A} + \mathbb{F} = C(X)$. Suppose $\mathscr{A} \neq C(X)$. Then $C(X)/\mathscr{A}$ is one dimensional; thus $\mathscr{A}^{\perp}$ is one dimensional (Theorem 2.2). Let $\mu \in \mathscr{A}^{\perp}$, $\|\mu\| = 1$. If $f \in \mathscr{A}$, then $f\mu \in \mathscr{A}^{\perp}$; hence there is an α in $\mathbb{F}$ such that $f\mu = \alpha\mu$. This implies that each f in $\mathscr{A}$ is constant on the support of μ. But the functions in $\mathscr{A}$ separate the points of X. Hence the support of μ is a single point x_0 and so $\mathscr{A}^{\perp} = \{\beta\delta_{x_0}: \beta \in \mathbb{F}\}$. Thus $\mathscr{A} = {}^{\perp}\mathscr{A}^{\perp} = \{f \in C(X): f(x_0) = 0\}$. ∎

There are many examples of subalgebras of $C(X)$ that separate the points of X, contain the constants, but are not necessarily closed under complex conjugation. Indeed, a subalgebra of $C(X)$ having these properties is called a *uniform algebra* or *function algebra* and their study forms a separate area of mathematics (Gamelin [1969]). One example (the most famous) is obtained by letting X be a subset of $\mathbb{C}$ and letting $\mathscr{A} = R(X) \equiv$ the uniform closure of rational functions with poles off X.

Let $x_0, x_1 \in X$, $x_0 \neq x_1$, and let $\mathscr{A} \equiv \{f \in C(X): f(x_0) = f(x_1)\}$. Then $\mathscr{A}$ is a uniformly closed subalgebra of $C(X)$, contains the constant functions, and is closed under conjugation. In a certain sense this is the worst that can happen if the only hypothesis of the Stone–Weierstrass Theorem that does not hold is that $\mathcal{A}$ fails to separate the points of X (see Exercise 4).

If X is only assumed to be locally compact, then the story is similar.

8.3. Corollary. *If X is locally compact and $\mathcal{A}$ is a closed subalgebra of $C_0(X)$ such that*

(a) *for each x in X there is an f in $\mathcal{A}$ such that $f(x) \neq 0$;*
(b) *$\mathcal{A}$ separates the points of X;*
(c) *$\bar{f} \in \mathcal{A}$ whenever $f \in \mathcal{A}$;*

then $\mathcal{A} = C_0(X)$.

PROOF. Let X_∞ = the one point compactification of X and identify $C_0(X)$ with $\{f \in C(X_\infty): f(\infty) = 0\}$. So $\mathcal{A}$ becomes a subalgebra of $C(X_\infty)$. Now apply Corollary 8.2. The details are left to the reader. ∎

What are the extreme points of the unit ball of $M(X)$? The characterization of these extreme points as well as the extreme points of the set $P(X)$ of probability measures on X is given in the next theorem. [A *probability measure* is a positive measure μ such that $\mu(X) = 1$.]

8.4. Theorem. *If X is compact, then the set of extreme points of* ball $M(X)$ *is*

$$\{\alpha\delta_x: |\alpha| = 1 \text{ and } x \in X\}.$$

The set of extreme points of $P(X)$, the probability measures on X, is

$$\{\delta_x: x \in X\}.$$

PROOF. It is left as an exercise for the reader to show that if $x \in X$, δ_x is an extreme point of $P(X)$ and $\alpha\delta_x$ is an extreme point of ball $M(X)$ (Exercise 3).

It will now be shown that if μ is an extreme point of $P(X)$, then μ is an extreme point of ball $M(X)$. Thus the first part of the theorem implies the second. Suppose μ is an extreme point of $P(X)$ and $\nu_1, \nu_2 \in$ ball $M(X)$ such that $\mu = \frac{1}{2}(\nu_1 + \nu_2)$. Then $1 = \|\mu\| \leq \frac{1}{2}(\|\nu_1\| + \|\nu_2\|) \leq 1$; hence $\|\nu_1\| + \|\nu_2\| = 2$ and so $\|\nu_1\| = \|\nu_2\| = 1$. Also, $1 = \mu(X) = \frac{1}{2}(\nu_1(X) + \nu_2(X))$. Now $|\nu_1(X)|, |\nu_2(X)| \leq 1$ and 1 is an extreme point of $\{\alpha \in \mathbb{F}: |\alpha| \leq 1\}$. Hence for $k = 1, 2$, $\|\nu_k\| = \nu_k(X) = 1$. By Exercise III.7.2, $\nu_k \in P(X)$ for $k = 1, 2$. Since $\mu \in \text{ext } P(X)$, $\mu = \nu_1 = \nu_2$. So μ is an extreme point of ball $M(X)$. Thus it suffices to prove the first part of the theorem.

Suppose that μ is an extreme point of ball $M(X)$ and let K be the support of μ. That is,

$$K = X \setminus \cup\{V: V \text{ is open and } |\mu|(V) = 0\}.$$

Hence $|\mu|(X \setminus K) = 0$ and $\int f d\mu = \int_K f d\mu$ for every f in $C(X)$. It will be shown that K is a singleton set.

Fix x_0 in K and suppose there is a second point x in K, $x \neq x_0$. Let U and V be open subsets of X such that $x_0 \in U$, $x \in V$, and $\operatorname{cl} U \cap \operatorname{cl} V = \square$. By Urysohn's Lemma there is an f in $C(X)$ such that $0 \leq f \leq 1$, $f(y) = 1$ for y in $\operatorname{cl} U$, and $f(y) = 0$ for y in $\operatorname{cl} V$. Consider the measures $f\mu$ and $(1 - f)\mu$. Put $\alpha = \|f\mu\| = \int |f| \, d|\mu| = \int f \, d|\mu|$. Then $\alpha = \int f \, d|\mu| \leq \|\mu\| = 1$ and $\alpha = \int f \, d|\mu| \geq |\mu|(U) > 0$ since U is open and $U \cap K \neq \square$. Also, $1 - \alpha = 1 - \int f \, d|\mu| = \int (1 - f) \, d|\mu| = \|(1 - f)\mu\|$ and so $1 - \alpha \geq \int_V (1 - f) \, d|\mu| = |\mu|(V) > 0$ since $x \in K$. Hence $0 < \alpha < 1$.

But $f\mu/\alpha$ and $(1 - f)\mu/(1 - \alpha) \in$ ball $M(X)$ and

$$\mu = \alpha \left[\frac{f\mu}{\alpha} \right] + (1 - \alpha) \left[\frac{(1 - f)\mu}{1 - \alpha} \right].$$

Since μ is an extreme point of ball $M(X)$ and $\alpha \neq 0$, $\mu = f\mu/\alpha$. This can only happen if $f \equiv \alpha < 1$ a.e. $[\mu]$. But $f \equiv 1$ on U and $|\mu|(U) > 0$, a contradiction. Hence $K = \{x_0\}$.

Since the only measures whose support can be the singleton set $\{x_0\}$ have the form $\alpha \delta_{x_0}$, α in $\mathbb{F}$, the theorem is proved. ∎

EXERCISES

1. Suppose that $\mathscr{A}$ is a subalgebra of $C(X)$ that separates the points of X and $1 \in \mathscr{A}$. Show that if $x_1, \ldots, x_n$ are distinct points in X and $\alpha_1, \ldots, \alpha_n \in \mathbb{F}$, there is an f in $\mathscr{A}$ such that $f(x_j) = \alpha_j$ for $1 \leq j \leq n$.

2. Give the details of the proof of Corollary 8.3.

3. If X is compact, show that for each x in X, δ_x is an extreme point of $P(X)$ and $\alpha \delta_x$, $|\alpha| = 1$, is an extreme point of ball $M(X)$.

4. Let X be compact and let $\mathscr{A}$ be a closed subalgebra of $C(X)$ such that $1 \in \mathscr{A}$ and $\mathscr{A}$ is closed under conjugation. Define an equivalence relation $\sim$ on X by declaring $x \sim y$ if and only if $f(x) = f(y)$ for all f in $\mathscr{A}$. Let $X/\sim$ be the corresponding quotient space and let $\pi: X \to X/\sim$ be the natural map. Give $X/\sim$ the quotient topology. (a) Show that if $f \in \mathscr{A}$, then there is a unique function $\pi^*(f)$ in $C(X/\sim)$ such that $\pi^*(f) \circ \pi = f$. (b) Show that $\pi^*: \mathscr{A} \to C(X/\sim)$ is an isometry. (c) Show that π^* is surjective. (d) Show that $\mathscr{A} = \{f \in C(X): f(x) = f(y)$ whenever $x \sim y\}$.

5. (This exercise requires Exercise IV.4.7.) Let X be completely regular and topologize $C(X)$ as in Example IV.1.5. If $\mathscr{A}$ is a closed subalgebra of $C(X)$ such that $1 \in \mathscr{A}$, $\mathscr{A}$ separates the points of X, and $\bar{f} \in \mathscr{A}$ whenever $f \in \mathscr{A}$, then $\mathscr{A} = C(X)$.

6. Let X, Y be compact spaces and show that if $f \in C(X \times Y)$ and $\varepsilon > 0$, then there are functions $g_1, \ldots, g_n$ in $C(X)$ and $h_1, \ldots, h_n$ in $C(Y)$ such that $|f(x, y) - \sum_{k=1}^n g_k(x) h_k(y)| < \varepsilon$ for all (x, y) in $X \times Y$.

7. Let $\mathscr{A}$ be the uniformly closed subalgebra of $C_b(\mathbb{R})$ generated by $\sin x$ and $\cos x$. Show that $\mathscr{A} = \{f \in C_b(\mathbb{R}): f(t) = f(t + 2\pi)$ for all t in $\mathbb{R}\}$.

8. If K is a compact subset of $\mathbb{C}$, $f \in C(K)$, and $\varepsilon > 0$, show that there is a polynomial $p(z, \bar{z})$ in z and $\bar{z}$ such that $|f(z) - p(z, \bar{z})| < \varepsilon$ for all z in K.

§9*. The Schauder Fixed-Point Theorem

Fixed-point theorems hold a fascination for mathematicians and they are very applicable to a variety of mathematical and physical situations. In this section and the next two such theorems are presented.

The results of this section are different from the rest of this book in an essential way. Although we will continue to look at convex subsets of Banach spaces, the functions will not be assumed to be linear or affine. This is a small part of nonlinear functional analysis.

To begin with, recall the following classical result whose proof can be found in any algebraic topology book. (Also see Dugundji [1966].)

9.1.   Brouwer's Fixed-Point Theorem. *If* $1 \le d < \infty$, $B =$ *the closed unit ball of* $\mathbb{R}^d$, *and* $f: B \to B$ *is a continuous map, then there is a point* x *in* B *such that* $f(x) = x$.

9.2.   Corollary. *If* K *is a nonempty compact convex subset of a finite-dimensional normed space* $\mathscr{X}$ *and* $f: K \to K$ *is a continuous function, then there is a point* x *in* K *such that* $f(x) = x$.

PROOF. Since $\mathscr{X}$ is isomorphic to either $\mathbb{C}^d$ or $\mathbb{R}^d$, it is homeomorphic to either $\mathbb{R}^{2d}$ or $\mathbb{R}^d$. So it suffices to assume that $\mathscr{X} = \mathbb{R}^d$, $1 \le d < \infty$. If $K = \{x \in \mathbb{R}^d: \|x\| \le r\}$, then the result is immediate from Brouwer's Theorem (Exercise). If K is any compact convex subset of $\mathbb{R}^d$, let $r > 0$ such that $K \subseteq B \equiv \{x \in \mathbb{R}^d: \|x\| \le r\}$. Let $\phi: B \to K$ be the function defined by $\phi(x) =$ the unique point y in K such that $\|x - y\| = \text{dist}(x, K)$ (I.2.5). Then ϕ is continuous (Exercise) and $\phi(x) = x$ for each x in K. (In topological parlance, K is a retract of B.) Hence $f \circ \phi: B \to K \subseteq B$ is continuous. By Brouwer's Theorem, there is an x in B such that $f(\phi(x)) = x$. Since $f \circ \phi(B) \subseteq K$, $x \in K$. Hence $\phi(x) = x$ and $f(x) = x$.   ∎

Schauder's Fixed-Point Theorem is a generalization of the preceding corollary to infinite-dimensional spaces.

9.3.   Definition. If $\mathscr{X}$ is a normed space and $E \subseteq \mathscr{X}$, a function $f: E \to \mathscr{X}$ is said to be *compact* if f is continuous and $\text{cl } f(A)$ is compact whenever A is a bounded subset of E.

If E is itself a compact subset of $\mathscr{X}$, then every continuous function from E into $\mathscr{X}$ is compact.

The following lemma will be needed in the proof of Schauder's Theorem.

9.4.   Lemma. *If* K *is a compact subset of the normed space* $\mathscr{X}$, $\varepsilon > 0$, *and* A *is a finite subset of* K *such that* $K \subseteq \bigcup\{B(a; \varepsilon): a \in A\}$, *define* $\phi_A: K \to \mathscr{X}$ *by*

$$\phi_A(x) = \frac{\sum\{m_a(x)a: a \in A\}}{\sum\{m_a(x): a \in A\}},$$

*where $m_a(x) = 0$ if $\|x - a\| \geq \varepsilon$ and $m_a(x) = \varepsilon - \|x - a\|$ if $\|x - a\| \leq \varepsilon$.
Then ϕ_A is a continuous function and*

$$\|\phi_A(x) - x\| < \varepsilon$$

for all x in K.

PROOF. Note that for each a in A, $m_a(x) \geq 0$ and $\Sigma\{m_a(x): a \in A\} > 0$
for all x in K. So ϕ_A is well defined on K. The fact that ϕ_A is continuous
follows from the fact that for each a in A, $m_a: K \to [0, \varepsilon]$ is continuous.
(Verify!)

If $x \in K$, then

$$\phi_A(x) - x = \frac{\Sigma\{m_a(x)[a - x]: a \in A\}}{\Sigma\{m_a(x): a \in A\}}.$$

If $m_a(x) > 0$, then $\|x - a\| < \varepsilon$. Hence

$$\|\phi_A(x) - x\| \leq \frac{\Sigma\{m_a(x)\|a - x\|; a \in A\}}{\Sigma\{m_a(x); a \in A\}} < \varepsilon.$$

This concludes the proof. ∎

9.5.   The Schauder Fixed-Point Theorem. *Let E be a closed bounded convex
subset of a normed space $\mathscr{X}$. If $f: E \to \mathscr{X}$ is a compact map such that
$f(E) \subseteq E$, then there is an x in E such that $f(x) = x$.*

PROOF. Let $K = \operatorname{cl} f(E)$, so $K \subseteq E$. For each positive integer n let A_n be a
finite subset of K such that $K \subseteq \cup\{B(a; 1/n): a \in A_n\}$. For each n let
$\phi_n = \phi_{A_n}$ as in the preceding lemma. Now the definition of ϕ_n clearly implies
that $\phi_n(K) \subseteq \operatorname{co}(K) \subseteq E$ since E is convex; thus $f_n \equiv \phi_n \circ f$ maps E into
E. Also, Lemma 9.4 implies

9.6 $\|f_n(x) - f(x)\| < 1/n$ for x in E.

Let $\mathscr{X}_n$ be the linear span of the set A_n and put $E_n = E \cap \mathscr{X}_n$. So $\mathscr{X}_n$ is a
finite-dimensional normed space, E_n is a compact convex subset of $\mathscr{X}_n$, and
$f_n: E_n \to E_n$ (Why?) is continuous. By Corollary 9.2, there is a point x_n in
E_n such that $f_n(x_n) = x_n$.

Now $\{f(x_n)\}$ is a sequence in the compact set K, so there is a point x_0
and a subsequence $\{f(x_{n_j})\}$ such that $f(x_{n_j}) \to x_0$. Since $f_{n_j}(x_{n_j}) = x_{n_j}$,
(9.6) implies

$$\|x_{n_j} - x_0\| \leq \|f_{n_j}(x_{n_j}) - f(x_{n_j})\| + \|f(x_{n_j}) - x_0\|$$

$$\leq \frac{1}{n_j} + \|f(x_{n_j}) - x_0\|.$$

Thus $x_{n_j} \to x_0$. Since f is continuous, $f(x_0) = \lim f(x_{n_j}) = x_0$. ∎

There is a generalization of Schauder's Theorem where $\mathscr{X}$ is only assumed
to be a LCS. See Dunford and Schwartz [1958], p. 456.

EXERCISE

1. Let $E = \{x \in l^2(\mathbb{N}): \|x\| \le 1\}$ and for x in E define $f(x) = ((1 - \|x\|^2),$ $x(1), x(2), \ldots)$. Show that $f(E) \subseteq E$, f is continuous, and f has no fixed points.

§10*. The Ryll–Nardzewski Fixed-Point Theorem

This section begins by proving a fixed-point theorem that in addition to being used to prove the result in the title of this section has some interest of its own. Recall that a map T defined from a convex set K into a vector space is said to be *affine* if $T(\Sigma \alpha_j x_j) = \Sigma \alpha_j T(x_j)$ when $x_j \in K$, $\alpha_j \ge 0$, and $\Sigma \alpha_j = 1$.

10.1. The Markov–Kakutani Fixed-Point Theorem. *If K is a nonempty compact convex subset of a LCS $\mathscr{X}$ and $\mathscr{F}$ is a family of continuous affine maps of K into itself that is abelian, then there is an x_0 in K such that $T(x_0) = x_0$ for all T in $\mathscr{F}$.*

PROOF. If $T \in \mathscr{F}$ and $n \ge 1$, define $T^{(n)}: K \to K$ by

$$T^{(n)} = \frac{1}{n} \sum_{k=0}^{n-1} T^k.$$

If S and $T \in \mathscr{F}$ and $n, m \ge 1$, then it is easy to check that $S^{(n)}T^{(m)} = T^{(m)}S^{(n)}$. Let $\mathscr{K} = \{T^{(n)}(K): T \in \mathscr{F}, n \ge 1\}$. Each set in $\mathscr{K}$ is compact and convex. If $T_1, \ldots, T_p \in \mathscr{F}$ and $n_1, \ldots, n_p \ge 1$, then the commutativity of $\mathscr{F}$ implies that $T_1^{(n_1)} \cdots T_p^{(n_p)}(K) \subseteq \cap_{j=1}^{p} T_j^{(n_j)}(K)$. This says that $\mathscr{K}$ has the finite intersection property and hence there is an x_0 in $\cap \{B: B \in \mathscr{K}\}$. It is claimed that x_0 is the desired common fixed point for the maps in $\mathscr{F}$.

If $T \in \mathscr{F}$ and $n \ge 1$, then $x_0 \in T^{(n)}(K)$. Thus there is an x in K such that

$$x_0 = T^{(n)}(x) = \frac{1}{n}\left[x + T(x) + \cdots + T^{n-1}(x)\right].$$

Using this equation for x_0, it follows that

$$T(x_0) - x_0 = \frac{1}{n}\left[T(x) + \cdots + T^{(n)}(x)\right]$$

$$-\frac{1}{n}\left[x + T(x) + \cdots + T^{n-1}(x)\right]$$

$$= \frac{1}{n}\left[T^n(x) - x\right]$$

$$\in \frac{1}{n}\left[K - K\right].$$

Now K is compact and so $K - K$ is also. If U is an open neighborhood of 0 in $\mathcal{X}$, there is an integer $n \geq 1$ such that $n^{-1}[K - K] \subseteq U$. Therefore $T(x_0) - x_0 \in U$ for every open neighborhood U of 0. This implies that $T(x_0) - x_0 = 0$. ∎

If p is a seminorm on $\mathcal{X}$ and $A \subseteq \mathcal{X}$, define the p-*diameter* of A to be the number

$$p\text{-diam } A \equiv \sup\{ p(x - y): x, y \in A \}.$$

10.2. Lemma. *If $\mathcal{X}$ is a LCS, K is a nonempty separable weakly compact convex subset of $\mathcal{X}$, and p is a continuous seminorm on $\mathcal{X}$, then for every $\varepsilon > 0$ there is a closed convex subset C of K such that:*

(a) $C \neq K$;
(b) $p\text{-diam}(K \setminus C) \leq \varepsilon$.

PROOF. Let $S = \{ x \in \mathcal{X}: p(x) \leq \varepsilon/4 \}$ and let $D =$ the weak closure of the set of extreme points of K. Note that $D \subseteq K$. By hypothesis there is a countable subset A of K such that $D \subseteq K \subseteq \bigcup\{ a + S: a \in A \}$. Now each $a + S$ is weakly closed. (Why?) Since D is weakly compact, there is an a in A such that $(a + S) \cap D$ has interior in the relative weak topology of D (Exercise 2). Thus, there is a weakly open subset W of $\mathcal{X}$ such that

10.3 $(a + S) \cap D \supseteq W \cap D \neq \square.$

Let $K_1 = \overline{\text{co}}(D \setminus W)$ and $K_2 = \overline{\text{co}}(D \cap W)$. Because K_1 and K_2 are compact and convex and $K_1 \cup K_2$ contains the extreme points of K, the Krein–Milman Theorem and Exercise 7.8 imply $K = \text{co}(K_1 \cup K_2)$.

10.4. Claim. $K_1 \neq K$.

In fact, if $K_1 = K$, then $K = \overline{\text{co}}(D \setminus W)$ so that $\text{ext } K \subseteq D \setminus W$ (Theorem 7.8). This implies that $D \subseteq D \setminus W$, or that $W \cap D = \square$, a contradiction to (10.3).

Now (10.3) implies that $K_2 \subseteq a + S$; so the definition of S implies that $p\text{-diam } K_2 \leq \varepsilon/2$. Let $0 < r \leq 1$ and define $f_r: K_1 \times K_2 \times [r, 1] \to K$ by $f_r(x_1, x_2, t) = tx_1 + (1 - t)x_2$. So f_r is continuous and $C_r \equiv f_r(K_1 \times K_2 \times [r, 1])$ is weakly compact and convex. (Verify!)

10.5. Claim. $C_r \neq K$ for $0 < r \leq 1$.

In fact, if $C_r = K$ and $e \in \text{ext } K$, then $e = tx_1 + (1 - t)x_2$ for some t, $r \leq t \leq 1$, x_j in K_j. Because e is an extreme point and $t \neq 0$, $e = x_1$. Thus $\text{ext } K \subseteq K_1$ and $K = K_1$, contradicting (10.4).

Let $y \in K \setminus C_r$. The definition of C_r and the fact that $K = \text{co}(K_1 \cup K_2)$ imply $y = tx_1 + (1 - t)x_2$ with x_j in K_j and $0 \leq t < r$. Hence $p(y - x_2)$

$= p(t(x_1 - x_2)) = tp(x_1 - x_2) \le rd$, where $d = p\text{-diam } K$. Therefore, if $y' = t'x_1' + (1 - t')x_2' \in K \setminus C_r$, then $p(y - y') \le p(y - x_2) + p(x_2 - x_2') + p(x_2' - y') \le 2rd + p\text{-diam } K_2 \le 2rd + \varepsilon/2$. Choosing $r = \varepsilon/4d$ and putting $C = C_r$, we have proved the lemma.   ∎

10.6. Definition. Let $\mathscr{X}$ be a LCS and let Q be a nonempty subset of $\mathscr{X}$. If $\mathscr{S}$ is a family of maps (not necessarily linear) of Q into Q, then $\mathscr{S}$ is said to be a *noncontracting family of maps* if for two distinct points x and y in Q,

$$0 \notin \text{cl}\{T(x) - T(y): T \in \mathscr{S}\}.$$

The next lemma has a straightforward proof whose discovery is left to the reader.

10.7. Lemma. *If $\mathscr{X}$ is a LCS, $Q \subseteq \mathscr{X}$, and $\mathscr{S}$ is a family of maps of Q into Q, then $\mathscr{S}$ is a noncontracting family if and only if for every pair of distinct points x and y in Q there is a continuous seminorm p such that*

$$\inf\{p(T(x) - T(y)): T \in \mathscr{S}\} > 0.$$

10.8. The Ryll–Nardzewski Fixed-Point Theorem. *If $\mathscr{X}$ is a LCS, Q is a weakly compact convex subset of $\mathscr{X}$, and $\mathscr{S}$ is a noncontracting semigroup of weakly continuous affine maps of Q into Q, then there is a point x_0 in Q such that $T(x_0) = x_0$ for every T in $\mathscr{S}$.*

PROOF. The proof begins by showing that every finite subset of $\mathscr{S}$ has a common fixed point.

10.9. Claim. *If $\{T_1, \ldots, T_n\} \subseteq \mathscr{S}$, then there is an x_0 in Q such that $T_k x_0 = x_0$ for $1 \le k \le n$.*

Put $T_0 = (T_1 + \cdots + T_n)/n$; so $T_0: Q \to Q$ and T_0 is weakly continuous and affine. By (10.1), there is an x_0 in Q such that $T_0(x_0) = x_0$. It will be shown that $T_k(x_0) = x_0$ for $1 \le k \le n$. In fact, if $T_k(x_0) \ne x_0$ for some k, then by renumbering the T_k, it can be assumed that there is an integer m such that $T_k(x_0) \ne x_0$ for $1 \le k \le m$ and $T_k(x_0) = x_0$ for $m < k \le n$. Let $T_0' = (T_1 + \cdots + T_m)/m$. Then

$$x_0 = T_0(x_0)$$
$$= \frac{1}{n}[T_1(x_0) + \cdots + T_m(x_0)] + \left(\frac{n - m}{n}\right)x_0.$$

Hence

$$T_0'(x_0) = \frac{1}{m}[T_1(x_0) + \cdots + T_m(x_0)]$$
$$= \frac{n}{m}\frac{1}{n}[T_1(x_0) + \cdots + T_m(x_0)]$$
$$= \frac{n}{m}\left[x_0 - \left(\frac{n - m}{n}\right)x_0\right]$$
$$= x_0.$$

Thus it may be assumed that $T_k(x_0) \neq x_0$ for all k, but $T_0(x_0) = x_0$. Make this assumption.

By Lemma 10.7, there is an $\varepsilon > 0$ and there is a continuous seminorm p on $\mathcal{X}$ such that for every T in $\mathcal{S}$ and $1 \leq k \leq n$,

10.10
$$p\big(T(T_k(x_0)) - T(x_0)\big) > \varepsilon.$$

Let $\mathcal{S}_1 =$ the semigroup generated by $\{T_1, T_2, \ldots, T_n\}$. So $\mathcal{S}_1 \subseteq \mathcal{S}$ and $\mathcal{S}_1 = \{T_{l_1} \cdots T_{l_m} : m \geq 1, 1 \leq l_j \leq n\}$. Thus $\mathcal{S}_1$ is a countable subsemigroup of $\mathcal{S}$. Put $K = \overline{\text{co}}\{T(x_0) : T \in \mathcal{S}_1\}$. Therefore K is a weakly compact convex subset of Q and K is separable. By Lemma 10.2, there is a closed convex subset C of K such that $C \neq K$ and $p\text{-diam}(K \setminus C) \leq \varepsilon$.

Since $C \neq K$, there is an S in $\mathcal{S}_1$ such that $S(x_0) \in K \setminus C$. Hence

$$S(x_0) = ST_0(x_0) = \frac{1}{n}\big[ST_1(x_0) + \cdots + ST_n(x_0)\big] \in K \setminus C.$$

Since C is convex, there must be a k, $1 \leq k \leq n$, such that $ST_k(x_0) \in K \setminus C$. But this implies that $p(S(T_k(x_0)) - S(x_0)) \leq p\text{-diam}(K \setminus C) \leq \varepsilon$, contradicting (10.10). This establishes Claim 10.9.

Let $\mathcal{F} =$ all finite nonempty subsets of $\mathcal{S}$. If $F \in \mathcal{F}$, let $Q_F = \{x \in Q : T(x) = x \text{ for all } T \text{ in } F\}$. By Claim 10.9, $Q_F \neq \square$ for every F in $\mathcal{F}$. Also, since each T in $\mathcal{S}$ is weakly continuous and affine, Q_F is convex and weakly compact. It is easy to see that $\{Q_F : F \in \mathcal{F}\}$ has the finite intersection property. Therefore, there is an x_0 in $\bigcap\{Q_F : F \in \mathcal{F}\}$. The point x_0 is the desired common fixed point for $\mathcal{S}$. ∎

The original reference for this theorem is Ryll–Nardzewski [1967]; the treatment here is from Namioka and Asplund [1967]. An application of this theorem is given in the next section.

EXERCISES

1. Was local convexity used in the proof of Theorem 10.1?

2. Show that if X is locally compact and $X = \bigcup_{n=1}^{\infty} F_n$, where each F_n is closed in X, then there is an integer n such that int $F_n \neq \square$. (Hint: Look at the proof of the Baire Category Theorem.)

§11*. An Application: Haar Measure on a Compact Group

In this section the operation on all semigroups and groups is denoted by multiplication.

11.1. Definition. A *topological semigroup* is a semigroup G that also is a topological space and such that the map $G \times G \to G$ defined by $(x, y) \mapsto xy$ is continuous. A *topological group* is a topological semigroup that is also a group such that the map $G \to G$ defined by $x \mapsto x^{-1}$ is continuous.

So a topological group is both a group and a topological space with a property that ties these two structures together.

11.2. Examples

(a) $\mathbb{N}$ and $\mathbb{R}_{\geq 0}$ are topological semigroups under addition.
(b) $\mathbb{Z}$, $\mathbb{R}$, and $\mathbb{C}$ are topological groups under addition.
(c) $\partial \mathbb{D}$ is a topological group under multiplication.
(d) If X is a topological space and $G = \{ f \in C(X): f(X) \subset \partial \mathbb{D} \}$, define $(fg)(x) = f(x)g(x)$ for f, g in G and x in X. Then G is a group. If G is given the topology of uniform convergence on X, G is a topological group.
(e) For $n \geq 1$, let $M_n(\mathbb{C})$ = the $n \times n$ matrices with entries in $\mathbb{C}$; $O(n) \equiv \{ A \in M_n(\mathbb{C}): A$ is invertible and $A^{-1} = A^* \}$; $SO(n) \equiv \{ A \in O(n)$: det $A = 1 \}$. If $M_n(\mathbb{C})$ is given the usual topology, $O(n)$ and $SO(n)$ are compact topological groups under multiplication.

There are many more examples and the subject is a self-sustaining area of research. Some good references are Hewitt and Ross [1963] and Rudin [1962].

11.3.   Definition. If S is a semigroup and $f: S \to \mathbb{F}$, then for every x in S define $f_x: S \to \mathbb{F}$ and $_x f: S \to \mathbb{F}$ by $f_x(s) = f(sx)$ and $_x f(s) = f(xs)$ for all s in S. If S is also a group, let $f^{\#}(s) = f(s^{-1})$ for all s in S.

11.4.   Theorem. *If G is a compact topological group, then there is a unique positive regular Borel measure m on G such that*

(a) $m(G) = 1$;
(b) *if U is a nonempty open subset of G, then $m(U) > 0$;*
(c) *if Δ is any Borel subset of G and $x \in G$, then $m(\Delta) = m(\Delta x) = m(x\Delta)$ $= m(\Delta^{-1})$, where $\Delta x \equiv \{ ax: a \in \Delta \}$, $x\Delta \equiv \{ xa: a \in \Delta \}$, and $\Delta^{-1} \equiv \{ a^{-1}: a \in \Delta \}$.*

The measure m is called the *Haar measure* for G. If G is locally compact, then it is also true that there is a positive Borel measure m on G satisfying (b) and such that $m(\Delta x) = m(\Delta)$ for all x in G and every Borel subset Δ of G. It is not necessarily true that $m(\Delta) = m(x\Delta)$, let alone that $m(\Delta) = m(\Delta^{-1})$ (see Exercise 4). The measure m is necessarily unbounded if G is not compact, so that (a) is not possible. Uniqueness, however, is still true in a modified form: if m_1, m_2 are two such measures, then $m_1 = \alpha m_2$ for some $\alpha > 0$.

By using the Riesz Representation Theorem for representing bounded linear functionals on $C(G)$, Theorem 11.4 is equivalent to the following.

11.5. Theorem. *If G is a compact topological group, then there exists a unique positive linear functional $I: C(G) \to \mathbb{F}$ such that*

(a) $I(1) = 1$;
(b) *if $f \in C(G)$, $f \geq 0$, and $f \neq 0$, then $I(f) > 0$;*
(c) *if $f \in C(G)$ and $x \in G$, then $I(f) = I(f_x) = I({}_x f) = I(f^{\#})$.*

Before proving Theorem 11.5, we need the following lemma. For a compact topological group G, if $x \in G$, define $L_x: M(G) \to M(G)$ and $R_x: M(G) \to M(G)$ by

$$\langle f, L_x(\mu)\rangle = \int {}_x f \, d\mu,$$

$$\langle f, R_x(\mu)\rangle = \int f_x \, d\mu$$

for f in $C(G)$ and μ in $M(G)$. Define $S_0: M(G) \to M(G)$ by

$$\langle f, S_0(\mu)\rangle = \int f^{\#} \, d\mu$$

for f in $C(G)$ and μ in $M(G)$. It is easy to check that L_x, R_x, and S_0 are linear isometries of $M(G)$ onto $M(G)$ (Exercise 5).

11.6. Lemma. *If G is a compact topological group, $\mu \in M(G)$, and $\rho: G \times G \to (M(G), \text{wk}^*)$ is defined by $\rho(x, y) = L_x R_y(\mu)$, then ρ is continuous. Similarly, if $\rho_0: G \times G \to (M(G), \text{wk}^*)$ is defined by $\rho_0(x, y) = S_0 L_x R_y(\mu)$, then ρ is continuous.*

PROOF. Let $f \in C(G)$ and let $\varepsilon > 0$. Then (Exercise 10) there is a neighborhood U of e (the identity of G) such that $|f(x) - f(y)| < \varepsilon$ whenever $xy^{-1} \in U$ or $x^{-1}y \in U$. Suppose $\{(x_i, y_i)\}$ is a net in $G \times G$ such that $(x_i, y_i) \to (x, y)$. Let i_0 be such that for $i \geq i_0$, $x_i x^{-1} \in U$ and $y_i^{-1} y \in U$. If $x \in G$, then $|f(x_i z y_i) - f(xzy)| \leq |f(x_i z y_i) - f(xzy_i)| + |f(xzy_i) - f(xzy)|$. But if $i \geq i_0$ and $z \in G$, $(x_i z y_i)(xzy_i)^{-1} = x_i x^{-1} \in U$ and $(xzy_i)^{-1}(xzy) = y_i^{-1} y \in U$. Hence $|f(x_i z y_i) - f(xzy)| < 2\varepsilon$ for $i \geq i_0$ and for all z in G. Thus $\lim_i \int f(x_i z y_i) \, d\mu(z) = \int f(xzy) \, d\mu(z)$. Since f was arbitrary, this implies that $\rho(x_i, y_i) \to \rho(x, y)\text{wk}^*$ in $M(G)$. The proof for ρ_0 is similar. ∎

PROOF OF THEOREM 11.5. If $e =$ the identity of G, then

11.7
$$\begin{cases} L_x R_y = R_y L_x \\ L_x L_y = L_{yx} \\ R_x R_y = R_{xy} \\ S_0^2 = L_e = R_e = \text{the identity on } M(G) \\ S_0 L_x R_y = L_{y^{-1}} R_{x^{-1}} S_0 \end{cases}$$

for x, y in G. Hence

$$\left(S_0 L_x R_y \right)\left(S_0 L_u R_v \right) = \left(L_{y^{-1}} R_{x^{-1}} S_0 \right)\left(S_0 L_u R_v \right)$$
$$= L_{y^{-1}} R_{x^{-1}} L_u R_v$$
$$= L_{y^{-1}} L_u R_{x^{-1}} R_v$$
$$= L_{uy^{-1}} R_{x^{-1}v}.$$

Hence if $S_1 =$ the identity on $M(G)$,

$$\mathscr{S} = \left\{ S_i L_x R_y : i = 0,1;\ x, y \in G \right\}$$

is a group of surjective linear isometries of $M(G)$. Let $Q =$ the probability measures on G; that is, $Q = \{ \mu \in M(G) : \mu \geq 0 \text{ and } \mu(G) = 1 \}$. So Q is a convex subset of $M(G)$ that is wk* compact. Furthermore, $T(Q) \subseteq Q$ for every T in $\mathscr{S}$.

11.8. Claim. If $\mu \in M(G)$ and $\mu \neq 0$, then $0 \notin$ the weak* closure of $\{ T(\mu) : T \in \mathscr{S} \}$.

In fact, Lemma 11.6 implies that $\{ T(\mu) : T \in \mathscr{S} \}$ is weak* closed. Since each T in $\mathscr{S}$ is an isometry, $T(\mu) \neq 0$ for every T in $\mathscr{S}$.

By Claim 11.8, $\mathscr{S}$ is a noncontracting family of affine maps of Q into itself. Moreover, if $T = S_0 L_x R_y$ and $\{ \mu_i \}$ is a net in Q such that $\mu_i \to \mu(\text{wk*})$, then for every f in $C(G)$, $\langle f, T(\mu_i) \rangle = \int f(xs^{-1}y)\, d\mu_i(s) \to \int f(xs^{-1}y)\, d\mu = \langle f, T(\mu) \rangle$. So each T in $\mathscr{S}$ is wk* continuous on Q. By the Ryll–Nardzewski Fixed-Point Theorem, there is a measure m in Q such that $T(m) = m$ for all T in $\mathscr{S}$.

By definition, (a) holds. Also, for any x in G and f in $C(G)$, $\int f(xs)\, dm(s) = \langle f, L_x(m) \rangle = \int f\, dm$. By similar equations, (c) holds. Now suppose $f \in C(G)$, $f \geq 0$, and $f \neq 0$. Then there is an $\varepsilon > 0$ such that $U = \{ x \in G : f(x) > \varepsilon \}$ is nonempty. Since U is open, $G = \bigcup\{ Ux : x \in G \}$, and G is compact, there are $x_1, x_2, \ldots, x_n$ in G such that $G \subseteq \bigcup_{k=1}^n Ux_k$. (Why is Ux open?) Define $g_k(x) = f(xx_k^{-1})$ and put $g = \sum_{k=1}^n g_k$. Then $g \in C(G)$ and $\int g\, dm = \sum_{k=1}^n \int g_k\, dm = n \int f\, dm$ by (c). But for any x in G there is an x_k such that $xx_k^{-1} \in U$; hence $g(x) \geq g_k(x) = f(xx_k^{-1}) > \varepsilon$.

Thus

$$\int f \, dm = \frac{1}{n} \int g \, dm \geq \varepsilon/n > 0.$$

This proves (b).

To prove uniqueness, let μ be a probability measure on G having properties (a), (b), and (c). If $f \in C(G)$ and $x \in G$, then $\int f \, d\mu = \int_x f \, d\mu$. Hence

$$\int f \, d\mu = \int \left[\int f(y) \, d\mu(y) \right] dm(x)$$

$$= \int \left[\int f(xy) \, d\mu(y) \right] dm(x)$$

$$= \int \left[\int f(xy) \, dm(x) \right] d\mu(y)$$

$$= \int \left[\int f(x) \, dm(x) \right] d\mu(y)$$

$$= \int f \, dm.$$

Hence $\mu = m$. ∎

For further information on Haar measure see Nachbin [1965].

What happens if G is only a semigroup? In this case L_x and R_x may not be isometries, so $\{L_x R_y : x, y \in G\}$ may not be noncontractive. However, there are measures for some semigroups that are invariant (see Exercise 7). For further reading see Greenleaf [1969].

EXERCISES

1. Let G be a group and a topological space. Show that G is a topological group if and only if the map of $G \times G \to G$ defined by $(x, y) \mapsto x^{-1}y$ is continuous.

2. Verify the statements in (11.2).

3. Show that Theorems (11.4) and (11.5) are equivalent.

4. Let G be a locally compact group. If m is a regular Borel measure on G, show that any two of the following properties imply the third: (a) $m(\Delta x) = m(\Delta)$ for every Borel set Δ and every x in G; (b) $m(x\Delta) = m(\Delta)$ for every Borel set Δ and every x in G; (c) $m(\Delta) = m(\Delta^{-1})$ for every Borel set Δ.

5. Show that the maps S_0, L_x, R_x are linear isometries of $M(G)$ onto $M(G)$.

6. Prove (11.7).

7. Let S be an abelian semigroup and show that there is a positive linear functional $L: l^\infty(S) \to \mathbb{F}$ such that (a) $L(1) = 1$, (b) $L(f_x) = L(f)$ for every f in $l^\infty(S)$.

8. Show that if S and L are as in Exercise 7, and S is infinite, then $L(f) = 0$ whenever $\{s \in S: f(s) \neq 0\}$ is finite.

9. If $S = \mathbb{N}$, what does Exercise 7 say about Banach limits?

10. If G is a compact group, $f: G \to \mathbb{F}$ is a continuous function, and $\varepsilon > 0$, show that there is a neighborhood U of the identity in G such that $|f(x) - f(y)| < \varepsilon$ whenever $xy^{-1} \in U$. (Note that this says that every continuous function on a compact group is uniformly continuous.)

11. If G is a locally compact group and $f \in C_b(G)$, let $\mathcal{O}(f) \equiv$ the closure of $\{f_x: x \in G\}$ in $C_b(G)$. Let $AP(G) = \{f \in C_b(G): \mathcal{O}(f)$ is compact$\}$. Functions in $AP(G)$ are called *almost periodic*. (a) Show that every periodic function in $C_b(\mathbb{R})$ belongs to $AP(\mathbb{R})$. (b) If G is compact, show that $AP(G) = C(G)$. (c) Show that if $f \in C_b(\mathbb{R})$, then $f \in AP(\mathbb{R})$ if and only if for every $\varepsilon > 0$ there is a positive number T such that in every interval of length T there is a number p such that $|f(x) - f(x + p)| < \varepsilon$ for all x in $\mathbb{R}$. (d) If G is not compact, then the only function in $AP(G)$ having compact support is the zero function. (e) Prove that there is a bounded linear functional $L: AP(G) \to \mathbb{F}$ such that $L(1) = 1$, $L(f) \geq 0$ if $f \geq 0$, and $L(f_x) = L(f)$ for all f in $AP(G)$ and x in G.

§12*. The Krein–Smulian Theorem

Let A be a convex subset of a Banach space $\mathscr{X}$. If A is weakly closed, then for every $r > 0$, $A \cap \{x \in \mathscr{X}: \|x\| \leq r\}$ is weakly closed; this is clear since each of the sets in the intersection is weakly closed. But the converse of this is also true: if A is convex and $A \cap \{X \in \mathscr{X}: \|x\| \leq r\}$ is weakly closed for every $r > 0$, then A is weakly closed. In fact, because A is convex it suffices to prove that A is norm closed (Corollary 1.5). If $\{x_n\} \subseteq A$ and $\|x_n - x_0\| \to 0$, then there is a constant r such that $\|x_n\| \leq r$ for all n. By hypothesis, $A \cap \{x \in \mathscr{X}: \|x\| \leq r\}$ is weakly closed and hence norm closed. Thus $x_0 \in A$.

Now let A be a convex subset of $\mathscr{X}^*$, $\mathscr{X}$ a Banach space. If $A \cap \{x^* \in \mathscr{X}^*: \|x^*\| \leq r\}$ is weak-star closed for every $r > 0$, is A weak-star closed? If $\mathscr{X}$ is reflexive, then this is the same question that was asked and answered affirmatively in the preceding paragraph. If $\mathscr{X}$ is not reflexive, then the preceding argument fails since there are norm-closed convex subsets of $\mathscr{X}^*$ that are not weak-star closed. (Example: let $x^{**} \in \mathscr{X}^{**} \setminus \mathscr{X}$ and consider $A = \ker x^{**}$.) Nevertheless, even though the argument fails, the statement is true.

12.1. The Krein–Smulian Theorem. *If $\mathscr{X}$ is a Banach space and A is a convex subset of $\mathscr{X}^*$ such that $A \cap \{x^* \in \mathscr{X}^*: \|x^*\| \leq r\}$ is weak-star closed for every $r > 0$, then A is weak-star closed.*

To prove this theorem, two lemmas are needed.

12.2.   Lemma. *If $\mathscr{X}$ is a Banach space, $r > 0$, and $\mathscr{F}_r$ is the collection of all finite subsets of $\{x \in \mathscr{X}: \|x\| \leq r^{-1}\}$, then*

$$\cap\{F^0: F \in \mathscr{F}_r\} = \{x^* \in \mathscr{X}^*: \|x^*\| \leq r\}.$$

PROOF. Let $E = \cap\{F^0: F \in \mathscr{F}_r\}$; it is easy to see that $r(\text{ball } \mathscr{X}^*) \subseteq E$. If $x^* \notin r(\text{ball } \mathscr{X}^*)$, then there is an x in ball $\mathscr{X}$ such that $|\langle x, x^* \rangle| > r$. Hence $|\langle r^{-1}x, x^* \rangle| > 1$ and $x^* \notin E$.   ∎

12.3.   Lemma. *If A and $\mathscr{X}$ satisfy the hypothesis of the Krein–Smulian Theorem and, moreover, $A \cap \text{ball } \mathscr{X}^* = \square$, then there is an x in $\mathscr{X}$ such that*

$$\text{Re}\langle x, x^* \rangle \geq 1$$

for all x^* in A.

PROOF. The proof begins by showing that there are finite subsets $F_0, F_1, \ldots$ of $\mathscr{X}$ such that

12.4 $\begin{cases} \text{(i)   } nF_n \subseteq \text{ball } \mathscr{X}; \\ \text{(ii) } n(\text{ball } \mathscr{X}^*) \cap \cap_{k=0}^{n-1} F_k^0 \cap A = \square. \end{cases}$

To establish (12.4) use induction as follows. Let $F_0 = (0)$. Suppose that $F_0, \ldots, F_{n-1}$ have been chosen satisfying (12.4) and set $Q = [(n + 1)\text{ball } \mathscr{X}^*] \cap \cap_{k=0}^{n-1} F_k^0 \cap A$. Note that Q is wk* compact. So if $Q \cap F^0 \neq \square$ for every finite subset F of $n^{-1}\text{ball } \mathscr{X}$, then $\square \neq Q \cap \cap\{F^0: F$ is a finite subset of $n^{-1}(\text{ball } \mathscr{X})\} = Q \cap [n(\text{ball } \mathscr{X}^*)]$ by the preceding lemma. This contradicts (12.4ii). Therefore there is a finite subset F_n of $n^{-1}(\text{ball } \mathscr{X})$ such that $Q \cap F_n^0 = \square$. This proves (12.4).

If $\{F_n\}_{n=1}^\infty$ satisfies (12.4), then $A \cap \cap_{n=1}^\infty F_n^0 = \square$. Arrange the elements of $\cup_{n=1}^\infty F_n$ in a sequence and denote this sequence by $\{x_n\}$. Note that $\lim\|x_n\| = 0$. Thus if $x^* \in \mathscr{X}^*$, $\{\langle x_n, x^* \rangle\} \in c_0$. Define $T: \mathscr{X}^* \to c_0$ by $T(x^*) = \{\langle x_n, x^* \rangle\}$. It is easy to see that T is linear (and bounded, though this fact is unnecessary). Hence $T(A)$ is a convex subset of c_0. Also, from the construction of $\{x_n\} = \cup_{n=1}^\infty F_n$, for each x^* in A, $\|T(x^*)\| = \sup_n|\langle x_n, x^* \rangle| > 1$. That is, $T(A) \cap \text{ball } c_0 = \square$. Thus Theorem III.3.7 applies to the sets $T(A)$ and $\text{int}[\text{ball } c_0]$ and there is an f in $l^1 = c_0^*$ and an α in $\mathbb{R}$ such that $\text{Re}\langle \phi, f \rangle < \alpha \leq \text{Re}\langle T(x^*), f \rangle$ for every ϕ in $\text{int}[\text{ball } c_0]$ and x^* in A. That is,

12.5 $$\text{Re} \sum_{n=1}^\infty \phi(n)f(n) < \alpha \leq \text{Re} \sum_{n=1}^\infty \langle x_n, x^* \rangle f(n)$$

for every ϕ in c_0 with $\|\phi\| < 1$ and for every x^* in A. Replacing f by $f/\|f\|$ and α by $\alpha/\|f\|$, it is clear that it may be assumed that (12.5) holds with $\|f\| = 1$. If $\phi \in c_0$, $\|\phi\| < 1$, let $\mu \in \mathbb{F}$ such that $|\mu| = 1$ and $\langle \mu\phi, f \rangle = |\langle \phi, f \rangle|$. Applying this to (12.5) and taking the supremum over all ϕ in int[ball c_0] gives that $1 \leq \text{Re} \sum_{n=1}^{\infty} \langle x_n, x^* \rangle f(n)$ for all x^* in A. But $f \in l^1$ so $x = \sum_{n=1}^{\infty} f(n) x_n \in \mathcal{X}$ and $1 \leq \text{Re} \langle x, x^* \rangle$ for all x^* in A. ∎

Where was the completeness of $\mathcal{X}$ used in the preceding proof?

PROOF OF THE KREIN–SMULIAN THEOREM. Let $x_0^* \in \mathcal{X}^* \setminus A$; it will be shown that $x_0^* \notin wk^* - \text{cl } A$. It is easy to see that A is norm closed. So there is an $r > 0$ such that $\{ x^* \in \mathcal{X}^*: \|x^* - x_0^*\| \leq r \} \cap A = \square$. But this implies that ball $\mathcal{X}^* \cap [r^{-1}(A - x_0^*)] = \square$. With this it is easy to see that $r^{-1}(A - x_0^*)$ satisfies the hypothesis of the preceding lemma. Therefore there is an x in $\mathcal{X}$ such that $\text{Re} \langle x, x^* \rangle \geq 1$ for all x^* in $r^{-1}(A - x_0^*)$. In particular, $0 \notin wk^* - \text{cl}[r^{-1}(A - x_0^*)]$ and hence $x_0^* \notin wk^* - \text{cl } A$. ∎

12.6. Corollary. *If $\mathcal{X}$ is a Banach space and $\mathcal{Y}$ is a linear manifold in $\mathcal{X}^*$, then $\mathcal{Y}$ is weak-star closed if and only if $\mathcal{Y} \cap$ ball $\mathcal{X}^*$ is weak-star closed.*

12.7. Corollary. *If $\mathcal{X}$ is a separable Banach space and A is a convex subset of $\mathcal{X}^*$ that is weak-star sequentially closed, then A is weak-star closed.*

PROOF. Because $\mathcal{X}$ is separable, $r(\text{ball } \mathcal{X}^*)$ is weak-star metrizable for every $r > 0$ (Theorem 5.1). So if A is weak-star sequentially closed, $A \cap [r(\text{ball } \mathcal{X}^*)]$ is weak-star closed for every $r > 0$. Hence the Krein–Smulian Theorem applies. ∎

This last corollary is one of the most useful forms of the Krein-Smulian Theorem. To show that a convex subset A of $\mathcal{X}^*$ is weak-star closed it is not necessary to show that every weak-star convergent net from A has its limit in A; it suffices to prove this for sequences.

12.8. Corollary. *If $\mathcal{X}$ is a separable Banach space and $F: \mathcal{X}^* \to \mathbb{F}$ is a linear functional, then F is weak-star continuous if and only if F is weak-star sequentially continuous.*

PROOF. By Theorem IV.3.1, F is wk* continuous if and only if ker F is wk* closed. This corollary is, therefore, a direct consequence of the preceding one. ∎

There is a misinterpretation of the Krein–Smulian Theorem that the reader should be warned about. If A is a weak-star closed convex subset of ball $\mathcal{X}^*$, let $\mathcal{M} = \bigcup \{ rA: r > 0 \}$. It is easy to see that $\mathcal{M}$ is a linear manifold, but it does not follow that $\mathcal{M}$ is weak-star closed. What is true is the following.

12.9. Theorem. *Let $\mathcal{X}$ be a Banach space and let A be a weak-star closed subset of $\mathcal{X}^*$. If $\mathcal{Y} =$ the linear span of A, then $\mathcal{Y}$ is norm closed in $\mathcal{X}^*$ if and only if $\mathcal{Y}$ is weak-star closed.*

The proof will not be presented here. The interested reader can consult Dunford and Schwartz [1958], p. 429.

There is a method for finding the weak-star closure of a linear manifold that is quite useful despite its seemingly bizarre appearance. Let $\mathscr{X}$ be a Banach space and let $\mathscr{M}$ be a linear manifold in $\mathscr{X}^*$. For each ordinal number α define a linear manifold $\mathscr{M}_\alpha$ as follows. Let $\mathscr{M}_1 = \mathscr{M}$. Suppose α is an ordinal number and $\mathscr{M}_\beta$ has been defined for each ordinal $\beta < \alpha$. If α has an immediate predecessor, $\alpha - 1$, let $\mathscr{M}_\alpha$ be the weak-star sequential closure of $\mathscr{M}_{\alpha-1}$. If α is a limit ordinal and has no immediate predecessor, let $\mathscr{M}_\alpha = \bigcup\{\mathscr{M}_\beta : \beta < \alpha\}$. In each case $\mathscr{M}_\alpha$ is a linear manifold in $\mathscr{X}^*$ and $\mathscr{M}_\beta \subseteq \mathscr{M}_\alpha$ if $\beta \le \alpha$.

12.10. Theorem. *If $\mathscr{X}$ is a separable Banach space, $\mathscr{M}$ is a linear manifold in $\mathscr{X}^*$, and $\mathscr{M}_\alpha$ is defined as above for every ordinal number α, then $\mathscr{M}_\Omega$ is the weak-star closure of $\mathscr{M}$, where Ω is the first uncountable ordinal. Moreover, there is an ordinal number $\alpha < \Omega$ such that $\mathscr{M}_\alpha = \mathscr{M}_\Omega$.*

PROOF. By Corollary 12.7 it suffices to show that $\mathscr{M}_\Omega$ is weak-star sequentially closed. Let $\{x_n^*\}$ be a sequence in $\mathscr{M}_\Omega$ such that $x_n^* \to x^*$ (wk*). Since $\mathscr{M}_\Omega = \bigcup\{\mathscr{M}_\alpha : \alpha < \Omega\}$, for each n there is an $\alpha_n < \Omega$ such that $x_n^* \in \mathscr{M}_{\alpha_n}$. But $\alpha = \sup_n \alpha_n < \Omega$. Hence $x_n^* \in \mathscr{M}_\alpha$ for all n; thus $x \in \mathscr{M}_{\alpha+1} \subseteq \mathscr{M}_\Omega$ and $\mathscr{M}_\Omega$ is weak-star closed.

To see that $\mathscr{M}_\Omega = \mathscr{M}_\alpha$ for some $\alpha < \Omega$, let $\{x_n^*\}$ be a countable wk* dense subset of ball $\mathscr{M}_\Omega$. For each n there is an α_n such that $x_n^* \in \mathscr{M}_{\alpha_n}$. Put $\alpha = \sup_n \alpha_n$. So $\{x_n^*\} \subset$ ball $\mathscr{M}_\alpha$. Put ball $\mathscr{M}_\Omega$ is a compact metric space in the weak-star topology, so $\{x_n^*\}$ is wk* sequentially dense in ball $\mathscr{M}_\Omega$. Therefore ball $\mathscr{M}_\Omega \subseteq$ ball $\mathscr{M}_{\alpha+1}$ and $\mathscr{M}_\Omega = \mathscr{M}_{\alpha+1}$. ∎

When is $\mathscr{M}$ weak-star sequentially dense in $\mathscr{X}^*$? The following result of Banach answers this question.

12.11. Theorem. *If $\mathscr{X}$ is a separable Banach space and $\mathscr{M}$ is a linear manifold in $\mathscr{X}^*$, then the following statements are equivalent.*

(a) *$\mathscr{M}$ is weak-star sequentially dense in $\mathscr{X}^*$.*
(b) *There is a positive constant c such that for every x in $\mathscr{X}$,*

$$\|x\| \le \sup\{|\langle x, x^*\rangle| : x^* \in \mathscr{M}, \|x^*\| \le c\}.$$

(c) *There is a positive constant c such that if $x^* \in$ ball $\mathscr{X}^*$, there is a sequence $\{x_k^*\}$ in $\mathscr{M}$, $\|x_k^*\| \le c$, such that $x_k^* \to x^*$ (wk*).*

PROOF. It is clear that (c) implies (a). The proof will consist in showing that (a) implies (c) and that (b) and (c) are equivalent.

(a) $\Rightarrow$ (c): For each positive integer n, let $A_n = $ the wk* closure of $n(\text{ball } \mathscr{M})$. If $x^* \in \mathscr{X}^*$, let $\{x_k^*\}$ be a sequence in $\mathscr{M}$ such that $x_k^* \to x^*$ (wk*). By the PUB, there is an n such that $\|x_k^*\| \le n$ for all k. Hence

$x^* \in A_n$. That is, $\bigcup_{n=1}^{\infty} A_n = \mathscr{X}^*$. Clearly each A_n is norm closed, so the Baire Category Theorem implies that there is an A_n that has interior in the norm topology. Thus there is an x_0^* in A_n and an $r > 0$ such that $A_n \supseteq \{x^* \in \mathscr{X}^*: \|x^* - x_0^*\| \leq r\}$. Let $\{x_k^*\} \subseteq n(\text{ball } \mathscr{M})$ such that $x_k^* \to x_0^*$ (wk*). If $x^* \in \text{ball } \mathscr{X}^*$, then $x_0^* + rx^* \in A_n$; hence there is a sequence $\{y_k^*\}$ in $n(\text{ball } \mathscr{M})$ such that $y_k^* \to x_0^* + rx^*$ (wk*). Thus $r^{-1}(y_k^* - x_k^*) \to x^*$ (wk*) and $r^{-1}(y_k^* - x_k^*) \in c(\text{ball } \mathscr{M})$, where $c = 2n/r$ is independent of x^*.

(c) $\Rightarrow$ (b): If $x \in \mathscr{X}$, then Alaoglu's Theorem implies there is an x^* in ball $\mathscr{X}^*$ such that $\langle x, x^* \rangle = \|x\|$. By (c), there is a sequence $\{x_k^*\}$ in $c(\text{ball } \mathscr{M})$ such that $x_k^* \to x^*$ (wk*). Thus $\langle x_k^*, x \rangle \to \|x\|$ and (b) holds.

(b) $\Rightarrow$ (c): According to (b), ball $\mathscr{X} \supseteq {}^{\circ}[c(\text{ball } \mathscr{M})]$. Hence ball $\mathscr{X}^* = (\text{ball } \mathscr{X})^{\circ} \subseteq {}^{\circ}[c(\text{ball } \mathscr{M})]^{\circ}$. By (1.8), ${}^{\circ}[c(\text{ball } \mathscr{M})]^{\circ} = $ the weak-star closure of $c(\text{ball } \mathscr{M})$. But bounded subsets of $\mathscr{X}^*$ are weak-star metrizable (5.1) and hence (c) follows. ∎

EXERCISES

1. Suppose $\mathscr{X}$ is a normed space and that the only hyperplanes $\mathscr{M}$ in $\mathscr{X}^*$ such that $\mathscr{M} \cap$ ball $\mathscr{X}^*$ is weak-star closed are those that are weak-star closed. Prove that $\mathscr{X}$ is a Banach space.

2. (von Neumann) Let A be the subset of l^2 consisting of all vectors $\{x_{mn}: 1 \leq m < n < \infty\}$ where $x_{mn}(m) = 1$, $x_{mn}(n) = m$, and $x_{mn}(k) = 0$ if $k \neq m, n$. Show that $0 \in wk - \text{cl } A$ but no sequence in A converges weakly to 0.

3. Where were the hypotheses of the separability and completeness of $\mathscr{X}$ used in the proof of Theorem 12.11?

4. Let $\mathscr{X}$ be a separable Banach space. If $\mathscr{M}$ is a linear manifold in $\mathscr{X}^*$ give necessary and sufficient conditions that every functional in $wk^* - \text{cl } \mathscr{M}$ be the wk^* limit of a sequence from $\mathscr{M}$.

5. Let $\mathscr{X}$ be a normed space and let $\mathscr{T}$ be a locally convex topology on $\mathscr{X}$ such that ball $\mathscr{X}$ is $\mathscr{T}$-compact. Show that there is a Banach space $\mathscr{Y}$ such that $\mathscr{X}$ is isometrically isomorphic to $\mathscr{Y}^*$. (Hint: Let $\mathscr{Y} = \{x^* \in \mathscr{X}^*: x^*|\text{ball } \mathscr{X}$ is $\mathscr{T}$-continuous$\}$.)

§13*. Weak Compactness

In this section, two results are stated without proof. These results are among the deepest in the study of weak topologies.

13.1. The Eberlein–Smulian Theorem. *If $\mathscr{X}$ is a Banach space and $A \subseteq \mathscr{X}$, then the following statements are equivalent.*

(a) *Each sequence of elements of A has a subsequence that is weakly convergent.*

(b) *Each sequence of elements of A has a weak cluster point.*
(c) *The weak closure of A is weakly compact.*

The proof can be found in Dunford and Schwartz [1958], p. 430. The serious student should examine Chapter V of Dunford and Schwartz [1958] for several results not presented here as well as for some of the history behind the material of this chapter.

The following is an easy consequence of the Eberlein–Smulian Theorem.

13.2. Corollary. *If $\mathscr{X}$ is a Banach space and $A \subseteq \mathscr{X}$, then A is weakly compact if and only if $A \cap \mathscr{M}$ is weakly compact for every separable subspace $\mathscr{M}$ of $\mathscr{X}$.*

If $\mathscr{X}$ is a Banach space and A is a weakly compact subset of $\mathscr{X}$, then for each x^* in $\mathscr{X}^*$ there is an x_0 in A such that $|\langle x_0, x^* \rangle| = \sup\{|\langle x, x^* \rangle| : x \in A\}$. It is a rather deep fact due to R. C. James [1964a] that the converse is true.

13.3. James's Theorem. *If $\mathscr{X}$ is a Banach space and A is a closed convex subset of $\mathscr{X}$ such that for each x^* in $\mathscr{X}^*$ there is an x_0 in A with*

$$|\langle x_0, x^* \rangle| = \sup\{|\langle x, x^* \rangle| : x \in A\},$$

then A is weakly compact.

Another reference for a proof of this theorem as well as a number of other equivalent formulations of weak compactness and reflexivity is James [1964b]. Also, if $\mathscr{X}$ is only assumed to be a normed space in Theorem 13.2, the conclusion is false (see James [1971]).

The next result, presented with proof, is also called the Krein–Smulian Theorem and must not be confused with the theorem of the preceding section.

13.4. Krein–Smulian Theorem. *If $\mathscr{X}$ is a Banach space and K is a weakly compact subset of $\mathscr{X}$, then $\overline{\text{co}}(K)$ is weakly compact.*

PROOF. *Case 1*: $\mathscr{X}$ *is separable.* Endow K with the relative weak topology; so $M(K) = C(K)^*$. If $\mu \in M(K)$, define $F: \mathscr{X}^* \to \mathbb{F}$ by

$$F_\mu(x^*) = \int_K \langle x, x^* \rangle \, d\mu(x).$$

It is easy to see that F_μ is a bounded linear functional on $\mathscr{X}^*$ and $\|F_\mu\| \leq \|\mu\| \sup\{\|x\| : x \in K\}$.

13.5. Claim. $F_\mu: \mathscr{X}^* \to \mathbb{F}$ *is weak-star continuous.*

By (12.8) it suffices to show that F_μ is weak* sequentially continuous. Let $\{x_n^*\}$ be a sequence in $\mathscr{X}^*$ such that $x_n^* \to x^*$ (wk*). By the PUB, $M = \sup_n \|x_n^*\| < \infty$. Also, $\langle x, x_n^* \rangle \to \langle x, x^* \rangle$ for every x in K. By the Lebesgue Dominated Convergence Theorem, $F_\mu(x_n^*) = \int \langle x, x_n^* \rangle \, d\mu(x) \to F_\mu(X^*)$. So (13.5) is established.

By (1.3), $F_\mu \in \mathscr{X}$. That is, there is an x_μ in $\mathscr{X}$ such that $F_\mu(x^*) = \langle x_\mu, x^* \rangle$. Define $T: M(K) \to \mathscr{X}$ by $T(\mu) = x_\mu$.

13.6. Claim. $T: (M(K), \text{wk*}) \to (\mathscr{X}, \text{wk})$ is continuous.

In fact, this is clear. If $\mu_i \to 0$ weak* in $M(K)$, then for each x^* in $\mathscr{X}^*$, $x^*|K \in C(K)$. Hence $\langle T(\mu_i), x^* \rangle = \int \langle x, x^* \rangle \, d\mu_i(x) \to 0$.

Let $\mathscr{P} = $ the probability measures on K. By Alaoglu's Theorem $\mathscr{P}$ is weak* compact. Thus $T(\mathscr{P})$ is weakly compact and convex. However, if $x \in K$, $\langle T(\delta_x), x^* \rangle = \langle x, x^* \rangle$; that is, $T(\delta_x) = x$. So $T(\mathscr{P}) \supseteq K$. Hence $T(\mathscr{P}) \supseteq \overline{\text{co}}(K)$ and $\overline{\text{co}}(K)$ must be compact.

Case 2: $\mathscr{X}$ is arbitrary. Let $\{x_n\}$ be a sequence in $\overline{\text{co}}(K)$. So for each n there is a finite subset F_n of K such that $x_n \in \text{co}(F_n)$. Let $F = \bigcup_{n=1}^\infty F_n$ and let $\mathscr{M} = \bigvee F$. Then $K_1 = K \cap \mathscr{M}$ is weakly compact and $\{x_n\} \subseteq \overline{\text{co}}(K_1)$. Since $\mathscr{M}$ is separable, Case 1 implies that $\overline{\text{co}}(K_1)$ is weakly compact. By the Eberlein–Smulian Theorem, there is a subsequence $\{x_{n_k}\}$ and an x in $\overline{\text{co}}(K_1) \subseteq \overline{\text{co}}(K)$ such that $x_{n_k} \to x$. Thus $\overline{\text{co}}(K)$ is weakly compact. $\blacksquare$

EXERCISES

1. Prove Corollary 13.2.

2. If $\mathscr{X}$ is a Banach space and K is a compact subset of $\mathscr{X}$, prove that $\overline{\text{co}}(K)$ is compact.

3. In the proof of (13.4), if $\mathscr{P} = $ the probability measures on K, show that $T(\mathscr{P}) = \overline{\text{co}}(K)$.

4. Prove the Eberlein–Smulian Theorem in the setting of Hilbert space.

CHAPTER VI

Linear Operators on a Banach Space

As has been said before in this book, the theory of bounded linear operators on a Banach space has seen relatively little activity owing to the difficult geometric problems inherent in the concept of a Banach space. In this chapter several of the general concepts of this theory are presented. When combined with the few results from the next chapter, they constitute essentially the whole of the general theory of these operators.

We begin with a study of the adjoint of a Banach space operator. Unlike the adjoint of an operator on a Hilbert space (Section II.2), the adjoint of a bounded linear operator on a Banach space does not operate on the space but on the dual space.

§1. The Adjoint of a Linear Operator

Suppose $\mathcal{X}$ and $\mathcal{Y}$ are vector spaces and $T \colon \mathcal{X} \to \mathcal{Y}$ is a linear transformation. Let $\mathcal{Y}' = $ all of the linear functionals of $\mathcal{Y} \to \mathbb{F}$. If $y' \in \mathcal{Y}'$, then $y' \circ T \colon \mathcal{X} \to \mathbb{F}$ is easily seen to be a linear functional on $\mathcal{X}$. That is, $y' \circ T \in \mathcal{X}'$. This defines a map

$$T' \colon \mathcal{Y}' \to \mathcal{X}'$$

by $T'(y') = y' \circ T$. The first result shows that if $\mathcal{X}$ and $\mathcal{Y}$ are Banach spaces, then the map T' can be used to determine when T is bounded. Another equivalent formulation of boundedness is given by means of the weak topology.

1.1. **Theorem.** *If $\mathscr{X}$ and $\mathscr{Y}$ are Banach spaces and $T: \mathscr{X} \to \mathscr{Y}$ is a linear transformation, then the following statements are equivalent.*

(a) *T is bounded.*

(b) *$T'(\mathscr{Y}^*) \subseteq \mathscr{X}^*$.*

(c) *$T: (\mathscr{X}, \text{weak}) \to (\mathscr{Y}, \text{weak})$ is continuous.*

PROOF. (a) $\Rightarrow$ (b): If $y^* \in \mathscr{Y}^*$, then $T'(y^*) \in \mathscr{X}'$; it must be shown that $T'(y^*) \in \mathscr{X}^*$. But $|T'(y^*)(x)| = |y^* \circ T(x)| = |\langle T(x), y^* \rangle| \le \|T(x)\| \|y^*\| \le \|T\| \|y^*\| \|x\|$. So $T'(y^*) \in \mathscr{X}^*$.

(b) $\Rightarrow$ (c): If $\{x_i\}$ is a net in $\mathscr{X}$ and $x_i \to 0$ weakly, then for y^* in $\mathscr{Y}^*$, $\langle T(x_i), y^* \rangle = T'(y^*)(x_i) \to 0$ since $T'(y^*) \in \mathscr{X}^*$. Hence $T(x_i) \to 0$ weakly in $\mathscr{Y}$.

(c) $\Rightarrow$ (b): If $y^* \in \mathscr{Y}^*$, then $y^* \circ T: \mathscr{X} \to \mathbb{F}$ is weakly continuous by (c). Hence $T'(y^*) = y^* \circ T \in \mathscr{X}^*$ by (V.1.2).

(b) $\Rightarrow$ (a): Let $y^* \in \mathscr{Y}^*$ and put $x^* = T'(y^*)$. So $x^* \in \mathscr{X}^*$ by (b). So if $x \in$ ball $\mathscr{X}$, $|\langle T(x), y^* \rangle| = |\langle x, x^* \rangle| \le \|x^*\|$. That is, $\sup\{|\langle T(x), y^* \rangle|: x \in$ ball $\mathscr{X}\} < \infty$. Hence $T(\text{ball } \mathscr{X})$ is weakly bounded; by the PUB, $T(\text{ball } \mathscr{X})$ is norm bounded and so $\|T\| < \infty$. ∎

The preceding result is useful, though strictly speaking it is not necessary for the purpose of defining the adjoint of an operator A in $\mathscr{B}(\mathscr{X}, \mathscr{Y})$, which we now turn to. If $A \in \mathscr{B}(\mathscr{X}, \mathscr{Y})$ and $y^* \in \mathscr{Y}^*$, then $y^* \circ A = A'(y^*) \in \mathscr{X}^*$. This defines a map $A^*: \mathscr{Y}^* \to \mathscr{X}^*$, where $A^* = A'|\mathscr{Y}^*$. Hence

1.2 $$\langle x, A^*(y^*) \rangle = \langle A(x), y^* \rangle$$

for x in $\mathscr{X}$ and y^* in $\mathscr{Y}^*$. A^* is called the *adjoint* of A.

Before exploring the concept let's see how this compares with the definition of the adjoint of an operator on Hilbert space given in § II.2. There is a difference, but only a small one. When $\mathscr{H}$ is identified with $\mathscr{H}^*$, the dual space of $\mathscr{H}$, the identification is not linear but conjugate linear (if $\mathbb{F} = \mathbb{C}$). The isometry $h \mapsto L_h$ of $\mathscr{H}$ onto $\mathscr{H}^*$, where $L_h(f) = \langle f, h \rangle$, satisfies $L_{\alpha h} = \bar{\alpha} L_h$. Thus the definition of A^* given in (1.2) above is not the same as the adjoint of an operator on Hilbert space, since in (1.2) A^* is defined on $\mathscr{Y}^*$ and not some conjugate-linear isomorphic image of it. In particular, if the definition (1.2) is applied to a matrix A acting on $\mathbb{C}^d$ considered as a Banach space, its adjoint corresponds to the transpose of A. If $\mathbb{C}^d$ is considered as a Hilbert space, then the matrix of A^* is the conjugate transpose of the matrix of A. This difference will not confuse us but it will serve to explain minor differences that will appear in the treatment of the two types of adjoints. The first of these occurs in the next result.

1.3. **Proposition.** *If $\mathscr{X}$ and $\mathscr{Y}$ are Banach spaces, $A, B \in \mathscr{B}(\mathscr{X}, \mathscr{Y})$, and $\alpha, \beta \in \mathbb{F}$, then $(\alpha A + \beta B)^* = \alpha A^* + \beta B^*$.*

Note the absence of conjugates. The proof is left to the reader.

If $A \in \mathscr{B}(\mathscr{X}, \mathscr{Y})$, then it is easy to see that $A^* \in \mathscr{B}(\mathscr{Y}^*, \mathscr{X}^*)$. In fact, if $y^* \in$ ball $\mathscr{Y}^*$ and $x \in$ ball $\mathscr{X}$, then $|\langle x, A^*y^* \rangle| = |\langle Ax, y^* \rangle| \le \|Ax\| \le \|A\|$. Hence $\|A^*y^*\| \le \|A\|$ if $y^* \in$ ball $\mathscr{Y}^*$, so that $\|A^*\| \le \|A\|$. This implies that $(A^*)^* \equiv A^{**}$ can be defined,

$$A^{**}: \mathscr{X}^{**} \to \mathscr{Y}^{**},$$
$$\langle A^{**}x^{**}, y^* \rangle = \langle x^{**}, A^*y^* \rangle$$

for x^{**} in $\mathscr{X}^{**}$ and y^* in $\mathscr{Y}^*$.

Suppose $x \in \mathscr{X}$ and consider x as an element of $\mathscr{X}^{**}$ via the natural embedding of $\mathscr{X}$ into its double dual. What is $A^{**}(x)$? For y^* in $\mathscr{Y}^*$,

$$\langle A^{**}(x), y^* \rangle = \langle x, A^*y^* \rangle$$
$$= \langle Ax, y^* \rangle.$$

That is, $A^{**}|\mathscr{X} = A$. This is the first part of the next proposition.

1.4. Proposition. *If $\mathscr{X}$ and $\mathscr{Y}$ are Banach spaces and $A \in \mathscr{B}(\mathscr{X}, \mathscr{Y})$, then:*

(a) $A^{**}|\mathscr{X} = A$;
(b) $\|A^*\| = \|A\|$;
(c) *if A is invertible, then A^* is invertible and $(A^*)^{-1} = (A^{-1})^*$;*
(d) *if $\mathscr{Z}$ is a Banach space and $B \in \mathscr{B}(\mathscr{Y}, \mathscr{Z})$, then $(BA)^* = A^*B^*$.*

PROOF. Part (a) was proved above. It was also shown that $\|A^*\| \le \|A\|$. Thus $\|A^{**}\| \le \|A^*\|$. So if $x \in$ ball $\mathscr{X}$, then (a) implies that $\|Ax\| = \|A^{**}x\| \le \|A^{**}\| \le \|A^*\|$. Hence $\|A\| \le \|A^*\|$.

The remainder of the proof is left to the reader. ∎

1.5.   Example. Let (X, Ω, μ) and M_ϕ: $L^p(\mu) \to L^p(\mu)$ be as in Example III.2.2. If $1 \le p < \infty$ and $1/p + 1/q = 1$, then M_ϕ^*: $L^q(\mu) \to L^q(\mu)$ is given by $M_\phi^* f = \phi f$. That is, $M_\phi^* = M_\phi$.

1.6.   Example. Let K and k be as in Example III.2.3. If $1 \le p < \infty$ and $1/p + 1/q = 1$, then K^*: $L^q(\mu) \to L^q(\mu)$ is the integral operator with kernel $k^*(x, y) \equiv k(y, x)$.

1.7.   Example. Let X, Y, τ, and A be as in Example III.2.4. Then A^*: $M(Y) \to M(X)$ is given by

$$(A^*\mu)(\Delta) = \mu(\tau^{-1}(\Delta))$$

for every Borel subset Δ of X and every μ in $M(Y)$.

Compare (1.5) and (1.6) with (II.2.8) and (II.2.9) to see the contrast between the adjoint of an operator on Banach space with the adjoint of a Hilbert space operator.

1.8.   Proposition. *If $A \in \mathscr{B}(\mathscr{X}, \mathscr{Y})$, then $\ker A^* = (\operatorname{ran} A)^\perp$ and $\ker A = {}^\perp(\operatorname{ran} A^*)$.*

The proof of this useful result is similar to that of Proposition II.2.19 and is left to the reader.

This enables us to prove the converse of Proposition 1.4c.

1.9. Proposition. *If $A \in \mathcal{B}(\mathcal{X}, \mathcal{Y})$, then A is invertible if and only if A^* is invertible.*

PROOF. In light of (1.4c) it suffices to assume that A^* is invertible and show that A is invertible. By the Open Mapping Theorem, there is a constant $c > 0$ such that $A^*(\text{ball } \mathcal{Y}^*) \supseteq \{x^* \in \mathcal{X}^* : \|x^*\| \leq c\}$. So if $x \in \mathcal{X}$, then

$$\|Ax\| = \sup\{|\langle Ax, y^*\rangle| : y^* \in \text{ball } \mathcal{Y}^*\}$$
$$= \sup\{|\langle x, A^*y^*\rangle| : y^* \in \text{ball } \mathcal{Y}^*\}$$
$$\geq \sup\{|\langle x, x^*\rangle| : x^* \in \mathcal{X}^* \quad \text{and} \quad \|x^*\| \leq c\}$$
$$= c\|x\|.$$

Thus $\ker A = (0)$ and ran A is closed. (Why?) On the other hand, $(\text{ran } A)^\perp = \ker A^* = (0)$ since A^* is invertible. Thus ran A is also dense. This implies that A is surjective and thus invertible. ∎

This section concludes with the following useful result that seems to be somewhat unfamiliar to parts of the mathematical community.

1.10. Theorem. *If $\mathcal{X}$ and $\mathcal{Y}$ are Banach spaces and $A \in \mathcal{B}(\mathcal{X}, \mathcal{Y})$, then the following statements are equivalent.*

(a) *ran A is closed.*
(b) *ran A^* is weak* closed.*
(c) *ran A^* is norm closed.*

PROOF. It is clear that (b) implies (c), so it will be shown that (a) implies (b) and (c) implies (a). Before this is done, it will be shown that it suffices to prove the theorem under the additional hypothesis that A is injective and has dense range.

Let $\mathcal{Z} = \text{cl}(\text{ran } A)$. Thus $A: \mathcal{X} \to \mathcal{Z}$ induces a bounded linear map $B: \mathcal{X}/\ker A \to \mathcal{Z}$ defined by $B(x + \ker A) = Ax$. If $Q: \mathcal{X} \to \mathcal{X}/\ker A$ is the natural map, the diagram

$$\mathcal{X} \xrightarrow{\quad A \quad} \mathcal{Z} \hookrightarrow \mathcal{Y}$$
$$Q \searrow \qquad \nearrow B$$
$$\mathcal{X}/\ker A$$

commutes. (Why is B bounded?) It is easy to see that B is injective and that B has dense range. In fact, ran B = ran A, so ran A is closed if and only if ran B is closed. Let's examine $B^*: \mathcal{Z}^* \to (\mathcal{X}/\ker A)^*$. By (V.2.2), $(\mathcal{X}/\ker A)^* = (\ker A)^\perp = \text{wk}^*\text{cl}(\text{ran } A^*) \subseteq \mathcal{X}^*$ by (1.8). Also by (V.2.3), since $\mathcal{Z} \leq \mathcal{Y}$, $\mathcal{Z}^* = \mathcal{Y}^*/\mathcal{Z}^\perp = \mathcal{Y}^*/(\text{ran } A)^\perp = \mathcal{Y}^*/\ker A^*$ by (1.8). Thus,

$$B^*: \mathcal{Y}^*/\ker A^* \to (\ker A)^\perp.$$

1.11. Claim. $B^*(y^* + \ker A^*) = A^*y^*$ for all y^* in $\mathcal{Y}^*$.

To see this, let $x \in \mathcal{X}$ and $y^* \in \mathcal{Y}^*$. Making the appropriate identifications as in (V.2.2) and (V.2.3) gives $\langle x + \ker A, B^*(y^* + \ker A^*)\rangle = \langle B(x + \ker A), y^* + \ker A^*\rangle = \langle Ax, y^* + (\operatorname{ran} A)^\perp\rangle = \langle Ax, y^*\rangle = \langle x, A^*y^*\rangle = \langle x + {}^\perp(\operatorname{ran} A^*), A^*y^*\rangle = \langle x + \ker A, A^*y^*\rangle$. Since x was arbitrary, (1.11) is established.

Note that Claim 1.11 implies that $\operatorname{ran} B^* = \operatorname{ran} A^*$. Hence $\operatorname{ran} A^*$ is weak* (resp., norm) closed if and only if $\operatorname{ran} B^*$ is weak* (resp., norm) closed.

This discussion shows that the theorem is equivalent to the analogous theorem in which there is the additional hypothesis that A is injective and has dense range. It is assumed, therefore, that $\ker A = (0)$ and $\operatorname{cl}(\operatorname{ran} A) = \mathcal{Y}$.

(a) $\Rightarrow$ (b): Since $\operatorname{ran} A$ is closed, the additional hypothesis implies that A is bijective. By the Inverse Mapping Theorem, $A^{-1} \in \mathcal{B}(\mathcal{Y}, \mathcal{X})$. Hence A^* is invertible (1.4c). Since A^* is invertible, $\operatorname{ran} A^* = \mathcal{X}^*$ and hence is weak* closed.

(c) $\Rightarrow$ (b): Since $\operatorname{ran} A$ is dense in $\mathcal{Y}$, $\ker A^* = (\operatorname{ran} A)^\perp$ (1.8) $= (0)$. Thus $A^*: \mathcal{Y}^* \to \operatorname{ran} A^*$ is a bijection. Since $\operatorname{ran} A^*$ is norm closed, it is a Banach space. By the Inverse Mapping Theorem, there is a constant $c > 0$ such that $\|A^*y^*\| \geq c\|y^*\|$ for all y^* in $\mathcal{Y}^*$.

To show that $\operatorname{ran} A^*$ is weak* closed, the Krein–Smulian Theorem (V.12.6) will be used. Thus suppose $\{A^*y_i^*\}$ is a net in $\operatorname{ran} A^*$ with $\|A^*y_i^*\| \leq 1$ such that $A^*y_i^* \to x^* \, \sigma(\mathcal{X}^*, \mathcal{X})$ for some x^* in $\mathcal{X}^*$. Thus $\|y_i^*\| \leq c^{-1}$ for all y_i^*. By Alaoglu's Theorem there is a y^* in $\mathcal{Y}^*$ such that $y_i^* \xrightarrow[\text{cl}]{} y^* \, \sigma(\mathcal{Y}^*, \mathcal{Y})$. Thus (1.1c), $A^*y_i^* \xrightarrow[\text{cl}]{} A^*y^* \, \sigma(\mathcal{X}^*, \mathcal{X})$, and so $x^* = A^*y^* \in \operatorname{ran} A^*$. By (V.12.6), $\operatorname{ran} A^*$ is weak* closed.

(b) $\Rightarrow$ (a): Since $\operatorname{ran} A^*$ is weak* closed, $\operatorname{ran} A^* = (\ker A)^\perp = \mathcal{X}^*$. Also, $\ker A^* = (\operatorname{ran} A)^\perp = (0)$ since A has dense range. Thus A^* is a bijection and is thus invertible. By Proposition 1.9, A is invertible and thus has closed range. ∎

EXERCISES

1. Prove Proposition 1.3.

2. Complete the proof of Proposition 1.4.

3. Verify the statement made in (1.5).

4. Verify the statement made in (1.6).

5. Verify the statement made in (1.7).

6. Let $1 \leq p < \infty$ and define $S: l^p \to l^p$ by $S(\alpha_1, \alpha_2, \ldots) = (0, \alpha_1, \alpha_2, \ldots)$. Compute S^*.

7. Let $A \in \mathcal{B}(c_0)$ and for $n \geq 1$, define e_n in c_0 by $e_n(n) = 1$ and $e_n(m) = 0$ for $m \neq n$. Put $\alpha_{mn} = (Ae_n)(m)$ for $m, n \geq 1$. Prove: (a) $M \equiv \sup_m \sum_{n=1}^{\infty} |\alpha_{mn}| < \infty$; (b) for every n, $\alpha_{mn} \to 0$ as $m \to \infty$. Conversely, if $\{\alpha_{mn}: m, n \geq 1\}$ are scalars satisfying (a) and (b), then

$$(Ax)(m) = \sum_{n=1}^{\infty} \alpha_{mn} x(n)$$

defines a bounded operator A on c_0 and $\|A\| = M$. Find A^*.

8. Let $A \in \mathcal{B}(l^1)$ and for $n \geq 1$ define e_n in l^1 by $e_n(n) = 1$, $e_n(m) = 0$ for $m \neq n$. Put $\alpha_{mn} = (Ae_n)(m)$ for $m, n \geq 1$. Prove: (a) $M \equiv \sup_n \sum_{m=1}^{\infty} |\alpha_{mn}| < \infty$; (b) for every m, $\sup_n |\alpha_{mn}| < \infty$. Conversely, if $\{\alpha_{mn}: m, n \geq 1\}$ are scalars satisfying (a) and (b), then

$$(Af)(n) = \sum_{m=1}^{\infty} \alpha_{mn} f(m)$$

defines a bounded operator A on l^1 and $\|A\| = M$. Find A^*.

9. (F. F. Bonsall)   Let $\mathcal{X}$ be a Banach space, Z a nonempty set, and $u: Z \to \mathcal{X}$. If there are positive constants M_1 and M_2 such that (i) $\|u(z)\| \leq M_1$ for all z in Z and (ii) for every x^* in $\mathcal{X}$, $\sup \{| < u(z), x^* > |: z \in Z\} \geq M_2 \|x^*\|$; then for every x in $\mathcal{X}$ there is an f in $l^1(Z)$ such that $(*)x = \Sigma\{f(z)u(z): z \in Z\}$ and $M_2 \inf \|f\|_1 \leq \|x\| \leq M_1 \inf \|f\|_1$, where the infimum is taken over all f in $l^1(Z)$ such that $(*)$ holds. (Hint: define $T: l^1(Z) \to \mathcal{X}$ by $Tf = \Sigma\{f(z)u(z): z \in Z\}$.)

10. (F. F. Bonsall)   Let m be normalized Lebesgue measure on $\partial\mathbb{D}$ and for $|z| < 1$ and $|w| = 1$ let $p_z(w) = (1 - |z|^2)/|1 - \bar{z}w|^2$. So p_z is the Poisson kernel. Show that if $f \in L^1(m)$, then there is a sequence $\{z_n\} \subseteq \mathbb{D}$ and a sequence $\{\lambda_n\}$ in l^1 such that $(*)f = \Sigma_{n=0}^{\infty} \lambda_n p_{z_n}$. Moreover, $\|f\|_1 = \inf \Sigma_{n=1}^{\infty} |\lambda_n|$, where the infimum is taken over all $\{\lambda_n\}$ in l^1 such that $(*)$ holds. (Hint: use Exercise 9.)

§2*. The Banach–Stone Theorem

As an application of the adjoint of a linear map, the isometries between spaces of the form $C(X)$ and $C(Y)$ will be characterized. Note that if X and Y are compact spaces, $\tau: Y \to X$ is continuous map, and $Af = f \circ \tau$ for f in $C(X)$, then (III.2.4) A is a bounded linear map and $\|A\| = 1$. Moreover, A is an isometry if and only if τ is surjective. If A is a surjective isometry, then τ must be a homeomorphism. Indeed, suppose A is a surjective isometry; it must be shown that τ is injective. If $y_0, y_1 \in Y$ and $y_0 \neq y_1$, then there is a g in $C(Y)$ such that $g(y_0) = 0$ and $g(y_1) = 1$. Let $f \in C(X)$ such that $Af = g$. Thus $f(\tau(y_0)) = g(y_0) = 0$ and $f(\tau(y_1)) = 1$. Hence $\tau(y_0) \neq \tau(y_1)$.

So if $\tau\colon Y \to X$ is a homeomorphism and $\alpha\colon Y \to \mathbb{F}$ is a continuous function, with $|\alpha(y)| \equiv 1$, then $T\colon C(X) \to C(Y)$ defined by $(Tf)(y) = \alpha(y)f(\tau(y))$ is a surjective isometry. The next result gives a converse to this.

2.1. The Banach–Stone Theorem. *If X and Y are compact and $T\colon C(X) \to C(Y)$ is a surjective isometry, then there is a homeomorphism $\tau\colon Y \to X$ and a function α in $C(Y)$ such that $|\alpha(y)| = 1$ for all y and*

$$(Tf)(y) = \alpha(y)f(\tau(y))$$

for all f in $C(X)$ and y in Y.

PROOF. Consider $T^*\colon M(Y) \to M(X)$. Because T is a surjective isometry, T^* is also. (Verify.) Thus T^* is a weak* homeomorphism of ball $M(Y)$ onto ball $M(X)$ that distributes over convex combinations. Hence (Why?)

$$T^*(\text{ext}[\text{ball } M(Y)]) = \text{ext}[\text{ball } M(X)].$$

By Theorem V.8.4 this implies that for every y in Y there is a unique $\tau(y)$ in X and a unique scalar $\alpha(y)$ such that $|\alpha(y)| = 1$ and

$$T^*(\delta_y) = \alpha(y)\delta_{\tau(y)}.$$

By the uniqueness, $\alpha\colon Y \to \mathbb{F}$ and $\tau\colon Y \to X$ are well-defined functions.

2.2. Claim. $\alpha\colon Y \to \mathbb{F}$ is continuous.

If $\{y_i\}$ is a net in Y and $y_i \to y$, then $\delta_{y_i} \to \delta_y$ weak* in $M(Y)$. Hence $\alpha(y_i)\delta_{\tau(y_i)} = T^*(\delta_{y_i}) \to T^*(\delta_y) = \alpha(y)\delta_{\tau(y)}$ weak* in $M(X)$. In particular, $\alpha(y_i) = \langle 1, T^*(\delta_{y_i}) \rangle \to \langle 1, T^*(\delta_y) \rangle = \alpha(y)$, proving (2.2).

2.3. Claim. $\tau\colon Y \to X$ is a homeomorphism.

As in the proof of (2.2), if $y_i \to y$ in Y, then $\alpha(y_i)\delta_{\tau(y_i)} \to \alpha(y)\delta_{\tau(y)}$ weak* in $M(X)$. Also, $\alpha(y_i) \to \alpha(y)$ in $\mathbb{F}$ by (2.2). Thus $\delta_{\tau(y_i)} = \alpha(y_i)^{-1}[\alpha(y_i)\delta_{\tau(y_i)}] \to \delta_{\tau(y)}$. By (V.6.1) this implies that $\tau(y_i) \to \tau(y)$, so that $\tau\colon Y \to X$ is continuous.

If $y_1, y_2 \in Y$ and $y_1 \neq y_2$, then $\overline{\alpha(y_1)\delta_{y_1}} \neq \overline{\alpha(y_2)\delta_{y_2}}$. Since T^* is injective, it is easy to see that $\tau(y_1) \neq \tau(y_2)$ and so τ is one-to-one. If $x \in X$, then the fact that T^* is surjective implies that there is a μ in $M(Y)$ such that $T^*\mu = \delta_x$. It must be that $\mu \in \text{ext}[\text{ball } M(X)]$ (Why?), so that $\mu = \beta\delta_y$ for some y in Y and β in $\mathbb{F}$ with $|\beta| = 1$. Thus $\delta_x = T^*(\beta\delta_y) = \beta\alpha(y)\delta_{\tau(y)}$. Hence $\beta = \overline{\alpha(y)}$ and $\tau(y) = x$. Therefore $\tau\colon Y \to X$ is a continuous bijection and hence must be a homeomorphism (A.2.8). This establishes (2.3).

If $f \in C(X)$ and $y \in Y$, then $T(f)(y) = \langle Tf, \delta_y \rangle = \langle f, T^*\delta_y \rangle = \langle f, \alpha(y)\delta_{\tau(y)} \rangle = \alpha(y)f(\tau(y))$. ■

§3. Compact Operators

The following definition generalizes the concept of a compact operator from a Hilbert space to a Banach space.

3.1. Definition. If $\mathscr{X}$ and $\mathscr{Y}$ are Banach spaces and $A: \mathscr{X} \to \mathscr{Y}$ is a linear transformation, then A is *compact* if cl $A(\text{ball } \mathscr{X})$ is compact in $\mathscr{Y}$.

The reader should become reacquainted with Section II.4.

It is easy to see that compact operators are bounded.

For operators on a Hilbert space the following concept is equivalent to compactness, as will be seen.

3.2. Definition. If $\mathscr{X}$ and $\mathscr{Y}$ are Banach spaces and $A \in \mathscr{B}(\mathscr{X}, \mathscr{Y})$, then A is *completely continuous* if for any sequence $\{x_n\}$ in $\mathscr{X}$ such that $x_n \to x$ weakly it follows that $\|Ax_n - Ax\| \to 0$.

3.3. Proposition. *Let $\mathscr{X}$ and $\mathscr{Y}$ be Banach spaces and let $A \in \mathscr{B}(\mathscr{X}, \mathscr{Y})$.*

(a) *If A is a compact operator, then A is completely continuous.*

(b) *If $\mathscr{X}$ is reflexive and A is completely continuous, then A is compact.*

PROOF. (a) Let $\{x_n\}$ be a sequence in $\mathscr{X}$ such that $x_n \to 0$ weakly. By the PUB, $M = \sup_n \|x_n\| < \infty$. Without loss of generality, it may be assumed that $M \leq 1$. Hence $\{Ax_n\} \subseteq \text{cl } A(\text{ball } \mathscr{X})$. Since A is compact, there is a subsequence $\{x_{n_k}\}$ and a y in $\mathscr{Y}$ such that $\|Ax_{n_k} - y\| \to 0$. But $x_{n_k} \to 0$ (wk) and $A: (\mathscr{X}, \text{wk}) \to (\mathscr{Y}, \text{wk})$ is continuous (1.1c). Hence $Ax_{n_k} \to A(0) = 0$ (wk). Thus $y = 0$. Since 0 is the unique cluster point of $\{Ax_n\}$ and this sequence is contained in a compact set, $\|Ax_n\| \to 0$.

(b) First assume that $\mathscr{X}$ is separable; so $(\text{ball } \mathscr{X}, \text{wk})$ is a compact metric space. So if $\{x_n\}$ is a sequence in ball $\mathscr{X}$ there is an x in $\mathscr{X}$ and a subsequence $\{x_{n_k}\}$ such that $x_{n_k} \to x$ weakly. Since A is completely continuous, $\|Ax_{n_k} - Ax\| \to 0$. Thus $A(\text{ball } \mathscr{X})$ is sequentially compact; that is, A is a compact operator.

Now let $\mathscr{X}$ be arbitrary and let $\{x_n\} \subseteq \text{ball } \mathscr{X}$. If $\mathscr{X}_1 =$ the closed linear span of $\{x_n\}$, then $\mathscr{X}_1$ is separable and reflexive. If $A_1 = A|\mathscr{X}_1$, then $A_1: \mathscr{X}_1 \to \mathscr{Y}$ is easily seen to be completely continuous. By the first paragraph, A_1 is compact. Thus $\{Ax_n\} = \{A_1 x_n\}$ has a convergent subsequence. Since $\{x_n\}$ was arbitrary, A is a compact operator. ∎

The fact that in the proof of (3.3b), $A(\text{ball } \mathscr{X})$ was shown to be compact, and hence closed, is a consequence of the reflexivity of A.

By Proposition V.5.2, every operator in $\mathscr{B}(l^1)$ is completely continuous. However, there are noncompact operators in $\mathscr{B}(l^1)$ (for example, the identity operator).

There has been relatively little study of completely continuous operators that I am aware of. Most of the effort has been devoted to the study of compact operators and this is the direction we now pursue.

3.4. Schauder's Theorem. *If $A \in \mathcal{B}(\mathcal{X}, \mathcal{Y})$, then A is compact if and only if A^* is compact.*

PROOF. Assume A is a compact operator and let $\{y_n^*\}$ be a sequence in ball $\mathcal{Y}^*$. It must be shown that $\{A^*y_n^*\}$ has a norm convergent subsequence or, equivalently, a cluster point in the norm topology. By Alaoglu's Theorem, there is a y^* in ball $\mathcal{Y}^*$ such that $y_n^* \xrightarrow{\text{cl}} y^*$ (weak*). It will be shown that $A^*y_n^* \xrightarrow{\text{cl}} A^*y^*$ in norm.

Let $\varepsilon > 0$ and fix $N \geq 1$. Because $A(\text{ball } \mathcal{X})$ has compact closure, there are vectors $y_1, \ldots, y_m$ in $\mathcal{Y}$ such that $A(\text{ball } \mathcal{X}) \subseteq \bigcup_{k=1}^m \{y \in \mathcal{Y}: \|y - y_k\| < \varepsilon/3\}$. Since $y_n^* \xrightarrow{\text{cl}} y^*$ (weak*), there is an $n \geq N$ such that $|\langle y_k, y^* - y_n^* \rangle| < \varepsilon/3$ for $1 \leq k \leq m$. Let x be an arbitrary element in ball $\mathcal{X}$ and choose y_k such that $\|Ax - y_k\| < \varepsilon/3$. Then

$$|\langle x, A^*y^* - A^*y_n^* \rangle| = |\langle Ax, y^* - y_n^* \rangle|$$
$$\leq |\langle Ax - y_k, y^* - y_n^* \rangle| + |\langle y_k, y^* - y_n^* \rangle|$$
$$\leq 2\|Ax - y_k\| + \varepsilon/3 < \varepsilon.$$

Thus $\|A^*y - A^*y_n^*\| \leq \varepsilon$.

For the converse, assume A^* is compact. By the first half of the proof, $A^{**}: \mathcal{X}^{**} \to \mathcal{Y}^{**}$ is compact. It is easy to check that $A = A^{**}|\mathcal{X}$ is compact. ∎

For Banach spaces $\mathcal{X}$ and $\mathcal{Y}$, $\mathcal{B}_0(\mathcal{X}, \mathcal{Y})$ denotes the set of all compact operators from $\mathcal{X}$ into $\mathcal{Y}$; $\mathcal{B}_0(\mathcal{X}) = \mathcal{B}(\mathcal{X}, \mathcal{X})$.

3.5. Proposition. *Let $\mathcal{X}, \mathcal{Y}$, and $\mathcal{Z}$ be Banach spaces.*

(a) $\mathcal{B}_0(\mathcal{X}, \mathcal{Y})$ *is a closed linear subspace of* $\mathcal{B}(\mathcal{X}, \mathcal{Y})$.
(b) *If* $K \in \mathcal{B}_0(\mathcal{X}, \mathcal{Y})$ *and* $A \in \mathcal{B}(\mathcal{Y}, \mathcal{Z})$, *then* $AK \in \mathcal{B}_0(\mathcal{X}, \mathcal{Z})$.
(c) *If* $K \in \mathcal{B}_0(\mathcal{X}, \mathcal{Y})$ *and* $A \in \mathcal{B}(\mathcal{Z}, \mathcal{X})$, *then* $KA \in \mathcal{B}_0(\mathcal{Z}, \mathcal{Y})$.

The proof of (3.5) is left as an exercise.

3.6. Corollary. *If $\mathcal{X}$ is a Banach space, $\mathcal{B}_0(\mathcal{X})$ is a closed two-sided ideal in the algebra $\mathcal{B}(\mathcal{X})$.*

Let $\mathcal{B}_{00}(\mathcal{X}, \mathcal{Y})$ = the bounded operators $T: \mathcal{X} \to \mathcal{Y}$ for which ran T is finite dimensional. Operators in $\mathcal{B}_{00}(\mathcal{X}, \mathcal{Y})$ are called operators with *finite rank*. It is easy to see that $\mathcal{B}_{00}(\mathcal{X}, \mathcal{Y}) \subseteq \mathcal{B}_0(\mathcal{X}, \mathcal{Y})$ and by (3.5a) the closure of $\mathcal{B}_{00}(\mathcal{X}, \mathcal{Y})$ is contained in $\mathcal{B}_0(\mathcal{X}, \mathcal{Y})$. Is $\mathcal{B}_{00}(\mathcal{X}, \mathcal{Y})$ dense in $\mathcal{B}_0(\mathcal{X}, \mathcal{Y})$?

It was shown in (II.4.4) that if $\mathcal{H}$ is a Hilbert space, then $\mathcal{B}_0(\mathcal{H})$ is indeed the closure of $\mathcal{B}_{00}(\mathcal{H})$. Note that the ability to find an orthonormal basis in a Hilbert space played a significant role in the proof of this theorem. There is a concept of a basis for a Banach space called a Schauder basis. Any Banach space $\mathcal{X}$ with a Schauder basis has the property that $\mathcal{B}_{00}(\mathcal{X})$ is dense in $\mathcal{B}_0(\mathcal{X})$. Enflo [1973] gave an example of a separable reflexive Banach space $\mathcal{X}$ for which $\mathcal{B}_{00}(\mathcal{X})$ is not dense in $\mathcal{B}_0(\mathcal{X})$, and, hence, has no Schauder basis. Davie [1973] and [1975] have simplifications of Enflo's proof. For the classical Banach spaces, however, every compact operator is the limit of a sequence of finite-rank operators.

The remainder of this section is devoted to proving that for X compact, $\mathcal{B}_{00}(C(X))$ is dense in $\mathcal{B}_0(C(X))$. This begins with material that may be familiar to many readers but will be presented for those who are unacquainted with it.

3.7. Definition. If X is completely regular and $\mathcal{F} \subseteq C(X)$, then $\mathcal{F}$ is *equicontinuous* if for every $\varepsilon > 0$ and for every x_0 in X there is a neighborhood U of x_0 such that $|f(x) - f(x_0)| < \varepsilon$ for all x in U and for all f in $\mathcal{F}$.

Note that for a single function f in $C(X)$, $\mathcal{F} = \{f\}$ is equicontinuous. The concept of equicontinuity states that one neighborhood works for all f in $\mathcal{F}$.

3.8. The Arzela–Ascoli Theorem. *If X is compact and $\mathcal{F} \subseteq C(X)$, then $\mathcal{F}$ is totally bounded if and only if $\mathcal{F}$ is bounded and equicontinuous.*

PROOF. Suppose $\mathcal{F}$ is totally bounded. It is easy to see that $\mathcal{F}$ is bounded. If $\varepsilon > 0$, then there are $f_1, \ldots, f_n$ in $\mathcal{F}$ such that $\mathcal{F} \subseteq \bigcup_{k=1}^n \{f \in C(X): \|f - f_k\| < \varepsilon/3\}$. If $x_0 \in X$, let U be an open neighborhood of x_0 such that for $1 \le k \le n$ and x in U, $|f_k(x) - f_k(x_0)| < \varepsilon/3$. If $f \in \mathcal{F}$, let f_k be such that $\|f - f_k\| < \varepsilon/3$. Then for x in U,

$$
\begin{aligned}
|f(x) - f(x_0)| &\le |f(x) - f_k(x)| + |f_k(x) - f_k(x_0)| \\
&\quad + |f_k(x_0) - f(x_0)| \\
&< \varepsilon.
\end{aligned}
$$

Hence $\mathcal{F}$ is equicontinuous.

Now assume that $\mathcal{F}$ is equicontinuous and $\mathcal{F} \subseteq \text{ball } C(X)$. Let $\varepsilon > 0$. For each x in $\mathcal{X}$, let U_x be an open neighborhood of x such that $|f(x) - f(y)| < \varepsilon/2$ for f in $\mathcal{F}$ and y in U_x. Now $\{U_x : x \in X\}$ is an open covering of X. Since X is compact, there are points $x_1, \ldots, x_n$ in X such that $X = \bigcup_{j=1}^n U_{x_j}$.

Let $\{\alpha_1, \ldots, \alpha_m\} \subseteq \mathbb{D}$ such that $\text{cl } \mathbb{D} \subseteq \bigcup_{k=1}^m \{\alpha: |\alpha - \alpha_k| < \varepsilon/2\}$. Let $B = $ all ordered n-tuples of scalars $(\beta_1, \ldots, \beta_n)$ such that $\{\beta_1, \ldots, \beta_n\} \subseteq \{\alpha_1, \ldots, \alpha_m\}$. (So B has m^n elements.) Let $\phi_1, \ldots, \phi_n$ be a partition of unity subordinate to the cover $\{U_{x_1}, \ldots, U_{x_n}\}$ (V.6.5). For $b = (\beta_1, \ldots, \beta_n)$ in B, let $g_b = \sum_{j=1}^{n} \beta_j \phi_j$.

3.9. Claim. $\mathscr{F} \subseteq \bigcup_{b \in B} \{f: \|f - g_b\| < \varepsilon\}$.

Note that (3.9) implies that $\mathscr{F}$ is totally bounded.

For f in $\mathscr{F}$, $\{f(x_1), \ldots, f(x_n)\} \subseteq \text{cl } \mathbb{D}$. Pick $b = (\beta_1, \ldots, \beta_n)$ in B such that $|\beta_j - f(x_j)| < \varepsilon/2$ for $1 \leq j \leq n$. If $x \in X$, then $\sum_j \phi_j(x) = 1$ and so

$$|f(x) - g_b(x)| = \left| f(x) - \sum_{j=1}^{n} \beta_j \phi_j(x) \right|$$

$$= \left| \sum_{j=1}^{n} [f(x) - \beta_j] \phi_j(x) \right|$$

$$\leq \sum_{j=1}^{n} |f(x) - \beta_j| |\phi_j(x)|.$$

Now if $\phi_j(x) > 0$, $x \in U_j$ and so $|f(x) - \beta_j| \leq |f(x) - f(x_j)| + |f(x_j) - \beta_j| < \varepsilon$. Hence $|f(x) - g_b(x)| < \varepsilon$ for all x in X. That is, $\|f - g_b\| < \varepsilon$. ∎

3.10.   Corollary. *If X is compact and $\mathscr{F} \subseteq C(X)$, then $\mathscr{F}$ is compact if and only if $\mathscr{F}$ is closed, bounded, and equicontinuous.*

3.11.   Theorem. *If X is compact, then $\mathscr{B}_{00}(C(X))$ is dense in $\mathscr{B}_0(C(X))$.*

PROOF. Let $T \in \mathscr{B}_0(C(X))$. Thus $T(\text{ball } C(X))$ is bounded and equicontinuous by the Arzela–Ascoli Theorem. If $\varepsilon > 0$ and $x \in X$, let U_x be an open neighborhood of x such that $|(Tf)(x) - (Tf)(y)| < \varepsilon$ for all f in ball $C(X)$ and y in U_x. Let $\{x_1, \ldots, x_n\} \subseteq X$ such that $X \subseteq \bigcup_{j=1}^{n} U_{x_j}$. Let $\{\phi_1, \ldots, \phi_n\}$ be a partition of unity subordinate to $\{U_{x_1}, \ldots, U_{x_n}\}$. Define T_ε: $C(X) \to C(X)$ by

$$T_\varepsilon f = \sum_{j=1}^{n} (Tf)(x_j) \phi_j.$$

Since ran $T_\varepsilon \subseteq \bigvee \{\phi_1, \ldots, \phi_n\}$, $T_\varepsilon \in \mathscr{B}_{00}(C(X))$.

If $f \in \text{ball } C(X)$ and $x \in X$, then

$$|(T_\varepsilon f)(x) - (Tf)(x)| = \left| \sum_{j=1}^{n} [(Tf)(x_j) - (Tf)(x)] \phi_j(x) \right|$$

$$\leq \sum_{j=1}^{n} |(Tf)(x_j) - (Tf)(x)| \phi_j(x)$$

$$< \varepsilon$$

by an argument like the one used to prove (3.9). ∎

If X is locally compact, then the operators on $C_0(X)$ of finite rank are dense in $\mathscr{B}_0(C_0(X))$. See Exercise 18.

EXERCISES

1. If $\mathscr{X}$ is reflexive and $A \in \mathscr{B}(\mathscr{X}, \mathscr{Y})$, show that $A(\text{ball } \mathscr{X})$ is closed in $\mathscr{Y}$.

2. Prove Proposition 3.5.

3. If $A \in \mathscr{B}_0(\mathscr{X}, \mathscr{Y})$, show that cl[ran A] is separable.

4. If $A \in \mathscr{B}_0(\mathscr{X}, \mathscr{Y})$ and ran A is closed, show that ran A is finite dimensional.

5. If $A \in \mathscr{B}_0(\mathscr{X})$ and A is invertible, show that dim $\mathscr{X} < \infty$.

6. Let (X, Ω, μ) be a finite measure space, $1 < p < \infty$, and $1/p + 1/q = 1$. If $k: X \times X \to \mathbb{F}$ is an $\Omega \times \Omega$-measurable function such that $\sup\{\int |k(x, y)|^q \, d\mu(y): x \in X\} < \infty$, then $(Kf)(x) = \int k(x, y) f(y) \, d\mu(y)$ defines a compact operator on $L^p(\mu)$.

7. Let (X, Ω, μ) be an arbitrary measure space, $1 < p < \infty$, and $1/p + 1/q = 1$. If $k: X \times X \to \mathbb{F}$ is an $\Omega \times \Omega$-measurable function such that $M = [(\int |k(x, y)|^p \, d\mu(x))^{q/p} \, d\mu(y)]^{1/q} < \infty$ and if $(Kf)(x) = \int k(x, y) f(y) \, d\mu(y)$, then $K \in \mathscr{B}_0(L^p(\mu))$ and $\|K\| \le M$.

8. Let X be a compact space and let μ be a positive Borel measure on X. Let $T \in \mathscr{B}(L^p(\mu), C(X))$ where $1 < p < \infty$. Show that if $A: L^p(\mu) \to L^p(\mu)$ is defined by $Af = Tf$, then A is compact.

9. (B. J. Pettis) If $\mathscr{X}$ is reflexive and $T \in \mathscr{B}(\mathscr{X}, l^1)$, then T is a compact operator. Also, if $\mathscr{Y}$ is reflexive and $T \in \mathscr{B}(c_0, \mathscr{Y})$, T is compact.

10. If X is compact and $\{f_1, \ldots, f_n, g_1, \ldots, g_n\} \subseteq C(X)$, define $k(x, y) = \sum_{j=1}^n f_j(x) g_j(y)$ for $x, y \in X$. Let μ be a regular Borel measure on X and put $Kf(x) = \int k(x, y) f(y) \, d\mu(y)$. Show that $K \in \mathscr{B}(C(X))$ and K has finite rank.

11. If X is compact, $k \in C(X \times X)$, and μ is a regular Borel measure on X, show that $Kf(x) = \int k(x, y) f(y) \, d\mu(y)$ defines a compact operator on $C(X)$.

12. Let (X, Ω, μ) be a σ-finite measure space and for ϕ in $L^\infty(\mu)$ let $M_\phi: L^p(\mu) \to L^p(\mu)$ be the multiplication operator defined in Example III.2.2. Give necessary and sufficient conditions on (X, Ω, μ) and ϕ for M_ϕ to be compact.

13. Let $\tau: [0, 1] \to [0, 1]$ be continuous and define $A: C[0, 1] \to C[0, 1]$ by $Af = f \circ \tau$. Give necessary and sufficient conditions on τ for A to be compact.

14. Let $A \in \mathscr{B}(c_0)$ and let (α_{mn}) be the corresponding matrix as in Exercise 1.7. Give necessary and sufficient conditions on (α_{mn}) for A to be compact.

15. Let $A \in \mathscr{B}(l^1)$ and let (α_{mn}) be the corresponding matrix as in Exercise 1.8. Give a necessary and sufficient condition on (α_{mn}) for A to be compact.

16. If (X, d) is a compact metric space and $F \subseteq C(X)$, show that $\mathscr{F}$ is equicontinuous if and only if for every $\varepsilon > 0$ there is a $\delta > 0$ such that $|f(x) - f(y)| < \varepsilon$ whenever $d(x, y) < \delta$ and $f \in \mathscr{F}$.

17. If X is locally compact and $\mathscr{F} \subseteq C_0(X)$, show that $\mathscr{F}$ is totally bounded if and only if (a) $\mathscr{F}$ is bounded; (b) $\mathscr{F}$ is equicontinuous; (c) for every $\varepsilon > 0$ there is a compact subset K of X such that $|f(x)| < \varepsilon$ for all f in $\mathscr{F}$ and x in $X \setminus K$.

18. If X is locally compact and $A \in \mathscr{B}_0(C(X))$, then there is a sequence $\{A_n\}$ of finite-rank operators such that $\|A_n - A\| \to 0$.

19. Let $\mathscr{X}$ be a Banach space and suppose there is a net $\{F_i\}$ of finite-rank operators on $\mathscr{X}$ such that (a) $\sup_i \|F_i\| < \infty$; (b) $\|F_i x - x\| \to 0$ for all x in $\mathscr{X}$. Show that if $A \in \mathscr{B}_0(\mathscr{X})$, then $\|F_i A - A\| \to 0$ and hence there is a sequence $\{A_n\}$ of finite-rank operators on $\mathscr{X}$ such that $\|A_n - A\| \to 0$.

20. Let $1 \leq p \leq \infty$ and let (X, Ω, μ) be a σ-finite measure space. If $A \in \mathscr{B}_0(L^p(\mu))$, show that there is a sequence $\{A_n\}$ of finite-rank operators such that $\|A_n - A\| \to 0$. (Hint: Use Exercise 19.)

21. Let X be compact and let $\mathscr{U}$ be the collection of all pairs (C, F) where $C = \{U_1, \ldots, U_n\}$ is a finite open cover of X and $F = \{x_1, \ldots, x_n\} \subseteq X$ such that $x_j \in U_j$ for $1 \leq j \leq n$. If (C_1, F_1) and $(C_2, F_2) \in \mathscr{U}$, define $(C_1, F_1) \leq (C_2, F_2)$ to mean: (a) C_2 is a refinement of C_1; that is, each member of C_2 is contained in some member of C_1. (b) $F_1 \subseteq F_2$. If $\alpha = (C, F) \in \mathscr{U}$ let $\{\phi_1, \ldots, \phi_n\}$ be a partition of unity subordinate to C. If $F = \{x_1, \ldots, x_n\}$, define $T_\alpha: C(X) \to C(X)$ by

$$(T_\alpha f)(x) = \sum_{j=1}^{n} f(x_j) \phi_j(x).$$

Then: (a) $T_\alpha \in \mathscr{B}_{00}(C(X))$; (b) $\|T_\alpha\| = 1$; (c) $(\mathscr{U}, \leq)$ is a directed set and $\{T_\alpha: \alpha \in \mathscr{U}\}$ is a net; (d) $\|T_\alpha f - f\| \to 0$ for each f. Now apply Exercise 19 to obtain a new proof of Theorem 3.11.

§4. Invariant Subspaces

4.1. Definition. If $\mathscr{X}$ is a Banach space and $T \in \mathscr{B}(\mathscr{X})$, an *invariant subspace* for T is a closed linear subspace $\mathscr{M}$ of $\mathscr{X}$ such that $Tx \in \mathscr{M}$ whenever $x \in \mathscr{M}$. $\mathscr{M}$ is nontrivial if $\mathscr{M} \neq (0)$ or $\mathscr{X}$. Lat T = the collection of all invariant subspaces for T. If $\mathscr{A} \subseteq \mathscr{B}(\mathscr{X})$, then Lat $\mathscr{A} = \bigcap\{\text{Lat } T: T \in \mathscr{A}\}$.

This generalizes the corresponding concept of invariant subspace for an operator on Hilbert space (II.3.5). Note that the idea of a reducing subspace for an operator on a Hilbert space has no generalization to Banach spaces since there is no concept of an orthogonal complement in Banach spaces.

4.2. Proposition.

(a) If $\mathscr{M}_1, \mathscr{M}_2 \in \text{Lat } T$, then $\mathscr{M}_1 \vee \mathscr{M}_2 \equiv \text{cl}(\mathscr{M}_1 + \mathscr{M}_2) \in \text{Lat } T$ and $\mathscr{M}_1 \wedge \mathscr{M}_2 \equiv \mathscr{M}_1 \cap \mathscr{M}_2 \in \text{Lat } T$.

(b) *If* $\{\mathcal{M}_i: i \in I\} \subseteq \text{Lat}\,T$, *then* $\bigvee\{\mathcal{M}_i: i \in I\}$, *the closed linear span of* $\bigcup_i \mathcal{M}_i$, *and* $\bigwedge\{\mathcal{M}_i: i \in I\} \equiv \bigcap_i \mathcal{M}_i$ *belong to* $\text{Lat}\,T$.

The proof of this proposition is left as an exercise. The proposition, however, does justify the use of the symbol "Lat" to denote the collection of invariant subspaces. With the operations $\vee$ and $\wedge$, $\text{Lat}\,T$ is a lattice (a) that is complete (b). Moreover, $\text{Lat}\,T$ has a largest element, $\mathcal{X}$, and a smallest element, (0).

The main question is: does $\text{Lat}\,T$ have any elements besides (0) and $\mathcal{X}$? In other words, does T have a nontrivial invariant subspace? C. J. Read [1984] has given an example of a bounded operator on l^1 that has no nontrivial invariant subspaces. This deep work does not completely settle the matter. Which Banach spaces $\mathcal{X}$ have the property that there is a bounded operator on $\mathcal{X}$ with no nontrivial invariant subspaces? If $\mathcal{X}$ is reflexive, is $\text{Lat}\,T$ nontrivial for every T in $\mathcal{B}(\mathcal{X})$? The question is unanswered even if $\mathcal{X}$ is a Hilbert space. However, for certain specific operators and classes of operators it has been shown that the lattice of invariant subspaces is not trivial. In this section it will be shown that any compact operator has a nontrivial invariant subspace. This will be obtained as a corollary of a more general result of V. Lomonosov. But first some examples.

4.3. Example. If $\mathcal{X}$ is a finite-dimensional space over $\mathbb{C}$ and $T \in \mathcal{B}(\mathcal{X})$, then $\text{Lat}\,T$ is not trivial. In fact, let $\mathcal{X} = \mathbb{C}^d$ and let $T =$ a matrix. Then $p(z) = \det(T - zI)$ is a polynomial of degree d. Hence it has a zero, say α. If $\det(T - \alpha I) = 0$, then $(T - \alpha I)$ is not invertible. But in finite-dimensional spaces this means that $T - \alpha I$ is not injective. Thus $\ker(T - \alpha I) \neq (0)$. Let $\mathcal{M} \leq \ker(T - \alpha I)$ such that $\mathcal{M} \neq (0)$. If $x \in \mathcal{M}$, then $Tx = \alpha x \in \mathcal{M}$, so $\mathcal{M} \in \text{Lat}\,T$.

4.4. Example. If $T = \begin{bmatrix} 0 & -1 \\ 1 & 0 \end{bmatrix}$ on $\mathbb{R}^2$, then $\text{Lat}\,T$ is trivial. Indeed, if $\text{Lat}\,T$ is not trivial, there is a one-dimensional space $\mathcal{M}$ in $\text{Lat}\,T$. Let $\mathcal{M} = \{\alpha e: \alpha \in \mathbb{R}\}$. Since $\mathcal{M} \in \text{Lat}\,T$, $Te = \lambda e$ for some λ in $\mathbb{R}$. Hence $T^2 e = T(Te) = \lambda Te = \lambda^2 e$. But $T^2 = -I$, so $-e = \lambda^2 e$ and it must be that $\lambda^2 = -1$ if $e \neq 0$. But this cannot be if λ is real.

If $d \geq 3$, however, and $T \in \mathcal{B}(\mathbb{R}^d)$, then $\text{Lat}\,T$ is not trivial (Exercise 6).

4.5. Example. If $V: L^2[0,1] \to L^2[0,1]$ is the Volterra operator, $Vf(x) = \int_0^x f(t)\,dt$, and $0 \leq \alpha \leq 1$, put $\mathcal{M}_\alpha = \{f \in L^2[0,1]: f(t) = 0 \text{ for } 0 \leq t \leq \alpha\}$. Then $\mathcal{M}_\alpha \in \text{Lat}\,V$. Moreover, it can be shown that $\text{Lat}\,V = \{\mathcal{M}_\alpha: 0 \leq \alpha \leq 1\}$. (See Donoghue [1957], and Radjavi and Rosenthal [1973], p. 68).

4.6. Example. If $S: l^p \to l^p$ is defined by $S(\alpha_1, \alpha_2, \ldots) = (0, \alpha_1, \alpha_2, \ldots)$, and $\mathcal{M}_n = \{x \in l^p: x(k) = 0 \text{ for } 1 \leq k \leq n\}$, then $\mathcal{M}_n \in \text{Lat}\,S$.

4.7. Example. Let (X, Ω, μ) be a σ-finite measure space and for ϕ in $L^{\infty}(\mu)$ let M_{ϕ} denote the multiplication operator on $L^{p}(\mu)$, $1 \leq p \leq \infty$. If $\Delta \in \Omega$, let $\mathcal{M}_{\Delta} = \{ f \in L^{p}(\mu) : f = 0 \text{ a.e. } [\mu] \text{ off } \Delta \}$. Then for each ϕ in $L^{\infty}(\mu)$, $M_{\Delta} \in \operatorname{Lat} M_{\phi}$.

It is a difficult if not impossible task to determine all the invariant subspaces of a specific operator. The Volterra operator and the shift operator are examples where all the invariant subspaces have been determined. But there are multiplication operators M_{ϕ} for which there is no characterization of $\operatorname{Lat} M_{\phi}$ as well as some M_{ϕ} for which such a characterization has been achieved. One such example follows: let $\mu = $ Lebesgue area measure on $\mathbb{D}$ and let $(Af)(z) = zf(z)$ for f in $L^{2}(\mu)$. There is no known characterization of $\operatorname{Lat} A$.

It is necessary at this point to return to the geometry of Banach spaces to prove the following classical theorem.

4.8. Mazur's Theorem. *If $\mathcal{X}$ is a Banach space and K is a compact subset of $\mathcal{X}$, then $\overline{\operatorname{co}}(K)$ is compact.*

PROOF. It suffices to show that $\overline{\operatorname{co}}(K)$ is totally bounded. Let $\varepsilon > 0$ and choose $x_{1}, \ldots, x_{n}$ in K such that $K \subseteq \bigcup_{j=1}^{n} B(x_{j}; \varepsilon/4)$. Put $C = \operatorname{co}\{ x_{1}, \ldots, x_{n} \}$. It is easy to see that C is compact. Hence there are vectors $y_{1}, \ldots, y_{m}$ in C such that $C \subseteq \bigcup_{i=1}^{m} B(y_{i}; \varepsilon/4)$. If $w \in \overline{\operatorname{co}}(K)$, there is a z in $\operatorname{co}(K)$ with $\|w - z\| < \varepsilon/4$. Thus $z = \sum_{p=1}^{l} \alpha_{p} k_{p}$, where $k_{p} \in K$, $\alpha_{p} \geq 0$, and $\sum \alpha_{p} = 1$. Now for each k_{p} there is an $x_{j(p)}$ with $\|k_{p} - x_{j(p)}\| < \varepsilon/4$. Therefore

$$\left\| z - \sum_{p=1}^{l} \alpha_{p} x_{j(p)} \right\| = \left\| \sum_{p=1}^{l} \alpha_{p} (k_{p} - x_{j(p)}) \right\|$$

$$\leq \sum_{p=1}^{l} \alpha_{p} \| k_{p} - x_{j(p)} \|$$

$$< \varepsilon/4.$$

But $\sum_{p} \alpha_{p} x_{j(p)} \in C$ so there is a y_{i} with $\| \sum_{p} \alpha_{p} x_{j(p)} - y_{i} \| < \varepsilon/4$. The triangle inequality now shows that $\operatorname{co}(K) \subseteq \bigcup_{i=1}^{m} B(y_{i}; \varepsilon)$ and so $\overline{\operatorname{co}}(K)$ is totally bounded. ∎

The next result is from Lomonosov [1973]. When it appeared it caused great excitement, both for the strength of its conclusion and for the simplicity of its proof. The proof uses Schauder's Fixed-Point Theorem (V.9.5).

4.9. Lomonosov's Lemma. *If $\mathcal{A}$ is a subalgebra of $\mathcal{B}(\mathcal{X})$ such that $1 \in \mathcal{A}$ and $\operatorname{Lat} \mathcal{A} = \{(0), \mathcal{X}\}$ and if K is a nonzero compact operator on $\mathcal{X}$, then there is an A in $\mathcal{A}$ such that $\ker(AK - 1) \neq 0$.*

PROOF. It may be assumed that $\|K\| = 1$. Fix x_0 in $\mathscr{X}$ such that $\|Kx_0\| > 1$ and put $S = \{x \in \mathscr{X}: \|x - x_0\| \leq 1\}$. It is easy to check that

4.10 $0 \notin S$ and $0 \notin \text{cl } K(S)$.

Now if $x \in \mathscr{X}$ and $x \neq 0$, $\text{cl}\{Tx: T \in \mathscr{A}\}$ is an invariant subspace for $\mathscr{A}$ (because $\mathscr{A}$ is an algebra) that contains the nonzero vector x (because $1 \in \mathscr{A}$). By hypothesis, $\text{cl}\{Tx: T \in \mathscr{A}\} = \mathscr{X}$. By (4.10) this says that for every y in $\text{cl } K(S)$ there is a T in $\mathscr{A}$ with $\|Ty - x_0\| < 1$. Equivalently,

$$\text{cl } K(S) \subseteq \bigcup_{T \in \mathscr{A}} \{y: \|Ty - x_0\| < 1\}.$$

Because $\text{cl } K(S)$ is compact, there are $T_1, \ldots, T_n$ in $\mathscr{A}$ such that

4.11 $\text{cl } K(S) \subseteq \bigcup_{j=1}^{n} \{y: \|T_j y - x_0\| < 1\}.$

For y in $\text{cl } K(S)$ and $1 \leq j \leq n$, let $a_j(y) = \max\{0, 1 - \|T_j y - x_0\|\}$. By (4.11), $\sum_{j=1}^{n} a_j(y) > 0$ for all y in $\text{cl } K(S)$. Define $b_j: \text{cl } K(S) \to \mathbb{R}$ by

$$b_j(y) = \frac{a_j(y)}{\sum\limits_{i=1}^{n} a_i(y)},$$

and define $\psi: S \to \mathscr{X}$ by

$$\psi(x) = \sum_{j=1}^{n} b_j(Kx) T_j Kx.$$

It is easy to see that $a_j: \text{cl } K(S) \to [0, 1]$ is a continuous function. Hence b_j and ψ are continuous.

If $x \in S$, then $Kx \in K(S)$. If $b_j(Kx) > 0$, then $a_j(Kx) > 0$ and so $\|T_j Kx - x_0\| < 1$. That is, $T_j Kx \in S$ whenever $b_j(Kx) > 0$. Since S is a convex set and $\sum_{j=1}^{n} b_j(Kx) = 1$ for x in S,

$$\psi(S) \subseteq S.$$

Note that $T_j K \in \mathscr{B}_0(\mathscr{X})$ for each j so that $\bigcup_{j=1}^{n} T_j K(S)$ has compact closure. By Mazur's Theorem, $\overline{\text{co}}(\bigcup_{j=1}^{n} T_j K(S))$ is compact. But this convex set contains $\psi(S)$ so that $\text{cl } \psi(S)$ is compact. That is, ψ is a compact map. By the Schauder Fixed-Point Theorem, there is a vector x_1 in S such that $\psi(x_1) = x_1$.

Let $\beta_j = b_j(Kx_1)$ and put $A = \sum_{j=1}^{n} \beta_j T_j$. So $A \in \mathscr{A}$ and $AKx_1 = \psi(x_1) = x_1$. Since $x_1 \neq 0$ (Why?), $\ker(AK - 1) \neq (0)$. ∎

4.12. Definition. If $T \in \mathscr{B}(\mathscr{X})$, then a *hyperinvariant subspace* for T is a subspace $\mathscr{M}$ of $\mathscr{X}$ such that $A\mathscr{M} \subseteq \mathscr{M}$ for every operator A in the commutant of T, $\{T\}'$; that is, $A\mathscr{M} \subseteq \mathscr{M}$ whenever $AT = TA$.

Note that every hyperinvariant subspace for T is invariant.

4.13. Lomonosov's Theorem. *If $T \in \mathcal{B}(\mathcal{X})$, T is not a multiple of the identity, and $TK = KT$ for some nonzero compact operator K, then T has a nontrivial hyperinvariant subspace.*

PROOF. Let $\mathcal{A} = \{T\}'$. We want to show that Lat $\mathcal{A} \neq \{(0), \mathcal{X}\}$. If this is not the case, then Lomonosov's Lemma implies that there is an operator A in $\mathcal{A}$ such that $\mathcal{N} = \ker(AK - 1) \neq (0)$. But $\mathcal{N} \in \text{Lat}(AK)$ and $AK|\mathcal{N}$ is the identity operator. Since $AK \in \mathcal{B}_0(\mathcal{X})$, $AK|\mathcal{N} \in \mathcal{B}_0(\mathcal{N})$. Thus dim $\mathcal{N} < \infty$. Since $AK \in \mathcal{A} = \{T\}'$, for any x in $\mathcal{N}$, $AK(Tx) = T(AKx) = Tx$; hence $T\mathcal{N} \subseteq \mathcal{N}$. But dim $\mathcal{N} < \infty$ so that $T|\mathcal{N}$ must have an eigenvalue λ. Thus $\ker(T - \lambda) = \mathcal{M} \neq (0)$. But $\mathcal{M} \neq \mathcal{X}$ since T is not a multiple of the identity. It is easy to check that $\mathcal{M}$ is hyperinvariant for T. ∎

4.14. Corollary. (Aronszajn-Smith [1954].). *If $K \in \mathcal{B}_0(\mathcal{X})$, then* Lat K *is nontrivial.*

The next result appeared in Bernstein and Robinson [1966], where it is proved using nonstandard analysis. Halmos [1966] gave a proof using standard analysis. Now it is an easy consequence of Lomonosov's Theorem.

4.15. Corollary. *If $\mathcal{X}$ is infinite dimensional, $A \in \mathcal{B}(\mathcal{X})$, and there is a polynomial in one variable, p, such that $p(A) \in \mathcal{B}_0(\mathcal{X})$, then* Lat A *is nontrivial.*

PROOF. If $p(A) \neq 0$, then Lomonosov's Theorem applies. If $p(A) = 0$, let $p(z) = \alpha_0 + \alpha_1 z + \cdots + \alpha_n z^n$, $\alpha_n \neq 0$. For $x \neq 0$, let $\mathcal{M} = \bigvee\{x, Ax, \ldots, A^{n-1}x\}$. Since $A^n = -\alpha_n^{-1}[\alpha_0 + \alpha_1 A + \cdots + \alpha_{n-1} A^{n-1}x]$, $\mathcal{M} \in \text{Lat } A$. Since $x \in \mathcal{M}$, $\mathcal{M} \neq (0)$; since dim $\mathcal{M} < \infty$, $\mathcal{M} \neq \mathcal{X}$. ∎

4.16. Corollary. *If $K_1, K_2 \in \mathcal{B}_0(\mathcal{X})$ and $K_1 K_2 = K_2 K_1$, then K_1 and K_2 have a common nontrivial invariant subspace.*

EXERCISES

1. Let $A, B, T \in \mathcal{B}(\mathcal{X})$ such that $TA = BT$. Show that graph $(T) \in \text{Lat}(A \oplus B)$.

2. Prove that $\mathcal{M} \in \text{Lat } T$ if and only if $\mathcal{M}^\perp \in \text{Lat } T^*$. What does the map $\mathcal{M} \mapsto \mathcal{M}^\perp$ of Lat T into Lat T^* do to the lattice operations?

3. Let $\{e_1, e_2, e_3\}$ be the usual basis for $\mathbb{F}^3$ and let $\alpha_1, \alpha_2, \alpha_3 \in \mathbb{F}$. Define T: $\mathbb{F}^3 \to \mathbb{F}^3$ by $Te_j = \alpha_j e_j$, $1 \leq j \leq 3$. (a) If $\alpha_1, \alpha_2, \alpha_3$ are all distinct, show that $\mathcal{M} \in \text{Lat } T$ if and only if $\mathcal{M} = \bigvee E$, where $E \subseteq \{e_1, e_2, e_3\}$. (b) If $\alpha_1 = \alpha_2 \neq \alpha_3$, show that $\mathcal{M} \in \text{Lat } T$ if and only if $\mathcal{M} = \mathcal{N} + \mathcal{L}$, where $\mathcal{N} \leq \bigvee\{e_1, e_2\}$ and $\mathcal{L} \leq \{\alpha e_3 : \alpha \in \mathbb{F}\}$.

4. Generalize Exercise 3 by characterizing Lat T, where T is defined by $Te_j = \alpha_j e_j$, $1 \le j \le d$, for any choice of scalars $\alpha_1, \ldots, \alpha_d$ and where $\{e_1, \ldots, e_d\}$ is the usual basis for $\mathbb{F}^d$.

5. Let $\{e_1, \ldots, e_d\}$ be the usual basis for $\mathbb{F}^d$, let $\{\alpha_1, \ldots, \alpha_{d-1}\} \subseteq \mathbb{F}$. If $Te_j = \alpha_j e_{j+1}$ for $1 \le j \le d - 1$ and $Te_d = 0$, find Lat T.

6. If $T \in \mathcal{B}(\mathbb{R}^d)$ and $d \ge 3$, show that T has a nontrivial invariant subspace.

7. Show that if $T \in \mathcal{B}(\mathcal{X})$ and $\mathcal{X}$ is not separable, then T has a nontrivial invariant subspace.

8. Give an example of an invertible operator T on a Banach space $\mathcal{X}$ and an invariant subspace $\mathcal{M}$ for T such that $\mathcal{M}$ is not invariant for T^{-1}.

9. Let $K \in \mathcal{B}_0(\mathcal{X})$ and show that if $\mathcal{C}$ is a maximal chain in Lat K, then $\mathcal{C}$ is a maximal chain in the lattice of all subspaces of $\mathcal{X}$.

§5. Weakly Compact Operators

5.1. Definition. If $\mathcal{X}$ and $\mathcal{Y}$ are Banach spaces, an operator T in $\mathcal{B}(\mathcal{X}, \mathcal{Y})$ is *weakly compact* if the closure of $T(\text{ball } \mathcal{X})$ is weakly compact.

Weakly compact operators are generalizations of compact operators, but the hypothesis is not sufficiently strong to yield good information about their structure.

Recall that in a reflexive Banach space the weak closure of any bounded set is weakly compact. Also, a bounded operator $T: \mathcal{X} \to \mathcal{Y}$ is continuous if both $\mathcal{X}$ and $\mathcal{Y}$ have their weak topologies (1.1). With these facts in mind, the proof of the next result becomes an easy exercise for the reader.

5.2. Proposition.

(a) *If either $\mathcal{X}$ or $\mathcal{Y}$ is reflexive, then every operator in $\mathcal{B}(\mathcal{X}, \mathcal{Y})$ is weakly compact.*

(b) *If $T: \mathcal{X} \to \mathcal{Y}$ is weakly compact and $A \in \mathcal{B}(\mathcal{Y}, \mathcal{Z})$, then AT is weakly compact.*

(c) *If $T: \mathcal{X} \to \mathcal{Y}$ is weakly compact and $B \in \mathcal{B}(\mathcal{Z}, \mathcal{X})$, then TB is weakly compact.*

This proposition shows that assuming that an operator is weakly compact is not that strong an assumption. For example, if $\mathcal{X}$ is reflexive, every operator in $\mathcal{B}(\mathcal{X})$ is weakly compact. In particular, every operator on a Hilbert space is weakly compact. So any theorem about weakly compact operators is a theorem about all operators on a reflexive space.

In fact, there is a degree of validity for the converse of this statement. In a certain sense, theorems about operators on reflexive spaces are also theorems about weakly compact operators. The precise meaning of this statement is the content of Theorem 5.4 below. But before we begin to prove this, a lemma is needed.

Let $\mathcal{Y}$ be a Banach space and let W be a bounded convex balanced subset of $\mathcal{Y}$. For $n > 1$ put $U_n = 2^n W + 2^{-n}\text{int}[\text{ball}\,\mathcal{Y}]$. Let $p_n = $ the gauge of U_n (IV.1.14). Because $U_n \supseteq 2^{-n}\text{int}[\text{ball}\,\mathcal{Y}]$, it is easy to check that p_n is a norm on $\mathcal{Y}$. In fact, p_n and $\|\cdot\|$ are equivalent norms. To see this note that if $\|y\| < 1$, then $2^{-n}y \in U_n$ so that $p_n(y) < 2^n$. Hence $p_n(y) \leq 2^n\|y\|$. Also, because W is bounded, U_n must be bounded; let $M > \sup\{\|y\|\colon y \in U_n\}$. So if $p_n(y) < 1$, $\|y\| < M$. Thus $\|y\| \leq Mp_n(y)$, and $\|\cdot\|$ and p_n are equivalent norms.

5.3. Lemma. *For a Banach space $\mathcal{Y}$ let W, U_n, and p_n be as above. Let $\mathcal{R} = $ the set of all y in $\mathcal{Y}$ such that $\|\|y\|\| \equiv [\sum_{n=1}^{\infty} p_n(y)^2]^{1/2} < \infty$. Then*

(a) $W \subseteq \{y\colon \|\|y\|\| < 1\}$;

(b) $(\mathcal{R}, \|\|\cdot\|\|)$ *is a Banach space and the inclusion map $A\colon \mathcal{R} \to \mathcal{Y}$ is continuous;*

(c) $A^{**}\colon \mathcal{R}^{**} \to \mathcal{Y}^{**}$ *is injective and $(A^{**})^{-1}(\mathcal{Y}) = \mathcal{R}$;*

(d) $\mathcal{R}$ *is reflexive if and only if $\text{cl}\,W$ is weakly compact.*

PROOF. (a) If $w \in W$, then $2^n w \in U_n$. Hence $1 > p_n(2^n w) = 2^n p_n(w)$, so $p_n(w) < 2^{-n}$. Thus $\|\|w\|\|^2 < \sum_n(2^{-n})^2 < 1$.

(b) Let $\mathcal{Y}_n = \mathcal{Y}$ with the norm p_n and put $\mathcal{X} = \oplus_2 \mathcal{Y}_n$ (III.4.4). Define $\Phi\colon \mathcal{R} \to \mathcal{X}$ by $\Phi(y) = (y, y, \dots)$. It is easy to see that Φ is an isometry, though it is clearly not surjective. In fact, $\text{ran}\,\Phi = \{(y_n) \in \mathcal{X}\colon y_n = y_m$ for all $n, m\}$. Thus $\mathcal{R}$ is a Banach space. Let $P_1 = $ the projection of $\mathcal{X}$ onto the first coordinate. Then $A = P_1 \circ \Phi$ and hence A is continuous.

(c) With the notation from the proof of (b), it follows that $\mathcal{X}^{**} = \oplus_2 \mathcal{Y}_n^{**}$ and $\Phi^{**}\colon \mathcal{R}^{**} \to \mathcal{X}^{**}$ is given by $\Phi^{**}(y^{**}) = (A^{**}y^{**}, A^{**}y^{**}, \dots)$. Now the fact that Φ is an isometry implies that Φ^* is surjective. (This follows in two ways. One is by a direct argument (see Exercise 2). Also, $\text{ran}\,\Phi^*$ is closed since $\text{ran}\,\Phi$ is closed (1.10), and $\text{ran}\,\Phi^*$ is dense since $^{\perp}(\text{ran}\,\Phi^*) = \ker\Phi = (0)$.) Hence $\ker\Phi^{**} = (\text{ran}\,\Phi^*)^{\perp} = (0)$; that is, Φ^{**} is injective. Therefore A^{**} is injective.

Now let $y^{**} \in A^{**-1}(\mathcal{Y})$. It follows that $\Phi^{**}y^{**} = x \in \mathcal{X}$. Let $\{y_i\}$ be a net in $\mathcal{R}$ such that $\|y_i\| \leq \|y^{**}\|$ for all i and $y_i \to y^{**}$ $\sigma(\mathcal{R}^{**}, \mathcal{R}^*)$ (V.4.1). Thus $\Phi^{**}(y_i) \to \Phi^{**}(y^{**})$ $\sigma(\mathcal{X}^{**}, \mathcal{X}^*)$. But $\Phi^{**}(y_i) = \Phi(y_i) \in \mathcal{X}$ and $\Phi^{**}(y^{**}) = x$. Hence $\Phi(y_i) \to x$ $\sigma(\mathcal{X}, \mathcal{X}^*)$. Since $\text{ran}\,\Phi$ is closed, $x \in \text{ran}\,\Phi$; let $\Phi(y) = x$. Then $0 = \Phi^{**}(y^{**} - y)$. Since Φ^{**} is injective, $y^{**} = y \in \mathcal{R}$.

(d) An argument using Alaoglu's Theorem shows that $A^{**}(\text{ball}\,\mathcal{R}^{**}) = $ the $\sigma(\mathcal{Y}^{**}, \mathcal{Y}^*)$ closure of $A(\text{ball}\,\mathcal{R})$. Put $C = A(\text{ball}\,\mathcal{R})$. Suppose $\text{cl}\,W$ is weakly compact. Now $C \subseteq 2^n\text{cl}\,W + 2^{-n}\text{ball}\,\mathcal{Y}^{**}$ and this set is $\sigma(\mathcal{Y}^{**}, \mathcal{Y}^*)$ compact. From the preceding paragraph, $A^{**}(\text{ball}\,\mathcal{R}^{**}) \subseteq$

$2^n \mathrm{cl}\, W + 2^{-n} \mathrm{ball}\, \mathcal{Y}^{**}$. Thus,

$$A^{**}(\mathrm{ball}\, \mathcal{R}^{**}) \subseteq \bigcap_{n=1}^{\infty} [2^n \mathrm{cl}\, W + 2^{-n} \mathrm{ball}\, \mathcal{Y}^{**}]$$

$$\subseteq \bigcap_{n=1}^{\infty} [\mathcal{Y} + 2^{-n} \mathrm{ball}\, \mathcal{Y}^{**}]$$

$$= \mathcal{Y}.$$

By (c), $\mathcal{R}^{**} = \mathcal{R}$ and $\mathcal{R}$ is reflexive.

Now assume $\mathcal{R}$ is reflexive; thus ball $\mathcal{R}$ is $\sigma(\mathcal{R}, \mathcal{R}^{*})$-compact. Therefore $C = A(\mathrm{ball}\, \mathcal{R})$ is weakly compact in $\mathcal{Y}$. By (a), $\mathrm{cl}\, W$ is weakly compact. ∎

The next theorem, as well as the preceding lemma, are from Davis, Figel, Johnson, and Pelczynski [1974].

5.4. Theorem. *If $\mathcal{X}, \mathcal{Y}$ are Banach spaces and $T \in \mathcal{B}(\mathcal{X}, \mathcal{Y})$, then T is weakly compact if and only if there is a reflexive space $\mathcal{R}$ and operators A in $\mathcal{B}(\mathcal{R}, \mathcal{Y})$ and B in $\mathcal{B}(\mathcal{X}, \mathcal{R})$ such that $T = AB$.*

PROOF. If $T = AB$, where A, B have the described form, then T is weakly compact by Proposition 5.2.

Now assume that T is weakly compact and put $W = T(\mathrm{ball}\, \mathcal{X})$. Define $\mathcal{R}$ as in Lemma 5.3. By (5.3d), $\mathcal{R}$ is reflexive. Let $A: \mathcal{R} \to \mathcal{Y}$ be the inclusion map. Note that if $x \in \mathrm{ball}\, \mathcal{X}$, then $Tx \in W$. Hence $2^n Tx \in U_n$ and so $1 > p_n(2^n Tx) = 2^n p_n(Tx)$. Thus $p_n(Tx) < 2^{-n}$ for x in ball $\mathcal{X}$. Hence if $\|x\| \leq 1$, $\|\|Tx\|\|^2 = \sum_n p_n(Tx)^2 < \sum 4^{-n} = c$. So $B: \mathcal{X} \to \mathcal{R}$ defined by $Bx = Tx$ is a bounded operator. Clearly $AB = T$. ∎

The preceding result can be used to prove several standard results from antiquity.

5.5. Theorem. *If $\mathcal{X}, \mathcal{Y}$ are Banach spaces and $T \in \mathcal{B}(\mathcal{X}, \mathcal{Y})$, the following statements are equivalent.*

(a) *T is weakly compact.*
(b) *$T^{**}(\mathcal{X}^{**}) \subseteq \mathcal{Y}$.*
(c) *T^{*} is weakly compact.*

PROOF. (a) $\Rightarrow$ (b): Let $\mathcal{R}$ be a reflexive space, $A \in \mathcal{B}(\mathcal{R}, \mathcal{Y})$, and $B \in \mathcal{B}(\mathcal{X}, \mathcal{R})$ such that $T = AB$. So $T^{**} = A^{**}B^{**}$. But $A^{**}: \mathcal{R} \to \mathcal{Y}^{**}$ since $\mathcal{R}^{**} = \mathcal{R}$. Hence $A^{**} = A$. Thus $T^{**} = AB^{**}$, and so $\mathrm{ran}\, T^{**} \subseteq \mathrm{ran}\, A \subseteq \mathcal{Y}$.

(b) $\Rightarrow$ (a): $T^{**}(\mathrm{ball}\, \mathcal{X}^{**})$ is $\sigma(\mathcal{Y}^{**}, \mathcal{Y}^{*})$ compact by Alaoglu's Theorem and the weak* continuity of T^{**}. By (b), $T^{**}(\mathrm{ball}\, \mathcal{X}^{**}) = C$ is $\sigma(\mathcal{Y}, \mathcal{Y}^{*})$

compact in $\mathcal{Y}$. Hence $T(\text{ball }\mathcal{X}) \subseteq C$ and must have weakly compact closure.

(c) $\Rightarrow$ (a): Let $\mathcal{S}$ be a reflexive space, $C \in \mathcal{B}(\mathcal{Y}^*, \mathcal{S})$, $D \in \mathcal{B}(\mathcal{S}, \mathcal{X}^*)$ such that $T^* = DC$. So $T^{**} = C^*D^*$, $D^*: \mathcal{X}^{**} \to \mathcal{S}^*$, and $C^*: \mathcal{S}^* \to \mathcal{Y}^{**}$. Put $\mathcal{R} = \text{cl } D^*(\mathcal{X})$ and $B = D^*|\mathcal{X}$; then $B: \mathcal{X} \to \mathcal{R}$ and $\mathcal{R}$ is reflexive. Let $A = C^*|\mathcal{R}$; so $A: \mathcal{R} \to \mathcal{Y}^{**}$. But if $x \in \mathcal{X}$, $ABx = C^*D^*x = T^{**}x = Tx \in \mathcal{Y}$. Thus $A: \mathcal{R} \to \mathcal{Y}$. Clearly $AB = T$.

(a) $\Rightarrow$ (c): Exercise. ∎

EXERCISES

1. Prove Proposition 5.2.

2. If $\mathcal{R}$ and $\mathcal{X}$ are Banach spaces and $\Phi: \mathcal{R} \to \mathcal{X}$ is an isometry, give an elementary proof that Φ^* is surjective.

3. Let $\mathcal{X}$ be a Banach space and recall the definition of a weakly Cauchy sequence (V.4.4). (a) Show that every bounded sequence in c_0 has a weakly Cauchy subsequence, but not every weakly Cauchy sequence in c_0 converges. (b) Show that if $T \in \mathcal{B}(c_0)$ and T is weakly compact, then T is compact.

4. Say that a Banach space $\mathcal{X}$ is *weakly compactly generated* (WCG) if there is a weakly compact subset K of $\mathcal{X}$ such that $\mathcal{X}$ is the closed linear span of K. Prove (Davis, Figel, Johnson, and Pelczynski, [1974]) that $\mathcal{X}$ is WCG if and only if there is a reflexive space and an injective bounded operator $T: \mathcal{R} \to \mathcal{X}$ such that $\text{ran } T$ is dense. (Hint: The Krein–Smulian Theorem (V.13.4) may be useful.)

5. If (X, Ω, μ) is a finite-measure space, $k \in L^\infty(X \times X, \Omega \times \Omega, \mu \times \mu)$, and $K: L^1(\mu) \to L^1(\mu)$ is defined by $(Kf)(x) = \int k(x, y)f(y)\,d\mu(y)$, show that K is weakly compact and K^2 is compact.

6. Let $\mathcal{Y}$ be a weakly sequentially complete Banach space. That is, if $\{y_n\}$ is a sequence in $\mathcal{Y}$ such that $\{\langle y_n, y^*\rangle\}$ is a Cauchy sequence in $\mathbb{F}$ for every y^* in $\mathcal{Y}^*$, then there is a y in $\mathcal{Y}$ such that $y_n \to y$ weakly [see (V.4.4)]. (a) If $T \in \mathcal{B}(\mathcal{X}, \mathcal{Y})$ and $x^{**} \in \mathcal{X}^{**}$ such that x^{**} is the $\sigma(\mathcal{X}^{**}, \mathcal{X}^*)$ limit of a sequence from $\mathcal{X}$, show that $T^{**}(x^{**}) \in \mathcal{Y}$. Let X be a compact space and put $\mathcal{F} =$ all subsets of X that are the union of a countable number of compact G_δ sets. Let $\mathcal{L} =$ the linear span of $\{\chi_F: F \in \mathcal{F}\}$ considered as a subset of $M(X)^* = C(X)^{**}$. (b) Show that if $T \in \mathcal{B}(C(X), \mathcal{Y})$, then $T^{**}(\mathcal{L}) \subseteq \mathcal{Y}$. (c) (Grothendieck [1953].) If $T \in \mathcal{B}(C(X), \mathcal{Y})$, then T is weakly compact. [Hint (Spain [1976]): Use James's Theorem [(V.13.3)].

CHAPTER VII

Banach Algebras and Spectral Theory for Operators on a Banach Space

The theory of Banach algebras is a large area in functional analysis with several subdivisions and applications to diverse areas of analysis and the rest of mathematics. Some monographs on this subject are by Bonsall and Duncan [1973] and C. R. Rickart [1960].

A significant change occurs in this chapter that will affect the remainder of this book. In order to prove that the spectrum of an element of a Banach algebra is nonvoid (Section 3), it is necessary to assume that the underlying field of scalars $\mathbb{F}$ is the field of complex numbers $\mathbb{C}$. It will be assumed from Section 3 until the end of this book that all vector spaces are over $\mathbb{C}$. This will also enable us to apply the theory of analytic functions to the study of Banach algebras and linear operators.

In this chapter only the rudiments of this subject are discussed. Enough, however, is presented to allow a treatment of the basics of spectral theory for operators on a Banach space.

§1. Elementary Properties and Examples

An *algebra* over $\mathbb{F}$ is a vector space $\mathscr{A}$ over $\mathbb{F}$ that also has a multiplication defined on it that makes $\mathscr{A}$ into a ring such that if $\alpha \in \mathbb{F}$ and $a, b \in \mathscr{A}$, $\alpha(ab) = (\alpha a)b = a(\alpha b)$.

1.1. Definition. A *Banach algebra* is an algebra $\mathscr{A}$ over $\mathbb{F}$ that has a norm $\|\cdot\|$ relative to which $\mathscr{A}$ is a Banach space and such that for all a, b in $\mathscr{A}$,

1.2 $$\|ab\| \le \|a\|\,\|b\|.$$

If $\mathscr{A}$ has an identity, e, then it is assumed that $\|e\| = 1$.

The fact that (1.2) is satisfied is not essential. If $\mathscr{A}$ is an algebra and has a norm relative to which $\mathscr{A}$ is a Banach space and is such that the map of $\mathscr{A} \times \mathscr{A} \to \mathscr{A}$ defined by $(a, b) \mapsto ab$ is continuous, then there is an equivalent norm on $\mathscr{A}$ that satisfies (1.2) (Exercise 1).

If $\mathscr{A}$ has an identity e, then the map $\alpha \mapsto \alpha e$ is an isomorphism of $\mathbb{F}$ into $\mathscr{A}$ and $\|\alpha e\| = |\alpha|$. So it will be assumed that $\mathbb{F} \subseteq \mathscr{A}$ via this identification. Thus the identity will be denoted by 1.

The content of the next proposition is that if $\mathscr{A}$ does not have an identity, it is possible to find a Banach algebra $\mathscr{A}_1$ that contains $\mathscr{A}$, that has an identity, and is such that dim $\mathscr{A}_1/\mathscr{A} = 1$.

1.3.   Proposition. *If $\mathscr{A}$ is a Banach algebra without an identity, let $\mathscr{A}_1 = \mathscr{A} \times \mathbb{F}$. Define algebraic operations on $\mathscr{A}_1$ by*

(i) $(a, \alpha) + (b, \beta) = (a + b, \alpha + \beta)$;
(ii) $\beta(a, \alpha) = (\beta a, \beta \alpha)$;
(iii) $(a, \alpha)(b, \beta) = (ab + \alpha b + \beta a, \alpha \beta)$.

Define $\|(a, \alpha)\| = \|a\| + |\alpha|$. Then $\mathscr{A}_1$ with this norm and the algebraic operations defined in (i), (ii), *and* (iii) *is a Banach algebra with identity $(0, 1)$ and $a \mapsto (a, 0)$ is an isometric isomorphism of $\mathscr{A}$ into $\mathscr{A}_1$.*

PROOF. Only (1.2) will be verified here; the remaining details are left to the reader. If $(a, \alpha), (b, \beta) \in \mathscr{A}_1$, then $\|(a, \alpha)(b, \beta)\| = \|(ab + \beta a + \alpha b, \alpha \beta)\| = \|ab + \beta a + \alpha b\| + |\alpha \beta| \leq \|a\| \|b\| + |\beta| \|a\| + |\alpha| \|b\| + |\alpha| |\beta| = \|(a, \alpha)\| \|(b, \beta)\|$. ∎

1.4.   Example. If X is a compact space, then $\mathscr{A} = C(X)$ is a Banach algebra if $(fg)(x) = f(x)g(x)$ whenever $f, g \in \mathscr{A}$ and $x \in X$. Note that $\mathscr{A}$ is abelian and has an identity (the constantly 1 function).

If X is completely regular and $\mathscr{A} = C_b(X)$, then $\mathscr{A}$ is also a Banach algebra. In fact, $C_b(X) \cong C(\beta X)$ (V.6) so that this is a special case of Example 1.4. Another special case is l^∞.

1.5.   Example. If X is a locally compact space, $\mathscr{A} = C_0(X)$ is a Banach algebra when the multiplication is defined pointwise as in the preceding example. $\mathscr{A}$ is abelian, but if X is not compact, $\mathscr{A}$ does not have an identity. If X_∞ is the one-point compactification of X, then $C(X_\infty) \supseteq C_0(X)$ and $C(X_\infty)$ is a Banach algebra with identity.

Note that c_0 is a special case of Example 1.5.

1.6.   Example. If (X, Ω, μ) is a σ-finite measure space and $\mathscr{A} = L^\infty(X, \Omega, \mu)$, then $\mathscr{A}$ is an abelian Banach algebra with identity if the operations are defined pointwise.

1.7.   Example. Let $\mathscr{X}$ be a Banach space and put $\mathscr{A} = \mathscr{B}(\mathscr{X})$. If multiplication is defined by composition, then $\mathscr{A}$ is a Banach algebra with identity, 1. If dim $\mathscr{X} \geq 2$, $\mathscr{A}$ is not abelian.

1.8.   Example. If $\mathscr{X}$ is a Banach space and $\mathscr{A} = \mathscr{B}_0(\mathscr{X})$, the compact operators on $\mathscr{X}$, then $\mathscr{A}$ is a Banach algebra without identity if dim $\mathscr{X} = \infty$. In fact, $\mathscr{B}_0(\mathscr{X})$ is an ideal of $\mathscr{B}(\mathscr{X})$.

Note that a special case of Example 1.7 occurs when $\mathscr{A} = M_n(\mathbb{F})$, the $n \times n$ matrices, where $\mathscr{A}$ is given the norm resulting when $M_n(\mathbb{F})$ is identified with $\mathscr{B}(\mathbb{F}^n)$.

1.9.   Example. Let G be a locally compact topological group and let $M(G) =$ all finite regular Borel measures on G. If $\mu, \nu \in M(G)$, define L: $C_0(G) \to \mathbb{F}$ by

$$L(f) = \int \int f(xy)\, d\mu(x)\, d\nu(y) = \int \int f(xy)\, d\nu(y)\, d\mu(x).$$

Then L is a linear functional on $C_0(G)$ and

$$|L(f)| \leq \int \int |f(xy)|\, d|\mu|(x)\, d|\nu|(y)$$

$$\leq \|f\| \, \|\mu\| \, \|\nu\|.$$

So $L \in C_0(G)^* = M(G)$. Define $\mu * \nu$ by $L(f) = \int f d\mu * \nu$ for f in $C_0(G)$. That is,

1.10 $$\int f d\mu * \nu = \int \int f(xy)\, d\mu(x)\, d\nu(y).$$

Note that $\|\mu * \nu\| = \|L\| \leq \|\mu\| \, \|\nu\|$. It follows that $M(G)$ is a Banach algebra with this definition of multiplication. The product $\mu * \nu$ is called the *convolution* of μ and ν.

Let $e =$ the identity of G and let $\delta_e =$ the unit point mass at e. If $f \in C_0(G)$, then

$$\int f d\mu * \delta_e = \int \int f(xy)\, d\mu(x)\, d\delta_e(y)$$

$$= \int f(xe)\, d\mu(x)$$

$$= \int f d\mu.$$

So $\mu * \delta_e = \mu$; similarly, $\delta_e * \mu = \mu$. Hence δ_e is the identity for $M(G)$.

If $x, y \in G$, then it is easy to check that $\delta_x * \delta_y = \delta_{xy}$ and $M(G)$ is abelian if and only if G is abelian.

1.11.   Example. Let G be a locally compact group and let $m =$ right Haar measure on G. That is, m is a non-negative regular Borel measure on G such that $m(U) > 0$ for every nonempty open subset U of G and $\int f(xy)\, dm(x)$ $= \int f(x)\, dm(x)$ for every f in $C_c(G)$ (the continuous functions $f: G \to \mathbb{F}$ with compact support). If G is compact, the existence of m was established in Section V.11. If G is not compact, m exists but its existence must be established by nonfunctional analytic methods (see Nachbin [1965]).

If $f, g \in L^1(m)$, let $\mu = fm$ and $\nu = gm$ as in the proof of (V.8.1). Then $\mu, \nu \in M(G)$ and $\|\mu\| = \|f\|_1$, $\|\nu\| = \|g\|_1$. In fact, the Radon–Nikodym Theorem makes it possible to identify $L^1(m)$ with a closed subspace of $M(G)$. Is it a closed subalgebra?

Let $\phi \in C_c(G)$. Then

$$\int \phi \, d\mu * \nu = \int \int \phi(xy) f(x) g(y) \, dm(x)\, dm(y)$$
$$= \int g(y) \left[\int \phi(xy) f(x) \, dm(x) \right] dm(y)$$
$$= \int g(y) \left[\int \phi(x) f(xy^{-1}) \, dm(x) \right] dm(y)$$
$$= \int \phi(x) \left[\int f(xy^{-1}) g(y) \, dm(y) \right] dm(x)$$
$$= \int \phi(x) h(x) \, dm(x),$$

where $h(x) = \int f(xy^{-1}) g(y) \, dm(y)$, x in G. It follows that $h \in L^1(m)$ (see Exercise 4). Thus $\mu * \nu = hm$, so $L^1(m)$ is a Banach subalgebra of $M(G)$. In fact, the preceding discussion enables us to define $f * g$ in $L^1(m)$ for f, g in $L^1(m)$ by

$$f * g(x) = \int f(xy^{-1}) g(y) \, dm(y).$$

The algebra $L^1(m)$ is denoted by $L^1(G)$.

It can be shown that $L^1(G)$ is abelian if and only if G is abelian and $L^1(G)$ has an identity if and only if G is discrete (in which case $L^1(G) = M(G)$—what is m?). This algebra is examined more closely in Section 9.

If $\{\mathscr{A}_i\}$ is a collection of Banach algebras, let $\oplus_0 \mathscr{A}_i \equiv \{a \in \prod_i \mathscr{A}_i : \text{for all } \varepsilon > 0, \{i: \|a(i)\| \geq \varepsilon\} \text{ is finite}\}$.

1.12. Proposition. *If* $\{\mathscr{A}_i\}$ *is a collection of Banach algebras,* $\oplus_0 \mathscr{A}_i$ *and* $\oplus_\infty \mathscr{A}_i$ *are Banach algebras.*

PROOF. Exercise.

EXERCISES

1. Let $\mathscr{A}$ be an algebra that is also a Banach space and such that if $a \in \mathscr{A}$, the maps $x \mapsto ax$ and $x \mapsto xa$ of $\mathscr{A} \to \mathscr{A}$ are continuous. Let $\mathscr{A}_1 = \mathscr{A} \times \mathbb{F}$ as in Proposition 1.3. If $a \in \mathscr{A}$, define $L_a: \mathscr{A}_1 \to \mathscr{A}_1$ by $L_a(x, \xi) = (ax + \xi a, 0)$.

Show that $L_a \in \mathcal{B}(\mathcal{A}_1)$ and if $\|\|a\|\| = \|L_a\|$, then $\|\| \cdot \|\|$ is equivalent to the norm of $\mathcal{A}$ and $\mathcal{A}$ with $\|\| \cdot \|\|$ is a Banach algebra.

2. Complete the proof of Proposition 1.3.

3. Verify the statements made in Examples (1.4) through (1.9) and (1.11).

4. Let G be a locally compact group. (a) If $\phi \in C_c(G)$ and $\varepsilon > 0$, show that there is an open neighborhood U of e in G such that $\|\phi_x - \phi_y\| < \varepsilon$ whenever $xy^{-1} \in U$. [Here $\phi_x(z) = \phi(xz)$.] (b) Show that if $f \in L^p(G)$, $1 \leq p < \infty$, and $\varepsilon > 0$, there is an open neighborhood U of e in G such that $\|f_x - f_y\|_p < \varepsilon$ whenever $xy^{-1} \in U$. (c) Show that if $f \in L^1(G)$ and $g \in L^\infty(G)$, $h(x) = \int f(xy^{-1})g(y)\, dm(y)$ defines a bounded continuous function $h: G \to \mathbb{F}$. (d) If $f, g \in L^1(G)$ and h is defined as in (c), show that $h \in L^1(G)$.

5. Prove Proposition 1.12.

6. Let $\{\mathcal{A}_i: i \in I\}$ be a collection of Banach algebras. (a) Show that $\oplus_0 \mathcal{A}_i$ is a closed ideal of $\oplus_\infty \mathcal{A}_i$. (b) Show that $\oplus_\infty \mathcal{A}_i$ has an identity if and only if each $\mathcal{A}_i$ has an identity. (c) Show that $\oplus_0 \mathcal{A}_i$ has an identity if and only if I is finite and each $\mathcal{A}_i$ has an identity.

7. If X, Y are completely regular, show that $C_b(X) \oplus_\infty C_b(Y)$ is isometrically isomorphic to $C_b(X \oplus Y)$, where $X \oplus Y$ is the disjoint union of X and Y.

8. If X and Y are locally compact, show that $C_0(X) \oplus_\infty C_0(Y)$ is isometrically isomorphic to $C_0(X \oplus Y)$.

9. Let $\{X_i: i \in I\}$ be a collection of locally compact spaces and let $X =$ the disjoint union of these spaces furnished with the topology $\{U \subseteq X: U \cap X_i$ is open in X_i for all $i\}$. Show that X is locally compact and $\oplus_0 C_0(X_i)$ is isometrically isomorphic to $C_0(X)$.

§2. Ideals and Quotients

If $\mathcal{A}$ is an algebra, a *left ideal* of $\mathcal{A}$ is a subalgebra $\mathcal{M}$ of $\mathcal{A}$ such that $ax \in \mathcal{M}$ whenever $a \in \mathcal{A}$, $x \in \mathcal{M}$. A *right ideal* of $\mathcal{A}$ is a subalgebra $\mathcal{M}$ such that $xa \in \mathcal{M}$ whenever $a \in \mathcal{A}$, $x \in \mathcal{M}$. A (*bilateral*) *ideal* is a subalgebra of $\mathcal{A}$ that is both a left ideal and a right ideal.

If $a \in \mathcal{A}$ and $\mathcal{A}$ has an identity 1, say that a is *left invertible* if there is an x in $\mathcal{A}$ with $xa = 1$. Similarly, define *right invertible* and *invertible* elements. If a is invertible and $x, y \in \mathcal{A}$ such that $xa = 1 = ay$, then $y = 1y = (xa)y = x(ay) = x1 = x$. So if a is invertible, there is a unique element a^{-1} such that $aa^{-1} = a^{-1}a = 1$.

If $\mathcal{M}$ is a left ideal in $\mathcal{A}$, $a \in \mathcal{M}$, and a is left invertible, then $\mathcal{M} = \mathcal{A}$. In fact, if $xa = 1$, then $1 \in \mathcal{M}$ since $\mathcal{M}$ is a left ideal. Thus for y in $\mathcal{A}$, $y = y1 \in \mathcal{M}$. This forms a link between ideals and invertibility.

In the case of a Banach algebra some bonuses occur due to the interplay of the norm and the algebra. The results of this section will be for Banach algebras with an identity. To discuss invertibility this is, of course, the only feasible setting. For Banach algebras without an identity some analogous results can be obtained, however, by a consideration of the algebra obtained by adjoining an identity (1.3). The concept of a modular ideal and a modular unit can also be employed (see Exercise 6).

The next proof is based on the geometric series.

2.1. Lemma. *If $\mathscr{A}$ is a Banach algebra with identity and $x \in \mathscr{A}$ such that $\|x - 1\| < 1$, then x is invertible.*

PROOF. Let $y = 1 - x$; so $\|y\| = r < 1$. Since $\|y^n\| \leq \|y\|^n = r^n$ (Why?), $\sum_{n=0}^{\infty}\|y^n\| < \infty$. Hence $z = \sum_{n=0}^{\infty}y^n$ converges in $\mathscr{A}$. If $z_n = 1 + y + y^2 + \cdots + y^n$,

$$z_n(1 - y) = (1 + y + \cdots + y^n) - (y + y^2 + \cdots + y^{n+1}) = 1 - y^{n+1}.$$

But $\|y^{n+1}\| \leq r^{n+1}$, so $y^{n+1} \to 0$ as $n \to \infty$. Hence $z(1 - y) = \lim z_n(1 - y) = 1$. Similarly, $(1 - y)z = 1$. So $(1 - y)$ is invertible and $(1 - y)^{-1} = z = \sum_0^{\infty}y^n$. But $1 - y = 1 - (1 - x) = x$. ∎

Note that completeness was used to show that $\sum y^n$ converges.

2.2. Theorem. *If $\mathscr{A}$ is a Banach algebra with identity, $G_l = \{a \in \mathscr{A}: a$ is left invertible$\}$, $G_r = \{a \in \mathscr{A}: a$ is right invertible$\}$, and $G = \{a \in \mathscr{A}: a$ is invertible$\}$, then G_l, G_r, and G are open subsets of $\mathscr{A}$. Also, the map $a \mapsto a^{-1}$ of $G \to G$ is continuous.*

PROOF. Let $a_0 \in G$ and let $b_0 \in \mathscr{A}$ such that $b_0a_0 = 1$. If $\|a - a_0\| < \|b_0\|^{-1}$, then $\|b_0a - 1\| = \|b_0(a - a_0)\| < 1$. By the preceding lemma, $x = b_0a$ is invertible. If $b = x^{-1}b_0$, then $ba = 1$. Hence $G_l \supseteq \{a \in \mathscr{A}: \|a - a_0\| < \|b_0\|^{-1}\}$ and G_l must be open. Similarly, G_r is open. Since $G = G_l \cap G_r$ (Why?), G is open.

To prove that $a \mapsto a^{-1}$ is a continuous map of $G \to G$, first assume that $\{a_n\}$ is a sequence in G such that $a_n \to 1$. Let $0 < \delta < 1$ and suppose $\|a_n - 1\| < \delta$. From the preceding lemma, $a_n^{-1} = (1 - (1 - a_n))^{-1} = \sum_{k=0}^{\infty}(1 - a_n)^k = 1 + \sum_{k=1}^{\infty}(1 - a_n)^k$. Hence

$$\|a_n^{-1} - 1\| = \left\| \sum_{k=1}^{\infty} (1 - a_n)^k \right\|$$

$$\leq \sum_{k=1}^{\infty} \|1 - a_n\|^k$$

$$< \delta/(1 - \delta).$$

If $\varepsilon > 0$ is given, then δ can be chosen such that $\delta/(1 - \delta) < \varepsilon$. So $\|a_n - 1\| < \delta$ implies $\|a_n^{-1} - 1\| < \varepsilon$. Hence $\lim a_n^{-1} = 1$.

Now let $a \in G$ and suppose $\{a_n\}$ is a sequence in G such that $a_n \to a$. Hence $a^{-1}a_n \to 1$. By the preceding paragraph, $a_n^{-1}a = (a^{-1}a_n)^{-1} \to 1$. Hence $a_n^{-1} = a_n^{-1}aa^{-1} \to a^{-1}$. ∎

Two facts surfaced in the preceding proofs that are worth recording for the future.

2.3. Corollary. *Let $\mathscr{A}$ be a Banach algebra with identity.*

(a) *If $\|a - 1\| < 1$, then $a^{-1} = \sum_{k=0}^{\infty}(1 - a)^k$.*
(b) *If $b_0 a_0 = 1$ and $\|a - a_0\| < \|b_0\|^{-1}$, then a is left invertible.*

A *maximal ideal* is a proper ideal that is contained in no larger proper ideal.

2.4. Corollary. *If $\mathscr{A}$ is a Banach algebra with identity, then*

(a) *the closure of a proper left, right, or bilateral ideal is a proper left, right, or bilateral ideal;*
(b) *a maximal left, right, or bilateral ideal is closed.*

PROOF. (a) Let $\mathscr{M}$ be a proper left ideal and let G_l be the set of left-invertible elements in $\mathscr{A}$. It follows that $\mathscr{M} \cap G_l = \square$. (See the introduction to this section.) Thus $\mathscr{M} \subseteq \mathscr{A} \setminus G_l$. By the preceding theorem, $\mathscr{A} \setminus G_l$ is closed. Hence cl $\mathscr{M} \subseteq \mathscr{A} \setminus G_l$; and thus cl $\mathscr{M} \neq \mathscr{A}$. It is easy to check that cl $\mathscr{M}$ is an ideal. The proof of the remainder of (a) is similar.

(b) If $\mathscr{M}$ is a maximal left ideal, cl $\mathscr{M}$ is a proper left ideal by (a). Hence $\mathscr{M} = $ cl $\mathscr{M}$ by maximality. ∎

If $\mathscr{A}$ does not have an identity, then $\mathscr{A}$ may contain some proper, dense ideals. For example, let $\mathscr{A} = C_0(\mathbb{R})$. Then $C_c(\mathbb{R})$, the continuous functions with compact support, is a dense ideal in $C_0(\mathbb{R})$. There is something that can be said, however (see Exercise 6).

2.5. Proposition. *If $\mathscr{A}$ is a Banach algebra with identity, then every proper left, right, or bilateral ideal is contained in a maximal ideal of the same type.*

The proof of the preceding proposition is an exercise in the application of Zorn's Lemma and is left to the reader. Actually, this is a theorem from algebra and it is not necessary to assume that $\mathscr{A}$ is a Banach algebra.

Let $\mathscr{A}$ be a Banach algebra and let $\mathscr{M}$ be a proper closed ideal. Note that $\mathscr{A}/\mathscr{M}$ becomes an algebra. Indeed, $(x + \mathscr{M})(y + \mathscr{M}) = xy + \mathscr{M}$ is a well-defined multiplication on $\mathscr{A}/\mathscr{M}$. (Why?)

2.6. Theorem. *If $\mathscr{A}$ is a Banach algebra and $\mathscr{M}$ is a proper closed ideal in $\mathscr{A}$, then $\mathscr{A}/\mathscr{M}$ is a Banach algebra. If $\mathscr{A}$ has an identity, so does $\mathscr{A}/\mathscr{M}$.*

PROOF. We have already seen that $\mathscr{A}/\mathscr{M}$ is a Banach space and, as was mentioned prior to the statement of the theorem, $\mathscr{A}/\mathscr{M}$ is an algebra. If $x, y \in \mathscr{A}$ and $u, v \in \mathscr{M}$, then $(x + u)(y + v) = xy + (xv + uy + uv) \in xy + \mathscr{M}$. Hence $\|(x + \mathscr{M})(y + \mathscr{M})\| = \|xy + \mathscr{M}\| \leq \|(x + u)(y + v)\| \leq \|x + u\| \, \|y + v\|$. Taking the infimum over all u, v in $\mathscr{M}$ gives that $\|(x + \mathscr{M})(y + \mathscr{M})\| \leq \|x + \mathscr{M}\| \, \|y + \mathscr{M}\|$. The remainder of the proof is left to the reader. ∎

It may be that $\mathscr{A}/\mathscr{M}$ has an identity even if $\mathscr{A}$ does not. For example, let $\mathscr{A} = C_0(\mathbb{R})$ and let $\mathscr{M} = \{\phi \in C_0(\mathbb{R}): \phi(x) = 0 \text{ when } |x| \leq 1\}$. If $\phi_0 \in C_0(\mathbb{R})$ such that $\phi_0(x) = 1$ for $|x| \leq 1$, then $\phi_0 + \mathscr{M}$ is an identity for $\mathscr{A}/\mathscr{M}$. In fact, if $\phi \in C_0(\mathbb{R})$, $(\phi\phi_0 - \phi)(x) = 0$ if $|x| \leq 1$. Hence $(\phi + \mathscr{M})(\phi_0 + \mathscr{M}) = \phi + \mathscr{M}$ (see Exercises 6 through 9).

EXERCISES

1. Let $\mathscr{A}$ be a Banach algebra and let $\mathscr{L}$ be all of the closed left ideals in $\mathscr{A}$. If $I_1, I_2 \in \mathscr{L}$, define $I_1 \vee I_2 \equiv \text{cl}(I_1 + I_2)$ and $I_1 \wedge I_2 = I_1 \cap I_2$. Show that with these definitions $\mathscr{L}$ is a complete lattice with a largest and a smallest element.

2. Let X be locally compact. For every open subset U of X, let $I(U) = \{\phi \in C_0(X): \phi = 0 \text{ on } X \setminus U\}$. Show that $U \mapsto I(U)$ is a lattice monomorphism of the collection of open subsets of X into the lattice of closed ideals of $C_0(X)$. (It is, in fact, surjective, but the proof of that should wait.)

3. Let (X, Ω, μ) be a σ-finite measure space and let I be an ideal in $L^\infty(X, \Omega, \mu)$ that is weak* closed. Show that there is a set Δ in Ω such that $I = \{\phi \in L^\infty(X, \Omega, \mu): \phi = 0 \text{ on } \Delta\}$.

4. Let $\mathscr{A} = \left\{ \begin{bmatrix} \alpha & 0 \\ \beta & \alpha \end{bmatrix} : \alpha, \beta \in \mathbb{F} \right\}$ and let $\mathscr{M} = \left\{ \begin{bmatrix} 0 & 0 \\ \beta & 0 \end{bmatrix} : \beta \in \mathbb{F} \right\}$. Show that $\mathscr{A}$ is a Banach algebra and $\mathscr{M}$ is a maximal ideal in $\mathscr{A}$.

5. Show that for $n \geq 1$, $M_n(\mathbb{C})$ has no nontrivial ideals. How about $M_n(\mathbb{R})$?

6. Let $\mathscr{A}$ be a Banach algebra but do not assume that $\mathscr{A}$ has an identity. If I is a left ideal of $\mathscr{A}$, say that I is a *modular left ideal* if there is a u in $\mathscr{A}$ such that $\mathscr{A}(1 - u) \equiv \{a - au: a \in \mathscr{A}\} \subseteq I$; call such an element u of $\mathscr{A}$ a *right modular unit* for I. Similarly, define *right modular ideals* and *left modular units*. Prove the following. (a) If u is a right modular unit for the left ideal I and $u \in I$, then $I = \mathscr{A}$. (b) Maximal modular left ideals are maximal left ideals. (c) If I is a proper modular left ideal, then I is contained in a maximal left ideal. (d) If I is a proper modular left ideal and u is a modular right unit for I, then $\|u - x\| \geq 1$ for all x in I and cl I is a proper modular left ideal. (e) Every maximal modular left ideal of $\mathscr{A}$ is closed.

7. Using the terminology of Exercise 6, let I be an ideal of $\mathscr{A}$. Show: (a) if u is a right modular unit for I and v is a left modular unit for I, then $u - v \in I$. (b) If I is closed, $\mathscr{A}/I$ has an identity if and only if there is a right modular unit and a left modular unit for I. Call an ideal I such that $\mathscr{A}/I$ has an identity a

modular ideal. An element u such that $u + I$ is an identity for $\mathscr{A}/I$ is called a *modular identity* for I.

8. If $\mathscr{A}$ is a Banach algebra, a net $\{e_i\}$ in $\mathscr{A}$ is called an *approximate identity* for $\mathscr{A}$ if $\sup_i \|e_i\| < \infty$ and for each a in $\mathscr{A}$, $e_i a \to a$ and $a e_i \to a$. Show that $\mathscr{A}$ has an approximate identity if and only if there is a bounded subset E of $\mathscr{A}$ such that for every $\varepsilon > 0$ and for every a in $\mathscr{A}$ there is an e in E with $\|ae - a\| + \|ea - a\| < \varepsilon$.

9. Show that if X is locally compact, then $C_0(X)$ has an approximate identity.

10. If $\mathscr{H}$ is a Hilbert space, show that $\mathscr{B}_0(\mathscr{H})$ has an approximate identity.

11. If G is a locally compact group, show that $L^1(G)$ (1.11) has an approximate identity. [Hint: Let $\mathscr{U}$ = all neighborhoods U of the identity e of G such that cl U is compact. Order $\mathscr{U}$ by reverse inclusion. For U in $\mathscr{U}$, let $f_U = m(U)^{-1}\chi_U$. Then $\{f_U: U \in \mathscr{U}\}$ is an approximate identity for $L^1(G)$.]

12. For $0 < r < 1$, let $P_r\colon \partial\mathbb{D} \to [0, \infty)$ be defined by $P_r(z) = \sum_{n=-\infty}^{\infty} r^{|n|} z^n$ (the *Poisson kernel*). Show that $\{P_r\}$ is an approximate identity for $L^1(\partial\mathbb{D})$ (under convolution).

13. If $\mathscr{H}$ is a Hilbert space and P is a finite-rank projection, show that $\mathscr{B}_0(\mathscr{H})P$ is a closed modular left ideal of $\mathscr{B}_0(\mathscr{H})$. What is the associated right modular unit?

14. Find the minimal closed proper left ideals of $M_n(\mathbb{F})$.

15. Find the minimal closed proper left ideals of $\mathscr{B}_0(\mathscr{H})$, $\mathscr{H}$ a Hilbert space. How about for $\mathscr{B}_0(\mathscr{X})$, $\mathscr{X}$ a Banach space?

16. What are the maximal modular left ideals of $\mathscr{B}_0(\mathscr{H})$, $\mathscr{H}$ a Hilbert space?

§3. The Spectrum

3.1. Definition. If $\mathscr{A}$ is a Banach algebra with identity and $a \in \mathscr{A}$, the spectrum of a, denoted by $\sigma(a)$, is defined by

$$\sigma(a) = \{\alpha \in \mathbb{F}: a - \alpha \text{ is not invertible}\}.$$

The *left spectrum*, $\sigma_l(a)$, is the set $\{\alpha \in \mathbb{F}: a - \alpha \text{ is not left invertible}\}$; the *right spectrum*, $\sigma_r(a)$, is defined similarly.

The *resolvent set* of a is defined by $\rho(a) = \mathbb{F} \setminus \sigma(a)$. The *left* and *right* resolvents of a are $\rho_l(a) = \mathbb{F} \setminus \sigma_l(a)$ and $\rho_r(a) = \mathbb{F} \setminus \sigma_r(a)$.

3.2. Example. Let X be compact. If $f \in C(X)$, then $\sigma(f) = f(X)$. In fact, if $\alpha = f(x_0)$, then $f - \alpha$ has a zero and cannot be invertible. So $f(X) \subseteq \sigma(f)$. On the other hand, if $\alpha \notin f(X)$, $f - \alpha$ is a nonvanishing continuous function on X. Hence $(f - \alpha)^{-1} \in C(X)$ and so $f - \alpha$ is invertible. Thus $\alpha \notin \sigma(f)$.

3.3. **Example.** If $\mathscr{X}$ is a Banach space and $A \in \mathscr{B}(\mathscr{X})$, then $\sigma(A) = \{\alpha \in \mathbb{F}$: either $\ker(A - \alpha) = (0)$ or $\mathrm{ran}(A - \alpha) \neq \mathscr{X}\}$. In fact, this means that $\rho(A) = \mathbb{F} \setminus \sigma(A) = \{\alpha \in \mathbb{F}: A - \alpha \text{ is bijective}\}$. If $\alpha \in \rho(A)$, there is an operator T in $\mathscr{B}(\mathscr{X})$ such that $T(A - \alpha) = (A - \alpha)T = 1$; clearly, $A - \alpha$ is bijective. On the other hand, if $A - \alpha$ is bijective, $(A - \alpha)^{-1} \in \mathscr{B}(\mathscr{X})$ by the Inverse Mapping Theorem.

3.4. **Example.** If $\mathscr{H}$ is a Hilbert and $A \in \mathscr{B}(\mathscr{H})$, then $\sigma_l(A) = \{\alpha \in \mathbb{F}: \inf\{\|(A - \alpha)h\|: \|h\| = 1\} = 0\}$. In fact, suppose $B \in \mathscr{B}(\mathscr{H})$ such that $B(A - \alpha) = 1$. If $\|h\| = 1$, then $1 = \|h\| = \|B(A - \alpha)h\| \leq \|B\| \|(A - \alpha)h\|$. So $\|(A - \alpha)h\| \geq \|B\|^{-1}$ whenever $\|h\| = 1$.

Conversely, suppose $\|(A - \alpha)h\| \geq \delta > 0$ whenever $\|h\| = 1$. Note that $\ker(A - \alpha) = (0)$. It will now be shown that $\mathrm{ran}(A - \alpha)$ is closed. In fact, assume that $(A - \alpha)f_n \to g$. Then $\delta\|f_n - f_m\| \leq \|(A - \alpha)(f_n - f_m)\| = \|(A - \alpha)f_n - (A - \alpha)f_m\|$. Thus $\{f_n\}$ is a Cauchy sequence. Let $f_n \to f$. Then $g = \lim(A - \alpha)f_n = (A - \alpha)f$; hence $g \in \mathrm{ran}(A - \alpha)$. Let $\mathscr{X} = \mathrm{ran}(A - \alpha)$; so $(A - \alpha): \mathscr{H} \to \mathscr{X}$ is a bijection. Thus $(A - \alpha)^{-1}: \mathscr{X} \to \mathscr{H}$ is bounded. Define $B: \mathscr{H} \to \mathscr{H}$ by letting $B(k + h) = (A - \alpha)^{-1}k$ when $k \in \mathscr{X}$ and $h \in \mathscr{X}^\perp$. Thus $B \in \mathscr{B}(\mathscr{H})$ and $B(A - \alpha) = 1$.

3.5. **Example.** If $\mathscr{A} = M_2(\mathbb{R})$ and $A = \begin{bmatrix} 0 & -1 \\ 1 & 0 \end{bmatrix}$, then $\sigma(A) = \square$. In fact, $A - \alpha$ is not invertible if and only if $0 = \det(A - \alpha) = \alpha^2 + 1$, which is impossible in $\mathbb{R}$.

The phenomenon of the last example does not occur if $\mathscr{A}$ is a Banach algebra over $\mathbb{C}$.

3.6. **Theorem.** *If $\mathscr{A}$ is a Banach algebra over $\mathbb{C}$ with an identity, then for each a in $\mathscr{A}$, $\sigma(a)$ is a nonempty compact subset of $\mathbb{C}$. Moreover, if $|\alpha| > \|a\|$, $\alpha \notin \sigma(a)$ and $z \mapsto (z - a)^{-1}$ is an $\mathscr{A}$-valued analytic function defined on $\rho(a)$.*

Before beginning the proof, a few words on vector-valued analytic functions are in order. If G is a region in $\mathbb{C}$ and $\mathscr{X}$ is a Banach space, define the derivative of $f: G \to \mathscr{X}$ at z_0 to be $\lim_{h \to 0} h^{-1}[f(z_0 + h) - f(z_0)]$ if the limit exists. Say that f is analytic if f has a continuous derivative on G. The whole theory of analytic functions transfers to this situation. The statements and proofs of such theorems as Cauchy's Integral Formula, Liouville's Theorem, etc., transfer verbatim. Also, $f: G \to \mathscr{X}$ is analytic if for each z_0 in G there is a sequence $x_0, x_1, x_2, \ldots$ in $\mathscr{X}$ such that $f(z) = \sum_{k=0}^{\infty}(z - z_0)^k x_k$ whenever $z \in B(z_0; r)$, where $r = \mathrm{dist}(z_0, \partial G)$. Moreover, the convergence is uniform on compact subsets of $B(z_0; r)$.

There is also a way of obtaining the vector-valued case as a consequence of the scalar-valued case (see Exercise 4).

PROOF OF THEOREM 3.6. If $|\alpha| > \|a\|$, then $\alpha - a = \alpha(1 - a/\alpha)$ and $\|a/\alpha\|$ < 1. By Corollary 2.3, $(1 - a/\alpha)$ is invertible. Hence $\alpha - a$ is invertible and so $\alpha \notin \sigma(a)$. Thus $\sigma(a) \subseteq \{\alpha \in \mathbb{C}: |\alpha| \leq \|a\|\}$ and $\sigma(a)$ is bounded.

Let G be the set of invertible elements of $\mathscr{A}$. The map $\alpha \mapsto (\alpha - a)$ is a continuous function of $\mathbb{C} \to \mathscr{A}$. Since G is open and $\rho(a)$ is the inverse image of G under this map, $\rho(a)$ is open. Thus $\sigma(a) = \mathbb{C} \setminus \rho(a)$ is compact.

Define $F: \rho(a) \to \mathscr{A}$ by $F(z) = (z - a)^{-1}$. In the identity $x^{-1} - y^{-1} = x^{-1}(y - x)y^{-1}$, let $x = (\alpha + h - a)$ and $y = (\alpha - a)$, where $\alpha \in \rho(a)$ and $h \in \mathbb{C}$ such that $h \neq 0$ and $\alpha + h \in \rho(a)$. This gives

$$\frac{F(\alpha + h) - F(\alpha)}{h} = \frac{(\alpha + h - a)^{-1}(-h)(\alpha - a)^{-1}}{h}$$

$$= -(\alpha + h - a)^{-1}(\alpha - a)^{-1}.$$

Since $(\alpha + h - a)^{-1} \to (\alpha - a)^{-1}$ as $h \to 0$, $F'(\alpha)$ exists and

$$F'(\alpha) = -(\alpha - a)^{-2}.$$

Clearly $F': \rho(a) \to \mathscr{A}$ is continuous, so F is analytic on $\rho(a)$.

From the first paragraph of the proof and Corollary 2.3, if $|z| > \|a\|$,

$$F(z) = \frac{1}{z}\left(1 - \frac{a}{z}\right)^{-1} = \frac{1}{z}\sum_{k=0}^{\infty}\left(\frac{a}{z}\right)^k.$$

Hence

$$\|F(z)\| \leq \frac{1}{|z|}\sum_{k=0}^{\infty}\left(\frac{\|a\|}{|z|}\right)^k$$

$$= \frac{1}{|z|}\frac{1}{1 - \|a\|/|z|}$$

$$= (|z| - \|a\|)^{-1}.$$

Thus $F(z) \to 0$ as $z \to \infty$. Therefore if $\rho(a) = \mathbb{C}$, F is an entire function that vanishes at ∞. By Liouville's Theorem F is constant. Since $F' \neq 0$, this is a contradiction. Thus $\rho(a) \neq \mathbb{C}$, or $\sigma(a) \neq \square$. ∎

Because the spectrum of an element of a complex Banach algebra is not empty, the following assumption is made.

Assumption. *Henceforward, all Banach spaces and all Banach algebras are over* $\mathbb{C}$.

3.7. Definition. If $\mathscr{A}$ is a Banach algebra with identity and $a \in \mathscr{A}$, the *spectral radius* of $a, r(a)$, is defined by

$$r(a) = \sup\{|\alpha|: \alpha \in \sigma(a)\}.$$

Because $\sigma(a) \neq \square$ and is bounded, $r(a)$ is well defined and finite; because $\sigma(a)$ is compact, this supremum is attained.

Let $\mathscr{A} = M_2(\mathbb{C})$ and let $A = \begin{bmatrix} 0 & 0 \\ 1 & 0 \end{bmatrix}$. Then $A^2 = 0$ and $\sigma(A) = \{0\}$; so $r(A) = 0$. So it is possible to have $r(A) = 0$ with $A \neq 0$.

3.8. Proposition. *If $\mathscr{A}$ is a Banach algebra with identity and $a \in \mathscr{A}$, $\lim \|a^n\|^{1/n}$ exists and*

$$r(a) = \lim \|a^n\|^{1/n}.$$

PROOF. Let $G = \{z \in \mathbb{C}: z = 0 \text{ or } z^{-1} \in \rho(a)\}$. Define $f: G \to \mathscr{A}$ by $f(0) = 0$ and for $z \neq 0$, $f(z) = (z^{-1} - a)^{-1}$. Since $(a - \alpha)^{-1} \to 0$ as $\alpha \to \infty$, f is analytic on G, and so f has a power series expansion. In fact, by Corollary 2.3, for $|z| < \|a\|^{-1}$,

$$f(z) = \sum_{n=0}^{\infty} a^n/(z^{-1})^{n+1} = z \sum_{n=0}^{\infty} z^n a^n.$$

From complex variable theory, this power series converges for $|z| < R \equiv \text{dist}(0, \partial G) = \text{dist}(0, \sigma(a)^{-1})$ (Here $\sigma(a)^{-1} = \{z^{-1}: z \in \sigma(a)\}$). Thus $R = \inf\{|\alpha|: \alpha^{-1} \in \sigma(a)\} = r(a)^{-1}$. Also, from the theory of power series, $R^{-1} = \limsup \|a^n\|^{1/n}$. Thus

$$r(a) = \limsup \|a^n\|^{1/n}.$$

Now if $\alpha \in \mathbb{C}$ and $n \geq 1$, $\alpha^n - a^n = (\alpha - a)(\alpha^{n-1} + \alpha^{n-2}a + \cdots + a^{n-1}) = (\alpha^{n-1} + \alpha^{n-2}a + \cdots + a^{n-1})(\alpha - a)$. So if $\alpha^n - a^n$ is invertible, $\alpha - a$ is invertible and $(\alpha - a)^{-1} = (\alpha^n - a^n)^{-1}(\alpha^{n-1} + \cdots + a^{n-1})$. So for α in $\sigma(a)$, $\alpha^n - a^n$ is not invertible for every $n \geq 1$. By Theorem 3.6, $|\alpha|^n \leq \|a^n\|$. Hence $|\alpha| \leq \|a^n\|^{1/n}$ for all $n \geq 1$ and α in $\sigma(a)$. So if $\alpha \in \sigma(a)$, $|\alpha| \leq \liminf \|a^n\|^{1/n}$. Taking the supremum over all α in $\sigma(a)$ gives that $r(a) \leq \liminf \|a^n\|^{1/n} \leq \limsup \|a^n\|^{1/n} = r(a)$. So $r(a) = \lim \|a^n\|^{1/n}$. ∎

3.9. Proposition. *Let $\mathscr{A}$ be a Banach algebra with identity and let $a \in \mathscr{A}$.*

(a) *If $\alpha \in \rho(a)$, then $\text{dist}(\alpha, \sigma(a)) \geq \|(\alpha - a)^{-1}\|^{-1}$.*

(b) *If $\alpha, \beta \in \rho(a)$, then*

$$(\alpha - a)^{-1} - (\beta - a)^{-1} = (\beta - \alpha)(\alpha - a)^{-1}(\beta - a)^{-1}$$

$$= (\beta - \alpha)(\beta - a)^{-1}(\alpha - a)^{-1}.$$

PROOF. (a) By Corollary 2.3, if $\alpha \in \rho(a)$ and $\|x - (\alpha - a)\| < \|(\alpha - a)^{-1}\|^{-1}$, x is invertible. So if $\beta \in \mathbb{C}$ and $|\beta| < \|(\alpha - a)^{-1}\|^{-1}$, $(\beta + \alpha - a)$ is invertible; that is, $\alpha + \beta \in \rho(a)$. Hence $\text{dist}(\alpha, \sigma(a)) \geq \|(\alpha - a)^{-1}\|^{-1}$.
(b) This follows by letting $x = \alpha - a$ and $y = \beta - a$ in the identity $x^{-1} - y^{-1} = x^{-1}(y - x)y^{-1} = y^{-1}(y - x)x^{-1}$. ∎

The identity in part (b) of the preceding proposition is called the *resolvent identity* and the function $\alpha \mapsto (\alpha - a)^{-1}$ of $\rho(a) \to \mathscr{A}$ is called the *resolvent* of a.

EXERCISES

1. Let S be the unilateral shift on l^2 (II.2.10). Show that S is left invertible but not right invertible.

2. If $\mathscr{A}$ is a Banach algebra with identity and $a \in \mathscr{A}$ and is nilpotent (that is, $a^n = 0$ for some n), then $\sigma(a) = \{0\}$.

3. Let (X, Ω, μ) be a σ-finite measure space and let $\mathscr{A} = L^\infty(X, \Omega, \mu)$ (1.6). If $\phi \in \mathscr{A}$, show that the following are equivalent: (a) $\alpha \in \sigma(\phi)$; (b) $0 = \sup\{\inf\{|\phi(x) - \alpha|: x \in X \setminus \Delta\}: \Delta \in \Omega \text{ and } \mu(\Delta) = 0\}$; (c) if $\varepsilon > 0$, $\mu(\{x \in X: |\phi(x) - \alpha| < \varepsilon\}) > 0$; (d) if ν is the measure defined on the Borel subsets of $\mathbb{C}$ by $\nu(\Delta) = \mu(\phi^{-1}(\Delta))$, then $\alpha \in$ the support of ν.

4. If G is an open subset of $\mathbb{C}$ and $f: G \to \mathscr{X}$ is a continuous function such that for each x^* in $\mathscr{X}^*$, $x^* \circ f: G \to \mathbb{C}$ is analytic, then f is analytic.

5. If $\mathscr{A}$ is a Banach algebra with identity, $\{a_n\} \subseteq \mathscr{A}$, $a_n \to a$, $\alpha_n \in \sigma(a_n)$, and $\alpha_n \to \alpha$, then $\alpha \in \sigma(a)$.

6. If $\mathscr{A}$ is a Banach algebra with identity and $r: \mathscr{A} \to [0, \infty)$ is the spectral radius, show that r is upper semicontinuous. If $a \in \mathscr{A}$ such that $r(a) = 0$, show that r is continuous at a.

7. If $\mathscr{A}$ is a Banach algebra with identity, $a, b \in \mathscr{A}$, and α is a nonzero scalar such that $(\alpha - ab)$ is invertible, show that $(\alpha - ba)$ is invertible and $(\alpha - ba)^{-1} = \alpha^{-1} + \alpha^{-1}b(\alpha - ab)^{-1}a$. Show that $\sigma(ab) \cup \{0\} = \sigma(ba) \cup \{0\}$ and give an example such that $\sigma(ab) \neq \sigma(ba)$.

§4. The Riesz Functional Calculus

Before coming to the main course of this section, it is necessary to have an appetizer from complex analysis. Many of these topics can be found in Conway [1978] with complete proofs. Only a few results are presented here.

If γ is a closed rectifiable curve in $\mathbb{C}$ and $a \notin \{\gamma\} \equiv \{\gamma(t): 0 \le t \le 1\}$, then the *winding number of γ about a* is defined to be the number

$$n(\gamma; a) = \frac{1}{2\pi i} \int_\gamma \frac{1}{z - a} \, dz.$$

The number $n(\gamma; a)$ is always an integer and is constant on each component of $\mathbb{C} \setminus \{\gamma\}$ and vanishes on the unbounded component of $\mathbb{C} \setminus \{\gamma\}$.

Let G be an open subset of $\mathbb{C}$ and let $\mathscr{X}$ be a Banach space. If $f: G \to \mathscr{X}$ is analytic and $x^* \in \mathscr{X}^*$, then $z \mapsto \langle f(z), x^* \rangle$ is analytic on G and its derivative is $\langle f'(z), x^* \rangle$. By Exercise 4 of the preceding section, if f: $G \to \mathscr{X}$ is a continuous function such that $z \mapsto \langle f(z), x^* \rangle$ is analytic for each x^* in $\mathscr{X}^*$, then f: $G \to \mathscr{X}$ is analytic. These facts will help in discussing and proving many of the results below.

If γ is a rectifiable curve in G and f is a continuous function defined in a neighborhood of $\{\gamma\}$ with values in $\mathscr{X}$, then $\int_\gamma f$ can be defined as for a scalar-valued f as the limit in $\mathscr{X}$ of sums of the form

$$\sum_j \left[\gamma(t_j) - \gamma(t_{j-1}) \right] f(t_j),$$

where $\{t_0, t_1, \ldots, t_n\}$ is a partition of $[0, 1]$. Hence $\int_\gamma f = \int_0^1 f(\gamma(t)) \, d\gamma(t) \in \mathscr{X}$. It is easy to see that for every x^* in $\mathscr{X}^*$, $\langle \int_\gamma f, x^* \rangle = \int_\gamma \langle f(\cdot), x^* \rangle$.

4.1. Cauchy's Theorem. *If $\mathscr{X}$ is a Banach space, G is an open subset of $\mathbb{C}$, f: $G \to \mathscr{X}$ is an analytic function, and $\gamma_1, \ldots, \gamma_m$ are closed rectifiable curves in G such that $\sum_{j=1}^m n(\gamma_j; a) = 0$ for all a in $\mathbb{C} \setminus G$, then $\sum_{j=1}^m \int_{\gamma_j} f = 0$.*

PROOF. If $x^* \in \mathscr{X}^*$, then $\langle \sum_{j=1}^m \int_{\gamma_j} f, x^* \rangle = \sum_{j=1}^m \int_{\gamma_j} \langle f(\cdot), x^* \rangle = 0$ by the scalar-valued version of Cauchy's Theorem. Hence $\sum_{j=1}^m \int_{\gamma_j} f = 0$. ∎

4.2. Cauchy's Integral Formula. *If $\mathscr{X}$ is a Banach space, G is an open subset of $\mathbb{C}$, f: $G \to \mathscr{X}$ is analytic, γ is a closed rectifiable curve in G such that $n(\gamma; a) = 0$ for every a in $\mathbb{C} \setminus G$, and $\lambda \in G \setminus \{\gamma\}$, then for every integer $k \geq 0$,*

$$n(\gamma; \lambda) f^{(k)}(\lambda) = \frac{k!}{2\pi i} \int_\gamma (z - \lambda)^{-(k+1)} f(z) \, dz.$$

4.3. Definition. A closed rectifiable curve γ is *positively oriented* if for every a in $G \setminus \{\gamma\}$, $n(\gamma; a)$ is either 0 or 1. In this case the *inside* of γ, denoted by ins γ, is defined by

$$\text{ins } \gamma \equiv \{ a \in \mathbb{C} \setminus \{\gamma\} : n(\gamma; a) = 1 \}.$$

The *outside* of γ, denoted by out γ, is defined by

$$\text{out } \gamma \equiv \{ a \in \mathbb{C} \setminus \{\gamma\} : n(\gamma; a) = 0 \}.$$

Thus $\mathbb{C} = \{\gamma\} \cup \text{ins } \gamma \cup \text{out } \gamma$.

A curve γ: $[0, 1] \to \mathbb{C}$ is *simple* if $\gamma(s) = \gamma(t)$ implies that either $s = t$ or $s = 0$ and $t = 1$. The Jordan Curve Theorem says that if γ is a simple closed rectifiable curve, then $\mathbb{C} \setminus \{\gamma\}$ has two components and $\{\gamma\}$ is the boundary of each. Hence $n(\gamma; a)$ takes on only two values and one of these must be 0; the other must be ± 1.

If $\Gamma = \{\gamma_1, \ldots, \gamma_m\}$ is a collection of closed rectifiable curves, then Γ is *positively oriented* if: (a) $\{\gamma_i\} \cap \{\gamma_j\} = \square$ for $i \neq j$; (b) for a in $\mathbb{C} \setminus \bigcup_{j=1}^m \{\gamma_j\}$, $n(\Gamma; a) \equiv \sum_{j=1}^m n(\gamma_j; a)$ is either 0 or 1. The *inside* of Γ, ins Γ, is defined by

$$\text{ins } \Gamma \equiv \{ a : n(\Gamma; a) = 1 \}.$$

The *outside* of Γ, out Γ, is defined by

$$\text{out } \Gamma \equiv \{ a \colon n(\Gamma; a) = 0 \}.$$

4.4. Proposition. *If G is an open subset of $\mathbb{C}$ and K is a compact subset of G, then there is a positively oriented system of curves $\Gamma = \{ \gamma_1, \ldots, \gamma_m \}$ in $G \setminus K$ such that $K \subseteq \text{ins } \Gamma$ and $\mathbb{C} \setminus G \subseteq \text{out } \Gamma$. The curves $\gamma_1, \ldots, \gamma_m$ can be found such that they are infinitely differentiable.*

The proof of this proposition can be found on p. 195 of Conway [1978], though some details are missing.

If $\Gamma = \{ \gamma_1, \ldots, \gamma_m \}$ and each γ_j is rectifiable, define

$$\int_\Gamma f = \sum_{j=1}^m \int_{\gamma_j} f$$

whenever f is continuous in a neighborhood of $\{\Gamma\}$.

Let $\mathscr{A}$ be a Banach algebra with identity and let $a \in \mathscr{A}$. One of the principal uses of Proposition 4.4 in this book will occur when $K = \sigma(a)$. If $f \colon G \to \mathbb{C}$ is analytic and $\sigma(a) \subseteq G$, we will define an element $f(a)$ in $\mathscr{A}$ by

4.5
$$f(a) = \frac{1}{2\pi i} \int_\Gamma f(z)(z - a)^{-1} \, dz$$

where Γ is as in Proposition 4.4 with $K = \sigma(a)$. But first it must be shown that (4.5) does not depend on the choice of Γ. That is, it must be shown that $f(a)$ is well defined.

4.6. Proposition. *Let $\mathscr{A}$ be a Banach algebra with identity, let $a \in \mathscr{A}$, and let G be an open subset of $\mathbb{C}$ such that $\sigma(a) \subseteq G$. If $\Gamma = \{ \gamma_1, \ldots, \gamma_m \}$ and $\Lambda = \{ \lambda_1, \ldots, \lambda_k \}$ are two positively oriented collections of curves in G such that $\sigma(a) \subseteq \text{ins } \Gamma$ and $\sigma(a) \subseteq \text{ins } \Lambda$ and if $f \colon G \to \mathbb{C}$ is analytic, then*

$$\int_\Gamma f(z)(z - a)^{-1} \, dz = \int_\Lambda f(z)(z - a)^{-1} \, dz.$$

PROOF. For $1 \leq j \leq k$, let $\gamma_{m+j} = \lambda_j^{-1}$; that is, $\gamma_{m+j}(t) = \lambda_j(1 - t)$ for $0 \leq t \leq 1$. If $z \notin G \setminus \sigma(a)$, then either $z \in \mathbb{C} \setminus G$ or $z \in \sigma(a)$. If $z \in \mathbb{C} \setminus G$, then $\sum_{j=1}^{m+k} n(\gamma_j; z) = n(\Gamma; z) - n(\Lambda; z) = 0 - 0 = 0$. If $z \in \sigma(a)$, then $\sum_{j=1}^{m+k} n(\gamma_j; z) = n(\Gamma; z) - n(\Lambda; z) = 1 - 1 = 0$. Thus $\Sigma \equiv \{ \gamma_j \colon 1 \leq j \leq m + k \}$ is a system of closed curves in $U = G \setminus \sigma(a)$ such that $n(\Sigma; z) = 0$ for all z in $\mathbb{C} \setminus U$. Since $z \mapsto f(z)(z - a)^{-1}$ is analytic on U, Cauchy's Theorem implies

$$0 = \int_\Sigma f(z)(z - a)^{-1} \, dz = \int_\Gamma f(z)(z - a)^{-1} \, dz - \int_\Lambda f(z)(z - a)^{-1} \, dz. \qquad \blacksquare$$

As was pointed out before, Proposition 4.6 implies that (4.5) gives a well-defined element $f(a)$ of $\mathscr{A}$ whenever f is analytic in a neighborhood of $\sigma(a)$. Let $\text{Hol}(a)$ = all of the functions that are analytic in a neighborhood of $\sigma(a)$. Note that $\text{Hol}(a)$ is an algebra where if $f, g \in \text{Hol}(a)$ and f and g have domains $D(f)$ and $D(g)$, then fg and $f + g$ have domain $D(f) \cap D(g)$. $\text{Hol}(a)$ is not, however, a Banach algebra.

4.7. The Riesz Functional Calculus. *Let $\mathscr{A}$ be a Banach algebra with identity and let $a \in \mathscr{A}$.*

(a) *The map $f \mapsto f(a)$ of $\text{Hol}(a) \to \mathscr{A}$ is an algebra homomorphism.*
(b) *If $f(z) = \sum_{k=0}^{\infty} \alpha_k z^k$ has radius of convergence $> r(a)$, then $f \in \text{Hol}(a)$ and $f(a) = \sum_{k=0}^{\infty} \alpha_k a^k$.*
(c) *If $f(z) \equiv 1$, then $f(a) = 1$.*
(d) *If $f(z) = z$ for all z, $f(a) = a$.*
(e) *If $f, f_1, f_2, \ldots$ are all analytic on G, $\sigma(a) \subseteq G$, and $f_n(z) \to f(z)$ uniformly on compact subsets of G, then $\|f_n(a) - f(a)\| \to 0$ as $n \to \infty$.*

PROOF. (a) Let $f, g \in \text{Hol}(a)$ and let G be an open neighborhood of $\sigma(a)$ on which both f and g are analytic. Let Γ be a positively oriented system of closed curves in G such that $\sigma(a) \subseteq \text{ins}\,\Gamma$. Let Λ be a positively oriented system of closed curves in G such that $(\text{ins}\,\Gamma) \cup \{\Gamma\} = \text{cl}(\text{ins}\,\Gamma) \subseteq \text{ins}\,\Lambda$. Then

$$f(a)g(a) = -\frac{1}{4\pi^2}\left[\int_\Gamma f(z)(z-a)^{-1}\,dz\right]\left[\int_\Lambda g(\zeta)(\zeta-a)^{-1}\,d\zeta\right]$$

$$= -\frac{1}{4\pi^2}\int_\Gamma\int_\Lambda f(z)g(\zeta)(z-a)^{-1}(\zeta-a)^{-1}\,d\zeta\,dz$$

[by (3.9b)] $$= -\frac{1}{4\pi^2}\int_\Gamma\int_\Lambda f(z)g(\zeta)\left[\frac{(z-a)^{-1}-(\zeta-a)^{-1}}{\zeta-z}\right]d\zeta\,dz$$

$$= -\frac{1}{4\pi^2}\int_\Gamma f(z)\left[\int_\Lambda \frac{g(\zeta)}{\zeta-z}\,d\zeta\right](z-a)^{-1}\,dz$$

$$+ \frac{1}{4\pi^2}\int_\Lambda g(\zeta)\left[\int_\Gamma \frac{f(z)}{\zeta-z}\,dz\right](\zeta-a)^{-1}\,d\zeta.$$

But for ζ on Λ, $\zeta \in \text{out}\,\Gamma$ and hence $\int_\Gamma[f(z)/(\zeta-z)]\,dz = 0$ (Cauchy's Theorem). If $z \in \{\Gamma\}$, then $z \in \text{ins}\,\Lambda$ and so $\int_\Lambda[g(\zeta)/(\zeta-z)]\,d\zeta = 2\pi i g(z)$. Hence

$$f(a)g(a) = \frac{1}{2\pi i}\int_\Gamma f(z)g(z)(z-a)^{-1}\,dz$$

$$= (fg)(a).$$

The proof that $(\alpha f + \beta g)(a) = \alpha f(a) + \beta g(a)$ is left to the reader.

(c) and (d). Let $f(z) = z^k$, $k \geq 0$. Let $\gamma(t) = R\exp(2\pi it)$, $0 \leq t \leq 1$, where $R > \|a\|$. So $\sigma(a) \subset \operatorname{ins}\gamma$, and hence

$$f(a) = \frac{1}{2\pi i}\int_\gamma z^k(z-a)^{-1}\,dz$$

$$= \frac{1}{2\pi i}\int_\gamma z^{k-1}\left(1 - \frac{a}{z}\right)^{-1}\,dz$$

$$= \frac{1}{2\pi i}\int_\gamma z^{k-1}\sum_{n=0}^\infty a^n/z^n\,dz,$$

since $\|a/z\| < 1$ for $|z| = R$. Since this infinite series converges uniformly for z on γ,

$$f(a) = \sum_{n=0}^\infty \left[\frac{1}{2\pi i}\int_\gamma \frac{1}{z^{n-k+1}}\,dz\right]a^n.$$

If $n \neq k$, then $z^{-(n-k+1)}$ has a primitive and hence $\int_\gamma z^{-(n-k+1)}\,dz = 0$. For $n = k$ this integral becomes $\int_\gamma z^{-1}\,dz = 2\pi i$. Hence $f(a) = a^k$.

(e) Let $\Gamma = \{\gamma_1,\ldots,\gamma_m\}$ be a positively oriented system of closed curves in G such that $\sigma(a) \subseteq \operatorname{ins}\Gamma$. Fix $1 \leq k \leq m$; then

$$\left\|\int_{\gamma_k} f_n(z)(z-a)^{-1}\,dz - \int_{\gamma_k} f(z)(z-a)^{-1}\,dz\right\|$$

$$= \left\|\int_0^1 [f_n(\gamma_k(t)) - f(\gamma_k(t))][\gamma_k(t) - a]^{-1}\,d\gamma_k(t)\right\|$$

$$\leq \int_0^1 |f_n(\gamma_k(t)) - f(\gamma_k(t))|\,\|[\gamma_k(t) - a]^{-1}\|\,d|\gamma_k|(t).$$

Now $t \mapsto \|[\gamma_k(t) - a]^{-1}\|$ is continuous on $[0,1]$ and hence bounded by some constant, say M. Thus

$$\left\|\int_{\gamma_k} f_n(z)(z-a)^{-1}\,dz - \int_{\gamma_k} f(z)(z-a)^{-1}\,dz\right\|$$

$$\leq M\|\gamma_k\|\max\{|f_n(z) - f(z)|: z \in \{\gamma_k\}\},$$

where $\|\gamma_k\|$ is the total variation (length) of γ_k. By hypothesis it follows that $\|f_n(a) - f(a)\| \to 0$ as $n \to \infty$.

(b) If $p(z) = \sum_{k=0}^n \alpha_k z^k$ is a polynomial, then (a), (c), and (d) combine to give that $p(a) = \sum_{k=0}^n \alpha_k a^k$. Now let $f(z) = \sum_{k=0}^\infty \alpha_k z^k$ have radius of convergence $R > r(a)$, the spectral radius of a. If $p_n(z) = \sum_{k=0}^n \alpha_k z^k$, $p_n(z) \to f(z)$ uniformly on compact subsets of $\{z: |z| < R\}$. By (e), $p_n(a) \to f(a)$. So (b) follows. ∎

The Riesz Functional Calculus is used in the study of Banach algebras and is especially useful in the study of linear operators on a Banach space

(Sections 6 and 7). Now our attention must focus on the basic properties of this functional calculus. The first such property is its uniqueness.

4.8.   Proposition. *Let $\mathscr{A}$ be a Banach algebra with identity and let $a \in \mathscr{A}$. Let τ: Hol$(a) \to \mathscr{A}$ be a homomorphism such that (a) $\tau(1) = 1$, (b) $\tau(z) = a$, (c) if $\{f_n\}$ is a sequence of analytic functions on an open set G such that $\sigma(a) \subseteq G$ and $f_n(z) \to f(z)$ uniformly on compact subsets of G, then $\tau(f_n) \to \tau(f)$. Then $\tau(f) = f(a)$ for every f in Hol(a).*

PROOF. The proof uses Runge's Theorem (III.8.1), but first it must be shown that $\tau(f) = f(a)$ whenever f is a rational function. If $n \geq 1$, $\tau(z^n) = \tau(z)^n = a^n$; hence $\tau(p) = p(a)$ for any polynomial p. Let q be a polynomial such that q never vanishes on $\sigma(a)$, so $1/q \in$ Hol(a). Also, $1 = \tau(1) = \tau(q \cdot q^{-1}) = \tau(q)\tau(q^{-1}) = q(a)\tau(q^{-1})$. Hence $q(a)$ is invertible and $q(a)^{-1} = \tau(q^{-1})$. But using the Riesz Functional Calculus, a similar argument shows that $q(a)^{-1} = (1/q)(a)$. Thus $\tau(q^{-1}) = (1/q)(a)$. Therefore if $f = p/q$, where p and q are polynomials and q never vanishes on $\sigma(a)$, $\tau(f) = \tau(p \cdot q^{-1}) = \tau(p)\tau(q^{-1}) = p(a)(1/q)(a) = f(a)$.

Now let $f \in$ Hol(a) and suppose f is analytic on an open set G such that $\sigma(a) \subseteq G$. By Runge's Theorem there are rational functions $\{f_n\}$ in Hol(a) such that $f_n(z) \to f(z)$ uniformly on compact subsets of G. By (iii) of the hypothesis, $\tau(f_n) \to \tau(f)$. But $\tau(f_n) = f_n(a)$ and $f_n(a) \to f(a)$ by (4.7e). Hence $\tau(f) = f(a)$. ∎

A fact that has been implicit in the manipulations involving the functional calculus is that $f(a)$ and $g(a)$ commute for all f and g in Hol(a). In fact, if τ: Hol$(a) \to \mathscr{A}$ is defined by $\tau(a) = f(a)$, then $f(a)g(a) = \tau(fg) = \tau(gf) = g(a)f(a)$. Still more can be said.

4.9.   Proposition. *If $a, b \in \mathscr{A}$, $ab = ba$, and $f \in$ Hol(a), then $f(a)b = bf(a)$.*

PROOF. An algebraic exercise demonstrates that $f(a)b = bf(a)$ if f is a rational function with poles off $\sigma(a)$. The general result now follows by Runge's Theorem. ∎

4.10. The Spectral Mapping Theorem. *If $a \in \mathscr{A}$ and $f \in$ Hol(a), then*

$$\sigma(f(a)) = f(\sigma(a)).$$

PROOF. If $\alpha \in \sigma(a)$, let $g \in$ Hol(a) such that $f(z) - f(\alpha) = (z - \alpha)g(z)$. If it were the case that $f(\alpha) \notin \sigma(f(a))$, then $(a - \alpha)$ would be invertible with inverse $g(a)[f(a) - f(\alpha)]^{-1}$. Hence $f(a) \in \sigma(f(a))$; that is, $f(\sigma(a)) \subseteq \sigma(f(a))$.

Conversely, if $\beta \notin f(\sigma(a))$, then $g(z) = [f(z) - \beta]^{-1} \in$ Hol(a) and so $g(a)[f(a) - \beta] = 1$. Thus $\beta \notin \sigma(f(a))$; that is, $\sigma(f(a)) \subseteq f(\sigma(a))$. ∎

This section closes with an application of the functional calculus that is typical.

4.11.   Proposition. *Suppose $a \in \mathcal{A}$ and $\sigma(a) = F_1 \cup F_2$, where F_1 and F_2 are disjoint nonempty closed sets. Then there is a nontrivial idempotent e in $\mathcal{A}$ such that*

(a) *if $ba = ab$, then $be = eb$;*
(b) *if $a_1 = ae$ and $a_2 = a(1 - e)$, then $a = a_1 + a_2$ and $a_1 a_2 = a_2 a_1 = 0$;*
(c) *$\sigma(a_1) = F_1 \cup \{0\}$, $\sigma(a_2) = F_2 \cup \{0\}$.*

PROOF. Let G_1, G_2 be disjoint open subsets of $\mathbb{C}$ such that $F_j \subset G_j$, $j = 1, 2$. Let Γ be a positively oriented system of closed curves in G_1 such that $F_1 \subseteq \operatorname{ins} \Gamma$, $F_2 \subseteq \operatorname{out} \Gamma$. If $f =$ the characteristic function of G_1, $f \in \operatorname{Hol}(a)$; let $e = f(a)$. Since $f^2 = f$, $e^2 = e$. Part (a) follows from (4.9).

Note that $e(1 - e) = 0 = (1 - e)e$. Hence (b) is immediate. Let $f_1(z) = zf(z)$, $f_2(z) = z(1 - f(z))$. It follows from (4.7a) that $a_j = f_j(a)$, $j = 1, 2$. Hence the Spectral Mapping Theorem implies that $\sigma(a_j) = f_j(\sigma(a)) = F_j \cup \{0\}$.   ∎

Part (c) of the preceding proposition has the somewhat unattractive conclusion that $\sigma(a_1) = F_1 \cup \{0\}$. It would be much neater if the conclusion were that $\sigma(a_1) = F_1$. This is, in a sense, the case. Since $a_1(1 - e) = 0$ and $1 - e \neq 0$, a_1 cannot be invertible. However, consider the algebra $\mathcal{A}_1 \equiv \mathcal{A}e$. It is left to the reader to show that $\mathcal{A}_1$ is a Banach algebra and e is the identity for $\mathcal{A}_1$. If a_1 is considered as an element of the algebra $\mathcal{A}_1$, then its spectrum as an element of $\mathcal{A}_1$ is F_1. This is an illustration of how the spectrum depends on the Banach algebra (the subject of the next section; also see Exercise 9).

EXERCISES

1. Let $\mathcal{A} = C(X)$, X compact (see Example 3.2). If $g \in C(X)$ and $f \in \operatorname{Hol}(g)$, show that $f(g) = f \circ g$.

2. Let a be a nilpotent element of $\mathcal{A}$. For f, g in $\operatorname{Hol}(a)$, give a necessary and sufficient condition on f and g that $f(a) = g(a)$.

3. Let $d \geq 1$ and let $A \in M_d(\mathbb{C})$. Give a necessary and sufficient condition on f in $\operatorname{Hol}(A)$ such that $f(a) = 0$. (Hint: Consider the Jordan canonical form for A.)

4. If $\mathcal{A}$ is a Banach algebra with identity, $a \in \mathcal{A}$, $f \in \operatorname{Hol}(a)$, and g is analytic in a neighborhood of $f(\sigma(a))$, then $g \circ f \in \operatorname{Hol}(a)$ and $g(f(a)) = g \circ f(a)$.

5. If $\mathcal{X}$ is a Banach space, $A \in \mathcal{B}(\mathcal{X})$, and $\mathcal{M} \leq \mathcal{X}$ such that $(A - \alpha)^{-1}\mathcal{M} \subseteq \mathcal{M}$ for all α in $\rho(A)$, show that $f(A)\mathcal{M} \subseteq \mathcal{M}$ whenever $f \in \operatorname{Hol}(A)$.

6. If $\mathcal{X}$ is a Banach space, $A \in \mathcal{B}(\mathcal{X})$, and $f \in \operatorname{Hol}(A)$, show that $f(A)^* = f(A^*)$. (See (6.1) below.)

7. If $\mathcal{H}$ is a Hilbert space, $A \in \mathcal{B}(\mathcal{H})$, and $f \in \text{Hol}(A)$, show that $f(A)^* = \tilde{f}(A^*)$, where $\tilde{f}(z) = \overline{f(\bar{z})}$. (See (6.1) below.)

8. If $\mathcal{H}$ is a Hilbert space, A is a normal operator on $\mathcal{H}$, and $f \in \text{Hol}(A)$, show that $f(A)$ is normal.

9. Let $\mathcal{X}$ be a Banach space and let $A \in \mathcal{B}(\mathcal{X})$. Show that if $\sigma(A) = F_1 \cup F_2$ where F_1, F_2 are disjoint closed subsets of $\mathbb{C}$, then there are topologically complementary subspaces $\mathcal{X}_1, \mathcal{X}_2$ of $\mathcal{X}$ such that (a) $B\mathcal{X}_j \subseteq \mathcal{X}_j$ $(j = 1, 2)$ whenever $BA = AB$; (b) if $A_j = A|\mathcal{X}_j$, $\sigma(A_j) = F_j$; (c) there is an invertible operator $R: \mathcal{X} \to \mathcal{X}_1 \oplus_1 \mathcal{X}_2$ such that $RAR^{-1} = A_1 \oplus A_2$.

10. Let $A \in M_d(\mathbb{C})$, $\sigma(A) = \{\alpha_1, \ldots, \alpha_n\}$, where $\alpha_i \neq \alpha_j$ for $i \neq j$. Show that for $1 \leq j \leq n$ there is a matrix A_j in $M_{d_j}(\mathbb{C})$ such that $\sigma(A_j) = \{\alpha_j\}$ and A is similar to $A_1 \oplus \cdots \oplus A_n$.

§5. Dependence of the Spectrum on the Algebra

If $\partial \mathbb{D} = \{z \in \mathbb{C}: |z| = 1\}$, let $\mathcal{B} = $ the uniform closure of the polynomials in $C(\partial \mathbb{D})$. (Here "polynomial" means a polynomial in z.) If $\mathcal{A} = C(\partial \mathbb{D})$, then the spectrum of z as an element of $\mathcal{A}$ is $\partial \mathbb{D}$ (Example 3.2). That is,

$$\sigma_{\mathcal{A}}(z) = \partial \mathbb{D}.$$

Now $z \in \mathcal{B}$ and so it has a spectrum as an element of this algebra; denote this spectrum by $\sigma_{\mathcal{B}}(z)$. There is no reason to believe that $\sigma_{\mathcal{B}}(z) = \sigma_{\mathcal{A}}(z)$. In fact, they are not equal.

5.1. Example. If $\mathcal{B} = $ the closure in $C(\partial \mathbb{D})$ of the polynomials in z, then $\sigma_{\mathcal{B}}(z) = \text{cl } \mathbb{D}$.

To see this first note that $\|z\| = 1$, so that $\sigma_{\mathcal{B}}(z) \subseteq \text{cl } \mathbb{D}$ by Theorem 3.6. If $|\lambda| \leq 1$ and $\lambda \notin \sigma_{\mathcal{B}}(z)$, there is an f in $\mathcal{B}$ such that $(z - \lambda)f = 1$. Note that this implies that $|\lambda| < 1$. Because $f \in \mathcal{B}$, there is a sequence of polynomials $\{p_n\}$ such that $p_n \to f$ uniformly on $\partial \mathbb{D}$. Thus for every $\varepsilon > 0$ there is a N such that for $m, n \geq N$, $\varepsilon > \|p_n - p_m\|_{\partial \mathbb{D}} \equiv \sup\{|p_n(z) - p_n(z)|: z \in \partial \mathbb{D}\}$. By the Maximum Principle, $\varepsilon > \|p_n - p_m\|_{\text{cl } \mathbb{D}}$ for $m, n \geq N$. Thus $g(z) = \lim p_n(z)$ is analytic on $\mathbb{D}$ and continuous on $\text{cl } \mathbb{D}$; also, $g|\partial \mathbb{D} = f$. By the same argument, since $p_n(z)(z - \lambda) \to 1$ uniformly on $\partial \mathbb{D}$, $p_n(z)(z - \lambda) \to 1$ uniformly on $\mathbb{D}$. Thus $g(z)(z - \lambda) = 1$ on $\mathbb{D}$. But $1 = g(\lambda)(\lambda - \lambda) = 0$, a contradiction. Thus, $\text{cl } \mathbb{D} \subseteq \sigma_{\mathcal{B}}(z)$.

Thus the spectrum not only depends on the element of the algebra, but also the algebra. Precisely how this dependence occurs is given below, but it can be said that the example above is typical, both in its statement and its proof, of the general situation. To phrase these results it is necessary to introduce the polynomially convex hull of a compact subset of $\mathbb{C}$.

5.2. Definition. If A is a set and $f: A \to \mathbb{C}$, define

$$\|f\|_A \equiv \sup\{|f(z)|: z \in A\}.$$

If K is a compact subset of $\mathbb{C}$, define the *polynomially convex hull* of K to be the set $\hat{K}$ given by

$$\hat{K} \equiv \{z \in \mathbb{C}: |p(z)| \le \|p\|_K \text{ for every polynomial } p\}.$$

The set K is *polynomially convex* if $K = \hat{K}$.

Note that the polynomially convex hull of $\partial \mathbb{D}$ is cl $\mathbb{D}$. This is, again, quite typical. If K is any compact set, then $\mathbb{C} \setminus K$ has a countable number of components, only one of which is unbounded. The bounded components are sometimes called the *holes* of K; a few pictures should convince the reader of the appropriateness of this terminology.

5.3. Proposition. *If K is a compact subset of $\mathbb{C}$, then $\mathbb{C} \setminus \hat{K}$ is the unbounded component of $\mathbb{C} \setminus K$. Hence K is polynomially convex if and only if $\mathbb{C} \setminus K$ is connected.*

PROOF. Let $U_0, U_1, \ldots$ be the components of $\mathbb{C} \setminus K$, where U_0 is unbounded. Put $L = \mathbb{C} \setminus U_0$; hence $L = K \cup \bigcup_{n=1}^{\infty} U_n$. Clearly $K \subseteq \hat{K}$. If $n \ge 1$, then U_n is a bounded open set and a topological argument implies $\partial U_n \subset K$. By the Maximum Principle $U_n \subseteq \hat{K}$. Thus, $L \subseteq \hat{K}$.

If $\alpha \in U_0$, $(z - \alpha)^{-1}$ is analytic in a neighborhood of L. By (III.8.5), there is a sequence of polynomials $\{p_n\}$ such that $\|p_n - (z - \alpha)^{-1}\|_L \to 0$. If $q_n = (z - \alpha)p_n$, then $\|q_n - 1\|_L \to 0$. Thus for large n, $\|q_n - 1\|_L < 1/2$. Since $K \subset L$ and $|q_n(\alpha) - 1| = 1$, this implies that $\alpha \notin \hat{K}$. Thus $\hat{K} \subseteq L$. ∎

5.4. Theorem. *If $\mathscr{A}$ and $\mathscr{B}$ are Banach algebras with identity such that $\mathscr{B} \subseteq \mathscr{A}$ and $a \in \mathscr{B}$, then*

(a) $\sigma_{\mathscr{A}}(a) \subseteq \sigma_{\mathscr{B}}(a)$ *and* $\partial \sigma_{\mathscr{B}}(a) \subseteq \partial \sigma_{\mathscr{A}}(a)$.
(b) $\sigma_{\mathscr{A}}(a)\hat{} = \sigma_{\mathscr{B}}(a)\hat{}$.
(c) *If G is a hole of $\sigma_{\mathscr{A}}(a)$, then either $G \subseteq \sigma_{\mathscr{B}}(a)$ or $G \cap \sigma_{\mathscr{B}}(a) = \square$.*
(d) *If $\mathscr{B}$ is the closure in $\mathscr{A}$ of all polynomials in a, then $\sigma_{\mathscr{B}}(a) = \sigma_{\mathscr{A}}(a)\hat{}$.*

PROOF. (a) If $\alpha \notin \sigma_{\mathscr{B}}(a)$, then there is a b in $\mathscr{B}$ such that $b(a - \alpha) = (a - \alpha)b = 1$. Since $\mathscr{B} \subseteq \mathscr{A}$, $\alpha \notin \sigma_{\mathscr{A}}(a)$. Now assume that $\lambda \in \partial \sigma_{\mathscr{B}}(a)$. Since int $\sigma_{\mathscr{A}}(a) \subseteq$ int $\sigma_{\mathscr{B}}(a)$, it suffices to show that $\lambda \in \sigma_{\mathscr{A}}(a)$. Suppose $\lambda \notin \sigma_{\mathscr{A}}(a)$; there is thus an x in $\mathscr{A}$ such that $x(a - \alpha) = (a - \alpha)x = 1$.

Since $\lambda \in \partial\sigma_{\mathscr{B}}(a)$, there is a sequence $\{\lambda_n\}$ in $\mathbb{C} \setminus \sigma_{\mathscr{B}}(a)$ such that $\lambda_n \to \lambda$. Let $(a - \lambda_n)^{-1}$ be the inverse of $(a - \lambda_n)$ in $\mathscr{B}$; so $(a - \lambda_n)^{-1} \in \mathscr{A}$. Since $\lambda_n \to \lambda$, $(a - \lambda_n) \to (a - \lambda)$. By Theorem 2.2, $(a - \lambda_n)^{-1} \to x$. Thus $x \in \mathscr{B}$ since $\mathscr{B}$ is complete. This contradicts the fact that $\lambda \in \sigma_{\mathscr{B}}(a)$.

(b) This is a consequence of (a) and the Maximum Principle.

(c) Let G be a hole of $\sigma_{\mathscr{A}}(a)$ and put $G_1 = G \cap \sigma_{\mathscr{B}}(a)$ and $G_2 = G \setminus \sigma_{\mathscr{B}}(a)$. So $G = G_1 \cup G_2$ and $G_1 \cap G_2 = \square$. Clearly G_2 is open. On the other hand, the fact that $\partial\sigma_{\mathscr{B}}(a) \subseteq \sigma_{\mathscr{A}}(a)$ and $G \cap \sigma_{\mathscr{A}}(a) = \square$ implies that $G_1 = G \cap \operatorname{int} \sigma_{\mathscr{B}}(a)$, so G_1 is open. Because G is connected, either G_1 or G_2 is empty.

(d) Let $\mathscr{B}$ be as in (d). From (a) and (b) it is known that $\sigma_{\mathscr{A}}(a) \subseteq \sigma_{\mathscr{B}}(a) \subseteq \sigma_{\mathscr{A}}(a)\hat{\ }$. Fix λ in $\sigma_{\mathscr{A}}(a)\hat{\ }$. If $\lambda \notin \sigma_{\mathscr{B}}(a)$, $(a - \lambda)^{-1} \in \mathscr{B} \subseteq \mathscr{A}$. Hence there is a sequence of polynomials $\{p_n\}$ such that $p_n(a) \to (a - \lambda)^{-1}$. Let $q_n(z) = (z - \lambda)p_n(z)$. Thus $\|q_n(a) - 1\| \to 0$. By the Spectral Mapping Theorem, $\sigma_{\mathscr{A}}(q_n(a)) = q_n(\sigma_{\mathscr{A}}(a))$. Thus, because $\lambda \in \sigma_{\mathscr{A}}(a)\hat{\ }$,

$$\|q_n(a) - 1\| \geq r(q_n(a) - 1)$$

$$= \sup\{|z - 1| : z \in \sigma_{\mathscr{A}}(q_n(a))\}$$

$$= \sup\{|q_n(w) - 1| : w \in \sigma_{\mathscr{A}}(a)\}$$

$$\geq |q_n(\lambda) - 1|$$

$$= 1.$$

This is a contradiction.   ∎

EXERCISES

1. If K is a compact subset of $\mathbb{C}$, let $P(K)$ be the closure of the polynomials in $C(K)$. Show that the identity map on polynomials extends to an isometric isomorphism of $P(K)$ onto $P(K\hat{\ })$.

2. If K is a compact subset of $\mathbb{C}$, let $R(K)$ be the closure in $C(K)$ of all rational functions with poles off K. If $f \in R(K)$, show that $\sigma_{R(K)}(f) = f(K)$. If $f \in P(K)$, show that $\sigma_{P(K)}(f) = f(K\hat{\ })$.

3. Let $\mathscr{A}, \mathscr{B}$ be as in Theorem 5.4. If $a \in \mathscr{B}$ and $\sigma_{\mathscr{B}}(a) \subseteq \mathbb{R}$, show that $\sigma_{\mathscr{B}}(a) = \sigma_{\mathscr{A}}(a)$.

4. Let $\mathscr{A}$ be a Banach algebra with identity and let $a \in \mathscr{A}$. If $G_1, G_2, \ldots$ are the holes of $\sigma_{\mathscr{A}}(a)$ and $1 \leq n_1 \leq n_2 \ldots$, show that there is a subalgebra $\mathscr{B}$ of $\mathscr{A}$ such that $a \in \mathscr{B}$ and $\sigma_{\mathscr{B}}(a) = \sigma_{\mathscr{A}}(a) \cup \bigcup_{k=1}^{\infty} G_{n_k}$.

5. If $\mathscr{A}$, $\mathscr{B}$, and a are as in Theorem 5.4, $\mathscr{A}$ is not abelian, and $\mathscr{B}$ is a maximal abelian subalgebra of $\mathscr{A}$, show that $\sigma_{\mathscr{A}}(a) = \sigma_{\mathscr{B}}(a)$.

6. If K is a nonempty compact subset of $\mathbb{C}$ that is polynomially convex, show that the components of $\operatorname{int} K$ are simply connected.

§6. The Spectrum of a Linear Operator

The proof of the first result is left as an exercise.

6.1. Proposition.

(a) *If $\mathscr{X}$ is a Banach space and $A \in \mathscr{B}(\mathscr{X})$, $\sigma(A^*) = \sigma(A)$.*
(b) *If $\mathscr{H}$ is a Hilbert space and $A \in \mathscr{B}(\mathscr{H})$, $\sigma(A^*) = \sigma(A)^*$, where for any subset Δ of $\mathbb{C}$, $\Delta^* \equiv \{\bar{z}: z \in \Delta\}$.*

In this section only results about operators on Banach spaces will be given. For the corresponding results about operators on a Hilbert space involving the adjoint, the reader is asked to supply the details. The preceding proposition should be kept in mind as a model of the probable differences.

In this section and the next $\mathscr{X}$ always denotes a Banach space over $\mathbb{C}$.

6.2. Definition.
If $A \in \mathscr{B}(\mathscr{X})$, the *point spectrum* of $A, \sigma_p(A)$, is defined by

$$\sigma_p(A) \equiv \{\lambda \in \mathbb{C}: \ker(A - \lambda) \neq (0)\}.$$

As in the case of operators on a Hilbert space, elements of $\sigma_p(A)$ are called *eigenvalues*. If $\lambda \in \sigma_p(A)$, vectors in $\ker(A - \lambda)$ are called *eigenvectors*; $\ker(A - \lambda)$ is called the *eigenspace* of A at λ.

6.3. Definition.
If $A \in \mathscr{B}(\mathscr{X})$, the *approximate point spectrum* of A, $\sigma_{ap}(A)$, is defined by

$$\sigma_{ap}(A) \equiv \{\lambda \in \mathbb{C}: \text{there is a sequence } \{x_n\} \text{ in } \mathscr{X}$$

$$\text{such that } \|x_n\| = 1 \text{ for all } n \text{ and } \|(A - \lambda)x_n\| \to 0\}.$$

Note that $\sigma_p(A) \subseteq \sigma_{ap}(A)$.

6.4. Proposition.
If $A \in \mathscr{B}(\mathscr{X})$ and $\lambda \in \mathbb{C}$, the following statements are equivalent.

(a) $\lambda \notin \sigma_{ap}(A)$.
(b) $\ker(A - \lambda) = (0)$ *and* $\operatorname{ran}(A - \lambda)$ *is closed.*
(c) *There is a constant $c > 0$ such that $\|(A - \lambda)x\| \geq c\|x\|$ for all x.*

PROOF. Clearly it may be assumed that $\lambda = 0$.
    (a) $\Rightarrow$ (c): Suppose (c) fails to hold; then for every n there is a vector x_n with $\|Ax_n\| \leq \|x_n\|/n$. If $y_n = x_n/\|x_n\|$, $\|y_n\| = 1$ and $\|Ay_n\| \to 0$. Hence $0 \in \sigma_{ap}(A)$.
    (c) $\Rightarrow$ (b): Suppose $\|Ax\| \geq c\|x\|$. Clearly $\ker A = (0)$. If $Ax_n \to y$, $\|x_n - x_m\| \leq c^{-1}\|Ax_n - Ax_m\|$, so $\{x_n\}$ is a Cauchy sequence. Let $x = \lim x_n$; therefore $Ax = y$ and $\operatorname{ran} A$ is closed.

(b) $\Rightarrow$ (a): Let $\mathscr{Y} = \operatorname{ran} A$; so $A: \mathscr{X} \to \mathscr{Y}$ is a continuous bijection. By the Inverse Mapping Theorem, there is a bounded operator $B: \mathscr{Y} \to \mathscr{X}$ such that $BAx = x$ for all x in $\mathscr{X}$. Thus if $\|x\| = 1$, $1 = \|BAx\| \leq \|B\| \|Ax\|$. That is, $\|Ax\| \geq \|B\|^{-1}$ whenever $\|x\| = 1$. Hence $0 \notin \sigma_{ap}(A)$. ∎

It may be that $\sigma_p(A)$ is empty, but it will be shown that $\sigma_{ap}(A)$ is never empty. The first statement follows from the next result (or from other examples that have been presented); the second statement will be proved later.

6.5. Proposition. *If $1 \leq p \leq \infty$, define $S: l^p \to l^p$ by $S(x_1, x_2, \dots) = (0, x_1, x_2, \dots)$. Then $\sigma(S) = \operatorname{cl} \mathbb{D}$, $\sigma_p(S) = \square$, and $\sigma_{ap}(S) = \partial \mathbb{D}$. Moreover, for $|\lambda| < 1$, $\operatorname{ran}(S - \lambda)$ is closed and $\dim[l^p/\operatorname{ran}(S - \lambda)] = 1$.*

PROOF. Let S_p be the shift on l^p. For $1 \leq p \leq \infty$, define $T_p: l^p \to l^p$ by $T_p(x_1, x_2, \dots) = (x_2, x_3, \dots)$. It is easy to check that for $1 \leq p < \infty$ and $1/p + 1/q = 1$, $S_p^* = T_q$. Since $\|S_p\| = 1$, $\sigma(S_p) \subseteq \operatorname{cl} \mathbb{D}$.

Suppose $x = (x_1, x_2, \dots) \in l^p$, $\lambda \neq 0$. If $S_p x = \lambda x$, $0 = \lambda x_1$, $x_1 = \lambda x_2, \dots$. Hence $0 = x_1 = x_2 = \cdots$. Since S_p is an isometry, $\ker S_p = (0)$. Thus $\sigma_p(S_p) = \square$.

Let $1 \leq p \leq \infty$ and $|\lambda| < 1$. Put $x_\lambda = (1, \lambda, \lambda^2, \dots)$. Then $\|x_\lambda\|_p^p = \sum_{n=0}^{\infty} |\lambda^p|^n < \infty$. Also, $T_p x_\lambda = (\lambda, \lambda^2, \dots) = \lambda x_\lambda$. Hence $\lambda \in \sigma_p(T_p)$ and $x_\lambda \in \ker(T_p - \lambda)$. If $1 \leq p < \infty$ and $1/p + 1/q = 1$, $T_q = S_p^*$; so $\mathbb{D} \subseteq \sigma(T_q) = \sigma(S_p)$. Also, $S_\infty = T_1^*$, so $\mathbb{D} \subseteq \sigma(S_\infty)$. Thus for all p, $\mathbb{D} \subseteq \sigma(S_p) \subseteq \operatorname{cl} \mathbb{D}$. Since $\sigma(S_p)$ is necessarily closed, $\sigma(S_p) = \operatorname{cl} \mathbb{D}$.

If $|\lambda| < 1$ and $x \in l^p$, $\|(S_p - \lambda)x\|_p = \|S_p x - \lambda x\|_p \geq |\|S_p x\|_p - |\lambda| \|x\|_p| = |\|x\|_p - |\lambda| \|x\|_p| = (1 - |\lambda|)\|x\|_p$. By (6.4), $\lambda \notin \sigma_{ap}(S_p)$. Hence $\sigma_{ap}(S_p) \subseteq \partial \mathbb{D}$. The fact that $\sigma_{ap}(S_p) = \partial \mathbb{D}$ follows from the next proposition (6.7).

Fix $|\lambda| < 1$, so $\operatorname{ran}(S_p - \lambda)$ is closed. If $1 \leq p < \infty$, then $\dim[l^p/\operatorname{ran}(S_p - \lambda)] = \dim[\operatorname{ran}(S_p - \lambda)^\perp]$ (Why?) $= \dim \ker(S_p^* - \lambda)$ (VI.1.8) $= \dim \ker(T_q - \lambda)$. Also, $\dim[l^\infty/\operatorname{ran}(S_\infty - \lambda)] = \dim[^\perp \operatorname{ran}(S_\infty - \lambda)]$ (Why?) $= \dim \ker(T_1 - \lambda)$. So to complete the proof it suffices to show that $\dim \ker(T_p - \lambda) = 1$ for $1 \leq p \leq \infty$. If $x \in l^p$ and $T_p x = \lambda x$, then $(x_2, x_3, \dots) = (\lambda x_1, \lambda x_2, \dots)$. So $x_{n+1} = \lambda x_n$ for all n. Thus $x_2 = \lambda x_1$; $x_3 = \lambda x_2 = \lambda^2 x_1, \dots$. Hence $x_{n+1} = \lambda^n x_1$. That is, if $x_\lambda = (1, \lambda, \lambda^2, \dots)$, then $x = x_1 x_\lambda$. Since it has already been shown that $x_\lambda \in \ker(T_p - \lambda)$, this completes the proof. ∎

6.6. Corollary. *If $1 \leq p \leq \infty$ and $T: l^p \to l^p$ is defined by $T(x_1, x_2, \dots) = (x_2, x_3, \dots)$, then $\sigma(T) = \operatorname{cl} \mathbb{D}$ and for $|\lambda| < 1$, $\ker(T - \lambda)$ is the one-dimensional space spanned by the vector $(1, \lambda, \lambda^2, \dots)$.*

The next result shows that if S is as in (6.5), then $\partial \mathbb{D} \subseteq \sigma_{ap}(S)$.

6.7. Proposition. *If $A \in \mathcal{B}(\mathcal{X})$, then $\partial\sigma(A) \subseteq \sigma_{ap}(A)$.*

PROOF. Let $\lambda \in \partial\sigma(A)$ and let $\{\lambda_n\} \subseteq \mathbb{C} \setminus \sigma(A)$ such that $\lambda_n \to \lambda$.

6.8. Claim. $\|(A - \lambda_n)^{-1}\| \to \infty$ as $n \to \infty$.

In fact, if the claim were false, then by passing to a subsequence if necessary, it follows that there is a constant M such that $\|(A - \lambda_n)^{-1}\| \le M$ for all n. Choose n sufficiently large that $|\lambda_n - \lambda| < M^{-1}$. Then $\|(A - \lambda) - (A - \lambda_n)\| < \|(A - \lambda_n)^{-1}\|^{-1}$. By (2.3b), this implies that $(A - \lambda)$ is invertible, a contradiction. This establishes (6.8).

Let $\|x_n\| = 1$ such that $\alpha_n \equiv \|(A - \lambda_n)^{-1}x_n\| > \|(A - \lambda_n)^{-1}\| - n^{-1}$, so $\alpha_n \to \infty$. Put $y_n = \alpha_n^{-1}(A - \lambda_n)^{-1}x_n$; hence $\|y_n\| = 1$. Now

$$(A - \lambda)y_n = (A - \lambda_n)y_n + (\lambda - \lambda_n)y_n$$
$$= \alpha_n^{-1}x_n + (\lambda - \lambda_n)y_n.$$

Thus $\|(A - \lambda)y_n\| \le \alpha_n^{-1} + |\lambda - \lambda_n|$, so that $\|(A - \lambda)y_n\| \to 0$ as $n \to \infty$. That is, $\lambda \in \sigma_{ap}(A)$. ∎

Let $A \in \mathcal{B}(\mathcal{X})$ and suppose Δ is a *clopen* subset of $\sigma(A)$; that is, Δ is a subset of $\sigma(A)$ that is both closed and relatively open. So $\sigma(A) = \Delta \cup (\sigma(A) \setminus \Delta)$. As in Proposition 4.11 (and Exercise 4.9),

6.9
$$E(\Delta) = E(\Delta; A) = \frac{1}{2\pi i}\int_\Gamma (z - A)^{-1}\, dz,$$

where Γ is a positively oriented Jordan system such that $\Delta \subseteq \text{ins}\,\Gamma$ and $\sigma(A) \setminus \Delta \subseteq \text{out}\,\Gamma$, is an idempotent. Moreover, $E(\Delta)B = BE(\Delta)$ whenever $AB = BA$ and if $\mathcal{X}_\Delta = E(\Delta)\mathcal{X}$, $\sigma(A|\mathcal{X}_\Delta) = \Delta$. If $\Delta = $ a singleton set $\{\lambda\}$, let $E(\lambda) = E(\{\lambda\})$ and $\mathcal{X}_\lambda = \mathcal{X}_{\{\lambda\}}$. Note that if λ is an isolated point of $\sigma(A)$, then $\{\lambda\}$ is a clopen subset of $\sigma(A)$.

6.10. Example. Let $\{\alpha_n\} \in l^\infty$, $1 \le p \le \infty$, and define $A\colon l^p \to l^p$ by $(Ax)(n) = \alpha_n x(n)$. Then $\sigma(A) = \text{cl}\{\alpha_n\}$ and $\sigma_p(A) = \{\alpha_n\}$. Let $e_n(k) = 0$ if $k \ne n$ and 1 if $k = n$. For each k, define $N_k = \{n \in \mathbb{N}; \alpha_n = \alpha_k\}$ and define $P_k\colon l^p \to l^p$ by $P_k x = \chi_{N_k} x$. If α_k is an isolated point of $\sigma(A)$, then $\{\alpha_k\}$ is a clopen subset of $\sigma(A)$ and $E(\{\alpha_k\}; A) = P_k$.

Suppose $A \in \mathcal{B}(\mathcal{X})$ and λ_0 is an isolated point in $\sigma(A)$. Hence $E(\lambda_0) = E(\lambda_0; A)$ is a well-defined idempotent. Also, λ_0 is an isolated singularity of the analytic function $z \mapsto (z - A)^{-1}$ on $\mathbb{C} \setminus \sigma(A)$. Perhaps the nature of this singularity (pole or essential singularity) will reveal something of the nature of λ_0 as an element of $\sigma(A)$. First it is helpful to get the precise form of the Laurent expansion of $(z - A)^{-1}$ about λ_0.

6.11. Lemma. *If λ_0 is an isolated point of $\sigma(A)$, then*

$$(z - A)^{-1} = \sum_{n=-\infty}^{\infty} (z - \lambda_0)^n A_n$$

for $0 < |z - \lambda_0| < r_0 = \text{dist}(\lambda_0, \sigma(A) \setminus \{\lambda_0\})$, where

$$A_n = \frac{1}{2\pi i} \int_\gamma (z - \lambda_0)^{-n-1} (z - A)^{-1} dz$$

for $\gamma = $ any circle centered at λ_0 with radius $< r_0$.

The proof follows the lines of the usual Laurent series development (Conway [1978]).

6.12. Proposition. *If λ_0 is an isolated point of $\sigma(A)$, then λ_0 is a pole of $(z - A)^{-1}$ of order n if and only if $(\lambda_0 - A)^n E(\lambda_0) = 0$ and $(\lambda_0 - A)^{n-1} E(\lambda_0) \neq 0$.*

PROOF. Let $(z - A)^{-1} = \sum_{n=-\infty}^\infty (z - \lambda_0)^n A_n$ as in (6.11). Now λ_0 is a pole of order n if and only if $A_{-n} \neq 0$ and $A_{-k} = 0$ for $k > n$. Let Γ be a positively oriented system of curves such that $\sigma(A) \setminus \{\lambda_0\} \subseteq \text{ins}\,\Gamma$ and $\lambda_0 \in \text{out}\,\Gamma$. Let γ be a circle centered at λ_0 and contained in out Γ. Let $e(z) \equiv 1$ in a neighborhood of $\gamma \cup \text{ins}\,\gamma$ and $e(z) \equiv 0$ in a neighborhood of $\Gamma \cup \text{ins}\,\Gamma$. So $e \in \text{Hol}(A)$ and $e(A) = E(\lambda_0)$. If $k \geq 1$,

$$A_{-k} = \frac{1}{2\pi i} \int_\gamma (z - \lambda_0)^{k-1} (z - A)^{-1} dz$$

$$= \frac{1}{2\pi i} \int_{\gamma + \Gamma} e(z)(z - \lambda_0)^{k-1} (z - A)^{-1} dz$$

$$= E(\lambda_0)(A - \lambda_0)^{k-1}$$

since $\sigma(A) \subseteq \text{ins}(\gamma + \Gamma) = \text{ins}\,\gamma \cup \text{ins}\,\Gamma$. The proposition now follows. ∎

6.13. Corollary. *If λ_0 is an isolated point of $\sigma(A)$ and is a pole of $(z - A)^{-1}$, then $\lambda_0 \in \sigma_p(A)$.*

In fact, the preceding result implies that if n is the order of the pole, then $(0) \neq (\lambda_0 - A)^{n-1} E(\lambda_0) \mathcal{X} \subseteq \ker(A - \lambda_0)$.

6.14. Example. A measurable function $k: [0,1] \times [0,1] \to \mathbb{C}$ is called a *Volterra kernel* if k is bounded and $k(x, y) = 0$ when $x < y$. If $1 \leq p \leq \infty$ and k is a Volterra kernel, define $V_k: L^p(0,1) \to L^p(0,1)$ by

$$V_k f(x) = \int_0^1 k(x, y) f(y) \, dy = \int_0^x k(x, y) f(y) \, dy.$$

Then $V_k \in \mathcal{B}(L^p)$ and $\|V_k\| \leq \|k\|_\infty$ (VI.2.3).
   If k, h are Volterra kernels and

$$(hk)(x, y) = \int_0^1 h(x, t) k(t, y) \, dt,$$

then hk is a Volterra kernel, $\|hk\|_\infty \le \|h\|_\infty \|k\|_\infty$, and $V_{hk} = V_h V_k$. Note that if $k(x, y)$ is the characteristic function of $\{(x, y) \in [0, 1] \times [0, 1] : y < x\}$, then V_k is the Volterra operator (II.1.7).

If k is a Volterra kernel, then

$$\sigma(V_k) = \{0\}.$$

Indeed, from the preceding paragraph it is known that $V_k^n = V_{k^n}$. This will be used to show that the spectral radius of V_k is 0.

6.15. Claim. $|k^n(x, y)| \le \dfrac{\|k\|_\infty^n}{(n-1)!}(x - y)^{n-1}$ for $y < x$.

This is proved by induction. Clearly it holds for $n = 1$. Suppose (6.15) is true for some $n \ge 1$. Then

$$|k^{n+1}(x, y)| = \left| \int_y^x k(x, t) k^n(t, y)\, dt \right|$$

$$\le \int_y^x |k(x, t)| |k^n(t, y)|\, dt$$

$$\le \|k\|_\infty \frac{\|k\|_\infty^n}{(n-1)!} \int_y^x (t - y)^{n-1}\, dt$$

$$\le \frac{\|k\|_\infty^{n+1}}{n!}(x - y)^n.$$

This establishes the claim.

From (6.15) it follows that

$$\|V_k^n\| \le \|k^n\|_\infty \le \frac{\|k\|_\infty^n}{(n-1)!}.$$

Therefore

$$\|V_k^n\|^{1/n} \le \|k\|_\infty [(n-1)!]^{-1/n}.$$

Since $[(n-1)!]^{-1/n} \to 0$ as $n \to \infty$, $r(V_k) = 0$. Thus $\square \ne \sigma(V_k) \subseteq \{\lambda \in \mathbb{C} : |\lambda| \le 0\}$; that is, $\sigma(V_k) = \{0\}$.

It is possible for $\ker V_k$ to be nontrivial. For example, if $k(x, y) = \chi_{(0, 1/2)}(y)$ when $y < x$ and 0 otherwise, then

$$V_k f(x) = \begin{cases} \int_0^x f(y)\, dy & \text{if } x \le \tfrac{1}{2}, \\[2mm] \int_0^{1/2} f(y)\, dy & \text{if } x \ge \tfrac{1}{2}. \end{cases}$$

So if $f(y) = 0$ for $0 \le y \le \tfrac{1}{2}$, $V_k f = 0$.

On the other hand, the Volterra operator V $[= V_k$ for $k(x, y) =$ the characteristic function of $\{(x, y): y < x\}]$ has $\ker V = (0)$. In fact, if $0 = Vf$, then for all x, $0 = \int_0^x f(y)\, dy$. Differentiating gives that $f = 0$.

Is there an analogy between V_k for a Volterra kernel k and a lower triangular matrix?

EXERCISES

1. Prove Proposition 6.1.

2. Show that for $\mathscr{X}$ a Banach space and A in $\mathscr{B}(\mathscr{X})$, $\sigma_l(A) = \sigma_r(A^*)$. What happens in a Hilbert space?

3. If $\mathscr{H}$ is a Hilbert space and K is a compact subset of $\mathbb{C}$, show that there is an A in $\mathscr{B}(\mathscr{H})$ such that $\sigma(A) = K$. Can A be found such that $\sigma(A) = \sigma_{ap}(A) = K$?

4. Let K be a compact subset of $\mathbb{C}$. Does there exist an operator A in $\mathscr{B}(C[0,1])$ such that $\sigma(A) = K$?

5. If $\mathscr{X}$ is a Banach space and $A \in \mathscr{B}(\mathscr{X})$, show that A is left invertible if and only if $\ker A = (0)$ and $\operatorname{ran} A$ is a closed complemented subspace of $\mathscr{X}$.

6. If $\mathscr{X}$ is a Banach space and $A \in \mathscr{B}(\mathscr{X})$, show that A is right invertible if and only if $\operatorname{ran} A = \mathscr{X}$ and $\ker A$ is a complemented subspace of $\mathscr{X}$.

7. If $\mathscr{X}$ is a Banach space and $T: \mathscr{X} \to \mathscr{X}$ is an isometry, then either $\sigma(T) \subseteq \partial \mathbb{D}$ or $\sigma(T) = \operatorname{cl} \mathbb{D}$.

8. Verify the statements made in Example 6.10.

9. Let $1 \le p \le \infty$ and suppose $0 < \alpha_1 \le \alpha_2 \cdots$ such that $r = \lim \alpha_n < \infty$. Define $A: l^p \to l^p$ by $A(x_1, x_2, \dots) = (0, \alpha_1 x_1, \alpha_2 x_2, \dots)$. Show that $\sigma(A) = \{ z \in \mathbb{C}: |z| \le r \}$ and $\sigma_{ap}(A) = \partial \sigma(A)$. If $|\lambda| < r$, then $\operatorname{ran}(A - \lambda)$ is closed and has codimension 1. Also, $\sigma_p(A) = \square$.

10. Verify the statements made in Example 6.14.

11. Let $1 \le p \le \infty$ and let (X, Ω, μ) be a σ-finite measure space. For ϕ in $L^\infty(\mu)$, define M_ϕ on $L^p(\mu)$ as in Example III.2.2. Find $\sigma(M_\phi)$, $\sigma_{ap}(M_\phi)$, and $\sigma_p(M_\phi)$.

12. If $A \in \mathscr{B}(\mathscr{X})$, $f \in \operatorname{Hol}(A)$, and $\lambda \in \sigma_p(A)$, does $f(\lambda) \in \sigma_p(f(A))$? If $\lambda \in \sigma_{ap}(A)$, does $f(\lambda) \in \sigma_{ap}(f(A))$? Is there a relation between $f(\sigma_{ap}(A))$ and $\sigma_{ap}(f(A))$?

13. If $A \in \mathscr{B}(\mathscr{X})$ say that a complex number λ has *finite index* if there is a positive integer k such that $\ker(A - \lambda)^k = \ker(A - \lambda)^{k+1}$; the *index* of λ, denoted by $\nu(\lambda)$ or $\nu_A(\lambda)$, is the smallest such integer k. (a) Show that if λ is an isolated point of $\sigma(A)$ and a pole of order n of $(z - A)^{-1}$, then $\nu(\lambda) = n$. (b) If $\nu(\lambda) < \infty$, show that $\ker(A - \lambda)^{\nu(\lambda)} = \ker(A - \lambda)^{\nu(\lambda)+k}$ for all $k \ge 0$. (c) If

$$\mathscr{X} = \mathbb{C}^n \text{ and } A = \begin{bmatrix} 0 & & & & \\ 1 & 0 & & & \\ & 1 & \cdot & & \\ & & \cdot & \cdot & \\ & & & \cdot & \\ & & & 1 & 0 \end{bmatrix}, \text{ then } \sigma(A) = \{0\} \text{ and } \nu(0) = n.$$

14. If V is the Volterra operator, show that 0 is an essential singularity of $(z - V)^{-1}$.

15. Let $\mathscr{A}$ be a Banach algebra with identity. If $a \in \mathscr{A}$, define $L_a, R_a \in \mathscr{B}(\mathscr{A})$ by $L_a(x) = ax$ and $R_a(x) = xa$. Show that $\sigma(L_a) = \sigma(R_a) = \sigma(a)$.

§7. The Spectral Theory of a Compact Operator

Recall that for a Banach space $\mathcal{X}$, $\mathcal{B}_0(\mathcal{X})$ is the algebra of all compact operators. This Banach algebra has no identity, so if $A \in \mathcal{B}_0(\mathcal{X})$, then $\sigma(A)$ refers to the spectrum of A as an element of $\mathcal{B}(\mathcal{X})$. Of course, if $\mathcal{A} = \mathcal{B}_0(\mathcal{X}) + \mathbb{C}$, then $\mathcal{A}$ is a Banach algebra with identity (Why?) and we could consider $\sigma_{\mathcal{A}}(A)$ for A in $\mathcal{B}_0(\mathcal{X})$. By Theorem 5.4, $\sigma(A) \subseteq \sigma_{\mathcal{A}}(A)$, $\partial\sigma_{\mathcal{A}}(A) \subseteq \sigma(A)$, and $\sigma(A)^{\hat{}} = \sigma_{\mathcal{A}}(A)^{\hat{}}$. Below, in Theorem 7.1, it will be shown that $\sigma(A)$ is a countable set and hence $\sigma(A) = \partial\sigma(A) = \sigma(A)^{\hat{}}$. Thus $\sigma(A) = \sigma_{\mathcal{A}}(A)$.

7.1. Theorem. (F. Riesz) *If* $\dim \mathcal{X} = \infty$ *and* $A \in \mathcal{B}_0(\mathcal{X})$, *then one and only one of the following possibilities occurs.*

(a) $\sigma(A) = \{0\}$.
(b) $\sigma(A) = \{0, \lambda_1, \ldots, \lambda_n\}$, *where for* $1 \le k \le n$, $\lambda_k \ne 0$, *each* λ_k *is an eigenvalue of* A, *and* $\dim \ker(A - \lambda_k) < \infty$.
(c) $\sigma(A) = \{0, \lambda_1, \lambda_2, \ldots\}$, *where for each* $k \ge 1$, λ_k *is an eigenvalue of* A, $\dim \ker(A - \lambda_k) < \infty$, *and, moreover,* $\lim \lambda_k = 0$.

The proof will use several lemmas. The first lemma was given in the case that $\mathcal{X}$ is a Hilbert space in Proposition II.4.14. The proof is identical and will not be repeated here.

7.2. Lemma. *If* $A \in \mathcal{B}_0(\mathcal{X})$, $\lambda \ne 0$, *and* $\ker(A - \lambda) = (0)$, *then* $\operatorname{ran}(A - \lambda)$ *is closed.*

The proof of the next lemma is like that of Corollary II.4.15.

7.3. Lemma. *If* $A \in \mathcal{B}_0(\mathcal{X})$, $\lambda \ne 0$, *and* $\lambda \in \sigma(A)$, *then either* $\lambda \in \sigma_p(A)$ *or* $\lambda \in \sigma_p(A^*)$.

7.4. Lemma. *If* $\mathcal{M} \le \mathcal{N} \le \mathcal{X}$, $\mathcal{M} \ne \mathcal{N}$, *and* $\varepsilon > 0$, *then there is a* y *in* $\mathcal{N}$ *such that* $\|y\| = 1$ *and* $\operatorname{dist}(y, \mathcal{M}) \ge 1 - \varepsilon$.

PROOF. Let $\delta(y) = \operatorname{dist}(y, \mathcal{M})$ for every y in $\mathcal{N}$. Now if $y_1 \in \mathcal{N}$, there is an x_0 in $\mathcal{M}$ such that $\delta(y_1) \le \|x_0 - y_1\| \le (1 + \varepsilon)\delta(y_1)$. Let $y_2 = y_1 - x_0$. Then $(1 + \varepsilon)\delta(y_2) = (1 + \varepsilon)\inf\{\|y_2 - x\| : x \in \mathcal{M}\} = (1 + \varepsilon)\inf\{\|y_1 - x_0 - x\| : x \in \mathcal{M}\} = (1 + \varepsilon)\delta(y_1)$ since $x_0 \in \mathcal{M}$. Thus $(1 + \varepsilon)\delta(y_2) > \|x_0 - y_1\| = \|y_2\|$. Let $y = \|y_2\|^{-1} y_2$. So $\|y\| = 1$, $y \in \mathcal{N}$, and if $x \in \mathcal{M}$, then

$$\|y - x\| = \|\|y_2\|^{-1} y_2 - x\|$$

$$= \|y_2\|^{-1}\|y_2 - \|y_2\|x\| > [(1 + \varepsilon)\delta(y_2)]^{-1}\|y_2 - \|y_2\|x\|$$

$$\ge (1 + \varepsilon)^{-1} > 1 - \varepsilon. \quad \blacksquare$$

If $\mathcal{M}$ and $\mathcal{N}$ are finite dimensional in the preceding lemma, then y can be chosen in $\mathcal{N}$ such that $\|y\| = 1$ and $\operatorname{dist}(y, \mathcal{M}) = 1$ (see Exercise 1).

7.5. Lemma. *If $A \in \mathcal{B}_0(\mathcal{X})$ and $\{\lambda_n\}$ is a sequence of distinct elements in $\sigma_p(A)$, then $\lim \lambda_n = 0$.*

PROOF. For each n let $x_n \in \ker(A - \lambda_n)$ such that $x_n \neq 0$. It follows that if $\mathcal{M}_n = \bigvee\{x_1, \ldots, x_n\}$, then $\dim \mathcal{M}_n = n$ (Exercise). Hence $\mathcal{M}_n \leq \mathcal{M}_{n+1}$ and $\mathcal{M}_n \neq \mathcal{M}_{n+1}$. By the preceding lemma there is a vector y_n in $\mathcal{M}_n$ such that $\|y_n\| = 1$ and $\operatorname{dist}(y_n, \mathcal{M}_{n-1}) > \frac{1}{2}$. Let $y_n = \alpha_1 x_1 + \cdots + \alpha_n x_n$. Hence

$$(A - \lambda_n)y_n = \alpha_1(\lambda_1 - \lambda_n)x_1 + \cdots + \alpha_{n-1}(\lambda_{n-1} - \lambda_n)x_{n-1} \in \mathcal{M}_{n-1}.$$

So if $n > m$,

$$A(\lambda_n^{-1}y_n) - A(\lambda_m^{-1}y_m) = \lambda_n^{-1}(A - \lambda_n)y_n - \lambda_m^{-1}(A - \lambda_m)y_m + y_n - y_m$$

$$= y_n - \left[y_m + \lambda_m^{-1}(A - \lambda_m)y_m - \lambda_n^{-1}(A - \lambda_n)y_n\right].$$

But the bracketed expression belongs to $\mathcal{M}_{n-1}$. Hence $\|A(\lambda_n^{-1}y_n) - A(\lambda_m^{-1}y_m)\| \geq \operatorname{dist}(y_n, \mathcal{M}_{n-1}) > \frac{1}{2}$. Therefore $A(\lambda_n^{-1}y_n)$ can have no convergent subsequence. But A is a compact operator so that if S is any bounded subset of $\mathcal{X}$, $\operatorname{cl} A(S)$ is compact. Thus it must be that $\{\lambda_n^{-1}y_n\}$ has no bounded subsequence. Since $\|y_n\| = 1$ for all n, it must be that $\|\lambda_n^{-1}y_n\| = |\lambda_n|^{-1} \to \infty$. That is, $0 = \lim \lambda_n$. ∎

PROOF OF THEOREM 7.1. The first step is to establish the following.

7.6. Claim. *If $\lambda \in \sigma(A)$ and $\lambda \neq 0$, then λ is an isolated point of $\sigma(A)$.*

In fact, if $\{\lambda_n\} \subseteq \sigma(A)$ and $\lambda_n \to \lambda$, then each λ_n belongs to either $\sigma_p(A)$ or $\sigma_p(A^*)$ (7.3). So either there is a subsequence $\{\lambda_{n_k}\}$ that is contained in $\sigma_p(A)$ or there is a subsequence contained in $\sigma_p(A^*)$. If $\{\lambda_{n_k}\} \subseteq \sigma_p(A)$, then Lemma 7.5 implies $\lambda_{n_k} \to 0$, a contradiction. If $\{\lambda_{n_k}\} \subseteq \sigma_p(A^*)$, then the fact that A^* is compact gives the same contradiction. Thus λ must be isolated if $\lambda \neq 0$.

7.7. Claim. *If $\lambda \in \sigma(A)$ and $\lambda \neq 0$, then $\lambda \in \sigma_p(A)$ and $\dim \ker(A - \lambda) < \infty$.*

By (7.6), λ is an isolated point of $\sigma(A)$ so that $E(\lambda)$ can be defined as in (6.9). Let $\mathcal{X}_\lambda = E(\lambda)\mathcal{X}$ and $A_\lambda = A|\mathcal{X}_\lambda$. By Exercise 4.9 [also see (4.11)], $\sigma(A_\lambda) = \{\lambda\}$. Thus A_λ is an invertible compact operator. By Exercise VI.3.5, $\dim \mathcal{X}_\lambda < \infty$. If $n = \dim \mathcal{X}_\lambda$, then $A_\lambda - \lambda$ is a nilpotent operator on an n-dimensional space. Thus $(A_\lambda - \lambda)^n = 0$. Let $\nu = $ the positive integer such that $(A_\lambda - \lambda)^\nu = 0$ but $(A_\lambda - \lambda)^{\nu-1} \neq 0$. Let $x \in \mathcal{X}_\lambda$ such that $0 \neq (A_\lambda - \lambda)^{\nu-1}x = y$; then $(A - \lambda)y = 0$. Thus $\lambda \in \sigma_p(A)$.

Also, $\ker(A - \lambda) \in \operatorname{Lat} A$ and $A|\ker(A - \lambda)$ is compact. But $Ax = \lambda x$ for all x in $\ker(A - \lambda)$, so $\dim \ker(A - \lambda) < \infty$.

Now for the *dénouement*. If $\dim \mathcal{X} = \infty$ and $A \in \mathcal{B}_0(\mathcal{X})$, then A cannot be invertible (Exercise VI.3.5). Thus $0 \in \sigma(A)$. If $\lambda \in \sigma(A)$ and $\lambda \neq 0$, then Claim 7.7 says that $\lambda \in \sigma_p(A)$ and dim ker$(A - \lambda) < \infty$. So if $\sigma(A)$ is finite, either (a) or (b) of (7.1) hold. If $\sigma(A)$ is infinite, then Claim 7.6 implies that $\sigma(A)$ is countable. So let $\sigma(A) = \{0, \lambda_1, \lambda_2, \dots\}$. By Lemma 7.5 and Claim 7.7, (c) holds. ∎

Part of the following surfaced in the proof of the theorem.

7.8. Corollary. *If $A \in \mathcal{B}_0(\mathcal{X})$ and $\lambda \in \sigma(A)$ with $\lambda \neq 0$, then λ is a pole of $(z - A)^{-1}$, ker$(A - \lambda) \subseteq E(\lambda)\mathcal{X}$, and dim $E(\lambda)\mathcal{X} < \infty$.*

PROOF. The only part of this corollary that did not appear in the preceding proof is the fact that ker$(A - \lambda) \subseteq E(\lambda)\mathcal{X}$.

Let $\Delta = \sigma(A) \setminus \{\lambda\}$, $\mathcal{X}_\Delta = E(\Delta)\mathcal{X}$, $A_\Delta = A|\mathcal{X}_\Delta$. By Exercise 4.9, $\sigma(A_\Delta) = \Delta$; so $A_\Delta - \lambda$ is invertible on $\mathcal{X}_\Delta$. If $x \in$ ker$(A - \lambda)$, then $x = E(\lambda)x + E(\Delta)x$. Hence $0 = (A - \lambda)x = (A - \lambda)E(\lambda)x + (A - \lambda)E(\Delta)x = (A_\lambda - \lambda)E(\lambda)x + (A_\Delta - \lambda)E(\Delta)x$. But $\mathcal{X}_\lambda$ and $\mathcal{X}_\Delta \in$ Lat A, so $(A_\lambda - \lambda)E(\lambda)x \in \mathcal{X}_\lambda$ and $(A_\Delta - \lambda)E(\Delta)x \in \mathcal{X}_\Delta$; since $\mathcal{X}_\lambda \cap \mathcal{X}_\Delta = (0)$, $0 = (A_\lambda - \lambda)E(\lambda)x = (A_\Delta - \lambda)E(\Delta)x$. But $A_\Delta - \lambda$ is invertible so $E(\Delta)x = 0$; that is, $x = E(\lambda)x \in \mathcal{X}_\lambda$. Hence ker$(A - \lambda) \subseteq \mathcal{X}_\lambda$. ∎

If k is a Volterra kernel (6.14), then V_k is a compact operator (Exercise VI.3.6) and $\sigma(V_k) = \{0\}$. So the first possibility of Theorem 7.1 can occur. If V is the Volterra operator, then $\sigma_p(V) = \square$.

Let V be the Volterra operator on $L^p(0,1)$, $1 < p < \infty$. If $\lambda_1, \dots, \lambda_n \in \mathbb{C}$, let $D: \mathbb{C}^n \to \mathbb{C}^n$ be defined by $D(z_1, \dots, z_n) = (\lambda_1 z_1, \dots, \lambda_n z_n)$. Then $A = V \oplus D$ on $L^p(0,1) \oplus \mathbb{C}^n$ is compact and $\sigma(A) = \{0, \lambda_1, \dots, \lambda_n\}$. So the second possibility of (7.1) occurs. If $\{\lambda_n\} \subseteq \mathbb{C}$ and $\lim \lambda_n = 0$, then define $D: l^p \to l^p$ $(1 \leq p \leq \infty)$ by $(Dx)(n) = \lambda_n x(n)$. If $A = V \oplus D$ on $L^p(0,1) \oplus l^p$, A is compact and $\sigma(A) = \{0, \lambda_1, \lambda_2, \dots\}$ (see Exercise 3).

The next result has a number of applications in the theory of integral equations.

7.9. The Fredholm Alternative. *If $A \in \mathcal{B}_0(\mathcal{X})$, $\lambda \in \mathbb{C}$, and $\lambda \neq 0$, then* ran$(A - \lambda)$ *is closed and* dim ker$(A - \lambda) =$ dim ker$(A - \lambda)^* < \infty$.

PROOF. It suffices to assume that $\lambda \in \sigma(A)$. Put $\Delta = \sigma(A) \setminus \{\lambda\}$, $\mathcal{X}_\lambda = E(\lambda)\mathcal{X}$, $\mathcal{X}_\Delta = E(\Delta)\mathcal{X}$, $A_\lambda = A|\mathcal{X}_\lambda$, and $A_\Delta = A|\mathcal{X}_\Delta$. Now $\lambda \notin \Delta = \sigma(A_\Delta)$, so $A_\Delta - \lambda$ is invertible. Thus ran$(A_\Delta - \lambda) = \mathcal{X}_\Delta$. Hence ran$(A - \lambda) = (A - \lambda)\mathcal{X}_\lambda + (A - \lambda)\mathcal{X}_\Delta =$ ran$(A_\lambda - \lambda) + \mathcal{X}_\Delta$. Since dim $\mathcal{X}_\lambda < \infty$, ran$(A - \lambda)$ is closed (III.4.3).

Also note that

$$\mathcal{X}/\text{ran}(A - \lambda) = (\mathcal{X}_\Delta + \mathcal{X}_\lambda)/[\text{ran}(A_\lambda - \lambda) + \mathcal{X}_\Delta]$$

$$\approx \mathcal{X}_\lambda/\text{ran}(A_\lambda - \lambda).$$

Since dim $\mathcal{X}_\lambda < \infty$, dim$[\mathcal{X}/\text{ran}(A - \lambda)] =$ dim $\mathcal{X}_\lambda -$ dim ran$(A_\lambda - \lambda) =$

$\dim \ker(A_\lambda - \lambda) = \dim \ker(A - \lambda) < \infty$ since $\ker(A - \lambda) \subseteq \mathcal{X}_\lambda$ (7.8). But $[\mathcal{X}/\mathrm{ran}(A - \lambda)]^* = [\mathrm{ran}(A - \lambda)]^\perp$ (III.10.2) $= \ker(A - \lambda)^*$. Hence $\dim \ker(A - \lambda) = \dim \ker(A - \lambda)^*$. ∎

7.10. Corollary. *If $A \in \mathcal{B}_0(\mathcal{X})$, $\lambda \in \mathbb{C}$, and $\lambda \neq 0$, then for every y in $\mathcal{X}$ there is an x in $\mathcal{X}$ such that*

7.11
$$(A - \lambda)x = y$$

if and only if the only vector x such that $(A - \lambda)x = 0$ is $x = 0$. If this condition is satisfied, then the solution to (7.11) is unique.

This corollary is a rephrasing of part of the Fredholm Alternative together with the fact that an operator has dense range if and only if its adjoint has a trivial kernel.

The applications of the Fredholm Alternative occur by taking the compact operator to be an integral operator.

EXERCISES

1. If $\mathcal{M}, \mathcal{N}$ are finite-dimensional subspaces of $\mathcal{X}$ and $\mathcal{M} \leq \mathcal{N}, \mathcal{M} \neq \mathcal{N}$, then there is a y in $\mathcal{N}$ such that $\|y\| = 1$ and $\mathrm{dist}(y, \mathcal{M}) = 1$.

2. Let $A \in \mathcal{B}(\mathcal{X})$ and let $\lambda_1, \ldots, \lambda_n$ be distinct points in $\sigma_p(A)$. If $x_k \in \ker(A - \lambda_k)$, $1 \leq k \leq n$, and $x_k \neq 0$, show that $\{x_1, \ldots, x_n\}$ is a linearly independent set.

3. Let $\mathcal{X}_1, \mathcal{X}_2, \ldots$ be Banach spaces and put $\mathcal{X} = \oplus_p \mathcal{X}_n$. Let $A_n \in \mathcal{B}(\mathcal{X}_n)$ such that $\sup_n \|A_n\| < \infty$ and define $A \colon \mathcal{X} \to \mathcal{X}$ by $A\{x_n\} = \{A_n x_n\}$. Show that $A \in \mathcal{B}(\mathcal{X})$ and $\|A\| = \sup_n \|A_n\|$. Show that $A \in \mathcal{B}_0(\mathcal{X})$ if and only if each $A_n \in \mathcal{B}_0(\mathcal{X})$ and $\lim \|A_n\| = 0$.

4. Suppose $A \in \mathcal{B}(\mathcal{X})$ and there is a polynomial p such that $p(A) \in \mathcal{B}_0(\mathcal{X})$. What can be said about $\sigma(A)$?

5. Suppose $A \in \mathcal{B}(\mathcal{X})$ and there is an entire function f such that $f(A) \in \mathcal{B}_0(\mathcal{X})$. What can be said about $\sigma(A)$?

6. With the terminology of Exercise 6.13, if $A \in \mathcal{B}_0(\mathcal{X})$, $\lambda \in \sigma(A)$, and $\lambda \neq 0$, what can be said about the index of λ?

§8. Abelian Banach Algebras

Recall that it is assumed that every Banach algebra is over $\mathbb{C}$. Also assume that all Banach algebras contain an identity.

A *division algebra* is an algebra such that every nonzero element has a multiplicative inverse. It may seem incongruous that the first theorem in this section allows the algebra to be nonabelian. However, the conclusion is that the algebra is abelian—and much more.

8.1. The Gelfand–Mazur Theorem. *If $\mathscr{A}$ is a Banach algebra that is also a division ring, then $\mathscr{A} = \mathbb{C}$ ($\equiv \{\lambda 1 : \lambda \in \mathbb{C}\}$).*

PROOF. If $a \in \mathscr{A}$, then $\sigma(a) \neq \square$. If $\lambda \in \sigma(a)$, then $a - \lambda$ has no inverse. But $\mathscr{A}$ is a division ring, so $a - \lambda = 0$. That is, $a = \lambda$. ∎

As a corollary of the preceding theorem, the algebra of quaternions, $\mathbb{H}$, is not a Banach algebra. That is, it is impossible to put a norm on $\mathbb{H}$ that makes it into a Banach algebra. Can you show this directly?

8.2. Proposition. *If $\mathscr{A}$ is an abelian Banach algebra and $\mathscr{M}$ is a maximal ideal, then there is a homomorphism h: $\mathscr{A} \to \mathbb{C}$ such that $\mathscr{M} = \ker h$. Conversely, if h: $\mathscr{A} \to \mathbb{C}$ is a nonzero homomorphism, then $\ker h$ is a maximal ideal. Moreover, this correspondence $h \mapsto \ker h$ between homomorphisms and maximal ideals is bijective.*

PROOF. If $\mathscr{M}$ is a maximal ideal, then $\mathscr{M}$ is closed (2.4b). Hence $\mathscr{A}/\mathscr{M}$ is a Banach algebra with identity. Let π: $\mathscr{A} \to \mathscr{A}/\mathscr{M}$ be the natural map. If $a \in \mathscr{A}$ and $\pi(a)$ is not invertible in $\mathscr{A}/\mathscr{M}$, then $\pi(\mathscr{A}a) = \pi(a)[\mathscr{A}/\mathscr{M}]$ is an ideal in $\mathscr{A}/\mathscr{M}$ that is proper. Let $I = \{b \in \mathscr{A} : \pi(b) \in \pi(\mathscr{A}a)\} = \pi^{-1}(\pi(\mathscr{A}a))$. Then I is a proper ideal of $\mathscr{A}$ and $\mathscr{M} \subseteq I$. Since $\mathscr{M}$ is maximal, $\mathscr{M} = I$. Thus $\pi(a\mathscr{A}) \subseteq \pi(I) = \pi(\mathscr{M}) = (0)$. That is, $\pi(a) = 0$. This says that $\mathscr{A}/\mathscr{M}$ is a field. By the Gelfand–Mazur Theorem $\mathscr{A}/\mathscr{M} = \mathbb{C}$ $= \{\lambda + \mathscr{M} : \lambda \in \mathbb{C}\}$. Define $\tilde{h}$: $\mathscr{A}/\mathscr{M} \to \mathbb{C}$ by $h(\lambda + \mathscr{M}) = \lambda$ and define h: $\mathscr{A} \to \mathbb{C}$ by $h = \tilde{h} \circ \pi$. Then h is a homomorphism and $\ker h = \mathscr{M}$.

Conversely, suppose h: $\mathscr{A} \to \mathbb{C}$ is a nonzero homomorphism. Then $\ker h = \mathscr{M}$ is a nontrivial ideal and $\mathscr{A}/\mathscr{M} \approx \mathbb{C}$. (Why?) So $\mathscr{M}$ is maximal. If h, h' are two nonzero homomorphisms and $\ker h = \ker h'$, then there is an α in $\mathbb{C}$ such that $h = \alpha h'$ (A.1.4). But $1 = h(1) = \alpha h'(1) = \alpha$, so $h = h'$. ∎

8.3. Corollary. *If $\mathscr{A}$ is an abelian Banach algebra and h: $\mathscr{A} \to \mathbb{C}$ is a homomorphism, then h is continuous.*

PROOF. Maximal ideals are closed (2.4b). ∎

The next result improves the preceding corollary a little. Remember that by (8.3) if h: $\mathscr{A} \to \mathbb{C}$ is a homomorphism, then $h \in \mathscr{A}^*$ (the Banach space dual of $\mathscr{A}$).

8.4. Proposition. *If $\mathscr{A}$ is abelian and h: $\mathscr{A} \to \mathbb{C}$ is a homomorphism, then $\|h\| = 1$.*

PROOF. By (8.3), $h \in \mathscr{A}^*$ so that $\|h\| < \infty$. Let $a \in \mathscr{A}$ and put $\lambda = h(a)$. If $|\lambda| > \|a\|$, then $\|a/\lambda\| < 1$. Hence $1 - a/\lambda$ is invertible. Let $b = (1 - a/\lambda)^{-1}$, so $1 = b(1 - a/\lambda) = b - ba/\lambda$. Since $h(1) = 1$, $1 = h(b - ba/\lambda)$

$= h(b) - h(b)h(a)/\lambda = h(b) - h(b) = 0$, a contradiction. Hence $\|a\| \geq$
$|\lambda| = |h(a)|$; so $\|h\| \leq 1$. Since $h(1) = 1$, $\|h\| = 1$.   ∎

8.5.   Definition. If $\mathscr{A}$ is an abelian Banach algebra, let Σ = the collection
of all nonzero homomorphisms of $\mathscr{A} \to \mathbb{C}$. Give Σ the relative weak*
topology that it has as a subset of $\mathscr{A}^*$. Σ with this topology is called the
maximal ideal space of $\mathscr{A}$.

8.6.   Theorem. *If $\mathscr{A}$ is an abelian Banach algebra, then its maximal ideal
space Σ is a compact Hausdorff space. Moreover, if $a \in \mathscr{A}$, then $\sigma(a) =$
$\Sigma(a) \equiv \{ h(a): h \in \Sigma \}$.*

PROOF. Since $\Sigma \subseteq$ ball $\mathscr{A}^*$, it suffices for the proof of the first part of the
theorem to show that Σ is weak* closed. Let $\{ h_i \}$ be a net in Σ and suppose
$h \in$ ball $\mathscr{A}^*$ such that $h_i \to h$ weak*. If $a, b \in \mathscr{A}$, then $h(ab) =$
$\lim_i h_i(ab) = \lim_i h_i(a)h_i(b) = h(a)h(b)$. So h is a homomorphism. Since
$h(1) = \lim_i h_i(1) = 1$, $h \in \Sigma$. Thus Σ is compact.

If $h \in \Sigma$ and $\lambda = h(a)$, then $a - \lambda \in \ker h$. So $a - \lambda$ is not invertible
and $\lambda \in \sigma(a)$; that is, $\Sigma(a) \subseteq \sigma(a)$. Now assume that $\lambda \in \sigma(a)$; so $a - \lambda$
is not invertible and, hence, $(a - \lambda)\mathscr{A}$ is a proper ideal. Let $\mathscr{M}$ be a
maximal ideal in $\mathscr{A}$ such that $(a - \lambda)\mathscr{A} \subseteq \mathscr{M}$. If $h \in \Sigma$ such that $\mathscr{M} =$
$\ker h$, then $0 = h(a - \lambda) = h(a) - \lambda$; hence $\sigma(a) \subseteq \Sigma(a)$.   ∎

Now it is time for an example. Here is one that is a little more than an
example. If X is compact and $x \in X$, let $\delta_x \colon C(X) \to \mathbb{C}$ be defined by
$\delta_x(f) = f(x)$. It is easy to see that δ_x is a homomorphism on the algebra
$C(X)$.

8.7.   Theorem. *If X is compact and Σ is the maximal ideal space of $C(X)$,
then the map $x \mapsto \delta_x$ is a homeomorphism of X onto Σ.*

PROOF. Let $\Delta \colon X \to \Sigma$ be defined by $\Delta(x) = \delta_x$. As was pointed out before,
$\Delta(X) \subseteq \Sigma$. It was shown in Proposition V.6.1 that $\Delta \colon X \to (\Delta(X), \text{weak}^*)$
is a homeomorphism. Thus it only remains to show that $\Delta(X) = \Sigma$. If
$h \in \Sigma$, then there is a measure μ in $M(X)$ such that $h(f) = \int f d\mu$ for all f
in $C(X)$. Also, $\|\mu\| = \|h\| = 1$ and $\mu(X) = \int 1 d\mu = h(1) = 1$. Hence $\mu \geq 0$
(Exercise III.7.2). Let $x \in \text{support}(\mu)$. It will be shown that $h = \delta_x$.

Let $\mathscr{M} = \{ f \in C(X): f(x) = 0 \}$. So $\mathscr{M}$ is a maximal ideal of $C(X)$.
Note that if it can be shown that $\ker h \subseteq \mathscr{M}$, then it must be that $\ker h = \mathscr{M}$
and so $h = \delta_x$. So let $f \in \ker h$. Because $\ker h$ is an ideal, $|f|^2 = \bar{f}f \in \ker h$.
Hence $0 = h(|f|^2) = \int |f|^2 d\mu$. Since $\mu \geq 0$ and $|f|^2 \geq 0$, it must be that
$f = 0$ a.e. $[\mu]$. Since f is continuous, $f \equiv 0$ on support (μ). In particular,
$f(x) = 0$ and so $f \in \mathscr{M}$.   ∎

It follows from the preceding theorem that the maximal ideals of $C(X)$
are all of the form $\{ f \in C(X): f(x) = 0 \}$ for some x in X.

8.8.   Definition. Let $\mathscr{A}$ be an abelian Banach algebra with maximal ideal space Σ. If $a \in \mathscr{A}$, then the *Gelfand transform* of a is the function $\hat{a}$: $\Sigma \rightarrow \mathbb{C}$ defined by $\hat{a}(h) = h(a)$.

8.9.   Theorem. *If $\mathscr{A}$ is an abelian Banach algebra with maximal ideal space Σ and $a \in \mathscr{A}$, then the Gelfand transform of a, $\hat{a}$, belongs to $C(\Sigma)$. The map $a \mapsto \hat{a}$ of $\mathscr{A}$ into $C(\Sigma)$ is a continuous homomorphism of $\mathscr{A}$ into $C(\Sigma)$ of norm 1 and its kernel is*

$$\bigcap \{ \mathscr{M} : \mathscr{M} \text{ is a maximal ideal of } \mathscr{A} \}.$$

Moreover, for each a in $\mathscr{A}$,

$$\|\hat{a}\|_{\infty} = \lim_{n \to \infty} \|a^n\|^{1/n}.$$

PROOF. If $h_i \rightarrow h$ in Σ, then $h_i \rightarrow h$ weak* in $\mathscr{A}^*$. So if $a \in \mathscr{A}$, $\hat{a}(h_i) = h_i(a) \rightarrow h(a) = \hat{a}(h)$. Thus $\hat{a} \in C(\Sigma)$.

Define $\gamma: \mathscr{A} \rightarrow C(\Sigma)$ by $\gamma(a) = \hat{a}$. If $a, b \in \mathscr{A}$, then $\gamma(ab)(h) = \widehat{ab}(h) = h(ab) = h(a)h(b) = \hat{a}(h)\hat{b}(h)$. Therefore $\gamma(ab) = \gamma(a)\gamma(b)$. It is easy to see that γ is linear, so γ is a homomorphism. Also, by (8.4), if $a \in \mathscr{A}$, $|\hat{a}(h)| = |h(a)| \leq \|a\|$; thus $\|\gamma(a)\|_{\infty} = \|\hat{a}\|_{\infty} \leq \|a\|$. So γ is continuous and $\|\gamma\| \leq 1$. Since $\gamma(1) = 1$, $\|\gamma\| = 1$.

Note that $a \in \ker \gamma$ if and only if $\hat{a} \equiv 0$; that is, $a \in \ker \gamma$ if and only if $h(a) = 0$ for each h in Σ. Thus $a \in \ker \gamma$ if and only if a belongs to every maximal ideal of $\mathscr{A}$.

Finally, by Theorem 8.6, if $a \in \mathscr{A}$, then $\|\hat{a}\|_{\infty} = \sup\{|\lambda|: \lambda \in \sigma(a)\}$. The last part of this theorem is thus a consequence of this observation and Proposition 3.8.   ∎

The homomorphism $a \mapsto \hat{a}$ of $\mathscr{A}$ into $C(\Sigma)$ is called the *Gelfand transform* of $\mathscr{A}$. The kernel of the *Gelfand* transform is called the *radical* of $\mathscr{A}$, rad $\mathscr{A}$. So

$$\text{rad } \mathscr{A} = \bigcap \{ \mathscr{M} : \mathscr{M} \text{ is a maximal ideal of } \mathscr{A} \}.$$

If X is compact and Σ, the maximal ideal space of $C(X)$, is identified with X as in Theorem 8.7, then the Gelfand transform $C(X) \rightarrow C(\Sigma)$ becomes the identity map.

If $\mathscr{A}$ is an abelian algebra, say that a in $\mathscr{A}$ is a *generator* of $\mathscr{A}$ if $\{ p(a): p$ is a polynomial$\}$ is dense in $\mathscr{A}$.

Recall that if $\tau: X \rightarrow Y$ is a homeomorphism, then $A: C(Y) \rightarrow C(X)$ defined by $Af = f \circ \tau$ is an isometric isomorphism (VI.2.1). Denote the relationship between A and τ by $A = \tau^{\#}$.

8.10.   Proposition. *If $\mathscr{A}$ is an abelian Banach algebra with identity and a is a generator of $\mathscr{A}$, then there is a homeomorphism $\tau: \Sigma \rightarrow \sigma(a)$ such that if $\gamma: \mathscr{A} \rightarrow C(\Sigma)$ is the Gelfand transform and p is a polynomial, then $\gamma(p(a)) = \tau^{\#}(p)$.*

PROOF. Define $\tau\colon \Sigma \to \sigma(a)$ by $\tau(h) = h(a)$. By Theorem 8.6 τ is surjective. It is easy to see that τ is continuous. To see that τ is injective, suppose $\tau(h_1) = \tau(h_2)$, so $h_1(a) = h_2(a)$. Hence $h_1(a^n) = h_2(a^n)$ for all $n \geq 0$. By linearity, $h_1(p(a)) = h_2(p(a))$ for every polynomial p. Since a is a generator for $\mathscr{A}$ and h_1 and h_2 are continuous on $\mathscr{A}$, $h_1 = h_2$, and τ is injective. Since Σ is compact, τ is a homeomorphism.

The remainder of the proposition follows from the fact that γ and $\tau^{\#}$ are homomorphisms. Hence $\gamma(p(a))(h) = p(\gamma(a))(h) = p(\hat{a})(h) = p(\hat{a}(h)) = p(\tau(h)) = \tau^{\#}(p)(h)$. ∎

8.11. Corollary. *If $\mathscr{A}$ has two generators, a_1 and a_2, then $\sigma(a_1)$ and $\sigma(a_2)$ are homeomorphic.*

The converse to (8.11) is not true. If $\mathscr{A} = C[-1, 1]$, then $f(x) = x$ defines a generator f for $\mathscr{A}$. If $g(x) = x^2$, then $\sigma(g) = g([-1, 1]) = [0, 1]$. So $\sigma(f)$ and $\sigma(g)$ are homeomorphic. However, g is not a generator for $\mathscr{A}$. In fact, the Banach algebra generated by g consists of the even functions in $C[-1, 1]$.

8.12. Example. If $V\colon L^2(0, 1) \to L^2(0, 1)$ is the Volterra operator and $\mathscr{A}$ is the closure in $\mathscr{B}(L^2(0, 1))$ of $\{p(V)\colon p$ is a polynomial in $z\}$, then $\mathscr{A}$ is an abelian Banach algebra and rad $\mathscr{A} = \mathrm{cl}\{p(V)\colon p$ is a polynomial in z and $p(0) = 0\}$. In other words, $\mathscr{A}$ has a unique maximal ideal, rad $\mathscr{A}$. In fact, if $\mathscr{B} = \mathscr{B}(L^2(0, 1))$, Theorem 5.4 implies that $\partial\sigma_{\mathscr{A}}(V) \subseteq \sigma_{\mathscr{B}}(V) \subseteq \sigma_{\mathscr{A}}(V)$. Since $\sigma_{\mathscr{B}}(V) = \{0\}$ (6.14), $\sigma_{\mathscr{A}}(V) = \{0\}$. The statement above now follows by Proposition 8.10.

8.13. Example. Let $\mathscr{A}$ be the closure in $C(\partial\mathbb{D})$ of the polynomials in z. If Σ is the maximal ideal space of $\mathscr{A}$, then Σ is homeomorphic to $\sigma_{\mathscr{A}}(z)$. (Here z is the function whose value at λ in $\partial\mathbb{D}$ is λ.) Now $\sigma_{\mathscr{A}}(z) = \mathrm{cl}\,\mathbb{D}$ as was shown in Example 5.1. If $f \in \mathscr{A}$, then the Maximum Modulus Theorem shows that f has a continuous extension to $\mathrm{cl}\,\mathbb{D}$ that is analytic in $\mathbb{D}$ [see (5.1)]. Also denote this extension by f. The proof of (8.10) shows that the continuous homomorphisms on $\mathscr{A}$ are of the form $f \mapsto f(\lambda)$ for some λ in $\mathrm{cl}\,\mathbb{D}$.

In the next section the Banach algebra $L^1(G)$ is examined for a locally compact abelian group and its maximal ideals are characterized.

EXERCISES

1. Let $\mathscr{A}$ be a Banach algebra with identity and let J be the smallest closed two-sided ideal of $\mathscr{A}$ containing $\{xy - yx\colon x, y \in \mathscr{A}\}$. J is called the *commutator ideal* of $\mathscr{A}$. (a) Show that $\mathscr{A}/J$ is an abelian Banach algebra. (b) If I is a closed ideal of $\mathscr{A}$ such that $\mathscr{A}/I$ is abelian, then $I \supseteq J$. (c) If $h\colon \mathscr{A} \to \mathbb{C}$ is a

homomorphism, then $J \subseteq \ker h$ and h induces a homomorphism $\tilde{h}: \mathscr{A}/J \to \mathbb{C}$ such that $\tilde{h} \circ \pi = h$, where $\pi: \mathscr{A} \to \mathscr{A}/J$ is the natural map. Hence $\|\tilde{h}\| = 1$. (d) Let Σ be the set of homomorphisms of $\mathscr{A} \to \mathbb{C}$ and let $\tilde{\Sigma}$ be the set of homomorphisms of $\mathscr{A}/J$. Show that the map $h \mapsto \tilde{h}$ defined in (c) is a homeomorphism of Σ onto $\tilde{\Sigma}$.

2. Using the terminology of Exercises 2.6 and 2.7, let $\mathscr{A}$ be an abelian Banach algebra without identity and show that if $\mathscr{M}$ is a maximal modular ideal, then there is a homomorphism $h: \mathscr{A} \to \mathbb{C}$ such that $\mathscr{M} = \ker h$. Conversely, if $h: \mathscr{A} \to \mathbb{C}$ is a nonzero homomorphism, then $\ker h$ is a maximal modular ideal. Moreover, the correspondence $h \mapsto \ker h$ is a bijection between homomorphisms and maximal modular ideals.

3. If $\mathscr{A}$ is an abelian Banach algebra and $h: \mathscr{A} \to \mathbb{C}$ is a homomorphism, then h is continuous and $\|h\| \leq 1$. If $\mathscr{A}$ has an approximate identity $\{e_i\}$ such that $\|e_i\| \leq 1$ for all i, then $\|h\| = 1$ (see Exercise 2.8).

4. Let $\mathscr{A}$ be an abelian Banach algebra and let Σ be the set of nonzero homomorphisms of $\mathscr{A} \to \mathbb{C}$. Show that Σ is locally compact if it has the relative weak* topology from $\mathscr{A}^*$ (Exercise 3).

5. With the notation of Exercise 4, assume that $\mathscr{A}$ has no identity and let $\mathscr{A}_1$ be the algebra obtained by adjoining an identity. For a in $\mathscr{A}$, let $\sigma(a)$ be the spectrum of a as an element of $\mathscr{A}_1$ and show that $\sigma(a) = \{h(a): h \in \Sigma\} \cup \{0\}$. Also, show that the maximal ideal space of $\mathscr{A}_1$, Σ_1, is the one-point compactification of Σ.

6. With the notation of Exercise 4, for each a in $\mathscr{A}$ define $\hat{a}: \Sigma \to \mathbb{C}$ by $\hat{a}(h) = h(a)$. Show that $\hat{a} \in C_0(\Sigma)$ and the map $a \mapsto \hat{a}$ of $\mathscr{A}$ into $C_0(\Sigma)$ is a contractive homomorphism with kernel $= \bigcap\{\mathscr{M}: \mathscr{M}$ is a maximal modular ideal of $\mathscr{A}\}$.

7. If X is locally compact, show that $x \mapsto \delta_x$ is a homeomorphism of X onto the maximal ideal space of $C_0(X)$.

8. Let X be locally compact and for each open subset U of X let $C_0(U) = \{f \in C_0(X): f(x) = 0$ for x in $X \setminus U\}$. Show that $C_0(U)$ is a closed ideal of $C_0(X)$ and that every closed ideal of $C_0(X)$ has this form. Moreover, the map $U \mapsto C_0(U)$ is a lattice isomorphism from the lattice of open subsets of X onto the lattice of ideals of $C_0(X)$.

9. With the notation of the preceding exercise, show that $C_0(U)$ is a modular ideal if and only if $X \setminus U$ is compact.

10. If $\mathscr{A}$ is an abelian Banach algebra and $a \in \mathscr{A}$, say that a is a *rational generator* of $\mathscr{A}$ if $\{f(a): f$ is a rational function with poles off $\sigma(a)\}$ is dense in $\mathscr{A}$. Show that if a is a rational generator of $\mathscr{A}$, then Σ is homeomorphic to $\sigma(a)$.

11. Verify the statements made in Example 8.12.

12. Say that $a_1, \ldots, a_n$ are generators of $\mathscr{A}$ if $\mathscr{A}$ is the smallest Banach algebra with identity that contains $\{a_1, \ldots, a_n\}$. Show that $a_1, \ldots, a_n$ are generators of $\mathscr{A}$ if and only if $\mathscr{A} = \mathrm{cl}\{p(a_1, \ldots, a_n): p$ is a polynomial in n complex variables $z_1, \ldots, z_n\}$, and if Σ is the maximal ideal space, then there is a

homeomorphism τ of Σ onto a compact subset K of $\mathbb{C}^n$ such that if p is a polynomial in n variables, then $\gamma(p(a_1, \ldots, a_n)) = \tau^{\#}(p)$.

13. Verify the statements made in Example 8.13.

14. (Zelazko [1968].) Let $\mathscr{A}$ be a Banach algebra and suppose $\phi\colon \mathscr{A} \to \mathbb{C}$ is a linear functional such that $\phi(a^2) = \phi(a)^2$ for all a in $\mathscr{A}$. Show that ϕ is a homomorphism.

15. Let $\mathscr{A}$ be an abelian Banach algebra with identity that is semisimple [that is, rad $\mathscr{A} = (0)$]. If $\|\cdot\|$ is the norm on $\mathscr{A}$ and $\|\cdot\|_1$ is another norm on $\mathscr{A}$ that also makes $\mathscr{A}$ into a Banach algebra, then these two norms are equivalent. (Hint: use the Closed Graph Theorem to show that the identity map $i\colon$ $(\mathscr{A}, \|\cdot\|) \to (\mathscr{A}, \|\cdot\|_1)$ is continuous.)

16. Let $\mathscr{A}$ be as in Example 8.13 and let $K = \{\phi \in \mathscr{A}^*\colon \phi(1) = \|\phi\| = 1\}$. Show that ext $K = \{\delta_z\colon |z| = 1\}$. (See (V.7).)

17. Show that $f(x) = \exp(\pi i x)$ is a generator of $C([0,1])$ but $g(x) = \exp(2\pi i x)$ is not.

§9*. The Group Algebra of a Locally Compact Abelian Group

If G is a locally compact abelian group and m is Haar measure on G, then $L^1(G) \equiv L^1(m)$ is a Banach algebra (Example 1.11), where for f, g in $L^1(G)$ the product $f * g$ is the convolution of f and g:

$$f * g(x) = \int_G f(xy^{-1})g(y)\, dy.$$

Note that dy is used to designate integration with respect to m rather than $dm(y)$. Because G is abelian, $L^1(G)$ is abelian. In fact, $g * f(x) = \int g(xy^{-1})f(y)\, dy$. If $y^{-1}x$ is substituted for y in this integral, the value of the integral does not change because Haar measure is translation invariant. Hence $g * f(x) = \int g(y)f(y^{-1}x)\, dy = \int g(y)f(xy^{-1})\, dy = f * g(x)$.

Let e denote the identity of G. If G is discrete, then $\delta_e \in L^1(G)$ and δ_e is an identity for $L^1(G)$. If G is not discrete, then $L^1(G)$ does not have an identity (Exercise 1).

Some examples of nondiscrete locally compact abelian groups are $\mathbb{R}^n$ and $\mathbb{T}^n$, where $\mathbb{T} =$ the unit circle $\partial \mathbb{D}$ in $\mathbb{C}$ with the usual multiplication. Note that $\mathbb{T}^\infty$ is also a compact abelian group while $\mathbb{R}^\infty$ fails to be locally compact. The Cantor set can be identified with the product of a countable number of copies of $\mathbb{Z}_2$ and is thus a compact abelian group. Indeed, the product of a countable number of finite sets (with the discrete topology) is homeomorphic to the Cantor set, so that the Cantor set has infinitely many nonisomorphic group structures.

For a topological group G, $L^1(G)$ is called the *group algebra* for G. If G is discrete, the algebraists talk of the *group algebra* over a field K as the set of all $f = \sum_{g \in G} a_g g$, where $a_g \in K$ and $a_g \neq 0$ for at most a finite number of g in G. If $K = \mathbb{C}$, this is the set of functions $f: G \to \mathbb{C}$ with finite support. Thus in the discrete case the group algebra of the algebraists can be identified with a dense manifold in $L^1(G) = l^1(G)$.

Unlike §V.11, if $f: G \to \mathbb{C}$ and $x \in G$, define $f_x: G \to \mathbb{C}$ by $f_x(y) = f(yx^{-1})$; so $f_x(y) = f(x^{-1}y)$ for G abelian. We want to examine the function $x \mapsto f_x$ of $G \to L^p(G)$, $1 \leq p < \infty$. To do this we first prove the following (see Exercise V.11.10).

9.1. Proposition. *If G is a topological group and $f: G \to \mathbb{C}$ is a continuous function with compact support, then for any $\varepsilon > 0$ there is a neighborhood U of e such that $|f(x) - f(y)| < \varepsilon$ whenever $x^{-1}y \in U$.*

PROOF. Let $\mathcal{U}$ be the collection of open neighborhoods U of e such that $U = U^{-1}$. Note that if V is any neighborhood of e, then $U = V \cap V^{-1} \in \mathcal{U}$ and $U \subseteq V$. Order $\mathcal{U}$ by reverse inclusion.

Suppose the result is false. Then there is an $\varepsilon > 0$ such that for every U in $\mathcal{U}$ there are points x_U, y_U in G with $x_U^{-1}y_U$ in U and $|f(x_U) - f(y_U)| \geq \varepsilon$. Note that either x_U or $y_U \in K \equiv$ support f. Since $U = U^{-1}$, we may assume that $x_U \in K$ for every U in $\mathcal{U}$. Now $\{x_U: U \in \mathcal{U}\}$ is a net in K. Since K is compact, there is a point x in K such that $x_U \xrightarrow{\text{cl}} x$. But $x_U^{-1}y_U \to e$. Since multiplication is continuous, $y_U = x_U(x_U^{-1}y_U) \xrightarrow{\text{cl}} x$. Therefore if W is any neighborhood of x, there is a U in $\mathcal{U}$ with $x_U, y_U \in W$. But f is continuous at x so W can be chosen such that $|f(x) - f(w)| < \varepsilon/2$ whenever $w \in W$. With this choice of W, $|f(x_U) - f(y_U)| < \varepsilon$, a contradiction. ∎

One can rephrase (9.1) by saying that continuous functions on a topological group that have compact support are uniformly continuous.

In the next result it is the case $p = 1$ which is of principal interest for us at this time. The proof of the general theorem is, however, no more difficult than this special case.

9.2. Proposition. *If G is a locally compact group, $1 \leq p < \infty$, and $f \in L^p(G)$, then the map $x \mapsto f_x$ is a continuous function from G into $L^p(G)$.*

PROOF. Fix f in $L^p(G)$, x in G, and $\varepsilon > 0$; it must be shown that there is a neighborhood V of x such that for y in V, $\|f_y - f_x\|_p < \varepsilon$. First note that there is a continuous function $\phi: G \to \mathbb{C}$ having compact support such that $\|f - \phi\|_p < \varepsilon/3$. Let $K = \text{spt } \phi$. Note that because Haar measure is translation invariant, for any y in G, $\|f_y - \phi_y\|_p = \|f - \phi\|_p < \varepsilon/3$. Now by Proposition 9.1, there is a neighborhood U of e such that $|\phi(y) - \phi(w)|$

$< \frac{1}{3}\varepsilon[2m(K)]^{-1/p}$ whenever $y^{-1}w \in U$. Put $V = Ux$. If $y \in V$, then

$$\|\phi_y - \phi_x\|_p^p = \int |\phi(zy^{-1}) - \phi(zx^{-1})|^p \, dz.$$

But $y = ux$ for some u in U, so $(zy^{-1})^{-1}(zx^{-1}) = yx^{-1} = u \in U$. Thus

$$\|\phi_y - \phi_x\|_p^p = \int_{Ky \cup Kx} |\phi(zy^{-1}) - \phi(zx^{-1})|^p \, dz$$

$$\leq \left(\frac{\varepsilon}{3}\right)^p [2m(K)]^{-1} m(Ky \cup Kx)$$

$$\leq \left(\frac{\varepsilon}{3}\right)^p.$$

Therefore if $y \in V$, $\|f_x - f_y\|_p \leq \|f_x - \phi_x\|_p + \|\phi_x - \phi_y\|_p + \|\phi_y - f_y\|_p < \varepsilon$. ∎

The aim of this section is to discuss the homomorphisms on $L^1(G)$ when G is abelian and to examine the Gelfand transform. There is a bit of a difficulty here since $L^1(G)$ does not have an identity when G is not discrete. If δ_e is the unit point mass at e, then δ_e is the identity for $M(G)$ and hence acts as an identity for $L^1(G)$. Nevertheless $\delta_e \notin L^1(G)$ if G is not discrete. All is not lost as $L^1(G)$ has an approximate identity (Exercise 2.8) of a nice type.

9.3. Proposition. *If $f \in L^1(G)$ and $\varepsilon > 0$, then there is a neighborhood U of e such that if g is a non-negative Borel function on G that vanishes off U and has $\int g(x)\,dx = 1$, then $\|f - f * g\|_1 < \varepsilon$.*

PROOF. By the preceding proposition, there is a neighborhood U of e such that $\|f - f_y\|_1 < \varepsilon$ whenever $y \in U$. If g satisfies the conditions, then $f(x) - f * g(x) = \int [f(x) - f(xy^{-1})]g(y)\,dy$ for all x. Thus,

$$\|f - f * g\|_1 = \int \left| \int_U [f(x) - f(xy^{-1})] g(y)\,dy \right| dx$$

$$\leq \int_U g(y) \int |f(x) - f(xy^{-1})|\,dx\,dy$$

$$= \int_U g(y)\|f - f_y\|_1\,dy$$

$$\leq \varepsilon. \quad ∎$$

9.4. Corollary. *There is a net $\{e_i\}$ of non-negative functions in $L^1(G)$ such that $\int e_i\,dm = 1$ for all i and $\|e_i * f - f\|_1 \to 0$ for all f in $L^1(G)$.*

PROOF. Let $\mathcal{U}$ be the collection of all neighborhoods of e and order $\mathcal{U}$ by reverse inclusion. Let $\mathcal{U} = \{U_i : i \in I\}$ where $i \leq j$ if and only if $U_j \subseteq U_i$. For each i in I put $e_i = m(U_i)^{-1}\chi_{U_i}$, so $e_i \geq 0$ and $\int e_i\,dm = 1$. If $f \in L^1(G)$

and $\varepsilon > 0$, let U_i be as in the preceding proposition. So if $j \geq i$, e_j satisfies the conditions on g in (9.3) and hence $\|f - f * e_j\|_1 < \varepsilon$. ∎

9.5. Corollary. *If h: $L^1(G) \to \mathbb{C}$ is a nonzero homomorphism, then h is bounded and $\|h\| = 1$.*

PROOF. The fact that h is bounded and $\|h\| \leq 1$ is Exercise 8.3. In light of the preceding corollary if $h(f) \neq 0$, $h(f) = \lim h(f * e_i) = h(f)\lim h(e_i)$. Hence $h(e_i) \to 1$. Since $\|e_i\| = 1$ for all i, $\|h\| = 1$. ∎

Even though Haar measure on most of the popular examples is σ-finite, this is not true in general. For example, if D is an uncountable discrete group, the Haar measure on D is counting measure and, hence, not σ-finite. Similarly, Haar measure on $D \times \mathbb{R}$ is not σ-finite. Nevertheless, it is true that $L^1(G)^* = L^\infty(G)$ for any locally compact group because (G, m) is an example of a decomposable measure space. This fact will be assumed here. The interested reader can consult Hewitt and Ross [1963].

9.6. Theorem. *If G is a locally compact abelian group and γ: $G \to \mathbb{T}$ is a continuous homomorphism, define $\hat{f}(\gamma)$ by*

9.7
$$\hat{f}(\gamma) = \int f(x)\gamma(x^{-1})\,dx$$

for every f in $L^1(G)$. Then $f \mapsto \hat{f}(\gamma)$ is a nonzero homomorphism on $L^1(G)$. Conversely, if h: $L^1(G) \to \mathbb{C}$ is a nonzero homomorphism, there is a continuous homomorphism γ: $G \to \mathbb{T}$ such that $h(f) = \hat{f}(\gamma)$.

PROOF. First note that if γ: $G \to \mathbb{T}$ is a homomorphism, $\gamma(xy) = \gamma(x)\gamma(y)$ and $\gamma(x^{-1}) = \gamma(x)^{-1} = \overline{\gamma(x)}$, the complex conjugate of $\gamma(x)$. If $f, g \in L^1(G)$, then

$$\widehat{f * g}(\gamma) = \int (f * g)(x)\gamma(x^{-1})\,dx$$

$$= \int \gamma(x^{-1}) \int f(xy^{-1})g(y)\,dy\,dx$$

$$= \int g(y)\gamma(y^{-1}) \left[\int f(xy^{-1})\gamma\big((xy^{-1})^{-1}\big)\,dx \right] dy.$$

But the invariance of the Haar integral gives that $\int f(xy^{-1})\gamma((xy^{-1})^{-1})\,dx = \int f(x)\gamma(x^{-1})\,dx$. Hence

$$\widehat{f * g}(\gamma) = \int g(y)\gamma(y^{-1}) \left[\int f(x)\gamma(x^{-1})\,dx \right] dy = \hat{f}(\gamma)\hat{g}(\gamma).$$

So $f \mapsto \hat{f}(\gamma)$ is a homomorphism. Since γ is continuous and $\gamma(G) \subseteq \mathbb{T}$, $\gamma \in L^\infty(G)$ and $\|\gamma\|_\infty = 1$. Thus $f \mapsto \hat{f}(\gamma)$ is not identically zero.

Now assume that h: $L^1(G) \to \mathbb{C}$ is a nonzero homomorphism. Since h is a bounded linear functional, there is a ϕ in $L^\infty(G)$ such that $h(f) =$

$\int f(x)\phi(x)\,dx$ and $\|\phi\|_\infty = \|h\| = 1$. If $f, g \in L^1(G)$, then $h(f * g) = \int (f * g)(x)\phi(x)\,dx = \int g(y)[\int f(xy^{-1})\phi(x)\,dx]\,dy = \int g(y)h(f_y)\,dy$. [Note that $y \mapsto h(f_y)$ is continuous scalar-valued function by Proposition 9.2.] But $h(f * g) = h(f)h(g) = \int g(y)h(f)\phi(y)\,dy$. So

$$0 = \int g(y)\big[h(f_y) - h(f)\phi(y)\big]\,dy$$

for every g in $L^1(G)$. But $y \mapsto h(f_y) - h(f)\phi(y)$ belongs to $L^\infty(G)$, so for any f in $L^1(G)$,

9.8 $h(f_y) = h(f)\phi(y)$

for almost all y in G. Pick f in $L^1(G)$ such that $h(f) \neq 0$. By (9.8), $\phi(y) = h(f_y)/h(f)$ a.e. But the right-hand side of this equation is continuous. Hence we may assume that ϕ is a continuous function. Thus for every f in $L^1(G)$, (9.8) holds everywhere.

In (9.8), replace y by xy and we obtain $h(f)\phi(xy) = h(f_{xy}) = h((f_x)_y)$. Now replace f in (9.8) by f_x to get $h(f_x)\phi(y) = h(f_{xy})$. Thus $h(f)\phi(xy) = h(f_x)\phi(y) = [h(f)\phi(x)]\phi(y)$. If $h(f) \neq 0$, this implies $\phi(xy) = \phi(x)\phi(y)$ for all x, y in G. Thus $\phi: G \to \mathbb{C}$ is a homomorphism and $|\phi(x)| \le 1$ for all x. But $1 = \phi(e) = \phi(x)\phi(x^{-1}) = \phi(x)\phi(x)^{-1}$ and $|\phi(x)|$, $|\phi(x)^{-1}| \le 1$. Hence $|\phi(x)| = 1$ for all x in G. If $\gamma(x) = \phi(x^{-1})$, then $\gamma: G \to \mathbb{T}$ is a continuous homomorphism and $h(f) = \hat{f}(\gamma)$ for all f in $L^1(G)$. ∎

Let Σ be the set of nonzero homomorphisms on $L^1(G)$, where G is assumed to be abelian (both here and throughout the rest of the chapter). So $\Sigma \subseteq$ ball $L^1(G)^*$. If $h \in$ ball $L^1(G)^*$ and $\{h_i\}$ is a net in Σ such that $h_i \to h$ weak*, then it is easy to see that h is multiplicative. Thus the weak* closure of $\Sigma \subseteq \Sigma \cup \{0\}$. Hence the relative weak* topology on Σ makes Σ into a locally compact Hausdorff space (see Exercise 8.4).

Let $\Gamma =$ all the continuous homomorphisms $\gamma: G \to \mathbb{T}$. By Theorem 9.6, Σ and Γ can be identified using formula (9.7). In fact, the map defined in (9.7) is the Gelfand transform when this identification is made. (Just look at the definitions.) Since Σ and Γ are identified and Σ has a topology, Γ can be given a topology. Thus Γ becomes a locally compact space with this topology. (For another description of the topology, see Exercise 6.) The functions in Γ are called *characters* and are sometimes denoted by $\Gamma = \hat{G}$ and called the *dual group*.

Also notice that in a natural way Γ is a group. If $\gamma_1, \gamma_2 \in \Gamma$, then $(\gamma_1\gamma_2)(x) \equiv \gamma_1(x)\gamma_2(x)$ and $\gamma_1\gamma_2 \in \Gamma$.

9.9. Proposition. *Γ is a locally compact abelian group.*

Clearly Γ is an abelian group and we know that Γ is a locally compact space. It must be shown that Γ is a topological group. To do this we first prove a lemma.

9.10. Lemma.

(a) *The map* $(x, \gamma) \mapsto \gamma(x)$ *of* $G \times \Gamma \to \mathbb{T}$ *is continuous.*
(b) *If* $\{\gamma_i\}$ *is a net in* Γ *and* $\gamma_i \to \gamma$ *in* Γ, *then* $\gamma_i(x) \to \gamma(x)$ *uniformly for* x
   belonging to any compact subset of G.

PROOF. First note that if $x \in G$ and $f \in L^1(G)$, then for every γ in Γ,

$$\hat{f}_x(\gamma) = \int f_x(y) \gamma(y^{-1}) \, dy$$

$$= \int f(yx^{-1}) \gamma(y^{-1}) \, dy$$

$$= \int f(z) \gamma(z^{-1} x^{-1}) \, dz$$

$$= \gamma(x^{-1}) \hat{f}(\gamma).$$

So if $\gamma_i \to \gamma$ in Γ and $x_i \to x$ in G,

$$|f(\gamma_i) \gamma_i(x_i) - \hat{f}(\gamma) \gamma(x)| = |\hat{f}_{x_i^{-1}}(\gamma_i) - \hat{f}_{x^{-1}}(\gamma)|$$

$$\le |\hat{f}_{x_i^{-1}}(\gamma_i) - \hat{f}_{x^{-1}}(\gamma_i)| + |\hat{f}_{x^{-1}}(\gamma_i) - \hat{f}_{x^{-1}}(\gamma)|.$$

But $|\hat{f}_{x_i^{-1}}(\gamma_i) - \hat{f}_{x^{-1}}(\gamma_i)| \le \|f_{x_i^{-1}} - f_{x^{-1}}\|_1 \to 0$ by (9.2). Because $f_{x^{-1}} \in$
$L^1(G)$, $\hat{f}_{x^{-1}}(\gamma_i) \to \hat{f}_{x^{-1}}(\gamma)$ since $\gamma_i \to \gamma$. Thus $\hat{f}(\gamma_i)\gamma_i(x_i) \to \hat{f}(\gamma)\gamma(x)$. If f
is chosen so that $\hat{f}(\gamma) \ne 0$, then because $\hat{f}(\gamma_i) \to \hat{f}(\gamma)$, there is an i_0 such
that $\hat{f}(\gamma_i) \ne 0$ for $i \ge i_0$. Therefore $\gamma_i(x_i) \to \gamma(x)$ and (a) is proven.

Now let K be a compact subset of G and let $\{\gamma_i\}$ be a net in Γ such that
$\gamma_i \to \gamma_0$. Suppose $\{\gamma_i(x)\}$ does not converge uniformly on K to $\gamma_0(x)$. Then
there is an $\varepsilon > 0$ such that for every i, there is a $j_i \ge i$ and an x_i in K such
that $|\gamma_{j_i}(x_i) - \gamma_0(x_i)| \ge \varepsilon$. Now $\{\gamma_{j_i}\}$ is a net and $\gamma_{j_i} \to \gamma_0$ (Exercise). Since
K is compact, there is an x_0 in K such that $x_i \xrightarrow{\text{cl}} x_0$. Now part (a)
implies that the map $(x, \gamma) \mapsto (\gamma(x), \gamma_0(x))$ of $G \times \Gamma$ into $\mathbb{T} \times \mathbb{T}$ is continuous. Since $(x_i, \gamma_{j_i}) \xrightarrow{\text{cl}} (x_0, \gamma_0)$ in $G \times \Gamma$, $(\gamma_{j_i}(x_i), \gamma_0(x_i)) \xrightarrow{\text{cl}}$
$(\gamma_0(x_0), \gamma_0(x_0))$. So for any i_0, there is an $i \ge i_0$ such that $|\gamma_{j_i}(x_i) - \gamma_0(x_0)|$
$< \varepsilon/2$ and $|\gamma_0(x_i) - \gamma_0(x_0)| < \varepsilon/2$. Hence $|\gamma_{j_i}(x_i) - \gamma_0(x_i)| < \varepsilon$, a contradiction. ∎

PROOF OF PROPOSITION 9.9. Let $\{\gamma_i\}, \{\lambda_i\}$ be nets in Γ such that $\gamma_i \to \gamma$
and $\lambda_i \to \lambda$. It must be shown that $\gamma_i \lambda_i^{-1} \to \gamma \lambda^{-1}$. Let $\phi \in C_c(G)$ and put
$K = \operatorname{spt} \phi$. Then $\hat{\phi}(\gamma_i \lambda_i^{-1}) = \int_K \phi(x) \gamma_i(x^{-1}) \lambda_i(x) \, dx$. By the preceding
lemma, $\gamma_i(x^{-1}) \to \gamma(x^{-1})$ and $\lambda_i(x) \to \lambda(x)$ uniformly for x in K. Thus
$\hat{\phi}(\gamma_i \lambda_i^{-1}) \to \hat{\phi}(\gamma \lambda^{-1})$. If $f \in L^1(G)$ and $\varepsilon > 0$, let $\phi \in C_c(G)$ such that
$\|f - \phi\|_1 < \varepsilon/3$. Then

$$|\hat{f}(\gamma_i \lambda_i^{-1}) - \hat{f}(\gamma \lambda^{-1})| < \frac{2\varepsilon}{3} + |\hat{\phi}(\gamma_i \lambda_i^{-1}) - \hat{\phi}(\gamma \lambda^{-1})|.$$

It follows that $\hat{f}(\gamma_i \lambda_i^{-1}) \to \hat{f}(\gamma \lambda^{-1})$ for every f in $L^1(G)$. Hence $\gamma_i \lambda_i^{-1} \to$
$\gamma \lambda^{-1}$ in Γ. ∎

Since Γ is a locally compact abelian group, it too has a dual group. Let $\hat{\Gamma}$ be this dual group. If $x \in G$, define $\rho(x): \Gamma \to \mathbb{T}$ by $\rho(x)(\gamma) = \gamma(x)$. It is easy to see that ρ is a homomorphism. It is a rather deep fact, entitled the Pontryagin Duality Theorem, that $\rho: G \to \hat{\Gamma}$ is a homeomorphism and an isomorphism. That is, G "is" the dual group of its dual group. The interested reader may consult Rudin [1962]. We turn now to some examples.

9.11. Theorem. *If $y \in \mathbb{R}$, then $\gamma_y(x) = e^{ixy}$ defines a character on $\mathbb{R}$ and every character on $\mathbb{R}$ has this form. The map $y \mapsto \gamma_y$ is a homeomorphism and an isomorphism of $\mathbb{R}$ onto $\hat{\mathbb{R}}$. If $y \in \mathbb{R}$ and $f \in L^1(\mathbb{R})$, then*

9.12 $$\hat{f}(\gamma_y) = \hat{f}(y) = \int_{-\infty}^{\infty} f(x) e^{-ixy} \, dx,$$

the Fourier transform of f.

PROOF. If $y \in \mathbb{R}$, then $|\gamma_y(x)| = 1$ for all x and $\gamma_y(x_1 + x_2) = \gamma_y(x_1)\gamma_y(x_2)$. So $\gamma_y \in \hat{\mathbb{R}}$. Also, $\gamma_{y_1+y_2}(x) = \gamma_{y_1}(x)\gamma_{y_2}(x)$. Hence $y \mapsto \gamma_y$ is a homomorphism of $\mathbb{R}$ into $\hat{\mathbb{R}}$.

Now let $\gamma \in \hat{\mathbb{R}}$. $\gamma(0) = 1$ so that there is a $\delta > 0$ such that $\int_0^\delta \gamma(x) \, dx = a \neq 0$. Thus

$$a\gamma(x) = \gamma(x) \int_0^\delta \gamma(t) \, dt$$

$$= \int_0^\delta \gamma(x + t) \, dt$$

$$= \int_x^{x+\delta} \gamma(t) \, dt.$$

Hence $\gamma(x) = a^{-1} \int_x^{x+\delta} \gamma(t) \, dt$. Because γ is continuous, the Fundamental Theorem of Calculus implies that γ is differentiable. Also,

$$\frac{\gamma(x + h) - \gamma(x)}{h} = \gamma(x) \left[\frac{\gamma(h) - 1}{h} \right].$$

So $\gamma'(x) = \gamma'(0)\gamma(x)$. Since $\gamma(0) = 1$ and $|\gamma(x)| = 1$ for all x, the elementary theory of differential equations implies that $\gamma = \gamma_y$ for some y in $\mathbb{R}$. This implies that $y \mapsto \gamma_y$ is an isomorphism of $\mathbb{R}$ onto $\hat{\mathbb{R}}$.

It is clear from (9.7) that (9.12) holds. From here it is easy to see that $y \mapsto \gamma_y$ is a homeomorphism of $\mathbb{R}$ onto $\hat{\mathbb{R}}$. ∎

So the preceding result says that $\mathbb{R}$ is its own dual group. Because of (9.12), the function $\hat{f}$ as defined in (9.7) is called the *Fourier transform* of f. The next result lends more weight to the use of this terminology.

9.13. Theorem. *If $n \in \mathbb{Z}$, define $\gamma_n: \mathbb{T} \to \mathbb{T}$ by $\gamma_n(z) = z^n$. Then $\gamma_n \in \hat{\mathbb{T}}$ and the map $n \mapsto \gamma_n$ is a homeomorphism and an isomorphism of $\mathbb{Z}$ onto $\hat{\mathbb{T}}$. If*

$n \in \mathbb{Z}$ and $f \in L^1(\mathbb{T})$, then

9.14
$$\hat{f}(\gamma_n) = \hat{f}(n) \equiv \frac{1}{2\pi} \int_0^{2\pi} f(e^{i\theta}) e^{-in\theta} \, d\theta.$$

PROOF. It is left to the reader to check that $\gamma_n \in \hat{\mathbb{T}}$ and $n \mapsto \gamma_n$ is an injective homomorphism of $\mathbb{Z}$ into $\hat{\mathbb{T}}$. If $\gamma \in \hat{\mathbb{T}}$, define $\sigma \colon \mathbb{R} \to \mathbb{T}$ by $\sigma(t) = \gamma(e^{it})$; it follows that $\sigma \in \hat{\mathbb{R}}$. By (9.11), $\sigma(t) = e^{iyt}$ for some y in $\mathbb{R}$. But $\sigma(t + 2\pi) = \sigma(t)$, so $e^{2\pi i y} = 1$. Hence $y = n \in \mathbb{Z}$. Thus $\gamma(e^{i\theta}) = \sigma(\theta) = e^{in\theta}$, $\gamma = \gamma_n$, and $n \mapsto \gamma_n$ is an isomorphism of $\mathbb{Z}$ onto $\hat{\mathbb{T}}$. Formula (9.14) is immediate from (9.7). The fact that $n \mapsto \gamma_n$ is a homeomorphism is left as an exercise. ∎

So $\hat{\mathbb{T}} = \mathbb{Z}$, a discrete group. This can be generalized.

9.15. Theorem. *If G is compact, $\hat{G}$ is discrete; if G is discrete, $\hat{G}$ is compact.*

PROOF. Put $\Gamma = \hat{G}$. If G is discrete, then $L^1(G)$ has an identity. Hence its maximal ideal space is compact. That is, Γ is compact.

Now assume that G is compact. Hence $\Gamma \subseteq L^1(G)$ since $m(G) = 1$. Suppose $\gamma \in \Gamma$ and $\gamma \neq$ the identity for Γ, then there is a point x_0 in G such that $\gamma(x_0) \neq 1$. Thus

$$\int \gamma(x) \, dx = \int \gamma(xx_0^{-1}x_0) \, dx$$

$$= \gamma(x_0) \int \gamma(xx_0^{-1}) \, dx$$

$$= \gamma(x_0) \int \gamma(x) \, dx,$$

since Haar measure is translation invariant. Since $\gamma(x_0) \neq 1$, this implies that

$$\int_G \gamma(x) \, dx = 0 \qquad \text{if } \gamma \neq 1.$$

Of course if $\gamma = 1$, $\int 1 \, dx = m(G) = 1$. So if $f = 1$ on G, $\hat{f} \in L^1(G)$ and $\hat{f}(\gamma) = \int \gamma(x^{-1}) \, dx = \chi_{\{1\}}(\gamma)$. Since $\hat{f}$ is continuous on Γ, $\{1\}$ is an open set. By translation, every singleton set in Γ is open and hence Γ is discrete. ∎

9.16. Theorem. *If $a \in \mathbb{T}$, define $\gamma_a \colon \mathbb{Z} \to \mathbb{T}$ by $\gamma_a(n) = a^n$. Then $\gamma_a \in \hat{\mathbb{Z}}$ and the map $a \mapsto \gamma_a$ is a homeomorphism and an isomorphism of $\mathbb{T}$ onto $\hat{\mathbb{Z}}$. If $a \in \mathbb{T}$ and $f \in L^1(\mathbb{Z}) = l^1(\mathbb{Z})$, then*

9.17
$$\hat{f}(\gamma_a) = \hat{f}(a) = \sum_{n=-\infty}^{\infty} f(n) a^{-n}.$$

PROOF. Again the proof that $a \mapsto \gamma_a$ is a monomorphism of $\mathbb{T} \to \hat{\mathbb{Z}}$ is left to the reader. If $\gamma \in \hat{\mathbb{Z}}$, let $\gamma(1) = a \in \mathbb{T}$. Also, $\gamma(n) = \gamma(1)^n = a^n$, so $\gamma = \gamma_a$.

Hence $a \mapsto \gamma_a$ is an isomorphism. It is easy to show that this map is continuous and hence, by compactness, a homeomorphism. ∎

For additional reading, consult Rudin [1962].

EXERCISES

1. Prove that if $L^1(G)$ has an identity, then G is discrete.

2. If $f \in L^\infty(G)$, show that $x \mapsto f_x$ is a continuous function from G into $(L^\infty(G), \text{wk}*)$.

3. Is there a measure μ on $\mathbb{R}$ different from Lebesgue measure such that for f in $L^1(\mu)$, $x \mapsto f_x$ is continuous? Is there a measure for which this map is discontinuous?

4. If $f \in C_0(G)$, show that $x \mapsto f_x$ is a continuous map from $G \to C_0(G)$.

5. If $f \in L^\infty(G)$ and f is uniformly continuous on G, show that $x \mapsto f_x$ is a continuous function from $G \to L^\infty(G)$. Is the converse true?

6. If K is a compact subset of G, $\gamma_0 \in \Gamma$, and $\varepsilon > 0$, let $U(K, \gamma_0, \varepsilon) = \{\gamma \in \Gamma : |\gamma(x) - \gamma_0(x)| < \varepsilon$ for all x in $K\}$. Show that the collection of all such sets is a base for the topology of Γ. (This says that the topology on Γ is the *compact-open topology*.)

7. Show that there is a discontinuous homomorphism $\gamma : \mathbb{R} \to \mathbb{T}$. If $\gamma : \mathbb{R} \to \mathbb{T}$ is a homomorphism that is a Borel function, show that γ is continuous.

8. If G is a compact abelian group, show that the linear span of Γ is dense in $C(G)$.

9. If G is a compact abelian group, show that Γ forms an orthonormal basis in $L^2(G)$.

10. If G is a compact abelian group, show that G is metrizable if and only if Γ is countable.

11. Let $\{G_\alpha\}$ be a family of compact abelian groups and $G = \Pi_\alpha G_\alpha$. If $\Gamma_\alpha = \hat{G}_\alpha$, show that the character group of G is $\{\{\gamma_\alpha\} \in \Pi_\alpha \Gamma_\alpha : \gamma_\alpha = e$ except for at most a finite number of $\alpha\}$.

CHAPTER VIII

C*-Algebras

A C^*-algebra is a particular type of Banach algebra that is intimately connected with the theory of operators on a Hilbert space. If $\mathcal{H}$ is a Hilbert space, then $\mathcal{B}(\mathcal{H})$ is an example of a C^*-algebra. Moreover, if $\mathcal{A}$ is any C^*-algebra, then it is isomorphic to a subalgebra of $\mathcal{B}(\mathcal{H})$ (see Section 5). Some of the general theory developed in this chapter will be used in the next chapter to prove the Spectral Theorem, which reveals the structure of normal operators.

A more thorough treatment of C^*-algebras is available in Arveson [1976] or Sakai [1971].

§1. Elementary Properties and Examples

If $\mathcal{A}$ is a Banach algebra, an *involution* is a map $a \mapsto a^*$ of $\mathcal{A}$ into $\mathcal{A}$ such that the following properties hold for a and b in $\mathcal{A}$ and α in $\mathbb{C}$:

(i) $(a^*)^* = a$;
(ii) $(ab)^* = b^*a^*$;
(iii) $(\alpha a + b)^* = \bar{\alpha}a^* + b^*$.

Note that if $\mathcal{A}$ has involution and an identity, then $1^* \cdot a = (1^* \cdot a)^{**} = (a^* \cdot 1)^* = (a^*)^* = a$; similarly, $a \cdot 1^* = a$. Since the identity is unique, $1^* = 1$. Also, for any α in $\mathbb{C}$, $\alpha^* = \bar{\alpha}$.

1.1. Definition. A C^*-*algebra* is a Banach algebra $\mathcal{A}$ with an involution such that for every a in $\mathcal{A}$,

$$\|a^*a\| = \|a\|^2.$$

1.2.   Example. If $\mathcal{H}$ is a Hilbert space, $\mathcal{A} = \mathcal{B}(\mathcal{H})$ is a C*-algebra where for each A in $\mathcal{B}(\mathcal{H})$, $A^* =$ the adjoint of A. (See Proposition II.2.7.)

1.3.   Example. If $\mathcal{H}$ is a Hilbert space, $\mathcal{B}_0(\mathcal{H})$ is a C*-subalgebra of $\mathcal{B}(\mathcal{H})$, though $\mathcal{B}_0(\mathcal{H})$ does not have an identity.

1.4.   Example. If X is a compact space, $C(X)$ is a C*-algebra where $f^*(x) = \overline{f(x)}$ for f in $C(X)$ and x in X.

1.5.   Example. If (X, Ω, μ) is a σ-finite measure space, $L^\infty(X, \Omega, \mu)$ is a C*-algebra where the involution is defined as in (1.4).

1.6.   Example. If X is locally compact but not compact, $C_0(X)$ is a C*-algebra without identity.

1.7.   Proposition. *If $\mathcal{A}$ is a C*-algebra and $a \in \mathcal{A}$, then $\|a^*\| = \|a\|$.*

PROOF. Note that $\|a\|^2 = \|a^*a\| \leq \|a^*\|\|a\|$; so $\|a\| \leq \|a^*\|$. Since $a = a^{**}$, substituting a^* for a in this inequality gives $\|a^*\| \leq \|a\|$.   ∎

1.8.   Proposition. *If $\mathcal{A}$ is a C*-algebra and $a \in \mathcal{A}$, then*

$$\|a\| = \sup\{\|ax\|: x \in \mathcal{A}, \|x\| \leq 1\}$$

$$= \sup\{\|xa\|: x \in \mathcal{A}, \|x\| \leq 1\}.$$

PROOF. Let $\alpha = \sup\{\|ax\|: x \in \mathcal{A}, \|x\| \leq 1\}$. Then $\|ax\| \leq \|a\|\|x\|$ for any x in $\mathcal{A}$; hence $\alpha \leq \|a\|$. If $x = a^*/\|a\|$, then $\|x\| = 1$ by the preceding proposition. For this x, $\|ax\| = \|a\|$, so $\alpha = \|a\|$. The proof of the other equality is similar.   ∎

This last proposition has an alternate formulation that is useful. If $a \in \mathcal{A}$, define $L_a: \mathcal{A} \to \mathcal{A}$ by $L_a(x) = ax$. By (1.8), $L_a \in \mathcal{B}(\mathcal{A})$ and $\|L_a\| = \|a\|$. If $\rho: \mathcal{A} \to \mathcal{B}(\mathcal{A})$ is defined by $\rho(a) = L_a$, then ρ is a homomorphism and an isometry. That is, $\mathcal{A}$ is isometrically isomorphic to a subalgebra of $\mathcal{B}(\mathcal{A})$. The map ρ is called the *left regular representation* of $\mathcal{A}$.

The left regular representation can be used to discuss the process of adjoining an identity to $\mathcal{A}$. Since $\mathcal{A}$ is isomorphic to a subalgebra $\mathcal{B}(\mathcal{A})$ and $\mathcal{B}(\mathcal{A})$ has an identity, why not just look at the subalgebra of $\mathcal{B}(\mathcal{A})$ generated by $\mathcal{A}$ and the identity operator? Why not, indeed. This is just what is done below.

1.9.   Proposition. *If $\mathcal{A}$ is a C*-algebra, then there is a C*-algebra $\mathcal{A}_1$ with an identity such that $\mathcal{A}_1$ contains $\mathcal{A}$ as an ideal. If $\mathcal{A}$ does not have an identity, then $\mathcal{A}_1/\mathcal{A}$ is one dimensional. If $\mathcal{C}$ is a C*-algebra with identity,* and v: $\mathscr{A} \to \mathscr{C}$ is a $*$-homomorphism, then v_1: $\mathscr{A}_1 \to \mathscr{C}$, defined by $v_1(a + \alpha) = v(a) + \alpha$ for a in $\mathscr{A}$ and α in $\mathbb{C}$, is a $*$-homomorphism.

PROOF. It may be assumed that $\mathscr{A}$ does not have an identity. Let $\mathscr{A}_1 = \{a + \alpha: a \in \mathscr{A}, \alpha \in \mathbb{C}\}$ ($a + \alpha$ is just a formal sum). Define multiplication and addition in the obvious way. Let $(a + \alpha)^* = a^* + \bar{\alpha}$ and define the norm on $\mathscr{A}_1$ by

$$\|a + \alpha\| = \sup\{\|ax + \alpha x\|: x \in \mathscr{A}, \|x\| \le 1\}.$$

Clearly, this is a norm on $\mathscr{A}_1$. It must be shown that $\|y^*y\| = \|y\|^2$ for every y in $\mathscr{A}_1$.

Fix a in $\mathscr{A}$ and α in $\mathbb{C}$. If $\varepsilon > 0$, then there is an x in $\mathscr{A}$ such that $\|x\| = 1$ and

$$\|a + \alpha\|^2 - \varepsilon < \|ax + \alpha x\|^2 = \|(x^*a^* + \bar{\alpha}x^*)(ax + \alpha x)\|$$

$$= \|x^*(a + \alpha)^*(a + \alpha)x\| \le \|(a + \alpha)^*(a + \alpha)\|.$$

Thus $\|a + \alpha\|^2 \le \|(a + \alpha)^*(a + \alpha)\|$.

It is left to the reader to prove that the norm on $\mathscr{A}_1$ makes $\mathscr{A}_1$ a Banach algebra. For the other inequality, note that $\|(a + \alpha)^*(a + \alpha)\| \le \|(a + \alpha)^*\|\|a + \alpha\|$. So the proof will be complete if it can be shown that $\|(a + \alpha)^*\| \le \|a + \alpha\|$. Now if $x, y \in \mathscr{A}$ and $\|x\|, \|y\| \le 1$, then $\|y(a + \alpha)^*x\| = \|ya^*x + \bar{\alpha}yx\| = \|x^*ay^* + \alpha x^*y^*\| = \|x^*(a + \alpha)y^*\| \le \|a + \alpha\|$. Taking the supremum over all such x, y gives the desired inequality.

It remains to prove the statement concerning the $*$-homomorphism v. Note that being a $*$-homomorphism means, besides being an algebra homomorphism, that $v(a^*) = v(a)^*$. The details are left to the reader. ∎

If $\mathscr{A}$ is a C^*-algebra with identity and $a \in \mathscr{A}$, then $\sigma(a)$, the spectrum of a, is well defined. If $\mathscr{A}$ does not have an identity, $\sigma(a)$ is defined as the spectrum of a as an element of the C^*-algebra $\mathscr{A}_1$ obtained in Proposition 1.9.

1.10. Definition. If $\mathscr{A}$ is a C^*-algebra and $a \in \mathscr{A}$, then (a) a is *hermitian* if $a = a^*$; (b) a is *normal* if $a^*a = aa^*$; (c) a is *unitary* if $a^*a = aa^* = 1$ (this only makes sense if $\mathscr{A}$ has an identity).

1.11. Proposition. *Let $\mathscr{A}$ be a C^*-algebra and let $a \in \mathscr{A}$.*

(a) *If a is invertible, then a^* is invertible and $(a^*)^{-1} = (a^{-1})^*$.*
(b) *$a = x + iy$ where x and y are hermitian elements of $\mathscr{A}$.*
(c) *If u is a unitary element of $\mathscr{A}$, $\|u\| = 1$.*
(d) *If $\mathscr{B}$ is a C^*-algebra and ρ: $\mathscr{A} \to \mathscr{B}$ is a $*$-homomorphism, then $\|\rho(a)\| \le \|a\|$.*
(e) *If $a = a^*$, then $\|a\| = r(a)$.*

PROOF. The proofs of (a), (b), and (c) are left as exercises.

(e) Since $a^* = a$, $\|a^2\| = \|a^*a\| = \|a\|^2$; by induction, $\|a^{2n}\| = \|a\|^{2n}$ for $n \geq 1$. That is, $\|a^{2n}\|^{1/2n} = \|a\|$ for $n \geq 1$. Hence $r(a) = \lim\|a^{2n}\|^{1/2n} = \|a\|$.

(d) A $*$-homomorphism ρ: $\mathscr{A} \to \mathscr{B}$ is an algebra homomorphism such that $\rho(a)^* = \rho(a^*)$ for all a in $\mathscr{A}$. If $\mathscr{A}$ has an identity, it is not assumed that $\rho(1) = $ the identity of $\mathscr{B}$. However, it is easy to see that $\rho(1)$ is the identity for $\text{cl}\,\rho(\mathscr{A})$. If $\mathscr{A}$ does not have an identity, then ρ can be extended to a $*$-homomorphism ρ_1: $\mathscr{A}_1 \to \mathscr{B}_1$ such that $\rho_1(1) = 1$ (1.9). Thus it suffices to prove the proposition under the additional assumption that $\mathscr{A}$ and $\mathscr{B}$ have identities and $\rho(1) = 1$.

If $x \in \mathscr{A}$, then it follows that $\sigma(\rho(x)) \subseteq \sigma(x)$ (Verify!) and hence $r(\rho(x)) \leq r(x)$. So, using part (e) and the fact that a^*a is hermitian, $\|\rho(a)\|^2 = \|\rho(a^*a)\| = r(\rho(a^*a)) \leq r(a^*a) = \|a^*a\| = \|a\|^2$. ∎

1.12. Proposition. *If $\mathscr{A}$ is a C^*-algebra and h: $\mathscr{A} \to \mathbb{C}$ is a homomorphism, then*:

(a) $h(a) \in \mathbb{R}$ *whenever* $a = a^*$;
(b) $h(a^*) = \overline{h(a)}$ *for all a in $\mathscr{A}$*;
(c) $h(a^*a) \geq 0$ *for all a in $\mathscr{A}$*;
(d) *if* $1 \in \mathscr{A}$ *and u is unitary, then* $|h(u)| = 1$.

PROOF. If $\mathscr{A}$ has no identity, extend h to $\mathscr{A}_1$ by letting $h(1) = 1$. Thus, we may assume that $\mathscr{A}$ has an identity. By Exercise VII.8.1, $\|h\| = 1$. If $a = a^*$ and $t \in \mathbb{R}$,

$$|h(a + it)|^2 \leq \|a + it\|^2 = \|(a + it)^*(a + it)\|$$

$$= \|(a - it)(a + it)\|$$

$$= \|a^2 + t^2\| \leq \|a\|^2 + t^2.$$

If $h(a) = \alpha + i\beta$, α, β in $\mathbb{R}$, then this yields

$$\|a\|^2 + t^2 \geq |\alpha + i(\beta + t)|^2$$

$$= \alpha^2 + (\beta + t)^2$$

$$= \alpha^2 + \beta^2 + 2\beta t + t^2;$$

hence $\|a\|^2 \geq \alpha^2 + \beta^2 + 2\beta t$ for all t in $\mathbb{R}$. If $\beta \neq 0$, then letting $t \to \pm\infty$, depending on the sign of β, gives a contradiction. Therefore $\beta = 0$ or $h(a) \in \mathbb{R}$. This proves (a).

Let $a = x + iy$, where x and y are hermitian. Since $h(x)$, $h(y) \in \mathbb{R}$ by (a) and $a^* = x - iy$, (b) follows. Also, $h(a^*a) = h(a^*)h(a) = |h(a)|^2 \geq 0$, so (c) holds. Finally, if u is unitary, $|h(u)|^2 = h(u^*)h(u) = h(u^*u) = h(1) = 1$. ∎

Note that part (b) of the preceding proposition implies that any homomorphism $h: \mathscr{A} \to \mathbb{C}$ is a ∗-homomorphism. This, coupled with (VII.8.6), gives the following corollary.

1.13. Corollary. *If $\mathscr{A}$ is an abelian C^*-algebra and a is a hermitian element of $\mathscr{A}$, then $\sigma(a) \subseteq \mathbb{R}$.*

This corollary is short-lived as the conclusion remains valid even if $\mathscr{A}$ is not abelian.

1.14. Proposition. *Let $\mathscr{A}$ and $\mathscr{B}$ be C^*-algebras with identities such that $\mathscr{A} \subseteq \mathscr{B}$. If $a \in \mathscr{A}$, then $\sigma_{\mathscr{A}}(a) = \sigma_{\mathscr{B}}(a)$.*

PROOF. First assume that a is hermitian and let $\mathscr{C} = C^*(a)$, the C^*-algebra generated by a and 1. So $\mathscr{C}$ is abelian. By Corollary 1.13 $\sigma_{\mathscr{C}}(a) \subseteq \mathbb{R}$. By Theorem VII.5.4, $\sigma_{\mathscr{A}}(a) \subseteq \sigma_{\mathscr{C}}(a) = \partial\sigma_{\mathscr{C}}(a) \subseteq \sigma_{\mathscr{A}}(a)$; so $\sigma_{\mathscr{A}}(a) = \sigma_{\mathscr{C}}(a) \subseteq \mathbb{R}$. By similar reasoning, $\sigma_{\mathscr{B}}(a) = \sigma_{\mathscr{C}}(a)$, and hence $\sigma_{\mathscr{A}}(a) = \sigma_{\mathscr{B}}(a)$.

Now let a be arbitrary. It suffices to show that if a is invertible in $\mathscr{B}$, a is invertible in $\mathscr{A}$. So suppose there is a b in $\mathscr{B}$ such that $ab = ba = 1$. Thus, $(a^*a)(bb^*) = (bb^*)(a^*a) = 1$. Since a^*a is hermitian, the first part of the proof implies a^*a is invertible in $\mathscr{A}$. But inverses are unique, so $bb^* = (a^*a)^{-1} \in \mathscr{A}$. Hence $b = b(b^*a^*) = (bb^*)a^* \in \mathscr{A}$. ∎

This result must, of course, be contrasted with Theorem VII.5.4.

EXERCISES

1. Verify the statements made in Examples 1.2 through 1.6.

2. Let $\mathscr{A} = \{f \in C(\mathrm{cl}\,\mathbb{D}): f$ is analytic in $\mathbb{D}\}$ and for f in $\mathscr{A}$ define f^* by $f^*(z) = \overline{f(\bar{z})}$. Show that $\mathscr{A}$ is a Banach algebra, $f^* \in \mathscr{A}$ when $f \in \mathscr{A}$, and $\|f^*\| = \|f\|$, but $\mathscr{A}$ is not a C^*-algebra.

3. If $\{\mathscr{A}_i: i \in I\}$ is a collection of C^*-algebras, show that $\oplus_{\infty}\mathscr{A}_i$ and $\oplus_0\mathscr{A}_i$ are C^*-algebras.

4. Let X be a locally compact space and let $\mathscr{A}$ be a C^*-algebra. If $C_b(X, \mathscr{A}) =$ the collection of bounded continuous functions from $X \to \mathscr{A}$, show that $C_b(X, \mathscr{A})$ is a C^*-algebra. Let $C_0(X, \mathscr{A}) =$ all of the continuous functions $f: X \to \mathscr{A}$ such that for every $\varepsilon > 0$, $\{x \in X: \|f(x)\| \geq \varepsilon\}$ is compact. Show that $C_0(X, \mathscr{A})$ is a C^*-algebra.

§2. Abelian C^*-Algebras and the Functional Calculus in C^*-Algebras

The next theorem is the basic result of this section. It will be used to develop a functional calculus for normal elements that extends the Riesz Functional Calculus.

2.1. Theorem. *If $\mathscr{A}$ is an abelian C^*-algebra with identity and Σ is its maximal ideal space, then the Gelfand transform γ: $\mathscr{A} \to C(\Sigma)$ is an isometric $*$-isomorphism of $\mathscr{A}$ onto $C(\Sigma)$.*

PROOF. By Theorem VII.8.9, $\|\hat{x}\|_\infty \leq \|x\|$ for every x in $\mathscr{A}$. But $\|\hat{x}\|_\infty$ is the spectral radius of x, so by (1.11e), $\|x\| = \|\hat{x}\|_\infty$ for every hermitian element x of $\mathscr{A}$. In particular, $\|x^*x\| = \|x^*x\|_\infty$ for every x in $\mathscr{A}$.

If $a \in \mathscr{A}$ and $h \in \Sigma$, then $\hat{a}^*(h) = h(a^*) = \overline{h(a)} = \overline{\hat{a}(h)}$. That is, $\widehat{a^*} = \overline{\hat{a}}$. Equivalently, $\gamma(a^*) = \gamma(a)^*$ since the involution on $C(\Sigma)$ is defined by complex conjugation. Thus, γ is a $*$-homomorphism. Also, $\|a\|^2 = \|a^*a\| = \|\widehat{a^*a}\|_\infty = \| |\hat{a}|^2 \|_\infty = \|\hat{a}\|_\infty^2$; therefore $\|a\| = \|\hat{a}\|_\infty$ and γ is an isometry.

Because γ is an isometry, it has closed range. To show that γ is surjective, therefore, it suffices to show that it has dense range. This is accomplished by applying the Stone–Weierstrass Theorem. Note that $\hat{1} = 1$, so $\gamma(\mathscr{A})$ is a subalgebra of $C(\Sigma)$ containing the constants. Because γ preserves the involution, $\gamma(\mathscr{A})$ is closed under complex conjugation. It remains to show that $\gamma(\mathscr{A})$ separates the points of Σ. But if h_1 and h_2 are distinct homomorphisms in Σ, they are distinct because there is an a in $\mathscr{A}$ such that $h_1(a) \neq h_2(a)$. Hence $\hat{a}(h_1) \neq \hat{a}(h_2)$. ∎

By combining the preceding theorem with Proposition 1.9 and Exercise VII.8.6, the following is obtained.

2.2. Corollary. *If $\mathscr{A}$ is an abelian C^*-algebra without identity and Σ is its maximal ideal space, then the Gelfand transform γ: $\mathscr{A} \to C_0(\Sigma)$ is an isometric $*$-isomorphism of $\mathscr{A}$ onto $C_0(\Sigma)$.*

In order to focus our attention on the key concepts and not be distracted by peripheral considerations, we now make the following.

Assumption. *All C^*-algebras that are considered have an identity.*

Let $\mathscr{B}$ be an arbitrary C^*-algebra and let a be a normal element of $\mathscr{B}$. So if $\mathscr{A} = C^*(a)$, the C^*-algebra generated by a (and 1), $\mathscr{A}$ is abelian. Hence $\mathscr{A} \cong C(\Sigma)$, where Σ is the maximal ideal space of $\mathscr{A}$. So by Theorem 2.1 if $f \in C(\Sigma)$, there is a unique element x of $\mathscr{A}$ such that $\hat{x} = f$. We want to think of x as $f(a)$ and thus define a functional calculus for normal elements of a C^*-algebra. To be useful, however, we should have a ready way of identifying Σ. Moreover, since $\mathscr{A} = C^*(a)$ and thus depends on a, it should be that Σ depends on a in a clear way. The idea embodied in Proposition VII.8.10 that Σ and $\sigma(a)$ are homeomorphic via a natural map is the key here, although (VII.8.10) is not directly applicable here since a is not a generator of $C^*(a)$ as a Banach algebra but only as a C^*-algebra. [If $a = a^*$, then a is a generator of $C^*(a)$ as a Banach algebra.] Nevertheless the result is true.

2.3. Proposition. *If $\mathscr{A}$ is an abelian C^*-algebra with maximal ideal space Σ and $a \in \mathscr{A}$ such that $\mathscr{A} = C^*(a)$, then the map $\tau \colon \Sigma \to \sigma(a)$ defined by $\tau(h) = h(a)$ is a homeomorphism. If $p(z, \bar{z})$ is a polynomial in z and $\bar{z}$ and $\gamma \colon \mathscr{A} \to C(\Sigma)$ is the Gelfand transform, then $\gamma(p(a, a^*)) = p \circ \tau$.*

The proof of this result follows, with a few variations, along the lines of the proof of Proposition VII.8.10 and is left to the reader.

If $\tau \colon \Sigma \to \sigma(a)$ is defined as in the preceding proposition, then $\tau^\# \colon C(\sigma(a)) \to C(\Sigma)$ is defined by $\tau^\#(f) = f \circ \tau$. Note that $\tau^\#$ is a $*$-isomorphism and an isometry, because τ is a homeomorphism. Note that $\mathscr{A} = C^*(a)$ is the closure of $\{ p(a, a^*) \colon p(z, \bar{z}) \text{ is a polynomial in } z \text{ and } \bar{z} \}$. Now such a polynomial $p(z, \bar{z})$ is, of course, a function on $\sigma(a)$. [Just evaluate the polynomial at any z in $\sigma(a)$.] The last part of (2.3) says that $\gamma(p(a, a^*)) = \tau^\#(p)$. We define a map $\rho \colon C(\sigma(a)) \to C^*(a)$ so that the following diagram commutes:

2.4

$$C^*(a) \xrightarrow{\;\;\gamma\;\;} C(\Sigma)$$

$$\rho \nwarrow \qquad \nearrow \tau^\#$$

$$C(\sigma(a))$$

Note that if $\mathscr{B}$ is any C^*-algebra and a is a normal element of $\mathscr{B}$, then $\mathscr{A} = C^*(a)$ is an abelian C^*-algebra contained in $\mathscr{B}$ and so (2.4) applies. Moreover, in light of Proposition 1.4, the spectrum of a does not depend on whether a is considered as an element of $\mathscr{A}$ or $\mathscr{B}$. The following definition is, therefore, unambiguous.

2.5. Definition. If $\mathscr{B}$ is a C^*-algebra with identity and a is a normal element of $\mathscr{B}$, let $\rho \colon C(\sigma(a)) \to C^*(a) \subseteq \mathscr{B}$ be as in (2.4). If $f \in C(\sigma(a))$, define

$$f(a) \equiv \rho(f).$$

The map $f \mapsto f(a)$ of $C(\sigma(a)) \to \mathscr{B}$ is called the *functional calculus for a.*

Note that if $p(z, \bar{z})$ is a polynomial in z and $\bar{z}$, then $\rho(p(z, \bar{z})) = p(a, a^*)$. In particular, $\rho(z^n \bar{z}^m) = a^n a^{*m}$ so that $\rho(z) = a$ and $\rho(\bar{z}) = a^*$. Also, $\rho(1) = 1$.

The properties of this functional calculus can be obtained from the fact that ρ is an isometric $*$-isomorphism of $C(\sigma(a))$ into $\mathscr{B}$—with one exception. How does this functional calculus compare with the Riesz Functional Calculus? If $f \in \text{Hol}(a)$, $f|\sigma(a) \in C(\sigma(a))$; so $f(a)$ has two possible interpretations. Or does it?

2.6. Theorem. *If $\mathscr{B}$ is a C^*-algebra and a is a normal element of $\mathscr{B}$, then the functional calculus has the following properties.*

(a) $f \mapsto f(a)$ *is a $*$-monomorphism.*
(b) $\|f(a)\| = \|f\|_\infty$.
(c) $f \mapsto f(a)$ *is an extension of the Riesz Functional Calculus.*

Moreover, the functional calculus is unique in the sense that if τ: $C(\sigma(a)) \to C^*(a)$ is a $*$-homomorphism that extends the Riesz Functional Calculus, then $\tau(f) = f(a)$ for every f in $C(\sigma(a))$.

PROOF. Let ρ: $C(\sigma(a)) \to C^*(a)$ be the map defined by $\rho(f) = f(a)$. From (2.4), (a) and (b) are immediate.

Let π: $\text{Hol}(a) \to \mathscr{A} \subseteq \mathscr{B}$ denote the map defined by the Riesz Functional Calculus. Since $\rho(z) = \pi(z) = a$, an algebraic manipulation gives that $\rho(f) = \pi(f)$ for every rational function f with poles off $\sigma(a)$. If $f \in \text{Hol}(a)$, then by Runge's Theorem there is a sequence $\{f_n\}$ of such rational functions such that $f_n(z) \to f(z)$ uniformly in a neighborhood of $\sigma(a)$. Thus $\pi(f_n) \to \pi(f)$. By (b), $\rho(f_n) \to \rho(f)$. Thus $\rho(f) = \pi(f)$.

To prove uniqueness, let τ: $C(\sigma(a)) \to \mathscr{B}$ be a $*$-homomorphism that extends the Riesz Functional Calculus. If $f \in C(\sigma(a))$, then there is a sequence $\{p_n\}$ of polynomials in z and $\bar{z}$ such that $p_n(z, \bar{z}) \to f(z)$ uniformly on $\sigma(a)$. But $\tau(p_n) = p_n(a, a^*)$, $\tau(p_n) \to \tau(f)$, and $p_n(a, a^*) \to f(a)$. Hence $\tau(f) = f(a)$. ∎

Because of the uniqueness statement in the preceding theorem, it is not necessary to remember the form of the functional calculus $f \mapsto f(a)$, but only the fact that it is an isometric $*$-monomorphism that extends the Riesz Functional Calculus. Indeed, by the uniqueness of the Riesz Functional Calculus, it suffices to have that $f \mapsto f(a)$ is an isometric $*$-monomorphism such that if $f(z) \equiv 1$, then $f(a) = 1$, and if $f(z) = z$, then $f(a) = a$. Any properties or applications of the functional calculus can be derived or justified using only these properties. There may, however, be an occasion when the precise form of the functional calculus [viz., (2.4)] facilitates a proof. There are also situations in which the definition of the functional calculus gets in the way of a proof and the properties in (2.6) give the clean way of applying this powerful tool.

2.7. Spectral Mapping Theorem. *If $\mathscr{A}$ is a C^*-algebra and a is a normal element of $\mathscr{A}$, then for every f in $C(\sigma(a))$,*

$$\sigma(f(a)) = f(\sigma(a)).$$

PROOF. Let ρ: $C(\sigma(a)) \to C^*(a)$ be defined by $\rho(f) = f(a)$. So ρ is a $*$-isomorphism. Hence $\sigma(f(a)) = \sigma(\rho(f)) = \sigma(f)$. But $\sigma(f) = f(\sigma(a))$ (VII.3.2). ∎

Once again (1.14) was used implicitly in the preceding proof.

EXERCISES

1. Prove a converse to Proposition 2.3. If K is a compact subset of $\mathbb{C}$, $C(K)$ is a singly generated C^*-algebra.

2. If $\mathscr{A}$ is an abelian C^*-algebra with a finite number of C^*-generators $a_1, \ldots, a_n$, then there is a compact subset X of $\mathbb{C}^n$ and an isometric $*$-isomorphism ρ: $\mathscr{A} \to C(X)$ such that $\rho(a_k) = z_k$, $1 \le k \le n$, where $z_k(\lambda_1, \ldots, \lambda_n) = \lambda_k$ (see Exercise VII.8.12).

3. A *Stonean space* is a compact space X such that the closure of every open subset of X is open. (a) Show that the Cantor set is a Stonean space. (b) Show that a compact space X is a Stonean space if and only if each connected subset of X is a singleton set. (c) Show that X is a Stonean space if and only if $C(X)$ is the closed linear span of its projections ($\equiv$ hermitian idempotents).

4. Using the terminology of Exercise 3, show that if (X, Ω, μ) is a σ-finite measure space, the maximal ideal space of $L^\infty(X, \Omega, \mu)$ is a Stonean space.

5. If $\mathscr{A}$ is a C^*-algebra with identity and $a = a^*$, show that $\exp(ia) = u$ is unitary. Is the converse true?

6. Let X be compact and fix a point x_0 in X. Let $\mathscr{A} = \{\{f_n\}: f_n \in C(X),$ $\sup_n \|f_n\| < \infty,$ and $\{f_n(x_0)\}$ is a convergent sequence$\}$. Show that $\mathscr{A}$ is an abelian C^*-algebra with identity and find its maximal ideal space.

7. If X is completely regular, then $C_b(X)$ is a C^*-algebra and its maximal ideal space is the Stone–Čech compactification of X.

§3. The Positive Elements in a C^*-Algebra

This section is an application of the functional calculus developed in the preceding section. The results here are very useful in the study of operators on a Hilbert space and they demonstrate the power of the functional calculus.

If $\mathscr{A}$ is a C^*-algebra, let Re$\mathscr{A}$ denote the hermitian elements of $\mathscr{A}$.

3.1. Definition. If $\mathscr{A}$ is a C^*-algebra and $a \in \mathscr{A}$, then a is *positive* if $a \in \mathrm{Re}\,\mathscr{A}$ and $\sigma(a) \subseteq [0, \infty)$. If a is positive, this is denoted by $a \geq 0$. Let $\mathscr{A}_+$ be the set of all positive elements of $\mathscr{A}$.

3.2. Example. If $\mathscr{A} = C(X)$, then f is positive in $\mathscr{A}$ if and only if $f(x) \geq 0$ for all x in X.

3.3. Example. If $\mathscr{A} = L^\infty(\mu)$ and $f \in L^\infty(\mu)$, then $f \geq 0$ if and only if $f(x) \geq 0$ a.e. $[\mu]$.

3.4. Proposition. *If $a \in \mathrm{Re}\,\mathscr{A}$, then there are unique positive elements u, v in $\mathscr{A}$ such that $a = u - v$ and $uv = vu = 0$.*

PROOF. Let $f(t) = \max(t, 0)$, $g(t) = -\min(t, 0)$. Then $f, g \in C(\mathbb{R})$ and $f(t) - g(t) = t$. Using the functional calculus, let $u = f(a)$ and $v = g(a)$. So u and v are hermitian and by the Spectral Mapping Theorem $u, v \geq 0$. Also, $u - v = f(a) - g(a) = a$ and $uv = vu = (gf)(a) = 0$ since $fg \equiv 0$.

To show uniqueness, let $u_1, v_1 \in \mathscr{A}_+$ such that $u_1 - v_1 = a$ and $u_1 v_1 = v_1 u_1 = 0$. Let $\{ p_n \}$ be a sequence of polynomials such that $p_n(0) = 0$ for all n and $p_n(t) \to f(t)$ uniformly on $\sigma(a)$. Hence $p_n(a) \to u$ in $\mathscr{A}$. But $u_1 a = a u_1$. So $u_1 p_n(a) = p_n(a) u_1$ for all n; hence $u_1 u = u u_1$. Similarly, it follows that a, u, v, u_1, and v_1 are pairwise commuting hermitian elements of $\mathscr{A}$. Let $\mathscr{B} =$ the C^*-algebra generated by a, u, v, u_1, and v_1; so $\mathscr{B}$ is abelian. Hence $\mathscr{B} \cong C(\Sigma)$ where Σ is the maximal ideal space of $\mathscr{B}$. The uniqueness now follows from the uniqueness statement for $C(\Sigma)$ (Exercise 1). ∎

The next result follows in a similar way.

3.5. Proposition. *If $a \in \mathscr{A}_+$ and $n \geq 1$, there is a unique element b in $\mathscr{A}_+$ such that $a = b^n$.*

The decomposition $a = u - v$ of a hermitian element a is sometimes called the orthogonal decomposition of a. The elements u and v are called the *positive* and *negative parts* of a and are denoted by $u = a_+$ and $v = a_-$. Note that $a_- \geq 0$.

If $a \in \mathscr{A}_+$, then the unique b obtained in (3.5) is called the *n*th *root* of a and is denoted by $b = a^{1/n}$. Note that if b is not assumed to be positive, it is not necessarily unique (see Exercise 5).

If X is compact and $f \in C(X)_+$, then notice that $|f(x) - t| \leq t$ for every real number $t \geq \|f\|$. Conversely, if $|f(x) - t| \leq t$ for some $t \geq \|f\|$, then $f(x) \geq 0$ for all x and so $f \geq 0$. These observations are behind some of the statements in the next result.

3.6. Theorem. *If $\mathscr{A}$ is a C^*-algebra and $a \in \mathscr{A}$, then the following statements are equivalent.*

(a) $a \geq 0$.
(b) $a = b^2$ for some b in $\mathrm{Re}\,\mathscr{A}$.

(c) $a = x^*x$ for some x in $\mathscr{A}$.

(d) $a = a^*$ and $\|t - a\| \leq t$ for all $t \geq \|a\|$.

(e) $a = a^*$ and $\|t - a\| \leq t$ for some $t \geq \|a\|$.

PROOF. It is clear that (b) implies (c) and (d) implies (e). By (3.5), (a) implies (b).

(e) $\Rightarrow$ (a): Since $a = a^*$, $C^*(a)$ is abelian. If $X = \sigma(a)$, $X \subseteq \mathbb{R}$ and $f \mapsto f(a)$ is a $*$-isomorphism of $C(X)$ onto $C^*(a)$. Using this isomorphism and (e), $|t - x| \leq t$ for some $t \geq \|a\| = \sup\{|s|: s \in X\}$ and all x in X. From the discussions preceding this theorem (with $f(x) = x$), $x \geq 0$ for all x in X. That is, $X = \sigma(a) \subseteq [0, \infty)$. Hence $a \geq 0$.

(a) $\Rightarrow$ (d): This proof follows the lines of the preceding paragraph and is left to the reader.

(c) $\Rightarrow$ (a): If $a = x^*x$ for some x in $\mathscr{A}$, then it is clear that $a = a^*$. Let $a = u - v$, where $u, v \geq 0$ and $uv = vu = 0$. It must be shown that $v = 0$. If $xv^{1/2} = b + ic$, where $b, c \in \mathrm{Re}\,\mathscr{A}$, then $(xv^{1/2})^*(xv^{1/2}) = (b - ic)(b + ic) = b^2 + c^2 + i(bc - cb)$. But also $(xv^{1/2})^*(xv^{1/2}) = v^{1/2}x^*xv^{1/2} = v^{1/2}(u - v)v^{1/2} = -v^2$. So the uniqueness of the Cartesian decomposition implies that $b^2 + c^2 = -v^2$ and $bc = cb$. Thus $\mathscr{B}$, the C^*-algebra generated by b and c, is abelian. Hence if $\lambda \in \sigma(b^2 + c^2)$, there is a homomorphism h: $\mathscr{B} \to \mathbb{C}$ such that $\lambda = h(b^2 + c^2) = h(b)^2 + h(c)^2$. Since $h(b), h(c) \in \mathbb{R}$ (1.12), $\lambda \geq 0$. Thus $b^2 + c^2 \in \mathscr{A}_+$. But $-(b^2 + c^2) = v^2$ and the same type of argument shows that $v^2 \in \mathscr{A}_+$. Thus $v^2 \in \mathscr{A}_+ \cap (-\mathscr{A}_+)$. By Proposition 3.7 below, $v^2 = 0$. Since $v \geq 0$, $v = 0$ (3.5). ∎

The next result will be proved only using the equivalence of (a), (d), and (e) from the preceding theorem.

3.7. Proposition. *If $\mathscr{A}$ is a C^*-algebra, then $\mathscr{A}_+$ is a closed cone.*

PROOF. Let $\{a_n\} \subseteq \mathscr{A}_+$ and suppose $a_n \to a$. Clearly $a \in \mathrm{Re}\,\mathscr{A}$. By (3.6d), $\|a_n - \|a_n\|\| \leq \|a_n\|$. Hence $\|a - \|a\|\| \leq \|a\|$, so by (3.6e), $a \geq 0$.

Clearly, $\lambda a \geq 0$ if $a \geq 0$ and $\lambda \geq 0$. Let $a, b \in \mathscr{A}_+$; it must be shown that $a + b \geq 0$. It suffices to assume that $\|a\|, \|b\| \leq 1$. But $\|1 - \frac{1}{2}(a + b)\| = \frac{1}{2}\|(1 - a) + (1 - b)\| \leq 1$ by (3.6d). So by (3.6e), $\frac{1}{2}(a + b) \geq 0$.

If $a \in \mathscr{A}_+ \cap (-\mathscr{A}_+)$, then $a = a^*$ and $\sigma(a) = \{0\}$. But $\|a\| = r(a)$ (1.11e). ∎

Now to look at one more example—a very important one.

3.8. Theorem. *If $\mathscr{H}$ is a Hilbert space and $A \in \mathscr{B}(\mathscr{H})$, then $A \geq 0$ if and only if $\langle Ah, h \rangle \geq 0$ for all h in $\mathscr{H}$.*

PROOF. If $A \geq 0$, then (3.6c) $A = T^*T$ for some T in $\mathscr{B}(\mathscr{H})$. Hence $\langle Ah, h \rangle = \|Th\|^2 \geq 0$. Conversely, suppose $\langle Ah, h \rangle \geq 0$ for all h in $\mathscr{H}$. By (II.2.12), $A = A^*$. It remains to show that $\sigma(A) \subseteq [0, \infty)$. If $h \in \mathscr{H}$ and

$\lambda < 0$, then

$$\|(A - \lambda)h\|^2 = \|Ah\|^2 - 2\lambda\langle Ah, h\rangle + \lambda^2\|h\|^2$$

$$\geq -2\lambda\langle Ah, h\rangle + \lambda^2\|h\|^2 \geq \lambda^2\|h\|^2$$

since $\lambda < 0$ and $\langle Ah, h\rangle \geq 0$. By (VII.6.4), $\lambda \notin \sigma_{ap}(A)$. But this implies that $A - \lambda$ is left invertible (Exercise VII.6.5). Since $A - \lambda$ is self-adjoint, $A - \lambda$ is also right invertible. Thus $\lambda \notin \sigma(A)$ and $A \geq 0$. ∎

3.9. Definition. If $\mathscr{A}$ is a C^*-algebra and $a, b \in \text{Re}\,\mathscr{A}$, then $a \leq b$ if $b - a \in \mathscr{A}_+$.

This ordering makes a C^*-algebra into a partially ordered vector space (over $\mathbb{C}$).

Note that if A and B are hermitian operators on the Hilbert space $\mathscr{H}$, then $A \leq B$ if and only if $\langle Ah, h\rangle \leq \langle Bh, h\rangle$ for all h in $\mathscr{H}$.

This section closes with an application of positivity to obtain the polar decomposition of an operator. If $\lambda \in \mathbb{C}$, then $\lambda = |\lambda|e^{i\theta}$ for some θ; this is the polar decomposition of λ. Can an analogy be found for operators? To answer this question we might first ask what is the analogy of $|\lambda|$ and $e^{i\theta}$ among operators. If $A \in \mathscr{B}(\mathscr{H})$, then the proper definition for $|A|$ would seem to be $|A| \equiv (A^*A)^{1/2}$ [see (3.5)]. How about an analogy of $e^{i\theta}$? Should it be a unitary operator? An isometry? For an arbitrary operator neither of these is correct. The following new class of operators is needed.

3.10. Definition. A *partial isometry* is an operator W such that for h in $(\ker W)^\perp$, $\|Wh\| = \|h\|$. The space $(\ker W)^\perp$ is called the *initial space* of W and the space $\text{ran}\,W$ is called the *final space* of W. See Exercises 15–20 for more on partial isometries.

3.11. Polar Decomposition. *If $A \in \mathscr{B}(\mathscr{H})$, then there is a partial isometry W with $(\ker A)^\perp$ as its initial space and $\text{cl}(\text{ran}\,A)$ as its final space such that $A = W|A|$. Moreover, if $A = UP$ where $P \geq 0$ and U is a partial isometry with initial and final spaces $(\ker A)^\perp$ and $\text{cl}(\text{ran}\,A)$, respectively, then $P = |A|$ and $U = W$.*

PROOF. If $h \in \mathscr{H}$, then $\|Ah\|^2 = \langle Ah, Ah\rangle = \langle A^*Ah, h\rangle = \langle |A|h, |A|h\rangle$. Thus

3.12 $$\|Ah\|^2 = \|\,|A|h\|^2.$$

Since $(\text{ran}\,A^*)^\perp = \ker A$, $\text{ran}\,A^*$ is dense in $(\ker A)^\perp$. If $f \in \text{ran}\,A^*$, $f = A^*g$ for g in $(\ker A^*)^\perp = \text{cl ran}\,A$. Therefore, $\{A^*Ak: k \in \mathscr{H}\}$ is dense in $\text{cl}[\text{ran}\,A^*] = (\ker A)^\perp$. But $A^*Ak = |A|^2k = |A|h$, where $h = |A|k$. That is, $\{|A|h: h \in \mathscr{H}\}$ is dense in $(\ker A)^\perp$. If $W: \text{ran}|A| \to \text{ran}\,A$ is defined by

3.13 $$W(|A|h) = Ah,$$

then (3.12) implies that W is a well-defined isometry. Thus W extends to an isometry $W: (\ker A)^{\perp} \to \mathrm{cl}(\mathrm{ran}\, A)$. If $Wh = 0$ for h in $\ker A$, W is a partial isometry. By (3.13), $W|A| = A$.

For the uniqueness, note that $A^*A = PU^*UP$. Now $U^*U = E \equiv$ the projection onto the initial space of U (Exercise 16), $(\ker A)^{\perp}$. But $\ker A \supseteq \ker P$, so $(\ker A)^{\perp} \subseteq (\ker P)^{\perp} = \mathrm{cl}(\mathrm{ran}\, P)$, since $P = P^*$. Hence $EP = P$. Thus $A^*A = PEP = P^2$. By the uniqueness of the positive square root, $P = |A|$. Since $A = U|A|$, $U|A|h = Ah = W|A|h$. That is, U and W agree on a dense subset of their common initial space. Hence $U = W$. ∎

EXERCISES

1. Prove the uniqueness statement in Proposition 3.4 for the case that $\mathscr{A}$ is abelian.

2. Prove Proposition 3.5.

3. Let $A \in \mathscr{B}(L^2(0,1))$ be defined by $(Af)(t) = tf(t)$. Show that $A \geq 0$ and find $A^{1/n}$.

4. Let (X, Ω, μ) be a σ-finite measure space, let $\phi \in L^{\infty}(X, \Omega, \mu)$, and define M_{ϕ} as in Theorem II.1.5. Show that $M_{\phi} \geq 0$ if and only if $\phi(x) \geq 0$ a.e. $[\mu]$. What is $M_{\phi}^{1/n}$? If $M_{\phi} \in \mathrm{Re}\,\mathscr{B}(\mathscr{H})$, find the positive and negative parts of M_{ϕ}.

5. Find an example of a positive operator on a Hilbert space that has a nonhermitian square root.

6. If $a \in \mathrm{Re}\,\mathscr{A}$, show that $|a| \equiv (a^2)^{1/2} = a_+ + a_-$.

7. If $a \in \mathscr{A}_+$, show that $x^*ax \in \mathscr{A}_+$ for every x in $\mathscr{A}$.

8. If $a, b \in \mathscr{A}, 0 \leq a \leq b$, and a is invertible, then b is invertible and $b^{-1} \leq a^{-1}$.

9. If $a, b \in \mathrm{Re}\, A$, $a \leq b$, and $ab = ba$, then $f(a) \leq f(b)$ for every increasing continuous function f on $\mathbb{R}$.

10. If $a \in \mathrm{Re}\,\mathscr{A}$ and $\|a\| \leq 1$, show that a is the sum of two unitaries. (Hint: First solve this for $\mathscr{A} = \mathbb{C}$.)

11. If $\alpha > 0$, define $f_{\alpha}: (-\alpha^{-1}, \infty) \to \mathbb{R}$ by $f_{\alpha}(t) = t/(1 + \alpha t) = \alpha^{-1}[1 - (1 + \alpha t)^{-1}]$. Show:
   (a) If $a \leq b$ in $\mathscr{A}$, $f_{\alpha}(a) \leq f_{\alpha}(b)$ for all $\alpha > 0$;
   (b) $f_{\alpha}(t) < \min\{t, \alpha^{-1}\}$ for $t > 0$;
   (c) $\lim_{\alpha \to 0} f_{\alpha}(t) = t$ uniformly on bounded intervals in $[0, \infty)$;
   (d) if $\alpha \leq \beta$, $f_{\alpha} \leq f_{\beta}$;
   (e) $f_{\alpha} \circ f_{\beta} = f_{\alpha+\beta}$;
   (f) $\lim_{\alpha \to \infty} \alpha f_{\alpha}(t) = 1$ uniformly on bounded intervals in $[0, \infty)$.

12. If $a, b \in \mathscr{A}_+$ and $a \leq b$, show that $a^{\beta} \leq b^{\beta}$ for $0 \leq \beta \leq 1$. ($a^{\beta} = f(a)$ where $f(t) = t^{\beta}$.) (Hint: Let f_{α} be as in Exercise 11 and show that $\int_0^{\infty} f_{\alpha}(t)\alpha^{-\beta}\, d\alpha = \gamma t^{\beta}$ where $\gamma > 0$. Use the definition of the improper integral and the functional calculus.)

13. Give an example of a C^*-algebra $\mathscr{A}$ and positive elements a, b in $\mathscr{A}$ such that $a \le b$ but $b^2 - a^2 \notin \mathscr{A}_+$.

14. Let $\mathscr{A} = \mathscr{B}(l^2)$, let $a = $ the unilateral shift on l^2, and let $b = a^*$. Show that $\sigma(ab) \ne \sigma(ba)$.

15. Let $W \in \mathscr{B}(\mathscr{H})$ and show that the following statements are equivalent: (a) W is a partial isometry; (b) W^* is a partial isometry; (c) W^*W is a projection; (d) WW^* is a projection; (e) $WW^*W = W$; (f) $W^*WW^* = W^*$.

16. If W is a partial isometry, show that W^*W is the projection onto the initial space of W and WW^* is the projection onto the final space of W.

17. If W_1, W_2 are partial isometries, define $W_1 \precsim W_2$ to mean that $W_1^*W_1 \le W_2^*W_2$, $W_1W_1^* \le W_2W_2^*$, and $W_2h = W_1h$ whenever h is in the initial space of W_1. Show that $\precsim$ is a partial ordering on the set of partial isometries and that a partial isometry W is a maximal element in this ordering if and only if either W or W^* is an isometry.

18. Using the terminology of Exercise 17, show that the extreme points of ball $\mathscr{B}(\mathscr{H})$ are the maximal partial isometries.

19. Find the polar decomposition of each of the following operators: (a) M_ϕ as defined in (II.1.5); (b) the unilateral shift; (c) the weighted unilateral shift $[A(x_1, x_2, \ldots) = (0, \alpha_1 x_1, \alpha_2 x_2, \ldots)$ for x in l^2 and $\sup_n |\alpha_n| < \infty]$ with non-zero weights; (d) $A \oplus \alpha$ (in terms of the polar decomposition of A).

20. Let $A \in \mathscr{B}(\mathscr{H})$ such that $\ker A = (0)$ and $A \ge 0$ and define S on $\mathscr{K} = \mathscr{H} \oplus \mathscr{H} \oplus \cdots$ by $S(h_1, h_2, \ldots) = (0, Ah_1, Ah_2, \ldots)$. Find the polar decomposition of S, $S = W|S|$, and show that $S = |S|W$.

21. Show that the parts of the polar decomposition of a normal operator commute.

22. If $A \in \mathscr{B}(\mathscr{H})$, show that there is a positive operator P and a partial isometry W such that $A = PW$. Discuss the uniqueness of P and W.

23. If A is normal and $\sigma(A) \cap \{re^{i\theta}: r \ge 0 \text{ and } \alpha \le \theta \le \beta\} = \square$, where $0 < \beta - \alpha < 2\pi$, show that the parts of the polar decomposition of A belong to $C^*(A)$.

24. Give an example of a normal operator A such that the partial isometry in the polar decomposition of A does not belong to $C^*(A)$.

§4*. Ideals and Quotients for C^*-Algebras

We begin with a basic result.

4.1. Proposition. *If I is a closed left or right ideal in the C^*-algebra $\mathscr{A}$, $a \in I$ with $a = a^*$, and if $f \in C(\sigma(a))$ with $f(0) = 0$, then $f(a) \in I$.*

PROOF. Note that if I is proper, then $0 \in \sigma(a)$ since a cannot be invertible. Since $\sigma(a) \subseteq \mathbb{R}$, the Weierstrass Theorem implies there is a sequence $\{p_n\}$

of polynomials such that $p_n(t) \to f(t)$ uniformly for t in $\sigma(a)$. Hence $p_n(0) \to f(0) = 0$. Thus $q_n(t) = p_n(t) - p_n(0) \to f(t)$ uniformly on $\sigma(a)$ and $q_n(0) = 0$ for all n. Thus $q_n(a) \in I$ and by the functional calculus, $\|q_n(a) - f(a)\| \to 0$. Hence $f(a) \in I$.   ∎

4.2.   Corollary. *If I is a closed left or right ideal, $a \in I$ with $a = a^*$, then a_+, a_-, $|a|$, and $|a|^{1/2} \in I$.*

Note that if I is a left ideal of $\mathscr{A}$, then $\{a^*: a \in I\}$ is a right ideal. Therefore a left ideal I is an ideal if $a^* \in I$ whenever $a \in I$.

4.3.   Theorem. *If I is a closed ideal in the C^*-algebra $\mathscr{A}$, then $a^* \in I$ whenever $a \in I$.*

PROOF. Fix a in I. Thus $a^*a \in I$ since I is an ideal. The idea is to construct a sequence $\{u_n\}$ of continuous functions defined on $[0, \infty)$ such that

4.4   (i) $u_n(0) = 0$ and $u_n(t) \geq 0$ for all t;

  (ii) $\|au_n(a^*a) - a\| \to 0$ as $n \to \infty$.

Note that if such a sequence $\{u_n\}$ can be constructed, then $u_n(a^*a) \geq 0$ and $u_n(a^*a) \in I$ by Proposition 4.1. Also, $u_n(a^*a)a^* \in I$ since I is an ideal and $\|u_n(a^*a)a^* - a^*\| = \|au_n(a^*a) - a\| \to 0$ by (ii). Thus $a^* \in I$ whenever $a \in I$. It remains to construct the sequence $\{u_n\}$.

Note that

$$\|au_n(a^*a) - a\|^2$$
$$= \|[au_n(a^*a) - a]^*[au_n(a^*a) - a]\|$$
$$= \|u_n(a^*a)a^*au_n(a^*a) - a^*au_n(a^*a) - u_n(a^*a)a^*a + a^*a\|.$$

If $b = a^*a$, then the fact that $bu_n(b) = u_n(b)b$ implies that $\|au_n(a^*a) - a\|^2 = \|f_n(b)\| \leq \sup\{|f_n(t)|: t \geq 0\}$, where $f_n(t) = tu_n(t)^2 - 2tu_n(t) + t = t[u_n(t) - 1]^2$. If $u_n(t) = nt$ for $0 \leq t \leq n^{-1}$ and $u(t) = 1$ for $t \geq n^{-1}$, then it is seen that $\sup\{|f_n(t)|: t \geq 0\} = 4/27n \to 0$ as $n \to \infty$; so (4.4) is satisfied.   ∎

Notice that the construction of the sequence $\{u_n\}$ satisfying (4.4) actually proves more. It shows that there is a "local" approximate identity. That is, the proof of the preceding theorem shows that the following holds.

4.5.   Proposition. *If $\mathscr{A}$ is a C^*-algebra and I is an ideal of $\mathscr{A}$, then for every a in I there is a sequence $\{e_n\}$ of positive elements in I such that:*

(a) $e_1 \leq e_2 \leq \cdots$ *and* $\|e_n\| \leq 1$ *for all n;*
(b) $\|ae_n - a\| \to 0$ *as $n \to \infty$.*

In the preceding proposition the sequence $\{e_n\}$ depends on the element $\{a\}$. It is also true that there is a positive increasing net $\{e_i\}$ in I such that

$\|e_i a - a\| \to 0$ and $\|ae_i - a\| \to 0$ for every a in I (see p. 36 of Arveson [1976]).

We turn now to an important consequence of Theorem 4.3.

4.6. Theorem. *If $\mathscr{A}$ is a C*-algebra and I is a closed ideal of $\mathscr{A}$, then for each $a + I$ in $\mathscr{A}/I$ define $(a + I)^* = a^* + I$. Then $\mathscr{A}/I$ with its quotient norm is a C*-algebra.*

To prove (4.6), a lemma is needed.

4.7. Lemma. *If I is an ideal in a C*-algebra $\mathscr{A}$ and $a \in \mathscr{A}$, then $\|a + I\| = \inf\{\|a - ax\|: x \in I, x \geq 0, \text{ and } \|x\| \leq 1\}$.*

PROOF. If $(\text{ball } I)_+ = \{x \in \text{ball } I: x \geq 0\}$, then clearly $\|a + I\| \leq \inf\{\|a - ax\|: x \in (\text{ball } I)_+\}$ since $aI \subseteq I$. Let $y \in I$ and let $\{e_n\}$ be a sequence in $(\text{ball } I)_+$ such that $\|y - ye_n\| \to 0$ as $n \to \infty$. Now $0 \leq 1 - e_n \leq 1$, so $\|(a + y)(1 - e_n)\| \leq \|a + y\|$. Hence

$$\|a + y\| \geq \liminf \|(a + y)(1 - e_n)\|$$
$$= \liminf \|(a - ae_n) + (y - ye_n)\|$$
$$= \liminf \|a - ae_n\|$$

since $\|y - ye_n\| \to 0$. Thus $\|a + y\| \geq \inf_n \|a - ae_n\| \geq \inf\{\|a - ax\|: x \in (\text{ball } I)_+\}$. Taking the infimum over all y in I gives the desired remaining inequality. $\blacksquare$

PROOF OF THEOREM 4.6. The only difficult part of this proof is to show that $\|a + I\|^2 = \|a^*a + I\|$ for every a in $\mathscr{A}$. If $x \in I$, then

$$\|a + x\|^2 = \|(a + x)^*(a + x)\|$$
$$= \|a^*a + a^*x + x^*a + x^*x\| \geq \inf\{\|a^*a + y\|: y \in I\}$$
$$= \|a^*a + I\|$$

since $a^*x + x^*a + x^*x \in I$ whenever $x \in I$ (4.3). On the other hand, the preceding lemma gives that

$$\|a + I\|^2 = \inf\{\|a - ax\|^2: x \in (\text{ball } I)_+\}$$
$$= \inf\{\|a(1 - x)\|^2: x \in (\text{ball } I)_+\}$$
$$= \inf\{\|(1 - x)a^*a(1 - x)\|: x \in (\text{ball } I)_+\}$$
$$\leq \inf\{\|a^*a(1 - x)\|: x \in (\text{ball } I)_+\}$$
$$= \inf\{\|a^*a - a^*ax\|: x \in (\text{ball } I)_+\}$$
$$= \|a^*a + I\|. \quad \blacksquare$$

If $\mathscr{A}, \mathscr{B}$ are C*-algebras with ideals I, J, respectively, and $\rho: \mathscr{A} \to \mathscr{B}$ is a *-homomorphism such that $\rho(I) \subseteq J$, then ρ induces a *-homomorphism $\tilde{\rho}$: $\mathscr{A}/I \to \mathscr{B}/J$ defined by $\tilde{\rho}(a + I) = \rho(a) + J$. In particular, if $I = \ker\rho$, then $\tilde{\rho}$: $\mathscr{A}/\ker\rho \to \mathscr{B}$ is a $*$-homomorphism and $\tilde{\rho} \circ \pi = \rho$, where π: $\mathscr{A} \to \mathscr{A}/\ker\rho$ is the natural map. Keep these facts in mind when reading the proof of the next result.

4.8.   Theorem. *If $\mathscr{A}, \mathscr{B}$ are C^*-algebras and ρ: $\mathscr{A} \to \mathscr{B}$ is a $*$-homomorphism, then $\|\rho(a)\| \leq \|a\|$ for all a and $\operatorname{ran}\rho$ is closed in $\mathscr{B}$. If ρ is a $*$-monomorphism, then ρ is an isometry.*

PROOF. The fact that $\|\rho(a)\| \leq \|a\|$ is a restatement of (1.11d). Now assume that ρ is a $*$-monomorphism. As in the proof of (1.11d), it suffices to assume that $\mathscr{A}$ and $\mathscr{B}$ have identities and $\rho(1) = 1$. (Why?)

If $a \in \mathscr{A}$ and $a = a^*$, then it is easy to see that $\rho(a) = \rho(a)^*$ and $\sigma(\rho(a)) \subseteq \sigma(a)$. If $\sigma(\rho(a)) \neq \sigma(a)$, there is a continuous function f on $\sigma(a)$ such that $f(t) = 0$ for all t in $\sigma(\rho(a))$ but f is not identically zero on $\sigma(a)$. Thus $f(\rho(a)) = 0$, but $f(a) \neq 0$. Let $\{p_n\}$ be polynomials such that $p_n(t) \to f(t)$ uniformly on $\sigma(a)$. Thus $p_n(a) \to f(a)$ and $p_n(\rho(a)) \to f(\rho(a)) = 0$. But $p_n(\rho(a)) = \rho(p_n(a)) \to \rho(f(a))$. Thus $\rho(f(a)) = f(\rho(a)) = 0$. Since ρ was assumed injective, $f(a) = 0$, a contradiction. Hence $\sigma(a) = \sigma(\rho(a))$ if $a = a^*$. Thus by (1.11e), $\|a\| = r(a) = r(\rho(a)) = \|\rho(a)\|$ if $a = a^*$. But then for arbitrary a, $\|a\|^2 = \|a^*a\| = \|\rho(a^*a)\| = \|\rho(a)^*\rho(a)\| = \|\rho(a)\|^2$ and ρ is an isometry.

To complete the proof let ρ: $\mathscr{A} \to \mathscr{B}$ be a $*$-homomorphism and let $\tilde{\rho}$: $\mathscr{A}/\ker\rho \to \mathscr{B}$ be the induced $*$-monomorphism. So $\tilde{\rho}$ is an isometry and hence $\operatorname{ran}\tilde{\rho}$ is closed. But $\operatorname{ran}\tilde{\rho} = \operatorname{ran}\rho$. ∎

We turn now to some specific examples of C^*-algebras and their ideals.

4.9.   Proposition. *If X is compact and I is a closed ideal of $C(X)$, then there is a closed subset F of X such that $I = \{f \in C(X): f(x) = 0$ for all x in $F\}$. Moreover, $C(X)/I$ is isometrically isomorphic to $C(F)$.*

PROOF. Let $F = \{x \in X: f(x) = 0$ for all f in $I\}$, so F is a closed subset of x. If $\mu \in M(X)$ and $\mu \perp I$, then $\int |f|^2 d\mu = 0$ for every f in I since $|f|^2 = f\bar{f} \in I$ whenever $f \in I$. Thus each f must vanish on the support of μ; hence $|\mu|(X \setminus F) = 0$. Conversely, if $\mu \in M(X)$ and the support of μ is contained in F, $\int f d\mu = 0$ for every f in I. Thus $I^{\perp} = \{\mu \in M(X): |\mu|(X \setminus F) = 0\}$. Since I is closed, $I = {}^{\perp}(I^{\perp}) = \{f \in C(X): f(x) = 0$ for all x in $F\}$. The remainder of the proof is left to the reader. ∎

4.10.   Proposition. *If I is a closed ideal of $\mathscr{B}(\mathscr{H})$, then $I \supseteq \mathscr{B}_0(\mathscr{H})$ or $I = (0)$.*

PROOF. Suppose $I \neq (0)$ and let T be a nonzero operator in I. Thus there are vectors f_0, f_1 in $\mathscr{H}$ such that $Tf_0 = f_1 \neq 0$. Let g_0, g_1 be arbitrary nonzero vectors in $\mathscr{H}$. Define A: $\mathscr{H} \to \mathscr{H}$ by letting $Ah = \|g_0\|^{-2}\langle h, g_0\rangle f_0$.

Then $Ag_0 = f_0$ and $Ah = 0$ if $h \perp g_0$. Define $B: \mathcal{H} \to \mathcal{H}$ by letting $Bh = \|f_1\|^{-2}\langle h, f_1 \rangle g_1$. So $Bf_1 = g_1$ and $Bh = 0$ if $h \perp f_1$. Thus $BTAh = 0$ if $h \perp g_0$ and $BTAg_0 = g_1$. Hence for any pair of nonzero vectors g_0, g_1 in $\mathcal{H}$ the rank-one operator that takes g_0 to g_1 and is zero on $[g_0]^\perp$ belongs to I. From here it easily follows that I contains all finite-rank operators. Since I is closed, $I \supseteq \mathcal{B}_0(\mathcal{H})$. ∎

It will be shown in (IX.4.2), after we have the spectral theorem, that if I is a closed ideal in $\mathcal{B}(\mathcal{H})$ and $\mathcal{H}$ is separable, then $I = (0)$, $\mathcal{B}_0(\mathcal{H})$, or $\mathcal{B}(\mathcal{H})$.

Exercises

1. Complete the proof of Proposition 4.9.

2. Show that $M_n(\mathbb{C})$ has no nontrivial ideals. Find all of the left ideals.

3. If α is an infinite cardinal number, let $I = \{ A \in \mathcal{B}(\mathcal{H}): \dim \mathrm{cl}(\mathrm{ran}\, A) \le \alpha \}$. Show that I_α is a closed ideal in $\mathcal{B}(\mathcal{H})$.

4. Let S be the unilateral shift on l^2. Show that $C^*(S) \supseteq \mathcal{B}_0(l^2)$ and $C^*(S)/\mathcal{B}_0(l^2)$ is abelian. Show that the maximal ideal space of $C^*(S)/\mathcal{B}_0(l^2)$ is homeomorphic to $\partial \mathbb{D}$.

5. If V is the Volterra operator on $L^2(0,1)$, show that $C^*(V) = \mathbb{C} + \mathcal{B}_0(L^2(0,1))$.

6. If $\mathcal{A}$ is a C^*-algebra, I is a closed ideal of $\mathcal{A}$, and $\mathcal{B}$ is a C^*-subalgebra of $\mathcal{A}$, show that the C^*-algebra generated by $I \cup \mathcal{B}$ is $I + \mathcal{B}$.

7. If $\mathcal{A}$ is a C^*-algebra and I and J are closed ideals in $\mathcal{A}$, show that $I + J$ is a closed ideal of $\mathcal{A}$.

§5*. Representations of C^*-Algebras and the Gelfand–Naimark–Segal Construction

5.1.   Definition. A *representation* of a C^*-algebra is a pair $(\pi, \mathcal{H})$, where $\mathcal{H}$ is a Hilbert space and $\pi: \mathcal{A} \to \mathcal{B}(\mathcal{H})$ is a $*$-homomorphism. If $\mathcal{A}$ has an identity, it is assumed that $\pi(1) = 1$. (The algebras considered in this book are assumed to have an identity. This proviso is given for the reader's convenience when consulting the literature.) Often $\mathcal{H}$ is deleted and we say that π is a representation.

5.2.   Example. If $\mathcal{H}$ is a Hilbert space and $\mathcal{A}$ is a C^*-subalgebra of $\mathcal{B}(\mathcal{H})$, then the inclusion map $\mathcal{A} \hookrightarrow \mathcal{B}(\mathcal{H})$ is a representation.

5.3.   Example. If n is any cardinal number and $\mathcal{H}$ is a Hilbert space, let $\mathcal{H}^{(n)}$ denote the direct sum of $\mathcal{H}$ with itself n times. If $A \in \mathcal{B}(\mathcal{H})$, then $A^{(n)}$ is the direct sum of A with itself n times; so $A^{(n)} \in \mathcal{B}(\mathcal{H}^{(n)})$ and

$\|A^{(n)}\| = \|A\|$. The operator $A^{(n)}$ is called the *inflation* of A. If $\pi: \mathscr{A} \to \mathscr{B}(\mathscr{H})$ is a representation, the *inflation* of π is the map $\pi^{(n)}: \mathscr{A} \to \mathscr{B}(\mathscr{H}^{(n)})$ defined by $\pi^{(n)}(a) = \pi(a)^{(n)}$ for all a in $\mathscr{A}$.

5.4. Example. If (X, Ω, μ) is a σ-finite measure space and $\mathscr{H} = L^2(\mu)$, then $\pi: L^\infty(\mu) \to \mathscr{B}(\mathscr{H})$ defined by $\pi(\phi) = M_\phi$ is a representation.

5.5. Example. If X is a compact space and μ is a positive Borel measure on X, then $\pi: C(X) \to \mathscr{B}(L^2(\mu))$ defined by $\pi(f) = M_f$ is a representation.

5.6. Definition. A representation π of a C^*-algebra $\mathscr{A}$ is *cyclic* if there is a vector e in $\mathscr{H}$ such that $\mathrm{cl}[\pi(\mathscr{A})e] = \mathscr{H}$; e is said to be a *cyclic vector* for the representation π.

Note that the representations in Examples 5.4 and 5.5 are cyclic (Exercises 2 and 3). Also, the identity representation $i: \mathscr{B}(\mathscr{H}) \to \mathscr{B}(\mathscr{H})$ is cyclic and every nonzero vector is a cyclic vector for this representation. If $\mathscr{A} = \mathbb{C} + \mathscr{B}_0(\mathscr{H})$, then the identity representation is cyclic. On the other hand, if $n \geq 2$, then the inflation $\pi^{(n)}$ of a representation of $C(X)$ is never cyclic (Exercise 4).

There is another way to obtain representations.

5.7. Definition. If $\{(\pi_i, \mathscr{H}_i): i \in I\}$ is a family of representations of $\mathscr{A}$, then the *direct sum* of this family is the representation $(\pi, \mathscr{H})$, where $\mathscr{H} = \oplus_i \mathscr{H}_i$ and $\pi(a) = \{\pi_i(a)\}$ for every a in $\mathscr{A}$.

Note that since $\|\pi_i(a)\| \leq \|a\|$ for every i (4.8), $\pi(a)$ is a bounded operator on $\mathscr{H}$. It is easy to check that π is a representation.

5.8. Example. Let X be a compact space and let $\{\mu_n\}$ be a sequence of measures on X. For each n let $\pi_n: C(X) \to \mathscr{B}(L^2(\mu_n))$ be defined by $\pi_n(f) = M_f$ on $L^2(\mu_n)$. Then $\pi = \oplus_n \pi_n$ is a representation. If the measures $\{\mu_n\}$ are pairwise mutually singular, then π is equivalent (below) to the representation $f \to M_f$ of $C(X) \to \mathscr{B}(L^2(\mu))$, where $\mu = \sum_{n=1}^\infty \mu_n / 2^n \|\mu_n\|$ (Exercise 5).

The concept of equivalence for representations is that of unitary equivalence. That is, two representations of a C^*-algebra $\mathscr{A}$, $(\pi_1, \mathscr{H}_1)$ and $(\pi_2, \mathscr{H}_2)$, are *equivalent* if there is an isomorphism $U: \mathscr{H}_1 \to \mathscr{H}_2$ such that $U\pi_1(a)U^{-1} = \pi_2(a)$ for every a in $\mathscr{A}$. The importance of cyclic representations arises from the fact, given in the next result, that every representation is equivalent to the direct sum of cyclic representations.

5.9. Theorem. *If π is a representation of the C^*-algebra $\mathscr{A}$, then there is a family of cyclic representations $\{\pi_i\}$ of $\mathscr{A}$ such that π and $\oplus_i \pi_i$ are equivalent.*

PROOF. Let $\mathscr{E} =$ the collection of all subsets E of nonzero vectors in $\mathscr{H}$ such that $\pi(\mathscr{A})e \perp \pi(\mathscr{A})f$ for e, f in E with $e \neq f$. Order $\mathscr{E}$ by inclusion. An application of Zorn's Lemma implies that $\mathscr{E}$ has a maximal element E_0. Let $\mathscr{H}_0 = \oplus\{cl[\pi(\mathscr{A})e]: e \in E_0\}$. If $h \in \mathscr{H} \ominus \mathscr{H}_0$, then $0 = \langle \pi(a)e, h \rangle$ for every a in $\mathscr{A}$ and e in E_0. So if $a, b \in \mathscr{A}$ and $e \in E_0$, $0 = \langle \pi(b^*a)e, h \rangle = \langle \pi(b)^*\pi(a)e, h \rangle = \langle \pi(a)e, \pi(b)h \rangle$. That is, $\pi(\mathscr{A})e \perp \pi(\mathscr{A})h$ for all e in E_0. Hence $E_0 \cup \{h\} \in \mathscr{E}$, contradicting the maximality of E_0. Therefore $\mathscr{H} = \mathscr{H}_0$.

For e in E_0 let $\mathscr{H}_e = cl[\pi(\mathscr{A})e]$. If $a \in \mathscr{A}$, clearly $\pi(a)\mathscr{H}_e \subseteq \mathscr{H}_e$. Since $a^* \in \mathscr{A}$ and $\pi(a)^* = \pi(a^*)$, $\mathscr{H}_e$ reduces $\pi(a)$. So if $\pi_e: \mathscr{A} \to \mathscr{B}(\mathscr{H}_e)$ is defined by $\pi_e(a) = \pi(a)|\mathscr{H}_e$, π_e is a representation of a. Clearly $\pi = \oplus\{\pi_e: e \in E_0\}$. ∎

In light of the preceding theorem, it becomes important to understand cyclic representations. To do this, let $\pi: \mathscr{A} \to \mathscr{B}(\mathscr{H})$ be a cyclic representation with cyclic vector e. Define $f: \mathscr{A} \to \mathbb{C}$ by $f(a) = \langle \pi(a)e, e \rangle$. Note that f is a bounded linear functional on $\mathscr{A}$ with $\|f\| \leq \|e\|^2$. Since $f(1) = \|e\|^2$, $\|f\| = \|e\|^2$. Moreover, $f(a^*a) = \langle \pi(a^*a)e, e \rangle = \langle \pi(a)^*\pi(a)e, e \rangle = \|\pi(a)e\|^2 \geq 0$.

5.10.   Definition. If $\mathscr{A}$ is a C^*-algebra, a linear functional $f: \mathscr{A} \to \mathbb{C}$ is *positive* if $f(a) \geq 0$ whenever $a \in \mathscr{A}_+$. A *state* on $\mathscr{A}$ is a positive linear functional on $\mathscr{A}$ of norm 1.

5.11.   Proposition. *If f is a positive linear functional on a C^*-algebra $\mathscr{A}$, then*

$$|f(y^*x)|^2 \leq f(y^*y)f(x^*x)$$

for every x, y in $\mathscr{A}$.

PROOF. If $[x, y] = f(y^*x)$ for x, y in $\mathscr{A}$, then $[\cdot, \cdot]$ is a semi-inner product on $\mathscr{A}$. The proposition now follows by the CBS inequality (I.1.4). ∎

5.12.   Corollary. *If f is a positive linear functional on the C^*-algebra $\mathscr{A}$, then f is bounded and $\|f\| = f(1)$.*

5.13.   Example. If X is a compact space, then the positive linear functionals on $C(X)$ correspond to the positive measures on X. The states correspond to the probability measures on X.

As was shown above, each cyclic representation gives rise to a positive linear functional. It turns out that each positive linear functional gives rise to a cyclic representation.

5.14. Gelfand–Naimark–Segal Construction. *Let $\mathscr{A}$ be a C^*-algebra with identity.*

(a) *If f is a positive linear functional on $\mathscr{A}$, then there is a cyclic representation $(\pi_f, \mathscr{H}_f)$ of $\mathscr{A}$ with cyclic vector e such that $f(a) = \langle \pi_f(a)e, e \rangle$ for all a in $\mathscr{A}$.*

(b) *If $(\pi, \mathscr{H})$ is a cyclic representation of $\mathscr{A}$ with cyclic vector e and $f(a) \equiv \langle \pi(a)e, e \rangle$ and if $(\pi_f, \mathscr{H}_f)$ is constructed as in (a), then π and π_f are equivalent.*

Before beginning the proof, it will be helpful if the theorem is examined when $\mathscr{A}$ is abelian. So let $\mathscr{A} = C(X)$ where X is compact. If f is a positive linear functional on $\mathscr{A}$, then there is a positive measure μ on X such that $f(\phi) = \int \phi \, d\mu$ for all ϕ in $\mathscr{A}$. The representation $(\pi_f, \mathscr{H}_f)$ is the one obtained by letting $\mathscr{H}_f = L^2(\mu)$ and $\pi_f(\phi) = M_\phi$, but let us look a little closer. One way to obtain $L^2(\mu)$ from $C(X)$ and μ is to let $\mathscr{L} = \{\phi \in C(X): \int |\phi|^2 \, d\mu = 0\}$. Note that $\mathscr{L}$ is an ideal in $C(X)$. Define an inner product on $C(X)/\mathscr{L}$ by $\langle \phi + \mathscr{L}, \psi + \mathscr{L} \rangle = \int \phi \bar{\psi} \, d\mu$. The completion of $C(X)/\mathscr{L}$ with respect to this inner product is $L^2(\mu)$.

To see part (b) in the abelian case, let $\pi: C(X) \to \mathscr{B}(\mathscr{H})$ be a cyclic representation with cyclic vector e. Let μ be the positive measure on X such that $\int \phi \, d\mu = \langle \pi(\phi)e, e \rangle = f(\phi)$. Now define $U_1: C(X) \to \mathscr{H}$ by $U_1(\phi) = \pi(\phi)e$. Note that U_1 is linear and has dense range. If $\mathscr{L}$ is as in the preceding paragraph and $\phi \in \mathscr{L}$, then $\|U_1(\phi)\|^2 = \langle \pi(\phi)e, \pi(\phi)e \rangle = \langle \pi(\phi^*\phi)e, e \rangle = \int |\phi|^2 \, d\mu = 0$. So $U_1 \mathscr{L} = 0$. Thus U_1 induces a linear map $U: C(X)/\mathscr{L} \to \mathscr{H}$ where $U(\phi + \mathscr{L}) = \pi(\phi)e$. If $\langle \phi + \mathscr{L}, \psi + \mathscr{L} \rangle \equiv \int \phi \bar{\psi} \, d\mu$, then $\langle U(\phi + \mathscr{L}), U(\psi + \mathscr{L}) \rangle = \langle \pi(\phi)e, \pi(\psi)e \rangle = \langle \pi(\phi \psi^*)e, e \rangle = \int \phi \bar{\psi} \, d\mu = \langle \phi + \mathscr{L}, \psi + \mathscr{L} \rangle$. Thus U extends to an isomorphism U from the completion of $\mathscr{A}/\mathscr{L} = L^2(\mu)$ onto $\mathscr{H}$. So $U: L^2(\mu) \to \mathscr{H}$ and if $\phi \in C(X)$ and we think of $C(X)$ as a (dense) subset of $L^2(\mu)$, $U\phi = \pi(\phi)e$. If $\phi, \psi \in C(X)$, then $UM_\phi \psi = U(\phi \psi) = \pi(\phi \psi)e = \pi(\phi)\pi(\psi)e = \pi(\phi)U(\psi)$; that is, $UM_\phi = \pi(\phi)U$ on a dense subset of $L^2(\mu)$ and, hence, $UM_\phi = \pi(\phi)U$ for every ϕ in $C(X)$. In other words, π is equivalent to the representation $\phi \mapsto M_\phi$.

PROOF OF THEOREM 5.14. Let f be a positive linear functional on $\mathscr{A}$ and put $\mathscr{L} = \{x \in \mathscr{A}: f(x^*x) = 0\}$. It is easy to see that $\mathscr{L}$ is closed in $\mathscr{A}$. Also if $a \in \mathscr{A}$ and $x \in \mathscr{L}$, then (5.11) implies that

$$f\big((ax)^*(ax)\big)^2 = f\big(x^*(a^*ax)\big)^2$$
$$\leq f(x^*x)f(x^*a^*aa^*ax)$$
$$= 0.$$

That is, $\mathscr{L}$ is a closed left ideal in $\mathscr{A}$. Now consider $\mathscr{A}/\mathscr{L}$ as a vector space. (Since $\mathscr{L}$ is only a left ideal, $\mathscr{A}/\mathscr{L}$ is not an algebra.) For x, y in $\mathscr{A}$, define

$$\langle x + \mathscr{L}, y + \mathscr{L} \rangle = f(y^*x).$$

It is left as an exercise for the reader to show that $\langle \cdot, \cdot \rangle$ is a well-defined inner product on $\mathscr{A}/\mathscr{L}$. Let $\mathscr{H}_f$ be the completion of $\mathscr{A}/\mathscr{L}$ with respect to the norm defined on $\mathscr{A}/\mathscr{L}$ by this inner product.

Because $\mathscr{L}$ is a left ideal of $\mathscr{A}$, $x + \mathscr{L} \mapsto ax + \mathscr{L}$ is a well-defined linear transformation on $\mathscr{A}/\mathscr{L}$. Also, $\|ax + \mathscr{L}\|^2 = \langle ax + \mathscr{L}, ax + \mathscr{L} \rangle = f(x^*a^*ax)$. Now if $\|a^*a\|$ is considered as an element of $\mathscr{A}$ (it is a multiple of the identity), then an appeal to the functional calculus for a^*a shows that $\|a^*a\| - a^*a \geq 0$. Hence (Exercise 3.7) $0 \leq x^*(\|a^*a\| - a^*a)x = \|a\|^2 x^*x - x^*a^*ax$; that is, $x^*a^*ax \leq \|a\|^2 x^*x$. Therefore $\|ax + \mathscr{L}\|^2 \leq \|a\|^2 f(x^*x) = \|a\|^2 \|x + \mathscr{L}\|^2$. Thus if $\pi_f(a) \colon \mathscr{A}/\mathscr{L} \to \mathscr{A}/\mathscr{L}$ is defined by $\pi_f(a)(x + \mathscr{L}) = ax + \mathscr{L}$, $\pi_f(a)$ is a bounded linear operator with $\|\pi_f(a)\| \leq \|a\|$. Hence $\pi_f(a)$ extends to an element of $\mathscr{B}(\mathscr{H}_f)$. It is left to the reader to verify that $\pi_f \colon \mathscr{A} \to \mathscr{B}(\mathscr{H}_f)$ is a representation.

Put $e = 1 + \mathscr{L}$ in $\mathscr{H}_f$. Then $\pi_f(\mathscr{A})e = \{a + \mathscr{L} \colon a \in \mathscr{A}\} = \mathscr{A}/\mathscr{L}$ which, by definition, is dense in $\mathscr{H}_f$. Thus e is a cyclic vector for π_f. [Also note that $\langle \pi_f(a)e, e \rangle = f(a)$.] This proves (a).

Now let $(\pi, \mathscr{H})$, e, and f be as in (b) and let $(\pi_f, \mathscr{H}_f)$ be the representation constructed. Let e_f be the cyclic vector for π_f so that $f(a) = \langle \pi_f(a)e_f, e_f \rangle$ for all a in $\mathscr{A}$. Hence $\langle \pi_f(a)e_f, e_f \rangle = \langle \pi(a)e, e \rangle$ for all a in $\mathscr{A}$. Define U on the dense manifold $\pi_f(\mathscr{A})e_f$ in $\mathscr{H}_f$ by $U\pi_f(a)e_f = \pi(a)e$. Note that $\|\pi(a)e\|^2 = \langle \pi(a)e, \pi(a)e \rangle = \langle \pi(a^*a)e, e \rangle = \langle \pi_f(a^*a)e_f, e_f \rangle = \|\pi_f(a)e_f\|^2$. This implies that U is well defined and an isometry. Thus U extends to an isomorphism of $\mathscr{H}_f$ onto $\mathscr{H}$. If $x, a \in \mathscr{A}$, then $U\pi_f(a)\pi_f(x)e_f = U\pi_f(ax)e_f = \pi(a)\pi(x)e = \pi(a)U\pi_f(x)e_f$. Thus $\pi(a)U = U\pi_f(a)$ so that π and π_f are equivalent. ∎

The Gelfand–Naimark–Segal construction is often called the GNS construction.

It is not difficult to show that if f is a positive linear functional on $\mathscr{A}$ and $\alpha > 0$, then the representations π_f and $\pi_{\alpha f}$ are equivalent (Exercise 8). So it is appropriate to only consider the cyclic representations corresponding to states. If $\mathscr{A}$ is a C^*-algebra, let $S_{\mathscr{A}}$ = the collection of all states on $\mathscr{A}$. Note that $S_{\mathscr{A}} \subseteq$ ball $\mathscr{A}^*$. $S_{\mathscr{A}}$ is called the *state space* of $\mathscr{A}$.

5.15. Proposition. *If $\mathscr{A}$ is a C^*-algebra with identity, then $S_{\mathscr{A}}$ is a weak* compact convex subset of $\mathscr{A}^*$ and if $a \in \mathscr{A}_+$, then $\|a\| = \sup\{f(a) \colon f \in S_{\mathscr{A}}\}$ and this supremum is attained.*

PROOF. Since $S_{\mathscr{A}} \subseteq$ ball $\mathscr{A}^*$, to show that $S_{\mathscr{A}}$ is weak* compact, it suffices to show that $S_{\mathscr{A}}$ is weak* closed. The reader can supply this proof using nets. Clearly $S_{\mathscr{A}}$ is convex.

If $\mathscr{A} = C(X)$ with X compact and $f \in C(X)_+$, then there is a point x in X such that $f(x) = \|f\|$. Thus $\|f\| = \int f\, d\delta_x = \sup\{ \int f\, d\mu : \mu \in$ (ball $M(X))_+\}$. If $\mathscr{A}$ is arbitrary and $a \in \mathscr{A}_+$, then $\|a\| \geq \sup\{ f(a) : f \in S_{\mathscr{A}}\}$. Also, from the argument in the abelian case, there is a state f_1 on $C^*(a)$ such that $f_1(a) = \|a\|$. If we can show that f_1 extends to a state f on $\mathscr{A}$, the proof is complete. That this can be done is a consequence of the next result. ∎

5.16. Proposition. *Let $\mathscr{A}$, $\mathscr{B}$ be C^*-algebras with $\mathscr{B} \subseteq \mathscr{A}$. If f_1 is a state on $\mathscr{B}$, then there is a state f on $\mathscr{A}$ such that $f|\mathscr{B} = f_1$.*

PROOF. Consider the real linear spaces $\mathrm{Re}\,\mathscr{A}$ and $\mathrm{Re}\,\mathscr{B}$. If $a \in \mathscr{A}_+$, then $a \leq \|a\|$ in $\mathscr{A}$. Since $1 \in \mathrm{Re}\,\mathscr{B}$, $\mathrm{Re}\,\mathscr{B}$ has an order unit. By Corollary III.9.12, if $f_1 \in S_{\mathscr{B}}$ there is a positive linear functional f on $\mathrm{Re}\,\mathscr{A}$ such that $f|\mathrm{Re}\,\mathscr{B} = f_1$. Since $e \in \mathscr{B}$, $f(1) = f_1(1) = 1$. Now let $f(a) = f((a + a^*)/2) + if((a - a^*)/2i)$ for an arbitrary a in $\mathscr{A}$. It follows that $f \in S_{\mathscr{A}}$ and $f|\mathscr{B} = f_1$. ∎

The next result says that every C^*-algebra is isomorphic to a C^*-algebra contained in $\mathscr{B}(\mathscr{H})$ for some $\mathscr{H}$. Thus each C^*-algebra "is" an algebra of operators.

5.17. Theorem. *If $\mathscr{A}$ is a C^*-algebra, then there is a representation $(\pi, \mathscr{H})$ of $\mathscr{A}$ such that π is an isometry. If $\mathscr{A}$ is separable, then $\mathscr{H}$ can be chosen separable.*

PROOF. Let F be a weak* dense subset of $S_{\mathscr{A}}$ and let $\pi = \oplus\{\pi_f : f \in F\}$, $\mathscr{H} = \oplus\{\mathscr{H}_f : f \in F\}$. Thus $\|a\|^2 \geq \|\pi(a)\|^2 = \sup_f\|\pi_f(a)\|^2$. If e_f is the cyclic vector for π_f, then $\|e_f\|^2 = \langle e_f, e_f \rangle = \langle \pi_f(1)e_f, e_f \rangle = f(1) = 1$. Hence $\|\pi_f(a)\|^2 \geq \|\pi_f(a)e_f\|^2 = \langle \pi_f(a^*a)e_f, e_f \rangle = f(a^*a)$, and $\|a\|^2 \geq \|\pi(a)\|^2 \geq \sup\{ f(a^*a) : f \in F\}$. Since F is weak* dense in $S_{\mathscr{A}}$, Proposition 5.15 implies $\sup\{ f(a^*a) : f \in F\} = \|a^*a\| = \|a\|^2$. Hence π is an isometry.

If $\mathscr{A}$ is separable, (ball $\mathscr{A}^*$, wk*) is a compact metric space (V.5.1). Hence $S_{\mathscr{A}}$ is weak* separable so that the set F of the preceding paragraph can be chosen to be countable. Now if $f \in F$, $\pi(\mathscr{A})f$ is a separable dense submanifold in $\mathscr{H}_f$ since $\mathscr{A}$ is separable. Thus $\mathscr{H}_f$ is separable. It follows that $\mathscr{H}$ is separable. ∎

Actually, more can be said if $\mathscr{A}$ is separable. In fact, every separable C^*-algebra has a cyclic representation that is isometric (Exercise 12).

EXERCISES

1. Let $\mathscr{A}$ be a C^*-algebra with identity and let $\pi: \mathscr{A} \to \mathscr{B}(\mathscr{H})$ be a $*$-homomorphism [but don't assume that $\pi(1) = 1$]. Let $P_1 = \pi(1)$. Show that P_1 is a projection and $\mathscr{H}_1 \equiv P_1\mathscr{H}$ reduces $\pi(\mathscr{A})$. If $\pi_1(a) = \pi(a)|\mathscr{H}_1$, show that $\pi_1: \mathscr{A} \to \mathscr{B}(\mathscr{H}_1)$ is a representation.

2. Show that the representation in Example 5.4 is a cyclic representation and find all of the cyclic vectors.

3. Show that the representation in Example 5.5 is a cyclic representation and find all the cyclic vectors.

4. If X is compact and μ is a positive measure on X, let $\pi_\mu\colon C(X) \to \mathcal{B}(L^2(\mu))$ be the representation defined in Example 5.5. If μ, ν are positive measures on X show that $\pi_\mu \oplus \pi_\nu$ is cyclic if and only if $\mu \perp \nu$. If $\mu \perp \nu$, then $\pi_\mu \oplus \pi_\nu$ is equivalent to $\pi_{\mu+\nu}$. Also, $\pi_\mu^{(n)}$ is not cyclic if $n \geq 2$.

5. Verify the statements in Example 5.8.

6. If $\mathcal{A} = \mathbb{C} + \mathcal{B}_0(\mathcal{H})$ and $\pi\colon \mathcal{A} \to \mathcal{B}(\mathcal{H})$ is the identity representation, show that $\pi^{(\infty)}$ is a cyclic representation.

7. Fix a Banach limit LIM on $l^\infty(\mathbb{N})$ and let $\mathcal{H}$ be a separable Hilbert space with an orthonormal basis $\{e_n\}$. Define $f\colon \mathcal{B}(\mathcal{H}) \to \mathbb{C}$ by $f(T) = \mathrm{LIM}\{\langle Te_n, e_n \rangle\}$. Show that f is a state on $\mathcal{B}(\mathcal{H})$. If π_f is the corresponding cyclic representation, show that $\ker \pi_f = \mathcal{B}_0(\mathcal{H})$. Hence π_f induces a cyclic representation of $\mathcal{B}(\mathcal{H})/\mathcal{B}_0(\mathcal{H})$ that is isometric. Is $\mathcal{H}_f$ separable?

8. If f is a positive linear functional on $\mathcal{A}$ and $\alpha \in (0, \infty)$, show that π_f and $\pi_{\alpha f}$ are equivalent representations.

9. If $a \in \mathcal{A}$, then $a \geq 0$ if and only if $f(a) \geq 0$ for every state f.

10. If $a \in \mathcal{A}$ and $a \neq 0$, then there is a state f on $\mathcal{A}$ such that $f(a) \neq 0$.

11. If f is a state on $\mathcal{A}$ and π_f is the corresponding representation, then π_f is injective if and only if $\{x \in \mathcal{A}\colon f(x^*x) = 0\} = (0)$.

12. If $\mathcal{A}$ is a separable C^*-algebra and $\{f_n\}$ is a countable weak* dense subset of $S_\mathcal{A}$, let $f = \sum_n 2^{-n} f_n$. Show that π_f is an isometry.

CHAPTER IX

Normal Operators on Hilbert Space

In this chapter the Spectral Theorem for normal operators on a Hilbert space is proved. This theorem is then used to answer a number of questions concerning normal operators. In fact, the Spectral Theorem can be used to answer essentially every question about normal operators.

§1. Spectral Measures and Representations of Abelian C^*-Algebras

Before beginning this section the reader should familiarize himself with the definitions and examples in (VIII.5.1) through (VIII.5.8).

In this section we want to focus our attention on representations of abelian C^*-algebras. The reason for this is that the Spectral Theorem and its generalizations can be obtained as a special case of such a theory. The idea is the following. Let N be a normal operator on $\mathcal{H}$. Then $C^*(N)$ is an abelian C^*-algebra and the functional calculus $f \mapsto f(N)$ is a $*$-isomorphism of $C(\sigma(N))$ onto $C^*(N)$ (VIII.2.6). Thus $f \mapsto f(N)$ is a representation $C(\sigma(N)) \to \mathcal{B}(\mathcal{H})$ of the abelian C^*-algebra $C(\sigma(N))$. A diagnosis of such representations yields the Spectral Theorem.

A representation $\rho: C(X) \to \mathcal{B}(\mathcal{H})$ is a $*$-homomorphism with $\rho(1) = 1$. Also, $\|\rho\| = 1$ (VIII.1.11d). If $f \in C(X)_+$, then $f = g^2$ where $g \in C(X)_+$; hence $\rho(f) = \rho(g)^2 = \rho(g)^*\rho(g) \geq 0$. So ρ is a positive map. One might expect, by analogy with the Riesz Representation Theorem, that $\rho(f) = \int f \, dE$ for some type of measure E whose values are operators rather than scalars. This is indeed the case. We begin by introducing these measures and defining the integral of a scalar-valued function with respect to one of them.

1.1.   Definition. If X is a set, Ω is a σ-algebra of subsets of X, and $\mathcal{H}$ is a Hilbert space, a *spectral measure* for $(X, \Omega, \mathcal{H})$ is a function $E: \Omega \to \mathcal{B}(\mathcal{H})$ such that:

(a) for each Δ in Ω, $E(\Delta)$ is a projection;
(b) $E(\square) = 0$ and $E(X) = 1$;
(c) $E(\Delta_1 \cap \Delta_2) = E(\Delta_1)E(\Delta_2)$ for Δ_1 and Δ_2 in Ω;
(d) if $\{\Delta_n\}_{n=1}^\infty$ are pairwise disjoint sets from Ω, then

$$E\left(\bigcup_{n=1}^\infty \Delta_n \right) = \sum_{n=1}^\infty E(\Delta_n).$$

A word or two concerning condition (d) in the preceding definition. If $\{E_n\}$ is a sequence of pairwise orthogonal projections on $\mathcal{H}$, then it was shown in Exercise II.3.5 that for each h in $\mathcal{H}$, $\sum_{n=1}^\infty E_n(h)$ converges in $\mathcal{H}$ to $E(h)$, where E is the orthogonal projection of $\mathcal{H}$ onto $\bigvee\{ E_n(\mathcal{H}): n \geq 1\}$. Thus it is legitimate to write $E = \sum_{n=1}^\infty E_n$. Now if $\Delta_1 \cap \Delta_2 = \square$, then (b) and (c) above imply that $0 = E(\Delta_1)E(\Delta_2) = E(\Delta_2)E(\Delta_1)$; that is, $E(\Delta_1)$ and $E(\Delta_2)$ have orthogonal ranges. So if $\{\Delta_n\}_1^\infty$ is a sequence of pairwise disjoint sets in Ω, the ranges of $\{E(\Delta_n)\}$ are pairwise orthogonal. Thus the equation $E(\bigcup_1^\infty \Delta_n) = \sum_1^\infty E(\Delta_n)$ in (d) has the precise meaning just discussed.

Another way to discuss this is by the introduction of two topologies that will also be of value later.

1.2.   Definition. If $\mathcal{H}$ is a Hilbert space, the *weak operator topology* (WOT) on $\mathcal{B}(\mathcal{H})$ is the locally convex topology defined by the seminorms $\{ p_{h,k}: h, k \in \mathcal{H}\}$ where $p_{h,k}(A) = |\langle Ah, k\rangle|$. The *strong operator topology* (SOT) is the topology defined on $\mathcal{B}(\mathcal{H})$ by the family of seminorms $\{ p_h: h \in \mathcal{H}\}$, where $p_h(A) = \|Ah\|$.

1.3.   Proposition. *Let $\mathcal{H}$ be a Hilbert space and let $\{A_i\}$ be a net in $\mathcal{B}(\mathcal{H})$.*

(a) $A_i \to A$ (WOT) *if and only if* $\langle A_i h, k\rangle \to \langle Ah, k\rangle$ *for all h, k in $\mathcal{H}$.*
(b) *If $\sup_i\|A_i\| < \infty$ and $\mathcal{T}$ is a total subset of $\mathcal{H}$, then $A_i \to A$ (WOT) if and only if* $\langle A_i h, k\rangle \to \langle Ah, k\rangle$ *for all h, k in $\mathcal{T}$.*
(c) $A_i \to A$ (SOT) *if and only if* $\|A_i h - Ah\| \to 0$ *for all h in $\mathcal{H}$.*
(d) *If $\sup_i\|A_i\| < \infty$ and $\mathcal{T}$ is a total subset of $\mathcal{H}$, then $A_i \to A$ (SOT) if and only if* $\|A_i h - Ah\| \to 0$ *for all h in $\mathcal{T}$.*
(e) *If $\mathcal{H}$ is separable, then the WOT and SOT are metrizable on bounded subsets of $\mathcal{B}(\mathcal{H})$.*

PROOF. The proofs of (a) through (d) are left as exercises. For (e), let $\{h_n\}$ be any countable total subset of ball $\mathcal{H}$. If $A, B \in \mathcal{B}(\mathcal{H})$, let

$$d_s(A, B) = \sum_{n=1}^{\infty} 2^{-n} \|(A - B)h_n\|,$$

$$d_w(A, B) = \sum_{m,n=1}^{\infty} 2^{-n-m} |\langle (A - B)h_n, h_m \rangle|.$$

Then d_s and d_w are metrics on $\mathcal{B}(\mathcal{H})$. It is left as an exercise to show that d_s and d_w define the SOT and WOT on bounded subsets of $\mathcal{B}(\mathcal{H})$. ∎

1.4. Example. Let (X, Ω, μ) be a σ-finite measure space. If $\phi \in L^{\infty}(\mu)$, let M_{ϕ} be the multiplication operator on $L^2(\mu)$. Then a net $\{\phi_i\}$ in $L^{\infty}(\mu)$ converges weak* to ϕ if and only if $M_{\phi_i} \to M_{\phi}$ (WOT). In fact, if $f, g \in L^2(\mu)$ and $\phi_i \to \phi$ weak* in $L^{\infty}(\mu)$, then $\langle M_{\phi_i} f, g \rangle = \int \phi_i f\bar{g}\, d\mu \to \int \phi f\bar{g}\, d\mu = \langle M_{\phi} f, g \rangle$ since $f\bar{g} \in L^1(\mu)$. Conversely, if $M_{\phi_i} \to M_{\phi}$ (WOT) and $f \in L^1(\mu)$, then $f = g_1 \bar{g}_2$, where $g_1, g_2 \in L^2(\mu)$. (Why?) So $\int \phi_i f\, d\mu = \langle M_{\phi_i} g_1, g_2 \rangle \to \langle M_{\phi} g_1, g_2 \rangle = \int \phi f\, d\mu$.

1.5. Example. If $\{E_n\}$ is a sequence of pairwise orthogonal projections on $\mathcal{H}$, then $\sum_1^{\infty} E_n$ converges (SOT) to the projection of $\mathcal{H}$ onto $\bigvee\{E_n(\mathcal{H}): n \geq 1\}$.

In light of (1.5), a spectral measure for $(X, \Omega, \mathcal{H})$ could be defined as a SOT-countably additive projection-valued measure.

1.6. Example. Let X be a compact set. $\Omega =$ the Borel subsets of X, $\mu = $ a measure on Ω, and $\mathcal{H} = L^2(\mu)$. For Δ in Ω, let $E(\Delta) = $ multiplication by χ_{Δ}, the characteristic function of Δ. E is a spectral measure for $(X, \Omega, \mathcal{H})$.

1.7. Example. If E is a spectral measure for $(X, \Omega, \mathcal{H})$, the *inflation*, $E^{(n)}$, of E, defined by $E^{(n)}(\Delta) = E(\Delta)^{(n)}$, is a spectral measure for $(X, \Omega, \mathcal{H}^{(n)})$.

1.8. Example. Let X be any set, $\Omega = $ all the subsets of X, $\mathcal{H} = $ any separable Hilbert space, and fix a sequence $\{x_n\}$ in X. If $\{e_1, e_2, \ldots\}$ is some orthonormal basis for $\mathcal{H}$, define $E(\Delta) = $ the projection onto $\bigvee\{e_n: x_n \in \Delta\}$. E is a spectral measure for $(X, \Omega, \mathcal{H})$.

The next lemma is useful in studying spectral measures as it allows us to prove things about spectral measures from known facts about complex-valued measures.

1.9. Lemma. *If E is a spectral measure for $(X, \Omega, \mathcal{H})$ and $g, h \in \mathcal{H}$, then*

$$E_{g,h}(\Delta) \equiv \langle E(\Delta)g, h \rangle$$

defines a countably additive measure on Ω with total variation $\leq \|g\| \|h\|$.

PROOF. That $\mu = E_{g,h}$ as defined above, is a countably additive measure is left for the reader to verify. If $\Delta_1, \ldots, \Delta_n$ are pairwise disjoint sets in Ω, let $\alpha_j \in \mathbb{C}$ such that $|\alpha_j| = 1$ and $|\langle E(\Delta_j)g, h\rangle| = \alpha_j \langle E(\Delta_j)g, h\rangle$. So $\Sigma_j |\mu(\Delta_j)| = \Sigma_j \alpha_j \langle E(\Delta_j)g, h\rangle = \langle \Sigma_j E(\Delta_j)\alpha_j g, h\rangle \leq \|\Sigma_j E(\Delta_j)\alpha_j g\| \|h\|$. Now $\{E(\Delta_j)\alpha_j g: 1 \leq j \leq n\}$ is a finite sequence of pairwise orthogonal vectors so that $\|\Sigma_j E(\Delta_j)\alpha_j g\|^2 = \Sigma_j \|E(\Delta_j)g\|^2 = \|E(\bigcup_{j=1}^n \Delta_j)g\|^2 \leq \|g\|^2$; hence $\Sigma_j |\mu(\Delta_j)| \leq \|g\| \|h\|$. Thus $\|\mu\| \leq \|g\| \|h\|$. ∎

It is possible to use spectral measures to define representations. The next result is crucial for this purpose. It tells us how to integrate with respect to a spectral measure.

1.10. Proposition. *If E is a spectral measure for $(X, \Omega, \mathcal{H})$ and $\phi: X \to \mathbb{C}$ is a bounded Ω-measurable function, then there is a unique operator A in $\mathcal{B}(\mathcal{H})$ such that if $\varepsilon > 0$ and $\{\Delta_1, \ldots, \Delta_n\}$ is an Ω-partition of X with $\sup\{|\phi(x) - \phi(x')|: x, x' \in \Delta_k\} < \varepsilon$ for $1 \leq k \leq n$, then for any x_k in Δ_k,*

$$\left\| A - \sum_{k=1}^n \phi(x_k) E(\Delta_k) \right\| < \varepsilon.$$

PROOF. Define $B(g, h) \equiv \int \phi \, dE_{g,h}$ for g, h in $\mathcal{H}$. By the preceding lemma it is easy to see that B is a sesquilinear form with $|B(g, h)| \leq \|\phi\|_\infty \|g\| \|h\|$. Hence there is a unique operator A such that $B(g, h) = \langle Ag, h\rangle$ for all g and h in $\mathcal{H}$.

Let $\{\Delta_1, \ldots, \Delta_n\}$ be an Ω-partition satisfying the condition in the statement of the proposition. If g and h are arbitrary vectors in $\mathcal{H}$ and $x_k \in \Delta_k$ for $1 \leq k \leq n$, then

$$\left| \langle Ag, h\rangle - \sum_{k=1}^n \phi(x_k)\langle E(\Delta_k)g, h\rangle \right|$$

$$= \left| \sum_{k=1}^n \int_{\Delta_k} [\phi(x) - \phi(x_k)] d\langle E(x)g, h\rangle \right|$$

$$\leq \sum_{k=1}^n \int_{\Delta_k} |\phi(x) - \phi(x_k)| d|\langle E(x)g, h\rangle|$$

$$\leq \varepsilon \int d|\langle E(x)g, h\rangle| \leq \varepsilon \|g\| \|h\|. \quad ∎$$

The operator A obtained in the preceding proposition is the *integral of ϕ with respect to E* and is denoted by

$$\int \phi \, dE.$$

Therefore if $g, h \in \mathcal{H}$ and ϕ is a bounded Ω-measurable function on X, the preceding proof implies that

1.11
$$\left\langle \left(\int \phi \, dE \right) g, h \right\rangle = \int \phi \, dE_{g,h}.$$

Let $B(X, \Omega)$ denote the set of bounded Ω-measurable functions ϕ: $X \to \mathbb{C}$ and let $\|\phi\| = \sup\{|\phi(x)|: x \in X\}$. It is easy to see that $B(X, \Omega)$ is a Banach algebra with identity. In fact, if $\phi^*(x) \equiv \overline{\phi(x)}$, then $B(X, \Omega)$ is an abelian C^*-algebra. The properties of the integral $\int \phi \, dE$ are summarized by the following result.

1.12. Proposition. *If E is a spectral measure for $(X, \Omega, \mathcal{H})$ and ρ: $B(X, \Omega)$ $\to \mathcal{B}(\mathcal{H})$ is defined by $\rho(\phi) = \int \phi \, dE$, then ρ is a representation of $B(X, \Omega)$.*

PROOF. It will only be shown that ρ is multiplicative; the remainder is an exercise. Let ϕ and $\psi \in C(X)$. Let $\varepsilon > 0$ and choose a Borel partition $\{\Delta_1, \ldots, \Delta_n\}$ of X such that $\sup\{|\omega(x) - \omega(x')|: x, x' \in \Delta_k\} < \varepsilon$ for $\omega = \phi, \psi$ or $\phi\psi$ and for $1 \le k \le n$. Hence, if $x_k \in \Delta_k$ $(1 \le k \le n)$,

$$\left\| \int \omega \, dE - \sum_{k=1}^{n} \omega(x_k) E(\Delta_k) \right\| < \varepsilon$$

for $\omega = \phi$, ψ, or $\phi\psi$. Thus, using the triangle inequality,

$$\left\| \int \phi\psi \, dE - \left(\int \phi \, dE \right) \left(\int \psi \, dE \right) \right\|$$

$$\le \varepsilon + \left\| \sum_{k=1}^{n} \phi(x_k)\psi(x_k) E(\Delta_k) - \left[\sum_{i=1}^{n} \phi(x_i) E(\Delta_i) \right]\left[\sum_{j=1}^{n} \psi(x_j) E(\Delta_j) \right] \right\|$$

$$+ \left\| \left[\sum_{i=1}^{n} \phi(x_i) E(\Delta_i) \right]\left[\sum_{j=1}^{n} \psi(x_j) E(\Delta_j) \right] - \left(\int \phi \, dE \right)\left(\int \psi \, dE \right) \right\|.$$

But $E(\Delta_i) E(\Delta_j) = E(\Delta_i \cap \Delta_j)$ and $\{\Delta_1, \ldots, \Delta_n\}$ is a partition. So the middle term in this sum is zero. Hence

$$\left\| \int \phi\psi \, dE - \left(\int \phi \, dE \right)\left(\int \phi \, dE \right) \right\|$$

$$\le \varepsilon + \left\| \left[\sum_{i=1}^{n} \phi(x_i) E(\Delta_i) \right]\left[\sum_{j=1}^{n} \psi(x_j) E(\Delta_j) - \int \psi \, dE \right] \right\|$$

$$+ \left\| \left[\sum_{i=1}^{n} \phi(x_i) E(\Delta_i) - \int \phi \, dE \right]\left[\int \psi \, dE \right] \right\| \le \varepsilon[1 + \|\phi\| + \|\psi\|].$$

Since ε was arbitrary, $\int \phi\psi \, dE = (\int \phi\psi \, dE)(\int \psi \, dE)$. ■

1.13. Corollary. *If X is a compact Hausdorff space and E is a spectral measure defined on the Borel subsets of X, then ρ: $C(X) \to \mathcal{B}(\mathcal{H})$ defined by $\rho(u) = \int u \, dE$ is a representation of $C(X)$.*

The next result is the main result of this section and it states that the converse to the preceding corollary holds.

1.14. Theorem. *If* $\rho: C(X) \to \mathscr{B}(\mathscr{H})$ *is a representation, there is a unique spectral measure E defined on the Borel subsets of X such that*

$$\rho(u) = \int u \, dE$$

for every u in $C(X)$.

PROOF. The idea of the proof is similar to the idea of the proof of the Riesz Representation Theorem for linear functionals on $C(X)$. We wish to extend ρ to a representation $\tilde{\rho}: B(X) \to \mathscr{B}(\mathscr{H})$, where $B(X)$ is the C^*-algebra of bounded Borel functions. The measure E of a Borel set Δ is then defined by letting $E(\Delta) = \tilde{\rho}(\chi_\Delta)$. In fact, it is possible to give a proof of the theorem patterned on the proof of the Riesz Representation Theorem. Here, however, the proof will use the Riesz Representation Theorem to simplify the technical details.

If $g, h \in \mathscr{H}$, then $u \mapsto \langle \rho(u)g, h \rangle$ is a linear functional on $C(X)$ with norm $\leq \|g\| \|h\|$. Hence there is a unique measure, $\mu_{g,h}$, in $M(X)$ such that

1.15 $$\langle \rho(u)g, h \rangle = \int u \, d\mu_{g,h}$$

for all u in $C(X)$. It is easy to verify that the map $(g, h) \mapsto \mu_{g,h}$ is sesquilinear (use uniqueness) and $\|\mu_{g,h}\| \leq \|g\| \|h\|$. Now fix ϕ in $B(X)$ and define $[g, h] = \int \phi \, d\mu_{g,h}$. Then $[\cdot, \cdot]$ is sesquilinear form and $\|[g, h]\| \leq \|\phi\| \|g\| \|h\|$. Hence there is a unique bounded operator A such that $[g, h] = \langle Ag, h \rangle$ and $\|A\| \leq \|\phi\|$ (II.2.2). Denote the operator A by $\tilde{\rho}(\phi)$. So $\tilde{\rho}: B(X) \to \mathscr{B}(\mathscr{H})$ is a well-defined function, $\|\tilde{\rho}(\phi)\| \leq \|\phi\|$, and for all g, h in $\mathscr{H}$,

1.16 $$\langle \tilde{\rho}(\phi)g, h \rangle = \int \phi \, d\mu_{g,h}.$$

1.17. Claim. $\tilde{\rho}: B(X) \to \mathscr{B}(\mathscr{H})$ is a representation and $\tilde{\rho}|C(X) = \rho$.

The fact that $\tilde{\rho}(u) = \rho(u)$ whenever $u \in C(X)$ follows immediately from (1.15) and (1.16). If $\phi \in B(X)$, consider ϕ as an element of $M(X)^*$ ($= C(X)^{**}$); that is, ϕ corresponds to the linear functional $\mu \mapsto \int \phi \, d\mu$. By Proposition V.4.1, $\{u \in C(X): \|u\| \leq \|\phi\|\}$ is $\sigma(M(X)^*, M(X))$ dense in $\{L \in M(X)^*: \|L\| \leq \|\phi\|\}$. Thus there is a net $\{u_i\}$ in $C(X)$ such that $\|u_i\| \leq \|\phi\|$ for all u_i and $\int u_i \, d\mu \to \int \phi \, d\mu$ for every μ in $M(X)$. If $\psi \in B(X)$, then $\psi\mu \in M(X)$ whenever $\mu \in M(X)$. Hence $\int u_i \psi \, d\mu \to \int \phi\psi \, d\mu$ for every ψ in $B(X)$ and μ in $M(X)$. By (1.16), $\tilde{\rho}(u_i\psi) \to \tilde{\rho}(\phi\psi)$ (WOT) for all ψ in $B(X)$. In particular, if $\psi \in C(X)$, then $\tilde{\rho}(\phi\psi) = \text{WOT} - \lim \tilde{\rho}(u_i\psi) =$

WOT $- \lim \rho(u_i)\rho(\psi) = \tilde{\rho}(\phi)\rho(\psi)$. That is,

$$\tilde{\rho}(\phi\psi) = \tilde{\rho}(\phi)\rho(\psi)$$

whenever $\phi \in B(X)$ and $\psi \in C(X)$. Hence $\tilde{\rho}(u_i\psi) = \rho(u_i)\tilde{\rho}(\psi)$ for any ψ in $B(X)$ and for all u_i. Since $\tilde{\rho}(u_i) \to \tilde{\rho}(\phi)$ (WOT) and $\tilde{\rho}(u_i\psi) \to \tilde{\rho}(\phi\psi)$ (WOT), this implies that

$$\tilde{\rho}(\phi\psi) = \tilde{\rho}(\phi)\tilde{\rho}(\psi)$$

whenever $\phi, \psi \in B(X)$.

The proof that $\tilde{\rho}$ is linear is immediate by (1.16). To see that $\tilde{\rho}(\phi)^* = \tilde{\rho}(\bar{\phi})$, let $\{u_i\}$ be the net obtained in the preceding paragraph. If $\mu \in M(X)$, let $\bar{\mu}$ be the measure defined by $\bar{\mu}(\Delta) = \overline{\mu(\Delta)}$. Then $\rho(u_i) \to \tilde{\rho}(\phi)$ (WOT) and so $\rho(u_i)^* \to \tilde{\rho}(\phi)^*$ (WOT). But $\int \bar{u}_i \, d\mu = \overline{\int u_i \, d\bar{\mu}} \to \overline{\int \phi \, d\bar{\mu}} = \int \bar{\phi} \, d\mu$ for every measure μ. Hence $\rho(\bar{u}_i) \to \tilde{\rho}(\bar{\phi})$. But $\rho(u_i)^* = \rho(\bar{u}_i)$ since ρ is a $*$-homomorphism. Thus $\tilde{\rho}(\phi)^* = \tilde{\rho}(\bar{\phi})$ and $\tilde{\rho}$ is a representation.

For any Borel subset Δ of X let $E(\Delta) \equiv \tilde{\rho}(\chi_\Delta)$. We want to show that E is a spectral measure. Since χ_Δ is a hermitian idempotent in $B(X)$, $E(\Delta)$ is a projection by (1.17). Since $\chi_\square = 0$ and $\chi_X = 1$, $E(\square) = 0$ and $E(X) = 1$. Also, $E(\Delta_1 \cap \Delta_2) = \tilde{\rho}(\chi_{\Delta_1 \cap \Delta_2}) = \tilde{\rho}(\chi_{\Delta_1}\chi_{\Delta_2}) = E(\Delta_1)E(\Delta_2)$. Now let $\{\Delta_n\}$ be a pairwise disjoint sequence of Borel sets and put $\Lambda_n = \bigcup_{k=n+1}^\infty \Delta_k$. It is easy to see that E is finitely additive so if $h \in \mathscr{H}$, then

$$\left\| E\left(\bigcup_{k=1}^\infty \Delta_k \right)h - \sum_{k=1}^n E(\Delta_k)h \right\|^2 = \langle E(\Lambda_n)h, E(\Lambda_n)h \rangle$$
$$= \langle E(\Lambda_n)h, h \rangle$$
$$= \langle \tilde{\rho}(\chi_{\Lambda_n})h, h \rangle$$
$$= \int \chi_{\Lambda_n} \, d\mu_{h,h}$$
$$= \sum_{k=n+1}^\infty \mu_{h,h}(\Delta_k) \to 0$$

as $n \to \infty$. Therefore E is a spectral measure.

It remains to show that $\rho(u) = \int u \, dE$. It will be shown that $\tilde{\rho}(\phi) = \int \phi \, dE$ for every ϕ in $B(X)$. Fix ϕ in $B(X)$ and $\varepsilon > 0$. If $\{\Delta_1, \ldots, \Delta_n\}$ is any Borel partition of X such that $\sup\{|\phi(x) - \phi(x')| : x, x' \in \Delta_k\} < \varepsilon$ for $1 \le k \le n$, then $\|\phi - \sum_{k=1}^n \phi(x_k)\chi_{\Delta_k}\|_\infty < \varepsilon$ for any choice of x_k in Δ_k. Since $\|\tilde{\rho}\| = 1$, $\varepsilon > \|\tilde{\rho}(\phi) - \sum_{k=1}^n \phi(x_k)E(\Delta_k)\|$. This implies that $\tilde{\rho}(\phi) = \int \phi \, dE$ for any ϕ in $B(X)$.

The proof of the uniqueness of E is left to the reader. ∎

EXERCISES

1. Prove Proposition 1.3.

2. Show that ball $\mathscr{B}(\mathscr{H})$ is WOT compact.

3. Show that $\operatorname{Re}\mathcal{B}(\mathcal{H})$ and $\mathcal{B}(\mathcal{H})_+$ are WOT and SOT closed.

4. If $L\colon \mathcal{B}(\mathcal{H}) \to \mathbb{C}$ is a linear functional, show that the following statements are equivalent: (a) L is SOT-continuous; (b) L is WOT-continuous; (c) there are vectors $h_1, \ldots, h_n, g_1, \ldots, g_n$ in $\mathcal{H}$ such that $L(A) = \sum_{j=1}^{n}\langle Ah_j, g_j\rangle$.

5. Show that a convex subset of $\mathcal{B}(\mathcal{H})$ is WOT closed if and only if it is SOT closed.

6. Verify the statement in Example 1.5.

7. Verify the statements made in Examples 1.6, 1.7, and 1.8.

8. For the spectral measures in (1.6), (1.7), and (1.8), give the corresponding representations.

9. If $\{E_i\}$ is a net of projections and E is a projection, show that $E_i \to E$ (WOT) if and only if $E_i \to E$ (SOT).

10. For the representation in (VIII.5.5), find the corresponding spectral measure.

11. In Example VIII.5.4, the representation is not quite covered by Theorem 1.14 since it is a representation of $L^\infty(\mu)$ and not $C(X)$. Nevertheless, this representation is given by a spectral measure defined on Ω. Find it.

12. Let X be a compact Hausdorff space and let $\{x_n\}$ be a sequence in X. Let $\{e_n\}$ be an orthonormal basis for $\mathcal{H}$ and for each u in $C(X)$ define $\rho(u)$ in $\mathcal{B}(\mathcal{H})$ by $\rho(u)e_n = u(x_n)e_n$. Show that ρ is a representation and find the corresponding spectral measure.

13. A representation $\rho\colon \mathcal{A} \to \mathcal{B}(\mathcal{H})$ is *irreducible* if the only projections in $\mathcal{B}(\mathcal{H})$ that commute with every $\rho(a)$, a in $\mathcal{A}$, are 0 and 1. Prove that if $\mathcal{A}$ is abelian and ρ is an irreducible representation of $\mathcal{A}$, then $\dim \mathcal{H} = 1$. Find the corresponding spectral measure.

14. Show that a representation $\rho\colon C(X) \to \mathcal{B}(\mathcal{H})$ is injective if and only if $E(G) \neq 0$ for every open set G, where E is the corresponding spectral measure.

15. Let $\{A_i\}$ be a net of hermitian operators on $\mathcal{H}$ and suppose that there is a hermitian operator T such that $A_i \leq T$ for all i. If $\{\langle A_i h, h\rangle\}$ is an increasing net in $\mathbb{R}$ for every h in $\mathcal{H}$, then there is a hermitian operator A such that $A_i \to A$ (WOT).

16. Show that there is a contraction $\tau\colon \mathcal{B}(\mathcal{H})^{**} \to \mathcal{B}(\mathcal{H})$ such that $\tau(T) = T$ for T in $\mathcal{B}(\mathcal{H})$. If $\rho\colon C(X) \to \mathcal{B}(\mathcal{H})$ is a representation, show that the map $\tilde{\rho}$ in the proof of Theorem 1.14 is given by $\tilde{\rho}(\phi) = \tau \circ \rho^{**}(\phi)$.

§2. The Spectral Theorem

The Spectral Theorem is a landmark in the theory of operators on a Hilbert space. It provides a complete statement about the nature and structure of normal operators. This accolade will be seen to hold when in Section 10 it is used to give a complete set of unitary invariants. Two operators A and B are *unitarily equivalent* if there is a unitary operator U such that $UAU^* = B$; in symbols, $A \cong B$. Using the Spectral Theorem, a (countable) set of objects are attached to a normal operator N on a (separable) Hilbert space. It is then shown that two normal operators are unitarily equivalent if and only if these objects are equal.

The Spectral Theorem for a normal operator N on a Hilbert space with $\dim \mathscr{H} = d < \infty$ says that N can be diagonalized. That is, if $\alpha_1, \ldots, \alpha_d$ are the eigenvalues of N (repeated as often as their multiplicities), then the corresponding eigenvectors $e_1, e_2, \ldots, e_d$ form an orthonormal basis for $\mathscr{H}$. In infinite-dimensional spaces a normal operator need not have eigenvalues. For example, let $N =$ multiplication by the independent variable on $L^2(0, 1)$. So an alternate formulation that can be generalized is desired.

Let N be normal on $\mathscr{H}$, $\dim \mathscr{H} = d < \infty$. Let $\lambda_1, \ldots, \lambda_n$ be the distinct eigenvalues of N and let E_k be the orthogonal projection of $\mathscr{H}$ onto $\ker(N - \lambda_k)$, $1 \le k \le n$. Then the Spectral Theorem says that

2.1
$$N = \sum_{k=1}^{n} \lambda_k E_k.$$

In this form a generalization is possible. Rather than discuss orthogonal projections on eigenspaces (which may not exist), the concept of a spectral measure is used; rather than the sum that appears in (2.1), an integral is used. It is worth mentioning that the finite-dimensional version is a corollary of the general theorem (see Exercise 4).

2.2. The Spectral Theorem. *If N is a normal operator, there is a unique spectral measure E on the Borel subsets of $\sigma(N)$ such that:*

(a) $N = \int z \, dE(z)$;
(b) *if G is a nonempty relatively open subset of $\sigma(N)$, $E(G) \ne 0$;*
(c) *if $A \in \mathscr{B}(\mathscr{H})$, then $AN = NA$ and $AN^* = N^*A$ if and only if $AE(\Delta) = E(\Delta)A$ for every Δ.*

PROOF. Let $\mathscr{A} = C^*(N)$, the C^*-algebra generated by N. So $\mathscr{A}$ is the closure of all polynomials in N and N^*. By Theorem VIII.2.6, there is an isometric isomorphism $\rho: C(\sigma(N)) \to \mathscr{A} \subseteq \mathscr{B}(\mathscr{H})$ given by $\rho(u) = u(N)$ (the functional calculus). By Theorem 1.14 there is a unique spectral measure E defined on the Borel subsets of $\sigma(N)$ such that $\rho(u) = \int u \, dE$ for all u in $C(\sigma(N))$. In particular, (a) holds since $N = \rho(z)$.

If G is a nonempty relatively open subset of $\sigma(N)$, there is a nonzero continuous function u on $\sigma(N)$ such that $0 \le u \le \chi_G$. Using Claim 1.17, one obtains that $E(G) = \tilde{\rho}(\chi_G) \ge \rho(u) \ne 0$; so (b) holds.

Now let $A \in \mathscr{B}(\mathscr{H})$ such that $AN = NA$ and $AN^* = N^*A$. It is not hard to see that this implies, by the Stone–Weierstrass Theorem, that $A\rho(u) = \rho(u)A$ for every u in $C(\sigma(N))$; that is, $Au(N) = u(N)A$ for all u

in $C(\sigma(N))$. Let $\Omega = \{\Delta: \Delta$ is a Borel set and $AE(\Delta) = E(\Delta)A\}$. It is left to the reader to show that Ω is a σ-algebra. If G is an open set in $\sigma(N)$, there is a sequence $\{u_n\}$ of positive continuous functions on $\sigma(N)$ such that $u_n(z) \uparrow \chi_G(z)$ for all z. Thus

$$
\begin{aligned}
\langle AE(G)g, h \rangle &= \langle E(G)g, A^*h \rangle \\
&= E_{g, A^*h}(G) \\
&= \lim \int u_n \, dE_{g, A^*h} \\
&= \lim \langle u_n(N)g, A^*h \rangle \\
&= \lim \langle Au_n(N)g, h \rangle \\
&= \lim \langle u_n(N)Ag, h \rangle \\
&= \langle E(G)Ag, h \rangle.
\end{aligned}
$$

So Ω contains every open set and, hence, it must be the collection of Borel sets. The converse is left to the reader. ∎

The unique spectral measure E obtained in the Spectral Theorem is called the *spectral measure* for N. An abbreviation for the Spectral Theorem is to say, "Let $N = \int \lambda \, dE(\lambda)$ be the *spectral decomposition* of N." If ϕ is a bounded Borel function on $\sigma(N)$, define $\phi(N)$ by

$$
\phi(N) \equiv \int \phi \, dE,
$$

where E is the spectral measure for N.

2.3. Theorem. *If N is a normal operator on $\mathcal{H}$ with spectral measure E and $B(\sigma(N))$ is the C^*-algebra of bounded Borel functions on $\sigma(N)$, then the map*

$$
\phi \mapsto \phi(N)
$$

is a representation of the C^*-algebra $B(\sigma(N))$. If $\{\phi_i\}$ is a net in $B(\sigma(N))$ such that $\int \phi_i \, d\mu \to 0$ for every μ in $M(\sigma(N))$, then $\phi_i(N) \to 0$ (WOT). This map is unique in the sense that if $\tau: B(\sigma(N)) \to \mathcal{B}(\mathcal{H})$ is a representation such that $\tau(z) = N$ and $\tau(\phi_i) \to 0$ (WOT) whenever $\{\phi_i\}$ is a net in $B(\sigma(N))$ such that $\int \phi_i \, d\mu \to 0$ for every μ in $M(\sigma(N))$, then $\tau(\phi) = \phi(N)$ for all ϕ in $B(\sigma(N))$.

PROOF. The fact that $\phi \mapsto \phi(N)$ is a representation is a consequence of Proposition 1.12. If $\{\phi_i\}$ is as in the statement, then the fact that $E_{g,h} \in M(\sigma(N))$ implies that $\phi_i(N) \to 0$ (WOT).

To prove uniqueness, let $\tau: B(\sigma(N)) \to \mathcal{B}(\mathcal{H})$ be a representation with the appropriate properties. Then $\tau(u) = u(N)$ if $u \in C(\sigma(N))$ by the uniqueness of the functional calculus for normal elements of a C^*-algebra (VIII.2.6). If $\phi \in B(\sigma(N))$, then Proposition V.4.1 implies that there is a net $\{u_i\}$ in $C(\sigma(N))$ such that $\|u_i\| \leq \|\phi\|$ for all u_i and $\int u_i \, d\mu \to \int \phi \, d\mu$ for every μ in $M(\sigma(N))$. Thus $u_i(N) \to \phi(N)$ (WOT). But $\tau(\phi) = $ WOT $- \lim \tau(u_i) = $ WOT $- \lim u_i(N)$; therefore $\tau(\phi) = \phi(N)$. ∎

It is worthwhile to rewrite (1.11) as

2.4
$$\langle \phi(N)g, h \rangle = \int \phi \, dE_{g,h}$$

for ϕ in $B(\sigma(N))$ and g, h in $\mathcal{H}$. If $\phi \in B(\mathbb{C})$, then the restriction of ϕ to $\sigma(N)$ belongs to $B(\sigma(N))$. Since the support of each measure $E_{g,h}$ is contained in $\sigma(N)$, (2.4) holds for every bounded Borel function ϕ on $\mathbb{C}$. This has certain technical advantages that will become apparent when we begin to apply (2.4).

Proposition 2.3 thus extends the functional calculus for normal operators. This functional calculus or, equivalently, the Spectral Theorem, will be exploited in this chapter. But right now we look at some examples.

2.5. Example. If μ is a regular Borel measure on $\mathbb{C}$ with compact support K, define N_μ on $L^2(\mu)$ by $N_\mu f = zf$ for each f in $L^2(\mu)$. It is easy to check that $N_\mu^* f = \bar{z}f$, and, hence, N_μ is normal.

(a) $\sigma(N_\mu) = K = $ support of μ. (Exercise.)
(b) If, for a bounded Borel function ϕ, we define M_ϕ on $L^2(\mu)$ by $M_\phi f = f$, then $\phi(N_\mu) = M_\phi$.

Indeed, this is an easy application of the uniqueness part of (2.3).

(c) If E is the spectral measure for N_μ, then $E(\Delta) = M_{\chi_\Delta}$.

Just note that $E(\Delta) = \chi_\Delta(N)$.

2.6. Example. Let (X, Ω, μ) be any σ-finite measure space and put $\mathcal{H} = L^2(X, \Omega, \mu)$. For ϕ in $L^\infty(\mu) \equiv L^\infty(X, \Omega, \mu)$, define M_ϕ on $\mathcal{H}$ by $M_\phi f = \phi f$.

(a) M_ϕ is normal and $M_\phi^* = M_{\bar\phi}$ (II.2.8).
(b) $\phi \mapsto M_\phi$ is a representation of $L^\infty(\mu)$ (VIII.5.4).
(c) If $\phi \in L^\infty(\mu)$, $\|\phi\|_\infty = \|M_\phi\|$ (II.1.5).
(d) Define the *essential range* of ϕ by

$$\text{ess-ran}(\phi) \equiv \cap \{\text{cl}(\phi(\Delta)): \Delta \in \Omega \text{ and } \mu(X \setminus \Delta) = 0\}.$$

Then $\sigma(M_\phi) = \text{ess-ran}(\phi)$. (This appears as Exercise VII.3.3, but a proof is given here.)

First assume that $\lambda \notin \text{ess-ran}(\phi)$. So there is a set Δ in Ω with $\mu(X \setminus \Delta) = 0$ and λ not in $\text{cl}(\phi(\Delta))$; thus, there is a $\delta > 0$ with $|\phi(x) - \lambda| \geq \delta$ for all x in Δ. If $\psi = (\phi - \lambda)^{-1}$, $\psi \in L^\infty(\mu)$ and $M_\psi = (M_\phi - \lambda)^{-1}$.

Conversely, assume $\lambda \in \text{ess-ran}(\phi)$. It follows that for every integer n there is a set Δ_n in Ω such that $0 < \mu(\Delta_n) < \infty$ and $|\phi(x) - \lambda| < 1/n$ for all x in Δ_n. Put $f_n = (\mu(\Delta_n))^{-1/2}\chi_{\Delta_n}$; so $f_n \in L^2(\mu)$ and $\|f_n\|_2 = 1$. However, $\|(M_\phi - \lambda)f_n\|^2 = (\mu(\Delta_n))^{-1}\int_{\Delta_n}|\phi - \lambda|^2 \, d\mu \le 1/n^2$, showing that $\lambda \in \sigma_{ap}(M_\phi)$.

(e) If E is the spectral measure for M_ϕ [so E is defined on the Borel subsets of $\sigma(M_\phi) = $ ess-ran$(\phi) \subseteq \mathbb{C}$], then for every Borel subset Δ of $\sigma(M_\phi)$, $E(\Delta) = M_{\chi_\phi^{-1}(\Delta)}$.

2.7. Proposition. *If for $k \ge 1$, N_k is a normal operator on $\mathcal{H}_k$ with* $\sup_k\|N_k\| < \infty$, E_k *is the spectral measure for N_k, and if $N = \oplus_{k=1}^\infty N_k$ on* $\mathcal{H} = \oplus_{k=1}^\infty \mathcal{H}_k$, *then:*

(a) $\sigma(N) = \text{cl}[\bigcup_{k=1}^\infty \sigma(N_k)]$;
(b) *if E is the spectral measure for N, $E(\Delta) = \oplus_{k=1}^\infty E_k(\Delta \cap \sigma(N_k))$ for every Borel subset Δ of $\sigma(N)$.*

PROOF. Exercise.

A historical account of the spectral theorem is an enormous undertaking by itself. One such account is Steen [1973]. You might also consult the notes in Dunford and Schwartz [1963] and Halmos [1951].

EXERCISES

Throughout these exercises, N is a normal operator on $\mathcal{H}$ with spectral measure E.

1. Show that $\lambda \in \sigma_p(N)$ if and only if $E(\{\lambda\}) \ne 0$. Moreover, if $\lambda \in \sigma_p(N)$, $E(\{\lambda\})$ is the orthogonal projection onto ker$(N - \lambda)$.

2. If Δ is a clopen subset of $\sigma(N)$, show that $E(\Delta)$ is the Riesz idempotent associated with Δ.

3. Prove Theorem II.5.1 and its corollaries by using the Spectral Theorem.

4. Prove Theorem II.7.6 and its corollaries by using the Spectral Theorem.

5. Obtain Theorem II.7.11 as a consequence of (2.3).

6. Verify the statements in Example 2.5.

7. Verify (2.6e).

8. Let A be a hermitian operator with spectral measure E on a separable space. For each real number t define a projection $P(t) = E(-\infty, t)$. Show:
   (a) $P(s) \le P(t)$ for $s \le t$;
   (b) if $t_n \le t_{n+1}$ and $t_n \to t$, $P(t_n) \to P(t)$ (SOT);
   (c) for all but a countable number of points t, $P(t_n) \to P(t)$ (SOT) if $t_n \to t$;
   (d) for f in $C(\sigma(A))$, $f(A) = \int_{-\infty}^\infty f(t) \, dP(t)$, where this integral is to be defined (by the reader) in the Riemann–Stieltjes sense.

9. Show that a normal operator N is (a) hermitian if and only if $\sigma(N) \subseteq \mathbb{R}$; (b) positive if and only if $\sigma(N) \subseteq [0, \infty)$; (c) unitary if and only if $\sigma(N) \subseteq \partial\mathbb{D}$.

10. Show that if $\mathcal{H}$ is separable, there are at most a countable number of points $\{z_n\}$ in $\sigma(N)$ such that $E(z_n) \neq 0$. By Exercise 1, these are the eigenvalues of N.

11. Show that $E(\sigma(N) \setminus \sigma_p(N)) = 0$ if and only if N is diagonalizable; that is, there is a basis for $\mathcal{H}$ consisting of eigenvectors for N.

12. Show that if $N = U|N|$ ($|N| = (N^*N)^{1/2}$) is the polar decomposition of N, $U = \phi(N)$ for some Borel function ϕ on $\phi(N)$. Hence $U|N| = |N|U$.

13. Show that $N = W|N|$ for some unitary W that is a function of N.

14. Prove that if A is hermitian, $\exp(iA)$ is unitary. Is the converse true?

15. Show that there is a normal operator M such that $M^2 = N$ and $M = \phi(N)$ for some Borel function ϕ. Is there only one such normal operator?

16. Define $N: L^2(\mathbb{R}) \to L^2(\mathbb{R})$ by $(Nf)(t) = f(t + 1)$. Show that N is normal and find its spectral decomposition.

17. Suppose that $N_1, \ldots, N_d$ are normal operators such that $N_j N_k^* = N_k^* N_j$ for $1 \leq j, k \leq d$. Show that there is a subset X of $\mathbb{C}^d$ and a spectral measure E defined on the Borel subsets of X such that $N_k = \int z_k \, dE(z)$ for $1 \leq k \leq d$ (z_k = the kth coordinate function)(see Exercise VIII.2.2).

18. If $N_1, \ldots, N_d$ are as in Exercise 17 and each is compact, show that there is a basis for $\mathcal{H}$ consisting of eigenvectors for each N_k. (This is the *simultaneous diagonalization* of $N_1, \ldots, N_d$.)

19. This exercise gives the properties of Hilbert–Schmidt operators (defined below).
    (a) If $\{e_i\}$ and $\{f_j\}$ are two orthonormal bases for $\mathcal{H}$ and $A \in \mathcal{B}(\mathcal{H})$, then

$$\sum_i \|Ae_i\|^2 = \sum_j \|Af_j\|^2 = \sum_i \sum_j |\langle Ae_i, f_j \rangle|^2.$$

(b) If $A \in \mathcal{B}(\mathcal{H})$ and $\{e_i\}$ is a basis for $\mathcal{H}$, define

$$\|A\|_2 = \left[\sum_i \|Ae_i\|^2 \right]^{1/2}.$$

By (a) $\|A\|_2$ is independent of the basis chosen and hence is well defined. If $\|A\|_2 < \infty$, A is called a *Hilbert–Schmidt operator*. $\mathcal{B}_2 = \mathcal{B}_2(\mathcal{H})$ denotes the set of all Hilbert–Schmidt operators. (c) $\|A\| \leq \|A\|_2$ for every A in $\mathcal{B}(\mathcal{H})$ and $\|\cdot\|_2$ is a norm on $\mathcal{B}_2$. (d) If $T \in \mathcal{B} = \mathcal{B}(\mathcal{H})$ and $A \in \mathcal{B}_2$, then $\|TA\|_2 \leq \|T\|\|A\|_2$, $\|A^*\|_2 = \|A\|_2$, and $\|AT\|_2 \leq \|A\|_2\|T\|$. (e) $\mathcal{B}_2$ is an ideal of $\mathcal{B}$ that contains $\mathcal{B}_{00}$, the finite-rank operators. (f) $A \in \mathcal{B}_2$ if and only if $|A| \equiv (A^*A)^{1/2} \in \mathcal{B}_2$; in this case $\|A\|_2 = \||A|\|_2$. (g) $\mathcal{B}_2 \subseteq \mathcal{B}_0$; moreover, if A is a compact operator and $\lambda_1, \lambda_2, \ldots$ are the eigenvalues of $|A|$, each repeated as often as its multiplicity, then $A \in \mathcal{B}_2(\mathcal{H})$ iff $\sum_{n=1}^\infty \lambda_n^2 < \infty$. In this case, $\|A\|_2 = (\sum \lambda_n^2)^{1/2}$. (h) If (X, Ω, μ) is a measure space and $k \in L^2(\mu \times \mu)$, let $K: L^2(\mu) \to L^2(\mu)$ be the integral operator with kernel k. Then $K \in \mathcal{B}_2(L^2(\mu))$ and $\|K\|_2 = \|k\|_2$ (see Proposition II.4.7 and Lemma II.4.8). (i) Interpret part (h) for a purely atomic measure space. More information on $\mathcal{B}_2$ is contained in the next exercise.

20. This exercise discusses trace-class operators (defined below) and assumes a knowledge of Exercise 19. $\mathcal{B}_1(\mathcal{H}) = \{AB: A \text{ and } B \in \mathcal{B}_2(\mathcal{H})\}$. Operators belonging to $\mathcal{B}_1(\mathcal{H})$ are called *trace-class operators* and $\mathcal{B}_1(\mathcal{H}) = \mathcal{B}_1$ is called the trace class. (a) If $A \in \mathcal{B}_1(\mathcal{H})$ and $\{e_i\}$ is a basis, then $\sum|\langle Ae_i, e_i\rangle| < \infty$. Moreover, the sum $\sum\langle Ae_i, e_i\rangle$ is independent of the choice of basis. (Hint: If $A = C^*B$, B, C in $\mathcal{B}_2$, show that $|\langle Ae_i, e_i\rangle| \equiv \frac{1}{2}(\|Be_i\|^2 + \|Ce_i\|^2)$.) (b) If $\{e_i\}$ is a basis for $\mathcal{H}$, define tr: $\mathcal{B}_1 \to \mathbb{C}$ by

$$\mathrm{tr}(A) = \sum_i \langle Ae_i, e_i\rangle.$$

By (a) the definition of tr(A) does not depend on the choice of a basis; tr(A) is called the *trace* of A. If dim $\mathcal{H} < \infty$, then tr(A) is precisely the sum of the diagonal terms of any matrix representation of A. (c) If $A \in \mathcal{B}(\mathcal{H})$, then the following are equivalent: (1) $A \in \mathcal{B}_1$; (2) $|A| = (A^*A)^{1/2} \in \mathcal{B}_1$; (3) $|A|^{1/2} \in \mathcal{B}_2$; (4) tr($|A|$) $< \infty$. (d) If $A \in \mathcal{B}_1$ and $T \in \mathcal{B}$, then AT and TA are in $\mathcal{B}_1$ and tr(AT) = tr(TA). Moreover, tr: $\mathcal{B}_1 \to \mathbb{C}$ is a positive linear functional such that if $A \in \mathcal{B}_1$, $A \geq 0$, and tr(A) = 0, then $A = 0$. (e) If $A \in \mathcal{B}_1$, define $\|A\|_1 \equiv$ tr($|A|$). If $A \in \mathcal{B}_1$ and $T \in \mathcal{B}$, show that $|\mathrm{tr}(TA)| \leq \|T\|\|A\|_1$. (f) $\|A\|_1 = \|A^*\|_1$ if $A \in \mathcal{B}_1$. (g) If $T \in \mathcal{B}$ and $A \in \mathcal{B}_1$, then $\|TA\|_1 \leq \|T\|\|A\|_1$ and $\|AT\|_1 \leq \|T\|\|A\|_1$. (h) $\|\cdot\|_1$ is a norm on $\mathcal{B}_1$. It is called the *trace norm*. (i) $\mathcal{B}_1$ is an ideal in $\mathcal{B}(\mathcal{H})$ that contains $\mathcal{B}_{00}$. (j) If $A \in \mathcal{B}_1$ and $\{e_i\}$ and $\{f_i\}$ are two bases for $\mathcal{H}$, then $\sum_i|\langle Ae_i, f_i\rangle| \leq \|A\|_1$. (d) $\mathcal{B}_1 \subseteq \mathcal{B}_0$. Also, if $A \in \mathcal{B}_0$ and $\lambda_1, \lambda_2, \ldots$ are the eigenvalues of $|A|$, each repeated as often as its multiplicity, then $A \in \mathcal{B}_1$ if and only if $\sum_{n=1}^{\infty}\lambda_n < \infty$. In this case, $\|A\|_1 = \sum_{n=1}^{\infty}\lambda_n$. (1) If A and $B \in \mathcal{B}_2$, define $(A, B) = \mathrm{tr}(B^*A)$. Then $(\cdot, \cdot)$ is an inner product on $\mathcal{B}_2$, $\|\cdot\|_2$ is the norm defined by this inner product, and $\mathcal{B}_2$ is $\|\cdot\|_2$ complete. In other words, $\mathcal{B}_2$ is a Hilbert space. (m) $(\mathcal{B}_1, \|\cdot\|_1)$ is a Banach space. (n) $\mathcal{B}_{00}$ is dense in both $\mathcal{B}_1$ and $\mathcal{B}_2$. (For more on these matters, see Ringrose [1971] and Schatten [1960].)

21. This exercise assumes a knowledge of Exercise 20. If $g, h \in \mathcal{H}$, let $g \otimes h$ denote the rank-one operator defined by $(g \otimes h)(f) = \langle f, h\rangle g$. (a) If $g, h \in \mathcal{H}$ and $A \in \mathcal{B}(\mathcal{H})$, tr($A(g \otimes h)$) = $\langle Ag, h\rangle$. (b) If $T \in \mathcal{B}_1$, then $\|T\|_1 = \sup\{|\mathrm{tr}(CT)|: C \in \mathcal{B}_0, \|C\| \leq 1\}$. (c) If $T \in \mathcal{B}_1$, define $L_T: \mathcal{B}_0 \to \mathbb{C}$ by $L_T(C) = \mathrm{tr}(TC) (= \mathrm{tr}(CT))$. Show that the map $T \mapsto L_T$ is an isometric isomorphism of $\mathcal{B}_1$ onto $\mathcal{B}_0^*$. (d) If $B \in \mathcal{B}$, define $F_B: \mathcal{B}_1 \to \mathbb{C}$ by $F_B(T) = \mathrm{tr}(BT)$. Show that $B \mapsto F_B$ is an isometric isomorphism of $\mathcal{B}$ onto $\mathcal{B}_1^*$. (e) If $L \in \mathcal{B}^*$ show that $L = L_0 + L_1$ where $L_0, L_1 \in \mathcal{B}^*$, $L_1(B) = \mathrm{tr}(BT)$ for some T in $\mathcal{B}_1$, and $L_0(C) = 0$ for every compact operator C. Show that $\|L\| = \|L_0\| + \|L_1\|$ and that L_0 and L_1 are unique.

22. Prove that if U is any unitary operator on $\mathcal{H}$, then there is a continuous function $u: [0, 1] \to \mathcal{B}(\mathcal{H})$ such that $u(t)$ is unitary for all t, $u(0) = U$, and $u(1) = 1$.

23. If N is normal, show that there is a sequence of invertible normal operators that converges to N.

§3. Star-Cyclic Normal Operators

Recall the definition of a reducing subspace and some of its equivalent formulations (Section II.3).

3.1.   Definition. A vector e_0 in $\mathcal{H}$ is a *star-cyclic vector* for A if $\mathcal{H}$ is the smallest reducing subspace for A that contains e_0. The operator A is *star cyclic* if it has a star-cyclic vector. A vector e_0 is *cyclic* for A if $\mathcal{H}$ is the smallest invariant subspace for A that contains e_0; A is *cyclic* if it has a cyclic vector.

3.2.   Proposition. (a) *A vector e_0 is a star-cyclic vector for A if and only if $\mathcal{H} = \operatorname{cl}\{Te_0: T \in C^*(A)\}$, where $C^*(A) =$ the C^*-algebra generated by A.* (b) *A vector e_0 is a cyclic vector for A if and only if $\mathcal{H} = \operatorname{cl}\{p(A)e_0: p = a$ polynomial $\}$.*

PROOF. Exercise.

Note that if e_0 is a star-cyclic vector for A, then it is a cyclic vector for the algebra $C^*(A)$.

3.3.   Proposition. *If A has either a cyclic or a star-cyclic vector, then $\mathcal{H}$ is separable.*

PROOF. It is easy to see that $C^*(A)$ and $\{p(A): p = $ a polynomial$\}$ are separable subalgebras of $\mathcal{B}(\mathcal{H})$. Now use (3.2). ∎

Let μ be a compactly supported measure on $\mathbb{C}$ and let N_μ be defined on $L^2(\mu)$ as in Example 2.5. If $K = $ support μ, then $C^*(N_\mu) = \{M_u: u \in C(K)\}$. Since $C(K)$ is dense in $L^2(\mu)$, it follows that 1 is a star-cyclic vector for N_μ. The converse of this is also true.

3.4.   Theorem. *A normal operator N is star-cyclic if and only if N is unitarily equivalent to N_μ for some compactly supported measure μ on $\mathbb{C}$. If e_0 is a star-cyclic vector for N, then μ can be chosen such that there is an isomorphism $V: \mathcal{H} \to L^2(\mu)$ with $Ve_0 = 1$ and $VNV^{-1} = N_\mu$. Under these conditions, V is unique.*

PROOF. If $N \cong N_\mu$, then we have already seen that N is star cyclic. So suppose that N has a star-cyclic vector e_0. If E is the spectral measure for N, put $\mu(\Delta) = \|E(\Delta)e_0\|^2 = \langle E(\Delta)e_0, e_0 \rangle$ for every Borel subset Δ of $\mathbb{C}$ (see Lemma 1.9). Let $K = $ support μ.

If $\phi \in B(K)$, then (2.4) implies

$$
\begin{aligned}
\|\phi(N)e_0\|^2 &= \langle \phi(N)e_0, \phi(N)e_0 \rangle \\
&= \langle |\phi|^2(N)e_0, e_0 \rangle \\
&= \int |\phi(z)|^2 \, d\langle E(z)e_0, e_0 \rangle \\
&= \int |\phi|^2 \, d\mu.
\end{aligned}
$$

So if $B(K)$ is considered as a submanifold of $L^2(\mu)$, $V\phi = \phi(N)e_0$ defines an isometry from $B(K)$ onto $\{\phi(N)e_0 : \phi \in B(K)\}$. But e_0 is a star-cyclic vector, so the range of V is dense in $\mathcal{H}$. Hence V extends to an isomorphism $V: L^2(\mu) \to \mathcal{H}$.

If $\phi \in B(K)$, then $VN_\mu V^{-1}(\phi(N)e_0) = VN_\mu(\phi) = V(z\phi) = N\phi(N)e_0$. Hence $VN_\mu V^{-1} = N$ on $\{\phi(N)e_0 : \phi \in B(K)\}$, which is dense in $\mathcal{H}$. So $VN_\mu V^{-1} = N$.

The proof of the uniqueness statement is an exercise. ∎

Any theorem about the operators N_μ is a theorem about star-cyclic normal operators. With this in mind, the next theorem gives a complete unitary invariant for star-cyclic normal operators. But first, a definition.

3.5. Definition. Two measures, μ_1 and μ_2, are *mutually absolutely continuous* if they have the same sets of measure zero; that is, $\mu_1(\Delta) = 0$ if and only if $\mu_2(\Delta) = 0$. This will be denoted by $[\mu_1] = [\mu_2]$. (The more standard notation in the literature is $\mu_1 \equiv \mu_2$, but this seems insufficient.) If $[\mu_1] = [\mu_2]$, then the Radon–Nikodym derivatives $d\mu_1/d\mu_2$ and $d\mu_2/d\mu_1$ are well defined. Say that μ_1 and μ_2 are *boundedly mutually absolutely continuous* if $[\mu_1] = [\mu_2]$ and the Radon–Nikodym derivatives are essentially bounded functions.

3.6. Theorem. $N_{\mu_1} \cong N_{\mu_2}$ *if and only if* $[\mu_1] = [\mu_2]$.

PROOF. Suppose $[\mu_1] = [\mu_2]$ and put $\phi = d\mu_1/d\mu_2$. So if $g \in L^1(\mu_1)$, $g\phi \in L^1(\mu_2)$ and $\int g\phi \, d\mu_2 = \int g \, d\mu_1$. Hence, if $f \in L^2(\mu_1)$, $\sqrt{\phi}f \in L^2(\mu_2)$ and $\|\sqrt{\phi}f\|_2 = \|f\|_2$; that is, $U: L^2(\mu_1) \to L^2(\mu_2)$ defined by $Uf = \sqrt{\phi}f$ is an isometry. If $g \in L^2(\mu_2)$, then $f = \phi^{-1/2}g \in L^2(\mu_1)$ and $Uf = g$; hence U is surjective and $U^{-1}g = \phi^{-1/2}g$ for g in $L^2(\mu_2)$. If $g \in L^2(\mu_2)$, then $UN_{\mu_1}U^{-1}g = UN_{\mu_1}\phi^{-1/2}g = Uz\phi^{-1/2}g = zg$, and so $UN_{\mu_1}U^{-1} = N_{\mu_2}$.

Now assume that $V: L^2(\mu_1) \to L^2(\mu_2)$ is an isomorphism such that $VN_{\mu_1}V^{-1} = N_{\mu_2}$. Put $\psi = V(1)$; so $\psi \in L^2(\mu_2)$. For convenience, put $N_j = N_{\mu_j}$, $j = 1, 2$. It is easy to see that $VN_1^k V^{-1} = N_2^k$ and $VN_1^{*k}V^{-1} = N_2^{*k}$.

Hence $Vp(N_1, N_1^*)V^{-1} = p(N_2, N_2^*)$ for any polynomial p in z and $\bar{z}$. Since $N_1 \cong N_2$, $\sigma(N_1) = \sigma(N_2)$; hence support $\mu_1 = $ support $\mu_2 = K$. By taking uniform limits of polynomials in z and $\bar{z}$, $Vu(N_1)V^{-1} = u(N_2)$ for u in $C(K)$. Hence for u in $C(K)$, $V(u) = Vu(N_1)1 = u(N_2)V1 = u\psi$. Because V is an isometry, this implies that $\int|u|^2 \, d\mu_1 = \int|u|^2|\psi|^2 \, d\mu_2$ for every u in $C(K)$. Hence $\int v \, d\mu_1 = \int v|\psi|^2 \, d\mu_2$ for v in $C(K)$, $v \geq 0$. By the uniqueness part of the Riesz Representation Theorem, $\mu_1 = |\psi|^2\mu_2$, so $\mu_1 \ll \mu_2$.

By using V^{-1} instead of V and reversing the roles of N_1 and N_2 in the preceding argument, it follows that $\mu_2 \ll \mu_1$. Hence $[\mu_1] = [\mu_2]$. ∎

EXERCISES

1. If μ is a compactly supported measure on $\mathbb{C}$ and $f \in L^2(\mu)$, f is a star-cyclic vector for N if and only if $\mu(\{x: f(x) = 0\}) = 0$.

2. Prove Proposition 3.2.

3. If μ_1 and μ_2 are compactly supported measures on $\mathbb{C}$, show that the following statements are equivalent: (a) μ_1 and μ_2 are boundedly mutually absolutely continuous; (b) there is an isomorphism $V: L^2(\mu_1) \to L^2(\mu_2)$ such that $VN_{\mu_1}V^{-1} = N_{\mu_2}$ and $VL^\infty(\mu_1) = L^\infty(\mu_2)$; (c) there is a bounded bijection $R: L^2(\mu_1) \to L^2(\mu_2)$ such that $Rp(z, \bar{z}) = p(z, \bar{z})$ for every polynomial in z and $\bar{z}$.

4. Show that if N is a star-cyclic normal operator and $\lambda \in \sigma_p(N)$, then $\dim\ker(N - \lambda) = 1$.

5. If N is diagonalizable and star cyclic and if $\sigma_p(N) = \{\lambda_1, \lambda_2, \dots\}$, show that N is unitarily equivalent to N_μ, where $\mu = \sum_{n=1}^\infty 2^{-n}\delta_{\lambda_n}$ (see Exercise 2.11).

6. Let N be a diagonalizable normal operator. Show that $N \cong M$ if and only if M is a diagonalizable normal operator, $\sigma_p(N) = \sigma_p(M)$, and $\dim\ker(N - \lambda) = \dim\ker(M - \lambda)$ for all λ. (Compare this with Theorem II.8.3.)

7. Let U be the bilateral shift on $l^2(\mathbb{Z})$. If e_0 is the vector in $l^2(\mathbb{Z})$ that has 1 in the zeroth place and zeros elsewhere, then e_0 is a star-cyclic vector for U. If μ is the compactly supported measure on $\mathbb{C}$ and $V: l^2(\mathbb{Z}) \to L^2(\mu)$ is the isomorphism such that $Ve_0 = 1$ and $VUV^{-1} = N_\mu$, then
(a) $\mu = m = $ normalized arc length on $\partial\mathbb{D}$;
(b) $V^{-1} = $ the Fourier transform on $L^2(m) = L^2(\partial\mathbb{D})$.

8. Suppose $N_1, \dots, N_d$ are normal operators such that $N_j N_k^* = N_k^* N_j$ for $1 \leq j, k \leq d$ and suppose there is a vector e_0 in $\mathcal{H}$ such that $\mathcal{H}$ is the only subspace of $\mathcal{H}$ containing e_0 that reduces each of the operators $N_1, \dots, N_d$. Show that there is a compactly supported measure μ on $\mathbb{C}^d$ and an isomorphism $V: \mathcal{H} \to L^2(\mu)$ such that $VN_k V^{-1}f = z_k f$ for f in $L^2(\mu)$ and $1 \leq k \leq d$ ($z_k = $ the kth coordinate function) (see Exercise 2.17).

§4. Some Applications of the Spectral Theorem

In this section a few diverse applications of the Spectral Theorem are presented. These will show the power and finesse of the Spectral Theorem as well as demonstrate some of the methods used to apply it. One result in this section (Theorem 4.6) is more than an application. Indeed, many regard this as the optimal statement of the Spectral Theorem.

If N is a normal operator and $N = \int z \, dE(z)$ is its spectral representation, then $\phi \mapsto \phi(N) \equiv \int \phi \, dE$ is a $*$-homomorphism of $B(\mathbb{C})$ into $\mathscr{B}(\mathscr{H})$. Thus, if $\phi, \psi \in B(\mathbb{C})$, $(\int \phi \, dE)(\int \psi \, dE) = \int \phi \psi \, dE$ and $\|\int \phi \, dE\| \leq \sup\{|\phi(z)|: z \in \sigma(N)\}$.

4.1. Proposition. *If N is a normal operator and $N = \int z \, dE(z)$, then N is compact if and only if for every $\varepsilon > 0$, $E(\{z: |z| > \varepsilon\})$ has finite rank.*

PROOF. If $\varepsilon > 0$, let $\Delta_\varepsilon = \{z: |z| > \varepsilon\}$ and $E_\varepsilon = E(\Delta_\varepsilon)$. Then

$$N - NE_\varepsilon = \int z \, dE(z) - \int z \chi_{\Delta_\varepsilon}(z) \, dE(z)$$

$$= \int z \chi_{\mathbb{C} \setminus \Delta_\varepsilon}(z) \, dE(z) = \phi(N)$$

where $\phi(z) = z\chi_{\mathbb{C} \setminus \Delta_\varepsilon}(z)$. Thus $\|N - NE_\varepsilon\| \leq \sup\{|z|: z \in \mathbb{C} \setminus \Delta_\varepsilon\} \leq \varepsilon$. If E_ε has finite rank for every $\varepsilon > 0$, then so does NE_ε. Thus $N \in \mathscr{B}_0(\mathscr{H})$.

Now assume that N is compact and let $\varepsilon > 0$. Put $\phi(z) = z^{-1}\chi_{\Delta_\varepsilon}(z)$; so $\phi \in B(\mathbb{C})$. Since N is compact, so is $N\phi(N)$. But $N\phi(N) = \int z z^{-1}\chi_{\Delta_\varepsilon}(z) \, dE(z) = E_\varepsilon$. Since E_ε is a compact projection, it must have finite rank. (Why?) ■

The preceding result could have been proved by using the fact that compact normal operators are diagonalizable and the eigenvalues must converge to 0.

4.2. Theorem. *If $\mathscr{H}$ is separable and I is an ideal of $\mathscr{B}(\mathscr{H})$ that contains a noncompact operator, then $I = \mathscr{B}(\mathscr{H})$.*

PROOF. If $A \in I$ and $A \notin \mathscr{B}_0(\mathscr{H})$, consider A^*A; let $A^*A = \int t \, dE(t)$ ($\sigma(A^*A) \subseteq [0, \infty)$). By the preceding proposition, there is an $\varepsilon > 0$ such that $P = E(\varepsilon, \infty)$ has infinite rank. But $P = (\int t^{-1}\chi_{(\varepsilon, \infty)}(t) \, dE(t))A^*A \in I$. Since $\mathscr{H}$ is separable, $\dim P\mathscr{H} = \dim \mathscr{H} = \aleph_0$. Let $U: \mathscr{H} \to P\mathscr{H}$ be a surjective isometry. It is easy to check that $1 = U^*PU$. But $P \in I$, so $1 \in I$. Hence $I = \mathscr{B}(\mathscr{H})$. ■

In Proposition VIII.4.10, it was shown that every nonzero ideal of $\mathscr{B}(\mathscr{H})$ contains the finite-rank operators. When combined with the preceding result, this yields the following.

4.3.   Corollary. *If $\mathcal{H}$ is separable, then the only nontrivial closed ideal of $\mathcal{B}(\mathcal{H})$ is the ideal of compact operators.*

The next proposition is related to Theorem VIII.5.9. Indeed, it is a consequence of it so that the proof will only be sketched.

Let N be a normal operator on $\mathcal{H}$ and for every vector e in $\mathcal{H}$ let $\mathcal{H}_e \equiv \{ N^{*k}N^j e : k, j \geq 0 \}$. So $\mathcal{H}_e$ is the smallest subspace of $\mathcal{H}$ that contains e and reduces N. Also, $N|\mathcal{H}_e$ is a star-cyclic normal operator.

4.4.   Proposition. *If N is a normal operator on $\mathcal{H}$, then there are reducing subspaces $\{ \mathcal{H}_i : i \in I \}$ for N such that $\mathcal{H} = \oplus_i \mathcal{H}_i$ and $N|\mathcal{H}_i$ is star cyclic.*

PROOF. Using Zorn's Lemma find a maximal set of vectors $\mathcal{E}$ in $\mathcal{H}$ such that if $e, f \in \mathcal{E}$ and $e \neq f$, then $\mathcal{H}_e \perp \mathcal{H}_f$. It follows that $\mathcal{H} = \oplus_e \mathcal{H}_e$.   ∎

4.5.   Corollary. *Every normal operator is unitarily equivalent to the direct sum of star-cyclic normal operators.*

By combining the preceding proposition with Theorem 3.4 on the representation of star-cyclic normal operators we can obtain the following theorem.

4.6.   Theorem. *If N is a normal operator on $\mathcal{H}$, then there is a measure space (X, Ω, μ) and a function ϕ in $L^\infty(X, \Omega, \mu)$ such that N is unitarily equivalent to M_ϕ on $L^2(X, \Omega, \mu)$.*

PROOF. If $\mathcal{M}$ is a reducing subspace for N then $N \cong N|\mathcal{M} \oplus N|\mathcal{M}^\perp$; thus $\sigma(N|\mathcal{M}) \subseteq \sigma(N)$. So if $\{N_i\}$ is a collection of star-cyclic normal operators such that $N \cong \oplus_i N_i$ (4.5), then $\sigma(N_i) \subseteq \sigma(N)$ for every N_i. By Theorem 3.4 there is a measure μ_i supported on $\sigma(N)$ such that $N_i \cong N_{\mu_i}$. Let $X_i = $ the support of μ_i and let $\Omega_i = $ the Borel subsets of X_i. Let $X = $ the disjoint union of $\{X_i\}$. Define Ω to be the collection of all subsets Δ of X such that $\Delta \cap X_i \in \Omega_i$ for all i. It is easy to check that Ω is a σ-algebra. If $\Delta \in \Omega$ let $\mu(\Delta) \equiv \Sigma_i \mu_i(\Delta \cap X_i)$; then (X, Ω, μ) is a measure space. If $f \in L^2(X, \Omega, \mu)$ then $f_i = f|X_i \in L^2(\mu_i)$. Moreover, the map $U : L^2(\mu) \to \oplus_i L^2(\mu_i)$ defined by $Uf = \oplus_i (f|X_i)$ is easily seen to be an isomorphism. Define $\phi : X \to \mathbb{C}$ by letting $\phi(z) = z$ if $z \in X_i$ ($\subseteq \mathbb{C}$); since $X_i \subseteq \sigma(N)$ for every i, ϕ is a bounded function. If G is an open subset of $\mathbb{C}$, $\phi^{-1}(G) \cap X_i = G \cap X_i \in \Omega_i$; hence ϕ is Ω-measurable. Therefore $\phi \in L^\infty(X, \Omega, \mu)$. It is left to the reader to check that $UM_\phi U^{-1} = \oplus_i N_{\mu_i} \cong N$.   ∎

4.7.   Proposition. *If $\mathcal{H}$ is separable, then the measure space in Theorem 4.6 is σ-finite.*

PROOF. This is true because if $L^2(X, \Omega, \mu)$ is separable, then (X, Ω, μ) must be σ-finite. Indeed, let $\mathcal{E}$ be a collection of pairwise disjoint sets from Ω

having nonzero measure. A computation shows that $\{(\mu(\Delta))^{-1/2}\chi_\Delta: \Delta \in \mathscr{E}\}$ are pairwise orthogonal vectors in $L^2(\mu)$. If $L^2(\mu)$ is separable, then $\mathscr{E}$ must be countable. Therefore (X, Ω, μ) is σ-finite. ∎

Of course if (X, Ω, μ) is finite it is not necessarily true that $L^2(\mu)$ is separable.

The next result will be useful later in this book and it also provides a different type of application of the Spectral Theorem.

4.8. Proposition. *If $\mathscr{A}$ is an SOT-closed C^*-subalgebra of $\mathscr{B}(\mathscr{H})$, then $\mathscr{A}$ is the norm closed linear span of the projections in $\mathscr{A}$.*

PROOF. If $A \in \mathscr{A}$, $A + A^*$ and $A - A^* \in \mathscr{A}$; hence $\mathscr{A}$ is the linear span of Re $\mathscr{A}$. Suppose $A \in$ Re $\mathscr{A}$ and $A = \int t\, dE(t)$. If $[a, b] \subseteq \mathbb{R}$, then there is a sequence $\{u_n\}$ in $C(\mathbb{R})$ such that $0 \le u_n \le 1$, $u_n(t) = 1$ for $a \le t \le b - n^{-1}$, $u_n(t) = 0$ for $t \le a - n^{-1}$ and $t \ge b$. Hence $u_n(t) \to \chi_{[a,b)}(t)$ as $n \to \infty$. If $h \in \mathscr{H}$, then

$$\|\{u_n(A) - E[a, b)\}h\|^2 = \int |u_n(t) - \chi_{[a,b)}(t)|^2\, dE_{h,h}(t) \to 0$$

by the Lebesgue Dominated Convergence Theorem. That is, $u_n(A) \to E[a, b)$ (SOT). Since $\mathscr{A}$ is SOT-closed, $E[a, b) \in \mathscr{A}$. Now let (α, β) be an open interval containing $\sigma(A)$. If $\varepsilon > 0$, then there is a partition $\{\alpha = t_0 < \cdots < t_n = \beta\}$ such that $|t - \sum_{k=1}^n t_k \chi_{[t_{k-1}, t_k)}(t)| < \varepsilon$ for t in $\sigma(A)$; hence $\|A - \sum_{k=1}^\infty t_k E[t_{k-1}, t_k)\| < \varepsilon$. Thus every self-adjoint operator in $\mathscr{A}$ belongs to the closed linear span of the projections in $\mathscr{A}$. ∎

EXERCISES

1. If N is a normal operator show that ran N is closed if and only if 0 is an isolated point of $\sigma(N)$.

2. Give an example of a non-normal operator A such that 0 is an isolated point of $\sigma(A)$ and ran A is closed. Give an example of a non-normal operator B such that ran B is closed and 0 is not an isolated point of $\sigma(B)$.

3. If $\mathscr{H}$ is a nonseparable Hilbert space find an example of a nontrivial closed ideal of $\mathscr{B}(\mathscr{H})$ that is different from $\mathscr{B}_0(\mathscr{H})$.

4. Let (X, Ω, μ) be the measure space obtained in the proof of Theorem 4.6 and show that $L^1(X, \Omega, \mu)^*$ is isometrically isomorphic to $L^\infty(X, \Omega, \mu)$.

5. Show that $\mathscr{H}$ is separable if and only if every collection of pairwise orthogonal projections in $\mathscr{B}(\mathscr{H})$ is countable.

6. If (X, Ω, μ) is a measure space, then (X, Ω, μ) is σ-finite if and only if every collection of pairwise orthogonal projections in $\{M_\phi: \phi \in L^\infty(\mu)\}$ is countable.

7. If $N = \int z\, dE(z)$ and $\varepsilon > 0$, show that ran $E(\{z: |z| > \varepsilon\}) \subseteq$ ran N.

8. Let $\mathscr{M}$ be a linear manifold in $\mathscr{H}$ and show that $\mathscr{M}$ has the property that $\mathscr{M}$ contains no closed infinite-dimensional subspaces if and only if whenever $A \in \mathscr{B}(\mathscr{H})$ and ran $A \subseteq \mathscr{M}$, then A is compact.

9. Show that the extreme points of $\{A \in \mathscr{B}(\mathscr{H}): 0 \le A \le 1\}$ are the projections.

§5. Topologies on $\mathscr{B}(\mathscr{H})$

In this section some results on the SOT and WOT on $\mathscr{B}(\mathscr{H})$ are presented. These results are necessary for understanding some of the results that are to follow in later sections and also for a proper comprehension of a number of other subjects in mathematics.

The first result appeared as Exercise 1.4.

5.1. Proposition. *If $L: \mathscr{B}(\mathscr{H}) \to \mathbb{C}$ is a linear functional, then the following statements are equivalent.*

(a) *L is SOT continuous.*

(b) *L is WOT continuous.*

(c) *There are vectors $g_1, \ldots, g_n, h_1, \ldots, h_n$ in $\mathscr{H}$ such that $L(A) = \sum_{k=1}^{n}\langle Ag_k, h_k \rangle$ for every A in $\mathscr{B}(\mathscr{H})$.*

PROOF. Clearly (c) implies (b) and (b) implies (a). So assume (a). By (IV.3.1f) there are vectors $g_1, \ldots, g_n$ in $\mathscr{H}$ such that

$$|L(A)| \le \sum_{k=1}^{n} \|Ag_k\| \le \sqrt{n}\left[\sum_{k=1}^{n}\|Ag_k\|^2\right]^{1/2}$$

for every A in $\mathscr{B}(\mathscr{H})$. Replacing g_k by $\sqrt{n}\, g_k$, it may be assumed that

$$|L(A)| \le \left[\sum_{k=1}^{n}\|Ag_k\|^2\right]^{1/2} \equiv p(A).$$

Now p is a seminorm and $p(A) = 0$ implies $L(A) = 0$. Let $\mathscr{K} = \text{cl}\{Ag_1 \oplus Ag_2 \oplus \cdots \oplus Ag_n: A \in \mathscr{B}(\mathscr{H})\}$; so $\mathscr{K} \subseteq \mathscr{H} \oplus \cdots \oplus \mathscr{H}$ (n times). Note that if $Ag_1 \oplus \cdots \oplus Ag_n = 0$, $p(A) = 0$, and hence, $L(A) = 0$. Thus $F(Ag_1 \oplus \cdots \oplus Ag_n) = L(A)$ is a well-defined linear functional on a dense manifold in $\mathscr{K}$. But

$$|F(Ag_1 \oplus \cdots \oplus Ag_n)| \le p(A) = \|Ag_1 \oplus \cdots \oplus Ag_n\|.$$

So F can be extended to a bounded linear functional F_1 on $\mathscr{H}^{(n)}$. Hence, there are vectors $h_1, \ldots, h_n$ in $\mathscr{H}$ such that

$$F_1(f_1 \oplus \cdots \oplus f_n) = \langle f_1 \oplus \cdots \oplus f_n, h_1 \oplus \cdots \oplus h_n \rangle$$

$$= \sum_{k=1}^{n}\langle f_k, h_k \rangle.$$

In particular, $L(A) = F(Ag_1 \oplus \cdots \oplus Ag_n) = \sum_{k=1}^{n}\langle Ag_k, h_k \rangle$. ∎

5.2.   Corollary. *If $\mathscr{C}$ is a convex subset of $\mathscr{B}(\mathscr{H})$, the* WOT *closure of $\mathscr{C}$ equals the* SOT *closure of $\mathscr{C}$.*

PROOF. Combine the preceding proposition with Corollary V.1.4.   ∎

When discussing the closure (WOT or SOT) of a convex set it is usually better to discuss the SOT. Shortly an "algebraic" characterization of the SOT closure of a subalgebra of $\mathscr{B}(\mathscr{H})$ will be given. But first recall (VIII.5.3) that if $1 \le n \le \infty$, $\mathscr{H}^{(n)}$ denotes the direct sum of $\mathscr{H}$ with itself n times ($\aleph_0$ times if $n = \infty$). If $A \in \mathscr{B}(\mathscr{H})$, $A^{(n)}$ is the operator on $\mathscr{H}^{(n)}$ defined by $A^{(n)}(h_1, \ldots, h_n) = (Ah_1, \ldots, Ah_n)$. If $\mathscr{S} \subset \mathscr{B}(\mathscr{H})$, $\mathscr{S}^{(n)} \equiv \{A^{(n)}: A \in \mathscr{S}\}$. It is rather interesting that the SOT closure of an algebra can be characterized using its lattice of invariant subspaces.

5.3.   Proposition. *If $\mathscr{A}$ is a subalgebra of $\mathscr{B}(\mathscr{H})$ containing 1, then the* SOT *closure of $\mathscr{A}$ is*

5.4 $\{B \in \mathscr{B}(\mathscr{H}): \text{ for every finite } n, \text{ Lat } \mathscr{A}^{(n)} \subseteq \text{Lat } B^{(n)}\}.$

PROOF. It is left as an exercise for the reader to show that if $B \in \text{SOT} - \text{cl } \mathscr{A}$, B belongs to the set (5.4). Now assume that B belongs to the set (5.4). Fix $f_1, f_2, \ldots, f_n$ in $\mathscr{H}$ and $\varepsilon > 0$. It must be shown that there is an A in $\mathscr{A}$ such that $\|(A - B)f_k\| < \varepsilon$ for $1 \le k \le n$.

Let $\mathscr{M} = \bigvee\{(Af_1, \ldots, Af_n): A \in \mathscr{A}\}$. Because $\mathscr{A}$ is an algebra, $\mathscr{M} \in \text{Lat } \mathscr{A}^{(n)}$; hence $\mathscr{M} \in \text{Lat } B^{(n)}$. Because $1 \in \mathscr{A}$, $(f_1, \ldots, f_n) \in \mathscr{M}$. Since $\{(Af_1, \ldots, Af_n); A \in \mathscr{A}\}$ is a dense manifold and $(Bf_1, \ldots, Bf_n) \in \mathscr{M}$, there is an A in $\mathscr{A}$ with $\varepsilon^2 > \sum_{k=1}^n \|(A - B)f_k\|^2$; hence $B \in \text{SOT} - \text{cl } \mathscr{A}$.   ∎

5.5.   Proposition. *The closed unit ball of $\mathscr{B}(\mathscr{H})$ is* WOT *compact.*

PROOF. The proof of this proposition follows along the lines of the proof of Alaoglu's Theorem. For each h in ball $\mathscr{H}$ let $X_h = $ a copy of ball $\mathscr{H}$ with the weak topology. Put $X = \Pi\{X_h: \|h\| \le 1\}$. If $A \in $ ball $\mathscr{B}(\mathscr{H})$ let $\tau(A) \in X$ defined by $\tau(A)_h = Ah$. Give X the product topology. Then τ: (ball $\mathscr{B}(\mathscr{H}), \text{WOT}) \to X$ is a continuous function and a homeomorphism onto its image (verify). Now show that $\tau(\text{ball } \mathscr{B}(\mathscr{H}))$ is closed in X. From here it follows that ball $\mathscr{B}(\mathscr{H})$ is WOT compact.   ∎

EXERCISES

1. Show that if $B \in \text{SOT} - \text{cl } \mathscr{A}$, then B belongs to the set defined in (5.4).

2. Show that $\mathscr{B}_{00}$ is SOT dense in $\mathscr{B}$.

3. If $\{A_k\}$ and $\{B_k\}$ are sequences in $\mathscr{B}(\mathscr{H})$ such that $A_k \to A(\text{WOT})$ and $B_k \to B(\text{SOT})$, then $A_k B_k \to AB(\text{WOT})$.

4. With the notation of Exercise 3, show that if $A_k \to A$(SOT), then $A_k B_k \to AB$(SOT).

5. Let S be the unilateral shift on $l^2(\mathbb{N})$ (II.2.10). Examine the sequences $\{S^k\}$ and $\{S^{*k}\}$ and their relation to Exercises 3 and 4.

6. (Halmos.) Fix an orthonormal basis $\{e_n: n \geq 1\}$ for $\mathcal{H}$. (a) Show that $0 \in$ weak closure of $\{\sqrt{n}\, e_n: n \geq 1\}$ (Halmos [1982], Solution 28). (b) Let $\{n_i\}$ be a net of integers such that $\sqrt{n_i}\, e_{n_i} \to 0$ weakly. Define $A_i f = \sqrt{n_i}\, \langle f, e_{n_i}\rangle e_{n_i}$ for f in $\mathcal{H}$. Show that $A_i \to 0$ (SOT) but $\{A_i^2\}$ does not converge to 0 (SOT).

§6. Commuting Operators

If $\mathcal{S} \subseteq \mathcal{B}(\mathcal{H})$, let $\mathcal{S}' \equiv \{A \in \mathcal{B}(\mathcal{H}): AS = SA \text{ for every } S \text{ in } \mathcal{S}\}$. $\mathcal{S}'$ is called the *commutant* of $\mathcal{S}$. It is not difficult to see that $\mathcal{S}'$ is always an algebra. Similarly, $\mathcal{S}'' \equiv (\mathcal{S}')'$ is called the *double commutant* of $\mathcal{S}$. This process can continue, but (thankfully) $\mathcal{S}''' = \mathcal{S}'$ (Exercise 1). In some circumstances, $\mathcal{S} = \mathcal{S}''$.

The problem of determining the commutant or double commutant of a single operator or a collection of operators leads to some exciting and interesting mathematics. The commutant is an algebraic object and the idea is to bring the force of analysis to bear in the characterization of this algebra.

We begin by examining the commutant of a direct sum of operators. Recall that if $\mathcal{H} = \mathcal{H}_1 \oplus \mathcal{H}_2 \oplus \cdots$ and $A_n \in \mathcal{B}(\mathcal{H}_n)$ for $n \geq 1$, then $A = A_1 \oplus A_2 \oplus \cdots$ defines a bounded operator on $\mathcal{H}$ if and only if $\sup_n \|A_n\| < \infty$; in this case $\|A\| = \sup_n \|A_n\|$. Also, each operator B on $\mathcal{H}$ has a matrix representation $[B_{ij}]$ where $B_{ij} \in \mathcal{B}(\mathcal{H}_j, \mathcal{H}_i)$.

6.1. Proposition. (a) *If $A = A_1 \oplus A_2 \oplus \cdots$ is a bounded operator on $\mathcal{H} = \mathcal{H}_1 \oplus \mathcal{H}_2 \oplus \cdots$ and $B = [B_{ij}] \in \mathcal{B}(\mathcal{H})$, then $AB = BA$ if and only if $B_{ij}A_j = A_i B_{ij}$ for all i, j.*

(b) *If $B = [B_{ij}] \in \mathcal{B}(\mathcal{H}^{(n)})$, $BA^{(n)} = A^{(n)}B$ if and only if $B_{ij}A = AB_{ij}$ for all i, j.*

The proof of this proposition is an easy exercise in matrix manipulation and is left to the reader.

6.2. Proposition. *If $A \in \mathcal{B}(\mathcal{H})$ and $1 \leq n \leq \infty$, then $\{A^{(n)}\}'' = \{B^{(n)}: B \in \{A\}''\} = \{\{A\}''\}^{(n)}$.*

PROOF. The second equality in the statement is a tautology and it is the first equality that forms the substance of the proposition. If $B \in \{A\}''$, then the preceding proposition implies that $B^{(n)} \in \{A^{(n)}\}''$. Now let $B \in \{A^{(n)}\}''$. To simplify the notation, assume $n = 2$. So $B \in \{A \oplus A\}''$; let $B = [B_{ij}]$,

$B_{ij} \in \mathcal{B}(\mathcal{H})$. Since $\begin{bmatrix} 0 & 1 \\ 0 & 0 \end{bmatrix} \in \{A \oplus A\}'$, matrix multiplication shows that $B_{11} = B_{22}$ and $B_{21} = 0$. Similarly, the fact that $\begin{bmatrix} 0 & 0 \\ 1 & 0 \end{bmatrix}$ commutes with $A \oplus A$ implies that $B_{12} = 0$. If $C = B_{11} \ (= B_{22})$, $B = C \oplus C$. If $T \in \{A\}'$, then $T \oplus T \in \{A \oplus A\}'$, so $B(T \oplus T) = (T \oplus T)B$. This shows that $C \in \{A\}''$. ∎

6.3. Corollary. *If* $\mathcal{S} \subseteq \mathcal{B}(\mathcal{H})$, $\{\mathcal{S}^{(n)}\}'' = \{\mathcal{S}''\}^{(n)}$.

Say that a subspace $\mathcal{M}$ of $\mathcal{H}$ reduces a collection $\mathcal{S}$ of operators if it reduces each operator in $\mathcal{S}$. By Proposition II.3.7, $\mathcal{M}$ reduces $\mathcal{S}$ if and only if the projection of $\mathcal{H}$ onto $\mathcal{M}$ belongs to $\mathcal{S}'$. This is important in the next theorem, due to von Neumann [1929].

6.4. The Double Commutant Theorem. *If* $\mathcal{A}$ *is a* C^*-*subalgebra of* $\mathcal{B}(\mathcal{H})$ *containing* 1, *then* SOT − cl $\mathcal{A}$ = WOT − cl $\mathcal{A} = \mathcal{A}''$.

PROOF. By Corollary 5.2, WOT − cl $\mathcal{A}$ = SOT − cl $\mathcal{A}$. Also, since $\mathcal{A}''$ is SOT closed (Exercise 2) and $\mathcal{A} \subseteq \mathcal{A}''$, SOT − cl $\mathcal{A} \subseteq \mathcal{A}''$.

It remains to show that $\mathcal{A}'' \subseteq$ SOT − cl $\mathcal{A}$. To do this Proposition 5.3 will be used.

Let $B \in \mathcal{A}''$, $n \geq 1$, and let $\mathcal{M} \in \text{Lat } \mathcal{A}^{(n)}$. It must be shown that $B^{(n)}\mathcal{M} \subseteq \mathcal{N}$. Because $\mathcal{A}$ is a C^*-algebra, so is $\mathcal{A}^{(n)}$. So the fact that $\mathcal{M} \in \text{Lat } \mathcal{A}^{(n)}$ and $A^{*(n)} \in \mathcal{A}^{(n)}$ whenever $A^{(n)} \in \mathcal{A}^{(n)}$ implies that $\mathcal{M}$ reduces $A^{(n)}$ for each A in $\mathcal{A}$. So if P is the projection of $\mathcal{H}^{(n)}$ onto $\mathcal{M}$, $P \in \{\mathcal{A}^{(n)}\}'$. But $B \in \mathcal{A}''$; so by Corollary 6.3, $B^{(n)} \in \{\mathcal{A}^{(n)}\}''$. Hence $B^{(n)}P = PB^{(n)}$ and $\mathcal{M} \in \text{Lat } B^{(n)}$. ∎

6.5. Corollary. *If* $\mathcal{A}$ *is a* SOT *closed* C^*-*subalgebra of* $\mathcal{B}(\mathcal{H})$ *and* $A \in \mathcal{B}(\mathcal{H})$ *such that* $A(P\mathcal{H}) \subseteq P\mathcal{H}$ *for every projection* P *in* $\mathcal{A}'$, *then* $A \in \mathcal{A}$.

PROOF. This uses, in addition to the Double Commutant Theorem, Proposition 4.8 as applied to $\mathcal{A}'$. Indeed, $\mathcal{A}'$ is a SOT closed C^*-algebra and hence it is the norm-closed linear span of its projections. So if $A \in \mathcal{B}(\mathcal{H})$ and $AP\mathcal{H} \subseteq P\mathcal{H}$ for every projection P in $\mathcal{A}'$, then $A(1 - P)\mathcal{H} \subseteq (1 - P)\mathcal{H}$ for every projection P in $\mathcal{A}'$. Thus $P\mathcal{H}$ reduces A and, hence, $AP = PA$. By (4.8), $A \in \mathcal{A}'' = \mathcal{A}$. ∎

6.6. Theorem. *If* (X, Ω, μ) *is a* σ-*finite measure space and* $\phi \in L^{\infty}(\mu)$, *define* M_{ϕ} *on* $L^2(\mu)$ *by* $M_{\phi}f = \phi f$. *If* $\mathcal{A}_{\mu} \equiv \{M_{\phi}: \phi \in L^{\infty}(\mu)\}$, *then* $\mathcal{A}_{\mu}' = \mathcal{A}_{\mu} = \mathcal{A}_{\mu}''$.

PROOF. It is easy to see that if $\mathcal{A} = \mathcal{A}'$, then $\mathcal{A} = \mathcal{A}''$. Since $\mathcal{A}_{\mu} \subseteq \mathcal{A}_{\mu}'$, it suffices to show that $\mathcal{A}_{\mu}' \subseteq \mathcal{A}_{\mu}$. So fix A in $\mathcal{A}_{\mu}'$; it must be shown that $A = M_{\phi}$ for some ϕ in $L^{\infty}(\mu)$.

Case 1: $\mu(X) < \infty$. Here $1 \in L^2(\mu)$; put $\phi = A(1)$. Thus $\phi \in L^2(\mu)$. If $\psi \in L^{\infty}(\mu)$, then $\psi \in L^2(\mu)$ and $A(\psi) = AM_{\psi}1 = M_{\psi}A1 = M_{\psi}\phi = \phi\psi$. Also, $\|\phi\psi\|_2 = \|A\psi\|_2 \leq \|A\|\|\psi\|_2$.

Let $\Delta_n = \{x \in X: |\phi(x)| \geq n\}$. Putting $\psi = \chi_{\Delta_n}$ in the preceding argument gives

$$\|A\|^2 \mu(\Delta_n) = \|A\|^2 \|\psi\|^2 \geq \|\phi\psi\|^2 = \int_{\Delta_n} |\phi|^2 \, d\mu \geq n^2 \mu(\Delta_n).$$

So if $\mu(\Delta_n) \neq 0$, $\|A\| \geq n$. Since A is bounded, $\mu(\Delta_n) = 0$ for some n; equivalently, $\phi \in L^\infty(\mu)$. But $A = M_\phi$ on $L^\infty(\mu)$ and $L^\infty(\mu)$ is dense in $L^2(\mu)$, so $A = M_\phi$ on $L^2(\mu)$.

Case 2: $\mu(X) = \infty$. If $\mu(\Delta) < \infty$, let $L^2(\mu|\Delta) = \{f \in L^2(\mu): f = 0$ off $\Delta\}$. For f in $L^2(\mu|\Delta)$, $Af = A\chi_\Delta f = \chi_\Delta Af \in L^2(\mu|\Delta)$. Let $A_\Delta =$ the restriction of A to $L^2(\mu|\Delta)$. By Case 1, there is a ϕ_Δ in $L^\infty(\mu|\Delta)$ such that $A_\Delta = M_{\phi_\Delta}$. Now if $\mu(\Delta_1) < \infty$ and $\mu(\Delta_2) < \infty$, $\phi_{\Delta_1}|\Delta_1 \cap \Delta_2 = \phi_{\Delta_2}|\Delta_1 \cap \Delta_2$ (Exercise).

Write $X = \bigcup_{n=1}^\infty \Delta_n$, where $\Delta_n \in \Omega$ and $\mu(\Delta_n) < \infty$. From the argument above, if $\phi(x) = \phi_{\Delta_n}(x)$ when $x \in \Delta_n$, ϕ is a well-defined measurable function on X. Now $\|\phi_\Delta\|_\infty = \|M_{\phi_\Delta}\|$ (II.1.5) $= \|A_\Delta\| \leq \|A\|$; hence $\|\phi\|_\infty \leq \|A\|$. It is easy to check that $A = M_\phi$. ∎

The next result will enable us to solve a number of problems concerning normal operators. It can be considered as a result that removes a technicality, but it is much more than that.

6.7. The Fuglede–Putnam Theorem. *If N and M are normal operators on $\mathcal{H}$ and $\mathcal{K}$, and $B: \mathcal{K} \to \mathcal{H}$ is an operator such that $NB = BM$, then $N^*B = BM^*$.*

PROOF. Note that it follows from the hypothesis that $N^k B = BM^k$ for all $k \geq 0$. So if $p(z)$ is a polynomial, $p(N)B = Bp(M)$. Since for a fixed z in $\mathbb{C}$, $\exp(izN)$ and $\exp(izM)$ are limits of polynomials in N and M, respectively, it follows that $\exp(izN)B = B\exp(izM)$ for all z in $\mathbb{C}$. Equivalently, $B = e^{-izN}Be^{izM}$. Because $\exp(X + Y) = (\exp X)(\exp Y)$ when X and Y commute, the fact that N and M are normal implies that

$$f(z) \equiv e^{-izN^*}Be^{izM^*}$$

$$= \bar{e}^{izN^*}e^{-izN}Be^{izM}e^{izM^*}$$

$$= e^{-i(zN^* + \bar{z}N)}Be^{i(\bar{z}M + zM^*)}.$$

But for every z in $\mathbb{C}$, $zN^* + \bar{z}N$ and $zM^* + \bar{z}M$ are hermitian operators. Hence $\exp[-i(zN^* + zN)]$ and $\exp[i(zM^* + zM)]$ are unitary (Exercise 2.14). Therefore $\|f(z)\| \leq \|B\|$. But $f: \mathbb{C} \to \mathcal{B}(\mathcal{K}, \mathcal{H})$ is an entire function. By Liouville's Theorem, f is constant.

Thus, $0 = f'(z) = -iN^*e^{-izN^*}BE^{izM^*} + ie^{-izN^*}BM^*e^{izM^*}$. Putting $z = 0$ gives $0 = -iN^*B + iBM^*$, whence the theorem. ∎

This theorem was originally proved in Fuglede [1950] under the assumption that $N = M$. As stated, the theorem was proved in Putnam [1951]. The proof given here is due to Rosenblum [1958]. Another proof is in Radjavi and Rosenthal [1973]. Berberian [1959] observed that Putnam's version can be derived from Fuglede's original theorem by the following matrix trick. If

$$L = \begin{bmatrix} N & 0 \\ 0 & M \end{bmatrix} \text{ and } A = \begin{bmatrix} 0 & B \\ 0 & 0 \end{bmatrix}$$

then L is normal on $\mathscr{H} \oplus \mathscr{H}$ and $LA = AL$. Hence $L^*A = AL^*$, and this gives Putnam's version.

6.8. Corollary. *If $N = \int z \, dE(z)$ and $BN = NB$, then $BE(\Delta) = E(\Delta)B$ for every Borel set Δ.*

PROOF. If $BN = NB$, then $BN^* = N^*B$; the conclusion now follows by The Spectral Theorem. ∎

The Fuglede–Putnam Theorem can be combined with some other results we have obtained to yield the following.

6.9. Corollary. *If μ is a compactly supported measure on $\mathbb{C}$, then*

$$\{ N_\mu \}' = \mathscr{A}_\mu \equiv \{ M_\phi \colon \phi \in L^\infty(\mu) \}.$$

PROOF. Clearly $\mathscr{A}_\mu \subseteq \{ N_\mu \}'$. If $A \in \{ N_\mu \}'$, then Theorem 6.7 implies $AN_\mu^* = N_\mu^*A$. By an easy algebraic argument, $AM_\phi = M_\phi A$ whenever ϕ is a polynomial in z and $\bar{z}$. By taking weak* limits of such polynomials, it follows that $A \in \mathscr{A}_\mu'$. By Theorem 6.6 $A \in \mathscr{A}_\mu$. ∎

Putnam applied his generalization of Fuglede's Theorem to show that similar normal operators must be unitarily equivalent. This has a formal generalization which is useful.

6.10. Proposition. *Let N_1 and N_2 be normal operators on $\mathscr{H}_1$ and $\mathscr{H}_2$. If $X \colon \mathscr{H}_1 \to \mathscr{H}_2$ is an operator such that $XN_1 = N_2 X$, then:*

(a) *cl(ran X) reduces N_2;*
(b) *ker X reduces N_1;*
(c) *If $M_1 = N_1|(\ker X)$ and $M_2 = N_2|\mathrm{cl}(\mathrm{ran}\, X)$, then $M_1 \cong M_2$.*

PROOF. (a) If $f_1 \in \mathscr{H}_1$, $N_2 Xf_1 = XN_1 f_1 \in \mathrm{ran}\, X$; so cl(ran X) is invariant for N_2. By the Fuglede–Putnam Theorem, $XN_1^* = N_2^* X$, so cl(ran X) is invariant for N_2^*.

(b) Exercise.

(c) Since $X(\ker X)^\perp \subseteq \mathrm{cl}(\mathrm{ran}\, X)$, part (c) will be proved if it can be shown that $N_1 \cong N_2$ when $\ker X = (0)$ and ran X is dense. So make these assumptions and consider the polar decomposition of X, $X = UA$. Because $\ker X = (0)$ and ran X is dense, A is a positive operator on $\mathscr{H}_1$ and $U \colon \mathscr{H}_1 \to \mathscr{H}_2$ is an isomorphism. Now $X^*N_2^* = N_1^* X^*$, so $X^*N_2 = N_1 X^*$. A calculation shows that $A^2 = X^*X \in \{N_1\}'$, so $A \in \{N_1\}'$. (Why?) Hence $N_2 UA = N_2 X = UAN_1 = UN_1 A$; that is, $N_2 U = UN_1$ on the range of A. But $\ker A = (0)$, so ran A is dense in $\mathscr{H}_1$. Therefore $N_2 U = UN_1$, or $N_2 = UN_1 U^{-1}$. ∎

6.11. Corollary. *Two similar normal operators are unitarily equivalent.*

The corollary appears in Putnam [1951], while Proposition 6.10 first appeared in Douglas [1969].

EXERCISES

1. If $\mathscr{S} \subseteq \mathscr{B}(\mathscr{H})$, show that $\mathscr{S}' = \mathscr{S}'''$.

2. If $\mathscr{S} \subseteq \mathscr{B}(\mathscr{H})$, show that $\mathscr{S}'$ is always a SOT closed subalgebra of $\mathscr{B}(\mathscr{H})$.

3. Prove Proposition 6.1.

4. Let $\mathscr{H}$ be a Hilbert space of dimension α and define $S: \mathscr{H}^{(\infty)} \to \mathscr{H}^{(\infty)}$ by $S(h_1, h_2, \ldots) = (0, h_1, h_2, \ldots)$. S is called the *unilateral shift of multiplicity* α. (a) Show that $A = [A_{ij}] \in \{\mathscr{S}\}'$ if and only if $A_{ij} = 0$ for $j > i$ and $A_{ij} = A_{i+1, j+1}$ for $i \geq j$. (b) Show that $A = [A_{ij}] \in \{\mathscr{S}\}''$ if and only if $A_{ij} = 0$ for $j > i$ and $A_{ij} = A_{i+1, j+1} = $ a multiple of the identity for $i \geq j$.

5. What is $\{N_\mu \oplus N_\mu\}'$? $\{N_\mu \oplus N_\mu\}''$?

6. If $\mathscr{A}$ is a subalgebra of $\mathscr{B}(\mathscr{H})$, show that $\mathscr{A}$ is a maximal abelian subalgebra of $\mathscr{B}(\mathscr{H})$ if and only if $\mathscr{A} = \mathscr{A}'$.

7. Find a non-normal operator that is similar to a normal operator. (Hint: Try dim $\mathscr{H} = 2$.)

8. Let μ be a compactly supported measure on $\mathbb{C}$ and let $\mathscr{K}$ be a separable Hilbert space. A function $f: \mathbb{C} \to \mathscr{K}$ is a Borel function if $f^{-1}(G)$ is a Borel set when G is weakly open in $\mathscr{K}$. Define $L^2(\mu, \mathscr{K})$ to be the equivalence classes of Borel functions $f: \mathbb{C} \to \mathscr{K}$ such that $\int \|f(x)\|^2 d\mu(x) < \infty$. Define $\langle f, g \rangle = \int \langle f(x), g(x) \rangle d\mu(x)$ for f and g in $L^2(\mu, \mathscr{K})$. (a) Show that $L^2(\mu, \mathscr{K})$ is a Hilbert space. Define N on $L^2(\mu, \mathscr{K})$ by $(Nf)(z) = zf(z)$. (b) Show that N is a normal operator and $\sigma(N) = $ support μ. Calculate N^*. (c) Show that $N \cong N_\mu^{(\alpha)}$, where $\alpha = \dim \mathscr{K}$. (d) Find $\{N\}'$. (Hint: Use 6.1.) (e) Find $\{N\}''$.

9. Let $\mathscr{H}$ be separable with basis $\{e_n\}$. Let A be the diagonal operator on $\mathscr{H}$ given by $Ae_n = \lambda_n e_n$, where $\sup_n |\lambda_n| < \infty$. Determine $\{A\}'$ and $\{A\}''$. Give necessary and sufficient conditions on $\{\lambda_n\}$ such that $\{A\}' = \{A\}''$.

10. Let $\mathscr{A}$ be a C^*-subalgebra of $\mathscr{B}(\mathscr{H})$ but do not assume that $\mathscr{A}$ contains the identity operator. Let $\mathscr{M} = \bigvee \{\text{ran } A: A \in \mathscr{A}\}$ and let $P = $ the projection of $\mathscr{H}$ onto $\mathscr{M}$. Show that SOT $- \text{cl } \mathscr{A} = \mathscr{A}''P = P\mathscr{A}''$.

§7. Abelian von Neumann Algebras

7.1.   Definition. A *von Neumann algebra* $\mathscr{A}$ is a C^*-subalgebra of $\mathscr{B}(\mathscr{H})$ such that $\mathscr{A} = \mathscr{A}''$.

Note that if $\mathscr{A}$ is a von Neumann algebra, then $1 \in \mathscr{A}$ and $\mathscr{A}$ is SOT closed. Conversely, if $1 \in \mathscr{A}$ and $\mathscr{A}$ is a SOT closed C^*-subalgebra of $\mathscr{B}(\mathscr{H})$, then $\mathscr{A}$ is a von Neumann algebra by the Double Commutant Theorem.

7.2.   Examples. (a) $\mathscr{B}(\mathscr{H})$ and $\mathbb{C}$ are von Neumann algebras.

(b) If (X, Ω, μ) is a σ-finite measure space, then $\mathscr{A}_\mu \equiv \{ M_\phi : \phi \in L^\infty(\mu) \}$ $\subseteq \mathscr{B}(L^2(\mu))$ is an abelian von Neumann algebra by Theorem 6.6. In fact, it is a maximal abelian von Neumann algebra.

It will be shown in this section that $\mathscr{A}_\mu$ is the only abelian von Neumann algebra up to a $*$-isomorphism. However, there are many others that are not unitarily equivalent to $\mathscr{A}_\mu$.

For $\mathscr{A}_j \subseteq \mathscr{B}(\mathscr{H}_j)$, $j \geq 1$, $\mathscr{A}_1 \oplus \mathscr{A}_2 \oplus \cdots$ is used to denote the l^∞ direct sum of $\mathscr{A}_1, \mathscr{A}_2, \ldots$. That is, $\mathscr{A}_1 \oplus \mathscr{A}_2 \oplus \cdots = \{ A_1 \oplus A_2 \oplus \cdots : A_j \in \mathscr{A}_j$ for $j \geq 1$ and $\sup_j \|A_j\| < \infty \}$. Note that $\mathscr{A}_1 \oplus \mathscr{A}_2 \oplus \cdots \subseteq \mathscr{B}(\mathscr{H}_1 \oplus \mathscr{H}_2 \oplus \cdots)$ and $\|A_1 \oplus A_2 \oplus \cdots\| = \sup_j \|A_j\|$.

7.3.   Proposition. (a) *If* $\mathscr{A}_1, \mathscr{A}_2, \ldots$ *are von Neumann algebras, then so is* $\mathscr{A}_1 \oplus \mathscr{A}_2 \oplus \cdots$. (b) *If* $\mathscr{A}$ *is a von Neumann algebra and* $1 \leq n \leq \infty$, *then* $\mathscr{A}^{(n)}$ *is a von Neumann algebra.*

PROOF. Exercise.

The proof of the next result is also an exercise.

7.4.   Proposition. *Let* $\mathscr{A}_j$ *be a von Neumann algebra on* $\mathscr{H}_j$, $j = 1, 2$. *If* $U:$ $\mathscr{H}_1 \to \mathscr{H}_2$ *is an isomorphism such that* $U\mathscr{A}_1 U^{-1} = \mathscr{A}_2$, *then* $U\mathscr{A}_1' U^{-1} = \mathscr{A}_2'$.

Now let (X, Ω, μ) be a σ-finite measure space and define $\rho: \mathscr{A}_\mu \to \mathscr{A}_\mu^{(2)}$ by $\rho(T) = T \oplus T$. Then ρ is a $*$-isomorphism. However, $\mathscr{A}_\mu$ and $\mathscr{A}_\mu^{(2)}$ are not *spatially isomorphic*. That is, there is no Hilbert space isomorphism $U:$ $L^2(\mu) \to L^2(\mu) \oplus L^2(\mu)$ such that $U\mathscr{A}_\mu U^{-1} = \mathscr{A}_\mu^{(2)}$. Why? One way to see that no such U exists is to note that $\mathscr{A}_\mu$ has a cyclic vector (give an example). However, $\mathscr{A}_\mu^{(2)}$ does not have a cyclic vector as shall be seen presently (Theorem 7.8).

7.5.   Definition. If $\mathscr{A} \subseteq \mathscr{B}(\mathscr{H})$ and $e_0 \in \mathscr{H}$, then e_0 is a *separating vector* for $\mathscr{A}$ if the only operator A in $\mathscr{A}$ such that $Ae_0 = 0$ is the operator $A = 0$.

If (X, Ω, μ) is a σ-finite measure space and $f \in L^2(\mu)$ such that $\mu(\{x \in X: f(x) = 0\}) = 0$ (Why does such an f exist?), then f is a separating vector for $\mathscr{A}_\mu$ as well as a cyclic vector. If $\mathscr{A} = \mathscr{B}(\mathscr{H})$, then no vector in $\mathscr{H}$ is separating for $\mathscr{A}$ while every nonzero vector is a cyclic vector. If $\mathscr{A} = \mathbb{C}$ and $\dim \mathscr{H} > 1$, then $\mathscr{A}$ has no cyclic vectors but every nonzero vector is separating for $\mathscr{A}$.

7.6.   Proposition. *If e_0 is a cyclic vector for $\mathscr{A}$, then e_0 is a separating vector for $\mathscr{A}'$.*

PROOF. If $T \in \mathscr{A}'$ and $Te_0 = 0$, then for every A in $\mathscr{A}$, $TAe_0 = ATe_0 = 0$. Since $\bigvee \mathscr{A}e_0 = \mathscr{H}$, $T = 0$. ∎

7.7.   Corollary. *If $\mathscr{A}$ is an abelian subalgebra of $\mathscr{B}(\mathscr{H})$, then every cyclic vector for $\mathscr{A}$ is a separating vector for $\mathscr{A}$.*

PROOF. Because $\mathscr{A}$ is abelian, $\mathscr{A} \subseteq \mathscr{A}'$. ∎

Since $\mathscr{B}(\mathscr{H})' = \mathbb{C}$, Proposition 7.6 explains some of the duality exhibited prior to (7.6). Also note that if (X, Ω, μ) is a finite-measure space, $1 \oplus 0$, $0 \oplus 1$, and $1 \oplus 1$ are all separating vectors for $\mathscr{A}_\mu^{(2)}$. Because $\mathscr{A}_\mu^{(2)} \neq (\mathscr{A}_\mu^{(2)})'$, the next theorem says that $\mathscr{A}_\mu^{(2)}$ has no cyclic vector.

Although it is easy to see that conditions (a) and (b) in the next result are equivalent, irrespective of any assumption on $\mathscr{H}$, the equivalence of the remaining parts to (a) and (b) is not true unless some additional assumption is made on $\mathscr{H}$ or $\mathscr{A}$ (see Exercise 5). We are content to assume that $\mathscr{H}$ is separable.

7.8.   Theorem. *Assume that $\mathscr{H}$ is separable and $\mathscr{A}$ is an abelian C*-subalgebra of $\mathscr{B}(\mathscr{H})$. The following statements are equivalent.*

(a) *$\mathscr{A}$ is a maximal abelian von Neumann algebra.*
(b) *$\mathscr{A} = \mathscr{A}'$.*
(c) *$\mathscr{A}$ has a cyclic vector and is SOT closed.*
(d) *There is a compact metric space X, a positive Borel measure μ with support X, and an isomorphism $U: L^2(\mu) \to \mathscr{H}$ such that $U\mathscr{A}_\mu U^{-1} = \mathscr{A}$.*

PROOF. The proof that (a) and (b) are equivalent is left as an exercise.

(b) $\Rightarrow$ (c): By Zorn's Lemma and the separability of $\mathscr{H}$, there is a maximal sequence of unit vectors $\{e_n\}$ such that for $n \neq m$, cl$[\mathscr{A}e_n] \perp$ cl$[\mathscr{A}e_m]$. It follows from the maximality of $\{e_n\}$ that $\mathscr{H} = \bigoplus_{n=1}^\infty$ cl$[\mathscr{A}e_n]$.

Let $e_0 = \sum_{n=1}^\infty e_n/\sqrt{2^n}$. Since $e_n \perp e_m$ for $n \neq m$, $\|e_0\|^2 = \Sigma 2^{-n} = 1$. Let P_n = the projection of $\mathscr{H}$ onto $\mathscr{H}_n = $ cl$[\mathscr{A}e_n]$. Clearly $\mathscr{A}$ leaves $\mathscr{H}_n$ invariant and so, since $\mathscr{A}$ is a *-algebra, $\mathscr{H}_n$ reduces $\mathscr{A}$. Thus $P_n \in \mathscr{A}' = \mathscr{A}$ and cl$[\mathscr{A}e_0] \supseteq$ cl$[\mathscr{A}P_ne_0] = $ cl$[\mathscr{A}e_n] = \mathscr{H}_n$. Therefore cl$[\mathscr{A}e_0] = \mathscr{H}$ and e_0 is a cyclic vector for $\mathscr{A}$.

(c) $\Rightarrow$ (d): Since $\mathscr{H}$ is separable, ball $\mathscr{A}$ is WOT metrizable and compact (1.3 and 5.5). By picking a countable WOT dense subset of ball $\mathscr{A}$ and letting $\mathscr{A}_1$ be the C^*-algebra generated by this countable dense subset, it follows that $\mathscr{A}_1$ is a separable C^*-algebra whose SOT closure is $\mathscr{A}$. Let X be the maximal ideal space of $\mathscr{A}_1$ and let $\rho: C(X) \to \mathscr{A}_1 \subseteq \mathscr{A} \subseteq \mathscr{B}(\mathscr{H})$ be the inverse of the Gelfand map. By Theorem 1.14 there is a spectral measure E defined on the Borel subsets of X such that $\rho(u) = \int u\, dE$ for u in $C(X)$. If $\phi \in B(X)$ and $\{u_i\}$ is a net in $C(X)$ such that $\int u_i\, d\nu \to \int \phi\, d\nu$ for every ν in $M(X)$, then $\rho(u_i) = \int u_i\, dE \to \int \phi\, dE$ (WOT). Thus $\{\int \phi\, dE: \phi \in B(X)\} \subseteq \mathscr{A}$ since $\mathscr{A}$ is SOT closed.

Let e_0 be a cyclic vector for $\mathscr{A}$ and put $\mu(\Delta) = \|E(\Delta)e_0\|^2 = \langle E(\Delta)e_0, e_0 \rangle$. Thus $\langle (\int \phi\, dE)e_0, e_0 \rangle = \int \phi\, d\mu$ for every ϕ in $B(X)$. Consider $B(X)$ as a linear manifold in $L^2(\mu)$ by identifying functions that agree a.e. $[\mu]$. If $\phi \in B(X)$, then

$$\left\| \left(\int \phi\, dE \right) e_0 \right\|^2 = \left\langle \left(\int \phi\, dE \right)^* \left(\int \phi\, dE \right) e_0, e_0 \right\rangle$$

$$= \int |\phi|^2\, d\mu.$$

This says two things. First, if $\phi = 0$ a.e. $[\mu]$, then $(\int \phi\, dE)e_0 = 0$. Hence $U: B(X) \to \mathscr{H}$ defined by $U\phi = (\int \phi\, dE)e_0$ is a well-defined map from the dense manifold $B(X)$ in $L^2(\mu)$ into $\mathscr{H}$. Second, U is an isometry. Since the domain and range of U are dense (Why?), U extends to an isomorphism $U: L^2(\mu) \to \mathscr{H}$.

If $\phi \in B(X)$ and $\psi \in L^\infty(\mu)$, then $UM_\psi\phi = U(\psi\phi) = (\int \psi\phi\, dE)e_0 = (\int \psi\, dE)(\int \phi\, dE)e_0 = (\int \psi\, dE)U\phi$. Hence $UM_\psi U^{-1} = \int \psi\, dE$ and $U\mathscr{A}_\mu U^{-1} \subseteq \mathscr{A}$. On the other hand, $U\mathscr{A}_\mu U^{-1}$ is a SOT closed C^*-subalgebra of $\mathscr{B}(\mathscr{H})$ that contains $UC(X)U^{-1} = \mathscr{A}_1$. (Why?) So $U\mathscr{A}_\mu U^{-1} = \mathscr{A}$.

Because $\mathscr{A}_1$ is separable, X is metrizable.

(d) $\Rightarrow$ (b): This is a consequence of Theorem 6.6 and Proposition 7.4. ∎

7.9. Corollary. *If $\mathscr{A}$ is an abelian C^*-subalgebra of $\mathscr{B}(\mathscr{H})$ and $\mathscr{H}$ is separable, then $\mathscr{A}$ has a separating vector.*

PROOF. By Zorn's Lemma, $\mathscr{A}$ is contained in a maximal abelian C^*-algebra, $\mathscr{A}_m$. It is easy to see that $\mathscr{A}_m$ must be SOT closed, so $\mathscr{A}_m$ is a maximal abelian von Neumann algebra. By the preceding theorem, there is a cyclic vector e_0 for $\mathscr{A}_m$. But (7.7) e_0 is separating for $\mathscr{A}_m$, and hence for any subset of $\mathscr{A}_m$. ∎

The preceding corollary may seem innocent, but it is, in fact, the basis for the next section.

1. Prove Proposition 7.3.

2. Prove Proposition 7.4.

3. Why are $\mathscr{A}_\mu$ and $\mathscr{A}_\mu^{(2)}$ not spatially isomorphic?

4. Show that if X is any compact metric space, there is a separable Hilbert space $\mathscr{H}$ and a $*$-monomorphism $\tau: C(X) \to \mathscr{B}(\mathscr{H})$. Find the spectral measure for τ.

5. Let $\mathscr{A}$ be an abelian C^*-subalgebra of $\mathscr{B}(\mathscr{H})$ such that $\mathscr{A}'$ contains no uncountable collection of pairwise orthogonal projections. Show that the following statements are equivalent: (a) $\mathscr{A}$ is a maximal abelian von Neumann algebra; (b) $\mathscr{A} = \mathscr{A}'$; (c) $\mathscr{A}$ has a cyclic vector and is SOT closed; (d) there is a finite-measure space (X, Ω, μ) and an isomorphism U; $L^2(\mu) \to \mathscr{H}$ such that $U\mathscr{A}_\mu U^{-1} = \mathscr{A}$.

6. Let $\{P_n\}$ be a sequence of commuting projections in $\mathscr{B}(\mathscr{H})$ and put $A = \sum_{n=1}^\infty 3^{-n}(2P_n - 1)$. Show that $C^*(A)$ is the C^*-algebra generated by $\{P_n\}$. (Do you see a connection between A and the Cantor–Lebesgue function?)

7. If $\mathscr{A}$ is an abelian von Neumann algebra on a separable Hilbert space $\mathscr{H}$, show that there is a hermitian operator A such that $\mathscr{A}$ equals the smallest von Neumann algebra containing A. (Hint: Let $\{P_n\}$ be a countable WOT dense subset of the set of projections in $\mathscr{A}$ and use Exercise 6. This proof is due to Rickart [1960], pp. 293–294.)

8. If X is a compact space, show that $C(X)$ is generated as a C^*-algebra by its characteristic functions if and only if X is totally disconnected. If A is as in Exercise 6, show that $\sigma(A)$ is totally disconnected.

9. If X and Y are compact spaces and $\tau: C(X) \to C(Y)$ is a homomorphism with $\tau(1) = 1$, show that there is a continuous function $\phi: Y \to X$ such that $\tau(u) = u \circ \phi$ for every u in $C(X)$. Show that τ is injective if and only if ϕ is surjective, and, in this case, τ is an isometry. Show that τ is surjective if and only if ϕ is injective.

10. Let X and Z be compact spaces, $Y = X \times Z$, and let $\phi: Y \to X$ be the projection onto the first coordinate. Define $\tau: C(X) \to C(Y)$ by $\tau(u) = u \circ \phi$. Describe the range of τ.

11. Adopt the notation of Exercise 9. Define an equivalence relation $\sim$ on Y by saying $y_1 \sim y_2$ if and only if $\phi(y_1) = \phi(y_2)$. Let $q: Y \to Y/\sim$ be the natural map and $q_*: C(Y/\sim) \to C(Y)$ the induced homomorphism. Show that there is a $*$-epimorphism $\rho: C(X) \to C(Y/\sim)$ such that the diagram

$$C(X) \xrightarrow{\quad\tau\quad} C(Y)$$
$$\rho \searrow \qquad \nearrow q^*$$
$$C(Y/\sim)$$

commutes. Find the corresponding injection $Y/\sim \to X$.

12. If X is any compact metric space, show that there is a totally disconnected compact metric space Y and a continuous surjection $\phi\colon Y \to X$. (Hint: Start by embedding $C(X)$ into $\mathscr{B}(\mathscr{H})$ and use Exercises 7 and 8.)

13. Show that every totally disconnected compact metric space is the continuous image of the Cantor ternary set. (Do this directly; do not try to use C^*-algebras.) Combine this with Exercise 12 to get that every compact metric space is the continuous image of the Cantor set.

14. If $\mathscr{A}$ is an abelian von Neumann algebra and X is its maximal ideal space, then X is a *Stonean space*; that is, if U is open in X, then cl U is open in X.

15. This exercise assumes Exercise 2.21 where it was proved that $\mathscr{B}_1^* = \mathscr{B}$. When referring to the weak* topology on $\mathscr{B} = \mathscr{B}(\mathscr{H})$, we mean the topology $\mathscr{B}$ has as the Banach dual of $\mathscr{B}_1$. (a) Show that on bounded subsets of $\mathscr{B}$ the weak* topology = WOT. (b) Show that a C^*-subalgebra of $\mathscr{B}(\mathscr{H})$ is a von Neumann algebra if and only if $\mathscr{A}$ is weak* closed. (c) Show that WOT and the weak* topology agree on abelian von Neumann algebras. (d) Give an example of a weak* closed subspace of $\mathscr{B}$ that is not WOT closed.

§8. The Functional Calculus for Normal Operators: The Conclusion of the Saga

In this section it will always be assumed that

all Hilbert spaces are separable.

Indeed, this assumption will remain in force for the rest of the chapter. This assumption is necessary for the validity of some of the results and minimizes the technical details in others.

If N is a normal operator on $\mathscr{H}$, let $W^*(N)$ be the von Neumann algebra generated by N. That is, $W^*(N)$ is the intersection of all of the von Neumann algebras containing N. Hence $W^*(N)$ is the WOT closure of $\{p(N, N^*)\colon p(z, \bar{z})$ is a polynomial in z and $\bar{z}\}$.

8.1. Proposition. *If N is a normal operator, then $W^*(N) = \{N\}'' \supseteq \{\phi(N)\colon \phi \in B(\sigma(N))\}$.*

PROOF. The equality results from combining the Double Commutant Theorem and the Fuglede–Putnam Theorem. If $\phi \in B(\sigma(N))$, $N = \int z\, dE(z)$, and $T \in \{N\}'$, then $T \in \{N, N^*\}'$ by the Fuglede–Putnam Theorem and $TE(\Delta) = E(\Delta)T$ for every Borel set Δ by the Spectral Theorem. Hence $T\phi(N) = \phi(N)T$ since $\phi(N) = \int \phi\, dE$. ∎

The purpose of this section is to prove that the containment in the preceding proposition is an equality. In fact, more will be proved. A measure μ whose support is $\sigma(N)$ will be found such that $\phi(N)$ is well defined if $\phi \in L^\infty(\mu)$ and the map $\phi \mapsto \phi(N)$ is a $*$-isomorphism of $L^\infty(\mu)$ onto $W^*(N)$. To find μ, Corollary 7.9 (which requires the separability of $\mathcal{H}$) is used.

By Corollary 7.9, $W^*(N)$, being an abelian von Neumann algebra, has a separating vector e_0. Define a measure μ on $\sigma(N)$ by

8.2 $$\mu(\Delta) = \langle E(\Delta)e_0, e_0 \rangle = \|E(\Delta)e_0\|^2.$$

8.3. Proposition. $\mu(\Delta) = 0$ if and only if $E(\Delta) = 0$.

PROOF. If $\mu(\Delta) = 0$, then $E(\Delta)e_0 = 0$. But $E(\Delta) = \chi_\Delta(N) \in W^*(N)$. Since e_0 is a separating vector, $E(\Delta) = 0$. The reverse implication is clear. ∎

8.4. Definition. A *scalar-valued spectral measure for N* is a positive measure μ on $\sigma(N)$ such that $\mu(\Delta) = 0$ if and only if $E(\Delta) = 0$; that is, μ and E are mutually absolutely continuous.

So Proposition 8.3 says that scalar-valued spectral measures exist. It will be shown (8.9) that every scalar-valued spectral measure is defined by (8.2) where e_0 is a separating vector for $W^*(N)$. In the process additional information is obtained about a normal operator and its functional calculus.

If $h \in \mathcal{H}$, let $\mu_h \equiv E_{h,h}$ and let $\mathcal{H}_h \equiv \mathrm{cl}[W^*(N)h]$. Note that $\mathcal{H}_h$ is the smallest reducing subspace for N that contains h. Let $N_h \equiv N|\mathcal{H}_h$. Thus N_h is a $*$-cyclic normal operator with $*$-cyclic vector h. The uniqueness of the spectral measure for a normal operator implies that the spectral measure for N_h is $E(\Delta)|\mathcal{H}_h$; that is, $\chi_\Delta(N_h) = \chi_\Delta(N)|\mathcal{H}_h = E(\Delta)|\mathcal{H}_h$. Thus Theorem 3.4 implies there is a unique isomorphism U_h: $\mathcal{H}_h \to L^2(\mu_h)$ such that $U_h h = 1$ and $U_h N_h U_h^{-1} f = zf$ for all f in $L^2(\mu_h)$. The notation of this paragraph is used repeatedly in this section.

The way to understand what is going on is to consider each N_h as a localization of N. Since N_h is unitarily equivalent to M_z on $L^2(\mu_h)$ we can agree that we thoroughly understand the local behavior of N. Can we put together this local behavior of N to understand the global behavior of N? This is precisely what is done in §10.

In the present section the objective is to show that if h is a separating vector for $W^*(N)$, then the functional calculus for N is completely determined by the functional calculus for N_h. The sense in which this "determination" is made is the following. If $A \in W^*(N)$, then the definition of $\mathcal{H}_h$ shows that $A\mathcal{H}_h \subseteq \mathcal{H}_h$. Since $A^* \in W^*(N)$, $\mathcal{H}_h$ reduces each operator in $W^*(N)$; thus $A|\mathcal{H}_h$ is meaningful. It will be shown that the map $A \to A|\mathcal{H}_h$ is a $*$-isomorphism of $W^*(N)$ onto $W^*(N_h)$ if h is a separating vector for $W^*(N)$. Since N_h is $*$-cyclic, Theorem 6.6 and Corollary 6.9 show how to determine $W^*(N_h)$.

We begin with a modest lemma.

8.5.   Lemma. *If $h \in \mathcal{H}$ and ρ_h: $W^*(N) \to W^*(N_h)$ is defined by $\rho_h(A) = A|\mathcal{H}_h$, then ρ_h is a $*$-epimorphism that is WOT-continuous. Moreover, if $\psi \in B(\sigma(N))$, then $\rho_h(\psi(N)) = \psi(N_h)$ and if $A \in W^*(N)$, then there is a ϕ in $B(\sigma(N_h))$ such that $\rho_h(A) = \phi(N_h)$.*

PROOF. First let us see that ρ_h maps $W^*(N)$ into $W^*(N_h)$. If $p(z, \bar{z})$ is a polynomial in z and $\bar{z}$, then $\rho_h[p(N, N^*)] = p(N_h, N_h^*)$ as an algebraic manipulation shows. If $\{p_i\}$ is a net of such polynomials such that $p_i(N, N^*) \to A(\text{WOT})$, then for f, g in $\mathcal{H}_h$, $\langle p_i(N, N^*)f, g \rangle \to \langle Af, g \rangle$; thus $p_i(N_h, N_h^*) \to \rho_h(A)(\text{WOT})$ and so $\rho_h(A) \in W^*(N_h)$. It is left as an exercise for the reader to show that ρ_h is a $*$-homomorphism. Also, the preceding argument can be used to show that ρ_h is WOT continuous.

If $\psi \in B(\sigma(N))$, there is a net $\{p_i(z, \bar{z})\}$ of polynomials in z and $\bar{z}$ such that $\int p_i \, d\nu \to \int \psi \, d\nu$ for every ν in $M(\sigma(N))$. (Why?) Since $\sigma(N_h) \subseteq \sigma(N)$ (Why?), $\int p_i \, d\eta \to \int \psi \, d\eta$ for every η in $M(\sigma(N_h))$. Therefore $p_i(N, N^*) \to \psi(N)(\text{WOT})$ and $p_i(N_h, N_h^*) \to \psi(N_h)(\text{WOT})$. But $\rho_h(p_i(N, N^*)) = p_i(N_h, N_h^*)$ and $\rho_h(p_i(N, N^*)) \to \rho_h(\psi(N))$; hence $\rho_h(\psi(N)) = \psi(N_h)$.

Let U_h: $\mathcal{H}_h \to L^2(\mu_h)$ be the isomorphism such that $U_h h = 1$ and $U_h N_h U_h^{-1} = N_{\mu_h}$. If $A \in W^*(N)$ and $A_h = \rho_h(A)$, then $A_h N_h = N_h A_h$; thus $U_h A_h U_h^{-1} \in \{N_{\mu_h}\}'$. By Corollary 6.9, there is a ϕ in $B(\sigma(N_h))$ such that $U_h A_h U_h^{-1} = M_\phi$. It follows (How?) that $A_h = \phi(N_h)$.

Finally, to show that ρ_h is surjective note that if $B \in W^*(N_h)$, then (use the argument in the preceding paragraph) $B = \psi(N_h)$ for some ψ in $B(\sigma(N_h))$. Extend ψ to $\sigma(N)$ by letting $\psi = 0$ on $\sigma(N) \setminus \sigma(N_h)$. Then $\psi(N) \in W^*(N)$ and $\rho_h(\psi(N)) = \psi(N_h) = B$.   ∎

8.6.   Lemma. *If $e \in \mathcal{H}$ such that μ_e is a scalar-valued spectral measure for N and if ν is a positive measure on $\sigma(N)$ such that $\nu \ll \mu_e$, then there is an h in $\mathcal{H}$ such that $\nu = \mu_h$.*

PROOF. This proof is just an application of the Radon–Nikodym Theorem once certain identifications are made; namely, $f = [d\nu/d\mu_e]^{1/2} \in L^2(\mu_e)$, so put $h = U_e^{-1}f$. Hence $h \in \mathcal{H}_e$. For any Borel set Δ, $\nu(\Delta) = \int \chi_\Delta \, d\nu = \int \chi_\Delta f\!\!f \, d\mu_e = \langle M_{\chi_\Delta}f, f \rangle = \langle U_e^{-1}M_{\chi_\Delta}f, U_e^{-1}f \rangle = \langle E(\Delta)h, h \rangle = \mu_h(\Delta)$.   ∎

8.7.   Lemma. $W^*(N) = \{\phi(N): \phi \in B(\sigma(N))\}$.

PROOF. Let $\mathcal{A} = \{\phi(N): \phi \in B(\sigma(N))\}$. Hence $\mathcal{A}$ is a $*$-algebra and $\mathcal{A} \subseteq W^*(N)$ by Proposition 8.1. Since $N \in \mathcal{A}$ it suffices to prove that $\mathcal{A}$ is WOT closed. Let $\{\phi_i\}$ be a net in $B(\sigma(N))$ such that $\phi_i(N) \to A$ (WOT); so $A \in W^*(N)$. By (8.5) $\phi_i(N_h) \to A|\mathcal{H}_h(\text{WOT})$ for any h in $\mathcal{H}$. Also, by Lemma 8.5, for every h in $\mathcal{H}$ there is a ϕ_h in $B(\mathbb{C})$ such that $A|\mathcal{H}_h = \phi_h(N_h)$. Fix a separating vector e for $W^*(N)$; hence μ_e is a scalar-valued spectral measure for N.

If $h \in \mathcal{H}$, then the fact that $\phi_i(N_h) \to \phi_h(N_h)(\text{WOT})$ implies $\phi_i \to \phi_h$ weak* in $L^\infty(\mu_h)$. Also, $\phi_i \to \phi_e$ weak* in $L^\infty(\mu_e)$. But $\mu_h \ll \mu_e$ so that

$d\mu_h/d\mu_e \in L^1(\mu_e)$; hence for any Borel set Δ

$$\int_\Delta \phi_i \, d\mu_h = \int_\Delta \phi_i \frac{d\mu_h}{d\mu_e} \, d\mu_e \rightarrow \int_\Delta \phi_e \, d\mu_h.$$

But also

$$\int_\Delta \phi_i \, d\mu_h \rightarrow \int_\Delta \phi_h \, d\mu_h.$$

So $0 = \int_\Delta (\phi_e - \phi_h) \, d\mu_h$ for every Borel set Δ. Therefore $\phi_h = \phi_e$ a.e. $[\mu_h]$. But if $g \in \mathcal{H}_h$, then $\langle \phi_h(N_h)g, g \rangle = \langle \phi_h(N)g, g \rangle = \int \phi_h \, d\mu_g = \int \phi_e \, d\mu_g$ since $\mu_g \ll \mu_h$. Thus $\langle \phi_h(N_h)g, g \rangle = \langle \phi_e(N_h)g, g \rangle$; that is, $\phi_h(N_h) = \phi_e(N_h)$. In particular, $Ah = \phi_h(N_h)h = \phi_e(N_h)h = \phi_e(N)h$. Since h was arbitrary, $A = \phi_e(N)$.   ∎

8.8. Corollary. *If $\rho_h\colon W^*(N) \rightarrow W^*(N_h)$ is the $*$-epimorphism of Lemma 8.5, then* $\ker \rho_h = \{\phi(N)\colon \phi = 0 \text{ a.e. } [\mu_h]\}$.

8.9. Theorem. *If N is a normal operator and $e \in \mathcal{H}$, the following statements are equivalent.*

(a) *e is a separating vector for $W^*(N)$.*
(b) *μ_e is a scalar-valued spectral measure for N.*
(c) *The map $\rho_e\colon W^*(N) \rightarrow W^*(N_e)$ defined in (8.5) is a $*$-isomorphism.*
(d) *$\{\phi \in B(\sigma(N))\colon \phi(N) = 0\} = \{\phi \in B(\sigma(N))\colon \phi = 0 \text{ a.e. } [\mu_e]\}$.*

PROOF. (a) $\Rightarrow$ (b): Proposition 8.3.

(b) $\Rightarrow$ (c): By Lemma 8.5, ρ_e is a $*$-epimorphism. By Corollary 8.8, $\ker \rho_e = \{\phi(N)\colon \phi = 0 \text{ a.e. } [\mu_e]\}$. But if $\phi = 0$ a.e. $[\mu_e]$, (b) implies that $\phi = 0$ off a set Δ such that $E(\Delta) = 0$. Thus $\phi(N) = \int_\Delta \phi \, dE = 0$.

(c) $\Rightarrow$ (d): Combine (c) with Corollary 8.8.

(d) $\Rightarrow$ (a): Suppose $A \in W^*(N)$ and $Ae = 0$. By Lemma 8.7, there is a ϕ in $B(\sigma(N))$ such that $\phi(N) = A$. Thus, $0 = \|Ae\|^2 = \langle A^*Ae, e \rangle = \int |\phi|^2 \, d\mu_e$. So $\phi = 0$ a.e. $[\mu_e]$. By (d), $A = 0$.   ∎

These results can now be combined to yield the final statement of the functional calculus for normal operators.

8.10. The Functional Calculus for a Normal Operator. *If N is a normal operator on the separable Hilbert space $\mathcal{H}$ and μ is a scalar-valued spectral measure for N, then there is a well-defined map $\rho\colon L^\infty(\mu) \rightarrow W^*(N)$ given by the formula $\rho(\phi) = \phi(N)$ such that*

(a) *ρ is a $*$-isomorphism and an isometry;*
(b) *$\rho\colon (L^\infty(\mu), \text{weak}^*) \rightarrow (W^*(N), \text{WOT})$ is a homeomorphism.*

PROOF. Let e be a separating vector such that $\mu = \mu_e$ [by (8.6) and (8.9)]. If $\phi \in B(\sigma(N))$ and $\phi = 0$ a.e. $[\mu]$, then $\phi(N) = 0$ by (8.9d); so $\rho(\phi) = \phi(N)$

is a well-defined map. It is left to the reader to show that ρ is a $*$-homomorphism. By Lemma 8.7, ρ is surjective. Also, if $\rho(\phi) = \phi(N) = 0$, then $\phi = 0$ a.e. $[\mu]$ by (8.9d). Thus ρ is a $*$-isomorphism. By (VIII.4.8) ρ is an isometry. (A proof avoiding (VIII.4.8) is possible—it is left as an exercise.) This proves (a).

Let $\{\phi_i\}$ be a net in $L^\infty(\mu)$ and suppose that $\phi_i(N) \to 0(\text{WOT})$. If $f \in L^1(\mu)$ and $f \geq 0$, $f\mu \ll \mu = \mu_e$. By Lemma 8.6 there is a vector h such that $f\mu = \mu_h$. Thus $\int \phi_i f \, d\mu = \int \phi_i \, d\mu_h = \langle \phi_i(N)h, h \rangle \to 0$. Thus $\phi_i \to 0$ (weak$*$) in $L^\infty(\mu)$. This proves half of (b); the other half is left as an exercise. ∎

8.11. The Spectral Mapping Theorem. *If N is a normal operator on a separable space and μ is a scalar-valued spectral measure for N and if $\phi \in L^\infty(\mu)$, then $\sigma(\phi(N)) =$ the μ-essential range of ϕ.*

PROOF. Use (8.10) and the fact (2.6) that the μ-essential range of ϕ is the spectrum of ϕ as an element of $L^\infty(\mu)$. ∎

8.12. Proposition. *Let N, μ, ϕ be as in (8.11). If $N = \int z \, dE$, then $\mu \circ \phi^{-1}$ is a scalar-valued spectral measure for $\phi(N)$ and $E \circ \phi^{-1}$ is its spectral measure.*

PROOF. Exercise.

EXERCISES

1. What is a scalar spectral measure for a diagonalizable normal operator?

2. Let N_1 and N_2 be normal operators with scalar spectral measures μ_1 and μ_2. What is a scalar spectral measure for $N_1 \oplus N_2$?

3. Let $\{e_n\}$ be an orthonormal basis for $\mathcal{H}$ and put $\mu(\Delta) = \sum_{n=1}^\infty 2^{-n}\|E(\Delta)e_n\|^2$. Show that μ is a scalar spectral measure for N.

4. Give an example of a normal operator on a nonseparable space which has no scalar-valued spectral measure.

5. Prove that the map ρ in (8.10) is an isometry without using (VIII.4.8).

6. Prove Proposition 8.12.

7. Show that if μ and ν are compactly supported measures on $\mathbb{C}$, the following statements are equivalent: (a) $N_\mu \oplus N_\nu$ is $*$-cyclic; (b) $W^*(N_\mu \oplus N_\nu) = W^*(N_\mu) \oplus W^*(N_\nu)$; (c) $\mu \perp \nu$.

8. If M and N are normal operators with scalar spectral measures μ and ν, respectively, show that the following are equivalent: (a) $W^*(M \oplus N) = W^*(M) \oplus W^*(N)$; (b) $\{M \oplus N\}' = \{M\}' \oplus \{N\}'$; (c) there is no operator A such that $MA = AN$ other than $A = 0$; (d) $\mu \perp \nu$.

9. If M and N are normal operators, show that $C^*(M \oplus N) = C^*(M) \oplus C^*(N)$ if and only if $\sigma(M) \cap \sigma(N) = \square$.

10. Give an example of two normal operators M and N such that $W^*(M \oplus N) = W(M) \oplus W^*(N)$ but $C^*(M \oplus N) \neq C^*(M) \oplus C^*(N)$. In fact, find M and N such that $W^*(M \oplus N)$ splits, but $\sigma(M) = \sigma(N)$.

11. If U is the bilateral shift and V is any unitary operator, show that $W^*(U \oplus V) = W^*(U) \oplus W^*(V)$ if and only if V has a spectral measure that is singular to arc length on $\partial \mathbb{D}$.

12. If $\mathscr{A}$ is an abelian von Neumann algebra on a separable space, show that there is a compactly supported measure μ on $\mathbb{R}$ such that $\mathscr{A}$ is $*$-isomorphic to $L^\infty(\mu)$. (Hint: Use Exercise 7.7.)

13. (This exercise assumes a knowledge of Exercise 2.21). Let $N = \int z \, dE(z)$ be a normal operator with scalar-valued spectral measure μ and define $\alpha \colon \mathscr{B}_1(\mathscr{H}) \to L^1(\mu)$ by $\alpha(T)(\Delta) = \operatorname{tr}(TE(\Delta))$. Show that α is a surjective contraction. What is α^*? [$L^1(\mu)$ is identified, via the Radon–Nikodym Theorem, with the set of complex-valued measures that are absolutely continuous with respect to μ.]

§9. Invariant Subspaces for Normal Operators

Remember that we continue to assume that all Hilbert spaces are separable.

Every normal operator on a Hilbert space of dimension of at least 2 has a nontrivial invariant subspace. This is an easy consequence of the Spectral Theorem. Indeed, if $N = \int z \, dE(z)$, $E(\Delta)\mathscr{H}$ is a reducing subspace for every Borel set Δ.

If $A \in \mathscr{B}(\mathscr{H})$, $\mathscr{M}$ is a linear subspace of $\mathscr{H}$, and P is the orthogonal projection of $\mathscr{H}$ onto $\mathscr{M}$, then $\mathscr{M}$ reduces A if and only if $P \in \{A\}'$. Also, $\mathscr{M} \in \operatorname{Lat} A$ (= the lattice of invariant subspaces for A) if and only if $AP = PAP$. Since the spectral projections of a normal operator belong to $W^*(N)$, they are even more than reducing.

9.1. Definition. An operator A is *reductive* if every invariant subspace for A reduces A. Equivalently, A is reductive if and only if $\operatorname{Lat} A = \operatorname{Lat} A^*$.

Thus, every self-adjoint operator is reductive. Every normal operator on a finite-dimensional space is reductive. More generally, every normal compact operator is reductive (Ando [1963]). However, the bilateral shift is not reductive. Indeed, if U is the bilateral shift on $l^2(\mathbb{Z})$, $\mathscr{H} = \{f \in l^2(\mathbb{Z}) \colon f(n) = 0 \text{ if } n < 0\} \in \operatorname{Lat} U$, but $\mathscr{H} \notin \operatorname{Lat} U^*$. Wermer [1952] first studied reductive normal operators and characterized the reductive unitary operators. A first step towards characterizing the reductive normal operators will be taken here. The final step has been taken but it will not be viewed in this book. The result is due to Sarason [1972]. Also see Conway [1981], § VII.5.

9.2. Definition. If μ is a compactly supported measure on $\mathbb{C}$, $P^\infty(\mu)$ denotes the weak* closure of the polynomials in $L^\infty(\mu)$.

Because the support of μ is compact, every polynomial, when restricted to that support, belongs to $L^{\infty}(\mu)$.

For any operator A, let $W(A)$ denote the WOT closed subalgebra of $\mathscr{B}(\mathscr{H})$ generated by A; that is, $W(A)$ is the WOT closure in $\mathscr{B}(\mathscr{H})$ of the polynomials in A. The next result is an immediate consequence of The Functional Calculus for Normal Operators.

9.3. Theorem. *If N is a normal operator and μ is a scalar-valued spectral measure for N, then the functional calculus, when restricted to $P^{\infty}(\mu)$, is an isometric isomorphism $\rho: P^{\infty}(\mu) \to W(N)$ and a weak*-WOT homeomorphism. Also, $\rho(z) = N$.*

9.4. Definition. An operator A is *reflexive* if whenever $B \in \mathscr{B}(\mathscr{H})$ and $\mathrm{Lat}\, A \subseteq \mathrm{Lat}\, B$, then $B \in W(A)$.

It is easy to see that if $B \in W(A)$, then $\mathrm{Lat}\, A \subseteq \mathrm{Lat}\, B$ (Exercise). An operator is reflexive precisely when it has sufficiently many invariant subspaces to characterize $W(A)$. For a survey of reflexive operators and some related topics, see Radjavi and Rosenthal [1973].

9.5. Theorem. (*Sarason* [1966].). *Every normal operator is reflexive.*

PROOF. Suppose N is normal and $\mathrm{Lat}\, N \subseteq \mathrm{Lat}\, A$. If P is a projection in $\{N\}'$, then $P\mathscr{H}$ and $(P\mathscr{H})^{\perp} \in \mathrm{Lat}\, N \subseteq \mathrm{Lat}\, A$, so $AP = PA$. By Corollary 6.5, $A \in W^{*}(N)$. Let μ be a scalar-valued spectral measure for N. By Theorem 8.10, there is a ϕ in $L^{\infty}(\mu)$ such that $A = \phi(N)$. By Theorem 9.3, it must be shown that $\phi \in P^{\infty}(\mu)$.

Now let's focus our attention on a special case: Assume that N is $*$-cyclic; thus $N = N_{\mu}$. Suppose $f \in L^{1}(\mu)$ and $\int f\psi\, d\mu = 0$ for every ψ in $P^{\infty}(\mu)$. If it can be shown that $\int f\phi\, d\mu = 0$, then the Hahn–Banach Theorem implies that $\phi \in P^{\infty}(\mu)$. This is the strategy we follow. Let $f = g\bar{h}$ for some g, h in $L^{2}(\mu)$. Put $\mathscr{M} = \bigvee\{z^{k}g: k \geq 0\}$. Clearly $\mathscr{M} \in \mathrm{Lat}\, N$, so $\mathscr{M} \in \mathrm{Lat}\, A = \mathrm{Lat}\, \phi(N) = \mathrm{Lat}\, M_{\phi}$. Hence $\phi g \in \mathscr{M}$. But $0 = \int z^{k}f\, d\mu = \int z^{k}g\bar{h}\, d\mu = \langle N^{k}g, h \rangle$ for all $k \geq 0$; hence $h \perp \mathscr{M}$. Thus $0 = \langle \phi g, h \rangle = \int \phi g\bar{h}\, d\mu = \int \phi f\, d\mu$, and $\phi \in P^{\infty}(\mu)$.

Now we return to the general case. By Theorem 8.9 there is a separating vector e for $W^{*}(N)$ such that $\mu(\Delta) = \|E(\Delta)e\|^{2}$, where $N = \int z\, dE(z)$. Let $\mathscr{K} = \bigvee\{N^{*k}N^{j}e: k, j \geq 0\}$. Clearly $\mathscr{K}$ reduces N and $N|\mathscr{K}$ is $*$-cyclic. Hence $\mathscr{K}$ reduces A and $A|\mathscr{K} = \phi(N|\mathscr{K})$. By the preceding paragraph $\phi \in P^{\infty}(\mu)$. ∎

An immediate consequence of the preceding theorem is the first step in the characterization of reductive normal operators.

9.6. Corollary. *If N is a normal operator and μ is a scalar spectral measure for N, then N is reductive if and only if $P^{\infty}(\mu) = L^{\infty}(\mu)$.*

PROOF. If $\mathcal{M} \in \text{Lat } N$, then $\mathcal{M} \in \text{Lat } \phi(N)$ for every ϕ in $P^{\infty}(\mu)$. So if $P^{\infty}(\mu) = L^{\infty}(\mu)$, $\bar{z} \in P^{\infty}(\mu)$ and, hence, $\mathcal{M} \in \text{Lat } N^*$ whenever $\mathcal{M} \in \text{Lat } N$.

Now suppose N is reductive. This means that $\text{Lat } N \subseteq \text{Lat } N^*$. By the preceding theorem, this implies $N^* \in W(N)$; equivalently, $\bar{z} \in P^{\infty}(\mu)$. Since $P^{\infty}(\mu)$ is an algebra, every polynomial in z and $\bar{z}$ belongs to $P^{\infty}(\mu)$. By taking weak* limits this implies that $P^{\infty}(\mu) = L^{\infty}(\mu)$. ∎

The preceding corollary fails to be a good characterization of reductive normal operators since it only says that one difficult problem is equivalent to another. A way is needed to determine when $P^{\infty}(\mu) = L^{\infty}(\mu)$. This is what was done in Sarason [1972].

Are there any reductive operators that are not normal? This natural and seemingly innocent question has much more to it than meets the eye. Dyer, Pedersen, and Porcelli [1972] have shown that this question has an affirmative answer if and only if every operator on a Hilbert space has a nontrivial invariant subspace.

EXERCISES

1. (Ando [1963].) Use Corollary 9.6 to show that every compact normal operator is reductive.

2. Determine all of the invariant subspaces of a compact normal operator.

3. (Rosenthal [1968].) Show that every reductive compact operator is normal.

§10. Multiplicity Theory for Normal Operators: A Complete Set of Unitary Invariants

Throughout this section only separable Hilbert spaces are considered.

When are two normal operators unitarily equivalent? The answer to this question must be given in the following way: to each normal operator we must attach a collection of objects such that two normal operators are unitarily equivalent if and only if the two collections are equal (or equivalent). Furthermore, it should be easier to verify that these collections are equivalent than to verify that the normal operators are equivalent. This is contained in the following result due to Hellinger [1907]. Note that it generalizes Theorem 3.6.

10.1.   Theorem. (a) *If N is a normal operator, then there is a sequence of measures $\{\mu_n\}$ (possibly finite) on $\mathbb{C}$ such that $\mu_{n+1} \ll \mu_n$ for all n and*

10.2 $N \cong N_{\mu_1} \oplus N_{\mu_2} \oplus \cdots .$

(b) *If N and $\{\mu_n\}$ are as in (a) and $M \cong N_{\nu_1} \oplus N_{\nu_2} \oplus \cdots$, where $\nu_{n+1} \ll \nu_n$ for all n, then $N \cong M$ if and only if $[\mu_n] = [\nu_n]$ for all n.*

The proof of this theorem requires several lemmas. Before beginning, we will examine a couple of false starts for a proof. This will cause us to arrive at the correct strategy for a proof and show us the necessity for some of the lemmas.

Let $N = \int z\, dE(z)$. If $e \in \mathcal{H}$ and $\mathcal{H}_e = \mathrm{cl}[W^*(N)e]$, then $N|\mathcal{H}_e$ is a $*$-cyclic normal operator. An application of Zorn's Lemma and the separability of $\mathcal{H}$ produces a maximal sequence $\{e_n\}$ in $\mathcal{H}$ such that $\mathcal{H}_{e_n} \perp \mathcal{H}_{e_m}$. By the maximality of e_n, $\mathcal{H} = \oplus_n \mathcal{H}_{e_n}$. If $N_n = N|\mathcal{H}_{e_n}$, $N_n = N_{\mu_n}$, where $\mu_n(\Delta) = \|E(\Delta)e_n\|^2$; thus $N \cong \oplus_n N_{\mu_n}$. The trouble here is that μ_{n+1} is not necessarily absolutely continuous with respect to μ_n. Just using Zorn's Lemma to produce the sequence $\{\mu_n\}$ eliminates any possibility of having $\{\mu_n\}$ canonical and producing the unitary invariant desired for normal operators. Let's try again.

Note that if $\mu_{n+1} \ll \mu_n$ for all n in (10.2), then $\mu_n \ll \mu_1$ for all n. This in turn implies that μ_1 is a scalar-valued spectral measure for N. Using Lemma 8.6 we are thus led to choose μ_1 as follows. Let e_1 be a separating vector for $W^*(N)$; this exists by Corollary 7.9 and the separability of $\mathcal{H}$. Put $\mu_1(\Delta) = \|E(\Delta)e_1\|^2$. If $\mathcal{H}_1 = \mathrm{cl}[W^*(N)e_1]$, then $N|\mathcal{H}_1 \cong N_{\mu_1}$. Let $N_2 \equiv N|\mathcal{H}_1^\perp$; so N_2 is normal. A pair of easy exercises shows that the spectral measure E_2 for N_2 is given by $E_2(\Delta) = E(\Delta)|\mathcal{H}_1^\perp$ and $W^*(N_2) = W^*(N)|\mathcal{H}_1^\perp$ ($\equiv \{A \in \mathcal{B}(\mathcal{H}_1): A = T|\mathcal{H}_1^\perp$ for some T in $W^*(N)\}$). Let e_2 be a separating vector for $W^*(N_2)$ and put $\mu_2(\Delta) = \|E_2(\Delta)e_2\|^2$. By the easy exercises above, $\mu_2(\Delta) = \|E(\Delta)e_2\|^2$, so that $\mu_2 \ll \mu_1$, and $\mathcal{H}_2 \equiv \mathrm{cl}[W^*(N)e_2] = \mathrm{cl}[W^*(N_2)e_2] \leq \mathcal{H}_1^\perp$. Also, $N|\mathcal{H}_2 \cong N_{\mu_2}$.

Continuing in this way produces a sequence of vectors $\{e_n\}$ such that if $\mathcal{H}_n = \mathrm{cl}[W^*(N)e_n]$ and $\mu_n(\Delta) = \|E(\Delta)e_n\|^2$, then $\mathcal{H}_n \perp \mathcal{H}_m$ for $n \neq m$, $\mu_{n+1} \ll \mu_n$, and $N|\mathcal{H}_n \cong N_{\mu_n}$. The difficulty here is that $\mathcal{H}$ is not necessarily equal to $\oplus_n \mathcal{H}_n$ so that N and $\oplus_n N_{\mu_n}$ cannot be proved to be unitarily equivalent. (Actually, N and $\oplus_n N_{\mu_n}$ are unitarily equivalent, but to show this we need the force of Theorem 10.1. See Exercise 2.) The following provides us with a look at an example to see what can go wrong.

10.3.   Example. For $n \geq 1$ let $\mu_n =$ Lebesgue measure on $[0, 1 + 2^{-n}]$ and let $\mu_\infty =$ Lebesgue measure on $[0, 1]$. Put $N = \oplus_{n=1}^\infty N_{\mu_n} \oplus N_{\mu_\infty}$. If the process of the preceding paragraph is followed, it might be that vectors $\{e_n\}$ that are chosen are the vectors with a 1 in the $L^2(\mu_n)$ coordinate and zeros elsewhere. Thus the spaces $\{\mathcal{H}_n\}$ are precisely the spaces $\{L^2(\mu_n)\}$ and $[\oplus_1^\infty \mathcal{H}_n]^\perp = L^2(\mu_\infty)$.

(Nevertheless, $N \cong \oplus_1^\infty N_{\mu_n}$. Indeed, let $\nu_n = \mu_n|[1, 1 + 2^{-n}]$; so $\mu_n = \mu_\infty + \nu_n$ and $\mu_\infty \perp \nu_n$. Thus $N_{\mu_n} \cong N_{\nu_n} \oplus N_{\mu_\infty}$. Therefore

$$N = \oplus_1^\infty N_{\mu_n} \oplus N_{\mu_\infty}$$

$$\cong \oplus_1^\infty N_{\nu_n} \oplus N_{\mu_\infty}^{(\infty)} \oplus N_{\mu_\infty}$$

$$\cong \oplus_1^\infty N_{\nu_n} \oplus N_{\mu_\infty}^{(\infty)}$$

$$\cong \oplus_1^\infty N_{\mu_n}.)$$

After an examination of the statement of Theorem 10.1, it becomes clear that some procedure like the one outlined in the paragraph preceding Example 10.3 should be used. It only becomes necessary to modify this procedure so that the vectors $\{e_n\}$ can be chosen in such a way that $\mathcal{H} = \oplus_1^\infty \mathcal{H}_n$. For example, let, as above, e_1 be a separating vector for $W^*(N)$ and let $\{f_n\}$ be an orthonormal basis for $\mathcal{H}$ such that $f_1 = e_1$. We now want to choose the vectors $\{e_n\}$ such that $\{f_1, \ldots, f_n\} \subseteq \mathcal{H}_1 \oplus \cdots \oplus \mathcal{H}_n$. In this way we will meet success. The vital link here is the next result.

10.4. Proposition. *If N is a normal operator on $\mathcal{H}$ and $e \in \mathcal{H}$, then there is a separating vector e_0 for $W^*(N)$ such that $e \in \mathrm{cl}[W^*(N)e_0]$.*

PROOF. Let f_0 be any separating vector for $W^*(N)$, let E be the spectral measure for N, let $\mu(\Delta) = \|E(\Delta)f_0\|^2$, and put $\mathcal{G} = \mathrm{cl}[W^*(N)f_0]$. Write $e = g_1 + h_1$, where $g_1 \in \mathcal{G}$ and $h_1 \in \mathcal{G}^\perp$.

Let $\eta(\Delta) = \|E(\Delta)h_1\|^2$ and let $\mathcal{L} = \mathrm{cl}[W^*(N)h_1]$. Hence $\eta \ll \mu$, N is reduced by both $\mathcal{L}$ and $\mathcal{G}$, and $\mathcal{L} \perp \mathcal{G}$. Moreover, $N|\mathcal{G} \cong N_\mu$ and $N|\mathcal{L} \cong N_\eta$. Now the fact that $\eta \ll \mu$ implies that there is a Borel set Δ such that $[\eta] = [\mu|\Delta]$. (Why?) Hence $N|\mathcal{L} \cong N_\nu$ if $\nu = \mu|\Delta$ (Theorem 3.6). Let $U: \mathcal{G} \oplus \mathcal{L} \to L^2(\mu) \oplus L^2(\nu)$ be an isomorphism such that $U(N|\mathcal{G} \oplus \mathcal{L})U^{-1} = N_\mu \oplus N_\nu$. Since $e = g_1 + h_1 \in \mathcal{G} \oplus \mathcal{L}$, let $Ue = g \oplus h$. Because h_1 is a $*$-cyclic vector for $N|\mathcal{L}$, $h(z) \neq 0$ a.e. $[\nu]$.

This reduces the proof of this proposition to proving the next lemma. ∎

10.5. Lemma. *Let μ be a compactly supported measure on $\mathbb{C}$, Δ a Borel subset of the support of μ, and put $\nu = \mu|\Delta$. If $N = N_\mu \oplus N_\nu$ on $L^2(\mu) \oplus L^2(\nu)$ and $g \oplus h \in L^2(\mu) \oplus L^2(\nu)$ such that $h(z) \neq 0$ a.e. $[\nu]$, then there is an f in $L^2(\mu)$ such that $f \oplus h$ is a separating vector for $W^*(N)$ and $g \oplus h \in \mathrm{cl}[W^*(N)(f \oplus h)]$.*

PROOF. Define $f(z) = g(z)$ for z in Δ and $f(z) = 1$ for z not in Δ. Put $\mathcal{H} = \mathrm{cl}[W^*(N)(f \oplus h)] = \mathrm{cl}\{\phi f \oplus \phi h : \phi \in L^\infty(\mu)\}$ since μ is a scalar-valued spectral measure for N. If $\Delta' = $ the complement of Δ, then note that $\phi \chi_{\Delta'} \oplus 0 = \phi \chi_{\Delta'}(f \oplus h) \in \mathcal{H}$ for all ϕ in $L^\infty(\mu)$. Hence $L^2(\mu|\Delta') \oplus 0 \subseteq \mathcal{H}$. This implies that $(1 - g)\chi_{\Delta'} \oplus 0 \in \mathcal{H}$, so $g \oplus h = f \oplus h - (1 - g)\chi_{\Delta'} \oplus 0 \in \mathcal{H}$.

On the other hand, if $\phi \in L^\infty(\mu)$ and $0 = \phi f \oplus \phi h$, then $\phi f = \phi h = 0$ a.e. $[\mu]$. Since $h(z) \neq 0$ a.e. $[\nu]$, $\phi(z) = 0$ a.e. $[\mu]$ on Δ. But for z in Δ', $f(z) = 1$; hence $\phi(z) = 0$ a.e. $[\mu]$ on Δ'. Thus, $f \oplus h$ is a separating vector for $W^*(N)$. ∎

PROOF OF THEOREM 1.10 (a): Let e_1 be a separating vector for $W^*(N)$ and let $\{f_n\}$ be an orthonormal basis for $\mathscr{H}$ such that $f_1 = e_1$. Put $\mathscr{H}_1 = \mathrm{cl}[W^*(N)e_1]$, $\mu_1(\Delta) = \|E(\Delta)e_1\|^2$, and $N_2 = N|\mathscr{H}_1^\perp$. Let $f_2' =$ the orthogonal projection of f_2 onto $\mathscr{H}_1^\perp$. By Proposition 10.4 there is a separating vector e_2 for $W^*(N_2)$ such that $f_2' \in \mathrm{cl}[W^*(N_2)e_2] \equiv \mathscr{H}_2$. Note that $\mathscr{H}_2 = \mathrm{cl}[W^*(N)e_2]$ and $\{f_1, f_2\} \subseteq \mathscr{H}_1 \oplus \mathscr{H}_2$. Put $\mu_2(\Delta) = \|E(\Delta)e_2\|^2$. Now continue by induction. ∎

Now for part (b) of Theorem 10.1. If $[\mu_n] = [\nu_n]$ (the notation is that of Theorem 10.1) for every n, then $N_{\mu_n} \cong N_{\nu_n}$ by Theorem 3.6. Therefore $N \cong M$. Thus it is the converse that causes difficulties. So assume that $N \cong M$. If $M \in \mathscr{B}(\mathscr{K})$, $U: \mathscr{H} \to \mathscr{K}$ is an isomorphism such that $UNU^{-1} = M$, and e_1 is a separating vector for $W^*(N)$, then $Ue_1 = f_1$ is easily seen to be a separating vector for $W^*(M)$. Since μ_1 and ν_1 are scalar-valued spectral measures for N and M, respectively, it follows that $[\mu_1] = [\nu_1]$; thus $N_{\mu_1} \cong N_{\nu_1}$ by Theorem 3.6. However, here is the difficulty—the isomorphism that shows that $N_{\mu_1} \cong N_{\nu_1}$ may not be related to U; that is, if $\mathscr{H} = \oplus_n \mathscr{H}_n$, $\mathscr{K} = \oplus_1^\infty \mathscr{K}_n$, where $N|\mathscr{H}_n \cong N_{\mu_n}$ and $M|\mathscr{K}_n \cong N_{\nu_n}$, then $N|\mathscr{H}_1 \cong M|\mathscr{K}_1$, but we do not know that $U\mathscr{H}_1 = \mathscr{K}_1$. Thus we want to argue that because $N \cong M$ and $N|\mathscr{H}_1 \cong M|\mathscr{K}_1$, then $N|\mathscr{H}_1^\perp \cong M|\mathscr{K}_1^\perp$. In this way we can prove (10.1b) by induction. This step is justified by the following.

10.6. Proposition. *If N, A, and B are normal operators, N is $*$-cyclic, and $N \oplus A \cong N \oplus B$, then $A \cong B$.*

PROOF. Let $N \in \mathscr{B}(\mathscr{K})$, $A \in \mathscr{B}(\mathscr{H}_A)$, $B \in \mathscr{B}(\mathscr{H}_B)$, and let $U: \mathscr{K} \oplus \mathscr{H}_A \to \mathscr{K} \oplus \mathscr{H}_B$ be an isomorphism such that $U(N \oplus A)U^{-1} = N \oplus B$. Now U can be written as a 2×2 matrix,

$$U = \begin{bmatrix} U_{11} & U_{12} \\ U_{21} & U_{22} \end{bmatrix},$$

where $U_{11}: \mathscr{K} \to \mathscr{K}$, $U_{12}: \mathscr{H}_A \to \mathscr{K}$, $U_{21}: \mathscr{K} \to \mathscr{H}_B$, $U_{22}: \mathscr{H}_A \to \mathscr{H}_B$. Expressing $N \oplus A$ and $N \oplus B$ as

$$\begin{bmatrix} N & 0 \\ 0 & A \end{bmatrix} \text{ and } \begin{bmatrix} N & 0 \\ 0 & B \end{bmatrix}$$

respectively, the equation $U(N \oplus A) = (N \oplus B)U$ becomes

10.7
$$\begin{bmatrix} U_{11}N & U_{12}A \\ U_{21}N & U_{22}A \end{bmatrix} = \begin{bmatrix} NU_{11} & NU_{12} \\ BU_{21} & BU_{22} \end{bmatrix}.$$

Similarly, $U(N \oplus A)^* = (N \oplus B)^* U$ becomes

10.7*
$$\begin{bmatrix} U_{11}N^* & U_{12}A^* \\ U_{21}N^* & U_{22}A^* \end{bmatrix} = \begin{bmatrix} N^*U_{11} & N^*U_{12} \\ B^*U_{21} & B^*U_{22} \end{bmatrix}.$$

Parts of the preceding equations will be referred to as $(10.7)_{ij}$ and $(10.7)^*_{ij}$, $i, j = 1, 2$.

An examination of the equations $U^*U = 1$ and $UU^* = 1$, written in matrix form, yields the equations

10.8
$$\begin{cases} \text{(a)} \;\; U_{11}^*U_{12} + U_{21}^*U_{22} = 0 & (\text{on } \mathcal{H}_A) \\ \text{(b)} \;\; U_{11}^*U_{11}^* + U_{12}U_{12}^* = 1 & (\text{on } \mathcal{K}) \\ \text{(c)} \;\; U_{21}U_{11}^* + U_{22}U_{12}^* = 0 & (\text{on } \mathcal{K}). \end{cases}$$

Now equation $(10.7)_{22}$ and Proposition 6.10 imply that $(\ker U_{22})^\perp$ reduces A, $\mathrm{cl}(\operatorname{ran} U_{22}) = (\ker U_{22}^*)^\perp$ reduces B, and

10.9
$$A|(\ker U_{22})^\perp \cong B|(\ker U_{22}^*)^\perp .$$

What about $A|\ker U_{22}$ and $B|\ker U_{22}^*$? If they are unitarily equivalent, then $A \cong B$ and we are done. If $h \in \ker U_{22} \subseteq \mathcal{H}_A$, then

$$\begin{bmatrix} U_{11} & U_{12} \\ U_{21} & U_{22} \end{bmatrix} \begin{bmatrix} 0 \\ h \end{bmatrix} = \begin{bmatrix} U_{12}h \\ 0 \end{bmatrix}.$$

Since U is an isometry, it follows that U_{12} maps $\ker U_{22}$ isometrically onto a closed subspace of $\mathcal{K}$. Put $\mathcal{M}_1 = U_{12}(\ker U_{22})$. Equations $(10.7)_{12}$ and $(10.7)^*_{12}$ and the fact that $\ker U_{22}$ reduces A imply that $\mathcal{M}_1$ reduces N. Thus, the restriction of U_{12} to $\ker A_{22}$ is the required isomorphism to show that

10.10
$$A|\ker U_{22} \cong N|\mathcal{M}_1.$$

Similarly, U_{21}^* maps $\ker U_{22}^* = (\operatorname{ran} U_{22})^\perp$ isometrically onto $\mathcal{M}_2 = U_{21}^*(\ker U_{22}^*)$, $\mathcal{M}_2$ reduces N, and

10.11
$$B|\ker U_{22}^* \cong N|\mathcal{M}_2.$$

Note that if $\mathcal{M}_1 = \mathcal{M}_2$, then (10.9), (10.10), and (10.11) show that $A \cong B$. Could it be that $\mathcal{M}_1$ and $\mathcal{M}_2$ are equal?

If $h \in \ker U_{22}$, then (10.8a) implies that $U_{11}^*U_{12}h = -U_{21}^*U_{22}h = 0$. Hence $\mathcal{M}_1 = U_{12}(\ker U_{22}) \subseteq \ker U_{11}^*$. On the other hand, if $f \in \ker U_{11}^*$, then (10.8b) implies $f = (U_{11}U_{11}^* + U_{12}U_{12}^*)f = U_{12}U_{12}^*f$. But by (10.8c), $U_{22}U_{12}^*f = -U_{21}U_{11}^*f = 0$, so $U_{12}^*f \in \ker U_{22}$. Hence $f \in U_{12}(\ker U_{22})$. Thus,

$$\mathcal{M}_1 = \ker U_{11}^*.$$

Similarly,

$$\mathcal{M}_2 = \ker U_{11}.$$

Until this point we have not used the fact that N is $*$-cyclic. Equation $(10.7)_{11}$ implies that $U_{11} \in \{N\}'$. By Corollary 6.9 this implies that U_{11} is normal. Hence, $\ker U_{11}^* = \ker U_{11}$, or $\mathcal{M}_1 = \mathcal{M}_2$. ∎

If the hypothesis in the preceding proposition that N is $*$-cyclic is deleted, the conclusion is no longer valid. For example, let N and A be the identities on separable infinite-dimensional spaces and let B be the identity on a finite-dimensional space. Then $N \oplus A \cong N \oplus B$, but A and B are not equivalent. However, the requirement that N be $*$-cyclic can be replaced by another, even when N, A, and B are not assumed to be normal. For the details see Kadison and Singer [1957].

The proof of Theorem 10.1(b) is now a straightforward argument as outlined before the statement of Proposition 10.6. The details are left to the reader.

If μ and ν are measures and $\nu \ll \mu$, then there is a Borel set Δ such that $[\nu] = [\mu | \Delta]$. Using this fact, Theorem 10.1 can be restated as follows.

10.12.   Corollary. (a) *If N is a normal operator with scalar spectral measure μ, then there is a decreasing sequence $\{\Delta_n\}$ of Borel subsets of $\sigma(N)$ such that $\Delta_1 = \sigma(N)$ and*

$$N \cong N_\mu \oplus N_{\mu|\Delta_2} \oplus N_{\mu|\Delta_3} \oplus \cdots .$$

(b) *If M is another normal operator with scalar spectral measure ν and if $\{\Sigma_n\}$ is a decreasing sequence of Borel subsets of $\sigma(M)$ such that $M \cong N_\nu \oplus N_{\nu|\Sigma_2} \oplus N_{\nu|\Sigma_3} \oplus \cdots$, then $N \cong M$ if and only if (i) $[\mu] = [\nu]$ and (ii) $\mu(\Delta_n \setminus \Sigma_n) = 0 = \mu(\Sigma_n \setminus \Delta_n)$ for all n.*

10.13.   Example. Let μ be Lebesgue measure on $[0, 1]$ and let μ_n be Lebesgue measure on $[1/n + 1, 1/n]$ for $n \geq 1$. (So $\mu = \Sigma \mu_n$.) Let $N = N_{\mu_1} \oplus N_{\mu_2}^{(2)} \oplus N_{\mu_3}^{(3)} \oplus \cdots$. The direct sum decomposition of N that appears in Corollary 10.12 is obtained by letting $\Delta_n = [0, 1/n]$, $n \geq 1$. Then $N \cong N_\mu \oplus N_{\mu|\Delta_2} \oplus N_{\mu|\Delta_3} \oplus \cdots$.

What does Theorem 10.1 say for normal operators on a finite-dimensional space? If $\dim \mathcal{H} < \infty$, there is an orthonormal basis $\{e_n\}$ for $\mathcal{H}$ consisting of eigenvectors for N. Observe that N is cyclic if and only if each eigenvalue has multiplicity 1. So each summand that appears in (10.2) must operate on a subspace of $\mathcal{H}$ that contains only one basis element e_n per eigenvalue. Moreover, since μ_1 is a scalar spectral measure for N, it must be that the first summand in (10.2) contains one basis element for each eigenvalue for N. Thus, if $\sigma(N) = \{\lambda_1, \lambda_2, \ldots, \lambda_n\}$, where $\lambda_i \neq \lambda_j$ for $i \neq j$, then (10.2) becomes

10.14 $N \cong D_1 \oplus D_2 \oplus \cdots \oplus D_m,$

where $D_1 = \operatorname{diag}(\lambda_1, \lambda_2, \ldots, \lambda_n)$ and, for $k \geq 2$, D_k is a diagonalizable operator whose diagonal consists of one, and only one, of each of the eigenvalues of N having multiplicity at least k.

There is another decomposition for normal operators that furnishes a complete set of unitary invariants and has a connection with the concept of multiplicity. For normal operators on a finite-dimensional space, this decomposition takes on the following form.

Let Λ_k = the eigenvalues of N having multiplicity k. So for λ in Λ_k, $\dim \ker(N - \lambda) = k$. If $\Lambda_k = \{\lambda_j^{(k)}: 1 \le j \le m_k\}$, let N_k be the diagonalizable operator on a km_k dimensional space whose diagonal contains each $\lambda_j^{(k)}$ repeated k times. So $N \cong N_1 \oplus N_2 \oplus \cdots \oplus N_p$, if $\sigma(N) = \Lambda_1 \cup \cdots \cup \Lambda_p$. Now $\sigma(N_k) = \Lambda_k$ and each eigenvalue of N_k has multiplicity k. Thus $N_k \cong A_k^{(k)}$, where A_k is a diagonalizable operator on an m_k dimensional space with $\sigma(A_k) = \Lambda_k$. Thus

10.15 $$N \cong A_1 \oplus A_2^{(2)} \oplus \cdots \oplus A_p^{(p)},$$

and $\sigma(A_i) \cap \sigma(A_j) = \square$ for $i \ne j$.

Now the big advantage of the decomposition (10.15) is that it permits a discussion of $\{N\}'$. Because the spectra of the operators A_k are disjoint,

$$\{N\}' = \{N_1\}' \oplus \{N_2\}' \oplus \cdots \oplus \{N_p\}'.$$

(Why?) If $\mathcal{H}_j^{(k)} = \ker(N - \lambda_j^{(k)})$, then $\dim \mathcal{H}_j^{(k)} = k$ and $\oplus_{j=1}^{m_k} \mathcal{H}_j^{(k)} = $ the domain of N_k. Since $\lambda_i^{(k)} \ne \lambda_j^{(k)}$ for $i \ne j$,

$$\{N_k\}' = \mathcal{B}(\mathcal{H}_1^{(k)}) \oplus \cdots \oplus \mathcal{B}(\mathcal{H}_{n_k}^{(k)}),$$

and each $\mathcal{B}(\mathcal{H}_j^{(k)})$ is isomorphic to the $k \times k$ matrices.

The decomposition of an arbitrary normal operator that is analogous to decomposition (10.15) for finite-dimensional normal operators is contained in the next result. The corresponding discussion of the commutant will follow this theorem.

10.16. Theorem. *If N is a normal operator, then there are mutually singular measures $\mu_\infty, \mu_1, \mu_2, \ldots$ (some of which may be zero) such that*

$$N \cong N_{\mu_\infty}^{(\infty)} \oplus N_{\mu_1} \oplus N_{\mu_2}^{(2)} \oplus \cdots.$$

If M is another normal operator with corresponding measures $\nu_\infty, \nu_1, \nu_2, \ldots$, then $N \cong M$ if and only if $[\mu_n] = [\nu_n]$ for $1 \le n \le \infty$.

PROOF. Let μ be a scalar spectral measure for N and let $\{\Delta_n\}$ be the sequence of Borel subsets of $\sigma(N)$ obtained in Corollary 10.12. Put $\Sigma_\infty = \cap_{n=1}^\infty \Delta_n$ and $\Sigma_n = \Delta_n \setminus \Delta_{n+1}$ for $1 \le n < \infty$; let $\mu_n = \mu|\Sigma_n$, $1 \le n < \infty$. Put $\nu_n = \mu|\Delta_n$, $1 \le n < \infty$. Now $\Delta_n = \Sigma_\infty \cup (\Delta_n \setminus \Delta_{n+1}) \cup (\Delta_{n+1} \setminus \Delta_{n+2}) \cup \cdots = \Sigma_\infty \cup \Sigma_n \cup \Sigma_{n+1} \cup \cdots$. Hence $\nu_n = \mu_\infty + \mu_n + \mu_{n+1} + \cdots$ and the measures $\mu_\infty, \mu_n, \mu_{n+1}, \ldots$ are pairwise singular. Hence $N_{\nu_n} \cong N_{\mu_\infty}$

$\oplus N_{\mu_n} \oplus N_{\mu_{n+1}} \cdots$. Combining this with Corollary 10.12 gives

$$N \cong N_{\nu_1} \oplus N_{\nu_2} \oplus N_{\nu_3} \oplus \cdots$$

$$\cong \left(N_{\mu_\infty} \oplus N_{\mu_1} \oplus N_{\mu_2} \oplus \cdots \right) \oplus \left(N_{\mu_\infty} \oplus N_{\mu_2} \oplus N_{\mu_3} \oplus \cdots \right)$$

$$\oplus \left(N_{\mu_\infty} \oplus N_{\mu_3} \oplus N_{\mu_4} \oplus \cdots \right) \oplus \cdots$$

$$\cong N_{\mu_\infty}^{(\infty)} \oplus N_{\mu_1} \oplus N_{\mu_2}^{(2)} \oplus N_{\mu_3}^{(3)} \oplus \cdots .$$

The proof of the uniqueness part of the theorem is left to the reader. ∎

Note that the form of the normal operator presented in Example 10.13 is the form of the operator given in the conclusion of the preceding theorem.

Now to discuss $\{N\}'$. Fix a compactly supported measure μ on $\mathbb{C}$ and let $\mathcal{H}_n$ be an n-dimensional Hilbert space, $1 \leq n \leq \infty$. Define a function $f: \mathbb{C} \to \mathcal{H}_n$ to be a Borel function if $z \mapsto \langle f(z), g \rangle$ is a Borel function for each g in $\mathcal{H}_n$. If $f: \mathbb{C} \to \mathcal{H}_n$ is a Borel function and $\{e_j\}$ is an orthonormal basis for $\mathcal{H}_n$, then $\|f(z)\|^2 = \Sigma_j |\langle f(z), e_j \rangle|^2$, so $z \to \|f(z)\|^2$ is a Borel function. Let $L^2(\mu; \mathcal{H}_n)$ be the space of all Borel functions $f: \mathbb{C} \to \mathcal{H}_n$ such that $\|f\|^2 \equiv \int \|f(z)\|^2 \, d\mu(z) < \infty$, where two functions agreeing a.e. $[\mu]$ are identified. If f and $g \in L^2(\mu; \mathcal{H}_n)$, $\langle f, g \rangle \equiv \int \langle f(z), g(z) \rangle \, d\mu(z)$ defines an inner product on $L^2(\mu; \mathcal{H}_n)$. It is not difficult to show that $L^2(\mu; \mathcal{H}_n)$ is a Hilbert space.

10.17. Proposition. *If N is multiplication by z on $L^2(\mu; \mathcal{H}_n)$, then $N \cong N_\mu^{(n)}$.*

PROOF. Let $\{e_j: 1 \leq j \leq n\}$ be an orthonormal basis for $\mathcal{H}_n$ and define $U: L^2(\mu; \mathcal{H}_n) \to L^2(\mu)^{(n)}$ by $Uf = (\langle f(\cdot), e_1 \rangle, \langle f(\cdot), e_2 \rangle, \ldots)$. Then U is an isomorphism and $UNU^{-1} = N_\mu^{(n)}$. The details are left to the reader. ∎

Combining the preceding proposition with Proposition 6.1(b), we can find $\{N\}'$; namely, $\{N_\mu^{(n)}\}' = $ all matrices (T_{ij}) on $\mathcal{B}(L^2(\mu)^{(n)})$ such that $T_{ij} \in \{N_\mu\}'$ for all i, j. By Corollary 6.9, $\{N_\mu^{(n)}\}' = $ all matrices $(M_{\phi_{ij}})$ that belong to $\mathcal{B}(L^2(\mu)^{(n)})$, such that $\phi_{ij} \in L^\infty(\mu)$. Now the idea is to use Proposition 10.17 to bring this back to $\mathcal{B}(L^2(\mu; \mathcal{H}_n))$ and describe $\{N\}'$.

A function $\phi: \mathbb{C} \to \mathcal{B}(\mathcal{H}_n)$ is defined to be a Borel function if for each f and g in $\mathcal{H}_n$, $z \mapsto \langle \phi(z)f, g \rangle$ is a Borel function. If $\{f_j\}$ is a countable dense subset of the unit ball of $\mathcal{H}_n$, $\|\phi(z)\| = \sup\{|\langle \phi(z)f_i, f_j \rangle|: 1 \leq i, j < \infty\}$, so $z \mapsto \|\phi(z)\|$ is a Borel function. Let $L^\infty(\mu; \mathcal{B}(\mathcal{H}_n))$ be the equivalent classes of bounded Borel functions from $\mathbb{C}$ into $\mathcal{B}(\mathcal{H}_n)$ furnished with the μ-essential supremum norm.

If $\phi \in L^\infty(\mu; \mathcal{B}(\mathcal{H}_n))$ and $f \in L^2(\mu; \mathcal{H}_n)$, let $f(z) = \Sigma f_j(z)e_j$, where $\{e_j\}$ is an orthonormal basis for $\mathcal{H}_n$ and $f_j(z) = \langle f(z), e_j \rangle$, so $\Sigma |f_j(z)|^2 = \|f(z)\|^2$. Thus $\phi(z)f(z) = \Sigma_j f_j(z)\phi(z)e_j$. So for any e in $\mathcal{H}_n$, $\langle \phi(z)f(z), e \rangle = \Sigma f_j(z)\langle \phi(z)e_j, e \rangle$ is a Borel function. It is easy to check that $\phi f \in L^2(\mu; \mathcal{H}_n)$ and $\|\phi f\| \leq \|\phi\|_\infty \|f\|$. Let M_ϕ: $L^2(\mu; \mathcal{H}_n) \to L^2(\mu; \mathcal{H}_n)$ be defined by $M_\phi f = \phi f$. Combined with the preceding remarks, the following result can be shown to hold. (The proof is left to the reader.)

10.18. **Proposition.** *If N is multiplication by z on $L^2(\mu; \mathcal{H}_n)$, then*

$$\{N\}' = \{M_\phi: \phi \in L^\infty(\mu; \mathcal{B}(\mathcal{H}_n))\}.$$

Also, $\|M_\phi\| = \|\phi\|_\infty$ for every ϕ in $L^\infty(\mu; \mathcal{B}(\mathcal{H}_n))$.

The next lemma is a consequence of Proposition 6.10 and the fact that unitarily equivalent normal operators have mutually absolutely continuous scalar-valued spectral measures.

10.19. **Lemma.** *If N_1 and N_2 are normal operators with mutually singular scalar spectral measures and $XN_1 = N_2 X$, then $X = 0$.*

Using the observation made prior to Corollary 6.8, the preceding lemma implies that $\{N_1 \oplus N_2\}' = \{N_1\}' \oplus \{N_2\}'$ whenever N_1 and N_2 are as in the lemma.

The next theorem of this section can be proved by piecing together Theorem 10.16 and the remaining results of this section. The details are left to the reader.

10.20. **Theorem.** *If N is a normal operator on $\mathcal{H}$, there are mutually singular measures $\mu_\infty, \mu_1, \mu_2, \ldots$ and an isomorphism*

$$U: \mathcal{H} \to L^2(\mu_\infty; \mathcal{H}_\infty) \oplus L^2(\mu_1) \oplus L^2(\mu_2; \mathcal{H}_2) \oplus \cdots$$

such that

$$UNU^{-1} = N_\infty \oplus N_1 \oplus N_2 \oplus \cdots$$

where N_n = multiplication by z on $L^2(\mu_n; \mathcal{H}_n)$. Also,

$$\{N_\infty \oplus N_1 \oplus N_2 \oplus \cdots\}' = L^\infty(\mu_\infty; \mathcal{B}(\mathcal{H}_\infty)) \oplus L^\infty(\mu_1)$$
$$\oplus L^\infty(\mu_2; \mathcal{B}(\mathcal{H}_2)) \oplus \cdots.$$

Using the notation of the preceding theorem, if μ is a scalar-valued spectral measure for N, then there are pairwise disjoint Borel sets $\Delta_\infty, \Delta_1, \ldots$ such that $[\mu_n] = [\mu|\Delta_n]$. Define a function m_N: $\mathbb{C} \to \{0, 1, \ldots \infty\}$ by letting $m_N = \infty\chi_{\Delta_\infty} + \chi_{\Delta_1} + 2\chi_{\Delta_2} + \cdots$. As it stands the definition of m_N depends on the choice of the sets $\{\Delta_n\}$ as well as N. However, any two choices of the sets $\{\Delta_n\}$ differ from one another by sets of μ-measure zero. The function m_N is called the *multiplicity function* for N. Note that m_N is a Borel function.

If m: $\mathbb{C} \to \{\infty, 0, 1, 2, \ldots\}$ is a Borel function and μ is a compactly supported measure such that $\mu(\{z: m(z) = 0\}) = 0$, let $\Delta_n = \{z: m(z) = n\}$, $n = \infty, 1, 2, \ldots$. If $N_n = N_{\mu|\Delta_n}$, then $N = N_\infty^{(\infty)} \oplus N_1 \oplus N_2^{(2)} \oplus \cdots$ is a normal operator whose spectral measure is μ and whose multiplicity function agrees with m a.e. $[\mu]$.

10.21. Theorem. *Two normal operators are unitarily equivalent if and only if they have the same scalar-valued spectral measure μ and their multiplicity functions are equal a.e. $[\mu]$.*

There is some notation that is used by many and we should mention its connection with what we have just finished. Suppose $m: \mathbb{C} \to \{\infty, 0, 1, 2, \ldots\}$ is a Borel function and μ is a compactly supported measure on $\mathbb{C}$ such that $\mu(\{z: m(z) = 0\}) = 0$. If $z \in \mathbb{C}$ let $\mathcal{H}(z)$ be a Hilbert space of dimension $m(z)$. The *direct integral* of the spaces $\mathcal{H}(z)$, denoted by $\int \mathcal{H}(z) \, d\mu(z)$, is precisely the space

$$L^2(\mu|\Delta_\infty; \mathcal{H}_\infty) \oplus L^2(\mu|\Delta_1) \oplus L^2(\mu|\Delta_2; \mathcal{H}_2) \oplus \cdots,$$

where $\Delta_n = \{z: m(z) = n\}$ and $\dim \mathcal{H}_n = n$. If $\phi: \mathbb{C} \to \mathcal{B}(\mathcal{H}_\infty) \cup \mathcal{B}(\mathbb{C}) \cup \mathcal{B}(\mathcal{H}_2) \cup \cdots$ such that $\phi(z) \in \mathcal{B}(\mathcal{H}_n)$ when $z \in \Delta_n$, $\phi: \Delta_n \to \mathcal{B}(\mathcal{H}_n)$ is a Borel function, and there is a constant M such that $\|\phi(z)\| \leq M$ a.e. $[\mu]$, then $\int \phi(z) \, d\mu(z)$ denotes the operator $M_{\phi|\Delta_\infty} \oplus M_{\phi|\Delta_1} \oplus \cdots$ as in (10.20). Although the direct integral notation is quite suggestive, one must revert to the notation of (10.20) to produce proofs.

REMARKS. There are several sources for multiplicity theory. Most begin by proving Theorem 10.16. This is done for nonseparable spaces in Halmos [1951] and Brown [1974]. Another source is Arveson [1976], where the theory is set in the context of C^*-algebras which is its proper milieu. Also, Arveson shows that the theory can be applied to some non-normal operators. The details of this more general multiplicity theory are carried out in Ernest [1976] as part of a more general classification scheme. Another source for multiplicity theory is Dunford and Schwartz [1963].

By Theorem 4.6, every normal operator is unitarily equivalent to a multiplication operator M_ϕ on $L^2(X, \Omega, \mu)$ for some measure space (X, Ω, μ). The scalar-valued spectral measure for M_ϕ is $\mu \circ \phi^{-1}$. What is the multiplicity function for M_ϕ? One is tempted to say that $m_{M_\phi}(z) =$ the number of points in $\phi^{-1}(z)$. This is not quite correct. The answer can be found in Abrahamse and Kriete [1973]. Also, Abrahamse [1978] contains a survey of spectral multiplicity for normal operators treated from this point of view.

EXERCISES

1. Let A and B be operators on $\mathcal{H}$ and $\mathcal{K}$, respectively. Let $\mathcal{H}_0$ and $\mathcal{K}_0$ be reducing subspaces for A and B and suppose that $A \cong B|\mathcal{K}_0$ and $B \cong A|\mathcal{H}_0$. Show that $A \cong B$.

2. Let $\mu_1, \mu_2, \ldots$ be compactly supported measures on $\mathbb{C}$ such that $\mu_{n+1} \ll \mu_n$ for all n. Show that if M is any normal operator whose spectral measure is

absolutely continuous with respect to each μ_n, then $N_{\mu_1} \oplus N_{\mu_2} \oplus \cdots \cong (N_{\mu_1} \oplus N_{\mu_2} \oplus \cdots) \oplus M$.

3. If $\mu = $ Lebesgue measure on $[0, 1]$, show that $N_\mu \cong N_\mu^p$ for $0 < p < \infty$.

4. Let $\mu = $ Lebesgue measure on $[0, 1]$ and characterize the functions ϕ in $L^\infty(\mu)$ such that $N_\mu \cong \phi(N_\mu)$.

5. Let $\mu = $ area measure on $\mathbb{D}$ and show that N_μ and N_μ^2 are not unitarily equivalent.

6. Let $\mu = $ Lebesgue measure on $[0, 1]$ and let $\nu = $ Lebesgue measure on $[-1, 1]$. Show that $N_\nu^2 \cong N_\mu \oplus N_\mu$. How about N_ν^3?

7. Let μ be Lebesgue measure on $\mathbb{R}$ and $N = $ multiplication by $\sin x$ on $L^2(\mu)$. Find the decompositions of N obtained in Theorems 10.1 and 10.16.

8. If μ is Lebesgue measure on $\mathbb{R}$ and $N = $ multiplication by e^{ix} on $L^2(\mu)$, show that $N \cong N_m^{(\infty)}$ where $m = $ arc length measure on $\partial\mathbb{D}$.

9. Define $U: L^2(\mathbb{R}) \to L^2(\mathbb{R})$ by $(Uf)(t) = f(t - 1)$. Show that U is unitary and find its scalar-valued spectral measure and multiplicity function.

10. Represent N as in Theorem 10.1 and find the corresponding representation for $N \oplus N = N^{(2)}$; for $N^{(3)}$, for $N^{(\infty)}$. (Are you surprised by the result for $N^{(\infty)}$?)

11. Prove the results and solve the exercises from § II.8.

12. Let N be a normal operator and show that $N \cong N^{(2)}$ if and only if there is a $*$-cyclic normal operator M such that $N \cong M^{(\infty)}$. What does this say about the multiplicity function for N?

13. Let (X, Ω, μ) be a measure space such that $L^2(\mu)$ is separable, let $\phi \in L^\infty(\mu)$, and let $N = M_\phi$ on $L^2(\mu)$. Find the decompositions of N obtained in Theorems 10.1 and 10.16.

14. Let μ be a compactly supported measure on $\mathbb{C}$, ϕ a bounded Borel function on $\mathbb{C}$, and suppose $\{\Delta_n\}$ are pairwise disjoint Borel sets such that ϕ is one-to-one on each Δ_n and $\mu(\mathbb{C} \setminus \bigcup_{n=1}^\infty \Delta_n) = 0$. Let $\phi_n = \phi\chi_{\Delta_n}$ and $\mu_n = \mu \circ \phi_n^{-1}$ for $n \geq 1$. Prove that M_ϕ on $L^2(\mu)$ is unitarily equivalent to $\bigoplus_{n=1}^\infty N_{\mu_n}$.

CHAPTER X

Unbounded Operators

It is unfortunate for the world we live in that all of the operators that arise naturally are not bounded. But that is indeed the case. Thus it is important to study such operators.

The idea here is not to study an arbitrary linear transformation on a Hilbert space. In fact, such a study is the province of linear algebra rather than analysis. The operators that are to be studied do possess certain properties that connect them to the underlying Hilbert space. The properties that will be isolated are inspired by natural examples.

All Hilbert spaces in this chapter are assumed separable.

§1. Basic Properties and Examples

The first relaxation in the concept of operator is not to assume that the operators are defined everywhere on the Hilbert space.

1.1. Definition. If $\mathcal{H}, \mathcal{K}$ are Hilbert spaces, a *linear operator* $A: \mathcal{H} \to \mathcal{K}$ is a function whose domain of definition is a linear manifold, dom A, in $\mathcal{H}$ and such that $A(\alpha f + \beta g) = \alpha A f + \beta A g$ for f, g in dom A and α, β in $\mathbb{C}$. A is *bounded* if there is a constant $c > 0$ such that $\|Af\| \leq c\|f\|$ for all f in dom A.

Note that if A is bounded, then A can be extended to a bounded linear operator on cl[dom A] and then extended to $\mathcal{H}$ by letting A be 0 on (dom $A)^{\perp}$. So unless it is specified to the contrary, a bounded operator will always be assumed to be defined on all of $\mathcal{H}$.

If A is a linear operator from $\mathcal{H}$ into $\mathcal{K}$, then A is also a linear operator from cl[dom A] into $\mathcal{K}$. So we will often only consider those A such that dom A is dense in $\mathcal{H}$; such an operator A is said to be *densely defined*. $\mathcal{B}(\mathcal{H})$ still denotes the bounded operators defined on $\mathcal{H}$.

If A, B are linear operators from $\mathcal{H}$ into $\mathcal{K}$, then $A + B$ is defined with dom$(A + B) = $ dom $A \cap$ dom B. If $B: \mathcal{H} \to \mathcal{K}$ and $A: \mathcal{K} \to \mathcal{L}$, then AB is a linear operator from $\mathcal{H}$ into $\mathcal{L}$ with dom$(AB) = B^{-1}(\text{dom } A)$.

1.2. Definition. If A, B are operators from $\mathcal{H}$ into $\mathcal{K}$, then A is an *extension* of B if dom $B \subseteq$ dom A and $Ah = Bh$ whenever $h \in$ dom B. In symbols this is denoted by $B \subseteq A$.

Note that if $A \in \mathcal{B}(\mathcal{H})$, then the only extension of A is itself. So this concept is only of value for unbounded operators.

If $A: \mathcal{H} \to \mathcal{K}$, the *graph* of A is the set

$$\text{gra } A \equiv \{ h \oplus Ah \in \mathcal{H} \oplus \mathcal{K} : h \in \text{dom } A \}.$$

It is easy to see that $B \subseteq A$ if and only if gra $B \subseteq$ gra A.

1.3. Definition. An operator $A: \mathcal{H} \to \mathcal{K}$ is *closed* if its graph is closed in $\mathcal{H} \oplus \mathcal{K}$. An operator is *closable* if it has a closed extension. Let $\mathscr{C}(\mathcal{H}, \mathcal{K})$ = the collection of all closed densely defined operators from $\mathcal{H}$ into $\mathcal{K}$. Let $\mathscr{C}(\mathcal{H}) = \mathscr{C}(\mathcal{H}, \mathcal{H})$. (It should be emphasized that the operators in $\mathscr{C}(\mathcal{H}, \mathcal{K})$ are densely defined.)

When is a subset of $\mathcal{H} \oplus \mathcal{K}$ a graph of an operator from $\mathcal{H}$ into $\mathcal{K}$? If $\mathcal{G} = $ gra A for some $A: \mathcal{H} \to \mathcal{K}$, then $\mathcal{G}$ is a submanifold of $\mathcal{H} \oplus \mathcal{K}$ such that if $k \in \mathcal{K}$ and $0 \oplus k \in \mathcal{G}$, then $k = 0$. The converse is also true. That is, suppose that $\mathcal{G}$ is a submanifold of $\mathcal{H} \oplus \mathcal{K}$ such that if $k \in \mathcal{K}$ and $0 \oplus k \in \mathcal{G}$, then $k = 0$. Let $\mathcal{D} = \{ h \in \mathcal{H}:$ there exists a k in $\mathcal{K}$ with $h \oplus k$ in $\mathcal{G} \}$. If $h \in \mathcal{D}$ and $k_1, k_2 \in \mathcal{K}$ such that $h \oplus k_1, h \oplus k_2 \in \mathcal{G}$, then $0 \oplus (k_1 - k_2) = h \oplus k_1 - h \oplus k_2 \in \mathcal{G}$. Hence $k_1 = k_2$. That is, for every h in $\mathcal{D}$ there is a unique k in $\mathcal{K}$ such that $h \oplus k \in \mathcal{G}$; denote k by $k = Ah$. It is easy to check that A is a linear map and $\mathcal{G} = $ gra A. This gives an internal characterization of graphs that will be useful in the next proposition.

1.4. Proposition. *An operator $A: \mathcal{H} \to \mathcal{K}$ is closable if and only if* cl[gra A] *is a graph.*

PROOF. Let cl[gra A] be a graph. That is, there is an operator $B: \mathcal{H} \to \mathcal{K}$ such that gra $B = $ cl[gra A]. Clearly gra $A \subseteq$ gra B, so A is closable.

Now assume that A is closable; that is, there is a closed operator $B: \mathcal{H} \to \mathcal{K}$ with $A \subseteq B$. If $0 \oplus k \in$ cl[gra A], $0 \oplus k \in$ gra B and hence $k = 0$. By the remarks preceding this proposition, cl[gra A] is a graph. ∎

If A is closable, call the operator whose graph is cl[gra A] the *closure* of A.

1.5.   Definition. If $A: \mathcal{H} \to \mathcal{K}$ is densely defined, let

$$\text{dom } A^* = \{ k \in \mathcal{K}: h \mapsto \langle Ah, k \rangle \text{ is a bounded linear}$$

$$\text{functional on dom } A \}.$$

Because dom A is dense in $\mathcal{H}$, if $k \in \text{dom } A^*$, then there is a unique vector f in $\mathcal{H}$ such that $\langle Ah, k \rangle = \langle h, f \rangle$ for all h in dom A. Denote this unique vector f by $f = A^*k$. Thus

$$\langle Ah, k \rangle = \langle h, A^*k \rangle$$

for h in dom A and k in dom A^*.

1.6.   Proposition. If $A: \mathcal{H} \to \mathcal{K}$ is a densely defined operator, then:

(a) A^* is a closed operator;
(b) A^* is densely defined if and only if A is closable;
(c) if A is closable, then its closure is $(A^*)^* \equiv A^{**}$.

Before proving this, a lemma is needed which will also be useful later.

1.7.   Lemma. If $A: \mathcal{H} \to \mathcal{K}$ is densely defined and $J: \mathcal{H} \oplus \mathcal{K} \to \mathcal{K} \oplus \mathcal{H}$ is defined by $J(h \oplus k) = (-k) \oplus h$, then J is an isomorphism and

$$\text{gra } A^* = [J \text{ gra } A]^{\perp}$$

PROOF. It is clear that J is an isomorphism. To prove the formula for gra A^*, note that gra $A^* = \{ k \oplus A^*k \in \mathcal{K} \oplus \mathcal{H}: k \in \text{dom } A^* \}$. So if $k \in$ dom A^* and $h \in \text{dom } A$,

$$\langle k \oplus A^*k, J(h \oplus Ah) \rangle = \langle k \oplus A^*k, -Ah \oplus h \rangle$$

$$= -\langle k, Ah \rangle + \langle A^*k, h \rangle = 0.$$

Thus gra $A^* \subseteq [J \text{ gra } A]^{\perp}$. Conversely, if $k \oplus f \in [J \text{ gra } A]^{\perp}$, then for every h in dom A, $0 = \langle k \oplus f, -Ah \oplus h \rangle = -\langle k, Ah \rangle + \langle f, h \rangle$, so $\langle Ah, k \rangle = \langle h, f \rangle$. By definition $k \in \text{dom } A^*$ and $A^*k = f$. ∎

PROOF OF PROPOSITION 1.6. The proof of (a) is clear from Lemma 1.7. For the remainder of the proof notice that because the map J in (1.7) is an isomorphism, $J^* = J^{-1}$ and so $J^*(k \oplus h) = h \oplus (-k)$.

(b) Assume A is closable and let $k_0 \in (\text{dom } A^*)^{\perp}$. We want to show that $k_0 = 0$. Thus $k_0 \oplus 0 \in [\text{gra } A^*]^{\perp} = [J \text{ gra } A]^{\perp \perp} = \text{cl}[J \text{ gra } A] = J[\text{cl}(\text{gra } A)]$. So $0 \oplus -k_0 = J^*(k_0 \oplus 0) \in J^*J[\text{cl}(\text{gra } A)] = \text{cl}(\text{gra } A)$. But because A is closable, cl(gra A) is a graph; hence $k_0 = 0$. For the converse, assume dom A^* is dense in $\mathcal{K}$. Thus $A^{**} \equiv (A^*)^*$ is defined. By (a), A^{**} is a closed operator. It is easy to see that $A \subseteq A^{**}$, so A has a closed extension.

(c) Note that by Lemma 1.7 gra $A^{**} = [J^* \text{gra } A^*]^{\perp} = [J^*[J \text{ gra } A]^{\perp}]^{\perp}$. But for any linear manifold $\mathcal{M}$ and any isomorphism J, $(J\mathcal{M})^{\perp} = J(\mathcal{M}^{\perp})$.

Hence $J^*[(J\mathcal{M})^\perp] = \mathcal{M}^\perp$ and, thus, $[J^*[J\mathcal{M}]^\perp]^\perp = \mathcal{M}^{\perp\perp} = \text{cl }\mathcal{M}$. Putting $\mathcal{M} = \text{gra } A$ gives that $\text{gra } A^{**} = \text{cl gra } A$. ∎

1.8. Corollary. *If $A \in \mathcal{C}(\mathcal{H}, \mathcal{K})$, then A^* is densely defined and $A^{**} = A$.*

1.9. Example. Let $e_0, e_1, \ldots$ be an orthonormal basis for $\mathcal{H}$ and let $\alpha_0, \alpha_1, \ldots$ be complex numbers. Define $\mathcal{D} = \{h \in \mathcal{H}: \sum_0^\infty |\alpha_n \langle h, e_n \rangle|^2 < \infty\}$ and let $Ah = \sum_0^\infty \alpha_n \langle h, e_n \rangle e_n$ for h in $\mathcal{D}$. Then $A \in \mathcal{C}(\mathcal{H})$ with $\text{dom } A = \mathcal{D}$. Also, $\text{dom } A^* = \mathcal{D}$ and $A^*h = \sum_0^\infty \bar{\alpha}_n \langle h, e_n \rangle e_n$ for all h in $\mathcal{D}$.

1.10. Example. Let (X, Ω, μ) be a σ-finite measure space and let $\phi: X \to \mathbb{C}$ be an Ω-measurable function. Let $\mathcal{D} = \{f \in L^2(\mu): \phi f \in L^2(\mu)\}$ and define $Af = \phi f$ for all f in $\mathcal{D}$. Then $A \in \mathcal{C}(L^2(\mu))$, $\text{dom } A^* = \mathcal{D}$, and $A^*f = \bar{\phi} f$ for f in $\mathcal{D}$.

1.11. Example. Let $\mathcal{D} = $ all functions $f: [0, 1] \to \mathbb{C}$ that are absolutely continuous with $f' \in L^2(0, 1)$ and such that $f(0) = f(1) = 0$. $\mathcal{D}$ includes all polynomials p with $p(0) = p(1) = 0$. So the uniform closure of $\mathcal{D}$ is $\{f \in C[0, 1]: f(0) = f(1) = 0\}$. Thus $\mathcal{D}$ is dense in $L^2(0, 1)$. Define A: $L^2(0, 1) \to L^2(0, 1)$ by $Af = if'$ for f in $\mathcal{D}$. To see that A is closed, suppose $\{f_n\} \subseteq \mathcal{D}$ and $f_n \oplus if_n' \to f \oplus g$ in $L^2 \oplus L^2$. Let $h(x) = -i\int_0^x g(t)\, dt$; so h is absolutely continuous. Now using the Cauchy-Schwarz inequality we get that $|f_n(x) - h(x)| = |\int_0^x [f_n'(t) + ig(t)]\, dt| \leq \|f_n' + ig\|_2 = \|if_n' - g\|_2$. Thus $f_n(x) \to h(x)$ uniformly on $[0, 1]$. Since $f_n \to f$ in $L^2(0, 1)$, $f(x) = h(x)$ a.e. So we may assume that $f(x) = -i\int_0^x g(t)\, dt$ for all x. Therefore f is absolutely continuous and $f_n(x) \to f(x)$ uniformly on $[0, 1]$; thus $f(0) = f(1) = 0$ and $f' = -ig \in L^2(0, 1)$. So $f \in \mathcal{D}$ and $f \oplus g = f \oplus if' \in \text{gra } A$; that is, $A \in \mathcal{C}(L^2(0, 1))$.

Note that $\{f': f \in \mathcal{D}\} = \{h \in L^2(0, 1): \int_0^1 h(x)\, dx = 0\} = [1]^\perp$.

Claim. $\text{dom } A^* = \{g: g \text{ is absolutely continuous on } [0, 1], g' \in L^2(0, 1)\}$ and for g in $\text{dom } A^*$, $A^*g = ig'$.

In fact, suppose $g \in \text{dom } A^*$ and let $h = A^*g$. Put $H(x) = \int_0^x h(t)\, dt$. Using integration by parts, for every f in $\mathcal{D}$, $i\int_0^1 f'\bar{g} = \langle Af, g \rangle = \langle f, h \rangle = \int_0^1 f\bar{h} = \int_0^1 f(x)\, d\overline{H(x)} = -\int_0^1 f'(x)\overline{H(x)}\, dx$; that is, $\langle f', -ig \rangle = \langle f', -H \rangle$ for all f in $\mathcal{D}$. Thus $H - ig \in \{f': f \in \mathcal{D}\}^\perp = [1]^{\perp\perp}$; hence $H - ig = c$, a constant function. Thus $g = ic - iH$ so that g is absolutely continuous and $g' = -ih \in L^2$. Also note that $A^*g = h = ig'$. The other inclusion is left to the reader.

1.12. Example. Let $\mathscr{E} = \{f \in L^2(0, 1): f \text{ is absolutely continuous}, f' \in L^2,$ and $f(0) = f(1)\}$. Define $Bf = if'$ for f in $\mathscr{E}$. As in (1.11), $B \in \mathcal{C}(L^2(0, 1))$ and $\text{ran } B = [1]^\perp$.

Claim. $\text{dom } B^* = \mathscr{E}$ and $B^*g = ig'$ for g in $\mathscr{E}$.

Let $g \in \operatorname{dom} B^*$. Put $h = B^*g$ and $H(x) = \int_0^x h(t)\, dt$. As in (1.11), $H(0) = H(1) = 0$ and for every f in $\mathscr{E}$, $i\int_0^1 f'\bar{g} = -\int_0^1 f'\overline{H}$. Hence $0 = \int_0^1 (if'\bar{g} + f'\overline{H}) = \int_0^1 if'\overline{(g + iH)}$. Thus $g + iH \perp \operatorname{ran} B$ and so $g + iH = c$, a constant function. Thus $g = c - iH$ is absolutely continuous, $g' = -ih \in L^2$, and $g(0) = g(1) = c$. Thus $g \in \mathscr{E}$ and $B^*g = h = ig'$. The other inclusion is left to the reader.

The preceding two examples illustrate the fact that the calculation of the adjoint depends on the domain of the operator, not just the formal definition of the operator. Note the fact that the next result generalizes (II.2.19).

1.13. Proposition. *If* $A: \mathscr{H} \to \mathscr{K}$ *is densely defined, then*

$$(\operatorname{ran} A)^{\perp} = \ker A^*.$$

If A *is also closed, then*

$$(\operatorname{ran} A^*)^{\perp} = \ker A.$$

PROOF. If $h \perp \operatorname{ran} A$, then for every f in $\operatorname{dom} A$, $0 = \langle Af, h \rangle$. Hence $h \in \operatorname{dom} A^*$ and $A^*h = 0$. The other inclusion is clear. By Corollary 1.8, if $A \in \mathscr{C}(\mathscr{H}, \mathscr{K})$, $A^{**} = A$. So the second equality follows from the first. ∎

1.14. Definition. If $A: \mathscr{H} \to \mathscr{K}$ is a linear operator, A is *boundedly invertible* if there is a bounded linear operator $B: \mathscr{K} \to \mathscr{H}$ such that $AB = 1$ and $BA \subseteq 1$.

Note that if $BA \subseteq 1$, then BA is bounded on its domain. Call B the (*bounded*) *inverse* of A.

1.15. Proposition. *Let* $A: \mathscr{H} \to \mathscr{K}$ *be a linear operator.*

(a) *A is boundedly invertible if and only if* $\ker A = (0)$, $\operatorname{ran} A = \mathscr{K}$, *and the graph of A is closed.*
(b) *If A is boundedly invertible, its inverse is unique.*

PROOF. (a) Let B be the bounded inverse of A. So $\operatorname{dom} B = \mathscr{K}$. Since $BA \subseteq 1$, $\ker A = (0)$; since $AB = 1$, $\operatorname{ran} A = \mathscr{K}$. Also, $\operatorname{gra} A = \{h \oplus Ah: h \in \operatorname{dom} A\} = \{Bk \oplus k: k \in \mathscr{K}\}$. Since B is bounded, $\operatorname{gra} A$ is closed. Conversely, if A has the stated properties, $Bk = A^{-1}k$ for k in $\mathscr{K}$ is a well-defined operator on $\mathscr{K}$. Because $\operatorname{gra} A$ is closed, $\operatorname{gra} B$ is closed. By the Closed Graph Theorem, $B \in \mathscr{B}(\mathscr{K}, \mathscr{H})$.

(b) This is an exercise. ∎

1.16. Definition. If $A: \mathscr{H} \to \mathscr{H}$ is a linear operator, $\rho(A)$, the *resolvent set* for A, is defined by $\rho(A) = \{\lambda \in \mathbb{C}: \lambda - A$ is boundedly invertible$\}$. The *spectrum* of A is the set $\sigma(A) = \mathbb{C} \setminus \rho(A)$:

It is easy to see that if $A: \mathscr{H} \to \mathscr{H}$ is a linear operator and $\lambda \in \mathbb{C}$, $\operatorname{gra} A$ is closed if and only if $\operatorname{gra}(A - \lambda)$ is closed. So if A does not have closed graph, $\sigma(A) = \Box$. Even if A has closed graph, it is possible that $\sigma(A)$ is empty (see Exercise 10). However, some of the other properties of the spectrum hold. The proof of the next result is left to the reader.

1.17. Proposition. *If A: $\mathcal{H} \to \mathcal{H}$ is a linear operator, then $\sigma(A)$ is closed and $z \mapsto (z - A)^{-1}$ is an analytic function on $\rho(A)$.*

Note that if A is defined as in Example 1.9, then $\sigma(A) = \mathrm{cl}\{\alpha_n\}$. Hence it is possible for $\sigma(A)$ to equal any closed subset of $\mathbb{C}$.

1.18. Proposition. *Let $A \in \mathscr{C}(\mathcal{H})$.*

(a) $\lambda \in \rho(A)$ *if and only if* $\ker(A - \lambda) = (0)$ *and* $\mathrm{ran}(A - \lambda) = \mathcal{H}$.
(b) $\sigma(A^*) = \{\bar{\lambda}: \lambda \in \sigma(A)\}$ *and for λ in $\rho(A)$, $(A - \lambda)^{*-1} = [(A - \lambda)^{-1}]^*$.*

PROOF. Exercise.

EXERCISES

1. If A, B, and AB are densely defined linear operators, show that $(AB)^* \supseteq B^*A^*$.

2. Verify the statements in Example 1.9.

3. Verify the statements in Example 1.10.

4. Define an unbounded weighted shift and determine its adjoint.

5. Verify the statements in Example 1.11.

6. If $\mathcal{H}$ is infinite dimensional, show that there is a linear operator A: $\mathcal{H} \to \mathcal{H}$ such that gra A is dense in $\mathcal{H} \oplus \mathcal{H}$. What does this say about dom A^*?

7. Let $\mathscr{D}$ be the set of absolutely continuous functions f such that $f' \in L^2(0,1)$. Let $Df = f'$ for f in $\mathscr{D}$ and let $(Af)(x) = xf(x)$ for f in $L^2(0,1)$. Show that $DA - AD \subseteq 1$.

8. If $\mathscr{A}$ is a Banach algebra with identity, show that there are no elements a, b in $\mathscr{A}$ such that $ab - ba = 1$. (Hint: compute $a^n b - ba^n$.)

9. Prove Proposition 1.18.

10. Define A: $L^2(\mathbb{R}) \to L^2(\mathbb{R})$ by $(Af)(x) = \exp(-x^2)f(x - 1)$ for all f in $L^2(\mathbb{R})$. (a) Show that $A \in \mathscr{B}(L^2(\mathbb{R}))$. (b) Find $\|A^n\|$ and show that $r(A) = 0$ so that $\sigma(A) = \{0\}$. (c) Show that A is injective. (d) Find A^* and show that ran A is dense. (e) Define $B = A^{-1}$ with dom $B = $ ran A and show that $B \in \mathscr{C}(L^2(\mathbb{R}))$ with $\sigma(B) = \Box$.

11. If $A \in \mathscr{C}(\mathcal{H})$, show that $A^*A \in \mathscr{C}(\mathcal{H})$. Show that $-1 \notin \sigma(A^*A)$ and that if $B = (1 + A^*A)^{-1}$, $\|B\| \leq 1$.

12. If B is the bounded operator obtained in Exercise 11, show that $C = AB$ is also bounded and $\|C\| \leq 1$.

§2. Symmetric and Self-Adjoint Operators

A correct introduction to this section consists in a careful examination of Examples 1.11 and 1.12 in the preceding section. In (1.11) we saw that the operator A seemed to be inclined to be self-adjoint, but dom A^* was different from dom A so we could not truly say that $A = A^*$. In (1.12), $B = B^*$ in any sense of the concept of equality. This points out the distinction between symmetric and self-adjoint operators that it is necessary to make in the theory of unbounded operators.

2.1. Definition. An operator A: $\mathcal{H} \to \mathcal{H}$ is *symmetric* if A is densely defined and $\langle Af, g \rangle = \langle f, Ag \rangle$ for all f, g in dom A.

The proof of the next proposition is left to the reader.

2.2. Proposition. *If A is densely defined, the following statements are equivalent.*

(a) *A is symmetric.*
(b) *$(Af, f) \in \mathbb{R}$ for all f in* dom A.
(c) *$A \subseteq A^*$.*

If A is symmetric, then the fact that $A \subseteq A^*$ implies dom A^* is dense. Hence A is closable by Proposition 1.6.

It is easy to check that the operators in Examples 1.11 and 1.12 are symmetric.

2.3. Definition. A densely defined operator A: $\mathcal{H} \to \mathcal{H}$ is *self-adjoint* if $A = A^*$.

Let us emphasize that the condition that $A = A^*$ in the preceding definition carries with it the requirement that dom $A = $ dom A^*. Now clearly every self-adjoint operator is symmetric, but the operator A in Example 1.11 shows that there are symmetric operators that are not self-adjoint. If, however, an operator is bounded, then it is self-adjoint if and only if it is symmetric. The operator B in Example 1.12 is an unbounded self-adjoint operator and Examples 1.9 and 1.10 can be used to furnish additional examples of unbounded self-adjoint operators.

Note that Proposition 1.6 implies that a self-adjoint operator is necessarily closed.

2.4. Proposition. *Suppose A is a symmetric operator on $\mathcal{H}$.*

(a) *If* ran A *is dense, then A is injective.*
(b) *If $A = A^*$ and A is injective, then* ran A *is dense and A^{-1} is self-adjoint.*

(c) *If* dom $A = \mathcal{H}$, *then* $A = A^*$ *and* A *is bounded.*
(d) *If* ran $A = \mathcal{H}$, *then* $A = A^*$ *and* $A^{-1} \in \mathcal{B}(\mathcal{H})$.

PROOF. The proof of (a) is trivial and (b) is an easy consequence of (1.13) and some manipulation.

(c) We have $A \subseteq A^*$. If dom $A = \mathcal{H}$, then $A = A^*$ and so A is closed. By the Closed Graph Theorem $A \in \mathcal{B}(\mathcal{H})$.

(d) If ran $A = \mathcal{H}$, then A is injective by (a). Let $B = A^{-1}$ with dom $B =$ ran $A = \mathcal{H}$. If $f = Ag$ and $h = Ak$, with g, k in dom A, then $\langle Bf, h \rangle = \langle g, Ak \rangle = \langle Ag, k \rangle = \langle f, k \rangle = \langle f, Bh \rangle$. Hence B is symmetric. By (c), $B = B^* \in \mathcal{B}(\mathcal{H})$. By (b), $A = B^{-1}$ is self-adjoint. ∎

We now will turn our attention to the spectral properties of symmetric and self-adjoint operators. In particular, it will be seen that symmetric operators can have nonreal numbers in their spectrum, though the nature of the spectrum can be completely diagnosed (2.8). Self-adjoint operators, however, must have real spectra. The next result begins this spectral discussion.

2.5. Proposition. *Let A be a symmetric operator and let $\lambda = \alpha + i\beta$, α and β real numbers.*

(a) *For each f in* dom A, $\|(A - \lambda)f\|^2 = \|(A - \alpha)f\|^2 + \beta^2\|f\|^2$.
(b) *If $\beta \neq 0$,* ker$(A - \lambda) = (0)$.
(c) *If A is closed and $\beta \neq 0$,* ran$(A - \lambda)$ *is closed.*

PROOF. Note that

$$\|(A - \lambda)f\|^2 = \|(A - \alpha)f - i\beta f\|^2$$
$$= \|(A - \alpha)f\|^2 + 2\operatorname{Re} i\langle(A - \alpha)f, \beta f\rangle + \beta^2\|f\|^2.$$

But

$$\langle(A - \alpha)f, \beta f\rangle = \beta\langle Af, f\rangle - \alpha\beta\|f\|^2 \in \mathbb{R},$$

so (a) follows. Part (b) is immediate from (a). To prove (c), note that $\|(A - \lambda)f\|^2 \geq \beta^2\|f\|^2$. Let $\{f_n\} \subseteq$ dom A such that $(A - \lambda)f_n \to g$. The preceding inequality implies that $\{f_n\}$ is a Cauchy sequence in $\mathcal{H}$; let $f = \lim f_n$. But $f_n \oplus (A - \lambda)f_n \in$ gra$(A - \lambda)$ and $f_n \oplus (A - \lambda)f_n \to f \oplus g$. Hence $f \oplus g \in$ gra$(A - \lambda)$ and so $g = (A - \lambda)f \in$ ran$(A - \lambda)$. This proves (c). ∎

2.6. Lemma. *If $\mathcal{M}, \mathcal{N}$ are closed subspaces of $\mathcal{H}$ and $\mathcal{M} \cap \mathcal{N}^{\perp} = (0)$, then* dim $\mathcal{M} \leq$ dim $\mathcal{N}$.

PROOF. Let P be the orthogonal projection of $\mathcal{H}$ onto $\mathcal{N}$ and define T: $\mathcal{M} \to \mathcal{N}$ by $Tf = Pf$ for f in $\mathcal{M}$. Since $\mathcal{M} \cap \mathcal{N}^{\perp} = (0)$, T is injective. If $\mathcal{L}$ is a finite-dimensional subspace of $\mathcal{M}$, dim $\mathcal{L} =$ dim $T\mathcal{L} \leq$ dim $\mathcal{N}$. Since $\mathcal{L}$ was arbitrary, dim $\mathcal{M} \leq$ dim $\mathcal{N}$. ∎

2.7.   Theorem. *If A is a closed symmetric operator, then* $\dim \ker(A^* - \lambda)$ *is constant for* $\operatorname{Im} \lambda > 0$ *and constant for* $\operatorname{Im} \lambda < 0$.

PROOF. Let $\lambda = \alpha + i\beta$, α and β real numbers and $\beta \neq 0$.

Claim. If $|\lambda - \mu| < |\beta|$, $\ker(A^* - \mu) \cap [\ker(A^* - \lambda)]^\perp = (0)$.

Suppose this is not so. Then there is an f in $\ker(A^* - \mu) \cap [\ker(A^* - \lambda)]^\perp$ with $\|f\| = 1$. By (2.5c), $\operatorname{ran}(A - \bar{\lambda})$ is closed. Hence $f \in [\ker(A^* - \lambda)]^\perp = \operatorname{ran}(A - \bar{\lambda})$. Let $g \in \operatorname{dom} A$ such that $f = (A - \bar{\lambda})g$. Since $f \in \ker(A^* - \mu)$,

$$0 = \langle (A^* - \mu)f, g \rangle = \langle f, (A - \bar{\mu})g \rangle$$
$$= \langle f, (A - \bar{\lambda} + \bar{\lambda} - \bar{\mu})g \rangle$$
$$= \|f\|^2 + (\lambda - \mu)\langle f, g \rangle.$$

Hence $1 = \|f\|^2 = |\lambda - \mu| \, |\langle f, g \rangle| \leq |\lambda - \mu| \, \|g\|$. But (2.5a) implies that $1 = \|f\| = \|(A - \bar{\lambda})g\| \geq |\beta| \, \|g\|$; so $\|g\| \leq |\beta|^{-1}$. Hence $1 \leq |\lambda - \mu| \, \|g\| \leq |\lambda - \mu| \, |\beta|^{-1} < 1$ if $|\lambda - \mu| < |\beta|$. This contradiction establishes the claim.

Combining the claim with Lemma 2.6 gives that $\dim \ker(A^* - \mu) \leq \dim \ker(A^* - \lambda)$ if $|\lambda - \mu| < |\beta| = |\operatorname{Im} \lambda|$. Note that if $|\lambda - \mu| < \frac{1}{2}|\beta|$, then $|\lambda - \mu| < |\operatorname{Im} \mu|$, so that the other inequality also holds. This shows that the function $\lambda \mapsto \dim \ker(A^* - \lambda)$ is locally constant on $\mathbb{C} \setminus \mathbb{R}$. A simple topological argument demonstrates the theorem. ∎

2.8.   Theorem. *If A is a closed symmetric operator, then one and only one of the following possibilities occurs*:

(a) $\sigma(A) = \mathbb{C}$;
(b) $\sigma(A) = \{\lambda \in \mathbb{C}: \operatorname{Im} \lambda \geq 0\}$;
(c) $\sigma(A) = \{\lambda \in \mathbb{C}: \operatorname{Im} \lambda \leq 0\}$;
(d) $\sigma(A) \subseteq \mathbb{R}$.

PROOF. Let $H_\pm = \{\lambda \in \mathbb{C}: \pm \operatorname{Im} \lambda > 0\}$. By (2.5) for λ in $H_\pm$, $A - \lambda$ is injective and has closed range. So if $A - \lambda$ is surjective, $\lambda \in \rho(A)$ (2.4d). But $[\operatorname{ran}(A - \lambda)]^\perp = \ker(A^* - \bar{\lambda})$. So the preceding theorem implies that either $H_\pm \subseteq \sigma(A)$ or $H_\pm \cap \sigma(A) = \square$. Since $\sigma(A)$ is closed, if $H_\pm \subseteq \sigma(A)$, then either $\sigma(A) = \mathbb{C}$ or $\sigma(A) = \operatorname{cl} H_\pm$. If $H_\pm \cap \sigma(A) = \square$, $\sigma(A) \subseteq \mathbb{R}$. ∎

2.9.   Corollary. *If A is a closed symmetric operator, the following statements are equivalent.*

(a) *A is self-adjoint.*
(b) $\sigma(A) \subseteq \mathbb{R}$.
(c) $\ker(A^* - i) = \ker(A^* + i) = 0$.

PROOF. If A is symmetric, every eigenvalue of A is real (Exercise 1). So if $A = A^*$ and Im $\lambda \neq 0$, $\ker(A^* - \lambda) = \ker(A - \lambda) = (0)$. By Theorem 2.8, $\sigma(A) \subseteq \mathbb{R}$, so (a) implies (b).

If $\sigma(A) \subseteq \mathbb{R}$, $\ker(A^* \pm i) = [\operatorname{ran}(A \pm i)]^{\perp} = \mathscr{H}^{\perp} = (0)$. Hence (b) implies (c).

If (c) holds, then this, combined with (2.5) and (1.15a), implies $A \pm i$ is boundedly invertible. By (1.18), $A^* \pm i$ is boundedly invertible. Let $h \in \operatorname{dom} A^*$. Then there is an f in dom A such that $(A + i)f = (A^* + i)h$. But $A^* + i \supseteq A + i$, so $(A^* + i)f = (A^* + i)h$. But $A^* + i$ is injective, so $h = f \in \operatorname{dom} A$. Thus $A = A^*$. ∎

2.10. Corollary. *If A is a closed symmetric operator and $\sigma(A)$ does not contain $\mathbb{R}$, then $A = A^*$.*

It may have occurred to the reader that a symmetric operator A fails to be self-adjoint because its domain is too small and that this can be rectified by merely increasing the size of the domain. Indeed, if A is the symmetric operator in Example 1.11, then the operator B of Example 1.12 is a self-adjoint extension of A. However, the general situation is not always so cooperative.

Fix a symmetric operator A and suppose B is a symmetric extension of A: $A \subseteq B$. It is easy to verify that $B^* \subseteq A^*$. Since $B \subseteq B^*$, we get $A \subseteq B \subseteq B^* \subseteq A^*$. Thus every symmetric extension of A is a restriction of A^*.

2.11. Proposition. (a) *A symmetric operator has a maximal symmetric extension.* (b) *Maximal symmetric extensions are closed.* (c) *A self-adjoint operator is a maximal symmetric operator.*

PROOF. Part (a) is an easy application of Zorn's Lemma. If A is symmetric, $A \subseteq A^*$ and so A is closable. The closure of a symmetric operator is symmetric (Exercise 3), so part (b) is immediate. Part (c) is a consequence of the comments preceding this proposition. ∎

2.12. Definition. Let A be a closed symmetric operator. The *deficiency subspaces* of A are the spaces

$$\mathscr{L}_+ = \ker(A^* - i) = [\operatorname{ran}(A + i)]^{\perp},$$

$$\mathscr{L}_- = \ker(A^* + i) = [\operatorname{ran}(A - i)]^{\perp}.$$

The *deficiency indices* of A are the numbers $n_\pm = \dim \mathscr{L}_\pm$.

It is possible for any pair of deficiency indices to occur (see Exercise 6).

In order to study the closed symmetric extensions of a symmetric operator we also introduce the spaces

$$\mathcal{K}_+ = \{ f \oplus if : f \in \mathcal{L}_+ \},$$
$$\mathcal{K}_- = \{ g \oplus ig : g \in \mathcal{L}_- \}.$$

So $\mathcal{K}_\pm \leq \mathcal{H} \oplus \mathcal{H}$. Notice that $\mathcal{K}_\pm$ are contained in gra A^* and are the portions of the graph of A^* that lie above $\mathcal{L}_\pm$. The next lemma will indicate why the deficiency subspaces are so named.

2.13. Lemma. *If A is a closed symmetric operator,*

$$\text{gra } A^* = \text{gra } A \oplus \mathcal{K}_+ \oplus \mathcal{K}_-.$$

PROOF. Let $f \in \mathcal{L}_+$ and $h \in \text{dom } A$. Then

$$\langle h \oplus Ah, f \oplus if \rangle = \langle h, f \rangle - i \langle Ah, f \rangle$$
$$= -i \langle (A + i)h, f \rangle$$
$$= 0$$

since $\mathcal{L}_+ = [\text{ran}(A + i)]^\perp$. The remainder of the proof that gra A, $\mathcal{K}_+$, and $\mathcal{K}_-$ are pairwise orthogonal is left to the reader. Since it is clear that gra $A \oplus \mathcal{K}_+ \oplus \mathcal{K}_- \subseteq$ gra A^*, it remains to show that this direct sum is dense in gra A^*.

Let $h \in \text{dom } A^*$ and assume $h \oplus A^*h \perp$ gra $A \oplus \mathcal{K}_+ \oplus \mathcal{K}_-$. Since $h \oplus A^*h \perp$ gra A, for every f in dom A, $0 = \langle h \oplus A^*h, f \oplus Af \rangle = \langle h, f \rangle + \langle A^*h, Af \rangle$. So $\langle A^*h, Af \rangle = -\langle h, f \rangle$ for every f in dom A. This implies that $A^*h \in \text{dom } A^*$ and $A^*A^*h = -h$. Therefore $(A^* - i)(A^* + i)h = (A^*A^* + 1)h = 0$. Thus $(A^* + i)h \in \mathcal{L}_+$. Reversing the order of these factors also shows that $(A^* - i)h \in \mathcal{L}_-$. But if $g \in \mathcal{L}_+$, $0 = \langle h \oplus A^*h, g \oplus ig \rangle = \langle h, g \rangle - i \langle A^*h, g \rangle = -i \langle (A^* + i)h, g \rangle$. Since g can be taken equal to $(A^* + i)h$, we get that $(A^* + i)h = 0$, or $h \in \mathcal{L}_-$. Similarly, $h \in \mathcal{L}_+$. So $h \in \mathcal{L}_+ \cap \mathcal{L}_- = (0)$. ∎

2.14. Definition. If A is a closed symmetric operator and $\mathcal{M}$ is a linear manifold in dom A^*, then $\mathcal{M}$ is A-*symmetric* if $\langle A^*f, g \rangle = \langle f, A^*g \rangle$ for all f, g in $\mathcal{M}$. Call such a manifold A-*closed* if $\{ f \oplus A^*f : f \in \mathcal{M} \}$ is closed in $\mathcal{H} \oplus \mathcal{H}$.

So $\mathcal{M}$ is both A-symmetric and A-closed precisely when $A^*|\mathcal{M}$, the restriction of A^* to $\mathcal{M}$, is a closed symmetric operator; if $\mathcal{M} \supseteq \text{dom } A$, then $A^*|\mathcal{M}$ is a closed symmetric extension of A.

2.15. Lemma. *If A is a closed symmetric operator on $\mathcal{H}$ and B is a closed symmetric extension of A, then there is an A-closed, A-symmetric submanifold $\mathcal{M}$ of $\mathcal{L}_+ + \mathcal{L}_-$ such that*

2.16 $$\text{gra } B = \text{gra } A + \text{gra}(A^*|\mathcal{M}).$$

Conversely, if $\mathcal{M}$ is an A-closed, A-symmetric manifold in $\mathcal{L}_+ + \mathcal{L}_-$, then there is a closed symmetric extension B of A such that (2.16) holds.

PROOF. If the A-closed, A-symmetric manifold $\mathcal{M}$ in $\mathcal{L}_+ + \mathcal{L}_-$ is given, let $\mathcal{D} = \text{dom } A + \mathcal{M}$. Since $\mathcal{D} \subseteq \text{dom } A^*$, $B = A^*|\mathcal{D}$ is well defined. Let $f = f_0 + f_1$, $g = g_0 + g_1$, f_0, g_0 in dom A and f_1, g_1 in $\mathcal{M}$. Then

$$\langle A^*f, g \rangle = \langle A^*f_0 + A^*f_1, g_0 + g_1 \rangle$$
$$= \langle Af_0, g_0 \rangle + \langle Af_0, g_1 \rangle + \langle A^*f_1, g_0 \rangle + \langle A^*f_1, g_1 \rangle.$$

Using the A-symmetry of $\mathcal{M}$, the symmetry of A, and the definition of A^* we get

$$\langle A^*f, g \rangle = \langle f_0, Ag_0 \rangle + \langle f_0, A^*g_1 \rangle + \langle f_1, Ag_0 \rangle + \langle f_1, A^*g_1 \rangle$$
$$= \langle f, A^*g \rangle.$$

So $B = A^*|\mathcal{D}$ is symmetric. Note that gra $A \perp \text{gra}(A^*|\mathcal{M})$ in $\mathcal{H} \oplus \mathcal{H}$. Since both of these spaces are closed, gra B, given by (2.16), is closed.

Now let B be any closed symmetric extension of A. As discussed before, $A \subseteq B \subseteq A^*$; so gra $A \subseteq \text{gra } B \subseteq \text{gra } A^* = \text{gra } A \oplus \mathcal{K}_+ \oplus \mathcal{K}_-$. Let $\mathcal{G} = \text{gra } B \cap (\mathcal{K}_+ \oplus \mathcal{K}_-)$ and let $\mathcal{M} =$ the set of first coordinates of elements in $\mathcal{G}$. Clearly, $\mathcal{M}$ is a manifold in $\mathcal{L}_+ + \mathcal{L}_-$ and $\mathcal{M} \subseteq \text{dom } B$. Hence for f, g in $\mathcal{M}$, $\langle A^*f, g \rangle = \langle Bf, g \rangle = \langle f, Bg \rangle = \langle f, A^*g \rangle$. So $\mathcal{M}$ is A-symmetric. Clearly, $\text{gra}(A^*|\mathcal{M}) = \mathcal{G}$, so $\mathcal{M}$ is A-closed. If $h \oplus Bh \in \text{gra } B$, let $h \oplus Bh = (f \oplus Af) + k$ when $f \in \text{dom } A$ and $k \in \mathcal{K}_+ \oplus \mathcal{K}_-$. Since $A \subseteq B$, $k \in \text{gra } B$; so $k \in \mathcal{G}$. This shows that (2.16) holds. ∎

2.17. Theorem. *Let A be a closed symmetric operator. If W is a partial isometry with initial space in $\mathcal{L}_+$ and final subspace in $\mathcal{L}_-$, let*

2.18 $$\mathcal{D}_W = \{ f + g + Wg : f \in \text{dom } A, g \in \text{initial } W \}$$

and define A_W on $\mathcal{D}_W$ by

2.19 $$A_W(f + g + Wg) = Af + ig - iWg.$$

Then A_W is a closed symmetric extension of A. Conversely, if B is any closed symmetric extension of A, then there is a unique partial isometry W such that $B = A_W$ as in (2.19).

If W is such a partial isometry and W has finite rank, then

$$n_\pm(A_W) = n_\pm(A) - \dim(\text{ran } W).$$

PROOF. Let W be a partial isometry with initial space I_+ in $\mathcal{L}_+$ and final space I_- in $\mathcal{L}_-$. Define $\mathcal{D}_W$ and A_W as in (2.18) and (2.19). Let $\mathcal{M} = \{ g + Wg : g \in I_+ \}$; so $\mathcal{M}$ is a manifold in $\mathcal{L}_+ + \mathcal{L}_-$. If $g, h \in I_+$, then $\langle Wg, Wh \rangle = \langle g, h \rangle$. Hence $\langle A^*(g + Wg), h + Wh \rangle = \langle A^*g, h \rangle + \langle A^*g, Wh \rangle + \langle A^*Wg, h \rangle + \langle A^*Wg, Wh \rangle$. Since $g \in \ker(A^* - i)$ and Wg

$\in \ker(A^* + i)$,

$$\langle A^*(g + Wg), h + Wh \rangle = i\langle g, h \rangle + i\langle g, Wh \rangle - i\langle Wg, h \rangle - i\langle Wg, Wh \rangle$$
$$= i\langle g, Wh \rangle - i\langle Wg, h \rangle.$$

Similarly, $\langle g + Wg, A^*(h + Wh) \rangle = i\langle g, Wh \rangle - i\langle Wg, h \rangle$, so that $\mathcal{M}$ is A-symmetric. If $\{g_n\} \subseteq I_+$ and $(g_n + Wg_n) \oplus (ig_n - iWg_n) \to f \oplus h$ in $\mathcal{H} \oplus \mathcal{H}$, then $2ig_n = i(g_n + Wg_n) + (ig_n - iWg_n) \to if + h$ and $2iWg_n = i(g_n + Wg_n) - (ig_n - iWg_n) \to if - h$. If $g = (2i)^{-1}(if + h)$, then $f = g + Wg$ and $h = ig - iWg$. Hence $\mathcal{M}$ is A-closed. By Lemma 2.15, A_W is a closed symmetric extension of A.

To prove that $n_+(A_W) = n_+(A) - \dim I_+$, let $f \in \operatorname{dom} A$, $g \in I_+$. Then

$$(A_W + i)(f + g + Wg) = (A + i)f + ig - iWg + ig + iWg$$
$$= (A + i)f + 2ig.$$

Thus $\operatorname{ran}(A_W + i) = \operatorname{ran}(A + i) \oplus I_+$, and so $n_+(A_W) = \dim[\operatorname{ran}(A_W + i)]^\perp = \dim \mathcal{L}_+ \ominus I_+ = n_+(A) - \dim I_+$. Similarly, $n_-(A_W) = n_-(A) - \dim I_-= n_-(A) - \dim I_+$.

Now let B be a closed symmetric extension of A. By Lemma 2.15 there is an A-symmetric, A-closed manifold $\mathcal{M}$ in $\mathcal{L}_+ + \mathcal{L}_-$ such that $\operatorname{gra} B = \operatorname{gra} A + \operatorname{gra}(A^*|\mathcal{M})$. If $f \in \mathcal{M}$, let $f = f^+ + f^-$, where $f^\pm \in \mathcal{L}_\pm$; put $I_+ = \{f^+ : f \in \mathcal{M}\}$. Since $\mathcal{M}$ is A-symmetric, $0 = \langle A^*f, f \rangle - \langle f, A^*f \rangle = 2i\langle f^+, f^+ \rangle - 2i\langle f^-, f^- \rangle$; hence $\|f^+\| = \|f^-\|$ for all f in $\mathcal{M}$. So if $Wf^+ = f^-$ whenever $f = f^+ + f^- \in \mathcal{M}$ and if I_+ is closed, W is a partial isometry and (2.18) and (2.19) are easily seen to hold. It remains to show that I_+ is closed. Suppose $\{f_n\} \subseteq \mathcal{M}$ and $f_n^+ \to g^+$ in $\mathcal{L}_+$. Since $\|f_n^- - f_m^-\| = \|f_n^+ - f_m^+\|$, there is a g^- in $\mathcal{L}_-$ such that $f_n^- \to g^-$. Clearly $f_n \to g^+ + g^- = g$. Also, $A^*f_n^\pm = \pm if_n^\pm \to \pm ig^\pm$. It follows that $g \oplus A^*g \in \operatorname{cl} \operatorname{gra}(A^*|\mathcal{M}) = \operatorname{gra}(A^*|\mathcal{M})$; thus $g^+ \in I^+$. ∎

2.20. Theorem. *Let A be a closed symmetric operator with deficiency indices* $n_\pm$.

(a) *A is self-adjoint if and only if $n_+ = n_- = 0$.*
(b) *A has a self-adjoint extension if and only if $n_+ = n_-$. In this case the set of self-adjoint extensions is in natural correspondence with the set of isomorphisms of $\mathcal{L}_+$ onto $\mathcal{L}_-$.*
(c) *A is a maximal symmetric operator that is not self-adjoint if and only if either $n_+ = 0$ and $n_- > 0$ or $n_+ > 0$ and $n_- = 0$.*

PROOF. Part (a) is a rephrasing of Corollary 2.9. For (b), $n_+ = n_-$ if and only if $\mathcal{L}_+$ and $\mathcal{L}_-$ are isomorphic. But this is equivalent to stating that there is a partial isometry on $\mathcal{H}$ with initial and final spaces $\mathcal{L}_+$ and $\mathcal{L}_-$, respectively. Part (c) follows easily from the preceding theorem. ∎

2.21. Example. Let A and $\mathcal{D}$ be as in Example 1.11; so A is symmetric. The operator B of Example 1.12 is a self-adjoint extension of A. Let us determine all self-adjoint extensions of A. To do this it is necessary to determine $\mathcal{L}_\pm$. Now $f \in \mathcal{L}_\pm$ if and only if $f \in \operatorname{dom} A^*$ and $\pm if = A^*f =$

if ', so $\mathscr{L}_+ = \{ \alpha e^{\pm x} \colon \alpha \in \mathbb{C} \}$. Hence $n_+ = 1$. Also, the isomorphisms of $\mathscr{L}_+$ onto $\mathscr{L}_-$ are all of the form $W_\lambda e^x = \lambda e^{-x}$ where $|\lambda| = 1$. If $|\lambda| = 1$, let

$$D_\lambda \equiv \{ f + \alpha e^x + \lambda \alpha e^{-x} \colon \alpha \in \mathbb{C}, f \in \mathscr{D} \},$$

$$A_\lambda (f + \alpha e^x + \lambda \alpha e^{-x}) = if' + \alpha i e^x - i\lambda \alpha e^{-x},$$

if $f \in \mathscr{D}$, $\alpha \in \mathbb{C}$.

According to Theorem 2.17, $\{ (A_\lambda, \mathscr{D}_\lambda) \colon |\lambda| = 1 \}$ are all of the self-adjoint extensions of A. The operator B of Example 1.12 is the extension A_1.

For more information on symmetric operators and the relation of the problem of finding self-adjoint extensions to physical problems, see Reed and Simon [1975] from which much of the present development is taken.

EXERCISES

1. If A is symmetric, show that all of the eigenvalues of A are real.

2. If A is symmetric and λ, μ are distinct eigenvalues, show that $\ker(A - \lambda) \perp \ker(A - \mu)$.

3. Show that the closure of a symmetric operator is symmetric.

4. Let $\mathscr{D} = \{ f \in L^2(0, \infty) \colon$ for every $c > 0$, f is absolutely continuous on $[0, c]$, $f(0) = 0$, and $f' \in L^2(0, \infty) \}$. Define $Af = if'$ for f in $\mathscr{D}$. Show that A is a densely defined closed operator and find $\operatorname{dom} A^*$. Show that A is symmetric with deficiency indices $n_+ = 0$ and $n_- = 1$.

5. Let $\mathscr{E} = \{ f \in L^2(-\infty, 0) \colon$ for every $c < 0$, f is absolutely continuous on $[c, 0]$, $f(0) = 0$, and $f' \in L^2(-\infty, 0) \}$. Define $Af = if'$ for f in $\mathscr{E}$. Show that A is a densely defined closed operator and find $\operatorname{dom} A^*$. Show that A is symmetric with deficiency indices $n_+ = 1$, $n_- = 0$.

6. If k, l are any nonnegative integers or ∞, show that there is a closed symmetric operator A with $n_+ = k$ and $n_- = l$. (Hint: Use Exercises 4 and 5.)

7. Let $C_c^2(0, 1)$ be all twice continuously differentiable functions on $(0, 1)$ with compact support and let $Af = -f''$ for f in $C_c^2(0, 1)$. Show that the closure of A is a densely defined symmetric operator and determine all of its self-adjoint extensions.

8. If $A \in \mathscr{C}(\mathscr{H})$, show that A^*A is self-adjoint (see Exercise 1.11).

§3. The Cayley Transform

Consider the Möbius transformation

$$M(z) = \frac{z - i}{z + i}.$$

It is immediate that $M(0) = -1$, $M(1) = -i$, and $M(\infty) = 1$. Thus M

maps the upper half plane onto $\mathbb{D}$ and $M(\mathbb{R} \cup \infty) = \partial\mathbb{D}$. So if A is self-adjoint, $M(A)$ should be unitary. Suppose A is symmetric; does $M(A)$ make sense? What is $M(A)$?

To answer these questions, we should first investigate the meaning of $M(A)$ if A is symmetric. We want to define $M(A)$ as $(A - i)(A + i)^{-1}$. As was seen in the last section, however, $\operatorname{ran}(A + i)$ is not necessarily all of $\mathcal{H}$ if A is not self-adjoint. In fact, $(\operatorname{ran}(A + i))^{\perp} = \mathcal{L}_{+}$ and $(\operatorname{ran}(A - i))^{\perp} = \mathcal{L}_{-}$, the deficiency spaces for A. However (2.5), if A is closed and symmetric, $\operatorname{ran}(A \pm i)$ is closed. Also, realize that if $w = M(z)$, then $z = M^{-1}(w) = i(1 + w)/(1 - w)$.

3.1. Theorem. (a) *If A is a closed densely defined symmetric operator with deficiency subspaces $\mathcal{L}_{\pm}$, and if $U: \mathcal{H} \to \mathcal{H}$ is defined by letting $U = 0$ on $\mathcal{L}_{+}$ and*

3.2
$$U = (A - i)(A + i)^{-1}$$

on $\mathcal{L}_{+}^{\perp}$, then U is a partial isometry with initial space $\mathcal{L}_{+}^{\perp}$, final space $\mathcal{L}_{-}^{\perp}$, and such that $(1 - U)(\mathcal{L}_{+}^{\perp})$ is dense in $\mathcal{H}$.

(b) *If U is a partial isometry with initial and final spaces $\mathcal{M}$ and $\mathcal{N}$, respectively, and such that $(1 - U)\mathcal{M}$ is dense in $\mathcal{H}$, then*

3.3
$$A = i(1 + U)(1 - U)^{-1}$$

is a densely defined closed symmetric operator with deficiency subspaces $\mathcal{L}_{+} = \mathcal{M}^{\perp}$ and $\mathcal{L}_{-} = \mathcal{N}^{\perp}$.

(c) *If A is given as in (a) and U is defined by (3.2), then A and U satisfy (3.3). If U is given as in (b) and A is defined by (3.3), then A and U satisfy (3.2).*

PROOF. (a) By (2.5c), $\operatorname{ran}(A \pm i)$ is closed and so $\mathcal{L}_{\pm}^{\perp} = \operatorname{ran}(A \pm i)$. By (2.5b), $\ker(A + i) = (0)$, so $(A + i)^{-1}$ is well defined on $\mathcal{L}_{+}^{\perp}$. Moreover, $(A + i)^{-1}\mathcal{L}_{+}^{\perp} \subseteq \operatorname{dom} A$ so that U defined by (3.2) makes sense and gives a well-defined operator. If $h \in \mathcal{L}_{+}^{\perp}$, then $h = (A + i)f$ for a unique f in dom A. Hence $\|Uh\|^2 = \|(A - i)f\|^2 = (2.5a) \|Af\|^2 + \|f\|^2 = \|(A + i)f\|^2 = \|h\|^2$. Hence U is a partial isometry, $(\ker U)^{\perp} = \mathcal{L}_{+}^{\perp}$, and $\operatorname{ran} U = \mathcal{L}_{-}^{\perp}$. Once again, if $f \in \operatorname{dom} A$ and $h = (A + i)f$, then $(1 - U)h = h - (A - i)f = (A + i)f - (A - i)f = 2if$. So $(1 - U)\mathcal{L}_{+}^{\perp} = \operatorname{dom} A$ and is dense in $\mathcal{H}$.

(b) Now assume that U is a partial isometry as in (b). It follows that $\ker(1 - U) = (0)$. In fact, if $f \in \ker(1 - U)$, then $Uf = f$; so $\|f\| = \|Uf\|$ and hence $f \in \operatorname{initial} U$. Since U^*U is the projection onto initial U, $f = U^*Uf = U^*f$; so $f \in \ker(1 - U^*) = \operatorname{ran}(1 - U)^{\perp} \subseteq [(1 - U)\mathcal{M}]^{\perp} = (0)$ by hypothesis. Thus $f = 0$ and $1 - U$ is injective.

Let $\mathcal{D} = (1 - U)\mathcal{M}$ and define $(1 - U)^{-1}$ on $\mathcal{D}$. Because $1 - U$ is bounded, $\operatorname{gra}(1 - U)^{-1}$ is closed. If A is defined as in (3.3), it follows that A is a closed densely defined operator. If $f, g \in \mathcal{D}$, let $f = (1 - U)h$ and

$g = (1 - U)k$, $h, k \in \mathcal{M}$. Hence

$$\langle Af, g \rangle = i\langle (1 + U)h, (1 - U)k \rangle$$
$$= i[\langle h, k \rangle + \langle Uh, k \rangle - \langle h, Uk \rangle - \langle Uh, Uk \rangle].$$

Since $h, k \in \mathcal{M}$, $\langle Uh, Uk \rangle = \langle h, k \rangle$; hence $\langle Af, g \rangle = i[\langle Uh, k \rangle - \langle h, Uk \rangle]$. Similarly, $\langle f, Ag \rangle = -i\langle (1 - U)h, (1 + U)k \rangle = -i[\langle h, Uk \rangle - \langle Uh, k \rangle] = \langle Af, g \rangle$. Hence A is symmetric.

Finally, if $h \in \mathcal{M}$ and $f = (1 - U)h$, then $(A + i)f = Af + if = i(1 + U)h + i(1 - U)h = 2ih$. Thus $\operatorname{ran}(A + i) = \mathcal{M}$. Similarly, $(A - i)f = i(1 + U)h - i(1 - U)h = 2Uh$, so that $\operatorname{ran}(A - i) = \operatorname{ran} U = \mathcal{N}$.

(c) Suppose A is as in (a) and U is defined as in (3.2). If $g \in (1 - U)\mathcal{L}_+^\perp$, put $g = (1 - U)h$, where $h \in \mathcal{L}_+^\perp = \operatorname{ran}(A + i)$. Hence $h = (A + i)f$ for some f in dom A. Thus $g = h - Uh = (A + i)f - (A - i)f = 2if$; so $f = -\frac{1}{2}ig$. Also,

$$i(1 + U)(1 - U)^{-1}g = i(1 + U)h$$
$$= i[h + Uh]$$
$$= i[(A + i)f + (A - i)f]$$
$$= 2iAf$$
$$= Ag.$$

Therefore (3.3) holds.

The proof of the remainder of (c) is left to the reader. ■

3.4. Definition. If A is a densely defined closed symmetric operator, the partial isometry U defined by (3.2) is called the *Cayley transform* of A.

3.5. Corollary. *If A is a self-adjoint operator and U is its Cayley transform, then U is a unitary operator with $\ker(1 - U) = (0)$. Conversely, if U is a unitary with $1 \notin \sigma_p(U)$, then the operator A defined by (3.3) is self-adjoint.*

PROOF. If A is a densely defined symmetric operator, then A is self-adjoint if and only if $\mathcal{L}_\pm = (0)$. A partial isometry is a unitary operator if and only if its initial and final spaces are all of $\mathcal{H}$. This corollary is now seen to follow from Theorem 3.1. ■

One use of the Cayley transform is to study self-adjoint operators by using the theory of unitary operators. Indeed, the preceding results say that there is a bijective correspondence between self-adjoint operators and the set of unitary operators without 1 as an eigenvalue.

EXERCISES

1. If U is a partial isometry, show that the following statements are equivalent: (a) $\ker(1 - U) = (0)$; (b) $\ker(1 - U^*) = (0)$; (c) $\operatorname{ran}(1 - U)$ is dense; (d) $\operatorname{ran}(1 - U^*)$ is dense.

2. Let U be a partial isometry with initial and final spaces $\mathcal{M}$ and $\mathcal{N}$, respectively. Show that the following statements are equivalent: (a) $(1 - U)\mathcal{M}$ is dense; (b) $(1 - U^*)\mathcal{N}$ is dense; (c) $\ker(U^* - U^*U) = (0)$; (d) $\ker(U - UU^*) = (0)$.

3. Find a partial isometry U such that $\ker(1 - U) = (0)$ but $(1 - U)(\ker U)^\perp$ is not dense.

4. If A is a densely defined closed symmetric operator and B and C are the operators defined in Exercises 1.11 and 1.12, then the Cayley transform of A is $(C - iB)(C + iB)^{-1}$.

5. Find the Cayley transform of the operator in Example 1.9 when each α_n is real.

6. Find the Cayley transform of the operator in Example 1.10 when ϕ is real valued.

7. Let S be the unilateral shift of multiplicity 1 and find the symmetric operator A such that S is the Cayley transform of A.

8. Let $U = S^*$, where S is the unilateral shift of multiplicity 1. Is U the Cayley transform of a symmetric operator A? If so, find it.

§4. Unbounded Normal Operators and the Spectral Theorem

If A is self-adjoint, the classical way to obtain the spectral decomposition of A is to let U be the Cayley transform of A, obtain the spectral decomposition of U, and then use the inverse Cayley transform to translate this back to a decomposition for A. There is a spectral theorem for unbounded normal operators, however, and the Cayley transform is not applicable here.

In this section the approach is to prove the spectral theorem for normal operators by using that theorem for the bounded case. The spectral theorem for self-adjoint operators is then only a special case.

4.1. Definition. A linear operator N on $\mathcal{H}$ is *normal* if N is closed, densely defined, and $N^*N = NN^*$.

Note that the equation $N^*N = NN^*$ that appears in Definition 4.1 implicitly carries the condition that dom $N^*N =$ dom NN^*. The operators in Examples 1.9 and 1.10 are normal and every self-adjoint operator is normal. Examining Example 1.9 it is easy to see that for a normal operator it is not necessarily the case that dom $N^*N =$ dom N.

Parts of the next result have appeared in various exercises in this chapter, but a complete proof is given here.

4.2. Proposition. *If* $A \in \mathscr{C}(\mathcal{H})$, *then*

(a) $1 + A^*A$ *has a bounded inverse defined on all of* $\mathcal{H}$.
(b) *If* $B = (1 + A^*A)^{-1}$, *then* $\|B\| \leq 1$ *and* $B \geq 0$.
(c) *The operator* $C = A(1 + A^*A)^{-1}$ *is a contraction.*

(d) *A*A is self-adjoint.*
(e) $\{h \oplus Ah: h \in \text{dom } A^*A\}$ *is dense in* gra A.

PROOF. Define $J: \mathcal{H} \oplus \mathcal{H} \to \mathcal{H} \oplus \mathcal{H}$ by $J(h \oplus k) = (-k) \oplus h$. By Lemma 1.7, gra $A^* = [J \text{ gra } A]^\perp$. So if $h \in \mathcal{H}$, there are f in dom A and g in dom A^* such that $0 \oplus h = J(f \oplus Af) + g \oplus A^*g = (-Af) \oplus f + g \oplus A^*g$. Hence $0 = -Af + g$, or $g = Af$; also, $h = f + A^*g = f + A^*Af = (1 + A^*A)f$. Thus $\text{ran}(1 + A^*A) = \mathcal{H}$.

Also, for f in dom A^*A, $Af \in$ dom A^* and $\|f + A^*Af\|^2 = \|f\|^2 + 2\|Af\|^2 + \|A^*Af\|^2 \geq \|f\|^2$. Hence $\ker(1 + A^*A) = (0)$. Thus $(1 + A^*A)^{-1}$ exists, is closed, and is defined on all of $\mathcal{H}$. It must be that $(1 + A^*A)^{-1} \in \mathcal{B}(\mathcal{H})$ (1.15). This proves (a).

It was shown that $\|(1 + A^*A)f\| \geq \|f\|$ whenever $f \in$ dom A^*A. If $h = (1 + A^*A)f$ and $B = (1 + A^*A)^{-1}$, then this implies that $\|Bh\| \leq \|h\|$. Hence $\|B\| \leq 1$. In addition, $\langle Bh, h \rangle = \langle f, (1 + A^*A)f \rangle = \|f\|^2 + \|Af\|^2 \geq 0$, so (b) holds.

Put $C = A(1 + A^*A)^{-1} = AB$; if $f \in$ dom A^*A and $(1 + A^*A)f = h$, then $\|Ch\|^2 = \|Af\|^2 \leq \|(1 + A^*A)f\|^2 = \|h\|^2$ by the argument used to prove (a). Hence $\|C\| \leq 1$, so (c) is proved.

Now to prove (e). Since A is closed, it suffices to show that no nonzero vector in gra A is orthogonal to $\{h \oplus Ah: h \in \text{dom } A^*A\}$. So let $g \in$ dom A and suppose that for every h in dom A^*A,

$$
\begin{aligned}
0 &= \langle g \oplus Ag, h \oplus Ah \rangle \\
&= \langle g, h \rangle + \langle Ag, Ah \rangle \\
&= \langle g, h \rangle + \langle g, A^*Ah \rangle \\
&= \langle g, (1 + A^*A)h \rangle.
\end{aligned}
$$

So $g \perp \text{ran}(1 + A^*A) = \mathcal{H}$; hence $g = 0$.

To prove (d), note that (e) implies that dom A^*A is dense. Now let $f, g \in$ dom A^*A; so $f, g \in$ dom A and $Af, Ag \in$ dom A^*. Hence $\langle A^*Af, g \rangle = \langle Af, Ag \rangle = \langle f, A^*Ag \rangle$. Thus A^*A is symmetric. Also, $1 + A^*A$ has a bounded inverse. This implies two things. First, $1 + A^*A$ is closed, and so A^*A is closed. Also, $-1 \notin \sigma(A^*A)$ so that by Corollary 2.10, A^*A is self-adjoint. ∎

4.3. Proposition. *If N is a normal operator, then* dom $N =$ dom N^* *and* $\|Nf\| = \|N^*f\|$ *for every f in* dom N.

PROOF. First observe that if $h \in$ dom $N^*N =$ dom NN^*, then $Nh \in$ dom N^* and $N^*h \in$ dom N. Hence $\|Nh\|^2 = \langle N^*Nh, h \rangle = \langle NN^*h, h \rangle = \|N^*h\|^2$. Now if $f \in$ dom N, (4.2e) implies that there is a sequence $\{h_n\}$ in dom N^*N such that $h_n \oplus Nh_n \to f \oplus Nf$; so $\|Nh_n - Nf\| \to 0$. But from the first part of this proof, $\|N^*h_n - N^*h_m\| = \|Nh_n - Nh_m\|$. So there is a g in $\mathcal{H}$ such that $N^*h_n \to g$. Thus $h_n \oplus N^*h_n \to f \oplus g$. But N^* is closed; thus $f \in$ dom N^* and $g = N^*f$. So dom $N \subseteq$ dom N^* and $\|Nf\| = \lim\|Nh_n\| = \lim\|N^*h_n\| = \|N^*f\|$.

On the other hand, N^* is normal (Why?), and so dom $N^* \subseteq$ dom N^{**} $=$ dom N. ∎

4.4.   Lemma. *Let $\mathcal{H}_1$, $\mathcal{H}_2$, ... be Hilbert spaces and let $A_n \in \mathcal{B}(\mathcal{H}_n)$ for all $n \geq 1$. If $\mathcal{D} = \{(h_n) \in \bigoplus_n \mathcal{H}_n: \sum_{n=1}^\infty \|A_n h_n\|^2 < \infty\}$ and A is defined on $\mathcal{H} = \bigoplus_n \mathcal{H}_n$ by $A(h_n) = (A_n h_n)$ whenever $(h_n) \in \mathcal{D}$, then $A \in \mathcal{C}(\mathcal{H})$. A is a normal operator if and only if each A_n is normal.*

PROOF. Since $\mathcal{H}_n \subseteq \mathcal{D}$ for each n, $\mathcal{D}$ is dense in $\mathcal{H}$. Clearly A is linear. If $\{h^{(j)}\} \subseteq \operatorname{dom} A$ and $h^{(j)} \oplus A h^{(j)} \to h \oplus g$ in $\mathcal{H} \oplus \mathcal{H}$, then for each n, $h_n^{(j)} \oplus A_n h_n^{(j)} \to h_n \oplus g_n$. Since A_n is bounded, $A_n h_n = g_n$. Hence $\sum_n \|A h_n\|^2 = \sum \|g_n\|^2 = \|g\|^2 < \infty$; so $h \in \operatorname{dom} A$. Clearly $Ah = g$, so $A \in \mathcal{C}(\mathcal{H})$.

It is left to the reader to show that $\operatorname{dom} A^* = \{(h_n) \in \mathcal{H}: \sum_{n=1}^\infty \|A_n^* h_n\|^2 < \infty\}$ and $A^*(h_n) = (A_n^* h_n)$ when $(h_n) \in \operatorname{dom} A^*$. From this the rest of the lemma easily follows.   ∎

If (X, Ω) is a measurable space and $\mathcal{H}$ is a Hilbert space, recall the definition of a spectral measure E for $(X, \Omega, \mathcal{H})$ (IX.1.1). If $h, k \in \mathcal{H}$, let $E_{h,k}$ be the complex-valued measure given by $E_{h,k}(\Delta) = \langle E(\Delta)h, k \rangle$ for each Δ in Ω.

Let $\phi: X \to \mathbb{C}$ be an Ω-measurable function and for each n let $\Delta_n = \{x \in X: n - 1 \leq |\phi(x)| < n\}$. So $\chi_{\Delta_n}\phi$ is a bounded Ω-measurable function. Put $\mathcal{H}_n = E(\Delta_n)\mathcal{H}$. Since $\bigcup_{n=1}^\infty \Delta_n = X$ and the sets $\{\Delta_n\}$ are pairwise disjoint, $\bigoplus_{n=1}^\infty \mathcal{H}_n = \mathcal{H}$. If $E_n(\Delta) = E(\Delta \cap \Delta_n)$, E_n is a spectral measure for $(X, \Omega, \mathcal{H}_n)$. Also, $\int \phi \, dE_n$ is a normal operator on $\mathcal{H}_n$. Define

4.5 $$\mathcal{D}_\phi \equiv \left\{ h \in \mathcal{H}: \sum_{n=1}^\infty \left\| \left(\int \phi \, dE_n \right) E(\Delta_n) h \right\|^2 < \infty \right\}.$$

By Lemma 4.4, $N_\phi: \mathcal{H} \to \mathcal{H}$ given by

4.6 $$N_\phi h = \sum_{n=1}^\infty \left(\int \phi \, dE_n \right) E(\Delta_n) h$$

for h in $\mathcal{D}_\phi$ is a normal operator. The operator N_ϕ is also denoted by

$$N_\phi = \int \phi \, dE.$$

4.7.   Theorem. *If E is a spectral measure for $(X, \Omega, \mathcal{H})$, $\phi: X \to \mathbb{C}$ is an Ω-measurable function, and $\mathcal{D}_\phi$ and N_ϕ are defined as in (4.5) and (4.6), then:*

(a) $\mathcal{D}_\phi = \{h \in \mathcal{H}: \int |\phi|^2 \, dE_{h,h} < \infty\}$;
(b) *for h in $\mathcal{D}_\phi$ and f in $\mathcal{H}$, $\phi \in L^1(|E_{h,f}|)$ with*

4.8 $$\int |\phi| \, d|E_{h,f}| \leq \|f\| \left(\int |\phi|^2 \, dE_{h,h} \right)^{1/2},$$

4.9 $$\left\langle \left(\int \phi \, dE \right) h, f \right\rangle = \int \phi \, dE_{h,f}.$$

PROOF. Using the ∗-homomorphic properties associated with a spectral measure (IX.1.12), one obtains

$$\left\| \left(\int \phi \, dE_n \right) E(\Delta_n) h \right\|^2 = \left\langle \left(\int \chi_{\Delta_n} \phi \, dE \right)^* \left(\int \chi_{\Delta_n} \phi \, dE \right) h, h \right\rangle$$

$$= \int_{\Delta_n} |\phi|^2 \, dE_{h,h}.$$

From here, (a) is immediate.

Now let $h \in \mathscr{D}_\phi$, $f \in \mathscr{H}$. By the Radon–Nikodym Theorem, there is an Ω-measurable function u such that $|u| \equiv 1$ and $|E_{h,f}| = u E_{h,f}$, where $|E_{h,f}|$ is the variation for $E_{h,f}$. Let $\phi_n = \sum_{k=1}^n \chi_{\Delta_k} \phi$; so ϕ_n is bounded as is $u\phi_n$. Thus

$$\int |\phi_n| \, d|E_{h,f}| = \int |\phi_n| u \, dE_{h,f}$$

$$= \left\langle \left(\int |\phi_n| u \, dE \right) h, f \right\rangle$$

$$\leq \|f\| \left\| \left(\int |\phi_n| u \, dE \right) h \right\|.$$

But

$$\left\| \left(\int |\phi_n| u \, dE \right) h \right\|^2 = \left\langle \left(\int |\phi_n| u \, dE \right) h, \left(\int |\phi_n| u \, dE \right) h \right\rangle$$

$$= \left\langle \left(\int |\phi_n|^2 \, dE \right) h, h \right\rangle$$

$$= \int |\phi_n|^2 \, dE_{h,h}$$

$$\leq \int |\phi|^2 \, dE_{h,h}.$$

Combining this with the preceding inequality gives that $\int |\phi_n| \, d|E_{h,f}| \leq \|f\| (\int |\phi|^2 \, dE_{h,h})^{1/2}$ for all n. Letting $n \to \infty$ gives (4.8). Since ϕ_n is bounded, $\langle (\int \phi_n \, dE) h, f \rangle = \int \phi_n \, dE_{h,f}$. If $h \in \mathscr{D}_\phi$ and $f \in \mathscr{H}$, then (4.8) and the Lebesgue Dominated Convergence Theorem imply that $\int \phi_n \, dE_{h,f} \to \int \phi \, dE_{h,f}$ as $n \to \infty$. But

$$\left(\int \phi_n \, dE \right) h = \left(\int \phi \, dE \right) E \left(\bigcup_{j=1}^n \Delta_j \right) h$$

$$= E \left(\bigcup_{j=1}^n \Delta_j \right) \left(\int \phi \, dE \right) h.$$

Since $E(\bigcup_{j=1}^n \Delta_j) \to E(X) = 1$ (SOT) as $n \to \infty$. $\langle (\int \phi_n \, dE) h, f \rangle \to \langle (\int \phi \, dE) h, f \rangle$ as $n \to \infty$. This proves (4.9). ■

Note that as a consequence of (4.7) dom N_ϕ and the definition of N_ϕ do not depend on the choice of the sets $\{\Delta_n\}$ as would seem to be the case from (4.5) and (4.6).

4.10. Theorem. *If (X, Ω) is a measurable space, $\mathcal{H}$ is a Hilbert space, and E is a spectral measure for $(X, \Omega, \mathcal{H})$, let $\Phi(X, \Omega)$ be the algebra of all Ω-measurable functions ϕ: $X \to \mathbb{C}$ and define ρ: $\Phi(X, \Omega) \to \mathscr{C}(\mathcal{H})$ by $\rho(\phi) = \int \phi\, dE$. Then for ϕ, ψ in $\Phi(X, \Omega)$:*

(a) $\rho(\phi)^* = \rho(\bar{\phi})$;
(b) $\rho(\phi\psi) \supseteq \rho(\phi)\rho(\psi)$ *and* $\mathrm{dom}(\rho(\phi)\rho(\psi)) = \mathscr{D}_\psi \cap \mathscr{D}_{\phi\psi}$;
(c) *If ψ is bounded,* $\rho(\phi)\rho(\psi) = \rho(\psi)\rho(\phi) = \rho(\phi\psi)$;
(d) $\rho(\phi)^*\rho(\phi) = \rho(|\phi|^2)$.

The proof of this theorem is left as an exercise.

4.11. The Spectral Theorem. *If N is a normal operator on $\mathcal{H}$, then there is a unique spectral measure E defined on the Borel subsets of $\mathbb{C}$ such that:*

(a) $N = \int z\, dE(z)$;
(b) $E(\Delta) = 0$ *if* $\Delta \cap \sigma(N) = \square$;
(c) *if U is an open subset of $\mathbb{C}$ and $U \cap \sigma(N) \neq \square$, then $E(U) \neq 0$;*
(d) *if $A \in \mathscr{B}(\mathcal{H})$ such that $AN \subseteq NA$ and $AN^* \subseteq N^*A$, then $A(\int\phi\,dE) \subseteq (\int\phi\,dE)A$ for every Borel function ϕ on $\mathbb{C}$.*

Before launching into the proof, a few words motivating the proof are appropriate. Suppose a spectral measure E defined on the Borel subsets of $\mathbb{C}$ is given and let $N = \int z\, dE(z)$. It is not difficult to see that if $0 \le a < b < \infty$ and Δ is the annulus $\{z: a \le |z| \le b\}$, then $\mathcal{H}_\Delta = E(\Delta)\mathcal{H} = \{h \in \mathrm{dom}\, N: h \in \mathrm{dom}\, N^n \text{ for all } n \text{ and } a^n\|h\| \le \|N^n h\| \le b^n\|h\|\}$. $\mathcal{H}_\Delta$ is a closed subspace of $\mathcal{H}$ that reduces N and $N|\mathcal{H}_\Delta$ is bounded. The idea behind the proof is to write $\mathbb{C}$ as the disjoint union of annuli $\{\Delta_j\}$ such that for each Δ_j there is a reducing subspace $\mathcal{H}_{\Delta_j}$ for N with $N_j \equiv N|\mathcal{H}_{\Delta_j}$ bounded, and, moreover, such that $\mathcal{H} = \bigoplus_j \mathcal{H}_{\Delta_j}$. Once this is done the Spectral Theorem for bounded normal operators can be applied to each N_j and direct sums of these can be formed to obtain the spectral measure for N.

So we would like to show that for the annulus $\{z: a \le |z| \le b\}$, $\{h \in \mathrm{dom}\, N: h \in \mathrm{dom}\, N^n \text{ for all } n \text{ and } a^n\|h\| \le \|N^n h\| \le b^n\|h\|\}$ is a reducing subspace for N. To facilitate this, we will use the operator $B = (1 + N^*N)^{-1}$ which is a positive contraction (4.2). To understand what is done below note that $z \mapsto (1 + |z|^2)^{-1}$ maps $\mathbb{C}$ onto $(0, 1]$ and $a \le |z| \le b$ if and only if $(1 + a^2)^{-1} \ge (1 + |z|^2)^{-1} \ge (1 + b^2)^{-1}$.

4.12. Lemma. *If N is a normal operator, $B = (1 + N^*N)^{-1}$, and $C = N(1 + N^*N)^{-1}$, then $BC = CB$ and $(1 + N^*N)^{-1}N \subseteq C$.*

PROOF. From (4.2), B and C are contractions and $B \geq 0$. It will first be shown that $(1 + N^*N)^{-1}N \subseteq C$; that is, $BN \subseteq NB$. If $f \in \text{dom } BN$, then $f \in \text{dom } N$. Let $g \in \text{dom } N^*N$ such that $f = (1 + N^*N)g = Bg$. Then $N^*Ng \in \text{dom } N$; hence $Ng \in \text{dom } NN^* = \text{dom } N^*N$. Thus $Nf = Ng + NN^*Ng = (1 + N^*N)Ng$. Therefore $BNf = B(1 + N^*N)Ng = Ng$. But $NBf = Ng$, so $BN = NB$ on dom N. Thus $BN \subseteq NB$.

If $h \in \mathcal{H}$, let $f \in \text{dom } N^*N$ such that $h = (1 + N^*N)f$. So $BCh = BNBh = BNf = NBf = NBBh = CBh$. Hence $BC = CB$. ∎

4.13. Lemma. *With the same notation as in Lemma 4.12, if $B = \int t \, dP(t)$ is its spectral representation, $1 > \delta > 0$, and Δ is a Borel subset of $[\delta, 1]$, then $\mathcal{H}_\Delta = P(\Delta)\mathcal{H} \subseteq \text{dom } N$, $\mathcal{H}_\Delta$ is invariant for both N and N^*, and $N|\mathcal{H}_\Delta$ is a bounded normal operator with $\|N|\mathcal{H}_\Delta\| \leq [(1 - \delta)/\delta]^{1/2}$.*

PROOF. If $h \in \mathcal{H}_\Delta$, then because $P(\Delta) = \chi_\Delta(B)$, $\|Bh\|^2 = \langle B^2 P(\Delta)h, h \rangle = \int_\Delta t^2 \, dP_{h,h} \geq \delta^2 \|h\|^2$. So $B|\mathcal{H}_\Delta$ is invertible and there is a g in $\mathcal{H}_\Delta$ such that $h = Bg$. But ran $B = \text{dom}(1 + N^*N) \subseteq \text{dom } N$. Hence $h \in \text{dom } N$; that is, $\mathcal{H}_\Delta \subseteq \text{dom } N$.

Let $h \in \mathcal{H}_\Delta$ and again let $g \in \mathcal{H}_\Delta$ such that $h = Bg$. Hence $Nh = NBg = Cg$. By Lemma 4.12, $BC = CB$; so by (IX.2.2), $P(\Delta)C = CP(\Delta)$. Since $g \in \mathcal{H}_\Delta$, $Nh = Cg \in \mathcal{H}_\Delta$. Note that if $M = N^*$ and $B_1 = (1 + M^*M)^{-1}$, then $B_1 = B$. From the preceding argument $N^*\mathcal{H}_\Delta = M\mathcal{H}_\Delta \subseteq \mathcal{H}_\Delta$. It easily follows that $N|\mathcal{H}_\Delta$ is normal.

Finally, if $h \in \mathcal{H}_\Delta$, then

$$\begin{aligned}
\|Nh\|^2 &= \langle N^*Nh, h \rangle \\
&= \langle [(N^*N + 1) - 1]h, h \rangle \\
&= \int_\delta^1 (t^{-1} - 1) \, dP_{h,h}(t) \leq \|h\|^2 (1 - \delta)/\delta.
\end{aligned}$$

Hence $\|N|\mathcal{H}_\Delta\| \leq [(1 - \delta)/\delta]^{1/2}$. ∎

PROOF OF THE SPECTRAL THEOREM. Let $B = (1 + N^*N)^{-1}$ and $C = N(1 + N^*N)^{-1}$ as in Lemma 4.12. Let $B = \int_0^1 t \, dP(t)$ be the spectral decomposition of B and put $P_n = P(1/(n + 1), 1/n]$ for $n \geq 1$. Since $\ker B = (0) = P(\{0\})$, $\sum_{n=1}^\infty P_n = 1$. Let $\mathcal{H}_n = P_n \mathcal{H}$. By Lemma 4.13, $\mathcal{H}_n \subseteq \text{dom } N$, $\mathcal{H}_n$ reduces N, and $N_n \equiv N|\mathcal{H}_n$ is a bounded normal operator with $\|N_n\| \leq n^{1/2}$. Also, if $h \in \mathcal{H}_n$, $(1 + N_n^*N_n)Bh = B(1 + N_n^*N_n)h = h$; that is,

$$B|\mathcal{H}_n = (1 + N_n^*N_n)^{-1}.$$

Thus if $\lambda \in \sigma(N_n)$, $(1 + |\lambda|^2)^{-1} \in \sigma(B|\mathcal{H}_n) \subseteq [1/(n + 1), 1/n]$. Thus $\sigma(N_n) \subseteq \{z \in \mathbb{C} : (n - 1)^{1/2} \leq |z| \leq n^{1/2}\} \equiv \Delta_n$. Let $N_n = \int z \, dE_n(z)$ be the spectral decomposition of N_n. For any Borel subset Δ of $\mathbb{C}$, let $E(\Delta)$ be defined by

4.14
$$E(\Delta) = \sum_{n=1}^\infty E_n(\Delta \cap \Delta_n).$$

Note that $E_n(\Delta \cap \Delta_n)$ is a projection with range in $\mathcal{H}_n$. Since $\mathcal{H}_n \perp \mathcal{H}_m$ for $n \neq m$, (4.14) defines a projection in $\mathcal{B}(\mathcal{H})$. (Technically $E(\Delta)$ should be defined by $E(\Delta) = \sum_{n=1}^{\infty} E_n(\Delta \cap \Delta_n) P_n$. But this technicality does not add anything to understanding.)

To show that E is a spectral measure, it is clear that $E(\mathbb{C}) = 1$ and $E(\square) = 0$. If Λ_1 and Λ_2 are Borel subsets of $\mathbb{C}$, then

$$E(\Lambda_1 \cap \Lambda_2) = \sum_{n=1}^{\infty} E_n(\Lambda_1 \cap \Lambda_2 \cap \Delta_n)$$

$$= \sum_{n=1}^{\infty} E_n(\Lambda_1 \cap \Delta_n) E_n(\Lambda_2 \cap \Delta_n).$$

Again, the fact that the spaces $\{\mathcal{H}_n\}$ are pairwise orthogonal implies

$$E(\Lambda_1 \cap \Lambda_2) = \left(\sum_{n=1}^{\infty} E_n(\Lambda_1 \cap \Delta_n) \right) \left(\sum_{n=1}^{\infty} E_n(\Lambda_2 \cap \Delta_n) \right)$$

$$= E(\Lambda_1) E(\Lambda_2).$$

If $h \in \mathcal{H}$, then $\langle E(\Delta)h, h \rangle = \sum_{n=1}^{\infty} \langle E_n(\Delta \cap \Delta_n)h, h \rangle$. So if $\{\Lambda_j\}_{j=1}^{\infty}$ are pairwise disjoint Borel sets,

$$\left\langle E\left(\bigcup_{j=1}^{\infty} \Lambda_j \right)h, h \right\rangle = \sum_{n=1}^{\infty} \left\langle E_n\left(\left(\bigcup_{j=1}^{\infty} \Lambda_j \right) \cap \Delta_n \right)h, h \right\rangle$$

$$= \sum_{n=1}^{\infty} \sum_{j=1}^{\infty} \langle E_n(\Lambda_j \cap \Delta_n)h, h \rangle.$$

Since each term in this double summation is positive, the order of summation can be reversed. Thus

$$\left\langle E\left(\bigcup_{j=1}^{\infty} \Lambda_j \right)h, h \right\rangle = \sum_{j=1}^{\infty} \sum_{n=1}^{\infty} \langle E_n(\Lambda_j \cap \Delta_n)h, h \rangle$$

$$= \sum_{j=1}^{\infty} \langle E(\Lambda_j)h, h \rangle.$$

So $E(\bigcup_{j=1}^{\infty} \Lambda_j) = \sum_{j=1}^{\infty} E(\Lambda_j)$; therefore E is a spectral measure.

Let $M = \int z \, dE(z)$ be defined as in Theorem 4.7. Thus $\mathcal{H}_n \subseteq \text{dom } M$ and by the Spectral Theorem for bounded operators, $Mh = N_n h = Nh$ if $h \in \mathcal{H}_n$. If h is any vector in dom M, $h = \sum_{1}^{\infty} h_n$, $h_n \in \mathcal{H}_n$, and $\sum_{1}^{\infty} \|Nh_n\|^2 < \infty$. Thus $h \in \text{dom } N$ and $Nh = Mh$. This proves (a).

4.15. Claim.
$$\sigma(N) = \text{cl}\left[\bigcup_{n=1}^{\infty} \sigma(N_n) \right].$$

It is left to the reader to show that $\bigcup_{n=1}^{\infty} \sigma(N_n) \subseteq \sigma(N)$. Since $\sigma(N)$ is closed, this proves half of (4.15). If $\lambda \notin \text{cl}[\bigcup_{n=1}^{\infty} \sigma(N_n)]$, then there is a $\delta > 0$ such that $|\lambda - z| \geq \delta$ for all z in $\bigcup_{n=1}^{\infty} \sigma(N_n)$. Thus $(N_n - \lambda)^{-1}$ exists and

$\|(N_n - \lambda)^{-1}\| \leq \delta^{-1}$ for all n. Thus $A = \oplus_{n=1}^{\infty}(N_n - \lambda)^{-1}$ is a bounded operator. It follows that $A = (N - \lambda)^{-1}$, so $\lambda \notin \sigma(N)$.

By (4.15) if $\Delta \cap \sigma(N) = \square$, $\Delta \cap \sigma(N_n) = \square$ for all n. Thus $E_n(\Delta) = 0$ for all n. Hence $E(\Delta) = 0$ and (b) holds.

If U is open and $U \cap \sigma(N) \neq \square$, then (4.15) implies $U \cap \sigma(N_n) \neq \square$ for some n. Since $E_n(U) \neq 0$, $E(U) \neq 0$ and (c) is true.

Now let $A \in \mathscr{B}(\mathscr{H})$ such that $AN \subseteq NA$ and $AN^* \subseteq N^*A$. Thus $A(1 + N^*N) \subseteq (1 + N^*N)A$. It follows that $AB = BA$. By the Spectral Theorem for bounded operators, A commutes with the spectral projections of B. In particular, each $\mathscr{H}_n$ reduces A and if $A_n \equiv A|\mathscr{H}_n$, then $A_n N_n = N_n A_n$. Hence $A_n \phi(N_n) = \phi(N_n)A_n$ for any bounded Borel function ϕ. The remaining details are left to the reader. ∎

The Fuglede–Putnam Theorem holds for unbounded normal operators (Exercise 8), so that the hypothesis in part (d) of the Spectral Theorem can be weakened to $AN \subseteq NA$.

4.16. Definition. If N is a normal operator on $\mathscr{H}$, then a vector e_0 is a *star-cyclic* vector for N if for all positive integers k and l, $e_0 \in \mathrm{dom}(N^{*k}N^l)$ and $\mathscr{H} = \bigvee\{N^{*k}N^l e_0 : k, l \geq 0\}$.

4.17. Example. Let μ be a finite measure on $\mathbb{C}$ such that every polynomial in z and $\bar{z}$ belongs to $L^2(\mu)$. Let $\mathscr{D}_{\mu} = \{f \in L^2(\mu): zf \in L^2(\mu)\}$ and define $N_{\mu}f = zf$ for f in $\mathscr{D}_{\mu}$. Then N_{μ} is a normal operator and 1 is a star-cyclic vector for N_{μ}.

Note that $d\mu(z) = e^{-|z|}d\,\mathrm{Area}(z)$ is a measure satisfying the conditions of (4.17).

4.18. Theorem. *If N is a normal operator on $\mathscr{H}$ with a star-cyclic vector e_0, then there is a finite measure μ on $\mathbb{C}$ such that every polynomial in z and $\bar{z}$ belongs to $L^2(\mu)$ and there is an isomorphism $W: \mathscr{H} \to L^2(\mu)$ such that $We_0 = 1$ and $WNW^{-1} = N_{\mu}$.*

The proof is similar to the proof of Theorem IX.3.4 and is left to the reader.

4.19. Theorem. *If N is a normal operator on the separable Hilbert space $\mathscr{H}$, then there is a σ-finite measure space (X, Ω, μ) and an Ω-measurable function ϕ such that N is unitarily equivalent to M_{ϕ} on $L^2(\mu)$.*

The proof of Theorem 4.19 is only sketched. Write N as the (unbounded) direct sum of bounded normal operators $\{N_n\}$. By Theorem IX.4.6, there is a σ-finite measure space (X_n, Ω_n, μ_n) and a bounded Ω_n-measurable function ϕ_n such that $N_n \cong M_{\phi_n}$. Let $X =$ the disjoint union of $\{X_n\}$ and let

$\Omega = \{\Delta \subseteq X: \Delta \cap X_n \in \Omega_n \text{ for every } n\}$. If $\Delta \in \Omega$, let $\mu(\Delta) = \sum_1^\infty \mu_n(\Delta \cap X_n)$. Let $\phi: X \to \mathbb{C}$ be defined by $\phi(x) = \phi_n(x)$ if $x \in X_n$. Then ϕ is Ω-measurable and $N \cong M_\phi$ on $L^2(X, \Omega, \mu)$.

EXERCISES

1. Prove Theorem 4.10.

2. Show that if A is a symmetric operator that is normal, then A is self-adjoint.

3. Using the notation of Theorem 4.10, what is $\sigma(\int \phi \, dE)$?

4. If Δ_n and E_n are as in the proof of the Spectral Theorem, show that $E_n(\Delta_{n+1}) = E_n(\Delta_{n-1}) = 0$.

5. Use the Spectral Theorem to show that if $0 < a \le b < \infty$, $\Delta = \{z \in \mathbb{C}: a \le |z| \le b\}$, and $N = \int z \, dE(z)$ is the spectral decomposition of the normal operator N, then $E(\Delta)\mathcal{H} = \{h \in \text{dom } N: a^n\|h\| \le \|N^n h\| \le b^n\|h\| \text{ for all } n \ge 1\}$.

6. State and prove a polar decomposition for operators in $\mathcal{C}(\mathcal{H}, \mathcal{K})$.

7. If A is self-adjoint, prove that $\exp(iA)$ is unitary.

8. (Fuglede–Putnam Theorem.) If N, M are normal operators and A is a bounded operator such that $AN \subseteq MA$, then $AN^* \subseteq M^*A$.

9. Prove Theorem 4.18.

10. If μ_1, μ_2 are finite measures on $\mathbb{C}$ and N_{μ_1}, N_{μ_2} are defined as in Example 4.17, show that $N_{\mu_1} \cong N_{\mu_2}$ iff $[\mu_1] = [\mu_2]$.

11. Fill in the details of the proof of Theorem 4.19.

§5. Stone's Theorem

If A is a self-adjoint operator on $\mathcal{H}$, then $\exp(iA)$ is a unitary operator (Exercise 4.7). Hence $U(t) = \exp(itA)$ is unitary for all t in $\mathbb{R}$. The purpose of this section is not to investigate the individual operators $\exp(itA)$, but rather the entire collection of operators $\{\exp(itA): t \in \mathbb{R}\}$. In fact, as the first theorem shows, $U: \mathbb{R} \to$ unitaries on $\mathcal{H}$ is a group homomorphism with certain properties. Stone's Theorem provides a converse to this; every such homomorphism arises in this way.

5.1. Theorem. *If A is self-adjoint and $U(t) = \exp(itA)$ for t in $\mathbb{R}$, then*

(a) $U(t)$ *is unitary;*
(b) $U(s + t) = U(s)U(t)$ *for all s in $\mathbb{R}$;*
(c) *if $h \in \mathcal{H}$, then $\lim_{s \to t} U(s)h = U(t)h$;*
(d) *if $h \in \text{dom } A$, then*

5.2 $$\lim_{t \to 0} \frac{1}{t}[U(t)h - h] = iAh;$$

(e) *if $h \in \mathcal{H}$ and $\lim_{t \to 0} t^{-1}[U(t)h - h]$ exists, then $h \in \text{dom } A$.*

PROOF. As was mentioned, part (a) is an exercise. Since $\exp(itx)\exp(isx) = \exp(i(s + t)x)$ for all x in $\mathbb{R}$, (b) is a consequence of the functional calculus for normal operators [(4.10) and (4.11)]. Also note that $U(0)U(t) = U(t)$, so that $U(0) = 1$.

(c) If $h \in \mathscr{H}$, then $\|U(t)h - U(s)h\| = \|U(t - s + s)h - U(s)h\| = $ [by (b)] $\|U(s)[U(t - s)h - h]\| = \|U(t - s)h - h\|$ since $U(s)$ is unitary. Thus (c) will be shown if it is proved that $\|U(t)h - h\| \to 0$ as $t \to 0$. If $A = \int_{-\infty}^{\infty} x \, dE(x)$ is the spectral decomposition of A, then

$$\|U(t)h - h\|^2 = \int_{-\infty}^{\infty} |e^{itx} - 1|^2 \, dE_{h,h}(x).$$

Now $E_{h,h}$ is a finite measure on $\mathbb{R}$; for each x in $\mathbb{R}$, $|e^{itx} - 1|^2 \to 0$ as $t \to 0$; and $|e^{itx} - 1|^2 \le 4$. So the Lebesgue Dominated Convergence Theorem implies that $U(t)h \to h$ as $t \to 0$.

(d) Note that $t^{-1}[U(t) - 1] - iA = f_t(A)$, where $f_t(x) = t^{-1}[\exp(itx) - 1] - ix$. So if $h \in \mathrm{dom}\, A$,

$$\left\| \frac{1}{t}[U(t)h - h] - iAh \right\|^2 = \|f_t(A)h\|^2$$

$$= \int_{-\infty}^{\infty} \left| \frac{e^{itx} - 1}{t} - ix \right|^2 dE_{h,h}(x).$$

As $t \to 0$, $t^{-1}[e^{itx} - 1] - ix \to 0$ for all x in $\mathbb{R}$. Also, $|e^{is} - 1| \le |s|$ for all real numbers s (Why?), hence $|f_t(x)| \le |t|^{-1}|e^{itx} - 1| + |x| \le 2|x|$. But $|x| \in L^2(E_{h,h})$ by Theorem 4.7(a). So again the Lebesgue Dominated Convergence Theorem implies that (5.2) is true.

(e) Let $\mathscr{D} = \{h \in \mathscr{H}: \lim_{t \to 0} t^{-1}[U(t)h - h]$ exists in $\mathscr{H}\}$. For h in $\mathscr{D}$, let Bh be defined by

$$Bh = -i \lim_{t \to 0} \frac{U(t)h - h}{t}.$$

It is easy to see that $\mathscr{D}$ is a linear manifold in $\mathscr{H}$ and B is linear on $\mathscr{D}$. Also, by (d), $B \supseteq A$ so that B is densely defined. Moreover, if $h, g \in \mathscr{D}$, then

$$\langle Bh, g \rangle = -i \lim_{t \to 0} \left\langle \frac{U(t)h - h}{t}, g \right\rangle.$$

By (b) and the fact that each $U(t)$ is unitary, it follows that $U(t)^* = U(t)^{-1} = U(-t)$. Hence

$$\langle Bh, g \rangle = -i \lim_{t \to 0} \left\langle h, \frac{U(-t)g - g}{t} \right\rangle$$

$$= \lim_{t \to 0} \left\langle h, -i\left[\frac{U(-t)g - g}{-t} \right] \right\rangle$$

$$= \langle h, Bg \rangle.$$

Hence B is a symmetric extension of A. Since self-adjoint operators are maximal symmetric operators (2.11), $B = A$ and $\mathcal{D} = \operatorname{dom} A$.   ■

Inspired by the preceding theorem, the following definition is made.

5.3. Definition. A *strongly continuous one-parameter unitary group* is a function $U: \mathbb{R} \to \mathcal{B}(\mathcal{H})$ such that for all s and t in $\mathbb{R}$ (a) $U(t)$ is a unitary operator; (b) $U(s + t) = U(s)U(t)$; (c) if $h \in \mathcal{H}$, then $U(t)h \to U(t_0)h$ as $t \to t_0$.

Note that by Theorem 5.1, if A is self-adjoint, then $U(t) = \exp(itA)$ defines a strongly continuous one-parameter unitary group.

Also, $U(0) = 1$ and $U(-t) = U(t)^{-1}$, so that $\{U(t): t \in \mathbb{R}\}$ is indeed a group. Property (c) also implies that $U: \mathbb{R} \to (\mathcal{B}(\mathcal{H}), \mathrm{SOT})$ is continuous. By Exercise 1, if U is only assumed to be WOT-continuous, then U is SOT-continuous. However, this condition can be relaxed even further as the following result of von Neumann [1932] shows.

5.4.   Theorem. *If $\mathcal{H}$ is separable, $U: \mathbb{R} \to \mathcal{B}(\mathcal{H})$ satisfies conditions* (a) *and* (b) *of Definition 5.3, and if for all h, g in $\mathcal{H}$ the function $t \mapsto \langle U(t)h, g \rangle$ is Lebesgue measurable, then U is a strongly continuous one-parameter unitary group.*

PROOF. If $0 < a < \infty$ and $h, g \in \mathcal{H}$, then $t \mapsto \langle U(t)h, g \rangle$ is a bounded measurable function on $[0, a]$ and hence

$$\int_0^a |\langle U(t)h, g \rangle|\, dt \leq a\|h\|\|g\|.$$

Thus

$$h \mapsto \int_0^a \langle U(t)h, g \rangle\, dt$$

is a bounded linear function on $\mathcal{H}$. Therefore there is a g_a in $\mathcal{H}$ such that

5.5 $$\langle h, g_a \rangle = \int_0^a \langle U(t)h, g \rangle\, dt$$

and $\|g_a\| \leq a\|g\|$.

Claim. $\{g_a: g \in \mathcal{H}, a > 0\}$ is total in $\mathcal{H}$.

In fact, suppose $h \in \mathcal{H}$ and $h \perp \{g_a: g \in \mathcal{H}, a > 0\}$. Then by (5.5), for every $a > 0$ and every g in $\mathcal{H}$,

$$0 = \int_0^a \langle U(t)h, g \rangle\, dt.$$

Thus for every g in $\mathcal{H}$, $\langle U(t)h, g \rangle = 0$ a.e. on $\mathbb{R}$. Because $\mathcal{H}$ is separable there is a subset Δ of $\mathbb{R}$ having measure zero such that if $t \notin \Delta$, $\langle U(t)h, g \rangle$

$= 0$ whenever g belongs to a preselected countable dense subset of $\mathscr{H}$. Thus $U(t)h = 0$ if $t \notin \Delta$. But $\|h\| = \|U(t)h\|$, so $h = 0$ and the claim is established.

Now if $s \in \mathbb{R}$,

$$\langle h, U(s)g_a \rangle = \langle U(-s)h, g_a \rangle$$
$$= \int_0^a \langle U(t-s)h, g \rangle \, dt$$
$$= \int_{-s}^{a-s} \langle U(t)h, g \rangle \, dt.$$

Thus $\langle h, U(s)g_a \rangle \to \langle h, g_a \rangle$ as $s \to 0$. By the claim and the fact that the group is uniformly bounded, $U \colon \mathbb{R} \to (\mathscr{B}(\mathscr{H}), \text{WOT})$ is continuous at 0. By the group property, $U \colon \mathbb{R} \to (\mathscr{B}(\mathscr{H}), \text{WOT})$ is continuous. Hence U is SOT-continuous (Exercise 1). ∎

We now turn our attention to the principal result of this section, Stone's Theorem, which states that the converse of Theorem 5.1 is valid. Note that if $U(t) = \exp(itA)$ for a self-adjoint operator A, then part (d) of Theorem 5.1 instructs us how to recapture A. This is the route followed in the proof of Stone's Theorem, proved in Stone [1932].

5.6. Stone's Theorem. *If U is a strongly continuous one-parameter unitary group, then there is a self-adjoint operator A such that $U(t) = \exp(itA)$.*

PROOF. Begin by defining $\mathscr{D}$ to be the set of all vectors h in $\mathscr{H}$ such that $\lim_{t \to 0} t^{-1}[U(t)h - h]$ exists; since $0 \in \mathscr{D}$, $\mathscr{D} \neq \square$. Clearly $\mathscr{D}$ is a linear manifold in $\mathscr{H}$.

5.7. Claim. $\mathscr{D}$ is dense in $\mathscr{H}$.

Let $\mathscr{L} =$ all continuous functions ϕ on $\mathbb{R}$ such that $\phi \in L^1(0, \infty)$. Hence for any h in $\mathscr{H}$, $t \mapsto \phi(t)U(t)h$ is a continuous function of $\mathbb{R}$ into $\mathscr{H}$. Because $\|U(t)h\| = \|h\|$ for all t, a Riemann integral, $\int_0^\infty \phi(t)U(t)h \, dt$, can be defined and is a vector in $\mathscr{H}$. Put

5.8
$$T_\phi h = \int_0^\infty \phi(t)U(t)h \, dt.$$

It is easy to see that $T_\phi \colon \mathscr{H} \to \mathscr{H}$ is linear and bounded with $\|T_\phi\| \leq \int_0^\infty |\phi(t)| \, dt$. Similarly, for each ϕ in $\mathscr{L}$

5.9
$$S_\phi h = \int_0^\infty \phi(t)U(-t)h \, dt$$

defines a bounded operator on $\mathscr{H}$.

For any ϕ in $\mathscr{L}$ and t in $\mathbb{R}$,

$$U(t)T_\phi h = U(t)\int_0^\infty \phi(s)U(s)h\,ds$$

$$= \int_0^\infty \phi(s)U(t+s)h\,ds$$

$$= \int_t^\infty \phi(s-t)U(s)h\,ds.$$

Similarly,

$$U(t)S_\phi h = \int_{-t}^\infty \phi(s+t)U(-s)h\,ds.$$

Now let $\mathscr{L}^{(1)} =$ all ϕ in $\mathscr{L}$ that are continuously differentiable with ϕ' in $\mathscr{L}$. For ϕ in $\mathscr{L}^{(1)}$,

$$-\frac{i}{t}[U(t)-1]T_\phi h = -\frac{i}{t}\int_t^\infty \phi(s-t)U(s)h\,ds + \frac{i}{t}\int_0^\infty \phi(s)U(s)h\,ds$$

$$= -i\int_t^\infty \left[\frac{\phi(s-t)-\phi(s)}{t}\right]U(s)h\,ds$$

$$+ \frac{i}{t}\int_0^t \phi(s)U(s)h\,ds.$$

Now

$$\left\|\int_0^t \left[\frac{\phi(s-t)-\phi(s)}{t}\right]U(s)h\,ds\right\|$$

$$\leq \|h\|\sup\{|\phi(s-t)-\phi(s)| : 0 \leq s \leq t\}$$

$$\to 0$$

as $t \to 0$. Hence

$$\lim_{t\to 0}\int_t^\infty \left[\frac{\phi(s-t)-\phi(s)}{t}\right]U(s)h\,ds = -\int_0^\infty \phi'(s)U(s)h\,ds$$

$$= -T_{\phi'}h.$$

Since $s \mapsto \phi(s)U(s)h$ is continuous and $U(0) = 1$, the Fundamental Theorem of Calculus implies that

$$\lim_{t\to 0}\frac{1}{t}\int_0^t \phi(s)U(s)h\,ds = \phi(0)h.$$

Hence for ϕ in $\mathscr{L}^{(1)}$ and h in $\mathscr{H}$,

5.10 $$\lim_{t\to 0} -\frac{i}{t}[U(t)-1]T_\phi h = iT_{\phi'}h + i\phi(0)h.$$

Similarly, for ϕ in $\mathscr{L}^{(1)}$ and h in $\mathscr{H}$,

5.11 $$\lim_{t\to 0} -\frac{i}{t}[U(t)-1]S_\phi h = -iS_{\phi'}h - i\phi(0)h.$$

So (5.10) implies that

$$\mathcal{D} \supseteq \left\{ T_\phi h: \phi \in \mathcal{L}^{(1)} \text{ and } h \in \mathcal{H} \right\}.$$

But for every positive integer n there is a ϕ_n in $\mathcal{L}^{(1)}$ such that $\phi_n \geq 0$, $\phi_n(t) = 0$ for $t \geq 1/n$, and $\int_0^\infty \phi_n(t)\, dt = 1$ (Exercise 2). Hence

$$T_{\phi_n} h - h = \int_0^{1/n} \phi_n(t)[U(t) - 1] h\, dt$$

and so $\| T_{\phi_n} h - h \| \leq \sup\{\|U(t)h - h\|: 0 \leq t \leq 1/n\}$. Therefore $\| T_{\phi_n} h - h \| \to 0$ as $n \to \infty$ since U is strongly continuous. This says that $\mathcal{D}$ is dense.

For h in $\mathcal{D}$, define

5.12
$$Ah = -i \lim_{t \to 0} \frac{1}{t} [U(t) - 1] h.$$

5.13. Claim. A is symmetric.

The proof of this is left to the reader.

By (2.2c), A is closable; also denote the closure of A by A. According to Corollary 2.9, to prove that A is self-adjoint it suffices to prove that $\ker(A^* \pm i) = (0)$. Equivalently, it suffices to show that $\operatorname{ran}(A \pm i)$ is dense. It will be shown that there are operators $B_\pm$ such that $(A \pm i)B_\pm = 1$, so that $A \pm i$ is surjective.

Notice that according to (5.10),

$$(A + i)T_\phi = AT_\phi + iT_\phi = i(T_{\phi'} + T_\phi) + i\phi(0).$$

So taking $\phi(t) = -ie^{-t}$, $(A + i)T_\phi = 1$. According to (5.11),

$$(A - i)S_\psi = AS_\psi - iS_\psi = -i(S_{\psi'} + S_\psi) - i\psi(0).$$

Taking $\psi(t) = ie^{-t}$, $(A - i)S_\psi = 1$. Hence A is self-adjoint.

Put $V(t) = \exp(iAt)$. It remains to show that $V = U$. Let $h \in \mathcal{D}$. By Theorem 5.1(d),

$$s^{-1}[V(t + s) - V(t)]h = s^{-1}[V(s) - 1]V(t)h \to iAV(t)h;$$

that is, $V'(t)h = iAV(t)h$. Similarly,

$$s^{-1}[U(t + s) - U(t)]h = s^{-1}[U(s) - 1]U(t)h \to iAU(t)h.$$

So if $h(t) = U(t)h - V(t)h$, then $h: \mathbb{R} \to \mathcal{H}$ is differentiable and

$$h'(t) = iAU(t)h - iAV(t)h = iAh(t).$$

But

$$\frac{1}{s}\|h(t + s) - h(t)\|^2 = \left\langle \frac{h(t + s) - h(t)}{s}, h(t + s) - h(t) \right\rangle.$$

Thus $(d/dt)\|h(t)\|^2 = 0$ and so $\|h\|: \mathbb{R} \to \mathbb{R}$ is a constant function. But $h(0) = 0$, so $h(t) \equiv 0$. This says that $U(t)h = V(t)h$ for all h in $\mathcal{D}$ and all t in $\mathbb{R}$. Since $\mathcal{D}$ is dense, $U = V$. ∎

5.14. Definition. If U is a strongly continuous one-parameter unitary group, then the self-adjoint operator A such that $U(t) = \exp(itA)$ is called the *infinitesimal generator* of U.

By virtue of Stone's Theorem and Theorem 5.1, there is a one-to-one correspondence between self-adjoint operators and strongly continuous one-parameter unitary groups. Thus, it should be able to characterize certain properties of a group in terms of its infinitesimal generator and vice versa. For example, suppose the infinitesimal generator is bounded; what can be said about the group?

5.15. Proposition. *If U is a strongly continuous one-parameter unitary group with infinitesimal generator A, then A is bounded if and only if $\lim_{t \to 0} \| U(t) - 1 \| = 0$.*

PROOF. First assume that A is bounded. Hence $\| U(t) - 1 \| = \| \exp(itA) - 1 \| = \sup\{|e^{itx} - 1| : x \in \sigma(A)\} \to 0$ as $t \to 0$ since $\sigma(A)$ is compact.

Now assume that $\| U(t) - 1 \| \to 0$ as $t \to 0$. Let $0 < \varepsilon < \pi/4$; then there is a $t_0 > 0$ such that $\| U(t) - 1 \| < \varepsilon$ for $|t| < t_0$. Since $U(t) - 1 = \int_{\sigma(A)} (e^{itx} - 1) \, dE(t)$, $\sup\{|e^{itx} - 1| : x \in \sigma(A)\} = \| U(t) - 1 \| < \varepsilon$ for $|t| < t_0$. Thus for a small δ, $tx \in \bigcup_{n=-\infty}^{\infty} (2\pi n - \delta, 2\pi n + \delta) \equiv G$ whenever $x \in \sigma(A)$ and $|t| < t_0$. In fact, if ε is chosen sufficiently small, then δ is small enough that the intervals $\{(2\pi n - \delta, 2\pi n + \delta)\}$ are the components of G. If $x \in \sigma(A)$, $\{tx: 0 \le t < t_0\}$ is the interval from 0 to $t_0 x$ and is contained in G. Hence $tx \in (-\delta, \delta)$ for x in $\sigma(A)$ and $|t| < t_0$. In particular, $t_0 \sigma(A) \subseteq [-\delta, \delta]$ so $\sigma(A)$ is compact and A is bounded. ∎

Let μ be a positive measure on $\mathbb{R}$ and let $A_\mu f = xf$ for f in $\mathcal{D}_\mu = \{ f \in L^2(\mu): xf \in L^2(\mu) \}$. We have already seen that A is self-adjoint. Clearly $\exp(itA_\mu) = M_{e_t}$ on $L^2(\mathbb{R})$, where e_t is the function $e_t(x) = \exp(itx)$. This can be generalized a bit.

5.16. Proposition. *Let (X, Ω, μ) be a σ-finite measure space and let ϕ be a real-valued Ω-measurable function on X. If $A = M_\phi$ on $L^2(\mu)$ and $U(t) = \exp(itA)$, then $U(t) = M_{e_t}$, where $e_t(x) = \exp(it\phi(x))$.*

Since each self-adjoint operator on a separable Hilbert space can be represented as a multiplication operator (Theorem 4.19), the preceding proposition gives a representation of all strongly continuous one-parameter semigroups.

EXERCISES

1. If $U: \mathbb{R} \to \mathcal{B}(\mathcal{H})$ is such that $U(t)$ is unitary for all t, $U(s + t) = U(s)U(t)$ for all s, t, and $U: \mathbb{R} \to (\mathcal{B}(\mathcal{H}), \mathrm{WOT})$ is continuous, then U is SOT-continuous.

2. Show that for every integer n there is a continuously differentiable function ϕ_n such that both ϕ_n and $\phi_n' \in L^1(0, \infty)$, $\phi_n(t) = 0$ if $t \ge 1/n$, and $\int_0^\infty \phi_n(t) \, dt = 1$.

3. Prove Claim 5.13.

4. Adopt the notation from the proof of Stone's Theorem. Let $\phi, \psi \in \mathscr{L}$ and show
   (a) $T_\phi^* = S_{\bar\phi}$; (b) $T_\phi T_\psi = T_{\phi\psi}$, and $S_\phi S_\psi = S_{\phi\psi}$; (c) $T_\phi A \subseteq A T_\phi$.

5. Let U be a strongly continuous one-parameter unitary group with infinitesimal
   generator A. Suppose e is a nonzero vector in $\mathscr{H}$ such that $Ae = \lambda e$. What is
   $U(t)e$? Conversely, suppose there is a nonzero t such that $U(t)$ has an eigenvector. What can be said about A? $U(s)$?

§6. The Fourier Transform and Differentiation

Perhaps the best way to begin this section is by examining an example.

6.1. Example. Let $\mathscr{D} = \{f \in L^2(\mathbb{R}): f$ is absolutely continuous on every
bounded interval in $\mathbb{R}$ and $f' \in L^2(\mathbb{R})\}$. For f in $\mathscr{D}$, let $Af = if'$. Then A
is self-adjoint.

First let's show that A is symmetric. If $f \in \mathscr{D}$, note that $f(x) \to 0$ as
$x \to \pm\infty$ since $f \in L^2(\mathbb{R})$. So if $f, g \in \mathscr{D}$, $0 < a < \infty$,

$$i\int_{-a}^{a} f'(x)\overline{g(x)}\, dx = i\left[f(a)\overline{g(a)} - f(-a)\overline{g(-a)}\right] - i\int_{-a}^{a} f(x)\overline{g'(x)}\, dx.$$

Hence $\langle Af, g \rangle = \langle f, Ag \rangle$ and A is symmetric.

Now let $g \in \operatorname{dom} A^*$ and for $0 < a < \infty$ let $\mathscr{D}_a = \{f \in \mathscr{D}: f(x) = 0$ for
$|x| \geq a\}$. The proof that $g \in \operatorname{dom} A$ follows the lines of the argument used
in Example 1.11. In fact, let $h = A^*g$. So if $f \in \mathscr{D}$, then $\int f(x)\overline{h(x)}\, dx =
i\int f'(x)\overline{g(x)}\, dx$. Let $H(x) = \int_0^x h(t)\, dt$. Then using integration by parts we
get that for f in $\mathscr{D}_a$,

$$\int_{-a}^{a} f\bar h = \overline{H(a)}f(a) - \overline{H(-a)}f(-a) - \int_{-a}^{a} f'\overline{H}$$

$$= -\int_{-a}^{a} f'\overline{H}.$$

Therefore $\int_{-a}^{a} f'[\overline{H} - (\overline{ig})] = 0$ for every f in $\mathscr{D}_a$. As in (1.11), it follows
that $H - ig$ is constant on $[-a, a]$ and g is absolutely continuous. More-
over, $0 = H' - ig' = h - ig'$; hence $A^*g = h = ig'$. Thus $g \in \mathscr{D}$ and A is
self-adjoint.

If A is the differentiation operator in Example 6.1, what is the group
$U(t) = \exp(itA)$? Since A is not represented as a multiplication operator,
Proposition 5.16 cannot be applied. One could proceed to try and discover
the spectral measure for A. Since $A = \int x\, dE(x)$, $U(t) = \int e^{itx}\, dE(x)$. Or
one could be clever.

Later in this section it will be shown that if $\mathscr{F}: L^2(\mathbb{R}) \to L^2(\mathbb{R})$ is the
Fourier–Plancherel transform, then $\mathscr{F}$ is a unitary operator (6.17) and
$\mathscr{F}^{-1}A\mathscr{F} =$ the operator on $L^2(\mathbb{R})$ of multiplication by the independent

variable (6.18). Thus $\mathscr{F}^{-1}U(t)\mathscr{F}$ is multiplication by e^{ixt}. But it is possible to find $U(t)$ directly.

Recall that if $f \in \operatorname{dom} A$,

$$Af = -i \lim_{t \to 0} \frac{U(t)f - f}{t}.$$

So

$$f'(x) = \lim_{t \to 0} - \frac{(U(t)f)(x) - f(x)}{t}.$$

Being clever, one might guess that $(U(t)f)(x) = f(x - t)$.

6.2.   Theorem. *If A and $\mathscr{D}$ are as in Example 6.1 and $U(t) = \exp(itA)$, then $(U(t)f)(x) = f(x - t)$ for all f in $L^2(\mathbb{R})$ and x, t in $\mathbb{R}$.*

PROOF. Let $(V(t)f)(x) = f(x - t)$. It is easy to see that V is a strongly continuous one-parameter unitary group. Let B be the infinitesimal generator of V. It must be shown that $B = A$.

Note that $f \in \operatorname{dom} B$ if and only if $\lim_{t \to 0} t^{-1}(V(t)f - f)$ exists. Let $f \in C_c^{(1)}(\mathbb{R})$; that is, f is continuously differentiable and has compact support. Thus for $t > 0$,

$$\left[\frac{V(t)f - f}{t} \right](x) = \frac{f(x - t) - f(x)}{t} = -\frac{1}{t}\int_{x-t}^x f'(y)\,dy$$

and

$$\left| \frac{V(t)f(x) - f(x)}{t} + f'(x) \right| \le \frac{1}{t}\int_{x-t}^x |f'(x) - f'(y)|\,dy$$

$$\le \sup\{|f'(x) - f'(y)| : |x - y| \le t\}.$$

Because f' is continuous with compact support, f' is uniformly continuous. Let $K = \{x : \operatorname{dist}(x, \operatorname{spt} f') \le 2\}$. So K is compact. For $\varepsilon > 0$, let $\delta(\varepsilon) < 1$ be such that if $|x - y| < \delta(\varepsilon)$, then $|f'(x) - f'(y)| < \varepsilon$. Hence $\|t^{-1}[Vf - f] + f'\|_2 \le \varepsilon^2 |K|$, where $|K| = $ the Lebesgue measure of K. Thus $C_c^{(1)}(\mathbb{R}) \subseteq \operatorname{dom} B$ and $Bf = Af$ for f in $C_c^{(1)}(\mathbb{R})$. But if $f \in \operatorname{dom} A$, there is a sequence $\{f_n\} \subseteq C_c^{(1)}(\mathbb{R})$ such that $f_n \oplus Af_n \to f \oplus Af$ in $\operatorname{gra} A$ (Exercise 1). But $f_n \oplus Af_n = f_n \oplus Bf_n \in \operatorname{gra} B$, so $f \oplus Af \in \operatorname{gra} B$; that is, $A \subseteq B$. Since self-adjoint operators are maximal symmetric operators (2.11), $A = B$. ∎

To show that the Fourier transform demonstrates that M_x and $i\,d/dx$ are unitarily equivalent, we introduce the Schwartz space of rapidly decreasing functions.

6.3.   Definition. A function $\phi \colon \mathbb{R} \to \mathbb{R}$ is *rapidly decreasing* if ϕ is infinitely differentiable and for all integers $m, n \ge 0$,

6.4 $\|\phi\|_{m,n} \equiv \sup\{|x^m \phi^{(n)}(x)| : x \in \mathbb{R}\} < \infty.$

Let $\mathscr{S} = \mathscr{S}(\mathbb{R})$ be the set of all rapidly decreasing functions on $\mathbb{R}$.

Note that if $\phi \in \mathscr{S}$, then for all $m, n \geq 0$ there is a constant $C_{m,n}$ such that

$$|\phi^{(n)}(x)| \leq C_{m,n}|x|^{-m}.$$

Thus if p is any polynomial and $n \geq 0$, $|p(x)\phi^{(n)}(x)| \to 0$ as $|x| \to \infty$. In fact, this is equivalent to ϕ belonging to $\mathscr{S}$ (Exercise 3). Also note that if $\phi \in \mathscr{S}$, then $x^m\phi^{(n)} \in \mathscr{S}$ for all $m, n \geq 0$.

It is not difficult to see that $\| \cdot \|_{m,n}$ is a seminorm on $\mathscr{S}$. Also, $\mathscr{S}$ with all of these seminorms is a Frechet space (Exercise 2). The space $\mathscr{S}$ is sometimes called the *Schwartz space* after Laurent Schwartz.

6.5. Proposition. *If $1 \leq p \leq \infty$, $\mathscr{S} \subseteq L^p(\mathbb{R})$. If $1 \leq p < \infty$, $\mathscr{S}$ is dense in $L^p(\mathbb{R})$; $\mathscr{S}$ is weak-star dense in $L^\infty(\mathbb{R})$.*

PROOF. We already have that $\mathscr{S} \subseteq L^\infty(\mathbb{R})$. If $1 \leq p < \infty$ and $\phi \in \mathscr{S}$, then

$$\int_{-\infty}^{\infty} |\phi|^p \, dx = \int_{-\infty}^{\infty} (1 + x^2)^{-p}(1 + x^2)^p|\phi|^p \, dx$$

$$\leq \|(1 + x^2)^p|\phi|^p\|_\infty \int_{-\infty}^{\infty} (1 + x^2)^{-p} \, dx.$$

Since $(1 + x^2)^p \geq 1 + x^2$,

$$\|\phi\|_p \leq \pi^{1/p}\|(1 + x^2)\phi\|_\infty$$

$$\leq \pi^{1/p}[\|\phi\|_{0,0} + \|\phi\|_{2,0}].$$

Since $C_c^{(\infty)}(\mathbb{R}) \subseteq \mathscr{S}$, the density statements are immediate. ∎

6.6. Definition. If $f \in L^1(\mathbb{R})$, the *Fourier transform* of f is the function $\hat{f}$ defined by

$$\hat{f}(x) = \frac{1}{\sqrt{2\pi}} \int_{\mathbb{R}} f(t)e^{-ixt} \, dt.$$

Because $f \in L^1(\mathbb{R})$, this integral is well defined.

The interested reader may want to peruse §VII.9, where the Fourier transform is presented in the more general context of locally compact abelian groups. That section will not be assumed here.

Recall that if $f, g \in L^1$, then the convolution of f and g,

$$f * g(x) = (2\pi)^{-1/2} \int_{\mathbb{R}} f(x - t)g(t) \, dt,$$

belongs to $L^1(\mathbb{R})$ and $\|f * g\|_1 \leq \|f\|_1\|g\|_1$. It is also true that if $f \in L^p(\mathbb{R})$, $1 \leq p \leq \infty$, then $f * g \in L^p(\mathbb{R})$ and $\|f * g\|_p \leq \|f\|_p\|g\|_1$ (see Exercise 4).

6.7.   Theorem. (a) *If $f \in L^1(\mathbb{R})$, then $\hat{f}$ is a continuous function on $\mathbb{R}$ that vanishes at $\pm \infty$. Also, $\|\hat{f}\|_\infty \le \|f\|_1$.*
(b) *If $\phi \in \mathcal{S}$, $\hat{\phi} \in \mathcal{S}$. Also for $m, n \ge 0$,*

6.8 $$(ix)^m \left(\frac{d}{dx}\right)^n \hat{\phi} = \left[\left(\frac{d}{dx}\right)^m ((-ix)^n \phi)\right]^{\hat{}} .$$

(c) *If $f, g \in L^1(\mathbb{R})$, then $(f * g)^{\hat{}} = \hat{f}\hat{g}$.*
(*Note:* $[\ ]^{\hat{}} =$ *the Fourier transform of the function defined in the brackets.*)

PROOF. (a) The fact that $\hat{f}$ is continuous is an easy consequence of Lebesgue's Dominated Convergence Theorem; it is clear that $\|\hat{f}\|_\infty \le \|f\|_1$. For the other part of (a), let $f =$ the characteristic function of the interval (a, b). Then $\hat{f}(x) = i(2\pi)^{-1/2} x^{-1} [e^{-ixb} - e^{-ixa}] \to 0$ as $|x| \to \infty$. So $\hat{f}(x)$ vanishes at $\pm \infty$ if f is a linear combination of such characteristic functions. The result for a general f follows by approximation.

(b) It is convenient to introduce the notation $D\phi = \phi'$. Thus $D^n \phi = \phi^{(n)}$. Also in this proof, as in many others of this section, x will be used to denote the function whose value at t is t and it will also be used occasionally to denote the independent variable.

If $\phi \in \mathcal{S}$, then differentiation under the integral sign (Why is this justified?) gives

$$(D\hat{\phi})(y) = \frac{1}{\sqrt{2\pi}} \int_{-\infty}^{\infty} (-it) e^{-iyt} \phi(t)\, dt$$

$$= [(-ix)\phi]^{\hat{}}(y).$$

By induction we get that for $n \ge 0$,

(6.9) $$D^n \hat{\phi} = [(-ix)^n \phi]^{\hat{}}.$$

Using integration by parts,

$$(D\phi)^{\hat{}}(y) = \frac{1}{\sqrt{2\pi}} \int_{-\infty}^{\infty} e^{-iyt} \phi'(t)\, dt$$

$$= \frac{-1}{\sqrt{2\pi}} \int_{-\infty}^{\infty} \phi(t) \frac{d}{dt} [e^{-iyt}]\, dt$$

$$= \frac{iy}{\sqrt{2\pi}} \int_{-\infty}^{\infty} e^{-iyt} \phi(t)\, dt.$$

That is, $(D\phi)^{\hat{}} = (ix)\hat{\phi}$. By induction,

6.10 $$(D^n \phi)^{\hat{}} = (ix)^n \hat{\phi}$$

for all $n \ge 0$. Combining (6.9) and (6.10) gives (6.8).

By (6.8) if $m, n \geq 0$, then for ϕ in $\mathscr{S}$,

$$\|\hat{\phi}\|_{m,n} = \sup\{|x^m(D^n\hat{\phi})(x)|: x \in \mathbb{R}\}$$

$$= \sup\left\{\left|\frac{1}{\sqrt{2\pi}} \int_{-\infty}^{\infty} e^{-ixt}\left(\frac{d}{dt}\right)^m [(-it)^n\phi(t)]\, dt\right|: x \in \mathbb{R}\right\}$$

$$\leq \frac{1}{\sqrt{2\pi}} \int_{-\infty}^{\infty} \left|\left(\frac{d}{dt}\right)^m [t^n\phi(t)]\right| dt$$

$$< \infty$$

since $D^m(x^n\phi) \in L^1(\mathbb{R})$ (6.5).

(c) This is an easy exercise in integration theory and is left to the reader. ∎

The fact that $\hat{f}(x) \to 0$ as $|x| \to \infty$ is called the *Riemann–Lebesgue Lemma*.

The process now begins whereby it will be shown that the Fourier transform on $L^1 \cap L^2$ extends to a unitary operator on $L^2(\mathbb{R})$. Moreover, the adjoint of this unitary will be calculated and it will be shown that if $i\, d/dx$ is conjugated by this unitary, then the resulting self-adjoint operator is M_x.

Changing notation a little, let U_y [instead of $U(y)$] denote the translation operator. Moreover, think of U_y as operating on all of the L^p spaces, not just L^2, so $(U_y f)(x) = f(x - y)$ for f in $L^p(\mathbb{R})$. Also, let e_y be the function $e_y(x) = \exp(ixy)$.

6.11. Proposition. *If $f \in L^1(\mathbb{R})$ and $y \in \mathbb{R}$, then*

$$[U_y f]\hat{} = e_{-y}\hat{f},$$
$$[e_y f]\hat{} = U_y\hat{f}.$$

PROOF. If $f \in L^1(\mathbb{R})$,

$$[U_y f]\hat{}(x) = (2\pi)^{-1/2} \int [U_y f](t) e^{-ixt}\, dt$$

$$= (2\pi)^{-1/2} \int f(t - y) e^{-ixt}\, dt$$

$$= (2\pi)^{-1/2} \int f(s) e^{-ix(s+y)}\, ds$$

$$= e_{-y}(x)\hat{f}(x).$$

The proof of the other equation is left as an exercise. ∎

In the proof of the next lemma the fact that $\int_{-\infty}^{\infty} e^{-t^2}\, dt = \sqrt{\pi}$ is needed. Those who have never seen this can verify it by putting $I = \int_0^{\infty} e^{-x^2}\, dx$, noting that $I^2 = \int_0^{\infty}\int_0^{\infty} e^{-(x^2+y^2)}\, dx\, dy$, and using polar coordinates.

6.12.   Lemma. *If $\varepsilon > 0$ and $\rho_\varepsilon(t) = e^{-\varepsilon^2 t^2}$, then*

$$\hat{\rho}_\varepsilon(x) = \frac{1}{\varepsilon\sqrt{2}} e^{-x^2/4\varepsilon^2}.$$

PROOF. Note that $\rho_\varepsilon \in \mathscr{S}$. By (6.8), $D\hat{\rho}_\varepsilon = (-ix\rho_\varepsilon)^{\hat{}}$. Using integration by parts,

$$\begin{aligned}
(D\hat{\rho}_\varepsilon)(x) &= \frac{-i}{\sqrt{2\pi}} \int_{-\infty}^{\infty} e^{-\varepsilon^2 t^2} t e^{-ixt} \, dt \\
&= \frac{-i}{\sqrt{2\pi}} \left(\frac{-1}{2\varepsilon^2} \right) \int_{-\infty}^{\infty} e^{-ixt} \, d\left(e^{-\varepsilon^2 t^2} \right) \\
&= \frac{-i}{2\varepsilon^2 \sqrt{2\pi}} \int_{-\infty}^{\infty} e^{-\varepsilon^2 t^2} (-ix) e^{-ixt} \, dt \\
&= \frac{-x}{2\varepsilon^2} \hat{\rho}_\varepsilon(x).
\end{aligned}$$

Let $\psi_\varepsilon(x) = e^{-x^2/4\varepsilon^2}$. Then both $\hat{\rho}_\varepsilon$ and ψ_ε satisfy the differential equation $u'(x) = -(x/2\varepsilon^2)u(x)$. Hence $\hat{\rho}_\varepsilon = c\psi_\varepsilon$ for some constant c. But $\psi_\varepsilon(0) = 1$, and

$$\begin{aligned}
\hat{\rho}_\varepsilon(0) &= \frac{1}{\sqrt{2\pi}} \int_{-\infty}^{\infty} e^{-\varepsilon^2 t^2} \, dt \\
&= \frac{1}{\varepsilon\sqrt{2\pi}} \int_{-\infty}^{\infty} e^{-s^2} \, ds \\
&= \frac{1}{\varepsilon\sqrt{2\pi}} \sqrt{\pi} = \frac{1}{\varepsilon\sqrt{2}}. \qquad \blacksquare
\end{aligned}$$

6.13.   Proposition. *If $\psi \in L^1(\mathbb{R})$ such that $(2\pi)^{-1/2}\int_\mathbb{R} \psi(x) \, dx = 1$ and if for $\varepsilon > 0$, $\psi_\varepsilon(x) = \varepsilon^{-1}\psi(x/\varepsilon)$, then for every f in $C_0(\mathbb{R})$, $\psi_\varepsilon * f(x) \to f(x)$ uniformly on $\mathbb{R}$.*

PROOF. Note that $(2\pi)^{-1/2}\int\psi_\varepsilon(x) \, dx = 1$ for all $\varepsilon > 0$. Hence for any x in $\mathbb{R}$,

$$\begin{aligned}
\psi_\varepsilon * f(x) - f(x) &= (2\pi)^{-1/2} \int [f(x - t) - f(x)] \frac{1}{\varepsilon}\psi\left(\frac{t}{\varepsilon}\right) dt \\
&= (2\pi)^{-1/2} \int [f(x - s\varepsilon) - f(x)] \psi(s) \, ds.
\end{aligned}$$

Put $\omega(y) = \sup\{|f(x - y) - f(x)|: y \in \mathbb{R}\}$. Now f is uniformly continuous (Why?), so if $\varepsilon > 0$, then there is a $\delta > 0$ such that $\omega(y) < \varepsilon$ if $|y| < \delta$. Thus $\omega(y) \to 0$ as $|y| \to 0$. Moreover, the inequality above implies

$$\|\psi_\varepsilon * f - f\|_\infty \leq (2\pi)^{-1/2} \int \omega(s\varepsilon)|\psi(s)| \, ds.$$

Since $\psi \in L^1(\mathbb{R})$, the Lebesgue Dominated Convergence Theorem implies that $\|\psi_{\varepsilon_k} * f - f\|_\infty \to 0$ whenever $\varepsilon_k \to 0$. This proves the proposition. $\blacksquare$

The next result is often called the *Multiplication Formula*. Remember that if $f \in L^1(\mathbb{R})$, $\hat{f} \in C_0(\mathbb{R})$. Hence $\hat{f}g \in L^1(\mathbb{R})$ when both f and $g \in L^1(\mathbb{R})$.

6.14. Theorem. *If $f, g \in L^1(\mathbb{R})$, then*

$$\int_{\mathbb{R}} \hat{f}(x)g(x)\, dx = \int_{\mathbb{R}} f(x)\hat{g}(x)\, dx.$$

PROOF. The proof is an easy consequence of Fubini's Theorem. In fact, if $f, g \in L^1(\mathbb{R})$, then

$$\int \hat{f}(x)g(x)\, dx = \int \left[\frac{1}{\sqrt{2\pi}} \int f(t)e^{-ixt}\, dt \right] g(x)\, dx$$

$$= \int f(t) \left[\frac{1}{\sqrt{2\pi}} \int g(x)e^{-ixt}\, dx \right] dt$$

$$= \int f(t)\hat{g}(t)\, dt. \qquad \blacksquare$$

6.15. Inversion Formula. *If $\phi \in \mathcal{S}$, then*

$$\phi(x) = \frac{1}{\sqrt{2\pi}} \int_{-\infty}^{\infty} \hat{\phi}(t)e^{ixt}\, dt.$$

PROOF. Let $\rho_\varepsilon(x) = e^{-\varepsilon^2 x^2}$ and put $\psi(x) = \hat{\rho}_1(x)$. Then by Lemma 6.12 $\psi_\varepsilon(x) = \varepsilon^{-1}\psi(x/\varepsilon) = \hat{\rho}_\varepsilon(x)$. Also,

$$(2\pi)^{-1/2} \int \psi(x)\, dx = (2\pi)^{-1/2} \int_{-\infty}^{\infty} 2^{-1/2} e^{-x^2/4}\, dx = 1.$$

So $\psi_\varepsilon * h(x) \to h(x)$ uniformly for any h in $C_0(\mathbb{R})$. If $\phi \in \mathcal{S}$, put $f = \phi$ and $g = e_x \rho_\varepsilon$ in (6.14). By Proposition 6.11 and Lemma 6.12, $\hat{g} = U_x \hat{\rho}_\varepsilon = U_x \psi_\varepsilon$. Thus

$$\frac{1}{\sqrt{2\pi}} \int_{-\infty}^{\infty} \hat{\phi}(t)e^{itx}e^{-\varepsilon^2 t^2}\, dt = \frac{1}{\sqrt{2\pi}} \int_{-\infty}^{\infty} \phi(t)\psi_\varepsilon(t-x)\, dt$$

$$= \phi * \psi_\varepsilon(x)$$

$$\to \phi(x)$$

as $\varepsilon \to 0$. The Lebesgue Dominated Convergence Theorem implies the left-hand side converges to $(2\pi)^{-1/2}\int \hat{\phi}(t)e^{ixt}\, dt$ and the theorem is proved. $\blacksquare$

In many ways the next result is a rephrasing of the preceding theorem.

6.16. Theorem. *If $\mathscr{F}: \mathcal{S} \to \mathcal{S}$ is defined by $\mathscr{F}\phi = \hat{\phi}$, $\mathscr{F}$ is a bijection with*

$$(\mathscr{F}^{-1}\phi)(x) = \frac{1}{\sqrt{2\pi}} \int_{-\infty}^{\infty} \phi(t)e^{ixt}\, dt.$$

Moreover, if $\mathscr{S}$ is given the topology induced by the seminorms $\{\| \cdot \|_{m, n}:$ $m, n \geq 0\}$ that were defined in (6.4), $\mathscr{F}$ is a homeomorphism.

PROOF. By (6.7b), $\mathscr{F}\mathscr{S} \subseteq \mathscr{S}$. The preceding theorem says that $\mathscr{F}$ is bijective and gives the formula for $\mathscr{F}^{-1}$. The proof of the topological statement is left to the reader. ∎

6.17. Plancherel's Theorem. *If $\phi \in \mathscr{S}$, then $\|\phi\|_2 = \|\hat{\phi}\|_2$ and the Fourier transform $\mathscr{F}$ extends to a unitary operator on $L^2(\mathbb{R})$.*

PROOF. Let $\phi \in \mathscr{S}$ and put $\psi(x) = \overline{\phi(-x)}$. So $\rho = \phi * \psi \in L^1(\mathbb{R})$ and $\hat{\rho} = \hat{\phi}\hat{\psi}$. An easy calculation shows that $\hat{\psi} = \overline{\hat{\phi}}$; hence $\hat{\rho} = |\hat{\phi}|^2$. Also, the Inversion Formula shows that $\rho(0) = (2\pi)^{-1/2}\int\hat{\rho}(x)\, dx = (2\pi)^{-1/2}$ $\int|\hat{\phi}(x)|^2\, dx$. Thus

$$\int |\hat{\phi}(x)|^2\, dx = (2\pi)^{1/2}\rho(0)$$

$$= (2\pi)^{1/2}\phi * \psi(0)$$

$$= \int \phi(x)\psi(0 - x)\, dx$$

$$= \int |\phi(x)|^2\, dx.$$

So if $\mathscr{S}$ is considered as a subspace of $L^2(\mathbb{R})$, $\mathscr{F}$, the Fourier transform, is an isometry on $\mathscr{S}$. By Proposition 6.5 and the preceding theorem, $\mathscr{F}$ extends to a unitary operator on $L^2(\mathbb{R})$. ∎

Warning! The content of the Plancherel Theorem is that the Fourier transform extends to an isometry. The formula for this isometry is not given by the formula for the Fourier transform. Indeed, this formula does not make sense when f is not an L^1 function. However, the same symbol, $\mathscr{F}$, will be used to denote this unitary operator on $L^2(\mathbb{R})$. For emphasis it is called the Plancherel transform.

6.18. Theorem. *Let A be the operator on $L^2(\mathbb{R})$ given by $Af = i\, d/dx$ and let M be the operator defined by $Mf = xf$. If $\mathscr{F}: L^2(\mathbb{R}) \to L^2(\mathbb{R})$ is the Plancherel Transform, then $\mathscr{F}\mathrm{dom}\, M = \mathrm{dom}\, A$ and*

$$\mathscr{F}^{-1}A\mathscr{F} = M.$$

PROOF. The fact that $A\mathscr{F} = \mathscr{F}M$ on $\mathscr{S}$ is an immediate consequence of Theorem 6.7(b). Since $\mathscr{S}$ is dense in both dom A and dom M, the rest of the result follows (with some work—give the details). ∎

Fourier analysis is a subject unto itself. One source is Stein and Weiss [1971]; another is Reed and Simon [1975].

EXERCISES

1. If $\mathscr{D}$ is as in Example 6.1, show that for every f in $\mathscr{D}$ there is a sequence $\{f_n\}$ in $C_c^{(1)}(\mathbb{R})$ such that $f_n \to f$ and $f_n' \to f'$ in $L^2(\mathbb{R})$.

2. Show that the Schwartz space $\mathscr{S}$ with the seminorms $\{\|\cdot\|_{m,n}: m, n \geq 0\}$ is a Frechet space.

3. If ϕ is infinitely differentiable on $\mathbb{R}$, show that $\phi \in \mathscr{S}$ if and only if for every integer $n \geq 0$ and every polynomial p, $\phi^{(n)}(x)p(x) \to 0$ as $|x| \to \infty$.

4. If $f \in L^p(\mathbb{R})$, $1 \leq p \leq \infty$, and $g \in L^1(\mathbb{R})$, show that $f * g \in L^p(\mathbb{R})$ and $\|f * g\|_p \leq \|f\|_p \|g\|_1$. (Hint: See Dunford and Schwartz [1958], p. 530, Exercise 13 for a generalization of Minkowski's Inequality.)

5. If ψ and ψ_ε are as in Proposition 6.13 and $f \in L^p(\mathbb{R})$, $1 \leq p < \infty$, show that $\|f * \psi_\varepsilon - f\|_p \to 0$ as $\varepsilon \to 0$. If $f \in L^\infty(\mathbb{R})$, show that $f * \psi_\varepsilon \to f$ (weak*).

6. If $f \in L^1(\mathbb{R})$ and $\hat{f} \in L^1(\mathbb{R})$, show that $f(x) = (2\pi)^{-1/2}\int_{\mathbb{R}} \hat{f}(t)e^{ixt}\, dt$ a.e.

7. If $\mathscr{F}: L^2(\mathbb{R}) \to L^2(\mathbb{R})$ is the Plancherel Transform and $f \in L^2(\mathbb{R})$, show that $(\mathscr{F}^{-1}f)(x) = (\mathscr{F}f)(-x)$.

8. Show that $\mathscr{F}^4 = 1$ but $\mathscr{F}^2 \neq 1$. What does this say about $\sigma(\mathscr{F})$?

9. Find the Fourier transform of the Hermite polynomials. What do you think?

§7. Moments

To understand this section, the preceding two sections are unnecessary.

Let μ be a positive Borel measure on $\mathbb{R}$ such that $\int |t|^n \, d\mu(t) = m_n < \infty$ for every $n \geq 0$. The numbers $\{m_n\}$ are called the *moments* of μ in analogy with the corresponding concept from mechanics. The central problem here, called the *Hamburger moment problem*, is to characterize those sequences of numbers that are moment sequences. Just as self-adjoint operators are connected to measures, the theory of self-adjoint operators is connected to the solution of this moment problem.

7.1. Theorem. *If $\{m_n: n \geq 0\}$ is a sequence of real numbers, the following statements are equivalent.*

(a) *There is a positive regular Borel measure μ on $\mathbb{R}$ such that $\int |t|^n \, d\mu(t) < \infty$ for all $n \geq 0$ and $m_n = \int t^n \, d\mu(t)$.*

(b) *If $\alpha_0, \dots, \alpha_n \in \mathbb{C}$, then $\sum_{j,k=0}^n m_{j+k}\alpha_j\bar{\alpha}_k \geq 0$.*

(c) *There is a self-adjoint operator A and a vector e such that $e \in \text{dom } A^n$ for all n and $m_n = \langle A^n e, e \rangle$ for all $n \geq 0$.*

Before proving this theorem, a preliminary result is needed. This result is useful in many other situations and is one of the standard ways to show that a symmetric operator has a self-adjoint extension.

7.2. Proposition. *Let T be a symmetric operator on $\mathcal{H}$ and suppose there is a function J: $\mathcal{H} \to \mathcal{H}$ having the following properties:*

(a) *J is conjugate linear (that is, $J(h + g) = Jh + Jg$ and $J(\alpha h) = \bar{\alpha} Jh$);*
(b) *$J^2 = 1$;*
(c) *J is continuous;*
(d) *$J \operatorname{dom} T \subseteq \operatorname{dom} T$ and $TJ \subseteq JT$.*

Then T has a self-adjoint extension.

PROOF. First note that if $h \in \operatorname{dom} T$, then $Jh \in \operatorname{dom} T$ and $h = J(Jh)$. Hence $J \operatorname{dom} T = \operatorname{dom} T$ and $JT = TJ$.

Let $h \in \mathcal{H}$ and define L: $\mathcal{H} \to \mathbb{C}$ by $L(f) = \langle h, Jf \rangle$. Since J is conjugate linear, L is a linear functional. By (c), L is continuous. Thus there is a unique vector h^* in $\mathcal{H}$ such that $L(f) = \langle f, h^* \rangle$. Let $J^*h = h^*$. Thus J^*: $\mathcal{H} \to \mathcal{H}$ and

7.3 $\langle f, J^*h \rangle = \langle h, Jf \rangle.$

It is clear that J^* is additive. If $\alpha \in \mathbb{C}$, then $\langle f, J^*(\alpha h) \rangle = \langle \alpha h, Jf \rangle = \alpha \langle f, J^*h \rangle = \langle f, \bar{\alpha} J^*h \rangle$. Thus J^* is conjugate linear. Since $J^2 = 1$, it follows that $J^{*2} = 1$.

Let $h \in \operatorname{dom} T^*$ and $f \in \operatorname{dom} T$. Then $\langle TJf, h \rangle = \langle Jf, T^*h \rangle = \langle J^*T^*h, f \rangle$ by (7.3). But also by (d), $\langle TJf, h \rangle = \langle JTf, h \rangle = \langle J^*h, Tf \rangle$. So $\langle J^*T^*h, f \rangle = \langle J^*h, Tf \rangle$ for all h in $\operatorname{dom} T^*$ and f in $\operatorname{dom} T$. But this says that $J^*h \in \operatorname{dom} T^*$ whenever $h \in \operatorname{dom} T^*$ and, furthermore, $T^*J^*h = J^*T^*h$. Since $J^{*2} = 1$, it follows that $J^* \operatorname{dom} T^* = \operatorname{dom} T^*$ and $J^*T^* = T^*J^*$.

Now let $h \in \ker(T^* \pm i)$. Then $T^*J^*h = J^*T^*h = J^*(\pm ih) = \mp iJ^*h$. Thus $J^* \ker(T^* \pm i) \subseteq \ker(T^* \mp i)$. Since $J^{*2} = 1$, $J^* \ker(T^* \pm i) = \ker(T^* \mp i)$. But J^* is injective. Indeed, if $J^*h = 0$, then $h = J^*(J^*h) = 0$. Thus the deficiency indices of T are equal. By Theorem 2.20, T has a self-adjoint extension. ∎

PROOF OF THEOREM 7.1. (a) *implies* (b). If $\alpha_0, \ldots, \alpha_n \in \mathbb{C}$, then

$$\sum_{j,k=0}^{n} m_{j+k} \alpha_j \bar{\alpha}_k = \int \sum_{j,k=0}^{n} \alpha_j \bar{\alpha}_k t^{j+k} \, d\mu(t)$$

$$= \int \left(\sum_{j=0}^{n} \alpha_j t^j \right) \left(\sum_{k=0}^{n} \bar{\alpha}_k t^k \right) d\mu(t)$$

$$= \int \left| \sum_{k=0}^{n} \alpha_k t^k \right|^2 d\mu(t) \geq 0.$$

(b) *implies* (c). Let $\mathcal{H}_0$ = the collection of all finitely nonzero sequences of complex numbers $\{\alpha_n : n \geq 0\}$. That is, $\{\alpha_0, \alpha_1, \ldots\} \in \mathcal{H}_0$ if $\alpha_n \in \mathbb{C}$ for all $n \geq 0$ and $\alpha_n = 0$ for all but a finite number of values of n. If

$x = \{\alpha_n\}$, $y = \{\beta_n\} \in \mathcal{H}_0$ define $[x, y]$ by

7.4
$$[x, y] \equiv \sum_{j,k=0}^{\infty} m_{j+k}\alpha_j\bar{\beta}_k.$$

It is easy to see that $\mathcal{H}_0$ is a vector space and (7.4) defines a semi-inner product on $\mathcal{H}_0$. In fact, it is routine that $[\cdot, \cdot]$ is sesquilinear and condition (b) implies that $[x, x] \geq 0$ for all x in $\mathcal{H}_0$.

Let $\mathcal{K}_0 = \{x \in \mathcal{H}_0 : [x, x] = 0\}$ and let $\mathcal{H}_1$ be the quotient vector space $\mathcal{H}_0/\mathcal{K}_0$. If $h = x + \mathcal{K}_0$ and $f = y + \mathcal{K}_0 \in \mathcal{H}_1$, then

7.5
$$\langle h, f \rangle \equiv [x, y]$$

can be verified to be a well-defined inner product on $\mathcal{H}_1$. Let $\mathcal{H}$ be the Hilbert space obtained by completing $\mathcal{H}_1$ with respect to the norm defined by the inner product (7.5).

Now to define some operators. If $x = \{\alpha_n\} \in \mathcal{H}_0$, let $T_0x = \{0, \alpha_0, \alpha_1, \dots\}$. It is easy to check that T_0 is a linear transformation on $\mathcal{H}_0$. Also, if $x = \{\alpha_n\}$, $y = \{\beta_n\} \in \mathcal{H}_0$, let $T_0x = \{\gamma_n\}$. So $\gamma_0 = 0$ and $\gamma_n = \alpha_{n-1}$ if $n \geq 1$. Hence

$$\begin{aligned}
[T_0x, y] &= \sum_{j,k=0}^{\infty} m_{j+k}\gamma_j\bar{\beta}_k \\
&= \sum_{\substack{j=1 \\ k=0}}^{\infty} m_{j+k}\alpha_{j-1}\bar{\beta}_k \\
&= \sum_{j,k=0}^{\infty} m_{j+k+1}\alpha_j\bar{\beta}_k \\
&= \sum_{\substack{j=0 \\ k=1}}^{\infty} m_{j+k}\alpha_j\bar{\beta}_{k-1} \\
&= [x, T_0y].
\end{aligned}$$

In particular, if $x \in \mathcal{K}_0$, then the preceding equation and the CBS inequality imply that

$$\begin{aligned}
|[T_0x, T_0x]| = |[T_0^2x, x]| &\leq [T_0^2x, T_0^2x][x, x] \\
&= 0.
\end{aligned}$$

Hence $T_0\mathcal{K}_0 \subseteq \mathcal{K}_0$. Thus T_0 induces a linear transformation T on $\mathcal{H}_1$ defined by $T(x + \mathcal{K}_0) = T_0x + \mathcal{K}_0$. It follows that $\langle Th, f \rangle = \langle h, Tf \rangle$ for all h, f in $\mathcal{H}_1$. Since $\mathcal{H}_1$ is, by definition, dense in $\mathcal{H}$, T is a densely defined symmetric operator on $\mathcal{H}$. Now to show that T has a self-adjoint extension.

Define $J_0 : \mathcal{H}_0 \to \mathcal{H}_0$ by $J_0(\{\alpha_n\}) = \{\bar{\alpha}_n\}$. It is easy to see that J_0 is conjugate linear and $J_0^2 = 1$. Also, $J_0T_0 = T_0J_0$. An easy calculation shows that $[J_0x, J_0y] = [x, y]$ for all x, y in $\mathcal{H}_0$. So $J_0\mathcal{K}_0 \subseteq \mathcal{K}_0$ and J_0 induces a conjugate linear function $J_1 : \mathcal{H}_1 \to \mathcal{H}_1$ defined by $J_1(x + \mathcal{K}_0) = J_0x + \mathcal{K}_0$. It follows that $J_1T = TJ_1$, $J_1^2 = 1$, and $\|J_1h\| = \|h\|$ for all h in $\mathcal{H}_1$. Thus

J_1 extends to a conjugate linear $J: \mathcal{H} \to \mathcal{H}$ such that $J^2 = 1$ and $\|Jh\| = \|h\|$ for all h in $\mathcal{H}$. Hence J is continuous. Also, $J \operatorname{dom} T = J_1 \mathcal{H}_1 \subseteq \mathcal{H}_1 = \operatorname{dom} T$ and $TJ \subseteq JT$. By Proposition 7.2, T has a self-adjoint extension A.

Let $e_0 = \{1, 0, 0, \dots\} \in \mathcal{H}_0$. Hence $T_0^n e_0$ has a 1 in the nth place and zeros elsewhere. If $e = e_0 + \mathcal{H}_0$, then $e \in \operatorname{dom} T^n \subseteq \operatorname{dom} A^n$ for all $n \geq 0$. Also,

$$\langle A^n e, e \rangle = [T_0^n e_0, e_0] = m_n$$

for $n \geq 0$.

(c) *implies* (a). Let $\mathcal{L}$ be the closed linear span of $\{A^n e : n \geq 0\}$. For h in $\mathcal{L} \cap \operatorname{dom} A$, let $Bh = Ah$. It follows that B is a self-adjoint operator on $\mathcal{L}$ and e is a cyclic vector for B. By Theorem 4.18, there is a positive measure μ on $\mathbb{R}$ (because B is self-adjoint) such that $\int |t|^n \, d\mu(t) < \infty$ for every $n \geq 0$ and an isomorphism $W: \mathcal{L} \to L^2(\mu)$ such that $We = 1$ and $WBW^{-1} = M_t$. Thus

$$\int t^n \, d\mu(t) = \langle M_t^n 1, 1 \rangle$$

$$= \langle W^{-1} M_t^n 1, W^{-1} 1 \rangle$$

$$= \langle A^n e, e \rangle$$

$$= m_n. \quad \blacksquare$$

EXERCISES

1. (Stieltjes.) Let $\{m_n : n \geq 0\}$ be a sequence of real numbers and show that the following statements are equivalent. (a) There is a positive regular Borel measure μ on $[0, \infty)$ such that $m_n = \int t^n \, d\mu(t)$ for all $n \geq 0$. (b) If $\alpha_0, \dots, \alpha_n \in \mathbb{C}$, then $\sum_{j,k=0}^n m_{j+k} \alpha_j \bar{\alpha}_k \geq 0$ and $\sum_{j,k=0}^n m_{j+k+1} \alpha_j \bar{\alpha}_k \geq 0$. (c) There is a self-adjoint operator A with $\sigma(A) \subseteq [0, \infty)$ and a vector e in $\operatorname{dom} A^n$ for all $n \geq 0$ such that $m_n = \langle A^n e, e \rangle$ for $n \geq 0$.

2. (Bochner.) Let $m: \mathbb{R} \to \mathbb{C}$ be a function and show that the following statements are equivalent. (a) There is a finite positive measure μ on $\mathbb{R}$ such that $m(t) = \int e^{ixt} \, d\mu(x)$ for all t in $\mathbb{R}$. (b) m is continuous and if $\alpha_0, \dots, \alpha_n \in \mathbb{C}$ and $t_0, \dots, t_n \in \mathbb{R}$, then $\sum_{j,k=0}^n m(t_j - t_k) \alpha_j \bar{\alpha}_k \geq 0$. (c) There is a strongly continuous one-parameter unitary group $U(t)$ and a vector e such that $m(t) = \langle U(t)e, e \rangle$ for all t. (Hint: Let $\mathcal{H}_0 = $ all functions $f: \mathbb{R} \to \mathbb{C}$ that vanish off a finite set.)

3. Let $\{m_n : n \in \mathbb{Z}\} \subseteq \mathbb{C}$ and show that the following statements are equivalent. (a) There is a positive measure μ on $\mathbb{D}$ such that $m_n = \int z^n \, d\mu(z)$ for all n in $\mathbb{Z}$. (b) If $\alpha_{-n}, \dots, \alpha_{-1}, \alpha_0, \alpha_1, \dots, \alpha_n \in \mathbb{C}$, then $\sum_{j,k=-n}^n m_{j-k} \alpha_j \bar{\alpha}_k \geq 0$. (c) There is a unitary operator U and a vector e such that $m_n = \langle U^n e, e \rangle$ for all n.

4. Show that the operator A that appears in the proof that (7.1b) implies (7.1c) is cyclic.

CHAPTER XI

Fredholm Theory

This chapter is entirely independent of the preceding one and only tangentially dependent on Chapters VIII and IX.

The purpose of this chapter is to study certain properties of operators on a Hilbert space that are invariant under compact perturbations. That is, we want to study properties of an operator A in $\mathcal{B}(\mathcal{H})$ that are also possessed by $A + K$ for every K in $\mathcal{B}_0(\mathcal{H})$. The correct view here is to consider this undertaking as a study of the quotient algebra $\mathcal{B}(\mathcal{H})/\mathcal{B}_0(\mathcal{H}) = \mathcal{B}/\mathcal{B}_0$—the *Calkin algebra*. Any property associated with an element of the Calkin algebra is a property associated with a coset of operators and vice versa. It is useful—indeed essential—to relate these properties to the way in which the operators act on the underlying Hilbert space.

§1. The Spectrum Revisited

In Section VII.6 we saw several properties of the spectrum of an operator on a Banach space. In particular, the concepts of point spectrum, $\sigma_p(A)$, and approximate point spectrum, $\sigma_{ap}(A)$, were explored. It was also shown (VII.6.7) that $\partial\sigma(A) \subseteq \sigma_{ap}(A)$. Recall that $\sigma_l(A)$ is the left spectrum of A and $\sigma_r(A)$ is the right spectrum of A.

1.1. Proposition. *If $A \in \mathcal{B}(\mathcal{H})$, the following statements are equivalent.*

(a) $\lambda \notin \sigma_{ap}(A)$; *that is,* $\inf\{\|(A - \lambda)h\|: \|h\| = 1\} > 0$.
(b) $\operatorname{ran}(A - \lambda)$ *is closed and* $\dim\ker(A - \lambda) = 0$.
(c) $\lambda \notin \sigma_l(A)$.

(d) $\bar{\lambda} \notin \sigma_r(A^*)$.
(e) $\text{ran}(A^* - \bar{\lambda}) = \mathcal{H}$.

PROOF. By Proposition VII.6.4, (a) and (b) are equivalent. Also, if $B \in \mathcal{B}(\mathcal{H})$, then $B(A - \lambda) = 1$ if and only if $(A^* - \bar{\lambda})B^* = 1$ so that (c) and (d) are easily seen to be equivalent.

  (b) *implies* (c). Let $\mathcal{M} = \text{ran}(A - \lambda)$ and define $T: \mathcal{H} \to \mathcal{M}$ by $Th = (A - \lambda)h$; then T is bijective. By the Open Mapping Theorem, $T^{-1}: \mathcal{M} \to \mathcal{H}$ is continuous. Define $B: \mathcal{H} \to \mathcal{H}$ by letting $B = T^{-1}$ on $\mathcal{M}$ and $B = 0$ on $\mathcal{M}^{\perp}$. Then $B \in \mathcal{B}(\mathcal{H})$ and $B(A - \lambda) = 1$. (Note that we used a property of Hilbert spaces here; see Exercise VII.6.5.)

  (d) *implies* (e). Since $\bar{\lambda} \notin \sigma_r(A^*)$, there is an operator C in $\mathcal{B}(\mathcal{H})$ such that $(A^* - \bar{\lambda})C = 1$. Hence $\mathcal{H} = (A^* - \bar{\lambda})C\mathcal{H} \subseteq \text{ran}(A^* - \bar{\lambda})$.

  (e) *implies* (a). Let $\mathcal{N} = \ker(A^* - \bar{\lambda})^{\perp}$ and define $T: \mathcal{N} \to \mathcal{H}$ by $Th = (A^* - \bar{\lambda})h$. Then T is bijective and hence invertible. Let $C: \mathcal{H} \to \mathcal{H}$ be defined by $Ch = T^{-1}h$. Then $C\mathcal{H} = \mathcal{N}$ and $(A^* - \bar{\lambda})C = 1$. Thus $C^*(A - \lambda) = 1$ so that if $h \in \mathcal{H}$, $\|h\| = \|C^*(A - \lambda)h\| \leq \|C^*\| \|(A - \lambda)h\|$. Hence $\inf\{\|(A - \lambda)h\|: \|h\| = 1\} \geq \|C^*\|^{-1}$. ∎

  If $\Delta \subseteq \mathbb{C}$, $\Delta^* \equiv \{\bar{\lambda}: \lambda \in \Delta\}$.

1.2. Corollary. *If $A \in \mathcal{B}(\mathcal{H})$, then $\partial\sigma(A) \subseteq \sigma_l(A) \cap \sigma_r(A) = \sigma_{ap}(A) \cap \sigma_{ap}(A^*)^*$.*

PROOF. The equality is immediate from the preceding theorem. In fact, $\sigma_l(A) = \sigma_{ap}(A)$ and $\sigma_r(A) = \sigma_l(A^*)^* = \sigma_{ap}(A^*)^*$. If $\lambda \in \partial\sigma(A)$, then (VII.6.7) $\lambda \in \sigma_{ap}(A)$. But $\bar{\lambda} \in \partial\sigma(A^*)$ so that $\bar{\lambda} \in \sigma_{ap}(A^*)$. ∎

  For normal elements there is less variety. The pertinent result is proved here in a more general setting than that of operators.

1.3. Proposition. *Let $\mathcal{A}$ be a C*-algebra with identity. If a is a normal element of $\mathcal{A}$, then the following statements are equivalent.*

(a) *a is invertible.*
(b) *a is left invertible.*
(c) *a is right invertible.*

PROOF. Assume that a is left invertible, so there is a b in $\mathcal{A}$ such that $ba = 1$. Thus for any x in $\mathcal{A}$, $\|x\| = \|bax\| \leq \|b\| \|ax\|$, and hence $\|ax\| \geq \|b\|^{-1}\|x\|$. In particular, this is true whenever $x \in C^*(a)$. Because a is normal, $C^*(a)$ is isomorphic to $C(K)$ where $K = \sigma(a)$ and where the isomorphism takes a into the function z ($z(w) = w$). The inequality above thus becomes: $\|zf\| \geq \|b\|^{-1}\|f\|$ for every f in $C(K)$. It must be shown that $0 \notin K$ ($= \sigma(a)$). If $0 \in K$, then for every integer n there is a function f_n in $C(K)$ such that $0 \leq f_n \leq 1$, $f_n(0) = 1$ and $f_n(z) = 0$ for z in K and

$|z| \geq n^{-1}$. Since $0 \in K$, $\|f_n\| = 1$. But $\|zf_n\| \leq 1/n$. This contradicts the inequality and so $0 \notin \sigma(a)$; that is, a is invertible.

The argument above shows that (b) implies (a). If a is right invertible, then a^* is left invertible. By the preceding argument a^* is invertible, and hence so is a. ∎

1.4. Proposition. *If N is a normal operator, then $\sigma(N) = \sigma_r(N) = \sigma_l(N)$. If λ is an isolated point of $\sigma(N)$, then $\lambda \in \sigma_p(N)$.*

PROOF. The first part of the proposition is immediate from the preceding result. If λ is an isolated point of $\sigma(N)$ and $N = \int z\, dE(z)$, then $0 \neq E(\{\lambda\}) = \ker(N - \lambda)$ (Exercise IX.2.1). ∎

EXERCISES

1. Let S be the unilateral shift of multiplicity 1 on $l^2(\mathbb{N})$ and find $\sigma_l(S)$ and $\sigma_r(S)$.

2. The *compression spectrum* of A, $\sigma_c(A)$, is defined by $\sigma_c(A) = \{\lambda \in \mathbb{C}: \text{ran}(A - \lambda)$ is not dense in $\mathcal{H}\}$. Show: (a) $\lambda \in \sigma_c(A)$ if and only if $\bar{\lambda} \in \sigma_p(A^*)$. (b) $\sigma_c(A) \subseteq \sigma_r(A)$, but this inclusion may be proper. (c) $\sigma_c(A)$ is not necessarily closed. (d) $\sigma(A) = \sigma_{ap}(A) \cup \sigma_c(A)$.

3. If $A \in \mathcal{B}(\mathcal{H})$ and $f \in \text{Hol}(A)$, then $f(\sigma_{ap}(A)) \subseteq \sigma_{ap}(f(A))$ and $f(\sigma_p(A)) \subseteq \sigma_p(f(A))$.

4. If f is a rational function with poles off $\sigma(A)$, show that $f(\sigma_{ap}(A)) = \sigma_{ap}(f(A))$ and $f(\sigma_p(A)) = \sigma_p(f(A))$. Give necessary and sufficient conditions on a function f in $\text{Hol}(A)$ that these equalities hold.

§2. The Essential Spectrum and Semi-Fredholm Operators

Let $\mathcal{B}/\mathcal{B}_0$ be the Calkin algebra and let $\pi: \mathcal{B} \to \mathcal{B}/\mathcal{B}_0$ be the natural map ($\mathcal{H}$ is being suppressed here). Since $\mathcal{B}_0$ is an ideal in $\mathcal{B}$, $\mathcal{B}/\mathcal{B}_0$ is a Banach algebra with identity.

2.1. Definition. If $A \in \mathcal{B}(\mathcal{H})$, then the *essential spectrum* of A, $\sigma_e(A)$, is the spectrum of $\pi(A)$ in $\mathcal{B}/\mathcal{B}_0$; that is, $\sigma_e(A) = \sigma(\pi(A))$. Similarly, the *left* and *right essential spectrum* of A are defined by $\sigma_{le}(A) \equiv \sigma_l(\pi(A))$ and $\sigma_{re}(A) \equiv \sigma_r(\pi(A))$, respectively.

2.2. Proposition. *Let $A \in \mathcal{B}(\mathcal{H})$.*

(a) $\sigma_{le}(A) \cup \sigma_{re}(A) = \sigma_e(A)$.
(b) $\sigma_{le}(A) = \sigma_{re}(A^*)^*$.
(c) $\sigma_{le}(A) \subseteq \sigma_l(A)$, $\sigma_{re}(A) \subseteq \sigma_r(A)$, and $\sigma_e(A) \subseteq \sigma(A)$.

(d) $\sigma_{le}(A)$, $\sigma_{re}(A)$, and $\sigma_e(A)$ are closed sets.
(e) If $K \in \mathcal{B}_0(\mathcal{H})$, $\sigma_{le}(A) = \sigma_{le}(A + K)$ and $\sigma_{re}(A) = \sigma_{re}(A + K)$.

PROOF. Parts (a), (b), (c), and (e) are trivial, and part (d) is a consequence of a general fact about Banach algebras. ■

In order to best appreciate and use the idea of essential spectrum, a better understanding of invertibility in $\mathcal{B}/\mathcal{B}_0$ is needed. The following terminology is traditional.

2.3. Definition. If $A \in \mathcal{B}(\mathcal{H})$, A is a *left Fredholm operator* if $\pi(A)$ is left invertible in $\mathcal{B}/\mathcal{B}_0$; A is a *right Fredholm operator* if $\pi(A)$ is right invertible in $\mathcal{B}/\mathcal{B}_0$; A is a *Fredholm operator* if $\pi(A)$ is invertible in $\mathcal{B}/\mathcal{B}_0$. Let $\mathcal{F}_l, \mathcal{F}_r, \mathcal{F}$ denote the set of left Fredholm, right Fredholm, and Fredholm operators. So $\mathcal{F} = \mathcal{F}_l \cap \mathcal{F}_r$. Operators in the set $\mathcal{SF} = \mathcal{F}_l \cup \mathcal{F}_r$ are called *semi-Fredholm operators*.

2.4. Proposition. The sets $\mathcal{F}_l, \mathcal{F}_r, \mathcal{F}$ are all open in $\mathcal{B}(\mathcal{H})$ and $A \in \mathcal{F}_l$ if and only if $A^* \in \mathcal{F}_r$.

PROOF. Each of these sets is the inverse image under π of an open subset of $\mathcal{B}/\mathcal{B}_0$. The other statement is trivial. ■

The next result, characterizing left Fredholm operators, is from Wolf [1959] and Fillmore, Stampfli, and Williams [1972].

2.5. Theorem. If $A \in \mathcal{B}(\mathcal{H})$, the following statements are equivalent.

(a) A is a left Fredholm operator.
(b) ran A is closed and dim ker $A < \infty$.
(c) There is no sequence $\{h_n\}$ of unit vectors in $\mathcal{H}$ such that $h_n \to 0$ weakly and $\lim\|Ah_n\| = 0$.
(d) There is no orthonormal sequence $\{e_n\}$ in $\mathcal{H}$ such that $\lim\|Ae_n\| = 0$.
(e) There is a $\delta > 0$ such that $\{h \in \mathcal{H} : \|Ah\| \leq \delta\|h\|\}$ contains no infinite-dimensional manifold.
(f) If the positive operator $(A^*A)^{1/2} = \int_0^\infty t\, dE(t)$, then there is a $\delta > 0$ such that $E[0, \delta]\mathcal{H}$ is finite dimensional.
(g) If $K \in \mathcal{B}_0(\mathcal{H})$, then dim ker$(A + K) < \infty$.

PROOF. (a) *implies* (b). According to (a) there is a bounded operator B such that $\pi(B)\pi(A) = 1$; that is, $\pi(BA - 1) = 0$. Hence $BA = 1 + K$ for some compact operator K. But ker $A \subseteq$ ker $BA =$ ker$(1 + K)$. Since the eigenspaces corresponding to nonzero eigenvalues of compact operators are finite dimensional, dim ker $A < \infty$. Also, the Fredholm Alternative (VII.7.9) implies ran $BA =$ ran$(K + 1)$ is closed. Hence there is a constant $c > 0$ such that for $h \perp$ ker(BA), $\|BAh\| \geq c\|h\|$. Thus if $h \in [$ker $BA]^\perp$, $c\|h\| \leq \|B\|\|Ah\|$, or $\|Ah\| \geq (c/\|B\|)\|h\|$. This implies that $A([$ker $BA]^\perp)$ is closed. But ran $A = A($ker $BA]^\perp) + A($ker $BA)$. Since $A($ker $BA)$ is finite dimensional, ran A is closed.

(b) *implies* (c). Let P be the projection of $\mathcal{H}$ onto $(\ker A)^{\perp}$. Since $\operatorname{ran} A$ is closed and A is a bijective map of $(\ker A)^{\perp}$ onto $\operatorname{ran} A$, there is a bounded operator B on $\mathcal{H}$ such that $B = 0$ on $(\operatorname{ran} A)^{\perp}$ and $BAh = h$ for h in $(\ker A)^{\perp}$. Thus $BA = P$. Now $1 - P$ has finite rank. So if $\{h_n\}$ is a sequence of unit vectors such that $h_n \to 0$ weakly, then $\|h_n - Ph_n\| \to 0$. But $1 = \|h_n\|^2 = \|(1 - P)h_n\|^2 + \|Ph_n\|^2 = \|(1 - P)h_n\|^2 + \|BAh_n\|^2$. Hence $\|BAh_n\| \to 1$ so that $\liminf \|Ah_n\| > 0$.

(c) *implies* (d). Orthonormal sequences converge weakly to zero.

(d) *implies* (e). If (e) is false, then for every positive integer n there is an infinite-dimensional subspace $\mathcal{M}_n$ such that $\|Ah\| \le (1/n)\|h\|$ for all h in $\mathcal{M}_n$. Let e_1 be a unit vector in $\mathcal{M}_1$. Suppose $e_1, \dots, e_n$ are orthonormal vectors such that $e_k \in \mathcal{M}_k$, $1 \le k \le n$. Let E be the projection of $\mathcal{H}$ onto $\bigvee\{e_1, \dots, e_n\}$. If $\mathcal{M}_{n+1} \cap [e_1, \dots, e_n]^{\perp} = (0)$, then E is injective on $\mathcal{M}_{n+1}$. Since $\dim \mathcal{M}_{n+1} = \infty$ and $\dim \operatorname{ran} E < \infty$, this is impossible. Thus there is a unit vector e_{n+1} in $\mathcal{M}_{n+1}$ such that $e_{n+1} \perp \{e_1, \dots, e_n\}$. The orthonormal sequence $\{e_n\}$ shows that (d) does not hold.

(e) *implies* (f). Let $|A| = \int t\, dE(t)$ and let $\delta > 0$. If $h \in E[0, \delta]\mathcal{H}$, then

$$
\begin{aligned}
\|Ah\|^2 &= \langle A^*Ah, h \rangle \\
&= \langle |A|^2 h, h \rangle \\
&= \int_0^{\delta} t^2\, dE_{h,h}(t) \le \delta^2 E_{h,h}[0, \delta] \\
&= \delta^2 \|h\|^2.
\end{aligned}
$$

So $E[0, \delta]\mathcal{H} \subseteq \{h: \|Ah\| \le \delta\|h\|\}$. By (e) there is a $\delta > 0$ such that $E[0, \delta]\mathcal{H}$ is finite dimensional.

(f) *implies* (a). Let $\mathcal{M}_{\delta} = \{E[0, \delta]\mathcal{H}\}^{\perp}$. Now $|A|$ maps $\mathcal{M}_{\delta}$ bijectively onto $\mathcal{M}_{\delta}$. In fact, the inverse of $|A|: \mathcal{M}_{\delta} \to \mathcal{M}_{\delta}$ is $(\int_{\delta}^{\infty} t^{-1}\, dE(t))|\mathcal{M}_{\delta}$. Let $A = U|A|$ be the polar decomposition of A. Since $\mathcal{M}_{\delta} \subseteq \operatorname{ran}|A| = \operatorname{initial} U$, U maps $\mathcal{M}_{\delta}$ isometrically onto some closed subspace $\mathcal{L}$ of $\operatorname{ran} A$. Let $V = $ the inverse of U on $\mathcal{L}$ and $V = 0$ on $\mathcal{L}^{\perp}$; that is, $V|\mathcal{L}^{\perp} = 0$ and $V|\mathcal{L} = (U|\mathcal{M}_{\delta})^{-1}$. Hence V is a partial isometry. Let $B_1 = \int_{\delta}^{\infty} t^{-1}\, dE(t)$ and put $B = B_1 V$. If $h \in \mathcal{M}_{\delta}$, then $BAh = B_1 VU|A|h = h$. If $h \in \mathcal{M}_{\delta}^{\perp} = E[0, \delta]h$, $|A|h \in \mathcal{M}_{\delta}^{\perp}$ and so $U|A|h \perp \mathcal{L}$; thus $BAh = 0$. Hence $BA = E(\delta, \infty) = 1 - E[0, \delta]$. Since $E[0, \delta]$ has finite rank, $\pi(B)\pi(A) = 1$.

(a) *implies* (g). If K is a compact operator, $\pi(A) = \pi(A - K)$. Thus $\pi(A - K)$ has a left inverse in the Calkin algebra. Since (a) implies (b), $\dim \ker(A - K) < \infty$.

(g) *implies* (d). Suppose (d) does not hold. So there is an orthonormal sequence $\{e_n\}$ such that $\|Ae_n\| \to 0$. By passing to a subsequence if necessary, it may be assumed that $\sum_{n=1}^{\infty} \|Ae_n\|^2 < \infty$. Thus for any h in $\mathcal{H}$,

$$
\begin{aligned}
\sum |\langle h, e_n \rangle| \|Ae_n\| &\le \left[\sum |\langle h, e_n \rangle|^2\right]^{1/2} \left[\sum \|Ae_n\|^2\right]^{1/2} \\
&\le C\|h\|,
\end{aligned}
$$

where $C = [\sum \|Ae_n\|^2]^{1/2}$. Thus $Kh = \sum_{n=1}^{\infty} \langle h, e_n \rangle Ae_n$ defines a bounded operator. Moreover, if $K_n h = \sum_{j=1}^{n} \langle h, e_j \rangle A e_j$, it is easy to see that $\| K_n - K \| \to 0$. Thus K is compact. But $(A - K)e_n = 0$ for every n, so $\dim \ker(A - K) = \infty$. ■

Each of the parts of the preceding theorem can be used to give a statement equivalent to the fact that a point belongs to the left essential spectrum. Only one of these statements will receive such a translation.

2.6. Corollary.

(a) $\lambda \in \sigma_{le}(A)$ if and only if $\dim \ker(A - \lambda) = \infty$ or $\operatorname{ran}(A - \lambda)$ is not closed.

(b) $\lambda \in \sigma_{re}(A)$ if and only if $\dim[\operatorname{ran}(A - \lambda)]^{\perp} = \infty$ or $\operatorname{ran}(A - \lambda)$ is not closed.

PROOF. Part (a) is straightforward. Part (b) follows immediately from the facts that $\sigma_{re}(A) = \sigma_{le}(A^*)^*$ and that $\operatorname{ran}(A - \lambda)$ is closed if and only if $\operatorname{ran}(A - \lambda)^*$ is closed (VI.1.10). ■

In order to prove part (b) of the preceding corollary it is not necessary to quote Theorem VI.1.10. For operators on a Hilbert space it is possible to give a direct proof that is easier than the Banach space case (see Exercise 2).

The reader should compare Corollary 2.6 and Proposition 1.1.

2.7. Proposition. If $A \in \mathcal{B}(\mathcal{H})$, then $\sigma_{ap}(A) = \sigma_{le}(A) \cup \{\lambda \in \sigma_p(A): \dim \ker(A - \lambda) < \infty\}$.

PROOF. If $\lambda \in \sigma_{ap}(A)$, then (1.1) either $\operatorname{ran}(A - \lambda)$ is not closed or $\ker(A - \lambda) \neq 0$. If $\operatorname{ran}(A - \lambda)$ is not closed or if $\dim \ker(A - \lambda) = \infty$, then $\lambda \in \sigma_{le}(A)$ by (2.6). The other inclusion is left to the reader. ■

2.8. Proposition. If N is a normal operator and $\lambda \in \sigma(N)$, then $\operatorname{ran}(N - \lambda)$ is closed if and only if λ is an isolated point of $\sigma(N)$.

PROOF. Assume λ is an isolated point of $\sigma(N)$; thus $X = \sigma(N) \setminus \{\lambda\}$ is a closed subset of $\sigma(N)$. If $N = \int z \, dE(z)$ and $\mathcal{H}_1 = E(X)\mathcal{H}$, then $\mathcal{H}_1$ reduces N and $\sigma(N|\mathcal{H}_1) = X$. Hence $(N - \lambda)\mathcal{H}_1$ is closed. Since $\mathcal{H}_1^{\perp} = \ker(N - \lambda)$, $\operatorname{ran}(N - \lambda) = (N - \lambda)\mathcal{H}_1$; hence $N - \lambda$ has closed range.

Now assume that $\lambda \in \sigma(N)$ but λ is not an isolated point. Then there is a strictly decreasing sequence $\{r_n\}$ of positive real numbers such that $r_n \to 0$ and such that each open annulus $A_n = \{z: r_{n+1} < |z - \lambda| < r_n\}$ has nonempty intersection with $\sigma(N)$. Thus $E(A_n)\mathcal{H} \neq (0)$; let e_n be a unit vector in $E(A_n)\mathcal{H}$. Then $e_n \perp \ker(N - \lambda) (= E(\{\lambda\})\mathcal{H})$ and

$$\|(N - \lambda)e_n\|^2 = \int_{A_n} |z - \lambda|^2 \, dE_{e_n, e_n}(z) \leq r_n^2 \to 0.$$

That is, $\inf\{\|(N - \lambda)h\|: \|h\| = 1,\ h \perp \ker(N - \lambda)\} = 0$ and so, by the Open Mapping Theorem, $N - \lambda$ does not have closed range. ■

2.9. Proposition. *If N is a normal operator, then $\sigma_e(N) = \sigma_{le}(N) = \sigma_{re}(N)$ and $\sigma(N) \setminus \sigma_e(N) = \{\lambda \in \sigma(N): \lambda$ is an isolated eigenvalue of N having finite multiplicity$\}$.*

PROOF. The first part follows by applying Proposition 1.3 to the Calkin algebra. If λ is an isolated point of $\sigma(N)$, then $\operatorname{ran}(N - \lambda)$ is closed by the preceding proposition. So if $\dim\ker(N - \lambda) < \infty$, $\lambda \notin \sigma_{le}(N) = \sigma_e(N)$ by Corollary 2.6. Conversely, if $\lambda \in \sigma(N) \setminus \sigma_e(N)$, then $\operatorname{ran}(N - \lambda)$ is closed and $\dim\ker(N - \lambda) < \infty$. By the preceding proposition, λ is an isolated point of $\sigma(N)$. Thus λ is an eigenvalue of finite multiplicity. ■

It is also worthwhile, before proceeding, explicitly to reformulate Theorem 2.5 to give a characterization of Fredholm and semi-Fredholm operators. The proof is left to the reader.

2.10. Proposition.

(a) *An operator A is a Fredholm operator if and only if $\operatorname{ran} A$ is closed and both $\ker A$ and $\ker A^* = (\operatorname{ran} A)^\perp$ are finite dimensional.*

(b) *An operator A is a semi-Fredholm operator if and only if $\operatorname{ran} A$ is closed and either $\ker A$ or $(\operatorname{ran} A)^\perp$ is finite dimensional.*

2.11. Example. Let G be a bounded region in $\mathbb{C}$ and, to avoid pathologies, assume $\partial G = \partial[\operatorname{cl} G]$. Let $\mathscr{H} = L_a^2(G)$ (I.1.10) and define $S\colon \mathscr{H} \to \mathscr{H}$ by $(Sf)(z) = zf(z)$. Then $\sigma(S) = \operatorname{cl} G$, $\sigma_e(S) = \sigma_{le}(S) = \sigma_{re}(S) = \partial G = \sigma_{ap}(S)$, $\sigma_p(S) = \square$, and for λ in G, $\operatorname{ran}(S - \lambda)$ is closed and $\dim[\operatorname{ran}(S - \lambda)]^\perp = 1$.

To show that these statements are true, begin by proving:

2.12   If $\lambda \in G$, $\operatorname{ran}(S - \lambda) = \{f \in L_a^2(G): f(\lambda) = 0\}$

In fact, if $h \in L_a^2(G)$, then $[(S - \lambda)h](z) = (z - \lambda)h(z)$ so that $f = (z - \lambda)h$ vanishes at λ. Conversely, suppose $f \in L_a^2(G)$ and $f(\lambda) = 0$; then $f(z) = (z - \lambda)h(z)$ for some analytic function h on G. It must be shown that $h \in L_a^2(G)$. Let $r > 0$ such that $D = \{z: |z - \lambda| \le r\} \subseteq G$. Then

$$\int\int |h|^2 = \int\int_D |h|^2 + \int\int_{G\setminus D} |h|^2.$$

Now $\int\int_D |h|^2 < \infty$ since h is bounded on D. For z in $G \setminus D$, $|h(z)| = |f(z)|/|z - \lambda| \le r^{-1}|f(z)|$. Hence

$$\int\int_{G\setminus D} |h|^2 \le r^{-2} \int\int_G |f|^2 < \infty.$$

Thus $h \in L_a^2(G)$ and $f = (S - \lambda)h$. This proves (2.12).

Using Corollary I.1.12, $f \mapsto f(\lambda)$ is a bounded linear functional on $L_a^2(G)$ whenever $\lambda \in G$. By (2.12), ran$(S - \lambda)$ is the kernel of this linear functional and hence is closed.

Because G is bounded, the constant functions belong to $L_a^2(G)$. So if $f \in L_a^2(G)$, $f = [f - f(\lambda)] + f(\lambda)$ and $f - f(\lambda) \in \text{ran}(S - \lambda)$. Thus $L_a^2(G) = \text{ran}(S - \lambda) + \mathbb{C}$. Therefore $\dim[\text{ran}(S - \lambda)]^{\perp} = \dim[L_a^2(G)/\text{ran}(S - \lambda)] = 1$ when $\lambda \in G$.

If $\lambda \in G$, then $S - \lambda$ is not surjective; hence $G \subseteq \sigma(S)$. If $\lambda \notin \text{cl} G$, then $(z - \lambda)^{-1}$ is a bounded analytic function on G. If $Af = (z - \lambda)^{-1} f$, then A is a bounded operator on $L_a^2(G)$ and it is easy to check that $A(S - \lambda) = (S - \lambda)A = 1$. Thus $\sigma(S) \subseteq \text{cl} G$. Combining these two containments, we get $\sigma(S) = \text{cl} G$.

From Proposition 2.10 we have that $S - \lambda$ is a Fredholm operator whenever $\lambda \in G$; thus $G \cap \sigma_e(S) = \square$. So $\sigma_e(S) \subseteq \partial G = \partial[\text{cl} G]$. If $\lambda \in \partial G$, then $\lambda \in \partial\sigma(S)$; thus $\lambda \in \sigma_{ap}(S)$ (1.2). Since $\ker(S - \lambda) = (0)$, $\text{ran}(S - \lambda)$ is not closed. Thus $\partial G \subseteq \sigma_{le}(S) \cap \sigma_{re}(S)$. This proves that $\sigma_e(S) = \sigma_{le}(S) = \sigma_{re}(S) = \partial G = \sigma_{ap}(S)$.

EXERCISES

1. Give a direct proof that (b) implies (a) in Theorem 2.5.

2. If $A \in \mathcal{B}(\mathcal{H})$ and ran A is closed, prove that ran A^* is closed without using Theorem VI.1.10. [Hint: Show that there is a bounded operator B on $\mathcal{H}$ such that $BA = $ the projection of $\mathcal{H}$ onto $(\ker A)^{\perp}$.]

3. (Putnam [1968].) If $A \in \mathcal{B}(\mathcal{H})$, $\lambda \in \partial\sigma(A)$, and λ is not an isolated point of $\sigma(A)$, then ran$(A - \lambda)$ is not closed. Give an example of an operator A such that 0 is an isolated point of $\sigma(A)$ but ran A is not closed.

4. Let G be a bounded region in $\mathbb{C}$ such that $\partial G = \partial[\text{cl} G]$ and let ϕ be a function that is analytic in a neighborhood of cl G. Define A: $L_a^2(G) \to L_a^2(G)$ by $Af = \phi f$. Find all of the parts of the spectrum of A.

5. Let S be the unilateral shift and show that $\sigma(S) = \sigma_r(S) = \text{cl} \mathbb{D}$, $\sigma_l(S) = \sigma_{le}(S) = \sigma_{re}(S) = \partial \mathbb{D}$, $\sigma_p(S) = \square$, and for $|\lambda| < 1$, ran$(S - \lambda)$ is closed with $\dim[\text{ran}(S - \lambda)]^{\perp} = 1$.

6. Let S be the unilateral shift and put $A = S \oplus S^*$. Find the parts of the spectrum of A.

7. Let S be the unilateral shift and put $A = S^{(\infty)}$. Show that $\sigma(A) = \sigma_e(A) = \sigma_{re}(A) = \text{cl} \mathbb{D}$ and $\sigma_{le}(A) = \partial \mathbb{D}$.

8. Let $A, B, C \in \mathcal{B}(\mathcal{H})$ and define X: $\mathcal{H}^{(2)} \to \mathcal{H}^{(2)}$ by the matrix $X = \begin{bmatrix} A & B \\ 0 & C \end{bmatrix}$.
   (a) Show that if $A \in \mathcal{F}$, then $X \in \mathcal{F}$ if and only if $C \in \mathcal{F}$. (b) If $A \in \mathcal{F}$, show that $X \in \mathcal{SF}$ if and only if $C \in \mathcal{SF}$. (c) Suppose $A, C \in \mathcal{SF}$ with dim ker $A = \infty$ and dim ker $C^* = \infty$. Show that $0 \in \sigma_{le}(X) \cap \sigma_{re}(X)$.

9. (Fillmore, Stampfli, and Williams [1972].) If $\lambda \in \sigma_{le}(A)$, then there is a projection P, having infinite rank, such that $\pi(A - \lambda)\pi(P) = 0$.

10. (Fillmore, Stampfli, and Williams [1972].) Let $A \in \mathcal{B}(\mathcal{H})$. (a) If A has a cyclic vector e, show that $\dim\{Ae, A^2e, \ldots\}^{\perp} \leq 1$. (b) Let $\lambda \in \sigma_{le}(A^*)$. If $\varepsilon > 0$, let f_1, f_2 be orthonormal vectors such that $\|(A^* - \lambda)f_j\| < \varepsilon$ for $j = 1, 2$ and let $P =$ the projection onto $V\{f_1, f_2\}$. Put $B = \bar{\lambda}P + (1 - P)A$. Show that $\|B - A\| < 2\varepsilon$. (c) Show that the noncyclic operators are dense in $\mathcal{B}(\mathcal{H})$.

§3. The Fredholm Index

The author would like to acknowledge that James P. Williams made available to him a set of unpublished notes on the Fredholm index which formed the basis of this section.

If A is a semi-Fredholm operator, define the (*Fredholm*) *index* of A, ind A, by

3.1
$$\text{ind } A = \dim \ker A - \dim(\operatorname{ran} A)^{\perp}$$

$$= \dim \ker A - \dim \ker A^*.$$

Note that ind $A \in \mathbb{Z} \cup \{\pm\infty\}$ and it is necessary for either $\ker A$ or $\ker A^*$ to be finite dimensional in order for (3.1) to make sense. For ind A to be well defined, it is not necessary that ran A be closed (the other part of the characterization of semi-Fredholm operators), but this property will be used in a critical way when the properties of the index are established. The main result of this section is the following.

3.2. Theorem. *If the set of semi-Fredholm operators, $\mathcal{SF}$, has the relative norm topology from $\mathcal{B}(\mathcal{H})$ and $\mathbb{Z} \cup \{\pm\infty\}$ has the discrete topology, then* ind: $\mathcal{SF} \to \mathbb{Z} \cup \{\pm\infty\}$ *is continuous. Moreover, if $A \in \mathcal{SF}$ and $K \in \mathcal{B}_0(\mathcal{H})$, then* ind $A = \text{ind}(A + K)$.

One of the uses of Theorem 3.2 is in the study of various integral and differential equations. More recently it has been used to study a variety of approximation questions in $\mathcal{B}(\mathcal{H})$ as well as several connections between topology and operator theory.

Before proving Theorem 3.2, which will require a few lemmas, we will examine some additional properties of the index and a few examples.

First observe that the Fredholm Alternative (VII.7.9) is an easy consequence of Theorem 3.2. Indeed, if $\lambda \in \mathbb{C}$, $\lambda \neq 0$, then the operator λ is invertible and so ind$(\lambda) = 0$. If $K \in \mathcal{B}_0(\mathcal{H})$, then (3.2) implies that ind$(\lambda - K) = 0$. Thus $\dim\ker(\lambda - K) = \dim[\operatorname{ran}(\lambda - K)]^{\perp}$ and we have the Fredholm Alternative.

3.3.   Proposition.

(a) *If $A \in \mathcal{SF}$, then $A^* \in \mathcal{SF}$ and ind $A = -$ ind A^*.*
(b) *If N is normal and $N \in \mathcal{SF}$, then $N \in \mathcal{F}$ and ind $N = 0$.*

PROOF. (a) is clear. If N is normal and $N \in \mathcal{SF}$, then $N \in \mathcal{F}_l \cup \mathcal{F}_r$. By
Proposition 1.3, $N \in \mathcal{F}$. Also, $\|N^*h\| = \|Nh\|$, so $\ker N = \ker N^*$. Hence
ind $N = 0$.   ∎

It is a good thing to keep in mind that if $A \in \mathcal{SF}$ and ind A is a finite
number, then $A \in \mathcal{F}$ since both $\ker A$ and $\ker A^*$ must be finite dimensional.

The continuity statement in Theorem 3.2 has an easy interpretation.
Because $\mathcal{SF}$ is an open set, its components are open. Since $\mathbb{Z} \cup \{\pm \infty\}$
has the discrete topology, the continuity of the index is equivalent to the
statement that the index is constant on the components of $\mathcal{SF}$. This is
quite useful in applications.

One of the uses of the index is to examine $\text{ind}(A - \lambda)$ for all λ for which
this makes sense. When does it make sense? It must be that $A - \lambda \in \mathcal{SF}$
and this is true precisely when $\lambda \notin \sigma_{le}(A) \cap \sigma_{re}(A)$. The next result is a
consequence of (3.2).

3.4.   Proposition. *If $A \in \mathcal{B}(\mathcal{H})$, then $\text{ind}(A - \lambda)$ is constant on the components of $\mathbb{C} \setminus \sigma_{le}(A) \cap \sigma_{re}(A)$. If λ is an isolated point of $\sigma(A)$ and $\lambda \notin
\sigma_{le}(A) \cap \sigma_{re}(A)$, then $\text{ind}(A - \lambda) = 0$.*

PROOF. The map $\lambda \mapsto A - \lambda$ is a continuous map of $\mathbb{C} \setminus \sigma_{le}(A) \cap \sigma_{re}(A)$
into $\mathcal{SF}$. So the first part of the proposition follows from the preceding
remarks. If λ is an isolated point of $\sigma(A)$ and $\lambda \notin \sigma_{le}(A) \cap \sigma_{re}(A)$, then
there is a sequence $\{\lambda_n\}$ in $\mathbb{C} \setminus \sigma(A)$ such that $\lambda_n \to \lambda$. Thus $\text{ind}(A - \lambda_n)$
$\to \text{ind}(A - \lambda)$. Since $\text{ind}(A - \lambda_n) = 0$ for all n, the result follows.   ∎

3.5.   Example. Let G be a bounded region in $\mathbb{C}$ such that $\partial G = \partial[\text{cl } G]$ and
define S: $L^2 a(G) \to L^2_a(G)$ by $Sf = zf$. Then $\sigma_{le}(S) \cap \sigma_{re}(S) = \partial G$ and
$\text{ind}(S - \lambda) = -1$ for λ in G. If $\lambda \notin \text{cl } G$, $S - \lambda$ is invertible.

In fact, in Example 2.11 it was shown that $\partial G = \sigma_{le}(S) = \sigma_{re}(S) = \sigma_e(S)$,
$\sigma_p(S) = \square$, and $\dim[\text{ran}(S - \lambda)]^\perp = 1$ for λ in G.

3.6.   Example. Let S be the unilateral shift on l^2. Then $\sigma_{le}(S) \cap \sigma_{re}(S) =
\partial \mathbb{D}$ and $\text{ind}(S - \lambda) = -1$ for $|\lambda| < 1$.

In Proposition VII.6.5 it was shown that $\sigma(S) = \text{cl } \mathbb{D}$, $\sigma_p(S) = \square$, and
$\sigma_{ap}(S) = \partial \mathbb{D}$. Thus for $|\lambda| = 1$, $\text{ran}(S - \lambda)$ is not closed and hence $\partial \mathbb{D} \subseteq
\sigma_{le}(S) \cap \sigma_{re}(S)$. Also, if $|\lambda| < 1$, it was shown that $\text{ran}(S - \lambda)$ is closed and
$\dim[\text{ran}(S - \lambda)]^\perp = 1$. This implies that $\partial \mathbb{D} = \sigma_{le}(S) \cap \sigma_{re}(S)$ and $\text{ind}(S
- \lambda) = -1$ for λ in $\mathbb{D}$.

3.7.   Proposition. *If* $A, B \in \mathscr{F}$, *then* $A \oplus B \in \mathscr{F}$ *and* $\mathrm{ind}\, A \oplus B = \mathrm{ind}\, A$
$+\, \mathrm{ind}\, B$.

PROOF. Exercise.

Using this proposition and the preceding examples, more examples can be manufactured. Here is an interesting one.

3.8.   Example. Let S be the unilateral shift on l^2 and put $A = S^* \oplus S$. Then $\sigma_{le}(A) \cap \sigma_{re}(A) = \partial \mathbb{D}$, $\sigma(A) = \mathrm{cl}\, \mathbb{D}$, and for $|\lambda| < 1$, $\mathrm{ind}(A - \lambda) = 0$.

One of the most important properties of the index is contained in the next result. Note that Theorem 3.2 is not used in its proof so it can be used in the proof of (3.2).

3.9.   Theorem. *If* $A, B \in \mathscr{F}$, *then* $AB \in \mathscr{F}$ *and* $\mathrm{ind}\, AB = \mathrm{ind}\, A + \mathrm{ind}\, B$.

PROOF. Since $\pi(\mathscr{F})$ is the group of invertible elements of $\mathscr{B}/\mathscr{B}_0$, it is clear that $AB \in \mathscr{F}$ whenever A and $B \in \mathscr{F}$.

Clearly $\ker B \subseteq \ker AB$. Also, if $h \in \ker AB$, then $Bh \in \ker A \cap \mathrm{ran}\, B$. In fact, B maps $\ker AB$ onto $\ker A \cap \mathrm{ran}\, B$. Thus B induces a bijection of $\ker AB/\ker B$ onto $\ker A \cap \mathrm{ran}\, B$ and so

3.10 $\dim \ker AB = \dim \ker B + \dim[\ker A \cap \mathrm{ran}\, B]$.

(Note that because $A, B \in \mathscr{F}$, all of the dimensions that appear in (3.10) are finite integers.)

Since $\ker A$ is finite dimensional, there is a finite-dimensional subspace $\mathscr{M}$ of $\ker A$ such that $\mathscr{M} \cap [\ker A \cap \mathrm{ran}\, B] = (0)$ and $\ker A = \mathscr{M} + \ker A \cap \mathrm{ran}\, B$. Hence

3.11 $\dim \ker A = \dim \mathscr{M} + \dim[\ker A \cap \mathrm{ran}\, B]$.

It must be that $\mathscr{M} \cap \mathrm{ran}\, B = (0)$. In fact, $\mathscr{M} \cap \mathrm{ran}\, B = \mathscr{M} \cap [\ker A \cap \mathrm{ran}\, B] = (0)$ since $\mathscr{M} \le \ker A$. Because $\dim \mathscr{M} < \infty$, $\mathscr{M} + \mathrm{ran}\, B$ is closed (III.4.3). Let $Q =$ the projection of $\mathscr{H}$ onto $(\mathscr{M} + \mathrm{ran}\, B)^\perp = \mathscr{M}^\perp \cap \mathrm{ran}\, B^\perp$ and let $T = Q|\mathrm{ran}\, B^\perp$; so T is surjective. If $h \in \ker T$, then $h \in \mathrm{ran}\, B^\perp$ and $0 = Th = Qh$; thus $h \in \ker Q = (\mathscr{M} + \ker B)^{\perp\perp} = \mathscr{M} + \ker B$. Since $h \in \mathrm{ran}\, B^\perp$, this implies that $\dim \ker T = \dim \mathscr{M}$. Since all of the spaces are finite dimensional, we have

3.12 $\dim \mathrm{ran}\, B^\perp = \dim \mathscr{M} + \dim[\mathscr{M}^\perp \cap \mathrm{ran}\, B^\perp]$.

Now note that

$$A(\mathrm{ran}\, B) \cap A(\mathscr{M}^\perp \cap \mathrm{ran}\, B^\perp) = (0).$$

In fact, if $f \in A(\mathrm{ran}\, B) \cap A(\mathscr{M}^\perp \cap \mathrm{ran}\, B^\perp)$, then $f = Ah = Ag$, where $h \in \mathrm{ran}\, B$ and $g \in \mathscr{M}^\perp \cap \mathrm{ran}\, B^\perp$. Thus $A(h - g) = 0$, so $h - g \in \ker A = \mathscr{M} + \ker A \cap \mathrm{ran}\, B$. Let $h - g = m + k$, where $m \in \mathscr{M}$ and $k \in \ker A$

$\cap$ ran B. Therefore $\langle h - g, g \rangle = \langle m + k, g \rangle = \langle m, g \rangle + \langle k, g \rangle = 0$. Hence $0 = \langle h, g \rangle = \langle g, g \rangle = \|g\|^2$, so $f = 0$.

Now we show that

$$\text{ran } A = A(\text{ran } B) + A(\mathcal{M}^\perp \cap \text{ran } B^\perp).$$

In fact, $\mathcal{H} = (\mathcal{M} + \text{ran } B) \oplus (\mathcal{M} + \text{ran } B)^\perp = (\mathcal{M} + \text{ran } B) \oplus (\mathcal{M}^\perp \cap \text{ran } B^\perp)$. Since $\mathcal{M} \leq \ker A$, we get the desired equality.

Now an argument like that used to obtain (3.12), coupled with the fact that $A(\text{ran } B) = \text{ran } AB$, gives

3.13 $\dim \text{ran } AB^\perp = \dim \text{ran } A^\perp + \dim[\mathcal{M}^\perp \cap \text{ran } B^\perp]$.

We can now put the pieces of the puzzle together. Indeed, first using (3.13) and (3.10), we get

$$
\begin{aligned}
\text{ind } AB &= \dim \ker AB - \dim \text{ran } AB^\perp \\
&= \dim \ker B + \dim[\ker A \cap \text{ran } B] \\
&\quad - \{\dim \text{ran } A^\perp + \dim[\mathcal{M}^\perp \cap \text{ran } B^\perp]\} \\
(3.11) \quad &= \dim \ker B + \{\dim \ker A - \dim \mathcal{M}\} \\
&\quad - \dim \text{ran } A^\perp - \dim[\mathcal{M}^\perp \cap \text{ran } B^\perp] \\
&= \text{ind } A + \dim \ker B - \{\dim \mathcal{M} + \dim[\mathcal{M}^\perp \cap \text{ran } B^\perp]\} \\
(3.12) \quad &= \text{ind } A + \text{ind } B. \quad \blacksquare
\end{aligned}
$$

3.14. Corollary. *If $A \in \mathcal{F}$ and R is an invertible operator, then $RAR^{-1} \in \mathcal{F}$ and* $\text{ind } RAR^{-1} = \text{ind } A$.

We now begin to prove Theorem 3.2. If $A \in \mathcal{B}(\mathcal{H})$, define

$$\gamma(A) \equiv \inf\{\|Ah\| : \|h\| = 1, h \perp \ker A\}.$$

3.15. Proposition. *If $A \in \mathcal{B}(\mathcal{H})$, then $\gamma(A) = \sup\{\gamma > 0 : \|Ah\| \geq \gamma\|h\|$ for all $h \perp \ker A\} = \inf\{\|Ah\|/\|h\| : h \notin \ker A\}$.*

The proof of this proposition is left as an exercise.

3.16. Proposition. *Let $A \in \mathcal{B}(\mathcal{H})$.*

(a) *$\gamma(A) > 0$ if and only if $\text{ran } A$ is closed.*
(b) *$\gamma(A) = \gamma(A^*)$.*

PROOF. The proof of (a) has appeared several times in this book under different guises. The proof here is left to the reader. To see (b), let $h \perp \ker A$. Then $\|A^*Ah\| = \||A|\,|A|h\| = \||A|\,|A|h\|$. But $|A|h \in \text{cl ran } A^*$ (Why?) $= \ker A^\perp$. Hence the definition of $\gamma(A)$ implies that $\|A^*Ah\| \geq \gamma(A)\||A|h\| = \gamma(A)\|Ah\|$; that is, $\|A^*f\| \geq \gamma(A)\|f\|$ for every f in $\text{ran } A$. Since $\text{ran } A$ is dense in $(\ker A^*)^\perp$, $\gamma(A^*) \geq \gamma(A)$. But $A = A^{**}$, so $\gamma(A) \geq \gamma(A^*)$. $\quad \blacksquare$

From here we get the following consequence.

3.17.   Corollary. *If* $A \in \mathscr{B}(\mathscr{H})$, *then* ran A *is closed if and only if* ran A^* *is closed.*

3.18.   Lemma. *If* $\mathscr{M}, \mathscr{N} \le \mathscr{H}$ *and* dim $\mathscr{M} >$ dim $\mathscr{N}$, *then there is a vector* m *in* $\mathscr{M}$ *such that* $\|m\| = $ dist$(m, \mathscr{N})$.

PROOF. Let P be the projection of $\mathscr{H}$ onto $\mathscr{M}$, so dim $P(\mathscr{N}) \le$ dim $\mathscr{N} <$ dim $\mathscr{M}$. Thus $P(\mathscr{N})$ is a proper subspace of $\mathscr{M}$; let $m \in \mathscr{M} \cap P(\mathscr{N})^{\perp}$. If $n \in \mathscr{N}$, then $0 = \langle Pn, m \rangle = \langle n, Pm \rangle = \langle n, m \rangle$, so $m \perp \mathscr{N}$. Hence $\|m\| = $ dist$(m, \mathscr{N})$. ■

3.19.   Lemma. *If* $h \in \mathscr{H}$, *then* $\gamma(A)$dist$(h, \ker A) \le \|Ah\|$.

PROOF. Let P be the projection of $\mathscr{H}$ onto $\ker A^{\perp}$; then $\|Ph\| = $ dist$(h, \ker A)$. Hence $\|Ah\| = \|APh\| \ge \gamma(A)\|Ph\| = \gamma(A)$dist$(h, \ker A)$. ■

The next result has some interest by itself as well as being a major stepping stone to the proof of Theorem 3.2. If the role of $\gamma(A)$ in the next and subsequent propositions impresses the reader as somewhat mysterious, reflect that if A is invertible, then $\gamma(A) = \|A^{-1}\|^{-1}$ (Exercise 7). Now in Corollary VII.2.3, it was shown that if $\mathscr{A}$ is a Banach algebra, $a_0 \in \mathscr{A}$, and $b_0 a_0 = 1$, then $a + b$ is left invertible whenever $\|b\| < \|b_0\|^{-1}$. Of course, a similar result holds for right-invertible elements. The number $\gamma(A)$ is trying to play the role of the reciprocal of the norm of a one-sided inverse.

For example, if A is left invertible, then ran A is closed and $\ker A = (0)$; hence $A \in \mathscr{SF}$. The next result implies that if $\|B\| < \gamma(A)$, then $A + B$ is left invertible.

3.20.   Proposition. *If* $A \in \mathscr{SF}$ *and* $B \in \mathscr{B}(\mathscr{H})$ *such that* $\|B\| < \gamma(A)$, *then* $A + B \in \mathscr{SF}$ *and*:

(a) dim $\ker(A + B) \le$ dim $\ker A$;

(b) dim ran$(A + B)^{\perp} \le$ dim ran $A^{\perp}$.

PROOF. First note that because $A \in \mathscr{SF}$, $\gamma(A) > 0$.

If $h \in \ker(A + B)$ and $h \ne 0$, then $Ah = -Bh$. By Lemma 3.19, $\gamma(A)$dist$(h, \ker A) \le \|Bh\| \le \|B\|\|h\| < \gamma(A)\|h\|$. Thus dist$(h, \ker A) < \|h\|$ for every nonzero vector h in $\ker(A + B)$. By Lemma 3.18, (a) holds.

Since $\|B\| = \|B^*\|$ and $\gamma(A) = \gamma(A^*)$, (a) implies that dim $\ker(A^* + B^*) \le$ dim $\ker A^*$. But this inequality is equivalent to (b).

It remains to prove that ran$(A + B)$ is closed. Since $A \in \mathscr{SF}$, either dim $\ker A < \infty$ or dim $\ker A^* < \infty$. Suppose dim $\ker A < \infty$. It will be shown that $A + B \in \mathscr{F}_l$ by using Theorem 2.5(e) and showing that if $\delta = \gamma(A) - \|B\|$, then $\{h: \|(A + B)h\| < \delta\|h\|\}$ contains no infinite-dimensional manifold. Indeed, if it did, it would contain a finite-dimensional subspace $\mathcal{M}$ with $\dim \mathcal{M} > \dim \ker A$. By Lemma 3.18 there is a vector h in $\mathcal{M}$ with $\|h\| = \operatorname{dist}(h, \ker A)$. Now $\|(A + B)h\| < \delta\|h\|$, so Lemma 3.19 implies that $\gamma(A)\|h\| = \gamma(A)\operatorname{dist}(h, \ker A) \leq \|Ah\| \leq \|(A + B)h\| + \|Bh\| < (\delta + \|B\|)\|h\| = \gamma(A)\|h\|$, a contradiction. Thus $A + B \in \mathcal{F}_l$ and so $\operatorname{ran}(A + B)$ is closed.

If $\dim \ker A = \infty$, then $\dim \ker A^* < \infty$. The argument of the preceding paragraph gives that $\operatorname{ran}(A^* + B^*)$ is closed. By Corollary 3.17, $\operatorname{ran}(A + B)$ is closed. ∎

3.21. Proposition. *If $A \in \mathcal{SF}$ and $\operatorname{ind} A \leq 0$, then there is a finite-rank operator F such that $\ker(A + F) = (0)$ and $\operatorname{ind}(A + F) = \operatorname{ind} A$.*

PROOF. Since $0 \geq \operatorname{ind} A = \dim \ker A - \dim \operatorname{ran} A^{\perp}$, $\dim \ker A < \infty$ and $\dim \ker A \leq \dim \operatorname{ran} A^{\perp}$. Let $\{e_1, \ldots, e_n\}$ be an orthonormal basis for $\ker A$ and let $\{f_1, \ldots, f_n\}$ be orthonormal vectors in $\operatorname{ran} A^{\perp}$. Define F: $\mathcal{H} \to \mathcal{H}$ by $Fh = \sum_{j=1}^n \langle h, e_j \rangle f_j$. Thus F is a finite-rank partial isometry with initial $F = \ker A$ and final $F \leq \operatorname{ran} A^{\perp}$.

If $h \in \ker(A + F)$, then $Ah = -Fh$, hence $Ah \in \operatorname{ran} A \cap \operatorname{ran} A^{\perp}$. So $0 = Ah = Fh$; that is, $h \in \ker A = \operatorname{initial} F$. So $\|h\| = \|Fh\| = 0$, and, therefore, $\ker(A + F) = 0$. Also, since $\operatorname{ran} F \leq \operatorname{ran} A^{\perp}$, and initial $F = \ker A$, $\operatorname{ran}(A + F) = \operatorname{ran} A \oplus \operatorname{ran} F$. Thus $\operatorname{ind}(A + F) = -\dim \operatorname{ran}(A + F)^{\perp} = -\dim \mathcal{H} \ominus [\operatorname{ran} A \oplus \operatorname{ran} F] = -\dim \operatorname{ran} A^{\perp} + \dim \operatorname{ran} F = \operatorname{ind} A$. ∎

3.22. Corollary. *If A is invertible and $K \in \mathcal{B}_0$, then $\operatorname{ind}(A + K) = 0$.*

PROOF. By considering $A^* + K^*$ if necessary, it suffices to assume that $\operatorname{ind}(A + K) \leq 0$. By the preceding proposition, there is a finite-rank operator F such that $\ker(A + K + F) = (0)$ and $\operatorname{ind}(A + K) = \operatorname{ind}(A + K + F)$. Let $L = K + F$. Since $A + L = A(1 + A^{-1}L)$, $\ker(1 + A^{-1}L) = (0)$; thus $-1 \notin \sigma_p(A^{-1}L)$. But $A^{-1}L \in \mathcal{B}_0$, so $1 + A^{-1}L$ is invertible. By Theorem 3.9, $\operatorname{ind}(A + K) = \operatorname{ind}(A + L) = \operatorname{ind} A(1 + A^{-1}L) = \operatorname{ind} A + \operatorname{ind}(1 + A^{-1}L) = 0$. ∎

3.23. Corollary. *If $A \in \mathcal{F}$, then the following statements are equivalent.*

(a) $\operatorname{ind} A = 0$.
(b) *There is a compact operator K such that $A + K$ is invertible.*
(c) *There is a finite-rank operator F such that $A + F$ is invertible.*

PROOF. (a) *implies* (c). By Proposition 3.21 there is a finite-rank operator F such that $\ker(A + F) = (0)$ and $\operatorname{ind}(A + F) = \operatorname{ind} A = 0$. Hence $\dim \operatorname{ran}(A + F)^{\perp} = \dim \ker(A + F) = 0$ and $A + F$ is invertible.

(c) *implies* (b). Clear.

(b) *implies* (a). Apply Corollary 3.22 to $A + K$. ∎

3.24. Proposition. *If* $A \in \mathscr{SF}$, $\ker A = (0)$, *and* $B \in \mathscr{B}(\mathscr{H})$ *with* $\|B\| < \gamma(A)$, *then* $\ker(A + B) = (0)$ *and* $\operatorname{ind}(A + B) = \operatorname{ind} A$.

PROOF. By Proposition 3.20, $A + B \in \mathscr{SF}$, $\ker(A + B) = (0)$, and $\dim \operatorname{ran}(A + B)^{\perp} \leq \dim \operatorname{ran} A^{\perp}$. It remains to show that $\dim \operatorname{ran}(A + B)^{\perp} \geq \dim \operatorname{ran} A^{\perp}$.

Let $n \geq 1$ such that $\gamma(A) - \|B\| > \|B\| n^{-1}$. For $0 \leq k \leq n - 1$, $\|(1 - k/n)B\| < \gamma(A)$. So Proposition 3.20 implies that $A + (1 - k/n)B \in \mathscr{SF}$ and is injective. So if $h \in \mathscr{H}$, $\|h\| = 1$, then

$$\left\| \left[A + \left(1 - \frac{k}{n} \right) B \right] h \right\| \geq \|Ah\| - \left(1 - \frac{k}{n} \right) \|Bh\|$$

$$\geq \gamma(A) - \|B\| > 0.$$

Thus $\gamma(A + (1 - k/n)B) \geq \gamma(A) - \|B\| > \| - (1/n)B\|$. Again, applying (3.20) to $A + (1 - k/n)B$ and $-(1/n)B$, we have that

$$A + \left(1 - \frac{k}{n} \right) B - \frac{1}{n} B = A + \left(1 - \frac{k + 1}{n} \right) B \in \mathscr{SF}$$

and

$$\dim \operatorname{ran} \left(A + \left(1 - \frac{k + 1}{n} \right) B \right)^{\perp} \leq \dim \operatorname{ran} \left(A + \left(1 - \frac{k}{n} \right) B \right)^{\perp}$$

for $0 \leq k \leq n - 1$. Looking at these n inequalities and noticing that the left-hand side for $k = n - 1$ is $\dim \operatorname{ran} A^{\perp}$ and that the right side for $k = 0$ is $\dim \operatorname{ran}(A + B)^{\perp}$, we get that $\dim \operatorname{ran} A^{\perp} \leq \dim \operatorname{ran}(A + B)^{\perp}$. ∎

3.25. Lemma. *If* $A \in \mathscr{SF}$ *and* F *is a finite-rank operator, then* $\operatorname{ind}(A + F) = \operatorname{ind} A$.

PROOF. If $\operatorname{ind} A = \pm \infty$, then either $\ker A$ or $\operatorname{ran} A^{\perp}$ is infinite dimensional. Because F has finite rank, the same is true of $A + F$. Thus $\operatorname{ind} A = \operatorname{ind}(A + F)$. Therefore it may be assumed that $\operatorname{ind} A$ is finite; that is, it may be assumed that A is a Fredholm operator. The proof is by cases.

Case 1: $\ker F^{\perp} \subseteq \ker A$. Hence $\ker A^{\perp} \subseteq \ker F = \mathscr{N}$. So $\operatorname{ran} A = A(\ker A^{\perp}) \subseteq A\mathscr{N} = (A + F)\mathscr{N} \subseteq \operatorname{ran}(A + F)$. This implies that $\operatorname{ran} A^{\perp} \supseteq \operatorname{ran}(A + F)^{\perp}$ and therefore

3.26 $\quad \dim \operatorname{ran} A^{\perp} = \dim \operatorname{ran}(A + F)^{\perp} + \dim[\operatorname{ran}(A + F) \ominus \operatorname{ran} A]$.

Also, $\operatorname{ran} A + \operatorname{ran} F = A(\ker A^{\perp}) + F(\ker F^{\perp}) = \operatorname{ran}(A + F)$ since $\ker A \supseteq \ker F^{\perp}$. Since $\operatorname{ran} A \subseteq \operatorname{ran}(A + F)$, $\operatorname{ran}(A + F) \ominus \operatorname{ran} A$ and $\operatorname{ran}(A + F)/\operatorname{ran} A$ are isomorphic as vector spaces. Also, the natural map of $\operatorname{ran}(A + F)$ onto $\operatorname{ran}(A + F)/\operatorname{ran} A$ when restricted to $\operatorname{ran} F$ remains surjective. Thus $\operatorname{ran} F/\operatorname{ran} F \cap \operatorname{ran} A$ and $\operatorname{ran}(A + F)/\operatorname{ran} A$ are isomorphic as vector spaces. Combining these isomorphisms gives

$$\dim[\operatorname{ran}(A + F) \ominus \operatorname{ran} A] = \dim \operatorname{ran} F - \dim[\operatorname{ran} F \cap \operatorname{ran} A].$$

If we combine this with (3.26), we obtain

3.27 $\dim \operatorname{ran} A^{\perp} = \dim \operatorname{ran}(A + F)^{\perp}$

$$+ \dim \operatorname{ran} F - \dim[\operatorname{ran} F \cap \operatorname{ran} A].$$

Since we want to show that $\operatorname{ind} A = \operatorname{ind}(A + F)$, formula (3.27) demonstrates how $\dim \operatorname{ran} A^{\perp}$ and $\dim \operatorname{ran}(A + F)^{\perp}$ differ. Now we must see how $\dim \ker A$ and $\dim \ker(A + F)$ differ.

Note that $A^{-1}(\operatorname{ran} F) = \ker A \oplus [A^{-1}\operatorname{ran} F \cap \ker A^{\perp}]$. Hence

3.28 $\dim \ker A = \dim A^{-1}(\operatorname{ran} F) - \dim[A^{-1}\operatorname{ran} F \cap \ker A^{\perp}].$

But A is injective on $\ker A^{\perp}$ and $A[A^{-1}\operatorname{ran} F \cap \ker A^{\perp}] = \operatorname{ran} F \cap \operatorname{ran} A$. Thus

3.29 $\dim[A^{-1}\operatorname{ran} F \cap \ker A^{\perp}] = \dim[\operatorname{ran} F \cap \operatorname{ran} A].$

Also, $(A + F)[A^{-1}\operatorname{ran} F] = \operatorname{ran} F$. If $h \in \ker(A + F)$, then $Ah = -Fh$, so that $h \in [A^{-1}\operatorname{ran} F]$; that is, $\ker(A + F) \subseteq [A^{-1}\operatorname{ran} F]$. Hence $\ker((A + F)|[A^{-1}\operatorname{ran} F]) = \ker(A + F)$ and so

$$\dim[A^{-1}\operatorname{ran} F] = \dim \ker(A + F) + \dim \operatorname{ran} F.$$

If we combine this formula and (3.29) with formula (3.28), we obtain

3.30 $\dim \ker A = \dim \ker(A + F)$

$$+ \dim \operatorname{ran} F - \dim[\operatorname{ran} F \cap \operatorname{ran} A].$$

Combining (3.27) and (3.30), it is clear that $\operatorname{ind} A = \operatorname{ind}(A + F)$.

Case 2: $\operatorname{ran} F \subseteq \operatorname{ran} A^{\perp}$. Hence $\ker F^{*\perp} = \operatorname{ran} F \subseteq \ker A^{*}$. So Case 1 implies that $\operatorname{ind} A = -\operatorname{ind} A^{*} = -\operatorname{ind}(A^{*} + F^{*}) = \operatorname{ind}(A + F)$.

Case 3: $\ker F^{\perp} \subseteq \ker A^{\perp}$ and $\operatorname{ran} F \subseteq \operatorname{ran} A$. Let A_1 and F_1 be the operators defined from $\ker A^{\perp}$ into $\operatorname{ran} A$ by letting them be the restrictions of A and F to $\ker A^{\perp}$. We want to apply Corollary 3.22 to A_1 and F_1. In fact, A_1: $\ker A^{\perp} \to \operatorname{ran} A$ is invertible, but there is a bit of a difficulty here since A_1 does not map a Hilbert space into itself. But this can be overcome since $\ker A^{\perp}$ and $\operatorname{ran} A$ are isomorphic Hilbert spaces. (Why?) The details are left to the reader. By Corollary 3.22, $\operatorname{ind}(A_1 + F_1) = 0$. We now want to relate these dimensions to the corresponding dimensions for A and $A + F$.

Since $\ker A \subseteq \ker F$, $\ker A \subseteq \ker(A + F)$. Thus $\ker(A_1 + F_1) = \ker(A + F) \ominus \ker A$. Therefore,

3.31 $\dim \ker A = \dim \ker(A + F) - \dim \ker(A_1 + F_1).$

Also, since $\ker(A + F)^{\perp} \subseteq \ker A^{\perp}$, $\operatorname{ran}(A_1 + F_1) = (A_1 + F_1)\ker A^{\perp} = \operatorname{ran}(A + F)$. Hence $\operatorname{ran}(A_1 + F_1)^{\perp} = \operatorname{ran} A \ominus \operatorname{ran}(A + F)$. So $\dim \operatorname{ran}(A_1 + F_1)^{\perp} = \dim \operatorname{ran} A - \dim \operatorname{ran}(A + F) = \dim \operatorname{ran}(A + F)^{\perp}$ $\dim(\operatorname{ran} A)^{\perp}$. (Why?) Therefore

$$\dim(\operatorname{ran} A)^{\perp} = \dim \operatorname{ran}(A + F)^{\perp} - \dim \operatorname{ran}(A_1 + F_1)^{\perp}.$$

Combining this equation with (3.31) gives that $\operatorname{ind} A = \operatorname{ind}(A + F) - \operatorname{ind}(A_1 + F_1) = \operatorname{ind}(A + F)$.

Case 4: The general case. Let $A \in \mathcal{F}$ and let F be a finite-rank operator. Let P be the projection of $\mathcal{H}$ onto $\ker A^{\perp}$ and let Q be the projection of $\mathcal{H}$ onto ran A. So QFP is a finite-rank operator and $\ker QFP \supseteq \ker A$. Hence $(\ker QFP)^{\perp} \subseteq \ker A^{\perp}$ and clearly ran $QFP \subseteq$ ran A. By Case 3,

$$\operatorname{ind} A = \operatorname{ind}(A + QFP).$$

Also, $(1 - Q)FP$ is a finite-rank operator and $\operatorname{ran}(1 - Q)FP \subseteq \operatorname{ran} A^{\perp} \subseteq [\operatorname{ran}(A + QFP)]^{\perp}$. So Case 2 implies

$$\operatorname{ind}(A + QFP) = \operatorname{ind}(A + QFP + (1 - Q)FP)$$
$$= \operatorname{ind}(A + FP).$$

But $F(1 - P)$ has finite rank and $[\ker F(1 - P)]^{\perp} \subseteq \ker A \subseteq \ker(A + FP)$. So Case 1 implies that

$$\operatorname{ind}(A + FP) = \operatorname{ind}(A + FP + F(1 - P))$$
$$= \operatorname{ind}(A + F). \quad \blacksquare$$

PROOF OF THEOREM 3.2. The continuity of the index is the first order of business. Let $A \in \mathcal{SF}$ and assume that ind $A \leq 0$. It must be shown that there is a $\delta > 0$ such that if $C \in \mathcal{SF}$ and $\|A - C\| < \delta$, then ind $A = \operatorname{ind} C$.

By Proposition 3.21 there is a finite-rank operator F such that $\ker(A + F) = 0$ and ind $A = \operatorname{ind}(A + F)$. Let $\delta = \gamma(A + F)$. By Proposition 3.24, if $\|C - A\| < \delta$, then $\operatorname{ind}(A + F) = \operatorname{ind}(C + F)$. But Lemma 3.25 implies that ind $C = \operatorname{ind}(C + F)$; thus ind $A = \operatorname{ind} C$ if $\|A - C\| < \delta$. If ind $A \geq 0$, then the preceding argument shows that the index is continuous at A^{*}. It follows that it is continuous at A.

If K is a compact operator, let $\{F_n\}$ be a sequence of finite-rank operators such that $\|F_n - K\| \rightarrow 0$. By the first part of the proof, $\operatorname{ind}(A + F_n) \rightarrow \operatorname{ind}(A + K)$. But $\operatorname{ind}(A + F_n) = \operatorname{ind} A$ by Lemma 3.25. Hence ind $A = \operatorname{ind}(A + K)$. $\quad \blacksquare$

For a more detailed treatment of the index applicable to unbounded operators on a Banach space, see pp. 229–244 of Kato [1966].

EXERCISES

1. Prove Proposition 3.7.

2. Verify the statements made in Example 3.8.

3. If S is the unilateral shift, show that for every $\varepsilon > 0$ there is a rank-one operator F with $\|F\| < \varepsilon$ such that $\sigma(S^{*} \oplus S + F) = \partial \mathbb{D}$.

4. Let G be an open connected subset of $\sigma(A) \setminus \sigma_{le}(A) \cup \sigma_{re}(A)$ and suppose $\lambda_0 \in G$ such that $\operatorname{ind}(A - \lambda_0) = 0$. Show that there is a finite-rank operator F such that $A + F - \lambda_0$ is invertible. Show that $A + F - \lambda$ is invertible for every λ in G.

5. If $A \in \mathscr{B}(\mathscr{H})$ and ran A is closed, show that ran $A^{(\infty)}$ is closed. If $A \in \mathscr{S}\mathscr{F}$ and ker $A = (0)$, show that $A^{(\infty)} \in \mathscr{S}\mathscr{F}$ and ind $A^{(\infty)} = -\infty$ or 0.

6. Prove Proposition 3.15.

7. If A is invertible, show that $\gamma(A) = \|A^{-1}\|^{-1}$.

8. Let $A \in \mathscr{F}$ and suppose f is analytic in a neighborhood of $\sigma(A)$ and does not vanish on $\sigma_e(A)$. Show that $f(A) \in \mathscr{F}$ and find ind $f(A)$.

9. Let S be the unilateral shift and let f be an analytic function in a neighborhood of cl $\mathbb{D}$ such that $f(z) \neq 0$ if $|z| = 1$. Let $\gamma(t) = f(\exp(2\pi it))$, $0 \le t \le 1$. Show that $\sigma_e(f(S)) = f(\partial\mathbb{D}) = \{\gamma(t): 0 \le t \le 1\}$ and that if $\lambda \notin f(\partial\mathbb{D})$, ind$(f(S) - \lambda) = -n(\gamma; \lambda)$, where $n(\gamma; \lambda) = $ the winding number of γ about λ. Moreover, show that if ind$(f(S) - \lambda) = 0$, then $\lambda \notin \sigma(f(S))$.

10. Let S be the operator defined in Example 2.11 where $G = \mathbb{D}$. Show that there is a compact operator K such that $S + K$ is unitarily equivalent to the unilateral shift.

11. Does the unilateral shift have a square root?

12. Show that for every n in $\mathbb{Z} \cup \{\pm\infty\}$ there is an operator A in $\mathscr{S}\mathscr{F}$ such that ind $A = n$.

13. If $A \in \mathscr{S}\mathscr{F}$, then for every $n \ge 1$, $A^n \in \mathscr{S}\mathscr{F}$ and ind $A^n = n(\text{ind } A)$.

§4. The Components of $\mathscr{S}\mathscr{F}$

Since the index is continuous on $\mathscr{S}\mathscr{F}$ and assumes every possible value (Exercise 3.12), $\mathscr{S}\mathscr{F}$ cannot be connected. What are its components?

Note that because $\mathscr{S}\mathscr{F}$ is an open subset of a Banach space, its components are arcwise connected (Exercise IV.1.24).

4.1. Theorem. *If $A, B \in \mathscr{S}\mathscr{F}$, then A and B belong to the same component of $\mathscr{S}\mathscr{F}$ if and only if* ind $A = $ ind B.

Half of this theorem is easy. For the other half we first prove a lemma.

4.2. Lemma. *If $A \in \mathscr{S}\mathscr{F}$ and* ind $A = 0$, *then there is a path $\gamma: [0, 1] \to \mathscr{S}\mathscr{F}$ such that $\gamma(0) = 1$ and $\gamma(1) = A$.*

PROOF. By Corollary 3.23 there is a finite-rank operator F such that $A + F$ is invertible. If $\gamma(t) = A + tF$, $\gamma(0) = A$, $\gamma(1) = A + F$, and $\gamma(t) \in \mathscr{S}\mathscr{F}$ for all t. Thus we may assume that A is invertible.

Let $A = U|A|$ be the polar decomposition of A. Because A is invertible, U is a unitary operator and $|A|$ is invertible. Using the Spectral Theorem, $U = \exp(iB)$ when B is hermitian. Also, since $0 \notin \sigma(|A|)$, $|A| = \int_{[\delta, r]} x \, dE(x)$, where $0 < \delta < r = \|A\|$. Define $\gamma: [0, 1] \to \mathscr{B}(\mathscr{H})$ by

$$\gamma(t) = e^{itB} \int_{[\delta, r]} x^t \, dE(x) = e^{itB}|A|^t.$$

It is easy to check that γ is continuous, $\gamma(0) = 1$, and $\gamma(1) = A$. Also, each $\gamma(t)$ is invertible so $\gamma(t) \in \mathscr{SF}$. ∎

PROOF OF THEOREM 4.1. First assume that $A, B \in \mathscr{F}$ and ind $A =$ ind B. So there is an operator C such that $CB = 1 + K$ for some compact operator K. Thus $C \in \mathscr{F}_r$ and ind $C = -$ind $B = -$ind A. Hence $AC \in \mathscr{F}$ and ind $AC = 0$. By the preceding lemma there is a path $\gamma: [0,1] \to \mathscr{SF}$ such that $\gamma(0) = 1$ and $\gamma(1) = AC$. Put $\rho(t) = \gamma(t)B - tAK$. Because $AK \in \mathscr{B}_0$, $\rho(t) \in \mathscr{F}$ for all t in $[0,1]$. Also, $\rho(0) = B$ and $\rho(1) = ACB - AK = A(1 + K) - AK = A$.

Now assume that ind $A = -\infty$; so $\dim(\operatorname{ran} A)^{\perp} = \infty$ and $\dim \ker A < \infty$. Let F be a finite-rank operator such that $\ker(A + tF) = 0$ for $t \neq 0$. (Why does F exist?) This path shows that we may assume that $\ker A = (0)$. Let V be any isometry such that $\dim(\operatorname{ran} V)^{\perp} = \infty$ and consider the polar decomposition $A = U|A|$ of A. Since $A \in \mathscr{SF}$ and $\ker A = (0)$, $|A|$ is invertible and U is an isometry. Also, $\operatorname{ran} U = \operatorname{ran} A$, so (Exercise 4) there is a unitary operator W such that $WUW^* = V$. Let $\gamma: [0,1] \to \mathscr{F}$ such that $\gamma(0) = |A|$ and $\gamma(1) = 1$ and let $\rho: [0,1] \to \mathscr{F}$ such that $\rho(0) = 1$ and $\rho(1) = W$. Then $\sigma(t) = \rho(t)U\gamma(t)\rho(t)^*$ defines a path $\sigma: [0,1] \to \mathscr{SF}$ (Why?) such that $\sigma(0) = A$ and $\sigma(1) = V$. Similarly, if ind $B = -\infty$, there is a path connecting B to V; so A and B belong to the same component of $\mathscr{SF}$.

If ind $A =$ ind $B = +\infty$, apply the preceding paragraph to A^* and B^*. ∎

4.3. Corollary. *The component of the identity in $\mathscr{F}$, $\mathscr{F}_0$, is a normal subgroup of $\mathscr{F}$ and $\mathscr{F}/\mathscr{F}_0$ is an infinite cyclic group.*

PROOF. By Theorem 3.9, ind $\mathscr{F} \to \mathbb{Z}$ is a group homomorphism and it is surjective (Exercise 3.12). By Theorem 4.1, $\ker(\text{ind}) = \mathscr{F}_0$. ∎

EXERCISES

1. Let G be any topological group and let G_0 be the component of the identity. Show that G_0 is a normal subgroup of G.

2. What are the components of the set of invertible elements in $C(\partial \mathbb{D})$?

3. If $S =$ the unilateral shift, what are the components of the set of invertible elements of $C^*(S)$?

4. If V and U are isometries and $\dim(\operatorname{ran} V)^{\perp} = \dim(\operatorname{ran} U)^{\perp}$, then there is a unitary W such that $WUW^* = V$.

5. Find the components of the set of partial isometries. Find the unitary equivalence classes of the set of partial isometries.

§5. A Finer Analysis of the Spectrum

In this section we will examine the spectrum and the index more closely. For example, if $A \in \mathscr{SF}$, then Theorem 3.2 implies that there is a $\delta > 0$ such that if $\|B - A\| < \delta$, then $B \in \mathscr{SF}$ and ind B = ind A. How does dim ker B differ from dim ker A? In general a lot cannot be said; the next result is the best that can be said.

5.1. Proposition. *If $A \in \mathscr{SF}$ and either ker $A = (0)$ or ran $A = \mathscr{H}$, then there is a $\delta > 0$ such that if $\|B - A\| < \delta$, then dim ker B = dim ker A and dim ran B = dim ran A.*

PROOF. By Proposition 3.20 and Theorem 3.2 there is a $\delta > 0$ such that if $\|B - A\| < \delta$, then ind A = ind B, dim ker $B \leq$ dim ker A, and dim ran $B^{\perp}$ $\leq$ dim ran $A^{\perp}$. Since one of these dimensions for A is 0, the proposition is proved. ∎

If both ker A and ran A are nonzero, then there are semi-Fredholm operators B that are arbitrarily close to A such that dim ker $B <$ dim ker A (see Exercise 1). In fact, just about anything that can go wrong here does go wrong. However, dim ker$(A - \lambda)$ does behave rather nicely as a function of λ.

5.2. Theorem. *If $\lambda \notin \sigma_{le}(A) \cap \sigma_{re}(A)$, then there is a $\delta > 0$ such that dim ker$(A - \mu)$ and dim ran$(A - \mu)^{\perp}$ are constant for $0 < |\mu - \lambda| < \delta$.*

PROOF. We may assume that $\lambda = 0$, so $A \in \mathscr{SF}$. It follows that $A^n \in \mathscr{SF}$ for every $n \geq 1$ (Exercise 3.13). Hence ran $A^n = \mathscr{M}_n$ is closed. Let $\mathscr{M} = \bigcap_{n=1}^{\infty} \mathscr{M}_n$. Note that $\mathscr{M}_{n+1} \subseteq \mathscr{M}_n$ and $A\mathscr{M}_n = \mathscr{M}_{n+1}$; hence $A\mathscr{M} \subseteq \mathscr{M}$. Let $B = A|\mathscr{M}$.

Claim. $B\mathscr{M} = \mathscr{M}$.

If $h \in \mathscr{M}$, then $h \in$ ran A and there is a unique vector f in (ker $A)^{\perp}$ such that $Af = h$. Now $h \in \mathscr{M}_{n+1} = A\mathscr{M}_n = A(\mathscr{M}_n \ominus$ ker $A)$, so there is a vector f_n in $\mathscr{M}_n \ominus$ ker A such that $Af_n = h$. But the uniqueness of f implies that $f = f_n \in \mathscr{M}_n$ for every n. Hence $f \in \mathscr{M}$ and $h = Af = Bf \in B\mathscr{M}$.

Thus $B \in \mathscr{SF}$ and ind B = dim ker B. By (3.2) and (3.20) there is a $\delta > 0$ such that if $|\mu| < \delta$, then dim ker$(B - \mu) \leq$ dim ker B, dim ran$(B - \mu)^{\perp} = 0$, and ind$(B - \mu)$ = ind B. Thus dim ker$(B - \mu)$ = dim ker B for $|\mu| < \delta$. Also, choose δ such that ind$(A - \mu)$ = ind A for $|\mu| < \delta$.

On the other hand, if $\mu \neq 0$, then ker$(A - \mu) \subseteq \mathscr{M}$. In fact, if $h \in$ ker$(A - \mu)$, then $A^n h = \mu^n h$, so that $h = A^n(\mu^{-n}h) \in \mathscr{M}_n$ for every n. Thus for $0 < |\mu| < \delta$, dim ker$(A - \mu)$ = dim ker$(B - \mu)$ = dim ker B; that is, dim ker$(A - \mu)$ is constant for $0 < |\mu| < \delta$. Since ind$(A - \mu)$ is constant for these values of μ, dim ran$(A - \mu)^{\perp}$ is also constant. ∎

The next result is from Putnam [1968].

5.3. Theorem. *If* $\lambda \in \partial\sigma(A)$, *then either* λ *is an isolated point of* $\sigma(A)$ *or* $\lambda \in \sigma_{le}(A) \cap \sigma_{re}(A)$.

PROOF. Suppose $\lambda \in \sigma(A)$ but λ is neither an isolated point of $\sigma(A)$ nor a point of $\sigma_{le}(A) \cap \sigma_{re}(A)$. It must be shown that $\lambda \in \text{int } \sigma(A)$. In fact, since $\lambda \notin \sigma_{le}(A) \cap \sigma_{re}(A)$, $A - \lambda \in \mathcal{SF}$. Let $\delta > 0$ such that $A - \mu \in \mathcal{SF}$, $\dim \ker(A - \mu) = \dim \ker(A - \lambda)$, and $\dim \text{ran}(A - \mu)^{\perp} = \dim \text{ran}(A - \lambda)^{\perp}$ for $0 < |\mu - \lambda| < \delta$. Since $\lambda \in \sigma(A)$, at least one of $\ker(A - \lambda)$ and $\text{ran}(A - \mu)^{\perp}$ differs from (0). Hence $\mu \in \sigma(A)$ for $|\mu - \lambda| < \delta$. ∎

What happens if λ is an isolated point of $\sigma(A)$?

5.4. Proposition. *If* λ *is an isolated point of* $\sigma(A)$, *the following statements are equivalent.*

(a) $\lambda \notin \sigma_{le}(A) \cap \sigma_{re}(A)$.
(b) λ *is a pole of the function* $z \mapsto (z - A)^{-1}$.
(c) *The Riesz idempotent* $E(\lambda)$ *has finite rank.*
(d) $A - \lambda \in \mathcal{F}$ *and* $\text{ind}(A - \lambda) = 0$.

PROOF. Exercise 3.

If $n \in \mathbb{Z} \cup \{\pm\infty\}$ and $A \in \mathcal{B}(\mathcal{H})$, define
$$P_n(A) \equiv \{\lambda \in \sigma(A): A - \lambda \in \mathcal{SF} \text{ and } \text{ind}(A - \lambda) = n\}.$$
So for $n \neq 0$, $P_n(A)$ is an open subset of the plane; the set $P_0(A)$ consist of an open set together with some isolated points of $\sigma(A)$. In fact, Proposition 5.4 can be used to show that $P_0(A)$ contains precisely the isolated points of $\sigma(A)$ for which the Riesz idempotent has finite rank. The proof of the next result is easy.

5.5. Proposition. *If* $A \in \mathcal{B}(\mathcal{H})$, *then* $\sigma_e(A) = [\sigma_{le}(A) \cap \sigma_{re}(A)] \cup P_{+\infty}(A) \cup P_{-\infty}(A)$.

5.6. Definition. If $A \in \mathcal{B}(\mathcal{H})$, then the *Weyl spectrum* of A, $\sigma_w(A)$, is defined by
$$\sigma_w(A) = \cap\{\sigma(A + K): K \in \mathcal{B}_0\}.$$

Note that since $\sigma_e(A + K) = \sigma_e(A)$ for every compact operator, $\sigma_w(A)$ is nonempty and $\sigma_e(A) \subseteq \sigma_w(A)$. The way to think of the Weyl spectrum is that it is the largest part of the spectrum of A that remains unchanged under compact perturbations. It is clear that $\sigma_w(A) = \sigma_w(A + K)$ for every K in $\mathcal{B}_0$, but it is not so clear that $\sigma_w(A) \subseteq \sigma(A)$. The following result of Schechter [1965] gives this and some more.

5.7. Theorem. *If* $A \in \mathcal{B}(\mathcal{H})$, *then* $\sigma_w(A) = \sigma_e(A) \cup \bigcup_{n \neq 0} P_n(A)$.

PROOF. Clearly $X \equiv \sigma_e(A) \cup \bigcup_{n \neq 0} P_n(A) \subseteq \sigma_w(A)$. Now suppose $\lambda \notin X$. Then $A - \lambda \in \mathcal{F}$ and $\text{ind}(A - \lambda) = 0$. By Corollary 3.23 there is a finite-rank operator F such that $A + F - \lambda$ is invertible. Hence $\lambda \notin \sigma(A + F)$ so that $\lambda \notin \sigma_w(A)$. ∎

So for every operator A in $\mathcal{B}(\mathcal{H})$ there is a *spectral picture* for A (a term coined in Pearcy [1978]). There are the open sets $\{P_n(A): 0 < |n| \leq \infty\}$, the set $P_0(A) = G_0 \cup D$ where D consists of isolated points λ for which $\dim E(\lambda) = n_\lambda < \infty$, and there is the remainder of $\sigma(A)$, which is the set $\sigma_{le}(A) \cap \sigma_{re}(A)$. The next result is due to Conway [1977].

5.8. Proposition. *Let K be a compact subset of* $\mathbb{C}$, *let* $\{G_n: -\infty \leq n \leq \infty\}$ *be open subsets of K (some possibly empty), let D be a subset of the set of isolated points of K, and for each λ in D let $n_\lambda \in \{1, 2, \ldots\}$. Then there is an operator A on $\mathcal{H}$ such that $\sigma(A) = K$, $P_n(A) = G_n$ for $0 < |n| \leq \infty$, $P_0(A) = G_0 \cup D$, and $\dim E(\lambda) = n_\lambda$ for every λ in D.*

We prove only a special case of this result; the general case is left to the reader. Let K be any compact subset of $\mathbb{C}$ and let G be an open subset of K. Put $H = \text{int}[\text{cl } G]$; so $G \subseteq H$, but it may be that $H \neq G$. However, $\partial H = \partial[\text{cl } H]$. Let $Tf = zf$ for f in $L_a^2(H)$, so $H = P_{-1}(T)$, $\sigma(T) = \text{cl } H$, and $\sigma_{le}(T) \cap \sigma_{re}(T) = \partial H = \partial[\text{cl } G]$. Let $\{\lambda_k\}$ be a countable dense subset of $K \setminus G$ and let N be the diagonalizable normal operator with $\sigma_p(N) = \{\lambda_k\}$ and such that $\dim \ker(N - \lambda_k) = \infty$ for each λ_k. If $0 < n \leq \infty$ and $A = N \oplus T^{(n)}$, then $\sigma(A) = K$, $P_{-n}(A) = G$, and $K \setminus G = \sigma_{le}(A) \cap \sigma_{re}(A)$.

EXERCISES

1. Let $A \in \mathscr{SF}$ and suppose that $\ker A \neq (0)$ and $\text{ran } A^\perp \neq (0)$. Show that for every $\delta > 0$ there is an operator B in $\mathscr{SF}$ such that $\|B - A\| < \delta$, $\dim \ker B < \dim \ker A$, and $\dim \text{ran } B^\perp < \dim \text{ran } A^\perp$.

2. If $A \in \mathscr{SF}$, show that there is a $\delta > 0$ such that $\dim \ker(A - \mu) = \dim \ker A$ and $\dim \text{ran}(A - \mu)^\perp = \dim \text{ran } A^\perp$ for $|\mu| < \delta$ if and only if $\ker A \subseteq \text{ran } A^n$ for every $n \geq 1$.

3. Prove Proposition 5.4.

4. Prove Proposition 5.5.

5. If $\lambda \in \partial P_n(A)$ and $n \neq 0$, show that $\text{ran}(A - \lambda)$ is not closed. What happens if $n = 0$?

6. (Stampfli [1974].) If $A \in \mathcal{B}(\mathcal{H})$, then there is a K in $\mathcal{B}_0(\mathcal{H})$ such that $\sigma(A + K) = \sigma_w(A)$.

7. Prove Proposition 5.8.

Preliminaries

As was stated in the Preface, the prerequisites for understanding this book are a good course in measure and integration theory and, as a corequisite, analytic function theory. In this and the succeeding appendices an attempt is made to fill in some of the gaps and standardize some notation. These sections are not meant to be a substitute for serious study of these topics.

In Section 1 of this appendix some results from infinite-dimensional linear algebra are set forth. Most of this is meant as review. Proposition 1.4, however, seems to be a fact that is not stressed or covered in courses but that is used often in functional analysis. Section 2 on topology is presented mainly to discuss nets. This topic is often not covered in the basic courses and it is especially useful in discussing various ideas and proving results in functional analysis.

§1. Linear Algebra

Let $\mathscr{X}$ be a vector space over $\mathbb{F} = \mathbb{R}$ or $\mathbb{C}$. A subset E of $\mathscr{X}$ is *linearly independent* if for any finite subset $\{e_1, \ldots, e_n\}$ of E and for any finite set of scalars $\{\alpha_1, \ldots, \alpha_n\}$, if $\sum_{k=1}^n \alpha_k e_k = 0$, then $\alpha_1 = \cdots = \alpha_n = 0$. A *Hamel basis* is a maximal linearly independent subset of $\mathscr{X}$.

1.1. Proposition. *If E is a linearly independent subset of $\mathscr{X}$, then E is a Hamel basis if and only if every vector x in $\mathscr{X}$ can be written as $x = \sum_{k=1}^n \alpha_k e_k$ for scalars $\alpha_1, \ldots, \alpha_n$ and $\{e_1, \ldots, e_n\} \subseteq E$.*

PROOF. Suppose E is a basis and $x \in \mathscr{X}$, $x \notin E$. Then $E \cup \{x\}$ is not linearly independent. Thus there are $\alpha_0, \alpha_1, \ldots, \alpha_n$ in $\mathbb{F}$ and $e_1, \ldots, e_n$ in E

such that $0 = \alpha_0 x + \alpha_1 e_1 + \cdots + \alpha_n e_n$, with $\alpha_0 \neq 0$. (Why?) Thus $x = \sum_{k=1}^{n}(-\alpha_k/\alpha_0)e_k$.

Conversely, if $\mathscr{X}$ is the linear span of E, then for every x in $\mathscr{X} \setminus E$, $E \cup \{x\}$ is not linearly independent. Thus E is a basis. ∎

1.2. Proposition. *If E_0 is a linearly independent subset of $\mathscr{X}$, then there is a basis E that contains E_0.*

PROOF. Use Zorn's Lemma.

A *linear functional* on $\mathscr{X}$ is a function $f: \mathscr{X} \to \mathbb{F}$ such that $f(\alpha x + \beta y) = \alpha f(x) + \beta f(y)$ for x, y in $\mathscr{X}$ and α, β in $\mathbb{F}$. If $\mathscr{X}$ and $\mathscr{Y}$ are vector spaces over $\mathbb{F}$, a *linear transformation* from $\mathscr{X}$ into $\mathscr{Y}$ is a function $T: \mathscr{X} \to \mathscr{Y}$ such that $T(\alpha_1 x_1 + \alpha_2 x_2) = \alpha_1 T(x_1) + \alpha_2 T(x_2)$ for x_1, x_2 in $\mathscr{X}$ and α_1, α_2 in $\mathbb{F}$.

If $A, B \subseteq \mathscr{X}$, then $A + B \equiv \{a + b: a \in A, b \in B\}$; $A - B \equiv \{a - b: a \in A, b \in B\}$. For α in $\mathbb{F}$ and $A \subseteq \mathscr{X}$, $\alpha A \equiv \{\alpha a: a \in A\}$. If $\mathscr{M}$ is a *linear manifold* in $\mathscr{X}$ (that is, $\mathscr{M} \subseteq \mathscr{X}$ and $\mathscr{M}$ is also a vector space with the same operations defined on $\mathscr{X}$), then define $\mathscr{X}/\mathscr{M}$ to be the collection of all the subsets of $\mathscr{X}$ of the form $x + \mathscr{M}$. A set of the form $x + \mathscr{M}$ is called a *coset* of $\mathscr{M}$. Note that $(x + \mathscr{M}) + (y + \mathscr{M}) = (x + y) + \mathscr{M}$ and $\alpha(x + \mathscr{M}) = \alpha x + \mathscr{M}$ since $\mathscr{M}$ is a linear manifold. Hence $\mathscr{X}/\mathscr{M}$ becomes a vector space over $\mathbb{F}$. It is called the *quotient space* of $\mathscr{X}$ mod $\mathscr{M}$.

Define $Q: \mathscr{X} \to \mathscr{X}/\mathscr{M}$ by $Q(x) = x + \mathscr{M}$. It is easy to see that Q is a linear transformation. It is called the *quotient map*.

If $T: \mathscr{X} \to \mathscr{Y}$ is a linear transformation,

$$\ker T \equiv \{x \in \mathscr{X}: Tx = 0\},$$

$$\operatorname{ran} T \equiv \{Tx: x \in \mathscr{X}\};$$

$\ker T$ is the *kernel* of T and $\operatorname{ran} T$ is the *range* of T. If $\operatorname{ran} T = \mathscr{Y}$, T is *surjective*; if $\ker T = (0)$, T is *injective*. If T is both injective and surjective, then T is *bijective*. It is easy to see that the natural map $Q: \mathscr{X} \to \mathscr{X}/\mathscr{M}$ is surjective and $\ker Q = \mathscr{M}$.

Suppose now that $T: \mathscr{X} \to \mathscr{Y}$ is a linear transformation and $\mathscr{M}$ is a linear manifold in $\mathscr{X}$. We want to define a map $\hat{T}: \mathscr{X}/\mathscr{M} \to \mathscr{Y}$ by $\hat{T}(x + \mathscr{M}) = Tx$. But $\hat{T}$ may not be well defined. To ensure that it is we must have $Tx_1 = Tx_2$ if $x_1 + \mathscr{M} = x_2 + \mathscr{M}$. But $x_1 + \mathscr{M} = x_2 + \mathscr{M}$ if and only if $x_1 - x_2 \in \mathscr{M}$, and $Tx_1 = Tx_2$ if and only if $x_1 - x_2 \in \ker T$. So $\hat{T}$ is well defined if $\mathscr{M} \subseteq \ker T$. It is easy to check that if $\hat{T}$ is well defined, $\hat{T}$ is linear.

1.3. Proposition. *If $T: \mathscr{X} \to \mathscr{Y}$ is a linear transformation and $\mathscr{M}$ is a linear manifold in $\mathscr{X}$ contained in $\ker T$, then there is a linear transformation $\hat{T}: \mathscr{X}/\mathscr{M} \to \mathscr{Y}$ such that the diagram*

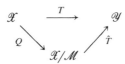

commutes.

The preceding proposition is especially useful if $\mathcal{M} = \ker T$. In that case $\hat{T}$ is injective.

The last proposition of this section will be quite helpful in the book.

1.4. Proposition. *Let* $f, f_1, \ldots, f_n$ *be linear functionals on* $\mathcal{X}$. *If* $\ker f \supseteq \bigcap_{k=1}^n \ker f_k$, *then there are scalars* $\alpha_1, \ldots, \alpha_n$ *such that* $f = \sum_{k=1}^n \alpha_k f_k$ *(that is,* $f(x) = \sum_{k=1}^n \alpha_k f_k(x)$ *for every* x *in* $\mathcal{X}$).

PROOF. It may be assumed without loss of generality that for $1 \leq k \leq n$,

$$\bigcap_{j \neq k} \ker f_j \neq \bigcap_{j=1}^n \ker f_j.$$

(Why?). So for $1 \leq k \leq n$, there is a y_k in $\bigcap_{j \neq k} \ker f_j$ such that $y_k \notin \bigcap_{j=1}^n \ker f_j$. So $f_j(y_k) = 0$ for $j \neq k$, but $f_k(y_k) \neq 0$. Let $x_k = [f_k(y_k)]^{-1} y_k$. Hence $f_k(x_k) = 1$ and $f_j(x_k) = 0$ for $j \neq k$.

Now let f be as in the statement of the proposition and put $\alpha_k = f(x_k)$. If $x \in \mathcal{X}$, let $y = x - \sum_{k=1}^n f_k(x) x_k$. Then $f_j(y) = f_j(x) - \sum_{k=1}^n f_k(x) f_j(x_k) = 0$. By hypothesis, $f(y) = 0$. Thus

$$0 = f(x) - \sum_{k=1}^n f_k(x) f(x_k)$$

$$= f(x) - \sum_{k=1}^n \alpha_k f_k(x);$$

equivalently, $f = \sum_{k=1}^n \alpha_k f_k$. ∎

§2. Topology

In this book all topological spaces are assumed to be Hausdorff.

This section will review some of the concepts and results using *nets* as this idea is frequently used in the text.

A *directed set* is a partially ordered set $(I, \leq)$ such that if $i_1, i_2 \in I$, then there is an i_3 in I such that $i_3 \geq i_1$ and $i_3 \geq i_2$. A good example of a directed set is to let $(X, \mathcal{T})$ be a topological space and for a fixed x_0 in X let $\mathcal{U} = \{U \text{ in } \mathcal{T}: x_0 \in U\}$. If $U, V \in \mathcal{U}$, define $U \geq V$ if $U \subseteq V$ (so bigger is smaller). $\mathcal{U}$ is said to be *ordered by reverse inclusion*. Another example is found if S is any set and $\mathcal{F}$ is the collection of all finite subsets of S. Define $F_1 \geq F_2$ in $\mathcal{F}$ if $F_1 \supseteq F_2$ (bigger means bigger). Here $\mathcal{F}$ is said to be *ordered by inclusion*. Both of these examples are used frequently in the text.

A *net* in X is a pair $((I, \leq), x)$, where $(I, \leq)$ is a directed set and x is a function from I into X. Usually we will write x_i instead of $x(i)$ and will use the phrase "let $\{x_i\}$ be a net in X."

Note that $\mathbb{N}$, the natural numbers, is a directed set, so every sequence is a net. If $(X, \mathcal{T})$ is a topological space, $x_0 \in X$, and $\mathcal{U} = \{U \text{ in } \mathcal{T}: x_0 \in U\}$, then let $x_U \in U$ for every U in $\mathcal{U}$. So $\{x_U: U \in \mathcal{U}\}$ is a net in X.

2.1. Definition. If $\{x_i\}$ is a net in a topological space X, then $\{x_i\}$ *converges* to x_0 (in symbols, $x_i \to x_0$ or $x_0 = \lim x_i$) if for every open subset U of X such that $x_0 \in U$, there is an $i_0 = i_0(U)$ such that $x_i \in U$ for $i \geq i_0$. The net *clusters* at x_0 (in symbols, $x_i \xrightarrow{\text{cl}} x_0$) if for every i_0 and for every open neighborhood U of x_0, there exists an $i \geq i_0$ such that $x_i \in U$.

These notions generalize the corresponding concepts for sequences. Also, if $x_i \to x_0$, then $x_i \xrightarrow{\text{cl}} x_0$. Note that the net $\{x_U: U \in \mathcal{U}\}$ defined just prior to the definition converges to x_0. This is a very important example of a convergent set.

2.2. Proposition. *If X is a topological space and $A \subseteq X$, then $x \in \text{cl } A$ (closure of A) if and only if there is a net $\{a_i\}$ in A such that $a_i \to x$.*

PROOF. Let $\mathcal{U} = \{U: U \text{ is open and } x \in U\}$. If $x \in \text{cl } A$, then for each U in $\mathcal{U}$ there is a point a_U in $A \cap U$. If $U_0 \in \mathcal{U}$, then $a_U \in U_0$ for every $U \geq U_0$; therefore $x = \lim a_U$. Conversely, if $\{a_i\}$ is a net in A and $a_i \to x$, then each U in $\mathcal{U}$ contains a point a_i and $a_i \in A \cap U$. Thus $x \in \text{cl } A$. ∎

2.3. Proposition. *If $A \subseteq X$, $\{a_i\}$ is a net in A, and $a_i \xrightarrow{\text{cl}} x$, then $x \in \text{cl } A$.*

PROOF. Exercise.

There is a concept of a subnet of a net and with this concept it is possible to prove that if a net clusters at a point x, then there is a subnet that converges to x. The concept of a subnet is, however, somewhat technical and is not what you might at first think it should be. Since this concept is not used in this book, the interested reader is referred to Kelley [1955]. It might also be appropriate to mention that a topological space is Hausdorff if and only if each convergent net has a unique limit point.

2.4. Proposition. *If X and Y are topological spaces and $f: X \to Y$, then f is continuous at x_0 if and only if $f(x_i) \to f(x_0)$ whenever $x_i \to x_0$.*

PROOF. First assume that f is continuous at x_0 and let $\{x_i\}$ be a net in X such that $x_i \to x_0$ in X. If V is open in Y and $f(x_0) \in V$, then there is an open set U in X such that $x_0 \in U$ and $f(U) \subseteq V$. Let i_0 be such that $x_i \in U$ for $i \geq i_0$. Hence $f(x_i) \in V$ for $i \geq i_0$. This says that $f(x_i) \to f(x_0)$.

Let $\mathcal{U} = \{U: U \text{ is open in } X \text{ and } x_0 \in U\}$. Suppose f is not continuous at x_0. Then there is an open subset V of Y such that $f(x_0) \in V$ and

$f(U) \setminus V \neq \square$ for every U in $\mathcal{U}$. Thus for each U in $\mathcal{U}$ there is a point x_U in U with $f(x_U) \notin V$. But $\{x_U\}$ is a net in X with $x_U \to x_0$ and clearly $\{f(x_U)\}$ cannot converge to $f(x_0)$. ∎

2.5. Proposition. *If $f: X \to Y$, f is continuous at x_0, and $\{x_i\}$ is a net in X that clusters at x_0, then $\{f(x_i)\}$ clusters at $f(x_0)$.*

PROOF. Exercise.

2.6. Proposition. *Let $K \subseteq X$. Then K is compact if and only if each net in K has a cluster point in K.*

PROOF. Suppose that K is compact and let $\{x_i: i \in I\}$ be a net in K. For each i let $F_i = \mathrm{cl}\{x_j: j \geq i\}$, so each F_i is a closed subset of K. It will be shown that $\{F_i: i \in I\}$ has the finite-intersection property. In fact, since I is directed, if $i_1, \ldots, i_n \in I$, then there is an $i \geq i_1, \ldots, i_n$. Thus $F_i \subseteq \bigcap_{k=1}^n F_{i_k}$ and $\{F_i\}$ has the finite-intersection property. Because K is compact, there is an x_0 in $\bigcap_i F_i$. But if U is open with x_0 in U and $i_0 \in I$, the fact that $x_0 \in \mathrm{cl}\{x_i: i \geq i_0\}$, implies there is an $i \geq i_0$ with x_i in U. Thus $x_i \xrightarrow{\mathrm{cl}} x_0$.

Now assume that each net in K has a cluster point in K. Let $\{K_\alpha: \alpha \in A\}$ be a collection of relatively closed subsets of K having the finite-intersection property. If $\mathscr{F} =$ the collection of all finite subsets of A, order $\mathscr{F}$ by inclusion. By hypothesis, if $F \in \mathscr{F}$, there is a point x_F in $\bigcap\{K_\alpha: \alpha \in F\}$. Thus $\{x_F\}$ is a net in K. By hypothesis, $\{x_F\}$ has a cluster point x_0 in K. Let $\alpha \in A$, so $\{\alpha\} \in \mathscr{F}$. Thus if U is any open set containing x_0 there is an F in $\mathscr{F}$ such that $\alpha \in F$ and $x_F \in U$. Thus $x_F \in U \cap K_\alpha$; that is, for each α in A and for every open set U containing x_0, $U \cap K_\alpha \neq \square$. Since K_α is relatively closed, $x_0 \in K_\alpha$ for each α in A. Thus $x_0 \in \bigcap_\alpha K_\alpha$ and K must be compact. ∎

The next result is used repeatedly in this book.

2.7. Proposition. *If X is compact, $\{x_i\}$ is a net in X, and x_0 is the only cluster point of $\{x_i\}$, then the net $\{x_i\}$ converges to x_0.*

PROOF. Let U be an open neighborhood of x_0 and let $J = \{j \in I: x_j \notin U\}$. If $\{x_i\}$ does not converge to x_0, then for every i in I there is a j in J such that $j \geq i$. In particular, J is also a directed set. Hence $\{x_j: j \in J\}$ is a net in the compact set $X \setminus U$. Thus it has a cluster point y_0. But the property of J mentioned before implies that y_0 is also a cluster point of $\{x_i: i \in I\}$, contradicting the assumption. Thus $x_i \to x_0$. ∎

The next result is rather easy, but it will be used so often that it should be explicitly stated and proved.

2.8. Proposition. *If $f: X \to Y$ is bijective and continuous and X is compact, then f is a homeomorphism.*

PROOF. If F is a closed subset of X, then F is compact. Thus $f(F)$ is compact in Y and hence closed. Since f maps closed sets to closed sets, f^{-1} is continuous. ∎

Note that the Hausdorff property was used in the preceding proof when we said that a compact subset of Y is closed.

In the study of functional analysis it is often the case that the mathematician is presented with a set that has two topologies. It is useful to know how properties of one topology relate to the other and when the two topologies are, in fact, one.

If X is a set and $\mathcal{T}_1, \mathcal{T}_2$ are two topologies on X, say that $\mathcal{T}_2$ is *larger* or *stronger* than $\mathcal{T}_1$ if $\mathcal{T}_2 \supseteq \mathcal{T}_1$; in this case you may also say that $\mathcal{T}_1$ is *smaller* or *weaker*. In the literature there is also an unfortunate nomenclature for these concepts; the words "finer" and "coarser" are used.

The following result is easy to prove (it is an exercise) but it is enormously useful in discussing a set with two topologies.

2.9.   Lemma. *If $\mathcal{T}_1, \mathcal{T}_2$ are topologies on X, then $\mathcal{T}_2$ is larger than $\mathcal{T}_1$ if and only if the identity map $i: (X, \mathcal{T}_2) \to (X, \mathcal{T}_1)$ is continuous.*

2.10.   Proposition. *Let $\mathcal{T}_1, \mathcal{T}_2$ be topologies on X and assume that $\mathcal{T}_2$ is larger than $\mathcal{T}_1$.*

(a) *If F is $\mathcal{T}_1$-closed, F is $\mathcal{T}_2$-closed.*
(b) *If $f: Y \to (X, \mathcal{T}_2)$ is continuous, then $f: Y \to (X, \mathcal{T}_1)$ is continuous.*
(c) *If $f: (X, \mathcal{T}_1) \to Y$ is continuous, then $f: (X, \mathcal{T}_2) \to Y$ is continuous.*
(d) *If K is $\mathcal{T}_2$-compact, then K is $\mathcal{T}_1$-compact.*
(e) *If X is $\mathcal{T}_2$-compact, then $\mathcal{T}_1 = \mathcal{T}_2$.*

PROOF. (b) Note that $f: Y \to (X, \mathcal{T}_1)$ is the composition of $f: Y \to (X, \mathcal{T}_2)$ and $i: (X, \mathcal{T}_2) \to (X, \mathcal{T}_1)$ and use Lemma 2.9.
(d) Use Lemma 2.9.
(e) Use Lemma 2.9 and Proposition 2.8.
   The remainder of the proof is an exercise. ∎

APPENDIX B

The Dual of $L^p(\mu)$

In this section we will prove the following which appears as III.5.5 and III.5.6 in the text.

Theorem. *Let* (X, Ω, μ) *be a measure space, let* $1 \le p < \infty$, *and let* $1/p + 1/q = 1$. *If* $g \in L^q(\mu)$, *define* $F_g \colon L^p(\mu) \to \mathbb{F}$ *by*

$$F_g(f) = \int fg \, d\mu.$$

If $1 < p < \infty$, *the map* $g \mapsto F_g$ *defines an isometric isomorphism of* $L^q(\mu)$ *onto* $L^p(\mu)^*$. *If* $p = 1$ *and* (X, Ω, μ) *is* σ-*finite*, $g \mapsto F_g$ *is an isometric isomorphism of* $L^\infty(\mu)$ *onto* $L^1(\mu)^*$.

PROOF. If $g \in L^q(\mu)$, then Hölder's Inequality implies that $|F_g(f)| \le \|f\|_p \|g\|_q$ for all f in $L^p(\mu)$. Hence $F_g \in L^p(\mu)^*$ and $\|F_g\| \le \|g\|_q$. Therefore $g \mapsto F_g$ is a linear contraction. It must be shown that this map is surjective and an isometry. Assume $F \in L^p(\mu)^*$.

  Case 1: $\mu(X) < \infty$. Here $\chi_\Delta \in L^p(\mu)$ for every Δ in Ω. Define $\nu(\Delta) = F(\chi_\Delta)$. It is easy to see that ν is finitely additive. If $\{\Delta_n\} \subseteq \Omega$ with $\Delta_1 \supseteq \Delta_2 \supseteq \cdots$ and $\bigcap_{n=1}^\infty \Delta_n = \square$, then

$$\|\chi_{\Delta_n}\|_p = \left[\int |\chi_{\Delta_n}|^p \, d\mu \right]^{1/p}$$

$$= \mu(\Delta_n)^{1/p} \to 0.$$

Hence $\nu(\Delta_n) \to 0$ since F is bounded. It follows by standard measure theory that ν is a countably additive measure. Moreover, if $\mu(\Delta) = 0$, $\chi_\Delta = 0$ in $L^p(\mu)$; hence $\nu(\Delta) = 0$. That is, $\nu \ll \mu$. By the Radon–Nikodym

Theorem there is an Ω-measurable function g such $\nu(\Delta) = \int_\Delta g \, d\mu$ for every Δ in Ω; that is, $F(\chi_\Delta) = \int \chi_\Delta g \, d\mu$ for every Δ in Ω. It follows that

B.1 $$F(f) = \int fg \, d\mu$$

for every simple function f.

B.2. Claim. $g \in L^q(\mu)$ and $\|g\|_q \leq \|F\|$.

Note that once this claim is proven, the proof of Case 1 is complete. Indeed, (B.2) says that $F_g \in L^p(\mu)^*$ and since F and F_g agree on a dense subset of $L^p(\mu)$ (B.1), $F = F_g$. Also, $\|g\|_q \leq \|F\| = \|F_g\| = \|g\|_q$.

To prove (B.2), let $t > 0$ and put $E_t = \{ x \in X : |g(x)| \leq t \}$. If $f \in L^p(\mu)$ such that $f = 0$ off E_t, then there is a sequence $\{f_n\}$ of simple functions such that for every n, $f_n = 0$ off E_t, $|f_n| \leq |f|$, and $f_n(x) \to f(x)$ a.e. $[\mu]$. (Why?) Thus $|(f_n - f)g| \leq 2t|f|$ and $\int |f| \, d\mu = \int |f| \cdot 1 \, d\mu \leq \|f\|_p \mu(X)^{1/q} < \infty$. By the Lebesgue Dominated Convergence Theorem, $F(f_n) = \int f_n g \, d\mu \to \int fg \, d\mu$. Also, $|f_n - f|^p \leq 2^p |f|^p$, so $\|f_n - f\|_p \to 0$; thus $F(f_n) \to F(f)$. Combining these results we get that for any $t > 0$ and any f in $L^p(\mu)$ that vanishes off E_t, (B.1) holds.

Case 1a: $1 < p < \infty$. So $1 < q < \infty$. Let $f = \chi_{E_t} |g|^q/g$, where $g(x) \neq 0$, and put $f(x) = 0$ when $g(x) = 0$. If $A = \{ x : g(x) \neq 0 \}$, then

$$\int |f|^p \, d\mu = \int_{E_t \cap A} \frac{|g|^{pq}}{|g|^p} \, d\mu = \int_{E_t} |g|^q \, d\mu$$

since $pq - p = q$. Therefore

$$\int_{E_t} |g|^q \, d\mu = \int fg \, d\mu = F(f) \leq \|F\| \|f\|_p = \|F\| \left[\int_{E_t} |g|^q \, d\mu \right]^{1/p}.$$

Thus

$$\|F\| \geq \left[\int_{E_t} |g|^q \, d\mu \right]^{1 - 1/p} \geq \left[\int_{E_t} |g|^q \, d\mu \right]^{1/q}.$$

Letting $t \to \infty$ gives that $\|g\|_q \leq \|F\|$.

Case 1b: $p = 1$. So $q = \infty$. For $\varepsilon > 0$ let $A = \{ x : |g(x)| > \|F\| + \varepsilon \}$. For $t > 0$ let $f = \chi_{E_t \cap A} \bar{g}/|g|$. Then $\|f\|_1 = \mu(A \cap E_t)$, and so

$$\|F\| \mu(A \cap E_t) \geq \int fg \, d\mu = \int_{A \cap E_t} |g| \, d\mu \geq (\|F\| + \varepsilon)\mu(A \cap E_t).$$

Letting $t \to \infty$ we get that $\|F\| \mu(A) \geq (\|F\| + \varepsilon)\mu(A)$, which can only be if $\mu(A) = 0$. Thus $\|g\|_\infty \leq \|F\|$.

Case 2: (X, Ω, μ) is arbitrary. Let $\mathscr{E} = $ all of the sets E in Ω such that $\mu(E) < \infty$. For E in Ω let $\Omega_E = \{ \Delta \in \Omega : \Delta \subseteq E \}$ and define $(\mu|E)(\Delta) = \mu(\Delta)$ for Δ in Ω_E. Put $L^p(\mu|E) = L^p(E, \Omega_E, \mu|E)$ and notice that $L^p(\mu|E)$

can be identified in a natural way with the functions in $L^p(X, \Omega, \mu)$ that vanish off E. Make this identification and consider the restriction of F: $L^p(\mu) \to \mathbb{F}$ to $L^p(\mu|E)$; denote the restriction by F_E: $L^p(\mu|E) \to \mathbb{F}$. Clearly F_E is bounded and $\|F_E\| \leq \|F\|$ for every E in $\mathscr{E}$.

By Case 1, for every E in $\mathscr{E}$ there is a g_E in $L^q(\mu|E)$ such that for f in $L^p(\mu|E)$,

B.3
$$F(f) = \int_E f g_E \, d\mu \quad \text{and} \quad \|g_E\|_q \leq \|F\|.$$

If $D, E \in \mathscr{E}$, then $L^p(\mu|D \cap E)$ is contained in both $L^p(\mu|D)$ and $L^p(\mu|E)$. Moreover, $F_D|L^p(\mu|D \cap E) = F_E|L^p(\mu|D \cap E) = F_{D \cap E}$. Hence $g_D = g_E = g_{D \cap E}$ a.e. $[\mu]$ on $D \cap E$. Thus, a function g can be defined on $\cup\{E: E \in \mathscr{E}\}$ by letting $g = g_E$ on E; put $g = 0$ off $\cup\{E: E \in \mathscr{E}\}$. A difficulty arises here in trying to show that g is measurable.

Case 2a: $1 < p < \infty$. Put $\sigma = \sup\{\|g_E\|_q: E \in \mathscr{E}\}$; so $\sigma \leq \|F\| < \infty$. Since $\|g_D\|_q \leq \|g_E\|_q$ if $D \subseteq E$, there is a sequence $\{E_n\}$ in $\mathscr{E}$ such that $E_n \subseteq E_{n+1}$ for all n and $\|g_{E_n}\|_q \to \sigma$. Let $G = \cup_{n=1}^\infty E_n$. If $E \in \mathscr{E}$ and $E \cap G = \square$, then $\|g_{E \cup E_n}\|_q^q = \|g_E\|_q^q + \|g_{E_n}\|_q^q \to \|g_E\|_q^q + \sigma^q$; thus $g_E = 0$. Therefore $g = 0$ off G and clearly g is measurable. Moreover, $g \in L^q(\mu)$ with $\|g\|_q = \sigma$.

If $f \in L^p(\mu)$, then $\{x: f(x) \neq 0\} = \cup_{n=1}^\infty D_n$ where $D_n \in \mathscr{E}$ and $D_n \subseteq D_{n+1}$ for all n. Thus $\chi_{D_n} f \to f$ in $L^p(\mu)$ and so $F(f) = \lim F(\chi_{D_n} f) = (\text{B.3})$ $\lim \int_{D_n} gf \, d\mu = \int gf \, d\mu$. Thus $F = F_g$ and $\|F\| = \|F_g\| = \|g\|_q \leq \sigma \leq \|F\|$.

Case 2b: $p = \infty$ and (X, Ω, μ) is σ-finite. This is left to the reader. ∎

EXERCISE

Look at the proof of the theorem and see if you can represent $L^1(X, \Omega, \mu)^*$ for an arbitrary measure space.

The Dual of $C_0(X)$

The purpose of this section is to show that the dual of $C_0(X)$ is the space of regular Borel measures on X and to put this result, and the accompanying definitions, in the context of complex-valued measures and functions.

Let X be any set and let Ω be a σ-algebra of subsets of X; so (X, Ω) is a *measurable space*. If μ is a countably additive function defined on Ω such that $\mu(\square) = 0$ and $0 \leq \mu(\Delta) \leq \infty$ for all Δ in Ω, call μ a *positive measure* on (X, Ω); (X, Ω, μ) is called a *measure space*.

If (X, Ω) is a measurable space, a *signed measure* is a countably additive function μ defined on Ω such that $\mu(\square) = 0$ and μ takes its values in $\mathbb{R} \cup \{\pm\infty\}$. (Note: μ can assume only one of the values $\pm\infty$.) It is assumed that the reader is familiar with the following result.

C.1. Hahn–Jordan Decomposition. *If μ is a signed measure on (X, Ω), then $\mu = \mu_1 - \mu_2$, where μ_1 and μ_2 are positive measures, and $X = E_1 \cup E_2$, where $E_1, E_2 \in \Omega$, $E_1 \cap E_2 = \square$, $\mu_1(E_2) = 0 = \mu_2(E_1)$. The measures μ_1 and μ_2 are unique and the sets E_1 and E_2 are unique up to sets of $\mu_1 + \mu_2$ measure zero.*

A *measure* (or *complex-valued measure*) is a complex-valued function μ defined on Ω that is countably additive and such that $\mu(\square) = 0$. Note that μ does not assume any infinite values. If μ is a measure, then $(\operatorname{Re}\mu)(\Delta) \equiv \operatorname{Re}(\mu(\Delta))$ is a signed measure, as is $(\operatorname{Im}\mu)(\Delta) \equiv \operatorname{Im}(\mu(\Delta))$; hence $\mu = \operatorname{Re}\mu + i \operatorname{Im}\mu$. Applying (C.1) to $\operatorname{Re}\mu$ and $\operatorname{Im}\mu$ we get

C.2
$$\mu = (\mu_1 - \mu_2) + i(\mu_3 - \mu_4)$$

where μ_j $(1 \leq j \leq 4)$ are positive measures, $\mu_1 \perp \mu_2$ (μ_1 and μ_2 are *mutually singular*) and $\mu_3 \perp \mu_4$. (C.2) will also be called the Hahn–Jordan decomposition of μ.

C.3.   Definition. If μ is a measure on (X, Ω) and $\Delta \in \Omega$, define the *variation* of μ, $|\mu|$, by

$$|\mu|(\Delta) = \sup\left\{ \sum_{j=1}^{m} |\mu(E_j)| : \{E_j\}_1^m \text{ is a measurable partition of } \Delta \right\}.$$

C.4.   Proposition. *If μ is a measure on (X, Ω), then $|\mu|$ is a positive finite measure on (X, Ω). If μ is a signed measure, $|\mu|$ is a positive measure. If (C.2) is satisfied, then $|\mu|(\Delta) \leq \sum_{k=1}^{4} \mu_k(\Delta)$; if μ is a signed measure, then $|\mu| = \mu_1 + \mu_2$.*

PROOF. Clearly $|\mu|(\Delta) \geq 0$. Let $\{\Delta_n\}$ be pairwise disjoint measurable sets and let $\Delta = \bigcup_{n=1}^{\infty} \Delta_n$. If $\varepsilon > 0$, then there is a measurable partition $\{E_j\}_{j=1}^{m}$ of Δ such that $|\mu|(\Delta) - \varepsilon < \sum_{j=1}^{m} |\mu(E_j)|$. Hence

$$|\mu|(\Delta) - \varepsilon \leq \sum_{j=1}^{m} \left| \sum_{n=1}^{\infty} \mu(E_j \cap \Delta_n) \right|$$

$$\leq \sum_{n=1}^{\infty} \sum_{j=1}^{m} |\mu(E_j \cap \Delta_n)|.$$

But $\{E_j \cap \Delta_n\}_{j=1}^{m}$ is a partition of Δ_n, so $|\mu|(\Delta) - \varepsilon \leq \sum_{n=1}^{\infty} |\mu|(\Delta_n)$. Therefore $|\mu|(\Delta) \leq \sum_{n=1}^{\infty} |\mu|(\Delta_n)$. For the reverse inequality we may assume that $|\mu|(\Delta) < \infty$. It follows that $|\mu|(\Delta_n) < \infty$ for every n. (Why?) Let $\varepsilon > 0$ and for each $n \geq 1$ let $\{E_1^{(n)}, \ldots, E_{m_n}^{(n)}\}$ be a partition of Δ_n such that $\sum_j |\mu(E_j^{(n)})| > |\mu|(\Delta_n) - \varepsilon/2^n$. Then

$$\sum_{n=1}^{N} |\mu|(\Delta_n) < \sum_{n=1}^{N} \left[\frac{\varepsilon}{2^n} + \sum_j |\mu(E_j^{(n)})| \right]$$

$$\leq \varepsilon + \sum_{n=1}^{N} \sum_j |\mu(E_j^{(n)})|$$

$$\leq \varepsilon + |\mu|(\Delta).$$

Letting $N \to \infty$ and $\varepsilon \to 0$ gives that $\sum_1^{\infty} |\mu|(\Delta_n) \leq |\mu|(\Delta)$.

Clearly $|\mu(\Delta)| \leq \sum_{k=1}^{4} \mu_k(\Delta)$, so $|\mu| \leq \sum_{k=1}^{4} \mu_k$. It is left to the reader to show that $|\mu| = \mu_1 + \mu_2$ if μ is a signed measure. Since $\mu_1, \mu_2, \mu_3, \mu_4$ are all finite, $|\mu|$ is finite if μ is complex-valued.   ∎

C.5.   Definition. If μ is a measure on (X, Ω) and ν is a positive measure on (X, Ω), say that μ is *absolutely continuous* with respect to ν ($\mu \ll \nu$) if $\mu(\Delta) = 0$ whenever $\nu(\Delta) = 0$. If ν is complex-valued, $\mu \ll \nu$ means $\mu \ll |\nu|$.

C.6.   Proposition. *Let μ be a measure and ν a positive measure on (X, Ω). The following statements are equivalent.*

(a) $\mu \ll \nu$.
(b) $|\mu| \ll \nu$.
(c) *If (C.2) holds, $\mu_k \ll \nu$ for $1 \le k \le 4$.*

PROOF. Exercise.

The Radon–Nikodym Theorem can now be proved for complex-valued measures μ by using (C.6) and applying the usual theorem to the real and imaginary parts of μ. The details are left to the reader.

C.7.   Radon–Nikodym Theorem. *If (X, Ω, ν) is a σ-finite measure space and μ is a complex-valued measure on (X, Ω) such that $\mu \ll \nu$, then there is a unique complex-valued function f in $L^1(X, \Omega, \nu)$ such that $\mu(\Delta) = \int_\Delta f\,d\nu$ for every Δ in Ω.*

The function f obtained in (C.7) is called the *Radon–Nikodym derivative* of μ with respect to ν and is denoted by $f = d\mu/d\nu$.

C.8.   Theorem. *Let (X, Ω, ν) be a σ-finite measure space and let μ be a complex-valued measure on (X, Ω) such that $\mu \ll \nu$ and let $f = d\mu/d\nu$.*

(a) *If $g \in L^1(X, \Omega, |\mu|)$, then $gf \in L^1(X, \Omega, \nu)$ and $\int g\,d\mu = \int gf\,d\nu$.*
(b) *For Δ in Ω, $|\mu|(\Delta) = \int_\Delta |f|\,d\nu$.*

PROOF. Part (a) follows from the corresponding result for signed measures by using (C.2) and a similar decomposition for f.

To prove (b), let $\{E_j\}$ be a measurable partition of Δ. Then

$$\sum_j |\mu(E_j)| \le \sum_j \int_{E_j} |f|\,d\nu = \int_\Delta |f|\,d\nu.$$

For the reverse inequality, let $g(x) = \overline{f(x)}/|f(x)|$ if $x \in \Delta$ and $f(x) \ne 0$; let $g(x) = 0$ otherwise. Let $\{g_n\}$ be a sequence of Ω-measurable simple functions such that $g_n(x) = 0$ off Δ, $|g_n| \le |g| \le 1$, and $g_n(x) \to g(x)$ a.e. $[\nu]$. Thus $fg_n \to |f|\chi_\Delta$ a.e. $[\nu]$. Also, $|fg_n| \le |f|\chi_\Delta$ and $f\chi_\Delta \in L^1(\nu)$ [see (C.2)]. By the Lebesgue Dominated Convergence Theorem, $\int fg_n\,d\nu \to \int_\Delta |f|\,d\nu$. If $g_n = \sum_j \alpha_j \chi_{E_j}$, where $\{E_j\}$ is a partition of Δ and $|\alpha_j| \le 1$, then $|\int fg_n\,d\nu| = |\int g_n\,d\mu| = |\sum_j \alpha_j \mu(E_j)| \le |\mu|(\Delta)$. Thus $\int_\Delta |f|\,d\nu \le |\mu|(\Delta)$. ∎

One way of phrasing (C.8b) is that $|d\mu/d\nu| = d|\mu|/d\nu$. The next result is left to the reader.

C.9.   Corollary. *If μ is a complex-valued measure on (X, Ω), then there is an Ω-measurable function f on X such that $|f| = 1$ a.e. $[|\mu|]$ and $\mu(\Delta) = \int_\Delta f\,d|\mu|$ for each Δ in Ω.*

C.10. Definition. Let X be a locally compact space and let Ω be the smallest σ-algebra of subsets of X that contains the open sets. Sets in Ω are called *Borel sets*. A positive measure μ on (X, Ω) is a *regular Borel measure* if (a) $\mu(K) < \infty$ for every compact subset K of X; (b) for any E in Ω, $\mu(E) = \sup\{\mu(K): K \subseteq E$ and K is compact$\}$; (c) for any E in Ω, $\mu(E) = \inf\{\mu(U): U \supseteq E$ and U is open$\}$. If μ is a complex-valued measure on (X, Ω), μ is a regular Borel measure if $|\mu|$ is. Let $M(X) = $ all of the complex-valued regular Borel measures on X. Note that $M(X)$ is a vector space over $\mathbb{C}$. For μ in $M(X)$, let

C.11 $$\|\mu\| \equiv |\mu|(X).$$

C.12. Proposition. (C.11) *defines a norm on* $M(X)$.

PROOF. Exercise.

C.13. Lemma. *If* $\mu \in M(X)$, *define* $F_\mu: C_0(X) \to \mathbb{C}$ *by* $F_\mu(f) = \int f\, d\mu$. *Then* $F_\mu \in C_0(X)^*$ *and* $\|F_\mu\| = \|\mu\|$.

PROOF. If $f \in C_0(X)$, then $|F_\mu(f)| \leq \int |f|\, d|\mu| \leq \|f\| \|\mu\|$. Hence $F_\mu \in C_0(X)^*$ and $\|F_\mu\| \leq \|\mu\|$.

To show equality, let f_0 be a Borel function such that $|f_0| = 1$ a.e. $[|\mu|]$ and $\mu(\Delta) = \int_\Delta f_0\, d|\mu|$. By Lusin's Theorem, if $\varepsilon > 0$, there is a continuous function ϕ on X with compact support such that $\int |\phi - \bar{f}_0|\, d|\mu| < \varepsilon$ and $\|\phi\| \leq \sup|f_0(x)| = 1$. Thus $\|\mu\| = \int f_0 \bar{f}_0\, d|\mu|$ (C.8a) $= \int \bar{f}_0\, d\mu = |\int \bar{f}_0\, d\mu| \leq |\int (\bar{f}_0 - \phi)\, d\mu| + |\int \phi\, d\mu| \leq \varepsilon + |F_\mu(\phi)| \leq \varepsilon + \|F_\mu\|$. Hence $\|\mu\| \leq \|F_\mu\|$. ∎

C.14. Corollary. (a) *If* U *is an open subset of* X *and* $\mu \in M(X)$, *then* $|\mu|(U) = \sup\{|\int \phi\, d\mu|: \phi \in C_c(X)$, $\mathrm{spt}\,\phi \subseteq U$, *and* $\|\phi\| \leq 1\}$. (b) *If* $\mu \geq 0$, $\mu(K) = \inf\{\int \phi\, d\mu: \phi \in C_0(X)$ *and* $\phi \geq \chi_K\}$.

PROOF. (a) If U is given the relative topology from X, U is locally compact. Let ν be the restriction of μ to U. Then (a) becomes a restatement of (C.13) for the space U together with the fact that $C_c(U)$ is norm dense in $C_0(U)$.

(b) If $\phi \geq \chi_K$, then because μ is positive, $\int \phi\, d\mu \geq \mu(K)$. Thus $\mu(K) \leq \alpha \equiv \inf\{\int \phi\, d\mu: \phi \in C_0(X)$ and $\phi \geq \chi_K\}$. Using the regularity of μ, for every integer n there is an open set U_n such that $K \subseteq U_n$ and $\mu(U_n \setminus K) < n^{-1}$. Let $\psi_n \in C_c(X)$ such that $0 \leq \psi_n \leq 1$, $\psi_n = 1$ on K, and $\psi_n = 0$ off U_n. Thus $\psi_n \geq \chi_K$ and so $\alpha \leq \int \psi_n\, d\mu \leq \mu(U_n) < \mu(K) + n^{-1}$. ∎

The next step in the process of representing bounded linear functionals on $C_0(X)$ by measures is to associate with each such functional a positive functional. If $\mu \in M(X)$, then the next lemma would associate with the functional F_μ the positive functional $I = F_{|\mu|}$.

C.15. Lemma. *If $F: C_0(X) \to \mathbb{C}$ is a bounded linear functional, then there is a unique linear functional $I: C_0(X) \to \mathbb{C}$ such that if $f \in C_0(X)$ and $f \geq 0$, then*

C.16 $I(f) = \sup\{|F(g)|: g \in C_0(X) \text{ and } |g| \leq f\}.$

Moreover $\|I\| = \|F\|$.

PROOF. Let $C_0(X)_+$ be the positive functions in $C_0(X)$ and for f in $C_0(X)_+$ define $I(f)$ as in (C.16). If $\alpha > 0$, then clearly $I(\alpha f) = \alpha I(f)$ if $f \in C_0(X)_+$. Also, if $g \in C_0(X)$ and $|g| \leq f$, then $|F(g)| \leq \|F\|\|g\| \leq \|F\|\|f\|$. Hence $I(f) \leq \|F\|\|f\| < \infty$.

Now we will show that $I(f_1 + f_2) = I(f_1) + I(f_2)$ whenever $f_1, f_2 \in C(X)_+$. If $\varepsilon > 0$, let $g_1, g_2 \in C_0(X)$ such that $|g_j| \leq f_j$ and $|F(g_j)| > I(f_j) - \frac{1}{2}\varepsilon$ for $j = 1, 2$. There are complex numbers β_j, $j = 1, 2$, with $|\beta_j| = 1$ and $F(g_j) = \beta_j|F(g_j)|$. Thus

$$I(f_1) + I(f_2) < \varepsilon + |F(g_1)| + |F(g_2)|$$
$$= \varepsilon + \bar{\beta}_1 F(g_1) + \bar{\beta}_2 F(g_2)$$
$$= \varepsilon + |F(\bar{\beta}_1 g_1 + \bar{\beta}_2 g_2)|.$$

But $|\bar{\beta}_1 g_1 + \bar{\beta}_2 g_2)| \leq |g_1| + |g_2| \leq f_1 + f_2$. Hence $I(f_1) + I(f_2) \leq \varepsilon + I(f_1 + f_2)$. Since ε was arbitrary, we have half of the desired equality.

For the other half of the equality, let $g \in C_0(X)$ such that $|g| \leq f_1 + f_2$ and $I(f_1 + f_2) < |F(g)| + \varepsilon$. Let $h_1 = \min(|g|, f_1)$ and $h_2 = |g| - h_1$. Clearly $h_1, h_2 \in C_0(X)_+$, $h_1 \leq f_1$, $h_2 \leq f_2$, and $h_1 + h_2 = |g|$. Define $g_j: X \to \mathbb{C}$ by

$$g_j(x) = \begin{cases} 0 & \text{if } g(x) = 0, \\[2mm] \dfrac{h_j(x)\overline{g(x)}}{|g(x)|} & \text{if } g(x) \neq 0. \end{cases}$$

It is left to the reader to verify that $g_j \in C_0(X)$ and $g_1 + g_2 = |g|$. Hence

$$I(f_1 + f_2) < |F(g_1) + F(g_2)| + \varepsilon$$
$$\leq |F(g_1)| + |F(g_2)| + \varepsilon$$
$$\leq I(f_1) + I(f_2) + \varepsilon.$$

Now let $\varepsilon \to 0$.

If f is a real-valued function in $C_0(X)$, then $f = f_1 - f_2$ where $f_1, f_2 \in C_0(X)_+$. If also $f = g_1 - g_2$ for some g_1, g_2 in $C_0(X)_+$, then $g_1 + f_2 = f_1 + g_2$. By the preceding argument $I(g_1) + I(f_2) = I(f_1) + I(g_2)$. Hence if we define $I: \operatorname{Re} C_0(X) \to \mathbb{R}$ by $I(f) = I(f_1) - I(f_2)$ where $f = f_1 - f_2$ with f_1, f_2 in $C_0(X)_+$, I is well defined. It is left to the reader to verify that I is $\mathbb{R}$-linear.

If $f \in C_0(X)$, then $f = f_1 + if_2$, where $f_1, f_2 \in \operatorname{Re} C_0(X)$. Let $I(f) = I(f_1) + iI(f_2)$. It is left to the reader to show that $I: C_0(X) \to \mathbb{C}$ is a linear functional.

To prove that $\|I\| = \|F\|$, first let $f \in C_0(X)$ and put $I(f) = \alpha |I(f)|$ where $|\alpha| = 1$. Hence $\bar{\alpha}f = f_1 + if_2$, where $f_1, f_2 \in \operatorname{Re} C_0(X)$. Thus $|I(f)| = \bar{\alpha}I(f) = I(f_1) + iI(f_2)$. Since $|I(f)|$ is a positive real number, $I(f_2) = 0$ and $I(f_1) = |I(f)|$. But $f_1 = \operatorname{Re}(\bar{\alpha}f) \leq |\bar{\alpha}f| = |f|$. Hence

$$|I(f)| \leq I(|f|).$$

From here we get, as in the beginning of this proof, that $\|I\| \leq \|F\|$. For the other half, if $\varepsilon > 0$, let $f \in C_0(X)$ such that $\|f\| \leq 1$ and $\|F\| < |F(f)| + \varepsilon$. Thus $\|F\| < I(|f|) + \varepsilon \leq \|I\| + \varepsilon$. ∎

C.17. Theorem. *If I: $C_0(X) \to \mathbb{C}$ is a bounded linear functional such that $I(f) \geq 0$ whenever $f \in C_0(X)_+$, then there is a positive measure ν in $M(X)$ such that $I(f) = \int f\,d\nu$ for every f in $C_0(X)$ and $\|I\| = \nu(X)$.*

The proof of this is an involved construction. Inspired by Corollary C.14, one defines $\nu(U)$ for an open set U by

$$\nu(U) = \sup\{ I(\phi): \phi \in C_c(X)_+, \phi \leq 1, \operatorname{spt}\phi \subseteq U \}.$$

Then for any Borel set E, let

$$\nu(E) = \inf\{ \nu(U): E \subseteq U \quad \text{and} \quad U \text{ is open}\}.$$

It now must be shown that ν is a positive measure and $I(f) = \int f\,d\nu$. For the details see (12.36) in Hewitt and Stromberg [1975] or §56 in Halmos [1974]. Indeed, Theorem C.17 is often called the Riesz Representation Theorem.

C.18. Riesz Representation Theorem. *If X is a locally compact space and $\mu \in M(X)$, define F_μ: $C_0(X) \to \mathbb{C}$ by*

$$F_\mu(f) = \int f\,d\mu.$$

Then $F_\mu \in C_0(X)^*$ and the map $\mu \mapsto F_\mu$ is an isometric isomorphism of $M(X)$ onto $C_0(X)^*$.

PROOF. The fact that $\mu \mapsto F_\mu$ is an isometry is the content of Lemma C.13. It remains to show that $\mu \mapsto F_\mu$ is surjective. Let $F \in C_0(X)^*$ and define I: $C_0(X) \to \mathbb{C}$ as in Lemma C.15. By Theorem C.17, there is a positive measure ν in $M(X)$ such that $I(f) = \int f\,d\nu$ for all f in $C_0(X)$. If $f \in C_0(X)$, then the definition of I implies that $|F(f)| \leq I(|f|) = \int |f|\,d\nu$. Thus, $f \mapsto F(f)$ defines a bounded linear functional on $C_0(X)$ considered as a linear manifold in $L^1(\nu)$. Now $C_0(X)$ is dense in $L^1(\nu)$ (Why?), so F has a unique extension to a bounded linear functional on $L^1(\nu)$. By Theorem B.1 there is a function ϕ in $L^\infty(\nu)$ such that $F(f) = \int f\phi\,d\nu$ for every f in $C_0(X)$ and $\|\phi\|_\infty \leq 1$. Let $\mu(E) = \int_E \phi\,d\nu$ for every Borel set E. Then $\mu \in M(X)$ and by Theorem C.8(a), $F(f) = \int f\,d\mu$; that is, $F = F_\mu$. ∎

Bibliography

M. B. Abrahamse [1978]. Multiplication operators. In *Hilbert Space Operators*. New York: Springer-Verlag Notes, Vol. 693, pp. 17–36.

M. B. Abrahamse and T. L. Kriete [1973]. The spectral multiplicity function of a multiplication operator. *Indiana J. Math.*, **22**, 845–857.

T. Ando [1963]. Note on invariant subspaces of a compact normal operator. *Archiv. Math.*, **14**, 337–340.

N. Aronszajn and K. T. Smith [1954]. Invariant subspaces of completely continuous operators. *Ann. Math.*, **60**, 345–350.

W. Arveson [1976]. *An Invitation to C*-algebras*. New York: Springer-Verlag.

S. Banach [1955]. *Theorie des opérations linéaires*. New York: Chelsea Publ. Co.

S. K. Berberian [1959]. Note on a theorem of Fuglede and Putnam. *Proc. Amer. Math. Soc.*, **10**, 175–182.

A. R. Bernstein and A. Robinson [1966]. Solution of an invariant subspace problem of K. T. Smith and P. R. Halmos. *Pacific J. Math.*, **16**, 421–431.

A. Beurling [1949]. On two problems concerning linear transformations in Hilbert space. *Acta. Math.*, **81**, 239–255.

F. F. Bonsall and J. Duncan [1973]. *Complete Normed Algebras*. Berlin: Springer-Verlag.

N. Bourbaki [1967]. *Espaces Vectoriels Topologiques*. Paris: Hermann.

A. Brown [1974]. A version of multiplicity theory. *Topics in Operator Theory*. Math. Surveys A.M.S., Vol. 13, 129–160.

R. C. Buck [1958]. Bounded continuous functions on a locally compact space. *Michigan Math. J.*, **5**, 95–104.

L. Carleson [1966]. On convergence and growth of partial sums of Fourier series. *Acta Math.*, **116**, 135–157.

M. D. Choi [1983]. Tricks or treats with the Hilbert matrix. *Amer. Math. Monthly*, **90**, 301–312.

J. A. Clarkson [1936]. Uniformly convex spaces. *Trans. Amer. Math. Soc.*, **40**, 396–414.

J. B. Conway [1977]. Every spectral picture is possible. *Notices Amer. Math. Soc.*, **24**, p. A431.

J. B. Conway [1978]. *Functions of One Complex Variable*. New York: Springer-Verlag.

J. B. Conway [1981]. *Subnormal Operators*. Boston: Pitman.

R. Courant and D. Hilbert [1953]. *Methods of Mathematical Physics*. New York: Interscience.

A. M. Davie [1973]. The approximation problem for Banach spaces. *Bull. London Math. Soc.*, **5**, 261–266.

A. M. Davie [1975]. The Banach approximation problem. *J. Approx. Theory*, **13**, 392–394.

W. J. Davis, T. Figel, W. B. Johnson, and A. Pelczynski [1974]. Factoring weakly compact operators. *J. Functional Anal.*, **17**, 311–327.

L. de Branges [1959]. The Stone-Weierstrass Theorem. *Proc. Amer. Math. Soc.*, **10**, 822–824.

W. F. Donoghue [1957]. The lattice of invariant subspaces of a completely continuous quasi-nilpotent transformation. *Pacific J. Math.*, **7**, 1031–1035.

R. G. Douglas [1969]. On the operator equation $S * XT = X$ and related topics. *Acta. Sci. Math. (Szeged)*, **30**, 19–32.

J. Dugundji [1966]. *Topology*. Boston: Allyn and Bacon.

N. Dunford and J. Schwartz [1958]. *Linear Operators. I*. New York: Interscience.

N. Dunford and J. Schwartz [1963]. *Linear Operators. II*. New York: Interscience.

J. Dyer, E. Pedersen, and P. Porcelli [1972]. An equivalent formulation of the invariant subspace conjecture. *Bull. Amer. Math. Soc.*, **78**, 1020–1023.

P. Enflo [1973]. A counterexample to the approximation problem in Banach spaces. *Acta Math.*, **130**, 309–317.

J. Ernest [1976]. Charting the operator terrain. *Memoirs Amer. Math. Soc.*, Vol. 71.

P. A. Fillmore, J. G. Stampfli, and J. P. Williams [1972]. On the essential numerical range, the essential spectrum, and a problem of Halmos. *Acta Sci. Math. (Szeged)*, **33**, 179–192.

B. Fuglede [1950]. A commutativity theorem for normal operators. *Proc. Nat. Acad. Sci.*, **36**, 35–40.

T. W. Gamelin [1969]. *Uniform Algebras*. Englewood Cliffs: Prentice-Hall.

L. Gillman and M. Jerison [1960]. *Rings of Continuous Functions*. Princeton: Van Nostrand. Reprinted by Springer-Verlag, New York.

F. Greenleaf [1969]. *Invariant Means on Topological Groups*. New York: Van Nostrand.

A. Grothendieck [1953]. Sur les applications linéaires faiblement compactes d'espaces du type $C(K)$. *Canadian J. Math.*, **5**, 129–173.

P. R. Halmos [1951]. *Introduction to Hilbert Space and the Theory of Spectral Multiplicity*. New York: Chelsea Publ. Co.

P. R. Halmos [1961]. Shifts on Hilbert spaces. *J. Reine Angew. Math.*, **208**, 102–112.

P. R. Halmos [1963]. What does the spectral theorem say? *Amer. Math. Monthly*, **70**, 241–247.

P. R. Halmos [1966]. Invariant subspaces of polynomially compact operators. *Pacific J. Math.*, **16**, 433–437.

P. R. Halmos [1974]. *Measure Theory*. New York: Springer-Verlag.

P. R. Halmos [1982]. *A Hilbert Space Problem Book, Second ed.* New York: Springer-Verlag.

P. R. Halmos and V. Sunder [1978]. *Bounded Integral Operators on L^2 Spaces*. New York: Springer-Verlag.

E. Hellinger [1907]. Die Orthogonal invarianten quadratischer Formen. Inaugural Dissertation, Göttingen.

H. Helson [1964]. *Invariant Subspaces*. New York: Academic Press.

E. Hewitt and K. A. Ross [1963]. *Abstract Harmonic Analysis, I*. New York: Springer-Verlag.

E. Hewitt and K. A. Ross [1970]. *Abstract Harmonic Analysis, II*. New York: Springer-Verlag.

E. Hewitt and K. Stromberg [1975]. *Real and Abstract Analysis*. New York: Springer-Verlag.

R. A. Hunt [1967]. On the convergence of Fourier series, orthogonal expansions and their continuous analogies. *Proc. of Conference at Edwardsville, Ill.* Southern Illinois Univ. Press, pp. 235–255.

R. C. James [1951]. A non-reflexive Banach space isometric with its second conjugate space. *Proc. Nat. Acad. Sci. USA*, **37**, 174–177.

R. C. James [1964a]. Weakly compact sets. *Trans. Amer. Math. Soc.*, **113**, 129–140.

R. C. James [1964b]. Weak compactness and reflexivity. *Israel J. Math.*, **2**, 101–119.

R. C. James [1971]. A counterexample for a sup theorem in normed spaces. *Israel J. Math.*, **9**, 511–512.

P. Jordan and J. von Neumann [1935]. On inner products in linear, metric spaces. *Ann. Math.*, **(2)36**, 719–723.

R. V. Kadison and I. M. Singer [1957]. Three test problems in operator theory. *Pacific J. Math.*, **7**, 1101–1106.

T. Kato [1966]. *Perturbation Theory for Linear Operators*. New York: Springer-Verlag.

J. L. Kelley [1955]. *General Topology*. New York: Springer-Verlag.

J. L. Kelley [1966]. Decomposition and representation theorems in measure theory. *Math. Ann.*, **163**, 89–94.

G. Köthe [1969]. *Topological Vector Spaces, I*. New York: Springer-Verlag.

C. Léger [1968]. Convexes compacts et leurs points extrêmaux. *Comptes Rend. Acad. Sci. Paris*, **267**, 92–93.

J. Lindenstrauss [1967]. On complemented subspaces of *m*. *Israel J. Math.*, **5**, 153–156.

J. Lindenstrauss and L. Tzafriri [1971]. On complemented subspaces problem. *Israel J. Math.*, **9**, 263–269.

V. Lomonosov [1973]. On invariant subspaces of families of operators commuting with a completely continuous operator. *Funkcional. Anal. i Prilozen*, **7**, 55–56.

F. J. Murray [1937]. On complementary manifolds and projections in spaces L_p and l_p. *Trans. Amer. Math. Soc.*, **41**, 138–152.

L. Nachbin [1965]. *The Haar Integral*. Princeton: D. Van Nostrand.

I. Namioka and E. Asplund [1967]. A geometric proof of Ryll-Nardzewski's fixed point theorem. *Bull. Amer. Math. Soc.*, **73**, 443–445.

J. von Neumann [1929]. Zur algebra der funktional-operatoren und theorie der normalen operatoren. *Math. Ann.*, **102**, 370–427.

J. von Neumann [1932]. Über einen Satz von Herrn M. H. Stone. *Ann. Math.*, **(2)33**, 567–573.

C. M. Pearcy [1978]. Some recent developments in operator theory. CBMS Series No. 36. Providence: American Mathematical Society.

A. Pelczynski [1960]. Projections in certain Banach spaces. *Studia Math.*, **19**, 209–228.

R. S. Phillips [1940]. On linear transformations. *Trans. Amer. Math. Soc.*, **48**, 516–541.

C. R. Putnam [1951]. On normal operators in Hilbert space. *Amer. J. Math.*, **73**, 357–362.

C. R. Putnam [1968]. The spectra of operators having resolvents of first-order growth. *Trans. Amer. Math. Soc.*, **133**, 505–510.

H. Radjavi and P. Rosenthal [1973]. *Invariant Subspaces*. New York: Springer-Verlag.

C. J. Read [1984]. A solution to the invariant subspace problem. *Bull. London Math. Soc.*, **16**, 337–401.

M. Reed and B. Simon [1975]. *Methods of Modern Mathematical Physics II*. New York: Academic.

M. Reed and B. Simon [1979]. *Methods of Modern Mathematical Physics III. Scattering Theory*. New York: Academic.

C. Rickart [1960]. *General Theory of Banach Algebras*. Princeton: D. Van Nostrand.

J. R. Ringrose [1971]. *Compact Non-self-adjoint Operators*. New York: Van Nostrand-Reinhold.

A. P. Robertson and W. Robertson [1966]. *Topological Vector Spaces*. Cambridge University.

M. Rosenblum [1958]. On a theorem of Fuglede and Putnam. *J. London Math. Soc.*, **33**, 376–377.

P. Rosenthal [1968]. Completely reducible operators. *Proc. Amer. Math. Soc.*, **19**, 826–830.

W. Rudin [1962]. *Fourier Analysis on Groups*. New York: Interscience.

C. Ryll-Nardzewski [1967]. On fixed points of semigroups of endomorphisms of linear spaces. *Fifth Berkeley Sympos. Math. Statist. and Prob., Vol. II: Contrib. to Prob. Theory, Part I*. Berkeley: University of California, pp. 55–61.

S. Sakai [1956]. A characterization of W^*-aglebras. *Pacific J. Math.*, **6**, 763–773.

S. Sakai [1971]. *C^*-algebras and W^*-algebras*. New York: Springer-Verlag.

D. Sarason [1966]. Invariant subspaces and unstarred operator algebras. *Pacific J. Math.*, **17**, 511–517.

D. Sarason [1972]. Weak-star density of polynomials. *J. Reine Angew. Math.*, **252**, 1–15.

D. Sarason [1974]. Invariant subspaces. *Topics in Operator Theory*. Math. Surveys, Vol. **13**, 1–47. Providence: American Mathematical Society.

H. H. Schaeffer [1971]. *Topological Vector Spaces*. New York: Springer-Verlag.

R. Schatten [1960]. *Norm Ideals of Completely Continuous Operators*. Berlin: Springer-Verlag.

M. Schechter [1965]. Invariance of the essential spectrum. *Bull. Amer. Math. Soc.*, **71**, 365–367.

B. Simon [1979]. *Trace Ideals and Their Applications*. London Math. Soc. Lecture Notes, Vol. *35*. Cambridge University.

A. Sobczyk [1941]. Projection of the space (m) on its subspace (c_0). *Bull. Amer. Math. Soc.*, **47**, 938–947.

P. G. Spain [1976]. A generalisation of a theorem of Grothendieck. *Quart. J. Math.*, **(2)27**, 475–479.

J. G. Stampfli [1974]. Compact perturbations, normal eigenvalues, and a problem of Salinas. *J. London Math. Soc.*, **9**, 165–175.

L. A. Steen [1973]. Highlights in the history of spectral theory. *Amer. Math. Monthly*, **80**, 359–381.

E. M. Stein and G. Weiss [1971]. *Introduction to Fourier Analysis on Euclidean Spaces*. Princeton University.

M. H. Stone [1932]. On one-parameter unitary groups in Hilbert space. *Ann. Math.*, **(2)33**, 643–648.

J. Wermer [1952]. On invariant subspaces of normal operators. *Proc. Amer. Math. Soc.*, **3**, 276–277.

R. J. Whitley [1966]. Projecting m onto c_0. *Amer. Math. Monthly*, **73**, 285–286

F. Wolf [1959]. On the invariance of the essential spectrum under a change of boundary conditions of partial differential operators. *Indag. Math.*, **21**, 142–147.

W. Zelazko [1968]. A characterization of multiplicative linear functionals in complex Banach algebras. *Studia Math.*, **30**, 83–85.

List of Symbols

Index

Graduate Texts in Mathematics

continued from page ii

Home Buying Kit

7th Edition

by Eric Tyson, MBA, and Ray Brown

A Wiley Brand

Home Buying Kit For Dummies®, 7th Edition

Published by: **John Wiley & Sons, Inc.**, 111 River Street, Hoboken, NJ 07030-5774, www.wiley.com

Copyright © 2020 by Eric Tyson and Ray Brown

Media and software compilation copyright © 2020 by John Wiley & Sons, Inc. All rights reserved.

Published simultaneously in Canada

For general information on our other products and services, please contact our Customer Care Department within the U.S. at 877-762-2974, outside the U.S. at 317-572-3993, or fax 317-572-4002. For technical support, please visit www.wiley.com/techsupport.

Wiley publishes in a variety of print and electronic formats and by print-on-demand. Some material included with standard print versions of this book may not be included in e-books or in print-on-demand. If this book refers to media such as a CD or DVD that is not included in the version you purchased, you may download this material at http://booksupport.wiley.com. For more information about Wiley products, visit www.wiley.com.

Library of Congress Control Number: 2020932608.

ISBN: 978-1-119-67479-5

ISBN 978-1-119-67481-8 (ebk); ISBN 978-1-119-67482-5 (ebk)

Manufactured in the United States of America

SKY10031413_111121

Contents at a Glance

Table of Contents

Introduction

Welcome to *Home Buying Kit For Dummies*, 7th Edition!

For about the cost of a couple of movie tickets or your monthly Netflix subscription, you can quickly and easily discover how to save thousands — perhaps even tens of thousands — of dollars the next time you buy a home.

How can we make such a claim? Simple. Each of us has spent decades personally advising thousands of people like you about home purchases and other important financial decisions. We've seen how ignorance of basic concepts and practices translates into money-draining mistakes. And we know that many of these mistakes are both needless and avoidable.

REMEMBER

No one is born knowing how to buy a home. Everyone who'd like to buy a home must learn how to do it. Unfortunately, too many people get a crash course in the school of hard knocks — and learn by making costly mistakes at their own expense.

We know you're not a dummy. You've already demonstrated an interest in discovering more about home buying by selecting this book, which can help you make smart moves and avoid financial land mines.

In the event that you're still wondering whether to buy this book, consider that buying a home may well be the largest purchase that you ever make. Buying a home can send shock waves through your personal finances and may even cause a sleepless night or two. Purchasing a home is a major financial step and a life event for most people. It certainly was for us when we bought our first homes. You owe it to yourself to do things right.

About This Book: The Eric Tyson/Ray Brown Difference

We know that many home-buying books are competing for your attention. Here are several other compelling reasons why this is the best book for you:

- » **It's in plain English.** Because we work with real people and answer real questions, our information is current, and we have a great deal of experience in explaining things. This experience can put you firmly in control of the home-buying process (rather than having it control you).

- » **It's objective.** We're not trying to sell you an expensive newsletter, seminar, or some real estate product you don't need. Our goal is to make you as knowledgeable as possible before you purchase a home. We even explain why you may *not* want to buy a home. We're not here to be real estate cheerleaders.

- » **It's holistic.** When you purchase a home, that purchase affects your ability to save money and accomplish other important financial goals. We help you understand how best to fit your home acquisition into the rest of your personal-finance plan.

- » **It's a reference.** You can read this book from cover to cover if you want. However, we know you're busy and you likely don't desire to become a real estate expert, so each portion of the book stands on its own. You can read it piecemeal to address your specific questions and immediate concerns.

Icons Used in This Book

Sprinkled throughout this book are cute little icons to help reinforce and draw attention to key points or to flag stuff that you can skip.

TIP

This bull's-eye notes key strategies that can improve your real estate deal and, in some cases, save you lots of moola. Think of these as helpful little paternalistic hints we would whisper in your ear if we were close enough to do so!

WARNING

Numerous land mines await novice as well as experienced home buyers. This explosive symbol marks those mines, and then we tell you how to sidestep them.

INVESTIGATE

Occasionally, we suggest that you do more research or homework. Don't worry: We tell you exactly what you need to do.

REMEMBER

"If I've told you once, I've told you a thousand times. . . ." Remember good old Mom and Dad? From time to time, we tell you something quite important and perhaps repeat ourselves. Just so you don't forget the point, this icon serves as a little nag to bring back those childhood memories.

TECHNICAL STUFF

Some of you are curious and have time to spare. Others are busy and just want to know the essentials. This geeky icon points out tidbits and information that you don't really have to know, but understanding this stuff can make you more self-confident and proud!

CHECK IT OUT

Want to calculate your monthly expenses? Or have a handy-dandy list available when you interview Realtors and home inspectors? Wanna see what your bank may include on a mortgage application? These along with lots of other important stuff can be found online at www.dummies.com/go/homebuyingkit7e.

Beyond the Book

In addition to the material in the print or e-book you're reading right now, this product also comes with some access-anywhere goodies on the web. Check out the free Cheat Sheet at www.dummies.com/cheatsheet/homebuyingkit for 20 home-buying tips and a monthly mortgage payment calculator.

Also, go online to www.dummies.com/go/homebuyingkit7e to access the "kit" part of this book. You'll find a variety of useful forms, including many of the lists and applications that we show in the book. You can print out the application forms and fill them in, just to ensure you have all the information you need. And you can also print out the lists of questions for potential Realtors and property inspectors so you're prepared when interviewing them.

Here's a list of what you'll find at www.dummies.com/go/homebuyingkit7e:

Table 1-1	Figuring future rent
Table 2-1	Your spending, now and after purchasing a home
Table 3-1	Monthly mortgage payment calculator
Chapter 3	Estimated homeownership expenses
Chapter 3	1040 Schedule A

Where to Go from Here

Odds are you're not quite ready to bolt over to the nearest bank and take out a mortgage — and we don't suggest that you blindly call the first Realtor you find online. It's up to you where you go from here, but if you're just beginning to think about buying your first home, we recommend that you read this book straight through, cover to cover, to maximize your home-buying savvy. But the A-to-Z approach isn't necessary — if you feel pretty confident in your knowledge of certain areas, pick other ones that you're most interested in by either skimming this book's table of contents or by relying on the well-crafted index at the back of the book.

1

Getting Started with Buying a Home

Chapter 1

Deciding Whether to Buy

E very month, week, and day, we buy things large and small: lunch, a new pair of shoes, and every now and then, a car.

Most people buy things without doing much comparison shopping, but instead draw upon their past experiences. When the counter help at the nearby coffeehouse is friendly and you like the brew, you go back for more.

Sometimes purchases lead you by association to related purchases. You get a pet cat or dog, for instance, and buying a collar and pet toys may naturally follow. By the same token, you buy a home, and before long you have a new sofa and dining table.

You end up being really happy with some items you purchase. Others fall short of your expectations . . . or worse. When the items in question don't cost you much, it's no big deal. Perhaps you return them or simply don't buy more in the future. But when it comes to buying a home, that kind of sloppy shopping can lead to financial and emotional disaster.

If you're not willing to invest time, and if you don't work with and heed the advice of the best people, you can end up overpaying for a home you hate. Our goals in this book are simple: to ensure that you're happy with the home you buy, that you get the best deal you can, and that owning the home helps you accomplish your financial goals.

Weighing the Advantages of Owning versus Renting

Nearly everyone seems to have an opinion about buying a home. People in the real estate business — including agents, lenders, property inspectors, and other related people — endorse homeownership. Of course, why wouldn't they? Their livelihoods depend upon it! Therein lies one fundamental problem of nearly all home-buying books written by people who have a vested interest in convincing their readers to buy a home.

REMEMBER

Homeownership isn't for everyone. One of our objectives in this chapter is to help you determine whether home buying is right for you.

Consider the case of Peter, who thought that owning a home was the best financial move he could make. What with tax write-offs and living in a place while it made money for him, he thought how could he lose? Peter envied his colleagues at work who'd seemingly made piles of money with property they bought years ago. Peter was a busy man and didn't have time to research other ways to invest his money.

Unfortunately, Peter bought a place that stretched his budget and required lots of attention and maintenance. Adding insult to injury, Peter went to graduate school clear across the country (something he knew he was likely to do at the time he bought) three years after he purchased. During these three years of his ownership, home prices dropped 10 percent in Peter's neighborhood. So after paying the expenses of sale and closing costs, Peter ended up losing his entire down payment when he sold.

Conversely, some people who continue to rent should buy. In her late 20s, Melody didn't want to buy a home, because she didn't like the idea of settling down. Her monthly rent seemed so cheap compared with the sticker prices on homes for sale.

As it always does, time passed. Melody's 20s turned into 30s, which melted into 40s and then 50s, and she was still renting. Her rent skyrocketed to eight times what it was when she first started renting. She fearfully looked ahead to escalating rental rates in the decades when she hoped to be retired.

Ownership advantages

Most people should eventually buy homes, but not everyone and not at every point in their lives. To decide whether now's the time for you to buy, consider the advantages of buying and whether they apply to you.

Owning should be less expensive than renting

You probably didn't appreciate it growing up, but in addition to the diaper changes, patience during potty training, help with homework, bandaging of bruised knees, and countless meals, your folks made sure that you had a roof over your head. Most of us take shelter for granted, unless we don't have it or are confronted for the first time with paying for it ourselves.

Remember your first apartment when you graduated from college or when your folks finally booted you out? That place probably made you appreciate the good deal you had before — even those cramped college dormitories may have seemed more attractive!

But even if you pay several hundred to a thousand dollars or more per month in rent, that expense may not seem so steep if you happen to peek at a home for sale. In most parts of the United States, we're talking about a big number — $150,000, $225,000, $350,000, or more for the sticker price. (Of course, if you're a higher-income earner, you may think that you can't find a habitable place to live for less than a half-million dollars, especially if you live in costly places such as New York City, Boston, Chicago, Los Angeles, or San Francisco.)

TIP

Here's a guideline that may change the way you view your seemingly cheap monthly rent. To figure out the price of a home you can buy for approximately the same monthly cost as your current rent, simply do the following calculation:

Take your monthly rent and multiply by 200, and you come up with the purchase price of a home.

$ _____ per month × 200 = $ _____

Example: $ 1,000 × 200 = $200,000

So, in the preceding example, if you were paying rent of $1,000 per month, you would pay approximately the same amount per month to own a $200,000 home (factoring in modest tax savings). Now your monthly rent doesn't sound quite so cheap compared with the cost of buying a home, does it? (Note that in Chapter 3 we show you how to accurately calculate the total costs of owning a home.)

WARNING

Even more important than the cost *today* of buying versus renting is the cost in the *future.* As a renter, your rent is fully exposed to increases in the cost of living, also known as *inflation.* A reasonable expectation for annual increases in your rent is 4 percent per year. Figure 1-1 shows what happens to a $1,000 monthly rent at just 4 percent annual rental inflation.

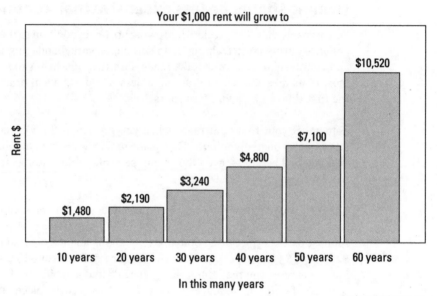

Your $1,000 rent will grow to

FIGURE 1-1:
The skyrocketing
cost of renting.

© *John Wiley & Sons, Inc.*

When you're in your 20s or 30s, you may not be thinking or caring about your golden years, but look what happens to your rent over the decades ahead with just modest inflation! Then remember that paying $1,000 rent per month now is the equivalent of buying a home for $200,000. Well, in 40 years, with 4 percent inflation per year, your $1,000-per-month rent will balloon to $4,800 per month. That's like buying a house for $960,000!

**CHECK IT
OUT**

In our example, we picked $1,000 for rent to show you what happens to that rent with a modest 4 percent annual rate of inflation. To see what may happen to your current rent at that rate of inflation (as well as at a slightly higher one), simply complete Table 1-1. (You can also access Table 1-1 online at www.dummies.com/go/ homebuyingkit7e.)

REMEMBER

If you're middle-aged or retired, you may not plan on having 40 to 60 years ahead of you. On the other hand, don't underestimate how many more years of housing you'll need. U.S. health statistics indicate that at age 50, you have a life expectancy of 30+ more years, and at age 65, 20+ more years (women on average tend to live a few years longer).

TABLE 1-1 **Figuring Future Rent**

Your Current Monthly Rent	Multiplication Factor to Determine Rent in Future Years at 4 Percent Annual Inflation Rate	Projected Future Rent
$_____	× 1.48	= $_____ in 10 years
$_____	× 2.19	= $_____ in 20 years
$_____	× 3.24	= $_____ in 30 years
$_____	× 4.80	= $_____ in 40 years
$_____	× 7.11	= $_____ in 50 years
$_____	× 10.52	= $_____ in 60 years
Your Current Monthly Rent	Multiplication Factor to Determine Rent in Future Years at 6 Percent Annual Inflation Rate	Projected Future Rent
$_____	× 1.79	= $_____ in 10 years
$_____	× 3.21	= $_____ in 20 years
$_____	× 5.74	= $_____ in 30 years
$_____	× 10.29	= $_____ in 40 years
$_____	× 18.42	= $_____ in 50 years
$_____	× 32.99	= $_____ in 60 years

Although the cost of purchasing a home generally increases over the years, after you purchase a home, the bulk of your housing costs aren't exposed to inflation if you use a fixed-rate mortgage to finance the purchase. As we explain in Chapter 6, a *fixed-rate mortgage* locks your mortgage payment in at a fixed amount (as opposed to an adjustable-rate mortgage payment that fluctuates in value with changes in interest rates). Therefore, only the comparatively smaller property taxes, insurance, and maintenance expenses will increase over time with inflation. (In Chapter 3, we cover in excruciating detail what buying and owning a home costs.)

REMEMBER

You're always going to need a place to live. And over the long term, inflation has almost always been around. Even if you must stretch a little to buy a home today, in the decades ahead, you'll be glad you did. The financial danger with renting long term is that *all* your housing costs (rent) increase over time. We're not saying that everyone should buy because of inflation, but we do think that if you're not going to buy, you should be careful to plan your finances accordingly. We discuss the pros and cons of renting later in this chapter.

USES FOR THE WEALTH YOU BUILD UP IN YOUR HOME

Over the many years that you're likely to own it, your home should become an important part of your financial *net worth* — that is, the difference between your *assets* (financial things of value that you own, such as bank accounts, retirement accounts, stocks, bonds, mutual funds, and so on) and your *liabilities* (debts). Why? Because homes generally increase in value over the decades while you're paying down the loan used to buy the home. Remember, we're talking about the long term here — decades, not just a few years. The housing market goes through downturns — the late 2000s being the most recent down period — but the long-term trend has always been higher.

Even if you're one of those rare people who own a home but don't see much *appreciation* (increase in the home's value) over the decades of your adult ownership, you still benefit from the monthly forced savings that result from paying down the remaining balance due on your mortgage. Older folks can tell you that owning a home free and clear of a mortgage is a joy.

All that home *equity* (the difference between the market value of a home and the outstanding loan on the home) can help your personal and financial situation in a number of ways. If, like most people, you hope to someday retire but (also like most people) saving doesn't come easily, your home's equity can help supplement your other sources of retirement income.

How can you tap into your home's equity? Here are three main ways:

- Some people choose to *trade down* — that is, to move to a less costly home in retirement. Sell your home for $500,000, replace it with one that costs $300,000, and you've freed up $200,000. Subject to certain requirements, you can sell your home and realize up to $250,000 in tax-free profits if you're single, or $500,000 if married. (See Chapter 17 to find out more about this homeownership tax break.)

- Another way to tap your home's equity is through borrowing. Your home's equity may be an easily tapped and low-cost source of cash (the interest you pay is generally tax-deductible — see Chapter 3).

- Some retirees also consider what's called a *reverse mortgage.* Under this arrangement, the lender sends you a monthly check that you can spend however you want. Meanwhile, a debt balance (that will be paid off when the property is finally sold) is built up against the property.

What can you do with all this home equity? Help pay for your children's college education, start your own business, remodel your home, use it toward your retirement, or whatever!

You can make your house your own

Think back to all the places you ever rented, including the rental in which you may currently be living. For each unit, make a list of the things you really didn't like that you would have changed if the property were yours: ugly carpeting, yucky exterior paint job, outdated appliances that didn't work well, and so on.

Although we know some tenants who actually do some work on their own apartments, we don't generally endorse this approach because it takes your money and time but financially benefits the building's owner. If, through persistence and nagging, you can get your landlord to make the improvements and repairs at her expense, great! Otherwise, you're out of luck or cash!

When you own your own place, however, you can do whatever you want to it. Want hardwood floors instead of ugly, green shag carpeting? Tear it out. Love neon-orange carpeting and pink exterior paint? You can add it!

WARNING

In your zest and enthusiasm to buy a place and make it your own, be careful of two things:

>> **Don't make the place too weird.** You'll probably want or need to sell your home someday, and the more outrageous you make it, the fewer buyers it will appeal to — and the lower the price it will likely fetch. If you don't mind throwing money away or are convinced that you can find a buyer with similarly (ahem) sophisticated tastes, be as unusual as you want. If you do make improvements, focus on those that add value: a deck addition for an outdoor living area, updated kitchens and bathrooms, and so on.

>> **Beware of running yourself into financial ruin.** Changing, improving, remodeling, or whatever you want to call it costs money. We know many home buyers who neglect other important financial goals (such as saving for retirement and their kids' college costs) in order to endlessly renovate their homes. Others rack up significant debts that hang like financial weights over their heads. In the worst cases, homes become money pits that cause owners to build up high-interest consumer debt as a prelude to bankruptcy or foreclosure.

You avoid unpleasant landlords

A final (and not inconsequential) benefit of owning your own home is that you don't have to subject yourself to the whims of an evil landlord. Much is made among real estate investors of the challenges of finding good tenants. As a tenant, perhaps you've already discovered that finding a good landlord isn't easy, either.

REMEMBER

The fundamental problem with some landlords is that they're slow to fix problems and make improvements. The best (and smartest) landlords realize that keeping the building shipshape helps attract and keep good tenants and maximizes rents and profits. But to some landlords, maximizing profits means being stingy with repairs and improvements.

When you own your home, the good news is that you're generally in control — you can get your stopped-up toilet fixed or your ugly walls painted whenever and however you like. No more hassling with unresponsive, obnoxious landlords. The bad news is that you're responsible for paying for and ensuring completion of the work. Even if you hire someone else to do it, you still must find competent contractors and oversee their work, neither of which is an easy responsibility.

Another risk of renting is that landlords may decide to sell the building and put you out on the street. You should ask your prospective landlords whether they have plans to sell. Some landlords won't give you a truthful answer, but the question is worth asking if this issue is a concern to you.

TIP

One way to avoid being jilted by a wayward landlord is to request that the lease contract guarantee you the right to renew your annual lease for a certain number of years, even with a change in building ownership. Unless landlords are planning on selling, and perhaps want to be able to boot you out, they should be delighted with a request that shows you're interested in staying a while. Also, by knowing if and when a landlord desires to sell, you may be able to be the buyer!

Renting advantages

Buying and owning a home throughout most of your adult life makes good financial and personal sense for most people — but not all people and not at all times. Renting works better for some people. The benefits of renting are many:

>> **Simplicity:** Yes, searching for a rental unit that meets your needs can take more than a few days (especially if you're in a tight rental market), but it should be a heck of a lot easier than finding a place to buy. When you buy, you must line up financing, conduct inspections, and deal with myriad other issues that renters never have to face. When you do it right, finding and buying a good home can be a time-consuming pain in the posterior.

>> **Convenience:** After you find and move into your rental, your landlord is responsible for the never-ending task of property maintenance and upkeep. Buildings and appliances age, and bad stuff happens: Fuses blow, plumbing backs up, heaters break in the middle of winter, roofs spring leaks during record-breaking rainfalls, trees come crashing down during windstorms. The list goes on and on and on. As a renter, you can kick back in the old recliner

with your feet up, a glass of wine in one hand and the remote control in the other, and say, "Ahhhhh, the joys of *not* being part of the landed gentry!"

» **Flexibility:** If you're the footloose and fancy-free type, you dislike feeling tied down. With a rental, as long as your lease allows (and most leases don't run longer than a year), you can move on. As a homeowner, if you want to move, you must deal with the significant chores of selling your home or finding a tenant to rent it.

» **Increased liquidity:** Unless you're the beneficiary of a large inheritance or work at a high-paying job, you'll probably be financially stretched when you buy your first home. Coming up with the down payment and closing costs usually cleans out most people's financial reserves. In addition, when you buy a home, you must meet your monthly mortgage payments, property taxes, insurance, and maintenance and repair expenses. As a renter, you can keep your extra cash to yourself, and budgeting is also easier without the upkeep-expense surprises that homeowners enjoy, such as the sudden urge to replace a leaking roof or old furnace.

TIP

You don't need to buy a home to cut your taxes. Should you have access to a retirement account such as a 401(k), 403(b), or SEP-IRA plan (see Chapter 2), you can slash your taxes while you save and invest your extra cash *as a renter*. So saving on taxes shouldn't be the sole motivation for you to buy a home.

» **Better diversification:** Many homeowners who are financially stretched have the bulk of their wealth tied up in their homes. As a renter, you can invest your money in a variety of sound investments, such as stocks, bonds, and perhaps your own small business. You can even invest a small amount of money in real estate through stocks or mutual funds if you want (see Chapter 16). Over the long term, the stock market has produced comparable rates of return to investing in the real estate market.

» **Maybe lower cost:** If you live in an area where home prices have rocketed ahead much faster than rental rates, real estate may be overpriced and not a good buy. In Chapter 4, we explain how to compare the cost of owning to the cost of renting in your area and how to spot a potentially overpriced real estate market.

REMEMBER

Renting should also be cheaper than buying if you expect to move soon. Buying and selling property costs big bucks. With real estate agent commissions, loan fees, title insurance, inspections, and all sorts of other costs, your property must appreciate approximately 15 percent just for you to break even and recoup these costs. Therefore, buying property that you don't expect to hold onto for at least three (and preferably five or more) years doesn't make much economic sense. Although you may sometimes experience appreciation in excess of 15 percent over a year or two, most of the time, you won't. If you're counting on such high appreciation, you're setting yourself up for disappointment.

The Pitfalls of the Rent-versus-Buy Decision

When you're considering purchasing a home, you can do lots of reflecting, crunch plenty of numbers, and conduct copious research to help you with your decision. We encourage these activities and show you how to do them in later chapters.

In reality, we know that many people are tempted to jump into making a decision about buying or continuing to rent without setting all their ducks in a row. At a minimum, we want to keep you from making common costly mistakes, so in the following sections, we go over the biggies to avoid.

Renting because it seems cheaper

When you go out to look at homes on the market *today*, the sticker prices are typically in the hundreds of thousands of dollars. Your monthly rent seems dirt-cheap by comparison.

REMEMBER

You must compare the *monthly* cost of homeownership with the monthly cost of renting. And you must factor in the tax savings you'll realize from homeownership tax deductions. (We show you how to make these calculations in Chapter 3.) But you must also think about the future. Just as your educational training affects your career prospects and income-earning ability for years to come, your rent-versus-buy decision affects your housing costs — not just this year, but also for years and decades to come.

Fretting too much over job security

Being insecure about your job is natural. Most people are — even corporate executives, superstar athletes, and movie stars. And buying a home seems like such a permanent thing to do. Job-loss fears can easily make you feel a financial noose tightening around your neck when you sit down to sign a contract to purchase a home.

Although a few people have real reasons to worry about losing their jobs, the reality is that the vast majority of people shouldn't worry about job loss. We don't mean to say that you *can't* lose your job — almost anyone can, in reality. Just remember that within a reasonable time, your skills and abilities will allow you to land back on your feet in a new, comparable position. We're not career experts, but we've witnessed many folks bounce back in just this way.

TIP

When losing your job is a high likelihood, and especially if you'd have to relocate for a new job, consider postponing the purchase of a home until your employment situation stabilizes. (If you haven't demonstrated a recent history of stable employment, most mortgage lenders won't want to lend you money anyway — see Chapters 6 and 7.) When you must move to find an acceptable or desirable job, selling your home and then buying another one can cost you thousands, if not tens of thousands, of dollars in transaction fees.

Buying when you expect to move soon

People move for many reasons other than job loss. You may want to move soon to advance your career, to be nearer to (or farther from!) relatives, to try living somewhere new, or just to get away from someplace old.

Unless you're planning to hold onto your home and convert it to a rental when you move, buying a home rarely makes sound financial sense when you expect to move within three years. Ideally, stay put for at least five years.

Succumbing to pushy salespeople

When you buy a house, you're the one who'll be coming home to it day after day — and you're the one on the hook for all the expenses. Don't ever forget these facts when you plunge into the thick of possibly purchasing a home. If you have lingering doubts about buying a home, apply the brakes.

WARNING

Many people involved in home-buying transactions have a vested interest in getting you to buy. They may push you to buy sooner (and buy more) than you intend to or can afford, given your other financial goals and obligations. The reasons: Many people who make their living in the real estate trade get paid only if and when you buy, and the size of their earnings depends upon how much you spend. In Chapter 9, we show you how to put together the best team to assist you in making a decision rather than push you into making a deal.

Ignoring logistics

Sometimes, when looking at homes, you can lose your perspective on big-picture issues. After months of searching, Frederick finally found a home that met his needs for both space and cost. He bought the home and moved in on a Saturday. Come Monday morning, Frederick hopped in his car and spent the next hour commuting. At the end of his workday, it was the same thing coming home. He was tired and grumpy when he arrived home Monday evening, and after making dinner for himself, he soon had to hit the hay to rise early enough to do it all over again on Tuesday.

Initially, Frederick hoped that the trying traffic was an aberration that would go away — but no such luck. In fact, on many days, his commute was worse than an hour each way. Frederick grew to hate his commute, his job, and his new home.

REMEMBER

When you buy a home, you're also buying the commute, the neighborhood, its amenities, and all the other stuff that comes along for the literal and figurative ride. Understand these issues *before* you buy. In the end, after 18 months of commuter purgatory, Frederick sold his home and went back to renting much closer to his job. Forgetting to consider what the commute from a home to his job would entail was an expensive lesson for Frederick. Don't make the same mistake Frederick made; take your time and consider all the important factors about the home you're thinking about purchasing.

Overbuying

Many first-time home buyers discover that their desires outstrip their budgets. Nelson and his wife, Laura, had good jobs and together made in excess of $150,000 per year. They got used to buying what they desired — they ate at fancy restaurants, took luxury vacations, and otherwise indulged themselves.

When it came time to purchase a home, they spent the maximum amount and borrowed the maximum amount that the mortgage person told them they could. After the home purchase, Laura got pregnant and eventually left her job to spend more time at home. With the high homeownership expenses, kid costs, and reduced household income, Nelson and Laura soon found themselves struggling to pay their monthly bills and started accumulating significant credit-card debts. Ultimately, they ended up filing bankruptcy.

REMEMBER

Either you own the home, or it owns you. Get your finances in order and understand how much you can truly afford to spend on a home before you buy (see Chapters 2 and 3).

Underbuying

Remember in the story *Goldilocks and the Three Bears* how Goldilocks had difficulty finding porridge to her liking? In one case, it was too cold, and in another, too hot. Well, just as you can overbuy when selecting a home, you can underbuy. That's what Nathan and Rebecca did when they bought their first home. They believed in living within their means — a good thing — but they took it to an extreme.

Nathan and Rebecca bought a home whose cost was far below the maximum amount they could have afforded. They borrowed $70,000 when they could have afforded to borrow three times that amount. They knew when they bought the home that they'd want to move to a bigger house within just a few years. Although this made the real estate agents and lenders happy, all the costs of buying and then selling soon after gobbled a huge chunk of Nathan and Rebecca's original down payment.

Buying because it's a grown-up thing to do

Peer pressure can be subtle or explicit. Some people even impose pressure on themselves. Buying a home is a major milestone and a tangible display of financial maturity and success. If your friends, siblings, and co-workers all seem to be homeowners, you may sometimes feel as though you're being a tad juvenile by not jumping on the same train.

Everyone has different needs, but not everyone should own a home, and certainly not at every point in his adult life. Besides, although they may never admit it, some homeowning friends and colleagues are jealous of you and other financially footloose and fancy-free renters.

A study even supports the notion that the life of a typical renter is, in some respects, better than that of the average homeowner. Peter Rossi and Eleanor Weber of the University of Massachusetts Social and Demographic Research Institute conducted a survey of thousands of people. Here are some of their findings:

>> Homeowners are less social, on average, than renters — spending less time with friends, neighbors, and co-workers.

>> Homeowners spend more time on household chores.

>> Perhaps for the preceding reasons, renters have more sex and less marital discord and cope better with parenting than homeowners do!

Buying because you're afraid that escalating prices will lock you out

From time to time, particular local real estate markets experience rapidly escalating prices. During such times, some prospective buyers panic, often with encouragement from those with a vested interest in converting prospective renters to buyers. Escalating housing prices make some renters feel left out of the party. Booming housing prices get all sorts of publicity, including from gloating homeowners clucking over their equity.

Never in the history of the real estate business have prices risen so high as to price vast numbers of people out of the market. In fact, patient buyers who can wait out a market that has increased sharply in value are often rewarded with steadying and, in some cases, declining prices (witness what happened in the late 2000s). Although you won't be locked out of the market forever, you should keep in mind that if you postpone buying for many years, you'll likely be able to buy less home for your money thanks to home prices increasing faster than the rate of inflation.

Misunderstanding what you can afford

When you make a major decision, be it personal or financial, it's perfectly natural and human to feel uncomfortable if you're flying by the seat of your pants and don't have enough background. With a home purchase, if you haven't examined your overall financial situation and goals, you're just guessing how much you should be spending on a home.

Again, the vested-interest folks won't generally bring this issue to your attention — partly because of their agendas and motivations, but also because it's not what they're trained and expert at doing. Look in the mirror to see the person who can help you with these important issues. (Chapter 2 walks you through all the important personal financial considerations you should explore before you set out on your buying expedition.)

Chapter 2

Getting Your Financial House in Order

When you're shopping for a home, you're the person best suited to look out for your overall interests. The people involved in typical real estate deals (such as real estate agents, bankers, loan brokers, and the like) are there to get their jobs done. It's *not* within their realm of responsibility to worry about how the real estate purchase fits with the rest of your personal finances and how best to arrange your finances before and after purchasing a home. This chapter explains how you can address these important issues.

WARNING

In the great history of home buying, many people have bought real estate without first getting their finances in order, setting some goals, and dealing with problems — and they've often paid dearly for this oversight. What are the consequences of plunging headlong into a home purchase before you're financially ready? For starters, you can end up paying tens of thousands of dollars more in taxes and interest over the years ahead. In the worst cases, we've witnessed the financial ruin of intelligent, hardworking people who end up over their heads in debt (and in some situations, even in foreclosure and bankruptcy). We want you to be happy and financially successful in your home — *so please read this chapter!*

Surveying Your Spending

Even if your income and spending fluctuate, you may have developed a basic spending routine. Every month, you earn a particular income and then spend most of, all of, or perhaps even more than what you earn on the necessities (and the not-so-necessary things) of life.

TIP

When you want to buy a home, saving is one area where it pays to be above average. Consistently saving more than 5 to 10 percent of your income can help turn you from a renter into a financially able and successful homeowner. Why? For two important reasons:

>> To purchase a home, you need to accumulate a decent chunk of money for the down payment and closing costs. True, wealthy relatives may help you out, but counting on their generosity is foolhardy. The attached strings may make such a gift or loan undesirable. If you're like most people, you probably don't have any wealthy relatives anyhow.

>> After you buy a home, your total monthly expenses will probably increase. So if you had trouble saving before the purchase, your finances are really going to be squeezed postpurchase. This will further handicap your ability to accomplish other important financial goals, such as accumulating money for retirement. If you don't take advantage of tax-sheltered retirement accounts, you'll miss out on thousands (if not tens of thousands) of dollars in valuable tax benefits. We discuss the importance and value of funding retirement accounts later in this chapter.

Gathering the data

One of the single most important things that you can and should do before you head out to purchase a home is to examine where (and on what) you're currently spending your money. Completing these financial calisthenics enables you to see what portion of your current income you're saving. Having a handle on your current budget also enables you to see how a given home purchase will fit within the budget or destroy it!

TIP

Review your spending data from at least a three-month span to determine how much you spend in a typical month on various things — such as rent, clothing, income taxes, haircuts, cellphone and streaming plans, and everything else (see Table 2-1; you can access Table 2-1 online at www.dummies.com/go/homebuying kit7e). If your spending fluctuates greatly throughout the year, you may need to analyze and average for 6 (or even 12) months to get an accurate sense of your spending behavior.

TABLE 2-1

Your Spending, Now and After Purchasing a Home

Item	Current Monthly Average ($)	Expected Monthly Average with Home Purchase ($)
Income	_____	_____
Taxes		
Social Security	_____	_____
Federal	_____	_____
State and local	_____	_____
Housing Expenses		
Rent	_____	n/a
Mortgage	n/a	_____
Property taxes	n/a	_____
Homeowners/renters insurance	_____	_____
Gas/electric/oil	_____	_____
Water/garbage	_____	_____
Phone/cellphone	_____	_____
Cable/satellite TV/streaming plans	_____	_____
Furniture/appliances	_____	_____
Maintenance/repairs	_____	_____
Food and Eating		
Supermarket	_____	_____
Restaurants and takeout	_____	_____
Transportation		
Gasoline	_____	_____
Maintenance/repairs	_____	_____
State registration fees	_____	_____
Tolls and parking	_____	_____
Bus or subway fares	_____	_____

(continued)

TABLE 2-1 *(continued)*

Item	Current Monthly Average ($)	Expected Monthly Average with Home Purchase ($)
Appearance		
Clothing	_____	_____
Shoes	_____	_____
Jewelry (watches, earrings)	_____	_____
Dry cleaning	_____	_____
Haircuts	_____	_____
Makeup	_____	_____
Other	_____	_____
Debt Repayments		
Credit/charge cards	_____	_____
Auto loans	_____	_____
Student loans	_____	_____
Other	_____	_____
Fun Stuff		
Entertainment (movies, concerts)	_____	_____
Vacation and travel	_____	_____
Gifts	_____	_____
Hobbies	_____	_____
Pets	_____	_____
Health club or gym	_____	_____
Other	_____	_____
Advisors		
Accountant	_____	_____
Attorney	_____	_____
Financial advisor	_____	_____
Healthcare		
Physicians and hospitals	_____	_____
Drugs	_____	_____

Item	Current Monthly Average ($)	Expected Monthly Average with Home Purchase ($)
Dental and vision	_____	_____
Therapy	_____	_____
Insurance		
Auto	_____	_____
Health	_____	_____
Life	_____	_____
Disability	_____	_____
Educational Expenses		
Courses	_____	_____
Books	_____	_____
Supplies	_____	_____
Kids		
Day care	_____	_____
Toys	_____	_____
Child support	_____	_____
Charitable Donations/Offerings	_____	_____
Other		
_____	_____	_____
_____	_____	_____
_____	_____	_____
_____	_____	_____
_____	_____	_____
Total Spending	_____	_____
Amount Saved	_____	_____

(subtract from income at the beginning of this table)

TRIMMING THE FAT FROM YOUR BUDGET

Most people planning to buy a home need to reduce their spending in order to accumulate enough money for the down payment and closing costs and to create enough slack in their budget to afford the extra costs of homeownership. (Increasing your income is another strategy, but that's usually more difficult to do.) Where you decide to make cuts in your budget is a matter of personal preference — but unless you're independently wealthy or a spendthrift, cut you must.

First, get rid of any and all consumer debt — such as that on credit cards and auto loans. Ridding yourself of such debt as soon as possible is vital to your long-term financial health. Consumer debt is as harmful to your financial health as smoking is to your personal health. Borrowing through consumer loans encourages you to live beyond your means and do the opposite of saving — call it "dis-saving" (or *deficit financing,* as those in Washington, D.C., say). The interest rates on consumer debt are high, and unlike the interest on a mortgage, the interest on consumer debt isn't tax-deductible, so you bear the full brunt of its cost.

Should you have accessible savings to pay down your consumer debts, by all means use those savings. You're surely paying a higher interest rate on such debt than you're earning from interest on your savings. Plus, interest on your savings is taxable. Just be sure you have access to sufficient emergency money through family or other means.

If you lack the savings to make your high-cost debts disappear, start by refinancing your high-cost credit-card debt onto cards with lower interest rates. Then work at reducing your spending to free up cash to pay down these debts as quickly as possible. And if you have a tendency to run up credit-card balances, consider getting rid of your credit cards and obtaining a Visa or MasterCard debit card. These debit cards look like credit cards and are accepted the same as credit cards by merchants, but they function like checks. When you make a purchase with a debit card, the money is deducted from your checking account within a day or two.

Trim unnecessary items from your budget. Even if you're not a high-income earner, some of the things you spend your money on aren't necessities. Although everyone needs food, shelter, clothing, and healthcare, people spend a great deal of additional money on luxuries and nonessentials. Even some of what we spend on the "necessity" categories is partly for luxury.

Purchase products and services that offer value. High quality doesn't have to cost more. In fact, higher-priced products and services are sometimes inferior to lower-cost alternatives.

Finally, buy in bulk. Most items are cheaper per unit when you buy them in larger sizes or volumes. Wholesale superstores such as Costco and Sam's Club and chain discount stores like Target offer family sizes and competitive pricing. Also watch for sales at local grocery and discount stores.

Financial software packages, such as Quicken, can help with the task of tracking and analyzing your spending, but old-fashioned paper and pencil work fine, too. What you need to do is assemble information that shows what you typically spend your money on. Access your checking account records, credit- and charge-card bills and transactions, online bill payment and banking summary, job pay stub, and your most recent tax return.

Whether you use our handy-dandy table or your own software isn't important. What does matter is that you capture the bulk of your spending. But you don't need to account for 100 percent of your spending and track every last penny (or even every last dollar). You're not designing an airplane or performing a financial audit for a major accounting firm here!

TIP

As you collect your spending data and consider your home purchase, think about how that purchase will affect and change your spending and ability to save. For example, as a homeowner, if you live farther away from your job than you did when you rented, how much will your transportation expenses increase? Also note that in Chapter 3, we walk you through estimating homeownership expenses, such as property taxes, insurance, maintenance, and the like.

Analyzing your spending numbers

Tabulating your spending is only half the battle on the path to fiscal fitness and a financially successful home purchase. You must *do* something with and about the personal spending information that you collect.

Here, in order of likelihood, are the possible outcomes of your spending analysis:

>> **You spend too much.** When most people examine their spending for the first time, they're somewhat horrified at *how much* they spend overall and *for what* specific things. Perhaps you had no idea that your café latte addiction is setting you back $100 per month or that you spend $400 per month on eating out.

TIP

Your challenge is to decide where to make reductions or cutbacks. (Check out the nearby sidebar, "Trimming the fat from your budget.") Everybody who has enough discretionary income to buy this book has fat in her budget (some have much more than others). For most people to reach their financial goals, they must save at least 10 percent of their pretax income. But how much you should be saving depends on what your goals are and how aggressive and successful an investor you are. If, for example, you want to retire early and don't have much put away yet, you may need to save much more than 10 percent per year to reach your goal.

TIP

>> **You save just right.** You may be one of those people who has mapped out a financial path and is right on track. Great! However, just as a cue ball sends a neatly racked set of billiard balls into disarray, buying a home can disrupt even the most organized and on-track budgets.

Reviewing what your budget may look like with a home in the picture is important. So if you haven't already done so, complete Table 2-1 to analyze your current spending and project how it may look after a home purchase.

>> **You save a lot.** Perhaps you're one of those rare sorts who save more than necessary. If so, you not only may be able to skip doing a budget but also may be able to stretch the amount you spend and borrow when buying a home. But even if you've made your financial plans and are saving more than enough, you still may want to complete Table 2-1 to ensure that your financial train doesn't get derailed.

Reckoning Your Savings Requirements

Not only do most people not know how much they're currently saving; even more people don't know how much they should be saving. You should know these amounts *before* you buy a home.

How much you should be saving likely differs from how much your neighbors and co-workers should be saving, because each person has a different situation, different resources, and different goals. Focus on your own situation.

Setting some goals

Most people find it enlightening to see how much they need to save to accomplish particular goals. Wanting to retire someday is a common goal. The challenge is that in your 20s and 30s, it's difficult to have more clearly defined goals — such as knowing that you want to retire at age 58 and move to New Mexico, where you'll join a shared-housing community and buy a home that currently costs $200,000. Not to worry — you don't need to know exactly when, where, and how you want to retire.

But you do want to avoid nasty surprises. When Peter and Nancy hit their 40s, they came to the painful realization that retirement was a long way off because they were still working off consumer debts and trying to initiate a regular savings program. Now they're confronted with a choice: having to work into their 70s to achieve their retirement goals or settling for a much less comfortable lifestyle in retirement.

THE WISE USE OF CREDIT

Just because borrowing on credit cards bears a high cost doesn't mean that all credit is bad for you. Borrowing money for long-term purposes can make sense if you borrow for sound, wealth-building investments. Borrowing money for a real estate purchase, for a small business, or for education can pay dividends down the road. *Note:* The amount that you borrow should be reasonable in comparison to your income and ability to repay.

When you borrow for investment purposes, you may earn tax benefits as well. With a home purchase, for example, home mortgage interest and property taxes are generally tax-deductible (as we discuss in Chapter 3).

If you own a business, you may deduct the interest expenses on loans that you take out for business purposes. Interest incurred through borrowing against your securities (stock and bond) investments (through so-called *margin loans*) is deductible against your investment income for the year.

In fact, you can even make wise use of short-term credit on your credit cards to make your money work harder for you. For example, you can use your credit cards for the convenience that they offer, not for their credit feature. When you pay your bill in full and on time during each monthly billing cycle, you've had free use of the money that you owed from the credit-card charges that you made during the previous month. (See Chapter 5 for details on how to use your positive credit experiences to obtain the best possible mortgage.)

If retirement isn't one of your goals, terrific! Should you want (and be able) to continue working throughout your 60s, 70s, and 80s, you don't need to accumulate the significant savings that others must in order to be loafing during those golden years. But counting on being able to keep working throughout your lifetime is risky — you don't know what the job market or your personal health may be like later in life.

Retirement-savings accounts and a dilemma

Prior tax reforms have taken away some previously available tax write-offs, except for one of the best and most accessible write-offs: funding a retirement-savings plan. Money that you contribute to an employer-based retirement plan — such as a 401(k) or a 403(b) — or to a self-employed plan, such as a SEP-IRA, is generally tax-deductible. This saves you both federal and state income taxes in the year for which the contribution is made. Additionally, all your money in these accounts compounds over time without taxation. (*Note:* The Roth IRA retirement accounts are unique in offering no up-front tax break but allowing the tax-free withdrawal of investment earnings subject to eligibility requirements.) These tax-reduction accounts are one of the best ways to save your money and make it grow.

The challenge for most people is keeping their spending down to a level that allows them to save enough to contribute to these terrific tax-reduction accounts. Suppose that you're currently spending all your income (a very American thing to do) and that you want to be able to save 10 percent of your income. Thanks to the tax savings that you'll net from funding your retirement account, if you're able to cut your spending by just 7.5 percent and put those savings into a tax-deductible retirement account, you'll actually be able to reach your 10 percent target.

WARNING

Generally speaking, when you contribute money to a retirement account, the money isn't accessible to you unless you pay a penalty. So if you're accumulating down-payment money for the purchase of a home, putting that money into a retirement account is generally a bad idea. Why? Because when you withdraw money from a retirement account, you not only owe current income taxes but also hefty penalties (10 percent of the amount withdrawn must go to the IRS, plus you must pay whatever penalty your state assesses).

So the dilemma is that you can save outside of retirement accounts and have access to your down-payment money but pay much more in taxes, or you can fund your retirement accounts and gain tax benefits but lack access to the money for your home purchase.

TIP

You can handle this dilemma in two ways. See whether your employer allows borrowing against retirement-savings-plan balances. And if you have an Individual Retirement Account (either a standard IRA or a Roth IRA), you're allowed to withdraw up to $10,000 (lifetime maximum) toward a home purchase so long as you haven't owned a home for the past two years. Tapping into a Roth IRA is a better deal because the withdrawal is free from income tax as long as the Roth account is at least five years old. Although a standard IRA has no such time restriction, withdrawals are taxed as income, so you'll net only the after-tax amount of the withdrawal toward your down payment.

Because most of us have limited discretionary dollars, we must decide what our priorities are. Saving for retirement and reducing your taxes are important, but when you're trying to save to purchase a home, some or most of your savings need to be outside a tax-sheltered retirement account. Putting your retirement savings on the back burner for a short time to build up your down-payment cushion is okay. Be careful, though, to purchase a home that offers enough slack in your budget to fund your retirement accounts after the purchase. Do the budget exercise in Table 2-1, earlier in this chapter!

Other reasons to save

Wanting to have the financial resources to retire someday is hardly the only reason to save. Most people have several competing reasons to squirrel away money. Here are some other typical financial objectives or goals that motivate people (or should be motivating them) to save money. We tell you how to fit each goal into your home-purchasing desires and your overall personal financial situation:

TIP

>> **Emergency reserve:** You simply can't predict when and exactly what impact a job loss, death in the family, accident, or unexpectedly large expense may have on you and your family. That's why it's a good idea to have an easily accessible and safe reservoir of money that you can tap should the need arise.

Make sure you have access to at least three months' worth of expenses (perhaps even six months' worth if you have a highly unstable job and volatile income). Ideally, you should keep this money in a money market fund because such funds offer you both high yields and liquidity. The major mutual fund companies (such as Vanguard, Fidelity, and T. Rowe Price) offer money funds with competitive yields, check-writing privileges, and access to other good investments. (See Chapter 3 to find out more about these funds and how you can use them for investing your down-payment money.) Alternatively, a bank savings account can work, but it will likely offer a lower yield. Should you have benevolent relatives who are willing to zap you some dough in a flash, they may serve as your emergency reserve as well.

WARNING

>> **Educational expenses:** If you have little cherubs at home, you want the best for them, and that typically includes a good college education. So when the first cash gifts start rolling in from Grandma and Grandpa, many a new parent establishes an investment account in the child's name.

Your best intentions can come back to haunt you, however, when Junior applies to enter college. All things being equal, the more you have available in your nonretirement accounts and in your child's name, the less financial aid your child will qualify for. (By financial aid, we mean all types of assistance, including grants and loans that aren't based on need.) Unless you're wealthy or are sure that you can afford to pay for the full cost of a college education for your kids, think long and hard before putting money in your child's name. Although it may sound selfish, you actually do yourself and your child a financial favor by taking full advantage of opportunities to fund your retirement accounts. Remember, too, that one of the advantages of being a homeowner is that you can borrow against your home's equity to help pay for your child's college expenses.

>> **Startup business expenses:** Another reason to save money is if you hope to start or purchase a business someday. When you have sufficient equity in your home, you can borrow against that equity to fund the business. But you may desire to accumulate a separate investment pool to fund your business.

REMEMBER

No matter what your personal and financial goals are, you're likely going to need to save a decent amount of money to achieve them. Consider what your goals are and how much you need to save to accomplish those goals, especially for retirement. Get your finances in order before you decide how much you can really afford to spend on a home. Otherwise, you may end up being a financial prisoner to your home.

Protecting Yourself, Your Dependents, and Your Assets

Not carrying proper insurance is potentially disastrous — both to yourself and your dependents. We're not talking about homeowners insurance here. (Heck, we haven't even explained how to find a home or get a loan yet! We get to homeowners insurance in Chapter 13.)

You need proper insurance protection for yourself personally, as well as for your assets. Sure, you can take your chances and hope that you never contract a dreaded disease, get into a horrible auto accident, or suffer some other misfortune or bad luck. But misfortune and bad luck usually come knocking without a warning.

Trust us when we say that we're optimistic, positive thinkers. However — and this is a big *however* — we know more than a few folks who got themselves (or their families or both) into major financial trouble after purchasing a home because they neglected to obtain proper insurance.

WARNING

Here are a few cautionary tales:

>> Steve bought a home and then found out from his doctor that he had multiple sclerosis. Steve ultimately had to cut back on work, and because he lacked proper long-term-disability insurance and now earned much lower work income, he was forced to sell his home at a large loss due to a soft real estate market in his area.

>> Mary owned a home in California and, despite the known risk of earthquakes, didn't purchase earthquake coverage. "It's so expensive, and besides, the insurance companies won't be able to meet the claims in a major quake. Government assistance will help," she said. Mary's home was a total loss in an earthquake, and although the government made a loan, it didn't *pay for* the loss — ultimately, the money came out of Mary's pocket.

>> Maggie and Donald were living a charmed life in the New England countryside with their two children, a white farmhouse, and a dog and a cat — until

Maggie came down with cancer. She left her job, which placed some strain on the family finances. After much treatment, Maggie died. Donald and the kids were forced to move because Maggie lacked proper life insurance.

>> Michelle had a walkway in disrepair. Unfortunately, one day an older man tripped and severely injured himself. To make a long story short, after lengthy legal proceedings, the settlement in favor of the man was significant enough to force Michelle to sell her home. A good chunk of the settlement money came out of Michelle's pocket because she lacked sufficient liability insurance.

Now, we're not about to try to tell you that insurance would have made these situations come out totally fine or that these are common occurrences. Insurance generally can't prevent most major medical problems, keep a person from dying, or stop someone from suing you. However, proper insurance can protect you and your family from the adverse and severe financial consequences of major problems. The right kind of insurance can make the difference between keeping versus losing your home, and it can help you and your family maintain your standard of living.

Wanting to skip insurance is tempting. After all, insurance costs you your hard-earned, after-tax dollars, and unlike a meal out, a vacation, or a new stereo, insurance has no up-front, tangible benefit.

REMEMBER

You hope you won't need to use insurance, but if you do need it, you're glad it's there to protect you and, in some cases, to protect your dependents. Buy too little insurance and it won't protect you and yours against a real catastrophe. So you need the right amount of coverage that balances good protection against cost.

Insuring yourself

Before you buy a home, get your insurance protection (for yourself and for your valuable assets) in order. Not doing so is the financial equivalent of driving down the highway in an old subcompact car at 90 miles per hour without a seat belt. You should purchase sufficient protection to prevent a financial catastrophe.

Disability insurance

Your ability to produce income should be insured. During your working years, your future income-earning ability is likely your most valuable asset — far more valuable than a car or even your home.

Long-term-disability insurance replaces most of your lost income in the event that a disability prevents you from working. Major disabilities are usually the result of accidents or medical problems — occurrences that, of course, can't be predicted. Even if you don't have financial dependents, you probably need disability coverage. Unless you're quite wealthy and no longer need to work for income, aren't *you* financially dependent upon your paycheck? Although many larger companies offer long-term-disability insurance, many small-company employees and self-employed people have no coverage — a risky proposition.

Life insurance

When you have dependents, you may also need life insurance protection. The question to ask yourself and your family is how they'd fare financially if you died and they no longer had your income. If your family is dependent on your income and you want them to be able to maintain their current standard of living in your absence, you need life insurance.

Term life insurance, like most other forms of insurance, is pure insurance protection and is the best type of insurance for the vast majority of people. The amount of coverage you buy should be based upon how many years' *worth* of your income you desire to provide your family in the event of your passing.

WARNING

Insurance brokers love to sell *cash-value life insurance* (also known as *whole* or *universal* life insurance) because of the hefty commissions they can earn by selling this type of insurance. (These commissions, of course, come out of your pocket.) Some mortgage lenders lobby you to buy the mortgage life insurance that they sell. Skip both these options. Mortgage life insurance is simply overpriced term insurance, and cash-value life insurance generally combines overpriced life insurance with a relatively low-return investment account.

Health insurance

REMEMBER

In addition to disability and life insurance, everyone should have a comprehensive health insurance policy. Even if you're in good health, you never know when an accident or illness may happen. Medical bills can mushroom into tens or hundreds of thousands of dollars in no time. Don't be without comprehensive health insurance.

Insuring your assets

As your wealth builds over the years (ideally, at least in part, from the increasing value of the home that we help you buy), so does the risk of losing — or facing a lawsuit arising from — your valuable assets. For example, you should have comprehensive insurance on your home and car(s). If your home burns to the ground, a comprehensive homeowners insurance policy would pay for the cost of rebuilding the home. Likewise, if your car is totaled in an accident, auto insurance should pay to replace the car.

WILLS, LIVING TRUSTS, AND ESTATE PLANNING

Although some of us don't like to admit it or even think about it, we're all mortal. Because of the way our legal and tax systems work, it's often beneficial to have legal documents in place specifying important details such as what should be done with your assets (including your home) when you die.

A *will* is the most basic of such documents, and for most people, particularly those who are younger or don't have great assets, the only critical one. Through a will, you can direct to whom your assets will go upon your death, as well as who will serve as guardian for your minor children. In the absence of a will, state law dictates these important issues.

Along with your will, also consider signing a *living will* and a *medical power of attorney*. These documents help your doctor and family members make important decisions regarding your healthcare, should you be unable to make those decisions for yourself.

Even a will and supporting medical and legal documents may not be enough to get your assets to your desired heirs, as well as minimize taxes and legal fees. When you hold significant assets (such as a home and business) outside tax-sheltered retirement accounts, in most states, those assets must be *probated* — which is the court-administered process for implementing your will. Attorneys' probate fees can run quite high — up to 5 percent of the value of the probated assets. Establishing and placing your home and other assets in a *living trust* can eliminate much of the hassle and cost of probate.

Finally, if your *net worth* (assets minus liabilities) exceeds $11.58 million, under current tax law, upon your death the federal (and perhaps your state's) government will levy significant estate taxes (numerous states have far lower limits). Estate planning can help minimize the portion of your estate subject to such taxation. One simple but powerful estate-planning strategy is to give money to your desired heirs to reduce your taxable estate. (If your relatives are in the fortunate position of having great wealth, they may give, free of tax, up to $15,000 yearly under current tax law to as many recipients as they want. If they give you $15,000, you can use this money toward your home's down payment.)

Wills, living trusts, and estate planning are nothing more than forms of insurance. Remember that it takes both time and money to generate these documents, and the benefits may be a long time off, so don't get carried away with doing too many of these things before you're older and have significant assets. Read the latest edition of *Personal Finance After 50 For Dummies* (Wiley), which Eric coauthored with Bob Carlson, to find out more about estate planning. Eric also enthusiastically recommends Ray's excellent book, *Winning the Endgame: A Guide to Aging Wisely and Dying Well*.

TIP

With all types of insurance that you purchase, take the highest deductible that you can comfortably afford. The *deductible* represents the amount of money that you must pay out of your own pocket when you have a loss for which you file a claim. High deductibles help keep the cost of your coverage low and also eliminate the hassle associated with filing small claims.

Along with buying insurance to cover the replacement costs for loss of or damage to your valuable assets, you should purchase adequate liability insurance for those assets. Both homeowners insurance and auto insurance come with liability protection. Make sure that you carry liability coverage for at least twice the value of your *net worth* (assets minus liabilities).

In addition to the liability protection that comes with auto and homeowners insurance, you may purchase a supplemental liability insurance policy known as an *umbrella* or *excess liability policy*. Purchased in increments of $1,000,000, this coverage can protect people with larger net worth. Note that this coverage doesn't protect against lawsuits arising from your work.

Invest in Yourself

Last but not least, in your zest to build your financial empire and buy ever bigger and more expensive homes, don't forget your best investment: you. Don't run your life and body into the ground by working horrendous hours just to afford what you consider your dream home.

In addition to investing in your health, your family, and your friends, invest in educating yourself and taking charge of your finances. If you need more help with assessing your current financial health; reducing your spending, your taxes, and your debts; and mapping out an overall financial plan (including dealing with your investments and insurance), be smart and pick up a copy of the latest edition of Eric's *Personal Finance For Dummies* (Wiley).

Chapter **3**

What Can You Afford to Buy?

When you're in the market for a car, the auto salesperson will eventually ask, "What is your budget?" or "How much can you afford to spend on a car?" Of course, they hope that a large number rolls off your tongue. If you're like many car buyers, you may be likely to say something along the lines of, "I'm not really sure."

Many car buyers today finance the purchase — so they allow a banker or other lender to determine how much car they can afford. Such determinations are based on a buyer's income and other debt obligations.

But here's where most people get confused. When a lender says you qualify to borrow, say, $30,000 for a car purchase, this doesn't mean you can *afford* to spend that much on a car. What the lender is effectively saying to you is, "Based on what little I know about your situation and the fact that I can't control your future behavior, this is the maximum amount that I think is a prudent risk for my organization to lend to you."

The lending organization normally requires a certain down payment to protect itself against the possibility that you may default on the loan. Should you default on an auto loan, for example, the lender has to send the repo man out to take away and sell your car. This process takes time and money, and the lender will surely get less for the car than the amount that you paid for it.

Much of the same logic applies to a home purchase with one important difference — over time, your home should hopefully appreciate in value whereas a car most definitely will not.

In this chapter, we help you determine what you can comfortably afford to spend on a home as well as how to calculate how much a particular home is likely to cost you.

Lenders Can't Tell You What You Can Afford

WARNING

Ultimately, a lender doesn't care about you, your financial situation, or your other needs as long as it has protected its financial interests. This is true whether you're borrowing to buy a car or a home. The lender doesn't know or care whether, for example, you're

>> Falling behind in saving for retirement

>> Wanting to save money for other important financial goals, such as starting or buying your own small business

>> Parenting a small army of kids (or facing steep private-schooling costs)

>> Lacking proper personal insurance protection

And therein lies the problem of making your decision about how much home (or car) you can afford to buy on the basis of how much money a lender is willing to lend you. That's what Walter and Susan did. They set out to purchase a home when Walter's business was booming. They were making in excess of $200,000 per year.

Walter and Susan really wanted to buy the biggest and best house that they could afford. When they met with their friendly neighborhood banker, he was more than willing to show them how they could borrow $900,000 by getting an adjustable-rate mortgage. (You can read all about these mortgages in Chapter 6. We'll simply tell you here that because some adjustable mortgages start out at an artificially low "teaser" interest rate, they enable you to qualify to borrow a good deal more than would be the case with a traditional, fixed-rate mortgage.)

Walter and Susan bought their dream home with an adjustable-rate $900,000 mortgage. Within a few years, Walter and Susan's dream home turned into the Nightmare on Oak Street. Their mortgage became a financial noose around their necks.

When blessed with young children, Walter and Susan didn't want to work such crazy long hours; yet they were forced to do so to meet their gargantuan mortgage payments. The initial payments on their adjustable-rate mortgage were low, but they ballooned gigantically as the loan's interest rate increased.

The financial strain led to personal strain as Walter and Susan had frequent arguments about money and childcare. We know of others who stretched themselves the same way that Walter and Susan did. Many of them continue slaving away long hours in jobs they don't like and making other unnecessary sacrifices, such as limiting the time they spend with family, in order to make their housing payments. Some end up divorcing, due in part to the financial strains. Others default on their loans and lose their homes and their good credit.

REMEMBER

People at all income levels, even the affluent, can get into trouble and overextend themselves by purchasing more house than they can afford and by taking on more debt than they can comfortably handle. Just because a lender or real estate agent says you're eligible for, or can qualify for, a certain size loan doesn't mean that's what you can afford given your personal financial situation. Lenders *can't* tell you what you can afford — they can tell you *only* the maximum that they'll lend to you.

The Cost of Buying and Owning a Home

Before you set out in search of your dream home, one of the single most important questions you should answer is, "What can I afford to spend on a home?" To answer that question intelligently, you first need to understand what your financial goals are, what it will take to achieve them, and where you are today. If you haven't yet read Chapter 2, now's the time (unless you're 100 percent sure that your personal finances are in tiptop shape). In the following sections, we dig into the costs of buying and owning a home.

Mortgage payments

In Chapter 6, we discuss selecting the best type of mortgage that fits your particular circumstances. In the meantime, you must still confront mortgages (with our assistance) because mortgages undoubtedly constitute the biggest component of the total cost of owning a home.

Start with the basics: A *mortgage* is a loan you take out to buy a home. A mortgage allows you to purchase a $200,000 home even though you have far less money than that to put toward the purchase.

With few exceptions, mortgage loans in the United States are typically repaid over a 15- or 30-year time span. Almost all mortgages require monthly payments. Here's how a mortgage works. Suppose that you're purchasing a $200,000 home and that (following our sage advice, appearing later in this chapter) you have dili-gently saved a 20 percent ($40,000, in this example) down payment. Thus, you're in the market for a $160,000 mortgage loan.

You sit down with a mortgage lender who asks you to complete a volume of paper-work (we navigate you through that morass in Chapter 7) that dwarfs the stack required for your annual income tax return. Just when you think the worst is over (after the paperwork blizzard subsides), the lender proceeds to give you an even bigger headache by talking about the literally hundreds of mortgage permutations and options.

Don't worry — we can help you cut through the clutter! Imagine, for a moment, a simple world where the mortgage lender offers you only two mortgage options: a 15-year fixed-rate mortgage and a 30-year fixed-rate mortgage (*fixed-rate* simply means that the interest rate on the loan stays fixed and level over the life of the loan). Here's what your monthly payment would be under each mortgage option:

$160,000, 15-year mortgage @ 4.75 percent = $1,245 per month

$160,000, 30-year mortgage @ 5.00 percent = $859 per month

As we discuss in Chapter 6, the interest rate is typically a little bit lower on a 15-year mortgage versus a 30-year mortgage because shorter-term loans are a little less risky for lenders. Note how much higher the monthly payment is on the 15-year mortgage than on the 30-year mortgage. Your payments must be higher for the 15-year mortgage because you're paying off the same size loan 15 years faster.

But don't let the higher monthly payments on the 15-year loan cause you to forget that at the end of 15 years, your mortgage payments disappear, whereas with the 30-year mortgage, you still have 15 more years' worth of monthly payments to go. So although you have a higher required monthly payment with the 15-year mort-gage, check out the difference in the total payments and interest on the two mort-gage options:

Mortgage Option	Total Payments	Total Interest
15-year mortgage	$224,064	$64,064
30-year mortgage	$309,312	$149,312

Note: In case you're curious about how we got the total interest amount, we simply subtracted the amount of the loan repaid ($160,000) from the "Total Payments." Also, the monthly payment numbers previously cited, as well as these total payments and interest numbers, are rounded off, so if you try multiplying 180 or 360 by the monthly payment numbers, you won't get answers identical to these numbers.

With the 30-year mortgage (compared with the 15-year mortgage), because you're borrowing the money over 15 additional years, it shouldn't come as a great surprise that (with a decent-size mortgage loan like this one) you end up paying more than $80,000 additional interest. The 30-year loan isn't necessarily inferior; for example, its lower payments may better allow you to accomplish other important financial goals, such as saving in a tax-deductible retirement account. (See Chapter 6 for more information about 15-year versus 30-year mortgages.)

In the early years of repaying your mortgage, nearly all your mortgage payment goes toward paying interest on the money that you borrowed. Not until the later years of your mortgage do you begin to rapidly pay down your loan balance, as shown in Figure 3-1.

TECHNICAL STUFF

As interest rates increase, so does the time required to pay off half the loan. For example, at a 10 percent interest rate, paying off half of a given loan takes almost 24 years of the loan's 30-year term.

Lender's limits

Because we've personally seen the financial consequences of people borrowing too much (yet still staying within the boundaries of what mortgage lenders allow), you won't hear us saying in this section that lenders can tell you the amount you can afford to spend on a home. They can't. All mortgage lenders can do is tell you their criteria for approving and denying mortgage applications and calculating the maximum that you're eligible to borrow. (For the inside scoop on lenders and their limits, see the first section of this chapter.)

Mortgage lenders tally up your monthly *housing expense,* the components of which they consider to be

> Mortgage payment (PI for principal and interest) + Property taxes (T for taxes) + Insurance (I for insurance) = Lender's definition of housing expense (PITI is the common acronym)

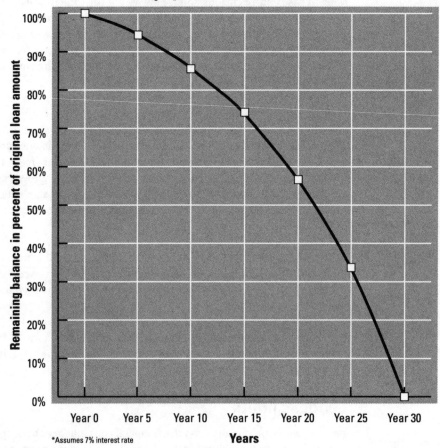

Repayment of a 30-Year Loan*

Y-axis: Remaining balance in percent of original loan amount

X-axis: Years — Year 0, Year 5, Year 10, Year 15, Year 20, Year 25, Year 30

*Assumes 7% interest rate

© John Wiley & Sons, Inc.

FIGURE 3-1: It takes many years into a mortgage to begin making real progress at repaying the amount originally borrowed. In this case, paying off half the loan balance takes nearly 22 years.

For a given property that you're considering buying, a mortgage lender calculates the housing expense and normally requires that it not exceed a certain percentage (typically around 40 percent or so) of your monthly before-tax *(gross)* income. (Some lenders allow the percentage to go a bit higher.) So, for example, if your monthly gross income is $6,000, your lender will not allow your expected monthly housing expense to exceed $2,400 (if the lender is using 40 percent). When you're self-employed and complete IRS Form 1040, Schedule C, mortgage lenders use your after-tax expenses *(net)* income, from the bottom line of Schedule C.

WARNING

Now, if you've been paying attention thus far in this chapter, you should smell something terribly wrong with such a simplistic, one-number-fits-all approach. This housing-expense ratio completely ignores almost all your other financial goals, needs, and obligations. It also ignores utilities, maintenance, and remodeling expenses, which can gobble up a lot of a homeowner's money.

About the only other financial considerations a lender takes into account (besides your income) are your other debts. Specifically, mortgage lenders examine the required monthly payments for student loans, an auto loan, credit-card bills, and other debts. In addition to the percentage of your income lenders allow for housing expenses, lenders typically allow an additional 5 percent of your monthly income to go toward other debt repayments. Thus, your monthly housing expense and monthly repayment of nonhousing debts can total up to, but generally be no more than, 45 percent.

STRETCHING MORE THAN LENDERS ALLOW

Sometimes, prospective home buyers feel that they can handle more debt than lenders will allow. Such home buyers may seek to borrow more money from family or fib on their mortgage application about their income. (Self-employed people have the greatest opportunity to do this.) Such behavior isn't unlike the shenanigans of some teenagers who drive above the speed limit, drink and smoke forbidden things, or stay out past curfew and sneak in the back door.

Although a few of these teenagers get away with such risky behavior, others end up in trouble academically or psychologically (or worse). The same is true of homeowners who stretch themselves financially thin to buy a more costly property. Some survive just fine, but others end up in financial, legal, and emotional trouble.

And, increasingly, home buyers who lie on their mortgage applications are getting caught. How? When you're ready to close on your loan, lenders can (and often do) ask you to sign a form authorizing them to request a copy of your income tax return from the IRS. This allows the lender to validate your income. (See Chapter 7 for more details.)

So although we say that the lender's word isn't the gospel as to how much home you can truly afford, we will go on record as saying that telling the truth on your mortgage application is the only way to go (and prevents you from committing perjury and fraud). Telling the truth is not only honest but also helps keep you from getting in over your head financially. Bankers don't want you to default on your loan, and you shouldn't want to take the risk of doing so either.

Should you have consumer debt, be sure to read Chapter 2. Suffice it to say here that you should get out (and stay out) of consumer debt. Consumer debt has a high cost, and unlike the interest on a mortgage loan, the interest on consumer debt isn't tax-deductible. And consumer debt handicaps your ability to qualify for and pay back your mortgage. *Consumer debt is the financial equivalent of cancer.*

Figuring the size of your mortgage payments

Calculating the size of your mortgage payment, after you know the amount you want to borrow, is simple. The hard part for most people is determining how much they can afford to borrow. If you already know how large a monthly mortgage payment you can afford, terrific! Go to the head of the class. Suppose you work through your budget in Chapter 2 and calculate that you can afford to spend $1,500 per month on housing. Determining the exact amount of mortgage that allows you to stay within this boundary is a little challenging, because the housing cost you figure that you can afford ($1,500, in our example) is made up of several components. Lucky for you, we cover each of these components in this chapter, including mortgage payments, property taxes, insurance, and maintenance. (Note that although lenders don't care about maintenance expenses in figuring what you can afford to buy, you shouldn't overlook this significant expense.)

As you change the amount that you're willing to spend on a home, the size of the mortgage you choose to take out also usually changes, but so do the other property cost components. So you may have to play with the numbers a bit to get them to work out just right. You may pick a certain home and then figure the property taxes, insurance, maintenance, and the like. When you tally everything up, you may find that the total comes in above or below your desired target ($1,500, in our example). Obviously, if you come out a little high, you need to cut back a bit and choose a slightly less-costly property and/or get a smaller mortgage.

Using Table 3-1, you can calculate the size of your mortgage payments based on the amount you want to borrow, the loan's interest rate, and the length (in years) the mortgage payments last. To determine the monthly payment on a mortgage, simply multiply the relevant number from Table 3-1 by the size of your mortgage expressed in (divided by) thousands of dollars. For example, if you take out a $150,000, 30-year mortgage at 5.50 percent, multiply 150 by 5.68 (from Table 3-1) to arrive at an $852 monthly payment. (You can access Table 3-1 online at www.dummies.com/go/homebuyingkit7e.)

TABLE 3-1

Monthly Mortgage Payment Calculator

Interest Rate	15-Year Mortgage	30-Year Mortgage
3	6.90	4.21
3⅛	6.96	4.28
3¼	7.02	4.35
3⅜	7.08	4.42
3½	7.14	4.49
3⅝	7.21	4.56
3¾	7.27	4.63
3⅞	7.33	4.70
4	7.40	4.77
4⅛	7.46	4.85
4¼	7.52	4.92
4⅜	7.59	4.99
4½	7.65	5.07
4⅝	7.71	5.14
4¾	7.78	5.22
4⅞	7.84	5.29
5	7.91	5.37
5⅛	7.98	5.45
5¼	8.04	5.53
5⅜	8.11	5.60
5½	8.18	5.68
5⅝	8.24	5.76
5¾	8.31	5.84
5⅞	8.38	5.92
6	8.44	6.00
6⅛	8.51	6.08
6¼	8.58	6.16

(continued)

TABLE 3-1 *(continued)*

Interest Rate	15-Year Mortgage	30-Year Mortgage
6⅜	8.65	6.24
6½	8.72	6.33
6⅝	8.78	6.41
6¾	8.85	6.49
6⅞	8.92	6.57
7	8.99	6.66
7⅛	9.06	6.74
7¼	9.13	6.83
7⅜	9.20	6.91
7½	9.28	7.00
7⅝	9.35	7.08
7¾	9.42	7.17
7⅞	9.49	7.26
8	9.56	7.34
8⅛	9.63	7.43
8¼	9.71	7.52
8⅜	9.78	7.61
8½	9.85	7.69
8⅝	9.93	7.78
8¾	10.00	7.87
8⅞	10.07	7.96
9	10.15	8.05
9⅛	10.22	8.14
9¼	10.30	8.23
9⅜	10.37	8.32
9½	10.45	8.41
9⅝	10.52	8.50
9¾	10.60	8.60

Interest Rate	15-Year Mortgage	30-Year Mortgage
9⅞	10.67	8.69
10	10.75	8.78
10⅛	10.83	8.87
10¼	10.90	8.97
10⅜	10.98	9.06
10½	11.06	9.15
10⅝	11.14	9.25
10¾	11.21	9.34
10⅞	11.29	9.43
11	11.37	9.53
11¼	11.53	9.72
11½	11.69	9.91
11¾	11.85	10.10
12	12.01	10.29
12¼	12.17	10.48
12½	12.17	10.48

CHECK IT OUT

Use this handy-dandy workspace (reproduced throughout this chapter and available online at www.dummies.com/go/homebuyingkit7e) to track your estimated homeownership expenses, starting with the mortgage payment:

Item	Estimated Monthly Expense
Mortgage payment	$
Property taxes	+ $
Insurance	+ $
Improvements, maintenance, and other	+ $
Homeownership expenses (pretax)	= $
Tax savings	– $
Homeownership expenses (after tax benefits)	= $

Property taxes

If you live and breathe, escaping taxes is darn near impossible. When you buy and own a home, your local government (typically through what's called a county tax collector's office or an equivalent for your local town) sends you an annual, lump-sum bill for property taxes. Receiving this bill and paying it are never much fun because most communities bill you just once or twice per year. And some home-owners find it aggravating to be paying so much in property taxes on top of all the federal and state income and sales taxes they pay. In case you're wondering, property taxes go toward expenses of the local community, such as the public schools and snow plowing (for those of us foolish enough to live where the winters are cold). Especially in higher-cost areas with few retail and commercial properties paying taxes, residential property taxes can be quite significant.

Should you make a small down payment (typically defined as less than 20 percent of the purchase price), many lenders insist on property tax and insurance *impound accounts.* These accounts require you to pay your property taxes and insurance to the lender each month along with your mortgage payment.

INVESTIGATE

Property taxes are typically based on the value of a property. Although an average property tax rate is about 1.5 to 2.0 percent of the property's purchase price per year, you should understand what the exact rate is in your area. Call the tax collector's office (you can find the phone number in the government pages section of your local phone directory under such headings as "Tax Collector," "Treasurer," or "Assessor"; or enter one of those terms and the name of the municipality where you live into a search engine) in the town where you're contemplating buying a home and ask what the property tax rate is and what additional fees and assessments may apply.

Be careful to make sure that you're comparing apples with apples when comparing communities and their property taxes. For example, some communities may nickel-and-dime you for extra assessments for services that are included in the standard property tax bills of other communities.

WARNING

Real estate listings, which are typically prepared by real estate agents, may list what the current property owner is paying in taxes. But relying on such data to understand what your real estate taxes will be if you buy the property can be financially dangerous. The current owner's taxes may be based on an outdated and much lower property valuation. Just as it's dangerous to drive forward by looking in the rearview mirror of your car, you shouldn't buy a property and bud-get for property taxes based on the current owner's taxes. Your property taxes (if you buy the home) will be recalculated based on the price you pay for the property.

Item	Estimated Monthly Expense
Mortgage payment	$
Property taxes	+ $
Insurance	+ $
Improvements, maintenance, and other	+ $
Homeownership expenses (pretax)	= $
Tax savings	– $
Homeownership expenses (after tax benefits)	= $

Insurance

When you purchase a home, your mortgage lender almost surely won't allow you to close the purchase until you demonstrate that you have proper homeowners insurance. Lenders aren't being paternalistic, but self-interested. You see, if you buy the home and make a down payment of, say, 20 percent of the purchase price, the lender is putting up the other 80 percent of the purchase price. So if the home burns to the ground and is a total loss, the lender has more invested financially than you do. In most states, your home is the lender's security for the loan.

Some lenders, in years past, learned the hard way that some homeowners may not care about losing their homes. In some cases, where homes were total losses, homeowners with little financial stake in the property and insufficient insurance coverage simply walked away from the problem and left the lender with the financial mess. Because of cases like this, almost all lenders today require you to purchase *private mortgage insurance* (PMI) if you put down less than 20 percent of the purchase price when you buy. (We discuss PMI further later in this chapter, in the section titled "The 20 percent solution.")

When you buy a home, you should want to protect your investment in the property (as well as cover the not-so-inconsequential cost of replacing your personal property, if it's ever damaged or stolen). In short order, your clothing, furniture, kitchen appliances, and beer-can collection can tally up to a lot of dollars to replace.

TIP

When you purchase homeowners insurance, you should buy the most comprehensive coverage that you can and take the highest deductible that you can afford to help minimize the cost. In Chapter 13, we explain how to do all that. To estimate what homeowners insurance may cost you, we suggest you contact some of the insurers we recommend in Chapter 13. Explain to them what type and price range of properties you're considering buying in which communities (zip codes), and they should be able to give you a ballpark monthly cost estimate for insurance. Calling insurance agents now also enables you to begin to evaluate which insurers offer the service and coverage you desire when the time comes to actually buy your dream home.

Just as you should do when you shop for a car, get quotes on insuring properties as you evaluate them, or ask current owners what they pay for their coverage. (Just remember that some homeowners overpay or don't buy the right kind of protection, so don't take what they pay as gospel.) If you overlook insurance costs until after you agree to buy a property, you can be in for a rude awakening.

Item	Estimated Monthly Expense
Mortgage payment	$
Property taxes	+ $
Insurance	+ $
Improvements, maintenance, and other	+ $
Homeownership expenses (pretax)	= $
Tax savings	– $
Homeownership expenses (after tax benefits)	= $

Maintenance and other costs

As a homeowner, you *must* make your mortgage and property tax payments. If you don't, you'll eventually lose your home. Homes also require maintenance over the years. You must do some kinds of maintenance (repairs, for example) at a certain time. You never know precisely when you may need to fix an electrical problem, patch a leaking roof, or replace the washer and dryer — until the problem rears its ugly head, which is why maintenance is difficult to budget for. (Painting and other elective improvements can take place at your discretion.)

TIP

As a rule of thumb, expect to spend about 1 percent of your home's purchase price each year on maintenance. So, for example, if you spend $150,000 on a home, you should budget about $1,500 per year (or about $125 per month) for maintenance. Although some years you may spend less, other years you may spend more. When your home's roof goes, for example, replacing it may cost you several years' worth of your budgeted maintenance expenses. With some types of housing, such as condominiums, you actually pay monthly dues into a homeowners association, which takes care of the maintenance for the complex. In that case, you're responsible for maintaining only the interior of your unit. Before you buy such a unit, check with the association to see what the dues are and whether any new assessments are planned for future repairs. (See Chapter 8 for more information.)

In addition to necessary maintenance, you should be aware (and beware) of what you may spend on nonessential home improvements. This *Other* category can really get you into trouble. Advertisements, your neighbors, and your co-workers can all entice you into blowing big bucks on new furniture, endless remodeling projects, landscaping, and you name it.

REMEMBER

Budget for these nonessentials; otherwise, your home can become a money pit by causing you to spend too much, not save enough, and (possibly) go into debt via credit cards and the like. (We cover the other dangers of over-improvement in Chapter 8.) Unless you're a terrific saver, can easily accomplish your savings goal, and have lots of slack in your budget, be sure not to overlook this part of your home-expense budget.

The amount you expect to spend on improvements is just a guess. It depends on how *finished* the home is that you buy and on your personal tastes and desires. Consider your previous spending behavior and the types of projects you expect to do as you examine potential homes for purchase.

Item	Estimated Monthly Expense
Mortgage payment	$
Property taxes	+ $
Insurance	+ $
Improvements, maintenance, and other	+ $
Homeownership expenses (pretax)	= $
Tax savings	– $
Homeownership expenses (after tax benefits)	= $

The tax benefits of homeownership

CHECK IT OUT

One of the benefits of homeownership is that the IRS and most state governments allow you to deduct, within certain limits, mortgage interest and property taxes when you file your annual income tax return. When you file your Federal IRS Form 1040, the mortgage interest and property taxes on your home are itemized deductions on Schedule A (see Figure 3-2, which is available online at www.dummies.com/go/homebuyingkit7e).

Up through 2017, mortgage interest and property tax payments for your home were generally tax-deductible on Schedule A of IRS Form 1040 except for the limitation on the mortgage interest deduction being "limited" to $1,000,000 of debt. There was no limit on property tax deductions.

SCHEDULE A
(Form 1040 or 1040-SR)
(Rev. January 2020)
Department of the Treasury
Internal Revenue Service (99)

Itemized Deductions

▶ Go to *www.irs.gov/ScheduleA* for instructions and the latest information.
▶ Attach to Form 1040 or 1040-SR.

Caution: If you are claiming a net qualified disaster loss on Form 4684, see the instructions for line 16.

OMB No. 1545-0074

2019

Attachment
Sequence No. 07

Name(s) shown on Form 1040 or 1040-SR

Your social security number

Medical and Dental Expenses	**Caution:** Do not include expenses reimbursed or paid by others.		
	1 Medical and dental expenses (see instructions)	1	
	2 Enter amount from Form 1040 or 1040-SR, line 8b **2**		
	3 Multiply line 2 by 7.5% (0.075)	3	
	4 Subtract line 3 from line 1. If line 3 is more than line 1, enter -0-		4
Taxes You Paid	5 State and local taxes.		
	a State and local income taxes or general sales taxes. You may include either income taxes or general sales taxes on line 5a, but not both. If you elect to include general sales taxes instead of income taxes, check this box . . . . ▶ ☐	5a	
	b State and local real estate taxes (see instructions)	5b	
	c State and local personal property taxes	5c	
	d Add lines 5a through 5c	5d	
	e Enter the smaller of line 5d or $10,000 ($5,000 if married filing separately)	5e	
	6 Other taxes. List type and amount ▶ _____	6	
	7 Add lines 5e and 6		7
Interest You Paid **Caution:** Your mortgage interest deduction may be limited (see instructions).	8 Home mortgage interest and points. If you didn't use all of your home mortgage loan(s) to buy, build, or improve your home, see instructions and check this box . . . . ▶ ☐		
	a Home mortgage interest and points reported to you on Form 1098. See instructions if limited	8a	
	b Home mortgage interest not reported to you on Form 1098. See instructions if limited. If paid to the person from whom you bought the home, see instructions and show that person's name, identifying no., and address . . . . ▶ _____	8b	
	c Points not reported to you on Form 1098. See instructions for special rules	8c	
	d Mortgage insurance premiums (see instructions)	8d	
	e Add lines 8a through 8d	8e	
	9 Investment interest. Attach Form 4952 if required. See instructions .	9	
	10 Add lines 8e and 9		10
Gifts to Charity **Caution:** If you made a gift and got a benefit for it, see instructions.	11 Gifts by cash or check. If you made any gift of $250 or more, see instructions	11	
	12 Other than by cash or check. If you made any gift of $250 or more, see instructions. You **must** attach Form 8283 if over $500 . . .	12	
	13 Carryover from prior year	13	
	14 Add lines 11 through 13		14
Casualty and Theft Losses	15 Casualty and theft loss(es) from a federally declared disaster (other than net qualified disaster losses). Attach Form 4684 and enter the amount from line 18 of that form. See instructions		15
Other Itemized Deductions	16 Other—from list in instructions. List type and amount ▶ _____		16
Total Itemized Deductions	17 Add the amounts in the far right column for lines 4 through 16. Also, enter this amount on Form 1040 or 1040-SR, line 9		17
	18 If you elect to itemize deductions even though they are less than your standard deduction, check this box . . . . ▶ ☐		

For Paperwork Reduction Act Notice, see the Instructions for Forms 1040 and 1040-SR. Cat. No. 17145C **Schedule A (Form 1040 or 1040-SR) 2019**

Source: U.S. Internal Revenue Service

FIGURE 3-2: Itemize mortgage interest and property tax deductions on Schedule A of your 1040.

Effective 2018, due to the Tax Cuts and Jobs Act, the tax benefits of home owner-ship were further limited, especially for those buying more expensive properties subject to higher property tax bills. The biggest change is that property taxes com-bined with your state income tax are now limited to a $10,000 annual deduction. And, the mortgage interest deduction now may be claimed on up to $750,000 of mortgage debt (or $1,000,000 if indebtedness was incurred prior to December 16, 2017), which obviously doesn't affect most home buyers.

Just because mortgage interest and property taxes are allowable deductions on your income tax return, don't think that the government is literally paying for these items for you. Consider that when you earn a dollar of income and must pay income tax on that dollar, you don't pay the entire dollar back to the government in taxes. Your tax bracket (see Table 3-2) determines the amount of taxes you pay on that dollar.

TABLE 3-2   # 2020 Federal Income Tax Brackets and Rates

Federal Tax Rate	For Single Individuals, Taxable Income Over	For Married Individuals Filing Joint Returns, Taxable Income Over	For Heads of Households, Taxable Income Over
10%	$0	$0	$0
12%	$9,875	$19,750	$14,100
22%	$40,125	$80,250	$53,700
24%	$85,525	$171,050	$85,500
32%	$163,300	$326,600	$163,300
35%	$207,350	$414,700	$207,350
37%	$518,400	$622,050	$518,400

Because of the new tax bill, determining the tax savings you may realize from homeownership has become much more complicated. Here's a shortcut that works reasonably well in determining your tax savings in homeownership: Multiply your federal-tax rate (which we explain in a moment) by the portion of your property taxes up to $10,000 when combined with your annual state income tax payments, and the portion of your mortgage payment on up to $750,000 of mortgage debt.

TIP

Even if you're under the $750,000 threshold, not all your mortgage payment is tax-deductible — only the portion of the mortgage payment that goes toward interest. Technically, you pay federal and state taxes, so you should consider your state tax savings as well when calculating your homeownership tax savings. However, to keep things simple and still get a reliable estimate, simply multiply your mortgage payment and property taxes by your *federal* income tax rate. This shortcut works well because the small portion of your mortgage payment that isn't deductible (because it's for the loan repayment) approximately offsets the overlooked state tax savings.

If you want to more accurately determine how homeownership may affect your tax situation, get out your tax return and try plugging in some reasonable numbers to estimate how your taxes will change. You can also speak with a tax advisor.

Item	Estimated Monthly Expense
Mortgage payment	$
Property taxes	+ $
Insurance	+ $
Improvements, maintenance, and other	+ $
Homeownership expenses (pretax)	= $
Tax savings	– $
Homeownership expenses (after tax benefits)	= $

TECHNICAL STUFF

The deductibility of the mortgage interest on up to $750,000 borrowed covers debt on both your primary residence and a second residence. (Buying and maintaining two homes is an expensive proposition and something few people can afford, so don't get any silly ideas from our mentioning this tax tidbit!)

Congratulations! You've totaled what your dream home should cost you on a monthly basis after factoring in the tax benefits of homeownership. Don't forget to plug these expected homeownership costs into your current monthly spending plans (see Chapter 2) to make sure you can afford to spend this much on a home and still accomplish your financial goals.

Closing Costs

On the day when a home becomes yours officially, known as *closing day,* many people (in addition to the seller) will have their hands in your wallet. Myriad one-time closing costs can leave you poorer or send you running to your relatives for financial assistance.

We don't want you to be unable to close your home purchase or be forced to get down on your hands and knees and beg for money from your mother-in-law. (Not only is such groveling hard on your ego, but also, she may expect grandchildren pronto.) Advance preparation for the closing costs saves your sanity and your finances.

Here are some typical closing costs (listed from those that are usually largest to those that are typically tiniest) and how much to budget for each (exact fees vary by property cost and location).

>> **Loan origination fees (points) and other loan charges:** These fees and charges range from nothing to 3 percent of the amount borrowed. Lenders generally charge all sorts of fees for things such as appraising the property, pulling your credit report, preparing loan documents, and processing your application, as well as charging a loan origination fee, which may be 1 or 2 percent of the loan amount. If you're strapped for cash, you can get a loan that has few or no fees; however, such loans have substantially higher interest rates over their lifetimes. As Chapter 12 explains, you may be able to cut a deal with the seller to pay these loan-closing costs.

>> **Escrow fees:** Escrow fees range from several hundred to over a thousand dollars, based on your home's purchase price. These fees cover the cost of handling all the purchase-related documents and funds. We explain escrows in much more detail in Chapters 9 and 14.

>> **Homeowners insurance:** This insurance typically costs several hundred to a thousand-plus dollars per year, depending on your home's value and how much coverage you want. As we discuss earlier in this chapter, you can't get a mortgage unless you prove to the lender that you have adequate homeowners insurance coverage. Promising to get this coverage isn't enough; lenders usually insist that you pay the first year's premium on said insurance policy at closing.

>> **Title insurance:** This insurance typically costs several hundred to a thousand dollars, depending on the home's purchase price. Lenders require that you purchase title insurance when you buy your home to make sure you have clear, marketable title to the property. Among other things, *title insurance* protects you and the lender against the remote possibility that the person selling you the home doesn't legally own it. We discuss title insurance in detail in Chapter 13.

>> **Property taxes:** These taxes typically cost several hundred to a couple thousand dollars and are based on the home's purchase price and the date that escrow closes. At the close of escrow, you may have to reimburse the sellers for any property taxes that they paid in advance. For example, suppose that (before they sell their home to you) the sellers have already paid their property taxes through June 30. If the sale closes on April 30, you owe the sellers two months' property taxes — the tax collector won't refund the property taxes already paid for May and June.

>> **Legal fees:** These fees range anywhere from nothing to hundreds of dollars. In some Eastern states, lawyers are routinely involved in real estate purchases. In most states, however, lawyers aren't needed for home purchases as long as

the real estate agents use standard, fill-in-the-blank contracts. Such contracts have the advance input and blessing of the legal eagles.

>> **Inspections:** Inspection fees can run from $200 to $1,000 (depending on the property's size and the scope of the inspection). As we explain in Chapter 13, you should never, ever consider buying a home without inspecting it. Because you're likely not a home-inspection expert, you benefit from hiring someone who inspects property as a full-time job. Sometimes you simply pay these costs directly; other times you pay these costs at closing.

>> **Private mortgage insurance (PMI):** Should you need it, this insurance can cost you several hundred dollars — or more — annually. As we explain in the next section, if you put less than 20 percent down on a home, some mortgage lenders require that you take out private mortgage insurance. This type of insurance protects the lender in the event that you default on the loan. At closing, you need to pay anywhere from a couple months' premiums to more than a year's premium in advance. If you can, avoid this cost by making a 20 percent down payment.

>> **Prepaid loan interest:** Lenders charge up to 30 days' interest on your loan to cover the interest that accrues from the date your loan is funded (usually, one business day before the escrow closes) up to 30 days prior to your first regularly scheduled loan payment. How much interest you actually have to pay depends on the timing of your first loan payment. If you're smart, and we know you are, you can work out this timing with the lender so you don't have to pay any advance loan interest.

TIP

To avoid paying three useless days of interest charges, *never schedule your escrow to close on a Monday.* Should you close on a Monday, the lender has to put your mortgage funds into escrow the preceding Friday. As a result, you're charged interest on your loan for Friday, Saturday, and Sunday even though you won't own the home until escrow closes on Monday. (This little tip more than pays for this book all by itself. Don't you feel smart now?)

>> **Recording:** The fee to record the deed and mortgage usually runs about $50.

>> **Overnight/courier fees:** These fees usually cost $50 or less. Remember the times when you sent something via the U.S. Postal Service to a destination that you could have driven to in less than a few hours, and it took them the better part of a week to get it there? Well, lenders and other players in real estate deals know that these snags can occur without warning, and because they don't want to derail your transaction or cost themselves money, they often send stuff the fastest way they can. And why not — it's your money!

>> **Notary:** Notary fees run from $10 to $20 *per signature per buyer.* At the close of escrow, you sign all sorts of important documents pledging your worldly possessions and firstborn child, should you renege on your mortgage. Therefore, you need to have your signature verified by a notary so everybody in the transaction knows that you really are who you say you are.

As you can see, closing costs can mount up in a hurry. In a typical real estate deal, closing costs total 2 to 5 percent of the property's purchase price. Thus, you shouldn't ignore them in figuring the amount of money you need to close the deal. Having enough to pay the down payment on your loan isn't sufficient.

TIP

When you're short of cash and hot to buy a home sooner rather than later, you can take out a mortgage with no out-of-pocket fees and points (see Chapter 6) and try to negotiate with the property seller to pay other closing costs (see Chapter 12). Expect to pay a higher mortgage interest rate for a low-up-front-fee loan. And all other things being equal, expect to pay a higher purchase price (with a corre-spondingly bigger mortgage) to entice the seller to pay your other closing costs. Also, don't blindly accept all the closing costs come closing time. (In Chapter 14, we explain the importance of auditing your closing statement.)

Accumulating the Down Payment

Jeremy went house hunting and soon fell in love with a home. Unfortunately, after he found his dream home, he soon discovered all the loan-documentation require-ments and the extra fees and penalties he would have to pay for having a small down payment. Ultimately, he couldn't afford to buy the home that he desired because he hadn't saved enough. "If I had known, I would have started saving much sooner — I thought that saving for the future was something you did when you turned middle-aged," he told Eric.

REMEMBER

We don't want you to be surprised when you finally set out to purchase a home. That's why now, in the comfort of your rental, commuter train, or bus (or any-where else you may be reading this book), we'd like you to consider the following:

>> How much money you should save for the down payment and closing costs for the purchase of your home

>> Where your down-payment money is going to come from

>> How you should invest this money while you're awaiting the purchase and closing

The 20 percent solution

Ideally, when buying a home you should have enough money accumulated for a down payment of 20 percent of the property's purchase price. Why 20 percent and not 10 or 15 or 25 or 30 percent? Twenty percent down is the magic number because it's generally a big enough cushion to protect lenders from default. Suppose, for example, that a buyer puts only 10 percent down, property values drop 5 percent, and the buyer defaults on the loan. When the lender forecloses — *after* paying a real estate commission, transfer tax, and other expenses of sale — the lender will be in the hole. Lenders don't like losing money. They've found that they're far less likely to lose money on mortgages where the borrower has put up a down payment of at least 20 percent of the property's value. (Unfortunately, lenders and the folks in Washington forgot this fact, which led to the late 2000s real estate market problems and high levels of foreclosures.)

If, like most people, you plan to borrow money from a bank or other mortgage lender, be aware that almost all require you to obtain (and pay for) private mortgage insurance (PMI) if your down payment is less than 20 percent of the property's purchase price. Although PMI typically adds several hundred dollars annually to your loan's cost, it protects the lender financially if you default. Should you buy an expensive home — into the hundreds-of-thousands-of-dollars price range — PMI can add $1,000 or more, annually, to your mortgage bill. (When you make a down payment of less than 20 percent, you can also expect worse loan terms, such as higher up-front fees and/or a higher ongoing interest rate on a mortgage.)

REMEMBER

PMI isn't a permanent cost. Your need for PMI vanishes when you can prove that you have at least 20 percent *equity* (home value minus loan balance outstanding) in the property. The 20 percent can come from loan paydown, appreciation, improvements that enhance the property's value, or any combination thereof. Note also that to remove PMI, most mortgage lenders require that an appraisal be done — at your expense.

Note: If you have (or expect to have) the 20 percent down payment and enough money for the closing costs, skip the next section and go to the section on how to invest your down-payment money.

Ways to buy with less money down

Especially if you're just starting to save or are still paying off student loans or worse — digging out from consumer debt — saving 20 percent of a property's purchase price as a down payment plus closing costs can seem like a financial mountain.

Don't panic, and don't give up. Here's a grab bag filled with time-tested ways to overcome this seemingly gargantuan obstacle:

>> **Boost your savings rate.** Say that you want to accumulate $30,000 for your home purchase, and you're saving just $100 per month. At this rate, it will take you nearly two decades to reach your savings goal! However, if you can boost

your savings rate by $300 per month, you should reach your goal in about five years.

TIP

Being efficient with your spending is always a good financial habit, but saving faster is a *necessity* for nearly all prospective home buyers. Without benevolent, loaded relatives or other sources for a financial windfall, you're going to need to accumulate money the old-fashioned way that millions of other home buyers have done in the past: by gradually saving it. Most people have fat in their budgets. Start by reading Chapter 2 for ways to assess your current spending and boost your savings rate.

>> **Set your sights lower.** Twenty percent of a big number is a big number, so it stands to reason that 20 percent of a smaller number is a smaller number. If the down payment and closing costs needed to purchase a $300,000 home are stretching you, scale back to a $240,000 or $200,000 home, which should slash your required cash for the home purchase by about 20 to 33 percent.

>> **Check out low-down-payment loan programs.** Some lenders offer low-down-payment mortgage programs where you can put down, say, 10 percent of the purchase price. To qualify for such programs, you generally must have excellent credit, have two to three months' worth of reserves for your housing expenses, and purchase private mortgage insurance (PMI). In addition to the extra expense of PMI, expect to get worse loan terms — higher interest rates and more up-front fees — with such low-money-down loans. Check with local lenders and real estate agents in your area. The best low-down-payment loan is the FHA purchase program. If you are a veteran, get a VA loan.

TIP

Unless you're chomping at the bit to purchase a home, take more time and try to accumulate a larger down payment. However, if you're the type of person who has trouble saving and may never save a 20 percent down payment, buying with less money down may be your best option. In this situation, be sure to shop around for the best loan terms.

>> **Access retirement accounts.** Some employers allow you to borrow against your retirement-savings plan. Just be sure you understand the repayment rules so you don't get tripped up and forced to treat the withdrawal as a taxable distribution. You're allowed to make penalty-free withdrawals from Individual Retirement Accounts for a first-time home purchase (see Chapter 2).

>> **Get family help.** Your folks or grandparents may like, perhaps even love, to help you with the down payment and closing costs for your dream home. Why would they do that? Well, perhaps they had financial assistance from family when they bought a home, way back when. Another possibility is that they have more money accumulated for their future and retirement than they may need. If they have substantial assets, holding onto all these assets until their death can trigger estate or inheritance taxes depending upon the state they live in. A final reason they may be willing to lend you money is that they're bank-and-bond-type investors and are earning paltry returns.

If your parents or grandparents (or other family members, for that matter) broach the topic of giving or lending you money for a home purchase, go ahead and discuss the matter. But in many situations, you (as the prospective home buyer) may need to raise the issue first. Some parents just aren't comfortable bringing up the topic of money or may be worried that you'll take their offer in the wrong way.

>> **Look into seller financing.** Some sellers don't need all the cash from the sale of their property when the transaction closes escrow. These sellers may be willing to offer you a second mortgage to help you buy their property. In fact, they often advertise that they're willing to assist with financing. Seller financing is usually due and payable in five to ten years. This gives you time to build up equity or save enough to refinance into a new, larger, 80 percent conventional mortgage before the seller's loan comes due.

Be cautious about seller financing. Some sellers who offer property with built-in financing are trying to dump a house that has major defects. It's also possible that the house may be priced far above its fair market value. Before accepting seller financing, make sure the property doesn't have fatal flaws (have a thorough inspection conducted, as we discuss in Chapter 13) and is priced competitively. Also be sure that the seller financing interest rate is as low as or lower than the rate you can obtain through a traditional mortgage lender.

>> **Get partners.** With many things in life, there is strength in numbers. You may be able to get more home for your money and may need to come up with less up-front cash if you find partners for a multiunit real estate purchase. For example, you can find one or two other partners and go in together to purchase a duplex or triplex.

Before you go into a partnership to buy a building, be sure to consider all the "what ifs." What if one of you wants out after a year? What if one of you fails to pay the pro-rata share of expenses? What if one of you wants to remodel and the other doesn't? And so forth. Have a lawyer prepare a co-ownership agreement that explicitly delineates how issues like these will be dealt with. Otherwise, you can face some major disagreements down the road, even if you go in together with friends or people you think you know well. We cover the pros and cons of partnerships in Chapter 8.

Where to invest the down payment

As with all informed investing decisions, which investment(s) you consider for money earmarked for your down payment should be determined by how soon you need the money back. The longer the time frame during which you can invest, the more growth–oriented and riskier (that is, more *volatile*) an investment you may consider. Conversely, when you have a short time frame — five years or less — during which you can invest, choosing volatile investments is dangerous.

When the stock market is rising, some folks are tempted to keep down-payment money in stocks. After all, when you're getting returns of 10, 15, 20 percent or more annually, you'll reach your down-payment savings goal far more quickly. Greedier investors lusting after high-flying technology and Internet stocks that seem to double in value every 90 days hope to quickly parlay their small savings for a shack into a money mountain for a mansion.

BUYING A HOME WITH "NO MONEY DOWN"

More than a few books written by (and high-priced seminars led by) real estate "gurus" claim that not only can you buy property with no money down but also that you can make piles of money doing so. A generation ago, this way of thinking was popularized by Robert Allen in his book *Nothing Down*.

Allen says that the key to buying property with no money down is to find a seller who's a don't-wanter — that is, someone who "will do anything to get rid of his property." Why would someone be that desperate? Well, perhaps the person is in financial trouble because of a job loss, an overextension of credit, or a major illness.

Perhaps when more people used to live in smaller, tight-knit communities where everyone supported one another, this type of vulture capitalism may not have flourished. But in these times, Allen says, a don't-wanter can offer you the most favorable mortgage terms, such as a low down payment and interest rate.

How do you find such downtrodden souls who are just waiting for you to take advantage of them? According to Allen's estimates, 10 percent of the sellers in the real estate market are don't-wanters. Simply call people who have property listed for sale in the newspaper, or place ads yourself saying that you'll buy in a hurry.

In our experience, finding homes that can be bought with no money down isn't easy to do. If you can find such a desperate seller, be aware that the property may have major flaws. If the property were a good one, logic dictates that the seller wouldn't have to sell under such lousy terms. Should you have the patience to hunt around and sift through perhaps hundreds of properties to find a good one available with seller financing at no money down, be our guest. Just don't expect the task to be easy or all that lucrative. Better to look for good properties and low-down-payment lender financing and to start saving a healthy down payment so you can qualify for a better loan.

Investing down-payment money in stocks is a dangerous strategy. Your expected home purchase may be delayed for years due to a sinking investment portfolio. Stocks are a generally inappropriate investment for down-payment money you expect to tap within the next five years. More aggressive individual stocks should have an even longer time horizon — ideally, seven to ten or more years. Consider what happened to the home-buying dreams of folks who foolishly parked their home-down-payment money in the stock market before and during the severe stock market decline of the early 2000s and the steep decline of 2007–08.

Investments for five years or less

Most prospective home buyers aren't in a position to take many risks with their down-payment money. The sooner you expect to buy, the less risk you should take. Unless you don't expect to buy for at least five years, you shouldn't even consider investing in more growth-oriented investments, such as stocks.

Although it may appear boring, the first (and likely best) place for accumulating your down-payment money is in a money market mutual fund. As with bank savings accounts, money market mutual funds don't put your principal at risk — the value of your original investment (principal) doesn't fluctuate. Rather, you simply earn interest on the money that you've invested. Money market funds invest in supersafe investments, such as Treasury bills, bank certificates of deposit, and *commercial paper* (short-term IOUs issued by the most creditworthy corporations).

If you really want to save through a bank, shop, shop, shop around. Smaller savings and loans and credit unions tend to offer more competitive yields than do the larger banks that spend gobs on advertising and have branches all around. Remember, more overhead means lower yields for your money.

In addition to higher yields, the best money market funds offer check writing (so you can easily access your money) and come in tax-free versions. If you're in a higher income tax bracket, a tax-free money market fund may allow you to earn a higher effective yield than a money fund that pays taxable interest. (*Note:* You pay tax only on money invested outside tax-sheltered retirement accounts.) When you're in a high tax bracket (refer to Table 3-2 earlier in this chapter), you should come out ahead by investing in tax-free money market funds. If you reside in a state with high income taxes, consider a state money market fund, which pays interest that's free of both federal and state tax.

The better money market funds also offer telephone exchange and redemption and automated, electronic exchange services with your bank account. Automatic investment comes in handy for accumulating your down payment for a home purchase. Once per month, for example, you can have money zapped from your bank account into your money market fund.

REMEMBER

Because a particular type of money market fund (general, Treasury, or tax-free municipal) is basically investing in the same securities as its competitors, opt for a fund that keeps lean-and-mean expenses. A money fund's operating expenses, which are deducted before payment of dividends, are the major factor in determining its yield. As with the high overhead of bank branches, the higher a money fund's operating expenses, the lower its yield. We recommend good ones in this section.

When you're not in a high federal-tax bracket, and you're not in a high state-tax bracket (that is, you pay less than 5 percent in state taxes), consider the following taxable money market funds for your home-down-payment money:

>> Fidelity Government Cash Reserves ($0 to open)

>> Vanguard Prime Money Market ($3,000 to open)

You can invest in a money market fund that invests in U.S. Treasury money market funds, which have the backing of the U.S. federal government — for what that's worth! From a tax standpoint, because U.S. Treasuries are state-tax-free but federally taxable, U.S. Treasury money market funds are appropriate when you're not in a high federal-tax bracket but you are in a high state-tax bracket (5 percent or higher). Should you choose to invest in a money market fund that invests in the U.S. Treasury, consider these:

>> Fidelity Government Money Market ($0 to open)

>> USAA Treasury Money Market ($3,000 to open)

>> Vanguard Federal Money Market ($3,000 to open)

Municipal (also known as *muni*) money market funds invest in short-term debt (meaning that it matures within the next few years) issued by state and local governments. A municipal money market fund, which pays you federally tax-free dividends, invests in munis issued by state and local governments throughout the country. A state-specific municipal fund invests in state and local government-issued munis for one state, such as New York. So if you live in New York and buy a New York municipal fund, the dividends on that fund are generally free of both federal and New York state taxes.

So how do you decide whether to buy a nationwide or state-specific municipal money market fund? Federal-tax-free-only money market funds are appropriate when you're in a high federal-tax bracket but not a high state-tax bracket (less than 5 percent). Your state may not have good (or any) state-tax-free money market funds available. If you live in any of those states, you're likely best off with one of the following national money market funds:

>> Fidelity Municipal Money Market Fund ($0 to open)

>> T. Rowe Price Summit Municipal Money Market ($25,000 to open)

>> Vanguard Tax-Exempt Money Market ($3,000 to open)

State-tax-free money market funds are appropriate when you're in a high federal-tax bracket and a high state-tax bracket (5 percent or higher). Contact fund companies listed in the sidebar "Getting in touch with mutual fund companies" to see whether they offer a money fund for your state.

GETTING IN TOUCH WITH MUTUAL FUND COMPANIES

Most mutual fund companies don't have many (or any) local branch offices. Generally, this fact helps mutual fund companies keep their expenses low so they can pay you greater yields on their money market funds.

So how do you deal with an investment company without a location near you? Simple: You open and maintain your mutual fund account via the fund's toll-free phone line, the mail, or the web. Some fund providers may have branch offices.

Here's how to reach, by phone or online, the major fund companies we recommend:

- Fidelity: 800-544-8888; www.fidelity.com

- T. Rowe Price: 800-638-5660; www.troweprice.com

- USAA: 800-382-8722; www.usaa.com

- Vanguard: 800-662-7447; www.vanguard.com

Investments for more than five years

Should you expect to hold onto your home-down-payment money for more than five years, you can comfortably consider riskier investments, such as longer-term bonds, as well as more conservative stocks. Eric covers these investments and many others in the latest editions of his books *Investing For Dummies* and *Mutual Funds For Dummies* (both published by Wiley).

SHORT-TERM BONDS AND BOND FUNDS

You may be thinking, "Three to five years is an awfully long time to keep my money dozing away in a money market fund."

Well, yes and no. During some time periods, investors who buy bonds maturing in five years get very little in the way of extra yield versus what they can get in a good money market fund. During other periods, three-year to five-year bonds yield a good deal more interest than money market funds yield.

Whenever you invest in bonds that won't mature soon, you take on risk. First is the risk that the bond issuer may fall into financial trouble between the time you buy the bond and the time it's due to mature. Second is the risk that interest rates in general can greatly increase. If the latter happens, typically caused by unexpected inflation, you may end up holding a bond that pays you less interest than the rate of inflation.

Most of the time, bonds that mature in a few years should produce a slightly higher rate of return for you than a money market or savings account. However, if you invest in such bonds, recognize that you may end up earning the same as (or perhaps even less than) you would have earned had you stuck with a money market fund. Rising interest rates can deflate the value of an investment in bonds.

Invest in bonds only if you expect to hold them for at least three to five years. If you want to invest in individual bonds and you're not in a high tax bracket, consider Treasury bonds, which don't require monitoring of credit risk — that is, unless the U.S. government slips into default! Also look at the yield on bank certificates of deposit. You may also consider some high-quality, short-term bond mutual funds that invest in — you guessed it — short-term bonds. A solid one is Vanguard's Short-Term Investment-Grade.

If you're in a high tax bracket, a tax-free money market fund is hard to beat. Some federal-tax-free bond funds to peruse include Vanguard's Short-Term Tax-Exempt and Vanguard's Limited-Term Tax-Exempt funds. Good, double-tax-free, short-term bond funds just don't exist.

Chapter **4**

Why Home Prices Rise and Fall

I f you're contemplating the purchase of a home, you may be concerned about the future direction of home prices. After all, who wants to buy a home and then see prices drop? Conversely, who in their right mind wouldn't love to jump into the real estate market before prices head higher? To understand what drives home prices, you must examine what drives the supply of, and demand for, homes.

REMEMBER

As we discuss what causes home prices to rise and fall in this chapter, please keep the following in mind: When it comes to buying and owning a home, don't get too hung up on the current state of your local market. If you take the perspective that after you buy a home, you're likely to own one for many decades, worrying about timing your purchase is generally not worth the trouble. *Timing* — that is, buying when prices are at rock bottom and getting out when you think that home values are cresting — is extraordinarily difficult to do. We know people who started waiting for lower home prices a generation ago — and they're still waiting!

What Drives Real Estate Markets and Prices?

Predicting what's going to happen with real estate prices in a particular neighborhood, town, region, or state over the next one, two, three, or more years isn't easy. Ultimately, the demand for and prices of homes in an area are driven largely by the economic health and vitality of that area. With an increase in jobs, particularly ones that pay well, comes a greater demand for housing.

REMEMBER

If you first buy a home when you're in your 20s, 30s, or even your 40s, you'll likely end up being a homeowner for several decades or more. Over such a lengthy time, the real estate markets in which you have your money invested will surely experience more than a few ups and downs. History shows that real estate prices experience more and bigger ups than downs over the long term (check out Figure 4-1 to see what we mean), so don't fret about the cloudiness of your real estate crystal ball.

Month	EHS Median Price ($)
2015.12	223,200
2016.03	221,400
2016.06	247,700
2016.09	235,200
2016.12	233,200
2017.03	236,600
2017.06	263,300
2017.09	247,600
2017.12	246,500
2018.03	249,800
2018.06	273,800
2018.09	256,900
2018.12	254,700
2019.03	259,700
2019.06	285,300
2019.09	271,500
2019.12	274,500

FIGURE 4-1: Median existing home sale prices in the United States since 1975.

© John Wiley & Sons, Inc.

UNDERSTANDING THE LATE 2000s HOUSING PRICE SLIDE

In the late 2000s, most local real estate markets suffered significant price declines. The extent of each market's decline depended on a number of factors. Areas that declined the most in price were characterized by tremendous overbuilding and rampant specula-tion during the run-up to the mid-2000s. This surge in home buying in the early to mid-2000s happened in large part due to higher-risk loans (for example, low or no down payment required) and/or loans that didn't require borrowers to document their employment income.

Although somewhat of an esoteric point, we feel it's worth mentioning that the decline in median home values was exacerbated in the late 2000s by a surge of so-called dis-tressed property sales such as foreclosures and short sales. These tended to be lower-priced properties and helped pull down the median numbers.

Some folks on the sidelines (renters) wonder who in their right mind would willingly buy a home after witnessing such a steep price decline. Why not wait longer for even lower prices?

Well, no one rings a bell at the precise time that local real estate prices bottom out. And, by the time it's obvious that the real estate market is recovering, mortgage interest rates and home prices rise and make homes less affordable.

That said, you may be ambivalent about buying a home at particular times in your life. Perhaps you're not sure you'll stay put for more than three to five years. The shorter the time period you expect to hold onto your home, the more important it is to be careful about when you buy. Thus, part of your home–buying decision may hinge on whether current home prices in your area offer you a good value. Even if you expect to stay put for a while, understanding what causes home prices to rise and fall and knowing ways to maximize your chances of getting a good buy can also be worth your while. This chapter helps you grasp these points.

If you're going to buy a home, you're making a significant investment — perhaps the single biggest investment you've ever made. You can do a mountain of research to decide what, where, and when to buy.

In the rest of this chapter, we explain what to look for, from an investment standpoint, both in a community and in the property that you buy. Some of the information that we provide requires you to think like an investor. Of course, for many people, buying a home to live in is different from buying an investment property to rent out.

Note: We discuss different types of properties (single-family homes, condominiums, and the like) and their investment desirability in Chapter 8.

Jobs, glorious jobs

A home provides shelter from the elements and a place to store and warehouse your consumer possessions. Because houses cost money to buy and maintain — and you're likely not a descendant of the Rockefellers, the Gettys, or Bill Gates — you need an ongoing source of money in order to afford your home. Where does this money usually come from? A job.

Okay, you may call it your *career* or (even better) one of your *passions*. But if you're like most folks, you work to pay your bills. And a home and its accompanying expenses are one of the biggest sources of expenses that people have (hence, one of the reasons we end up working so many decades as adults)!

REMEMBER

So it stands to reason that the demand for housing and the ability to pay for housing are deeply affected by the abundance and quality of jobs in a community or area. From an investment perspective, an ideal area where homes appreciate well in value has the following characteristics:

>> **Job growth:** The fact that an area has hundreds of thousands or millions of jobs means little if the number of jobs is shrinking. The New York City metropolitan area had millions of jobs yet experienced declining real estate prices in the early 1990s and late 2000s because of a deteriorating job base. Job *creation* and well-paying jobs is the lifeblood of a healthy local real estate market.

TIP

Check out the unemployment situation and examine how the jobless rate has changed in recent years. Good signs are a declining unemployment rate and increasing job growth.

>> **Job diversity:** If a community is reliant on a paper manufacturer and an underwear maker for half of all its jobs, you ought to be wary of buying a home there. Should both of these companies do poorly, the real estate market will follow. This type of scenario has played out in smaller communities that were badly hurt when large manufacturers and military bases lost many employees.

>> **Job quality:** All jobs aren't created equal. Which area do you think has faster-appreciating real estate prices: an area with more high-paying jobs in growth industries (such as technology) or an area that's producing mostly low-pay, low-skill jobs (such as jobs in fast-food joints)? As with food, entertainment, and sex, quality is just as important as (if not more important than) quantity. When most of the jobs in a community come from slow-growing or

shrinking employment sectors (such as farms, small retailers, shoe and apparel manufacturers, and government), real estate prices are unlikely to rise quickly in the years ahead. On the other hand, areas with a preponderance of high-growth industries (such as technology) should have a greater chance of experiencing faster price appreciation.

TIP

So how can you get your hands on data that gives you this type of perspective? The U.S. Bureau of Labor Statistics compiles employment and unemployment data for metropolitan areas and counties. You can find this department's treasure-trove of data at www.bls.gov, or try visiting a good local library or chamber of commerce. A real estate agent also may be able to help you track down the data.

Available housing

Although jobs create the demand for housing, the amount of housing available — both existing and new — is the supply side of the supply-and-demand equation. Even though jobs are being created, housing values may be stagnant if an overabundance of available housing exists. Conversely, a relatively low employment growth rate in an area with a housing shortage can trigger significant real estate price increases.

INVESTIGATE

Start by examining how well the existing supply of housing is being utilized. Vacancy rates, which measure how much or little demand there is for existing rental units, are a useful indicator to investigate. The *vacancy rate* is calculated simply by dividing the number of empty (unrented) rental units by the total number of rental units available. So, for example, if 50 rental units are vacant in Happy Valley, Tennessee, and 1,000 total units are available, the vacancy rate is 5 percent (50 divided by 1,000).

A low vacancy rate (under 5 percent) is generally a good indicator of future real estate price appreciation. When the vacancy rate is low and declining, more competition for few available rental units exists (or will soon exist). This competition tends to drive up rental rates, making renting more expensive and less attractive.

On the other hand, high vacancy rates indicate an excess supply of rentals, which tends to depress rents as landlords scramble to find tenants. All things being equal, high (more than 7 to 10 percent) and increasing vacancy rates are generally a bad sign for real estate sellers and prices but good for prospective home buyers.

INVESTIGATE

In addition to checking out vacancy rates, which tell you how well the existing housing supply is being used, smart real estate investors look at what's happening with building permits. In order to build new housing, a permit is required. The trend in the number of building permits can tell you how fast or how slowly the supply of real estate properties may be changing in the future.

WARNING

A significant increase in the number of permits being issued, such as what happened in some areas in the early to mid-2000s, can be a red flag because it may signal a future glut of housing. Such increases often happen after a sustained rise in housing prices in an area. As prices reach a premium level, builders race to bring new housing to market to capitalize on the high prices.

Conversely, depressed prices or a high cost of building can lead to little new housing being developed. Eventually, this trend should bode well for local real estate prices.

The supply of housing is also determined in part by the amount of land available to develop. Unless you think that houseboats or landfill sites are the waves of the future, you'll agree that land is needed to build housing. A limited supply of land generally bodes well for long-term real estate price appreciation in an area. Thus, real estate has appreciated very well over the decades (and is expensive today) in areas such as Manhattan, San Francisco, Hawaii, Hong Kong, and Tokyo, which are surrounded by water. Conversely, home prices tend to rise slowly in areas with vast tracts of developable land.

Inventory of homes for sale and actual sales

Just as scads of developable land and a barrage of new buildings place a lid on potential real estate price increases in the future, so do escalating numbers of properties listed for sale. Local associations of real estate agents, through their Multiple Listing Service (MLS), typically track the total number of *listings* (employment agreements between a property owner and a real estate agent). Properties that are "for sale by owner" (that is, without an agent) aren't included in this total, but such unlisted sales tend to follow the same trends as property listed with agents.

In a normal real estate market, the number of homes listed for sale stays at a relatively constant level as new homes come on the market and other ones sell. But as property prices start to reach high levels and some real estate owners/investors seek to cash in and invest elsewhere, the *listing inventory* (number of listings) can increase significantly. When home prices reach a high level relative to the cost of renting (see the next section), increasing numbers of potential buyers choose to rent. Buyer interest also may dry up because of significantly rising mortgage interest rates, or an economic slowdown or downturn (as happened in the late 2000s).

An increase in the number of newly listed houses for sale and a high inventory of unsold homes are two signs that soft home prices likely lie ahead. With many options to choose among, prospective buyers can be picky about what they buy. This competition among many sellers for a few buyers is what begins to exert downward pressure on prices and can create a *buyer's market* — a market in which prices are soft because of supply far exceeding demand.

A decreasing number of new listings and a low inventory of properties listed for sale bode well for home price increases. Fewer listings, multiple purchase offers, rapid sales, and offer prices higher than some list prices indicate that the demand from buyers exceeds the supply of property listed for sale — a *seller's market*.

When the local economy is strong and housing isn't expensive compared with rental rates, more renters elect to (and can afford to) purchase, thus increasing sales activity. If you're a seller, you're in heaven. As a buyer, you may be frustrated by dealing with constant price increases, losing homes in multiple-offer situations, or being beaten by other bidders in the race to a new listing.

INTEREST RATES AND HOME PRICES

Because the biggest expense of owning a home generally is the monthly mortgage payment, the level of interest rates on mortgages has a big impact on home prices. As interest rates drop, so can payments on mortgages of a given size.

Consider a $100,000, 30-year, fixed-rate mortgage. If the interest rate is 6 percent, the monthly mortgage payment is $600. At an interest rate of 10 percent, the mortgage payment balloons to $878.

It's certainly true that low interest rates enable more renters to become homeowners. So, you may think that declining interest rates would cause home prices to rise and, conversely, that increasing interest rates would lead to falling home prices. That this isn't the way the world operates is proved by the fact that even though interest rates trended higher in the late 1990s, home prices were rising briskly in many communities nationwide. Also, many parts of the United States in the late 2000s experienced falling home prices at the same time that interest rates were declining.

Clearly, other factors do influence home prices, especially the health of the local and national economies and consumer confidence. And although low interest rates make housing more affordable, low rates also make building more housing at lower costs possible. A larger supply of housing tends to dampen housing price increases.

What's the lesson of this story? Don't try to time your housing purchase based on what's happening, or what you expect to happen, with interest rates. The future change in your home's value can very well surprise you.

The rental market

Rental rates provide a useful indicator of the demand for housing. When the demand for rental housing exceeds the supply of rental housing and the local economy continues to grow, rents generally increase. This situation is a plus for future home price increases. As the cost of renting increases, purchasing a home looks all the more attractive to renters who are on the fence and are considering buying.

The trend in rents and the absolute level of rents won't tell you all that you need to know. Suppose that a two-bedroom, one-bath, 1,100-square-foot home in a decent neighborhood in your town is renting for $1,200 per month. So what? What you also need to know is how this rental cost compares with the cost of purchasing and owning the same home.

TIP

Compare the cost of renting a given home with the cost of owning it. Such a comparison is effectively what current renters do when they weigh the costs of buying a home and leaving their landlord behind. Comparing the cost of owning a home with the cost of renting that same property serves as a reality check on home prices.

To make a fair comparison between renting and owning, you must compare the monthly cost of renting with the monthly cost of owning. If you compare the cost of renting a home for $1,200 per month with the sticker price of buying that same home for $250,000, you're comparing apples with oranges. That $250,000 is the total purchase price of the home, not your monthly cost of owning it.

And when you calculate homeownership costs, you must also factor in tax benefits. Your biggest homeownership expenses — mortgage interest and property taxes — are generally tax-deductible (see Chapter 3), subject to limitations.

REMEMBER

Comparing the cost of owning a home with the cost of renting that same home is a simple yet powerful indicator of whether real estate in an area is overpriced, underpriced, or priced just right. Buying is generally safer (and a good value) when it costs about the same as renting. However, in some particularly desirable and in-demand communities, homeownership almost always costs more than renting. What's a reasonable premium? You won't find a simple answer, but if the monthly cost of owning is pushing past 20 to 30 percent more than the monthly cost of renting, be cautious.

After you purchase a home, you'll probably end up owning a home for decades to come in different areas throughout your working years. So, don't worry about timing your first home purchase. Trying to time your purchase has more importance if you may be moving in fewer than five years. In that case, be careful to avoid buying in an overheated market. The level of real estate prices compared with rents, the state of the job market, and the number of home listings for sale are useful indicators of the housing market's health.

REAL ESTATE GET-RICH-QUICK SCHEMES

Scores of books have been written and high-priced seminars conducted claiming to have the real estate investing approach that can "beat the system." Often, these promoters claim that you can become a multimillionaire through buying *distressed* property — property with financial, legal, or physical problems. One suggested strategy is to buy property on which a seller has defaulted or is about to default. Or how about buying a property in someone's estate through probate court? Maybe you'd like to try your hand at investing in a property that has been condemned or has toxic-waste contamination!

Getting a "good buy" and purchasing a problem property at a discount larger than the cost of fixing the property are possible. But such opportunities aren't easy to find as sellers of such properties are often unwilling to sell at a large enough discount to leave you sufficient profit. If you don't know how to thoroughly and correctly evaluate the problems of a property, you can end up overpaying. You may even get stuck with a property that has incurable defects such as poor location, excessive noise, or no backyard. (We tell you how to separate dumps from diamonds-in-the-rough fixer-upper properties in Chapter 8.)

In some cases, the strategies that these real estate gurus advocate involve taking advantage of people's lack of knowledge. For example, some people don't know that they can protect the equity in their home through filing personal bankruptcy. When you can find sellers in such dire financial straits, you may be able to get bargain buys on their homes.

Other methods of getting a good buy take a great deal of time and digging. Some involve cold-calling property owners to see whether they're interested in selling. Although you may eventually find a good candidate this way, when you factor in the value of your time, these deals appear far less attractive.

Tax rules and changes

How home purchases and ownership are treated under federal (and state) income tax rules impacts real estate values. Tax laws can and do change, as they did in late 2017 with the Tax Cuts and Jobs Act that took effect in 2018. As discussed in Chapter 3, the biggest change from that bill was to reduce the deductibility of home mortgage interest and property taxes. Under this new law, mortgage interest deductions are now limited to the first $750,000 of debt, down from $1,000,000 previously. The amount of property taxes combined with state income taxes that may be deducted is now limited to $10,000 annually (previously, there was no limit).

At the time these changes took effect, the so-called standard deduction that folks could take on their federal income tax returns was approximately doubled. This combined with the curtailed deductions allowed on Schedule A have led to more people taking the standard deduction, which is a time-saver. Some taxpayers, however, who had high property and state income taxes lost some of those write-offs. (That didn't necessarily mean such folks paid higher total federal income tax bills, because federal income tax rates were reduced across the board effective 2018.)

The overall impact of these recent tax law changes is that it reduced the value of real estate deductions, which means that it made real estate a bit more expensive to home buyers and homeowners. This was particularly true in areas with high home prices, high property taxes, and high state income taxes — think areas in and around New York City, San Francisco, Los Angeles, Chicago, Boston, and so on. All things being equal, you'd expect home prices, especially on more expensive properties, to take a hit from these recent changes, and that is indeed what we have seen.

Tax rules will continue to evolve in the years and decades ahead. Because legislation is required and existing stakeholders defend their positions, change is generally slow in coming and usually incremental and not major.

How to Get a Good Buy in Any Market

What if you have to (or want to) buy in a seller's market? Or you're simply frightened that you're going to overpay in any market because you're a home-buying novice? No one likes to be taken. And most folks like to feel or believe that they're getting a bargain.

Many times, when you purchase products and services through businesses (especially retailers), the sellers like to tell you how much of a discount or markdown they're offering you:

60 PERCENT OFF!

GOING-OUT-OF-BUSINESS SALE — EVERYTHING MUST GO!

SAVE UP TO $3 PER POUND!

SPRING CLEARANCE SALE!

Some home sellers and (more often) their agents like to use the same type of advertising. The following examples are from actual home-for-sale ads:

HUGE PRICE REDUCTION!

PRICE SLASHED $20,000!

REDUCED! BACK ON THE MARKET — OWNER MUST SELL!

TIP

Whenever you see these types of ads, rather than thinking, "Gee, that sounds like a good deal," you should be thinking, "That home must have been overpriced before, and/or it probably has a significant defect."

Now, we're not trying to tell you that you can't get a *deal* (in other words, buy a home for less than it's worth). But doing so isn't easy, and finding just the right situation takes a great deal of work. For most people, not overpaying — in other words, paying fair market value (which isn't necessarily the asking price) — is a good objective. (See Chapter 10 to find out more about determining home values.)

Read our following suggestions for finding a good buy even if you're willing to pay fair market value. These ideas can help prevent you from overpaying.

Seek hidden opportunities to add value

The easiest problems to correct are cosmetic. Some sellers and their agents are lazy and don't even bother to clean a property. Painting, tearing up dingy carpeting, refinishing hardwood floors, replacing outdated cabinets and appliances, and installing new landscaping need not be difficult projects. And such changes can make some properties look much better.

A somewhat more complicated way to add value is to identify properties not being fully used or developed according to their zoning. Sometimes you can make more productive use of a property. For example, you may be able to convert a duplex to two separate condominiums. Some single-family residences may incorporate a rental unit if local zoning allows. Perhaps you can add another level to create a panoramic view. A good real estate agent, a contractor, and the local planning office in the town or city in which you're looking at property should be able to help you identify properties for which the use can be changed.

REMEMBER

Identifying, evaluating, buying, and fixing up a property take valuable time and energy. If you have a talent for finding hidden opportunities and are willing to invest the time required to coordinate the fix-up work, by all means try your hand and money at it! Just be sure to be realistic when you assess whether the major defects can be corrected, how much money you may need to spend to improve the property, and how much value your improvements can really add. Also, be sure to hire a competent property inspector (see Chapter 13).

Buy when others are scared to buy

When the economy hits the skids and unemployment rises, the mood is somber and gloomy, and the number of home purchases usually drops. Prices tend to fall as well. This situation, such as the conditions brought on by the financial and credit crisis of the late 2000s, can signal a great time to step up and buy. Buy when homes are "on sale" and when you don't have to compete with many other buyers. Buy when you can have your pick of a larger inventory of homes for sale.

TIP

Few people feel comfortable buying an investment that has gone down in value, especially when things look bleak. (For some perverse psychological reason, though, many of us love shopping for bargains in retail stores.) Here are several signs that a soft real estate market is beginning to firm up:

>> The monthly cost of owning a home approximates (or is less than) the monthly cost of renting a similar property. One of the beauties of a major real estate price decline is that it can bring homeownership costs back in line with rent costs. That's exactly what happened in many previously overpriced areas during the late 2000s real estate price slump.

>> The inventory of homes listed for sale starts to fall from its peak as home sales pick up.

>> The rental market tightens (as evidenced by increasing rents and a low vacancy rate). Another good sign is that little new housing is being built.

>> The job market improves. Remember that jobs fuel the demand for housing. Home prices tend to rebound when employment increases. Watch for a decrease in the unemployment rate in your region.

REMEMBER

Despite lower home prices, an improving economy, and tightening rental and homes-for-sale inventories, prospective home buyers generally show far less interest in buying a home when things still look bleak. It takes courage to buy during those times when headlines and news reports trumpet more layoffs. Keep a level head and take advantage of buying opportunities when they occur. Years down the road, you may be glad that you did.

Find a motivated seller

When you take your time and peruse enough properties, you eventually cross paths with a property owner who really needs or wants to sell. The owner may need to relocate to another part of the country for a job, or perhaps the owner is trading up to a larger home and needs the cash from the sale of the current one to buy the new home. Sometimes a property owner simply can't afford to own and maintain a home any longer because of personal financial troubles.

TIP

Whatever the reason, buying a home at or below its fair market value is far easier if the seller is what we call *motivated*. How do you find a motivated seller? Simple: You ask questions! The number of prospective buyers who are too shy to (or don't think to) ask why a seller is selling is amazing. Many sellers will be honest, and more than a few real estate agents (especially those who love to talk) have loose lips and share plenty of details. But you gotta ask!

Buy during slow periods

Most local real estate markets go through predictable busy and slow periods like clockwork. Just as it makes good sense to buy when the overall real estate market in an area is depressed, it can be beneficial to buy during those typical slow periods.

For example, far fewer prospective buyers tend to be looking for homes during the holiday season in the dead of winter. In most markets, the period from Thanksgiving through January or February tends to be quite slow. The colder the region in which you live, the later into the new year this slow period typically lasts. In the blustery and snowy northernmost regions of the United States, the real estate market doesn't really start to pick up until April. In locales that are sunny and warm year-round, such as Florida and Southern California, home sales start to pick up early in the new year.

Another typically slow period is during the summer months of July and August. Many people take vacations then, and those families who want to buy in time for the next school year have likely already bought. The oppressive heat in the southern regions also keeps people indoors and near their air conditioners and iced tea.

The advantage of looking during the slow periods is that you have far less competition from other potential buyers. If you can find a motivated seller, you may really be able to negotiate a great deal without the intrusion of other potential buyers.

We're not saying, however, to look only during slow periods of the year, or that you can definitely get a good buy during these times. You must be realistic. In most markets, fewer properties are for sale during the slow periods. Smart sellers get their properties sold during the more active periods in the spring and fall. Also, be aware that a good portion of those properties on the market during slow periods may be the unwanted leftovers.

Become a great negotiator

Getting a good buy can be as simple as being a good negotiator. Good negotiation skills may enable you to buy a property at less than fair market value, especially if you find a seller who needs to sell soon.

Your negotiating position is also better if you're in a situation where you don't *have* to buy. The more patient you are and the more willing you are to walk away if you don't get a good deal, the better able you'll be to negotiate a good buy. Having a backup property in mind (or remembering that many other properties are out there for you to buy) also helps. See Chapter 12 to discover how to be a world-class negotiator.

Buy in a good neighborhood

If you buy a home in a desirable area, you should have a better chance at making a good investment. We explain how to find out whether a neighborhood is good in Chapter 8.

When you buy good real estate and hold it for the long term, you should earn a decent return on your investment. Over the long haul, having bought a property at a small discount becomes an insignificant issue. You'll make money from owning a home (and perhaps other real estate investments) as the overall real estate market appreciates.

Financing 101

2

Find out how to obtain and boost your credit score.

Understand the different types of mortgages and how to select the best one for your situation.

Tackle common mortgage problems.

Chapter **5**

Understanding and Improving Your Credit Score

You can't play the home-buying game if you can't pay. And most people can't pay without a mortgage.

When you apply for a loan, lenders try to determine your credit risk level. If they loan you money, what are the odds that you'll pay them back on time? To understand your credit risk, most lenders look at your credit score. Your score influences the credit that's available to you and the terms of any mortgage that lenders offer you.

Most lenders also use a number of other facts to make credit decisions. They usually look at the amount of debt you can reasonably handle given your income, your employment history, and your credit history. Based on their perception of this information, as well as their specific underwriting policies, lenders may extend credit to you even if your score is low, or decline your request for credit even if your score is high. But your chances for getting approved at the best possible loan terms improve when you have a good score.

REMEMBER

Understanding your credit score can help you manage your credit health. By knowing how lenders evaluate your credit risk, you can take action to lower your credit risk — and thus raise your score — over time. A better score may mean better loan options for you.

Follow the tips in this chapter to manage your credit score efficiently. Improving your score can help you

>> Get better credit offers

>> Lower your interest rates

>> Speed credit approvals

The Record You Can't Ignore: Your Credit Report

You've probably heard about credit reports and realize that they can make or break your application for a loan. However, these reports are mysterious to most people because they don't know what goes into a credit report, nor do they understand what they can do to improve their credit score. Wonder no more . . . .

What your credit history comprises

Credit reports tell a lender how well you manage your finances. Your report details your credit history as it has been reported to the credit-reporting agency by lenders who've extended credit to you. Your credit report lists

>> What types of credit you use

>> The length of time your accounts have been open

>> Whether you've paid your bills on time

It tells lenders how much credit you've used and whether you're seeking new sources of credit. It gives lenders a broader view of your credit history than one bank's own customer records (unless all your previous credit has been drawn from that one bank).

What goes into your credit report

REMEMBER

Although each credit-reporting agency formats and reports information differently, all credit reports contain basically the same kinds of information:

>> **Identifying information:** Your name, address, Social Security number, date of birth, and employment information are used to identify you. These factors aren't used to calculate your score, however. Updates to this information come from information you supply to lenders.

>> **Trade lines:** These are your credit accounts. Lenders report each account you've established with them. They report the type of account (credit card, auto loan, mortgage, and so on), the date you opened the account, your credit limit or loan amount, the account balance, your payment history, and when you closed the account.

>> **Inquiries:** When you apply for a loan, you authorize your lender to ask for a copy of your credit report. This is how inquiries appear on your credit report. The inquiries section contains a list of everyone who has accessed your credit report within the past two years. The report you see lists both *voluntary* inquiries, spurred by your own requests for credit, and *involuntary* inquiries, such as when lenders order your report so as to make you a preapproved credit offer in the mail.

>> **Public record and collection items:** Credit-reporting agencies also collect public record information from state and county courts and information on overdue debt from collection agencies. Public record information includes bankruptcies, foreclosures, suits, wage attachments, liens, and judgments.

Why you should check your credit report

If your credit report contains errors, it may be incomplete or contain information about someone else. This typically happens because

>> You applied for credit under slightly different names (Robert Jones, Bob Jones, and so on).

>> Someone made a clerical error in reading or entering name or address information from a handwritten application.

>> Someone gave an inaccurate Social Security number, or the lender misread the number.

>> Loan or credit-card payments were inadvertently applied to the wrong account.

INVESTIGATE

Derogatory information on your credit report may force you into a mortgage with a higher interest rate and fees or cause your mortgage application to be denied. If you find an error, the credit-reporting agency must investigate and respond to you within 30 days. If you're in the process of applying for a loan, immediately notify your lender of any incorrect information in your report. See the last section in this chapter for info on how to contact the credit-reporting agencies to obtain your credit report.

The Most Popular Kid on the Block: FICO Scores

The credit score most lenders use today was developed by Fair Isaac Corporation and is called a *FICO score.* FICO scores range from a low of 300 to a maximum of 850. The higher your FICO score, the lower the potential risk you pose for lenders. FICO scores are available to lenders at the three major credit-reporting agencies: Equifax, Experian, and TransUnion.

Although FICO scores are the most commonly used credit risk scores in the United States, lenders may use other scores to evaluate your credit risk. These include

>> **Application risk scores:** Many lenders use scoring systems that include the FICO score but also consider information from your loan application.

>> **Customer risk scores:** A lender may use these scores to make credit decisions about its current customers. Also called *behavior scores,* these scores generally consider the FICO score along with information on how you've paid that lender in the past.

>> **Other credit bureau scores:** These scores may evaluate your credit report differently than FICO scores, and in some cases a higher score may mean more risk, not less risk, as with FICO scores.

TIP

If you've been turned down for credit, the federal Equal Credit Opportunity Act (ECOA) gives you the right to find out why within 30 days. You're also entitled to a free copy of your credit report within 60 days, which you can request from the credit-reporting agencies. If your credit score was a primary part of the lender's decision, the lender will use the score reasons to explain why you didn't qualify for the credit.

How scores work — the short version

Each credit score is calculated by a mathematical equation that evaluates many types of information from your credit report at that agency. By comparing this information with the patterns in hundreds of thousands of past credit reports, the score identifies your level of estimated future credit risk.

For a FICO score to be calculated from your credit report, the report must contain at least one account that has been open for six months or longer. In addition, the report must contain at least one account that has been updated in the past six months. This ensures that enough recent information is in your report to calculate a score.

Your score can change whenever your credit report changes. But your score probably won't change a lot from one month to the next unless the information in your credit report changes substantially.

REMEMBER

Although a bankruptcy or late payment can quickly lower your score, improving your score takes time. That's why it's a good idea to check your score (especially if you have reason to be concerned about your credit history) at least six months before applying for a mortgage. That gives you time to take corrective action if needed. If you're actively working to improve your score, you should check it quarterly or even monthly to review changes.

FICO SCORES CAN DIFFER AMONG BUREAUS

FICO (previously known as Fair Isaac) makes the FICO scores as consistent as possible among the three credit-reporting agencies. If your information is identical at all three credit-reporting agencies, your scores from all three should be within a few points of one another.

But sometimes your FICO score may be quite different at each of the three credit-reporting agencies. The way lenders and other businesses report information to these agencies sometimes results in different information being in your credit report at the different agencies. The agencies may also report the same information in different ways. Even small differences in the information at the three credit-reporting agencies can affect your scores.

Because lenders may review your score and credit report from any one of the three credit-reporting agencies, check your credit report at all three to make sure each is correct (see the last section in this chapter to find out how to do that).

But no score, no matter how high or low, says whether you'll be a "good" or "bad" customer. Although many lenders use FICO scores to help them make lending decisions, each lender also has its own strategy, including the level of risk it finds acceptable for a given type of loan. There is no single minimum score used by all lenders.

How a FICO score assesses your credit history — the long version

The FICO score evaluates several categories of information: your payment history, the amount you owe, the length of your credit history, new credit you've acquired, types of credit you have in use, and the number of credit queries. Some, as you'd expect, are more important than others. It's important to note that

» **A score considers all these categories of information, not just one or two.** No one piece of information or factor alone determines your score.

» **The importance of any factor depends on the overall information in your credit report.** A given factor may be more important for some people than for others who have a different credit history. In addition, as the information in your credit report changes, so does the importance of any factor in determining your score. That's why it's impossible to say exactly how important any single factor is in determining your score — even the levels of importance shown in the following sections are for the general population and differ for different credit profiles.

» **Your FICO score looks only at information in your credit report.** Lenders often also look at other things when making a credit decision, including your income, how long you've worked at your present job, and the kind of credit you're requesting.

» **Your score considers both positive and negative information in your credit report.** Late payments lower your score, but establishing or reestablishing a good track record of making payments on time raises your score.

» **Raising your score is a bit like getting in shape.** It takes time, and there is no quick fix. In fact, quick-fix efforts can backfire. The best advice is to manage credit responsibly over time.

The percentages in Figure 5-1 are based on the importance of the five categories for the general population. For particular groups — for example, people who haven't been using credit long — the importance of these categories may be different.

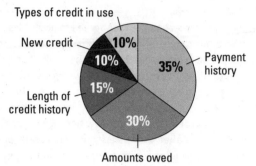

Types of credit in use

New credit

10%

10%

Length of credit history

15%

35%

Payment history

30%

Amounts owed

© John Wiley & Sons, Inc.

FIGURE 5-1:
How a credit score breaks down.

The following sections offer a complete look at the information that goes into a FICO score and how you can improve your score in each area. For a visual graphic of what contributes to your credit score, see Figure 5-1.

Your payment history

The first thing any lender wants to know is whether you've paid past credit accounts on time. Your track record for repaying creditors affects roughly 35 percent of your score.

Late payments aren't an automatic score-killer. An overall good credit picture can outweigh one or two instances of, say, late credit-card payments. On the other hand, having no late payments in your credit report doesn't mean you automatically get a great score. Some 60 to 65 percent of credit reports show no late payments at all. Your payment history is just one piece of information used in calculating your score.

In the area of payments, your score takes into account

>> **Payment information on many types of accounts:** These include credit cards such as Visa, MasterCard, American Express, and Discover; credit cards from stores where you do business; *installment loans* (loans such as a mortgage on which you make regular payments); and finance-company accounts.

>> **Public record and collection items:** These items include reports of events such as bankruptcies, foreclosures, suits, wage attachments, liens, and judgments. They're considered quite serious, although older items and items with small amounts count less than more recent items or those with larger amounts. Bankruptcies stay on your credit report for seven to ten years, depending on the type.

A foreclosure, a *short sale* (where the lender agrees to accept less than the total amount due on a mortgage), or a *deed in lieu of foreclosure* (in which a borrower deeds his property to the lender, who then sells it and uses the proceeds of the sale to repay the mortgage) all lower your credit score by about the same amount. If you've had the misfortune to go through one of these negative credit events, be advised that they're considered extremely serious delinquencies. Don't expect to get a new home loan with favorable terms for five to seven years after your foreclosure, short sale, or deed in lieu of foreclosure. Check with your lender or mortgage broker regarding your specific situation.

A property is *underwater* if the owner owes more on the mortgage than the house is worth. There are degrees of underwater. For instance, a property may be wading-pool deep or bottom-of-the-ocean submerged. Some owners who can afford to pay their mortgage feel it's pointless to do so because their property will never recover the value it has lost. A *strategic default* occurs when a borrower who has the income to make the loan payments voluntarily decides not to, which forces the lender to foreclose. Lenders frown upon this practice big time, as you may expect. Credit agencies treat it as an extremely serious delinquency, no different than any other foreclosure.

>> **Details on late or missed payments** *(delinquencies)* **and public record and collection items:** The FICO score considers how late such payments were, how much was owed, how recently they occurred, and how many there are. As a rule, a 60-day late payment isn't as damaging as a 90-day late payment. A 60-day late payment made just a month ago, however, penalizes you more than a 90-day late payment from five years ago.

>> **How many accounts show no late payments:** A good track record on most of your credit accounts increases your credit score.

How to improve your FICO score:

>> **Pay your bills on time.** Delinquent payments and collections can have a major negative impact on your score.

>> **If you've missed payments, get current and stay current.** The longer you pay your bills on time, the better your score.

>> **If you recently went through a foreclosure, short sale, or deed in lieu of foreclosure, the strategy for improving your credit score is the same.** Most important, stay current on your bills. Pay down your credit-card balances, and keep future balances low. Open new credit accounts only when absolutely needed. Your score will gradually improve if you stick to this strategy.

>> **If you're having trouble making ends meet, get help.** This doesn't improve your score immediately, but if you can begin to manage your credit and pay

on time, your score gets better over time. See our advice in Chapter 2 for credit-problem-solving strategies.

TECHNICAL STUFF

Paying off or closing an account doesn't remove it from your credit report. The score still considers this information because it reflects your past credit pattern.

The amount you owe

About 30 percent of your score is based on your current debt. Having credit accounts and owing money on them doesn't mean you're a high-risk borrower who'll receive a low score. However, owing a great deal of money on many accounts can indicate that a person is overextended and is more likely to make some payments late or not at all. Part of the science of scoring is determining how much is too much for a given credit profile.

In the area of debts, your score takes into account

>> **The amount owed on all accounts.** Note that even if you pay off your credit cards in full every month, your credit report may show a balance on those cards. The total balance on your last statement is generally the amount that will show in your credit report.

>> **The amount owed on all accounts and on different types of accounts.** In addition to the overall amount you owe, the score considers the amount you owe on specific types of accounts, such as credit cards and installment loans.

>> **Whether you show a balance on certain types of accounts.** In some cases, having a small balance without missing a payment shows that you've managed credit responsibly. On the other hand, closing unused credit accounts that show zero balances and that are in good standing doesn't raise your score.

>> **How many accounts have balances.** A large number can indicate higher risk of overextension.

>> **How much of the total credit line is being used on credit cards and other** *revolving credit* **(carrying a debt balance month to month) accounts.** Someone closer to "maxing out" on many credit cards may have trouble making payments in the future.

>> **How much of installment loan accounts is still owed compared with the original loan amounts.** For example, if you borrowed $10,000 to buy a car and you've paid back $2,000, you owe (with interest) more than 80 percent of the original loan. Paying down installment loans is a good sign that you're able and willing to manage and repay debt.

TIP

How to improve your FICO score:

>> **Keep balances low on credit cards and other revolving credit.** High outstanding debt can adversely affect a score.

>> **Pay off debt.** The most effective way to improve your score in this area is by paying down your revolving credit.

>> **Don't close unused credit cards as a short-term strategy to raise your score.** Generally, this doesn't work. In fact, it may *lower* your score. Late payments associated with old accounts won't disappear from your credit report if you close the account. Long-established accounts show that you have a longer history of managing credit, which is a good thing. And having available credit that you don't use doesn't lower your score. You may have reasons other than your score to shut down old credit-card accounts that you don't use, but don't do it in hopes of getting a better score.

>> **Don't open new credit-card accounts that you don't need just to increase your available credit.** This approach can backfire and actually lower your score.

The length of your credit history

How established is your credit history? About 15 percent of your score is based on this area. In general, a longer credit history will increase your score. However, even people who haven't been using credit long may get high scores, depending on how the rest of the credit report looks.

In this area, your score takes into account

>> **How long your credit accounts have been established, in general.** The score considers both the age of your oldest account and an average age of all your accounts.

>> **How long specific credit accounts have been established.** Your score reflects how long you've been handling different types of credit. Your score increases over time, for example, if you show that you can manage both a credit card and an installment loan such as a mortgage or car loan. This category doesn't have a specified minimum or maximum amount of time, but the longer you responsibly manage your credit, the higher your credit score.

>> **How long it has been since you used certain revolving credit accounts.** The score assesses revolving credit accounts such as credit cards differently if they haven't been used for a long time. An inactive credit-card account, no matter how old, is given less weight in your credit score than active accounts.

TIP

To improve your FICO score in this category, don't open a lot of new accounts too rapidly if you've been managing credit for only a short time. New accounts lower your average account age, which has a larger effect on your score if you don't have a lot of other credit information. Also, rapid account buildup can look risky if you're a new credit user.

New credit you've acquired

Taking on a lot of new debt affects your score, too. About 10 percent of your score is based on new credit and credit applications. Any credit less than a year old is considered new.

TECHNICAL STUFF

Applying for several new credit cards or accounts in a short period represents more risk — especially for people who don't have a long-established credit history. However, FICO scores do a good job of distinguishing between a search for *many* new credit accounts and rate shopping for *one* new account.

In the area of new credit, your score takes into account

>> **How many new accounts you have.** The score looks at how many new accounts you have by type of account (for example, how many newly opened credit-card accounts you have). It also may look at how many of your accounts are new accounts.

>> **How long it has been since you opened a new account.** Again, the score looks at this by type of account.

>> **How many recent requests for credit you've made.** This is indicated by inquiries to the credit-reporting agencies. Inquiries remain on your credit report for two years, although FICO scores consider only inquiries from the last 12 months. The scores have been carefully designed to count only those inquiries that truly impact credit risk.

>> **The length of time since lenders made credit report inquiries.** The older the lender inquiries, the better. Inquiries more than a year old are ignored by the FICO score. In this case, being ignored is good.

>> **Whether you have a good recent credit history following past payment problems.** Reestablishing credit and making payments on time after a period of late payment behavior both help raise a score over time.

How to improve your FICO score:

>> **Do your rate shopping for a specific loan within a focused period of time.** FICO scores distinguish between a search for a single loan and a search for many new credit lines, in part by the length of time over which inquiries occur. For more on this topic, see the section "The number of credit inquiries," later in this chapter.

>> **Reestablish your credit history if you've had problems.** Opening new accounts responsibly and paying them off on time will raise your score in the long term.

It's okay to request and check your own credit report and FICO score. This won't affect your score as long as you order your credit report directly from the credit-reporting agency or through an organization authorized to provide credit reports to consumers. See the last section in this chapter, "Getting Hold of Your Report and Score," for details on where you can obtain your credit score.

Types of credit you have in use

The credit mix usually isn't a key factor in determining your score, but it's given more weight if your credit report doesn't have a lot of other information on which to base a score. About 10 percent of your score is based on this category.

In this area, your score takes into account

>> **What kinds of credit accounts you have.** Your score considers your mix of credit cards, retail accounts, installment loans, finance-company accounts, and mortgage loans. Don't feel obligated to have one of each.

>> **How many of each type of credit account you have.** The score looks at the total number of accounts you have. How many is too many varies depending on the credit type. You don't need to have one of each type. Don't open credit accounts you don't intend to use just to hype up your total.

How to improve your FICO score:

>> **Apply for and open new credit accounts only as needed.** Don't open accounts just to have a better credit mix — it probably won't raise your score.

>> **Have credit cards, but manage them responsibly.** In general, having credit cards and installment loans (and making timely payments) raises your score.

>> **Note that closing an account doesn't make it go away.** A closed account still shows up on your credit report and may be included in the score.

The number of credit inquiries

A search for new credit can mean greater credit risk. This is why the FICO score counts inquiries — those requests a lender makes for your credit report or score when you apply for credit.

FICO scores consider inquiries very carefully because not all inquiries are related to credit risk. You should note three things about credit inquiries:

>> **Inquiries don't affect scores very much.** For most people, one additional credit inquiry takes less than 5 points off their FICO score. However, inquiries can have a greater impact if you have few accounts or a short credit history. Large numbers of inquiries also mean greater risk: People with six inquiries or more on their credit reports are eight times more likely to declare bankruptcy than people with no inquiries on their reports.

>> **Many kinds of inquiries aren't counted at all.** The score doesn't count it when you order your credit report or credit score from a credit-reporting agency. Also, the score doesn't count requests lenders make for your credit report or score in order to make you a preapproved credit offer or to review your account with them, even though you may see these inquiries on your credit report. Requests that are marked as coming from employers aren't counted either.

>> **The score looks for *rate shopping*.** Looking for a mortgage or an auto loan may cause multiple lenders to request your credit report, even though you're looking for only one loan. To compensate for this, the score counts multiple inquiries in any 45-day period as just one inquiry. In addition, the score ignores *all* inquiries made in the 30 days prior to scoring. So if you find a loan within 30 days, the inquiries won't affect your score while you're rate shopping.

What FICO scores ignore

FICO scores consider a wide range of information on your credit report. However, they don't consider

>> Your race, color, religion, national origin, sex, or marital status

>> U.S. law prohibits credit scoring from considering any receipt of public assistance, or the exercise of any consumer right under the Consumer Credit Protection Act.

>> Your age

>> Your salary, occupation, title, employer, date employed, or employment history

Lenders may consider this information, however:

>> Where you live

>> Any interest rate being charged on a particular credit card or other account

>> Any items reported as child/family support obligations or rental agreements

>> Certain types of inquiries (requests for your credit report or score)

    The score does *not* count any requests you make, any requests from employers, or any requests lenders make without your knowledge.

>> Any information not found in your credit report

>> Any information not proved to be predictive of future credit performance

Why your score is what it is

When a lender receives your FICO score, up to four *score reasons* are also delivered. These are the top reasons why your score wasn't higher. If the lender rejects your request for credit and your FICO score was part of the reason, these score reasons can help the lender tell you why.

TIP

These score reasons can be more useful to you than the score itself. They help you determine whether your credit report may contain errors and how you may improve your credit score. However, if you already have a high FICO score (for example, in the mid-700s or higher), some of the reasons may not be as helpful, because they may be marginal factors related to less important categories, such as your length of credit history, new credit, and types of credit in use.

Getting Hold of Your Report and Score

REMEMBER

Getting and retaining a copy of your personal credit report is a good idea. Because your personal credit report contains a history of your use (and abuse!) of credit, it's important that you're aware of what it contains and whether the information is accurate.

If you're applying for a mortgage, you can ask at that time for a copy of your credit report and your credit scores — after all, you're paying for them! You should also know that lenders are required to give you a copy of your credit report and credit scores without charge if they turn you down for a loan.

There's no substitute for knowing your credit scores *before* applying for a mortgage. Hindsight is golden, but it won't allay your disappointment if you don't get the loan you want. Learning your scores at least six months before you apply gives you time to improve them before your lender sees them.

TIP

Congress gave each of us the right to get one free credit report annually from each of the three big credit bureaus — Experian (888-397-3742; www.experian.com), Equifax (800-685-1111; www.equifax.com), and TransUnion (800-888-4213; www.transunion.com). Credit bureaus are also generally required to provide you with a free copy of your report if you were denied credit, employment, housing, or insurance over the most recent 60 days because of information in their credit files. You may also purchase a copy of your credit report from any of the three big bureaus (generally for less than $10). At the least, we encourage you to get your free reports annually. Only the report is free, however. You still have to pay to find out what your credit score is.

TIP

It's easy to check your FICO score and to find out specific things that you can do to raise it. At www.myFICO.com you can get your FICO scores for a fee. Information you receive includes

» Your current FICO scores from all three credit bureaus

» Your credit reports on which your FICO score is based

» An explanation of your score, the positive and negative factors behind it, and how lenders view your credit risk

» A FICO score simulator you can use to see how specific actions, such as paying off all your card balances, would affect your score

» Specific tips on what you can do to improve your FICO score over time

In addition, you can see current information on the average interest rates for home loans for different FICO score ranges.

WARNING

The sale of credit score monitoring products to consumers by credit bureaus and banks has become a big business. Don't confuse them with your FICO score. FICO scores are currently the only credit scores accepted by Fannie Mae, Freddie Mac, and the Federal Housing Administration (FHA), the big government organizations that play a central role in the home loan industry. That's why your lender will probably use FICO scores, too.

FREE CREDIT SCORE OR SCAM?

Over the years, we've seen lots of websites come and go, including those claiming to give you something for nothing. Years ago, FreeCreditScore.com flooded television and other media outlets with their attention grabbing and musical advertising (featuring pirates and slovenly dressed young adults). Eric noticed their ads (how could you not!) and was asked by folks about the "offer" of getting your credit score for "free."

When you watched their ads, it was hard to discern what they were actually selling. Their name and little jingle were all about getting your "free credit score" of course. But, what were you really getting?

By signing up on their website at the time, you could indeed get your Experian calculated credit score for "free." (The big credit bureau Experian owned and operated the website.) Obscured, however, was the fact that what FreeCreditScore.com (FCS) was actually selling was a credit monitoring service for $14.99 per month — which works out to $179.88 per year! The commercial said, "Offer applies with enrollment in freecreditscore.com," but fails to say what that means or that it even costs anything.

Seven days after giving you your "free" Experian credit score (which isn't what lenders use), FCS began charging those who signed up for their free credit score $14.99 per month through your credit card until you discontinued the service. And, they didn't make it easy to quit the service once you signed up. You couldn't do so online — you had to call a customer service number.

These deceptive practices landed Experian in trouble with the Federal Trade Commission which sued and fined them and they also got in trouble with the Florida Attorney General. Today, there are websites offering free credit scores. But, the free scores they are giving away aren't really worth much because they are credit bureau credit scores and not FICO scores.

Chapter **6**

Selecting a Mortgage

Most of us need to take out a mortgage to buy a home for the simple reason that doing so is the only way we can afford a home that meets our needs. This chapter helps all nonwealthy folks comprehend mortgages and then choose one. (If you *are* wealthy and have a great deal of money to put into a property, this chapter can also help you decide how much of your loot to put into your home purchase.)

Getting a Grasp on Mortgage Basics

What exactly is a mortgage? A *mortgage* is nothing more than a loan you obtain to close the gap between the cash you have for a down payment and the purchase price of the home you're buying. Homes in your area may cost $170,000, $470,000, $770,000, or more. No matter — most people don't have that kind of spare cash in their piggy banks.

Mortgages require that you make monthly payments to repay your debt. The mortgage payments comprise *interest*, which is what the lender charges for use of the money you borrow, and *principal*, which is repayment of the original amount you borrow.

As noted in Chapter 3, the lender may also insist that you establish an *impound account* if your cash down payment is less than 20 percent of the purchase price. Impound accounts require that you pay your monthly property taxes and insurance to the lender with your mortgage payment. Your lender puts impound account funds into a special escrow, and then makes these payments on your behalf when they're due. Escrows are covered in detail in Chapter 14.

Figuring out how to select a mortgage to meet your needs ensures that you'll be a happy homeowner for years to come. You also need to understand how to get a good deal when shopping around for a mortgage because your mortgage is typically the biggest monthly expense of homeownership (and perhaps of your entire household budget).

Suppose you borrow $200,000 (and contribute $50,000 from your savings as a 20 percent down payment) for the purchase of your $250,000 dream palace. If you borrow that $200,000 with a 30-year, fixed-rate mortgage at 5 percent, you end up paying a whopping $186,640 in interest charges alone over the life of your loan.

So that you don't spend any more than you need to on your mortgage, and so that you get the mortgage that best meets your needs, the time has come to get on with the task of understanding the mortgage options out there.

Fixed or Adjustable? That Is the Interest(ing) Question

Like some other financial and investment products, many different mortgage options are available for your choosing. The variations can be significant or trivial, expensive or less costly.

You will note throughout this chapter that two fundamentally different types of mortgages exist: fixed-rate mortgages and adjustable-rate mortgages. The chief differences between these two main types of mortgages are how their interest rate is determined and whether it can change.

Distinguishing fixed-rate mortgages from adjustables

Before adjustable-rate mortgages came into being, only fixed-rate mortgages existed. Usually issued for 15- or 30-year periods, *fixed-rate mortgages* (as the name suggests) have interest rates that are *fixed* (unchanging) during the entire life of the loan.

With a fixed-rate mortgage, the interest rate stays the same, and your monthly mortgage payment amount never changes. No surprises, no uncertainty, and no anxiety for you over interest-rate changes and changes in your monthly payment. If you like the predictability of your favorite TV show airing at the same time, you'll probably like fixed-rate mortgages.

SUBPRIME MORTGAGES AND THE LATE 2000s HOUSING SLUMP

During the late 1990s and continuing into the early to mid-2000s, mortgage lending was extended to more and more people who previously wouldn't have qualified for home loans. With little or no down payments and poor credit scores (that's the subprime reference), in the past these borrowers would have had to save more and improve their credit scores before qualifying for a home loan. In other words, they would have had to delay a desired home purchase until they upgraded their financial situation.

Thanks to government backing and incentives, subprime mortgages mushroomed, and more renters were able to borrow money and purchase homes. As long as real estate prices continued rising, jobs were plentiful, and the economy was strong, things worked out fine.

Unfortunately, during the mid-2000s, home prices began flattening, and then in the late 2000s prices started trending down, especially in those housing markets most heavily populated with subprime borrowers. On some of these loans, monthly mortgage payments jumped significantly higher after an initial period, and borrowers couldn't handle them financially. The economy soured, and layoffs and unemployment rose, which further pressured stretched homeowners. And as housing prices dropped more than 20 percent in particular markets, some of these subprime borrowers who bought with little or even no money down simply chose to walk away from homes worth quite a bit less than the outstanding mortgage balance.

Real estate market downturns aren't enjoyable for homeowners. No one likes to lose money, even on paper. And the late 2000s real estate market decline was a nasty decline for most parts of the country. But as with previous declines, this also passed. Home prices rose again. And lenders returned to more sensible lending criteria — expecting folks to make down payments and have decent credit or pay much, much more if they didn't.

On the other hand, *adjustable-rate mortgages* (ARMs for short) have an interest rate that varies (or *adjusts*). The interest rate on an ARM typically adjusts every 6 to 12 months, but it may change as frequently as every month.

As we discuss later in this chapter, the interest rate on an ARM is primarily determined by what's happening overall to interest rates. When interest rates are generally on the rise, odds are that your ARM will experience increasing rates, thus increasing the size of your mortgage payment. Conversely, when interest rates fall, ARM interest rates and payments generally fall.

If you like change, you may think that adjustable-rate mortgages sound good. Change is what makes life interesting, you say. Please read on, because even if you believe that variety is the spice of life, you may not like the financial variety and spice of adjustables!

Looking at hybrid loans

If only the world were so simple that only pure fixed-rate and pure adjustable-rate loans were available. But one of the rewards of living in a capitalistic society is that you often have a wide array of choices. Enter *hybrid loans* (or what lenders sometimes call *intermediate ARMs*). Such loans start out like a fixed-rate loan — the initial rate may be fixed for three, five, seven, or even ten years (often called 3/1, 5/1, 7/1, or 10/1 ARMs) — and then the loan converts to an ARM, usually adjusting every 12 months thereafter. Some hybrid loans adjust more frequently or, conversely, may only have one adjustment.

TIP

If you expect to keep your loan less than ten years, you may find 7/1 or 10/1 ARMs attractive. They offer greater stability than the shorter term ARMs as well as generally lower initial interest rates than 30-year, fixed-rate mortgages. However, if you don't pay off your 7/1 or 10/1 while the loan interest rate is fixed, remember that your interest rate and loan payments may go up a lot when the ARM adjusts.

Starting out risky: Interest-only mortgages

As housing prices mushroomed in already-high-cost urban areas and surrounding desired suburbs in the early 2000s, lenders responded with yet another twist and began pushing interest-only loans. Unlike a traditional mortgage, *interest-only mortgages* entice borrowers with artificially low payments in their early years. This is possible because the initial payments simply consist of interest (hence the name of the loans) with no repayment of principal.

At some predetermined point down the road (three, five, seven, or ten years), repayment of principal begins, and the monthly loan payment amount takes a significant jump of 7, 10, or 15 percent or more. And therein lies the major problem with these loans; some borrowers may be financially unprepared for the much higher payments. In the worst cases, these loans are actually balloon loans in disguise (see the nearby sidebar) carrying all the additional warts of balloon loans.

We're not big fans of interest-only mortgages. We urge you to completely understand all the terms and conditions before agreeing to take one.

BALLOON LOANS

One type of mortgage, known as a *balloon loan,* appears at first blush to be somewhat like a hybrid loan. The interest rate is fixed, for example, for five, seven, or ten years. However — and this is a big *however* — at the end of this time period, the full loan balance becomes due. In other words, you must pay off the *entire* loan.

Borrowers are attracted to balloon loans for the same reason that they're attracted to hybrid or ARM loans — because balloon loans start at a lower interest rate than do fixed-rate mortgages. Buyers are sometimes seduced into such loans during high-interest-rate periods or when they can't qualify for or afford the payments of a traditional mortgage.

We don't like balloon loans because they can blow up in your face. You may become trapped without a mortgage if you're unable to *refinance* (obtain a new mortgage to replace the old loan) when the balloon loan comes due. You may have problems refinancing if, for example, you lose your job, your income drops, the value of your property declines and the appraisal comes in too low to qualify you for a new loan, or interest rates increase and you can't qualify for a new loan at those higher rates.

In the real estate trade, balloon loans are also called *bullet loans.* Why? If the loan comes due during a period of high mortgage rates, a deep recession, or a credit crisis, industry people say that it's like getting hit by a bullet.

Remember that refinancing a mortgage is *never* a sure thing. Taking a balloon loan may be a financially hazardous short-term solution to your long-term financing needs.

The one circumstance under which we say that it's okay to *consider* a balloon loan is if you absolutely must have a particular property and the balloon loan is your one and only mortgage option. If that's the case, you should also be as certain as you can be that you'll be able to refinance when the balloon loan comes due. If you have family members who can step in to help with the refinancing, either by cosigning or by loaning you the money themselves, that's a big backup plus. Oh, and if you *must* take out a balloon loan, get as long a term as possible, ideally for no less than seven years (ten years is preferable), and always start the refinancing process at least 8 to 12 months before your balloon loan is due.

Making the fixed/adjustable decision

So how do you choose whether to take a fixed-rate or an adjustable-rate loan? Is it as simple as a personality test?

REMEMBER

In this section, we talk you through the pros and cons of your mortgage options, but as we do, please keep one very important fact in mind: In the final analysis, the mortgage that's best for you hinges on your personal and financial situation now and in the future. *You* are the one who's best positioned to make the call as to whether a fixed or an adjustable loan better matches your situation and desires.

Fixed-rate mortgages

It stands to reason that because the interest rate doesn't vary with a fixed-rate mortgage, the advantage of this type of mortgage is that you always know what your monthly payment is going to be. Thus, budgeting and planning are easier.

You'll pay a premium, in the form of a higher interest rate (generally ½ to ¾ percent), to get a lender to commit to lending you money over many years at a fixed rate. The longer the mortgage lender agrees to accept a fixed interest rate, the more risk that lender is taking. A lender who agrees to loan you money, for example, over 30 years at 6 percent will be weeping if interest rates skyrocket (as they did in the early 1980s) to the 15-plus percent level.

In addition to paying a premium interest rate when you take out the loan, another potential drawback to fixed-rate loans is that if interest rates fall significantly after you take out your mortgage, you face the risk of being stranded with your costly mortgage. That can happen if (because of deterioration in your financial situation or a decline in your property's value) you don't qualify to *refinance* (get a new loan to replace the old one). Even if you do qualify to refinance, doing so takes time and usually costs money for a new appraisal, loan fees, and title insurance. This may be 2 to 3 percent or more of the loan amount. A $200,000 loan, for example, may cost $4,000 to $6,000 or more to refinance.

Here are a couple other possible minor drawbacks to be aware of with some fixed-rate mortgages:

>> If you sell your house before paying off your fixed-rate mortgage, your buyers probably won't be able to assume that mortgage. The ability to pass your loan on to the next buyer (in real estate talk, the next buyer assumes your loan) can be useful if you're forced to sell during a rare period of ultra-high interest rates, such as occurred in the early 1980s.

>> Fixed-rate mortgages sometimes have prepayment penalties (explained in the sidebar "Avoid loans with prepayment penalties").

AVOID LOANS WITH PREPAYMENT PENALTIES

Some mortgages come with a provision that penalizes you for paying off the loan balance faster. Such penalties can amount to as much as several percentage points of the amount of the mortgage balance that's paid off early.

When you pay off a mortgage early because you sold the property or because you want to refinance the loan to take advantage of lower interest rates, a few lenders won't enforce their loan's prepayment penalties as long as they get to make the new mortgage. Even so, your hands are tied financially unless you go through the same lender. FHA, Fannie Mae, Freddie Mac, and VA loans do not have prepayment charges.

Many states place limits on the duration and amount of prepayment penalty that lenders may charge for mortgages made on owner-occupied residential property. The only way to know whether a loan has a prepayment penalty is to ask and to carefully review the federal truth-in-lending disclosure and the promissory note the mortgage lender provides you. We think you should avoid such loans. (Many so-called *no-points* loans have prepayment penalties.)

Adjustable-rate mortgages

Fixed-rate mortgages aren't your only option. Mortgage lenders were intelligent enough to realize that they couldn't foresee how much future interest rates would rise or fall; thus, adjustable-rate mortgages (*adjustables* or ARMs for short) were born.

Although some adjustables are more volatile than others, all are similar in that they *fluctuate* (or float) with the market level of interest rates. If the interest rate fluctuates, so does your monthly payment. And therein lies the risk: Because a mortgage payment is likely to be a big monthly expense for you, an adjustable-rate mortgage that's adjusting upward may wreak havoc with your budget.

Given all the trials, tribulations, and challenges of life as we know it, you may rightfully ask, "Why would anyone choose to accept an adjustable-rate mortgage?" Well, people who are stretching themselves — such as some first-time buyers or those *trading up* to a more expensive home — may financially force themselves into accepting adjustable-rate mortgages. Because an ARM starts out at a lower interest rate, such a mortgage enables you to *qualify* to borrow more. As we discuss in Chapter 3, just because you can qualify to borrow more doesn't mean you can *afford* to borrow that much, given your other financial goals and needs.

Some home buyers who can qualify for either an adjustable-rate or a fixed-rate mortgage of the same size have a choice and choose the fluctuating adjustable-rate mortgage. Why? Because they may very well save themselves money, in the form of smaller total interest charges, with an adjustable-rate loan rather than a fixed-rate loan.

Because you accept the risk of a possible increase in interest rates, mortgage lenders cut you a little slack. The *initial interest rate* (also sometimes referred to as the *teaser rate*) on an adjustable should be less than the initial interest rate on a comparable fixed-rate loan. In fact, an ARM's interest rate for the first year or two of the loan is generally lower than the rate on a fixed-rate mortgage.

Another advantage of an ARM is that if you purchase your home during a time of high interest rates, you can start paying your mortgage with the artificially depressed initial interest rate. Should interest rates subsequently decline, you can enjoy the benefits of lower rates without refinancing.

Another situation when adjustable-rate loans have an advantage over their fixed-rate brethren is when interest rates decline and you don't qualify to refinance your mortgage to reap the advantage of lower rates. The good news for homeowners who are unable to refinance and who have an ARM is that they usually capture many of the benefits of the lower rates. With a fixed-rate loan, you must refinance to realize the benefits of a decline in interest rates.

WARNING

The downside to an adjustable-rate loan is that if interest rates in general rise, your loan's interest and monthly payment will likely rise, too. During most time periods, if rates rise more than 1 or 2 percent and stay elevated, the adjustable-rate loan is likely to cost you more than a fixed-rate loan.

Before you make the final choice between a fixed-rate mortgage and an adjustable-rate mortgage, read the following two sections.

What would rising interest rates do to your finances?

WARNING

Far too many home buyers, especially first-timers, take out an adjustable-rate mortgage because doing so allows them to stretch and borrow more to buy a more expensive home. Although some of this overborrowing is caused by the spendthrift "I gotta have it today" attitude, overborrowing is also encouraged by some real estate and mortgage salespeople. After all, these salespeople's income, in the form of a commission, is a function of the cost of the home that you buy and the size of the mortgage that you take on. Resist the temptation to overspend on your first home. Buy what you can afford.

WHEN TO CONSIDER HYBRID LOANS

If you want more stability in your monthly payments than you can get with a regular adjustable, and you expect to keep your loan for no more than five to ten years, a *hybrid* (or intermediate ARM) loan with a fixed interest rate for the loan's first five to ten years, which is explained earlier in this chapter, may be the best loan for you.

The longer the initial interest rate stays locked in, the higher it will be, but the initial rate of a hybrid ARM is almost always lower than the interest rate on a 30-year, fixed-rate mortgage. However, because the initial rate of hybrid loans is locked in for a longer period of time than the 6-month or 1-year term of regular ARMs, hybrid ARMs have higher initial interest rates than regular ARM loans.

During periods when little difference exists between short-term and long-term interest rates, the interest-rate savings with a hybrid or regular adjustable (versus a fixed-rate loan) are minimal (less than 1 percent). In fact, during certain times, the initial interest rate on a 7- or 10-year hybrid was exactly the same as on a 30-year, fixed-rate loan. During such periods, fixed-rate loans offer the best overall value.

To evaluate hybrids, weigh the likelihood that you'll move before the initial loan interest rate expires. For example, with a 7-year hybrid, if you're saving, say, 0.5 percent per year versus the 30-year, fixed-rate mortgage but you're quite sure that you'll move within seven years, the hybrid will probably save you money. On the other hand, if there's a reasonable chance that you'll stay put for more than seven years and you don't want to face the risk of rising payments after seven years, you should opt for a 30-year, fixed-rate mortgage instead.

If you haven't already done so, read and digest Chapters 2 and 3 to understand how much you can really afford to spend on a home, given your other financial needs, commitments, and goals.

INVESTIGATE

When considering an ARM, you absolutely, positively must understand what rising interest rates (and, therefore, a rising monthly mortgage payment) would do to your personal finances. Consider taking an ARM only if you can answer all the following questions in the affirmative:

>> Is your monthly budget such that you can afford higher mortgage payments and still accomplish other financial goals that are important to you, such as saving for retirement?

>> Do you have an emergency reserve (equal to at least six months of living expenses) that you can tap into to make the potentially higher monthly mortgage payments?

>> Can you afford the highest payment allowed on the adjustable-rate mortgage?

The mortgage lender can tell you the highest possible monthly payment, which is the payment you would owe if the interest rate on your ARM went to the lifetime interest-rate cap allowed on the loan.

>> If you're stretching to borrow near the maximum the lender allows or an amount that will test the limits of your budget, are your job and income stable? If you're depending on financial help from your spouse, is your spouse's job and income secure?

>> If you expect to have children in the future, have you considered that your household expenses will rise and your income may fall with the arrival of those little bundles of joy?

>> Can you handle the psychological stress of changing interest rates and mortgage payments?

If you're fiscally positioned to take on the financial risks inherent to an adjustable-rate mortgage, by all means consider taking one — we're not trying to talk you into a fixed-rate loan. The odds are with you to save money, in the form of lower interest charges and payments, with an ARM. Your interest rate starts lower (and generally stays lower, if the overall level of interest rates doesn't change). Even if rates do go up, as they are sometimes prone to do, they'll surely come back down. So if you can stick with your ARM through times of high and low interest rates, you should still come out ahead.

Also recognize that although ARMs do carry the risk of a fluctuating interest rate, almost all adjustable-rate loans limit, or *cap,* the rise in the interest rate allowed on your loan. We certainly wouldn't recommend that you take an ARM without caps. Typical caps are 2 percent per year and 6 percent over the life of the loan. (We cover ARM interest-rate caps in detail later in this chapter.)

REMEMBER

Consider an adjustable-rate mortgage only if you're financially and emotionally secure enough to handle the maximum possible payments over an extended period of time. ARMs work best for borrowers who take out smaller loans than they're qualified for or who consistently save more than 10 percent of their monthly income. If you do choose an ARM, make sure you have a significant cash cushion that's accessible in the event that rates go up. Don't take an adjustable just because the initially lower interest rate allows you to afford a more expensive home. Better to buy a home that you can afford with a fixed-rate mortgage. (And don't forget hybrid loans if you want a loan with more payment stability but aren't willing to pay the premium of a long-term, fixed-rate loan.)

How long do you expect to stay in the home or hold the mortgage?

This is the single most important question you need to answer. As we explain earlier in this chapter, a mortgage lender takes more risk when lending money at a fixed rate of interest for many (15 to 30) years. Lenders charge you a premium, in the form of a higher interest rate than what the ARM starts at, for the interest-rate risk that they assume with a fixed-rate loan.

If you don't plan or expect to stay in your home for a long time, you should consider an ARM. Saving money on interest charges for most adjustables is usually guaranteed in the first two to three years because an ARM starts at a lower interest rate than a fixed-rate loan does. If you're reasonably certain that you'll hold onto your home for fewer than five years, you should come out ahead with an adjustable. However, you should also ask yourself why you're going to all the trouble and expense of buying a home that you expect to sell so soon.

TIP

If you expect to hold onto your home and mortgage for a long time — more than five years — a fixed-rate loan may make more sense, especially when you're not in a position to withstand the fluctuating monthly payments that come with an ARM.

When you're in the intermediate area (expecting to stay seven to ten years, for example), consider the hybrid loans we discuss earlier in this chapter.

If you're still stuck on the fence, go with the fixed-rate loan. A fixed-rate loan is financially safer than an ARM.

Deciding on your loan's life: 15 years or 30?

After you decide which type of mortgage — fixed or adjustable — you want, you may think that your mortgage quandaries are behind you. Unfortunately, they're not. You also need to make another important choice — typically between a 15-year and a 30-year mortgage. (Not all mortgages come in just 15- and 30-year varieties. You may run across some 20- and 40-year versions, but that won't change the issues we're about to tackle.)

When you're stretching to buy the home you want, you may also be forcing yourself to take the longer-term, 30-year mortgage. Doing so isn't necessarily bad and, in fact, has advantages.

The main advantage that a 30-year mortgage has over its 15-year peer is that it has lower monthly payments that free up more of your monthly income for other purposes, including saving for other important financial goals (such as retirement). You may want to have more money so you aren't a financial prisoner to your home and can have a life! A 30-year mortgage has lower monthly payments because you have a longer period to repay it (which translates into more payments). A fixed-rate, 30-year mortgage with an interest rate of 5 percent, for example, has payments that are approximately 32 percent lower than those on a comparable 15-year mortgage.

What if you can afford the higher payments that a 15-year mortgage requires? You shouldn't necessarily take it. What if, instead of making large payments on the 15-year mortgage, you make smaller payments on a 30-year mortgage and put that extra money to productive use?

NO ONE CAN PREDICT WHERE INTEREST RATES ARE HEADED

All you rational readers out there are probably commenting that the choice between an adjustable-rate mortgage and a fixed-rate mortgage is simple. All you need to know in order to make a decision is the direction of interest rates. It's only logical. When interest rates are about ready to rise, a fixed-rate mortgage is favorable. Lock in a low rate and smile smugly when interest rates skyrocket.

Conversely, if you think that rates are going to stay the same or drop, you want an ARM. Some real estate books that we've read even go so far as to say that your own personal interest-rate forecast should determine whether to take an ARM or fixed-rate mortgage! "Interest-rate forecasts should be the major factor in deciding whether or not to get an ARM," argues one such book.

Now, we know you're inquisitive and smart (after all, you bought our book), but you're *not* going to figure out which way rates are headed. The movement of interest rates isn't logical, and you certainly can't predict it. If you could, you would make a fortune investing in bonds, interest-rate futures, and options.

Even the money-management pros who work with interest rates and bonds as a full-time job can't consistently predict interest rates. Witness the fact that bond-fund managers at mutual fund companies have a tough time beating the buy-and-hold bond-market indexes. If bond-fund managers could foresee where rates were headed, they could easily beat the averages by trading into and out of bonds when they foresee interest-rate changes on the horizon.

TIP

If you do, indeed, make productive use of that extra money, the 30-year mortgage may be for you. A terrific potential use for that extra dough is to contribute it to a tax-deductible retirement account that you have access to. Contributions that you add to employer-based 401(k) and 403(b) plans (and self-employed SEP-IRAs) not only give you an immediate reduction in taxes but also enable your investment to compound, tax-deferred, over the years ahead. Everyone with employment income may also contribute to an Individual Retirement Account (IRA). However, your IRA contributions may not be immediately tax-deductible if your (or your spouse's) employer offers a retirement account or pension plan.

TIP

If you exhaust your options for contributing to all the retirement accounts that you can, and if you find it challenging to save money anyway, the 15-year mortgage may offer you a good forced-savings program.

When you elect to take a 30-year mortgage, you retain the flexibility to pay it off faster if you so choose. (Just be sure to avoid those mortgages that have a prepayment penalty, as we explain earlier in this chapter.) Constraining yourself with the 15-year mortgage's higher monthly payments does carry a risk. Should you fall on tough financial times, you may not be able to meet the required mortgage payments.

Finding a Fixed-Rate Mortgage

If you decide, based on our advice and selection criteria, to go with a fixed-rate loan, great! You shouldn't be disappointed. You'll have the peace of mind that comes with stable mortgage payments. And because fixed-rate loans have fewer options, they're a good deal easier to compare than adjustable-rate loans.

However, we don't want to give you the false impression that fixed-rate loans are as simple to shop for as carbonated beverages. Unfortunately, because of the hundreds of lenders that offer such loans and the seemingly endless number of extra fees and expenses that lenders tack onto loans, you need to put on your smart-consumer hat and sharpen your No. 2 pencil.

Be sure you understand the following sections before you attempt to choose the best fixed-rate loan to meet your needs.

The all-important interest rate

If you've ever borrowed money, you know that lenders aren't charities. Lenders make money by charging you, in the form of interest, for the use of their money.

Lenders normally quote the *rate of interest* as a percentage per year of the amount borrowed. You may be familiar with rates of interest if you've ever borrowed money through student loans, credit cards, or auto loans. In these cases, lenders may have charged you 8, 10, 12, or perhaps even 18 percent or more for the privilege of using their money. Similarly, mortgage lenders also quote you an annual interest rate.

You've shopped for other products and services online and by phone, so you may as well get online and on the horn and call lenders, as well. The first one you find may be offering a fixed-rate loan at an interest rate of 5.0 percent. Then you find another lender to try to beat that rate, and that lender says, "Sure, we can get you into a fixed-rate loan at 4.75 percent."

WARNING

Be careful of lenders promising you what you want right off the bat. Remember that no reputable lender can commit to an interest rate without reviewing your complete loan application and credit score.

The finer points of points

WARNING

If you blindly choose a 4.75 percent loan, you can be making a very expensive mistake. You have an idea of what we mean if you've heard the expression "Don't judge a book by its cover." You shouldn't judge a mortgage solely by its interest rate, either. You must also understand the points and many other loan fees that the lender charges.

Just as peanut butter goes with jelly (unless you have a nut allergy), the interest rate on a mortgage should go together, in your mind, with the points on the loan. We aren't talking about the kind of points that a basketball player tallies during a game for each successful shot. Points on a mortgage cost you money.

Points are up-front interest. Lenders charge points as a way of being paid for the work and expense of processing and approving your mortgage. When you buy a home, the points are tax-deductible — you get to claim them as an itemized expense on Schedule A of your IRS Form 1040 (see Chapter 3). When you refinance, in contrast, the points must be spread out for tax purposes and deducted over the life of the new loan.

Lenders quote points as a percentage of the mortgage amount and may require you to pay them at the time you close on your home purchase and begin the lengthy process of repaying your loan. One *point* is equal to 1 percent of the amount that you're borrowing. For example, if a lender says the loan being proposed to you has two points, that simply means you must pay 2 percent of the loan amount as points. On a $200,000 loan, for example, two points cost you $4,000. That's not chump change!

The interest rate on a fixed-rate loan has an inverse relationship to that loan's points. When you're able to (or desire to) pay more points on a mortgage, the lender should reduce the ongoing interest rate. This reduction may be beneficial to you if you have the cash to pay more points and want to lower the interest rate you'll pay year after year. If you expect to hold onto the home and mortgage for many years, the lower the interest rate, the better.

Conversely, if you want to (or need to) pay fewer points (perhaps because you're cash-constrained when you make the home purchase), you can pay a higher ongoing interest rate. The shorter the time that you expect to hold onto the mortgage, the more sense this strategy of paying less now makes. Again, the key question is, how long do you intend to keep your home?

Don't get suckered into believing that "no-point" loans are a good deal. There are no free lunches in the real estate world. Remember the points/interest-rate trade-off: If you pay less in points, the ongoing interest rate is higher. So if a loan has zero points, it must have a higher interest rate. This doesn't necessarily mean the loan is better or worse than comparable loans from other lenders. However, it has been our experience that lenders who aggressively push no-point loans aren't the most competitive lenders in terms of overall pricing. No-point loans make sense only when you're really tight on cash for your home purchase and expect not to hold onto the home and mortgage for the long term.

Take a look at a couple of specific mortgage options to understand the points/interest-rate trade-off. Suppose you want to borrow $150,000. One lender quotes you 5.25 percent on a 30-year, fixed-rate loan and charges one point (1 percent). Another lender quotes 5.75 percent and doesn't charge any points. Which offer is better? The answer depends mostly on how long you plan to keep the loan.

The 5.25-percent loan costs $828 per month compared with $875 per month for the 5.75-percent mortgage. You can save $47 per month with the 5.25-percent loan, but you have to pay $1,500 in points to get it.

To find out which loan is better for you, divide the cost of the points by the monthly savings ($1,500 divided by $47 equals 32). This gives you the number of months (in this case, 32) it will take you to recover the cost of the points. The 5.25-percent loan costs 0.5 percent less in interest annually than the 5.75-percent loan. Year after year, the 5.25-percent loan saves you 0.5 percent. But because you have to pay one point up-front on the 5.25-percent mortgage, it will take you about 32 months to earn back the savings to cover the cost of that point. So if you expect to keep the loan more than 32 months, go with the 5.25-percent, one-point option. If you don't plan to keep the loan for 32 months, choose the no-points loan.

TIP

To make a fair comparison of mortgages from different lenders, have the lenders provide interest-rate quotes at the *same* point level. Ask the mortgage contenders, for example, to tell you what their fixed-rate mortgage interest rate would be at one point. Also, make sure the loans are of the same term — for example, 30 years. Ask the lenders to provide a written, side-by-side comparison that lists all fees, charges, interest rates, and points on a single sheet of paper. A good lender will provide those details for you.

Other lender fees

You may think that because you're paying points with your mortgage, you don't have to pay any other up-front fees. Well, think again. There's no shortage of up-front loan-processing charges for you to investigate when making mortgage comparisons. If you don't understand the fee structure, you may end up with a high-cost loan or come up short of cash when the time comes to close on your home purchase.

TIP

Ask each lender whose services you're seriously considering for a written itemization of all these "other" charges. To reduce your chances of throwing money away on a mortgage for which you may not qualify, ask the lender whether your application may be turned down for some reason. For example, disclose any potential problems with the property that were discovered during inspections of the property (see Chapter 13 for details on inspections).

Just as some lenders have no-point mortgages, some lenders also have *no-fee mortgages*. If a lender is pitching a no-fee loan, odds are that the lender will charge you more in other ways, namely a much higher interest rate on your loan.

WARNING

No-point, no-fee mortgages can be very expensive. Ask the lender for a detailed, written comparison in advance of your different loan options so you know the true mortgage costs over the life of the loan.

Application and processing fees

Lenders generally charge $500 to $1,200 up-front as an *application* or *processing fee*. This charge is mainly to ensure that you're serious about wanting a loan from them and to compensate them in the event that your loan is rejected. Lenders want to cover their costs to keep from losing money on loan applications that don't materialize into actual loans. A few lenders don't charge this fee, or if they do, they return it if you take their loan.

Credit report

Your credit report tells a lender how responsibly you've dealt with prior loans. Did you pay all your previous loans back (and on time)? Credit reports don't cost a great deal, but you can expect to pay about $75 to $125 for the lender to obtain a complete, current copy of yours.

TIP

If you know you have blemishes on your credit report, address those problems *before* you apply for your mortgage. Otherwise, you're wasting your time and money. Don't apply for a loan that you know will be denied. Flip to Chapter 5 for tips on improving your credit score.

Appraisal

Mortgage lenders want an independent assessment to ensure that the property that you're buying is worth approximately what you agree to pay — that's the job of an appraiser. Why would the lender care? Simple — because the lender is likely loaning you a large portion of the property's purchase price. If you overpay and home values decline or you end up in financial trouble, you may be willing to walk away from the property and leave the lender holding the bag. Federal law now has strict standards for all independent, licensed appraisers to protect home buyers.

The cost of an appraisal varies with the size, complexity, and value of property. Expect to pay $300 to $500 for an appraisal of most modestly priced, average-type properties. You may have to pay up to $1,500 for a more expensive or unique home.

Arriving at the Absolute Best Adjustable

If you're the calm and collected type of person who isn't prone to panicking, can stomach interest-rate volatility, and have decided based on our sage advice to go with an adjustable-rate mortgage (ARM), you need to understand a bit more in order to choose a good one. Adjustables are more complicated to evaluate and select than fixed-rate mortgages are.

In addition to understanding points and other loan fees that we cover in the preceding section on fixed-rate loans, you'll be bombarded with such jargon as *margins, caps,* and *indexes.* Numbers geeks can easily spend hundreds of hours comparing different permutations of ARMs and determining how they may behave in different interest-rate environments.

REMEMBER

Unlike with a fixed-rate mortgage, precisely determining the amount of money a particular ARM is going to cost you isn't possible. As with choosing a home to buy, selecting an ARM that meets your needs and budget involves compromising and deciding what's important to you. So here's your crash course in understanding ARMs.

Where an ARM's interest rate comes from

WARNING

Most ARMs start at an artificially low interest rate. Selecting an ARM based on this rate is likely to be a huge mistake because you won't be paying this low rate for long, perhaps for just 6 to 12 months — or maybe even just 1 month! Lenders and mortgage brokers are like many other salespeople; they like to promote something that will catch your attention and get you thinking you're going to get a great deal. That's why lenders and brokers are most likely to tell you first about the low teaser rate. Don't be fooled.

REMEMBER

The starting rate on an ARM isn't anywhere near as important as what the future interest rate is going to be on the loan. How the future interest rate on an ARM is determined is the single most important feature for you to understand when evaluating an ARM.

All ARMs that we've ever seen are based on an equation that includes an index and margin, the two of which are added together to determine and set the future interest rate on the loan. Before we go further, please be sure you understand these terms:

>> **Index:** The index is a measure of interest rates that the lender uses as a reference. For example, the 6-month bank certificate of deposit index is used as a reference for many mortgages. Suppose the going rate on 6-month CDs is approximately 2 percent. The index theoretically indicates how much it costs the bank to take in money that it can then lend. Most commonly used indexes can be found in financial publications such as *The Wall Street Journal*.

>> **Margin:** The margin, which is set by lenders, is the lenders' profit (or markup) on the money they intend to lend. Most loans have margins of around 2.5 percent, but the exact margin depends on each individual lender and the index that lender is using. Some lenders have margins as high as 4 percent. When you compare loans that are tied to the same index and are otherwise the same, the loan with the lower margin is better (lower cost) for you.

>> **Interest rate:** The interest rate is the sum of the index and the margin. It's what you will pay (subject to certain limitations) on your loan.

Putting it all together, in our example of the 6-month CD index at 2 percent plus a margin of 2.5 percent, we get an interest rate sum of 4.5 percent. This figure is known as the *fully indexed rate*. If this loan starts out at 3.5 percent, for example, the fully indexed rate tells you what interest rate this ARM would increase to if the market level of interest rates, as measured by the CD index, stays constant. Never take an ARM unless you understand this important concept of the fully indexed rate:

Index + Margin = Interest Rate

WARNING

Many mortgage lenders know that more than a few borrowers focus on an ARM's initial interest rate and ignore the margin and the index that determine the loan rate. Take our advice and look at an ARM's starting rate *last*. Begin to evaluate an ARM by understanding what index it's tied to and what margin it has. The sections that follow explain common ARM indexes.

Treasury bills

The U.S. federal government is the largest borrower in the universe as we know it, so it should come as no surprise that at least one ARM index is based on the interest rate that the government pays on some of this pile of debt. The most commonly used government interest rate indexes for ARMs are for 6-month and 12-month Treasury bills.

The Treasury-bill indexes tend to be among the faster-moving ones around. In other words, they respond quickly to market changes in interest rates.

Certificates of deposit

Certificates of deposit (CDs) are interest-bearing bank investments that lock you in for a specific period of time. Adjustable-rate mortgages are usually tied to the average interest rate that banks are paying on 6-month CDs.

As with Treasury bills, CDs tend to move rapidly with overall changes in interest rates. However, CD rates tend to move up a bit more slowly when rates rise because profit-minded bankers like to drag their feet when paying more interest to depositors. Conversely, CD rates tend to come down quickly when rates decline so bankers can maintain their profits.

The 11th District Cost of Funds Index

The 11th District Cost of Funds Index (also known as COFI, pronounced like the caffeinated brew that some people drink in the morning) is published monthly by the Federal Home Loan Bank Board. This index shows the monthly weighted average cost of savings, borrowings, and advances for its member banks located in California, Arizona, and Nevada (the 11th District). Because the COFI is a moving average of the rates that bankers have paid depositors over recent months, it tends to be a relatively stable index.

REMEMBER

An ARM tied to a slower-moving index, such as the 11th District Cost of Funds Index, has the advantage of increasing more slowly when interest rates are on the upswing. On the other hand, you have to be patient to benefit from falling interest rates when rates are on the decline. The 11th District is slow to fall when interest rates overall decline.

Because ARMs tied to the 11th District Cost of Funds Index are slower to rise when overall interest rates rise, they generally begin at a higher rate of interest than do ARMs tied to faster-moving indexes.

The London Interbank Offered Rate Index

Okay, now for a more unusual index. The *London Interbank Offered Rate Index (LIBOR)* is an average of the interest rates that major international banks charge one another to borrow U.S. dollars in the London money market. Like the U.S. Treasury and CD indexes, LIBOR tends to move and adjust quite rapidly to changes in interest rates.

Why do we need an *international* interest-rate index? Well, foreign investors buy American mortgages as investments, and not surprisingly, these investors like ARMs tied to an easy-to-locate index that they understand and are more familiar with.

How often does the interest rate adjust?

Lenders usually adjust the interest rates on their ARMs every 6 or 12 months, using the mortgage-rate formula we discuss earlier in this section. Some loans adjust monthly. (Monthly adjustments are usually a red flag for negative-amortization loans; we explain why to stay away from these loans in the later sidebar, "Avoid adjustables with negative amortization.") In advance of each adjustment, the mortgage lender should send you a notice spelling out how the new rate is calculated according to the agreed-upon terms of your ARM.

REMEMBER

The less often your loan adjusts, the less financial risk you're accepting. In exchange for taking less risk, the mortgage lender normally expects you to pay a higher initial interest rate.

Limits on interest-rate adjustments

Despite the fact that an ARM has a system for calculating future interest rates (by adding the margin to the loan index), bankers limit how great a change can occur in the actual rate you pay. These limits, also known as *rate caps,* affect each future adjustment of an ARM's rate following the end of the initial rate period.

Periodic adjustment caps limit the maximum rate change, up or down, allowed at each adjustment. For ARMs that adjust at 6-month intervals, the adjustment cap is usually 1 percent. ARMs that adjust more than once annually generally restrict the maximum rate change allowed over the entire year, as well. This *annual rate cap* is usually 2 percent.

Finally, almost all adjustables come with *lifetime caps*. You should never take on an ARM without a lifetime cap. These caps limit the highest rate allowed over the entire life of the loan. ARMs commonly have lifetime caps of 5 to 6 percent higher than the initial rate.

REMEMBER

When you take on an ARM, be sure you can handle the maximum possible payment allowed, should the interest rate on the ARM rise to the lifetime cap. If, for example, the ARM's start rate is 5 percent and the lifetime cap is 6 percent, you can be paying 11 percent interest in just three short years in the worst possible case. On a $200,000 loan, for example, that's an increase of $832 per month. Ouch!

AVOID ADJUSTABLES WITH NEGATIVE AMORTIZATION

Some ARMs cap the increase of your monthly payment but not the increase of the interest rate. The size of your mortgage payment may not reflect all the interest you actually owe on your loan. So rather than paying the interest that's owed and paying off some of your loan balance every month, you may end up paying some (but not all) the interest you owe. Thus, the extra unpaid interest you still owe is added to your outstanding debt.

As you make mortgage payments over time, the loan balance you still owe is gradually reduced in a process called *amortizing the mortgage*. The reverse of this process (that is, increasing the size of your loan balance) is called *negative amortization*.

Liken negative amortization to paying only the minimum payment required on a credit card bill. You continue accumulating additional interest on the balance as long as you make only the minimum monthly payment. However, doing this with a mortgage defeats the purpose of your borrowing an amount that fits your overall financial goals (see Chapter 3).

Some lenders try to hide the fact that the ARM they're pitching you has negative amortization. How can you avoid negative-amortization loans? Simple — ask! And carefully review all written loan documents.

Also be aware that negative amortization pops up more often on mortgages that lenders consider risky to make. If you're having trouble finding lenders willing to offer you a mortgage, be especially careful.

Last but not least, realize that many lenders won't put additional financing in the form of second mortgages or home-equity lines of credit behind a negative-amortization loan. They fear that if the negative-amortization loan amount grows too large, it can swallow all the equity in the property.

Locating the Best, Lowest-Cost Lenders

For those of you out there who abhor shopping, we have some bad news. Unless you enjoy throwing away thousands of dollars, you need to shop around for the best deal on a mortgage. Think of it as "dialing for dollars" (or "surfing for dollars," for you web users).

Whether you do the footwork on your own or hire someone competent to help you doesn't matter. But you must make sure that this comparison shopping gets done.

Suppose you're in the market for a 30-year, $100,000 mortgage. If, through persistent and wise shopping, you're able to obtain a mortgage that is, for example, 0.5 percent per year lower in interest charges than you otherwise would have gotten, you'll save about $11,000 over the life of the loan. You can double those savings for a $200,000 mortgage.

WARNING

Although we encourage you to find the lowest-cost lenders, we must first issue a caution: Should someone offer you a deal that's much better than any other lender's, be skeptical and suspicious. Such a lender may be baiting you with a loan that doesn't exist or with one for which you can't qualify, and then you'll get stuck with a higher-cost loan if you don't have time to apply for another mortgage elsewhere.

Shopping on your own

Most areas have no shortage of mortgage lenders. Although having a large number of options to choose from is good for competition, so many alternatives can also make shopping a headache.

Many different types of companies offer mortgages today. The most common mortgage *originators* (as they're known in the business) are banks, savings and loan associations, and mortgage bankers.

"Who cares?" you ask. Well, mortgage bankers do only mortgages, and the best ones offer very competitive interest rates on a broad selection of mortgage programs, including FHA, Fannie Mae, Freddie Mac, and VA, as well as state and local Housing Agency loans. Mortgage bankers can also be flexible and competitive with loan fees.

Smaller banks and savings and loans can have good deals as well. As for the big banks whose names are drilled into your head from advertisements, they usually don't offer the best interest rates and only allow you to choose from their bank's loan programs.

As you begin your mortgage safari, you don't have to go it completely alone. If you've done a good job selecting a real estate agent to help you with your home purchase, for example, the agent should be able to rattle off a short list of good lenders and mortgage brokers (see the following section) in the area. Just remember to compare these lenders' loans and rates with those of some other mortgage lenders that you find on your own.

WARNING

Otherwise-good real estate agents may send you to lenders that don't necessarily offer the best mortgage interest rates. Some real estate agents may not be up to date with who has the best loans, or they may not be into shopping around. Others may have simply gotten comfortable doing business with certain lenders or gotten client referrals from said lenders previously.

The Internet offers yet another method for tapping into companies in a particular line of work. In Chapter 11, we offer plenty of advice and recommendations for websites that can assist you with mortgage shopping. Family, friends, and co-workers are also a good source of referrals, especially if their loan experience is recent.

Another way to find lenders is to look for tables of selected lenders' interest rates in the Sunday real estate section of the larger area newspapers. However, don't assume that such tables contain the best lenders in your area. In fact, many of these tables are sent to newspapers for free by firms that distribute information to mortgage brokers. Nonetheless, you can use these tables as a starting point by calling the lenders that list the best rates (realizing, of course, that rates can change daily and that the rates you see in the paper may not accurately reflect what's currently available).

TRAITS OF GOOD LENDERS

Yes, hundreds of mortgage lenders are out there. However, not anywhere near that many mortgage lenders are *good* lenders. Real estate agents and others in the real estate trade, as well as other borrowers whom you know, can serve as useful references for steering you toward the top-notch lenders and away from the losers. (To make sure you get unbiased recommendations, ask your agent whether she'll be paid a referral fee by any of the lenders being recommended.) As you solicit input from others and begin to interview lenders, seek to find lenders with the following traits:

- **Straightforward:** Good loan agents explain their various loan programs in plain English, without using double talk or jargon. They provide easy-to-read, written comparisons of their various loan programs that show all interest rates, costs, fees, and loan terms. They help you compare their loans with their competitors' loans.

(continued)

(continued)

Run as fast as you can from mortgage officers who talk down to you and try to snow you with lots of confusing lingo.

- **Approve locally:** Good lenders approve your loan locally. They don't send your loan application to an out-of-town loan committee, where you're transformed from a living, breathing human being into an inanimate loan number. Good lenders use appraisers who are familiar with the local real estate market and have experience appraising the types of properties that are commonly sold locally. Good lenders actively work with you and your real estate agent to get loan approval within your time frame.

- **Market savvy:** Good lenders understand the type of property you want to buy. Here's another big advantage of local loan approval: No deal-breaking, last-minute loan denials unexpectedly arise because you inadvertently run afoul of some obscure institutional policy.

  This type of snafu generally occurs when a mortgage broker tries to find the loan with the lowest interest rate currently being offered anywhere in the universe. Finding the loan is relatively easy. Getting the money, on the other hand, is nearly impossible, because you don't know what bizarre quirks lie buried deep in the loan documents' fine print.

  These quirky loans usually apply to absolutely pristine property. For example, an out-of-state lender once approved a loan subject to having all corrective work completed and the house painted inside and out prior to close of escrow. Given that the loan was approved on Monday and the sale was scheduled to close Friday, there was no way the work could be completed in four days. If possible, use a local lender.

- **Competitive:** Good lenders are competitive. Don't be afraid to ask the lender that you like best to match the interest rate of the lowest-priced lender you find. At worst, the lender will turn your rate request down. At best, you'll get the lender you want *and* the loan terms you want. Loan rates and charges *are* negotiable. Good lenders are flexible.

- **Detail-oriented:** Good lenders meet contract deadlines. They approve and fund loans on time. Your agent knows which lenders deliver on their promises and which don't. Talk's cheap. You need action, not empty promises. Missed deadlines may squash your purchase. Good lenders keep you and your real estate agent informed each step of the way.

- **National mortgage license:** The federally mandated National Mortgage Licensing Law went into effect January 2011. Loan officers originating residential mortgages are now required to complete a series of relevant financial courses, take a written examination, and pass a detailed background and credit check prior to being issued a license. Ask loan officers you are considering working with to show you their license. Many have not completed the licensing process and under this law are not permitted to do residential loans. In addition, many states mandate that loan officers obtain a state license.

Working with a mortgage broker

Mortgage brokers are middlemen, independent of banks or other financial institutions that have money to lend. They can do the mortgage shopping for you. But first get a list of their five most recent clients so you can do a reference check.

TIP

If your credit history and ability to qualify for a mortgage are questionable, a good licensed mortgage broker can help polish and package your application and steer you to the few lenders that may make you a loan. Brokers can also assist if lenders don't want to make loans on unusual properties that you're interested in buying. Many lenders don't like dealing with co-ops and tenancies-in-common (see Chapter 8), borrowers with credit problems, or situations where a home buyer seeks to borrow most (90 percent or more) of a property's value.

Mortgage brokers typically tell you that they can get you the best loan deal by shopping among many lenders. They may further argue that another benefit of using their service is that they can explain the multitude of loan choices, help you select a loan, and assist with the morass of paperwork that's (unfortunately) required to get a loan.

Good mortgage brokers can deliver on most of these promises, and for this service, they receive a cut of the amount that you borrow — typically, 0.5 to as much as 2 percent on smaller loans. Not cheap, but given what a headache finding and closing on a good mortgage can be, hiring a mortgage broker may be just what the financial doctor ordered.

If you're going to work with a mortgage broker, please keep in mind that such brokers are in the business of "selling" mortgages and derive a commission from this work, just as do stockbrokers who sell stock and car salespeople who sell cars. A difference, though, is that the interest rate and points that you pay to get most mortgages through a broker are the same as what you would pay a lender directly. Lenders reason that they can afford to share their normal fees with an outside mortgage broker who isn't employed by the bank. After all, if you got the loan directly from the bank, you'd have to work with and take up more of the time of one of the bank's own mortgage employees.

However, some lenders, including those with the lowest rates, don't market through mortgage brokers. And sometimes a loan obtained through a mortgage broker can end up costing you more than if you had obtained it directly from the lender — for example, if the mortgage broker is taking a big commission for himself.

The commission that the mortgage broker receives from the lender isn't set in stone and is completely negotiable, especially on larger loans. On a $100,000 loan, a 1 percent commission amounts to $1,000. The same commission rate on a $300,000 loan results in a $3,000 cut for the broker, even though this three-times-larger loan doesn't take up three times as much of the mortgage broker's time. You have every right to inquire of the mortgage broker what his take is. Don't become overwhelmed with embarrassment; remember, it's your money, and you have every right to know this information! Ask — and don't hesitate to negotiate.

TIP

Be sure to get all costs and fees in writing prior to committing to take a loan. Always ask whether the mortgage broker is receiving a *yield spread premium* from the lender. This is an undisclosed fee that you pay for with a higher interest rate on your loan. The Federal Good Faith Estimates Law requires that you receive this detailed estimate within three days after the mortgage broker gets your loan application. Be sure to read it. Ask questions if there's anything in the estimate that you don't understand. The estimate isn't carved in stone — it's negotiable.

INVESTIGATE

In addition to understanding and negotiating a commission with the mortgage broker, get answers to the following questions when choosing a mortgage broker:

>> **How many lenders does the broker do business with, and how does the broker keep up to date with new lenders and loans that may be better?** Some mortgage brokers, out of habit and laziness, send all their business to just a few lenders and don't get you the best deals. Ask brokers which lenders have approved the broker to represent them.

>> **How knowledgeable is the broker about the loan programs, and does the broker have the patience to explain all a loan's important features?** The more lenders a mortgage broker represents, the less likely the broker is to know the nuances of each and every loan. Be especially wary of a salesperson who aggressively pushes certain loan programs and glosses over or ignores explaining the important points we discuss in this chapter for evaluating particular mortgages.

WARNING

All the advice that we give for selecting a good lender applies doubly for choosing a good mortgage broker. Some brokers, for example, have been known to push programs with outrageous interest rates and points, which, not too surprisingly, entail big commissions for them. This problem occurs most frequently with borrowers who have questionable credit or other qualification problems.

Also head for cover if your mortgage broker pushes you toward balloon and negative-amortization loans (discussed earlier in this chapter). Balloon loans, which become fully due and payable several years after you get them, are dangerous because you may not be able to get new financing and can be forced to sell the property.

LOAN PREQUALIFICATION AND PREAPPROVAL

When you're under contract to buy a property, having your mortgage application denied (after waiting several weeks) may cause you to lose the property after having spent hundreds of dollars on loan fees and property inspections. Even worse, you may lose the home that you've probably spent countless hours searching for and a great deal of emotional energy to secure. Some house sellers won't be willing to wait or may need to sell quickly. If the sellers have other buyers waiting in the wings, you've likely lost the property.

How could you have avoided this heartache? Well, you may hear some people in the real estate business, particularly real estate agents and mortgage brokers, advocate that you go through mortgage prequalification or preapproval.

Prequalification is an informal discussion between borrower and lender. The lender provides an opinion of the loan amount that you can borrow based solely on what you, the borrower, tell the lender. The lender doesn't verify anything and isn't bound to make the loan when you're ready to buy. Prequalification is of little value to you and your real estate agent.

Preapproval is a much more rigorous process, which is why we prefer it if you have any reason to believe that you'll have difficulty qualifying for the loan you desire. Loan preapproval is based on documented and verified information regarding your likelihood of continued employment, your income, your liabilities, and the cash you have available to close on a home purchase. The only thing the lender can't preapprove is the property you intend to buy because, of course, you haven't found it yet.

Going through the preapproval process is a sign of your seriousness to house sellers — it places sort of a Good Borrowing Seal of Approval on you. A lender's preapproval letter is considerably stronger than a prequalification letter. In a multiple-offer situation where more than one prospective buyer bids on a home at the same time, buyers who've been preapproved for a loan have an advantage over buyers who haven't been proved creditworthy.

Lenders don't charge for prequalification. Given the extra work involved, some lenders do charge for preapproval (perhaps a few hundred dollars). Other lenders, however, offer free preapprovals to gain borrower loyalty. Don't choose a lender just because the lender doesn't charge for preapproval. That lender may not have the best loan terms.

If you do choose to get preapproved with a lender that charges for it, be sure that you're soon going to go through with a home purchase. Otherwise, you'll have thrown good money down the drain.

TIP

If you're on the fence about using a mortgage broker, take this simple test: If you're the type of person who dreads shopping and waits until the last minute to buy a gift, a good mortgage broker can probably help and save you money. A competent mortgage broker can be of greatest value to those who don't bother shopping around for a good deal or who may be shunned by most lenders.

Even if you plan to shop on your own, talking to a mortgage broker may be worthwhile. At the very least, you can compare what you find with what brokers say they can get for you.

WARNING

Be aware, though, that some brokers only tell you what you want to hear — that they can beat your best find. Later, you may discover that the broker isn't able to deliver when the time comes. If you find a good deal on your own and want to check with a mortgage broker to see what she has to offer, it may be wise not to tell the broker the terms of the best deal you've found. If you do, more than a few brokers will always come up with something that they say can beat it.

INVESTIGATE

When a mortgage broker quotes you a really good deal (you'll know this if you've shopped a little yourself), ask who the lender is. Most brokers refuse to reveal this information until you pay the few hundred dollars to cover the appraisal and credit report. In most cases, you can check with the actual lender to verify the interest rate and points that the broker quoted you and make sure you're eligible for the loan. (In some cases, lenders don't market loans directly to the public.)

Should you discover, in calling the lender directly, that the lender doesn't offer such attractive terms to its customers, don't leap to the conclusion that the mortgage broker lied to you. In rare cases, a mortgage broker may offer you a slightly better deal than what you could have gotten on your own.

TIP

If the broker is playing games to get your business, charging the broker's up-front fee on your credit card allows you to dispute the charge and get your money back.

SHOULD YOU APPLY FOR MORE THAN ONE MORTGAGE?

When you applied to college or for your last job, you likely didn't apply only to your first choice. You probably had a backup or two or three. Thus, when the time comes to apply for a mortgage, you may be tempted to apply to more than one mortgage lender. The advantage — if one lender doesn't deliver, you have a backup to . . . well, fall back on.

However, we believe that if you do your homework and pick a good lender with a reputation for competitive low rates, quality service, playing straight, and meeting borrowers' expectations, applying for more than one mortgage isn't necessary on most properties. When you apply for a second loan, you must pay for additional application fees, credit report charges, and a new appraisal because you can't transfer your appraisal to the second lender. You'll also spend more time and effort completing extra paperwork.

Applying to more than one mortgage lender makes more sense in special situations where you run a greater risk for having your loan application denied. The first case is when you have credit problems. Read Chapter 2 to whip your finances into shape before you embark on the home-buying journey; read Chapter 7 for tips on completing your loan application in a way that will make lenders salivate.

The second circumstance under which it makes sense to apply to more than one mortgage lender is when you want to buy a physically or legally "difficult" property. It's impossible, of course, to know in advance all the types of property idiosyncrasies that will upset a particular lender. In fact, both of your authors, earlier in our homeownership days, were denied mortgages because our prospective homes had quirks that a particular lender didn't care for. Minimize your chances for negative surprises by asking your agent and property inspector in advance whether any aspects of the property may give a lender cause for concern.

If you apply for two loans, tell both lenders that you're applying elsewhere. When the second lender pulls your credit report, the first lender's recent inquiry will show up. (Less-than-candid borrowers almost always get caught this way.) Also tell both lenders that you're sincerely interested — just as you would tell all prospective employers.

Chapter **7**

Mortgage Quandaries, Conundrums, and Forms

U nderstanding and selecting a mortgage (the subject of Chapter 6) isn't all that difficult a project after you cut through the jargon and know how to think about your overall financial situation and goals.

Unfortunately, when you apply for a mortgage, obstacles may get in your way. In this chapter, we show you how to conquer these irritating and sometimes not-so-trivial challenges. We also answer your queries about other perplexing (and, in some cases, desirable) alternatives you may have.

In the last section of this chapter, we explain how to complete those dreaded mortgage-application forms.

Overcoming Common Mortgage Problems

Few things in life are more frustrating than not being able to have something you really want, especially if you perceive, rightly or wrongly, that most other people you know have it. If you want to buy a home and you can't finance the purchase, odds are that your dream will have to be put on hold.

REMEMBER

Don't despair if obstacles stand in your way. You may have to exhibit a bit more patience than usual, but we've never met anyone who was determined to buy a home but wasn't able to overcome credit or other problems. This chapter shows you how to get the financing you need and deserve!

Insufficient income

Your desired mortgage lender may reject your loan application if you appear to be stretching yourself too thin financially. Don't get angry — the lender may be doing you a huge favor by keeping you from buying a home that will prevent you from saving money and achieving other financial goals that may be important to you over time. (For more on this topic, read Chapters 2 and 3 about getting your financial house in order and determining how much home you really can afford.)

TIP

If you *know* you can afford the home that you have your sights set on, here are some keys to getting your loan approved:

>> **Be patient.** When you have a low income (for example, if you're self-employed and have been deducting everything but the kitchen sink as a business expense), you may need to wait a year or two so you can demonstrate a higher income.

>> **Put more money down.** The likelihood of getting lender approval goes up if you put 25 to 30 percent down.

>> **Get a cosigner.** You always knew you'd hit your parents up again someday for help and favors. If your folks are in good financial shape, they may be able to cosign a loan to help you qualify. A financially solvent sibling, rich aunt, or wealthy pal can do the same.

WARNING

Be sure to consider the financial and nonfinancial ramifications of having a relative or buddy cosign a loan with you. If you default on the loan or make payments late, you'll besmirch not only your credit history but also your cosigner's. At a minimum, have a frank discussion about such issues before you enter into such an arrangement, and be sure to write up a loan agreement with your benevolent cosigner.

Debt and credit problems

When you seek to take out a mortgage, lenders examine your credit history, which is detailed in your personal credit report (see Chapter 5 for details on credit reports and scores). Lenders also analyze your current debts and liabilities, which you provide on your mortgage application. Your current debts and credit history can produce a number of red flags that may make lenders skittish about lending you money. This section tells you how to deal with the typical problems that concern lenders.

Credit report boo-boos

As you may know, creditors can report your loan delinquencies and defaults to credit bureaus. These blemishes will show up on your personal credit report.

TIP

Here's our suggested plan of attack for dealing with such problems:

>> **Be proactive.** If you know your credit report includes warts and imperfections, write a concise explanation (via email so you have a record and copy of it) to the lender explaining the flaws. For example, maybe you were late on some of your loan payments once because you were out of the country and didn't get your bills processed in time. Or perhaps you lost your job unexpectedly and fell behind in your payments until you located new employment.

>> **Shop around for understanding and flexible lenders.** Some lenders are more sympathetic to the fact that you're human and have sometimes erred. As you interview lenders, inquire whether your previous credit blemishes may pose a problem. You may also consider enlisting the services of a mortgage broker, who may be more accustomed to dealing with loan problems.

>> **Look to the property seller for a loan.** Property sellers who are interested in playing lender can also be flexible. Surprisingly, some won't even check your credit report. Those who check your credit report may be more willing than banks and other mortgage lenders to forgive past problems, especially if you're financially healthy and strong today.

>> **Fight and correct errors.** Credit-reporting agencies and creditors who report information to the agencies make mistakes. Unfortunately, in the financial world you're guilty until you can convince the credit agencies that you're innocent. Start by identifying the erroneous information on your credit report. Should the erroneous information pertain to an account you never had, tell the credit bureau to examine the possibility that the derogatory information belongs on someone else's report.

If the bad data *is* for one of your accounts but a creditor (for example, First Usury Bank, from which you obtained an auto loan) has made an error, you'll likely have to hound such a creditor until it instructs the credit bureau to fix the mistake. To get these sorts of errors corrected, you must be persistent and patient. By law, the credit bureaus are supposed to respond to your inquiry within 30 days. Dispute processes are generally done through online forms and submissions. You can contact the credit agencies at the phone number listed on your personal credit report. Should you get the runaround from the front-line customer-service representatives you talk to, ask to speak with a supervisor or manager until you get satisfaction. If that technique doesn't work, contact your local Better Business Bureau and file a complaint. You're also allowed to enter a statement of contention on your credit report so prospective creditors, such as mortgage lenders, that pull your credit report can see your side of the story. But your best strategy is to have the disparaging information removed from your credit report.

>> **Get a cosigner.** As we suggest earlier, a cosigner, such as a relative, can also help deal with credit problems that are knocking out your loan application.

>> **Save more, and build a better track record.** If you can continue to rent, buying yourself some more time may do the trick. Why rush buying if lenders avoid you like the plague and reject you, or only offer loans with ultrahigh interest rates? Spend a couple of years saving more money and keeping a clean credit record, and you'll eventually have lenders chasing you for your business!

TIP

If you're having problems getting approved, sit down with your loan officer and make a list of the items you must rectify to get an approval. Instead of trying to guess what's wrong, you'll have a checklist of everything you need to correct.

CREDIT (FICO) SCORES

Lenders generally use a credit-scoring system to help them streamline credit application processing. The most commonly used credit-scoring system by mortgage lenders is the FICO score. FICO is short for the company that developed this system — Fair Isaac Corporation.

Higher scores mean that the borrower is far more likely to make timely and complete payments when borrowing money. In other words, such borrowers are the types that lenders prefer making loans to and are generally said to be "low risk."

Credit scores such as the FICO score are determined by using information in your credit report. Because there are three major credit-rating agencies (Equifax, Experian, and TransUnion), you can have a FICO score from each of their reports. (We discuss credit reports and how to obtain them for free in Chapter 5.)

In the event that you're turned down for a mortgage or other loan, request a copy of your credit report and an explanation from the lender for the specific reasons that you were denied credit. You can improve your credit score over time by addressing lenders' concerns and by using the solutions we present elsewhere in this chapter for dealing with common mortgage-application problems.

The best defense against being turned down is to examine, understand, and even improve your credit score *before* applying for a mortgage. Please see Chapter 5 for all the details.

Excess debt

If you're turned down for a mortgage because of excess debt (such as on credit cards and auto loans), be grateful. The lender has actually done you a favor! Over the long term, such debt is a serious drag on your ability to save money and live within your income.

TIP

Should you have the cash available to pay off some or all of the debt, we emphatically urge you to do so. Mortgage lenders sometimes make this a condition of funding a mortgage, especially when you have significant debts or are on the margin of qualifying for the loan that you desire. If you lack sufficient cash to pay down the debt and buy the type of home you want, choose among the following options:

>> **Set your sights more realistically.** Buy a less expensive home for which you can qualify for a mortgage.

>> **Go on a financial diet.** Your best bet for getting rid of consumer debt is to take a hard look at your spending (see Chapter 2) and identify where you can make cuts. Use your savings to pay down the debt. Also explore boosting your employment income.

>> **Get family help.** Another potential option is to have your family help you, either by cosigning your loan or by lending or giving you money to pay down your high-interest debt.

Lack of down payment

Saving money in America, where *everything* is considered a necessity at one time or another, can be a chore. Should you lack sufficient money for a down payment, turn to Chapter 3 for suggestions about how to get financing.

Dealing with Appraisal Problems

Your loan application may be sailing smoothly through the loan-approval channels — thanks to your sterling (or at least acceptable) financial condition — and then, all of a sudden (like in a *Batman* episode):

POW!!! BANG!!! THUMP!!! KABOOM!!!

The property that you've fallen in love with isn't worth what you agreed to pay for it, at least according to the *appraiser* — the person who values property for lenders. You may be shocked, dismayed, and perhaps even frightened that the appraiser has given such a low estimate. What course of action you should take depends on which of the following three issues caused the low appraisal.

You've overpaid

Appraisals don't often come in lower than the contract's purchase price. When they do, more often than not they're low because you (and perhaps your real estate agent) overestimated what the home is worth. If this is the case, be grateful that the appraiser has provided you a big warning that you're about to throw away money, perhaps thousands of dollars, if you go through with paying the price specified in your purchase contract. It's also possible that the appraised value is low because the home needs a new roof, new foundation, or other major structural repairs. (We cover property inspections in Chapter 13.)

TIP

Because you obviously like the property (after all, you made an offer to buy it), use the appraisal as a tool to either renegotiate a lower purchase price with the seller or get a credit from the seller to do necessary repairs. If the seller won't play, move on to other properties. Be aware that even if the seller agrees to a credit, your lender may not. Most lenders cap the amount of credits the buyer can receive. If they reject the credit the seller is willing to give, asking the seller to reduce the purchase price may be the only way to move forward. Also reevaluate your agent's knowledge of property values and motivations — consider finding a new agent if the present agent prodded you into overpaying. (We provide guidance on finding an outstanding agent in Chapter 9.)

The appraiser doesn't know your area

If you and your agent know local property values and have seen comparable homes that fully justify the price you agreed to pay, it's possible that the appraiser simply doesn't know local property values. One clue that this is the case is if the appraiser doesn't normally appraise homes in your area. Another clue is if the comparable properties that the appraiser chose aren't good, representative comparisons. We get into exactly what is and isn't a comparable property in Chapter 10.

TIP

If you have reason to believe that the appraiser may be off base, express your concern to the mortgage lender you're using. Also, request a copy of the appraisal, which you're entitled to. The lender should be able to shed some light on the appraiser's background and experience with valuing homes in your area. Sometimes you can have a reappraisal done without an additional charge.

The appraiser/lender is sandbagging you

The least likely explanation for a low appraisal is that your mortgage lender may have come in with a low appraisal to get out of doing a loan that she feels is undesirable. In the business, this trick is called *sandbagging.*

Lenders that use in-house appraisals are best able to torpedo loans that they don't want to make. Why, you may reasonably wonder, would lenders sandbag you on a loan for which they've willingly accepted a loan application? Remember that the eager front-line mortgage person at the bank or the mortgage broker who placed your loan with the lender likely works on commission and isn't the person who makes mortgage-approval decisions at the lending company.

TIP

Should you suspect that your loan is being sandbagged, request a copy of your appraisal. If comparable sales data show that the appraisal is low, confront your lender on this issue and see what she has to say about it. If you get the runaround and no satisfaction, ask for a full refund of your loan application and appraisal fees and take your business to another lender. You may also consider filing a complaint with the state organization that regulates mortgage lenders in your area.

Those Darn Mortgage Forms

When you finally get to the part of your home purchase where you're applying for a mortgage, you're likely to become so sick of paperwork that you'll yearn for a paperless society. You may be interested in knowing that lenders (especially online lenders, which we discuss in detail in Chapter 11) are moving to a more computer-driven (and less pen-and-paper-oriented) mortgage-application process. No matter; you're still going to have to provide a great deal of personal and financial information.

In this section, we review the forms that you're commonly asked to complete in the mortgage-application process. If you're working with a skilled person at the mortgage lender's firm or mortgage-brokerage firm you've chosen, that person can help you navigate most of this dreaded paperwork.

But we know you probably have some questions about what kinds of information you're required to provide versus information you don't have to provide. You also may be uncomfortable revealing certain, how shall we say, less-than-flattering facts about your situation — facts that you feel may jeopardize your qualifications for a mortgage. Finally, no matter how good the mortgage person that you're working with is, the burden is still on you to pull together many facts, figures, and documents. So here we are, right by your side, to coach and cajole you along the way.

The laundry list of required documents

Many mortgage lenders provide you an incredibly lengthy list of documents that they require with mortgage applications. One quick look at the list is enough to make most prospective home buyers continue renting! The following list includes the typical documents that mortgage lenders will request from you:

>> Payroll records, typically for the most recent 30 consecutive days

>> Two most recent years' W-2 forms

>> Two most recent years' federal income tax returns

>> Signed IRS Form 4506-T Request for Transcript of Tax Return

>> Year-to-date profit-and-loss statement and current balance sheet if you're self-employed

>> Copies of past two months' bank, money market, and investment account statements

>> Recent statements for all outstanding mortgages

>> Copy of current declarations pages for homeowners insurance policies in force

>> Home purchase contract (if you're financing to buy a home)

>> Rental agreements for all rental properties

>> Divorce decrees

>> Federal corporate tax returns for the past two years

>> Partnership federal tax returns for the past two years

>> Partnership K-1s for the last two years

>> Condo or homeowners association documentation — such as CC&Rs (covenants, conditions, and restrictions); bylaws; articles of incorporation; budget; reserve study; current regular assessment amount owed, if any; special assessments due now or approved for the future, if any; and contact name, address, and phone number

>> Title report, abstract, and survey

>> Property inspection report and pest control inspection report (if you're buying a home)

>> Gift letter — the source of any funds being used toward your down payment must sign indicating that these funds are a "gift" and are not required to be repaid

>> Receipts for deposits (if you're buying a home)

But don't despair. This list must cover all possible situations, so some of the items won't apply to you.

Most of the items on this laundry list are required to prove and substantiate your current financial status to the mortgage lender and, subsequently, to other organizations that may buy your loan in the future. Payroll records, tax returns, and bank and investment account statements help document your income and assets. Lenders assess the risk of lending you money and determine how much they can lend you based on these items.

If you're wondering why lenders can't take you on your word about the personal and confidential financial facts and figures, remember that some people don't tell the truth. Even though we know you're an honest person, lenders have no way of knowing who's honest and who isn't. The unfortunate consequence is that lenders have to treat all their applicants as though they aren't honest.

Even though lenders require all this documentation, some buyers still falsify information. Worse yet, some mortgage brokers, in their quest to close more loans and earn more commissions, even coach buyers to lie to qualify for a loan. One example of how people cheat: Some self-employed people create bogus tax returns with inflated incomes. Although a few people have gotten away with such deception, we strongly discourage this wayward path — it's fraudulent.

WARNING

Falsifying loan documents is committing perjury and fraud, and isn't in your best interests. Besides the obvious legal objections, you can end up with more mortgage debt than you can really afford. Plus, mortgage lenders can catch you in your lies. How? Well, some mortgage lenders have you sign a document (at the time you close on your home purchase or at the time of your loan application) that allows them to request *directly from the IRS* a copy of the actual return you filed with the IRS. Form 4506 grants the lender permission to get a copy of your tax return (Form 4506 is available online at www.dummies.com/go/homebuyingkit7e). Another document that the lender may spring on you is Form 8821 (Form 8821 is accessible online at www.dummies.com/go/homebuyingkit7e), which asks the IRS to confirm specific information and is more likely to be sent in by a lender to verify your financial information as reported for tax purposes (see Figure 7-1). You typically get these documents at closing. You can refuse to sign them — but then again, the lender can refuse to make you a loan!

When you can't qualify for a desired mortgage without resorting to trickery, getting turned down is for your own good. Lenders have criteria to ensure that you'll be able to repay the money that you borrow and that you don't get in over your head.

Form **4506**

(March 2019)

Department of the Treasury
Internal Revenue Service

Request for Copy of Tax Return

▶ Do not sign this form unless all applicable lines have been completed.
▶ Request may be rejected if the form is incomplete or illegible.
▶ For more information about Form 4506, visit www.irs.gov/form4506.

OMB No. 1545-0429

Tip. You may be able to get your tax return or return information from other sources. If you had your tax return completed by a paid preparer, they should be able to provide you a copy of the return. The IRS can provide a **Tax Return Transcript** for many returns free of charge. The transcript provides most of the line entries from the original tax return and usually contains the information that a third party (such as a mortgage company) requires. See **Form 4506-T, Request for Transcript of Tax Return,** or you can quickly request transcripts by using our automated self-help service tools. Please visit us at IRS.gov and click on "Get a Tax Transcript..." or call 1-800-908-9946.

1a Name shown on tax return. If a joint return, enter the name shown first.

1b First social security number on tax return, individual taxpayer identification number, or employer identification number (see instructions)

2a If a joint return, enter spouse's name shown on tax return.

2b Second social security number or individual taxpayer identification number if joint tax return

3 Current name, address (including apt., room, or suite no.), city, state, and ZIP code (see instructions)

4 Previous address shown on the last return filed if different from line 3 (see instructions)

5 If the tax return is to be mailed to a third party (such as a mortgage company), enter the third party's name, address, and telephone number.

Caution: If the tax return is being mailed to a third party, ensure that you have filled in lines 6 and 7 before signing. Sign and date the form once you have filled in these lines. Completing these steps helps to protect your privacy. Once the IRS discloses your tax return to the third party listed on line 5, the IRS has no control over what the third party does with the information. If you would like to limit the third party's authority to disclose your return information, you can specify this limitation in your written agreement with the third party.

6 **Tax return requested.** Form 1040, 1120, 941, etc. and all attachments as originally submitted to the IRS, including Form(s) W-2, schedules, or amended returns. Copies of Forms 1040, 1040A, and 1040EZ are generally available for 7 years from filing before they are destroyed by law. Other returns may be available for a longer period of time. Enter only one return number. If you need more than one type of return, you must complete another Form 4506. ▶ _____

Note: If the copies must be certified for court or administrative proceedings, check here . . . . ☐

7 **Year or period requested.** Enter the ending date of the year or period, using the mm/dd/yyyy format. If you are requesting more than eight years or periods, you must attach another Form 4506.

_____   _____   _____   _____

_____   _____   _____   _____

8 **Fee.** There is a $50 fee for each return requested. **Full payment must be included with your request or it will be rejected. Make your check or money order payable to "United States Treasury." Enter your SSN, ITIN, or EIN and "Form 4506 request" on your check or money order.**

a Cost for each return . . . . $ 50.00

b Number of returns requested on line 7 . . . .

c Total cost. Multiply line 8a by line 8b . . . . $

9 If we cannot find the tax return, we will refund the fee. If the refund should go to the third party listed on line 5, check here . . . . . . ☐

Caution: Do not sign this form unless all applicable lines have been completed.

Signature of taxpayer(s). I declare that I am either the taxpayer whose name is shown on line 1a or 2a, or a person authorized to obtain the tax return requested. If the request applies to a joint return, at least one spouse must sign. If signed by a corporate officer, 1 percent or more shareholder, partner, managing member, guardian, tax matters partner, executor, receiver, administrator, trustee, or party other than the taxpayer, I certify that I have the authority to execute Form 4506 on behalf of the taxpayer. **Note:** This form must be received by IRS within 120 days of the signature date.

☐ Signatory attests that he/she has read the attestation clause and upon so reading declares that he/she has the authority to sign the Form 4506. See instructions.

Phone number of taxpayer on line 1a or 2a

Sign Here

▶ Signature (see instructions)   Date

▶ Title (if line 1a above is a corporation, partnership, estate, or trust)

▶ Spouse's signature   Date

For Privacy Act and Paperwork Reduction Act Notice, see page 2.   Cat. No. 41721E   Form **4506** (Rev. 3-2019)

Source: U.S. Internal Revenue Service

FIGURE 7-1:
These documents may be waiting for you in the lender's loan papers at closing.

Form 8821

(Rev. January 2018)

Department of the Treasury
Internal Revenue Service

Tax Information Authorization

▶ Go to *www.irs.gov/Form8821* for instructions and the latest information.

▶ **Don't sign this form unless all applicable lines have been completed.**

▶ **Don't use Form 8821 to request copies of your tax returns
or to authorize someone to represent you.**

OMB No. 1545-1165

For IRS Use Only

Received by:

Name _____

Telephone _____

Function _____

Date _____

1 Taxpayer information. Taxpayer must sign and date this form on line 7.

Taxpayer name and address	Taxpayer identification number(s)
	Daytime telephone number Plan number (if applicable)

2 Appointee. If you wish to name more than one appointee, attach a list to this form. **Check here if a list of additional appointees is attached** ▶ ☐

Name and address

CAF No. _____

PTIN _____

Telephone No. _____

Fax No. _____

Check if new: Address ☐   Telephone No. ☐   Fax No. ☐

3 Tax Information. Appointee is authorized to inspect and/or receive confidential tax information for the type of tax, forms, periods, and specific matters you list below. See the line 3 instructions.

☐ By checking here, I authorize access to my IRS records via an Intermediate Service Provider.

(a) Type of Tax Information (Income, Employment, Payroll, Excise, Estate, Gift, Civil Penalty, Sec. 4980H Payments, etc.)	(b) Tax Form Number (1040, 941, 720, etc.)	(c) Year(s) or Period(s)	(d) Specific Tax Matters

4 Specific use not recorded on Centralized Authorization File (CAF). If the tax information authorization is for a specific use not recorded on CAF, check this box. See the instructions. If you check this box, skip lines 5 and 6 . . . . . . ▶ ☐

5 Disclosure of tax information (you **must** check a box on line 5a or 5b unless the box on line 4 is checked):

a If you want copies of tax information, notices, and other written communications sent to the appointee on an ongoing basis, check this box . . . . ▶ ☐

Note. Appointees will no longer receive forms, publications, and other related materials with the notices.

b If you don't want any copies of notices or communications sent to your appointee, check this box . . . . ▶ ☐

6 Retention/revocation of prior tax information authorizations. If the line 4 box is checked, skip this line. If the line 4 box isn't checked, the IRS will automatically revoke all prior Tax Information Authorizations on file unless you check the line 6 box and attach a copy of the Tax Information Authorization(s) that you want to retain. . . . . ▶ ☐

To revoke a prior tax information authorization(s) without submitting a new authorization, see the line 6 instructions.

7 Signature of taxpayer. If signed by a corporate officer, partner, guardian, partnership representative, executor, receiver, administrator, trustee, or party other than the taxpayer, I certify that I have the authority to execute this form with respect to the tax matters and tax periods shown on line 3 above.

▶ **IF NOT COMPLETE, SIGNED, AND DATED, THIS TAX INFORMATION AUTHORIZATION WILL BE RETURNED.**

▶ **DON'T SIGN THIS FORM IF IT IS BLANK OR INCOMPLETE.**

Signature _____   Date _____

Print Name _____   Title (if applicable) _____

For Privacy Act and Paperwork Reduction Act Notice, see instructions.   Cat. No. 11596P   Form **8821** (Rev. 1-2018)

FIGURE 7-1:
(continued)

Permissions to inspect your finances

CHECK IT OUT

In order for a mortgage lender to make a proper assessment of your current financial situation, the lender needs to request detailed documentation. Thus, mortgage lenders or brokers ask you to sign a form (like the one shown in Figure 7-2) authorizing and permitting them to make such requests of your employer, the financial institutions that you do business with, and so on. (This form is available online at www.dummies.com/go/homebuyingkit7e.)

RELEASE OF AUTHORIZATION

I/We hereby authorize ComUnity Lending to verify any information necessary in connection with an F.H.A., V.A., Conventional or Second Trust Deed/Equity Line loan application, including but not limited to the following:

              1. *Credit History*
              2. *Employment Records*
              3. *Bank Accounts*
              4. *Mortgage History*

Authorization is further granted to ComUnity Lending to use a photostatic copy of my/our signature (s) below, to obtain information regarding any of the aforementioned items.

_____   _____
APPLICANT (Borrower)   SOCIAL SECURITY NUMBER

_____   _____
APPLICANT (Borrower)   SOCIAL SECURITY NUMBER

                                                 CLU/CR

FIGURE 7-2: This type of form grants permission to your mortgage lender or broker to verify and document the financial facts of your life.

© John Wiley & Sons, Inc.

REMEMBER

As we recommend in numerous places throughout this book, before you agree to do business with a lender, you should get, *in writing*, the lender's estimate of what your out-of-pocket expenditures will be in order to close on your home loan. The good news for you is that lenders are required by law to provide, within three days of your application, what's called a *Good Faith Estimate* of closing costs after you initiate a mortgage with them (see Figure 7-3). You can access this worksheet online at www.dummies.com/go/homebuyingkit7e.

The Uniform Residential Loan Application

This is the big enchilada, the whole cannoli, or whatever you want to call it. Mortgage lenders and brokers throughout this vast country use the *Uniform Residential Loan Application* to collect vital data about home purchases and proposed loans. Many lenders use this standardized document, known in the mortgage trade as *Form 1003*, because they sell their mortgages to investors.

GOOD FAITH ESTIMATE - BORROWER'S SETTLEMENT COSTS

This list gives an estimate of most of the charges you will have to pay at the settlement of your loan. The figures shown, as estimates, are subject to change. The figures shown are computed based on the sales price and financing indicated.

The numbers listed on the left-handed column of this form correspond to the line number on the HUD-1 form, which will be used in conjunction with the settlement of your loan.

THIS FORM DOES NOT COVER ALL ITEMS YOU WILL BE REQUIRED TO PAY IN CASH AT SETTLEMENT; FOR EXAMPLE, DEPOSITS IN ESCROW FOR REPAIRS OR PEST WORK, YOU MAY BE REQUIRED TO PAY OTHER ADDITIONAL AMOUNTS AT SETTLEMENT.

This estimate was prepared for _____ on the date of _____

for the purchase /refi of _____

(property address)

Sales Price / Value .		
1st Mortgage		
2nd Mortgage		
Total Financing		
Down Payment	(-)
Financed VA Funding Fee/MIP		
Total Financing including VA Funding Fee / MIP .		

NONRECURRING CLOSING COSTS

801	Origination Fee Paid To Lender	%
802	Discount Points Paid To Lender (Govt. Pts.)	%
803	Appraisal Fee	
817	Inspection Fee (442)	
804	Credit Report Fee	
805	Appraisal Review Fee	
806	Document Preparation	
808	Processing Fee Paid To Lender	
809	Underwriting Fee	
905	VA Funding Fee.	
810	Courier Fee	
811	Other _____	
813	Flood Certification .	
815	Wire Transfer Fee .	
816	Warehouse	
1101	Settlement or Closing (Escrow Fee)	
1102	Title Misc. Fees	
1104	(a) Title Insurance Lender's (ALTA)	
1105	(b) Title Insurance Owner's (CLTA)	
1103	Notary Fees	
1201	Recording Fees	
1106	Tax Service Fee	
1301	Pest Inspection	
1202	City Transfer Tax (3.30 x 1,000 of S.P. -split 50-50)	
	Subtotal Nonrecurring Closing Costs	

ITEMS PAID WHILE ACTING AS A BROKER

821	Commission Paid To Broker .	%
	Rebate Fee (Premium Pricing and/or Servicing Released Premium) Paid To Mortgage Broker	
825	Processing Fee Paid To Mortgage Broker	
	Subtotal Nonrecurring Closing Costs.	

RECURRING CLOSING COSTS OR PREPAID EXPENSES

	_____ Months Taxes	
	_____ Months Insurance	
902	PMI Premium (1st Year)	
	1 Month [ ] PMI 1 Month [ ] MMI	
901*	Interest _____ days	
	Subtotal: Prepaid Expenses	

TOTAL CASH REQUIRED: . . . . .TOTAL

Less monies advanced (escrow deposit)

Total cash required at closing

*This interest calculation represents the greatest amount of interest you could be required to pay at settlement. The actual amount will be determined by which day of the month your settlement is conducted. To determine the amount you will have to pay, multiply the number of days remaining in the month in which you settle times $ _____, which is the daily interest charge for your loan.

I hereby acknowledge receipt of a copy of this estimate.

Date: ___/___/___   Amended Date: ___/___/___

Borrower _____   Borrower _____

Borrower _____   Borrower _____

Prepared By: Prepared By:
4/5/93

Type of Program	
P & I @ _____ %	
P/MMI	
P & I (2nd)	
Taxes	
Insurance	
HOA Dues	
Total	

CLU/GFE

FIGURE 7-3:
Here's an estimated closing costs worksheet.

Some mortgage lenders may toss you a Form 1003 and expect you to return it to them completed. Other lenders and brokers help you fill out the form or even go so far as to complete it for you.

TIP

If you let someone fill out the *Uniform Residential Loan Application* on your behalf, know that you're still responsible for making sure that the information on the form is accurate and truthful. Also, be aware that in their sales efforts, some mortgage lenders and brokers may invite you to their offices or invite themselves to your home or office to complete this form for you or with you. Although we have no problem with good service, we do want you to keep in mind that you're not beholden or obligated to any lenders or brokers, even if they offer to come over and wash your car and provide you a pedicure! It's your money and your home purchase, so shop around for a good loan or mortgage broker. (Also, keep a copy of all the forms you complete for one mortgage lender to save time should you decide to apply to another lender.)

CHECK IT
OUT

If, like most people, you take the first whack at completing this form yourself, we trust that you'll find the upcoming sections (in which we walk you through the major items on this application) useful. You can access the Uniform Residential Loan Application online at www.dummies.com/go/homebuyingkit7e.

I. Borrower information

This section of the loan application (see Figure 7-4) is where you begin to provide personal information about yourself as well as any co-borrower you are buying with or currently own the property with. You will also detail your employment and income sources in this lengthy and important section.

The lender also wants to know where you've been living recently. If you've been in your most recent housing situation for at least two years, you need not list your prior residence. Lenders are primarily looking for stability here. Most lenders also request a letter from your landlord to verify that you've paid your rent in a timely fashion.

If you've moved frequently in recent years, most lenders check with previous landlords. If your application is on the borderline, good references can tip the scales in your favor. If you've paid the amount you owed on time, you should be fine. If you haven't, you should explain yourself, either by separate letter to the lender or in the blank space on page four of the application.

To be completed by the **Lender:**
Lender Loan No./Universal Loan Identifier _____ Agency Case No. _____

Uniform Residential Loan Application

Verify and complete the information on this application. If you are applying for this loan with others, each additional Borrower must provide information as directed by your Lender.

Section 1: Borrower Information. This section asks about your personal information and your income from employment and other sources, such as retirement, that you want considered to qualify for this loan.

1a. Personal Information

Name *(First, Middle, Last, Suffix)*

Social Security Number _____–____–_____
(or Individual Taxpayer Identification Number)

Alternate Names – *List any names by which you are known or any names under which credit was previously received (First, Middle, Last, Suffix)*

Date of Birth
(mm/dd/yyyy)
____/____/_____

○ **U.S. Citizen**
○ **Permanent Resident Alien**
○ **Non-Permanent Resident Alien**

○ I am applying for **individual credit.**
○ I am applying for **joint credit.** Total Number of Borrowers: _____
  Each Borrower intends to apply for joint credit. *Your initials:* _____

List Name(s) of Other Borrower(s) Applying for this Loan
(First, Middle, Last, Suffix)

Marital Status
○ Married
○ Separated
○ Unmarried*
*Single, Divorced, Widowed, Civil Union, Domestic Partnership, Registered Reciprocal Beneficiary Relationship

Dependents *(not listed by another Borrower)*
Number _____
Ages _____

Contact Information
Home Phone (____) ____–_____
Cell Phone   (____) ____–_____
Work Phone   (____) ____–_____   **Ext.**_____
Email _____

Current Address
Street _____ Unit # _____
City _____ State____ ⊡ Zip _____ Country _____
How Long at Current Address? _____ Years _____ Months   ○ Own   ○ Rent ($ _____ /month)   ○ No primary housing expense

If at Current Address for LESS than 2 years, list Former Address ☐ *Does not apply*
Street _____ Unit # _____
City _____ State____ ⊡ Zip _____ Country _____
How Long at Former Address? _____ Years _____ Months   ○ Own   ○ Rent ($ _____ /month)   ○ No primary housing expense

Mailing Address – *if different from Current Address* ☐ *Does not apply*
Street _____ Unit # _____
City _____ State____ ⊡ Zip _____ Country _____

Military Service – Did you (or your deceased spouse) ever serve, or are you currently serving, in the United States Armed Forces? ○ **NO** ○ **YES**
If YES, check all that apply: ☐ Currently serving on active duty with projected expiration date of service/tour ____ / _____ *(mm/yyyy)*
  ☐ Currently retired, discharged, or separated from service
  ☐ Only period of service was as a non-activated member of the Reserve or National Guard
  ☐ Surviving spouse

FIGURE 7-4: The first part of Section I summarizes your personal information.

Your recent work history is important to a mortgage lender (see Figure 7-5). Unless you're financially independent (wealthy) already, your lender knows that your employment income determines your ability to meet your monthly housing costs. As with your prior housing situation, lenders are seeking borrowers with stability, which can help push a marginal application through the loan-approval channels. To a lender, recent stability of the borrower in where they live, where they work, and the source and amount of their income, translates into predictability or future stability, which results in lower risk.

If you've held your recent job for at least the past two years, that's the only position you need to list (unless you currently work more than one job, in which case you should list all current jobs separately). Otherwise, you must list your prior employment to cover the past two-year period.

1b. Current Employment/Self Employment and Income   ☐ *Does not apply*

Employer or Business Name _____   Phone (___) ___ – _____

Address _____

City _____   State ___ ☐ Zip _____

Position or Title _____

Start Date ___ /_____ *(mm/yyyy)*

How long in this line of work? _____ Years _____ Months

☐ Check if you are the Business   ○ I have an ownership share of less than 25%.
   Owner or Self-Employed   ○ I have an ownership share of 25% or more.

Check if this statement applies:
☐ I am employed by a family member, property seller, real estate agent, or other party to the transaction.

Monthly Income (or Loss)
$ _____

Gross Monthly Income		
Base	$_____	/month
Overtime	$_____	/month
Bonus	$_____	/month
Commission	$_____	/month
Military Entitlements	$_____	/month
Other	$_____	/month
TOTAL	$_____	**/month**

1c. IF APPLICABLE, Complete Information for Additional Employment/Self Employment and Income   ☐ *Does not apply*

Employer or Business Name _____   Phone (___) ___ – _____

Address _____

City _____   State ___ ☐ Zip _____

Position or Title _____

Start Date ___ /_____ *(mm/yyyy)*

How long in this line of work? _____ Years _____ Months

☐ Check if you are the Business   ○ I have an ownership share of less than 25%.
   Owner or Self-Employed   ○ I have an ownership share of 25% or more.

Check if this statement applies:
☐ I am employed by a family member, property seller, real estate agent, or other party to the transaction.

Monthly Income (or Loss)
$ _____

Gross Monthly Income		
Base	$_____	/month
Overtime	$_____	/month
Bonus	$_____	/month
Commission	$_____	/month
Military Entitlements	$_____	/month
Other	$_____	/month
TOTAL	$_____	**/month**

1d. Previous Employment/Self-Employment and Income ONLY IF your Current Employment is LESS than 2 years.   ☐ *Does not apply*

Employer or Business Name _____

Address _____

City _____   State ___ ☐ Zip _____

Position or Title _____

Start Date ___ /_____ *(mm/yyyy)*   End Date ___ /_____ *(mm/yyyy)*

☐ Check if you were the Business Owner or Self-Employed

Previous Gross Monthly Income
$ _____

FIGURE 7-5:
The middle part of Section I details your employment.

We know that, in this ever-changing economy, some people change jobs fairly frequently and not always out of personal choice. Perhaps you held a position for only a few months (or less) and feel it would make your loan application look more attractive to simply leave the position off your forms. Others, who have had gaps in their employment, either because they took advantage of changing jobs to engage in other activities or because it took some time to find a suitable new position, may be tempted to discreetly gloss over gaps in employment.

TIP

What do we advise? Well, our overall perspective is that a mortgage application is somewhat like a resume. You should absolutely present your information in a positive and truthful way. It's better to show the employment gap than to have the lenders uncover it, which they can do because they often ask for the dates of your employment when verifying information with your employers.

As for leaving off a short-term or part-time job, the choice is up to you. This section of the application doesn't state that you must list every position.

Remember that lenders don't mind some job-hopping. If they see frequent job changes, then the *prospects for continued employment* section of the *verification-of-employment request* that your current employer receives from the lender will be more important.

This section of the application also asks that you list the monthly income from prior jobs. You don't provide your monthly income from your current job here because it's provided in the next section of the application.

You may also wonder (and be concerned about) why the lender wants your current and previous employers' phone numbers. Shortly before your loan is ready to close, the lender may call your current employer to verify that you're still employed, but verification of employment is usually done through current payroll documentation and W-2s.

It's highly unlikely that lenders will call your previous employers unless they need to verify an outstanding question about employment dates or similar information.

If you have other income sources, such as child support or alimony, list and describe them in the last portion of this section (see Figure 7-6). The more income you can list, the more likely you are to qualify for a mortgage with the most favorable terms for you. (To ensure that you "get credit" for child support or alimony payments received, the borrower must be able to prove receipt of the funds. This requirement can get sticky because the ex-spouse may be asked to provide a canceled check. See, there is a good reason to remain on good terms with your "ex".)

1e. Income from Other Sources ☐ ***Does not apply***

Include income from other sources below. Under Income Source, choose from the sources listed here:

• Alimony	• Child Support	• Interest and Dividends	• Mortgage Differential Payments	• Royalty Payments	• Unemployment Benefits
• Automobile Allowance	• Disability	• Notes Receivable	• Retirement (*e.g., Pension, IRA*)	• Separate Maintenance	• VA Compensation
• Boarder Income	• Foster Care	• Public Assistance		• Social Security	• Other
• Capital Gains	• Housing or Parsonage	• Mortgage Credit Certificate		• Trust	

NOTE: *Reveal alimony, child support, separate maintenance, or other income ONLY IF you want it considered in determining your qualification for this loan.*

Income Source – *use list above*	**Monthly Income**
	$
	$
	$
Provide TOTAL Amount Here	**$**

FIGURE 7-6: The last part of Section I details your other (nonemployment) income sources.

II. Financial information — Assets and liabilities

In Section II (see Figure 7-7), you present your personal balance sheet, which summarizes your assets and liabilities. Your assets would include bank checking and savings accounts, money market funds, retirement accounts, and other accounts you have.

Liabilities are any loans or debts you have outstanding. The more of these financial obligations you have, the more unwilling a mortgage lender is to lend you a large amount of money.

Section 2: Financial Information — Assets and Liabilities. This section asks about things you own that are worth money and that you want considered to qualify for this loan. It then asks about your liabilities (or debts) that you pay each month, such as credit cards, alimony, or other expenses.

2a. Assets – Bank Accounts, Retirement, and Other Accounts You Have

Include all accounts below. Under Account Type, choose from the account types listed here:

- Checking
- Savings
- Money Market
- Certificate of Deposit
- Mutual Fund
- Stocks
- Stock Options
- Bonds
- Retirement (e.g., 401k, IRA)
- Bridge Loan Proceeds
- Individual Development Account
- Trust Account
- Cash Value of Life Insurance (used for the transaction)

Account Type – use list above		Financial Institution	Account Number	Cash or Market Value
	▾			$
	▾			$
	▾			$
	▾			$
	▾			$
			Provide TOTAL Amount Here	$

2b. Other Assets You Have   ☐ Does not apply

Include all other assets below. Under Asset Type, choose from the asset types listed here:

- Earnest Money
- Proceeds from Sale of Non-Real Estate Asset
- Proceeds from Real Estate Property to be sold on or before closing
- Sweat Equity
- Employer Assistance
- Rent Credit
- Secured Borrowed Funds
- Trade Equity
- Unsecured Borrowed Funds
- Other

Asset Type – use list above		Cash or Market Value
	▾	$
	▾	$
	▾	$
	Provide TOTAL Amount Here	$

2c. Liabilities – Credit Cards, Other Debts, and Leases that You Owe   ☐ Does not apply

List all liabilities below (except real estate) and include deferred payments. Under Account Type, choose from the types listed here:
- Revolving (e.g., credit cards)   • Installment (e.g., car, student, personal loans)   • Open 30-Day (balance paid monthly)   • Lease (not real estate)   • Other

Account Type – use list above	Company Name	Account Number	Unpaid Balance	To be paid off at or before closing	Monthly Payment
▾			$	☐	$
▾			$	☐	$
▾			$	☐	$
▾			$	☐	$
▾			$	☐	$

2d. Other Liabilities and Expenses   ☐ Does not apply

Include all other liabilities and expenses below. Choose from the types listed here:
- Alimony   • Child Support   • Separate Maintenance   • Job Related Expenses   • Other

	Monthly Payment
▾	$
▾	$
▾	$

FIGURE 7-7: Section II asks for what you have and what you owe.

If you have the cash available to pay off high-cost consumer loans, such as credit card and charge card balances and auto or other consumer loans, consider doing so now (note the box you can check off indicating you intend to pay off certain loans before closing). These debts generally carry high interest rates that aren't tax-deductible, and they hurt your chances of qualifying for a mortgage (see Chapters 2 and 3 for an explanation of this matter).

Note (in the last part of this section) that you're to list child support and alimony payments that you make as well as out-of-pocket expenses related to your job if you aren't self-employed. These monthly expenses are like debts in the sense that they require monthly feeding.

III. Financial information — Real estate

Section III (see Figure 7-8) continues over onto page four and includes space for the details of real estate you already own. If you still own a home and are trying to sell it or have an offer on it but that transaction is pending, you should list that property in this section. This is where you would also include any second home or rental income properties.

If you make a profit from such holdings, that profit can help your chances of qualifying for other mortgages. Conversely, *negative cash flow* (property expenses exceeding income) from rentals reduces the amount that a mortgage lender will lend you. Most mortgage lenders want a copy of your tax return (and possibly copies of your rental agreements with tenants) to substantiate the information you enter in this space. Investment real estate is often owned in separate legal entities for tax and reduced liability purposes, so be prepared to provide your personal tax return, including the IRS form 1065 (Schedule K-1) for each of your real estate holdings.

Section 3: Financial Information — Real Estate. This section asks you to list all properties you currently own and what you owe on them. ☐ *I do not own any real estate*

3a. Property You Own If you are refinancing, list the property you are refinancing FIRST.

Address

Street _____ Unit # _____ City _____ State ___ ☐ Zip _____

Property Value	Status: Sold, Pending Sale, or Retained	Monthly Insurance, Taxes, Association Dues, etc. Not Included in Mortgage Payment	For Investment Property Only	
			Monthly Rental Income	For LENDER to Calculate: Net Monthly Rental Income
$	☐	$	$	$

Mortgage Loans on this Property   ☐ *Does not apply*

Creditor Name	Account Number	Monthly Mortgage Payment	Unpaid Balance	To be paid off at or before closing	Type: FHA, VA, Conventional, USDA-RD, Other	Credit Limit (if applicable)
		$	$	☐	☐	$
		$	$	☐	☐	$

3b. IF APPLICABLE, Complete Information for Additional Property   ☐ *Does not apply*

Address

Street _____ Unit # _____ City _____ State ___ ☐ Zip _____

Property Value	Status: Sold, Pending Sale, or Retained	Monthly Insurance, Taxes, Association Dues, etc. Not Included in Mortgage Payment	For Investment Property Only	
			Monthly Rental Income	For LENDER to Calculate: Net Monthly Rental Income
$	☐	$	$	$

Mortgage Loans on this Property   ☐ *Does not apply*

Creditor Name	Account Number	Monthly Mortgage Payment	Unpaid Balance	To be paid off at or before closing	Type: FHA, VA, Conventional, USDA-RD, Other	Credit Limit (if applicable)
		$	$	☐	☐	$
		$	$	☐	☐	$

FIGURE 7-8:
Section III has you list real estate you already own.

Net rental income (which lenders now calculate) refers to the difference between your rental real estate's monthly rents and expenses (excluding depreciation). *Rental property* is any property that you've bought for the purpose of renting it out. Therefore, *Net rental income* is the profit or loss that you make each month on rental property (excluding depreciation). If you've recently purchased the rental property, the lender counts only 75 percent of the current rent that you're collecting. If you've held your rental property long enough to complete a tax return, most lenders use the profit or loss (excluding depreciation) reported on your tax return.

IV. Loan and property information

The information you include in the *Loan Purpose* section (see Figure 7-9) indicates to the lender whether you plan to use the mortgage to buy a home, refinance an existing loan, or something else like construct a new home. Mortgage lenders and the investors who ultimately buy these types of mortgages want to know why you want to borrow money. So, in this section, you provide the address and the property value (usually the purchase price if you are buying the home).

Section 4: Loan and Property Information. This section asks about the loan's purpose and the property you want to purchase or refinance.

4a. Loan and Property Information

Loan Amount $ _____ **Loan Purpose** ○ Purchase   ○ Refinance   ○ Other _____

Property Address Street _____
Unit # _____ City _____ State ___ ⊡ Zip _____
County _____ Number of Units _____ **Property Value** $ _____

Occupancy ○ Primary Residence   ○ Second Home   ○ Investment Property   ○ FHA Secondary Residence

1. **Mixed-Use Property.** If you will occupy the property, will you set aside space within the property to operate your own business? *(e.g., daycare facility, medical office, beauty/barber shop)*   ○ NO ○ YES
2. **Manufactured Home.** Is the property a manufactured home? *(e.g., a factory built dwelling built on a permanent chassis)*   ○ NO ○ YES

4b. Other New Mortgage Loans on the Property You are Buying or Refinancing   ☐ *Does not apply*

Creditor Name	Lien Type	Monthly Payment	Loan Amount/ Amount to be Drawn	Credit Limit *(if applicable)*
	○ First Lien ○ Subordinate Lien	$	$	$

4c. Rental Income on the Property You Want to Purchase   **For Purchase Only**   ☐ *Does not apply*

Complete if the property is a 2-4 Unit Primary Residence or an Investment Property	Amount
Expected Monthly Rental Income	$
For LENDER to Calculate: Expected Net Monthly Rental Income	$

4d. Gifts or Grants You Have Been Given or Will Receive for this Loan   ☐ *Does not apply*

Include all gifts and grants below. Under Source, choose from the sources listed here:
· Relative   · Employer   · Community Nonprofit   · State Agency   · Other
· Unmarried Partner   · Religious Nonprofit   · Federal Agency   · Local Agency

Asset Type *(Cash Gift, Gift of Equity, Grant)*		Source – use list above		Cash or Market Value
	⊡ ○ Deposited ○ Not Deposited		⊡	$
	⊡ ○ Deposited ○ Not Deposited		⊡	$

FIGURE 7-9:
Section IV details the purpose of your loan and how you plan on using the property.

The lender also wants to know whether the property is your primary or secondary residence or is an investment property (you will also in this section estimate expected rental income). Your answers to these questions determine which loans your property is eligible for and the terms of the loans. From a lender's perspective, construction loans (which are usually short-term loans) and investment-property loans are riskier and therefore cost you more than other loans.

WARNING

You may be tempted (and some mortgage brokers and lender representatives have also been tempted) to lie on this part of the mortgage application to obtain more favorable loan terms. Be aware that lenders can — and sometimes do — challenge you to prove that you're going to live in the property if they suspect otherwise. Even after closing on a purchase and their loan, lenders have been known to ask for proof — such as a copy of a utility bill in your name — that you're living in the property. Some lenders have even been known to send a representative around to knock on the borrower's door to see who's living in the property. (Lenders only really require proof of owner-occupancy at the time of the loan closing and possibly a short time thereafter. But they don't care if months later the borrower relocates and uses the property as an income property.)

To ensure that the money for your down payment and closing costs isn't coming from another loan that may ultimately overburden your ability to repay the money they're lending you, mortgage lenders want to know the source of funds for your down payment and closing costs. Ideally, lenders want to see the down payment and closing costs coming from your personal savings. Tell the truth — lenders have many ways to trip you up in your lies here. For example, they ask to see the last several months of your bank or investment account statements to verify, for example, that someone else, such as a benevolent relative, didn't give you the money last week.

TIP

If you're receiving money from a relative as a gift to be used toward the down payment, have the gift giver write a short note (your broker/lender can provide a standardized gift letter) confirming that the money is indeed a present that you do *not* have to repay. Lenders are often suspicious that such payments are loans that must be repaid and that will add to your debt burden and risk of default. But in recent years it has become more difficult for younger people to buy homes, especially in high-cost areas, and it has become more common for parents and relatives to offer "no repayment required" funds to family members seeking to purchase their first condo or starter home.

V. Declarations

Figure 7-10 shows what this section looks like. Many of these questions are potential red flags to lenders or designed to see if you are giving answers consistent with the rest of your application.

If one of your answers in this section is likely to raise a red flag, then be sure to attach an explanation that honestly provides the lender with the background and details, so they understand if any negative situations were likely unusual or one-time events. Often showing that when you were faced with unexpected challenges in life you dealt with them head-on, without making excuses, can be a positive aspect in the overall evaluation of your loan application.

Section 5: Declarations. This section asks you specific questions about the property, your funding, and your past financial history.

5a. About this Property and Your Money for this Loan	
A. Will you occupy the property as your primary residence?	○ NO ○ YES
If YES, have you had an ownership interest in another property in the last three years?	○ NO ○ YES
If YES, complete (1) and (2) below:	
(1) What type of property did you own: primary residence (PR), FHA secondary residence (SR), second home (SH), or investment property (IP)?	▾
(2) How did you hold title to the property: by yourself (S), jointly with your spouse (SP), or jointly with another person (O)	▾
B. If this is a Purchase Transaction: Do you have a family relationship or business affiliation with the seller of the property?	○ NO ○ YES
C. Are you borrowing any money for this real estate transaction (*e.g., money for your closing costs or down payment*) or obtaining any money from another party, such as the seller or realtor, that you have not disclosed on this loan application?	○ NO ○ YES
If YES, what is the amount of this money?	$ _____
D. 1. Have you or will you be applying for a mortgage loan on another property (not the property securing this loan) on or before closing this transaction that is not disclosed on this loan application?	○ NO ○ YES
2. Have you or will you be applying for any new credit (*e.g., installment loan, credit card, etc.*) on or before closing this loan that is not disclosed on this application?	○ NO ○ YES
E. Will this property be subject to a lien that could take priority over the first mortgage lien, such as a clean energy lien paid through your property taxes (*e.g., the Property Assessed Clean Energy Program*)?	○ NO ○ YES

5b. About Your Finances	
F. Are you a co-signer or guarantor on any debt or loan that is not disclosed on this application?	○ NO ○ YES
G. Are there any outstanding judgments against you?	○ NO ○ YES
H. Are you currently delinquent or in default on a federal debt?	○ NO ○ YES
I. Are you a party to a lawsuit in which you potentially have any personal financial liability?	○ NO ○ YES
J. Have you conveyed title to any property in lieu of foreclosure in the past 7 years?	○ NO ○ YES
K. Within the past 7 years, have you completed a pre-foreclosure sale or short sale, whereby the property was sold to a third party and the Lender agreed to accept less than the outstanding mortgage balance due?	○ NO ○ YES
L. Have you had property foreclosed upon in the last 7 years?	○ NO ○ YES
M. Have you declared bankruptcy within the past 7 years? If YES, identify the type(s) of bankruptcy: ☐ Chapter 7 ☐ Chapter 11 ☐ Chapter 12 ☐ Chapter 13	○ NO ○ YES

FIGURE 7-10: Declare yourself in Section V; watch out for the red flags.

VI. Acknowledgments and agreements

If you haven't been honest on this form, here's your last chance to rethink what you're doing (see Figure 7-11). Remember, these days with the advent of "big data" and electronic records for everything we do financially, there is no such thing as getting away with misleading or lying in your loan application. The best policy is full and complete disclosure and transparency. Lenders know that almost everyone has one or more issues with their credit or employment or living situations, so be up-front and offer any details that may give the lender a comfort level with what happened and why you are a good risk for the pending loan.

Section 6: Acknowledgments and Agreements. This section tells you about your legal obligations when you sign this application.

Acknowledgments and Agreements

I agree to, acknowledge, and represent the following statements to:
- The Lender (this includes the Lender's agents, service providers and any of their successors and assigns); AND
- Other Loan Participants (this includes any actual or potential owners of a loan resulting from this application (the "Loan"), or acquirers of any beneficial or other interest in the Loan, any mortgage insurer, guarantor, any servicers or service providers of the Loan, and any of their successors and assigns).

By signing below, I agree to, acknowledge, and represent the following statements about:

(1) The Complete Information for this Application
- The information I have provided in this application is true, accurate, and complete as of the date I signed this application.
- If the information I submitted changes or I have new information before closing of the Loan, I must change and supplement this application or any real estate sales contract, including providing any updated/supplemented real estate sales contract.
- For purchase transactions: The terms and conditions of any real estate sales contract signed by me in connection with this application are true, accurate, and complete to the best of my knowledge and belief. I have not entered into any other agreement, written or oral, in connection with this real estate transaction.
- The Lender and Other Loan Participants may rely on the information contained in the application before and after closing of the Loan.
- Any intentional or negligent misrepresentation of information may result in the imposition of:
  (a) civil liability on me, including monetary damages, if a person suffers any loss because the person relied on any misrepresentation that I have made on this application, and/or
  (b) criminal penalties on me including, but not limited to, fine or imprisonment or both under the provisions of federal law (18 U.S.C. §§ 1001 et seq.).

(2) The Property's Security
- The Loan I have applied for in this application will be secured by a mortgage or deed of trust which provides the Lender a security interest in the property described in this application.

(3) The Property's Appraisal, Value, and Condition
- Any appraisal or value of the property obtained by the Lender is for use by the Lender and Other Loan Participants.
- The Lender and Other Loan Participants have not made any representation or warranty, express or implied, to me about the property, its condition, or its value.

(4) Electronic Records and Signatures
- The Lender and Other Loan Participants may keep any paper record and/or electronic record of this application, whether or not the Loan is approved.
- If this application is created as (or converted into) an "electronic application", I consent to the use of "electronic records" and "electronic signatures" as the terms are defined in and governed by applicable federal and/or state electronic transactions laws.
- I intend to sign and have signed this application either using my: (a) electronic signature; or (b) a written signature and agree that if a paper version of this application is converted into an electronic application, the application will be an electronic record, and the representation of my written signature on this application will be my binding electronic signature.
- I agree that the application, if delivered or transmitted to the Lender or Other Loan Participants as an electronic record with my electronic signature, will be as effective and enforceable as a paper application signed by me in writing.

(5) Delinquency
- The Lender and Other Loan Participants may report information about my account to credit bureaus. Late payments, missed payments, or other defaults on my account may be reflected in my credit report and will likely affect my credit score.
- If I have trouble making my payments I understand that I may contact a HUD-approved housing counseling organization for advice about actions I can take to meet my mortgage obligations.

(6) Use and Sharing of Information
I understand and acknowledge that the Lender and Other Loan Participants can obtain, use, and share the loan application, a consumer credit report, and related documentation for purposes permitted by applicable laws.

Borrower Signature _____ Date (mm/dd/yyyy) ____/____/_____

Borrower Signature _____ Date (mm/dd/yyyy) ____/____/_____

FIGURE 7-11:
Honesty counts in Section VI.

If you've had a mortgage broker (or anyone else) help you with this application, be sure to review the answers she provided for accuracy before you sign the agreement. Now is the time to ask yourself questions (and to review your responses) to ensure that you've presented your information in a positive-but-truthful light.

VII. Demographic information

You may refuse to answer the questions in this section (see Figure 7-12) if you want to — this information isn't required.

Section 7: Demographic Information. This section asks about your ethnicity, sex, and race.

Demographic Information of Borrower

The purpose of collecting this information is to help ensure that all applicants are treated fairly and that the housing needs of communities and neighborhoods are being fulfilled. For residential mortgage lending, federal law requires that we ask applicants for their demographic information (ethnicity, sex, and race) in order to monitor our compliance with equal credit opportunity, fair housing, and home mortgage disclosure laws. You are not required to provide this information, but are encouraged to do so. **The law provides that we may not discriminate** on the basis of this information, or on whether you choose to provide it. However, if you choose not to provide the information and you have made this application in person, federal regulations require us to note your ethnicity, sex, and race on the basis of visual observation or surname. The law also provides that we may not discriminate on the basis of age or marital status information you provide in this application.
Instructions: You may select one or more "Hispanic or Latino" origins and one or more designations for "Race." If you do not wish to provide some or all of this information, select the applicable check box.

Ethnicity
☐ Hispanic or Latino
   ☐ Mexican   ☐ Puerto Rican   ☐ Cuban
   ☐ Other Hispanic or Latino – *Enter origin:*

      Examples: Argentinean, Colombian, Dominican, Nicaraguan, Salvadoran, Spaniard, etc.
☐ Not Hispanic or Latino
☐ I do not wish to provide this information

Sex
☐ Female
☐ Male
☐ I do not wish to provide this information

Race
☐ American Indian or Alaska Native – *Enter name of enrolled or principal tribe:* _____
☐ Asian
   ☐ Asian Indian   ☐ Chinese   ☐ Filipino
   ☐ Japanese   ☐ Korean   ☐ Vietnamese
   ☐ Other Asian – *Enter race:* _____
      Examples: Hmong, Laotian, Thai, Pakistani, Cambodian, etc.
☐ Black or African American
☐ Native Hawaiian or Other Pacific Islander
   ☐ Native Hawaiian   ☐ Guamanian or Chamorro   ☐ Samoan
   ☐ Other Pacific Islander – *Enter race:*

      Examples: Fijian, Tongan, etc.
☐ White
☐ I do not wish to provide this information

To Be Completed by Financial Institution (*for application taken in person*):

	NO	YES
Was the ethnicity of the Borrower collected on the basis of visual observation or surname?	○	○
Was the sex of the Borrower collected on the basis of visual observation or surname?	○	○
Was the race of the Borrower collected on the basis of visual observation or surname?	○	○

The Demographic Information was provided through:

☐ Face-to-Face Interview (*includes Electronic Media w/ Video Component*)   ☐ Telephone Interview   ☐ Fax or Mail   ☐ Email or Internet

FIGURE 7-12:
Section VII is used to track discrimination.

The federal government tracks the ethnicity and gender of borrowers to see (among other things) whether lenders discriminate against certain people.

Other typical documents

All mortgage lenders and brokers have their own, individualized package of documents for you to complete. Some documents are standard because they are federally mandated. Covering all these forms here is certainly beyond the scope of this book — and most people's attention span. What follows are some of the other common forms that you're likely to encounter from your mortgage lender or broker.

Your right to receive a copy of appraisal

It wasn't always the case, but you now have the right to receive a copy of the appraisal report. That borrowers didn't always have this right is a bit absurd — after all, you're the one who's paying for the appraisal!

CHECK IT OUT

To make sure you know you have this right, the government requires that mortgage lenders and brokers present you with the document in Figure 7-13. You can access the form online at www.dummies.com/go/homebuyingkit7e.

EQUAL CREDIT OPPORTUNITY ACT
(REGULATION B)

RIGHT TO RECEIVE A COPY OF APPRAISAL

You have the right to a copy of the appraisal report used in connection with your application for credit. If you wish a copy, please write to us at the mailing address provided. We must receive your request no later than 90 days after we notify you of the action taken on your credit application, or you withdraw your application. In your letter you must provide us with your name, the address of the subject property, your current address, and the loan number assigned to your transaction.

I (We) have read and understand the aforementioned conditions regarding my right to receive a copy of our appraisal and acknowledge receipt to a copy of this disclosure.

_____   _____
Applicant   Date

_____   _____
Applicant   Date

FIGURE 7-13:
Exercise your right to your property's appraisal — ask for a copy.

TIP

Despite the fact that the notice tells you to make your request in writing, try making the request verbally to save yourself time. Then, if your request is ignored, go to the hassle of submitting a written request for your appraisal (within 90 days of the rendering of a decision to approve or reject your loan). Appraisals are good to have in your files — you never know when an appraisal may come in handy. At the very least, you can see what properties were used as comparables to yours in order to discover how good or bad the appraisal is.

Equal Credit Opportunity Act

Another form you'll probably see is one that discloses that it's a matter of federal law that a mortgage lender may not reject your loan because of any nonfinancial personal characteristic, such as race, sex, marital status, age, and so forth. You also don't have to disclose income that you receive as a result of being divorced (although we think that doing so is in your best interest because such income may help get your loan approved).

TIP

If you have reason to believe that a mortgage lender is discriminating against you, contact and file a complaint with your state's department of real estate or the government division that regulates mortgage lenders in your state. And start hunting around for a better, more ethical lender.

3
Property, Players, and Prices

Consider your options for where and what to buy.

Get tips on putting together a top-flight team of real estate professionals.

Understand how to determine what a given property is worth.

Surf the best internet websites and resources.

IN THIS CHAPTER

» Examining the relationship among location, value, and good neighborhoods

» Maximizing your investment

» Evaluating detached versus attached homes (condos and cooperative apartments)

» Understanding the rewards and risks of fixer-uppers, short sales, and foreclosures

» Considering the pros and cons of co-ownership

Chapter **8**

Where and What to Buy

What's your idea of the perfect car, the perfect job, and the perfect way to spend a day? Would you have said the same things ten years ago? Probably not. Perfection is a moving target — it changes as you change.

Where *the* perfect home is concerned, there's no such thing. For one thing, few people have the financial resources to afford what they think is the perfect home. Even if you're among the fortunate few with bucks to burn, it's still highly unlikely that one home will be perfect for you from birth to earth. The home that's great in your 20s when you're footloose and fancy-free probably won't cut it when you're in your 40s if you're married or raising a family. Fast-forward another 20 years to when you're nearing retirement. You may want or need to move to a smaller home that's easier to maintain.

Don't fret. Even though no single home stays perfect forever, this chapter shows you how to profitably achieve sequential perfection in your homes. And because moving is expensive, we also show you how to minimize the number of times you buy and sell.

You probably know someone who's lost money on a house sale. We're sure you don't plan to be the next victim of a capricious real estate market such as the one that hit during the late 2000s. Getting a bargain when you buy a home is a fine objective, but don't stop there. Don't you also want your home to appreciate in value while you own it?

REMEMBER

The best time to think about how much you'll get for your house when you sell it is before you buy it. Never let your enthusiasm for a house blind you to its flaws. Before you buy, try to look at the property through the eyes of the *next* potential buyer. Anything that disturbs you about the house or neighborhood will probably also bother the next buyer.

We're not suggesting that you should plan to sell your house immediately after buying it. For all we know, you'll live happily ever after in the home you're about to purchase. Then again, an unforeseen life change, such as a job transfer or family expansion, may force you to sell. If that happens, making a profit can take some of the sting out of moving day.

Appreciation is handy for a lot more than just increasing your net worth. Given that your home increases in value over time, you may someday find that this *equity* (the difference between market value and the mortgage you owe) can help you accomplish a multitude of important financial and personal goals. You can use the money any way you want — add to your retirement, help pay your kids' college education, start your own business, or take the Orient Express from London to Venice to celebrate your 25th wedding anniversary. Nest eggs are extremely versatile financial tools — and they're cholesterol free!

In a world filled with uncertainties, no one can guarantee that your home will increase in value. However, buying a good property in a desirable neighborhood tremendously increases your odds of making money. This maxim holds true whether the real estate market is strong or weak when you sell.

Property prices aren't static. They rise and fall because of such factors as the local job market, the supply of and demand for available housing and rental units, interest rates, and annual cycles of strong versus weak market activity. (For more on this subject, turn to Chapter 4.) Most of these things are beyond your control and ability to predict. But this doesn't mean that your financial destiny as a homeowner is a total fluke of fate. On the contrary, you control three important factors that greatly affect your home's value:

>> How much you pay for your home

>> Where your home is located

>> What home you buy

WARNING

The number-one controllable factor is how much you pay for your home. If you grossly overpay for your house when you buy it, you'll be extremely lucky to make a profit when you sell. That's why we devote Chapter 10 to making sure you know exactly how to spot well-priced properties and avoid overpriced turkeys.

This chapter focuses on the other two crucial factors under your control: where and what you buy.

Location, Location, Value

If you're wildly wealthy, you can afford to live anywhere you darn well please. The rest of us, however, have somewhat more limited budgets. Even so, unless you're foraging at the bottom of the housing food chain, you have many choices on places to spend your money. Where you ultimately decide to buy is up to you.

You've probably already heard that the three most important things you should look for when buying a home are "location, location, location." That axiom is largely true. People buy neighborhoods every bit as much as houses. In good times and bad, folks pay a premium to live in better neighborhoods. Conversely, rotten neighborhoods ravage home values. You'd have trouble selling the Taj Mahal if it were surrounded by junkyards and chicken farms.

But simply stating that the secret of making money in real estate when you buy is "location, location, location" is akin to saying you'll make a fortune in the stock market if you buy low and sell high. It takes more than glittering generalities to make money. You need specifics.

First off, we don't agree that the three most important factors are location, location, location. Besides, we don't see much point in repeating ourselves three times — it's not like you're a complete idiot or something! *Value* — what you get for your money — is important too.

If, for example, everyone knows that Elegant Estates is the *best* neighborhood in town, you'll pay a hefty premium to live there. And although Elegant Estates is currently king of the hill and may stay that way forever, keep in mind that this particular neighborhood has no place to go except downhill.

Other neighborhoods, ones that aren't held in such high esteem right now, may eventually improve what they offer home buyers and ultimately experience far greater property-value appreciation. Buying a home in a good location, though important, shouldn't be your sole home-shopping criterion. If you want to buy a home that is a good investment, you must look for good value. We explain how to do that in this chapter.

Characteristics of good neighborhoods

Good neighborhoods, like beauty, are in the eyes of the beholder. For example, being near excellent schools is important if you have young children. If, conversely, you're ready to retire, buying in a peaceful area with outdoor activities may appeal to you, and being next to a noisy junior high school is your nightmare! Neither neighborhood may suit you if you're the footloose and fancy-free type. Your ideal neighborhood is probably a singles' condo complex downtown so you can be near the action day or night.

Personal preferences aside, all good neighborhoods have the following characteristics:

>> **Economic health:** Nothing kills property values faster than a forest of "For Sale" signs precipitated by layoffs, foreclosures, or short sales where mortgage lenders agree to accept less money than is owed on properties to avoid going through the lengthy and costly foreclosure process. See Chapter 4 for ways to evaluate the employers and job markets in the various communities where you're contemplating buying a home.

>> **Amenities:** Amenities are special features of a neighborhood that make it an attractive, desirable place to live. Wide streets bordered by stately oak trees, lush green parks, ocean views, quiet cul-de-sacs, parking, and proximity to schools, churches, shopping, restaurants, transportation, playgrounds, tennis courts, and beaches are prime examples of amenities that add value to a neighborhood. Of course, few people can afford to buy in a neighborhood that has all these amenities, but the more of these perks a neighborhood has, the better.

>> **Quality schools:** You may not care how good or bad the local schools are if you don't have school-age children. However, unless you're buying in a remote, retirement, or vacation-type community, you had better believe that when you're ready to sell your house, most prospective buyers with kids will be deeply concerned about the school system. But you should care about the quality of nearby schools for more than just resale value, because good schools produce better kids, and that clearly affects the quality of life in the community.

TIP

Don't rely on test scores or someone's opinion when assessing school quality; visit the schools and speak with parents and teachers to get a handle on the schools in an area. We aim you toward websites chock-full of school information in Chapter 11.

>> **Low crime rates:** Most folks today are concerned with crime — and well they should be, given that crime rates in many parts of America are too high. As with schools, don't rely on hearsay or isolated news reports. Communities compile crime statistics, generally by neighborhood. Call the local police department, visit its website, or check the town's reference library to get the facts.

>> **Stability:** Some communities are in a constant state of flux. "Out with the old and in with the new" is their motto. Imagine what would happen to property values if a junkyard were replaced by a beautiful park. How about the reverse — an ugly, multistory, concrete parking garage appears where there was once a beautiful park? Check with the local planning department and a good real estate agent for the inside scoop on proposed developments in neighborhoods that you're considering.

>> **Pride of ownership:** A home's cost has no bearing on the amount of pride its owners take in it. Drive through any neighborhood, posh or modest, and you see in a flash whether the folks who live there are proud of their homes. A neighborhood filled with beautifully maintained homes and manicured lawns shouts pride of ownership.

WARNING

Property values sag when homeowners no longer take pride in their property. Avoid declining neighborhoods that display the red flags of dispirited owners — poorly kept houses, junk-filled yards, abandoned cars on the street, many absentee owners renting houses, and high rates of vandalism, crime, short sales, and foreclosures. Neighborhood deterioration is a blight that spreads from one house to another.

Selecting your best neighborhood

You may get lucky and find the neighborhood of your dreams right away. You're far more likely, however, to end up evaluating the strengths and weaknesses of several neighborhoods while trying to decide which one to favor with your purchase. If you're on a budget — and most people are — you may have to compromise and make tradeoffs.

Suppose that one neighborhood has the schools you like, the second is closest to your office (which would save you an hour a day commuting), and the third neighborhood is in a town with a delightful beach. They're all good neighborhoods, so your decision isn't easy.

The following sections offer ways to research and select the best neighborhood *for you*.

Prioritize your needs

Buying a home when you have budgetary constraints involves making tradeoffs. For example, if you want to live in the town with great schools and parks, you'll probably have to settle for a smaller home than you would if you bought in a more average community. When push comes to shove and you have to choose a place to live, you must decide what is most important to you.

Research

As we say earlier in this chapter, in the section "Characteristics of good neighborhoods," you should examine the health of the local economy, area amenities such as parks and entertainment, school quality, and crime rates before you buy a home. So where can you find this wealth of information?

>> **Tap local resources.** Check the local library. The local chamber of commerce is another excellent source of information.

>> **Talk to people who live in the neighborhoods.** Who knows more about a neighborhood than folks who live in it? In addition to asking how they feel about their neighborhood, see what residents say about the other neighborhoods you're considering. If you can spark some neighborhood rivalry, you'll get the dirt about the other neighborhoods' lousy weather, parking problems, unfriendly or snobby owners, and so on. Renters are also a great source of information. Because they don't have a wad of cash invested in a home, renters are generally candid about the shortcomings of a neighborhood. Last but not least, drive or walk through the neighborhoods at various times of the day and evening to make sure that their charm stays on 24 hours a day.

>> **Get days-on-market (DOM) statistics from your real estate agent.** DOM statistics indicate how long the average house in an area takes to sell after it goes on the market. As a rule, the faster property sells, the more likely it is to sell close to full asking price. Quick sales indicate strong buyer demand, which is nice to have when you're ready to sell.

>> **Get help from a professional.** Ask a real estate agent, lender, or appraiser to compare the upside potential of home values in each neighborhood. As Chapter 9 explains, home buying is a team sport. Get an analysis of each neighborhood's present and future property values from full-time real estate people.

TIP

Neither real estate agents nor lenders charge for opinions of value. They both, however, have a vested interest in selling you something. Appraisers, on the other hand, have no ax to grind. True, appraisers charge to analyze neighborhood property values and pricing trends. But if you're going to spend hundreds of thousands of dollars for a home, paying an additional few hundred dollars to get an unbiased, professional analysis of a neighborhood's property values may be money well spent.

>> **Go online.** Several real estate websites provide local community data and information. See Chapter 11 for recommended sites and surfing strategies.

Fundamental Principles for Selecting Your Home

Good news. It doesn't matter whether you buy a log cabin, Cape Cod colonial, French provincial, Queen Anne Victorian, or California ranch-style house. You can make money on any property by following three fundamental principles to select the home you buy. As you read the following guidelines, remember that they're not hard-and-fast rules — exceptions do exist.

The principle of progression: Why to buy one of the cheaper homes on the block

An appraiser can tell you that the *principle of progression* states that property of lesser value is enhanced by proximity to better properties. English translation, please? Buy one of the cheaper homes on the block, because the more expensive houses all around yours pull up the value of your home.

For instance, suppose that your agent shows you a house that just came on the market in a neighborhood you like. At $225,000, it's one of the least expensive homes you've seen in the area. The agent says that the other homes around it sell for anywhere from $275,000 to $300,000. You start to salivate.

INVESTIGATE

Don't whip out your checkbook yet. Do a little homework first. Find out why this house is so cheap. If the right things are wrong with it, write up the offer. If the wrong things are wrong with it, move on to the next property.

Curable defects

If a house is a bargain because it has defects that aren't too difficult or expensive to correct, go for it. For example, maybe the house is an ugly duckling that just needs a paint job, landscaping, and some other minor cosmetic touches in order to be transformed into a swan. Perhaps it's the only two-bedroom house on the block, but it has a large storage area that you can convert into a third bedroom for not more than $15,000. For $240,000 ($225,000 for the house plus $15,000 to add the bedroom), you're living in a $275,000 to $300,000 neighborhood. Such a deal!

Problems like these are *curable defects* — property deficiencies you can cure by upgrading, repairing, or replacing the defects relatively inexpensively. Painting, modernizing a bathroom, installing new counters and cabinets in the kitchen, and upgrading an electrical system are some examples of curable defects.

Depressed property values are another type of curable defect. A high number of short sales or foreclosures in a neighborhood drives property values down. One person's misfortune is another's opportunity. Some investors specialize in purchasing distressed properties at rock-bottom prices. These investors then rehab the houses and rent them out with an eye toward selling them for a profit when property values improve.

Incurable defects

WARNING

If a house has major problems, it's not a bargain at any price. Who'd want a house located next to a garbage dump? Or what about a *really* ugly home? Just because the seller made a fortune in the sausage business doesn't mean that you (or anyone else) would want to live in a house built in the shape of a giant hot dog. Maybe the house is cheap because a contractor says it's a wreck about ready to fall down, and you'd have to spend at least $125,000 on a new roof, a new foundation, new plumbing, and complete rewiring.

Enormous deficiencies like these are called *incurable defects.* They aren't economically feasible to correct. You can't fix the fact that a house is poorly located. Nor does it typically make sense to pay $225,000 for the hot-dog house so you can tear it down and build a new home (unless that's what comparable vacant lots sell for). By the same token, if you pay $225,000 for the wreck and then pour in another $125,000 on corrective work, you'll have the dubious honor of owning the most expensive house in the neighborhood.

Don't get us wrong. All rehabs aren't bad. We go into more detail about fixer-uppers later in this chapter.

The benefits of renovating cheaper homes

The less-expensive houses on the block are also the least risky ones to renovate, thanks to the principle of progression. For example, suppose that you just paid $225,000 for a house that needs a major rehab. Your construction project is located smack-dab in the middle of a neighborhood of $300,000 homes.

TIP

The difference between your purchase price and the value of the surrounding homes approximately defines the most you should consider spending on a rehab.

In the preceding example, you should spend no more than $75,000 to bring your home up to the prevailing standard set by the other houses. Of course, this is assuming that you can afford to spend that kind of money (see Chapter 2) and that you have the time and patience to coordinate the rehab work or do it yourself. As long as you improve the property wisely and stay within your budget, you'll probably get most or all the rehab money back when you sell the property.

REMEMBER

Use the principle of progression in conjunction with location, location, value. Buying one of the better less-expensive homes in a good neighborhood enhances your likelihood of property appreciation in the years ahead.

The principle of regression: Why not to buy the most expensive house on the block

You guessed it. The *principle of regression* is the economic opposite of the principle of progression.

If you buy the most expensive house on the block, the principle of regression punishes you when you sell. The lower value of all the other homes around you brings down your home's value.

WARNING

If an evil spirit whispers in your ear that you should buy the most expensive house on the block to flaunt your high status in life, go to an exorcist immediately. Don't succumb to the blandishments of the evil spirit unless the probability of losing money when you sell fills you with joy. Satisfy your ego — and make a wiser investment — by purchasing one of the less-expensive homes in a better neighborhood.

The most expensive house on the block is also the worst candidate for remodeling. Suppose that you buy a $300,000 home in a neighborhood of $200,000 houses. From an appraiser's perspective, the home already sticks out like a financial sore thumb. Spending another $50,000 to add a fancy new kitchen to what is already the most expensive house on the block further compounds your problem.

That new kitchen almost certainly won't increase your home's value to $350,000. No one can dispute the fact that you spent $50,000 on the kitchen if you have the receipts to prove your expenditures. But folks who buy $350,000 homes generally want to be surrounded by other homes worth as much as, or more than, the one they're buying.

Homes are like cups. When you fill a cup too full, it overflows. By the same token, when you make excessive improvements to your house (based on sale prices of comparable homes in the neighborhood), the money you spend on the rehab goes down the financial drain. This phenomenon is called *overimproving a property*.

Even if you buy the least expensive house in the neighborhood, you can overimprove it if you spend too much fixing it up. The best time to guard against overimproving your house is *before* you do the work.

REMEMBER

If you'll end up with the most expensive house on the block when you finish a project, don't do the project.

The principle of conformity: Why unusual is usually costly

The principles of progression and regression deal with economic conformity. If you want to maximize your chances for future appreciation of the home you buy, your home should also conform in size, age, condition, and style to the other homes in your neighborhood. That's the *principle of conformity*.

This principle doesn't mean that your home has to be an identical clone of every other house on the block. It should, however, stay within the prevailing standards of your neighborhood. For example:

>> **Size:** Your home shouldn't dwarf the other houses on the block, or vice versa. If your home is smaller than surrounding houses, use the principle of progression as a guide to bring it into size-conformity with the other houses and increase your home's value. If, conversely, you have a three-bedroom home in a neighborhood of two- and three-bedroom homes, adding a large fourth bedroom to your house would violate the principle of regression.

>> **Age:** You almost never see an older home in the midst of a tract of modern new homes. However, every now and then you find a brand-new home incongruously plunked in the midst of older homes. A modern home typically looks out of place in a neighborhood of gracious, older homes. Even if you get a terrific deal on the price, the modern home's lack of conformity with other homes on the block will probably come back to haunt you when you attempt to sell it.

>> **Condition:** The physical condition of your house has a tremendous impact on its value. Not surprisingly, your home loses value if it's a dilapidated dump compared with the rest of the houses on the block.

WARNING

Ironically, having your home in far nicer condition than other houses in the neighborhood isn't wise either. Even if your home conforms to all the other houses in size, age, and style, you still overimprove your home if the quality of materials, workmanship, and appliances in your home greatly exceeds the prevailing neighborhood quality standards.

>> **Style:** The architectural style of the house you buy isn't critical — as long as it conforms to the prevailing architectural style of other homes in the neighborhood. From an investment standpoint, for example, you don't want to buy the only Queen Anne Victorian in a block filled with Pennsylvania Dutch Colonial houses, or vice versa. Nor should you buy a three-story home when all the surrounding houses are one story high.

TIP

Your home doesn't have to be a bland, boring replica of every other house on the block. You can follow the principle of conformity and still express your individuality by the way you landscape, paint, and furnish your home. You know you've done well when people use words like "tasteful" and "exquisite" to describe your home. On the other hand, your decorating motif is a problem if folks refer to your house as "weird" or "eccentric."

Defining Home Sweet Home

What exactly is a home? When you come right down to it, home is an elusive concept. Everyone knows, for example, that home is where the heart is. That's fine and good if you're a romantic but not too helpful if you're a home buyer.

Up until now, we've loosely used the terms "home" and "house" to mean any place where you live or want to live. Under that definition, everything from a studio apartment in Manhattan to a grass hut on a Hawaiian beach qualifies as a home. Now, however, precision matters. We're about to focus on the specific types of property you're most likely to buy: detached homes, condominiums, and cooperative apartments. Each of these options offers homeowners distinct financial and personal advantages and disadvantages that you must understand to make a wise buying decision.

Detached residences

If you were raised in a big city, your mental image of home is probably an apartment in a multistory steel-and-concrete building, an attached brownstone, or some other type of row house. If, on the other hand, you grew up in a small town, when someone says "home" you most likely visualize a brick or wood-frame residence with a white picket fence, a garden, and a swingset in the yard.

TECHNICAL STUFF

To distinguish the kind of home you see in areas of abundantly cheap land from condos, co-ops, and other types of property that folks call home, the correct terminology for the white-picket-fence type property is *detached single-family dwelling.* The operative word is "detached," because such homes aren't attached to any of the surrounding properties. Now that you're properly dazzled by the depth and breadth of our knowledge, we'll just call these "homes" or "houses" like everyone else does.

Detached homes, like cars, come in two basic types: *new* and *used.*

New homes

If you're the type of person who'd never think of buying a used car because you like the new-car smell and don't like buying someone else's problems, you may feel the same way about new homes. They have some very appealing advantages:

>> **A properly constructed new home is built to satisfy today's buyers.** Choosing a new home produced by a reputable builder of high-quality properties gives you the peace of mind of knowing that your home doesn't contain asbestos, lead-based paints, formaldehyde, or other hazardous or toxic substances. Furthermore, you can rest assured that your new home complies with current (and more stringent) federal, state, and local building, fire, safety, and environmental codes. Of course, you have no guarantee that future years won't uncover more hazards!

>> **A properly constructed new home should be cheaper than a used home to operate and maintain.** Operating expenses are minimal because a new home should incorporate the latest technology in energy-efficient heating and cooling systems, modern plumbing and electrical service, energy-efficient appliances, and proper insulation levels. And with a quality new home, your initial maintenance expenses are practically nonexistent because everything is new — roof, appliances, interior and exterior paint, carpets, and so on. Other than changing the light bulbs, what's to fix?

>> **A properly designed new home won't force you to adjust your lifestyle to its limitations.** On the contrary, new homes have enough wall and floor outlets to accommodate all your high-tech goodies — microwave oven; espresso machine; satellite TV and cable outlets; hair dryers; electric razors; electric toothbrushes; and home-office gear such as computers, monitors, printers, broadband internet connections, and so on. No unsightly, hazardous tangle of extension cords for you.

INVESTIGATE

New homes are only as good as the developers who build them. Visit several of the developer's older projects. See with your own eyes how well the developments have weathered over the years. Ask homeowners in older developments whether they'd buy another new home from the same developer. See what kinds of problems, if any, they've had with their homes over the years. Inquire whether the builder closed the sale on time and honored all contractual commitments, including the completion of any unfinished construction work, on time. Also find out whether the developer amicably fixed defects that occurred or whether homeowners had to take legal action to get problems corrected. Ask real estate agents how much homes in the developments have appreciated in value over time and how that compares with other homes in the general area.

As you may expect, new homes also have some disadvantages. To wit:

>> **What you see usually isn't what you get.** You see a professionally decorated, exquisitely furnished, beautifully landscaped model home. You buy a bare-bones, unfinished house where nearly everything — appliances, carpets, window coverings, painting, fireplace finishes, landscaping, and so on — is an extra that isn't included in the base price. Developers often spend tens of thousands of dollars lavishly decorating model homes. Unwary new home buyers can spend small fortunes trying to duplicate the look of model homes. When touring a model home, ask the salesperson to explain exactly what is and isn't included in the no-frills base price.

>> **Prices are less negotiable.** Developers maintain price integrity to protect the value of their unsold inventory of homes and to sustain appraised values for loan purposes. In fact, a developer who cuts prices is warning you that the project is floundering. Rather than reduce their asking prices, developers bargain with you by throwing in free extras or giving you upgrades (for instance, more expensive grades of carpet, better appliances, or granite kitchen counters rather than Formica) in lieu of a price reduction.

WARNING

Some developers attract buyers by pricing bare-bones houses very close to their actual cost and then make substantial profits on extras and upgrades. If, upon doing some comparison shopping, you find that these items are outrageously overpriced, don't purchase them from the developer. Instead, buy the bare-bones house and purchase extras from outside suppliers.

>> **On a price-per-square-foot basis, new homes are usually more expensive than used ones.** No surprise. Land, labor, and material costs are higher today than they were years ago, when the used homes were built. And don't forget that you're buying a home without any wear and tear.

>> **New homes in more developed areas are generally built in spots previously considered undesirable or unbuildable.** It's the old "first-come, first-served" principle. Earlier developments got better sites. Today's developers take whatever land is available — steep hillsides, flood plains, and land located far away from the central business area. Ten or twenty years from now, today's so-called lousy sites will be considered prime areas for new building — it's all relative.

>> **New homes may have hidden operating costs.** Developments with extensive amenities usually charge the homeowners dues to cover operating and maintenance expenses of common areas such as swimming pools, tennis courts, exercise facilities, clubhouses, and the like. Some homeowners associations charge each owner the same annual fee. Others prorate dues based on the home's size or purchase price — the larger or more expensive your home, the higher your dues. If the development has a homeowners

association, find out how its dues are structured and what your dues would be. Also find out what rules (called *covenants*) govern what you can do with your home as part of the development. Some covenants limit the colors you can use when painting the house, what additions you can make to the property, whether you can rent the property, and so on. Although meant to maintain high property values, some of these rules can create problems later as you seek to adapt your property to your changing lifestyle. For more detailed information about the important documents associated with homeowners associations and covenants, be sure to see the section "Attached residences" later in this chapter.

WARNING

Sometimes, homeowners-association dues are set artificially low to camouflage the true cost of living in the development. When that happens, sooner or later homeowners get slugged with a special assessment to repaint the clubhouse, resurface the tennis courts, or whatever. Make sure that the homeowners association in the neighborhood you're considering has adequate reserves and that its dues accurately reflect actual operating and maintenance costs. Also check to see whether the historic rate of increase in dues has been reasonable and is in line with the current overall inflation rate, which you can probably determine by asking your lender.

>> **You may have to use the developer's real estate agent to represent you.** Developers always have their own sales staff and their own purchase contracts. Some developers, however, let you be represented by an outside real estate agent, which is called *broker cooperation*. Others insist that you use their agent. This isn't a negotiable item. If you don't like it, your option is to walk away without buying a home.

If you've fallen in love with a new home but the developer won't cooperate with outside agents, we recommend that you pay for an independent appraisal to get an unbiased opinion of the home's value. You also want to have a real estate lawyer of your own choosing review your contract. (See Chapter 9 for how to find one and what he can do for you.)

REMEMBER

Just because a home is brand-spanking new doesn't mean that it's flawless. People build homes. People are human. To err is human — that's why you hear the expression *human error*. Moreover, builders work for profit and may be tempted to cut corners to maximize their short-term profits, not to mention that some builders simply aren't very good. Also keep in mind that a completed new home is only as good as the sum of its parts. If construction took place during the rainy season, damp wood may develop mildew or even mold before being nailed into place and covered with sheetrock. Thus, even a brand-new, never-been-lived-in home should be *thoroughly* inspected from foundation to roof by a professional property inspector to discover possible human errors before you purchase it. We cover property inspections in Chapter 13.

Used homes

Perhaps you're wondering why we classify all homes as being either new or used. Why not "new and old" rather than "new and used"? Because *old* isn't a precise term. How old is old? Is a home built more than 25 years ago old? Or should the cutoff be homes constructed over 50 years ago? If homes built more than 50 years ago are old, what should we call homes built 100 or 200 years ago — decrepit? *Used*, on the other hand, merely means that someone owned the home before you did. (Considering how expensive homes are, you may prefer to call the place you purchase a "previously owned" home. If that makes you feel better, go right ahead.)

Regardless of what you choose to call them, used homes have many commendable features:

TIP

WARNING

>> **Used homes are generally less expensive than new homes.** As a rule, folks who bought houses years ago paid less for their homes than developers charge to build comparable new homes today. Furthermore, at any given time, more used homes are on the market than new homes. Good old competition holds down the price of used homes.

>> **Asking prices of used homes are generally much more negotiable than asking prices of new homes.** Sellers of used homes don't have to protect the property values of an entire development. They typically just want to get their money and move on to life's next great adventure.

>> **Used homes are usually located in well-established, proven neighborhoods.** With a used home, you don't have to wonder what the neighborhood will be like in a few years when it's fully developed. Just look around, and you can see exactly what kind of schools, transportation, shopping, entertainment, and other amenities you have.

>> **Used homes have been field tested.** By the time you buy a used home, its previous owners have usually discovered and corrected most of the problems that developed over time due to settling, structural defects, and construction flaws. You won't have to guess how well the home will age over the years — you can see it with your own eyes.

No matter how well a home ages, you should still have it thoroughly inspected (inside and out) by qualified professionals before you buy it. The last owners may not have had the time, desire, or money to fix problems. They may also not have been aware of hidden problems. Be sure that the home meets today's building codes; doesn't have environmental, health, or safety hazards; is well insulated; and so on. Never try to save money on home inspections just because the house looks fine to you. The only exception to this stern admonishment is if you happen to be a professional property inspector. (See Chapter 13 for more about home inspections and inspectors.)

>> **Used homes are "done" properties.** When you buy a used home, you generally don't have to go through the hassle and expense of buying and installing carpets, window coverings, and light fixtures; finishing off the fireplace; planting a lawn; landscaping the grounds; building fences and patios; installing sprinkler systems; and the like. The work is already done (unless the used home is a major rehab project), and everything is generally included in the purchase price.

>> **Buying a used home may be the only way to get the architectural style, craftsmanship, or construction materials you want.** What if your heart is set on owning an authentic 1800s New England farmhouse or a Queen Anne Victorian? Perhaps you want plaster walls, parquet floors, stained-glass windows, or some other kind of materials or craftsmanship that is unaffordable, if not impossible to find, in new homes. If that's the case, buy a used home.

Like new homes, used homes have some disadvantages:

>> **Used homes are generally more expensive than new homes to operate and maintain.** Some used homes have been retrofitted with energy-efficient heating and cooling systems. Even so, a used home with 12-foot-high ceilings will always be more expensive to heat than a new home with 9-foot-high ceilings. By the same token, the older a used home's roof, gutters, plumbing system, furnace, water heater, appliances, and so on, the sooner you'll need to repair or replace them.

TIP

Before buying a used home, ask the seller for copies of the past two years' utility bills (gas, electric, water, and sewer) so you can see for yourself exactly how much it costs to operate the house. If the utility bills are horrendous, ask your property inspector about the cost of making the house more energy efficient.

>> **Used homes generally have some degree of functional obsolescence.** Examples of functional obsolescence due to outdated floor plans or design features are things like the lack of a master bedroom, one bathroom in a three-bedroom house, no garage, inadequate electrical service, and no central heating or air conditioning. How much functional obsolescence is too much? That depends on you. What we think is charming, you may consider an uninhabitable disaster. We deal with extreme functional obsolescence in the fixer-upper section later in this chapter.

>> **Wonderful used homes are sometimes located in less-than-wonderful neighborhoods.** You may be attracted to an utterly charming older home in a lousy neighborhood. Despite how much you think you'd love living in it, don't forget that you'll have to travel through the undesirable surrounding area every time you want to get in and out of your dream house.

WARNING

Even though you may be able to ignore gang wars and graffiti on every wall, will prospective buyers be equally tolerant when you're ready to sell? Remember: "location, location, value." No matter how stunning the property or how great the deal you're offered on it, don't buy someone else's problem.

Attached residences

If you can't accept the rules and regulations that would, of necessity, be imposed on you by communal living, don't read any further. You're much too free a spirit to be happy owning an attached residence.

But if you're willing to put up with the constraints of communal living to get the economic and lifestyle goodies associated with it, read on. You may be pleasantly surprised.

Condominiums

What type of property offers first-time buyers their most affordable housing option and gives empty-nesters who own detached homes an ideal lifestyle alternative for their golden years? If you said "condominiums," go to the head of the class.

Some folks think that a *condominium* is a type of building. They're wrong. The kind of building in which a condo is located doesn't matter. Condos can be apartments in a Chicago high-rise or split-level townhouses in Dallas or Victorian flats in San Francisco. What makes a condo a condo is the way its *ownership* is structured.

THE INVESTMENT VALUE OF DETACHED HOMES

Americans have always had a deep-seated love for detached homes. Like spawning salmon returning to the stream where they were born, many people are inexorably drawn to the same kind of house they grew up in when it's their turn to buy a home. Even if you didn't grow up in a detached home, you may covet one because TV shows and advertisements have drilled into your head that such homes are desirable and a sign of success.

Buyer demand for detached homes makes them good investments. Compared with attached residences, such as condominiums and cooperative apartments, detached homes tend to hold their value better in weak markets and appreciate more rapidly in strong markets. Ask a local real estate agent for a comparison of property-value appreciation in detached versus attached residences, and you can see what we mean.

First, a quick break for today's foreign-language lesson. In Latin, *con* means "with," and *dominium* means "ownership." Put the two words together, and you get *condominium,* which translates to "ownership with others." You'll definitely dazzle your pals "con" that etymology trivia tidbit.

Suppose you buy a condo in a Chicago high-rise. You have a mortgage, property taxes, and a fancy deed suitable for framing to prove that you own unit 603, one of 100 condos in that building. So far, owning a condominium is pretty much like owning a detached home that floats in the sky.

When you buy a detached home, an invisible line runs along the border of your property to separate what belongs to you from what belongs to your neighbors. When you purchase a condo, on the other hand, generally your property line is the interior surfaces (walls, floors, ceilings, windows, and doors) of your unit. In other words, with a condo, you get a deed to the air inside your unit and everything filling it — carpeting, window coverings, and all.

Air and interior improvements aren't all you own. You and the other condo owners in the condominium complex share ownership of the *land* on which the project is located and the high-rise *building* that contains your individual units. Thus, all of you own a portion of the roof, exterior building walls, and foundation — as well as a chunk of the garage, elevators, lobby, hallways, swimming pool, tennis courts, exercise facilities, and so on. All the parts of the complex beyond the individual units are known as *common areas* because you own them *in common* with all the other condo owners.

If you buy a condo, you automatically become a member of the project's homeowners association. You don't have to attend the meetings unless you want to, but you must pay homeowners-association dues. The dues cover common-area operating and maintenance expenses for everything from staff salaries, chlorinating the pool, lighting the lobby, and garbage collection to fire insurance for the building. In most places, a portion of your dues goes into a reserve fund to cover inevitable repairs and replacements, such as painting the building occasionally and replacing the roof.

INVESTIGATE

Before buying a condo, find out exactly what percentage of joint ownership you'd have in the entire condominium complex. That amount establishes how much you'll be assessed for monthly homeowners-association dues and what percentage you'll pay of a special assessment that may be imposed on owners to cover unforeseen common-area expenses. It also determines how many votes you'd have in earthshaking matters affecting the complex, such as whether to paint the building aqua or tangerine, whether to repair the existing treadmill in the health club or buy a new one, and so on.

Condominiums use several different methods to establish the ownership percentages.

>> The simplest method is to give each owner an equal share of ownership in the entire development. Thus, each owner has one vote and pays an equal amount of the monthly dues and any special assessments.

>> If the ownership percentage is based on the size or market value of the condo, people who own the larger or more expensive units have more say in what happens in the complex than do owners of the smaller or less-expensive condos. However, the heavy hitters also have accordingly higher monthly homeowners-association dues and pay a larger percentage of special assessments.

WHY A CONDO?

Given their complexity, why do folks buy condominiums? Why doesn't everyone stick to simple, straightforward detached homes? Here's why:

>> **Attached residences increase your buying power.** Compare the price of a two-bedroom condo with a two-bedroom detached single-family dwelling in the same neighborhood. On the basis of livable square footage, condos generally sell for at least 20 to 30 percent less than comparable detached homes. Owning your very own roof, foundation, and plot of land is much more expensive than sharing these costs with a bunch of other owners.

For some would-be buyers, the choice is either buying a condo that meets their living-space needs or continuing to rent. Economic necessity explains why the path to the American dream for nearly one out of five first-time real estate buyers is condominium ownership. There's buying power in numbers.

>> **Attached residences generally cost less to maintain than detached homes.** Suppose that you're one of 100 condo owners in a Chicago high-rise. Unlike the owner of a detached home, who has to pay the entire cost of maintenance expenses such as installing a new roof or getting an exterior paint job, you can split these maintenance expenses with the other 99 owners. Although replacing the high-rise's roof, for example, costs more in absolute terms than replacing the roof of a detached single-family home, the cost per owner should be less. There's economy in numbers.

>> **Attached residences have amenities that you can't otherwise afford.** How many people do you know who own detached single-family homes with tennis courts, swimming pools, and fancy exercise clubs? Most homeowners can't afford expensive goodies like these. But when the cost is shared among all the owners in a large condo complex, the impossible dream is suddenly your hedonistic reality. There's luxury in numbers.

CONDOS NEED TO BE INSPECTED, TOO

When you buy a condo, you must inspect the entire building — not just your unit. As Chapters 9 and 13 explain, you need a professional property inspector on your real estate team because the structural and mechanical condition of a property greatly affects its value. What's the condition of expensive common-area components such as the roof, heating and cooling systems, plumbing and electrical systems, elevators, foundation, and the like? Are amenities such as tennis courts, swimming pool, and health facilities in good shape? Because you're buying part of all the common areas in addition to your individual unit, you need a professional's opinion of the entire complex's condition.

Check the building's soundproofing by asking other owners whether they're bothered by noises emanating from units above, below, or beside their unit. The building has a ventilation problem if you can smell other people's cooking odors in your unit or the hallways. If you discover that expensive repairs or replacements are necessary and the condominium's reserve fund doesn't have anywhere near enough money to cover the anticipated costs, don't buy a unit in this complex. Sooner or later, the owners can expect a special assessment and/or a big dues increase.

>> **Attached residences are ideal homes for some empty-nesters.** As you near retirement, you may find yourself rattling around in a detached single-family home like a little ol' pea in a great big empty pod. Perhaps a two-bedroom condo in a building with no maintenance hassles and a doorman who'll forward your mail while you're off on one of your frequent vacations would solve all your problems. There's lifestyle in numbers.

CONDO DRAWBACKS

Like detached homes, condos aren't for everyone. Judge for yourself how much the following drawbacks may affect you:

>> **Condominiums offer less privacy.** Shared walls mean you can hear others more easily. Noise pollution is one of the biggest problems with condos and the one area that prospective condo buyers frequently overlook. Visit the unit at different times of the day and different days of the week to listen for noise. Talk to owners of condominiums in the complex to see whether they're bothered by noise pollution. If possible, spend a few hours or an evening in a unit. Be sure to turn off the easy-listening music that real estate agents may have playing during your tour of the unit.

TIP

As a rule, the fewer common walls you share with neighbors, the more privacy you have in your unit. That's one reason corner units sell for a premium. And if your unit is on the top floor, you won't have people walking on your ceiling (unless there's a roof deck, of course). The ultimate in condo privacy, if you can afford it, is a top-floor corner unit.

» **Condominiums are legally complex.** Prior to buying your condo, you should receive copies of three extremely important documents: a Master Deed or Declaration of Covenants, Conditions, and Restrictions (CC&Rs); the homeowners-association bylaws; and the homeowners-association budget. (See the nearby sidebar, "Condominium documents," for more details.) Read these documents from cover to cover.

TECHNICAL STUFF

CONDOMINIUM DOCUMENTS

If you're the type of person who only wants to know what time it is, skip this. If, on the other hand, you're fascinated by how watches are made, you'll love this sidebar. It explains how condominiums are created and operated.

A condo project is born when the project developer records a map and a condominium plan along with a Declaration of Covenants, Conditions, and Restrictions (CC&Rs) in the county recorder's office, which officially makes this information a matter of public record for all the world to see. CC&Rs establish the condominium by creating a home-owners association, stipulating how the condominiums' maintenance and repairs will be handled, and regulating what can and can't be done to individual units and the condo-miniums' common areas. A similar procedure is sometimes used with new develop-ments for detached housing.

Bylaws keep the condominium functioning smoothly. They describe in minute detail the homeowners association's powers, duties, and operation. The bylaws also cover such nitty-gritty items as how the homeowners-association officers are elected and grant the association the right to levy assessments on individual condo owners.

Last, but far from least, the developer creates a budget. Unlike the government, the condominium's budget can't (theoretically, at least) operate in the red. The current bud-get establishes how much the condominium expects to spend this year to operate and maintain itself. Condo owners also receive an annual statement of income and expenses showing precisely how last year's dues were spent and spelling out the condo-minium's current financial condition.

WARNING

The CC&Rs, bylaws, and budget are legally binding on all condo owners. Even though they're bulky, bloated, and boring, you must read them very, very, very carefully. If you have questions about what these documents mean, or if you don't understand how they affect you, consult a real estate lawyer. And as long as we're talking about legal stuff, find out from your agent or the homeowners association whether the condominium is either currently involved in litigation or plans to be in the foreseeable future. Lawsuits are expensive.

>> **Condominiums are financially complex.** As a prospective owner, check the current operating budget. Be sure that it realistically covers building maintenance costs, staff salaries, utilities, garbage collection, insurance premiums, and other normal operating expenses. If the budget is too low, prepare to get slugged with a massive dues increase sooner or later. By the same token, make sure that the budget includes adequate reserve funds to provide for predictable major expenses such as occasional exterior paint jobs and new roofs. How much is adequate? Three to five percent of the condominium's gross operating budget is generally considered a minimally acceptable reserve. If the reserve fund is too low, you're in danger of getting a special assessment in the event of a financial emergency.

INVESTIGATE

We recommend that you review the past several years' operating budgets and financial statements for indicators of poor fiscal management. Here are some red flags to look out for:

- **Frequent, large homeowners-association dues increases.** Dues shouldn't be increasing annually much faster than the current rate of inflation.

- **Special assessments that wouldn't have been necessary if the association had an adequate reserve fund.** When discussing the budget and reserve fund, find out whether any dues increases or special assessments are anticipated in the near future to make up operating deficits or cover the cost of a major project.

WARNING

- **Too many homeowners who are delinquent in paying their dues.** Operating expenses continue unabated regardless of whether all the owners pay their dues. You can bet that homeowners in foreclosure or mired in a short sale probably aren't paying their dues. Too many distressed properties in a condo complex are the fiscal equivalent of a field of red flags flying.

>> **Some condominium rules are overly restrictive.** People who live in proximity to one another need a smattering of rules to maintain order and keep life blissful. Too many rules, however, can turn your condo into a prison. For example, the condominium may have rules specifying what kind of floor and window coverings you must have in your unit, rules regulating the type or number of pets you can have in your unit, rules limiting your ability to rent

your unit to someone else, rules forbidding you to make any alterations or improvements to your unit, rules limiting when or how often you can entertain in your unit, and so on. Before you buy, read the CC&Rs and bylaws carefully to find out exactly what kind of usage restrictions they contain. Some of these same restrictions can apply in new detached housing developments, as discussed previously.

If you discover that the condominium (or the new housing development) has restrictions you don't like, don't buy the unit. Trying to modify CC&Rs or bylaws to eliminate restrictions after you've bought a unit is usually an expensive exercise in frustration and futility. We know you have far better things to do with your life than waste a big chunk of it haggling with condominium associations and their lawyers.

WARNING

Prudent rental restrictions are good. Ideally, all units in the complex will be owner-occupied. If some owners *occasionally* let friends use their units or rent the units for a week or two while they're on vacation, no big deal. However, if most of the units are owned by absentee investors who rent them to an endless parade of partying strangers, that's bad if you happen to have difficulty sleeping with loud music blaring late at night. You may also have trouble getting a mortgage in a complex with too many renters.

>> **Brand-new condominium developments have the same advantages and disadvantages as new detached homes, compounded by a condo's added legal and economic complexity.** If you haven't yet read the section about new detached homes earlier in this chapter, now's the time to do so. All our cautionary statements about new detached homes also apply to new condominiums. Like new detached homes, new condo projects are as good or bad as the developers who build them and the lawyers who create them. Because any new project, by definition, doesn't have a track record yet, you must visit earlier projects done by the same developer to see how well they've aged and how satisfied the condo owners are.

WARNING

Some unscrupulous developers of new condominium projects purposely lowball monthly operating costs to deceive prospective purchasers into thinking that living there costs less than it really does. These developers pay a portion of the monthly expenses out of their own pockets to keep project costs artificially low. The economic ax falls when the developer turns the project over to the homeowners association, which is soon forced to jack up the dues to cover actual operating expenses. When projected operating costs look too good to be true, they probably are. Compare the new project's projected operating expenses with the actual operating expenses of a comparable established project.

>> **Where condominium parking and storage are concerned, the obvious isn't.** For example, does your condo deed include a deeded garage or parking space that only you can use, or is parking on a first-come, first-served basis?

Do parking privileges cost extra, or is parking part of the monthly dues? What about provisions for guest parking or a parking area for boats or trailers? Do you have a deeded storage area located outside your unit? If so, where is it? If you need even more storage, is any available, and how much does it cost? You're much better off getting answers to these questions before, rather than after, you buy.

» **Some older buildings that have been converted from apartments into condominiums have functional obsolescence problems.** Older buildings frequently have excellent detailing and craftsmanship. However, they also often have outdated heating and cooling systems, and may lack elevators, which are mighty handy if, for instance, you're carrying groceries or suitcases up several flights of stairs. If you're buying a condo in an older building, find out whether utilities are individually metered or lumped into the monthly homeowners-association dues. Does your unit have a thermostat to control its heating and air conditioning, or is the heating and cooling system centrally controlled?

WARNING

If utilities are part of the monthly dues, other condo owners have no incentive to economize by moderating their use of heat or air conditioning. If you're frugal, you just end up subsidizing owners who aren't. By the same token, in a building with central heating and cooling, your climate choices may be limited to roasting in the winter and freezing in the summer. Even if you can live with utility overcharges and personal discomfort, these factors may deter future buyers from purchasing when you try to sell your unit.

» **Size can be a problem.** Large condo complexes usually have a cold, impersonal, hotel-like feeling. And as a rule, people who live in large complexes tend not to pay much attention to finances and day-to-day operating details because the homeowners association hires professional property managers to run things for the owners. There are, however, a couple of offsetting advantages to owning a condo in a large complex. If, for instance, several owners in a 100-unit complex fail to pay their monthly dues, it's not the end of the world financially. What's more, socially speaking, the odds of regularly running into an owner you detest diminish as the complex increases in size.

TIP

Don't buy into a small condominium complex unless you enjoy intimate relations with your neighbors. Carefully size up the other owners. Be sure that they're the kind of folks you can trust to carry their fair share of the load financially and operationally. In a small condo, you actively participate in the homeowners association because you must. Every vote has an immediate impact on your finances and the quality of your life. You don't have to love the other owners. BUT (note the big "but") if some or all of them are the type of people you'll be unable to get along with, don't buy the unit.

After reading the disadvantages of condo ownership, you may think that only a fool would buy a condo. Not true. We know plenty of content condo owners who'd never consider buying a detached dwelling. In our attempt to protect you, we sometimes go a little overboard on the cautionary side of things. We do so with your best interests at heart.

TIP

Condominiums make the most sense for folks who don't want operating and maintenance hassles (remembering that you'll still have the *expense*), want to maximize their bang for the buck spent on living space, and don't need a private yard. Buying a condo for a few years while you save enough money to purchase a detached home usually doesn't make economic sense. Given the expenses of buying and selling a condo, combined with its probable lack of decent appreciation, you're probably better off waiting to buy a detached home if you think you can do so within five years.

Cooperative apartments

The two most common types of attached residences are condominiums and cooperative apartments, which are usually called co-ops. You can't tell which is which by looking at the building or individual units. Like condominiums, what makes a co-op a co-op is its legal status.

You'll be delighted to know that most of the pros and cons of condominium ownership also apply to co-ops, so you don't have to read a ton of new stuff. (If you haven't read the previous section on condos, do so now.) In the following sections, we focus on the three ways in which condos and co-ops differ: the definition of legal ownership, management, and your financing options.

DEFINITION OF LEGAL OWNERSHIP: DEED VERSUS STOCK

When you buy a condo, you get a deed to your unit. When you buy a co-op, you get a stock certificate (to prove that you own a certain number of shares of stock in the cooperative corporation) and a *proprietary lease,* which entitles you to occupy the apartment you bought. The corporation owns the building and has the deed in its name as, for example, the 10 West Eighty-Sixth Street Corporation. Thus, you're simultaneously a co-owner of the building (via your stock ownership) and a tenant in the building you co-own.

WARNING

In most cooperatives, shares are allocated based on how big a unit is and what floor it's on. Thus, a top-floor apartment usually has more shares than a ground-floor unit of the same size. The more shares you have, the greater your influence in the co-op because each share gives you one vote. Unfortunately, power has a price. Your proportionate share of the cooperative's total maintenance expenses is based on the number of shares you own in the corporation. If you own a great many shares, your monthly expenses will be disproportionately high. And when you're ready to sell, your unusually high monthly expenses may reduce your unit's value.

MANAGEMENT: HOMEOWNERS ASSOCIATION VERSUS BOARD OF DIRECTORS

If you've always fantasized about being the chairman of the board, here's your chance: Buy a co-op apartment, and work your way up the corporate ladder. Because your unit is in a building owned by a corporation, it's governed by a board of directors elected by you and the other owners. Nomenclature aside, just like the homeowners association in a condominium, the board of directors is responsible for the cooperative's day-to-day operations and finances.

BUYING AND SELLING CO-OPS IS OFTEN CHALLENGING

Buying and selling co-ops is usually a lot more difficult than buying and selling condos. Most cooperatives stipulate that individual owners can't sell or otherwise transfer their stock or proprietary leases without the express consent of either the board of directors or a majority of owners.

Prospective buyers generally must provide several letters of reference regarding their sterling character and Rock of Gibraltar creditworthiness. In addition, they may have to submit to a personal grilling by the board of directors. Given that the owners live in close proximity to one another and depend on one another financially, having the ability to screen out party animals, deadbeats, and the like is reasonable as long as that power isn't misused to unfairly discriminate against buyers.

Even so, some buyers find the approval process extremely intrusive and strenuously object to giving strangers their financial statements. The approval process also tends to slow the sale of co-op units on the market.

Owning a co-op is a two-edged sword. As a co-op owner, you have much more control over who your neighbors will (or won't) be than do condo owners. Unfortunately, that control cuts both ways. When you try to sell your unit, people you consider perfect buyers may be turned down by the co-op because your neighbors think that the prospective buyers would entertain too much or can't carry the load financially. Giving up the right to sell your co-op to the highest bidder may be too high a price to pay for the right to choose your neighbors.

FINANCING YOUR PURCHASE

Securing a mortgage to purchase your co-op may be difficult. Many lenders flat-out refuse to accept shares of stock in a cooperative corporation as security for a mortgage. Conversely, some co-ops absolutely won't permit any individual financing over and above the mortgage the corporation has on the building as a whole. These co-ops believe that one proof of creditworthiness is your ability to pay all cash for your unit.

WARNING

Unless you're richer than Midas, don't buy a co-op if only one or two lenders in your area make cooperative-apartment loans. Odds are you'll pay a higher interest rate because of the lack of lender competition and lender concerns about the greater risks of co-ops. Worse yet, what if these lenders stop making co-op mortgages and no other lenders take their place? You won't be able to sell your unit until you find an all-cash buyer (and they're few and far between) or until you have the financial resources to lend the money yourself to the next buyer.

Finding a Great Deal

If you're like most people, you're cursed with champagne taste and a beer budget. The homes you hunger for cost far more than you can afford. To buy one of these dream homes, you'd either have to get a really, really good deal or win the lottery.

Good deals *are* out there. The trick is knowing where to find them and how to evaluate them. Don't waste time looking at perfect houses if you're searching for a deal. People pay premium prices for perfection. The houses you find great deals on are imperfect properties — houses with either physical or financial problems. The deal you're offered is an inducement to tackle the problem. Whether the deal is ultimately better for you or for the seller is the question.

In Chapter 4, we discuss strategies for getting a good buy in any type of housing market. In the rest of this chapter, we cover special property situations that may be good deals or pigs in a poke.

Finding a fixer-upper

Fixer-uppers are run-down houses with physical problems. Real estate agents generally refer to fixer-uppers euphemistically as "needing work," "having great potential," or being a "handyman's special."

Fixer-uppers aren't very popular in sluggish real estate markets. Most buyers in such markets don't want to put up with the hassle or financial uncertainties associated with doing a major rehab. They prefer to buy houses in move-in condition. Such a house is a safe but passive investment. Because its potential has already been fully realized, the new owner can't do anything to significantly increase its value.

A fixer-upper, on the other hand, offers potentially larger rewards to folks who have the vision to see beyond the mess that is to the wonderful home that can be. A fixer-upper buyer must also have the financial resources and courage to tackle the risks. If you fit that profile, here's what you may be able to look forward to after you've transformed your ugly duckling into a swan:

>> You'll be living in a nicer home and a better neighborhood than you'd otherwise have been able to afford.

>> Instead of buying a home decorated in someone else's idea of good taste, your home will be done the way you like it.

>> You may have increased your home's fair-market value in excess of your out-of-pocket expenses for improvements you made.

For example, if you're handy, you can add thousands of dollars of value to a fixer-upper by doing labor-intensive jobs such as painting, wallpapering, and landscaping yourself. Sweat equity can pay big dividends.

WARNING

If, like us, you're mechanically challenged, forget sweat equity. It's less frustrating and cheaper in the long run to earn money doing what you do best and then using some of that money to hire competent contractors to do what *they* do best. Poor workmanship is a false economy; it looks awful and reduces property values. Doing the project well the first time is easier, faster, and ultimately less expensive than doing it badly yourself and then paying someone else to fix your mess. If you're one of those rare people who can do quality work yourself, by all means try your hand at it — just be realistic about the required time and costs.

Some fixer-uppers are easy to spot. They look like classic haunted houses — peeling paint, shutters falling off, overgrown yard, and so on. Things don't get any better on the inside. These houses may need everything from a good cleaning to electrical system and plumbing overhauls.

Other fixer-uppers, however, are much more subtle. Some older houses, condos, and co-ops, for example, may look fine at first glance but have functional obsolescence. They're livable, but they need improvements, such as adding master bedrooms, bathrooms, or garages, and upgrading their electrical systems to bring them up to today's more rigorous housing standards.

STRUCTURAL REPAIRS VERSUS RENOVATIONS

TIP

Work done on fixer-uppers falls into two broad categories: structural repairs and renovations.

- *Structural repairs* are changes you make to a property to bring it up to local health and safety standards. Such work can include foundation repairs, roof replacements, new electrical and plumbing-system installations, and so on — things that cost big bucks but add relatively little value to property. Ideally, you can get a credit from the seller to do some, if not all, of the necessary structural repairs. The less you have to take out of your pocket for corrective work, the more you have to spend on renovations.

- *Renovations* increase a fixer-upper's value by modernizing the home. Remodeling an old kitchen, installing a second bathroom, and adding a garage are a few examples of major structural renovations that make your home more functional, more pleasant to live in, and more valuable when you sell it.

- *Cosmetic renovations* (painting, carpeting, landscaping, and the like) also add value with far less expense and aggravation. The ideal fixer-uppers to buy are ones that look awful but simply need cosmetic fixes to look their best.

Digging for a diamond among the dumps

Finding the right fixer–upper isn't a matter of luck. On the contrary, it takes persistence, skill, and plain hard work. You spend lots of time tromping through properties; invest more precious time evaluating promising fixer–uppers that ultimately don't make sense economically; and then, just when you're ready to give up, you finally discover a diamond in the rough that you end up buying.

Here's how to separate diamonds from dumps:

>> **Read this book.** Everything you need to know is here. Pay special attention to the topics covered in this chapter (good neighborhoods; principles of progression, regression, and conformity; and used homes and condos), Chapter 10 (accurately determining fair-market value so you don't overpay), and Chapter 13 (property inspections). Also, be sure that you can financially afford all the necessary expenditures after the purchase for the fix-up work (see Chapter 2).

>> **Inspect the heck out of the fixer-upper before you buy it.** Every property should be carefully inspected prior to purchase. Fixer-uppers need even more scrutiny so you know precisely what you're getting yourself into. Make your purchase offer conditional upon your approval of the property inspections and satisfactory resolution of corrective-work issues you discover. You can find these clauses in Chapter 12.

>> **Get contractors' bids for structural repairs and renovations.** You can use contractors' bids as a negotiating tool to get a corrective-work credit or lower sales price from the sellers for structural repairs such as termite-damage repairs and a new roof. You should also get cost estimates for renovations such as bathroom modernization, new kitchen appliances and cabinets, central heating, and anything else required to bring the property up to date.

WARNING

If the bids and cost estimates you receive indicate that you'd end up with the most expensive house on the block, don't do the project. Fix-up work has three iron laws:

- It's always more disruptive than you expected.

- It always takes longer to finish than you planned.

- It always costs more than you estimated.

So if estimated fix-up costs would make the property the most expensive house on the block, by the time the work is finally completed, the actual costs will make it the most expensive house in the state!

TIP

Getting a loan is usually difficult if the cost of anticipated corrective-work repairs exceeds 3 percent of the property value, which is always the case with major fixer-uppers. However, a good real estate agent should know which lenders in your area specialize in fixer-upper loans. Given that one such lender finds you creditworthy and your project feasible, that lender may give you a mortgage to buy the property *and* a construction loan to make the improvements.

Final thoughts on fixer-uppers

Feeling somewhat overwhelmed by the risks associated with fixer-uppers is normal. Now you understand why most home buyers avoid them — they fear being sucked into a bottomless bog that utterly disrupts their lives and totally devours their savings.

WARNING

Most novice home buyers, especially first-time buyers, woefully underestimate the time and cost required to fix up homes. When all is said and done, nearly all people find that it would have cost them the same or less to buy a more finished home and avoid the headaches of doing or coordinating the renovations. Some folks have ended up in financial ruin and even divorced over the stresses of such renovations.

REMEMBER

If you like challenges and are willing to do a ton of extra detective work, remember these tips to maximize your chances of succeeding with a fixer-upper:

» Buy in the best neighborhood you can afford.

» Buy one of the cheaper houses on the best block.

» Make sure that the renovations will more than pay for themselves in increased property value. There are online resources that can help with this, like the Cost vs. Value Report by Remodeling Magazine (check it out at `www.remodeling.hw.net/cost-vs-value/2020/`). Here you will find cost estimates for many remodeling projects as well as the value those projects are likely to retain at resale.

» Make sure that the purchase price is low enough to allow you to do the corrective work and renovations without turning your property into the most expensive house on the block.

If the real estate gods play fair and square, whoever buys the exquisitely *finished* home you transformed from a dump will pay a bonus for your farsightedness to see the fixer-upper's potential, for your audacity to tackle the financial risk, and for your stamina to put up with the chaos and filth of a rehab. If (and only if) you select wisely, negotiate the price wisely, and renovate wisely, you'll enjoy years of blissful living in the wonderful home you created — and ideally make a fine profit to boot when you sell it.

Taking over a foreclosure

To get a mortgage, you give the lender the right to take your home away from you and sell it to pay the balance due on the mortgage if you

» Don't make your loan payments

» Don't pay your property taxes

» Let your homeowners insurance policy lapse

» Do anything else that financially endangers your home

The legal action to repossess a home and sell it is called a *foreclosure*.

Every year, hundreds of thousands of homes end up in foreclosure. Foreclosures in the late 2000s hit a high not seen in decades. Most foreclosures result from people overextended on debt, including mortgages whose payments ratchet higher than the borrowers are prepared to handle. In other cases, however, people fall on hard times — they lose a job, experience unexpected healthcare costs, suffer a death in the family, or go through a divorce. Finally, some borrowers who have little invested choose to walk away from properties that have declined in value.

BACK-ON-THE-MARKET PROPERTIES

When a house listed for sale receives an acceptable offer, the sellers usually tell their agent not to actively market the property or solicit other offers while they work with the buyers to satisfy the contract's terms and conditions of sale, such as property inspections and financing. If such a property comes back on the market (*BOM,* in real estate lingo), it means that the deal fell apart.

Property comes back on the market for many reasons. Perhaps the buyers couldn't qualify for a loan or got cold feet. Maybe the lender didn't think the house was worth as much money as the buyers were willing to pay for it and wouldn't approve their request for a loan. The far and away most common reason deals fall through, however, is that the buyers and sellers couldn't agree on how to handle the corrective work discovered during the inspections.

Ironically, the castle that all the buyers coveted when it was the newest listing on the market may turn into a "that old thing" pumpkin when it returns. Suddenly, suspicious buyers wonder what's wrong with the house. Real or imagined, that stigma of being a problem property repels a lot of people. They don't want to buy a house that someone else rejected. As a result of reduced buyer enthusiasm, BOM homes often sell for a lower price the second time around.

Don't categorically reject a house that comes back on the market. Find out why it's BOM. If the problem is related to property defects, ask the sellers to show you copies of the inspection reports. Given that the problems are correctable and that you can negotiate a good deal, the sellers' misfortune may be your good fortune. Sellers who've had a deal implode are frequently more willing to realistically negotiate on price and terms with the next buyer. If you apply the principles we cover in the fixer-upper section of this chapter, you can turn a BOM into a great deal.

You may have heard stories about people who got good deals buying foreclosures far below the property's appraised value. And in fact, some people who buy foreclosed property luck out. But for every lucky winner, many more people don't profit or, worse, actually lose money buying foreclosures.

WARNING

Buyer beware — foreclosures are generally legal and financial cesspools. Unless you have an expert on your team who can guide you through the entire foreclosure process from beginning to end, don't even think about buying a foreclosure at an auction.

If you buy a foreclosed home, you most likely also buy the previous owner's problems. Here's a list of risks to ponder:

>> **Physical:** Some homeowners react to the emotional devastation of a foreclosure with a scorched-earth attitude of "if we can't have it, we'll make darn sure that nobody else wants it." Before leaving, they take appliances, light fixtures, cabinets, sinks, toilets, and anything else of value. In extreme cases, they break windows, pour concrete down kitchen and bathroom drains, rip wiring out of walls, uproot shrubs, cut down trees, and do anything else they can think of to trash the property. What if you're the high bidder for a sabotaged house at an auction of foreclosed properties? Lucky you.

WARNING

Lenders usually won't let you inspect foreclosed properties prior to their auction. Nor can you make your offer to purchase subject to getting a loan. Lenders don't guarantee clear title to these properties, nor can you get title insurance to protect against undisclosed or undiscovered flaws in the chain of title or liens against the properties (see Chapter 13 for more on title insurance). The risk of buying a property at a foreclosure auction greatly exceeds the possible reward.

>> **Financial:** Depending on which state the house is located in, a foreclosure can take anywhere from four months to over a year to complete. Suppose that you get what appears to be a good deal from people who are actually selling partway through the foreclosure process to avoid the stigma of foreclosure. What if these people lie about how much they owe on their mortgage and property taxes? What if they don't tell you about unpaid homeowners-association fees, unrecorded mortgages, court judgments, or federal and state tax liens (outstanding tax bills) hanging over the house? One guess who's liable for debts secured by the property. Lucky you.

>> **Possession:** Suppose that after buying a foreclosure at an auction, you visit your new home and discover that the previous owners are still living in it with their last remaining possession — a shotgun. They have no intention of leaving peacefully. Who do you think will have the pleasure of evicting them? Lucky you.

TIP

Given possible sabotage by the previous owner, buying a foreclosure is never *entirely* safe. The least risky way to purchase a foreclosure is to buy a real estate owned (REO) property directly from a lender, government loan insurer, or other government agency that holds the title to the property because no one bought it at the foreclosure auction. Here's why:

>> Any recorded or undisclosed mortgages, court judgments, or tax liens on the house are either removed from the property or at least revealed to you prior to your purchase.

>> You can — absolutely must, in fact — have the house minutely scrutinized by professional property inspectors. Where foreclosures are concerned, you have to find out whether the previous owner left any hidden surprises for you. If we haven't scared you off yet, see Chapter 13 for the unique aspects of inspecting foreclosed property.

>> The price and terms of sale are negotiable. Even though foreclosures are normally listed at their appraised value, lenders may make allowances for corrective work by either reducing the price or giving you a credit to do the work. They'll also, as a rule, offer attractive loan terms (low cash down payments, no loan fees, and below-market interest rates) to get rid of these blighted properties quickly. After all, they're in the loan business — not property management.

WARNING

Think long and hard before buying a foreclosure. Even if you purchase the REO property directly from a lender, loan insurer, or government agency, you may be buying a house permeated from foundation to roof by shattered dreams. Such a house probably hasn't been given the best of care. Do your homework carefully, have the property thoroughly inspected, and understand *fully* what you're getting yourself into before you buy.

Seeking a short sale

Compared to acrimonious foreclosures where in the worst cases spiteful owners strip everything of any value from their house before maliciously trashing it, short sales are downright cordial. Short-sale owners may be *underwater* (owe more on the loan than their house is worth), but they're doing whatever they can to limit further damage to their credit rating by cooperating with the lender to maximize the property's sale price.

To that end, short-sale owners live in their house and continue to maintain it so it shows well. Depending on the severity of their financial problems, the owners may even make token loan payments. The lender, in turn, agrees to eat the loss by accepting the sale proceeds as full satisfaction of the debt. By working together, both parties avoid a far more odious foreclosure.

WARNING

Here are critical issues you should consider before beginning the process of buying a short-sale property:

>> **Short sales aren't screaming bargains.** Lenders typically price short-sale properties based on the sale prices of nondistressed property. They want to get as close as possible to fair-market value to minimize the loss on their mortgage. You may get a purchase price 5 to 10 percent below market

because short sales are considerably more time consuming, risky, and complex than a nondistressed property sale. However, don't count on getting an extraordinary deal.

>> **Short sales aren't short.** If you must complete your purchase quickly, forget doing a short sale. Even with knowledgeable, cooperative folks on both ends of the deal, short sales usually take at least 30 days for approval plus at least another 30 days to close the sale. If there's more than one lender and/or loan insurer involved, the process takes even longer. Worst of all, your offer remains contingent during that lengthy lender approval process. If the lenders receive a better offer before your offer has been approved, they'll drop yours like a hot potato. It doesn't matter that you made the first offer. It's not fair, but that's how the short-sale game is played.

>> **You can't do a short sale if you're related to or have a financial arrangement with the seller.** Buyer, seller, and agents must sign a form ensuring there are no pre-existing relationships or financial agreements. Lenders want to prevent secret deals where a buyer and seller agree to sell the house below its fair-market price and then kick back money to the seller under the table after close of escrow. This is collusion. This is fraud. You can go to jail. Enough said.

>> **Don't buy a short-sale property if you need financial help with closing costs or want a credit for repairs or corrective work.** By definition, short-sale sellers are short of cash, so don't expect financial help from them. And, whereas lenders infrequently offer to make minor repairs or throw a few bucks your way for closing costs, lenders aren't as willing or flexible in this area as traditional sellers.

>> **You can't make an offer contingent upon the sale of your home.** It's extremely unlikely that any lender will accept your offer if it hinges on selling your house to get the money to buy the short-sale property. This constraint limits the field of perspective short-sale purchasers to first-time buyers, folks who don't have to pull equity out of their present home to buy a property, and real estate investors.

>> **Short sales are sometimes "as is."** We recommend including an inspection contingency in your offer. Be advised that some lenders won't accept an offer subject to a property inspection. That means you must either spend hundreds of dollars having the property inspected before making your offer on the outside chance it will be accepted or forego an inspection. If you've read Chapter 13, you know we don't like buying a pig in a poke. If you haven't read Chapter 13, now's a fine time to do it.

>> **Don't make offers on multiple short-sale properties.** Unscrupulous get-rich-quick seminars recommend making offers on several short-sale properties at the same time. They encourage buyers to tie up multiple

properties, see which one works out best, and then cancel the other deals. Terrible idea! Real estate purchase agreements are binding contracts. They should never be entered into lightly. Such a practice is unfair to fiscally challenged short-sale sellers and can lead to severe financial penalties for capricious buyers.

If we haven't scared you off yet, here are questions you must have answered by the short-sale sellers or their agent prior to writing an offer:

INVESTIGATE

>> **How many loans are on the property and who are the lenders?** This tips you off regarding the short sale's complexity. The more lenders involved, the more potential problems. How quickly (or slowly) a bank bureaucracy digests the short sale and internal corporate negotiating strategies vary widely from lender to lender. Savvy real estate agents know which lenders are relatively easy to work with and, conversely, which ones are awful.

>> **Where are the sellers in the short-sale process?** Have the sellers just begun discussing a short sale with the lender? Has the lender notified the homeowners they've started the foreclosure process? If so, when? As noted in the foreclosure section, this sets a date when the property may be sold to the highest bidder or revert to the foreclosing lender. Is time your friend or your enemy? You must know.

>> **Who is negotiating the short sale with the lender — the seller, the seller's real estate agent, or a short-sale facilitator?** If a facilitator is involved, who pays for the facilitator's service? In some cases, the buyer is asked to pay that fee. This isn't illegal if properly disclosed in writing to all lenders involved. How much the facilitator charges and whether you want to pay the fee is another issue.

>> **Are there other liens on the property in addition to the mortgage?** As noted in the foreclosure section, there may be liens on the property for unpaid federal and state income taxes, unpaid property taxes, unpaid homeowners-association dues, mechanic's liens, and court judgments. You need to know what debts are on the property and who is responsible for paying them off prior to close of escrow.

TIP

Ah, the moment of truth. If a short sale seems feasible based on what you learned during your information-gathering phase, here are things you can do to improve the odds of having a successful transaction:

>> **Work with a real estate agent who understands the complexities of short sales.** As discussed in Chapter 9, you need an agent on your team with specialized training such as SFR (Short Sales and Foreclosures) or CDPE (Certified Distressed Property Expert). Ideally, your agent will also have

hands-on practical expertise gained by representing other buyers of short-sale properties.

>> **Get a letter from your lender specifying that you have been preapproved for a mortgage.** Not prequalified for a loan, *preapproved*. Short-sale lenders insist upon this. We cover loan preapprovals in Chapter 6.

>> **If possible, attach a short-sale addendum to your offer.** Standard purchase agreements don't have provisions for the unusual terms and conditions involved in a short sale. To correct that oversight, many states have addendums specifically written to deal with the nuances of short sales. We thoughtfully included a Short Sale Addendum and a Short Sale Information and Advisory form in Appendix B. Study them carefully to assist you in structuring your offer.

>> **Be realistic in your timing.** The short-sale lender will probably need 30 days or more to approve the contract. When it has been approved, you'll need at least another 30 days to close the sale.

>> **Don't be surprised if you get a counteroffer from the lender.** Remember — the lender isn't trying to dump the short-sale property. On the contrary, the lender is trying to obtain market value for it to limit its loss.

Pooling Your Resources: Ad Hoc Partnerships

As home prices have escalated in many densely populated parts of the country, so has the frequency of unrelated people forming ad hoc partnerships to buy houses. These couples aren't necessarily romantically involved. On the contrary, they're usually folks who decide that a good way of turning the American dream of owning a home into reality is to join forces as co-owners.

Types of residential co-ownership

One type of ad hoc co-ownership, known as *equity sharing*, involves outside investors who don't live in the property. The investor provides cash to buy a house, which the other co-owner lives in while the property (ideally) appreciates in value. After a specified period of time — say, five years — the co-owner who lives in the house has the option either to buy out the investor's share of the property or to sell the house and split the proceeds.

The other type of residential ad hoc co-ownership is one in which all co-owners live together in the jointly purchased property. By pooling their income and cash for a down payment with a significant other, friend, or relative, these individuals get a place to call home; a tax shelter; and, if the real estate gods are willing, a profit when they eventually sell and go their separate ways.

Live-in co-ownership arrangements are the real estate version of Siamese twins. People who live in ultraclose proximity to their partners every day have a personal relationship that is far more intense than in an equity-sharing co-ownership.

WARNING

We've seen live-in co-ownerships that turned out wonderfully. As time passed, the co-owners became even closer friends than they were before buying the house. Most, however, are no more than marriages of convenience that the parties suffer through solely to reap economic benefits. Bickering about things like whether to patch a leaky roof or get a new one, who left the sink full of dirty dishes (again!), whether to paint the living room purple or gold, and who gets the backyard for a party next Saturday can strain even the best of relationships.

THE CO-OWNERSHIP FROM HELL

Unfortunately, co-ownership of property occasionally can turn into unmitigated disasters. Irv's sad tale illustrates some of these pitfalls.

Irv and Sid, good pals for almost 25 years, bought a condo together. Irv used money inherited from his mother for the condo's down payment and closing costs. Sid, who had a much higher income than Irv but no cash, lived in the condo as his principal residence. Sid covered the monthly mortgage payments, property taxes, and homeowners-association dues, and also paid rent to Irv.

It was a perfect relationship. Sid got the tax deductions he needed, plus 25 percent of the appreciation when the condo sold. Sid's rent payments gave Irv a good return on the cash he'd invested, and he'd get the lion's share of the condo's appreciation when it sold. Irv and Sid were delighted with their arrangement.

All went well for nearly a year. Then, without warning, Sid filed bankruptcy.

Irv's rent payments stopped, of course. What's more, the bankruptcy court put a lien on the condo to tie up Sid's assets — and inadvertently tied up Irv's money as well. Worst of all, Irv and Sid didn't have a written co-ownership agreement describing their 75/25 equity split. Without something in writing, Irv was unable to prove this fact to the court's satisfaction.

Structuring a successful co-ownership

Too many co-ownership situations end up on the rocks unnecessarily. Why? The parties didn't anticipate problems related to co-ownership that arose. Ironically, most of these problems are foreseeable and avoidable. Proper planning prevents problems.

TIP

Before buying property with even your closest friends, you can do two things to greatly increase your odds for success:

>> **Give it a trial run.** If you're considering a live-in arrangement, we recommend that you live with your prospective co-owners for at least six months before buying a home together. You may discover that you have a major compatibility problem — for example, you may go to bed each night by nine, and your roomie may love to party into the wee hours of the morning. Ditto if you, Felix, insist on having everything in its place, and your partner-to-be, Oscar, uses the floor for a closet. Imagine the delights of co-ownership if you always pay your bills by the first of the month, and your partner's favorite sport is a spirited game of duck-the-bill-collector.

>> **Put it in writing.** We also recommend having a lawyer who handles residential real estate co-ownerships prepare a written "tenancy in common agreement" as soon as possible. Don't reinvent the wheel; let an experienced lawyer guide you and your prospective partner through all the foreseeable "what ifs" of owning property together.

Ideally, you should have the agreement drawn up well before you make an offer to purchase. Why the rush? To make sure you and your partner understand precisely how the co-ownership will operate and what your responsibilities to each other will be. The agreement should cover important issues such as the following:

>> **Financial arrangements:** This section of the agreement deals with the economics of buying, maintaining, and selling the property. It also specifies the tax deductibility of mortgage interest and property taxes for each partner if the partnership involves unequal financial contributions.

TIP

What happens if, for example, your co-owner suddenly dies or goes bankrupt? How you take title in the property is critical. (We cover this important issue in Chapter 14.) Should the co-owners give each other first right of refusal to buy the partner's share before it can be sold to an outsider? Planning for the unexpected sure beats reacting to a crisis. And last, your agreement must have equitable provisions for terminating the relationship.

>> **Dispute resolution:** What if you and your partner come to blows on a critically important issue like whether to plant daisies or roses along the side of the house? If only two co-owners are involved, how do you break tie votes? Anticipate disputes. Even the best of friends occasionally disagree — that's a fact of life. Provide a method (such as mediation or arbitration) to resolve the problems you can't work out between yourselves.

>> **Game plan:** If you and your partner intend to improve the property you purchase by, for example, remodeling the kitchen or converting a pair of flats to condominiums, your co-ownership agreement should be as specific as possible regarding the intended scope of work, project timing, cost, and so on. Plan now to prevent arguments later.

REMEMBER

Never rush into co-ownership — the economic consequences of a mistake may be devastating. Carefully weigh the pros and cons. Then get everything in writing with help from a lawyer experienced in covering all the "what if" situations just in case everything doesn't turn out as wonderfully as you hope.

Chapter 9

Assembling an All-Star Real Estate Team

Winston Churchill once characterized the former Soviet Union as "a riddle wrapped in a mystery inside an enigma." Churchill's apt description applies equally well to the home–buying process. If you're like most folks who are looking for a home, you're not an expert on property values, financing, or tax and real estate law. And when your life savings are on the line, ignorance isn't bliss. Not understanding what's involved in the process of buying a home can cost you big bucks and make you unhappy with the home that you ultimately purchase.

How can you find your way through the convoluted maze of constantly changing real estate market conditions, local laws, regulations, and tax codes? Where can you sign up for a crash course in home values? Even if you have the aptitude, how will you find the time to become an expert in so many fields?

You really can't become an expert in all aspects of buying a home (unless you're going to devote many hours and years to doing so), and you don't have to. In this chapter, we explain how to find competent experts who can help you buy a home and who won't charge you an arm and a leg for that help.

The Team Concept

Time and time again, we've seen smart people blunder into horrible situations while buying a home. More often than not, what got them into trouble was ignorance of something that they (or their advisors) should have known but didn't.

Strangely enough, knowing everything yourself isn't important. What is important is having good people on your team — people who know what you need to know so they can help you solve the problems that invariably arise.

Lining up the players

You don't have to become an instant expert in home values, mortgages, tax and real estate law, title insurance, escrows, pest-control work, and construction techniques to play the home-buying game well. You can choose to hire people who have mastered the skills you lack. Home buying is a team sport. Your job is to lead and coach the team, not play every position. After you've assembled a winning team, your players should give you solid advice so you can make brilliant decisions.

TIP

If cost were no object, you'd hire every competent expert you could get your hands on. But because you probably don't have an unlimited budget, you need to determine which experts are absolutely necessary and which tasks you can handle yourself. In this chapter, we explain which experts are generally worth hiring and which ones you can pass on. Ultimately, of course, you must determine how competent or challenged you feel with the various aspects of the home-buying process.

Here's a thumbnail sketch of the possible players on your team:

>> **You:** Always remember that you're the most important player on your team. In nearly every home purchase, something goes wrong — one of your players drops the ball or doesn't satisfy your needs. You have every right to politely, yet forcefully, insist that things be made right. Remember that you hire (and pay) the players on your team. They work for you. Bad players may see things the other way around — they want to believe (and want you to believe) that they're in charge. They may try to manipulate you to act in their interests rather than yours. Don't tolerate this. You're the boss — you can fire as well as hire.

>> **Real estate agent:** Because the home you're about to buy is probably the largest single investment you'll ever make, you must have someone on your team who knows property values. Your agent's primary mission is to help you find your dream home, tell you what the home is worth, and negotiate for it on your behalf.

- **Real estate broker:** Every state issues two kinds of real estate licenses: a salesperson's license and a broker's license. People with broker's licenses must satisfy much tougher educational and experience standards. If your real estate agent isn't an independent broker or the broker for a real estate office, he (or she) must be supervised by a broker who is responsible for everything that your agent does or fails to do. In a crisis, your transaction's success may depend upon backup support from your agent's broker.

- **Lender:** If you can't pay, you can't play. And because most folks can't pay all cash for their homes, you probably need a loan to buy your dream house. A good lender offers competitively priced loans and may even be able to help you select the best type of loan from the financial minefield of loan programs available today.

- **Property inspector:** A house's physical condition greatly affects its value. Have your dream home thoroughly inspected from roof to foundation before you purchase it to ensure you actually get what you think you're buying.

- **Escrow officer:** Mutual distrust is the underlying rule of every real estate deal. You and the seller need a neutral third party, an escrow officer, who handles funds and paperwork related to the transaction without playing favorites. The escrow officer is the home-buying game's referee.

- **Financial and tax advisors:** Before you buy a home, you should understand how the purchase fits into the context of your overall financial situation. You should address the issues of what your financial goals are and, given those goals, how much house you can afford. In Chapters 2 and 3, we explain how to do that.

- **Lawyer:** You may or may not need a lawyer on your team, depending on your contract's complexity, where your dream home is located, and your personal comfort level. The purchase agreement you sign when buying a home is a legally binding contract. If you have any questions about your contract's legality, put a lawyer who specializes in real estate law on your team.

REMEMBER

Odds are you won't win the game unless you have a winning team. But remember that your players are *advisors* — not decision makers. You're the boss and decision maker. The buck stops with you. After all, it's your money on the line.

Avoiding gratuitous advice

We'll say it again: Buying a home is a team sport. Successful transactions result from the coordinated efforts of many people — agents, brokers, lenders, property inspectors, escrow officers, tax advisors, and lawyers. Each player brings a different set of skills to the game and should make an important contribution to your team.

As long as your experts stick to what they know best, everything goes smoothly. Whenever one of your experts invades another expert's turf, however, war breaks out with a bang.

Unsolicited opinions related to property values are an example of one devastating type of gratuitous advice. Such opinions are usually volunteered by tax advisors or lawyers during a review of your transaction. Lawyers and tax advisors don't know property values. Making a lowball offer based on their bad advice, no matter how well intentioned, can blow the deal on your dream home.

Incredibly, some buyers foolishly *solicit* gratuitous advice in a misguided attempt to save a few bucks. "Why," buyers ask themselves, "hire a CPA for tax advice if we can get free tax advice from our agent? Why pay for legal advice from a lawyer if our escrow officer can give us free opinions about the best way for us to take title to our home?"

Why? Because if you're lucky, free advice from the wrong expert is worth exactly what you pay for it. Zip. Zero. Nada. Nothing.

If you're *unlucky*, free advice can be very expensive. The IRS, for example, shows no mercy if you make a mistake based on faulty advice. Ironically, this type of mistake usually ends up costing you far more than a lawyer or tax advisor would have charged you for correct advice.

Given the adverse consequences of bad advice, *good* experts don't offer guidance that they aren't qualified to give. If asked, they categorically refuse to give such advice. Instead, they redirect their clients to the proper experts. Good experts are wise enough to know what they don't know and humble enough to admit it. On a more selfish level, they don't want to get sued by their clients for giving lousy advice.

REMEMBER

Beware of experts who offer you gratuitous advice outside their fields of expertise.

Reeling in a Real Estate Agent

"What's it worth?"

The wrong answer to this question can cost you *big* bucks! Worse yet, there's no simple answer to this deceptively simple question, because home prices aren't precise. As Chapter 10 explains, you can't reduce home prices to a math problem where 2 plus 2 reassuringly equals 4 now and forevermore. Home prices aren't fixed — on the contrary, they slither all over the place.

REMEMBER

Houses sell for *fair market value*, which is whatever buyers offer and sellers accept. Fair market value isn't a specific number; it's a price *range*.

Suppose that you make an offer on a house worth about $250,000. If the seller has a better agent than you do, and you're desperate to buy, you may end up paying $275,000. On the other hand, if you're in no hurry to buy, and your agent is a good negotiator, you may be able to buy the home for $225,000. Home sale prices are often directly related both to the agent's knowledge of what comparable houses have sold for and to the agent's negotiating skills. Of course, other factors (such as the buyer's and seller's motivation, needs, and knowledge) are also important.

A good agent can be the foundation of your real estate team. An agent can help you find a home that meets your needs, negotiate for that home on your behalf, supervise property inspections, and coordinate the closing. Agents often have useful leads for mortgage loans. A good agent's negotiating skills and knowledge of property values can save you thousands of dollars.

Some people think that "good agent" is a contradiction in terms. These folks tell you that all agents have a hidden agenda: to make people buy more expensive homes than they can afford in order to fatten agents' commission checks.

Some agents *may* try to pressure you to buy sooner rather than later (and to pay more than you should) to fatten their own commissions. And unfortunately, many well-intentioned-but-inept agents are also out there. In the following sections, we explain how to avoid the bad agents — and how to sift through the masses of mediocre agents — in order to narrow the field down to good agents who are worthy of their commissions.

Types of agent relationships

Say that you've been working with an agent named Al who has been showing you property for several months. Yesterday, you finally found a home you like. The house seemed well priced, but you told Al to make a lowball offer anyhow. (What the heck. Everybody knows that prices are negotiable. If the sellers don't like the price, let them make a counteroffer.)

But what if your agent blew your cover? Suppose that Al told the sellers all your innermost secrets — such as how much cash you have for a down payment and how much you're *really* willing to pay for the house. Now the sellers can beat you at your own game. If you discovered what Al did to you (many victimized buyers don't), you'd undoubtedly feel hurt, betrayed, and pretty darn angry. You'd ask Al just whose side he's on.

The answer to that question has changed somewhat over the years. In decades past, buyers thought that they had agents who represented *them*. But in fact, they didn't. Back then, all agents were legally bound to be either agents or subagents *of the seller.*

TECHNICAL STUFF

Subagents, also called *cooperating agents,* work with one another through membership in a *Multiple Listing Service* (MLS). Agents use the MLS to promote their own real estate listings, and such agents offer to share their commissions with agents from other offices who actually sell the listed properties. As subagents of the seller, MLS participants are obliged to get top dollar for the sellers' properties.

Unfortunately, most buyers didn't know that the very agents who were working with them were actually representing the interests of the sellers. And the law didn't require agents to tell buyers which party they (the agents) actually represented before preparing offers on behalf of their buyers.

Times have changed somewhat. Some states (such as California) have adopted improved consumer-protection laws. These states passed laws that force agents to give both buyers and sellers a written disclosure regarding their duties as agents. The laws then allow buyers and sellers to select which type of relationship they want to have with their agents.

Home buyers and sellers can have three different types of relationships with real estate agents. We explain *dual agency* in the nearby sidebar "Dual agency and conflicts of interest." The other possible relationships are both types of *single agency,* which is when the agent represents only one of the two parties in the transaction:

>> **Seller's agent:** In this form of single agency, the agent works solely for the seller.

>> **Buyer's agent:** In this type of single agency, the agent works only for the buyer. A buyer's agent isn't an agent of the seller even if the buyer's agent gets a portion of the commission paid by the seller.

WARNING

Although single agency is an improvement over the old system, buyer's agents still suffer from the conflict of interest inherent in getting a commission that is tied to a percentage of the amount that a buyer spends for a property.

In rare cases, buyer's agents don't accept money from sellers. Instead, a buyer signs a contract to work exclusively with a buyer's agent, and the buyer pays the agent a retainer that is applied toward the fee owed when the buyer's agent finds the buyer a home. Depending on the contract provisions, the retainer may or may not be returned to the buyer if the buyer's agent fails to find the buyer a satisfactory property to purchase.

TIP

Here's a way to have the best of both worlds with a buyer's agent. This technique removes the buyer's agent's incentive to get you to spend more, yet it keeps you from paying a fee if you don't buy a home. Offer your buyer's agent a lump-sum commission plus a bonus if, *and only if,* the agent gets you a better buy. For example, if the agent typically receives 3 percent of a home's sale price, and you expect to buy a home for approximately $200,000, offer the agent a flat $5,000 commission plus an additional $100 bonus for every $1,000 below $200,000 the agent reduces the price for you, up to a maximum $6,000 commission.

Is your agent your ally or your enemy? Because laws regarding an agent's legal responsibility vary from state to state, know how the game is played in your state. Be sure you determine who your agent represents before you begin working together.

DUAL AGENCY AND CONFLICTS OF INTEREST

In certain transactions, an agent represents both the seller and the buyer. This type of representation is called *dual agency.*

Dual agency is the most confusing form of agency. Most people think that dual agency means that the exact same agent represents both the buyer and the seller. Such a situation is possible, but it's highly unusual and even more inadvisable. One agent can't possibly represent your best interests as a buyer and the seller's best interests at the same time.

In a more common kind of dual agency, the sale of a particular property involves two different agents who both work for the same real estate broker. Suppose that Sam Seller decides one sunny Sunday to list his house for sale with Sarah, an Acme Realty agent. Sarah smiles as she signs the agreement to represent Sam as the seller's exclusive agent.

Simultaneously, Betty Buyer bumps into Bob, who's also an Acme Realty agent, at a Sunday open house. Betty likes Bob's style and asks him to represent her exclusively as a buyer's agent. Bob enthusiastically agrees.

So far, so good. Sam has Sarah, his exclusive agent. Betty has Bob, her exclusive agent. Things get complicated later that afternoon when Bob shows Betty Buyer the house Sarah just listed for Sam Seller. Betty is bedazzled. She loves the house and tells Bob to write up an offer on it immediately.

When Betty decided to make an offer on Sam's house, the agency relationships that Betty and Sam had with their respective agents changed. Like it or not, Sarah suddenly represented both Sam and Betty. Similarly, Bob became the agent of both Betty and Sam.

(continued)

(continued)

Why? Even though two different agents are involved, both agents work for the same real estate broker, Acme Realty. As soon as Bob started to work on Betty's offer, Acme Realty represented the seller and the buyer of the same property. That's dual agency.

Dual agency probably won't be a problem if you end up working with an agent in a small office that has only a few agents. The odds that you'll buy a home listed by one of the other agents in your agent's office are slim. However, if the agent that you select works for a large brokerage operation with multiple offices and thousands of agents (such as Coldwell Banker), your odds of having to deal with dual agency skyrocket.

As a buyer, you must be on guard for two potential problems when confronted with dual agency. First, make sure your agent isn't sharing confidential information with any other agents at her real estate company. Second, watch out for agents who push their own company's listings because selling an in-house listing generates higher commissions for them.

Most states permit dual agency relationships as long as the agency status is disclosed to both the sellers and the buyers in advance, and both parties agree to it. Undisclosed dual agency can be used as grounds to have a purchase agreement revoked and usually permits the injured parties to seek recovery against the real estate agents.

How agents get paid

Real estate brokerage is an all-or-nothing business. As a rule, agents are paid a commission only when property sells. If the property doesn't sell, agents don't get paid.

WARNING

This payment method can create a conflict of interest between you and your agent. The payment method won't create a conflict of interest with *good* agents, because good agents put your best interests in front of their desire to get paid. You know that you're working with a bad agent, however, if the agent is more interested in quickly closing the sale and having you pay top dollar than in diligently educating you and getting you the best possible deal.

Allow us to answer your burning questions about real estate commissions:

> » **How much do real estate agents get in commissions?** Commissions are calculated as a percentage of the sale price. Depending on local custom, commissions on homes usually range from 4 to 7 percent of the sale price.

>> **Who pays the commission?** Typically, sellers. After all, sellers get money when property sells. Buyers rarely have much money left after making the down payment for their dream home and paying loan charges, property-inspection fees, homeowners-insurance premiums, moving costs, and the other expenses of purchase noted in Chapter 3. Because commission is part of the sales price, however, the effective cost of the commission comes out of both the buyer's and seller's pockets.

>> **Are commissions negotiable?** Absolutely. *Listing agreements* (the contracts that property owners sign with brokers to sell property) and purchase agreements usually state that commissions aren't fixed by law and may be negotiated between sellers and brokers.

>> **How is the commission distributed?** Suppose that a house sells for a nice, round $300,000. Assuming a 6 percent commission rate, the sale generates an $18,000 commission. That's a lot of money. At least it would be if it all went to one person, but commissions don't work that way as a rule.

Usually, the commission is divided in half at the close of escrow. The *listing broker,* who represents the sellers, gets half ($9,000, in our example) of the commission, and the other half ($9,000) goes to the *selling broker,* who represents the buyers.

TECHNICAL STUFF

If the selling or buying agent works for a broker, the broker typically gets a portion of the commission. The brokerage firm typically takes 30 to 50 percent of the commission, which leaves the agent 50 to 70 percent. In some firms, such as RE/MAX, agents pay a fixed monthly fee to their brokerage firm and end up keeping 80 to 90 percent of the commissions they bring into the firm. Agents who work on their own as independent brokers, of course, don't have to split their commissions with anyone.

Characteristics of good agents

Good agents can be male or female, and they come in a wide assortment of races, colors, creeds, and ages. All good agents, however, have the following character-istics that are beneficial to buyers:

>> **Good agents educate you.** Your agent knows the home-buying process and carefully explains each step so you *always* understand what's happening. Agents should be patient, not pushy. A good agent *never* uses your ignorance to manipulate you.

>> **Good agents don't make decisions for you.** Your agent *always* explains what your options are so *you* can make wise decisions regarding your best course of action.

WHERE AN AGENT'S TIME GOES

Some people think that real estate commissions are disproportionately large relative to the amount of work that agents do. That's a polite way of saying that agents are grossly overpaid.

Justifying a good agent's commission is easier if you understand what we call the Iceberg Theory. As you probably know, 90 percent of an iceberg's bulk is hidden underwater. You can't tell how big an iceberg is by the portion you see floating above the waterline. By the same token, you can't tell how much time agents spend working for you. Good agents spend at least nine hours working behind the scene for every hour spent in the presence of their clients.

Unfortunately, most buyers and sellers don't know this. Buyers and sellers think that commissions are excessive, given the relatively few hours they actually see their agents working for them.

Unlike lawyers and other professionals who bill clients by the hour, real estate agents don't itemize the time spent on a transaction from start to finish. If they did, you'd have a much better idea of where your agent's time goes.

Good real estate agents typically spend around 20 hours a week touring new properties and checking up on houses that have been on the market a while to see which houses are still available and which have had offers accepted on them. Agents do this legwork, week in and week out, to keep themselves current regarding what's on the market and how property values are changing.

After you select an agent, she starts targeting houses you may want to buy. Good agents screen several properties for each one they eventually show you, saving you the time of doing the screening yourself. Your agent spends time playing phone tag with listing agents, trying to get instructions about how to show properties, and scheduling showings. Then she spends more hours with you, touring houses and searching for your elusive dream home.

After you've found your dream home, your agent spends time preparing an offer to purchase, presenting the offer, and negotiating counteroffers with the seller's agent regarding the price and terms of sale. After the offer is accepted, a good agent spends more hours helping you with such things as securing a mortgage; coordinating transaction details with the seller's agent; providing information and paperwork to the escrow officer; going through the home with your various property inspectors; and reviewing mandated local, state, and federal disclosure statements from the sellers.

>> **Good agents tell you when they think that adding other experts (inspectors, lawyers, and the like) to your team is advisable.** Experts don't threaten a good agent. The agent's ego should always be secondary to the primary mission of serving you well.

>> **Good agents voluntarily restrict themselves geographically and by property type.** Your agent has ideally learned that trying to be all things to all people invariably results in mediocre service. Different communities can have radically different market conditions, laws, and restrictions. (For more information, see the nearby sidebar "Agents who work outside their areas of expertise are dangerous.")

>> **Good agents are full-time professionals, because serving you properly is a full-time job.** To reduce the financial impact of changing jobs, many people begin their real estate careers as part-timers, working as agents after normal business hours and weekends. That's fine for the agents, but not you.

TIP

One of the first questions you must ask any agent you're considering working with is "Are you a full-time agent?" Just as you wouldn't risk letting a part-time lawyer defend you, don't let a part-time agent represent you.

>> **Good agents have contacts.** Folks prefer doing business with people they know, respect, and trust. You can use your agent's working relationships with local lenders, property inspectors, lawyers, title officers, insurance agents, government officials, and other real estate agents. Good agents will refer you to highly skilled service providers who offer competitive pricing.

WARNING

Watch out for duplicitous agents with hidden agendas. Instead of referring you to the best possible service providers, these agents limit their recommendations to people who refer business to them or pay them a referral fee.

>> **Good agents have time.** Agents earn their living selling time, not houses. Success is a two-edged sword for busy agents. An agent who is already working with several other buyers and sellers probably won't have enough surplus time to serve you properly. Occasional scheduling conflicts are unavoidable. But if you often find your needs being neglected because your agent's time is overcommitted, get a new agent.

Selecting your agent

After you know the glittering generalities of a hypothetical good agent (see the previous section), you're ready to get down to the nitty-gritty specifics of choosing an agent of your very own. We strongly recommend that you interview at least three agents before selecting the lucky one.

WARNING

AGENTS WHO WORK OUTSIDE THEIR AREAS OF EXPERTISE ARE DANGEROUS

Many pitfalls await unwary buyers who trust agents who work outside their areas of expertise. Although extreme, here's a real-life example of a disaster caused by an agent who specialized in property located in Sonoma, a peaceful suburban town about 40 miles north of San Francisco.

The Sonoma agent represented her friend in the purchase of a small apartment building located in San Francisco. The buyer planned to convert the apartments to condominiums and then sell the condos individually at a profit.

Unfortunately, the Sonoma agent knew nothing about San Francisco's strict rent-control law or its equally strict condo-conversion ordinance. The intent of these laws is to discourage people from converting lower-rent apartments to upscale condos.

Had the building been converted to condos, the total proceeds from individual sales would've been less than the price the buyer had originally paid for the building, because of the restrictive nature of these two laws. The agent's negligence ultimately led the buyer to lose $125,000 when she later resold the building.

This agent not only made the mistake of working in "foreign territory," but she also failed to recommend that the buyer put a local real estate lawyer on her team to advise the buyer about these legal issues. The buyer could have sued her agent for malpractice. She didn't, because the agent was her "friend."

With friends like that, the buyer didn't need any enemies. If the Sonoma agent had been a true friend, she'd have referred the buyer to a good San Francisco agent.

Agents who go out of their area of geographical or property expertise do so because they're either greedy or just too darn inept to know better. Whatever the reason, avoid such agents like the plague.

Finding referral sources

If you have trouble finding three good agents to interview, here are some referral sources:

>> **Friends, business associates, and members of religious, professional, and social organizations to which you belong:** In short, anyone you know who's either house hunting or who owns a home in your target neighborhood can be a source of agent referrals. Don't just ask for names; find out why these folks liked their agents.

>> **Your employer:** The company you work for may have a relocation service that you can consult.

>> **Professionals in related fields:** Financial, tax, and legal advisors can be good agent-referral sources.

>> **The agent who sold your previous home:** If you're a homeowner who's moving into a new area, ask the agent who sold you your home to recommend a good agent in that area. Good agents network with one another.

>> **Sunday open houses:** While you're investigating the houses, check out the agents. These agents have already proved (by their open-house activity) that they work the neighborhood in which you want to buy.

Don't take any referral, even if it's from the Pope, as gospel. Most people who give referrals have limited or outdated experience with the recommended agent. Furthermore, the person making the referral is probably not a real estate expert.

Requesting an activity list

After you've identified at least three good agents, the fun begins. To avoid a misunderstanding, tell each agent that you plan to interview several agents before you select the one you'll work with. Ask each agent to bring to the interview a list of *every* property the agent listed or sold during the preceding 12 months. This list, called the *activity list*, is an extremely powerful analytical tool.

Here's what the activity list should include and how you should use the list during the interview (also available online at www.dummies.com/go/homebuyingkit7e):

>> **Property address:** Addresses help you zero in on the agent's geographical focus. See for yourself exactly how many properties the agent sold and listed in your target neighborhood(s). Eliminate agents who are focused outside your area *and* agents who have no geographical focus.

>> **Property type (house, condo, duplex, other):** You can use this information to determine whether the agent works on the kind of property you intend to buy. If, for example, an agent specializes in condos, and you want to buy a detached single-family home, you may have a problem. By the same token, if you want to buy a house in foreclosure or a short sale and the agent has never done a distressed property transaction, you have a problem. As noted in Chapter 8, distressed property requires a specialized skill set.

>> **Sales price:** Does the agent handle property in your price range? An agent who deals in *much* more or less expensive property than you expect to buy may not be the right agent for you. If, for example, you plan to spend $250,000, and the least expensive house the agent sold in the past year cost

$400,000, you have a mismatch. Such agents probably won't spend much time on you because they have bigger fish to fry.

>> **Date of sale:** Sales activity should be distributed fairly evenly throughout the year. If it isn't, find out why. A lack of recent sales activity may be because of illness or personal problems that may reduce the agent's effectiveness.

>> **Whom the agent represented — seller or buyer:** Seasoned agents work about half the time with buyers and the other half with sellers. Newer agents primarily work with buyers. Avoid agents who work primarily with sellers. These agents generally lack either the interest or aptitude to work effectively with buyers.

>> **Total dollar value of property sold during the preceding 12 months:** Comparing the three agents' grand-total property sales is a quick way to measure each agent's individual activity and success. There are, however, other equally important factors to consider when selecting your agent. You don't necessarily want a "top producer." These agents get to the top by listing and selling large quantities of property. They usually don't have the time or patience to do the hand-holding and education you may need, especially if you're a first-time buyer.

>> **Names and current phone numbers of sellers/buyers:** You'll use this later to spot-check references.

WARNING

Words whisper; actions thunder. The activity list transforms cheap chatter into solid facts. Good agents willingly give you their lists and encourage you to check client references. Bad agents don't want you talking to their unhappy victims. Eliminate from consideration any agent who won't give you a comprehensive activity list — she is trying to hide either a lack of sales or unhappy clients.

Interviewing agents

Begin each interview by spending a few minutes analyzing the agent's activity list (downloadable from online at www.dummies.com/go/homebuyingkit7e). After you've finished reviewing the list and had time to organize your thoughts, get answers to the following questions:

>> **Are you a full-time agent?** You should have asked this before inviting the agent to be interviewed. If you forgot, do it now. Don't work with part-time agents.

>> **Whom do you represent?** This topic gets back to the fundamental question of agency. Is the agent representing you exclusively, or is he a dual agent who represents both you and the seller? Be sure you know exactly whom your agent represents at all times.

>> **What can you tell me about your office?** Discuss office size, staff support, market specialization, and reputation. See whether the agent's broker is knowledgeable, is available to you if necessary, and is a good problem-solver. In a crunch, your transaction's success (or failure) may depend on the quality of backup support that you and the agent receive.

TIP

Don't put too much weight on the size of the agent's office. Some excellent agents work as sole practitioners, and other excellent agents prefer the synergism and support services of a huge office. Although larger offices tend to have more listings, no one office ever has a monopoly on the good listings. Quality of service is more important than quantity of agents or listings.

>> **How long have you been an agent?** You want an agent who keeps learning and growing. After five years in real estate, a good agent has five years' experience, whereas a mediocre agent has one year's experience five times. Time in the saddle is, by itself, no guarantee of competence.

>> **Do you have a salesperson's license or a broker's license?** An agent must satisfy more rigorous educational and field-sales experience requirements to get a broker's license. Many fine agents have only a salesperson's license throughout their entire careers. Although a broker's license isn't a guarantee of excellence, good agents often obtain a broker's license to improve their professional skills and to give themselves an advantage in agent-selection situations.

>> **Do you hold any professional designations? Have you taken any real estate classes recently? What do you read to keep current in your field?** Taking continuing-education courses and reading to stay abreast of changes in real estate brokerage are good signs. So is obtaining professional designations, such as the GRI (Graduate, Realtor Institute), SFR (Short Sales and Foreclosures), and CRS (Certified Residential Specialist) designations through the National Association of Realtors' study programs. A CDPE (Certified Distressed Property Expert) designation from the Charfen Institute is handy if you're seeking short sales or foreclosures. However, credentials in and of themselves are no guarantee of competence or ethics.

>> **What is your understanding of my home-buying needs?** You've probably already told the agent what type of property you want to buy, the neighborhood you want to live in, and how much you can spend. See whether the agent remembers what you said. If the agent doesn't remember, watch out. You need an agent who listens carefully to what you say.

WARNING

>> **What do you think of the other two agents (name them) whom I'm interviewing?** To encourage frankness, assure the agents that you won't repeat what they say to you. Good agents don't build themselves up by tearing down other agents. If all three agents are good ones, you won't hear any derogatory comments. However, if one of the agents (or the agent's firm)

has a bad reputation in the real estate community, the other two agents should tell you. Good or bad, the reputations of your agent and the agent's office rub off on you.

» **How many other buyers and sellers are you currently representing?** If, for example, the agent holds three listings open every weekend and is working with six other buyers to boot, where do you fit in? Although some scheduling conflicts are inevitable, you shouldn't have to contort your life to fit the agent's schedule. A good agent has time to accommodate your schedule.

» **Do you work in partnership with another agent or use assistants?** Some agents team up with another agent to handle buyers and sellers jointly. If this is the case, you must interview both agents. Other agents delegate time-consuming detail work to their assistants so they can focus on critical points in the transaction. If an agent relies on such assistants, be sure that the assistants are qualified and that you understand exactly how and when during the buying process the agent will work directly with you. You don't want to hire an agent only to find that you end up working most of the time with her assistant — whom you can't stand.

TIP

» **Is there anything I haven't asked about you or your firm that you think I should know?** Perhaps the agent is planning to change firms or is leaving next week to take an 80-day trip around the world. Maybe the agent's broker is going out of business. *This is the make-sure-that-I-find-out-everything-I-need-to-know-to-make-a-good-decision question.*

Checking agents' references

Here's your chance to profit from other people's mistakes, which is infinitely preferable to goofing up yourself. You should have activity lists with the names and phone numbers of every buyer and seller that the agents represented during the past 12 months. You can pick and choose the people you want to call instead of being restricted to a highly selective list of references who think that these agents are God's gift to real estate.

What's to prevent agents from culling out their worst transactions? Nothing. However, the more deals they delete, the less activity they have to show you — and the worse they look when you compare the agents' overall sales activity.

Suppose that each agent gives you a list containing 50 transactions. Assuming one buyer or seller for each transaction, 50 clients per agent times three agents interviewed equals 150 phone calls. You'd be on the phone forever!

Good news. You don't have to call each and every client to check references. You can get a pretty darn accurate picture of the agents by making as few as six calls per agent.

Here's a fast, easy way to get a representative sampling of client references (also available online at www.dummies.com/go/homebuyingkit7e):

1. **Because you're a buyer, ignore all references from sellers.**

   Doing so probably slices the list in half.

2. **Zero in on people who bought property similar in price, location, and property type to what you want to buy.**

3. **Call two of those representative buyers who purchased a home about 12 months ago, another two buyers who bought 6 months ago, and two buyers whose escrows closed most recently.**

   By spreading references over the past year, you can find out whether the agent's level of service has been consistently good.

INVESTIGATE

Now that you've identified which buyers to call, here's what to ask when you have them on the phone:

>> **Is the agent trustworthy? Honest? Did the agent follow through on promises?** Your agent can't be even the tiniest bit untrustworthy, dishonest, or unreliable. A negative answer to any of these questions is the kiss of death.

>> **Did the agent have enough time to serve you properly? Was the agent available as required to fit your schedule?** Occasional scheduling conflicts are okay. Frequent conflicts are absolutely, flat-out unacceptable.

>> **Did the agent explain everything that happened during the buying process clearly and in sufficient detail to satisfy you?** What one person thinks is sufficient detail may not be nearly enough information for another. You know which type of person you are — question agent references accordingly.

>> **Did the agent set realistic contract deadlines and meet or beat them?** "Time is of the essence" is a condition of every real estate contract. Contract time frames for obtaining a loan, completing property inspections, and the like are extremely important and must be strictly adhered to, or the deal goes belly-up. Good agents prepare well-written contracts with realistic time frames and then ensure that all deadlines are met on or before the due dates.

>> **Do the words *self-starter, committed,* and *motivated* describe the agent?** No one likes pushy people. But if you're under pressure to buy quickly, the last thing you want is a lethargic agent. You shouldn't have to jab your agent periodically with an electric prod to make sure he's still breathing. Find out how energetically the agent in question is prepared to work.

>> **Who found the home you bought — you or the agent?** This question is a double-check of the agent's market knowledge. Good agents know not only what's already on the market but also which houses will be coming on the market soon. You shouldn't have to find the house you buy — that's your agent's job.

>> **Did the agent negotiate a good price for your home?** See whether the agent's buyers *still* think they got a good deal. Good agents are frugal when spending their clients' money. Good agents use their knowledge of property values and their negotiating skills to make sure their clients pay the fair market value or less for the homes they buy. People who bought homes six months or a year ago can tell you how well their purchase prices have stood the test of time.

>> **Would you use the agent again?** This is the ultimate test of customer satisfaction. If someone says "no," find out why not. The negative answer may be due to a personality conflict between the buyer and the agent that won't bother you. On the other hand, the negative answer may reveal a horrendous flaw that you haven't yet discovered in the agent.

>> **Is there anything I haven't asked you about the agent or the agent's office that you think I should know?** You never know what you'll find out when you ask the famous catchall question.

Making your decision

REMEMBER

After analyzing all three agents' sales activity, interviewing the agents, and talking to their buyers, you have most of the facts you need to make an informed decision. Here are three final considerations to help you select the paragon of virtue that you need on your real estate team:

>> **Will you be proud having the agent represent you?** People who deal with your agent will form opinions of you based on their impressions of your agent. You can't afford to have anyone on your team who isn't a highly skilled professional.

>> **Do you communicate well with the agent?** Good agents make sure you completely understand everything they say. If you can't understand your agent, you're not stupid — the agent is a poor communicator.

>> **Do you enjoy the agent's personality?** Home buying is stressful, even for the coolest of cucumbers. You'll be sharing some extremely intense situations with your agent. Working with an agent you like may transform the home-buying process from a horrible experience into an exciting adventure — or at least a tolerable transaction.

BUYING WITHOUT AN AGENT

You may see a For Sale by Owner (also known as a FSBO — pronounced fiz-bo) or two during your home search. The sellers may even be friends, neighbors, or work colleagues. Because no real estate agent is involved on the selling side of a FSBO transaction, the sellers don't have to pay a commission. That shaves big bucks off their expenses of sale.

If the FSBO home meets your needs, and you found it yourself, you may rightfully wonder whether you need an agent to complete the deal. After all, if the sellers don't have to pay your agent's commission, they should be willing to sell you the home at a lower price.

Some home buyers have successfully purchased their dream homes without an agent. Others have made big boo-boos that way.

If you're a novice, using an agent usually makes sense. Consider the additional value that an agent brings to the transaction beyond finding you the property, such as negotiating; estimating market value; and helping coordinate property inspections, contingency removals, seller disclosures, financing, opening escrow, and myriad other details.

You may also consider asking an agent to represent you for less than the standard 3 percent commission because you found the property yourself. If you decide not to use an agent, consider hiring an attorney by the hour to review the contract and handle the transaction's important legal details.

Getting the most from your agent

After working so hard to find a great agent, it would be a shame to inadvertently ruin the relationship. Good buyer/agent relationships aren't accidental. Such relationships are based upon pillars of mutual loyalty and trust that develop over time.

Poor relationships, conversely, result from misconceptions of how the game is played. Some buyers act in what they think is their best interest, but they end up unintentionally harming themselves.

More isn't always better

One common fallacy is thinking that five agents are five times better than one agent. The theory sounds so logical. If you work with agents from a variety of offices, you can get better market coverage and first peek at the new listings that each office puts on the market. The more agents you work with, the better your chances of quickly finding your dream home.

Things don't work that way in the real world. When smart agents first meet you, they'll probably ask whether you're working with any other agents. These agents are trying to find out how much you know about the market (so they won't waste time showing you houses that you've already seen) and learn what you didn't like about the properties that you saw.

One good agent can quickly show you every home on the market that meets your price, neighborhood, size, and condition specifications. If none of the houses is what you want, good agents keep looking until the right home hits the market. Good agents don't limit their searches to houses listed by their offices. They investigate anything even remotely similar to what you want, regardless of which office listed the property. Whether you work with one agent or one hundred, you'll see the same houses.

Agents know they won't get paid if you don't buy. That risk comes with the job. What agents hate is losing a sale after months of hard work because they called you shortly *after* another agent called you about the same house. That risk is unnecessary. You're free, of course, to work with as many agents as you want. In fact, working with more than one agent makes sense if you're looking for a home in more than one geographic area. Don't be surprised, however, if good agents in the same area opt out of a horse race. Their odds of getting paid for their work increase dramatically when they spend their time on buyers who work exclusively with them. Loyalty begets loyalty.

WARNING

The risk of playing the field is rarely worth the reward. One loyal agent totally committed to finding you a home is infinitely better than five agents working for you as a last resort because they consider you just marginally better than Benedict Arnold. Like marriages, the best buyer/agent relationships are monogamous.

Your agent isn't the enemy

Another fallacy is viewing your agent as your adversary. True, you don't want to tell your innermost secrets to a loose-lipped agent who blithely blabs them to the seller or seller's agent. Some buyers think that the less their agent knows about them, the better. Such buyers believe that after agents know why they want to buy and how much cash they have, the agents will somehow magically manipulate them into spending far more than they can afford to spend for the home they eventually buy.

Not true. Good agents ask such questions because they need to be sure you're financially qualified in order to avoid wasting your time and theirs by showing you properties that you can't afford. If your agent knows that you're under deadline pressure to buy, she'll give your house hunt top priority.

REMEMBER

Good agents won't betray your trust. They know that if they take care of you, the commission takes care of itself. If you can't trust your agent, don't play cat-and-mouse games — *get a new agent.*

Ironically, smart agents fear you as much as or more than you fear them. They know you have the power to make or break their careers. If they please you, you'll be a source of glowing referrals for them. If they upset you, you'll tell everyone you know about the bad job they did.

Use the immense power of potential referrals to manage your relationship with the agent. If your agent does a lousy job, don't get mad — tell the world every gory detail of your rotten experience. Nothing ruins an agent's career faster than dissatisfied clients.

Bagging a Broker

Selecting a broker is easy. When you choose an agent, your agent's broker goes along for the ride. It's a package deal.

If your transaction rolls merrily along from the time your offer is accepted to the close of escrow, you probably never meet the broker. But if the engine begins to misfire and the wheels start coming off, one guess who you turn to for a quick repair job. Brokers are the invisible grease in problematic transactions.

All states issue two markedly different types of real estate licenses: one for salespeople (agents) and one for brokers. Agents who have broker's licenses must satisfy much more stringent educational and experience standards than agents with a salesperson's license do.

Your agent may have either type of license. Broker licensees have the option either to operate independently or to work for another broker. An agent who has a sales- person's license, on the other hand, *must* work under a broker's direct supervision so you have access to the broker's higher level of expertise should you need it.

When Harry Truman was president, he had a sign on his desk that read "The buck stops here." Like Truman, good brokers don't pass the buck. Here are some of their other characteristics:

>> **Excellent reputation:** The broker's image, good or bad, will be obvious from comments that you hear while checking agent references. You want the seller, lender, and all other people involved in your transaction working with you because of your broker's reputation, not in spite of it. Buying a home is hard

enough without the added burden of having to overcome guilt by association. If an agent's references disparage the agent's broker, dump the agent.

>> **Extensive business relationships:** Good brokers develop and maintain relationships with the people whom their office deals with — other brokers, lenders, title officers, city officials, and the like. This preexisting reservoir of goodwill is yours to use when the going gets rough. Brokers with strong business relationships can work near-miracles for you in a crisis.

>> **Strong problem-solving skills:** Participants in real estate transactions sometimes get highly emotional. When your life savings are on the line, you may occasionally lash out at your agent and the other players. Someone has to resolve the resulting quarrels and misunderstandings. That someone is the broker.

TIP

The broker's job is to help solve your problems. Call your broker into the game if your agent is stymied by a tough problem or if you're having trouble with your agent. Everything your agent does or fails to do is ultimately the broker's responsibility.

STREET-SMART VERSUS BOOK-SMART

Book-smart people have theoretical knowledge. They know only what should happen in a perfect world based upon what they've read. You're getting book-smart right now.

Street-smart people, conversely, know how things work in the real world. They learned the hard way through years of hands-on, practical experience. A good real estate broker is one of the most street-smart people on your home-buying team.

People learn very little from uneventful, routine transactions; they're like flying a plane on autopilot. Street smarts come from the sweaty-palm deals. Fortunately, most agents have only one or two of these gut-wrenching messes each year. When these turbulent transactions occur, the broker takes control of the plane, so to speak.

Because the broker participates directly or indirectly in every deal the office handles, your broker's practical experience is directly related to the number of agents in the office. A broker who manages a 25-agent office, for example, gets 25 years' of real estate experience per calendar year. Any broker who can survive five years of handling all the office's truly terrible transactions becomes a superb problem-solver out of sheer necessity.

Landing a Lender

Everyone thinks that buying a home is likely to be the largest single purchase you'll ever make. Unless you're an all-cash buyer, however, everyone is wrong. Here's why.

Suppose that your dream home's purchase price is $250,000. You make a 20 percent cash down payment of $50,000 and get a $200,000 fixed-rate loan at 5.0 percent interest from your friendly lender. Over the next 30 years, you conscientiously repay the loan with payments of about $1,074 a month. (We show you in Chapters 3 and 6 how to crunch these numbers for yourself.)

Your 360 monthly loan payments total approximately$387,000. If you originally borrowed $200,000, the additional $187,000 you paid is interest on your loan. The total interest charges over 30 years almost equal the $200,000 you borrowed to buy your home!

If you can't pay, you can't play. You need a good lender on your team to transform you from a home looker into a homeowner. By finding the right lender, you can save yourself big bucks over the life of the loan.

No one loan is right for everyone. A person fresh out of college who's struggling to buy a condo with 5 percent cash down has vastly different loan requirements than an older, cash-rich couple who put 50 percent cash down on a retirement cottage by using equity from the sale of their previous house.

Generations ago, finding the right loan was easy. You could get any kind of mortgage you wanted, as long as it was a 30-year, fixed-rate loan. All home loans were basically the same except for minor variations in loan fees and interest rates.

Those kinder, gentler days of yesteryear are long gone. Today, you're confronted by a bewildering array of fixed- and adjustable-rate mortgage (ARM) programs. In Chapter 6, we take the mystery out of securing a mortgage and selecting a lender.

Procuring Property Inspectors

A home's price is directly related to its physical condition. Homes in top shape sell for top dollar. Fixer-uppers sell at greatly reduced prices because whoever buys them must spend money on repairs to get them back into pristine condition.

Even if you're a rocket scientist, you can't know how much work a house may need just by looking at it. You can't see whether the roof leaks, the electrical system is shockingly defective, the plumbing is shot, the furnace's heat exchanger is cracked, the chimney is loose, or termites are feasting on the woodwork. Invisible defects like these cost major money to repair.

Because you don't want to inadvertently become the owner of a home with such expensive hidden problems, you need property inspectors on your team. None of the other players on your team — including real estate agents, lenders, and brokers — is qualified to advise you about a house's physical condition or the cost of necessary corrective work. In Chapter 13, we cover everything you need to know about property inspections and selecting property inspectors.

WARNING

Even though a home isn't quite as complicated as a space shuttle, it still has plenty of expensive systems that can go haywire. "Saving" money by forgoing inspections just because a home appears to be in good condition is a false economy. The vast majority of home problems aren't visible. Never buy a house that hasn't been thoroughly inspected from foundation to roof by a qualified inspector of your own choosing.

Electing an Escrow Officer

One common denominator crops up in most every real estate deal: mutual distrust. As a buyer, would you give the sellers your hard-earned money before every single condition of the sale is satisfied? Not likely. If your positions were reversed and you were the seller, would you give the buyers the *title* (ownership) to your house before you got their money? No way.

Deals would grind to a halt without something to bridge the gulf of mutual buyer-and-seller distrust. Even the simplest transaction involves myriad details that must be resolved to everyone's satisfaction before the sale can be completed. That's why real estate, like other team sports, has a referee.

Your *escrow officer* is the referee who keeps the game civilized. Strictly speaking, escrow officers aren't on anyone's team — they're neutral. They act as a disinterested third party for buyers and sellers without showing favoritism to either party.

TECHNICAL
STUFF

After you and the seller have a signed contract, all the documents, funds, and instructions related to your transaction are given to the escrow holder specified in your purchase agreement. We cover this process, known as *opening an escrow*, in detail in Chapter 14.

Buyers and sellers often select the escrow holder based on the recommendation of their real estate agents. Depending on where the property you're buying is located, local custom dictates whether your escrow is handled by a lawyer, bank, real estate broker, or the firm that issues the title-insurance policy.

Escrow fees range from a few hundred dollars to several thousand dollars and are based on your property's purchase price. Once again, local custom usually determines whether the buyer or the seller pays for the escrow or whether the escrow fees are split fifty-fifty. However, as we discuss in Chapter 12, this item is often negotiable.

Finding (or Forgoing) Financial and Tax Advisors

The real estate game is played with real money — your hard-earned cash. You've likely scrimped, saved, and done without in order to get the cash for your down payment. When you sell your house someday, its equity will probably be a major chunk of your net worth. Either way, buying or selling, you have big bucks on the line.

A home purchase has an enormous impact on your personal finances. Before you buy a home, you need to understand how a home purchase fits within the context of your overall finances and your other goals. Be sure to read Chapters 2 and 3, which deal with these important issues.

WARNING

You can elect to hire a financial advisor, but most such titled advisors aren't set up to handle home-buying questions objectively. The reason: Financial advisors stand to gain financially from the advice they render. Many so-called financial consultants get commissions from the investments they sell you. If this is the case, how motivated will they be to advise you to use your cash to buy a home rather than an investment from them?! Advisors who manage money on a fee (or a percentage) basis have the same conflict of interest.

If you're going to hire an advisor, use one who works by the hour and doesn't have a vested interested in your home-buying decision. Few financial advisors work on this basis. Although tax advisors are more likely to work on this basis, they tend to have a narrower-than-needed financial perspective. A competent tax advisor may be able to help you structure the home's purchase to maximize your tax benefits. For most transactions, however, a tax advisor is unnecessary.

If you want to hire a financial or tax advisor, interview several before you select one. Check with your agent, banker, lawyer, business associates, and friends for referrals. As is the case with selecting your agent, you should get client references from each tax advisor and call the references.

INVESTIGATE

Here's what to look for in a good financial or tax advisor:

>> **Is this a full-time job for the advisor?** The realm of personal finances and taxes is too vast for you to trust a part-timer. You need the services of a full-time professional.

>> **Does the advisor speak your language?** Good advisors can explain your financial alternatives in simple terms. If you don't understand exactly what the tax advisor is saying, ask for clarification. If you still don't understand, get another tax advisor. (See the nearby sidebar "Some experts give wrong advice.")

>> **Is the advisor objective?** Hire someone who works solely by the hour and doesn't have a vested interest in the advice he gives you about when to buy and how much to spend.

>> **What is the advisor's fee schedule?** Hourly fees vary widely. Don't pick someone strictly on a cost-per-hour basis. An advisor who's just beginning to practice, for example, may charge only half as much as one with 20 years' experience. If the rookie takes four hours to do what the old pro does in an hour, which advisor is more expensive in the long run? Furthermore, the quality of the seasoned veteran's advice may be superior to the quality of the novice's advice.

>> **Is the tax advisor a Certified Public Accountant (CPA) or Enrolled Agent (EA)?** These professional designations indicate that the tax advisor has satisfied special education and experience requirements and has passed a rigorous licensing exam. A CPA does general accounting and prepares tax returns. An EA focuses specifically on taxation. Only CPAs, EAs, and attorneys are authorized to represent you before the IRS in the event of an audit.

>> **Does the tax advisor have experience with real estate transactions?** Tax practice, like law or medicine, is an extremely broad field. The tax advisors that IBM uses (for example) are undoubtedly wonderful, but IBM's tax advisors aren't necessarily the best ones for you. You need a tax advisor whose clients have tax problems like yours.

TIP

The best advisors in the world can't do much to change the financial and tax consequences of a transaction *after* the deal is done. If you're going to consult advisors, do so *before* you make significant financial decisions. Plan your financial and tax situation instead of reacting to the consequences after the fact.

SOME EXPERTS GIVE WRONG ADVICE

Ray wasn't very sophisticated about tax advisors when he moved to San Francisco. Several people whom he worked with suggested that he use their tax advisor. He selected her based on their recommendations (plus the fact that her office was only two blocks away from his).

Ray knew he'd made a mistake when he went over to her office to review the tax return she'd prepared for him. He asked her how she had arrived at the itemized deductions for auto expenses and professional training. In both cases, they were much higher than the totals of the receipts Ray had given her.

She explained that these higher deductions reduced his tax bite and said no one would question the deductions because they were within acceptable IRS guidelines. Ray still didn't understand her concept of deducting more money than he had spent.

Ray's tax advisor rather impatiently went through the tax return again, speaking more slowly and using smaller words the second time. She concluded by saying he shouldn't worry about what she'd done because everyone overstates expenses.

Ray knew that she was speaking English because he recognized nearly all the words, but they seemed to be bouncing off his forehead without penetrating his brain. He was getting more and more frustrated. His tax advisor was getting later and later for her next appointment. In desperation, she finally gave Ray the return and sent him home like a schoolboy to "think it over."

He did.

Ray decided the problem wasn't that he was stupid. The problem was that he'd been given unethical advice. In her zeal to save Ray money, the tax advisor had falsified his deductions. That falsification was wrong. Worse, it was illegal. Ray solved the problem by getting a new tax advisor.

Never blindly follow the advice of experts because you're in awe of their expertise. Experts can be just as wrong as ordinary mortals.

Looking for Lawyers

Lawyers are like seat belts: You never know when you may need them. Your deal is rolling merrily along when out of nowhere — slam, bam, wham — you hit a legal pothole and end up in Sue City.

That real estate purchase agreement you sign is meant to be a legally binding contract between you and the seller. If you have any questions about the legality of your contract, get a lawyer on your team *pronto*. No one else on the team is qualified to give you legal advice.

Here's what determines whether you need a lawyer on your team:

>> **The location of the property you're buying:** In states such as California, lawyers rarely work on deals that involve only filling in the blanks on a standard, preprinted purchase agreement that has been previously reviewed and approved by members of the bar association. In other states, such as New York, however, lawyers routinely do everything from preparing purchase contracts to closing the escrow. Your agent knows the role that lawyers need to play in your locale.

>> **The complexity of your transaction:** You need a lawyer any time you get into a situation that isn't covered by a standard contract. Unless your agent is also a lawyer, she isn't qualified to do creative legal writing. Complicated issues, such as those that often arise from co-ownership agreements between unrelated people who buy property together, and the complex legal ramifications of taking title to your home should be handled by a lawyer. (We get into co-ownership agreements in Chapter 8 and taking title in Chapter 14.)

>> **When no agent is involved:** Say that you're buying a home that's being offered for sale directly by the owner. If neither you nor the seller has an agent, get a lawyer to prepare the contract and have the lawyer do the work that an agent would normally handle. Eliminating an agent doesn't eliminate the need for disclosures, inspections, contingency removals, and myriad other details involved in the home-buying process.

>> **To sleep at night:** You may have the world's easiest deal. Still, if you'd feel more comfortable having a lawyer review the contract, your peace of mind is certainly worth the cost of an hour or two of legal time.

Selecting your lawyer

If, for whatever reason, you decide you need a lawyer, interview several before making your selection. Law, like medicine, is highly specialized. A corporate attorney or the lawyer who handled your neighbor's divorce isn't the best choice for your real estate team. Get a lawyer who specializes in residential real estate transactions. Your agent and broker are excellent referral sources because they work with real estate lawyers all the time in their transactions.

A good lawyer:

>> **Is a full-time lawyer and licensed to practice law in your state:** Of course.

>> **Is local talent:** Real estate law, like real estate brokerage, is extremely provincial. The law varies not only from state to state but also from one area to another within the same state. Rent-control laws, condominium-conversion statutes, and zoning codes, for example, are usually passed by city or county governing agencies. A good local lawyer knows the laws and has working relationships with people who administer those laws in your area.

>> **Has a realistic fee schedule:** Lawyers' fees vary widely. A good lawyer gives you an estimate of how much handling your situation will cost. As with financial and tax advisors, the experience factor comes into play. Seasoned lawyers generally charge higher hourly fees than novice lawyers, but seasoned lawyers also tend to get a lot more done in an hour than inexperienced lawyers can. A low rate is no bargain if the novice is learning on your nickel.

>> **Has a good track record:** If your case may go to trial, find out whether the lawyer has courtroom experience. Some lawyers don't do trial work. Then ask about the lawyer's track record of wins versus losses. What good is a lawyer with a great deal of trial experience if that lawyer has never won a case?

>> **Is a deal-maker or a deal-breaker (whichever is appropriate):** Some lawyers are great at putting deals together. Others specialize in blowing them out of the water. Each skill is important. Good deal-makers, however, aren't always equally good deal-breakers, and vice versa. Depending on whether you want the lawyer to get you out of a deal or keep it together, be sure you have the right type of lawyer for your situation.

WARNING

If your lawyer's *only* solution to every problem is a lawsuit, you may be in the clutches of a deal-breaker who wants to run up big legal fees. Find another lawyer!

>> **Speaks your language:** Good lawyers explain your options clearly and concisely without resorting to incomprehensible legalese. Then they give you a *risk assessment* of your options to help you make a sound decision. For example, the lawyer may say that your first course of action will take longer but will give you a 90 percent chance of success, but the faster option gives you only a 50 percent chance of prevailing.

Getting the most out of a lawyer

Whoever said that an ounce of prevention is worth a pound of cure must have been thinking of lawyers. A two-hour preventive consultation with your lawyer is infinitely better than a two-month trial that may take place just because you "saved" money by avoiding a consulting fee.

TIP

If you're not sure whether you need a lawyer, Chapter 12 contains a clause you can put in your contract to get out of any deal that isn't approved by your lawyer. You don't actually need to get a lawyer if you use this clause; it just gives you the option to have the contract reviewed later by a lawyer if you want.

Good lawyers are strategists. Given adequate lead time, they can structure nearly any deal to your advantage. Conversely, if you bring wonderful lawyers into the game after the deal is done, all they can do is damage control. The best defense is a good offense.

REMEMBER

Beware of the *legal awe* factor. Some people hold lawyers in awe because their word is viewed as law. Disobey lawyers, they think, and you'll go to jail. Baloney. Don't blindly follow your lawyer's advice. If you don't understand the advice or if you disagree with it, question it. You may be correct, and the lawyer may be wrong. Lawyers are every bit as fallible as everyone else. In complicated and confusing situations, it may be worth getting a second opinion.

Chapter 10

What's It Worth?

Y ou see a home for sale. The asking price is $249,500. Is that charming cottage a steal or an overpriced turkey?

If you don't have the faintest idea, don't worry — that's normal. Most buyers don't know property values when they start hunting for a home. To become an educated buyer, you need to take time to familiarize yourself with property values. This chapter helps you understand property values so you can get your dream home for the best price.

Preparing to Tour an Endless Parade of Homes

When coauthor Ray began his real estate career, he spent dozens of hours each week looking at houses. Like all new agents, his appetite for property was boundless and indiscriminate — big houses, tiny condos, old property and new, houses in pristine condition or fixer-uppers, uptown, downtown, and midtown. If it had a roof and a For Sale sign, Ray toured the property inside and out.

Why? The best way to learn property values is to eyeball as many houses as possible and then monitor them until they sell. That's how agents educate themselves.

TIP

You don't need to see every house in town to get educated. A good agent can accelerate your learning curve by playing the real estate version of show and tell. You have to tour only houses that meet your specific wish list for budget, style, size, and neighborhood. After seeing no more than a dozen houses comparable to your dream home, you should be an educated buyer.

Don't be surprised if you're utterly confused after a day spent looking at property for sale. When you see six or seven houses in rapid succession, it's challenging to remember which one had the wonderful kitchen and which one had the huge backyard with a swing set. To make your property tours productive, follow these tips:

>> **Take notes.** You'll probably get a listing statement (those one-page, house-for-sale advertisements/marketing pieces), brochure, or Multiple Listing Service fact sheet describing each property you visit. To help you remember the house, make notes directly on your information sheet regarding distinguishing features such as a sunken living room, a crazy floor plan, or a location near a commuter rail stop.

>> **Review the tour.** After you finish for the day, discuss the houses you saw with your real estate agent (if you have one). If your memory is fuzzy about a property or two that you visited, your agent can probably fill in the details.

>> **Save the info sheets.** As you'll see when you read the "Determining Fair Market Value: Comparable Market Analysis" section later in this chapter, sale prices are mighty important negotiating tools. Ask your agent (or the listing agent, if you don't have an agent) to tell you when a house you toured sells and how much it sold for. Mark the sale price and date of sale on your info sheet for future reference.

The Three Elusive Components of Worth

Oscar Wilde said a cynic is someone who knows the price of everything and the value of nothing. In the real estate game, neither *cost* nor *price* is the same as *value*. When you understand what these words mean and how they differ, you can replace emotion with objectivity when looking at houses and during price negotiations after you finally make an offer. Out-facting people usually beats trying to out-argue them.

Value is a moving target

REMEMBER

Value is your opinion of what a particular home is worth to you, based on how you intend to use it now and in the future. Value isn't carved in stone; on the contrary, it's pretty darn elusive.

For one thing, opinions are subjective. We, your humble authors, may think that we resemble George Clooney and Brad Pitt. You, on the other hand, are of the opinion that we look like Boris Karloff and Bela Lugosi — in full monster makeup. No harm done, as long as we all realize that a big difference exists between subjective opinions and objective facts.

Furthermore, *internal factors* — things related to your personal situation — have a sneaky way of changing over time. Suppose that you currently place great value on a home with four bedrooms and a large, fenced-in backyard. The home must be located in a town with a good school system. Why? Because you have young children.

Twenty years from now, when the kids are grown and have moved out (you hope!), you may decide to sell the house. Why? Because you no longer need such a big home. Neither the house nor the school system changed — what changed were internal factors regarding your use for the property, and thus its value to you.

External factors are things outside your control that affect property values. If your commute time is cut in half because mass-transit rail service is extended into your neighborhood after you buy your home, your home's value may increase. If a garbage dump is built next door to you, you'll have a big problem getting top dollar for your house when you sell it.

TECHNICAL STUFF

The law of supply and demand is another external factor that affects value. If more people want to buy than sell, buyer competition drives home prices up. Conversely, if more people want to sell than buy, home prices drop. A high number of distressed property sales (foreclosures or short sales) in a neighborhood also drag down property values. See Chapter 4 for a complete explanation of all the factors that influence home prices.

Cost is yesterday

Cost measures past expenditures — for example, what the sellers paid when they bought their house. What the sellers originally paid or how much they spent fixing up the house after they bought it doesn't mean diddlypoo as far as a house's present or future value is concerned. That was then; this is now.

UNEDUCATED BUYERS ARE INADVERTENT LIARS

Coauthor Ray has only to look into a mirror to see the perfect prototype of a lying, uneducated buyer. Like all buyers, Ray honestly believed he was telling his agent the truth when he and his sweet wife, Annie B., began looking for a home in the wine country about 50 miles north of San Francisco. When they first met their agent, Beverly Mueller, Ray wasted no time establishing the ground rules of the house hunt.

"We don't need the Taj Mahal," Ray told Beverly. "I'm in the real estate army, not a civilian like your other buyers. Trust me when I tell you that $300,000 is the flat-out, absolute upper limit of what we'll spend."

"I understand, Ray. We'll only look at places under $300,000," Beverly said.

And look they did. Over the next several months, Beverly showed them every house in their price range on the market in the Sonoma Valley. Ray and Annie rejected each and every one. Either they liked the land and hated the house, or vice versa. They were ready to give up when they got lucky.

Ray and Annie found Woodpecker Haven, the home they ended up buying in Glen Ellen, thanks to Karen and Herman Isman, friends of theirs who were also working with Beverly. Karen and Annie drove from San Francisco to Glen Ellen together to see a house that Beverly thought the Ismans would like. It was love at first sight — not for Karen, but for Annie.

Why hadn't Beverly shown Ray and Annie the house? Because its asking price was $390,000 — far more than the $300,000 ceiling Ray imposed. Beverly's mistake was believing what Ray told her when they first met.

Why did Ray lie? He didn't intend to. Had he and Annie found what they wanted for less than $300,000, Ray would've been telling the truth. Only after three months of looking at property did it become clear that what Ray and Annie wanted to spend and what they wanted to live in were totally out of whack with market reality.

Many buyers hit the same wall sooner or later during their education process. Ray and Annie belatedly realized that they had to make trade-offs — either reduce their expectations to fit their budget or expand their budget until it satisfied their expectations.

That was when they became educated buyers. They were finally realistic enough to make tough decisions.

What Ray and Annie experienced happens to most people. For example, you ruefully decide that either the swimming pool or the family room has to go because you can't afford a home that has both and still buy in the neighborhood you like. Or perhaps you opt to keep your wish list intact and buy in a slightly less wonderful neighborhood. Something has to give when you're forced to confront reality.

The other alternative is expanding your buying power. Much as you'd like the security of a 30-year fixed-rate loan, you decide to get an adjustable-rate mortgage instead because the ARM allows you to qualify for a bigger loan. Much as you'd like to buy without financial assistance from your parents, you swallow your pride and ask them for a loan. Again, something has to give.

Groucho Marx once said he'd never belong to any club that would have him as a member. Paraphrasing Groucho, most folks would rather not settle for the house they can afford if, by stretching themselves a bit, they can buy their dream home.

For example, when home prices skyrocketed in most parts of both coasts during the latter half of the 1990s and into the mid-2000s, some buyers accused sellers of being greedy. "You paid $400,000 seven years ago. Now you're asking $850,000," they said. "If you get your price, you'll make an obscenely large profit."

"So what?" sellers replied compassionately. "If you don't want to pay our modest asking price, move out of the way so those nice buyers standing behind you can present their offer." In a hot seller's market, people who base their offering price on what sellers originally paid for property waste everyone's time.

However, the market doesn't always go in the same direction forever. In the early 1990s and again in the late 2000s, for example, prices declined in many areas. Sellers would've been ecstatic to find buyers willing to pay them what they'd paid five years earlier, when home prices peaked.

Price is what it's worth today

Sellers have *asking prices* on their houses. Buyers put *offering prices* in their contracts. Buyers and sellers negotiate back and forth to establish *purchase prices*. Today's purchase price is tomorrow's cost. Is the purchase price a good value? That depends.

You may get a bargain if you find a house owned by people who don't know property values or who must sell quickly because of an adverse life change such as divorce, job loss, or a death in the family. Folks who don't have time to sit around waiting for buyers willing to pay top dollar usually take a hit when they sell. Time is the seller's enemy and the buyer's pal.

If, however, you must buy quickly to relocate for a new job or to get your kids settled before school starts, watch out. You can overpay because you don't have enough time to search for a good deal.

REMEMBER

Cost is the past, price is the present, and value (like beauty) is in the eye of the beholder. What the sellers paid for their house years ago, or what they'd like to get for it today, doesn't matter. Don't squander your hard-earned money on an over-priced house to satisfy a seller's unrealistic fantasy.

Fair Market Value

Natural disasters aside, every home will sell at the right price. That price is defined as its *fair market value* (FMV) — the price a buyer will pay and a seller will accept for the house, given that neither buyer nor seller is under *duress*. Duress can come from life changes such as major health problems, divorce, or sudden job transfer, which put either the buyers or sellers under pressure to perform quickly. Distressed property sales are an extreme example of financial duress. If appraisers know that a sale is made under duress, they raise or lower the sale price accordingly to more accurately reflect the house's true fair market value.

Fair market value is more powerful than plain old *value.* As a buyer, you have an opinion of what the house is worth to you. The sellers have a separate, not neces-sarily equal (and probably higher) opinion of their home's value. These values are opinions, not facts. You can't bank opinions.

REMEMBER

Unlike value, fair market value is fact. It becomes a fact when buyers and sellers agree upon a *mutually acceptable price.* Just as it takes two to tango, it takes a buyer and a seller to make fair market value. Facts are bankable.

When fair market value isn't fair — need-based pricing

Whenever the real estate market gets all soft and mushy, many would-be sellers feel that fair market value isn't fair at all. "Why doesn't our house sell?" they ask. "Why can't we get our asking price? It's not fair."

"CAN'T SELL" VERSUS "WON'T SELL"

Two weeks of extraordinarily heavy winter rain years ago undermined the soil of a subdivision in the Anaheim Hills area of Los Angeles. After this drenching, homes in the 25-acre development began slipping downhill at the rate of about 1 inch a day.

Home foundations and swimming pools cracked. Streets and sidewalks buckled. Local authorities finally ordered everyone in the subdivision to evacuate their homes until the ground stabilized.

Unlike most frustrated sellers we know, these folks really *couldn't* sell their homes. Forces of nature beyond their control reduced their houses' value to zero. Other than salvage value, no market exists for unintentionally mobile homes.

Fortunately, most homeowners who claim that they "can't" sell their houses don't have this problem. They aren't disaster victims whose homes are suddenly rendered valueless by an act of God. On the contrary, they have buyers galore for their houses, as well as scads of lenders who'd make loans to those buyers.

If nothing is wrong with their houses, what's the problem? The homeowners. The problem isn't that they *can't* sell. These homeowners *won't* sell.

As long as homeowners choose not to accept what buyers are willing to pay for their houses, they won't sell — and those houses will remain on the market at their inflated asking prices. It's a self-fulfilling prophecy. As a prospective home buyer, beware of such greedy, unrealistic sellers.

REMEMBER

Don't let your highly developed sense of fair play make a sucker out of you. Sellers frequently confuse "fair" with "impartial." Despite its friendly name, fair market value isn't a warm, cuddly fairy godmother. On the contrary, it can be heartless and cruel. Need isn't a component of fair market value. Fair market value doesn't care about any of the following:

>> How much the sellers *need* because they overpaid for their house when they bought it

>> How much the sellers *need* to recover the money they spent fixing up their house after they bought it

>> How much money the sellers *need* to pay off their loan

>> How much money the sellers *need* from the sale to buy their next humble abode: Buckingham Palace

Here's why a seller's *need-based pricing* doesn't enter into fair market value. Suppose that two identical houses next door to each other are listed for sale. One house was purchased for $32,000 three decades ago. The other house sold a couple of years ago for $320,000, soon after home prices peaked in the area. The first home has no outstanding loan on it. The other still has a big mortgage.

Bill and Mary, who own the house purchased two years ago, *need* more money than Ed, owner of the house purchased thirty years ago. After all, they paid ten times as much as Ed for their house, and they owe the bank big bucks to pay off their mortgage.

Because the houses are basically identical in size, age, condition, and location, they have the same fair market value. Not surprisingly, they both sell for $275,000. That gives Ed a nice nest egg for retirement but barely pays off Mary and Bill's mortgage. Fair? Ed thinks so. Bill and Mary don't.

Fair market value is brutally impartial. It is what it is — not what buyers or sellers want it to be.

Median home prices versus fair market value

Some folks think that median sale prices for homes indicate fair market values. They don't.

TECHNICAL STUFF

Organizations such as the National Association of Realtors, the chamber of commerce, and private research firms generate *median sale-price statistics* by monitoring home sales in a specific geographic region such as a city, county, or state. One function of these organizations is to gather market-research data on home-sales activity.

There's nothing magical about the *median sale price.* It's simply the midpoint in a range of all the home sales for a reporting period. Half the sales during the reporting period fall above the median, and half fall below it. The median-price home, in other words, is the one exactly in the middle of the prices of all the houses that sold.

When this book went to press, the median sale price of a home in America was about $272,100, which tells you that half the homes in America sold for more than $272,100, and half sold for less than $272,100. Unfortunately, all you know about this hypothetical median-price home is its price.

You don't know how many bedrooms or baths the median-price home has. Nor do you know how many square feet of interior living space the house has, how old it is, or whether it has a garage or a yard. You don't even know where this elusive median-price house is located, other than that it's somewhere in the United States.

If median-price information is so vague, why bother with it? Because it tells you two important things:

>> **Price trends:** If the median price of a home in America was $221,800 ten years ago and is $272,100 now, you know home prices in general are rising. You don't know why median prices are going up, just that they are.

>> **Price relativity:** If the median-price home in Yakima, Washington, sells for $230,500 versus $661,300 for the median-price Honolulu home, you know that you'll get a much bigger bang for your housing buck in Yakima. Honolulu has many excellent qualities, but cheap housing isn't one of them.

Median-home-price statistics make interesting reading, but they aren't any more accurate for determining specific home values than median-income statistics are for determining how much you'll earn from your next employer. You need much more precise property-value information before you invest a major chunk of your life's savings in a home.

INVESTIGATE

You'll find a graph showing U.S. median home prices from 1975 to the present in Chapter 4. Note that they declined during the late 2000s because of a weak housing market and a large increase in foreclosures. Foreclosures distort regional statistics if counties with a large number of foreclosures are lumped together with other counties that have relatively few foreclosures but are in the same statistical market area. No single report gives a truly comprehensive snapshot of any market.

TIP

Chapter 3 can help you determine how much you can afford to spend on a home. You can find some websites in Chapter 11 that you can use to locate affordable areas by comparing median home prices on a town-by-town and neighborhood-by-neighborhood basis. When median-price statistics indicate that home prices are rising or falling sharply in an area, find out why by reading and talking to players on your real estate team, such as your agent.

Determining Fair Market Value: Comparable Market Analysis

Believe it or not, houses are like Red Delicious apples. Most houses are green and need more time on the real estate tree before they're ready to pick. A few are ripe for picking right now. The trick is knowing which is which, because houses don't turn red as they ripen.

That's one reason you must understand fair market value and know the asking prices and sale prices of houses comparable to the one you want to buy. Smart home buyers know which houses are green and which are ripe.

The basics of a helpful CMA

The best way to accurately determine a home's fair market value is to prepare a written *comparable market analysis* (CMA). A competent real estate agent can and should prepare a CMA for a home that you're interested in before you make your purchase offer. Every residential real estate office has its own CMA format. No matter how the information is presented to you, Tables 10-1 and 10-2 show you what good CMAs contain.

TABLE 10-1   **Sample CMA — "Recent Sales" Section**

Address	Date Sold	Sale Price	Bedrm/Bath	Parking	Condition	Remarks
210 Oak	04/30/20	$390,000	3/3	2 car	Very good	Best comp. Approx. same size and cond. as dream home (DH), slightly smaller lot. 1,867 sq. ft. $209/S.F.
335 Elm	02/14/20	$368,500	3/2	2 car	Fair	Busy street. Older baths. 1,805 sq. ft. $204/S.F.
307 Ash	03/15/20	$385,000	3/3	2 car	Good	Slightly larger than DH, but nearly same size and condition. Good comp. 1,850 sq. ft. $208/S.F.
555 Ash	01/12/20	$382,500	3/2.5	2 car	Excellent	Smaller than DH, but knockout renovation. 1,740 sq. ft. $220/S.F.
75 Birch	04/20/20	$393,000	3/3	3 car	Very good	Larger than DH, but location isn't as good. Superb landscaping. 1,910 sq. ft. $206/S.F.

These are facts. The CMA's "Recent Sales" section helps establish the fair market value of 220 Oak — your *dream home* that's currently on the market — by comparing it with *all* the other houses that:

>> Are located in the same neighborhood

>> Are approximately the same age, size, and condition

>> Have sold in the past six months

TABLE 10-2 **Sample CMA — "Currently for Sale" Section**

Address	Date Listed	Asking Price	Bedrm/Bath	Parking	Condition	Remarks
220 Oak (Dream Home)	04/25/20	$395,000	3/3	2 car	Very good	Quieter location than 123 Oak, good detailing, older kitchen. 1,880 sq. ft. $210/S.F.
123 Oak	05/01/20	$399,500	3/2	2 car	Excellent	High-end rehab. & priced accordingly. Done, done, done. 1,855 sq. ft. $215/S.F.
360 Oak	02/10/20	$375,000	3/2	1 car	Fair	Kitchen & baths need work, no fireplace. 1,695 sq. ft. $221/S.F.
140 Elm	04/01/20	$379,500	3/3	2 car	Good	Busy street, small rooms, small yard. 1,725 sq. ft. $220/S.F.
505 Elm	10/31/19	$425,000	2/2	1 car	Fair	Delusions of grandeur. Grossly overpriced! 1,580 sq. ft. $269/S.F.
104 Ash	04/17/20	$389,500	3/2.5	2 car	Very good	Great comp! Good floor plan, large rooms. Surprised it hasn't sold. 1,860 sq. ft. $209/S.F.
222 Ash	02/01/20	$419,500	3/2	1 car	Fair	Must have used 505 Elm as comp. Will never sell at this price. 1,610 sq. ft. $261/S.F.
47 Birch	03/15/20	$409,000	4/3.5	2 car	Good	Nice house, but over-improved for neighborhood. 2,005 sq. ft. $204/S.F.
111 Birch	04/25/20	$389,500	3/3	2 car	Very good	Gorgeous kitchen, no fireplace. 1,870 sq. ft. $208/S.F.

DISTRESSED PROPERTY SALES

The CMA should also include any distressed property sales. If foreclosures and short sales dominate the area, property values of all houses in the neighborhood are dragged lower. If, however, there have been only one or two distressed property sales in the neighborhood in the past six months, that's probably an aberration that shouldn't affect property values.

Comparing sale prices of owner-occupied, non-distressed properties to sales of bank-owned properties and short-sale properties can be difficult. When a bank takes a house back through foreclosure, the property is very often in poor condition. If that bank-owned property is sold as is, the sale price will be significantly lower than the sale price of an owner-occupied house that is the same age and size but in pristine condition. When doing an appraisal, appraisers adjust prices up or down to reflect the subject property's condition.

Short sales, in which a lender agrees to accept less than the outstanding loan balance to satisfy the debt to avoid going through the foreclosure process, present a different challenge. If, for example, you make an offer to purchase a non-distressed property, the seller usually responds to your offer within a few hours or, at worst, a couple of days. In a short sale, however, the seller doesn't have power to approve the sale. The lender must approve the sale. Accordingly, although the seller enters into contract with the buyer, the sale is conditioned upon the lender's approval of price and terms.

Banks generally move much, much more slowly because their representatives are over-loaded with cases. It's not unusual to wait 30 to 60 days for a response to your offer on a short-sale property — longer if more than one mortgage lender and private mortgage insurance are involved. The entire short-sale process from start to finish can take six months or, gasp, more. Appraisers make an adjustment if the buyer waited an inordinately long time before closing the sale.

These houses are called *comps,* which is short for *comparables.* Depending on when you began your house hunt, you probably haven't actually toured all the sold comps. No problem. A good real estate agent can show you listing statements for the houses you haven't seen, take you on a verbal tour of the properties, and explain how each one compares with your dream home.

TIP

Communicating well with your agent about subjective terms such as *large, lots of light, close to school,* and so on is critically important. You must understand precisely what the agent means when using such terms. Conversely, your agent must understand precisely what you want, need, and can afford.

If you and your agent were to analyze the sale comps in our example, you would find that houses comparable to the home you want to buy — 220 Oak, in Table 10-2 — are selling for slightly over $200 per square foot. Putting the sale prices into a price-per-square-foot basis makes comparisons much easier. As you can see in Table 10-2, anything that's way above or below the norm really leaps out at you.

The "Currently for Sale" section of the CMA compares your dream home (in this case, 220 Oak) with neighborhood comps that are *currently on the market*. These comps are included in the analysis to check price trends:

>> **If prices are falling:** Asking prices of houses on the market today will be lower than sale prices of comparable houses. This may be due to an increase of distressed property sales.

>> **If prices are rising:** You'll see higher asking prices today than for comps sold three to six months ago.

If you've been looking at houses in a specific area for a while, you've probably been in all the comps currently on the market in that area. You don't need anyone to tell you what you've seen with your own eyes. However, you do need an agent's help to compare the comps you've seen with comps you haven't seen, because some houses sold before you began your house hunt.

As Table 10-2 shows, your dream house appears to be priced very close to its fair market value based on the actual sale price of 210 Oak (in Table 10-1). Given that 220 Oak has 1,880 square feet, it's worth $392,920 at $209 per square foot. Factually establishing property value is easy when you know how.

REMEMBER

Your CMA must be comprehensive. It should include *all* comp sales in the past six months and *all* comps currently on the market. Getting an accurate picture of fair market values is more difficult if some parts of the puzzle are missing, especially in a neighborhood where homes don't sell frequently.

Like milk in your refrigerator, comps have expiration dates. Lenders usually won't accept houses that sold more than six months ago as comps. Their sale prices don't reflect current consumer confidence, business conditions, or mortgage rates. As a general rule, the older the comp, the less likely it is to represent today's fair market value.

Six months is generally accepted as long enough to have a good cross section of comp sales but short enough to have fairly consistent market conditions. But six months isn't carved in stone. If a major economic calamity occurred three months ago, for example, then six months is too long for a valid comparison. Conversely, if homes in a certain area rarely sell, you may need to examine comparable sales that occurred more than six months ago.

Sale prices are always given far more weight than asking prices when determining fair market value. Sellers can ask whatever they want for their houses; asking prices are sometimes fantasy. Sale prices are always facts — they indicate fair market value. The best proof of what a house is worth is its sale price. Don't guess — analyze the sale of comparable homes. Be sure that the comparable sales information factors in price reductions or large credits given for corrective work repairs (for example, a $5,000 credit from the sellers to the buyers to replace a broken furnace).

The flaws of CMAs

CMAs beat the heck out of median-price statistics for establishing fair market values, but even CMAs aren't perfect. We've seen people use exactly the same comps and arrive at very different opinions of fair market value. Discrepancies creep into the CMA process if you blindly compare comps without knowing all the following details of the subject properties:

>> **Wear and tear:** No two homes are the same after they've been lived in. Suppose that two identical tract homes are located next door to each other. One, owned by an older couple with no children or pets, is in pristine condition. The other, owned by a family with several small kids and several large dogs, resembles a federal disaster area. Your guess is as good as ours when figuring out how much it'll cost to repair the wear-and-tear damage in the second house. A good comparable analysis adjusts for this difference between the two homes.

>> **Site differences within a neighborhood:** Even though all the comps are in the same neighborhood, they aren't located on precisely the same plot of ground. How much is being located next to the beautiful park worth? How much will you pay to be seven minutes closer to the commuter-train stop? These value adjustments are a smidge less precise than brain surgery.

>> **Out-of-neighborhood comps:** Suppose that in the past six months, no homes were sold in the neighborhood where you want to live. Going into another neighborhood to find comps means that you and your agent must make value adjustments between two different neighborhoods' amenities (schools, shopping, transportation, and so on). Comparing different neighborhoods is far more difficult than making value adjustments within the same neighborhood.

INVESTIGATE

>> **Distressed property sales:** Foreclosures are easy to spot; short sales aren't. A house, for example, may come on the market as a normal owner-occupied property sale. However, sometime while it was being marketed or after an offer had been accepted but before the sale was completed, the house became a short sale because the owners didn't keep up their loan payments.

A good agent continually checks the Multiple Listing Service to determine whether any properties you're using as comps became short sales or bank-owned properties. If so, they may not be good comps because they sold under duress.

>> **Noncomp home sales:** What if five houses sold in the neighborhood in the past six months, but none of them were even remotely comparable in age, size, style, or condition to the house you want to buy? You and your agent must estimate value differences for three- versus four-bedroom homes, old versus new kitchens, small versus large yards, garage versus carport, and so on. If the home you want has a panoramic view and none of the other houses has any view at all, how much does the view increase the home's value? Guesstimates like these don't put astronauts on the moon.

These variables aren't insurmountable obstacles to establishing your dream home's fair market value. They do, however, greatly increase the margin of error when trying to determine a realistic offering price. You can minimize pricing problems created by these variables if you and/or your agent actually tour comparable homes inside and out.

WARNING

A valid comparison of your dream home to the other houses is impossible if you and your agent have only read about the comps in listing statements (brief data sheets about houses offered for sale) or seen the properties on a website. Here's why:

>> **Most listing statements are overblown to greater or lesser degrees.** You don't know how exaggerated the statement is if you haven't seen the house for yourself. You may consider the "large" master bedroom tiny. That "gourmet" kitchen's only distinction may be an especially fancy hot plate. The "sweeping" view from the living room may exist only if you're as tall as LeBron James. Of course, you won't know any of these things if you only read the houses' puff sheets instead of visiting them in person.

>> **Floor plans greatly affect a home's value.** Two houses, for example, may be approximately the same size, age, and condition, yet vary wildly in value. One house's floor plan flows beautifully from room to room; the rooms themselves are well proportioned with high ceilings. The other house doesn't work well because its floor plan is choppy and the ceilings are low. You can't tell which is which just by reading the two listing statements.

>> **Whoever controls the camera controls what you see.** Remember when viewing those stunning color photos or the video footage of a house advertised on a website that you're permitted to see only what the person who took the pictures wants you to see. You certainly won't get a peek at less desirable things, such as worn areas on the living room carpet or graffiti sprayed on the garage door of the house next door.

TIP

Eyeball. Eyeball. Eyeball. *Eyeballing* — personally touring houses and noting important details both inside and out with your own eyes — is the best way to decide which houses are true comps for your dream home.

Getting a Second Opinion: Appraisals versus CMAs

If you're in no rush to submit an offer and you're the suspicious type, you can double-check the opinion of value that you and your agent arrive at before making an offer on your dream home. You can pay several hundred dollars to get a professional appraisal of the house.

Getting an *unbiased* second opinion of value is always reassuring. An appraiser won't tell you what you want to hear just to make a sale. The appraiser isn't trying to sell you anything. Whether you buy the house or not, the appraiser gets paid.

WARNING

Unfortunately, the fact that the appraiser charges a fee regardless of whether you buy the house cuts both ways. Suppose that you and the sellers can't reach an agreement on price and terms of sale because the sellers are deluded. Even if your offer isn't accepted, you still get a bill from the appraiser. Paying for appraisals or property inspections before your offer is accepted generally isn't wise.

If you think a professional appraisal is vastly superior to your agent's opinion of value, think again. A good agent's CMA is usually as creditable as an appraisal. Conversely, if a professional appraisal is vastly superior because your agent is a lousy judge of property values, you should get a better agent (we show you how in Chapter 9).

In any given geographical area, appraisers usually don't eyeball nearly as many houses as agents who concentrate on that area. Appraisers aren't lazy; they use their time in other ways.

Formal appraisals are time-consuming. An appraiser inspects the property from foundation to attic, measures its square footage, makes detailed notes regarding everything from the quality of construction to the amount of wear and tear, photographs the house inside and out, photographs comps for the house being appraised, writes up the appraisal, and so on. Agents can tour 15 to 20 houses in the time it takes an appraiser to complete one appraisal.

Because touring properties is so time-consuming, because good agents are already doing the legwork, and because it's usually impossible to tour a home after the sale has been completed, appraisers frequently call agents to get information about houses the agents have listed or sold that may be comps. No matter how good an agent's description of the house is, however, personally touring the property is still best. Any appraisal's accuracy is reduced somewhat whenever the appraisal is based on comps the appraiser hasn't seen.

Agents also call one another about houses they haven't seen, so don't think that appraisers are the only ones who dial for info. However, remember that you're relying on your agent's local market knowledge to help you determine what a home is worth. If your agent hasn't seen most of the comps used in your CMA, get an agent who knows the market.

TIP

Unless you're pretty darn unsure about a property's value and willing to spend the money whether or not the deal goes through, don't waste money on a precontract appraisal.

Why Buyers and Sellers Often Start Far Apart

The average buyer may be brighter than the average seller. How else can you explain why buyers are generally so much more realistic about property prices?

It's not as though there are two different real estate markets: an expensive one for sellers and a cheap one for buyers. Sellers have access to exactly the same comps that buyers do. Yet buyers' initial offering prices tend to be far more realistic than sellers' initial asking prices. Why? Figure 10-1 may offer some insight into that question.

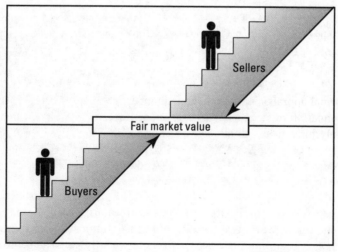

FIGURE 10-1:
How buyers and sellers approach fair market value.

© John Wiley & Sons, Inc.

BIDDING WARS

When house sellers select an agent, the interviewing process may go awry. Bidding wars often develop among the agents competing to list a seller's house for sale. The concept of fair market value is the bidding war's first victim. If you try to buy such a house, you could be the second victim. Here's what happens when a seller interviews agents:

"Thanks for giving me an opportunity to list your lovely house, Mr. and Mrs. Seller," the first agent says. "As you can see by reviewing the CMA, my fair market value analysis indicates that eight houses sold in your neighborhood in the past six months. Three of them were significantly larger than yours, so they can't be used as comps. The five houses comparable to yours in size, age, location, and condition sold in the range of $350,000 to $370,000. Based on their sale prices, I recommend an asking price of $375,000."

Now the second agent strikes. "Who cares what the comps indicate? Your house is painted a particularly attractive shade of turquoise, and your lawn is greener than the lawns of any of those other houses. I suggest starting with a nice, round asking price of $400,000."

Agent three knows that he probably won't get the listing unless he outbids the other two agents. "Our firm's internet marketing program is incredibly successful," he says, oozing confidence. "Through our relocation service and internet referrals, we'll undoubt-edly be able to find a buyer willing to pay $425,000."

This technique of successive agents giving ever-higher property valuations is known as *buying a listing*. Sellers, when confronted by the choice of market reality versus fantasy, often succumb to fantasy. They rationalize their decision by telling themselves that the highest bidding agent has the most faith in their house.

That's horse-hockey. If the sellers in this example select the highest bidder, it's because that agent dazzled them with the extra $50,000 they'd get by selecting him to sell their house. He told them what they wanted to hear. Greed triumphed over reason.

So who wins the bidding war? Not the folks who own the house. If their asking price has no basis in the real world, you won't purchase it. Neither will any other educated buyer.

How can you avoid becoming the victim of a bidding war? You know what we're going to say. Trust the comps to tell you what the house is worth. If the sellers won't listen to reason, move on. Comps don't fantasize. Neither should you.

Some people believe that the selfish interests of buyers and sellers force them to approach a house's fair market value from opposite directions. Buyers bring their offering price *up* to fair market value because they don't want to overpay. Sellers ratchet their price *down* to fair market value because they hate the thought of leaving any money on the table.

That's logical but simplistic. This reasoning still doesn't explain why many sellers initially tend to be so much more unrealistic than buyers.

The better you understand the warped thought processes of these sellers, the better you can handle their unreasonable objections to your eminently fair offer. To that end, here are the common causes of absurdly high asking prices.

Inept agents

Just because *you* use the information in Chapter 9 to select a great agent doesn't mean that everyone will. In fact, many people do a rotten job of picking an agent.

Perhaps the sellers' agent is an incompetent boob who doesn't know anything about property values. Those poor misguided sellers didn't put a smart asking price on their house because their agent gave them lousy advice.

When your agent discovers that the other agent is inept — either by the poor quality of the comps that the sellers' agent used to establish the asking price or by reputation (these things get around in the real estate brokerage community) — what can you do? That depends.

If the house has been on the market for a month or two and the sellers are open to reason, your agent's brilliant comps will prevail over their agent's fantasy pricing. The sellers will grasp the concept of fair market value and either gratefully accept your offer or make a modest counteroffer because your offering price is so logical, realistic, and fair.

You have a problem, however, if their house just came on the market. The sellers probably won't believe anything you and your agent say about the asking price being too high. They'll discount your opinion of their house's fair market value because they suspect that you're trying to steal their home. They'll be nearly as suspicious of a formal appraisal done by your hand-picked appraiser.

Time cures overpricing by inept agents. The longer the house stays on the market without selling, the more the sellers will doubt their own and their agent's opinion of value.

If the sellers' house isn't priced to sell and they won't listen to reason, move on to the next house on your list. No telling how long the sellers will take to get smart. Don't put your life on hold waiting for them to wise up. They may be very slow learners.

Unrealistic sellers

Some sellers get excellent pricing advice from agents — and choose to ignore it. Sellers attempting to sell without an agent often make the same mistake — they opt for the ever-popular need-based pricing method (which we describe earlier in this chapter) to set their asking price.

Sellers need time to accept that buyers don't care how much they paid for their house, how much they spent fixing it up, or how much they need to buy their next home. The sellers are stuck with these problems. The buyer isn't.

Unless an overpriced house has been on the market long enough to bring the sellers back to reality, move on. Most sellers aren't open to reason until they've tried their price for a couple of months or feel external pressure to sell. Trying to reason with such sellers prematurely is like trying to teach a pig how to whistle. Your time is wasted, and the pig gets upset.

SPOTTING OVERPRICED TURKEYS

Many agents show buyers overpriced houses, but their intention isn't to sell these houses. One of the tactics that smart agents adopt early in their careers is using OPTs (overpriced turkeys) to graphically demonstrate the value of well-priced homes.

Suppose an agent shows you a three-bedroom, two-bath house with a price tag of $299,500 and then takes you to an even more attractive four-bedroom, three-bath home in the same neighborhood *with the same asking price.* The agent doesn't have to say another word — the difference between price and value is glaringly obvious. The OPT makes the sale.

Here's another way to spot OPTs: They get lots of showings but no offers.

Chapter **11**

Tapping the Internet's Best Resources

No other personal financial transaction provokes more financial and emotional anxiety than buying a home. Most of us rightfully feel tremendous trepidation when buying something with upward of six digits in its sticker price. You have to confront deciding how much to spend, selecting a good real estate agent, negotiating on a property, choosing a mortgage, and handling the myriad closing details — all of which are mysterious, jargon-filled, and fraught with commission-hungry salespeople. (Ditto for the headaches when it comes time for you to sell.) Add in some time pressure for a job relocation or impending birth of a child (or two), and you have a surefire recipe for psychological stress.

Enter the internet. In this chapter, we highlight how you can use the internet to make yourself a more informed consumer and make your home search more efficient, effective, and perhaps a little less stressful. We name names and recommend for your consideration the best sites that we've reviewed. We also warn you about pitfalls you may encounter online.

Finding Useful Information

The number of real estate websites vastly exceeds those good enough to warrant a bookmark. In this section, we give some pointers so you can quickly head in the right direction when you start your search.

Get your feet wet at Realtor.com

Given the sheer number of sites among which to choose, one of the best places to start your realty web surfing is a comprehensive site that includes the most up-to-date listings of houses for sale and some useful residential real estate resources.

Our top pick for the best residential real estate website is www.realtor.com, sponsored by the National Association of Realtors (NAR). It has earned this high praise thanks to continued improvements.

Realtor.com has millions of listings of houses for sale from Multiple Listing Services (MLS) around the country. Most listings are quite detailed and usually include a map so you can see approximately where the property is located. Most listings also include numerous photos. Because this is the official website of the NAR, though, don't expect to find any For Sale by Owner listings here. Obviously, the goal of this site is to provide some resources to prospective home buyers (and sellers) in the hope of connecting you with a real estate agent. However, you may use this site as you like without providing any personal details or making a commitment to any agent.

This site also sports some useful resources to help you in your search for a specific town and neighborhood by accessing community statistics, including demographic data, public school information, and crime statistics. The site also includes some financing and moving resources.

Our chief complaint with this website is the sponsored content and ads. The good news, however, is that you can tap into this site's vast resources without having to commit to buy anything or work with anyone pitching his services.

Discover more at these sites

Several top real estate and government organizations offer useful information through their websites. So how do you know where and what to look for? Of course, we're going to tell you! Here are our top picks:

» **The Federal Emergency Management Agency** site (www.fema.gov) can help you see where various disasters (such as earthquakes, tornadoes, and floods) are most likely to strike, and it provides helpful educational resources on such topics as flood insurance and disaster preparation and prevention. The site also points you to useful resources should you or your loved ones need help someday.

» **The American Society of Home Inspectors** (www.ashi.org) can help you find an ASHI-certified inspector and teach you more about the home inspection process (which we cover in Chapter 13). You'll also find information and links to a variety of timely consumer protection issues, such as Consumer Product Safety Commission recalls and notices, water testing, and septic systems.

» **EricTyson.com** (www.erictyson.com), your humble coauthor, has a wonderful website that includes timely and thought-provoking articles on home buying and a variety of real estate topics, as well as all-important aspects of your personal finances.

» **The U.S. Department of Housing and Urban Development** (www.hud.gov) has an extensive website that includes HUD and other government agency listings of homes for sale. The site also has an excellent section for people with disabilities and their housing needs. Discrimination complaints (either as a buyer or renter) can be submitted to HUD.

» **The U.S. Department of Commerce's Bureau of Economic Analysis** (www.bea.gov) has a treasure-trove of state and local area economic data. If you want to review data to get more comfortable with the economic health of communities in which you may buy a home, this is the site for you!

Doing Some Preliminary Shopping

Our favorite thing about using the internet as a tool for hunting real estate is that you can quickly get an overview of what's going on. When it's convenient in your schedule, you can surf from website to website without salespeople pressuring you. Home buyers can use the internet to compare one community with another in such areas as school test scores, weather, and crime statistics. You can research many aspects of the buying process — including home prices, real estate agents, property inspectors, mortgages, and movers — without ever leaving your easy chair. With the click of a mouse, the internet can instantaneously whisk you *anywhere* in the world *any time* of the day or night. And you can expect to save money, too, because the web is supposed to cut out middlemen and their extra costs.

One thing we do want you to remember regarding the internet is this: Seeing is believing. Sitting in front of your computer screen in your current living quarters is simply no substitute for pounding the pavement and touring neighborhoods and properties. You can't talk to people you run into on the street, chat with school officials, or experience the quality and friendliness of local store owners through your computer.

Surveying homes for sale

Perhaps the internet's greatest contribution to the home-buying process is the ability it gives you to peruse many listings of homes for sale at your convenience. Especially if you're contemplating an out-of-state move and want a general idea of what you can get for your money, accessing home listings online is invaluable. The screening abilities of many sites with home listings sure beats scouring classified ads for a given area to find, say, all the houses in a particular community with at least three bedrooms and central air conditioning for under $300,000.

When you do a general web search (through Google, for example) for real estate in a given area (say Durham, North Carolina), the largest local and regional real estate firms' websites generally pop up on the first page or two of your search results. As we discuss in Chapter 9, the size of your prospective agent's firm has little bearing on their selection. But finding the larger realty company websites will get you in touch with more local information and listings.

Most real estate agents today will set you up to automatically receive email updates of newly listed homes for sale that meet specific criteria (for example, at least three bedrooms, at least 2,000 square feet, and so on). Just beware of agents who lazily use this instead of personally eyeballing properties and contacting you with precise feedback about given properties for sale that may meet your needs.

If you're not ready to work with an agent, start your home-search process on www. realtor.com, which allows you to search for homes that meet specific criteria.

If you're serious about actually buying a home you see promoted online, be aware that what you see isn't always what you can get. Even though one of the supposed virtues of the internet is how up-to-date it is, that's sometimes *not* the case with some sites listing homes for sale. Some websites contain outdated listings or listings with limited information. Nothing beats seeing the property in person.

Watching out for sites promoting foreclosures

With the declining real estate market in the late 2000s and the significant increase in foreclosures, many foreclosure-themed websites sprung up like weeds. Often, these sites hype the big money you can supposedly make buying foreclosures and downplay the major risks with buying and investing in foreclosed properties.

Among the various foreclosure websites, the biggest and most widely publicized is RealtyTrac, which is commonly cited as the main source for national and state foreclosure statistics. When you read and hear news reports about foreclosures increasing a certain percentage, RealtyTrac is typically the source of such data for the news media.

When Eric frequently saw RealtyTrac data, he wondered who these RealtyTrac people were and why they were collecting and publishing foreclosure data. After spending time on their website and doing some research, Eric realized that RealtyTrac's primary business is selling access to their foreclosure listings through a $49.95 monthly subscription fee.

But RealtyTrac seems to have disgruntled former customers with common complaints. The Better Business Bureau of the Southland in the Los Angeles area has logged hundreds of complaints against RealtyTrac.

Take the case of dissatisfied customer John Gassett, who signed up for RealtyTrac's "free trial" to access its foreclosure listings. Gassett was extremely disappointed with the prevalent inaccuracies he found in the local area listings he searched through, looking for potential foreclosure properties to purchase. "The information was very dated and very inaccurate. I was finding properties that had already sold six months ago."

Gassett's experience with RealtyTrac went from bad to worse when he attempted to cancel his subscription before racking up $49.95 in monthly fees. "Trying to even reach them was horrific . . . they do everything that they can to avoid interactions with customers," says Gassett. He contacted them by phone and sent numerous emails, all without success. After two months of credit-card charges, he had to step up his efforts by disputing the charges through his credit-card company. RealtyTrac failed to back down, so Gassett finally got his attorney (a relative willing to work pro bono) involved to get the monthly charges dropped and his account canceled.

Also, RealtyTrac actively and aggressively sells customer contact information to other companies. Gassett, for example, received numerous solicitations from real estate agents who bought his contact details from RealtyTrac.

Gassett ended up finding current and accurate foreclosure listings in local newspapers. Another potential way to find compilations of foreclosures is to do a web search using the name of a major city nearby and the word "foreclosure." For example, if you're interesting in looking for foreclosed properties outside of Pittsburgh, Pennsylvania, you can search "Pittsburgh PA foreclosures." But be aware that you'll find plenty of questionable sites in the search results. For more information about the realities and challenges of researching and buying foreclosures, see Chapter 8.

Sifting through school information

Even if you don't have school-age children of your own, you should investigate what's going on with local schools in areas where you're considering buying a home. If you do have school-age children, you should be even more motivated to collect information to assist you with the all-important decision of where to live.

The nonprofit website www.greatschools.org provides a useful starting point for collecting basic information about public schools in a given area. You can also see how a particular school and district compare with those in the rest of the nation. In addition, state departments of education often have useful (although often cumbersome-to-access) data on public schools. Also check out www.neighborhoodscout.com for useful school information, along with other community data.

REMEMBER

Please remember, of course, that getting data from a website is no substitute for visiting schools and talking to parents, administrators, and teachers. These latter steps take extra time and energy but often help folks make important decisions based on more complete information.

Perusing "best places" to live

Whenever a job relocation or family change prompts a move, it can raise the question, "Where's the best place to live?" What's best, of course, is in the eyes and desires of the beholder. One magazine, for example, claimed that Plymouth, Minnesota, is the best place to live, while another said it was Apex, North Carolina. Another publication gave the number one spot to Louisville, Colorado. With all due respect to those lovely towns and the people who call them home, we urge you not to accept any publication's or any person's rating of the supposed best place to live as being applicable to your situation.

Neighborhood Scout is a terrific site that allows you to select customized search criteria to find the best places that meet your needs. If you don't want to customize, you can begin your search with preset search criteria (which you may later modify) designed to appeal to common home-buying populations, such as "Families" or "Retirement Dream Areas."

THE INTERNET IS A SELLING TOOL

It may seem odd for us to include a tip about selling a house that you have yet to purchase. But sooner or later, most folks end up selling for one reason or another.

When it comes time to sell their home, many house sellers are tempted to cut real estate agents — and their typical 6 percent commission — out of the picture. Websites have made surprisingly little headway in helping consumers bypass agents by doing a For Sale by Owner — known in the trade as a FSBO.

Some sites offer scaled-back real estate agent services for a reduced commission. The communities typically served by such services tend to have higher-cost homes. Think of these realty brokers as discount brokers. With the reduced level of service comes a lower price. One site trumpets: "Sell with us and pay only 4.5 percent." To that we say, when it comes time to sell your house, should you choose to hire an agent, hire the best one you can, and negotiate the commission. You can have a great agent and a competitive commission to boot. Pick up a copy of the latest edition of our companion book, *House Selling For Dummies* (Wiley), to find out how.

Although the internet is presumed to threaten middlemen and their profit margins, that doesn't seem to be happening so far with the residential real estate buying and selling process. The fact that FSBO websites haven't really taken off is noteworthy. Remember — there's no substitute for pounding the pavement. Nothing, including the internet, beats actually seeing a home and the surrounding environment in person.

Familiarizing yourself with financing options

Many mortgage broker and lender websites provide rate quotes and offer to help you find a low-cost loan. The interactive features of some sites allow you to compare the total cost of loans, including points and fees, under different scenarios, such as how long you keep the loan and what happens to the interest rate on adjustable-rate mortgages. Interpreting these comparisons, however, requires a solid understanding of mortgage lingo and pricing. The worst sites are little more than glorified advertisement listings of mortgage lenders who pay a fee to be on the site.

TIP

Mortgage sites are best used to research the current mortgage marketplace rather than to actually apply for a loan, although you can do so on an increasing number of sites. The reason: Mortgage lending is still largely a locally driven business that varies based on the nuances of a local real estate market. Lynnea Key of Lynnea

Key Realty in San Francisco says, "I've had some near disasters with internet sites located out of the area that don't understand the local market here, where we've had multiple offers and overbids. I've had clients have to scramble at the last minute for a local lender. In multiple-offer situations, when I represent a prospective buyer, some listing agents are uneasy with mortgage preapproval letters from an internet site." (We discuss preapproval in Chapter 6.)

TIP

Mortgage data tracker and publisher HSH Associates (www.hsh.com) provides tons of information on mortgages and other types of loans through its website. If you're in the market for an adjustable-rate mortgage (ARM), you can check out the history of the interest-rate movements of various ARM indexes to see how quickly or slowly each index adjusts to overall changes in interest rates. The site also boasts useful articles and links on a variety of mortgages and selected home issues, such as environmental hazards in the home.

The Drawbacks of Searching for Houses in Cyberspace

The reality of real estate websites may fall short of what you expect. The following sections tell you what to watch out for.

Conflicts of interest

WARNING

Sadly, most websites are hardly bastions of objectivity or free of commercialism. On the contrary, internet sites often have an ax to grind or an agenda to push. They generally derive most, if not all, of their revenue from advertising. Thus, whenever you're online, always be on guard as to whether something is featured or promoted because it's truly the best or because a company or individual paid the owner of the website to be brought to your attention.

Bankruptcies

Most websites operated for commercial purposes will fail (as is the case with most small businesses in general). You should be careful when choosing to do business with any online company, especially smaller ones, those new to the real estate field, or those on the verge of running out of cash.

Misleading home-valuation tools

WARNING

Be careful when using the home-valuation tools that purport to help you determine the fair market value of a home or size up the fairness of your property taxes. You can find these tools on many general-interest real estate websites, as well as on sites such as Redfin.com, Trulia.com, and Zillow.com (these latter two are now both owned by Zillow Group). As a rule, these sites don't give you the kind of information you need to find and evaluate truly comparable properties. And they almost always end up being a referral resource for real estate agents who pay the site a fee for the referrals.

Now, with the click of your computer mouse, you can visit a multitude of websites that offer addresses, sale dates, and sale prices for houses that sold in recent years. You can look up towns where you may want to live and search by specific property, by street, or by price range. (*Note:* Not all communities' house-sales data can be found on such sites — in fact, public reporting of such information is prohibited in certain areas.)

These websites encourage prospective home buyers to use their invaluable sale price data to find areas they can afford and to determine what a particular house is worth. Existing homeowners can supposedly use the information to appeal unfairly high property-tax assessments.

If this sounds too good to be true, read Chapter 10 immediately to see the right way to value a home! Who cares whether 123 Main Street sold for $275,000 in April of last year? You need a lot more than an address, a sale price, and the date of sale to value a home properly. What about little details like size, age, condition, yard size, and so on?

Untrustworthy mortgage calculators

There's usually a significant difference between the amount the ubiquitous online mortgage calculators say you can borrow and how much you can actually *afford* to borrow. The dollar amount that a lender or online calculator comes up with is based solely on the ratio between your expected housing payments and income. Although this may satisfy a lender's concern that you won't default on a loan, such a simplistic calculation ignores your larger financial picture, such as how much (or little) you've put away for other long-term financial goals like retirement or college education for your children.

TIP

Take a hard look at your budget and goals before deciding how much you can afford to spend on a home; don't let some slick-looking online calculator decide this for you. Check out Chapters 2 and 3 for help.

4 Making the Deal

Chapter **12**

Negotiating Your Best Deal

When it comes to buying things, most Americans are lousy negotiators. Negotiation isn't part of our culture. We've been conditioned for generations to be docile buyers who pay whatever price is marked on a can of beans or a TV. Instead of negotiating with someone eyeball-to-eyeball to drive down the price, at best we comparison shop to find the store with the lowest price. (And many time-starved people don't even do that.)

Sure, we can negotiate when our back is to the wall. We haggle over expensive things like cars and dicker with the boss for a raise, but doing so makes us uncomfortable. We walk away from these encounters with the nagging suspicion that we came out on the short end of the deal — that someone else could've done better.

Realizing our nation's discomfort with negotiating, some car dealers have taken the haggling out of buying a car. Instead of using high-pressure sales tactics, these dealers post a sales price on the car — the *no-dicker sticker*. That's their price; take it or leave it. If you take it, you probably won't get the lowest price, but some people think that's a fair trade-off to avoid the unpleasantness of negotiating.

You won't find no-dicker stickers on homes. On the contrary, generally everything from the purchase price to the date that escrow closes is negotiable. Given today's high home prices in most of the densely populated parts of the United States, buying a home is the ultimate in high-stakes negotiating. Good negotiators come out of a home purchase smiling. Bad negotiators take it in the wallet.

Following the tips in this chapter will give you the negotiating advantage you so richly deserve throughout the home-buying process. And of course, these tips make getting the keys to your dream home faster, easier, and less expensive.

Understanding and Coping with Your Emotions

Emotion is an integral part of home buying. Real estate transactions are emotional roller coaster rides for everyone involved.

Sometimes, like San Francisco fog, emotion drifts into transactions so quietly that you hardly notice it. More often, however, it thunders into deals like a herd of elephants.

Examining the violent forces at work

Consider the violent forces acting on you during the home-buying process:

>> **You're dealing with people at their most primal level.** Shelter, food, and security are the three most basic necessities of life. Home is where the heart is. Your home is your castle. People become vicious when their homes are threatened. Speaking of primal urges, now you know why looking for a home is called house hunting.

>> **You're playing for large amounts of real money.** Whether this is your first home or your last, it's probably the largest purchase you've ever made. How much you pay for a home isn't the issue. When significant amounts of real money are at risk, the emotional intensity for you and the seller is just as great, whether the house you buy costs $250,000 or $2.5 million.

>> **You're probably going through a life change.** Buying a home would be plenty stressful if you only had to deal with seeking shelter and spending tons of money. Throw in a life change (such as marriage, divorce, birth, death, job change, or retirement), which is often the motivation to purchase a home, and you've created an emotional minefield.

REMEMBER

Because eliminating emotions from a home purchase is impossible, the next-best thing to do is recognize and manage them. By all means, share your concerns and frustrations in a productive way with your spouse or a friend who has purchased a home or, better yet, with a good therapist! Bottling up emotions isn't healthy or possible — the longer you stew, the worse the likely explosion is going to be. But the worst thing you can do is vent your frustrations and fears at other people in the transaction — especially the sellers. The folks who do the best job of controlling and properly directing their emotions generally end up getting the best deals.

Controlling yourself

TIP

Here are five techniques you can use to control your emotions during the home-buying negotiations:

>> **Put the transaction in perspective.** Which is worse: a failed home purchase or failed open-heart surgery? No matter how badly things go with your real estate transaction, keep reminding yourself that this isn't a life-or-death situation. Tomorrow is another day. The sun will rise again, roses will bloom again, birds will sing again, and children will laugh again. Life goes on. If worse comes to worst, the deal may die, but you'll live on to find another place that you can call home.

>> **Don't let time bully you.** Most life changes have predictable time frames. You have plenty of advance notice on marriages, births, retirements, and the like. Don't put yourself under needless pressure by procrastinating or by creating unnecessary, self-imposed deadlines. Allow yourself enough time to buy a home. Allocate time properly, and it will be your friend rather than your enemy.

>> **Maintain an emotional arm's length.** Keep your options open. Be ready to walk away from a potential house purchase if you can't reach a satisfactory agreement with the sellers on price and terms. Mentally condition yourself to the prospect that the deal may fall through. Houses are like buses: If you miss one, another will come along sooner or later.

>> **Accept uncertainty as a part of your transaction.** Much as you'd like to know everything about a property before making an offer on it, the game is played with incomplete information. You always have far more questions than answers at the beginning of a transaction. Don't worry; you'll be fine as long as you know what things you need to find out and get the answers in a timely manner during your transaction.

>> **Stay objective.** Use a comparable market analysis (CMA) to factually establish the fair market value of the home you want to buy (see Chapter 10). A good real estate agent can help you use this information to prepare an equitable offer. If you don't plan to use an agent, consider working with a real estate lawyer. Having someone to buffer you from your unavoidable emotional involvement is helpful if your composure starts to slip. Just make sure you work with professionals who are patient, not pushy, and who are committed to getting you the best deal. (Flip to Chapter 9 for tips on assembling a great real estate team.)

FIRST THINGS FIRST

Early in coauthor Ray's career, he worked with a buyer who insisted on having every question about a house answered before he'd submit an offer to purchase it. He wanted to structure a flawless offer. Because Ray didn't know any better, he went along with the plan. They spent several weeks fine-tuning the offering price by checking comparable home sales and getting quotes from contractors to do the necessary corrective work that had been discovered during the inspections ordered by a previous prospective buyer.

Unfortunately, Ray and the buyer got a hard lesson in accepting uncertainty because they overlooked one tiny detail; Ray's "buyer" didn't have a signed offer on the house. The seller got tired of their dithering around, endlessly gathering information, and sold the property to someone else.

If you're smart, you'll do what the successful buyer of this home did: Make a deal first. Condition your offer on getting all your questions answered while you have the house "tied up" with a contract.

That way, if everything goes well, you end up the proud owner of a wonderful home. If, however, you can't get a loan, or you don't like the findings of the inspection reports, you can either renegotiate the deal or bail out of the transaction and move on to a more promising home. In the meantime, however, you remove the property from the competition by getting your offer accepted.

Don't waste time getting answers to secondary questions until you answer the primary question: Can you and the seller agree on price and terms of sale? Failure to go for the commitment wastes time and money, and may cause you to lose the property.

The Art of Negotiating

Is negotiating like water or ice? If you said "water," go directly to the head of the class.

REMEMBER

Negotiating is fluid, not rigid. There's no one-size-fits-all *best* negotiating strategy that you can use in every home-buying situation. Good negotiators adjust their strategy based on a variety of factors, such as:

>> How well-priced a property is

>> How long it's been on the market

>> How motivated the sellers are

>> How motivated you are

>> Whether you're dealing from a position of strength (a buyer's market) or weakness (a seller's market)

"GOOD" DEPENDS ON YOUR PERSPECTIVE

Brace yourself. You may be shocked by the sellers' response to your offer to buy their home. From your perspective, you made a really good offer. They, on the other hand, may think your offer stinks.

Here, for example, is the perspective of first-time buyers who just blew their budget to smithereens making a $210,000 offer for a home listed at $239,000: "Honey, I'm so nervous. Do you think the sellers will accept our offer? I know their home costs a lot more than we planned to spend, but you know as well as I do that it's the best place we've seen in four months of looking. What's taking them so long to get back to us? The suspense is killing me."

And here's the perspective of the retired couple who got their $210,000 offer: "Calm down, dear. Your face is beet red. Remember your blood pressure. I'm sure that nice young couple didn't mean to insult us. And no matter what you say, I can't believe that they think we're doddering old fools who don't know how much our house is worth. They probably made the best offer they could. Please don't throw it away."

Two entirely different takes on the exact same offer. Buyers generally think that they're paying too much. Sellers usually think that they're giving their house away. When you're playing for real money, these conflicting perceptions fuel emotional fires that heat up the negotiating process.

Good negotiators, however, apply a few basic principles to every situation. If you understand these principles, you can greatly increase the odds of getting what you want.

Being realistic

Good negotiators understand that facts are the foundation of successful negotiation. If you want to become a good negotiator, you must see things as they are, rather than as you want them to be. Wishful thinking makes bad negotiation.

What's wishful thinking? A common wish in a rising real estate market, for example, is that you can pay yesteryear's price for today's home. Perhaps you saw a similar house offered for sale at a much lower price six months ago. You ignore the fact that prices have increased since then, which eliminates your chance of buying a home today at the old price. Another common (and generally unrealistic) wish is that you can afford to buy a home similar to the one that you were reared in.

How do you eliminate wishful thinking? By replacing fantasy with facts. Unfortunately, that's easier said than done, because we all inevitably get emotionally involved when we negotiate for something that we intensely desire. Even though that emotional involvement is part of human nature, allowing emotion to seep into a negotiation can cost you dearly.

The importance of objectivity

Unlike you and the seller, good real estate agents don't take things personally. The seller's agent, for example, won't be offended if your agent tells her that you hate the emerald-green paint in the kitchen and the red flocked wallpaper in the den. Your agent, by the same token, won't blow his cork if the seller's agent says that your offer is ridiculously low.

Agents find it easy to be objective. After all, they're not the ones who spent three long weekends painting the kitchen or months looking for just the right wallpaper to put in the den. Neither is it their life's savings on the negotiating table.

Good agents listen to what the market says a house is worth. They don't allow distracting details (such as how much the seller paid for the house ten years ago, or how little you can afford to spend for it today) to confuse negotiations. As you know if you've read Chapter 10, no correlation exists between these need-based issues and the current fair market value of a home.

Some folks think that agents have calculators for hearts. Not true. The good agents know that if they aren't coldly realistic about property values, the home won't sell, and they won't get paid.

The red flags in agent negotiations

If you follow our advice in Chapter 9 when selecting your real estate agent, you'll choose one who's a good negotiator.

WARNING

Doing a lousy job of selecting an agent can cost you big bucks. Bad agents don't know how to determine fair market values; as a result, you may pay too much for your home. And why should the bad agent care? After all, the more you pay, the more your agent makes, because agents' commissions are typically a percentage of the purchase price. If your agent pushes you to buy and can't justify the offering price by using comparable home sales, fire your agent and get a good one.

Good negotiators avoid making moral judgments. As long as the seller's position isn't illegal, it's neither immoral nor unfair. It's simply a negotiating position. Of course, agents are human. Sometimes, even the best agents *temporarily* lose their objectivity in the heat of battle. You know this has happened if your agent gets red in the face and starts accusing the other side of being unfair.

If your agent snaps out of the funk quickly, no problem. On the other hand, if your agent can't calm down, you've lost your emotional buffer. Agents who lose their professional detachment are incapable of negotiating well on your behalf.

TIP

No matter how satisfying it may be to go on an emotional rampage with your agent about the seller's utter lack of good taste, market knowledge, or scruples, getting angry won't get you the house. If your agent can't maintain a level head, ask your agent's broker (see Chapter 9) to negotiate for you, or get another agent.

Examining your negotiating style

Finding two people who have exactly the same negotiating style is as unlikely as finding two identical 200-year-old houses. All negotiating styles, however, boil down to variations on one of these two basic themes:

>> **Combative (I win, you lose):** These negotiators view winning only in the context of beating the other side. To them, negotiation is war. They take no prisoners.

>> **Cooperative (we both win):** These negotiators focus on solving problems rather than defeating opponents. Everyone involved in the transaction works together to find solutions that are satisfactory to both sides.

Which negotiating style is better? That depends on the kind of person you are, what your objectives are, and how much time you have.

Most folks opt for cooperation because they know that the world is round — what goes around nearly always comes back either to haunt you or to help you. Why fight battles in some weird game of mutually assured destruction if you can peacefully work together as allies to solve your common problems?

Combative negotiation is tolerated in a strong buyer's or seller's market. The operative word is *tolerated.* People grudgingly play an "I win, you lose" game when they have no alternative. However, in a balanced market that favors neither buyers nor sellers, combative negotiators are usually told, "I won't play your stupid game because I don't like your style."

FACT VERSUS OPINION

You and the sellers can use exactly the same facts (that is, recent sale prices of comparable houses) and yet reach entirely different opinions of fair market value. As we point out in Chapter 10, although houses may be comparable in terms of age, size, and condition, no two homes are identical after they've been lived in.

Furthermore, even though all the houses used in the comparable market analysis are in the same neighborhood, *site differences* (that is, proximity to schools, better view, bigger yard, and the like) usually affect individual property values. Last but not least, even though all the comparable houses were sold during the previous six months, property values can be affected by changes in mortgage rates and consumer confidence.

For example, your agent thinks that 123 Main Street, which sold two months ago for $280,000, is the best comparable *(comp)* for the house that you're trying to buy. The seller's agent agrees that it's a good comp but points out that this house has a two-car garage, whereas 123 Main Street has only a one-car garage. Your agent says that 123 Main Street has a larger kitchen with a breakfast nook and is two blocks closer to the park. The seller's agent says that the property you're considering has higher-quality kitchen cabinets and a new refrigerator, and is three blocks closer to the bus stop.

And so it goes. Everyone agrees on 123 Main Street's sale price and date of sale. These are *facts.* They're the same no matter who looks at them.

But how much value does a second garage space add to the home that you want to buy? Is being closer to the bus stop worth more to you than proximity to the park? Is an eat-in kitchen more or less valuable to you than fancy kitchen cabinets and a new refrigerator? The answers to these questions are *opinions* that are based on your value judgments. Another person would probably value the amenities somewhat differently.

Pricing isn't 100 percent scientific at this level of scrutiny. No two buyers are alike. Each buyer has different needs and, due to those differing needs, will reach different conclusions regarding opinions of value.

Cooperative negotiation, on the other hand, works well under all market conditions because its goal is to scratch everyone's itch. We all enjoy winning and hate losing. People sometimes cry when they're defeated. Problems, on the other hand, never cry when they're solved.

TIP

Unfortunately, some people are born competitors. The only cooperation they understand is the cooperation of a team that's working together to defeat its opponents. If you're a cooperative negotiator, here are two ways to protect yourself from combative negotiators:

>> **Try switching them from combative to cooperative by finding ways you both can win.** Shift their emphasis from beating you to solving the problem. "You want to sell. I want to buy. How can we do it?"

>> **If that fails, deep-six the deal.** If you keep negotiating, born competitors will strip the money from your bank account and the flesh from your bones. They confuse concessions with weakness. If, for example, you offer to split the difference, they'll take the 50 percent you give them as their just due and then go for the rest. They won't be happy until they thrill to a victory that's enhanced by your unconditional defeat. Life's too short to subject yourself to this kind of punishment. No matter how strong a seller's market you must contend with, you can find sellers who are cooperative negotiators — if you try.

Negotiating with finesse

TIP

Skillful negotiators get what they want through mutual agreement — not brute force. Brute force is crude, rude, ugly, and decidedly unfriendly. Here are some concepts that you may find useful for negotiating with finesse:

>> **Phones are for making appointments.** Never, never, never let your agent or lawyer present an offer or attempt to negotiate significant issues over the phone. Saying no over the phone is too easy for the sellers. Even if they agree with everything you want, they may change their minds by the time they actually have to sign the contract.

>> **Oral agreements are useless.** In our society, we have *written* contracts because people have notoriously selective memories. If you want your deal to be enforceable in a court of law, put everything about it in writing. Get into the habit of writing short, *dated* MFRs (Memos for Record) of important conversations (such as "June 2 — lender said we'd get 6.0 percent mortgage rate," "June 12 — sellers want to extend close of escrow a week," and so on).

Put these notes into your transaction file, just in case you need to refresh your memory. Heed the immortal words of Samuel Goldwyn: "A verbal agreement isn't worth the paper it's written on."

» **Deadline management is essential.** Real estate contracts are filled with deadlines for things like contingency removals, deposit increases, and (of course) the close of escrow. Failure to meet deadlines can have dreadful consequences. Your deal can fall apart — you can even get sued. Most deadlines, however, are flexible — if you handle them correctly. Suppose that you just found out that completing the property inspections will take longer than anticipated. *Immediately* contact the sellers to explain the reason for the delay and then get a *written* extension of the deadline. Reasonable delays can usually be accommodated if properly explained and promptly handled.

BRUTE FORCE VERSUS STYLE

Ray knows a real estate agent who's a superb technician. He's brilliant at determining a home's fair market value; writes flawless contracts; understands financing; and stays current on real estate laws, rules, and regulations. Technically, he's impeccable.

Unfortunately, he has no compassion. He's coldly perfect himself and expects equal perfection from everyone else. He's an ultra-hardball negotiator who neither gives nor expects mercy from his opponents. Some of his own clients don't even like him, but they all respect him because they know that he'll fight ruthlessly on their behalf.

Other agents hate working with this agent *because* he's such a brutal negotiator. They deal with him only when they have absolutely no alternative. If he represents a buyer in a multiple-offer situation, for example, his buyer's offer won't be selected if the selling agent can find any way to work with another buyer whose agent is less combative.

Ray Jones, a San Francisco agent who died several years ago, was this agent's exact opposite. Jones lacked technical polish, but he was kind, fair, and generous, and made folks smile with their hearts. His clients and other agents adored him. In a multiple-offer situation, his buyer's offer was either accepted or at least counteroffered, if possible. He made buying a home fun.

Think carefully when selecting an agent to represent you. People do business with you for only two reasons: because they have to or because they want to. The right agent can give you a negotiating advantage. The ruthless negotiator may make sense for buyers who are in no hurry to buy and who desire to get a good deal on a property. For others, such a piranha can be bad news.

The Negotiating Process

Negotiation is an ongoing process — a series of steps without a neatly defined beginning and end. Think of water flowing.

Each step in the negotiating process begins by gathering information. After you read this book, you'll understand the various aspects of buying a home. Then you can translate your information into action that generates more information that in turn leads to further action. And so it goes, until you're the proud owner of your dream home.

One way to begin the first action phase is to get your finances in order, get preapproved for a loan, and select an agent to work with you through the next information-gathering phase. You and your agent then investigate various neighborhoods and tour houses so you know what's on the market. You also figure out the difference between asking prices and fair market values. After you know what houses are really worth, you're ready to focus on the specific neighborhood you want to live in and begin seriously searching for your dream home.

Making an offer to purchase

After you find your dream home, you're ready for the next action step in the negotiating process: making an offer to purchase. No standard, universally accepted real estate purchase contract is used throughout the country. On the contrary, purchase contracts vary in length and terms from state to state and, within a state, from one locality to another.

We include the California Association of Realtors' *Residential Purchase Agreement* in Appendix A so you can see what a well-written, comprehensive residential real estate contract looks like. When you're ready to write an offer, your real estate agent or lawyer should provide a suitable contract for your area.

Real estate contracts are revised quite often because of such things as changes in real estate law and mandated seller disclosure requirements. A good agent or lawyer will use the most current version of the contract. Check the contract's revision date (usually noted in the bottom-left or -right corner of each page) to make sure you're not using a form just slightly newer than the Declaration of Independence.

WARNING

A carelessly worded, poorly thought-out offer can turn what should be a productive negotiation into an adversarial struggle between you and the sellers. Instead of working together to solve your common problem (that is, "you want to buy, and they want to sell — how can *we* each get what *we* want?"), you get sidetracked by issues that can't be resolved so early in the negotiating process.

Although buying a home can be a highly emotional experience, good offers defuse this potentially explosive situation by replacing emotion with facts. Buyers and sellers have feelings that can be hurt. Facts don't. That's why facts are the basis of successful negotiations.

All good offers have three things in common:

>> **Good offers are based on the sellers' most important concern: a realistic offering price.** You shouldn't pull the offering price out of thin air. Instead, base your offering price on houses (comparable to the seller's house in age, size, condition, and location) that have sold within the past six months. As Chapter 10 explains, sellers' asking prices are often fantasy. Actual sale prices of comparable houses are facts. *Focus on facts.*

>> **Good offers have realistic financing terms.** Your mortgage's interest rate, loan origination fee, and time allowed to obtain financing (explained in the upcoming section on contingencies) must be based on current lending conditions. Some offers get blown out of the water because a buyer's loan terms are unrealistic. *Focus on facts.*

TIP

If you've been prequalified or, better yet, preapproved for a loan (see Chapter 6), you or your agent should stress that advantage when you present your offer. This proves to the sellers that you're a creditworthy buyer who's ready, willing, and financially able to purchase their house.

>> **Good offers don't expect a blank check from the sellers.** Unless property defects are glaringly obvious, neither you nor the sellers will know whether any corrective work is needed at the time that your offer is initially submitted. Under these circumstances, it's smart to use property-inspection clauses (explained in the next section) that enable you to reopen negotiations regarding any necessary corrective work *after* you've received the inspection reports.

TIP

Remember that negotiation is an ongoing process. After the *action* of having your offer accepted, your property inspectors gather *information*. After they've determined what's actually required in the way of corrective work, you and the sellers can renew your negotiations *(action)* armed with hard facts *(information)*. This sequence beats wasting time and energy by arguing with the sellers about the cost to complete corrective work before any of you know the precise number of dollars needed to do the repairs. *Focus on facts.*

If the sellers agree with the price and terms contained in your offer, they'll sign it. Their agent should give you a signed copy of the offer immediately. When you actually receive a copy of the offer signed by the sellers, you have what's called a *ratified offer* (that is, a signed or accepted offer). This doesn't mean you own the house or it has been sold. All you can say for now is that a sale is pending.

Leaving an escape hatch: Contingencies

Even though the sellers have accepted your offer, it should contain extremely important escape clauses known as contingencies, which you cleverly built into the contract to protect yourself. A *contingency* is some specific future event that must be satisfied in order for the sale to go through. It gives you the right to pull out of the deal if that event fails to happen. If you don't remove a contingency, the sale falls apart, and your deposit money is usually returned.

These two contingencies appear in nearly every offer:

>> **Financing:** You can pull out of the deal if the loan specified in your contract isn't approved. Some contracts also have separate property appraisal contingencies that allow you to cancel the deal if a qualified appraiser does not think the home is worth what you've agreed to pay. As noted in Chapter 8, short sales require a special financing addendum. We thoughtfully included a short-sale addendum in Appendix B.

>> **Property inspections:** You can pull out of the deal if you don't approve the inspection reports or can't reach an agreement with the sellers about how to handle any necessary repairs.

Other standard contingencies give you the right to review and approve such things as a condominium's master deed, bylaws, and budget, as well as a property's title report. You can, if you want, make the deal contingent on events such as your lawyer's approval of the contract, or your parents' inspection of the house. As a rule, no *reasonable* contingency will be refused by the seller.

TIP

Don't go overboard with contingencies if you're competing for the property with several other buyers. Sellers, especially in strong real estate markets, don't like offers with lots of contingencies. From their perspective, the more contingencies in an offer, the more likely the deal is to fall apart. You must delicately balance the need to protect yourself with the compelling need to have your offer accepted. Keep your contingency time frames realistic but short. Resolve as many simple questions as possible before submitting the offer. For instance, if your parents insist on seeing the property you want to buy before they'll loan you money for a down payment, take them through the home before making your offer to eliminate that contingency.

WARNING

If you're considering making your offer subject to the sale of another house (such as the one you're living in now), don't do so if you're in a bidding war with other buyers. Check out the section "Negotiating from a position of weakness," later in this chapter, for reasons why including this type of contingency can cost you the house you're bidding on.

Here's a typical loan contingency:

> Conditioned [the magic word] upon buyer getting a 30-year, fixed-rate mortgage secured by the property in the amount of 80 percent of the purchase price. Said loan's interest rate shall not exceed 5.0 percent. Loan fees/points shall not exceed 2 percent of loan amount. If buyer can't obtain such financing within 30 days from acceptance of this offer, buyer must notify seller in writing of buyer's election to cancel this contract and have buyer's deposits returned.

If you want to see a more detailed financing contingency, read Paragraph 3 of the purchase contract in Appendix A of this book. We cover property inspections in Chapter 13.

The purchase agreement you sign is meant to be a legally binding contract. As we say in Chapter 9, it's wise to put a lawyer on your team immediately if you have *any* concerns about the legality of your contract. Even if you don't have a lawyer when you sign the contract, including the following clause in your offer may be prudent if you have legal questions:

> Conditioned upon my lawyer's review and approval of this contract within five days from acceptance.

Using this clause doesn't mean you actually have to hire a lawyer. It does, however, give you the option of having the contract reviewed later by a lawyer if you want. By the way, good contracts provide space to write in additional terms and conditions. In the contract in Appendix A, it's Paragraphs 5 and 6.

Include a provision in your contract that specifically states that contingencies must be removed in writing. Doing so should eliminate confusion between you and the sellers regarding whether a contingency has been satisfied. See Paragraph 14 of the contract in Appendix A for one way to handle this.

What good is a ratified offer filled with escape clauses? Well, a ratified offer (riddled with escape clauses or not) ties up the property. You don't have to worry about the owners selling the property to someone else while you're spending time and money inspecting it.

First get an agreement on the price and terms of sale — *then* get answers to all your other questions.

Getting a counteroffer

It's highly unlikely that the sellers will accept your offer as it's originally written. Even if they love your offering price, they'll probably tweak your offer here and there to make it acceptable to them. Sellers use *counteroffers* to fine-tune the price, terms, and conditions of offers they receive.

You'll be relieved to know that counteroffer forms are far less complicated than offer forms. Take a look at the *Buyer Counteroffer* in Figure 12-1, for example, and you'll see that it's only a one-page form. You can access a counteroffer form online at www.dummies.com/go/homebuyingkit7e.

Suppose you offer $275,000 for a home you like, and you ask to close escrow 30 days after the sellers accept your offer. Because they had the house listed at $289,500, the sellers think that your offering price is a mite low. Furthermore, they need six weeks to relocate.

Instead of rewriting your entire offer, they give you a counteroffer. It states that they're willing to accept all the terms and conditions of your offer except that they want $285,000 and six weeks after acceptance to close escrow.

The ball's in your court once again. You don't mind a six-week close of escrow, but you don't want to pay more than $280,000, so you give the sellers a *counter-counteroffer* to that effect.

Now only one bone of contention remains: the price. The sellers come back to you with a *firm* $284,000. You grudgingly respond at $281,000 and instruct your agent to make it clear to the sellers that you won't go any higher. Two can play the *firm* game. Negotiations now resemble the trench warfare of World War I.

If you really want the home, this phase of the game can be nerve-racking. You worry about another buyer making the sellers a better offer and stealing the house away while you're trying to get the price down that last $3,000. The sellers are equally concerned that they'll lose you by pushing too hard for the final $3,000. You don't want to pay a penny more than you have to. The sellers don't want to leave any money on the table.

You and the sellers are tantalizingly close to agreement on price. Your offering price and the sellers' asking price are both factually based on recent sales of comparable houses in the neighborhood. So why the deadlock? Because sometimes the same facts can lead to different conclusions (see the sidebar "Fact versus opinion," earlier in this chapter).

An equitable way to resolve this type of impasse is to split the difference fifty-fifty. If the sellers in our example use this technique, they'll come back to you with a $282,500 offer — down $1,500 from their *firm* asking price of $284,000 and up $1,500 from your *firm* offering price of $281,000. The mutual $1,500 concession equals less than 1 percent of the home's fair market value based on a $282,500 sale price. That's pinpoint accuracy in a real estate transaction.

CALIFORNIA ASSOCIATION OF REALTORS®

BUYER COUNTEROFFER
(C.A.R. Form BCO, 11/14)

Date _____

This is a counteroffer to the: ☐ Seller Counteroffer No.___, ☐ Seller Multiple Counteroffer No.___, or ☐ Other _____("Offer"),
dated _____, on property known as _____("Property"),
between _____ ("Buyer") and _____("Seller").

1. **TERMS:** The terms and conditions of the above referenced document are accepted subject to the following:
   A. **Paragraphs in the Offer that require initials by all parties, but are not initialed by all parties, are excluded from the final agreement unless specifically referenced for inclusion in paragraph 1C of this or another Counteroffer or an addendum.**
   B. **Unless otherwise agreed in writing, down payment and loan amount(s) will be adjusted in the same proportion as in the original Offer, but deposit amount(s) shall remain unchanged from the original Offer.**
   C. **OTHER TERMS:** _____
   _____
   _____
   _____
   _____
   _____
   _____
   _____
   _____
   _____
   _____

   D. **The following attached addenda are incorporated into this Buyer Counteroffer:** ☐ Addendum No. _____
   ☐ _____ ☐ _____

2. **EXPIRATION:** This Buyer Counteroffer shall be deemed revoked and the deposits, if any, shall be returned:
   A. Unless by 5:00pm on the third Day After the date it is signed in paragraph 3 (if more than one signature then, the last signature date)(or by ____ ☐ AM ☐ PM on _____ (date)) **(i)** it is signed in paragraph 4 by Seller and **(ii)** a copy of the signed Buyer Counteroffer is personally received by Buyer or _____, who is authorized to receive it.
   B. OR ☐ If Buyer withdraws it in writing (CAR Form WOO) anytime prior to Acceptance.

3. **OFFER: BUYER MAKES THIS COUNTEROFFER ON THE TERMS ABOVE AND ACKNOWLEDGES RECEIPT OF A COPY.**
   Buyer _____ Date _____
   Buyer _____ Date _____

4. **ACCEPTANCE: I/WE** accept the above Buyer Counteroffer **(If checked ☐ SUBJECT TO THE ATTACHED COUNTEROFFER)** and acknowledge receipt of a Copy.
   Seller_____Date_____Time_____AM/PM
   Seller_____Date_____Time_____AM/PM

CONFIRMATION OF ACCEPTANCE:

(____/____) (Initials) **Confirmation of Acceptance:** A Copy of Signed Acceptance was personally received by Buyer or Buyer's authorized agent as specified in paragraph 2A on (date) _____ at _____ AM/PM. **A binding Agreement is created when a Copy of Signed Acceptance is personally received by Buyer or Buyer's authorized agent whether or not confirmed in this document.**

Published and Distributed by:
REAL ESTATE BUSINESS SERVICES, LLC.
a subsidiary of the CALIFORNIA ASSOCIATION OF REALTORS®
525 South Virgil Avenue, Los Angeles, California 90020

BCO 11/14 (PAGE 1 OF 1) Print Date

EQUAL HOUSING OPPORTUNITY

BUYER COUNTER OFFER (BCO PAGE 1 OF 1)

FIGURE 12-1: A typical buyer counteroffer form.

TIP

Splitting the difference won't work in all situations. It is, however, a fair way to quickly resolve relatively small differences of opinion (a few percent or less of the home's price) so you can make a deal and get on with your life.

The Finer Points of Negotiating

A perfectly balanced market that favors neither buyer nor seller is rare. The market is almost always in a state of flux. As a result, the playing field usually tilts toward the buyer or seller.

Negotiating when the playing field isn't level

President Lyndon Johnson was a consummate politician. He'd cajole, promise, arm-twist, flatter, pressure, sweet-talk, threaten, jawbone, wheedle, bully, or horse-trade other politicians into supporting his legislation.

The late president's negotiating skills were legendary. Once, when accused of using somewhat unethical tactics to get the votes required to pass one of his Great Society programs, LBJ just shrugged. "Sorry you feel that way, son," he supposedly said. *"All I ever wanted was my unfair advantage."*

In a perfect world, you'd always have an unfair advantage. Unfortunately, the world is imperfect. No matter how good you are as a negotiator, sooner or later you'll have to negotiate from a position of weakness. The trick in these circumstances is to give yourself every possible advantage.

Buyer's and seller's markets

In the early 2000s, many home buyers complained bitterly about sellers taking unfair advantage of them. Given the seller's market at that time in many parts of the country, it wasn't unusual for owners of a well-priced house to receive multiple offers on it while their agent was still nailing up the For Sale sign. (Slight exaggeration, but you get the point.) Five years later, the hobnailed boot was on the other foot. Instead of a supply-demand imbalance, there was a demand-supply imbalance. The anguished screams now came from sellers who were complaining about buyers taking unfair advantage of *them.*

REMEMBER

The party in the weaker position always characterizes the market as "bad." Because you're a seeker of wisdom and truth, don't kid yourself. The market is, in reality, neither good nor bad. The market is impersonal. The market is the market. Moaning and groaning about unfair market dynamics won't help you if you're caught in a seller's market any more than complaining helps sellers caught in the viselike grip of a buyer's market.

Negotiating from a position of weakness

Newly listed homes that are priced to sell often generate multiple offers in a seller's market. But even when the market isn't a seller's market, a well-priced, attractive new listing may draw multiple offers.

WARNING

Unless you absolutely *must* have a particular home and price is no object, be careful about entering a bidding war. Such auctions can drive the price of a home above its fair market value. That situation is great for the seller, but it is financially deadly for you. We don't want you to overpay.

If you really want a home and you know that other offers will be made, here's how to improve your chances of winning in a multiple-offer situation:

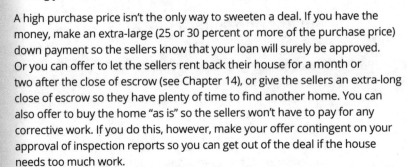

TIP

>> **Use comparable sales data to predetermine the upper limit of what you'll pay.** Don't get caught up by the excitement of a bidding war and let your emotions override your common sense. Be sure you read Chapter 10 and you know how to determine fair market value. Set no-matter-what limits on the amount you'll bid. Otherwise, you can grossly overpay.

>> **Put yourself in the sellers' position.** The sellers don't care how long you've been looking for a home or how little you can afford to pay. Faced with several offers, sellers select the offer that gives them the best combination of price, terms, and contingencies of sale. Find out what the sellers' needs are before making your offer. Their self-interest invariably prevails.

A high purchase price isn't the only way to sweeten a deal. If you have the money, make an extra-large (25 or 30 percent or more of the purchase price) down payment so the sellers know that your loan will surely be approved. Or you can offer to let the sellers rent back their house for a month or two after the close of escrow (see Chapter 14), or give the sellers an extra-long close of escrow so they have plenty of time to find another home. You can also offer to buy the home "as is" so the sellers won't have to pay for any corrective work. If you do this, however, make your offer contingent on your approval of inspection reports so you can get out of the deal if the house needs too much work.

>> **Make your best offer initially.** Buyers who win bidding contests, in the words of Civil War General Nathan Bedford Forrest, get there "firstest with the mostest." If you want the house, don't hold back in a multiple-offer situation: You may never get a second chance to make your best offer.

>> **Get preapproved for a loan.** Informed sellers worry about the financial strength of prospective buyers. They don't want to waste their time on buyers who can't qualify for a loan. All other things being equal, if you're preapproved for a loan (see Chapter 6), you should prevail over buyers whose financial

status is in doubt. And if you've been preapproved for a loan, you'll know *you* aren't wasting your time and money on a house that you may not qualify to buy.

>> **Don't make your offer subject to the sale of another house.** As discussed earlier in this chapter in the section on contingencies, if you own a house that you *must* sell in order to get the down payment for your new home, you're in trouble. You'll most likely be competing with other buyers who don't have that limitation. The sellers have enough problems selling their house without worrying about whether you can sell yours. Why should they take your offer if they can accept one without a subject-to-sale contingency in it? Offers made subject to the sale of another house get no respect in a multiple-offer situation.

TIP

>> **If you must sell in order to buy, put your old house on the market before seriously looking for a new home.** Ideally, you'll have a ratified offer on your old house before making an offer to buy a new place. Then, even with a subject-to-sale clause, your negotiating position will be much stronger. And you won't waste time worrying about how much money you'll have when and if your house sells. Stipulate a long close of escrow on the old house and the right to rent it back for several months after the sale so you'll have adequate lead time to buy your new home.

Spotting fake sellers

Why would anyone want to be a fake seller? That some people would knowingly waste their time and money on an exercise in futility is absurd.

The key word is *knowingly.* All sellers start out thinking that they're sincere. As the quest for a buyer continues, however, circumstances ultimately prove that some sellers are phony.

Fake sellers cleverly mimic genuine sellers. Like real sellers, counterfeit sellers sign listing agreements, have For Sale signs in their yards, advertise in newspapers, and have open houses on Sundays. They outwardly appear to be the real McCoy. If you don't know how to detect fake sellers, you'll waste your precious time, energy, and money by fruitlessly negotiating to buy a house that isn't really for sale.

Identifying bogus sellers is ridiculously easy once you know how. Here are five simple tests you can use to spot the fakes.

Are the sellers realistic?

The number-one reason that houses don't sell is that they have unrealistic asking prices. When people categorically state that they "can't" sell a grossly overpriced house, they expose themselves as fakes. What they're actually saying is that they refuse to accept the market's opinion of what their house is worth. People who won't listen to reason aren't sellers — they're property owners masquerading as sellers.

TIP

Real sellers may *inadvertently* overprice their homes initially. Unlike fake sellers, however, they eventually wise up. They know they have a problem if they get no offers (or only lowball offers). Authentic sellers accept the relevance of using recent sales of comparable houses in the neighborhood to establish their house's fair market value. Genuine sellers are realistic.

Are the sellers motivated?

Most folks don't sell their homes to generate commissions for real estate agents. Sellers are usually motivated by a life change, such as wedding bells, a job transfer, family expansion, retirement, or a death in the family. Perhaps the sellers are in contract to buy another home but can't complete the purchase until their house sells. Or their house may be in foreclosure. Real sellers always have a motive for selling.

In dire situations, such as an impending foreclosure or divorce, sellers often instruct their agents not to tell anyone why they're selling. If possible, however, find out why the house is being sold *before* making your offer. Knowing the sellers' motivation allows you to shape your offer's terms (that is, quick close of escrow, letting the sellers rent back the house after the sale, and the like) to fit the sellers' circumstances.

WARNING

Lack of motivation is a gigantic red flag. If the sellers or their agent say that they're testing the market, run as fast as you can in the opposite direction.

Do the sellers have a time frame?

Deadlines make things happen. Seller deadlines are often established by such things as when the twins are due, when school starts, when they have to begin new jobs in another city, when the escrow is due to close on the new home they're buying, and so on. Authentic sellers always have a deadline by which they must complete their sale.

Time is a powerful negotiating tool. If you aren't under pressure to buy and the sellers must sell immediately (if not sooner), time is your pal and their enemy. Conversely, if you have less than four weeks to find a place to live before the kids start school, the watch is on the other wrist. Ideally, you know the sellers' deadline, but they don't know yours. Most real negotiation occurs at the 11th hour, 59th minute, and 59th second of a 12-hour deadline.

SOMETIMES, HOMES SELL FOR MORE THAN THE ASKING PRICE

Amy was a buyer who knew precisely what she wanted. Her dream home didn't have to be large. It did, however, need a light and airy feeling, a gourmet kitchen, nice views, a beautiful garden, and a garage. She'd been house hunting a long time because she refused to settle for anything less than her dream home.

Amy had a good agent. When a house that met all of Amy's specifications was listed at $295,000, Amy and her agent were waiting at the front door on the first day the house was opened for inspection.

They weren't alone. The home was mobbed with drooling buyers and agents. Everything about the property, including its finely honed price, was flawless. The house was definitely priced to sell.

The listing agent told everyone that offers would be accepted in two days. Given the high level of buyer interest, Amy's agent knew there would be multiple offers. She suggested that Amy could probably beat the competition by offering $5,000 over the asking price. Based on the sale price of comparable houses in the neighborhood, the agent said the home was priced at (or perhaps slightly below) its fair market value. If all the other offers came in right at full asking price, Amy's $300,000 offer would stand out from the crowd.

Amy refused. Why, she reasoned, spend an extra $5,000 if she didn't have to? A full-price offer certainly wouldn't insult the sellers. If that wasn't enough money, Amy was sure that the sellers would give her a counteroffer.

She was wrong. The sellers didn't counter any of the many offers they received. Instead, they simply accepted the highest offer, which wasn't Amy's.

Amy took a calculated risk. She could've been right. In fact, we've seen multiple-offer situations in which not one of the offers was close to full asking price. Multiple offers are no guarantee that a house will sell at or over its asking price.

Each situation is different and must be evaluated on its own merits. And don't forget to look at the comparable sales data.

WARNING

You can be in deep trouble if you have a deadline and the sellers don't. If you reveal this information to the sellers, they may use your deadline to beat you to a pulp. Beware of procrastination. Don't let time bully you — and keep your deadlines to yourself.

Are the sellers forthright?

Genuine sellers are disarmingly candid about their house's physical, financial, and legal status. They know that withholding vital information endangers the sale and may lead to a lawsuit. Early disclosure of possible problems, on the other hand, gives everyone the lead time required to solve them. Real sellers don't have a "buyer beware" mind-set.

If you keep getting nasty surprises, you're working with fake sellers. Straightforward folks have only one defense against devious sellers who are playing an expensive, and possibly even devastating, game of "I've Got a Secret": Terminate the transaction.

Are the sellers cooperative?

Real sellers look for ways to make transactions go more smoothly. They work with you to solve problems rather than waste time trying to figure out who's to blame if something goes wrong. Genuine sellers have a let's-make-it-happen attitude. They're deal makers, not deal breakers.

WARNING

Inconsistent behavior is a red flag. If the sellers suddenly start missing contract deadlines or become strangely uncooperative, they may have lost their motivation to sell. Perhaps the wedding was postponed or the new job fell through. Whatever the reason, people sometimes switch from being real sellers to being fakes in mid-transaction. Find out why the sellers are acting strangely as soon as you notice the change, and you may be able to head off the problem. If you ignore the danger signs, you'll never know what hit you when the deal blows up in your face.

Lowballing

A *lowball* offer is one that's far below a property's actual fair market value. An example of a lowball offer is a $200,000 offer on a house that's worth every penny of $250,000.

Who makes lowball offers? Sometimes, it's a graduate from one of those scuzzy, get-rich-quick real estate seminars. Another lowball offer may come from somebody who's bottom-fishing for sellers in dire financial distress. More often, however, lowballing is a negotiating tactic used by people who state categorically, "No one ever pays full asking price. You always have to start low to end up with a fair price."

Those statements aren't true, of course. When you do your homework, you know the difference between well-priced properties and overpriced turkeys. (See Chapter 10 for a brush-up.)

Why lowballing is usually a bad idea

As we discuss earlier in this chapter, lowballing a well-priced house breaks the first rule of a good offer: Make a realistic offering price based on the sale price of comparable houses. Because skillful negotiators understand both sides of the issue, imagine that you're the seller of a house that's priced as close as humanly possible to its fair market value.

Several days after your house goes on the market, you receive an offer with an absurdly low purchase price. After the vein in your neck stops pounding, what conclusions can you form about the lowballing buyers?

>> **Taken in the best possible light, the buyers obviously haven't done their homework regarding comparable home sales.** Because they're grossly ignorant about fair market value, why should you try to educate them?

>> **Maybe the buyers think you don't know what your house is really worth and are trying to exploit your ignorance.** (That vein starts throbbing again.)

>> **Perhaps the buyers are trying to steal your house based on a mistaken impression that you're desperate to sell.** There's a name for critters that prey on misfortune: *vultures.*

None of these conclusions is at all favorable. As a seller, you'd probably make one of the following responses to buyers who lowballed your well-priced house:

>> **Let the buyers know their offer is totally unacceptable by having your agent return it with a message that you wouldn't sell your house to them if they were the last buyers on earth.** Why make a counteroffer to people who are either idiots or scoundrels?

>> **Make a full-price counteroffer.** To show your contempt for the buyers, you'll hardball them on each and every term and condition in their offer. (Two can play this game.)

REMEMBER

Buyers who lowball a well-priced property listed by sellers who can wait for a better offer destroy any chance of developing the mutual trust and sense of fair play on which cooperative negotiation is based. Bargaining is fine, but you must find a motivated seller and not aim too low. Starting at 25 percent below what the home is worth generally won't work unless the seller is desperate.

When low offers are justified

There's a huge difference between submitting an offer that's at the low end of a house's fair market value and lowballing. Suppose you offer $280,000 for a home listed at $299,500. You base your offering price on the fact that comparable houses in the neighborhood recently sold in the $280,000-to-$295,000 price range. You're at the low end of the range of fair market values. The sellers are at the high end. You're both being realistic.

If your offer is based on actual sales of comparable houses, it won't insult the seller. Such a low offer will, however, spark lively debate as both of you attempt to defend your respective prices. Coming in on the low side of a property's fair market value is fine as long as you have plenty of time to negotiate and reason to believe that the seller is motivated.

TIP

In situations like the preceding one, your best bet is to have an encyclopedic comparable market analysis and an agent who has *personally* eyeballed all the comps. Follow the guidelines that we discuss in Chapter 10.

A low offer is justified only when it isn't a *lowball* offer. Ironically, some sellers provoke low offers by their unwise pricing. These sellers insist on leaving room to negotiate in their price because they "know" that buyers never pay full asking price.

Sound familiar? This practice, unfortunately, becomes a self-fulfilling prophecy. When buyers who know property values make an offer on an overpriced house, their initial offering price is usually on the low side to give themselves room to negotiate. What goes around comes around.

Suppose that a house's fair market value is $300,000. If the sellers put this house on the market at $360,000 so they'll have a 20-percent negotiating cushion, and you offer $240,000 for the same reason, you and the sellers start out $120,000 apart. It takes a heap of extra negotiating to bridge a gap that big.

TIP

Don't play their silly game unless you have time to squander. Make your initial offer at the low end of the house's fair market value and see how the sellers respond to it. If they refuse to accept the hard evidence of recent comparable home sales in the neighborhood, don't waste valuable time trying to educate them. They aren't sellers yet — they're property owners masquerading as sellers. If you want the house, bide your time. Don't make your move until they wise up and lower their price, or their agent puts the word out that they're motivated sellers who won't turn down any reasonable offer.

Negotiating credits in escrow

Putting a "let's sell it" price on a house isn't always enough to get the house sold, especially in a buyer's (weak) market. Sellers often find that they have to give buyers money in the form of seller-paid financial concessions in order to close the deal. The two most common concessions are for nonrecurring closing costs and corrective work.

Nonrecurring closing costs

Some sellers come right out and tell you that they'll pay your nonrecurring closing costs if doing so will help put a deal together. *Nonrecurring closing costs* are one-time charges for such things as your appraisal, loan points, credit report, title insurance, and property inspections. If you've read Chapter 3, you know we're talking big bucks here. Closing costs can amount to 3 to 5 percent of the purchase price.

TIP

Even if the sellers don't offer to pay your nonrecurring closing costs, asking for this concession as one of the terms in your offer *usually* won't hurt. Two general exceptions to this rule are when it's a seller's (strong) market or when you're in a multiple-offer situation.

Here's how the credit works. Say you've signed a contract to buy a $250,000 house. You have $55,000 in cash, and the escrow officer has just told you that you'll have nonrecurring closing costs totaling 4 percent ($10,000) of the purchase price.

About now, you may be wondering, "Why not just reduce the purchase price to $240,000 instead of asking the sellers for a $10,000 credit?" After all, the sellers' net proceeds of the sale are the same either way, and simply reducing the purchase price is less complicated. Not to mention that because property taxes are often based on the purchase price, a lower purchase price will probably cut your annual tax bite.

The reason: If you're short of cash, as most buyers are, a credit is more helpful than a price reduction. If you have to pay $10,000 in closing costs, you won't have enough cash left to make a 20 percent ($48,000) down payment on your $240,000 home. With less than 20 percent down, your monthly loan costs increase because you have to pay a higher interest rate on your mortgage, plus private-mortgage-insurance costs. Neither will you have any cash left over for emergencies. Under these circumstances, you'd probably decide to buy a less expensive house.

Contrast that scenario with paying $250,000 for the house and getting a credit from the sellers for nonrecurring closing costs. After putting 20 percent ($50,000) cash down to get the loan with the lowest interest rate, you still have $5,000 in the bank thanks to the $10,000 credit. The credit makes the deal happen.

TIP

If you have plenty of cash, get a price reduction rather than a credit. In most areas, the lower your purchase price, the lower your annual property taxes. Just be aware that some less-than-scrupulous agents will lobby for the credit because a price reduction cuts into their commissions.

Corrective work

Typically, neither you nor the sellers know how much, if any, corrective work is needed when you submit your offer. Therefore, purchase contracts have provisions for additional negotiations regarding corrective-work credits *after* all the necessary inspections have been completed.

If the property inspectors find that little or no corrective work is required, you have little or nothing to negotiate. Suppose, however, that your inspectors discover the $250,000 house you want to buy needs $25,000 of corrective work for termite and dry-rot damage, foundation repairs, and a new roof. Big corrective-work bills can be deal killers.

TIP

Seeing is believing. We strongly recommend that you and the seller's agent be present, if possible, during property inspections so you both actually see the damage. And when you receive the inspection reports, use them as negotiating tools. Give the sellers copies of the reports for them to review before you meet with them to negotiate a corrective-work credit.

This is the moment of truth in most home sales. Sellers usually don't want to pay for the corrective work. Neither do you. The deal *will* fall through if this impasse can't be resolved.

At this point in the negotiations, it's critical that the sellers realize that the value of their house has just been reduced by the cost required to repair it. If comparable houses with no termite or dry-rot damage, with solid foundations, and with good roofs are selling for $250,000, the sellers' house is worth only $225,000 in its present condition. Given its reduced value, an 80 percent loan is $180,000 — not $200,000 based on a $250,000 fair market value. If you can borrow only $180,000 and the sellers refuse to reduce the selling price from $250,000 to $225,000, you have to drop out of the deal.

The sellers may refuse to pay for repairs found by inspectors that you've hired. The sellers may question the impartiality or validity of your inspection reports and order their own inspections to verify or refute yours. The sellers may even threaten to pull out of the contract if you don't back off on your demands.

Sellers who try to punish the messenger are usually making a big mistake. You didn't bring the damage with you when you came, and (luckily for you) you won't take it with you when you go. Like it or not, the sellers are stuck with it. If they drive you away, they may still have a legal obligation to tell other buyers what you've discovered. That disclosure will probably lower the price that any future buyer will pay for their house. All things considered, working things out with you will probably be faster (and no more expensive) than waiting for another buyer.

TECHNICAL STUFF

Lenders also participate in corrective-work problems. They get copies of inspection reports when borrowers tell them that a serious repair problem exists, when their appraisal indicates a property obviously needs major repairs, or when the purchase contract contains a credit for extensive repairs. Whenever the property's loan-to-value ratio exceeds 80 percent, lenders actively help buyers and sellers resolve corrective-work problems.

You can solve repair problems in a variety of ways:

>> **Ideally, the sellers leave enough money in escrow to cover the required corrective work with instructions for the escrow officer to pay the contractors as their work is completed.** This strategy has several advantages. You can supervise the work to be sure that it's done properly by contractors of your choice. The sellers don't have to suffer through having the work done while they're living in the house, and they don't have to incur any liability for the workmanship. Last but not least, the lender knows the work will be done.

>> **Alternatively, the lender withholds a portion of the full loan amount in a passbook savings account until the corrective work has been completed.** In cases involving major corrective work, the lender may refuse to fund the loan until the problems have been corrected.

>> **The sellers may give a credit for corrective work directly to buyers at the close of escrow.** Lenders usually don't approve of this approach, because it raises uncertainties about whether the corrective work will actually be completed. If it isn't, the security of the lender's loan is impaired.

TIP

You can make the sellers feel better by offering to get competitive bids on the work from several reputable, licensed contractors. As long as the lowest bidder will do a quality job in a timely fashion, you and the sellers benefit. This additional effort on your part shows the sellers that you don't want to get rich off their misfortune. All you want is what you thought you were buying in the first place: a well-maintained home with a good foundation and a roof that doesn't leak. Empathy is an excellent negotiating tactic.

Chapter **13**

Inspecting and Protecting Your Home

G iven how much houses cost today, it's idiotic not to have the home you plan to purchase carefully inspected before buying it. Skipping inspections to save a few bucks (relatively speaking) can be the most expensive mistake you ever make. Think of your biggest financial fiasco ever and multiply it by a hundred. That gives you some idea of the magnitude of the boo-boo you may make if you buy a home without first having it *thoroughly* inspected from foundation to roof.

Why are we so obsessed with property inspections? According to a study commissioned by *The Wall Street Journal* and conducted by housing economist Robert Sheehan, approximately two out of every five houses have at least one major defect. If the odds were two out of five that you'd get hit by a car the next time you walked across the street, you'd be pretty darn careful to inspect oncoming traffic before crossing!

Conducting Thorough Inspections

A home's physical condition greatly affects its value. You'd feel horrible if you paid top dollar for a home that you thought was in tiptop shape and discovered after you bought it that the house was riddled with expensive defects. Yet unless you're a professional property inspector, you probably don't have the faintest idea how much corrective work a house needs simply by looking at it.

Believe it or not, buying homes was even riskier a generation ago. The prevailing attitude then was extremely simple: "Buyer beware."

Today, fortunately, the situation has improved. Most states (but not all) now require that sellers and real estate agents make full, immediate disclosure to prospective buyers of all *known* mechanical, structural, and legal problems associated with owner-occupied residential property. If this trend continues, the time may come when the warning shifts to "Seller and real estate agent beware." (Real estate agents also have liability if they fail to disclose a known problem about a property.)

WARNING

Don't let your guard down. Even though the real estate market is a tad more consumer-friendly than it used to be, don't be lulled into a false sense of complacency. *Latent defects* — hidden problems that sellers and their agents aren't aware of regarding the home you're buying — can get you into a heap of budget-busting trouble after you complete your purchase.

If you haven't read Chapter 12 yet, take a quick look at the section about negotiating either a corrective-work credit or a price reduction. You can see how to use property inspections so they pay for themselves many times over.

All properties should be inspected

Overinspecting a house is much better than underinspecting it. Suppose you spend $350 to have the home you want to buy completely inspected by a qualified inspector, and you find out that nothing is wrong with it. Did you waste your money? Nope. You can sleep soundly, knowing that your home doesn't need any corrective work.

If, conversely, you skip the inspection to save $350 and later discover that your house needs $35,000 worth of repairs, you end up spending $100 in repairs for every dollar that you "saved." Such a deal! You may as well "save" money by not putting coins into parking meters and consider walloping parking tickets a normal driving expense!

Here's a list of properties that *must* be inspected prior to purchase:

>> **Used houses:** You're most likely to order inspections if your "new" home is someone else's used house. Obviously, the older the house, the greater the likelihood that you'll find defects in its mechanical and structural systems.

>> **New houses:** Even if you're buying a newly constructed, never-been-lived-in home, having it thoroughly inspected is wise. Just because the building is new doesn't guarantee that it was built properly. Believe it or not, brand-new houses often have construction flaws, sometimes major ones. Some home builders aren't competent, or they cut corners to save money and boost their profits.

>> **Condominiums:** You need an inspection before buying a condominium. Don't forget that when you buy a condo, you're also buying into the entire building in which your condo is located (see Chapter 8). As a co-owner of the building, you must pay your proportional share of the cost for corrective work required in common areas, such as the roof, heating system, or foundation.

>> **Townhouses, cooperative apartments, and all other forms of co-ownership property:** See the preceding bullet point about condominiums. Shared ownership doesn't get you off the hook. You still need property inspections.

WARNING

>> **Foreclosed property:** If you're thinking of buying a foreclosed home from a lender who got title to the property because no one purchased it at an auction, it's imperative that you have the house scrupulously scrutinized from top to bottom. You don't need a professional property inspector to tell you someone removed fixtures and appliances — you can see that with your own eyes. What you can't see is invisible stuff. Perhaps the copper wiring and pipes were pulled out of some walls and sold for scrap value, or maybe the previous owners poured acid down the kitchen and bathroom drains to sabotage the house. In the case of new construction that ends up in foreclosure, the inspector may, for example, discover wiring without junction boxes, or junction boxes without switches or fixtures attached to them. (See Chapter 8 for more details on foreclosures.)

REMEMBER

All properties need inspecting. Period. Inspect detached residences, attached residences, single-family dwellings, multifamily dwellings, condos, co-ops, townhouses, and anything else that has a foundation and a roof. If you're spending big bucks for a property, protect your investment by having it inspected.

The two types of defects: Patent and latent

Property defects come in two general categories: patent and latent.

>> *Patent defects* are right out in the open for the world to see. You don't need a professional property inspector to point out glaringly obvious stuff like water stains on the ceiling, cracks in the wall, or a flooded basement. You do, however, need a trained professional to tell you whether these defects are signs of major problems or merely inconsequential blemishes.

>> Latent defects can be even more financially devastating than patent defects because they're hidden. Like playing a high-priced game of hide-and-seek, you must find latent defects or literally pay the consequences.

Latent defects are out of sight — behind walls or concealed in inaccessible areas under the house or up in the attic, away from casual observation. Faulty wiring, termite damage, a cracked heat exchanger in the furnace, and health- and safety-code problems (such as lead in the water pipes and asbestos insulation) are some examples of latent physical flaws.

TIP

If you're considering a bank-owned foreclosed property, it may not be possible for your inspector to check water pressure, test for leaky pipes, and check pilot lights in the stove or furnace if water, gas, and electricity have been shut off. In that case, insist that the lender get the utilities turned on again so you can complete your inspection.

Legal blemishes, such as zoning violations and fraudulent title claims, illustrate another kind of latent defect that only experts can detect.

Patent defect red flags

You don't have to be a professional inspector to give property a basic once-over. Even the rankest amateur can check the water pressure and turn on water faucets to see whether they leak. Flipping light switches on and off and flushing the toilets to find out whether they work properly are easy but effective tests. Open the refrigerator to see whether it's cold inside. Turn on the stove's heating elements to see whether they get hot. You may be surprised how many defects you discover with these simple tests.

By the same token, you can spot the danger signs of possibly serious structural problems even if you've never had any special training, as long as you know what to look for as you walk through a property. Although we advocate that you hire a professional property inspector, here's a list of red flags that even a mechanically challenged home buyer should be able to spot:

>> **Cracks:** Check the property's foundation, interior walls, exterior retaining walls, fireplace, chimney, concrete floors (basement, garage, back porch, and the like), driveway, and sidewalk for large cracks. Any crack that you can stick a pencil into is a large crack. Watch for vertical cracks on any walls and long horizontal or diagonal cracks on exterior walls.

>> **Moisture:** Look for water stains on ceilings, walls, and floors. Feel basement walls for dampness. Sniff out the source of moldy smells. Check for drainage problems inside and out by looking for standing water. A sump pump in the basement or garage is a red flag waving to get your attention.

>> **Stickiness:** All doors (exterior, interior, garage, and cabinets) and windows should open and close easily.

>> **Looseness:** You shouldn't be able to see daylight around windows, doors, or skylights.

>> **Unevenness:** Floors shouldn't slope, and walls shouldn't bulge.

>> **Insects:** If the house you're buying is made of wood or wood and stucco, it may have problems with wood-destroying insects or organisms. Mud tubes along a house's foundation or in its basement are a sign of termite infestation. Look carefully at those areas of the property that come into contact with the earth — foundation, decks, garage, and fencing — for signs of decayed or rotted wood.

>> **Slides:** Check hillsides immediately behind the property to see whether they have netting on them or show evidence of recent earth or mudslides.

TIP

Before you have the property inspected, discuss any red flags you discover with your property inspector. Let the pro check them out to see whether they're major problems or only relatively minor flaws that you can quickly and inexpensively correct. A sticking front door, for example, can indicate either that the house has expensive foundation problems or simply that the door absorbed moisture because it wasn't properly sealed.

Types of property inspections

What inspections should you get to protect your investment? That depends on what area of the country you live in, how the building in question is constructed, and what you plan to do to the property after buying it. Here are the three most common types of inspections — which we recommend be done *after* you have an accepted offer to purchase but *before* removing your inspection contingencies (so you're able to negotiate the correction of problems discovered by the inspections, as we recommend in Chapter 12):

INVESTIGATE

>> **Prepurchase interior- and exterior-components inspection:** No matter whether you're buying a wood-frame cottage in the country or an urban condo in a 20-story steel-and-concrete building, you need a complete inspection of the property's interior and exterior. The inspection should cover such areas as the roof and gutters, plumbing, electrical work, heating and cooling systems, insulation, smoke detectors, kitchen, bathroom, and foundation. The inspection should also point out health, safety, and environmental hazards. This type of inspection usually takes several hours to complete and costs $300 to $600, depending on how large the property is and the inspection's length and degree of detail.

Don't be surprised if the property inspector recommends additional inspections. Good property inspectors are generalists who are trained to spot red flags. Like doctors who are general practitioners, good property inspectors refer their clients to specialists — such as roofers, structural engineers, and pest control inspectors — if they discover a problem beyond their scope of expertise. Property inspectors know you can't make good decisions unless you have the best possible information.

>> **Pest control inspection:** Warm climates, such as in the South and West, are a mixed blessing. You're not the only one who loves warm, balmy weather. So do termites, carpenter ants, powderpost beetles, dry rot, fungi, and other wood-munching infestations or infections. If these are a problem in your area, you also need a pest control inspection. These inspections generally cost $150 to $400.

TIP

Pest control inspections are very limited in scope — the inspectors check for property damage caused only by wood-destroying insects (infestations) and organisms (infections, such as dry rot and fungi). Although homes made of wood or wood and stucco are the wood destroyers' primary targets, even brick homes aren't safe. If you get a pest control inspection, it should be in addition to your prepurchase interior- and exterior-components inspection — not in lieu of it.

>> **Architect's or general contractor's inspection:** You need an architect or a general contractor on your team if you're buying a fixer-upper, intending to do corrective work, or planning a major property renovation, such as adding rooms or installing a new bathroom. The architect or general contractor can tell you whether what you want to do is structurally possible and meets local planning codes for such things as height restrictions and lot coverage. This inspector can also give you time and cost estimates for the project.

WARNING

Architects and general contractors usually don't charge for their initial property inspection because they're hoping to get your business. Although these people provide a valuable service, take their reports with a grain of salt. Don't expect them to give you a completely objective assessment as to whether you should buy the property because they'd probably love to do the work for you.

Inspecting inspectors

Unfortunately, some people who anoint themselves "home inspectors" have neither the background nor the training to do proper prepurchase home inspections. To compound the problem, most states don't certify, license, or regulate home inspectors. If you have a clipboard, a pickup truck, and a good "houseside manner," you too can be a home inspector nearly anywhere in the country.

WARNING

Worse yet, some contractors inspect houses and then do the corrective work that they discover during their own inspections. That situation ought to start a red flag waving in your mind. Unscrupulous contractors can — and do — manipulate this conflict of interest to their advantage by finding and creating work for themselves.

One way around this problem is to hire someone who only does inspections. A growing number of property inspectors are exactly that: professional property inspectors, not contractors. This distinction is more than just semantic. Performing property inspections requires a special expertise that not all contractors, engineers, and architects have.

Professional property inspectors are specifically trained to do inspections and only inspections; they make their living solely from inspection fees. They don't do corrective work, which eliminates the temptation to find unnecessary corrective work during their inspections.

Selecting your inspector

How can you find a qualified home inspector? Ask friends and business associates who've recently bought homes whom they used for their property inspections. Get a list of home inspectors from a real estate agent. Be careful, though, of inspectors who are popular with agents — that popularity *may* stem from not killing too many deals by going easy on their inspections. Also check the internet for "Building Inspection Services" or "Home Inspection Services." If several sources recommend the same inspector, you've probably found a good one.

Just because someone has a diploma issued by an organization with an impressive name doesn't mean that individual is a qualified property inspector. Unfortunately, some "professional" associations' only criterion for membership is the ability to pay an initiation fee and annual membership dues. No matter how fancy the diploma, don't be fooled by what can be a meaningless piece of paper.

The *American Society of Home Inspectors* (ASHI) is a professional association of independent home inspectors. Just because an inspector is an ASHI member doesn't guarantee that you'll get a good inspection, but it certainly increases the likelihood that you'll be working with a qualified professional. You can't just plunk down a membership fee and join. All ASHI-certified members have performed at least 250 property inspections and have passed two written proficiency exams as a prerequisite of membership. ASHI members must also adhere to ASHI's standards of practice, continuing education requirements, and code of ethics. To find members in your area, call ASHI at 800-743-2744 or visit its website at www.ashi.org.

We recommend that you interview several property inspectors prior to hiring one. Here are questions to help you select the best inspector (also available online at www.dummies.com/go/homebuyingkit7e):

>> **Are you a full-time, professional property inspector?** Only one answer is acceptable: Yes.

>> **What can you tell me about your company?** Discuss the company's size and how long it has been in business.

>> **Do you carry errors and omissions insurance?** Errors and omissions insurance covers the possibility that a property inspection can miss some problems. If an inspector makes an error that costs you big bucks, errors and omissions insurance can help make amends.

>> **How many inspections do you personally perform each year?** Although the average number of inspections varies from area to area, active inspectors usually conduct between 150 and 400 inspections per year. Find out whether the inspector works primarily in the area in which the property you want to have inspected is located and is thus familiar with *local* codes, *local* regulations, and *local* problems (such as floods, mudslides, earthquakes, tornadoes, and the like). Also, ask whether the inspector routinely inspects foreclosed property if you're considering buying a foreclosed home.

>> **Do you hold any special licenses or certifications?** Property inspectors usually have a background in some related field, such as construction, engineering, architecture, electricity, plumbing, or insurance claim adjusting.

This diversity adds extra insights to their inspections. Membership in ASHI or another trade association for property inspectors indicates at least a minimal knowledge of home inspection procedures.

>> **What's the scope of your prepurchase inspection?** Make sure that the inspection covers *all* the property's major structural and mechanical systems, inside and out, from foundation to roof. Anything less is unacceptable.

>> **How long will your inspection take?** Time actually spent at the site is an important consideration. This inspection isn't a race. It usually takes two or three hours to *thoroughly* inspect a condo or a home of average size.

>> **What type of report will I receive?** Verbal reports, like verbal contracts, are worthless. A boilerplate, checklist-type report is only marginally better. You must have a detailed description of your specific property's mechanical and structural condition. You need a narrative report, written in plain English, that clearly explains the implications of its findings.

TIP

Get a sample report from each inspector you interview. The best way to see whether a company writes good reports is to read one so you can draw your own conclusion. Figure 13-1 features a superficial inspection report; we include an example of a thorough inspection report in Appendix C so you know what a good one looks like.

>> **Do you mind if I tag along during your inspection?** Mind? On the contrary, good inspectors insist that you be present during the property inspection.

>> **Will your report include an estimate of the cost to do your recommended corrective work?** This is a trick question. If the inspector says yes, don't use the inspector. Good professional property inspectors do only inspections. They don't do corrective work. Neither do they solicit business for their friends. Good inspectors help you establish repair costs by referring you to three or four reputable contractors, roofers, electricians, and other repair people that you can contact for corrective-work quotes. Because there's usually more than one way to fix a defect, you have to decide how best to deal with a problem after you've consulted the appropriate repair people.

>> **How much does your inspection cost?** Unfortunately, this is generally the first question that buyers ask when shopping for a property inspector. This is no time to be penny-wise and pound-foolish. Watch out for unrealistically low "this week only" promotional fees that new inspectors may offer. Don't let green inspectors practice on you. Quality inspections cost more than quickie, one-size-fits-all, checklist inspections, because they're worth a lot more. Ultimately, because fees charged by good inspectors are usually pretty much the same (because of competitive pressure), you'll probably end up using the correct criteria to select your inspector: compatibility and competence.

INSPECTION REPORT

NAME _____) _____ ADDRESS _____ DATE 8 629/

COMMERCIAL/RESIDENTIAL/INCOME   FLOORS IN USE 3   NUMBER OF UNITS 3

DESCRIPTION: Wood Frame Stucco Front

BASEMENT/SUB-STRUCTURE	EX	GD	FR	PR
FOUNDATION				X
SILL PLATE				
WALLS/STUDS				
COLUMNS				
GIRDERS		X		
SUB-FLOOR		X		
FLOOR JOISTS				
SEISMIC BRACING				

INTERIOR	EX	GD	FR	PR
LATH & PLASTER		X		
SHEETROCK		X		
FLOORS:				
KITCHEN				
BATHROOM(S)				
GENERAL AREA			X	
DOORS		X		
STAIRS/HANDRAIL		X		

EXTERIOR	EX	GD	FR	PR
SIDING		X		
TRIM		X		
CAULKING				
FLASHING			X	
DRAINAGE/SLOPE				X
EARTH CLEARANCE				
PORCHES		X		
STAIRS				

PLUMBING	EX	GD	FR	PR
COPPER		X		
GALVANIZED				
MIXED				
DRAINS			X	
VENTS		X		
FIXTURES				
MAIN SERVICE - WATER				
MAIN SERVICE - GAS				

ROOF	EX	GD	FR	PR
TAR & GRAVEL		X		
ASPHALT SHINGLE				
ROLL ROOFING				
WOOD SHINGLE/SHAKE				
GUTTERS			X	
DOWNSPOUTS				X

HEATING SYSTEM	EX	GD	FR	PR
GRAVITY		X		
FORCED AIR				
SPACE:		X		
GAS				
ELECTRIC				
MIXED				

ELECTRICAL	EX	GD	FR	PR
110 ____ 220 ✓		X		
FUSES ____ BREAKERS ✓				

WATER HEATER	EX	GD	FR	PR
GAS		X		
ELECTRIC				
SIZE 50 GAL		X		

COMMENTS: 1) Romex wire Running Exposed in upper closet.
2) Wall fire protection is missing in closet behind Heater
3) Caulk is missing from Kitchen sink

BY. _____

SIGNED: _____

>> **Would you mind if I call some of your recent customers for references?**
Good property inspectors are happy to give you names and phone numbers of their satisfied customers. Bad inspectors may balk at providing references or direct you to people they know will say something positive about them. Be sure to check at least three references per inspector in the town where the property is located. Ask the references whether, after close of escrow, they discovered any major defects that their inspector missed and whether they'd use their inspector again.

Optimizing your inspection

Here are guidelines for getting the biggest bang out of the bucks that you invest in a prepurchase property inspection:

TIP

>> **Always make your offer to purchase a house subject to your review and approval of the inspection reports.** Doing so gives you the opportunity to either negotiate a credit or price reduction for corrective work that's discovered during the inspections or, if you want, get out of the deal. We cover this subject extensively in Chapter 12. As we note in Chapter 8, lenders generally won't let you inspect foreclosed property prior to an auction.

>> **See whether the sellers have any presale inspection reports that they ordered or any copies of inspection reports generated by previous prospective buyers.** If so, give the reports to your inspector to call attention to possible problem areas. Have your agent order a permit search on the property to find out whether electrical, plumbing, or other repairs or improvements have been made.

WARNING

Suppose that the sellers give you a presale inspection report that they ordered just before putting their house on the market. It says that their house is in perfect condition. You can save money by relying on their report instead of getting your own. Should you? No way. Never let the fox tell you how things are in the henhouse. Always pay for your own inspection by an inspector of your own choosing.

>> **Read your property inspector's report carefully.** If you don't see some defects listed in the report that your inspector specifically mentioned during the inspection, call the inspector to find out why. By the same token, don't be the least bit shy about calling your inspector to get a detailed explanation of *anything* you don't completely understand in the report.

>> **To minimize the cost of corrective repairs, get bids on the job from several reputable, licensed contractors.** Never try to save money by using unlicensed contractors to do the work without permits. Doing so is usually illegal, can create health and safety problems, and can adversely affect your home's resale value. Many states require that house sellers disclose to prospective purchasers the fact that work on the house was done without permits. If your state doesn't mandate this type of disclosure now, it probably will by the time that you're ready to sell your house.

>> **Use your property inspector during the contractor bidding process.** If the contractors have questions regarding items discussed in the inspection report, refer them to the report's author for clarification. For an additional fee, some property inspectors will help you evaluate bids you receive to do the corrective work.

Prepurchase property inspections are intended to give you a factual basis for negotiating the correction of big-ticket defects — not to nickel-and-dime sellers over credits for stained carpets and worn curtains. If your new home is someone else's used house, let your *offering price* reflect the home's reduced value due to normal wear-and-tear cosmetic defects.

WARNING

If your agent or the seller offers to pay for a *home warranty plan* or *home protection plan* (that is, a service contract that covers some of your home's major systems and appliances), it wouldn't be gracious of you to turn down a freebie. Never accept such a plan in lieu of an inspection, however, and don't use your own money to buy this type of plan for yourself. After spending $250 or so for the plan, you pay an additional $25 to $50 deductible each time you need someone to come out and look at a problem. Furthermore, these plans significantly limit how much they pay to correct major problems. Hiring a professional property inspector to inspect the home diligently and uncover all existing problems so you can negotiate their correction with the sellers is a better way to spend your money.

Don't expect your inspections to eliminate all future maintenance problems. In time, the garbage disposal will break. All roofs leak eventually. When these things happen, it isn't part of some hideous plot to defraud you.

REMEMBER

Anything in your home that can break or leak will break or leak, sooner or later. Repairs come with homeownership. After closing on your home purchase, normal upkeep is your responsibility — not the sellers'. They'll have repairs of their own to make to their new home.

SEEING AND READING ARE INFINITELY BETTER THAN JUST READING

You, your agent, and the seller's agent should join the inspector during the property inspection. Reading even the finest of inspection reports is, at best, a mediocre substitute for being at the property and looking at the defects with your own eyes. This may be your best opportunity to question the inspector about the ramifications of a defect and discuss various ways to correct problems. By seeing and talking about the defects, you gain a better understanding of why some defects are no big deal to fix, whereas others cost megabucks to repair.

From a negotiating standpoint, the sellers are more likely to accept the inspection report's findings if their agent is present when the inspection is performed. They'll know that the defects are real — their agent actually saw the defects and can point them out to the sellers before they get a copy of the inspection report. They'll know that a skilled professional inspected their house — not some stooge that you hired to defame their property so you could swindle them out of their hard-earned money. They'll know that even if they drive you away by hardballing you on the corrective work, they'll still be stuck with the problem of selling a defective house.

Even if the house is in perfect condition, you should know where certain things are. If you attend the inspection, your inspector can show you where to find important stuff like the furnace, water heater, and circuit breakers. The inspector should also show you where the emergency shutoff valves for the house's gas, electric, and water systems are. By attending the inspection, you learn much more about the house's care and maintenance than you'd ever pick up by reading the inspection report.

If it's flat-out impossible to be at the inspection because you're stuck in another city or must be at a command-performance business meeting, make sure to have someone you trust (your agent, a relative or friend, or someone equally trustworthy) at the inspection to act as your eyes and ears. Ask your surrogate to make an audio or video recording of the inspection, which you can use to supplement the inspection report. Watching a video isn't as good as personally being there, but it sure beats just reading a report. You can also call the inspector if you have questions about the report.

Last but not least, pay attention. Don't bring along a gaggle of kids, relatives, friends, business associates, painters, carpet suppliers, plumbers, electricians, or contractors who'll distract you from the job at hand: finding out everything you can about the property that you want to buy. Focus on the inspection.

Insuring Your Home

Nobody likes to spend money for insurance. But if something can cause you a financial catastrophe, you should insure against that risk. The point of insurance is that you spend a relatively small amount of money to protect against losing a great deal of money. For example, if your home burns to the ground and it's not insured, you can be out tens (if not hundreds) of thousands of dollars.

You shouldn't waste money insuring potentially small-dollar losses. Suppose that you mail a package that contains a gift worth $50. If the postal service loses it, you'll be bummed, but the loss won't be a financial catastrophe for you. You shouldn't waste your money on such insurance.

Here are the types of insurance that you need to have in place *before* you purchase your dream home.

Homeowners insurance

When you buy a home, most lenders require that you purchase homeowners insurance. Even if you're one of those rare people who can buy a home with cash without borrowing money, you should carry homeowners coverage. Why?

>> First, your home and the personal property (furniture, carpets, clothing and jewelry, computers, dishes, and the like) in your home would cost a small fortune to replace out of your own pocket.

>> Second, your home can lead to a lawsuit. If someone were injured or killed in your home, you could be sued for tens or hundreds of thousands of dollars, perhaps even a million dollars or more.

The following sections tell how to get the homeowners coverage you need.

The cost of rebuilding

If your home is destroyed, which most frequently happens from fires, your insurance policy should pay for the cost of rebuilding your home. The portion of your policy that takes care of this loss is the *dwelling coverage* section. The amount of this coverage should be equal to the cost of rebuilding the home you own. The cost to rebuild should be based on the square footage of your home. Your policy's dwelling coverage amount shouldn't be based on what you paid for the home or the amount of your mortgage. If you're buying a condominium or cooperative apartment, examine the coverage that your building's homeowners association carries.

TIP

Get a policy that includes a *guaranteed replacement cost* provision. This provision ensures that the insurance company will rebuild the home, even if the cost of construction is more than the policy coverage. If the insurance company underestimates your dwelling coverage, the company has to eat the difference.

Ask the insurers that you're speaking with how they define *guaranteed replacement cost coverage* — each insurer defines it differently. The most generous policies, for example, pay for the home's full replacement cost, no matter how much the replacement ends up costing. Other insurers set limits — for example, they agree to pay up to 120 percent of your policy's total dwelling coverage.

Lawsuit protection

Liability insurance protects you against lawsuits arising from bad things that happen to others while they're on your property. Suppose that a litigious passerby happens to slip on a banana peel that was left on your driveway. Or perhaps your second-floor deck collapses during a beer fest, and someone breaks a leg or two or worse.

TIP

Carry enough liability insurance to protect at least two times the value of your assets. Although the chances of being sued are remote, remember that if you're sued, the financial consequences can be staggering. In fact, if you have substantial assets (worth more than a couple hundred thousand dollars, for example) to protect, you might consider what's called an *umbrella*, or *excess liability policy*. Bought in increments of $1 million, this coverage adds to the liability coverage on your home and car(s). Check for such policies with your home and auto insurers.

Personal property protection

On a typical homeowners policy, the amount of personal property coverage is usually set at about 50 to 75 percent of the amount of dwelling coverage. If you're a condo or cooperative apartment owner, however, you generally need to choose a specific dollar amount for the personal property coverage you want.

Some policies come with *personal property replacement guarantees* that pay you for the replacement cost of an item rather than for the actual value of a used item at the time that it's damaged or stolen. If this feature isn't part of the standard policy sold by your insurer, you may want to purchase it as a *rider* (add-on provision), if such a rider is available.

TIP

If you ever need to file a claim, having documentation of your personal property helps. The simplest and fastest way to document your personal effects is to make a video of your belongings. Alternatively, you can maintain a file folder of receipts for major purchases and make a written inventory of your belongings. No matter how you document your belongings, be sure to place this documentation somewhere outside your home (and not in the vegetable garden). A list or video isn't going to do you much good if it's in your home and your house goes up in a puff of smoke during a fire or is irreparably damaged in a horrendous flood!

Where to get good coverage inexpensively

As with other types of insurance and other financial products, you must shop around. But we know you have better things to do with your time than shop, so here's a short list of companies that are known for offering high-quality, low-cost policies:

>> **Amica:** Although Amica does have good customer satisfaction, its prices are high in some areas. You can contact the company by calling 800-242-6422 or checking out the company's website at www.amica.com.

>> **Auto-Owners:** Call 517-323-1200, or visit www.auto-owners.com.

>> **Erie Insurance:** This company does business primarily in the Midwest and Mid-Atlantic. Call 800-458-0811 for a referral to a local agent, or visit the company's website at www.erieinsurance.com.

» **GEICO:** Call 800-841-2964 or visit www.geico.com.

» **Liberty Mutual:** Call 800-837-5254, or go to www.libertymutual.com.

» **Nationwide Mutual:** Call 877-669-6877, or see www.nationwide.com.

» **State Farm:** Call 844-803-1573, or visit www.statefarm.com.

» **USAA:** This company provides insurance for members of the military and their families. Call 800-531-8722 or visit its website at www.usaa.com to see whether you qualify.

You may have access to more specific information for your state. Many state insurance departments, which you can locate through the state government listings online, conduct surveys of insurers' prices and tabulate any complaints received. We also include information about some helpful websites in Chapter 11.

TIP

As you shop around, ask about special discounts for such things as homes with a security system or smoke-detection system, discounts for people who have multiple policies with the same insurer, and senior discounts.

Title insurance

Fast-forward to a time several months after you close escrow on the purchase of your dream home. Suppose we ask you to prove to us that you actually own the home.

"No problem," you say. You go to the safe deposit box where you keep all your important documents and pull out the fancy deed that the recorder's office mailed to you a couple weeks after you completed your purchase.

Sorry. That deed isn't proof positive.

For example, a man and his "wife" signed a deed that transferred their house's title to another couple. A few weeks later, the buyers were shocked to find that their deed wasn't valid because the real wife's signature had been forged. In fact, the real wife didn't even know that her husband had sold the property.

Title risks

In theory, you can go down to the local county recorder's office and find out who owns any piece of property in the county simply by checking the public record. In fact, all sorts of irregularities in the history of the various people who've owned the property since it was originally constructed can affect a property's title — irregularities that are difficult or impossible to find, no matter how diligently you comb the public records.

OTHER CATASTROPHES TO INSURE AGAINST

Depending on where the home you buy is located, it may be subjected to earthquakes, floods, hurricanes, mudslides, tornadoes, wildfires, or other bad stuff. Standard homeowners policies don't protect against all these vagaries, so you must secure additional riders.

Thousands of communities around the country are at risk for floods. Hence, if you live in one of these areas, you need to purchase a flood insurance rider. Check with prospective homeowners insurance providers. The federal government flood insurance program (800-638-6620; www.fema.gov/national-flood-insurance-program) can provide background information on the types of policies available through private insurance companies.

Earthquakes are another risk to insure against. In addition to California, parts of the Midwest (and even parts of the East Coast) have active fault lines.

Ask people in the area that you're considering moving to what the local risks are. The U.S. Geological Survey (check your local phone directory; www.usgs.gov) and the Federal Emergency Management Agency (800-358-9616; www.fema.gov) offer maps showing, respectively, earthquake and flood risks. Be aware and be informed.

Because the cost of earthquake and flood coverage is based on insurance companies' assessments of the risks of both your area and your property type, you should *not* decide whether to buy these riders based on only your perception of how small a risk a major quake or flood is. The risk is already built into the price.

You may be able to pay for much of the cost of earthquake or flood insurance by raising the deductibles on the main parts of both your homeowners insurance and the other insurance policies you carry. Remember — you can more easily afford the smaller claims than the big ones. If you think that flood or earthquake insurance is too costly, compare the coverage's costs with the expense that you'll incur to completely replace your home and personal property. Buy this insurance if you live in an area that has a chance of being affected by these catastrophes.

WARNING

Here are some causes of these hidden risks to titles:

>> **Secret spouses:** A seller may claim to be single when, in fact, he or she is secretly married in another state. Or perhaps the seller was divorced in a community property state where, through marriage, one spouse obtains a legal interest in property held individually by the other spouse. Whatever the reason, sometimes a present or former spouse no one knew about will show

up out of the blue and file a claim against the property. This explains why title company representatives are so infernally curious about your marital status. They must know whether you're single, married, divorced, or widowed to keep ownership records accurate.

>> **Undisclosed heirs:** When property owners die without wills, probate courts must decide who their rightful heirs are. Court decisions may not be binding on heirs who aren't notified of the proceeding. Even when there's a will, probate courts must sometimes settle questions concerning the will's interpretation. Undiscovered heirs sometimes magically appear and claim that they now own the property in question.

>> **Questionable competency:** Minors and people adjudged to be mentally incompetent can't enter into binding contracts unless their court-appointed guardians or conservators handle the transaction. If, for example, the seller was a minor or was mentally incompetent when a deed was signed, the transaction may be voidable or invalid.

>> **Goofs:** This is a highly technical, catchall category for human errors. It covers everything from clerks who overlook liens recorded against property (liens for unpaid federal and state income taxes or local property taxes, for example) and other important documents while doing title searches, to surveyors who incorrectly establish property boundaries. Honest mistakes create many title problems.

>> **Forgery and fraud:** As was the case with the fake wife, sellers are sometimes fraudulently impersonated. By the same token, signatures can be forged on documents. Escrow officers demand identification (that is, a photo ID, such as a driver's license issued within the past five years or a current passport) to *establish beyond a shadow of a doubt that you are who you claim to be.*

>> **Name confusion:** A lot of title problems result from people having names similar (or identical) to the buyer's name or seller's name. Even though you prove that you are who you claim you are, you also have to prove who you *aren't.* If you have a fairly common last name, you'll probably have to fill out a Statement of Information to help the title company distinguish you from other people with names like yours. If you have an ordinary name like Brown, Chen, Garcia, Gonzalez, Johnson, Jones, Lee, Miller, Nguyen, Williams, or the ever-popular Smith, expect to be asked to complete a Statement of Information.

TECHNICAL
STUFF

What type of information is requested in a Statement of Information? You (and your spouse if you're married) will have to provide your full name, Social Security number, date and year of birth, birthplace, date and place of marriage (if applicable), residence and employment information, previous marriages, and the like. This information is used to differentiate good old honest you from the legions of ne'er-do-wells out there with names similar to yours.

What title insurance does

Many people who buy homes spend hundreds of dollars for title insurance without really understanding what they're getting for their money. *Title insurance* assures homeowners and mortgage lenders that a property has a marketable *(valid)* title. If someone makes a claim that threatens your ownership of the home, the title insurance company protects you and the lender against loss or damage, according to the terms and provisions of your respective title insurance policies.

ACTUAL EXAMPLES OF TITLE INSURANCE PROBLEMS

Folks usually don't pay much attention to title insurance when they're buying a home. Most people get title insurance only because the lender won't give them a mortgage if they don't have it. But home buyers are mighty glad to have such a policy when a title problem rears its ugly head.

For example, a woman spent close to $10,000 to remodel an existing carport and shed and to build a fence along her property line after obtaining her neighbor's permission. So far, so good. Everything was fine for the woman until her neighbor sold his place several years later. The new owners had their property surveyed and discovered that her carport, shed, and fence extended about 2 feet onto their land. Instead of tearing everything down, which was the woman's first impulse after getting the bad news, she decided to file a claim on her title insurance policy.

The title company discovered that she was the victim of a faulty land survey. It solved the problem by buying from her new neighbors an *easement in perpetuity* to use the land that she had improved so she could leave everything (carport, shed, and fence) exactly where it was.

Another example involves a couple whose kitchen was destroyed by fire. The county building department said that to rebuild it, the couple would have to remove a preexisting carport that extended into a 5-foot lot setback. The previous owners had gotten all the necessary permits that were required to build the carport, but the local zoning laws had changed since the time the carport was built in 1970.

The clever couple knew what to do. They called their friendly title company representative. After investigating their problem, the title company paid a contractor $5,000 to remove the carport. The title company also paid the couple $19,000 to compensate them for the reduced value of their property because it no longer had covered parking.

Most of your title insurance premium pays for research to determine who legally owns the property that you want to buy and to find out whether any unpaid tax liens or judgments are recorded against it. Because title companies do a good job of eliminating title risks *before* folks buy property, only about 10 percent of the premium goes toward indemnifying homeowners against title claims *after* the close of escrow.

The title insurance premium that you pay at close of escrow is the one and only title insurance premium that you'll have to pay, *unless you refinance your mortgage.*

Title insurance deals with your risk of loss from *past* problems (such as unpaid property tax liens or forgery in the chain of title) that *may* exist at the time that your policy is issued. Because your policy covers the past, which is a fixed event, you pay only one title insurance premium *as long as you keep your original mortgage.*

TIP

If you refinance your mortgage, you have to get a new title insurance policy to protect the lender from title risks (such as income tax liens or property tax liens) that may have been recorded against your property between the time your previous policy was issued and the date of the refinance. If you refinance your loan, ask the title company whether you qualify for a *refinance rate* on the new title insurance policy. Most title companies will give you a big premium reduction — as much as 30 percent off their normal rates — if your previous policy was issued within five years of the new policy's issuance date.

Two kinds of title insurance

As a homeowner, you have a choice of two kinds of *owners* title insurance. Depending on the extent of the coverage that you desire, you can either get a standard coverage policy or an extended coverage policy.

>> *A standard title insurance policy* is less expensive than an extended policy because its coverage is more limited. Standard policies are limited to certain off-record risks (such as fraud in the chain of title, defective recordings, and competency), plus *recorded* (at the local county recorder's office) tax assessments, judgments, and other property defects that a search of public records can uncover.

>> *Extended title insurance policies* cover everything that standard policies do, plus they provide expanded coverage for off-record risks that can be discovered through a property inspection or by making inquiries of people in actual possession of the property, as well as defects such as *unrecorded* (never recorded at the county recorder's office) leases or contracts of sale. Only an extended title insurance policy would protect the homeowners in the faulty-land-survey and kitchen-fire examples in the nearby sidebar, "Actual examples of title insurance problems."

Title insurance costs vary greatly, depending on the geographic area in which your home is located, the home's purchase price, and the type of coverage you get. In addition to the owners policy that we recommend you purchase to protect your investment, most lenders will insist that you buy a policy to protect the mortgage lender against loss on the loan amount.

In some Eastern states, title companies are barred from doing title searches. If that prohibition exists in your area, you have to use a lawyer to handle your title search and escrow. In either case, shop around to see who offers the best combination of competitive premiums and good coverage.

WARNING

As we note in Chapter 8, foreclosure auctions are extremely risky for many reasons. Among other things, lenders won't guarantee that you'll get undisputed title to the property you purchase at an auction. Nor can you get title insurance to protect against undisclosed or undiscovered flaws in the chain of title or liens against auctioned property.

TIP

Local custom and practice determine who usually pays for title insurance. In some parts of the country, custom dictates that the buyer pays for it. In other areas, however, the seller pays the title insurance premium, or buyers and sellers split the cost fifty-fifty. As we point out in Chapter 12, the payment for title insurance is a negotiable item. Regardless of local custom, if you're in a strong buyer's market, the sellers may offer (or you can ask them) to pay your title insurance costs in order to put the deal together. If, conversely, you're bidding against several other buyers for a particularly desirable house, you'd be smart to sweeten your offer by paying for title insurance, even though local custom prescribes that sellers pay for it.

Chapter **14**

It Ain't Over till the Weight-Challenged Escrow Officer Sings

The big day draws near. Soon, if all goes well, you'll plunk down the balance of your down payment, sign on the dotted line, and pick up the keys to your dream home.

For most people, the final throes of buying a home involve elephantine incertitude, high anxiety, and flop sweats. You, however, are *not* most people. The tips you find in this chapter will soothe your fevered brow, smooth the yellow brick road to success, and make the endgame downright pleasant and enjoyable.

An Escrow Is a Good Thing

As soon as possible after you and the seller have a *ratified offer* (that is, a signed contract), all funds, documents, and instructions pertaining to your transaction should be delivered to a neutral third party: the *escrow holder* designated in your purchase agreement. The act of giving these funds, documents, and instructions to the escrow holder constitutes the *escrow*. Depending on the local custom in your area, a lawyer, an escrow firm, or a title company may handle the escrow. Buyers and sellers generally select an escrow holder based on recommendations from their agents. However, as with other companies you choose to do business with in your home-buying transaction, know that escrow fees and service quality vary.

Real estate deals are often characterized by mutual distrust. You and the seller need someone whom both of you can trust to hold the stakes while you two work through all the resolved and unresolved details in your contract. The escrow holder (also known as the *escrow officer*) is your referee — a neutral third party who shouldn't show any favoritism to either you or the seller.

Know thy escrow officer

Your escrow officer is responsible for preparing and reviewing papers related to the transfer of *title* — a legal document that stipulates ownership of the property. This includes getting the papers properly signed, delivered, and made a matter of public record; complying with your lender's funding instructions; ordering a title search (explained in Chapter 13); and accounting to you and the seller for your respective money. The escrow officer handles the nitty-gritty paperwork and money details.

When the escrow is opened, your contract will probably be filled with loopholes known as *contingencies* or *conditions of sale*. For example, your contract should be written so you can get out of the deal if you don't approve the property inspection reports, or if the seller can't give you clear title to the property, or if you can't get a loan. The escrow officer's job is to receive and follow your instructions. Don't instruct the escrow officer to give your money to the seller until you're *fully* satisfied that the seller has performed under the contract. Chapter 12 goes into great detail about contingencies.

Ideally, your escrow will go smoothly from start to close. If, however, the escrow officer ever gets conflicting instructions from you and the seller or lender, the escrow will stop dead in its tracks until the argument is resolved. What kind of conflicting instructions? Disputes about whether the purchase price includes an item of personal property (that is, a refrigerator, a fireplace screen, a light fixture, and the like) are always popular. So are disagreements about whether corrective work should be done before or after close of escrow.

TIP

Our friend Kip Oxman, a most excellent real estate attorney and broker, has a great saying that works wonders in dispute-resolution situations: "When all else fails, RTC." You can find the answers to most controversies if you **Read The Contract**. The real estate purchase agreement included in Appendix A is an example of an explicit contract that's intended to eliminate ambiguity.

Good escrow officers are worth their weight in gold in times of crisis when the shouting, tears, and threats of lawsuit begin. At moments like this, often only the escrow officer's incredible patience and crisis-mediation skills keep deals glued together.

TIP

Give yourself an unfair advantage by humanizing your escrow. Either call or visit your escrow officer at her office to introduce yourself. Ask whether she needs any additional information to make the escrow go faster and smoother. Some questions your escrow officer may ask include the following:

>> What's your middle name?

>> Where can you be reached during the day?

>> What's your insurance agent's name and phone number?

>> What's your Social Security number (so your deposit can be placed in an interest-bearing account)?

A little consideration and respect now will do wonders for you later if the escrow hits a rough patch.

Cover all the bases

To avoid truly horrible surprises, pay particular attention to the following three areas.

Closing costs

If you have a nice, orderly, sequential mind, you've undoubtedly read the preceding 13 chapters and know that we have a detailed itemization of closing costs in Chapter 3. If you're the kind of person who loves to skip around and sample random chapters that strike your fancy, we suggest that you read that section now, or the following tips won't make as much sense.

As soon as possible, get a rough idea of how much money you have to come up with at the close of escrow. Immediately after opening escrow, ask your lender or mortgage broker for a TILA RESPA Integrated Disclosure (TRID) Loan Estimate of your estimated closing costs (see Figure 14-1). Even though it may take several weeks to get actual costs for inspection fees, repair-work credits, homeowners insurance premiums, and the like, at least you'll have a preliminary number that you can fine-tune as additional information becomes available. Having the knowledge available in this preliminary statement beats getting hammered by unexpected closing costs a couple of days before the close of escrow. You can access the TRID Loan Estimate online at www.dummies.com/go/homebuyingkit7e.

FICUS BANK
4321 Random Boulevard • Somecity, ST 12340

Save this Loan Estimate to compare with your Closing Disclosure.

Loan Estimate

DATE ISSUED	2/15/2013
APPLICANTS	Michael Jones and Mary Stone
	123 Anywhere Street
	Anytown, ST 12345
PROPERTY	456 Somewhere Avenue
	Anytown, ST 12345
SALE PRICE	$180,000

LOAN TERM	30 years
PURPOSE	Purchase
PRODUCT	Fixed Rate
LOAN TYPE	☒ Conventional ☐ FHA ☐ VA ☐_____
LOAN ID #	123456789
RATE LOCK	☐ NO ☒ YES, until 4/16/2013 at 5:00 p.m. EDT

Before closing, your interest rate, points, and lender credits can change unless you lock the interest rate. All other estimated closing costs expire on 3/4/2013 at 5:00 p.m. EDT

Loan Terms

		Can this amount increase after closing?
Loan Amount	$162,000	**NO**
Interest Rate	3.875%	**NO**
Monthly Principal & Interest See Projected Payments below for your Estimated Total Monthly Payment	$761.78	**NO**
		Does the loan have these features?
Prepayment Penalty		**YES** • As high as **$3,240** if you pay off the loan during the first 2 years
Balloon Payment		**NO**

Projected Payments

Payment Calculation		Years 1-7	Years 8-30
Principal & Interest		$761.78	$761.78
Mortgage Insurance	+	82	+ —
Estimated Escrow Amount can increase over time	+	206	+ 206
Estimated Total Monthly Payment		**$1,050**	**$968**

		This estimate includes	In escrow?
Estimated Taxes, Insurance & Assessments Amount can increase over time	$206 a month	☒ Property Taxes ☒ Homeowner's Insurance ☐ Other:	YES YES
		See Section G on page 2 for escrowed property costs. You must pay for other property costs separately.	

Costs at Closing

Estimated Closing Costs	$8,054	Includes $5,672 in Loan Costs + $2,382 in Other Costs – $0 in Lender Credits. *See page 2 for details.*
Estimated Cash to Close	$16,054	Includes Closing Costs. *See Calculating Cash to Close on page 2 for details.*

Visit **www.consumerfinance.gov/mortgage-estimate** for general information and tools.

LOAN ESTIMATE   PAGE 1 OF 3 • LOAN ID # 123456789

FIGURE 14-1: TRID Loan Estimate. When your escrow closes, you will get a TRID Closing Disclosure.

BEWARE OF SCAMMERS AND WIRE FRAUD SCHEMES

Plenty of otherwise seemingly intelligent people of all ages have been falling for a horrible scheme. In recent years "wire fraud" has become a cottage industry for scammers and hackers.

As we explain in this chapter, just prior to closing, you the home buyer will need to get a good deal of money to the company (law firm, escrow company, or title company) handling the escrow and closing of your purchase. Typically, that happens by you wiring money from one of your accounts at a bank or other financial institution to an account for the firm handling the escrow. In the days leading up to the closing, those firms will provide you with wiring instructions.

Plans can change. One day, for example, you may open your email to find revised wiring instructions. The email is from a familiar looking email address at the company involved in your real estate transaction. You wire the funds but escrow doesn't receive them. Turns out the email, which looked official, was the work of a scammer. The email may even include a phone number you can call to verify the change in transfer instructions. If you call that number, it will, of course, be answered by the scammers. Or someone posing as an employee of that company calls you to pre-empt any concerns you may have.

Sadly, such scams are on the increase. In the worst cases, duped homebuyers have wired six-figure sums to what turned out to be offshore accounts of crooks. The money is gone. They will never get it back. Home purchases have fallen apart. Good folks' financial lives have been turned upside down and, in some cases, ruined.

To avoid such horrible misfortune, please remember to always use your phone. Call the person you know is handling your escrow to question and verify any details for sending them money, especially if you get an email or call with new and different instructions. Never place a call using a new number in an email. If you don't have the escrow company's verified information (such as on a business card), then look it up online.

TIP

Estimate the closing expenses on the high side. Overestimating expenses and finding, when actual costs come in, that you don't need as much money to close as you first expected is ideal. The sooner you put a box around your closing costs, the better. Don't react to the situation — control it.

TIP

If, like most folks, you must put additional money in escrow just prior to the close of the transaction, use a cashier's check or a money order, or have your funds wired directly to the escrow to prevent delays. Be sure you stay on top of your bank to ensure that the wire is expedited, because banks sometimes drop the ball.

Personal checks take time to clear, and credit cards don't cut it in escrows. If you have questions regarding what constitutes *good funds*, ask your escrow officer well in advance of the close of escrow. If your money is out of town, for example, in a money-market fund (such as is recommended in Chapter 3), check with your investment company about how you can wire money from your account to the escrow company and how long it will take so you are sure to meet all deadlines.

Preliminary report

Shortly after escrow is opened, you should receive an extremely important document: the *preliminary report* (or *prelim*) from your title company. This report shows who currently owns the property that you want to buy, as well as any money claims (such as mortgage liens, income-tax judgments, and property-tax assessments) that affect the property. Last but not least, the preliminary title report shows any third-party restrictions and interests — such as condominium covenants, conditions, and restrictions (CC&Rs) — and utility-company or private easements that limit your use of the property or some other claim about which you had no information.

REMEMBER

Your contract should be contingent upon your review and approval of the preliminary report. Look it over carefully. Ask your agent, escrow officer, title-company representative, or lawyer to explain anything in the report that you don't understand. Don't be shy — there's no such thing as a dumb question. You want to make sure your seller actually owns the property and there is enough money in the escrow to pay off all the debts that have nothing to do with you.

As per the purchase contract, you should have the right to *reasonably* disapprove of certain claims or restrictions that you don't want on the property and to ask the owner to clear them prior to the close of escrow. For example, asking the seller to pay off all debts secured by liens and judgments against the property is reasonable. Asking the seller of a condo to remove the CC&Rs would be unreasonable because, as noted in Chapter 8, the CC&Rs are an integral part of the property.

REMEMBER

A preliminary report is *not* title insurance. You can find more on the distinction between title insurance and a preliminary report in the title-insurance section of Chapter 13.

Closing Disclosure

You may believe that the most important piece of paper you get when escrow closes is the deed to your new home. From an accounting standpoint, however, the most important piece of paper is the TRID Closing Disclosure that you get from the escrow officer on the day that your escrow actually closes.

If you think of the escrow as a checking account, the final settlement statement is like your checkbook. It records all the money related to your home purchase that went through the escrow as either a credit or a debit:

>> **Credits:** Any money that you put into escrow (such as your initial deposit and down payment) appears as a credit to your account. You may also receive credits from the seller for such things as corrective-work repairs and property taxes. And, of course, your loan is a credit.

>> **Debits:** Funds paid out of escrow on your behalf are shown as debits. Your debits include modest and not-so-modest expenses, such as what you graciously paid the seller for your dream home, loan fees, homeowners insurance premiums, and property inspection fees.

INVESTIGATE

You meet with your escrow officer several days before close of escrow to sign the loan documents and other papers related to your home purchase. At that time, you'll receive a more precise TRID Closing Disclosure detailing what your closing costs are if the escrow closes as scheduled. Check this TRID Closing Disclosure *extremely* carefully, line by line and from top to bottom, to be absolutely certain that it accurately reflects your credits and debits.

Escrow officers are human — they sometimes make mistakes. So do other participants in the transaction who may have given the escrow officer incorrect information. And guess what — when mistakes turn up, whose favor do you think they're in? Probably not yours! It's your money on the table. Pay attention to detail. Review the closing statement and question whatever isn't clear or correct. You need not determine at the time you sign the loan documents precisely what's wrong with the closing statement. Take it home with you, and continue inspecting it and asking the various parties to the transaction to clarify anything you don't understand about it.

TIP

The TRID Closing Disclosure is extremely important. Keep a copy for your files — it will come in handy when the time comes to complete your annual income-tax return. As detailed in Chapter 3, some expenses (such as loan origination fees and property-tax payments) are tax-deductible. Furthermore, the TRID Closing Disclosure establishes your initial tax (cost) basis in the property. When you're ready to sell your property, you may owe capital-gains tax on any profit you make by selling the property for more than your cost basis (see Chapter 17 for more details).

'Tis the season: December escrows

As a rule, December is a slow month for home sales. A week or two before Thanksgiving, most buyers switch their attention from houses to holidays and family gatherings, and those buyers typically don't get back onto the home-buying track until around Super Bowl Sunday in mid-winter.

TIP

Here are two reasons that you may decide to buck the trend:

>> **Bargain hunting:** When the other buyers drop out of the market, you're the only game in town for sellers who must move soon, or for stubborn sellers who foolishly waited too long to get realistic about their asking price. If they must sell, sellers instruct their agents to put the word out that they're willing to deal. The magic phrase is, "Bring us an offer." If you're a lowballer looking for a deal, now's the time to make your move.

>> **Tax deductions:** What you get doesn't matter — what does matter is what you get to keep. Buying a home in December gives you tax deductions that you can use to reduce your federal and state income taxes in that calendar year. As we discuss in Chapter 3, owning a home gives you physical shelter and tax shelter. On your income taxes, you can, for example, write off your loan origination fee (points), mortgage interest, and property taxes that you pay prior to December 31.

Escrows are perverse creatures under even the best of circumstances. They're proof positive of Murphy's Law, which states that whatever can go wrong will — and *always* at the worst possible time. Experienced escrow officers know that nasty surprises can rear their ugly heads whenever you least expect them.

The list of potential surprises is unpleasantly long: missed deadlines, title glitches, problems paying off existing loans, changes in your loan's terms, insufficient funds to close escrow, funds not wired as promised, and so on.

December escrows are particularly perverse. Partying zaps your strength and reduces your effectiveness. People forget to sign papers before leaving on vacation. December 31 is an immutable deadline if you want to close this year for tax purposes. If you end up with a late December escrow, here are some things you (and your real estate agent) should do to make sure you meet your deadline:

>> **Stay in touch with your lender.** Lenders need copious documentation to substantiate loan applications. Be sure your lender has all the required documents as soon as possible. Lenders say that lack of follow-up on loan-document verification is the number one cause of escrow delays. In Chapter 7, we provide an extensive checklist of items — such as W-2s, tax returns, bank statements, and so on — that your lender may need you to provide to verify the information on your loan application.

>> **Don't leave any blank spaces on your loan application.** Draw a line through any section that doesn't apply to you. If you leave a section blank, the lender may assume that you forgot to complete it. And make a photocopy of everything you submit in case the originals get lost or you need to refer to the documents when the lender questions something you wrote.

>> **Stay in touch with your escrow officer.** Don't let your file get buried in a pile of pending escrows stuck on the corner of your escrow officer's desk. You or your agent should check with the escrow officer periodically to make sure things are going smoothly.

>> **Be available to sign your loan documents.** You may have only 24 to 48 hours after your loan package arrives at the escrow office to sign the documents and return them to the lender. A delay can cost you the loan.

>> **If you're leaving town for the holidays, tell your agent, lender, and escrow officer well in advance of your departure.** You can usually make special arrangements to close your escrow — no matter where you are — as long as people have advance warning and know how to reach you. The key to success is keeping everyone posted.

>> **Check the calendar.** Many offices are open only till noon on Christmas Eve and New Year's Eve. When Christmas Day and New Year's Day fall on Saturday or Sunday, office hours can really get crazy. Some businesses and public offices close on the preceding Friday, others close on the following Monday, and still others close on both Friday and Monday to give their employees a four-day holiday. Be sure to check the holiday office schedule of your agent, lender, escrow officer, and so on. Don't let a holiday office closing derail your deal.

WARNING

>> **Allow time between when you'd like to close and when you must close.** Give yourself maneuvering room to resolve last-minute problems that inevitably appear when you least expect them. Don't schedule your closing on the last business day of the year — you'll have no margin for error if you need to close by year's end.

Follow through

Engagements are to weddings what escrows are to buying houses. Just as wedding bells don't always ring for everyone who gets engaged, not all open escrows end in home purchases.

Many escrows could've been saved by applying a fundamental principle of winning tennis: Follow through. Tennis pros know that the game is more than simply making contact with the ball. Pros continue their swing "through the ball" after they hit it because they know the last part of the stroke is as important as the initial contact with the ball. If they don't follow through properly, the ball won't end up where they want it to go.

And so it is with real estate deals. Buyers, sellers, and agents often say that a house has been *sold* when the purchase contract is signed. *Not true!* Nothing was sold. The buyer and seller merely ratified an offer. *Big difference!*

REMEMBER

If you want to actually buy and move into the home — that is, close your escrow — everyone involved in your transaction must follow through on all the details. You won't be the proud owner of your dream house until the escrow officer says you are!

How You Take Title Is Vital

One of the most important decisions you make when buying a home is how you take title in the property. If you're unmarried, your choices are simpler because you take title as a sole owner. When two or more people co-own a property, however, the number of ways to take title multiplies dramatically.

How title is held is critically important. Each form of co-ownership has its own rainbow of advantages, disadvantages, tax consequences, and legal repercussions. You shouldn't make this decision in haste at an escrow office while signing your closing papers. Unfortunately, that's what usually happens.

What's the best form of co-ownership for you? That depends on your circumstances. Here are some forms of co-ownership and the advantages of each type.

Joint tenancy

Suppose that you and your spouse buy a house together as joint tenants. When your spouse dies 20 years from now, ownership of the house automatically transfers to you without going through probate. This feature of joint tenancy co-ownership is known as the *right of survivorship*.

Joint-tenancy benefits don't stop there. You also get a *stepped-up basis* on your spouse's half of the house. This stepped-up basis may save you big bucks on the capital-gains tax that you have to pay if you ever sell the house.

Here's how a stepped-up basis works. Say that you and your spouse paid $180,000 for the house when you bought it. Immediately after your spouse's death, the house is appraised at $300,000.

Your new cost-of-the-home basis for tax purposes is $240,000 ($90,000 for your half-share of the original purchase price, plus $150,000 for your spouse's half of the house at date of death) because no capital-gains tax applies to your spouse's $60,000 of appreciation in value.

TIP

Even though we use a married couple in our example, you need not be married to use joint tenancy co-ownership. However, a minimum of two people must co-own and each must possess an equal ownership interest in the property.

Community property

Only married couples can take title as community property. Compared with joint tenancy, an advantage of community property co-ownership is that both halves of your house get a stepped-up basis upon the death of your spouse. This gives you even bigger tax savings.

Using the same figures as the joint-tenancy example, as the surviving spouse, your cost basis is the full $300,000. Capital-gains tax is forgiven on every penny of appreciation in value between the date of purchase and your spouse's death.

Another advantage of community property co-ownership is the ability to will your share of the house to whomever you want. Because of the right of survivorship, this choice isn't possible when title is held as joint tenants.

TECHNICAL
STUFF

Seven states — Alaska, Arizona, California, Idaho, Louisiana, Nevada, New Mexico, Texas, and Wisconsin — plus Puerto Rico allow community property with rights of survivorship. This allows both the full step-up in basis of regular community property, plus the nonadministration of estates afforded by joint tenancy ownership.

Tenants-in-common

Holding title as tenants-in-common doesn't give you a stepped-up basis upon the death of a co-owner. This creates an obvious disadvantage from a tax standpoint.

Offsetting legal advantages exist, however, for unrelated persons who take title as tenants-in-common. Under this form of co-ownership, you generally have the right to will or sell your share of the property without permission of the co-owners. Furthermore, co-owners don't have to have equal ownership interests in the property — a nice feature for people who just want a small piece of the action.

Getting help drafting an agreement

TIP

If you're smart — and we know that you must be, or you wouldn't be reading this book — you and your co-owners should have a formal written agreement, prepared by a real estate lawyer, to cover situations likely to arise while you jointly own the property. (*Note:* Such agreements are generally used and advisable when individuals purchasing a property aren't a married couple.) Here's a recap of key provisions to include in your written agreement (you can find more on these items in the co-ownership section of Chapter 8):

>> Provisions to buy out a co-owner who has to sell when the other owners want to keep the property, including whether you want to avoid a forced sale if one wants to sell and the others want to hold

>> Provisions to prorate maintenance and repair costs among co-owners with unequal shares in the property

>> Provisions to resolve disputes regarding such things as what color to paint the house

>> Provisions for penalties if a co-owner can't cover his share of loan payments or property taxes

The preceding information isn't intended to be your definitive guide to the subtleties of real property title vesting. This chapter merely points out the most important issues that you should consider. Don't make a decision of this magnitude in haste, especially if your situation is unusual or complicated. In addition to deciding how to hold title, you should consider estate-planning issues, such as wills and potential trusts (see Chapter 2 to find out more).

Getting Possessive

The day your escrow closes is legally confusing. You don't own the home when the day begins at 12:01 a.m., but you're the owner of record when the day ends at midnight. Sometime during the day, the escrow officer gives the seller your money, notifies you that the deed has been recorded, and officially announces that you're now the proud owner of your dream home. Congratulations!

Moving day

When can you actually take possession of your home and move into it? That depends on the terms of your contract. Look at Paragraph 9 of the sample purchase contract in Appendix A to see an example of a Closing and Occupancy clause that specifies date and time of possession and delivery of keys from seller to buyer. Here are your usual options:

>> **Move in the same day that escrow closes.** This is fine if the sellers have already moved out. If, however, the sellers haven't moved yet and don't want

to deliver possession until they're absolutely, 100 percent certain that escrow has closed, you may have a logistical problem. For two moving vans to occupy exactly the same driveway at exactly the same time borders on the impossible. Moving into a house while someone else is moving out is something you'll never attempt more than once. You can find easier ways to go crazy.

>> **Move in the day after escrow closes.** We recommend this alternative if the sellers won't deliver possession until escrow closes. Let the sellers have the day that escrow closes as their moving day. After all, the sellers are still the owners until title transfers. Moving day is stressful, even under the best of circumstances. Why create unnecessary stress for yourself by trying to move in as the sellers are leaving?

After the sellers vacate but before your movers bring your belongings into the house, check your new home carefully for damage that may have been caused by the sellers' movers. When movers are involved, accidents can happen. We cover this problem in the next section.

TIP

Whether you move into your home the day that escrow closes or the following day, you start paying for utilities and homeowners insurance effective the day that escrow closes. Don't forget to coordinate resumption of utility services, if necessary, with the proper companies a couple of weeks prior to the scheduled close of escrow.

>> **Move in after a seller rent-back.** Sellers may remain in their house for several weeks after escrow closes while waiting to get into their new home. In that case, the buyer signs a separate *rent-back agreement* with the sellers, which becomes part of the purchase contract. The rent-back agreement covers such things as who pays for utilities and maintenance, what happens if property damage occurs, how much rent the sellers pay you, and what the penalties are if the sellers don't vacate the property on the date specified in the rent-back.

It's customary for the sellers to pay rent equal to what you're paying for principal and interest on your mortgage, plus property taxes and insurance so you don't have out-of-pocket expenses on what you pay to own the house during the term of their rental. The amount equaling *principal, interest, taxes, and insurance* (known as *PITI*) is prorated on a per-day basis from close of escrow until the sellers vacate.

Suppose that PITI is $50 per day, and the sellers expect to be out three weeks after escrow closes. You both instruct the escrow officer to hold *four* weeks' PITI in escrow to give you a cushion if the sellers encounter a delay in moving into their new place. When the sellers actually move out, you and the sellers jointly instruct the escrow officer to pay you PITI for the actual rental period and to refund to the sellers the unused portion of funds held in escrow.

If the home you're buying is vacant, you may be tempted to ask for permission to start fixing the house up before close of escrow. After all, painting or waxing floors, for example, is much easier and faster when the house is empty. *Don't do it.* If the deal falls through, you've spent your time and money fixing up someone else's house. If the house catches fire, you don't have insurance to cover your losses. The risk exceeds the reward. Instead, allow some time to do these tasks *after* escrow closes and *before* moving in.

Final verification of condition

Read the Final Verification of Condition clause in Paragraph 15 of the California Association of Realtors' purchase contract in Appendix A. If your state's contract doesn't have this type of clause in it, instruct your agent or lawyer to write such a clause into your contract.

INVESTIGATE

We urge you to inspect the property a few days (ideally, the day) before escrow closes to be sure it's still in the same general condition that it was when you signed the contract to buy it. What if the sellers knocked a big hole in the kitchen wall during a wild party? What if they forgot to water the lawn, and it turned into a rock garden? What if a sinkhole appeared smack-dab in the middle of the drive-way? The "what ifs" are endless.

You'll probably find that everything is fine. But if it isn't, you can order the escrow officer to stop the escrow while you resolve the problem. Such an action always gets the seller's and real estate agent's attention. If you and the seller can't work out a mutually satisfactory solution, you may have to kill the deal. Killing the deal is better than buying a problem. If you don't have a real estate agent, never attempt to kill a deal without first speaking with a real estate attorney.

Coping with Buyer's Remorse

Many home sellers are convinced that they left the family jewels on the table when they sold their houses. The notion that they "gave their house away" is called *seller's remorse.* Seller's remorse is painful, but it generally departs within a month or two after the sale. Sellers are lucky to have such an uncomplicated dementia.

If you're like most buyers, you'll experience the flip side of this nasty psychosis. *Buyer's remorse* is the sinking feeling that you paid way, way, waaaay too much for your new home.

Unfortunately, buyer's remorse is much more complex than seller's remorse. Buyer's remorse is compounded by many other anxieties — that you're getting the world's worst mortgage, that the bottom will fall out of property values in the years after you buy a home, that you'll lose your job, that your health will fail, and that your faithful dog will die.

REMEMBER

We're here to help you deal with fear of overpayment. Those other anxieties are absolutely normal reactions to the uncertainties most of us *initially* experience. They *will* go away. If it makes you feel any better, nearly all home buyers are traumatized by the same concerns while purchasing a home.

In time, you'll discover (as we and millions of others who've gone before you did) that you have a fine mortgage, property values are stable, your continued employment is secure, your health is great, and so is your dog's. Don't take our word for this. Do a little dialing for dollars to verify that you got a good loan, check the help-wanted ads for jobs like yours, discuss property values with neighbors, get a physical examination, and take your dog to the vet.

So what about the fear that you're paying waaaay too much? If the real estate gods love you, you'll get a light case of buyer's remorse that you can treat by taking a couple of aspirin. Some buyers, however, are so ravaged by it that they try to break their contract.

You can't deal with buyer's remorse until you accept it for what it is: raw, naked fear. You're afraid that you're overpaying for the house. That fear tears some buyers apart. The symptoms of typical, fear-driven buyer's remorse are easy to spot. After you've signed the contract to buy your dream home, you do one or more of the following:

>> **Read real estate listings online and in the real estate section of your local newspaper even more intently than you did before you signed the contract.** You're searching for similar or nicer houses with lower asking prices. (You forget that most houses read a lot better in ads than they eyeball when you tour them.)

>> **Spend Saturday and Sunday touring open houses.** Reading listings isn't enough for you. You pound the pavement, looking for better buys than you got. Seeing, after all, *is* believing. (Speaking of seeing, you may see the remorseful sellers making the rounds of the same houses that you're looking at, trying to find less-nice properties with bigger asking prices.)

>> **Discuss your purchase with friends, neighbors, business associates, and the guy standing behind you while you wait in line to buy movie tickets.** You ask anyone and everyone whether they think you're paying too much, even though 99.9 percent of the people you talk to don't have a clue about property values for homes similar to yours. (You accept as gospel any wild guess they make that confirms your suspicions.)

After going through these exercises *during* escrow and for a couple of months *after* the purchase (until you're emotionally and physically exhausted), you'll probably discover that your fears are groundless. Nothing's wrong or unusual about your concerns. What's wrong is letting these fears gnaw away at you secretly instead of openly confronting them.

TIP

Facts defeat fear. The faster you get the facts you need, the less you'll suffer.

As Chapter 10 explains, a home can have more than one correct price. Pricing and negotiation are arts, not precise sciences. Don't beat yourself up with *asking* prices. You're okay as long as your home's *purchase* price is in line with the *sale* prices of comparable houses.

If you follow the principles we cover in this book, you'll be just fine! Unlike many home buyers, you know how to get your finances in order before you buy, and you know how to determine what you can really afford to spend on a home. You know how to find a great neighborhood, a great property, a great mortgage, a great agent, and a great property inspector. You can spot an overpriced turkey and a good value. You know about property inspections and negotiating for repair of property defects. You know how to avoid nasty people and property surprises.

REMEMBER

Knowledge is power. After you've assimilated the advice in this book, you'll be extremely powerful. You have nothing to fear. Go for it!

5

The Part of Tens

Chapter **15**

Ten Financial "To Do's" After You Buy

lthough it may have seemed that the day would never come — here you are, a *homeowner*. A homeowner — can you believe it? Go ahead and pinch yourself!

Perhaps you're already correcting your friends who ask how it feels to be a homeowner. Most new owners say, "Well, the bank owns more of the house than I do." Actually, you own 100 percent of the property — you just owe the mortgage lender a bucketload of money! Trust us when we say that although it may seem like a lot of money now, it probably won't seem that way decades from now. You'll be glad then that you decided to buy rather than continue renting. What you owe today, you'll ideally own free and clear in 30 years, if not sooner.

If you think the hard part is over after you buy a home, you may be in for a surprise. Moving probably wasn't a picnic, but moving is just the beginning of your quest to transform your new slag heap into a beauteous home.

WARNING

As a new homeowner, you must sidestep the many solicitations that will be winging your way. Unfortunately, when you buy a home, you end up on mailing, email, and social media lists galore because your home purchase is a matter of public record. Some communities even publish home sales (complete with buyer and seller names and purchase price) in the local paper (including online), for goodness' sake!

This chapter can help you become a financially happy new homeowner and can help you avoid the pitfalls to which many new homeowners before you have fallen prey.

Stay on Top of Your Spending and Saving

After you buy and move into your home, if you're like most new homeowners, your furniture and other personal possessions seem to take on an even shabbier tinge than before. And because you're now living in the property, you soon discover aspects of it that you don't like as much as you did when you were looking at it from the outside as a prospective buyer. This is another bitter bite of *buyer's remorse*, a common affliction of new home buyers that we discuss in Chapter 14.

WARNING

Most home buyers can find unlimited furniture, appliances, and remodeling projects that quickly exhaust the incomes of even the rich and famous. Because of these spending temptations, more than a few home buyers end up not saving any of their hard-earned incomes. Some new homeowners even end up building credit-card (and other high interest) consumer debt because their spending outstrips their income.

Feeling a squeeze in the budget when you buy a home is perfectly normal. After all, your housing expenses are probably higher than they were when you were renting or living in a smaller, less expensive house. But that's all the more reason that you need to take a lean-and-mean approach to the rest of your budget and spending (see Chapter 2). You can also make your home more energy-efficient by doing some simple things such as adding insulation and installing water flow restrictors in faucets and shower heads. Also, use your home inspection report (see Chapter 13) to identify other opportunities for improvement.

Don't neglect saving toward important financial goals, such as retirement. And take your time transforming your new home into a veritable palace. Rejoice and take solace in the fact that you have a roof over your head, a warm and comfortable place to sleep, and adequate living space — things that many people around the world can only dream about.

Consider Electronic Mortgage Payments

Mortgage lenders want to be paid and to be paid on time. And you want to pay them on time. Late payments can cost you dearly — many mortgages have stipulations for penalties equal to about 5 percent of the mortgage payment amount if your payment is late. If your payment is one whole month late, a 5 percent penalty works out to an annualized interest rate in excess of 60 percent! Even being one day late can trigger this penalty. (And you thought that credit-card debt was costly to carry at 18 percent!) Late charges also show up as *derogatories* on your credit report.

TIP

Sign up for your mortgage lender's automatic-payment service to have your mortgage payment zapped electronically from your checking account to the lender on the same day each month. If your mortgage lender doesn't offer this service, establish it yourself through one of the many home-banking services, such as Fiserv (800-564-9184; www.fiserv.com), or through bill-payment software such as Quicken.

Rebuild Your Emergency Reserve

Most people clean out their emergency reserve (and then some) to scrape together enough cash to close on their home purchase. Ideally, you should have ready and available an emergency cash reserve equal to at least three months' worth of living expenses. If your employment is unstable and you lack family to lean on financially in a pinch, aim for six months' worth of living expenses. Keep the emergency money in a leading money market mutual fund (see Chapter 3 for details).

As with saving money to accomplish other important financial goals, rebuilding your emergency reserve requires you to go on a financial diet and spend less than you earn. Easier said than done, especially with all the tempting things to spend money on for your home. Avoid the malls, mail order catalogs, and home improvement stores until you're back on an even keel!

Ignore Solicitations for Mortgage Insurance

Soon after you move into your home — often within a matter of just weeks or months — you will be bombarded with solicitations offering you mortgage life insurance and mortgage disability insurance. Most of the solicitations come from your mortgage lender, but other solicitations may come from insurance firms that picked up on the publicly available information revealing that you recently bought your home.

WARNING

The fundamental problem with these insurance policies is that given the amount of insurance protection offered, they're usually grossly overpriced and don't provide the right amount of benefits. The size of your mortgage shouldn't necessarily determine the amount of life and disability insurance protection that you carry. If you need life insurance protection because you have dependents who rely on your income, buy low-cost, high-quality term insurance. Likewise, if you're dependent on your income, make sure you have proper long-term-disability insurance coverage. See Chapter 2 to find out more about satisfying your insurance needs.

Ignore Solicitations for Faster Payoff

Another type of solicitation that you may receive extols the virtues — thousands of dollars in interest savings — that you can reap if you pay off your mortgage faster. For a monthly fee, these services offer to turn your annual 12-monthly-payment mortgage into 26 biweekly payments, each of which is half of your current monthly payment. Thus, you'll be making 13 months' worth of mortgage payments every year instead of 12. Doing so will usually shave about 8 years off the repayment schedule of a 30-year mortgage.

These services have two problems:

» You're paying the service money for paying off your mortgage faster — something you can do without the service and its fees.

» Paying off your mortgage faster than necessary may not be in your best interest.

TIP

The question to ask yourself is what you would do with the extra money each month if you didn't pay off the mortgage faster. If you'd spend it on something frivolous that would provide only fleeting, superficial enjoyment, paying off your mortgage faster is probably a better use of the money. Likewise, if you're an older (or otherwise risk-averse) investor, you're unlikely to earn a high enough rate of return by investing your money to make it worth your while not to pay off your mortgage faster.

On the other hand, if you can instead put more money away into a tax-deductible retirement account, paying off your mortgage faster may actually cost you money rather than save you money. Neither is it wise to pay down your loan if doing so leaves you cash poor. Suppose you lose your job and take several months to find a new one. Or suppose your home needs a new roof, and you don't have the cash to pay for it. You should have at least three to six months' worth of living expenses in some readily available place such as a money market fund. If you don't, you may have to use high-interest-rate (and not tax-deductible) credit cards to pay for unexpected expenses.

Consider Protesting Your Tax Assessment

In most communities, real estate property taxes are based on an estimate of your home's value. If home prices have dropped since you bought your home, you may be able to appeal your assessment and enjoy a reduction in the property taxes that you're required to pay.

TIP

Contact your local assessor's office to inquire into the procedure for appealing your property taxes. Generally, the process involves providing comparable sales data in writing to the assessor to prove the reduced value of your home. If you need help with this exercise, contact the real estate agent who sold you the home. Just be aware that your agent may want to make you feel as though your home hasn't decreased as much in value in order to make you (and perhaps himself) feel better. Explain that you're trying to save money on your property taxes and need comps that sold for less than you paid for your house. See Chapter 10 if you need a quick refresher on establishing property values.

Refinance if Interest Rates Fall

In Chapters 6 and 7, we explain how to select a magnificent mortgage and provide many tips for getting the best mortgage deal you can. But after you're into the routine of making your mortgage payments, if you're like most people, staying on top of strategies to keep your mortgage costs to an absolute minimum is probably as high on your priority list as flossing your teeth three times a day.

Keep an eye on interest rates. See the mortgage shopping resources we recommend in Chapter 6, as well as some websites we recommend in Chapter 11 to efficiently assist you with that task. As you may already know, interest rates — like the weather — change. If interest rates decrease from where they were when you took out your mortgage, you may be able to refinance your mortgage and save yourself some money. *Refinancing* (as described in Chapter 6) simply means you take out another new (lower-cost) mortgage to replace your old (higher-cost) one.

TIP

If rates have dropped at least one full percentage point since you originally took out your loan, start to contemplate and assess refinancing. The key item to calculate is how many months it will take you to recoup the costs of refinancing (loan fees, title insurance, and the like). Suppose your favorite mortgage lender tells you that you can whack $150 off your monthly payment by refinancing. Sounds good, huh? Well, not so quick there, Poindexter. First, you won't save yourself $150 per month just because your payment drops by that amount — don't forget that you'll lose some tax write-offs if you have less mortgage interest to deduct.

To figure how much you'll really reduce your mortgage cost on an after-tax basis, take your tax rate (as delineated in Chapter 3) and decrease the monthly payment savings you expect from the refinance by that amount. If you're a moderate-income earner, suppose that between federal and state income taxes you're in the 28 percent tax bracket. So if your mortgage payment would drop by $150, and if you were to reduce that $150 by 28 percent (to account for the lost tax savings), your savings (on an after-tax basis) would actually be $108 per month.

Now, $108 per month is nothing to sneeze at, but you still must consider how much refinancing the loan will cost you. If the refinancing costs total, for example, $6,000, it will take you about 56 months ($6,000 divided by $108) to recover those costs. If you plan on moving within five years, refinancing won't save you money — it will actually cost you money. On the other hand, if it costs you just $3,000 to refinance, you can recover those costs within three years. If you expect to stay in your home for at least that long, refinancing is probably a good move.

Keep Receipts for All Improvements

Sooner or later, you'll spend money on your home. You should track and document some of what you spend money on for tax purposes to minimize the capital gain that you may owe tax on in the future. *Capital gain* simply means the difference between what you receive for the house when you sell it less what it cost you to buy the house — with one important modification. The IRS allows you to add the cost of improvements to the original cost of your home to calculate what's known as your *adjusted-cost basis*.

Capital gain = Net sale price – (Purchase price + Capital improvements)

For example, if you buy your home for $225,000, and over the years, it appreciates so that (after paying the costs of selling) your net selling price is $350,000, your capital gain is $125,000. Remember, though, that the IRS allows you to add the value of the capital improvements you make to your home to your purchase price.

Capital improvement is money you spend on your home that permanently increases its value and useful life — putting a new roof on your house, for example, rather than just patching the existing roof. So if you made $10,000 worth of improvements on the home you bought for $225,000, your capital gain would be reduced to $115,000. Money spent on maintenance, such as fixing a leaky pipe or replacing broken windows, is not added to your cost basis (see Chapter 3 for more details).

Before you sell your home, be sure to understand the tax consequences of such a transaction. As we discuss in Chapter 17, many homeowners are eligible to shelter a large chunk of their home's capital gain from taxation when the time comes for them to sell.

Ignore Solicitations to Homestead

Another pitch that you, as a new homeowner, may get in the mail is one offering to homestead your home if you pay the friendly firm anywhere from $50 to $100. *Homesteading* means protecting some of your home's equity from lawsuits. A firm may offer to file the appropriate (and quite simple) legal document to homestead.

If you live in a state where you need to take action to secure your homestead exemption, by all means do so. Just call the recorder's office and ask how to do it. The process is simple (and, in some states, unnecessary) and not worth paying a firm to do for you.

Take Time to Smell the Roses

Okay, so it's a cliché. But too often, people work, work, and work to afford a home and don't take the time to enjoy life, family, and friends (or even their home). If you buy a home that's within your financial means and you're resourceful and thrifty with your spending in the years that you live in it, your home shouldn't dictate your finances and your need to work. You should own the home. It shouldn't own you.

No one (that we're aware of) has ever said on her deathbed that she wished she had spent more time toiling away at work (and, therefore, less time with family, with friends, and for herself) so she could spend more money on her home.

Chapter **16**

Ten Things to Know When Investing in Real Estate

Both owning a home and paying down the mortgage on your home over the years should create *equity* — the difference between what your home is worth and what you owe on it. Even if the unlikely happens over the long term and your home doesn't appreciate in value, you'll build equity as you pay down your mortgage.

You can use the equity in your home in future years for a variety of important purposes, including (but not limited to) helping finance your retirement, paying for educational costs, and funding fun things such as traveling. In addition to owning your home, you can invest in real estate in other ways. In this chapter, we include our top ten tips and things you should know if you're going to invest in real estate.

TIP

If you want to invest in real estate or stocks, bonds, mutual funds, small businesses, and the like, first invest your time in finding out what makes such investments tick and in learning how you can make informed decisions that fit with your personal financial situation and goals. Pick up a copy of the latest editions of

Eric's *Investing For Dummies*, *Real Estate Investing For Dummies* (coauthored by Robert Griswold), and *Mutual Funds For Dummies* to find out more about investing.

Real Estate Is a Solid Long-Term Investment

Home values have always gone through up-and-down cycles. However, the long-term trend is up, and the rises usually are far greater than the subsequent declines, such as those that occurred in the late 2000s. (Stocks, as you may know, go through similar and often more violent cycles — witness the major stock market decline in the early and late 2000s.) So if you have a long-term (ideally, a decade or more) investing-time horizon, you should do just fine if you invest in real estate.

The average annual returns from investment real estate are comparable with those enjoyed by long-term stock market investors. The best time to buy well-located real estate is nearly always a decade or generation ago. A decade or generation from now, today's prices may look dirt-cheap.

Real Estate Investing Isn't for Everyone

If you're an impatient, busy person, investing in real estate probably isn't your ideal investment. For one thing, locating, negotiating, and closing on property can take a big chunk of your time if you want to buy good property at a competitive price. Then there's the chore (and the time sinkhole) of managing the property — you're responsible for everything from finding tenants to keeping the building clean and in good working order.

Even if you have the time to invest in real estate, you may also consider some other important aspects of your personal financial situation. As we discuss in Chapter 2, taking advantage of tax-deductible retirement accounts is vital to your long-term financial health and ability to retire. Buying investment real estate can prevent you from saving adequately in retirement accounts. Saving for your down payment can hamper your ability to fund these retirement accounts. And most of the properties that you buy will require additional out-of-pocket money in the early years.

REITs Are Good if You Loathe Being a Landlord

If you want to place some money in real estate but don't like the thought of being a landlord, consider *real estate investment trusts (REITs)*. REITs are managed by a company that pools your money with that of other investors to buy a variety of investment real estate properties that these trusts manage.

TIP

REITs trade on the major stock exchanges, and some mutual funds and exchange-traded funds also invest in REITs. Among the better REIT funds to consider are

>> Cohen & Steers Realty Shares: 800-330-7348; www.cohenandsteers.com

>> Fidelity Real Estate Investment Portfolio: 800-544-8888; www.fidelity.com

>> Vanguard Real Estate Index Fund: 800-662-7447; www.vanguard.com

Don't Invest in Limited Partnerships

WARNING

You should avoid real estate limited partnerships (LLPs) that are sold through securities brokerage firms. Securities brokers, often operating under the misleading titles of *financial consultant* or *financial advisor,* love to sell limited partnerships because of the hefty commissions that limited partnerships pay to the broker. The broker's take can be as high as 10 percent or more. Guess where this money comes from? If you said, "Out of my investment dollars," go to the head of the class!

In addition to the fatal flaw that only 90 cents on the dollar you invest actually go to work for you in the investment, broker-sold limited partnerships typically carry hefty annual operating fees ranging from 2 to 3 percent. So when you add it all up (or, we should say, after all these commissions and fees are subtracted from your hard-earned dollars), limited partnerships are destined to be poor investments for you. (And, to add insult to injury, selling an LLP can prove difficult and costly to do.)

Avoid Timeshare Condos and Vacation Homes

Another way that smart people lose a great deal of money when investing in real estate is through involvement in timeshare condominiums and vacation homes. The allure of both of these purchases is having a place to which you can escape for fun and relaxation.

With a timeshare, you're essentially buying the ownership of one week's use of a condo. Suppose that, for this privilege, you pay a one-time fee of $7,000. Although $7,000 may not sound like a lot, if one week costs $7,000, buying the entire year's rights to use the timeshare condo comes to more than $350,000. However, buying a similar condo in the area may set you back only about $125,000! So you're paying a *huge* markup on your week's ownership because of the costs of selling all those weeks and the need for the timeshare distributor to make a profit. (And you'll be on the hook for your share of the annual maintenance fees of a couple hundred dollars to more than a thousand dollars.)

TIP

A far better idea is to rent a condo that someone else owns — it's cheaper than owning, you can go to a different resort area each year (ski, beach, whatever), and you'll have no ownership and managing headaches. Or you can buy a condo outright, rent it out to others throughout the year, and reserve it for yourself for the week or two each year when you plan to take a vacation.

Another chilling thought: Timeshare condos are nearly impossible to sell. As we say in Chapter 8, the best time to think about selling a property is *before* you buy. Real estate is a relatively illiquid form of investment anyhow — why freeze your money solid in a timeshare condo?

Vacation homes present a different problem. Most people who purchase a vacation home use it for only a few weeks during the year. The rest of the time, the property is left vacant, creating a cash drain. Now, if you're affluent enough to afford this luxury, we're not going to stand in your way and discourage you from owning more than one home. But many people who buy vacation homes aren't wealthy enough to afford them.

REMEMBER

Before you buy a vacation property, examine your personal finances to determine whether you can still save enough each month after such a purchase to achieve your important financial goals, such as paying for higher education for your children or building your retirement nest egg. If you do buy a vacation home, consider buying a property that you can rent out during most of the time that you're not using it.

Residential Properties Are Your Best Investment Option

If you're going to invest in real estate, residential property is generally your safest and wisest investment:

>> First, such types of real estate are probably the most familiar to you because you've lived in (and perhaps bought) such properties already.

>> Second, residential property should be easier for you to manage and deal with on an ongoing basis. (Be sure you're knowledgeable about current rent-control laws — if any — in your community and how they affect the property you may buy.)

INVESTIGATE

Don't forget that being a landlord entails lots of responsibilities and occasional tricky situations. That's why we strongly recommend that you consult a good local real estate lawyer if you have *any* questions about legal issues connected with buying or operating your rental property (see Chapter 9).

Commercial or retail real estate has many financial and legal nuances that you probably haven't dealt with (and probably don't want to deal with). Business real estate also tends to be more volatile in value because it's more easily overbuilt, and the number of businesses, unlike the population of people in a locale, can shrink more quickly during poor economic conditions. We're not saying you should never try to invest in this more complicated type of property, but it's better for you to learn to swim in a backyard pool before leaping straight into a shark-infested sea.

Consider Fixer-Upper Income Property

Residential property that has curable defects (see Chapter 8) can be a good investment for people who can manage the repair and rehabilitation of the property. You must buy such property at an absolute rock-bottom price. To an investor, fixer-uppers can provide an income stream from rentals, as well as an appreciation in the property's value that can result from bringing the property back to its highest and best use.

TIP

Before you try your hand at buying and fixing up a property, talk to people in your local area who've already done so. Real estate agents and tax advisors can probably refer you to other like-minded investors. Ask these investors to explain all the work involved, what surprises they confronted, and whether they'd make similar purchases again if they had known what they do now.

Consider Converting Small Apartment Buildings to Condos

In real estate markets like those in densely populated urban areas, buying a multiunit building and converting the units to condominiums can be extremely profitable. To make such a transformation succeed, you need a good real estate lawyer and a good real estate agent who knows the value of apartments as condos. You also need a wise contractor who can help you estimate the costs of the work and the challenges involved in securing proper building permits.

TIP

As we recommend for those contemplating buying fixer-uppers, before you try to buy a multiunit building and attempt to convert it to condos, talk to other investors who've already done so. Also speak with your town's planning and building permit departments to see what sort of regulatory red tape awaits you. If the property is under rent control, be sure your lawyer investigates the ramifications of getting permission to do a condo conversion, which may include requirements such as giving a lifetime lease to handicapped and older tenants living in the building.

Consider the Property's Cash Flow

When you're considering the purchase of real estate for investment purposes, you must crunch some numbers. You don't need to remember any calculus (or even any high school algebra) to do these calculations — basic addition, subtraction, multiplication, and division will do.

TIP

To decide how much a specific property is worth (as a prospective buyer) and to understand the financial ramifications of your ownership of that property, calculate what's called the property's *cash flow*. You determine cash flow by summing the rental income that a property brings in on a monthly basis and then subtracting all the monthly expenses, such as the mortgage payment, property taxes, insurance, utility expenses that you (as the landlord) pay, repair and maintenance costs, advertising expenses, 5 percent (or more) vacancy factor, and so on. Be realistic and add up all the costs. If the current owners of the property you're buying are using it as investment real estate, ask them for a copy of Schedule E (Supplemental Income and Losses) from their income tax return.

Your Rental Losses Are Limited for Tax Purposes

TIP

If you purchase rental property that produces a negative cash flow, you should know before you buy whether you can claim that loss on your personal income tax return. If you're a high-income earner — making more than $100,000 per year — your ability to deduct rental losses may be limited. If you're a really high-income earner — making more than $150,000 per year — you may not be able to deduct any of your rental losses. Be sure to learn more about investing in real estate and related tax rules and strategies before heading down that road. Check out *Real Estate Investing For Dummies* by Eric Tyson and Robert S. Griswold (Wiley).

Chapter **17**

Ten Things to Consider When Selling Your House

When you own a home, the odds are extraordinarily high that someday you'll sell it. People who live their entire lives in their first home are rare. (Coauthor Eric's maternal grandparents accomplished this extraordinary feat — living more than six decades in the same home!)

Selling a house is generally somewhat less complicated than buying one. But just because selling a house may be easier than buying one doesn't mean that most people sell their houses properly.

If word gets out that you're considering selling your house, real estate agents will be attracted to you like hungry mosquitoes are to the only person on a desert island. And when you sell a house, the IRS and state tax authorities may be waiting to attempt to take a chunk of your profits, especially if you don't take the time to understand tax laws and how to make them work for you before you sell.

So in this chapter, we advise you about some important issues that you should weigh and ponder before you sell. And if you do decide to sell, we want you to do the best possible job of selling your house and avoiding the tax man (legally). In addition to reading this chapter, when it comes time to sell, check out our companion book: *Selling Your House For Dummies.*

Why Are You Selling?

Start with the basics. If you're contemplating selling, consider whether your reasons for selling are good ones. For example, who wouldn't like to live in a larger home with more amenities and creature comforts? But if you hastily put your home on the market in order to buy a bigger one, you may be making a major mistake. If your next, more expensive home stretches you too far financially, you may end up in ruin.

When you need to relocate for your job, or when you have a major life change, moving may be a necessity. Even so, you should weigh the pros and cons of keeping your property versus selling. Start this analytic process by reading the rest of this chapter.

Can You Afford to Buy the Next Home?

If you want to buy a more costly property, such a move is known in the real estate business as *trading up.* Doing an honest assessment of whether you can really afford to trade up is imperative. As we say in Chapter 3, no mortgage lender or real estate agent can objectively answer that question for you. Based on your income and down payment, the lender and agent can tell you the *most* that you can spend. They can't tell you what you can *afford* to spend and still accomplish your other financial and personal goals.

WARNING

One of the biggest mistakes that trade-up home buyers make is overextending themselves with debt to get into a more expensive property. The resulting impact on their budgets can be severe — no money may be left over for retirement savings, for educational expenses, or simply for having fun. In the worst cases, people have ended up losing their homes to foreclosure and bankruptcy when they suffered unexpected events, such as job losses or the deaths of spouses who had inadequate insurance.

Before you buy your next home, go through the same personal-finance exercises we advocate in Chapters 2 and 3. Get a handle on what you can really afford to spend on a home. Unless your income or assets have increased significantly since the time that you purchased your last home, you probably can't afford a significantly more expensive property. The most important issue for people to consider is how spending more money each month on a home will affect their ability to save for retirement.

What's It Worth?

When you're ready to sell your house, you'd better have a good understanding of what it's worth. You (and your agent, if you're using one) should analyze what comparable properties are currently selling for in your neck of the woods. For a discussion about comparable market analysis, see Chapter 10.

REMEMBER

If you need to sell your house without wasting a ton of time and energy, do what smart retailers do: Price it to sell. We're not advocating that you give your property away, so to speak, but we are suggesting that you avoid inflating your asking price to a point far above what the sales of comparable houses suggest that your house is worth.

You may be tempted, particularly when you're in no great hurry to sell, to grossly overprice your house in the hope that an uneducated buyer may pay you more than the property is really worth. The danger in this strategy is that you won't find a fool who will part with all that money for your overpriced property, and no one else will bid on it. Then, as you lower the price closer to what the house is really worth, prospective buyers may be wary of buying your property because of the extended length of time that it's been on the market. In the end, you may have a hard time getting 100 percent of what your house is really worth.

Have You Done Your Homework to Find a Good Real Estate Agent?

When most people are ready to sell their houses, they enlist the services of a real estate agent. Good agents can be worth their commission if they know how to prepare the property for sale, market it, and get it sold for top dollar. Unlike when you're a home buyer, your interests as a seller are aligned with a good agent's interests — the more you sell the property for, the more you net from the sale, and the more the agent gets paid.

Given how much homes actually cost (and how much they cost to sell and buy), you owe it to yourself to have a good agent representing you in the sale of your house. Be sure the agent you select isn't currently listing so many other properties for sale that she lacks enough time to properly service your listing. Also, the agent you worked with when you bought the home isn't necessarily the best agent to hire when you sell it. Different steps and expertise are required to sell (rather than buy) a house. See Chapter 9 for our guidance on finding a terrific agent.

Do You Have the Skills to Sell the House Yourself?

Although some property owners possess the skills and time needed to sell a house themselves, most don't. The carrot that may entice you to sell a house yourself is the avoidance of the 5 to 7 percent sales commission that agents ask for before they attempt to sell your property. Don't forget, however, that half of this commission goes to a buyer's agent. Because most buyers work with agents (partly because the agents' services appear to be at no cost to the buyers), you'll potentially save yourself only 2.5 to 3.5 percent of your property's final selling price by selling it without an agent on your side.

INVESTIGATE

Whether or not you sell the house yourself, interview several agents who've demonstrated that they know your neighborhood as a result of listing and selling properties in the area, and ask them to prepare a comparable market analysis for your house. Base your asking price on what comparable properties have sold for in the past six months. If you're shopping for an agent, also ask each one for an activity list of all the houses he has sold over the past 12 months so you can obtain references from property sellers who've worked with each agent.

Have You Properly Prepared the House for Sale?

The real work of selling a property begins before you ever formally place it on the market for sale or allow the first prospective buyer through the front door. Prepare your house for sale both inside and out. At a minimum, you should do the sort of cleanup work that you do before your parents (or perhaps the in-laws) visit — you know, scrambling around the house cleaning *everything* up (or at least tossing it under beds and into closets!).

If you'd like to take preparation to a higher level, consider staging (which we cover extensively in *Selling Your House For Dummies*). Just as stagehands set the stage for Broadway productions, smart sellers have their house staged to create a production designed to wow prospective purchasers. Professional stagers know how to emphasize the best features of a house and minimize the worst. You may not have to spend a fortune to stage your property. Small bucks spent wisely on staging can pay rich rewards.

TIP

But you have more to do than just running a vacuum (after you pick up the laundry from the floors) and washing the dishes. Have some good but brutally honest friends and prospective agents walk through the house with you to point out defects and flaws that won't cost you an arm and a leg to fix (for example, repairing leaky faucets or painting areas in need of new paint). Don't be defensive — take good notes! You should generally avoid major projects, such as kitchen renovations, room additions, and the like. Rarely will you get a high enough additional sales price to compensate you for the extra costs (and headaches) of these major projects (not to mention for the time that you spend coordinating or doing the work).

Do You Understand the House's Hot Buttons?

People don't buy homes — they buy a *hot button,* and the rest of the home goes with it. Hot buttons vary from home to home. Dynamite kitchens or baths, fireplaces, views, and gardens are often buyer turn-ons. Location is the hot button for people who *must* live in a certain neighborhood.

TIP

How can you determine your house's hot buttons? Think back to what appealed to you when you bought it. What you liked then will probably be the same hot buttons that will appeal to the next buyers. After you identify the hot buttons, emphasize them in your listing statement, multiple-listing description, and newspaper ads. Successful sellers know what the buyers will buy before they begin the marketing process.

What Are the Financial Ramifications of Selling?

Before you sell your house, you should understand the sale's financial consequences. For example, how much money will you spend on fix-up work? How much should you be netting from the sale in order to afford your next home?

REMEMBER

Unless you want to and can afford to be the proud owner of two homes, we advocate selling your current house before you commit to buying another. You can ask for a long close of escrow and a rent-back, if necessary, so you have time to close the sale of your next home without camping out on the street. Be sure about these things up-front so you won't have nasty surprises along the way or after you sell.

Do You Know the Rules for Capital-Gains Taxes on the Sale of a House?

Under current tax laws, most house sellers enjoy a significant tax break. Specifically, a large amount of *capital gains,* or profits, on the sale of a home are excluded from tax: up to $250,000 for single taxpayers and $500,000 for married couples filing jointly. (See Chapter 15 for information about calculating your profit.)

To qualify for this capital gains exclusion, the seller must have used the house as her principal residence for at least two of the previous five years. This requirement is reduced to one year for a person who has become physically or mentally unable to care for themselves. Also, time spent living in a nursing home or other healthcare facility counts toward this one-year requirement. You can use the capital gains exclusion no more than once every two years.

For the vast majority of house sellers out there, the current laws are a boon. Most people's house-sale profits don't come anywhere near the law's exclusion limits. However, there are a few homeowners out there — especially those who live in higher-cost areas and who've owned their homes for many years — whose gains exceed the $250,000 or $500,000 limits. If they sell, they'll owe tax on whatever profits exceed their applicable exclusion limits.

6 Appendixes

Appendix A

Sample Real Estate Purchase Contract

B ecause a real estate purchase contract is a legal document, your real estate agent or lawyer should provide you the appropriate contract form for your area and help you fill it out. As a rule, these contracts range from somewhat complex to quite complex and sometimes convoluted. Most purchase contracts include a warning that says something like this:

> "This is more than a receipt for money. It is intended to be a legally binding contract. Read it carefully."

Heed the warning!

Purchase contracts vary in length, complexity, and terms from state to state and, within a state, from one locality to another. This appendix includes a sample of the California Association of Realtors' real estate purchase contract. We chose California's contract because it's one of the most comprehensive residential real estate contracts around.

WARNING

Blank spaces are open invitations to confusion (at best) and deception (at worst)! Giving someone a contract with blank spaces above your signature is like giving someone a signed blank check. He can fill in whatever he wants over your signature, and you may have to pay. *Do not leave any spaces blank on your contract.*

Another important thing to check is the contract's *revision date* — usually located in the lower-left or -right corner of the page. Be sure you're working on the most recent version of the purchase contract.

CHECK IT OUT

See Chapter 12 for a more in-depth discussion of real estate purchase contracts. You can access the California Association of Realtors' sample real estate purchase contract online at www.dummies.com/go/homebuyingkit7e.

CALIFORNIA ASSOCIATION OF REALTORS®

CALIFORNIA RESIDENTIAL PURCHASE AGREEMENT AND JOINT ESCROW INSTRUCTIONS
(C.A.R. Form RPA, Revised 12/18)

Date Prepared: _____

1. **OFFER:**
   A. **THIS IS AN OFFER FROM** _____ ("Buyer").
   B. **THE REAL PROPERTY** to be acquired is _____, situated in
      _____ (City), _____ (County), California, _____ (Zip Code), Assessor's Parcel No. _____ ("Property").
   C. **THE PURCHASE PRICE** offered is _____ Dollars $ _____.
   D. **CLOSE OF ESCROW** shall occur on _____ (date)(or _____ **Days** After Acceptance).
   E. Buyer and Seller are referred to herein as the "Parties." Brokers are not Parties to this Agreement.

2. **AGENCY:**
   A. **DISCLOSURE:** The Parties each acknowledge receipt of a ☑ "Disclosure Regarding Real Estate Agency Relationships" (C.A.R. Form AD).
   B. **CONFIRMATION:** The following agency relationships are confirmed for this transaction:
      Seller's Brokerage Firm _____ License Number _____
      Is the broker of (check one): ☐ the seller; or ☐ both the buyer and seller. (dual agent)
      Seller's Agent _____ License Number _____
      Is (check one): ☐ the Seller's Agent. (salesperson or broker associate) ☐ both the Buyer's and Seller's Agent. (dual agent)

      Buyer's Brokerage Firm _____ License Number _____
      Is the broker of (check one): ☐ the buyer; or ☐ both the buyer and seller. (dual agent)
      Buyer's Agent _____ License Number _____
      Is (check one): ☐ the Buyer's Agent. (salesperson or broker associate) ☐ both the Buyer's and Seller's Agent. (dual agent)
   C. **POTENTIALLY COMPETING BUYERS AND SELLERS:** The Parties each acknowledge receipt of a ☑ "Possible Representation of More than One Buyer or Seller - Disclosure and Consent" (C.A.R. Form PRBS).

3. **FINANCE TERMS:** Buyer represents that funds will be good when deposited with Escrow Holder.
   A. **INITIAL DEPOSIT:** Deposit shall be in the amount of ....$ _____
      (1) Buyer Direct Deposit: Buyer shall deliver deposit directly to Escrow Holder by electronic funds transfer, ☐ cashier's check, ☐ personal check, ☐ other _____ within 3 business days after Acceptance (or _____);
      OR (2) ☐ Buyer Deposit with Agent: Buyer has given the deposit by personal check (or _____)
      to the agent submitting the offer (or to _____), made payable to
      _____. The deposit shall be held uncashed until Acceptance and then deposited with Escrow Holder within **3** business days after Acceptance (or _____).
      Deposit checks given to agent shall be an original signed check and not a copy.
      (Note: Initial and increased deposits checks received by agent shall be recorded in Broker's trust fund log.)
   B. **INCREASED DEPOSIT:** Buyer shall deposit with Escrow Holder an increased deposit in the amount of ....$ _____
      within _____ **Days** After Acceptance (or _____).
      If the Parties agree to liquidated damages in this Agreement, they also agree to incorporate the increased deposit into the liquidated damages amount in a separate liquidated damages clause (C.A.R. Form RID) at the time the increased deposit is delivered to Escrow Holder.
   C. ☐ **ALL CASH OFFER:** No loan is needed to purchase the Property. This offer is NOT contingent on Buyer obtaining a loan. Written verification of sufficient funds to close this transaction IS ATTACHED to this offer or ☐ Buyer shall, within **3 (or _____) Days** After Acceptance, Deliver to Seller such verification.
   D. **LOAN(S):**
      (1) **FIRST LOAN:** in the amount of ....$ _____
          This loan will be conventional financing **OR** ☐ FHA, ☐ VA, ☐ Seller financing (C.A.R. Form SFA), ☐ assumed financing (C.A.R. Form AFA), ☐ Other _____. This loan shall be at a fixed rate not to exceed _____% or, ☐ an adjustable rate loan with initial rate not to exceed _____%. Regardless of the type of loan, Buyer shall pay points not to exceed _____% of the loan amount.
      (2) ☐ **SECOND LOAN** in the amount of ....$ _____
          This loan will be conventional financing **OR** ☐ Seller financing (C.A.R. Form SFA), ☐ assumed financing (C.A.R. Form AFA), ☐ Other _____. This loan shall be at a fixed rate not to exceed _____% or, ☐ an adjustable rate loan with initial rate not to exceed _____%. Regardless of the type of loan, Buyer shall pay points not to exceed _____% of the loan amount.
      (3) **FHA/VA:** For any FHA or VA loan specified in 3D(1), Buyer has **17 (or ___) Days** After Acceptance to Deliver to Seller written notice (C.A.R. Form FVA) of any lender-required repairs or costs that Buyer requests Seller to pay for or otherwise correct. Seller has no obligation to pay or satisfy lender requirements unless agreed in writing. A FHA/VA amendatory clause (C.A.R. Form FVAC) shall be a part of this Agreement.
   E. **ADDITIONAL FINANCING TERMS:** _____
      _____
   F. **BALANCE OF DOWN PAYMENT OR PURCHASE PRICE** in the amount of................................$ _____
      to be deposited with Escrow Holder pursuant to Escrow Holder instructions.
   G. **PURCHASE PRICE (TOTAL):** ....$ _____

Buyer's Initials (_____)(_____)   Seller's Initials (_____)(_____)

© 2018, California Association of REALTORS®, Inc.

RPA REVISED 12/18 (PAGE 1 OF 10)   Print Date

CALIFORNIA RESIDENTIAL PURCHASE AGREEMENT (RPA PAGE 1 OF 10)

Property Address: _____ Date: _____

H. VERIFICATION OF DOWN PAYMENT AND CLOSING COSTS: Buyer (or Buyer's lender or loan broker pursuant to paragraph 3J(1)) shall, within **3 (or ___) Days** After Acceptance, Deliver to Seller written verification of Buyer's down payment and closing costs. (☐ Verification attached.)

I. APPRAISAL CONTINGENCY AND REMOVAL: This Agreement is (or ☐ is NOT) contingent upon a written appraisal of the Property by a licensed or certified appraiser at no less than the purchase price. Buyer shall, as specified in paragraph 14B(3), in writing, remove the appraisal contingency or cancel this Agreement within **17 (or ___) Days** After Acceptance.

J. LOAN TERMS:
(1) LOAN APPLICATIONS: Within **3 (or ___) Days** After Acceptance, Buyer shall Deliver to Seller a letter from Buyer's lender or loan broker stating that, based on a review of Buyer's written application and credit report, Buyer is prequalified or preapproved for any NEW loan specified in paragraph 3D. If any loan specified in paragraph 3D is an adjustable rate loan, the prequalification or preapproval letter shall be based on the qualifying rate, not the initial loan rate. (☐ Letter attached.)
(2) LOAN CONTINGENCY: Buyer shall act diligently and in good faith to obtain the designated loan(s). Buyer's qualification for the loan(s) specified above **is a contingency** of this Agreement unless otherwise agreed in writing. If there is no appraisal contingency or the appraisal contingency has been waived or removed, then failure of the Property to appraise at the purchase price does not entitle Buyer to exercise the cancellation right pursuant to the loan contingency if Buyer is otherwise qualified for the specified loan. Buyer's contractual obligations regarding deposit, balance of down payment and closing costs **are not contingencies** of this Agreement.
(3) LOAN CONTINGENCY REMOVAL:
Within **21 (or ___) Days** After Acceptance, Buyer shall, as specified in paragraph 14, in writing, remove the loan contingency or cancel this Agreement. If there is an appraisal contingency, removal of the loan contingency shall not be deemed removal of the appraisal contingency.
(4) ☐ NO LOAN CONTINGENCY: Obtaining any loan specified above is NOT a contingency of this Agreement. If Buyer does not obtain the loan and as a result does not purchase the Property, Seller may be entitled to Buyer's deposit or other legal remedies.
(5) LENDER LIMITS ON BUYER CREDITS: Any credit to Buyer, from any source, for closing or other costs that is agreed to by the Parties ("Contractual Credit") shall be disclosed to Buyer's lender. If the total credit allowed by Buyer's lender ("Lender Allowable Credit") is less than the Contractual Credit, then **(i)** the Contractual Credit shall be reduced to the Lender Allowable Credit, and **(ii)** in the absence of a separate written agreement between the Parties, there shall be no automatic adjustment to the purchase price to make up for the difference between the Contractual Credit and the Lender Allowable Credit.

K. BUYER STATED FINANCING: Seller is relying on Buyer's representation of the type of financing specified (including but not limited to, as applicable, all cash, amount of down payment, or contingent or non-contingent loan). Seller has agreed to a specific closing date, purchase price and to sell to Buyer in reliance on Buyer's covenant concerning financing. Seller shall pursue the financing specified in this Agreement. Seller has no obligation to cooperate with Buyer's efforts to obtain any financing other than that specified in the Agreement and the availability of any such alternate financing does not excuse Buyer from the obligation to purchase the Property and close escrow as specified in this Agreement.

4. SALE OF BUYER'S PROPERTY:
A. This Agreement and Buyer's ability to obtain financing are NOT contingent upon the sale of any property owned by Buyer.
OR B. ☐ This Agreement and Buyer's ability to obtain financing are contingent upon the sale of property owned by Buyer as specified in the attached addendum (C.A.R. Form COP).

5. ADDENDA AND ADVISORIES:
A. ADDENDA: ☐ Addendum # _____ (C.A.R. Form ADM)
☐ Back Up Offer Addendum (C.A.R. Form BUO) ☐ Court Confirmation Addendum (C.A.R. Form CCA)
☐ Septic, Well and Property Monument Addendum (C.A.R. Form SWPI)
☐ Short Sale Addendum (C.A.R. Form SSA) ☐ Other _____

B. BUYER AND SELLER ADVISORIES: ☑ Buyer's Inspection Advisory (C.A.R. Form BIA)
☐ Probate Advisory (C.A.R. Form PA) ☐ Statewide Buyer and Seller Advisory (C.A.R. Form SBSA)
☐ Trust Advisory (C.A.R. Form TA) ☐ REO Advisory (C.A.R. Form REO)
☐ Short Sale Information and Advisory (C.A.R. Form SSIA) ☐ Other _____

6. OTHER TERMS: _____
_____
_____
_____

7. ALLOCATION OF COSTS
A. INSPECTIONS, REPORTS AND CERTIFICATES: Unless otherwise agreed in writing, this paragraph only determines who is to pay for the inspection, test, certificate or service ("Report") mentioned; it **does not determine who is to pay for any work recommended or identified in the Report.**
(1) ☐ Buyer ☐ Seller shall pay for a natural hazard zone disclosure report, including tax ☐ environmental ☐ Other: _____
_____ prepared by _____
(2) ☐ Buyer ☐ Seller shall pay for the following Report _____
prepared by_____.
(3) ☐ Buyer ☐ Seller shall pay for the following Report _____
prepared by _____.

B. GOVERNMENT REQUIREMENTS AND RETROFIT:
(1) ☐ Buyer ☐ Seller shall pay for smoke alarm and carbon monoxide device installation and water heater bracing, if required by Law. Prior to Close Of Escrow ("COE"), Seller shall provide Buyer written statement(s) of compliance in accordance with state and local Law, unless Seller is exempt.
(2) (i) ☐ Buyer ☐ Seller shall pay the cost of compliance with any other minimum mandatory government inspections and reports

Buyer's Initials (_____)(_____)   Seller's Initials (_____)(_____)

RPA REVISED 12/18 (PAGE 2 OF 10)

CALIFORNIA RESIDENTIAL PURCHASE AGREEMENT (RPA PAGE 2 OF 10)

Property Address: _____ Date: _____

if required as a condition of closing escrow under any Law.

    (ii) ☐ Buyer ☐ Seller shall pay the cost of compliance with any other minimum mandatory government retrofit standards required as a condition of closing escrow under any Law, whether the work is required to be completed before or after COE.

    (iii) Buyer shall be provided, within the time specified in paragraph 14A, a copy of any required government conducted or point-of-sale inspection report prepared pursuant to this Agreement or in anticipation of this sale of the Property.

  C. ESCROW AND TITLE:

    (1) (a) ☐ Buyer ☐ Seller shall pay escrow fee _____

      (b) Escrow Holder shall be _____

      (c) The Parties shall, within **5 (or ___) Days** After receipt, sign and return Escrow Holder's general provisions.

    (2) (a) ☐ Buyer ☐ Seller shall pay for **owner's** title insurance policy specified in paragraph 13E _____.

      (b) Owner's title policy to be issued by _____

      (Buyer shall pay for any title insurance policy insuring Buyer's **lender**, unless otherwise agreed in writing.)

  D. OTHER COSTS:

    (1) ☐ Buyer ☐ Seller shall pay County transfer tax or fee _____

    (2) ☐ Buyer ☐ Seller shall pay City transfer tax or fee _____

    (3) ☐ Buyer ☐ Seller shall pay Homeowners' Association ("HOA") transfer fee _____

    (4) Seller shall pay HOA fees for preparing documents required to be delivered by Civil Code §4525.

    (5) ☐ Buyer ☐ Seller shall pay HOA fees for preparing all documents other than those required by Civil Code §4525.

    (6) Buyer to pay for any HOA certification fee.

    (7) ☐ Buyer ☐ Seller shall pay for any private transfer fee _____

    (8) ☐ Buyer ☐ Seller shall pay for _____

    (9) ☐ Buyer ☐ Seller shall pay for _____

    (10) ☐ Buyer ☐ Seller shall pay for the cost, not to exceed $ _____, of a standard (or ☐ upgraded) one-year home warranty plan, issued by _____, with the following optional coverages: ☐ Air Conditioner ☐ Pool/Spa ☐ Other: _____

    Buyer is informed that home warranty plans have many optional coverages in addition to those listed above. Buyer is advised to investigate these coverages to determine those that may be suitable for Buyer.

    OR ☐ **Buyer waives the purchase of a home warranty plan. Nothing in this paragraph precludes Buyer's purchasing a home warranty plan during the term of this Agreement.**

8. ITEMS INCLUDED IN AND EXCLUDED FROM SALE:

  A. NOTE TO BUYER AND SELLER: Items listed as included or excluded in the MLS, flyers or marketing materials are **not** included in the purchase price or excluded from the sale unless specified in paragraph 8 B or C.

  B. ITEMS INCLUDED IN SALE: Except as otherwise specified or disclosed,

    (1) All EXISTING fixtures and fittings that are attached to the Property;

    (2) EXISTING electrical, mechanical, lighting, plumbing and heating fixtures, ceiling fans, fireplace inserts, gas logs and grates, solar power systems, built-in appliances, window and door screens, awnings, shutters, window coverings, attached floor coverings, television antennas, satellite dishes, air coolers/conditioners, pool/spa equipment, garage door openers/remote controls, mailbox, in-ground landscaping, trees/shrubs, water features and fountains, water softeners, water purifiers, security systems/alarms and the following if checked: ☐ all stove(s), except _____; ☐ all refrigerator(s) except _____; ☐ all washer(s) and dryer(s), except _____.

    (3) The following additional items:_____

    (4) Existing integrated phone and home automation systems, including necessary components such as intranet and Internet-connected hardware or devices, control units (other than non-dedicated mobile devices, electronics and computers) and applicable software, permissions, passwords, codes and access information, are (☐ are NOT) included in the sale.

    (5) LEASED OR LIENED ITEMS AND SYSTEMS: Seller shall, within the time specified in paragraph 14A, **(i)** disclose to Buyer if any item or system specified in paragraph 8B or otherwise included in the sale is leased, or not owned by Seller, or specifically subject to a lien or other encumbrance, and **(ii)** Deliver to Buyer all written materials (such as lease, warranty, etc.) concerning any such item. Buyer's ability to assume any such lease, or willingness to accept the Property subject to any such lien or encumbrance, is a contingency in favor of Buyer and Seller as specified in paragraph 14B and C.

    (6) Seller represents that all items included in the purchase price, unless otherwise specified, **(i)** are owned by Seller and shall be transferred free and clear of liens and encumbrances, except the items and systems identified pursuant to 8B(5) and _____ , and **(ii)** are transferred without Seller warranty regardless of value.

  C. ITEMS EXCLUDED FROM SALE: Unless otherwise specified, the following items are excluded from sale: **(i)** audio and video components (such as flat screen TVs, speakers and other items) if any such item is not itself attached to the Property, even if a bracket or other mechanism attached to the component or item is attached to the Property; **(ii)** furniture and other items secured to the Property for earthquake purposes; and **(iii)** _____

_____. **Brackets attached to walls, floors or ceilings for any such component, furniture or item shall remain with the Property (or ☐ will be removed and holes or other damage shall be repaired, but not painted).**

9. CLOSING AND POSSESSION:

  A. Buyer intends (or ☐ does not intend) to occupy the Property as Buyer's primary residence.

  B. Seller-occupied or vacant property: Possession shall be delivered to Buyer: **(i)** at 6 PM or (_____ ☐ AM/☐ PM) on the date of Close Of Escrow; **(ii)** ☐ no later than ___ calendar days after Close Of Escrow; or **(iii)** ☐ at _____ ☐ AM/☐ PM on _____.

  C. Seller remaining in possession After Close Of Escrow: If Seller has the right to remain in possession after Close Of Escrow, **(i)** the Parties are advised to sign a separate occupancy agreement such as ☐ C.A.R. Form SIP, for Seller continued occupancy of less than 30 days, ☐ C.A.R. Form RLAS for Seller continued occupancy of 30 days or more; and **(ii)** the Parties are advised to consult with their insurance and legal advisors for information about liability and damage or injury to persons and personal and real property; and **(iii)** Buyer is advised to consult with Buyer's lender about the impact of Seller's occupancy on Buyer's loan.

  D. Tenant-occupied property: Property shall be vacant at least **5 (or ___) Days** Prior to Close Of Escrow, unless otherwise

Buyer's Initials (_____)(_____)   Seller's Initials (_____)(_____)

Property Address: _____ Date: _____

agreed in writing. **Note to Seller: If you are unable to deliver Property vacant in accordance with rent control and other applicable Law, you may be in breach of this Agreement.**

OR ☐ **Tenant to remain in possession** (C.A.R. Form TIP).

E. At Close Of Escrow: Seller assigns to Buyer any assignable warranty rights for items included in the sale; and Seller shall Deliver to Buyer available Copies of any such warranties. Brokers cannot and will not determine the assignability of any warranties.

F. At Close Of Escrow, unless otherwise agreed in writing, Seller shall provide keys, passwords, codes and/or means to operate all locks, mailboxes, security systems, alarms, home automation systems and intranet and Internet-connected devices included in the purchase price, and garage door openers. If the Property is a condominium or located in a common interest subdivision, Buyer may be required to pay a deposit to the Homeowners' Association ("HOA") to obtain keys to accessible HOA facilities.

10. **STATUTORY AND OTHER DISCLOSURES (INCLUDING LEAD-BASED PAINT HAZARD DISCLOSURES) AND CANCELLATION RIGHTS:**

A. **(1)** Seller shall, within the time specified in paragraph 14A, Deliver to Buyer: **(i)** if required by Law, a fully completed: Federal Lead-Based Paint Disclosures (C.A.R. Form FLD) and pamphlet ("Lead Disclosures"); and **(ii)** unless exempt, fully completed disclosures or notices required by sections 1102 et. seq. and 1103 et. seq. of the Civil Code ("Statutory Disclosures"). Statutory Disclosures include, but are not limited to, a Real Estate Transfer Disclosure Statement ("TDS"), Natural Hazard Disclosure Statement ("NHD"), notice or actual knowledge of release of illegal controlled substance, notice of special tax and/or assessments (or, if allowed, substantially equivalent notice regarding the Mello-Roos Community Facilities Act of 1982 and Improvement Bond Act of 1915) and, if Seller has actual knowledge, of industrial use and military ordnance location (C.A.R. Form SPQ or ESD).

(2) Any Statutory Disclosure required by this paragraph is considered fully completed if Seller has answered all questions and completed and signed the Seller section(s) and the Seller's Agent, if any, has completed and signed the Seller's Brokerage Firm section(s), or, if applicable, an Agent Visual Inspection Disclosure (C.A.R. Form AVID). Nothing stated herein relieves a Buyer's Brokerage Firm, if any, from the obligation to **(i)** conduct a reasonably competent and diligent visual inspection of the accessible areas of the Property and disclose, on Section IV of the TDS, or an AVID, material facts affecting the value or desirability of the Property that were or should have been revealed by such an inspection or **(ii)** complete any sections on all disclosures required to be completed by Buyer's Brokerage Firm.

(3) Note to Buyer and Seller: Waiver of Statutory and Lead Disclosures is prohibited by Law.

(4) Within the time specified in paragraph 14A, **(i)** Seller, unless exempt from the obligation to provide a TDS, shall, complete and provide Buyer with a Seller Property Questionnaire (C.A.R. Form SPQ); **(ii)** if Seller is not required to provide a TDS, Seller shall complete and provide Buyer with an Exempt Seller Disclosure (C.A.R. Form ESD).

(5) Buyer shall, within the time specified in paragraph 14B(1), return Signed Copies of the Statutory, Lead and other disclosures to Seller.

(6) In the event Seller or Seller's Brokerage Firm, prior to Close Of Escrow, becomes aware of adverse conditions materially affecting the Property, or any material inaccuracy in disclosures, information or representations previously provided to Buyer, Seller shall promptly provide a subsequent or amended disclosure or notice, in writing, covering those items. **However, a subsequent or amended disclosure shall not be required for conditions and material inaccuracies** of which Buyer is otherwise aware, or which are **disclosed in reports provided to or obtained by Buyer or ordered and paid for by Buyer.**

(7) If any disclosure or notice specified in paragraph 10A(1), or subsequent or amended disclosure or notice is Delivered to Buyer after the offer is Signed, Buyer shall have the right to cancel this Agreement within **3 Days** After Delivery in person, or **5 Days** After Delivery by deposit in the mail, or by an electronic record satisfying the Uniform Electronic Transactions Act (UETA), by giving written notice of cancellation to Seller or Seller's agent.

B. **NATURAL AND ENVIRONMENTAL HAZARD DISCLOSURES AND OTHER BOOKLETS:** Within the time specified in paragraph 14A, Seller shall, if required by Law: **(i)** Deliver to Buyer earthquake guide(s) (and questionnaire), environmental hazards booklet, and home energy rating pamphlet; **(ii)** disclose if the Property is located in a Special Flood Hazard Area; Potential Flooding (Inundation) Area; Very High Fire Hazard Zone; State Fire Responsibility Area; Earthquake Fault Zone; and Seismic Hazard Zone; and **(iii)** disclose any other zone as required by Law and provide any other information required for those zones.

C. **WITHHOLDING TAXES:** Within the time specified in paragraph 14A, to avoid required withholding, Seller shall Deliver to Buyer or qualified substitute, an affidavit sufficient to comply with federal (FIRPTA) and California withholding Law (C.A.R. Form AS or QS).

D. **MEGAN'S LAW DATABASE DISCLOSURE:** Notice: Pursuant to Section 290.46 of the Penal Code, information about specified registered sex offenders is made available to the public via an Internet Web site maintained by the Department of Justice at **www.meganslaw.ca.gov**. Depending on an offender's criminal history, this information will include either the address at which the offender resides or the community of residence and ZIP Code in which he or she resides. (Neither Seller nor Brokers are required to check this website. If Buyer wants further information, Broker recommends that Buyer obtain information from this website during Buyer's inspection contingency period. Brokers do not have expertise in this area.)

E. **NOTICE REGARDING GAS AND HAZARDOUS LIQUID TRANSMISSION PIPELINES:** This notice is being provided simply to inform you that information about the general location of gas and hazardous liquid transmission pipelines is available to the public via the National Pipeline Mapping System (NPMS) Internet Web site maintained by the United States Department of Transportation at **http://www.npms. phmsa.dot.gov/**. To seek further information about possible transmission pipelines near the Property, you may contact your local gas utility or other pipeline operators in the area. Contact information for pipeline operators is searchable by ZIP Code and county on the NPMS Internet Web site.

F. **CONDOMINIUM/PLANNED DEVELOPMENT DISCLOSURES:**

(1) SELLER HAS: 7 (or ___) Days After Acceptance to disclose to Buyer if the Property is a condominium, or is located in a planned development or other common interest subdivision (C.A.R. Form SPQ or ESD).

(2) If the Property is a condominium or is located in a planned development or other common interest subdivision, Seller has **3 (or ___) Days** After Acceptance to request from the HOA (C.A.R. Form HOA-IR): **(i)** Copies of any documents required by Law; **(ii)** disclosure of any pending or anticipated claim or litigation by or against the HOA; **(iii)** a statement containing the location and number of designated parking and storage spaces; **(iv)** Copies of the most recent 12 months of HOA minutes for regular and special meetings; **(v)** the names and contact information of all HOAs governing the Property (collectively, "CI Disclosures"); **(vi)** private transfer fees; **(vii)** Pet restrictions; and **(viii)** smoking restrictions. Seller shall itemize and Deliver

Buyer's Initials (_____)(_____)   Seller's Initials (_____)(_____)

RPA REVISED 12/18 (PAGE 4 OF 10)

CALIFORNIA RESIDENTIAL PURCHASE AGREEMENT (RPA PAGE 4 OF 10)

Property Address: _____ Date: _____

to Buyer all CI Disclosures received from the HOA and any CI Disclosures in Seller's possession. Buyer's approval of CI Disclosures is a contingency of this Agreement as specified in paragraph 14B(3). The Party specified in paragraph 7, as directed by escrow, shall deposit funds into escrow or direct to HOA or management company to pay for any of the above.

11. CONDITION OF PROPERTY: Unless otherwise agreed in writing: **(i)** the Property is sold (a) "AS-IS" in its PRESENT physical condition as of the date of Acceptance and (b) subject to Buyer's Investigation rights; **(ii)** the Property, including pool, spa, landscaping and grounds, is to be maintained in substantially the same condition as on the date of Acceptance; and **(iii)** all debris and personal property not included in the sale shall be removed by Close Of Escrow.

   A. Seller shall, within the time specified in paragraph 14A, DISCLOSE KNOWN MATERIAL FACTS AND DEFECTS affecting the Property, including known insurance claims within the past five years, and make any and all other disclosures required by law.

   B. Buyer has the right to conduct Buyer Investigations of the Property and, as specified in paragraph 14B, based upon information discovered in those investigations: **(i)** cancel this Agreement; or **(ii)** request that Seller make Repairs or take other action.

   C. Buyer is strongly advised to conduct investigations of the entire Property in order to determine its present condition. Seller may not be aware of all defects affecting the Property or other factors that Buyer considers important. Property improvements may not be built according to code, in compliance with current Law, or have had permits issued.

12. BUYER'S INVESTIGATION OF PROPERTY AND MATTERS AFFECTING PROPERTY:

   A. Buyer's acceptance of the condition of, and any other matter affecting the Property, is a contingency of this Agreement as specified in this paragraph and paragraph 14B. Within the time specified in paragraph 14B(1), Buyer shall have the right, at Buyer's expense unless otherwise agreed, to conduct inspections, investigations, tests, surveys and other studies ("Buyer Investigations"), including, but not limited to: **(i)** a general physical inspection; **(ii)** an inspection specifically for wood destroying pests and organisms. Any inspection for wood destroying pests and organisms shall be prepared by a registered Structural Pest Control company; shall cover the main building and attached structures; may cover detached structures; shall NOT include water tests of shower pans on upper level units unless the owners of property below the shower consent; shall NOT include roof coverings; and, if the Property is a unit in a condominium or other common interest subdivision, the inspection shall include only the separate interest and any exclusive-use areas being transferred, and shall NOT include common areas; and shall include a report ("Pest Control Report") showing the findings of the company which shall be separated into sections for evident infestation or infections (Section 1) and for conditions likely to lead to infestation or infection (Section 2); **(iii)** inspect for lead-based paint and other lead-based paint hazards; **(iv)** satisfy Buyer as to any matter specified in the attached Buyer's Inspection Advisory (C.A.R. Form BIA); **(v)** review the registered sex offender database; **(vi)** confirm the insurability of Buyer and the Property including the availability and cost of flood and fire insurance; and **(vii)** review and seek approval of leases that may need to be assumed by Buyer. Without Seller's prior written consent, Buyer shall neither make nor cause to be made: invasive or destructive Buyer Investigations, except for minimally invasive testing required to prepare a Pest Control Report; or inspections by any governmental building or zoning inspector or government employee, unless required by Law.

   B. Seller shall make the Property available for all Buyer Investigations. Buyer shall **(i)** as specified in paragraph 14B, complete Buyer Investigations and either remove the contingency or cancel this Agreement, and **(ii)** give Seller, at no cost, complete Copies of all such Investigation reports obtained by Buyer, which obligation shall survive the termination of this Agreement.

   C. Seller shall have water, gas, electricity and all operable pilot lights on for Buyer's Investigations and through the date possession is made available to Buyer.

   D. Buyer indemnity and seller protection for entry upon property: Buyer shall: **(i)** keep the Property free and clear of liens; **(ii)** repair all damage arising from Buyer Investigations; and **(iii)** indemnify and hold Seller harmless from all resulting liability, claims, demands, damages and costs. Buyer shall carry, or Buyer shall require anyone acting on Buyer's behalf to carry, policies of liability, workers' compensation and other applicable insurance, defending and protecting Seller from liability for any injuries to persons or property occurring during any Buyer Investigations or work done on the Property at Buyer's direction prior to Close Of Escrow. Seller is advised that certain protections may be afforded Seller by recording a "Notice of Non-Responsibility" (C.A.R. Form NNR) for Buyer Investigations and work done on the Property at Buyer's direction. Buyer's obligations under this paragraph shall survive the termination of this Agreement.

13. TITLE AND VESTING:

   A. Within the time specified in paragraph 14, Buyer shall be provided a current preliminary title report ("Preliminary Report"). The Preliminary Report is only an offer by the title insurer to issue a policy of title insurance and may not contain every item affecting title. Buyer's review of the Preliminary Report and any other matters which may affect title are a contingency of this Agreement as specified in paragraph 14B. The company providing the Preliminary Report shall, prior to issuing a Preliminary Report, conduct a search of the General Index for all Sellers except banks or other institutional lenders selling properties they acquired through foreclosure (REOs), corporations, and government entities. Seller shall within 7 Days After Acceptance, give Escrow Holder a completed Statement of Information.

   B. Title is taken in its present condition subject to all encumbrances, easements, covenants, conditions, restrictions, rights and other matters, whether of record or not, as of the date of Acceptance except for: **(i)** monetary liens of record (which Seller is obligated to pay off) unless Buyer is assuming those obligations or taking the Property subject to those obligations; and **(ii)** those matters which Seller has agreed to remove in writing.

   C. Within the time specified in paragraph 14A, Seller has a duty to disclose to Buyer all matters known to Seller affecting title, whether of record or not.

   D. At Close Of Escrow, Buyer shall receive a grant deed conveying title (or, for stock cooperative or long-term lease, an assignment of stock certificate or of Seller's leasehold interest), including oil, mineral and water rights if currently owned by Seller. Title shall vest as designated in Buyer's supplemental escrow instructions. THE MANNER OF TAKING TITLE MAY HAVE SIGNIFICANT LEGAL AND TAX CONSEQUENCES. CONSULT AN APPROPRIATE PROFESSIONAL.

   E. Buyer shall receive a CLTA/ALTA "Homeowner's Policy of Title Insurance", if applicable to the type of property and buyer. If not, Escrow Holder shall notify Buyer. A title company can provide information about the availability, coverage, and cost of other title policies and endorsements. If the Homeowner's Policy is not available, Buyer shall choose another policy, instruct Escrow Holder in writing and shall pay any increase in cost.

14. TIME PERIODS; REMOVAL OF CONTINGENCIES; CANCELLATION RIGHTS: The following time periods may only be extended, altered, modified or changed by mutual written agreement. Any removal of contingencies or cancellation under this paragraph by either Buyer or Seller must be exercised in good faith and in writing (C.A.R. Form CR or CC).

   A. SELLER HAS: 7 (or ___) Days After Acceptance to Deliver to Buyer all Reports, disclosures and information for which Seller

Buyer's Initials (_____)(_____)   Seller's Initials (_____)(_____)

Property Address: _____ Date: _____

is responsible under paragraphs 5, 6, 7, 8B(5), 10A, B, C, and F, 11A and 13A. If, by the time specified, Seller has not Delivered any such item, Buyer after first Delivering to Seller a Notice to Seller to Perform (C.A.R. Form NSP) may cancel this Agreement.

B. (1) **BUYER HAS: 17 (or ___) Days** After Acceptance, unless otherwise agreed in writing, to:
(i) complete all Buyer Investigations; review all disclosures, reports, lease documents to be assumed by Buyer pursuant to paragraph 8B(5), and other applicable information, which Buyer receives from Seller; and approve all matters affecting the Property; and (ii) Deliver to Seller Signed Copies of Statutory and Lead Disclosures and other disclosures Delivered by Seller in accordance with paragraph 10A.
(2) Within the time specified in paragraph 14B(1), Buyer may request that Seller make repairs or take any other action regarding the Property (C.A.R. Form RR). Seller has no obligation to agree to or respond to (C.A.R. Form RRRR) Buyer's requests.
(3) By the end of the time specified in paragraph 14B(1) (or as otherwise specified in this Agreement), Buyer shall Deliver to Seller a removal of the applicable contingency or cancellation (C.A.R. Form CR or CC) of this Agreement. However, if any report, disclosure or information for which Seller is responsible is not Delivered within the time specified in paragraph 14A, then Buyer has **5 (or ___) Days** After Delivery of any such items, or the time specified in paragraph 14B(1), whichever is later, to Deliver to Seller a removal of the applicable contingency or cancellation of this Agreement.
(4) **Continuation of Contingency:** Even after the end of the time specified in paragraph 14B(1) and before Seller cancels, if at all, pursuant to paragraph 14D, Buyer retains the right, in writing, to either (i) remove remaining contingencies, or (ii) cancel this Agreement based on a remaining contingency. Once Buyer's written removal of all contingencies is Delivered to Seller, Seller may not cancel this Agreement pursuant to paragraph 14D(1).
(5) **Access to Property:** Buyer shall have access to the Property to conduct inspections and investigations for **17 (or ___) Days** After Acceptance, whether or not any part of the Buyer's Investigation Contingency has been waived or removed.

C. ☐ **REMOVAL OF CONTINGENCIES WITH OFFER: Buyer removes the contingencies specified in the attached Contingency Removal form (C.A.R. Form CR). If Buyer removes any contingency without an adequate understanding of the Property's condition or Buyer's ability to purchase, Buyer is acting against the advice of Broker.**

D. **SELLER RIGHT TO CANCEL:**
(1) **Seller right to Cancel; Buyer Contingencies:** If, by the time specified in this Agreement, Buyer does not Deliver to Seller a removal of the applicable contingency or cancellation of this Agreement, then Seller, after first Delivering to Buyer a Notice to Buyer to Perform (C.A.R. Form NBP), may cancel this Agreement. In such event, Seller shall authorize the return of Buyer's deposit, except for fees incurred by Buyer.
(2) **Seller right to Cancel; Buyer Contract Obligations:** Seller, after first delivering to Buyer a NBP, may cancel this Agreement if, by the time specified in this Agreement, Buyer does not take the following action(s): (i) Deposit funds as required by paragraph 3A, or 3B or if the funds deposited pursuant to paragraph 3A or 3B are not good when deposited; (ii) Deliver a notice of FHA or VA costs or terms as required by paragraph 3D(3) (C.A.R. Form FVA); (iii) Deliver a letter as required by paragraph 3J(1); (iv) Deliver verification, or a satisfactory verification if Seller reasonably disapproves of the verification already provided, as required by paragraph 3C or 3H; (v) In writing assume or accept leases or liens specified in 8B5; (vi) Return Statutory and Lead Disclosures as required by paragraph 10A(5); or (vii) Sign or initial a separate liquidated damages form for an increased deposit as required by paragraphs 3B and 21B; or (viii) Provide evidence of authority to sign in a representative capacity as specified in paragraph 19. In such event, Seller shall authorize the return of Buyer's deposit, except for fees incurred by Buyer.

E. **NOTICE TO BUYER OR SELLER TO PERFORM:** The NBP or NSP shall: (i) be in writing; (ii) be signed by the applicable Buyer or Seller; and (iii) give the other Party at least **2 (or ___) Days** After Delivery (or until the time specified in the applicable paragraph, whichever occurs last) to take the applicable action. A NBP or NSP may not be Delivered any earlier than **2 Days** Prior to the expiration of the applicable time for the other Party to remove a contingency or cancel this Agreement or meet an obligation specified in paragraph 14.

F. **EFFECT OF BUYER'S REMOVAL OF CONTINGENCIES:** If Buyer removes, in writing, any contingency or cancellation rights, unless otherwise specified in writing, Buyer shall conclusively be deemed to have: (i) completed all Buyer Investigations, and review of reports and other applicable information and disclosures pertaining to that contingency or cancellation right; (ii) elected to proceed with the transaction; and (iii) assumed all liability, responsibility and expense for Repairs or corrections pertaining to that contingency or cancellation right, or for the inability to obtain financing.

G. **CLOSE OF ESCROW:** Before Buyer or Seller may cancel this Agreement for failure of the other Party to close escrow pursuant to this Agreement, Buyer or Seller must first Deliver to the other Party a demand to close escrow (C.A.R. Form DCE). The DCE shall: (i) be signed by the applicable Buyer or Seller; and (ii) give the other Party at least **3 (or ___) Days** After Delivery to close escrow. A DCE may not be Delivered any earlier than **3 Days** Prior to the scheduled close of escrow.

H. **EFFECT OF CANCELLATION ON DEPOSITS:** If Buyer or Seller gives written notice of cancellation pursuant to rights duly exercised under the terms of this Agreement, the Parties agree to Sign mutual instructions to cancel the sale and escrow and release deposits, if any, to the party entitled to the funds, less fees and costs incurred by that party. Fees and costs may be payable to service providers and vendors for services and products provided during escrow. Except as specified below, **release of funds will require mutual Signed release instructions from the Parties, judicial decision or arbitration award.** If either Party fails to execute mutual instructions to cancel escrow, one Party may make a written demand to Escrow Holder for the deposit (C.A.R. Form BDRD or SDRD). Escrow Holder, upon receipt, shall promptly deliver notice of the demand to the other Party. If, within 10 Days After Escrow Holder's notice, the other Party does not object to the demand, Escrow Holder shall disburse the deposit to the Party making the demand. If Escrow Holder complies with the preceding process, each Party shall be deemed to have released Escrow Holder from any and all claims or liability related to the disbursal of the deposit. Escrow Holder, at its discretion, may nonetheless require mutual cancellation instructions. **A Party may be subject to a civil penalty of up to $1,000 for refusal to sign cancellation instructions if no good faith dispute exists as to who is entitled to the deposited funds (Civil Code §1057.3).**

15. **FINAL VERIFICATION OF CONDITION:** Buyer shall have the right to make a final verification of the Property within **5 (or ___) Days** Prior to Close Of Escrow, NOT AS A CONTINGENCY OF THE SALE, but solely to confirm: (i) the Property is maintained pursuant to paragraph 11; (ii) Repairs have been completed as agreed; and (iii) Seller has complied with Seller's other obligations under this Agreement (C.A.R. Form VP).

16. **REPAIRS:** Repairs shall be completed prior to final verification of condition unless otherwise agreed in writing. Repairs to be

Buyer's Initials (_____)(_____)   Seller's Initials (_____)(_____)

performed at Seller's expense may be performed by Seller or through others, provided that the work complies with applicable Law, including governmental permit, inspection and approval requirements. Repairs shall be performed in a good, skillful manner with materials of quality and appearance comparable to existing materials. It is understood that exact restoration of appearance or cosmetic items following all Repairs may not be possible. Seller shall: **(i)** obtain invoices and paid receipts for Repairs performed by others; **(ii)** prepare a written statement indicating the Repairs performed by Seller and the date of such Repairs; and **(iii)** provide Copies of invoices and paid receipts and statements to Buyer prior to final verification of condition.

17. **PRORATIONS OF PROPERTY TAXES AND OTHER ITEMS:** Unless otherwise agreed in writing, the following items shall be PAID CURRENT and prorated between Buyer and Seller as of Close Of Escrow: real property taxes and assessments, interest, rents, HOA regular, special, and emergency dues and assessments imposed prior to Close Of Escrow, premiums on insurance assumed by Buyer, payments on bonds and assessments assumed by Buyer, and payments on Mello-Roos and other Special Assessment District bonds and assessments that are now a lien. The following items shall be assumed by Buyer WITHOUT CREDIT toward the purchase price: prorated payments on Mello-Roos and other Special Assessment District bonds and assessments and HOA special assessments that are now a lien but not yet due. Property will be reassessed upon change of ownership. Any supplemental tax bills shall be paid as follows: **(i)** for periods after Close Of Escrow, by Buyer; and **(ii)** for periods prior to Close Of Escrow, by Seller (see C.A.R. Form SPT or SBSA for further information). TAX BILLS ISSUED AFTER CLOSE OF ESCROW SHALL BE HANDLED DIRECTLY BETWEEN BUYER AND SELLER. Prorations shall be made based on a 30-day month.

18. **BROKERS:**
   A. **COMPENSATION:** Seller or Buyer, or both, as applicable, agree to pay compensation to Broker as specified in a separate written agreement between Broker and that Seller or Buyer. Compensation is payable upon Close Of Escrow, or if escrow does not close, as otherwise specified in the agreement between Broker and that Seller or Buyer.
   B. **SCOPE OF DUTY:** Buyer and Seller acknowledge and agree that Broker: **(i)** Does not decide what price Buyer should pay or Seller should accept; **(ii)** Does not guarantee the condition of the Property; **(iii)** Does not guarantee the performance, adequacy or completeness of inspections, services, products or repairs provided or made by Seller or others; **(iv)** Does not have an obligation to conduct an inspection of common areas or areas off the site of the Property; **(v)** Shall not be responsible for identifying defects on the Property, in common areas, or offsite unless such defects are visually observable by an inspection of reasonably accessible areas of the Property or are known to Broker; **(vi)** Shall not be responsible for inspecting public records or permits concerning the title or use of Property; **(vii)** Shall not be responsible for identifying the location of boundary lines or other items affecting title; **(viii)** Shall not be responsible for verifying square footage, representations of others or information contained in Investigation reports, Multiple Listing Service, advertisements, flyers or other promotional material; **(ix)** Shall not be responsible for determining the fair market value of the Property or any personal property included in the sale; **(x)** Shall not be responsible for providing legal or tax advice regarding any aspect of a transaction entered into by Buyer or Seller; and **(xi)** Shall not be responsible for providing other advice or information that exceeds the knowledge, education and experience required to perform real estate licensed activity. Buyer and Seller agree to seek legal, tax, insurance, title and other desired assistance from appropriate professionals.

19. **REPRESENTATIVE CAPACITY:** If one or more Parties is signing this Agreement in a representative capacity and not for him/herself as an individual then that Party shall so indicate in paragraph 31 or 32 and attach a Representative Capacity Signature Disclosure (C.A.R. Form RCSD). Wherever the signature or initials of the representative identified in the RCSD appear on this Agreement or any related documents, it shall be deemed to be in a representative capacity for the entity described and not in an individual capacity, unless otherwise indicated. The Party acting in a representative capacity **(i)** represents that the entity for which that party is acting already exists and **(ii)** shall Deliver to the other Party and Escrow Holder, within **3 Days** After Acceptance, evidence of authority to act in that capacity (such as but not limited to: applicable portion of the trust or Certification Of Trust (Probate Code §18100.5), letters testamentary, court order, power of attorney, corporate resolution, or formation documents of the business entity).

20. **JOINT ESCROW INSTRUCTIONS TO ESCROW HOLDER:**
   A. **The following paragraphs, or applicable portions thereof, of this Agreement constitute the joint escrow instructions of Buyer and Seller to Escrow Holder,** which Escrow Holder is to use along with any related counter offers and addenda, and any additional mutual instructions to close the escrow: paragraphs 1, 3, 4B, 5A, 6, 7, 10C, 13, 14G, 17, 18A, 19, 20, 26, 29, 30, 31, 32 and paragraph D of the section titled Real Estate Brokers on page 10. If a Copy of the separate compensation agreement(s) provided for in paragraph 18A, or paragraph D of the section titled Real Estate Brokers on page 10 is deposited with Escrow Holder by Broker, Escrow Holder shall accept such agreement(s) and pay out from Buyer's or Seller's funds, or both, as applicable, the Broker's compensation provided for in such agreement(s). The terms and conditions of this Agreement not set forth in the specified paragraphs are additional matters for the information of Escrow Holder, but about which Escrow Holder need not be concerned. Buyer and Seller will receive Escrow Holder's general provisions, if any, directly from Escrow Holder and will execute such provisions within the time specified in paragraph 7C(1)(c). To the extent the general provisions are inconsistent or conflict with this Agreement, the general provisions will control as to the duties and obligations of Escrow Holder only. Buyer and Seller will execute additional instructions, documents and forms provided by Escrow Holder that are reasonably necessary to close the escrow and, as directed by Escrow Holder, within **3 (or ____) Days,** shall pay to Escrow Holder or HOA or HOA management company or others any fee required by paragraphs 7, 10 or elsewhere in this Agreement.
   B. A Copy of this Agreement including any counter offer(s) and addenda shall be delivered to Escrow Holder within **3 Days** After Acceptance (or _____). Buyer and Seller authorize Escrow Holder to accept and rely on Copies and Signatures as defined in this Agreement as originals, to open escrow and for other purposes of escrow. The validity of this Agreement as between Buyer and Seller is not affected by whether or when Escrow Holder Signs this Agreement. Escrow Holder shall provide Seller's Statement of Information to Title company when received from Seller. If Seller delivers an affidavit to Escrow Holder to satisfy Seller's FIRPTA obligation under paragraph 10C, Escrow Holder shall deliver to Buyer a Qualified Substitute statement that complies with federal Law.
   C. Brokers are a party to the escrow for the sole purpose of compensation pursuant to paragraph 18A and paragraph D of the section titled Real Estate Brokers on page 10. Buyer and Seller irrevocably assign to Brokers compensation specified in paragraph 18A, and irrevocably instruct Escrow Holder to disburse those funds to Brokers at Close Of Escrow or pursuant to any other mutually executed cancellation agreement. Compensation instructions can be amended or revoked only with the written consent of Brokers. Buyer and Seller shall release and hold harmless Escrow Holder from any liability resulting from Escrow Holder's payment to Broker(s) of compensation pursuant to this Agreement.
   D. Upon receipt, Escrow Holder shall provide Seller and Seller's Broker verification of Buyer's deposit of funds pursuant to paragraph 3A and 3B. Once Escrow Holder becomes aware of any of the following, Escrow Holder shall immediately notify all

Buyer's Initials (_____)(_____)   Seller's Initials (_____)(_____)

Property Address: _____   Date: _____

Brokers: **(i)** if Buyer's initial or any additional deposit or down payment is not made pursuant to this Agreement, or is not good at time of deposit with Escrow Holder; or **(ii)** if Buyer and Seller instruct Escrow Holder to cancel escrow.

E. A Copy of any amendment that affects any paragraph of this Agreement for which Escrow Holder is responsible shall be delivered to Escrow Holder within 3 Days after mutual execution of the amendment.

21. REMEDIES FOR BUYER'S BREACH OF CONTRACT:

A. Any clause added by the Parties specifying a remedy (such as release or forfeiture of deposit or making a deposit non-refundable) for failure of Buyer to complete the purchase in violation of this Agreement shall be deemed invalid unless the clause independently satisfies the statutory liquidated damages requirements set forth in the Civil Code.

B. **LIQUIDATED DAMAGES: If Buyer fails to complete this purchase because of Buyer's default, Seller shall retain, as liquidated damages, the deposit actually paid. If the Property is a dwelling with no more than four units, one of which Buyer intends to occupy, then the amount retained shall be no more than 3% of the purchase price. Any excess shall be returned to Buyer.** Except as provided in paragraph 14G, release of funds will require mutual, Signed release instructions from both Buyer and Seller, judicial decision or arbitration award. **AT THE TIME OF ANY INCREASED DEPOSIT BUYER AND SELLER SHALL SIGN A SEPARATE LIQUIDATED DAMAGES PROVISION INCORPORATING THE INCREASED DEPOSIT AS LIQUIDATED DAMAGES (C.A.R. FORM RID).**

Buyer's Initials _____/_____   Seller's Initials _____/_____

22. DISPUTE RESOLUTION:

A. **MEDIATION:** The Parties agree to mediate any dispute or claim arising between them out of this Agreement, or any resulting transaction, before resorting to arbitration or court action through the C.A.R. Real Estate Mediation Center for Consumers (**www.consumermediation.org**) or through any other mediation provider or service mutually agreed to by the Parties. The Parties **also agree to mediate any disputes or claims with Broker(s), who, in writing, agree to such mediation prior to, or within a reasonable time after, the dispute or claim is presented to the Broker.** Mediation fees, if any, shall be divided equally among the Parties involved. If, for any dispute or claim to which this paragraph applies, any Party **(i)** commences an action without first attempting to resolve the matter through mediation, or **(ii)** before commencement of an action, refuses to mediate after a request has been made, then that Party shall not be entitled to recover attorney fees, even if they would otherwise be available to that Party in any such action. THIS MEDIATION PROVISION APPLIES WHETHER OR NOT THE ARBITRATION PROVISION IS INITIALED. **Exclusions from this mediation agreement are specified in paragraph 22C.**

B. **ARBITRATION OF DISPUTES:**

The Parties agree that any dispute or claim in Law or equity arising between them out of this Agreement or any resulting transaction, which is not settled through mediation, shall be decided by neutral, binding arbitration. The Parties also agree to arbitrate any disputes or claims with Broker(s), who, in writing, agree to such arbitration prior to, or within a reasonable time after, the dispute or claim is presented to the Broker. The arbitrator shall be a retired judge or justice, or an attorney with at least 5 years of residential real estate Law experience, unless the parties mutually agree to a different arbitrator. The Parties shall have the right to discovery in accordance with Code of Civil Procedure §1283.05. In all other respects, the arbitration shall be conducted in accordance with Title 9 of Part 3 of the Code of Civil Procedure. Judgment upon the award of the arbitrator(s) may be entered into any court having jurisdiction. Enforcement of this agreement to arbitrate shall be governed by the Federal Arbitration Act. Exclusions from this arbitration agreement are specified in paragraph 22C.

"NOTICE: BY INITIALING IN THE SPACE BELOW YOU ARE AGREEING TO HAVE ANY DISPUTE ARISING OUT OF THE MATTERS INCLUDED IN THE 'ARBITRATION OF DISPUTES' PROVISION DECIDED BY NEUTRAL ARBITRATION AS PROVIDED BY CALIFORNIA LAW AND YOU ARE GIVING UP ANY RIGHTS YOU MIGHT POSSESS TO HAVE THE DISPUTE LITIGATED IN A COURT OR JURY TRIAL. BY INITIALING IN THE SPACE BELOW YOU ARE GIVING UP YOUR JUDICIAL RIGHTS TO DISCOVERY AND APPEAL, UNLESS THOSE RIGHTS ARE SPECIFICALLY INCLUDED IN THE 'ARBITRATION OF DISPUTES' PROVISION. IF YOU REFUSE TO SUBMIT TO ARBITRATION AFTER AGREEING TO THIS PROVISION, YOU MAY BE COMPELLED TO ARBITRATE UNDER THE AUTHORITY OF THE CALIFORNIA CODE OF CIVIL PROCEDURE. YOUR AGREEMENT TO THIS ARBITRATION PROVISION IS VOLUNTARY."

"WE HAVE READ AND UNDERSTAND THE FOREGOING AND AGREE TO SUBMIT DISPUTES ARISING OUT OF THE MATTERS INCLUDED IN THE 'ARBITRATION OF DISPUTES' PROVISION TO NEUTRAL ARBITRATION."

Buyer's Initials _____/_____   Seller's Initials _____/_____

C. **ADDITIONAL MEDIATION AND ARBITRATION TERMS:**

(1) **EXCLUSIONS:** The following matters are excluded from mediation and arbitration: **(i)** a judicial or non-judicial foreclosure or other action or proceeding to enforce a deed of trust, mortgage or installment land sale contract as defined in Civil Code §2985; **(ii)** an unlawful detainer action; and **(iii)** any matter that is within the jurisdiction of a probate, small claims or bankruptcy court.

(2) **PRESERVATION OF ACTIONS:** The following shall not constitute a waiver nor violation of the mediation and arbitration provisions: **(i)** the filing of a court action to preserve a statute of limitations; **(ii)** the filing of a court action to enable the recording of a notice of pending action, for order of attachment, receivership, injunction, or other provisional remedies; or **(iii)** the filing of a mechanic's lien.

(3) **BROKERS:** Brokers shall not be obligated nor compelled to mediate or arbitrate unless they agree to do so in writing. Any Broker(s) participating in mediation or arbitration shall not be deemed a party to this Agreement.

23. SELECTION OF SERVICE PROVIDERS: Brokers do not guarantee the performance of any vendors, service or product providers ("Providers"), whether referred by Broker or selected by Buyer, Seller or other person. Buyer and Seller may select ANY Providers of their own choosing.

Buyer's Initials (_____)(_____)   Seller's Initials (_____)(_____)

CALIFORNIA RESIDENTIAL PURCHASE AGREEMENT (RPA PAGE 8 OF 10)

Property Address: _____ Date: _____

24. **MULTIPLE LISTING SERVICE ("MLS"):** Brokers are authorized to report to the MLS a pending sale and, upon Close Of Escrow, the sales price and other terms of this transaction shall be provided to the MLS to be published and disseminated to persons and entities authorized to use the information on terms approved by the MLS.

25. **ATTORNEY FEES:** In any action, proceeding, or arbitration between Buyer and Seller arising out of this Agreement, the prevailing Buyer or Seller shall be entitled to reasonable attorney fees and costs from the non-prevailing Buyer or Seller, except as provided in paragraph 22A.

26. **ASSIGNMENT:** Buyer shall not assign all or any part of Buyer's interest in this Agreement without first having obtained the separate written consent of Seller to a specified assignee. Such consent shall not be unreasonably withheld. Any total or partial assignment shall not relieve Buyer of Buyer's obligations pursuant to this Agreement unless otherwise agreed in writing by Seller (C.A.R. Form AOAA).

27. **EQUAL HOUSING OPPORTUNITY:** The Property is sold in compliance with federal, state and local anti-discrimination Laws.

28. **TERMS AND CONDITIONS OF OFFER:** This is an offer to purchase the Property on the above terms and conditions. The liquidated damages paragraph or the arbitration of disputes paragraph is incorporated in this Agreement if initialed by all Parties or if incorporated by mutual agreement in a counter offer or addendum. If at least one but not all Parties initial, a counter offer is required until agreement is reached. Seller has the right to continue to offer the Property for sale and to accept any other offer at any time prior to notification of Acceptance. The Parties have read and acknowledge receipt of a Copy of the offer and agree to the confirmation of agency relationships. If this offer is accepted and Buyer subsequently defaults, Buyer may be responsible for payment of Brokers' compensation. This Agreement and any supplement, addendum or modification, including any Copy, may be Signed in two or more counterparts, all of which shall constitute one and the same writing.

29. **TIME OF ESSENCE; ENTIRE CONTRACT; CHANGES:** Time is of the essence. All understandings between the Parties are incorporated in this Agreement. Its terms are intended by the Parties as a final, complete and exclusive expression of their Agreement with respect to its subject matter, and may not be contradicted by evidence of any prior agreement or contemporaneous oral agreement. If any provision of this Agreement is held to be ineffective or invalid, the remaining provisions will nevertheless be given full force and effect. Except as otherwise specified, this Agreement shall be interpreted and disputes shall be resolved in accordance with the Laws of the State of California. **Neither this Agreement nor any provision in it may be extended, amended, modified, altered or changed, except in writing Signed by Buyer and Seller.**

30. **DEFINITIONS:** As used in this Agreement:
    A. **"Acceptance"** means the time the offer or final counter offer is accepted in writing by a Party and is delivered to and personally received by the other Party or that Party's authorized agent in accordance with the terms of this offer or a final counter offer.
    B. **"Agreement"** means this document and any counter offers and any incorporated addenda, collectively forming the binding agreement between the Parties. Addenda are incorporated only when Signed by all Parties.
    C. **"C.A.R. Form"** means the most current version of the specific form referenced or another comparable form agreed to by the parties.
    D. **"Close Of Escrow"**, including "COE", means the date the grant deed, or other evidence of transfer of title, is recorded.
    E. **"Copy"** means copy by any means including photocopy, NCR, facsimile and electronic.
    F. **"Days"** means calendar days. However, after Acceptance, the last **Day** for performance of any act required by this Agreement (including Close Of Escrow) shall not include any Saturday, Sunday, or legal holiday and shall instead be the next Day.
    G. **"Days After"** means the specified number of calendar days after the occurrence of the event specified, not counting the calendar date on which the specified event occurs, and ending at 11:59 PM on the final day.
    H. **"Days Prior"** means the specified number of calendar days before the occurrence of the event specified, not counting the calendar date on which the specified event is scheduled to occur.
    I. **"Deliver", "Delivered"** or **"Delivery"**, unless otherwise specified in writing, means and shall be effective upon: personal receipt by Buyer or Seller or the individual Real Estate Licensee for that principal as specified in the section titled Real Estate Brokers on page 10, regardless of the method used (i.e., messenger, mail, email, fax, other).
    J. **"Electronic Copy"** or **"Electronic Signature"** means, as applicable, an electronic copy or signature complying with California Law. Buyer and Seller agree that electronic means will not be used by either Party to modify or alter the content or integrity of this Agreement without the knowledge and consent of the other Party.
    K. **"Law"** means any law, code, statute, ordinance, regulation, rule or order, which is adopted by a controlling city, county, state or federal legislative, judicial or executive body or agency.
    L. **"Repairs"** means any repairs (including pest control), alterations, replacements, modifications or retrofitting of the Property provided for under this Agreement.
    M. **"Signed"** means either a handwritten or electronic signature on an original document, Copy or any counterpart.

31. **EXPIRATION OF OFFER:** This offer shall be deemed revoked and the deposit, if any, shall be returned to Buyer unless the offer is Signed by Seller and a Copy of the Signed offer is personally received by Buyer, or by _____,
    who is authorized to receive it, by 5:00 PM on the third Day after this offer is signed by Buyer (or by ☐ _____ ☐AM/☐PM,
    on _____(date)).

☐ One or more Buyers is signing this Agreement in a representative capacity and not for him/herself as an individual. See attached Representative Capacity Signature Disclosure (C.A.R. Form RCSD-B) for additional terms.

Date _____ BUYER _____

(Print name) _____

Date _____ BUYER _____

(Print name) _____

☐ Additional Signature Addendum attached (C.A.R. Form ASA).

Buyer's Initials (_____)(_____)   Seller's Initials (_____)(_____)

RPA REVISED 12/18 (PAGE 9 of 10)

CALIFORNIA RESIDENTIAL PURCHASE AGREEMENT (RPA PAGE 9 OF 10)

Property Address: _____ Date: _____

32. ACCEPTANCE OF OFFER: Seller warrants that Seller is the owner of the Property, or has the authority to execute this Agreement. Seller accepts the above offer and agrees to sell the Property on the above terms and conditions. Seller has read and acknowledges receipt of a Copy of this Agreement, and authorizes Broker to Deliver a Signed Copy to Buyer.

☐ (If checked) SELLER'S ACCEPTANCE IS **SUBJECT TO ATTACHED COUNTER OFFER (C.A.R. Form SCO or SMCO) DATED:** _____.

☐ One or more Sellers is signing this Agreement in a representative capacity and not for him/herself as an individual. See attached Representative Capacity Signature Disclosure (C.A.R. Form RCSD-S) for additional terms.

Date _____ SELLER _____
(Print name) _____
Date _____ SELLER _____
(Print name) _____

☐ Additional Signature Addendum attached (C.A.R. Form ASA).

(___/___) **(Do not initial if making a counter offer.) CONFIRMATION OF ACCEPTANCE:** A Copy of Signed Acceptance was
(Initials) personally received by Buyer or Buyer's authorized agent on (date) _____ at _____
☐AM/☐PM. **A binding Agreement is created when a Copy of Signed Acceptance is personally received by Buyer or Buyer's authorized agent whether or not confirmed in this document. Completion of this confirmation is not legally required in order to create a binding Agreement; it is solely intended to evidence the date that Confirmation of Acceptance has occurred.**

REAL ESTATE BROKERS:
A. **Real Estate Brokers are not parties to the Agreement between Buyer and Seller.**
B. **Agency relationships are confirmed as stated in paragraph 2.**
C. If specified in paragraph 3A(2), Agent who submitted the offer for Buyer acknowledges receipt of deposit.
D. **COOPERATING (BUYER'S) BROKER COMPENSATION:** Seller's Broker agrees to pay Buyer's Broker and Buyer's Broker agrees to accept, out of Seller's Broker's proceeds in escrow, the amount specified in the MLS, provided Buyer's Broker is a Participant of the MLS in which the Property is offered for sale or a reciprocal MLS. If Seller's Broker and Buyer's Broker are not both Participants of the MLS, or a reciprocal MLS, in which the Property is offered for sale, then compensation must be specified in a separate written agreement (C.A.R. Form CBC). Declaration of License and Tax (C.A.R. Form DLT) may be used to document that tax reporting will be required or that an exemption exists.
E. **PRESENTATION OF OFFER:** Pursuant to Standard of Practice 1-7, if Buyer's Broker makes a written request, Seller's Broker shall confirm in writing that this offer has been presented to Seller.

Buyer's Brokerage Firm _____ DRE Lic. # _____
By _____ DRE Lic. # _____ Date _____
By _____ DRE Lic. # _____ Date _____
Address _____ City _____ State _____ Zip _____
Telephone _____ Fax _____ E-mail _____
Seller's Brokerage Firm _____ DRE Lic. # _____
By _____ DRE Lic. # _____ Date _____
By _____ DRE Lic. # _____ Date _____
Address _____ City _____ State _____ Zip _____
Telephone _____ Fax _____ E-mail _____

ESCROW HOLDER ACKNOWLEDGMENT:
Escrow Holder acknowledges receipt of a Copy of this Agreement, (if checked, ☐ a deposit in the amount of $ _____),
counter offer numbers _____ ☐ Seller's Statement of Information and _____
_____, and agrees to act as Escrow Holder subject to paragraph 20 of this Agreement, any
supplemental escrow instructions and the terms of Escrow Holder's general provisions.

Escrow Holder is advised that the date of Confirmation of Acceptance of the Agreement as between Buyer and Seller is _____

Escrow Holder _____ Escrow # _____
By _____ Date _____
Address _____
Phone/Fax/E-mail _____
Escrow Holder has the following license number # _____
☐ Department of Business Oversight, ☐ Department of Insurance, ☐ Department of Real Estate.

PRESENTATION OF OFFER: (_____)(_____) Seller's Broker presented this offer to Seller on _____ (date).
Broker or Designee Initials

REJECTION OF OFFER: (_____)(_____) No counter offer is being made. This offer was rejected by Seller on _____ (date).
Seller's Initials

Published and Distributed by:
REAL ESTATE BUSINESS SERVICES, LLC.
a subsidiary of the CALIFORNIA ASSOCIATION OF REALTORS®
525 South Virgil Avenue, Los Angeles, California 90020

Buyer Acknowledges that page 10 is part of this Agreement

(_____)(_____)
Buyer's Initials

RPA REVISED 12/18 (PAGE 10 of 10)

CALIFORNIA RESIDENTIAL PURCHASE AGREEMENT (RPA PAGE 10 OF 10)

The following form is copyrighted by the California Association of REALTORS®, ©2020, and is reprinted under a limited license with permission. Photocopying or any other reproduction, whether electronic or otherwise, is strictly prohibited.

Appendix **B**

Sample Short-Sale Addendum plus Short-Sale Information and Advisory

Why do you need a short-sale addendum? Because, as we note in Chapter 8, even the best standard real estate purchase contract isn't written to handle the unique terms and conditions involved in a short sale. A short-sale addendum is drafted specifically to deal with the nuances of short sales.

Everything we say in Appendix A about a real estate purchase contract applies equally to a short-sale addendum. It's a legal document. Your real estate agent or lawyer should provide the appropriate addendum form used in your area and help you fill it out. Don't leave any spaces blank on the addendum.

CHECK IT OUT

Like purchase contracts, short-sale addendums vary enormously in length, complexity, and terms from state to state and, within a state, from one locality to another. The forms included in this appendix are representative samples of well-written, comprehensive documents. You can access this short-sale addendum online at www.dummies.com/go/homebuyingkit7e.

**CALIFORNIA
ASSOCIATION
OF REALTORS®**

SHORT SALE ADDENDUM
(C.A.R. Form SSA, Revised 4/12)

This is an addendum to the ☐ California Residential Purchase Agreement, ☐ Counter Offer, ☐ Other _____
_____ ("Agreement"), dated _____,
on property known as _____ ("Property"),
between _____ ("Buyer") and
_____ ("Seller").

1. **SHORT SALE APPROVAL:**

   A. This Agreement is contingent upon Seller's receipt of and delivery to Buyer of written consent ("Short Sale Lenders' Consent") to the Agreement from all existing secured lenders and lienholders ("Short Sale Lenders"), by 5:00 P.M. no later than **45 (or** ☐ **_____) Days After Acceptance (or** ☐ on _____ (date) ("Short Sale Contingency Date"). If Buyer or Seller cancels this Agreement prior to the Short Sale Contingency Date, that party may be in breach of the Agreement unless the cancellation is made pursuant to some other paragraph in this addendum or in the Agreement, whether or not time periods in the Agreement have commenced.

   B. Short Sale Lenders' Consent means that all Short Sale Lenders shall collectively agree to reduce their respective loan balances by an amount sufficient to permit the proceeds from the sale of the Property to pay the existing balances on loans secured by the Property, real property taxes, brokerage commissions, closing costs, and other monetary obligations the Agreement requires Seller to pay at Close Of Escrow (including, but not limited to, escrow charges, title charges, documentary transfer taxes, prorations, retrofit costs, Homeowners Association Fees and Repairs) without requiring Seller to place any funds into escrow or have any continuing obligation to Short Sale Lenders.

   C. **(i)** Seller shall Deliver to Buyer a copy of Short Sale Lenders' Consent or term sheet(s) within 3 (or ☐ _____) Days After receipt by Seller. **(ii)** Seller's presentation to Buyer of Short Sale Lenders' Consent satisfying 1B removes the contingency in 1A.

   D. If by the Short Sale Contingency Date, **(i)** Seller has not received Short Sale Lenders' Consent satisfying 1B, Seller may in writing cancel this Agreement, or **(ii)** Buyer has not received a copy of Short Sale Lenders' Consent satisfying 1B, Buyer may cancel this Agreement in writing. In either case, Buyer shall be entitled to return of any remaining deposit delivered to escrow.

   E. Seller shall reasonably cooperate with existing Short Sale Lenders in the short sale process, but neither Seller nor Buyer is obligated to change the terms of their Agreement to satisfy Short Sale Lenders' consent or term sheet(s).

   F. If Short Sale Lenders' written consent or term sheet(s) provided to Seller require changes to the Agreement in order to satisfy the terms of 1B, **(i)** neither Buyer nor Seller shall be obligated to continue negotiations to satisfy any of the requirements of the term sheet(s) **(ii)** either party may in writing cancel this Agreement. And **(iii)** Seller is advised to seek legal, accounting and tax advice before agreeing to any such changes. If the Agreement is cancelled pursuant to this paragraph, Buyer shall be entitled to return of any remaining deposit delivered to escrow.

2. **TIME PERIODS.** Time periods in the Agreement for inspections, contingencies, covenants, and other obligations: **(i)** shall begin the Day After Seller delivers to Buyer Short Sale Lenders' Consent satisfying 1B. However, time periods for providing pre-approval/pre-qualification letters and verification of down payment and closing costs shall nonetheless begin as otherwise specified in the Agreement; or **(ii)** (if checked) ☐ shall begin as specified in the Agreement.

3. **BUYER'S DEPOSIT CHECK.** Buyer's deposit check shall be delivered to escrow within: **(i)** 3 business Days After Seller delivers to Buyer Short Sale Lenders' Consent satisfying 1B, or **(ii)** (if checked) ☐ as specified in the Agreement.

Buyer's Initials (_____)(_____)   Seller's Initials (_____)(_____)

SSA Revised 4/12 (PAGE 1 OF 2) Print Date

| Reviewed by _____ Date _____ |

SHORT SALE ADDENDUM (SSA PAGE 1 OF 2)

Property Address: _____ Date: _____

4. **NO ASSURANCE OF LENDER APPROVAL.** Buyer and Seller understand that Short Sale Lenders: **(i)** are not obligated to give consent to a short sale; **(ii)** may require Seller to forward any other offer received; and **(iii)** may give consent to other offers. Additionally, Short Sale Lenders may require that, in order to obtain their approval for a short sale, some terms of the Agreement, such as the Close of Escrow, be amended or that Seller sign a personal note or some other obligation for all or a portion of the amount of the secured debt reduction. Buyer and Seller do not have to agree to any of Short Sale Lenders' proposed terms. Buyer, Seller and Brokers do not have control over whether Short Sale Lenders will consent to a short sale, or control over any act, omission, or decision by any Short Sale Lender in the short sale process.

5. **BUYER AND SELLER COSTS.** Buyer and Seller acknowledge that each of them may incur costs in connection with rights or obligations under the Agreement. These costs may include, but are not limited to, payments for loan applications, inspections, appraisals, and other reports. Such costs will be the sole responsibility of the party incurring them if Short Sale Lenders do not consent to the transaction or either party cancels the transaction pursuant to the Agreement.

6. **OTHER OFFERS.** Unless otherwise agreed in writing, after Buyer's offer has been accepted by Seller, **(i)** Seller has the right to continue to market the Property for back-up offers; **(ii)** Seller has the right to accept back-up offers (C.A.R. Form BUO, Paragraph 1), and subject to Short Sale Lender(s) requirements present to Short Sale Lender(s) any accepted back-up offers that are received; and **(iii)** Seller shall notify buyer when any accepted back-up offers, are presented to Short Sale Lender(s).

7. **CREDIT, LEGAL AND TAX ADVICE.** Seller is informed that a short sale may have credit or legal consequences and may result in taxable income to Seller. **Seller is advised to seek advice from an attorney, certified public accountant or other expert regarding such potential consequences of a short sale.**

By signing below, Buyer and Seller each acknowledge that they have read, understand, accept and have received a copy of this Short Sale Addendum.

Date _____ Date _____

Buyer _____ Seller _____

Buyer _____ Seller _____

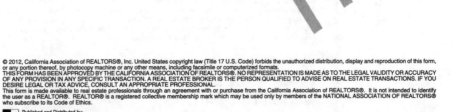

Published and Distributed by:
REAL ESTATE BUSINESS SERVICES, LLC.
a subsidiary of the CALIFORNIA ASSOCIATION OF REALTORS®
525 South Virgil Avenue, Los Angeles, California 90020

SSA Revised 4/12 (PAGE 2 OF 2)   Print Date

Reviewed by _____ Date _____

SHORT SALE ADDENDUM (SSA PAGE 2 OF 2)

CALIFORNIA
ASSOCIATION
OF REALTORS®

SHORT SALE INFORMATION AND ADVISORY
(C.A.R. Form SSIA, Revised 11/12)

Property Address: _____ Date: _____

This Short Sale Information and Advisory is intended to give general information regarding short sales, their potential impact, and the rights and responsibilities of the parties involved. It is not intended as legal advice for any particular property owner or buyer. Seller and Buyer should consult with their own professional advisors for legal, tax, credit and personal advice. Real estate brokers cannot and will not provide such advice.

1. **WHAT IS A SHORT SALE:** A short sale is the name used to describe a real estate transaction where the seller's lender(s) agree to allow the property owner to sell the property for less than the amount of the loan(s) secured by the property. The consent of a seller's lender(s) is necessary because without it there would not be enough money from the sale to pay off the lender(s) in full and to pay other costs of the sale. As a result, the lender's lien(s) would remain on title, and a seller would be unable to transfer title to a buyer free of monetary liens. (Properties that are worth less than the amount owed to the secured lender(s) are often referred to as being "underwater" or distressed properties).

2. **ALTERNATIVES TO A SHORT SALE:** Owners of distressed or underwater properties are faced with difficult choices that could have financial and emotional consequences. Any of the following or other alternatives potentially have negative tax or credit consequences, or both, for the owner:

   A. **A loan modification** is an arrangement between a borrower and a lender. It can involve a reduction in the interest rate on the loan, a deferment in payments on the loan, an extension of time to pay back the loan, a reduction in principal of the loan, a combination of these possibilities, or other changes to the repayment plan. A loan modification requires the consent of both lender and borrower.

   B. **A foreclosure** is a legal process through which the lender acquires title to the property from a borrower who has stopped making payments on a loan. The lender can foreclose whether or not the borrower consents.

   C. **A deed in lieu of foreclosure** occurs when the borrower offers to transfer the property to the lender, in lieu of the lender having to go through the foreclosure process, and the lender agrees to accept title to the property from the borrower and forgives the debt. A deed in lieu of foreclosure requires the consent of both lender and borrower.

   D. **Bankruptcy** is a legal action typically filed by a borrower to have debt(s) discharged. An automatic stay occurs as soon as a borrower files bankruptcy, staying all actions against the borrower. While a petition for bankruptcy can have the effect of delaying a foreclosure, it does not necessarily prevent a foreclosure from eventually occurring. No lender consent is required for a borrower to file bankruptcy.

3. **LENDER AGREEMENT TO SHORT SALE:** In order for a short sale to be completed, the lender(s), at a minimum, must agree to release the property from the lender(s) lien(s) to allow the sale. The lender(s) may or may not agree to reduce the amount owed to satisfy the debt. If not, the lender(s) may continue to pursue the borrower for payment of the balance of the debt. Prior to granting approval of the sale, the lender(s) may require the borrower to disclose all of the borrower's assets. They may require that the borrower liquidate other assets. They may require that the borrower sign an agreement to repay some or all of the debt at some later time. They may require that the borrower secure the unpaid debt with other assets owned by the borrower. Additionally, they will generally require that the transaction be arm's length, and that all terms of any benefit conferred on the seller be fully disclosed and that a seller cannot stay in the property following the sale. Finally, many first lien holders will limit the amount they will allow to be paid to a second lienholder, further complicating negotiations for the short sale. The lender will usually submit a "term sheet" to the borrower indicating the terms to which lender(s) will agree. If a seller and a buyer do not modify their contract to comply with the lender(s) terms, the lender(s) may not permit the short sale to proceed. Seller's are strongly advised to seek legal and tax advice regarding review of the term sheet. Brokers cannot and will not give legal or tax advice on the lender's term sheet or its effect on the Buyer and Seller's agreement or on the consequences to sellers and buyers should they proceed to close. There is no assurance that once the lender(s) have begun short sale negotiations, they will discontinue the foreclosure process.

4. **SELLER'S CONTINUING LIABILITY ON THE DEBT:** Many borrowers who attempt a short sale are concerned about whether the borrower is released from any further liability to the lender(s) or whether the lender can pursue the borrower for any unpaid balance of the debt. Some refer to continuing liability as a deficiency judgement. Generally speaking, a deficiency judgement is the right of a lender to pursue the borrower for the difference between the amount the lender receives and the amount the borrower owes on the debt. Deficiency judgements in California are prohibited in certain circumstances.

Buyer's Initials (_____)(_____)

Copyright © 2012, California Association of REALTORS®, Inc.

SSIA REVISED 11/12 (PAGE 1 OF 4)

Seller's Initials (_____)(_____)

| Reviewed by _____ Date _____ |

EQUAL HOUSING OPPORTUNITY

SHORT SALE INFORMATION AND ADVISORY (SSIA PAGE 1 OF 4)

Property Address: _____ Date: _____

   A. **Short Sale:**

     1. Beginning July 15, 2011, Code of Civil Procedure 580e provides that any lender who approves a short sale in writing is not permitted to seek or collect a deficiency against the borrower if the loan is secured by a Trust Deed on residential property containing 1-4 units. This law may not be waived. The July 15, 2011 law does not apply to: **(i)** lienholders on other types of property; or **(ii)** a borrower who has committed fraud or waste: or **(iii)** borrowers who are corporations, limited liability companies, or limited partnerships. Prior to this law coming into effect, from January 1, 2011 the restriction on lenders seeking deficiencies for approved short sales only applied to lenders holding a first trust deed on residential 1-4 units.

     2. For properties or borrowers to which CCP 580e does not apply, some lenders in negotiating a short sale will want the owner to sign a note for the balance of the unpaid principal. Other lenders will release the lien only, but not forgive the underlying debt. Some lenders will "reserve their rights." Thus, in these situations whether or not a lender retains the right to pursue a deficiency following a short sale becomes a negotiable term for each sale.

     3. Seller is encouraged to **(i)** obtain a written agreement from lender(s) or other applicable lien holders addressing whether and to what extent Seller will be released from any monetary or other claim, obligation, or liability upon approval of the short sale, and **(ii)** have that written agreement reviewed by an attorney, CPA or other appropriate professional of seller's choosing.

   B. **Foreclosure:**

     1. **Purchase Money Loans:** Loans given by lenders to purchase 1-4 unit properties, one of which will be occupied by the borrower, and seller-financed purchases are subject to "purchase money" anti-deficiency protection rules. Generally, this means that the lender cannot pursue the borrower for any deficiency after the property is **foreclosed** upon by the seller or lender, whether the seller or lender uses a non-judicial trustee sale or a judicial foreclosure. Refinancing a purchase money loan causes it to lose any purchase money protection it might have.

     2. **Trustee Sales:** If a lender **forecloses** by non-judicial trustee sale instead of by judicial foreclosure, that lender is barred from seeking any deficiency from the borrower after the trustee sale, even if the loan was not purchase money.

     3. **Refinanced Liens:** The anti-deficiency protections become much less clear for loans that are refinanced. Generally, loans that are refinanced lose their "purchase money" protection. Lenders extending refinances may be able to pursue a deficiency judgment against the borrower directly following a judicial foreclosure. However, beginning January 1, 2013 Borrowers who refinance a purchase money loan on owner-occupied residential property with 1-4 units, and do not take any cash out from the refinance receive the same anti-deficiency protection as if the refinance loan was a purchase money loan.

     4. **Junior Liens:** The anti-deficiency protections for Junior Lien holders are also somewhat unclear. Junior debt used to purchase the residence (such as 90/10 first and second) would have "purchase money" protection generally. However, junior liens that are refinanced or junior liens that are used to take out equity do not have "purchase money" protection. Such "non purchase money" junior lienholders may be able to pursue a deficiency judgement against the borrower directly after a Trustee's sale by a senior lienholder or after a judicial foreclosure by the junior lienholder. Although the law is not entirely clear, home equity loans (HELOCs) may fall into this category.

     5. **Other Liens:** Many other types of liens may be recorded on title including, without limitation, homeowners association liens, judgement liens, tax liens, and child support liens. Generally foreclosures by any lienholders senior to such liens do not protect the owner of the property from later legal action by the lienholder to collect on the obligation.

5. CREDIT AND TAX CONSEQUENCES:

   A. **CREDIT:** All of the owner's options discussed above will most likely have a negative impact on the owner's credit and on the owner's ability to finance or purchase property for some time. The credit impact and length of time the owner would have difficulty in obtaining a loan to purchase property again, or to be approved for any other credit transactions such as obtaining a credit card, leasing an apartment, or even to gain employment, varies. Lenders may view short sales and alternatives differently depending on their own underwriting guidelines and those established by governmental or quasi-governmental bodies. To find out more information about the impact to your credit score, go to www.myfico.com.

   B. **TAX:** With some exceptions, a reduction or discharge of a debt obligation by a lender results in income to the borrower. The income might not be taxable if the debt was being used to acquire, construct or substantially improve a borrower's principal residence. Another exception exists if the forgiveness of debt results from a situation where the lender is barred by law from collecting the debt, as in a foreclosure of purchase money debt. Insolvency and bankruptcy rules can also shield a borrower from forgiveness of debt income. Generally, when any debt is forgiven by a lender, they are required to provide the borrower a 1099 and it will be up to the borrower to make the proper claim on their tax return to avoid debt forgiveness income. Some of these rules are temporary, and state laws and federal laws differ. Broker has advised Seller that if Lender agrees to accept less than full payment, the difference may result in taxable income to Seller even though Seller does not receive any cash proceeds from the sale. Seller may also be taxed on the gain in value of the Property from the date of Seller's purchase to the date of sale, regardless of the amount of any existing loans/liens.

   C. **PROFESSIONAL ADVICE:** Seller is advised to discuss with an attorney, CPA or other professional of Seller's choosing before **(i)** accepting any offer to present to lender or **(ii)** agreeing to any changes requested by lender to an already accepted contract.

Buyer's Initials (_____)(_____)   Seller's Initials (_____)(_____)

SSIA 11/12 (PAGE 2 OF 4) | Reviewed by _____ Date _____ | **EQUAL HOUSING OPPORTUNITY**

SHORT SALE INFORMATION AND ADVISORY (SSIA PAGE 2 OF 4)

Property Address: _____ Date: _____

6. **POTENTIAL IMPROPRIETIES:** It is an unfortunate reality that many persons, including real estate licensees, mortgage lenders, and attorneys, among others, have taken advantage of owners of underwater or distressed properties. Some of the schemes present themselves as "rescues" of the homeowner, promising to let them stay in the property, to protect their credit, or to provide payments to them after closing, and usually outside of the escrow. Both the California Department of Real Estate (DRE) (http://www.dre.ca.gov) and the California Attorney General (http://www.ag.ca.gov) have issued written warnings of potential red flags in short sales and other rescue schemes. Some of these red flags are:

A. **No license:** The DRE believes that a real estate license is generally required to negotiate any short sale;

B. **MARS:** Short sale negotiators who do not represent a seller or buyer in a short sale are generally required to comply with the Mortgage Assistance Relief Services rules and provide required disclosures and notices to a seller.

C. **Up-front fees:** No real estate licensee can collect any up-front or advance fee without having first obtained a "no objection" letter from the DRE and no up-front fees may be taken for arranging a loan modification;

D. **Surcharges:** Charges by third parties that are not disclosed to the short sale lender and usually paid outside of escrow;

E. **3rd Party negotiations:** The licensing and fee requirements above apply whether the negotiation occurs through a Broker, representing a seller or a buyer in the transaction, or a 3rd party short sale negotiator. As with other real estate activity, short sale negotiator fees are negotiable and not set by law. The existence, fee and licensed status of any 3rd party short sale negotiator shall be disclosed to the lender and must be approved by the lender as part of the overall compensation to be paid in the short sale transaction.

F. **Straw buyers and house flipping:** Buyers misrepresent the value of the property to the short sale lender and flip the property to another buyer already in place;

G. **Other:** Other potential red flags include: guarantees to stop the foreclosure; instructions not to contact the lender; transfer of title prior to close (often to a trust) as a condition of negotiating with the lender; the buyer is an LLC; the buyer wants a power of attorney from the seller; and the buyer hires the third party negotiator or wants to negotiate directly with the lender.

While most of the activities on the above list on their face are not fraudulent, they serve as warning signs that the owner and the real estate agents involved should proceed with caution.

7. **BUYER CONSIDERATIONS:** Short sales are often difficult transactions taking considerably longer than a typical real estate transaction to complete. There is no guarantee that the lender or lenders will agree to the terms of the purchase offer or that they will respond in any timely fashion or even respond at all. There is no guarantee that a seller or a buyer will agree to any terms proposed by the lender as a condition of releasing the lien or the debt on the property. Buyers may expend money on inspections, loan applications, escrow fees and other costs that they will not be able to recover from anyone if the lender does not approve the transaction. Buyers may also have difficulty obtaining the return of their deposit in escrow, if a seller becomes noncommunicative during the short sale process. Generally, sellers also have the right to continue to give offers to their lender(s) even if they have a contract with an existing buyer. Brokers cannot give any assurances as to what will happen. Buyers are strongly cautioned that any undisclosed and unapproved payments to junior lienholders or to seller or to outside third party negotiators may be a form of lender fraud. Buyers are also strongly cautioned that writing offers on more than one short sale property with the intent to purchase only one such property could be a misrepresentation giving rise to legal claims by a seller including a claim for the buyer's deposit.

8. **NATIONAL MORTGAGE SETTLEMENT (SETTLEMENT) AND CALIFORNIA HOMEOWNER BILL OF RIGHTS:** In early 2012 California joined a national settlement agreement between five of the nation's largest lenders (Bank of America, JPMorgan Chase, Wells Fargo, Citigroup and All Financial) and most states ("the Settlement"). The Settlement obligates the lenders to, among other things, write down or refinance some loans, extinguish certain unpaid balances and provide transition assistance to some homeowners. Loans owned by Fannie Mae or Freddie Mac are not covered by the Settlement. In Fall of 2012, California enacted a series of laws, effective January 1, 2013, commonly and collectively referred to as the California Homeowner Bill of Rights ("Homeowners BOR"). These laws prohibit foreclosures while an approved short sale is pending or while a loan modification is in process or on appeal, as well as other requirements. Whether a borrower qualifies for any of the advantages of the Settlement or the Homeowner BOR requires an analysis of the borrower's loan and it applicability to either of those items. Such an analysis is beyond the scope of Brokers expertise. If a buyer or seller has questions about whether the borrower's loan is covered by either the Settlement or the Homeowner BOR, or how either of those items can affect a short sale transaction, that party should discuss the matter with a lawyer or accountant of their own choosing.

9. **BROKER ROLE:** A real estate broker cannot give legal or tax advice in connection with any of the options available to the borrower nor can the broker suggest what is the best course of action for the owner. Unfortunately, the owner is faced with extremely difficult choices having a lasting impact on the owner. Owners are strongly cautioned that they must seek legal and tax advice in what is not only a choice impacting taxes and credit, but also personal issues affecting the owner and often the owner's family. The broker's role is to assist the owner with the actual sale of the property in a short sale transaction, not to provide legal or tax advice or to guarantee the best possible outcome for the parties, or to assure a buyer that any particular transaction will be completed. Brokers do not, and cannot, assure that either a seller or a buyer will perform on their agreement or that the lender(s) will agree to any of the terms presented. Brokers are not a party to the contract between Buyer and Seller.

| Reviewed by _____ Date _____ |

SHORT SALE INFORMATION AND ADVISORY (SSIA PAGE 3 OF 4)

Property Address: _____ Date: _____

10. BROKER AUTHORITY: Seller authorizes Broker to: (1) market the Property for sale, (2) contact lenders concerning lender's approval of a short sale (C.A.R. Form ARC) and Seller agrees to give Broker any necessary information to negotiate with lenders, and (3) advertise in the MLS and other advertising medium that the property transfer, sales price and payment of commissions are subject to lender's approval. If lenders will not cooperate, Broker may cancel the listing agreement.

☐ Seller ☐ Buyer _____ Date _____

☐ Seller ☐ Buyer _____ Date _____

☐ Seller ☐ Buyer _____ Date _____

☐ Seller ☐ Buyer _____ Date _____

Real Estate Broker (Selling Firm) _____ DRE Lic # _____

By (Agent) _____ DRE Lic # _____ Date _____

Address _____ City _____ State _____ Zip _____

Telephone _____ Fax _____ Email _____

Real Estate Broker (Listing Firm) _____ DRE Lic # _____

By (Agent) _____ DRE Lic # _____ Date _____

Address _____ City _____ State _____ Zip _____

Telephone _____ Fax _____ Email _____

Published and Distributed by:
REAL ESTATE BUSINESS SERVICES, LLC.
a subsidiary of the CALIFORNIA ASSOCIATION OF REALTORS®
525 South Virgil Avenue, Los Angeles, California 90020

SSIA 11/12 (PAGE 4 of 4)

Reviewed by _____ Date _____

SHORT SALE INFORMATION AND ADVISORY (SSIA PAGE 4 of 4)

Appendix C

Example of a Good Inspection Report

I n Chapter 13, we show you an example of a lousy inspection report. This appendix gives an example of what you should expect to see in a good inspection report.

Like our sample, your report should paint a vivid word picture of the home you may purchase. The inspection report should be brimming with in-depth explanations — not merely a list of checkmarks, generic boilerplate, and hastily scribbled notes.

TIP

Get the most out of your inspection dollar. Find a professional inspection company that will thoroughly inspect the property's mechanical and structural systems inside and out, from foundation to roof, and present you a solid report on which you can make an informed home-buying decision. (See Chapter 13 for more advice on selecting a great home inspector.) You can access this inspection report online at www.dummies.com/go/homebuyingkit7e.

Warren Camp Inspection Services

P.O. Box 986, Arnold, CA 95223

(209) 795-7661

- **Inspection Date:** _____ XX, XXXX   • **Report Number:** XX - _____
- **Date of Report:** _____ XX, XXXX   • **Inspector:** Warren Camp, ASHI®
Certified Member, #732
- **Report:** Prepurchase inspection at ___ _____ Street, San Francisco
- **Dwelling Description:** Single-family dwelling
- **Present During Inspection:**   • **Weather:** No rain within past 10 days
  - **Buyer:** Red E. Toobuy
  - **Buyer's Agent:** Ken B. Elpful / *Manny Elpful and Associates*
  - **Seller's Agent:** A. Frank Lister / *Frank & Company, LLC*
  - **Others:** Bugzie O. Bliterate / *Nuke 'em, Heat 'em, Treat 'em Pest-Control Company*

- **The inspected unit was furnished at inspection.**
- **A structural/pest-control inspection report was not provided.**
- **The seller's disclosure form was not provided.**

As requested by the buyer's buyer's agent, this report is being prepared for the exclusive use of the buyer to accompany the on-site verbal presentation. In no way is it to be used by, nor are we obligated to review it with, any third parties. Because Warren Camp Inspection Services (WCIS) has not personally described the extent and nature of its findings to anyone but those present for the entire inspection, WCIS strongly discourages third parties from using of this report. Interested parties should arrange with WCIS for an inspection that meets their more individualized needs.

This report provides a professional opinion of general features and major deficiencies of the building andts systems at inspection. It does not necessarily analyze or report on adjacent properties, nor does it cover environmental/neighborhood concerns. It summarizes observations on components inspected in accordance with customary property-inspection standards. The scope of this inspection is limited only to items discussed. It is not technically exhaustive. Because certain findings are variable (separations and cracking lengths that increase in time, levelness and plumbness readings that may change over time, erosion and corrosion levels that do not remain static, and so on), no one should rely on any report findings for more than 60 days.

This is not a code compliance report; a home, product, or system guarantee of any kind; nor is it an evaluation of the property's saleability. It includes only items accessible to visual inspection; no furniture relocation, dismantling, demolition, or other manual handling, etc., would have occurred in its preparation. It does not fulfill the requirements set forth in California Civil Code Section 1102 as to the required disclosures of a transfer of real property.

The WCIS inspector explained to the client the two types of reports WCIS prepares. Rather than selecting the in-depth, narrative report with extensive recommendations, the client selected the present standard report. Findings and recommendations that would normally have been included in the extensive narrative report would be excluded from this report.

Please call WCIS with your questions.

Copyright © 2020 by WCIS

TABLE OF CONTENTS

CERTIFIED MEMBER

INTRODUCTION

The inspected property was a single-family dwelling. Most interior spaces were unfurnished. Low-voltage wiring, heat exchangers, gardens, fences, retaining walls, underground piping and storage tanks, and sprinklers are not included in the scope of this inspection report.

Warren Camp Inspection Service (WCIS) inspections are designed to meet or exceed recent "Standards of Practice" established by the American Society of Home Inspectors® (ASHI®) of which Warren Camp is a certified member. A copy of the Standards is available upon request.

For the most part, the building is a single-level, framed structure built over a crawl space. Built around 1955, the original structural work of this wood-framed building appeared customarily constructed. No unusual or extensive damage was apparent, however, several items need attention.

Portions of this single-family dwelling have been recently remodeled. Alteration of the plumbing and electrical systems, as well as several structural installations at the rear addition was made in a nonprofessional manner. These concerns are brought up in other sections of this report. Because many of the walls and ceilings were closed, it was not possible to ascertain the full extent of renovation. If more information about these altered areas is needed, (a) consult with a licensed structural engineer, (b) review copies of permits and remodeling contracts that may be made available, and (c) examine the seller's disclosure form.

The building interior and exterior were, for the most part, adequately maintained. But of course, *all* buildings have flaws. We'll discuss a number of these flaws, but we cannot discover and report on every one. This inspection and report is not technically exhaustive, and WCIS does not provide a thorough or fully detailed analysis of problem areas. With only a few hours to inspect the entire property, WCIS provides, at best, a professional opinion based upon experience. Inspector Warren Camp is not a licensed engineer or expert in every trade or craft. Only representative sample-checks of various exposed-to-view segments of this property were made. If additional items or conditions are found when repairs or improvements begin, call WCIS immediately before further work resumes.

All the main points of this report were fully discussed with Red E. Toobuy and his agent, Ken B. Elpful, at inspection. The following sections describe the findings discussed.

Repairs, corrections, and other follow-up items to consider (Note: Check-marked concerns are merely highlights of the inspector's findings. Read the entire report to fully appreciate this effort. Where you have interest, follow these and the following recommendations and have specialists address those discussed items that may not have been included in this inspection and report.)

> ✔ Check with the building inspection department about permits and inspections for any building construction, alterations, and additions.

EXTERIOR

Building Exterior

This building, with board-and-batten siding on the facade, and stucco on the balance, needs maintenance on the rear addition. Surfaces should be weathersealed in the not-too-distant future to prevent moisture entry. When references are made to the front, rear, left, and right, they are made facing the building from _____ Street.

Additional items not yet painted or waterproofed were found. They include five louvered wall vents, the garage door, one new entry door, and various windrow trim pieces.

At the garage rear, soil or pavement was close to or even with the foundation top. This condition can cause wood decay and deterioration. Because Warren Camp is not qualified as a structural/pest-control inspector, refer to a current report for findings and recommendations.

Wood-to-earth contacts were found at stairway posts and doorsills. Contacts encourage wood decay, entry of pests, and moisture retention. All contacts should be properly separated and appropriate grade levels maintained.

Rust deterioration was found on several exposed nails on the rear addition's roof eave trim. Prior to painting, these surfaces should either be fully prepared for paint or removed.

A few minor cracks in the stucco were evident at inspection. The cracking is likely due to material shrinkage or expansion, drought, seismic forces, or ordinary building movement. Cracks should be caulked and weathersealed. Contact a painting and waterproofing contractor to replenish and seal exterior surfaces to prevent water entry.

Stucco siding on the rear and side yard walls extended downward over the foundation and made contact with grade (ground covering). This is conducive to entry of wood-destroying pests or organisms from behind the stucco. As an upgrade, raise the base of stucco siding a few inches to expose the foundation or lower the grade level.

Windows and Doors

Aluminum windows on the rear addition appeared sound. WCIS suggests that exposed wooden frames, sills, and trim adjacent to the metal window frames be routinely maintained to prevent possible water entry. A sampling of these windows was operated at inspection and a number were openable to an acceptable degree.

Many of the original building's painted, wood-sash windows were unable to be opened when tested, causing some rooms to be without adequate ventilation. Further, with respect to fire egress and other emergency situations, it could be extremely difficult and dangerous to attempt exit through such inoperable closed windows. *This should be corrected immediately.* In addition, routine maintenance is recommended for exposed wooden sashes, adjacent window trim, and glazing putty to eliminate potential water seepage and extend lifespan.

The garage entry door was a tilt-up type without open-vent screens.

Its spring balances were not equipped with a safety device that prevents catapulting that might occur if the springs were to actually break under pressure. Contact a professional garage door installer for appropriate replacement/correction.

Pavement and Drainage

Excess changes in the height of pavement at the front walkway could be a trip hazard. Exercise caution in this area.

This building sits on a steep hillside. Erosive soil grooves were visible in a number of locations. Not being engineers of any kind, WCIS is unable to represent or evaluate this condition. Red E. Toobuy can contact qualified engineers regarding the stability of the building and hillside, as well as any past, present, or future soil embankment or ground or building movement.

Adequate soil drainage for Bay Area homes is imperative because soil types in this area swell when saturated and may damage a building's foundation. Grade at the front was noticeably sloped, likely providing adequate drainage away from the foundation during rainstorms.

A drainage pattern at the rear and side yards was not as easy to predict. Water entry is probable into the building subarea possibly because the soil and pavement was not significantly sloped away from the building but should be. Because calcification and/or staining were found

---

on the inside of a few foundation walls, grading and drainage should be monitored regularly and should be improved.

Moisture and drainage conditions vary with specific soil types, landscape/hardscape designs, and weather changes. Consequently, reporting on seepage and ponding conditions or making representations regarding soil stability cannot be made by this inspection company. Refer to a seller's disclosure and/or a soil engineer's report to learn of the possible presence of a subterranean French-drain system and to fully appreciate the potential for water entry— whether caused by light rains, natural springs, prolonged heavy rains, or other causes. Routinely keep all drains, patios, and walkways clean and well maintained.

Underground Piping

Understandably, inspection of inaccessible, underground perimeter drainage systems could not be inspected. Neither could WCIS inspect other underground devices such as conduits, gas and water piping, waste and vent lines, and so on, as well as under-slab components. Absolutely no testing of sewer lines is done by WCIS.

WCIS detected no outward signs of presently existing or previously placed underground fuel storage tanks (USTs) within the inspected areas (e.g., a fill spout, vent pipe, supply tubing and return line, or a fuse box labeled "oil burner"—typical indicators of USTs). Interested parties may wish to explore further since such exploration is not within the scope of ASHI® standard inspection practices.

- ✔ Free and maintain windows, trim, and hardware.
- ✔ Refer to a current structural/pest-control inspection report for findings and corrective recommendations.
- ✔ Paint or waterproof all raw materials.
- ✔ Repair needed items.
- ✔ Regrade/refinish landscape and hardscape surfaces in needed areas.
- ✔ Contact a soil engineer regarding erosion and hillside stability.

FOUNDATION

The foundation was only partially accessible because of low headroom throughout the subarea. No ratproofing membrane was yet placed beneath the family room, which might be a thoughtful consideration.

Visible foundation stem walls, as viewed from the subarea doorway and building exterior have been installed according to customary practiced standards.

Garage and crawl space legs were made of continuous concrete, which is often reinforced with internally placed steel bars that could not be examined or verified.

There was hairline or minor cracking on foundation sections of the family room addition. Such cracking in a building of this type and age is not uncommon and should be routinely monitored. If the cracking increases, or new cracks develop, contact a qualified engineer for an evaluation. Looking for any direct and current transference of foundation movement to adjacent finished walls, ceilings, and floors, no outward sign was detectable. It was not possible to determine if this cracking was a current condition. Determining whether foundations shift, settle, or rotate, or cracks will appear, or if existing cracks will extend further, is not within the scope of ASHI® inspections. Neither can WCIS predict the likelihood of future foundation failures, shifting, or settlement. If more information is needed, a qualified structural engineer, experienced in similar structures, should be contacted to fully inspect and evaluate findings on these and any other structural concerns, such as earthquake-preparedness measures.

✔ Routinely monitor the foundation.

STRUCTURAL FRAMING

Substructure

New and original framing was seen at the rear addition. The lower areas of partially accessible, exposed framing, were limited to portions of the crawl space. Framing, for the most part, was customary, with no visible sign of critical sags, cracks, deterioration, or movement.

Wood posts in the crawl space below the family room were in unsatisfactory condition. The bottoms of support posts beneath the center girder were not connected to embedded concrete piers. No fastening devices (screws, nails, or bolts) could be found. Corrective work is needed and would be easy to accomplish at the direction of a structural engineer.

The header supporting the garage access opening was some cause for concern because of the nonconforming size of fasteners used at each end. The header has also begun to rotate (shift) causing the fastening connections to weaken. Structurally, connections made between one structural member and another are essential. Post base-and-top connections, and beam connections to each other and to joists, were also minimal but could easily be supplemented.

Cripple-wall studs in the garage and subarea were tied together customarily, however, they were not yet benefited with supplemental fasteners or plywood shearwall panels known today to strengthen wooden structures located in earthquake country.

Many structural posts, beams, and studs had also not yet been retrofitted for earthquake preparedness. As a standard recommendation, these measures should be taken. All upgrade recommendations should come from a qualified, licensed, structural engineer.

Moisture staining was found on garage interior walls. These stains looked and felt dry, and, when tested with the biprobe electric moisture meter, accessible stains were dry. Because the cause of water staining, and the determination of its currentness, is difficult to determine, refer to a seller's disclosure statement to learn what efforts were made in each of these areas.

Rodent dropping was found throughout the subarea. Contact a pest-control company for an evaluation.

Portions of the subarea's ceiling were installed with thermal insulation, however, a calculated "energy inspection" is not within this industry's standards of practice. Several sections were loose or had fallen onto subarea soil. Corrective work is needed.

Main Structure

No evidence of *current* structural movement was noted during inspection of samplings of doors, windows, floors, walls, and ceilings. The tops of some doors were taper-cut to allow for wall shifting over the years. Any separations on walls, molding, or ceilings, or sagging or sloping of floors appeared to be the result of ordinary shifting and/or expansion within framing and supporting soils. In WCIS's opinion, the findings do not represent significant, current movement.

Attic Area

An attic had its access door in the hallway ceiling. No floor boards were yet sitting on attic joists making this area risky to traverse. In addition, thermal insulation covered many of the ceiling

---

joists. Consequently, only a limited visual inspection of the adjacent attic space was made from the access doorway.

Attic floor joists had runs of electrical wiring laid through and over their tops. Care must be taken whenever attic access is required.

Ceiling insulation was installed throughout much of the attic floor. Reporting on adequacy of building insulation is not within inspection-industry standards. Neither would we be able to examine or suspect any failures or hazards beneath or amid insulation. An electrical contractor could inspect the embedded wiring fully as a safety evaluation and provide a safety certificate.

Insulation baffles, required around most heat-producing elements, such as lighting fixtures and flues, were not readily visible and should be provided as needed.

Visually accessible roofing supports on the main building were customarily framed. A representative number of purlins (supporting members) and/or collar ties (members connecting two roof sections) were found. Unfortunately, the family room addition's attic was completely inaccessible so no inspection whatsoever was made.

There were beneficial vent openings at overhanging roof eaves. Ventilation was customarily provided and maintained.

> ✔ Hire a structural engineer to evaluate family room framing.
> ✔ Seismically retrofit structural posts, beams, plates, and studs for wind load and earth-quake preparedness.
> ✔ Correct the rotated header and refasten it appropriately.
> ✔ Refer to a current structural/pest-control inspection report for findings and corrective recommendations.
> ✔ Provide insulation baffles where needed and refasten fallen crawl space insulation.

FIRE SAFETY and SECURITY

Fire Safety

WCIS has some fire-safety concerns with this property. Garage wall surfaces, adjacent to habitable rooms, were not completely fire-resistant. Currently, there is risk of potential flame spread, as well as radon infiltration, into habitable spaces. Fully separate mechanical rooms from habitable rooms (e.g., by installing or patching all openings and separations with fire-resistant drywall, plaster, sheet metal, etc., or undertaking fire-resistant construction where appropriate). WCIS was unable to locate any smoke detectors or sprinklers in this area. A monitored alarm system with adequate smoke and heat detectors could be installed.

A few smoke detectors were installed in this building: on walls in three bedrooms and the ceiling of the common hallway. Because state and local codes change frequently, consult the building department for direction on optimal number, type, and location of units. Be certain to replace batteries every year with fresh batteries. Providing appropriately specified and located fire extinguishers also improves fire safety.

Security

The building's front door was equipped with a lock, deadbolt, and large glass pane. Glazing did not have a label certifying specification (e.g., tempered or safety). Door and window panes without safety glazing can be hazardous when broken, so current building codes require safety-labeled glass to minimize possible injury. Replacement is not customarily required; however, exercise caution and common sense in this area to prevent accidental breakage and possible bodily injury.

The front entry door lock requires minor adjustment for security as well as quick and easy operation.

It was a solid-core door, which is more resistant to breaking and entering, as well as to flame spread than a hollow-core door. (Not all solid-core doors are fire-resistant unless label-certified.)

The glass-and-flat-panel door from the garage to the side entry was a weak door offering little in the way of security.

Front and rear pedestrian garage doors were without at least a 1-inch-throw lock or deadbolt. At all exterior doors, deadbolts are the recommended auxiliary locking devices.

The side entrance (kitchen) deadbolt was a "double-keyed" type—a key for the inside as well as outside lock cylinder is required. If these keys are not easily accessible, emergency egress could be impossible, and bear serious safety consequences. Conversion to single-keyed bolts is easy, affordable, and should be considered. Contact a locksmith.

Glazing and Egress

Family room addition windows were installed close to the floor. Unfortunately, each glass pane lacked glazing labels certifying specification (e.g., tempered or safety). Replacement glazing, plastic film, or barrier installation may not now be required, but exercise caution and common sense in these areas until improvements are made.

Means of egress was a concern. When attempting to freely operate the fire-exit windows leading outside from each bedroom, the windows were painted shut and unable to open.

Interior and exterior lighting should be supplemented for overall security and safety.

And, as a reminder before taking possession of your new home, rekey all existing door-lock cylinders to improve overall security and provide peace of mind.

> ✔ Provide and install needed fire protection, separations, devices, and safety systems or components.
> ✔ Improve security and door-lock safety.
> ✔ Make fire-egress windows freely and fully openable.

PLUMBING

Water Supply

The main water-service shutoff valve was on the building's front wall. It was operable; no leakage was detected. A 3/4-inch copper waterline joined the building from the street. Visible domestic hot- and cold-water-supply lines were mostly made of copper.

WCIS found a combination of galvanized iron and copper waterlines at the front of the building. The seller might be able to provide information about the extent of copper piping replacement. Galvanized iron water piping is subject over time to corrosion from mineral build-up that can restrict water flow to fixtures. Corrosion may or may not be extensive and WCIS cannot predict at what rate this will occur.

Measured at the main valve, static pressure on the waterline was 89 pounds per square inch (PSI), which is a moderate-to-high level. Prescribed water-pressure-ratings are set at 55 to 65 PSI to prevent leakage from excessive pressures. A water-pressure regulator, pressure gauge on the incoming waterline, and routine water-pressure monitoring are always recommended.

Part of the hot-water piping in the subarea had no thermal insulation. Full insulation would reduce energy consumption and improve the hot-response time for each water fixture.

This area's copper water piping was also without proper or sufficient wall and ceiling fasteners. This omission might contribute to leakage or hammering noises in these lines, such as those WCIS detected in both bathrooms.

Vents, Drains, and Traps

Throughout this structure, visible waste and vent piping was made of cast iron. A 2-inch cast iron waste or vent line beneath the crawl space access door was cracked and deteriorated, requiring replacement of this piping in the not-too-distant future.

A waste/vent line in the garage was incomplete, lacking a cap or clean-out plug. Located to the right of the furnace, a plumber should simply install a proper cap or plug.

A number of drains were inspected and maintaining an effective water trap-seal.

Traps for both bathroom washbasins were a concern for WCIS because they were not fastened to a tailpiece and *each leaked*. Competent plumbing connections are essential. Trap replacement would be inexpensive and easy.

Gas Supply

The main gas-shutoff valve, located on the building's front exterior wall, was tight. If a shutoff valve is not now, or in cases of emergency, accessible or operable, the local utility company could be contacted for correction.

    ✔ Install a water-pressure regulator and gauge on the incoming water line.
    ✔ Strap water and waste lines securely to the structure.
    ✔ Replace cracked/deteriorated cast iron piping.
    ✔ Cap or plug the open waste/vent pipe.
    ✔ Replace defective traps on both bathroom washbasins.

WATER HEATER

The hot-water heater in the garage was a gas-fueled type that was operating during inspection. It had no fiberglass thermal jacket. A new thermal insulation blanket, equipped with a razor-cut, insulated access door placed directly over the heater's ID plate, should be installed.

The water heater also lacked adequate cross-strapping and restraining blocks designed to resist movement during an intense earthquake.

It was apparently a recently installed model. With a fiberglass tank, an identification plate indicated that this A. O. Smith appliance had a 40-gallon capacity, a setting of 38,000 BTUs, and a 40.4-gallon-per-hour recovery rating.

The tank bottom was free of rust. No leaks were evident.

A safety valve on water heater tops, referred to as a "temperature and pressure relief valve," is necessary for the safe operation of these appliances. The T&PR valve was properly located and a water overflow tube was connected to the valve according to accepted trade practices.

The shutoff valve on the cold-water supply piping was operational. No leakage was evident.

Hot-water piping immediately adjacent to the water heater had some thermal wrapping, however, an "energy inspection" is not within the scope of our inspections.

The drain valve at the base of the tank, when opened, showed minimal sludge deposits.

The gas-shutoff valve was difficult to operate and should be adjusted.

Fresh air needed for complete combustion was minimal in this area. Additional, continuous ventilation is suggested. Open air vents and windows provide such ventilation.

Gas-fueled heaters must always be vented safely. Visible portions of the exhaust vent flue were installed in a questionable and possibly unsafe condition (inaccessible portions of piping were not inspected). Flue connections were inappropriately "taped over" with asbestos-like materials. The mere presence of asbestos in a building material does not necessarily represent a health hazard. Many factors must be considered before making such determination (e.g., the percentage of asbestos make-up, exact type of asbestos, and current physical condition). Considering the age of this building, other asbestos-containing materials that may not be visibly detectable or identified in this report may be present. Contact specialty contractors to conduct lab-tests for asbestos presence and analysis, and if found to be positive, provide estimates for removal or encapsulation of these materials following the U.S. Environmental Protection Agency's standards of practice.

The flue was also stained, suggesting either leakage at the roofline or condensation from a lack of fresh air in this area. However, no moisture was evident in this area at inspection.

As a standard earthquake-preparedness consideration, some or all of the following installations should always be undertaken if not already present:
a. Flexible water-supply piping to water heaters
b. Fully functional seismic cross-strapping (see enclosed WCIS brochure)
c. Flexible gas-supply piping to heaters and all gas-fueled appliances

✔ Provide continuous fresh-air circulation.

LAUNDRY

A garage-area laundry area was no longer operational. No appliances were on location.

Air chambers on the water lines above the laundry sink had not yet been installed. Because they benefit the circulation of hot and cold water within these lines, air chambers should be installed by a qualified plumbing contractor before the sink is made operational.

The concrete-and-iron sink was adequately secured to the garage rear wall. Neither a sewer vent line above nor trap seal beneath this laundry tray had been installed. Contact a licensed plumbing contractor for such installation.

The gas-shutoff valve was tight and will require adjustment when connecting gas piping.

"Fresh air exchange" was minimal in this laundry area. Regularly opened windows aid such ventilation.

There was concern with the looseness of the 120-volt electric outlet next to the sink. This is a small repair job, and because looseness could compromise grounding protection, *a licensed electrician should promptly examine and fasten this outlet to its box.*

✔ Make the various recommended plumbing improvements when making the laundry area operational.
✔ Install a vent and trap for the sink and secure the loose electrical receptacle.

ELECTRICAL

Service and Main Disconnect
Electrical wiring for this building was fed from overhead and provided approximately 240 volts to the meter.

The main disconnect switch and panel on the right exterior wall had a 60-amp overcurrent protection device for the building. Gauge markings on the conductors were taped over but their size suggested #8 wire gauge. Ampacity (the service entrance capacity) was marginal, based on the building's current load-demands. To *ensure* adequate electrical size, or to increase it, contact a licensed electrician for a load calculation and evaluation.

The main electrical panel was fastened to the building exterior but was unprotected from the weather. Corrosion was found inside the panel and gutter chamber. Because it has been known to cause shorting, deteriorated parts should be replaced.

This panel was also extremely dirty, inside and out. Caution must be exercised in this location because foreign matter can allow arcing that can lead to shorting.

Subpanel Distribution

The building's main disconnect device was combined in a panel with other circuits. No other sub-panels were easily located or inspected.

Protected by circuit breakers, the combined main-disconnect-and-distribution panel had the following circuitry distribution:

> 1 @ 120-volt circuit at 15 amps
> 4 @ 120-volt circuits at 20 amps

This subpanel was not fully circuit-labeled but should be. It was, however, benefited by a closed-front protection cover.

"Double-tapping" (connecting two conductors to one circuit breaker) occurred within this panel. *Such wiring should be corrected immediately* because double tapping increases the possibility and frequency of tripping the overcurrent protection device.

As a part of regular property maintenance, all circuit breakers should be trip-tested, then re-set yearly, to insure that they are, and will remain, fully operational.

Grounding and Polarity

Of course, all electrical systems should be safely and properly grounded. An appropriately driven grounding rod was not easily located beneath the main panel. When sample testing outlets requiring adequate grounding, some had *little or no grounding protection.*

There was an "open ground" (ungrounded) condition in the living room, three bedrooms, and family room, which can be hazardous. *This should be corrected immediately for maximum personal safety.*

In a random sampling of receptacles, "reverse polarity" was present in a few locations (in both bathrooms and at each side of the kitchen sink). This condition, hazardous in certain instances, can be easily corrected and should be. What's more, the receptacle in the master bathroom was not protected with a ground-fault circuit interrupter (GFCI) device as expected.

Wiring

Electrical wiring for this building was comprised of original as well as supplemental wiring. Much of the exposed wiring was the Romex® type.

The following is only a sampling of wiring concerns and is not intended to take the place of an electrical contractor's findings:

a. A defective light switch for the kitchen's above-sink lighting fixture needs replacement.

The contents of this report have been prepared for the exclusive use of Red E. Toobuy. Reliance by others is prohibited.

Warren Camp Inspection Services

• Page 9

    b. Both bathroom light fixtures were extremely loose and this can compromise grounding protection. *Securely fasten each fixture immediately.*

    c. Some Romex wiring on garage walls was less than 8 feet high and unprotected, and this may permit damage to the cable. Exposed wiring should be piped into metal conduit or covered with approved protective material such as drywall.

    d. Extension cord wiring was inappropriately used as a substitute for permanent wiring in the master bedroom and should be properly wired if the fixture is to remain in use.

    e. Light fixtures in the garage and master bathroom were without lamps (bulbs). Because mistaken contact with hot sockets can be dangerous, replacement lamps should be provided in every such location.

In summary, safety concerns have arisen with interior wiring as well as the service entrance feeding the building. *Hire a licensed electrician as soon as possible* to examine the entire electrical system to its fullest and make corrections wherever needed.

    WCIS always recommends the installation of ground-fault circuit-interruption-type receptacles in kitchens, bathrooms, and other wet locations as an added safety measure. Provide where needed at the direction of a licensed electrician.

    ✔ Clean the main panel, replace deteriorated components therein, provide complete circuit identification labeling therein, and protect the panel from the elements.

    ✔ Remedy immediately all grounding, polarity, lighting, and wiring problems.

    ✔ Securely fasten loose switches, receptacles, and lighting fixtures.

HEATING
Heat Source Type and Condition

The Borg Warner brand gas-fueled, forced-air furnace in the garage rear had an estimated 64,000-BTU-input-capacity rating. Installed several years ago, it may be approaching the end of its "useful life." If a life-expectancy determination is needed, contact a heating specialist.

    The following heating concerns are noted:

    a. The gas-shutoff valve was tight and requires minor adjustment.

    b. Presence of a natural gas leak was detected at the furnace's front shelf. *Such leakage could be extremely hazardous and must be corrected immediately.* Contact a licensed plumber or a utility company representative for thorough testing and analysis.

    c. No service calendar was visible at inspection. A current record of scheduling visits would suggest maintenance history of the heating system.

    d. This appliance did not have benefit of pilotless ignition. A thermocouple device, which would shut off the gas supply if the pilot were not lit, would be a thoughtful pilot-safety upgrade.

    e. An electrical disconnect switch was mounted on the left side of the furnace to facilitate shutting down electrical power to this appliance for maintenance and repairs.

    f. No thermal insulation was presently installed on the return-air duct above the furnace. Because the duct is in an unconditioned or cold area, such omission lends itself to higher energy consumption than on insulated ducts.

    g. The warm-air ducts and plenum were wrapped with asbestos-like material that was crumbly. Contact specialty contractors to conduct a lab-test for asbestos presence and analysis here and around the house, and if found to be positive, provide estimates for removal or encapsulation of these materials following the U.S. Environmental Protection Agency's standards of practice.

Circulation and Ventilation

The return-air duct, flame ports, shelf, and furnace bottom were dirty. On the hall floor, the warm-air supply register and its interior were extremely dirty. These areas should be vacuumed promptly and regularly.

The filter box allows for a 14 x 25 x 1-inch filter. The filter was clean and properly installed. Furnace filters need to be changed every two to four months. Dirty filters actually block airflow to the heat exchanger causing it to overheat. Improper filter maintenance is a primary cause of premature cracking of a furnace's heat exchange components.

Oxygen sources necessary for complete combustion were minimal in the furnace chamber. Fresh-air entry was obstructed by boxes of personal belongings blocking the wall's ventilation screens. Remove obstructions and continually enable sufficient fresh-air exchange in this area.

Vent and Flue Piping

A cement-asbestos flue pipe (Transite®) was found. (See the discussion of asbestos elsewhere in this report.)

The flue was in a nonconforming location and may be dangerous to the building and occupants. Flues that do not pass the roofline should be properly extended upward or reinstalled.

Heat Exchanger

This heater's gas burners appeared to be out of balance with unusual flame characteristics known as "dancing flames." Unevenness is difficult for anyone but a heating contractor or utility company technician to analyze. Such a check-up should be made as soon as possible.

There was minor corrosion and pitting around the frontal entry of the heat exchanger area.

The firebox (heat exchanger) of this furnace separates and redirects hot air from ambient air, which it also warms and circulates. A full inspection of a heat exchanger is not possible without dismantling a furnace, which was not done by WCIS. There was also no access for an inspection mirror. Ask the local utility or a heating contractor to conduct a standard safety check of this and all gas appliances, supply lines, and flues, now and at every change of occupancy.

> ✔ Correct the gas leak at once.
> ✔ Loosen the shutoff valve.
> ✔ Address the nature and risk of asbestos-like material.
> ✔ Clean the ventilation and circulation components.
> ✔ Provide adequate fresh-air ventilation.
> ✔ Extend the flue top so it is sufficiently above the roofline.
> ✔ Ask the local utility or a heating contractor to activate the heater and conduct a
>   standard test and safety check of *all* gas-fueled appliances, supply lines, and flues.

INTERIOR
General Condition

Generally, walls, ceilings, and floors were adequately maintained. The inspection industry does not report on cosmetic details.

Windows, Doors, and Stairs

Because the operability of most windows has been affected by painting, fresh air availability has been diminished and should be increased by routinely opened. Repair as needed.

Windows in a number of rooms need additional attention. Some window locks and hardware need adjustment. In addition, at least three double-hung sashes in the master bedroom had broken wires. And one garage window had a cracked glass pane that should be replaced. Generally, any broken, deteriorated, and/or missing doors and windows, locks, and components, even though not specifically called out in this report, should be replaced or repaired.

Floors and Walls

Much of the family room flooring was carpeted. Uncovered hardwood flooring in the living room was in good condition and adequately maintained.

A hole was noticed on the hall wall. Apparently a missing door stop is responsible for this wall damage, which needs repair before repainting.

- ✔ Repair or adjust windows, doors, and hardware as needed.
- ✔ Install a missing doorstop device and repair the hole in the hall wall.

FIREPLACE

The living room fireplace had a sound firebox. Little cracking of bricks or mortar joints was detected. The firebox was empty.

Needing attention was the matter of cleanliness of the firebox, damper throat, and full extent of the chimney. The National Fire Prevention Association recommends that an in-depth inspection of the entire fireplace system be performed whenever there's a change of ownership of a home having a solid-fuel-burning fireplace. Thereafter, contact a professional chimney-sweep contractor to fully examine, repair, and clean all needed areas, as well as those that were not readily accessible for today's inspection, on a regular basis. This will insure continued safe and efficient fireplace operation.

The chimney flue had a cap on its top and it was the spark-arresting type. This protective ember screen was in satisfactory condition.

The fireplace damper door was operational and well fitted. However, neither a protective ember screen nor a glass-door assembly was presently in place at the firebox's outer hearth. Provide either type of protective barrier before lighting the next fire.

The wood mantle and breastplate, as well as the tiled outer hearth, were in good condition and well maintained.

- ✔ Provide a protective ember screen or glass-door assembly at the firebox's outer hearth.
- ✔ Hire a professional, full-service chimney-sweep inspector/contractor before activating the fireplace.

KITCHEN

The kitchen was well maintained. The sink, faucet, trap and drain, and shutoff valves were working when tested. Water pressure was adequate.

Leakage at the faucet ball needs *immediate correction.*

The electric garbage disposer was operational and functioned as expected. There was no unusual or excessive noise or vibration.

Inner surfaces of the dishwasher were empty and clean. It did not have an anti-siphoning device but was well secured to the underside of the counter. An anti-siphoning device, installed above the sink rim, prevents backflow of waste products into the clean dishwasher appliance if the sewer system were to become blocked. A licensed plumber should be contacted for this installation.

Stained hardwood cabinetry was in satisfactory condition; however, only a sampling of this kitchen's cabinet doors, drawers, and connections was made.

Plastic-laminated counters were in satisfactory condition, however, the backsplash to the right of the range was loose and needs securing and caulking. Joints in all counter, backsplash, and sink areas should be continuously sealed with a good quality, flexible caulk to help prevent moisture penetration.

A ducted exhaust fan in the overhead microwave appliance was operational. The exhaust fan filter was greasy, and the fan motor drew air weakly and may be grease-bound. For an efficient exchange of air, clean, repair, or replace components as needed.

Resilient vinyl flooring was recently installed and well maintained.

✔ Install an anti-siphoning device for the dishwasher.
✔ Secure the loose backsplash piece and apply caulking where needed.
✔ Clean, repair, or replace exhaust fan components.

BATHROOM

This building had two bathrooms that were recently remodeled. The sinks and faucets, traps and drains, and angle stops worked well when tested.

Water pressure was adequate; however, measurement is only a relative comparison rating. New owners should personally test each fixture to become familiar with each and make desired modifications.

Testing "dynamic water flow" (the running of two or more cold water fixtures concurrently) showed a noticeable drop in volume. Red E. Toobuy and his real-estate agent were told how to perform a "homeowner's dynamic water flow and temperature test" on each fixture to ascertain the risk of accidental scalding when cold faucets are activated while someone is taking a shower.

No evidence of significant or unusual deterioration was evident on visible drain lines and trap piping. Tested drains ran freely, however, water leakage was found at the guest bathroom sink drain. This leak needs *immediate repair and/or correction.*

Both toilets were secured and caulk-sealed to the floor. The guest bathroom's toilet seat was extremely loose and needs to be tightened.

Shower glass in the master bathroom did have a glazing label certifying composition (e.g., tempered or safety). Both tub and shower areas had well-fastened grab bars.

The guest bathroom ceiling fan drew air weakly and seemed to need cleaning, servicing, or replacement.

The guest bathroom's resilient floor covering had an open seam that needs adhesive and caulk.

✔ Correct the leaking guest bathroom sink drain.
✔ Fasten the loose toilet seat of the guest bathroom.
✔ Clean, service, or replace the exhaust fan.
✔ Secure the guest bathroom floor seam and apply caulk.

ROOFING

Accessibility

The roof was accessible by ladder. Inspector Warren Camp physically performed a full roof inspection. Only the general condition of visible roofing surfaces was observed. Watertesting of roof surfaces, membranes, chimneys, gutters, flashing, and so on, is not typically performed by home inspectors.

Membrane Type and Condition

Multiple layers of composition shingles appear to have been laid over this structure. The actual number or combined weight placed on structural members could not be determined. Multiple layers concern roofers and inspectors for different reasons: they create an uneven surface; retain moisture and/or gas vapors between membranes; may transfer decay to structural members in their contact; and may add excessive weight to the structure. Whenever multi-layered roofs receive their next membrane, all existing roofing materials should be torn off and discarded. Consider installing appropriately specified plywood sheathing at that time.

Although not fresh, and showing routine wear and tear due to exposure, the composition shingles on each roof slope appeared sound. There was little or no evidence of unusual or significant roof deterioration however moss growth was evident on the lower portion of the north-facing slope. Contact a roofing contractor to determine ways to eliminate this growth.

Debris was found on the front roof slope where the tall tree has been dropping leaves and branches over time. Roofing must be promptly and regularly cleaned and maintained.

Chimneys, Gutters, and Flashing

Step-shingle flashing is a quality feature. It was visible at the base of the fireplace chimney. Rust was observed on a portion of this flashing, suggesting further exploration and analysis by a sheet metal contractor.

Pipe vent and perimeter flashing were in satisfactory condition.

Sections of valley flashing were exposed. Overall, the condition was satisfactory.

Three gutter seams were noticeably rusted, especially on the unpainted interior face. In addition, two separated or missing gutter ends were found that need correction. Hire a sheet metal contractor to make needed repairs or replacements.

Gutters had collected organic debris from overhead trees. Keep gutters, downspouts, and all other drain openings free of debris for proper drainage throughout the year.

The downspout system was, for the most part, customarily installed. Unfortunately, a number of downspouts likely dump water directly onto foundation areas below, which can cause erosion and building settlement over time. Splash blocks or extenders can be placed at the base of such downspouts to divert collected water. As an option, see if a licensed plumbing contractor can connect downspout piping to an existing drain line.

Additional Concerns

Roofs are seldom, if ever, regularly inspected. Regardless of whether a WCIS roof inspection was made, roofing problems are often subtle and difficult to evaluate. Because property inspectors don't often have the hands-on training and accessibility roofers have, whenever questions of roofing adequacy arise, a licensed roofing contractor should be asked to provide a thorough inspection and evaluation.

Biennially, before the rainy season, roofs should be examined by a qualified roofing contractor, and routinely maintained.

✔ Clean the roof system of moss growth and tree dropping.
✔ Have the rusted step-shingle flashing analyzed by a sheet metal contractor.
✔ Repair or replace rusted gutters and separated/missing gutter end pieces.
✔ Extend or redirect downspout bottoms to divert rainwater away from the building foundation.

Seismic Map Evaluation Notations (an optional evaluation that was ordered by Red E. Toobuy)

Map #1 — Intensity of Ground Shaking During a Major Earthquake (having a Richter rating of 8.0 or higher): From "A" to "E," this property's location is rated "E" (the _least_ intense shaking rating in the city).

Map #2 — Potential Landslide Location: This building is _within_ such location. It's approximately three blocks from an active slide area.

Map #3 — Estimated Building Damage from a Major Quake: Seismologists anticipate _minimal_ damage to this building and from adjacent structures.

Map #4 — Potential Reservoir Failure: This building is _outside_ such location.

Map #5 — Geologic Makeup Beneath This Building: This building sits on unsheared Franciscan rock (designated KJU by geologists), which has the _highest stability_ rating in the city.

Map #6 — Liquefaction Potential: This building is _outside_ such location.

Map #7 — Subsidence Potential: This building is _outside_ such location.

Map #8 — Tsunami Potential: This building is _outside_ such location.

— • —

Thank you for calling Warren Camp, your ASHI-certified-member property inspector.

Additional articles/pamphlets provided:

All-Points Bulletin — a home remodeling and repair newsletter; a utility company pamphlet; published articles by Warren Camp about smoke detectors, asbestos, water intrusion, and GFCI electrical receptacles; and his year-round home-maintenance checklist.

Copies to:	delivery	mail	pick-up	fax	email
Buyer:	[ ]	[X]	[ ]	[ ]	[ ]
Buyer's Agent:	[ ]	[ ]	[X]	[ ]	[ ]
Seller's Agent:	[ ]	[ ]	[ ]	[ ]	[ ]

— • —

The contents of this report have been prepared for the exclusive use of Red E. Toobuy. Reliance by others is prohibited.

Warren Camp Inspection Services   • Page 15

CALIFORNIA ASSOCIATION OF REALTORS®

BUYER'S INSPECTION ADVISORY

(C.A.R. Form BIA, Revised 11/14)

Property Address _____ ("Property")

1. IMPORTANCE OF PROPERTY INVESTIGATION: The physical condition of the land and improvements being purchased is not guaranteed by either Seller or Brokers. You have an affirmative duty to exercise reasonable care to protect yourself, including discovery of the legal, practical and technical implications of disclosed facts, and the investigation and verification of information and facts that you know or that are within your diligent attention and observation. A general physical inspection typically does not cover all aspects of the Property nor items affecting the Property that are not physically located on the Property. If the professionals recommend further investigations, including a recommendation by a pest control operator to inspect inaccessible areas of the Property, you should contact qualified experts to conduct such additional investigations.

2. BROKER OBLIGATIONS: Brokers do not have expertise in all areas and therefore cannot advise you on many items, such as those listed below. If Broker gives you referrals to professionals, Broker does not guarantee their performance.

3. YOU ARE STRONGLY ADVISED TO INVESTIGATE THE CONDITION AND SUITABILITY OF ALL ASPECTS OF THE PROPERTY, INCLUDING BUT NOT LIMIITED TO THE FOLLOWING. IF YOU DO NOT DO SO, YOU ARE ACTING AGAINST THE ADVICE OF BROKERS.

    A. GENERAL CONDITION OF THE PROPERTY, ITS SYSTEMS AND COMPONENTS: Foundation, roof (condition, age, leaks, useful life), plumbing, heating, air conditioning, electrical, mechanical, security, pool/spa (cracks, leaks, operation), other structural and non-structural systems and components, fixtures, built-in appliances, any personal property included in the sale, and energy efficiency of the Property.

    B. SQUARE FOOTAGE, AGE, BOUNDARIES: Square footage, room dimensions, lot size, age of improvements and boundaries. Any numerical statements regarding these items are APPROXIMATIONS ONLY and have not been verified by Seller and cannot be verified by Brokers. Fences, hedges, walls, retaining walls and other barriers or markers do not necessarily identify true Property boundaries.

    C. WOOD DESTROYING PESTS: Presence of, or conditions likely to lead to the presence of wood destroying pests and organisms.

    D. SOIL STABILITY: Existence of fill or compacted soil, expansive or contracting soil, susceptibility to slippage, settling or movement, and the adequacy of drainage.

    E. WATER AND UTILITIES; WELL SYSTEMS AND COMPONENTS; WASTE DISPOSAL: Water and utility availability, use restrictions and costs. Water quality, adequacy, condition, and performance of well systems and components. The type, size, adequacy, capacity and condition of sewer and septic systems and components, connection to sewer, and applicable fees.

    F. ENVIRONMENTAL HAZARDS: Potential environmental hazards, including, but not limited to, asbestos, lead-based paint and other lead contamination, radon, methane, other gases, fuel oil or chemical storage tanks, contaminated soil or water, hazardous waste, waste disposal sites, electromagnetic fields, nuclear sources, and other substances, materials, products, or conditions (including mold (airborne, toxic or otherwise), fungus or similar contaminants).

    G. EARTHQUAKES AND FLOODING: Susceptibility of the Property to earthquake/seismic hazards and propensity of the Property to flood.

    H. FIRE, HAZARD AND OTHER INSURANCE: The availability and cost of necessary or desired insurance may vary. The location of the Property in a seismic, flood or fire hazard zone, and other conditions, such as the age of the Property and the claims history of the Property and Buyer, may affect the availability and need for certain types of insurance. Buyer should explore insurance options early as this information may affect other decisions, including the removal of loan and inspection contingencies.

    I. BUILDING PERMITS, ZONING AND GOVERNMENTAL REQUIREMENTS: Permits, inspections, certificates, zoning, other governmental limitations, restrictions, and requirements affecting the current or future use of the Property, its development or size.

    J. RENTAL PROPERTY RESTRICTIONS: Some cities and counties impose restrictions that limit the amount of rent that can be charged, the maximum number of occupants, and the right of a landlord to terminate a tenancy. Deadbolt or other locks and security systems for doors and windows, including window bars, should be examined to determine whether they satisfy legal requirements.

    K. SECURITY AND SAFETY: State and local Law may require the installation of barriers, access alarms, self-latching mechanisms and/or other measures to decrease the risk to children and other persons of existing swimming pools and hot tubs, as well as various fire safety and other measures concerning other features of the Property.

    L. NEIGHBORHOOD, AREA, SUBDIVISION CONDITIONS; PERSONAL FACTORS: Neighborhood or area conditions, including schools, law enforcement, crime statistics, registered felons or offenders, fire protection, other government services, availability, adequacy and cost of internet connections or other technology services and installations, commercial, industrial or agricultural activities, existing and proposed transportation, construction and development that may affect noise, view, or traffic, airport noise, noise or odor from any source, wild and domestic animals, other nuisances, hazards, or circumstances, protected species, wetland properties, botanical diseases, historic or other governmentally protected sites or improvements, cemeteries, facilities and condition of common areas of common interest subdivisions, and possible lack of compliance with any governing documents or Homeowners' Association requirements, conditions and influences of significance to certain cultures and/or religions, and personal needs, requirements and preferences of Buyer.

By signing below, Buyers acknowledge that they have read, understand, accept and have received a Copy of this Advisory. Buyers are encouraged to read it carefully.

Buyer _____   Buyer _____

Published and Distributed by:
REAL ESTATE BUSINESS SERVICES, INC.
a subsidiary of the California Association of REALTORS®
525 South Virgil Avenue, Los Angeles, California 90020

Reviewed by _____ Date _____

BIA REVISED 11/14 (PAGE 1 OF 1)   Print Date
BUYER'S INSPECTION ADVISORY (BIA PAGE 1 OF 1)

Appendix **D**

Glossary

The terms that appear in italic type within the definitions are defined in this glossary.

acceleration clause: Watch out for an *acceleration clause* in your mortgage contract. This provision gives the lender the right to demand payment of the entire outstanding balance if you miss a monthly payment, sell the property, or otherwise fail to perform as promised under the terms of your mortgage. (See also *due-on-sale clause.*) Ouch!

adjustable-rate mortgage (ARM): An *adjustable-rate mortgage* is a mortgage whose interest rate and monthly payments vary throughout its life. ARMs typically start with an unusually low interest rate (see *teaser rate*) that gradually rises over time. If the overall level of interest rates drops, as measured by a variety of indexes (see *index*), the interest rate of your ARM generally follows suit. Similarly, if interest rates rise, so does your mortgage's interest rate and monthly payment. Caps (see also *periodic cap* and *life cap*) limit the amount that the interest rate can fluctuate. Before you agree to an adjustable-rate mortgage, be sure that you can afford the highest payments that would result if the interest rate on your mortgage increased to the maximum allowed.

adjusted cost basis: For tax purposes, the *adjusted cost basis* is important when you sell your property because it allows you to determine what your profit or loss is. You can arrive at the adjusted cost basis by adding the cost of the *capital improvements* that you've made to the home to the price that you paid for the home. Capital improvements increase your property's value and its life expectancy.

adjustment period or **adjustment frequency:** This term has nothing to do with the first few weeks after you've broken up with your sweetheart; it refers to how often the *interest rate* for an *adjustable-rate mortgage* changes. Some adjustable-rate mortgages change every month, but more typically they have one or two adjustments per year. The less frequently your loan rate shifts, the less financial uncertainty you may have. But less frequent adjustments in your mortgage rate mean that you'll probably have a higher *teaser rate,* or initial interest rate. (The initial interest rate is also called the "start rate.")

annual percentage rate (APR): This figure states the total yearly cost of a mortgage as expressed by the actual rate of interest charged. The *APR* includes the base *interest rate, points,* and any other add-on loan fees and costs. Thus, the APR is invariably higher than the rate of interest that the lender quotes for the mortgage.

appraisal: You must pay for the mortgage lender to hire an appraiser to give an "opinion of value" (that is, the appraiser gives a measure of the market value) of the house you want to buy. This professional opinion helps protect the lender from lending you money on a home that's worth less than what you agree to pay for it. For typical homes, the *appraisal* fee is usually several hundred dollars.

appreciation/depreciation: *Appreciation* refers to the increase of a property's value. *Depreciation* (the reverse of appreciation) is when a property's value decreases.

arbitration of disputes: This is a method of solving contract disputes that generally, but not always, is less costly and faster than going to a court of law. In *arbitration,* buyers and sellers present their differences to a neutral arbitrator who, after hearing the evidence, makes a decision that resolves the disagreement. The arbitrator's decision is final and may be enforced as though it were a court judgment. Consult a real estate lawyer if you're ever a party to an arbitration. (Also see *mediation of disputes.*)

assessed value: The *assessed value* is the value of a property (according to your local county tax assessor) for the purpose of determining your *property tax.*

assumable mortgage: Some mortgages allow future buyers of your home to take over the remaining loan balance of your mortgage. If you need to sell your house but the *interest rates* currently offered by lenders are high, having an *assumable mortgage* may be handy. You may be able to offer the buyer your assumable loan at a lower interest rate than the current going interest rate. Most assumables are *adjustable-rate mortgages — fixed-rate, assumable mortgages* are nearly extinct these days because lenders realize that they lose a great deal of money on these types of mortgages when interest rates skyrocket.

balloon loans: These loans require level payments, just as a 15- or 30-year *fixed-rate mortgage* does. But well before their maturity date (the date when you pay them off) — typically, three to ten years after the start date — the loan's full remaining balance becomes due and payable. Although *balloon loans* can save you money because they charge a lower rate of interest relative to fixed-rate loans, balloon loans are dangerous. Being able to *refinance* a loan is never a sure thing. Beware of balloon loans!

bridge loan: If you find yourself in the inadvisable situation where you close on the purchase of a new home before you sell your old one, you may need a short-term *bridge loan*. Such loans enable you to borrow against the *equity* that's tied up in your old house until it sells. We say "bridge" because such a loan is the only thing keeping you above water financially during this period when you own two houses. Bridge loans are expensive compared with other alternatives, such as using a *cash reserve*, borrowing from family or friends, or using the proceeds from the sale of your current home. In most cases, you need the bridge loan for only a few months in order to tide you over until you sell your house. Thus, the loan fees can represent a high cost (about 10 percent of the loan amount) for such a short-term loan.

broker: A real estate *broker* is one level higher on the real estate professional totem pole than a *real estate agent* (or salesperson). Real estate agents can't legally work on their own — a broker must supervise them. To become a broker in most states, a real estate salesperson must have a number of years of full-time real estate experience, meet special educational requirements, and pass a state licensing exam. See also *real estate agent* and *Realtor*.

buydown: A *buydown* is a *VA loan* plan that's available only in some new housing developments and that targets veterans with low or modest incomes. Buydown simply means that a builder agrees to pay part of the home buyer's mortgage for the first few years. Sellers also sometimes do interest-rate buydowns to create attractive financing for buyers of their houses by paying lenders a predetermined amount of money so lenders will reduce their mortgage interest rates.

buyer's brokers: Historically, *real estate agents* and *brokers* worked only for sellers. Now they also work for buyers. The *buyer's broker* owes allegiance only to the buyer and doesn't have an agent relationship with the seller. Although this may seem to be an improvement for all the buyers in the world, don't be too ecstatic. Buyer's brokers are still paid on *commission* when you buy, so don't expect them to be supportive of you if you habitually lollygag. Also keep in mind that the higher the purchase price of the house, the more money the buyer's broker makes.

capital gain: For tax purposes, a *capital gain* is the profit that you make when you sell a home. If you buy a home for $175,000 and then (a number of years later) sell the house for $325,000, your capital gain is $150,000. A sizable amount of capital gain on a house sale is excluded from federal tax: up to $250,000 for qualifying single taxpayers and $500,000 for married couples filing jointly.

capital improvement: A *capital improvement* is money you spend on your home that permanently increases its value and useful life — putting a new roof on your house, for example, rather than just patching the existing roof.

caps: Two different types of caps for *adjustable-rate mortgages* exist. The *life cap* limits the highest or lowest *interest rate* that's allowed over the entire life of your mortgage. The *periodic cap* limits the amount that your interest rate can change in one *adjustment period*. A one-year *ARM*, for example, may have a start rate of 4 percent with a plus or minus 2 percent periodic adjustment cap and a 6 percent life cap. On a worst-case basis, the loan's interest rate would be 6 percent in the second year, 8 percent in the third year, and 10 percent (4 percent start rate plus the 6 percent life cap) forevermore, starting with the fourth year.

cash reserve: Most mortgage lenders require that home buyers have sufficient cash left over after closing on their home purchase to make the first two mortgage payments or to cover a financial emergency.

Certified Distressed Property Expert (CDPE): Real estate agents who successfully complete specialized training to help homeowners avoid foreclosure earn the *Certified Distressed Property Expert (CDPE)* designation from the Charfen Institute. If the owners can't afford to maintain their mortgages, a CDPE can advise them about other options such as mortgage modification or *deed in lieu of foreclosure*. CDPE agents also receive specialized short-sale training so they can facilitate transactions in which a lender agrees to accept less than what is owed on the mortgage upon sale of the property.

closing costs: After you pass every home-buying obstacle and reach the safe clearing in order to buy your home, one final potential land mine appears in the form of *closing costs*. These costs generally total 2 percent to 5 percent of the home's purchase price and are completely independent of (and in addition to) the *down payment*. Closing costs include such things as *points* (that is, the loan *origination fee* to cover the lender's administrative costs), an *appraisal* fee, a *credit report* fee, mortgage interest for the period between the closing date and the first mortgage payment, the *homeowners insurance* premium, *title insurance*, prorated *property taxes*, and recording and transferring charges. So when you're finally ready to buy, you need to have enough cash to pay all these costs in order to buy your dream home.

commission: The *commission* is the percentage of a house's selling price that's paid to the *real estate agent* and *broker*. Because most agents and brokers are paid by commission, understanding how the commission can influence the way that agents and brokers work is important for home buyers. Agents and brokers make money only when you make a purchase, and they make more money when you make a bigger purchase. Choose an agent carefully, and take your agent's advice with a grain of salt, because this inherent conflict of interest can often set an agent's visions and goals at odds with your visions and goals.

community property: Along with *joint tenancy* and *tenancy-in-common*, *community property* is a way that married couples may take title to real property. Community property offers two major advantages over joint tenancy and tenancy-in-common. First, community property ownership allows spouses to transfer interests, by a will or otherwise, to whomever they want. The second advantage of holding title to a home in community property is that the surviving spouse gets favorable tax treatment. The entire house's market value as of the spouse's date of death (such market value is also called the house's "stepped-up basis") is used rather than the house's original cost, which reduces the taxable profit (assuming that the home has appreciated in value) when the house is sold.

comparable market analysis (CMA): Buying a Toyota Prius from the first dealer you visit would be impulsive and foolish. You need to shop around to find out where the best deal on that type of car is. The same is true with home buying. If you're interested in buying a home, you need to find out how much money houses in the area have been selling for. You must identify "comparable" homes that have sold within the past six months, are in the immediate vicinity of the home you desire to purchase, and are as similar as possible to the one that you're interested in buying in terms of size, age, and condition. You must do the same thing for comparable houses currently on the market to see if prices are rising, flat, or falling. A written analysis of comparable houses currently being offered for sale and comparable houses sold in the past six months is called a *comparable market analysis (CMA)*.

conditions: See *contingencies*.

condominiums: *Condominiums* are housing units that are contained within a development area or within a single building in which you own your actual unit and a share of everything else in the development (lobby, parking areas, land, and the like, which are known as common areas). Condominiums are a less-expensive form of housing than single-family homes. For this reason, some people mistakenly view them as good starter houses. Unfortunately, condos generally don't increase in value as rapidly as single-family houses do, because the demand for them is lower than the demand for houses. And because condos are far easier for builders to develop than single-family homes, the supply of condos often exceeds the demand for them.

contingencies: *Contingencies* are conditions contained in almost all home purchase offers. The seller or buyer must meet or waive all contingencies before the deal can close. These conditions relate to such things as the buyer's review and approval of property inspections or the buyer's ability to get the mortgage financing that's specified in the contract.

convertible adjustable-rate mortgage: Unlike a conventional *adjustable-rate mortgage,* a *convertible adjustable-rate mortgage* gives you the opportunity to convert to a *fixed-rate mortgage,* usually between the 13th and 60th month of the loan. For this privilege, convertible adjustable-rate mortgage loans have a higher rate of interest than conventional adjustable-rate mortgages, and a conversion fee (which can range from a few hundred dollars to 1 percent or so of the remaining balance) is charged. Additionally, if you choose to convert to a fixed-rate mortgage, you'll pay a slightly higher rate than what you can get by shopping around for the best rates available at the time you convert.

cooperatives (co-ops): *Cooperatives* are apartment buildings where you own a share of a corporation whose main asset is the building that you live in. Each owner's rights to exclusive use of his apartment are spelled out in a lease in which the corporation is the landlord. In high-cost areas, cooperatives (like their cousins, *condominiums* and *townhouses*) are cheaper alternatives to buying single-family houses. Unfortunately, cooperatives also resemble their cousins in that they generally lag behind single-family homes in terms of *appreciation.* Co-ops are also, as a rule, harder to sell and obtain loans for than condominiums.

cosigner: If you have a checkered past in the credit world, you may need help securing a mortgage, even if you're financially stable. A friend or relative can come to your rescue by cosigning (which literally means being indebted for) a mortgage. A *cosigner* can't improve your *credit report* but can improve your chances of getting a mortgage. Cosigners should be aware, however, that cosigning for your loan adversely affects their future creditworthiness because your loan becomes what's known as a contingent liability against their borrowing power.

cost basis: See *adjusted cost basis.*

covenants, conditions, and restrictions (CC&Rs): *CC&Rs* establish a condominium by creating a homeowners association that stipulates how the *condominium*'s maintenance and repairs are handled and regulates what can and can't be done to individual units and the condominium's common areas. These restrictions may apply to lawn maintenance, window curtain colors, and the like. Some CC&Rs also put community decision-making rights into the hands of a homeowners association.

credit report: A *credit report* is a report that details credit you have taken out and how timely your repayments have been. This is the main report a lender uses to determine your creditworthiness. You must authorize the credit company to send the report to your lender and pay for the lender to obtain this report, which the lender uses to determine your ability to handle all forms of credit and to pay off loans in a timely fashion.

debt-to-income ratio: Before you go out home buying, you should determine what your price range is. Lenders generally figure that you shouldn't spend more than about 33 to 40 percent of your monthly income for your housing costs. The *debt-to-income ratio* measures your future monthly housing expenses — which include your proposed mortgage payment (debt), property tax, and insurance — in relation to your monthly income.

deed: A *deed* is the document that conveys title to real property. Before you receive the deed, the *title insurance* company must receive the mortgage company's payment and your payments for the *down payment* and *closing costs.* The title insurance company must also show that the seller holds clear and legal title to the property for which the title is being conveyed.

deed in lieu of foreclosure: Instead of *foreclosure,* which is generally costly and time-consuming for all parties, a *deed in lieu of foreclosure* is a voluntarily entered agreement whereby typically the borrower conveys ownership of the property in *default* to the lender to obtain a release from the lender from paying outstanding debt on that property.

default: *Default* is the failure to make your monthly mortgage payments on time. You're probably officially in default when you miss two or more monthly payments. Default also refers to other violations of the mortgage terms. Default can lead to *foreclosure* on your house.

delinquency: At first you're in *delinquency;* then you're in *default.* Delinquency occurs when a monthly mortgage payment isn't received by the due date.

distressed property: A *distressed property* is a mortgaged property in *default* that was either foreclosed upon by a lender or sold in a *short sale* to prevent the property from going into foreclosure. Chapter 8 delves into the subject of distressed property in distressing detail.

down payment: The *down payment* is the part of the purchase price that the buyer pays in cash, up-front, and doesn't finance with a mortgage. Generally, the larger the down payment, the better the deal you can get on a mortgage. You can usually get access to the best mortgage programs with a down payment of 20 percent of the home's purchase price.

due-on-sale clause: A *due-on-sale clause* contained in the mortgage entitles the lender to demand full payment of all money due on your loan when you sell or transfer title to the property.

earthquake insurance: Although the West Coast is often associated with earthquakes, other areas are also prone to earthquakes. An *earthquake insurance* rider on a homeowners policy pays to repair or rebuild your home if it's damaged in an earthquake. If you live in an area with earthquake risk, get earthquake insurance coverage!

equity: In the real estate world, *equity* refers to the difference between your home's market value and what you owe on it. For example, if your home is worth $250,000 and you have an outstanding mortgage of $170,000, your equity is $80,000.

escrow: *Escrow* isn't an exotic dish; it's the holding of important documents and money (related to the purchase/sale of a property) by a neutral third party (the escrow officer) prior to the transaction's close. After the seller accepts the buyer's offer, the buyer doesn't immediately move into the house. A period follows during which *contingencies* have to be met or waived. During this period, the escrow service holds the buyer's *down payment* and the buyer's and seller's documents related to the sale. "Closing escrow" means that the deal is completed. Among other duties, the escrow officer makes sure that the previous mortgage is paid off, your loan is funded, and the *real estate agent*s are paid.

Fannie Mae: See *Federal National Mortgage Association.*

Federal Home Loan Mortgage Corporation (FHLMC): The *FHLMC* (or *Freddie Mac*) is one of the best-known institutions in the secondary mortgage market. Freddie Mac buys mortgages from banks and other mortgage-lending institutions and, in turn, sells these mortgages to investors. These loan investments are generally considered safe because Freddie Mac buys mortgages only from companies that conform to its stringent mortgage regulations, and Freddie Mac guarantees the repayment of *principal* and interest on the mortgages that it sells.

Federal Housing Administration mortgages (FHA): *Federal Housing Administration mortgages* are marketed to people with modest means. The main advantage of these mortgages is that they require a small *down payment* (usually between 3 percent and 5 percent). FHA mortgages also offer a competitive *interest rate* — typically, 0.5 to 1 percent below the interest rates on other mortgages. The downside is that with an FHA mortgage, the buyer must purchase mortgage *default* insurance (see *private mortgage insurance*).

Federal National Mortgage Association (FNMA): The *FNMA* (or *Fannie Mae*) is one of the best-known institutions in the secondary mortgage market. Fannie Mae buys mortgages from banks and other mortgage-lending institutions and, in turn, sells them to investors. These loan investments are generally considered safe because Fannie Mae buys mortgages only from companies that conform to its stringent mortgage regulations, and Fannie Mae guarantees the repayment of *principal* and interest on the loans that it sells.

fixed-rate mortgage: The *fixed-rate mortgage* is the granddaddy of all mortgages. You lock into an *interest rate* (for example, 5.5 percent), and it never changes during the life (term) of your 15- or 30-year mortgage. Your mortgage payment is the same amount each and every month. Compare fixed-rate mortgages with *adjustable-rate mortgages.*

flood insurance: In federally designated flood areas, *flood insurance* is required. If even a remote chance exists that your area may flood, having flood insurance is prudent.

foreclosure: *Foreclosure* is the legal process in which the mortgage lender takes possession of and sells the property to attempt to satisfy indebtedness. When you *default* on a mortgage and the lender deems that you're incapable of making payments, you may lose your house to foreclosure. Being in default, however, doesn't necessarily lead to foreclosure. Some lenders are lenient and help you work out a solution if they see that your problems are temporary. The federal government also has several programs designed to help homeowners who are having problems paying their mortgages. Foreclosure is traumatic for the homeowner and expensive for the lender.

formula: The formula is how you calculate the *interest rate* for an *adjustable-rate mortgage.* Add the *margin* to the *index* to get the interest rate (margin + index = interest rate).

Freddie Mac: See *Federal Home Loan Mortgage Corporation.*

graduated-payment mortgage: A rare bird these days, the *graduated-payment mortgage* gives you the opportunity to cut your total interest costs. With a graduated-payment mortgage, your monthly payments are increased by a predetermined formula (for example, a 3 percent increase each year for seven years, after which time payments no longer fluctuate). If you expect to land a job that may allow you to make these higher payments, you may want to consider this option.

home equity loan: *Home equity loan* is technical jargon for what used to be called a *second mortgage.* With this type of loan, you borrow against the *equity* in your house. If used wisely, a home equity loan can help people pay off high-interest consumer debt, which is usually at a higher *interest rate* than a home equity loan and isn't tax-deductible; or a home equity loan can be used for other short-term needs, such as payments on a remodeling project.

home warranty plan: A *home warranty plan* is a type of insurance that covers repairs to specific parts of the home for a predetermined time period. Because home warranty plans typically cover small-potato items, such plans aren't worth buying. Instead, spend your money on a good *house inspection* before you buy the home to identify any major problems (electrical, plumbing, or structural).

homeowners insurance: Required and necessary. No ifs, ands, or buts about it — you need "dwelling coverage" that can cover the cost to rebuild your house. The liability insurance portion of this policy protects you against accidents that occur on your property. Another essential piece is the personal property coverage that pays to replace your lost worldly possessions and usually totals 50 to 75 percent of the dwelling coverage. Finally, get *flood insurance* or *earthquake insurance* if you're in an area susceptible to these natural disasters. As with other types of insurance, get the highest deductibles with which you're comfortable.

house inspection: Like *homeowners insurance,* we think that a *house inspection* is a necessity. The following should be inspected: overall condition of the property, inside and out; electrical, heating, and plumbing systems; foundation; roof; pest control and dry rot; and seismic/slide risk. A good house inspection can save you money by locating problems. With the inspection report in hand, you can ask the seller to either do repairs or reduce the purchase price. Hire your own inspector. Never be satisfied with a seller's inspection reports.

hybrid loans: Combining the features of *fixed-rate* and *adjustable-rate mortgage*s is the objective of *hybrid loans.* The initial *interest rate* for a hybrid loan may stay the same for the first three to ten years of the loan (as opposed to only six to twelve months for a standard adjustable-rate mortgage); then the interest rate adjusts biannually or annually. Remember that the longer the interest holds at the same initial rate, the higher the interest rate will be. These hybrid loans are best for people who plan to own their house for a short time (fewer than ten years) and who don't like the volatility of a typical adjustable-rate mortgage.

index: The *index* is the measure of the overall level of *interest rate*s that the lender uses as a reference to calculate the specific interest rate on an adjustable-rate loan. The index plus the *margin* is the *formula* for determining the interest rate on an *adjustable-rate mortgage.* One index used on some mortgages is the six-month Treasury bill. If the going rate for these Treasury bills is 3 percent and the margin is 2.5 percent, your interest rate would be 5.5 percent. Other common indexes used are certificates of deposit index, 11th District Cost of Funds index, and LIBOR index.

interest rate: Interest is what lenders charge you to use their money. Lenders generally charge higher rates of interest on higher-risk loans. For *fixed-rate mortgage*s, remember that the *interest rate* has a seesaw relationship with the *points.* A high number of points is usually associated with a lower interest rate, and vice versa. For an *adjustable-rate mortgage,* make sure that you understand the *formula* (the *index* plus the *margin*) that determines how the interest rate is calculated after the *teaser rate* expires.

interest-only mortgage: A mortgage that in its early years has the borrower making only interest payments. Typically a number of years after it starts, the payment jumps significantly as the borrower begins to make principal and interest payments.

investment property: Investment property is real estate that you don't use as your primary home but rather rent out to others. Real estate is a good long-term investment — it has produced returns similar to those from diversified stock portfolios over the years. In practice, investment in real estate is different from investment in stocks. You can also *leverage* your real estate investment — that is, you can make a profit on your investment as well as on borrowed money. Investing in real estate is time-intensive. You also need to be adept at managing people and money if you're to bear fruit with real estate investments. One drawback of *investment property* is that you can't shelter your investment-property profits in a retirement account the way you can shelter profits earned through stock investments.

joint tenancy: *Joint tenancy* is a form of co-ownership that gives each joint tenant equal interest and rights in the property, including the right of survivorship. At the death of one joint tenant, ownership automatically transfers to the surviving joint tenant(s). This form of ownership is good for married couples and often appropriate for unmarried people in a long-term relationship as long as they obtain title to the property at the same time. Some of the limitations of joint tenancy are, first, that each person must own an equal share of the house and, second, the right of survivorship terminates if one person transfers his interest to a third party.

late charge: A *late charge* is a fee that's charged if a mortgage payment is received late. A late charge can be steep — as much as 5 percent of the mortgage payment amount. Ouch! Get those payments in on time!

lease-option agreement: A *lease-option agreement* is a special contract used for a property that you can lease with an option to purchase for an agreed set price at a later date. These contracts usually require an up-front payment (called "option consideration") to secure the purchase option. The consideration is usually credited toward your *down payment* when you exercise your option to buy the home. An important factor in a lease-option agreement is what portion of the monthly rent payments (typically, one-third) is applied toward the purchase price if you buy. You'll usually pay a slightly higher rent because of the lease-option privilege. Lease-options can cause problems in the event of a default and should not be entered into without consulting an attorney.

leverage: *Leverage* refers to exerting great influence with little effort. Buying a house allows you to leverage your cash in two ways. Suppose that you make a 20 percent *down payment* on a $100,000 house — thus investing $20,000. The first leverage is that you control a $100,000 property with $20,000. If your house appreciates to a value of $120,000, you've made a $20,000 profit on a $20,000 investment — a 100 percent return thanks to leveraging. However, leverage works both ways, so if your house depreciates. . ..

lien: A *lien* is a legal claim against a property and an encumbrance on its title for the purpose of securing payment for work performed and money owed on account of loans, judgments, or claims. Liens are encumbrances on titles, and they need to be paid off before a property can be sold or title can be transferred to a subsequent buyer. The liens that have been recorded appear in a property's preliminary report.

life cap: The *life cap* determines the total amount that your *adjustable-rate mortgage interest rate* and monthly payment can fluctuate up or down during the loan's duration. The life cap is different from the *periodic cap,* which limits the extent to which your interest rate can change up or down in any one *adjustment period.*

liquidated damages: In most real estate contracts, buyers and sellers may agree at the beginning of the transaction regarding how much money one party, usually the seller, would receive if the other party were to violate the terms of the contract without good cause. *Liquidated damages* confine and define how much money the injured party may recover. Buyers, for example, generally limit their losses to the amount of their deposit. Discuss the advisability of using the liquidated damages provision with a lawyer or *real estate agent.*

lock-in: A *lock-in* is a mortgage lender's commitment and written agreement to guarantee a specified *interest rate* to the home buyer, provided that the loan closes within a set period of time. The lock-in also usually specifies the number of *points* to be paid at closing. Most lenders won't lock in your mortgage *interest rate* unless you've made an offer on the property and the property has been appraised. For the privilege of locking in the rate in advance of the loan closing, you may pay a slight interest-rate premium.

margin: The *margin* is the amount that's added to the *index* to calculate your *interest rate* for an *adjustable-rate mortgage.* Most loans have margins around 2.5 percent. Unlike the index (which constantly moves up and down), the margin never changes over the life of the loan.

mediation of disputes: *Mediation of disputes* is a fast, relatively inexpensive way to resolve simple contract disputes. In mediation, buyers and sellers present their differences to a neutral mediator who doesn't have the power to make a decision in favor of one side or the other. Instead, the mediator helps buyers and sellers work together to reach a mutually acceptable solution of their differences. It's probably in your best interest to mediate your problem before going to an arbitrator or suing in a court of law (also see *arbitration of disputes*).

mortgage broker: A *mortgage broker* is a person who can help you find a mortgage. Mortgage brokers buy mortgages wholesale from lenders and then mark the mortgages up (typically, from 0.5 to 1 percent) and sell them to buyers. A good mortgage broker is most helpful for people who won't shop around on their own for a mortgage, or for people who have blemishes on their *credit report*.

mortgage life insurance: *Mortgage life insurance* guarantees that the lender will receive its money in the event that you meet an untimely demise. Many people may try to convince you that you need this insurance to protect your dependents and loved ones. We recommend that you don't waste your time or money with this insurance! Mortgage life insurance is expensive. If you need life insurance, buy low-cost, high-quality term life insurance rather than mortgage life insurance.

Multiple Listing Service: A *Multiple Listing Service* (or *MLS*) is a *real estate agent*'s cooperative service that contains descriptions of most of the houses that are for sale. Real estate agents use this computer-based service to keep up with properties listed for sale by members of the Multiple Listing Service in their area.

negative amortization: Although it may sound like science-fiction jargon, *negative amortization* occurs when your outstanding mortgage balance increases despite the fact that you're making the required monthly payments. Negative amortization occurs with *adjustable-rate mortgage*s that cap the increase in your monthly payment but don't cap the *interest rate*. Therefore, your monthly payments don't cover all the interest that you actually owe. If you've ever watched your credit-card balance snowball as you make only the minimum monthly payment, you already have experience with this phenomenon. *Avoid loans with this feature!*

origination fee: See *points*.

partnership: A *partnership* is a combination of two or more people who come together to undertake an enterprise. Partnerships most often occur among people who have a business relationship and who buy the property as either a business asset or for investment purposes. If you are a partner in a partnership that owns real estate, you own your partnership interest, which is personal property, not real property. There are legal and tax consequences to ownership in this form. If you want to co-own real property, you need to own it as a tenant in common. See *tenancy-in-common*. If you intend to buy property with partners, have a real estate lawyer prepare a written partnership agreement for all the partners to sign before making an offer to purchase.

periodic cap: This cap limits the amount that the *interest rate* of an *adjustable-rate mortgage* can change up or down in one *adjustment period*. See also *caps*.

points: Also known as a loan's *origination fee*, points are interest charges paid up-front when you close on your loan. Points are actually a percentage of your total loan amount (one point is equal to 1 percent of the loan amount). For a $100,000 loan, one point costs you $1,000. Generally speaking, the more points that a loan has, the lower its *interest rate* should be. All the points that you pay on a purchase mortgage are deductible in the year that you pay them. If you *refinance* your mortgage, however, the points that you pay at the time you refinance must be amortized over the life of the loan. If you get a 30-year mortgage when you refinance, for example, you can deduct only one-thirtieth of the points on your taxes each year.

prepayment penalty: One advantage of most mortgages is that you can make additional payments to pay the loan off faster if you have the inclination and the money to do so. A *prepayment penalty* discourages you from doing this by penalizing you for early payments. Some states prohibit lenders from penalizing people who prepay their loans. Avoid mortgages that penalize prepayment!

principal: The *principal* is the amount that you borrow for a loan. If you borrow $100,000, your principal is $100,000. Each monthly mortgage payment consists of a portion of principal that must be repaid, plus the interest that the lender is charging you for use of the money. During the early years of your mortgage, your loan payment is primarily interest.

private mortgage insurance (PMI): If your *down payment* is less than 20 percent of your home's purchase price, you'll likely need to purchase *private mortgage insurance* (also known as "mortgage default insurance"). The smaller the down payment, the more likely a home buyer is to *default* on a loan. Private mortgage insurance can add hundreds of dollars per year to your loan costs. After the *equity* in your property increases to 20 percent, you no longer need the insurance. Don't confuse this insurance with *mortgage life insurance*.

probate sale: A *probate sale* is the sale of a home that occurs when a homeowner dies and the property is to be divided among inheritors or sold to pay debts. The executor of the estate organizes the probate sale, and a probate-court judge oversees the process. The highest bidder receives the property.

property tax: You have to pay *property tax* on the home you own. Annually, property tax averages 1 to 2 percent of a home's value, but property tax rates vary widely throughout this great land.

prorations: Certain items such as *property taxes* and homeowners-association dues are continuing expenses that must be prorated (distributed) between the buyers and sellers at close of *escrow*. If the buyers, for example, owe the sellers money for property taxes that the sellers paid in advance, the prorated amount of money due the sellers at close of escrow appears as a debit (charge) to the buyers and a credit to the sellers.

real estate agent: A *real estate agent* is the worker bee of real estate sales. Also called "salespeople," agents are supervised by a real estate *broker*. The state licenses agents; their pay is typically based totally on *commission*s generated by selling property.

real estate investment trust (REIT): A *real estate investment trust* is like a mutual fund of real estate investments. Such trusts invest in a collection of properties (from shopping centers to apartment buildings). REITs trade on the major stock exchanges. If you want to invest in real estate while avoiding the hassles inherent in owning property, real estate investment trusts may be the right choice for you.

real estate owned (REO): *Real estate owned (REO)* refers to property owned by a lender (typically a bank), government loan insurer, or other government agency because no one purchased it at a foreclosure auction. The opening bid is usually the outstanding loan amount due on the property. Because the property is used as security for the loan, the lender can legally repossess it if there are no interested bidders.

Realtor: A *Realtor* is a *real estate agent,* or *broker,* who belongs to the National Association of Realtors, a trade association whose members agree to its ways of doing business and code of ethics. The National Association of Realtors offers its members seminars and courses that deal with real estate topics.

refinance: *Refinance,* or "re-fi," is a fancy word for taking out a new mortgage loan (usually at a lower *interest rate*) to pay off an existing mortgage (generally at a higher interest rate). Refinancing is not automatic; neither is refinancing guaranteed. You have to apply for a new mortgage to replace the old one, provide all new documentation, and be reevaluated by the lender. Refinancing can be a hassle and expensive. Carefully weigh the costs and benefits of refinancing.

return on investment: The *return on investment* is the percentage of profit that you make on an investment. If you put $1,000 into an investment and one year later your account is worth $1,100, you've made a profit of $100. Your return is the profit ($100) divided by the initial investment ($1,000) — 10 percent. See also *leverage.*

reverse mortgage: A *reverse mortgage* enables elderly homeowners, typically those who are low on cash, to tap into their home's *equity* without selling their home or moving from it. Specifically, a lending institution makes a check out to you each month, and you can use the check as you want. This money is really a loan against the value of your home; because the money that you receive is a loan, the money is tax-free when you receive it. The downside of these loans is that they deplete your equity in your estate, the fees and *interest rates* tend to be on the high side, and some require repayment within a certain number of years.

second mortgage: A *second mortgage* is a mortgage that ranks after a first mortgage in priority of recording. In the event of a *foreclosure,* the proceeds from the home's sale go toward paying off the loans in the order in which they were recorded. You can have a third (or even a fourth) mortgage, but the farther down the line the mortgage is, the higher the risk of *default* on the mortgage — hence, the higher *interest rate* that you'll pay on the mortgage. See also *home equity loan.*

72-hour clause: The *72-hour clause,* also called a release clause, is commonly inserted into real estate purchase offers when the purchase of a home is contingent upon the sale of the buyer's current house. The seller accepts the buyer's offer but reserves the right to accept a better offer if one should happen to come along. However, the seller can't do this arbitrarily. If the seller receives an offer that he wants to accept, he must notify the buyer of that fact in writing. Then the buyer usually has 72 hours (though the allotted amount of time can vary) from the seller's notification to remove the contingency-of-sale clause and move on with the purchase; otherwise, the buyer's offer is wiped out.

shared-equity transaction: In a *shared-equity transaction,* a private investor contributes money toward the purchase of a house and subsequently shares *equity* as a co-owner. When the house is sold, the investor takes a share of the profit or loss. These shared-equity transactions can become fairly complicated because the investor co-owner and the resident co-owner may have conflicts of interest. For example, the investor co-owner may want to sell the property to make a profit, but the resident co-owner may want to stay put. If you intend to participate in an equity-sharing transaction, have a lawyer who works with residential real estate prepare a written co-ownership agreement for all parties to sign prior to purchasing the property.

short sale: Done with a property in *default,* to keep it from *foreclosure.* The lender agrees to accept the proceeds from selling the property as fully satisfying outstanding debt, even though the amount of that debt exceeds the sale proceeds.

strategic default: A *strategic default* occurs when borrowers who have sufficient income to make their mortgage payments decide not to do so, which forces the lender to foreclose. This usually happens when their property is so deeply *underwater* that the homeowners believe it won't recover the value it lost in the foreseeable future. For example, suppose they put $50,000 cash down (20 percent) on a $250,000 home that's worth no more than $125,000 today. Credit agencies treat a strategic default as an extremely serious delinquency, no different than any other foreclosure.

tax-deductible: *Tax-deductible* refers to payments that you may deduct against your federal and state taxable income. The interest portion of your mortgage payments, loan *points,* and *property taxes* are tax-deductible subject to limitations; your employment income is not!

teaser rate: Otherwise known as the initial *interest rate,* the *teaser rate* is the attractively low interest rate that most *adjustable-rate mortgages* start with. Don't be sucked into a mortgage because it has a low teaser rate. Look at the mortgage's *formula* (*index + margin* = interest rate) for a more reliable method of estimating the loan's future interest rate — the interest rate that will apply after the loan is "fully indexed."

tenancy-in-common: *Tenancy-in-common* is how all co-owners hold title. Community property and joint tenancy are co-ownerships with added rules and benefits. Co-owners don't need to own equal shares of the property they hold as tenants in common, and there is no right of survivorship. All co-owners should have a written agreement between and among them to make certain there are no disputes and each party knows what his or her rights and obligations are. Getting legal advice for any co-ownership is essential because the parties may want to address whether one co-owner can force the sale of the entire property.

title insurance: *Title insurance* covers the legal fees and expenses necessary to defend your title against claims that may be made against your ownership of the property. The extent of your coverage depends on whether you have an owner's standard coverage or extended coverage title insurance policy. To get a mortgage, you also have to buy a lender's title insurance policy to protect your lender against title risks. See Chapter 13.

top producers: People remark that 20 percent of all *real estate agents* account for 80 percent of all real estate sales. Be cautious. Why are those agents *top producers?* Some agents get to the top by being pushy and selling a great deal of property without patiently educating buyers — not the kind of agent that you want! If, however, the agent is a top producer because she works hard to meet the needs of her clients, being a top producer is a good thing.

townhouses: *Townhouse* is the decorative name for a row (or attached) home. It may or may not be a condominium or subject to an agreement regarding repair and use of elements servicing the townhouses. Townhouses are cheaper than single-family homes because they use common walls and roofs, thus saving land. In terms of investment appreciation potential, townhouses lie somewhere between single-family homes and *condominiums.*

underwater: A property is *underwater* when homeowners owe more on the mortgage than their house is worth. Underwater is a broad designation which covers everything from a mere toe dip under fair market value to submerged at the bottom of the economic ocean.

VA (Department of Veterans Affairs) loan: Congress passed the Serviceman's Readjustment Act, commonly known as the GI Bill of Rights, in 1944. One of its provisions enables the VA to help veterans and eligible people on active duty obtain mortgages on favorable terms (generally, 0.5 to 1 percent below the *interest rate* currently being charged on conventional loans) to buy primary residences. Like the *FHA,* the VA has no money of its own. It guarantees loans granted by conventional lending institutions that participate in VA mortgage programs.

Index

value
  defined, 229
  importance of, 159
  investment, 172
  seeking opportunities to add to houses, 77
verification-of-employment request, 144
voluntary inquiries, 85

W

Weber, Eleanor, 19
whole life insurance. *See* cash-value life insurance

wills, living, 35
Winning the Endgame (Brown), 35
wire fraud schemes, 313
wishful thinking, 264
workmanship, poor, 184

Y

yield spread premium, 124